High Frequency Over-the-Horizon Radar

Fundamental Principles, Signal Processing, and Practical Applications

Giuseppe Aureliano Fabrizio

I0894730

Mc Graw Hill Education

New York Chicago San Francisco
Lisbon London Madrid Mexico City
Milan New Delhi San Juan
Seoul Singapore Sydney Toronto

Library of Congress Cataloging-in-Publication Data

Fabrizio, Giuseppe Aureliano.
High frequency over-the-horizon radar : fundamental principles, signal processing, and practical applications / Giuseppe Aureliano Fabrizio. – First edition.
pages cm
Includes bibliographical references and index.
ISBN 9781265901233
1. Over-the-horizon radar. I. Title.
TK6592.O94F33 2013
621.3848–dc23 2013015637

McGraw-Hill Education books are available at special quantity discounts to use as premiums and sales promotions, or for use in corporate training programs. To contact a representative, please visit the Contact Us pages at www.mhprofessional.com.

High Frequency Over-the-Horizon Radar: Fundamental Principles, Signal Processing, and Practical Applications

1 2 3 4 5 6 7 8 9 0 DOC/DOC 10 9 8 7 6 5 4 3

ISBN 9781265901233
MHID 1265901236

Sponsoring Editor
Wendy Rinaldi

Editorial Supervisor
Janet Walden

Project Manager
Yashmita Hota, Cenveo® Publisher Services

Acquisitions Coordinator
Joya Anthony

Copy Editor
Cenveo Publisher Services

Proofreader
Cenveo Publisher Services

Indexer
Karin Arrigoni

Production Supervisor
Jean Bodeaux

Composition
Cenveo Publisher Services

Art Director, Cover
Jeff Weeks

High Frequency Over-the-Horizon Radar

About the Author

Giuseppe Aureliano Fabrizio received his B.E. with honors and Ph.D. degrees from the Department of Electrical and Electronic Engineering at Adelaide University, Australia, in 1992 and 2000. Since 1993, Dr. Fabrizio has been with the Defence Science and Technology Organization (DSTO), Australia, where he leads the EW and adaptive signal processing section of the high frequency radar branch. Dr. Fabrizio is responsible for the development and practical implementation of innovative and robust adaptive signal processing techniques to enhance the operational performance of modern OTH radar systems. He is a senior member of the IEEE and the main author of over 50 peer-reviewed journal and conference publications. Dr. Fabrizio is a corecipient of the prestigious M. Barry Carlton Award for the best paper published in the IEEE Transactions on Aerospace and Electronic Systems (AES) in a calendar year on two occasions—2003 and 2004. In 2007, he received the coveted DSTO Science Excellence award in recognition of his contributions to adaptive signal processing for the JORN OTH radar system. In the same year, he was granted a DSTO Defence Science Fellowship to pursue collaborative research at La Sapienza University in Rome, Italy. Dr. Fabrizio has delivered OTH radar tutorials at the 2008 IEEE Radar Conference, held in Rome, and at the 2010 IEEE International Radar Conference in Washington DC. He is an Australian representative on the IEEE International Radar Systems Panel, and is currently serving as VP of Education on the AESS Board of Governors. Dr. Fabrizio has collaborated with international defense agencies under the auspices of a Memorandum of Understanding (MoU) agreement and is the national representative to the NATO task group on Dynamic Waveform Diversity (SET-179). He has also formally collaborated with private industry, including Raytheon, Lockheed Martin, and BAE Systems, as well as numerous academic institutions, both in Australia and abroad. Dr. Fabrizio was selected as the recipient of the distinguished IEEE Fred Nathanson Memorial Radar Award in 2011 for his contributions to OTH radar and radar signal processing. He has written the first edition of *High Frequency Over-the-Horizon Radar: Fundamental Principles, Signal Processing, and Practical Applications* to complement the existing literature on radar and to provide a reference that may be built upon by future works on this subject.

In memory of my Father, and to my Family, with love.

Contents at a Glance

Color images for figures marked with this icon can be downloaded from www.mhprofessional.com/fabrizio-images

Contents

Color images for figures marked with this icon can be downloaded from www.mhprofessional.com/fabrizio-images

Part II Signal Description

Part III Processing Techniques

Preface

In contrast to the many professional texts available on a range of microwave radar topics, only one previous text has been exclusively devoted to the subject of high frequency (HF) over-the-horizon (OTH) radar. The text entitled *Fundamentals of Over-the-Horizon Radar* by A. A. Kolosov et al. was originally published in Russian (Radio i Svyaz, 1984) and subsequently translated into English by W. F. Barton (Artech House, Norwood, MA, 1987). Though this contribution continues to provide many valuable insights, the content has not been updated since the initial publication to capture the tremendous progress that has been made in OTH radar during the past two decades.

The chapter on OTH radar in the prestigious *Radar Handbook* edited by M. Skolnik and co-authored by J. M. Headrick and S. J. Anderson (3rd edition, McGraw-Hill, 2008) provides an excellent overview of the essential concepts, but the limited chapter length necessarily restricts the depth of treatment, perhaps most noticeably in the area of signal processing which has been a key element for enhancing the performance of modern OTH radar systems. Given the strong resurgence of interest in OTH surveillance systems from the international radar community, across the defense, commercial, and academic sectors, it is timely and appropriate to report the significant advances in OTH radar that have transpired over the last 20 years in a second book completely dedicated to this subject.

The primary objective of this book is to provide an up-to-date coverage of OTH radar systems, with the main contribution being a detailed description of the signal-processing models and techniques that have significantly advanced OTH radar state-of-the-art but are yet to receive in-depth attention in existing texts. The book aims to bridge this gap within a self-contained treatment that also describes the fundamental principles of OTH radar design and operation at a broader level for readers with no prior background in this area. It also strives to bring together a large amount of previously disconnected public domain literature on OTH radar and adaptive signal processing within a unified framework, drawing on a comprehensive citation list to clarify the relationship links between the numerous theoretical and experimental works published in these fields.

A distinguishing feature of this book is the prolific inclusion of experimental results to illustrate the practical application of processing techniques to data collected in the field by skywave and surface-wave OTH radar systems. It is anticipated that this aspect will benefit scientists and engineers wishing to gain a more detailed understanding of this technology, as well as radar practitioners and researchers with an interest in developing robust signal-processing algorithms for real-world systems. It is hoped that this book will also instill enthusiasm in younger scientists and engineers for OTH radar and the field of radar in general.

Acknowledgments

The opportunity to write and share this book with the international radar community is a great privilege and honor for the author, as it is to work alongside a remarkably talented team of scientists, engineers, and professionals at the Australian Defence Science and Technology Organization (DSTO). In particular, the author would like to express his gratitude to the Chief of the Intelligence, Surveillance and Reconnaissance Division (ISRD), Dr. Tony Lindsay, for granting permission to undertake this project. Dr. Gordon Frazer, Research Leader of HF Radar Branch, and Dr. Mike Turley, Head of Signal Processing and Propagation Group, are also gratefully acknowledged for supporting this text and the author over many years.

The former Chief of ISRD, Dr. Malcolm Golley, is warmly thanked for his faith in the author as a young engineer wishing to pursue a Ph.D. in the field of adaptive processing for OTH radar. Dr. Golley's support for this research activity and his inspirational leadership was instrumental to the author's professional development. Prof. Doug Gray, Prof. Yuri Abramovich, and Prof. Stuart Anderson are acknowledged with thanks for their guidance during these years. Indeed, much of the impetus behind Part II of the text has its foundations in research directions initially promoted by Prof. Abramovich.

The preparation of this book was actively encouraged by the author's colleague and dear friend Prof. Alfonso Farina. The long-standing and successful technical collaboration with Prof. Farina has been of great benefit to the author. The author is also grateful to Prof. Farina for hosting a defense science fellowship in Italy and for organizing the first tutorial on OTH radar as part of the IEEE Radar Conference series in Rome, 2008. This rare opportunity allowed the nucleus of the current text to be developed. Most important, the author acknowledges, with deep appreciation, the profound influence that Prof. Farina has had as his mentor in professional life.

This book has benefitted from material generated by many past and present ISRD staff members. The author is indebted to several colleagues within DSTO for providing expert peer-reviews of selected chapters, particularly the contributions received from Dr. Tony Lindsay, Dr. Mike Turley, Dr. Trevor Harris, Dr. Manuel Cervera, Mr. Mark Tyler, Dr. David Holdsworth, Dr. Justin Praschifka, and Dr. Lyndon Durbridge. A special acknowledgment is owed to Mr. Nick Spencer for proofreading some of the chapters. Mr. Brett Northey, Dr. David Netherway, and Dr. Andrew Heitmann are thanked for providing some of the environmental data. Ms. Kerry Barnes and Ms. Lee Hayes are also warmly thanked for their precious assistance in typing and checking some of the references.

The author is equally thankful for the constructive comments provided by a number of highly esteemed external reviewers. In particular, the author acknowledges the feedback kindly provided on individual chapters and technical areas by Dr. Ryan Riddolls (Defence Research and Development, Canada), Prof. Larry Marple (Georgia Tech Research Institute), Dr. David Emery (BAE Systems), Dr. Geoffrey San Antonio (Naval Research Laboratory), Dr. Fred Earl (National Systems), Dr. Ben Johnson (Lockheed Martin),

Dr. Jim Barnum (SRI International), Dr. Donald Barrick (Codar Ocean Sensors), and Dr. L. J. Nickisch (Northwest Research Associates).

The author and this text have benefitted significantly from interactions with colleagues in the wider radar and signal-processing communities. In particular, the author treasures the friendship and cooperation of Prof. Alex Gershman, Prof. Fulvio Gini, and Prof. Louis Scharf, which has led to the development of radar signal processing techniques described in this text. The author is also thankful to Prof. Hugh Griffiths, Dr. Braham Himed, Prof. Hermann Rohling, Dr. Bill Melvin, Prof. Don Sinnott, and Prof. Chris Baker for sharing their valuable knowledge and insights at international conferences and for their continued support.

The author kindly thanks the Editorial Director of McGraw-Hill Professional, Ms. Wendy Rinaldi, and all of the publishing team for the excellent manner in which this book has been managed and produced. Finally, the author wishes to express his loving gratitude to his wife, Lucrezia, for her unwavering support and patience, particularly at the time of starting her new life in Australia with newly born son, Leonardo. Without her help and encouragement, the text would not have been completed within the time frame needed to meet the publication schedule.

Abbreviations

2D	Two-dimensional
ACE	Adaptive coherence estimator
ACS	Auto-correlation sequence
ADC	Analog-to-digital converter
AF	Ambiguity function
AGW	Acoustic gravity wave
AIS	Automated identification system
AM	Amplitude-modulated
AMF	Adaptive matched filter
AR	Auto-regressive
APEX	Aperture extrapolation
ARD	Azimuth-range-Doppler
ARMA	Auto-regressive moving-average
ART	Analytical ray-tracing
ASD	Adaptive subspace detector
BANDEX	Bandwidth extrapolation
BMD	Ballistic missile defence
BSI	Blind system identification
BTL	Basic transmission loss
CA	Cell-averaging
CD	Coherently distributed
CH	Chain Home
CFAR	Constant false-alarm rate
CMD	Cross-modulation distortion
CME	Coronal mass ejection
CPI	Coherent processing interval
CR	Coordinate registration
CSF	Channel scattering function
CUT	Cell under test
CW	Continuous wave form
DARPA	Defense Advanced Research Projects Agency
DATEX	Data extrapolation
DDRx	Direct digital receiver
DFT	Discrete Fourier transform
DIR	Dwell illumination region
DOA	Direction of arrival
DOF	Degrees of freedom
DSTO	Defence Science and Technology Organization
EEZ	Exclusive Economic Zone
EUV	Extreme ultra-violet

FIR	Finite impulse response
FMCW	Frequency-modulated continuous waveform
FMS	Frequency management system
GAM	Generalized array manifold
GEMS	Generalized estimation of multipath signals
GFB	Go-fast boat
GLRT	Generalized likelihood-ratio test
GO	Greatest-of
GPS	Global positioning system
GSD	Generalized subspace detector
GWM	Generalized Watterson model
HF	High frequency
HFSW	High frequency surface-wave
HOS	Higher order statistics
HPF	High-pass filter
ID	Incoherently distributed
IDC	Ionospheric distortion correction
IF	Intermediate frequency
IID	Independent and identically distributed
IIR	Infinite impulse response
IMD	Intermodulation distortion
IMS	Integrated maritime surveillance
INE	Impulsive noise excision
I/Q	In-phase and quadrature
ITU	International Telecommunications Union
JFAS	Jindalee Facility Alice Springs
JORN	Jindalee Operational Radar Network
KRP	Known reference point
LCMV	Linearly constrained minimum variance
LFMCW	Linear frequency-modulated continuous waveform
LMS	Least mean square
LNA	Low-noise amplifier
LOS	Line-of-sight
LPDA	Log-periodic dipole array
LPF	low-pass filter
LRT	Likelihood-ratio test
MF	Matched filter
MIMO	Multiple-input multiple-output
ML	Maximum likelihood
MLE	Maximum likelihood estimate
MOM	Method-of-moments
MQP	Multi-segment quasi-parabolic
MSC	Magnitude squared coherence
MTI	Moving target indicator
MUF	Maximum useable frequency
MUSIC	Multiple signal classification
MVDR	Minimum variance distortionless response
NF	Noise figure

NHD	Non-homogeneity detector
NRL	Naval Research Laboratory
NRT	Numerical ray tracing
OI	Oblique incidence
OIS	Oblique incidence sounder
OS	Ordered statistics
OTH	Over the horizon
PBR	Passive bistatic radar
PCA	Polar cap absorption
PCD	Partially correlated distributed
PCL	Passive coherent location
PD	Probability of detection
PDF	Probability density function
PFA	Probability of false alarm
PRF	Pulse repetition frequency
PW	Pulse waveform
RAAF	Royal Australian Air Force
RAF	Royal Air Force
RCS	Radar cross section
RF	Radio frequency
RFI	Radio frequency interference
RMP	Recognized maritime picture
ROC	Receiver Operating Characteristic
ROTHR	Relocatable over-the-horizon radar
RTIM	Real-time ionospheric model
RX	Receiver
SAP	Spatial adaptive processing
SCM	Sample covariance matrix
SCR	Signal-to-clutter ratio
SCV	Sub-clutter visibility
SDR	Signal-to-disturbance ratio
SFDR	Spurious-free dynamic range
SHCR	Signal-to-hot clutter ratio
SIMO	Single-input multiple-output
SINR	Signal-to-interference plus noise ratio
SIRP	Spherically invariant random process
SMI	Sample matrix inverse
SNR	Signal-to-noise ratio
SO	Smallest-of
SOI	Signal of interest
SSL	Single site location
SSN	Sunspot number
STAP	Space-time adaptive processing
SVD	Singular value decomposition
SWF	Shortwave fadeout
TBP	Time-bandwidth product
TEC	Total electron content
TEP	Transequatorial propagation

TID	Traveling ionospheric disturbance
TkBD	Track-before-detect
TMF	Tapered matched filter
TSJ	Terrain-scattered jamming
TWERP	Twin-whip endfire receive pair
TX	Transmitter
UK	United Kingdom
ULA	Uniform linear array
UMP	Uniformly most powerful
US	United States
VI	Vertical incidence
VIS	Vertical incidence sounder
VSWR	Voltage standing wave ratio
WARF	Wide Aperture Research Facility
WFA	Wavefront analysis
WFG	Waveform generator
WRF	Waveform repetition frequency

CHAPTER 1

Introduction

Skywave over-the-horizon (OTH) radars operate in the high-frequency (HF) band (3–30 MHz) and exploit signal reflection from the ionosphere to detect and track airborne and surface targets at ranges an *order of magnitude* greater than is possible with conventional line-of-sight radars. More than half a century of international research and development in this area has resulted in the fielding of mature OTH radar systems capable of cost-effective early-warning surveillance over wide areas. In particular, the ability of OTH radar to persistently monitor remote geographical regions where microwave radar coverage is not feasible or convenient represents an important advantage of such systems.

The high performance achieved by state-of-the-art operational OTH radar systems is the outcome of a great deal of theoretical and experimental research in the areas of ionospheric propagation modeling, hardware system design, intelligent resource management, and digital signal processing. The knowledge gained and shared through joint programs of international collaboration has played a key role in the deployment of successful OTH radar systems worldwide.

OTH radars have evolved considerably in nearly all respects since the first systems emerged between the 1950s and 1970s, but the amount of progress that has transpired over the past four decades has been especially remarkable. This is mainly due to dramatic advances in key technology areas that have enabled the general capabilities of OTH radar systems to be enhanced significantly. For example, the availability of high-performance direct digital receivers coupled with the rapid improvement in computer capacity has allowed modern adaptive signal-processing techniques to increase OTH radar sensitivity in complex disturbance environments. Moreover, the improved performance resulting from the application of contemporary principles and techniques have notably augmented the operational value of OTH radar as a militarily useful surveillance tool.

An underlying reason for the long-standing interest in OTH radar technology is that the characteristics of such systems are not only unique but also complementary to those of many other surveillance sensors currently operating on maritime, airborne, and space-based platforms. The inherent advantages of OTH radar with respect to alternative surveillance systems are widely recognized to be of significant value for defense as well as civil applications, particularly in terms of the benefit that OTH radar can provide as an integrated element of a multi-layered sensor suite.

This chapter introduces the rudimentary principles of OTH radar in order to provide a preliminary appreciation of such systems. The main objective is to establish a "skeleton" framework connecting the most important features of OTH radar as a prelude to more detailed discussions that will follow in subsequent chapters. Specifically, the first section of this chapter starts with a brief description of the background and motivation that led to the development of OTH radar more than half a century ago. Two different types of systems, known as skywave and ground-wave OTH radar, are also distinguished in this section.[1] Although the term OTH radar refers to skywave systems throughout most of this book, the ground-wave OTH radar implementation, which operates by virtue of space-wave and surface-wave propagation, is also covered in a separate (self-contained) chapter.

The second section of this chapter describes the OTH radar concept of operation and the main challenges of operating in the HF environment. In addition, this section discusses the general characteristics of OTH radar systems and several practical applications (not limited to surveillance) that illustrate the various uses of OTH radar outputs for both military and civil purposes.

In the third section of this chapter, the nominal capabilities and limitations of an archetype OTH radar are discussed and contrasted with those of a typical microwave radar, which serves as a useful benchmark for comparisons. In particular, the radar equation is used as a vehicle for highlighting the different and often complementary properties of these two radar system designs. This theme is followed not only because the vast majority of radars are implemented at microwave frequencies, and radars synonymous with this frequency band represent the most popularly known systems, but also to clarify that OTH and conventional surveillance radars share many functions in common despite their substantial diversity in system architecture, operational characteristics, and nominal capabilities. The fourth and final section of this chapter summarizes the typical coverage, resolution, and accuracy that may be expected of a nominal OTH radar system.

Following the introduction, the core of the book is structured into three parts. The main topics covered in each part are summarized in a chapter-wise manner below to provide an overview of the book's layout at a glance. In essence, Part I describes a wide range of OTH radar principles and techniques at a fundamental level, Part II delves more deeply into the modeling of HF signals received by actual OTH radar systems based on experimental data analysis, while Part III is concerned with the application of advanced processing techniques to real and simulated OTH radar data. Although this organization is natural in the sense that later parts of the text build on earlier ones, the individual chapters are relatively self-contained and may be read in order of preference. The main body of the text is followed by the appendixes and bibliography.

- **Part I: Fundamental Principles** (Chapters 2–5) expands on the concepts presented in this chapter and explores a wide range of topics in more detail to provide a broad coverage of the essential OTH radar principles and techniques at a fundamental level for professionals with little or no prior background in this area. This commences with a basic description of the ionosphere and its properties as a propagation medium for HF radio waves in Chapter 2. The practical issues driving OTH radar system design and the key characteristics of the major subsystems are reviewed in Chapter 3. Chapter 4 discusses the different signal types received by OTH radar systems in the HF electromagnetic environment and explains the conventional signal and data-processing steps used for target detection, localization, and tracking. Chapter 5 is exclusively dedicated to the subject of HF surface-wave radar and explores a range of themes relevant to this specific OTH radar implementation.

[1]Ground-wave OTH radar is commonly referred to as HF surface-wave radar.

A brief summary of the development of OTH radar systems from a historical perspective is interwoven into the chapters, starting from the early work on radio-wave propagation via the ionosphere through to the implementation of modern skywave and surface-wave OTH radar systems in different countries around the world. This (non-exhaustive) background sheds light on some of the story of OTH radar, and serves to compare different systems in the context of actual programs, which lend further concreteness to the technical narrative.

- **Part II: Signal Description** (Chapters 6–9) deals with the mathematical modeling of narrowband HF signals received by OTH radar systems after ionospheric propagation in the dimensions of space and time. Specifically, Chapter 6 examines the ability of a deterministic signal description based on the wave-interference model to represent actual HF signals received over typical OTH radar coherent processing intervals in the order of a few seconds. For longer periods in the order of a few minutes, Chapter 7 investigates the ability of stationary statistical models to represent the space-time second-order statistics of the HF channel, which forms a basis for representing various signal types including clutter, interference, and target echoes. Different procedures for estimating the model parameters online using real data are presented in Chapter 8. Chapter 9 experimentally evaluates the validity of the multi-dimensional signal models and parameter estimation techniques for predicting adaptive beamformer performance in a practical interference rejection case study.

  As the understanding of HF signal characteristics is critical to the development of effective signal-processing techniques for OTH radar, this part of the text lays the foundation for the theoretical derivation and practical application of advanced processing methods discussed in Part III. While physical insights are often provided for the proposed signal models, it is emphasized that the main role of such models is to represent the characteristics of the digital data samples received by OTH radar systems from a signal-processing perspective.

- **Part III: Processing Techniques** (Chapters 10–13) leverages off the signal models developed in Part II to formulate robust adaptive signal-processing strategies appropriate for implementation in OTH radar systems. In particular, Chapter 10 considers the use of adaptive beamforming for the mitigation of interference with time-varying spatial structure over the OTH radar coherent processing interval. This chapter also describes the various factors that can potentially limit adaptive beamformer performance in real-world systems and some commonly used approaches to improve performance in practice. Chapter 11 motivates the use of space-time adaptive processing (STAP) in OTH radar and describes algorithms for mitigating multipath interference propagated via the ionosphere. Chapter 12 casts the target detection problem as a binary hypotheses test expressed directly in terms of the received data and derives adaptive algorithms based on the generalized likelihood-ratio test (GLRT) methodology. A variety of detection schemes exhibiting the desirable constant false-alarm rate (CFAR) property are described for different measurement models that account for useful signal uncertainty and the statistical heterogeneity of the disturbance (clutter and/or interference-plus-noise).

  The final chapter discusses a much less trodden research area as far as the radar field is traditionally concerned. Specifically, Chapter 13 deals with the problem of source and multipath separation for the case of uncooperative waveforms by means of blind signal processing (commonly applied in the communications field). It also explores the utility of such techniques for HF OTH passive coherent location (PCL) systems as well as in other applications. The key point of Chapter 13, and other chapters in this part of the text, is to illustrate the performance benefits that can be gained from using adaptive techniques in lieu of conventional processing by comparing results achieved on real and simulated data.

1.1 Background and Motivation

The "Chain Home" (CH) radar network installed on the east coast of Britain circa 1938 is widely regarded as the first military radar system to convincingly demonstrate its air-defense capabilities during wartime. The CH radars operated in the 20–30 MHz band because HF technology provided the only available means to generate sufficient power at the time. Unlike current skywave OTH radars, the CH system was designed to detect targets in the line of sight of the radar rather than at over-the-horizon ranges. In fact, echoes returned from very long distances by the ionosphere constituted "interference" for CH radar operators. The reader is referred to Neal (1985) for a comprehensive description of the CH operational radar system, which had a decisive influence during the Battle of Britain. Later during the second world war, radars exploiting frequencies in the UHF and microwave spectrum were successfully employed for line-of-sight surveillance applications.

1.1.1 Line-of-Sight Radars

The choice of frequency band has a profound influence on the physical characteristics and nominal capabilities of a radar system. The radar frequency bands extending from the HF to microwave spectrum are classified according to the International Telecommunications Union (ITU) letter designations in Table 1.1. The overwhelming majority of surveillance radar systems around the world are implemented at UHF or higher frequencies, since from a performance and cost perspective, these frequency bands are the most competitive for line-of-sight applications.

For example, airborne early-warning (AEW) radars operate in the UHF or L bands, multifunction radars used for surveillance and weapons fire-control on maritime platforms typically use the S and C bands, while X band and even higher frequencies may be exploited for missile defense and short-range missile guidance. In general, the use of higher frequencies provides greater resolution and accuracy along with the characteristic of physically smaller systems. On the other hand, radars operating at lower frequencies trade off these advantages for longer range capability and less susceptibility to weather clutter and shadowing effects.

It is reasonable to ask why surveillance radar systems for conventional line-of-sight applications were eventually implemented at UHF and higher frequencies in preference to the HF band when suitable technology became available. Figure 1.1 illustrates some of the compelling reasons for moving to higher frequencies, where the main technical advantages include: (1) greater useable bandwidths to provide fine range resolution in systems required to operate with a low fractional bandwidth, (2) the ability to form narrow (high-gain) beams using relatively small antenna apertures that can more readily

Band	Frequency	Wavelength	Example Radar Applications
HF	3–30 MHz	10–100 m	Over-the-horizon surveillance, ocean remote-sensing (skywave and ground-wave radars)
VHF	30–300 MHz	1–10 m	Very long-range air-surveillance, ground-penetrating radar, wind profilers
UHF	300–1000 MHz	0.3–1 m	Very long-range air-surveillance/airborne early-warning (AEW) radars (e.g., BMD)
L	1–2 GHz	15–30 cm	Long-range air-surveillance/AEW radars (maximum range coverage of ~500 km)
S	2–4 GHz	7.5–15 cm	Multifunction radar, terminal air traffic control (ATC) radar, marine radar
C	4–8 GHz	3.75–7.5 cm	Medium to short-range weapons fire-control, weather-mapping radar
X	8–12 GHz	2.5–3.75 cm	Airborne intercept and attack, missile defence radars, missile guidance
Ku	12–18 GHz	1.67–2.5 cm	Short-range seekers, maritime navigation radars (civil and military)
K	18–27 GHz	1.11–1.67 cm	Limited use (due to strong H_2O absorption)
Ka	27–40 GHz	0.75–1.11 cm	Very short-range seekers, airport surface movement detection radars

TABLE 1.1 ITU letter-band designations from the HF to microwave frequency region and some example civilian and defense radar applications. OTH radars operate in the HF band (3–30 MHz), while the majority of conventional (line-of-sight) surveillance radars operate at UHF and higher frequencies in the microwave spectrum. Note that radar systems are also implemented in lower and higher frequency bands than those listed in this table.

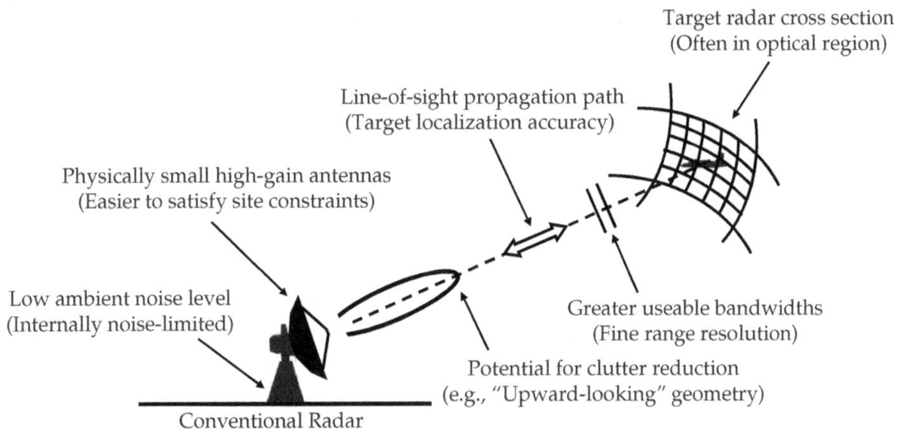

FIGURE 1.1 Summary of main (technical) reasons motivating the implementation of surveillance radars at UHF and microwave frequencies in preference to the HF band for line-of-sight applications when suitable technology became available.

satisfy site constraints, (3) target radar cross sections which lie in the optical region as opposed to the Rayleigh-resonance scattering regime, (4) considerably lower ambient noise levels that typically fall below the lowest achievable internal (thermal) noise floor limit of the receiver, (5) relatively stable and predictable (line-of-sight) propagation paths compared to ionospheric paths for more accurate target localization and tracking, and (6) potential for surface-clutter reduction at operational ranges by means of upward-looking radar geometries in certain practical applications. A comprehensive coverage of (ground-based and airborne) LOS surveillance radar systems can be found in several excellent texts, including the authoritative contributions of Skolnik (2008a), and Nathanson, Reilly, and Cohen (1999), among many others.

1.1.2 Coverage Limitations

Despite the significant advantages summarized in Figure 1.1, a fundamental limitation of conventional radar systems operating at microwave frequencies is that the range coverage is nominally restricted to regions where a line-of-sight (LOS) path exists between radar and target. In practice, the curvature of the Earth limits the range coverage of a microwave radar to ground distances that often do not extend much further than the geometrical horizon. In some cases, coverage may be limited to much shorter ranges when topographical features such as mountains effectively shadow targets from radar illumination. The performance of certain microwave radars may also be prone to meteorological effects, such as hail or rain, which has the potential to reduce target visibility due to weather clutter and increased signal attenuation.

The aforementioned limitations of conventional LOS radars are conceptually illustrated in Figure 1.2. Specifically, the top panel illustrates signal shadowing and attenuation effects, which may occur at relatively short ranges. On the other hand, the middle and bottom panels illustrate the obscuration of the LOS path from the radar to surface-vessels and aircraft targets due to the Earth's curvature for antennas mounted on ground-based and airborne radar platforms, respectively.

Under normal atmospheric conditions, microwave radar signals are bent down slightly toward the Earth due to refraction in the troposphere. This mechanism effectively extends the radio horizon beyond the geometrical horizon. For a ground-based radar operating at microwave frequencies, the increased distance of the radio horizon due to tropospheric refraction is often taken into account by assuming an effective Earth radius that is approximately four-thirds of the true value.[2] The nominal radio horizon limit may then be calculated as the geometrical horizon by simple trigonometry using this correction factor (Earth-radius multiplier value).

In unusual or anomalous atmospheric conditions, super-refraction of microwave frequency signals may at times extend the conventional radar coverage significantly beyond the nominal radio horizon. This can allow the detection of surface targets and low-altitude aircraft at ranges much greater than those expected under normal (globally averaged) atmospheric conditions. However, this propagation phenomenon, also known as "ducting," is neither frequent at certain locations/times nor controllable and thus

[2]The effective Earth radius model with multiplier $k = 4/3$ is appropriate for radar platform heights below approximately 1 km (where a linear variation of refractive index with height may be assumed) and radio frequencies above the HF band. The multiplier value tends to unity from above as the radar platform height increases above about 1 km and/or the radio frequency decreases below the VHF band.

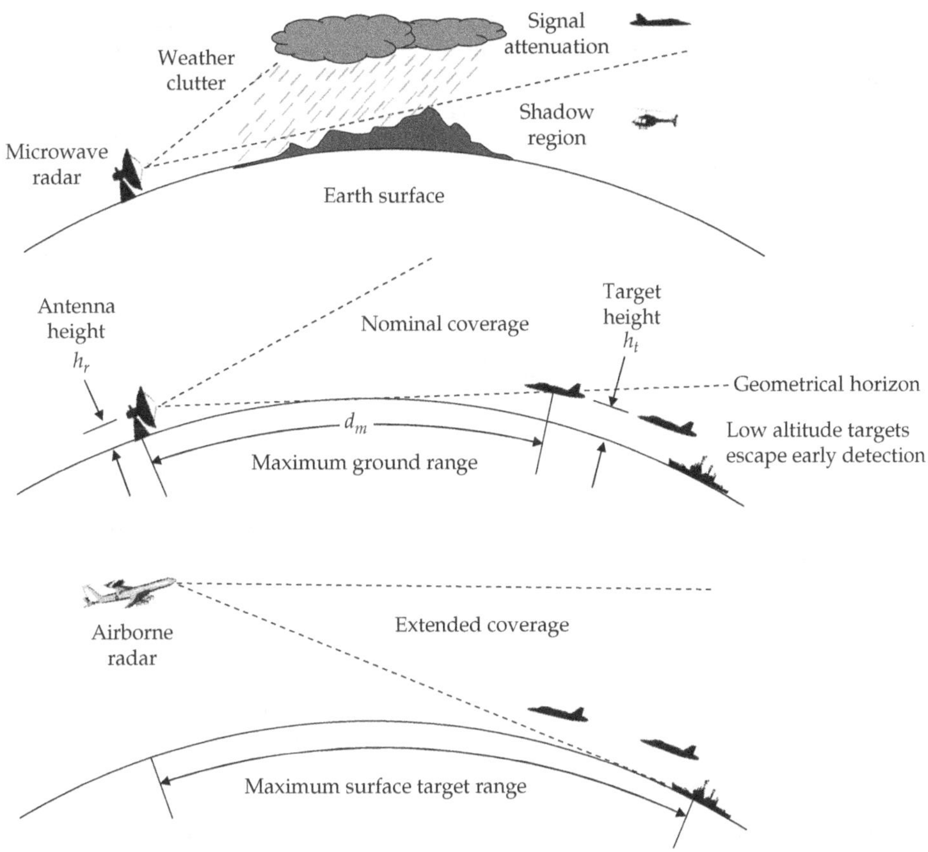

FIGURE 1.2 The middle and bottom panels notionally illustrate the nominal LOS range coverage limitation of a ground-based and airborne surveillance radar due to the Earth's curvature in terms of the maximum ground range to surface-vessels and aircraft targets over a smooth sphere. In practice, the coverage of conventional radar systems may be limited to much shorter ranges due to unfavorable meteorological conditions, which produce clutter returns and signal attenuation, and shadowing effects caused by topographical features such as mountains or valleys (top panel). © Commonwealth of Australia 2011.

cannot be relied upon to extend the range coverage of a conventional microwave radar beyond the nominal limitation in practice. Indeed, another type of anomalous propagation may decrease the effective range coverage with respect to that expected under normal atmospheric conditions by bending the signal away from the Earth in a process known as subrefraction.

With reference to the middle panel of Figure 1.2, the nominal range coverage limitation of a conventional LOS radar due to the Earth's curvature may be quantified in terms of the ground distance d_m. The radar platform is assumed to elevate the antenna to a height h_r above a smooth spherical Earth with an effective radius r_e that accounts for refraction under normal atmospheric conditions. The ground distance along the Earth's surface from a point vertically beneath the radar to the geometrical horizon is given by the simple arc-length formula $d = r_e \arccos\{r_e/(r_e + h_r)\}$. For a target at altitude h_t, the LOS path exists up to a maximum ground distance d_m given by Eqn. (1.1). At this ground range along the great-circle path, the radar and target are linked by a straight line that makes tangential contact with the Earth's surface. Any further increase in ground range causes the direct path to be lost due to the curvature of the Earth.

$$d_m = r_e[\arccos\{r_e/(r_e + h_r)\} + \arccos\{r_e/(r_e + h_t)\}] \tag{1.1}$$

As illustrated in the middle panel of Figure 1.2, a known limitation of LOS systems is that low-altitude targets can effectively escape early detection and may not become visible to a conventional microwave radar until they are potentially very close in range. As the heights h_r and h_t are (with few exceptions) much smaller than r_e, the arc-length distance is well approximated by the Pythagorean theorem $d_m \approx \sqrt{2h_r r_e + h_r^2} + \sqrt{2h_t r_e + h_t^2}$. Since $r_e \gg \max(h_r, h_t)$, this expression can in turn be approximated by the well-known maximum range formula for an LOS radar in Eqn. (1.2). This formula is appropriate for most practical surveillance radar scenarios, with space-based radar applications being a notable exception.

$$d_m \approx \sqrt{2h_r r_e} + \sqrt{2h_t r_e} \tag{1.2}$$

Figure 1.3a provides an indication of the maximum ground range d_m for an LOS radar system mounted on a ground-based or shipborne platform as a function of antenna height for relatively low-altitude targets. The curve labeled "geometrical horizon" is computed for the true Earth radius taken as $r = 6380$ km, while the curve labeled "radio horizon" assumes an effective Earth radius $r_e = kr$, where $k = 4/3$. The two remaining curves are computed for targets at heights of 20 and 80 m above the Earth's surface. For a radar antenna height below $h_r = 50$ m, surface targets ($h_t = 0$) at ground distances greater

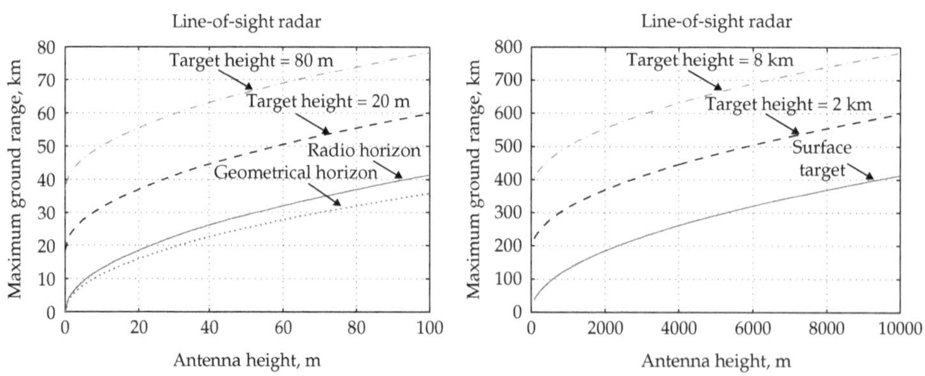

(a) Ground-based or shipborne radar.

(b) High-altitude airborne radar.

FIGURE 1.3 Nominal ground range coverage limitation of a line-of-sight radar system due to the Earth's curvature as a function of antenna height and target altitude.

than about $d_m = 30$ km fall outside the nominal range coverage of a conventional LOS radar. Ground-based or shipborne microwave radars are therefore susceptible to surface and low-altitude aircraft targets, which may exploit the Earth's curvature as a shield to avoid being illuminated by the radar signal.

Figure 1.3b provides an indication of the maximum nominal ground range of an airborne LOS radar coverage for surface and high-altitude targets at heights of 2 and 8 km. An airborne platform that elevates the radar antenna to a height of several kilometers can potentially extend the range coverage to perhaps a few hundred kilometers. Space-based sensors at heights of hundreds of kilometers may in principle increase the coverage to thousands of kilometers. However, the square-root dependence in Eqn. (1.2) implies that an order of magnitude increase in range coverage requires the antenna height to be raised by two orders of magnitude. Clearly, attempts to extend range coverage in this manner are met with an exponential rise in cost and complexity due to this inherent limitation of LOS technology.

1.1.3 Beyond the Horizon

The range coverage limitation of a conventional LOS radar becomes a problem when there is a requirement to continuously monitor remote area in regions near which it is not possible or convenient to permanently site an LOS surveillance system. This may include logistically inaccessible and sparsely populated regions where it is not viable to install infrastructure, or open areas of ocean far away from sovereign coastlines. Such situations can lead to undesirable surveillance gaps in space and/or time when attention is restricted to the sole use of LOS assets.

Mobile platforms with LOS surveillance capabilities can assist, but the deployment of such resources without real-time information to cue sensors to areas where suspicious activity is occurring can be ineffective and expensive. Territorially large nations such as Australia and the United States are required to regularly survey vast geographical areas and volumes of airspace in an efficient and reliable manner to provide a comprehensive real-time air and surface picture for military and civil uses. Surveillance solutions based exclusively on the deployment of a large number of LOS assets, each with a relatively circumscribed coverage area, may be financially prohibitive when very large and remote regions need to be routinely monitored.

Moreover, as the speed and stealth of threat technology increases in the modern era, surveillance systems capable of acquiring timely (early-warning) information on potential threats will remain of paramount importance to minimize the element of surprise. As far as the radar sensor is concerned, fulfilling the above-mentioned objectives in a cost-effective manner requires a fundamentally different technology that is not constrained by the LOS coverage limitation of conventional systems implemented at UHF or microwave frequency. Consequently, there has been a strong resurgence of interest in exploiting radar systems that operate at lower frequencies, more precisely in the HF band, where nature allows the line-of-sight restriction to be circumvented and the radar coverage to be extended tremendously.

An important property of the HF band (3–30 MHz) that is of great interest to the radar designer is the unique ability of HF signals to propagate over very long distances and illuminate the Earth's surface well beyond the horizon. As illustrated in Figure 1.4, OTH propagation of HF signals may occur via two different physical mechanisms known as skywave and surface-wave propagation. Skywave propagation refers to a mode in

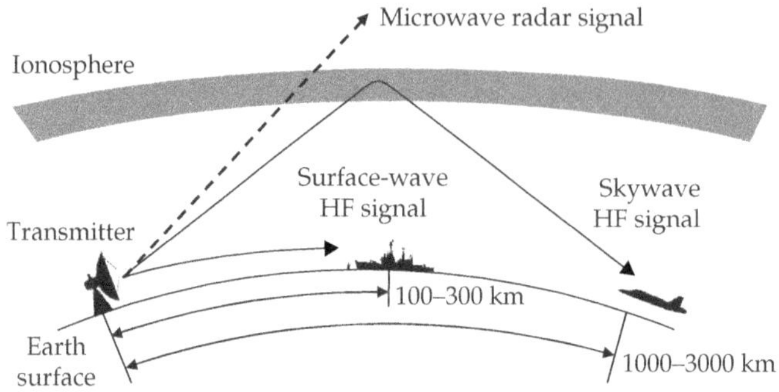

FIGURE 1.4 HF signals may propagate over-the-horizon by two physical mechanisms known as the skywave and surface-wave propagation modes, the former being essentially unique to the HF (3–30 MHz) band. The microwave radar signal has a much higher frequency which penetrates the ionosphere. © Commonwealth of Australia 2011.

which HF signals are "reflected" from regions of the upper atmosphere known as the ionosphere, where the concentration of free electrons within naturally formed ionized gas (plasma) reaches sufficiently high levels to significantly affect the propagation of HF radio waves. The ionosphere may reflect HF signals at one or more virtual heights between 100 and 600 km to illuminate the Earth's surface with significant power density at ground ranges of thousands of kilometers.

On the other hand, surface-wave propagation refers to a vertically polarized signal mode where a significant portion of HF radio-wave energy is able to propagate effectively over the conductive (saline) sea surface in a height region that extends down to the sea-air interface. In other words, this propagation mode effectively follows a height region immediately above the sea-air interface around the curvature of Earth to ranges of hundreds of kilometers. Surface-wave propagation should not be confused with anomalous propagation or ducting that occurs due to atmospheric refraction. The propagation mode in question depends on the electrical properties of the Earth's surface and would exist even in the absence of an atmosphere. The skywave mode is clearly of primary interest for very long-distance propagation. The surface-wave mode has interesting advantages for shorter but still OTH ranges that will be discussed in Chapter 5. Figure 1.4 also illustrates the "straight-line" propagation path of a microwave radar signal that penetrates the ionosphere due to its much higher frequency.

Below the HF band, medium frequency (MF) signals in the 0.3–3 MHz band experience high attenuation due to absorption in the lower altitude regions of the ionosphere, particularly over long-range paths during the day, while in even lower frequency bands (LF/VLF) there is barely sufficient bandwidth for more than a few voice channels. Above the HF band, the ionosphere rarely supports the reflection of VHF signals at frequencies much greater than about 30 MHz, although such signals may be diffusely scattered by irregularities in the ionosphere.

In essence, the lower and higher frequency bands adjacent to the HF band do not generally exhibit the same phenomenology that allows HF signals to propagate effectively well beyond the horizon for OTH radar purposes. Consequently, a considerable amount of attention in using the HF band for OTH radar emerged not long after the end of second world war. It was around this time that a combination of defense, commercial, and academic organizations in the United States commenced the early work on HF radar with the specific intent of achieving OTH target detection by virtue of skywave propagation.

By this stage skywave propagation over a one-way path had already been exploited for many years by short-wave communicators and for radio broadcasting. Indeed, the first long-distance "wireless" telegraphy signal was transmitted across the Atlantic ocean by Guglielmo Marconi on 12 December 1901. However, it was not until about 25 years later that the existence of the ionosphere and skywave propagation was experimentally proven. This was quite some time after the surface-wave mode was first (mistakenly) thought to be responsible for propagating Marconi's trans-Atlantic signals over a distance of about 3500 km from Poldhu on the Cornish coast of England to St. John's in Newfoundland, Canada.

From the surveillance radar perspective, the long distance propagation of HF signals via the skywave mode provides a means for the transmitter to illuminate surface or airborne targets well beyond the horizon and LOS, respectively. In contrast to one-way HF communication links or short-wave radio broadcasts, echoes of the radar signal scattered from the target then need to propagate back to a receiving system by the same physical mechanism to enable detection over a two-way propagation path.

The much longer range coverage as well as the immense surface areas and volumes of airspace that can be illuminated by a single HF radar system using skywave propagation compared to a microwave surveillance radar provided strong motivation to develop and assess the potential of OTH radar. The first ground-breaking experiments in skywave OTH radar began around the early 1950s at the US Naval Research Laboratory. In 1956, a definitive set of experiments conclusively demonstrated that such systems could succeed for long-range aircraft detection (Thomason 2003). The main threat that initially drove the design and procurement of operational skywave OTH radars for the US Navy was that of long- and medium-range bombers and missile carriers to the Battle Group at sea. As explained by Thomason (2003), a surveillance system that could detect targets over the entire theater of operations and provide timely information for tactical response was required to counter this threat.

After the pioneering achievements in the United States, a number of other countries including Australia, Russia (more precisely the former Soviet Union), China, and France developed skywave OTH radar systems for civil and military uses. In particular, the Australian HF radar program (Jindalee) commenced in the 1970s and has led to the development of three skywave OTH radars which today form part of the Jindalee Operational Radar Network (JORN). JORN is an operational system currently used by Australia's defense organization as a primary wide-area surveillance sensor.

1.2 OTH Radar Principles

Historically, skywave OTH radars have been developed for a variety of missions based on evolving system technologies and design perspectives, not to mention different fiscal constraints. For these reasons, past and present OTH radars exhibit a diverse range of characteristics. For a general introduction to the underlying principles regarding the operational concept, physical characteristics, and unique capabilities of such systems, it is convenient to restrict our attention to a nominal design which more or less typifies the essential features of several skywave OTH radars that are currently in use. With this purpose in mind, a skywave OTH radar implementation broadly exemplified by the most recent systems developed in the United States and Australia for the surveillance of air and surface targets serves as a suitable reference point for an initial description. Differences between a number of past and present skywave OTH radar system designs will be discussed in Chapter 3.

1.2.1 Operational Concept

The operational concept of a skywave OTH radar may be described in simple terms with reference to Figure 1.5. The transmit antenna radiates the HF radar signal as a directional beam that can be steered electronically in azimuth to search different sectors of the OTH radar coverage. To illuminate different range extents within the OTH radar coverage via skywave propagation, the transmit antenna typically has a broad vertical pattern that favours ray take-off angles from near-grazing elevation (~5 degrees) up to about 45 degrees. The emitted signal rays effectively propagate in straight lines in the lower atmosphere (ignoring tropospheric refraction) until they impinge at oblique angle on the bottom layer of the ionosphere, where the free electron density starts to be high enough

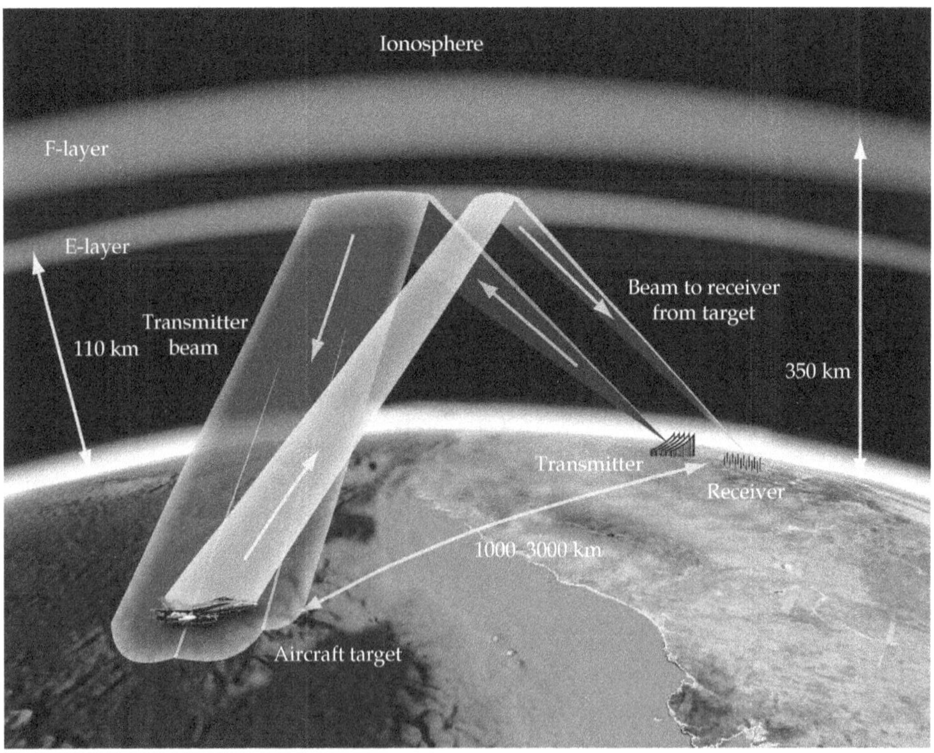

FIGURE 1.5 The principle of operation for a skywave OTH radar is inherently simple, but turning this concept into a reliable surveillance system creates many significant technical challenges. Note that the illustration shows E-layer propagation only, whereas the nominal range coverage of 1000–3000 km also accounts for F-layer propagation. © Commonwealth of Australia 2011.

to affect the propagation of HF radio waves. As the rays enter the ionosphere, they first encounter a free electron density that increases with height. Unlike microwave frequency signals, which continue to travel on a practically undeflected path and pass through the ionosphere, high-frequency signals are continuously refracted back toward the Earth due to the progressive change in free electron density, and hence radio refractive index, with altitude.

At one or more heights within the ionosphere, typically beyond 100 km and below 600 km, the electron density may be sufficiently high along the ray trajectory to cause total internal reflection. By this process, the reflected signal rays propagate downward on a refracted path and exit the ionosphere to illuminate a region of the Earth's surface and airspace above it well beyond the horizon. Indeed, the transmitted signals may be returned to Earth at ground distances of thousands of kilometers from the OTH radar. The actual curved path that the radar signal takes through the ionosphere and, hence, the range of the region which it illuminates on the Earth's surface depends on a number of critical factors, one of the most important being the signal frequency. Operators may therefore adjust the signal frequency and steer direction of the transmit antenna beam to selectively position the illuminated region in range and azimuth such that targets can be searched for over different geographical areas.

Any surface or airborne targets in the illuminated surveillance volume will intercept the incident signal power density and reradiate a portion of this intercepted power in all directions. A *minute fraction* of the power scattered by the target propagates back toward the highly sensitive OTH radar receiver via an ionospheric path similar to the one that illuminated the target. The receiver acquires the target skin-echo as a useful signal in addition to clutter signals, which include unwanted radar echoes backscattered from terrain or sea surfaces, ionized meteor trails, and ionospheric irregularities. Co-channel signals independent of the radar waveform that originate from other users of the HF band may also be received as interference at times. Finally, an OTH radar inevitably receives HF noise from atmospheric and galactic sources, which is more powerful than internal (thermal) noise and is collectively referred to as background noise.

In particular, the very powerful clutter returns completely obscure the much weaker target echoes, which cannot be distinguished directly from the raw data acquired by the OTH radar receive system. For this reason, signal processing is applied to the raw data such that very faint useful signals can be detected in a reliable manner. Among the various signal-processing steps, Doppler processing is essential to discriminate moving-target echoes from the competing clutter returns. Besides detecting the presence of targets within the surveillance volume, data processing is also applied to localize and track the position of targets over time. A geographic display of confirmed tracks that provide a human operator with a clear picture of air and surface target movements within the coverage area represents the final output of an OTH radar system.

The overall potential coverage of an OTH radar system nominally extends between 1000 and 3000 km in range, and depending on system design, it may span an arc of 60, 90, 180, or 360 degrees in azimuth. The range coverage of an OTH radar is an order of magnitude greater than that of a ground-based LOS system for the case of high-altitude targets within the troposphere. For the case of low-flying aircraft and surface vessels, the relative increase in detection range with respect to a ground-based conventional microwave radar actually approaches two orders of magnitude. This is because the "look-down" geometry of an OTH radar illuminates targets at all altitudes below the ionosphere.

An important property of HF signal propagation (diffraction effects coupled with the look-down geometry of a skywave path) is that it effectively eliminates (or vastly reduces) shadow regions in the coverage area, such that targets cannot exploit topographical features such as mountains or valleys to deliberately avoid illumination. In addition, the relatively long wavelengths used by OTH radars provides immunity to meteorological effects often experienced by microwave radar systems in the sense that signal attenuation and clutter due to precipitation is clearly not an issue in the HF band. Strictly speaking, OTH radar performance is not entirely independent of weather, as thunderstorm activity can significantly raise the noise level due to lightning discharges, and high sea-states can affect the detection of slow-moving targets against surface clutter.

The overall potential coverage area of a single OTH radar may be 6–12 million square kilometers. This vast area can be appreciated in Figure 1.6, which maps out the geographic coverage of the three Australian OTH radars forming the JORN network, located near Alice Springs (Northern Territory), Laverton (Western Australia), and Longreach (Queensland). A representative ground-based microwave radar coverage is shown for comparison in Figure 1.6. The shaded outer circle of the LOS coverage is for airborne targets at commercial-airline cruising altitudes, while the smaller inner circle is for surface and low-altitude targets. For reasons explained in a moment, an OTH radar cannot

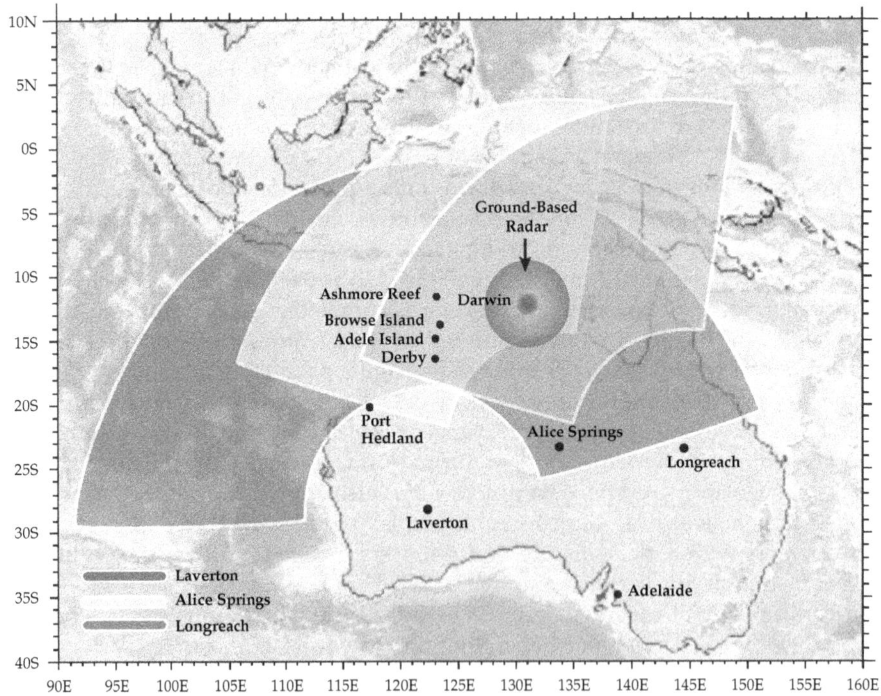

FIGURE 1.6 Geographic coverage of the Australian JORN OTH radar network with sites at Alice Springs (Northern Territory), Laverton (Western Australia), and Longreach (Queensland). The coverage of a ground-based microwave radar near Darwin is illustrated for comparison. The three JORN sites are controlled remotely from a coordination center near Adelaide, South Australia. © Commonwealth of Australia 2011.

monitor the entire potential coverage simultaneously. In general, a number of tasks (each composed of one or more surveillance regions) are concurrently scheduled on the radar time line to monitor a selected portion of the overall coverage for a designated period.

Depending on the type of mission and target class, which influence the demands placed on system resources, an operational OTH radar can detect and track targets in real time over areas ranging from a few tens of thousands to perhaps greater than one million square kilometers. Although the capital cost of an OTH radar is considerably higher than a conventional LOS radar, it provides outstandingly cost-effective surface and airspace surveillance per unit area and volume, respectively. Its chief advantage, however, resides in the capability to provide early-warning and wide-area surveillance over geographical areas where it is not possible or convenient to achieve persistent coverage through the deployment of microwave radar assets or other LOS sensors.

Figure 1.7 notionally depicts the overall coverage area of an OTH radar and a smaller surveillance region inside it known as the radar (or transmitter) footprint, which is illuminated simultaneously for a particular choice of signal carrier frequency and beam steer direction. The limitation of footprint size in range arises due to the finite range-depth of ionospheric propagation support at a fixed frequency, while the limitation in azimuth arises due to the confined width of the transmit antenna beam. The nominal range and cross-range dimensions of the OTH radar footprint will be quantified later. The main

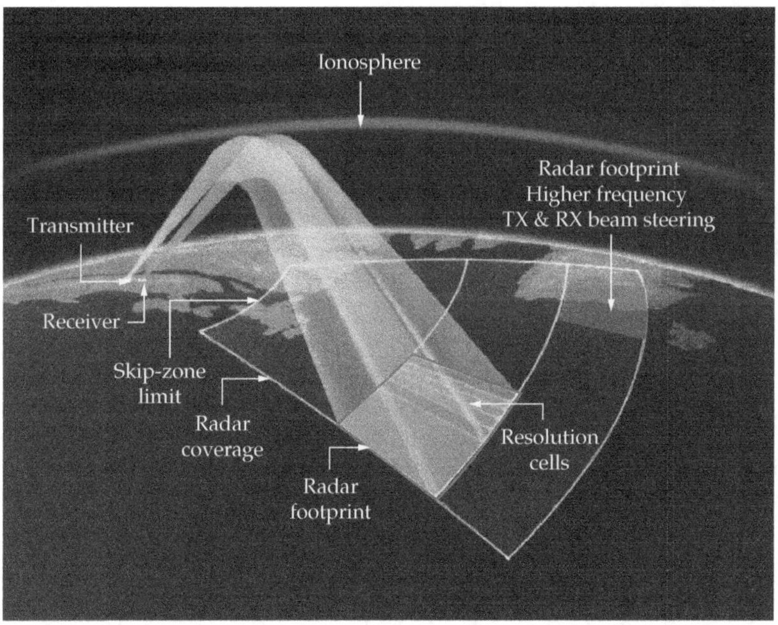

FIGURE 1.7 Notional diagram showing the overall coverage of a skywave OTH radar and the illumination of a smaller surveillance region within it known as the radar footprint. The concept is that the overall coverage cannot be illuminated simultaneously but is interrogated one surveillance region at a time by sequentially steering the radar footprint to different locations within the overall coverage. The radar dwells on each surveillance region during a coherent processing interval used for Doppler processing. © Commonwealth of Australia 2011.

point here is that it is necessary to search for targets in one surveillance region at a time by sequentially scanning the radar footprint in azimuth and range to illuminate different regions within the overall coverage.

Typically, an OTH radar concurrently monitors a number of active surveillance regions that cover the geographical areas of interest at a particular time. The radar dwells for a certain time to collect data on a particular surveillance region and then steps through to repeat the process for the other active surveillance regions in turn. In this way, an OTH radar revisits each of the active surveillance regions in a scheduled sequence. The size of each surveillance region is effectively bounded by the radar footprint dimensions. To emphasize that a surveillance region (or transmitter footprint) is simultaneously illuminated during a radar dwell, it is also commonly referred to as a dwell illumination region (DIR). The order in which the different DIRs are visited by the OTH radar is referred to as the scan sequence or scan policy. The area tiled by the set of DIRs scheduled concurrently on the OTH radar time-line may be interpreted as the total real-time coverage for target detection and tracking.

The number of DIRs which can be scheduled concurrently is limited by the coherent integration time deemed necessary for Doppler processing in each DIR and the revisit rate required on different DIRs for effective target tracking. As mentioned previously, these conflicting requirements almost invariably means that the operationally feasible real-time coverage of an OTH radar is less than the overall potential coverage. At this point, several questions arise. For instance, what factors determine the dimensions of the overall coverage, such as the minimum and maximum range, and what is the physical size of a DIR and the radar resolution cells within it? Other important questions are how many DIRs can be scheduled and how does the OTH radar scan the DIR over different regions of the coverage? Answers to these questions provide a preliminary indication of the nominal capabilities of OTH radar systems and will be discussed a little later in this chapter.

1.2.2 General Characteristics

Although the presence of the ionosphere allows OTH radars to detect and track targets at ranges an order of magnitude greater than conventional radar systems, this propagation medium is also the source of many uncertainties and difficulties in practice. Consequently, OTH radar performance is not solely determined by system design and operating parameters, but is also strongly influenced by the characteristics of the ionosphere as well as the prevailing HF electromagnetic (EM) environment. To continue our introduction, a few general properties about the ionosphere and HF signal environment are briefly described along with the main physical characteristics of typical OTH radar systems.

1.2.2.1 Propagation Medium

The naturally occurring ionized gas (plasma) in the ionosphere is created by radiation and energetic particles emitted from the sun. It is the sufficient concentration of free electrons formed in certain regions of the ionosphere that is responsible for returning HF signal energy to the Earth via the skywave propagation mode. The free electron concentration, more commonly referred to as electron density, can vary by over two orders of magnitude with height in the ionosphere. The ionization height profile also varies with time and geographic location. Despite there not being any direct radiation from the Sun at night, the ionosphere never completely disappears and always contains sufficient ionization to reflect HF signals for OTH radar operation.

Although the presence of the ionosphere may be regarded as certain, with gross characteristics that are often predictable to a certain extent, its detailed structure cannot be forecast accurately in practice. Besides this inherent unpredictability, the ionosphere also possesses a number of unique properties in its quality as a propagation medium for HF signals. Some of these differ markedly from the phenomenology considered in the microwave radar literature. A number of important characteristics of the ionosphere that influence HF signal propagation and hence OTH radar operation are identified below. More detailed descriptions of the ionosphere and its properties as a propagation medium for HF signals will be provided in Chapter 2.

In contrast to the LOS propagation channel at microwave frequencies, the ionosphere is a highly dynamic, spatially heterogeneous, and anisotropic propagation medium for HF signals with frequency-dependent characteristics that change significantly over the HF band. The operating frequency providing the most appropriate illumination for a particular surveillance region varies significantly with time of day (diurnal variation), month (seasonal variation), and over the 11-year solar cycle. Typically, the frequency needs to be adjusted every 10–30 minutes to track large-scale changes in the ionosphere, and perhaps more rapidly near the dawn and dusk terminators.

To provide an indication of the change in operating frequency, the optimum frequency to illuminate a particular surveillance region in the day may be twice that required at night. Moreover, the ratio of frequencies required to illuminate different ranges in the OTH radar coverage at a particular time may be 2 or 3 to 1. The operating frequency also needs to be modified for surveillance regions at the same range but in different directions, especially during the passage of a dawn or dusk terminator. The key point is that quality of skywave propagation for OTH radar target detection and tracking changes significantly as a function of frequency, time, and location.

In addition, the electron density in the ionosphere typically exhibits a horizontally stratified nature. This potentially allows an HF signal to be reflected from a number of physically distinct ionospheric regions or layers formed at different altitudes. In practice, the reflected HF signal is often composed of a superposition of multiple components or signal modes that propagate along different paths between two terminals on the Earth's surface. Four or more dominant modes with possibly quite different time delays, Doppler shifts, and directions of arrival can at times be present (and resolved by an OTH radar) for a single target. In general, multipath propagation conditions for HF signals reflected by the ionosphere are considerably more complex than the interference between a direct and surface-reflected path often considered in LOS radar applications.

The radio refractive index for HF signals in the ionosphere is anisotropic due to the influence of the Earth's magnetic field. This imposes a dependence of the radio refractive index on wave polarization and propagation direction. HF signal attenuation due to absorption may also be significant, particularly in a lower height region of the ionosphere known as the D-region (below about 90 km). Apart from this "normal" behavior, the ionosphere is also subject to traveling wave-like disturbances, geomagnetic storms, and a plethora of other phenomena which can dramatically effect its properties as a medium for radio wave propagation (Davies 1990). Some general characteristics of the HF skywave propagation channel are summarized in Table 1.2.

As alluded to earlier, effective OTH radar operation requires the signal frequency to be chosen in real time so as to optimize target detection performance in the surveillance

Characteristic	Skywave Propagation
Dynamic	Wide range of temporal scales: intra-CPI, mission lifetime, diurnal, seasonal, 11-year solar cycle
Heterogeneous	Small-scale irregularities, variability over radar region & coverage, magnetic latitude variability
Dispersive	Dispersive in frequency as well as in time-delay (group-range), Doppler, and ray angle-of-arrival
Anisotropic	Linear polarization splits into two elliptically polarized characteristic waves refracted differently
Multipath	Different layers in E- and F-regions give rise to multiple propagation modes over a skywave path
Absorption	D-layer absorption in the daytime may cause significant signal attenuation on long-range paths

TABLE 1.2 Some general characteristics of the HF skywave propagation channel that are relevant to OTH radar design and operation. These characteristics are discussed further in Chapter 2.

region. This requires an OTH radar to be *frequency agile* over wide portions of the HF spectrum. Carrier frequencies that can span two or more octaves may be needed for effective day/night operation over the entire OTH radar coverage. Since the properties of the ionosphere cannot be forecast to the high level of accuracy demanded by OTH radar systems, either using empirical databases or analytical models based upon them, a separate frequency management system (FMS) is required to provide the main OTH radar with real-time site-specific advice on propagation conditions. In addition to this distinguishing feature of OTH radar, spectral occupancy in the HF band needs to be continuously monitored to avoid interference among the large number of users.

The signal-to-noise ratio (SNR) criterion for frequency channel evaluation is critical for the detection of fast-moving targets that are often detected against noise rather than clutter after Doppler processing. Aircraft of fighter size or larger can be routinely detected by OTH radar. The ionospheric channel also imposes temporal modulations on the radar signal over time scales of fractions of a second that distort the amplitude and phase structure of the echo within the coherent processing interval. This phenomenon imposes Doppler shifts and spreads on the returned HF signal echoes. Doppler spectrum purity becomes the most important criterion for frequency channel evaluation in slow-moving target detection applications where useful signals need to be detected against clutter rather than noise. In favorable conditions, it is possible for an OTH radar to detect steel-hulled ships of ocean-going size.

1.2.2.2 System Characteristics

While the majority of conventional microwave radars are single-site systems that operate by using pulse waveforms in a truly monostatic configuration, many skywave OTH radar systems use continuous waveforms in a two-site architecture where the transmitter and receiver are separated to achieve isolation. The two-site configuration has its genesis in an early OTH radar known as the Wide Aperture Research Facility (WARF), which was built in central California by Stanford University staff and students in 1967 before it became a Stanford Research Institute (SRI) facility. This pioneering OTH radar arguably represents the first radar system to convincingly demonstrate continuous waveform operation based on a two-site architecture. Another legacy of the WARF, which is evident in many current OTH radars, is the use of a very wide aperture uniform linear

FIGURE 1.8 The Jindalee OTH radar transmit antenna. The two-band linear array consists of 16 (high band) and 8 (low band) uniformly spaced vertically polarized LPDA elements with an aperture of approximately 140 m. © Commonwealth of Australia 2011.

array of vertically polarized (twin-whip endfire-pair) antenna elements on receive and a relatively narrower aperture of vertically polarized log-periodic dipole array (LPDA) antenna elements on transmit (Barnum 1993). Figure 1.8 shows the Jindalee OTH radar transmit antenna located at Harts range, about 100 km north-east of Alice Springs in central Australia, while Figure 1.9 shows the Jindalee OTH radar receive antenna located at

FIGURE 1.9 The Jindalee OTH radar receive antenna consisting of a uniform linear array of 462 vertically polarized dual-fan antenna elements. The receive antenna has an aperture that is approximately 2.8-km long. © Commonwealth of Australia 2011.

Mt Everard, about 40 km north-west of Alice Springs. Although this is a bistatic OTH radar architecture, it is often referred to as a "quasi-monostatic" configuration because the distance between the transmit and receive sites (~100 km) is much smaller than the range to the radar coverage (1000–3000 km).

Microwave radar antennas typically have a major dimension in the order of a few meters, which makes it possible for such systems to be sited on various platforms. On the other hand, an OTH radar may have a receive antenna aperture that is 2–3 km long and a transmit antenna aperture that is 100–150 m long. OTH radars require very large areas of flat and open space at an electrically quiet location to site. A conventional microwave radar may be conveniently implemented as a single channel system with beam steering achieved by mechanical rotation of the entire antenna structure. On the other hand, OTH radars must operate using antenna systems based on multi-channel arrays that permit electronic beam steering on transmit and receive.

Besides these architectural differences, there is another important aspect that distinguishes OTH radar from its microwave counterpart. Specifically, an OTH radar system operates more effectively by using a dedicated network of geographically distributed auxiliary sensors to monitor the prevailing state of the ionosphere. This adjunct suite of sensors to the main radar, referred to previously as the FMS, also monitors channel occupancy so that frequency channels which are free of strong radio frequency interference (RFI) from other users can be identified as potential candidates for operation. The FMS is not only useful for providing the main radar with automated advice on optimum frequency selection for a particular task or mission, but also facilitates the conversion of target tracks formed in radar coordinates of angle-of-arrival and group-range to geographic position (latitude and longitude) in a process called *coordinate registration*.

Although an OTH radar needs to operate over most of the HF spectrum in response to changing propagation conditions, the radiated waveforms have a relatively small bandwidth and occupy very narrow bands of the spectrum at any particular time. Many two-site OTH radar systems transmit a repetitive linear frequency modulated continuous waveform (FMCW), or variant thereof, which has desirable ambiguity function properties while at the same time minimizing the level of out-of-band emissions. The radar signal illuminates the surveillance region over a coherent processing interval (CPI) or dwell time during which a number of linear FM pulses or sweeps are emitted and echoes received by the system. Further descriptions of the main radar subsystems, linear FMCW properties, and the FMS will be provided in Chapter 3.

1.2.2.3 HF Environment and Signal Processing

The flow chart in Figure 1.10 shows the different types of signals that may be received by an OTH radar. The left and right branches categorize signals as being either coherent or incoherent with the radar waveform, respectively. Coherent signals originate from the transmitted radar signal and may be further classified as clutter returns or useful signals (target echoes). In OTH radar, clutter returns arise due to backscatter from spatially extended regions of the Earth's surface, or backscatter from other passive sources that do not correspond to targets, such as echoes reflected from the transient ionization produced by meteor trails.

Incoherent signals are present irrespective of whether the radar transmitter is turned on or off and may be further divided into man-made interference and naturally occurring background noise. Naturally occurring background noise may either be of galactic origin (e.g., sun and other stars) or atmospheric origin (e.g., lightning discharges). On

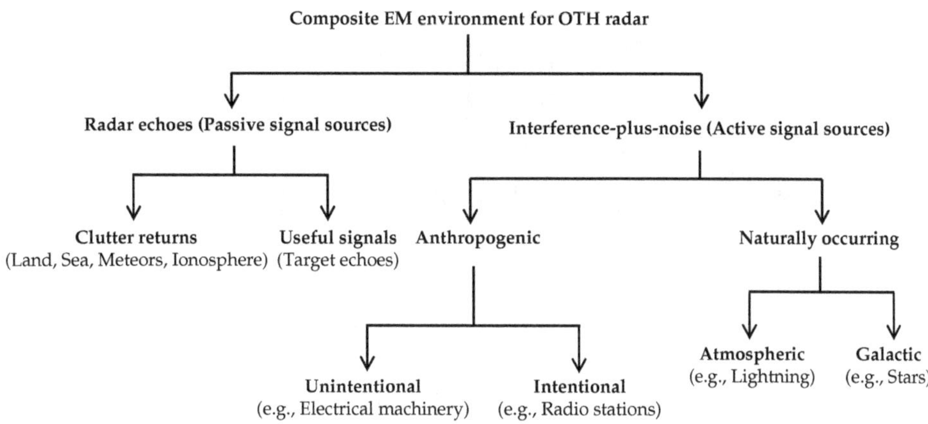

FIGURE 1.10 Composite HF signal environment for OTH radar. Relatively weak target echoes need to be detected against an additive mixture of clutter, interference, and noise, which collectively form the overall disturbance signal. © Commonwealth of Australia 2011.

the other hand, man-made interference may be regarded as unintentional (e.g., electrical machinery) or intentional (e.g., AM radio stations). OTH radars are in general permitted to use broad bands of the HF spectrum on a secondary basis. In exchange, such systems employ a policy of non-interference by selecting unoccupied frequency channels that are clear of other users in accordance with national standards and International Telecommunications Union (ITU) regulations. The properties of interference and noise received by an OTH radar are not only determined by the nature and geographic disposition of co-channel sources, but also the characteristics of the propagation medium (ionospheric circuits) that link the various interference and noise sources to the OTH radar receiver at the selected operating frequency. For the composite disturbance signal containing clutter, interference and noise, it is the residual level and distribution of unwanted signal energy over the target (azimuth-range-Doppler) search space at the *signal-processing output* that determines detection performance (i.e., after techniques to cancel unwanted signals have been applied) as opposed to the power of the various disturbance signals at the receiver input.

Effective mitigation of disturbance signals requires judicious signal processing and real-time frequency management to be considered jointly rather than separately in an OTH radar system. Traditionally, digital samples acquired by the receiver array during the radar dwell are range processed, beamformed, and Doppler processed. Ameliorative signal conditioning steps are also routinely performed to deal with transient disturbance signals arising from lightning discharges or meteor echoes. These rudimentary signal-processing steps will be described in Chapter 4, along with constant false-alarm rate processing, peak detection-estimation, tracking, and coordinate registration. Adaptive processing techniques will be discussed in Part III.

The importance of signal processing to the success of OTH radar may be illustrated using a simple example. In contrast to ground-based microwave radars, which may in some applications avoid strong clutter at operational ranges by means of upward-looking geometries, OTH radar target echoes must be detected in the presence of powerful clutter backscattered from large areas of the Earth's surface that are simultaneously illuminated at all ranges of interest. Although most HF signal energy tends to be scattered in the

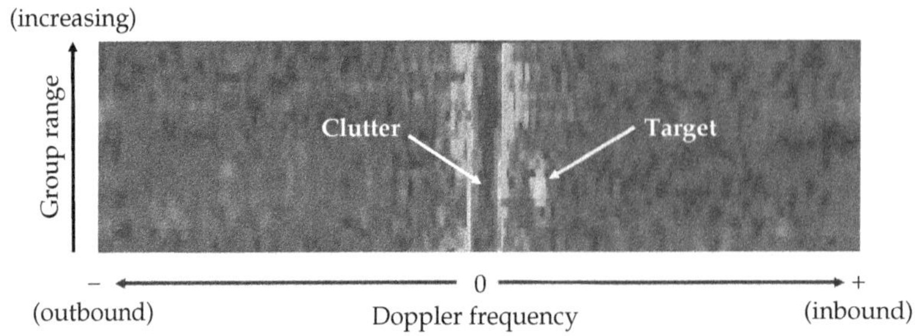

FIGURE 1.11 Intensity-modulated range-Doppler display at the output of a HF OTH radar beam steered in the direction of a real aircraft target. Doppler processing effectively separates the relatively faint echo reflected by the moving target (indicated in the display) from the surface-clutter echoes, which are concentrated near zero-Hertz Doppler frequency. This allows the inbound aircraft to be readily detected against background noise rather than the much more powerful clutter, which would have masked the target echo. Note that clutter backscattered from extended regions of the Earth's surface is present in all ranges (as indicated by the vertical "ridge" of finite width centred near zero-Hertz Doppler frequency), while the aircraft target is observed as a point scatterer and gives rise to an echo that is localized in range (i.e., along the vertical axis in the display). © Commonwealth of Australia 2011.

forward direction, the relatively large size of OTH radar resolution cells results in clutter returns that are perhaps 40 to 80 dB stronger than target echoes at the receiver. As the resolution cell size cannot be reduced beyond certain practical limits due to restrictions on the maximum bandwidth-aperture product of the system, Doppler processing is essential for target detection in OTH radar.

Fortunately, skywave HF signal paths are often sufficiently frequency-stable over time intervals in the order of seconds that the clutter energy returned from the Earth's surface is mostly confined to a small band of Doppler frequencies typically centered near zero Hertz and less than a few Hertz wide.[3] On the other hand, echoes from moving targets with a relative velocity v_r (defined as the negative of the echo path group-range rate) have a Doppler frequency shift given by the well-known expression $f_d = 2v_r f_c/c$, where f_c is the signal carrier frequency. Non-maneuvering aircraft targets typically produce echoes with a well-defined Doppler shift that often lies between 5 and 50 Hz, which is usually outside the clutter occupied region of the Doppler spectrum.

Importantly, the receiver needs to have sufficient dynamic range to faithfully preserve the spectral characteristics of strong clutter returns and much weaker target echoes. Providing this can be achieved, Doppler processing with an appropriately selected taper represents an effective means to isolate target echoes and clutter returns into different frequency bins. This ideally allows moving targets to be detected against a disturbance dominated by external background noise, which is normally higher in level than the internal receiver noise but much less powerful than the backscattered clutter. The ability of Doppler processing to reveal the presence of a target echo that would otherwise be submerged in clutter is illustrated by the real-data example in Figure 1.11.

[3]This is generally the case for HF signals reflected by the "quiet" mid-latitude ionosphere, while at low and high magnetic latitudes, the ionosphere is typically more disturbed and less frequency-stable.

1.2.3 Practical Applications

Skywave OTH radar systems are predominantly used for surveillance, where the primary mission is target detection and tracking. Under the umbrella of surveillance applications, the outputs of a skywave OTH radar may have defense and civil uses and different priorities may be accorded to broadly defined target classes such as aircraft and surface vessels. Besides surveillance missions, skywave OTH radars may also be engaged in remote-sensing applications, including oceanographic or ionospheric studies. In this case, useful signals are echoes from natural scatterers, while echoes from man-made objects (hard targets) represent clutter.

The following subsections of this chapter are mainly concerned with surveillance applications of OTH radar rather than remote sensing, mainly in the context of defense but also for civil purposes. Remote-sensing applications are briefly considered below, but will not be discussed in detail as this topic is rich enough to form the subject of an entire text in its own right. While surveillance and remote-sensing represent the two most common uses of skywave OTH radar, such technology may find other practical applications not limited to the ones discussed below.

1.2.3.1 Early-Warning Wide-Area Surveillance

OTH and microwave surveillance radars share many target types in common. Airborne targets of interest may include cruise or ballistic missiles, helicopters, private aircraft, fighter- and bomber-sized military aircraft as well as large commercial airliners. On the other hand, surface targets of interest may include land vehicles of varying size, small "go-fast" boats, ocean-going fishing boats, and large steel-hulled military ships such as patrol boats, cruisers, destroyers, and aircraft carriers. OTH and microwave radars also share similar surveillance functions, namely target detection, localization and tracking, while scanning the coverage to search for other targets, i.e., track-while-scan. An important distinction between the two systems is that OTH radars provide early-warning wide-area surveillance albeit at lower resolution and accuracy than conventional microwave radars.

The limited precision of an OTH radar generally means that such systems cannot be viewed as a complete solution to the airspace and maritime surveillance requirements of a nation. The value of OTH radar in defense applications needs to be measured in terms of the contribution it provides to national security as one element of a portfolio of surveillance assets that are managed as part of a comprehensive surveillance system. For example, the early-warning wide-area surveillance capability of an OTH radar may be utilized to cue higher precision surveillance and reconnaissance assets mounted on mobile platforms, such as patrol boats or airborne early-warning and control aircraft, to areas where abnormal activity has been detected. The ability to cue resources that have greater precision but more circumscribed coverage to targets of interest in real time by means of an integrated data distribution and command and control network leads to more efficient use of assets and possibly a reduction in the number of systems that need to be procured and maintained for an effective response to potential threats.

As mentioned before, the coverage area monitored in real-time by an OTH radar is typically organized into one or more tasks, where each task is in turn configured as one or more surveillance regions. The operating parameters of the OTH radar may be changed on a region-by-region basis. Operating parameters depend on the mission type and location of the surveillance region. Figure 1.12 illustrates a collection of representative tasks corresponding to four different mission types for a hypothetical OTH radar system. The different tasks are designated by the letters A to D. The radar dwells on (illuminates)

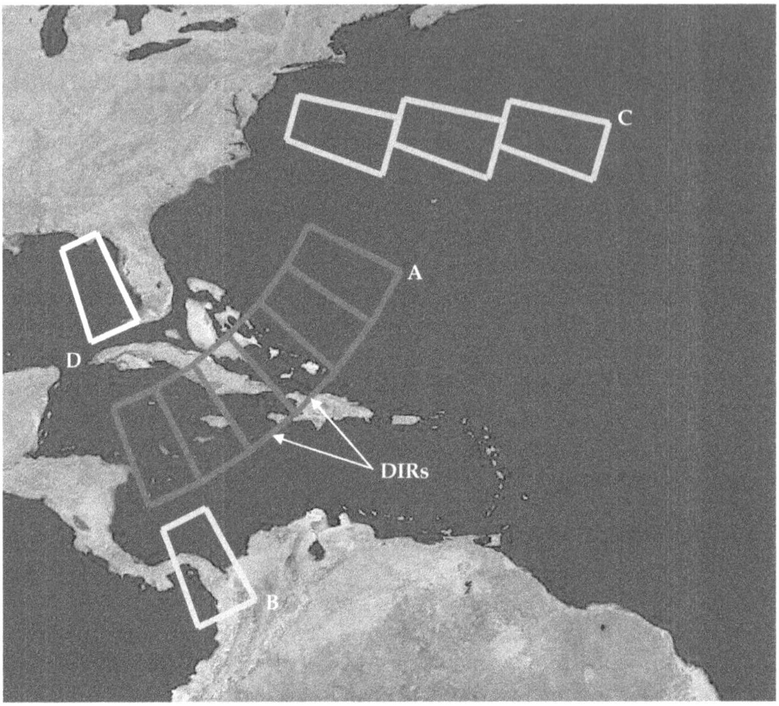

FIGURE 1.12 Examples of four OTH radar tasks labeled A to D. A particular task may contain one or more surveillance regions or DIRs revisited in a scheduled sequence. Different tasks and regions are usually configured with different operating frequencies to optimize skywave propagation to the geographical area of interest. Different radar waveform parameters are also used depending on the mission type. Multiple tasks may be interleaved on the radar time-line, but aircraft- and ship-detection tasks are often best performed at different times as they can be incompatible from a resource scheduling perspective. © Commonwealth of Australia 2011.

each DIR for the coherent processing interval and then steps through the different DIRs in a scheduled sequence. Multiple tasks may be interleaved on the radar time-line with selectable priorities, but care must be taken to ensure that the region revisit rate remains appropriate for each DIR to avoid a degradation in tracking performance.

While OTH radar tasks may be broadly classified as either air or surface tasks, their construction and use may have different tactical significance. Task A is commonly referred to as a "Barrier Task." This task is mainly intended for aircraft surveillance and provides coverage over a wide angular sector to detect targets moving through a certain range "barrier." When no other tasks are active, Task B may be referred to as a "Stare Task" because it provides continuous coverage focused on a single surveillance region. This type of task may be used to closely monitor airports, strategic shipping lanes, or missile launch sites, for example.

Task C may be referred to as a "Force Protection" task. It can be used to provide air route surveillance or navy fleet protection over a specified path. Task D is a remote sensing task that may be used for sea-state and surface-wind mapping or cyclone tracking. As this task is often demanding of the radar resource, it is normally performed in isolation of surveillance tasks. Air and surface tasks are often incompatible and performed separately due to the long dwell times needed for the latter and the high revisit rates needed for the former. Task selection not only depends on compatibility from a resource scheduling perspective, but also on the suitability of the prevailing propagation conditions.

The air and surface picture generated from the track output of an OTH radar may be used in several ways. As stated in Cameron (1995), OTH radar outputs may be used directly to support defense forces, either by aggregating surveillance information over a period of time to provide a knowledge-base of the normal pattern of activity in the region of interest, or during military operations, where real-time track information provides a live regional air and surface picture to assist military commanders with the assignment of missions and the tactical deployment of defense assets to best advantage. It is important to note that target classification beyond very broad target categories is difficult for OTH radar, as is target altitude estimation, although progress in the latter has been made. The surveillance capabilities of OTH radar are not only valuable for providing timely warning of critical events, but the known capacity for long-distance surveillance is also recognized as a deterrent to the escalation of conflict (Cameron 1995).

OTH radar outputs may also find civil uses for maritime and airspace control outside the range of conventional LOS radars. For example, OTH radar data may assist law enforcement agencies in the counter-drug offensive by monitoring strategic air and sea routes known to be favored by traffickers. More specifically, small private aircraft used to courier illegal drugs across international borders may be detected and tracked by OTH radar to cue interception and seizure initiatives by the Coast Guard or other government authorities (Ciboci 1998). OTH radars may also be used to detect illegal fishing activities in coastal waters and protect offshore assets such as oil production platforms by alerting patrols to the presence of unidentified maritime targets. Surface track data may also assist immigration and customs with border protection in detecting vessels used for "people smuggling," particularly on approaches to sparsely populated coastlines not covered by other sensors. OTH radars may also be used to assist with search and rescue operations.

1.2.3.2 Remote Sensing

As opposed to the detection and tracking of man-made targets, OTH radar systems may also be used to monitor the environment over regions that are vast and otherwise difficult to access. Perhaps the two main remote-sensing applications where HF radar techniques have significantly contributed to establishing climatological databases are in the study of the ionosphere as well as the mapping of sea-state and associated surface winds. Further details regarding the interaction of HF radio waves with the ionosphere and the ocean surface will appear in Chapters 2 and 5, respectively. Although such interactions are described with OTH radar surveillance applications in mind, much of the information is also pertinent to remote sensing. This section briefly discusses the two main HF radar remote-sensing applications. References are provided for readers interested in delving further.

In the mid-1920s, methods which represent the forerunners of modern HF radar were employed by Sir Edward Appleton in the United Kingdom and by G. Breit and M. Tuve in the United States to prove the existence of the ionosphere. More specifically, Appleton and his collaborators developed the ionosonde in 1924 and commenced ground-based soundings using frequency-modulated continuous waveforms (FMCW) to conclusively demonstrate the existence of an electrically charged layer, which Appleton named the E-layer. Appleton discovered a second reflecting layer in the following year, which he called the F-layer. At about the same time, Breit and Tuve employed HF sounding equipment using pulse waveforms to resolve and study the properties of echoes reflected from different layers of the ionosphere. Nowadays, radio probing of the ionosphere based on the radar principle is routinely performed with modern vertical and oblique incidence sounders to analyze the height structure and morphology of ionization in the upper atmosphere over a very wide range of spatial and temporal scales.

Physics-based models describing radio-wave propagation through the ionosphere may be used to invert the received HF echo traces to estimate ionization density height profiles at different times and locations. The collection of information from worldwide networks of sounding stations have contributed to creating empirical databases and models of ionospheric behavior as a function of time and geography. Such instruments are also used to study the formation and movement of electron density irregularities and traveling ionospheric disturbances. The detailed properties of signal propagation, such as multipath time-dispersion, amplitude fading depths/rates, and Doppler frequency shifts/spreads have been analyzed to guide the design of HF systems relying on skywave propagation. Specific information on the use of radio probing techniques to remotely gather data about ionospheric structure and dynamics can be found in Davies (1990).

The basis for remote sea-state sensing with HF radar is that the detailed structure of the sea echo Doppler spectrum contains significant information about the characteristics of the ocean surface and the associated surface-wind fields. In particular, the relationship between the directional wave-height spectrum of the sea surface and the structure of the Doppler spectrum of the scattered echo, formulated under relatively mild assumptions by Barrick (1972a), provides a theoretical framework for estimating and mapping oceanographic parameters using HF radar. Based on Barrick's physical model of the scattered field from the ocean surface, the sea echo Doppler spectrum in a particular radar resolution cell may be interpreted to provide spatially referenced oceanographic data. The characteristics of the associated surface-wind field are estimated from those of the sea-surface by inference.

Skywave and surface-wave HF radars have been used to estimate the ocean directional wave-height spectrum and to map surface-winds in speed and direction. Surface-wave systems are generally regarded as more suitable for extracting oceanographic parameters than skywave systems. The latter are relatively more susceptible to ionospheric contamination phenomena, whereas the coverage of the former is restricted to relatively shorter ranges. For a skywave OTH radar, the quality of data is largely determined by the severity of multipath propagation and the spectral contamination imposed on the signals by the ionosphere. The skywave mode also requires a zero-Hertz Doppler reference (e.g., land echoes) to estimate ocean surface currents due to Doppler shifts imposed by the ionosphere. More information regarding the application of HF radar to remote sensing of the ocean surface can be found in Anderson (1986) and references therein.

1.3 HF Radar Equation

A commonly adopted form of the OTH radar equation has signal-to-noise ratio (SNR) as the argument and may be expressed in the form of Eqn. (1.3) for a monostatic (or quasi-monostatic) system, as described in Skolnik (2008b). Based on the definitions of the various terms shown in Figure 1.13, it is observed that the OTH radar equation has

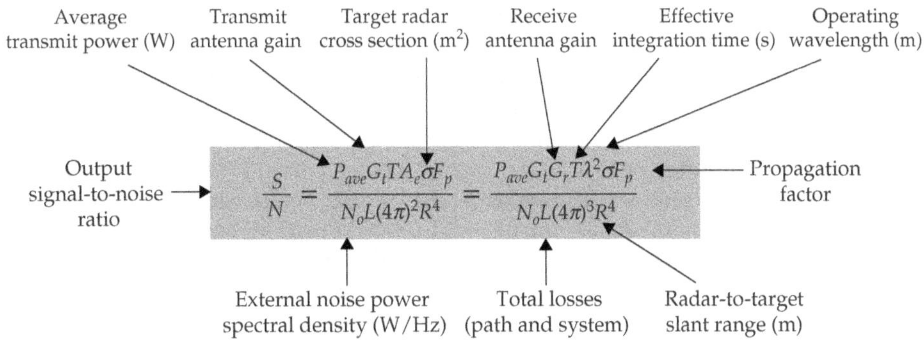

FIGURE 1.13 Definition of various terms in the noise-limited version of the OTH radar equation.

a similar *mathematical* form to the familiar (pulse-Doppler) microwave radar equation. However, significant differences between the two radar system types emerge when one delves into the detailed meaning and quantitative implications of the individual terms.

$$\frac{S}{N} = \frac{P_{ave} G_t G_r T \lambda^2 \sigma F_p}{N_o L (4\pi)^3 R^4} \tag{1.3}$$

This section discusses each term on the right-hand side of Eqn. (1.3) to further highlight the essential differences between OTH and microwave radar systems. Note that this (SNR) form of the radar equation is appropriate for providing an estimate of target detection performance in a *noise-limited* environment; that is, when the radial velocity of the target gives rise to a Doppler shift that is high enough to place the echo in a Doppler frequency region where noise dominates clutter.

The noise-limited version of the radar equation in Eqn. (1.3) is unsuitable for indicating detection performance in practical scenarios where the Doppler frequency of the target echo lies in a region of the spectrum where clutter dominates noise. In frequency-stable skywave propagation conditions, OTH radar clutter typically dominates noise over a relatively small band of Doppler frequencies (a few Hertz wide and centered near zero Hertz) such that ship and aircraft targets are usually detected against clutter and noise, respectively.

In unfavorable propagation conditions, dynamic irregularities in the ionosphere with speeds of hundreds of kilometers per hour can scatter strong spread-Doppler clutter that limits the detection of both slow- and fast-moving targets. In a clutter-limited environment, the signal-to-clutter ratio (SCR) becomes the primary measure of detection performance and an alternative form of the radar equation needs to be used. For convenience, we shall refer to Eqn. (1.3) in this chapter and discuss the SCR version of the HF radar equation in Chapter 5.

1.3.1 Slant Range

The term R in Eqn. (1.3) refers to the length of the one-way skywave signal path from the radar to the target via reflection from the ionosphere. This length is known as the *slant range* of the signal path, as opposed to the *ground range* measured along the Earth's surface. Apart from the difficulties of using the ionosphere as a propagation medium, serious design challenges are created by the sheer length of the OTH radar signal path. Specifically, Eqn. (1.3) shows that the SNR falls off with the *fourth* power of slant-range R and this results in extremely high spreading losses even when the radar coverage is restricted to one-hop ranges (i.e., those illuminated by a single ionospheric reflection).

This means that for an order of magnitude increase in path length R (compared to that of a conventional LOS radar), an OTH radar experiences a relative SNR loss of 40 dB due to spreading over a two-way path. To enable target detection over much longer path lengths, an OTH radar needs to compensate for this additional spreading loss. In other words, the 40-dB deficit in SNR relative to an LOS radar needs to be recouped by managing the OTH radar power budget with respect to other terms in the radar equation that are under the designer's control. As we shall see later in this section, it is such considerations which shape the physical and operational characteristics of an OTH radar system.

1.3.2 Transmit Power

The average power P_{ave} of an OTH radar transmit system may vary from about 10 kW up to 1 MW or higher. This large variation reflects the diversity of system designs, mission types, and performance requirements. A high transmit power improves SNR and hence radar sensitivity in a noise-limited environment. In practice, this mainly enhances detection performance for fast-moving (air) targets, particularly those with small RCS. An increase in P_{ave} is not expected to improve target detection performance in a clutter-limited environment. For slow-moving (surface) targets, P_{ave} may in principle be decreased without compromising detection performance until the point is reached where noise begins to dominate clutter over the target velocity search space.

Higher transmitting power is typically most beneficial for aircraft-detection missions, particularly at night when frequencies and target RCS are lower and atmospheric noise is higher. As slow-moving surface-vessels typically need to be detected against clutter, a higher transmit power will often not provide significant value (improve target detection performance) in ship-detection missions.

Operational OTH radars designed primarily for aircraft target detection and tracking typically transmit average power levels ranging from about 200 and 600 kW. Note that such values represent a 100-fold (20 dB) increase compared to the few kilowatts of average power delivered by a typical air traffic control (ATC) radar. According Eqn. (1.3), this provides a relative SNR gain of 20 dB for OTH radar, which (on its own) is not sufficient to fully offset the 40-dB relative loss in SNR due to spreading over a two-way path. Nevertheless, the much larger transmit power contributes significantly to reducing the 40-dB SNR deficit.

A coherent train of identical waveform pulses with amplitude-only, frequency or phase modulation may be used to operate OTH radars with co-located transmit and receive systems. While the monostatic radar configuration provides a number of operational and economic advantages relative to bistatic architectures, the need to use pulse waveforms with low duty-cycle reduces the average power that can be radiated by a practical transmitter with finite *peak*-power rating (handling capacity). Unit duty-cycle or continuous waveforms have therefore been preferred in several OTH radar designs to improve system sensitivity against noise. Besides making optimal use of the transmit resource from the SNR perspective, other advantages of continuous waveforms related to the control of out-of-band emissions will be described in Chapter 4.

Unit duty-cycle waveforms generally require the transmit and receive systems to be physically separated for effective operation. The choice to increase SNR through the use of such waveforms has in turn led to bistatic (or quasi-monostatic) OTH radar architectures. This design aspect distinguishes many OTH radars from the large majority of microwave radars. An inter-site separation of about 100 km over dry land typically provides sufficient isolation between the OTH radar transmitter and receiver due to the strong attenuation of the direct-path (ground-wave) signal. This configuration improves sensitivity against noise but it increases cost and precludes strictly reciprocal two-way propagation. For the quasi-monostatic geometry, the outbound and inbound propagation paths have closely spaced ionospheric control points and may often be considered nearly reciprocal due to the small bistatic angles involved (typically less than 5 degrees).

1.3.3 Antenna Gains

The OTH radar (transmit or receive) antenna gain is determined by the product of element gain and array gain. The array-gain component is related to the main lobe width of the beam and hence the spatial (angular) resolution of the antenna. A more detailed discussion of OTH radar antenna characteristics is provided in Chapter 3. This section introduces some basic properties of the OTH radar antennas in the context of Eqn. (1.3), noting that different antennas are used for transmit and receive in a bistatic system.

To achieve high gain and resolution, comparable to a conventional radar antenna, OTH radars require very large apertures because HF signal wavelengths are three orders of magnitude longer than those at microwave frequencies. For example, OTH radar receive antenna apertures may be up to 3 kilometers long, which is around 1000 times longer than the major dimension of a typical ATC microwave radar antenna. For practical reasons, OTH radar antennas are necessarily implemented as fixed multi-channel arrays of wire-type elements that enable beam steering to be performed electronically on both transmit and receive.

Traditionally, the OTH radar receive antenna array has a very wide aperture which is used to electronically (simultaneously) form a set of "finger" beams with high array-gain and spatial resolution to enhance target detection performance in noise- and clutter-limited environments, respectively, as well as to improve target localization accuracy for tracking. Depending on the operating frequency, the OTH radar receive antenna beam may have a gain of $G_r = 25$–35 dB and a (half-power) main lobe width of 0.2–2 degrees over the HF band. The transmit antenna typically has a lower gain of perhaps $G_t =$ 15–25 dB and a broader beam with a main lobe width of approximately 10–20 degrees. The relatively lower gain of the transmit antenna trades off SNR for an increase in coverage or coverage rate, since a broader beam allows a wider area to be simultaneously illuminated by the OTH radar using a single frequency (dwell).

OTH radars employing a linear antenna aperture for transmit and receive typically provide a coverage arc up to 90 degrees in azimuth. This may be extended to 180 or 360 degrees by using multiple ULAs with rotated boresight directions, or two-dimensional array geometries with antenna elements deployed on the ground (e.g., L-shaped or Y-shaped arrays). When no provision is made to independently steer the beam in elevation, the vertical radiation pattern of the transmit and receive antenna needs to be relatively broad. Vertical patterns providing high gain for elevation angles near grazing incidence up to about 40 degrees are typically required to ensure adequate range coverage as ionospheric conditions change.

HF antenna radiation patterns are strongly influenced by the electrical properties of the underlying ground. Ground mesh screens are usually installed under and in front of both the transmit and receive arrays to stabilize surface impedance and improve low elevation angle gain. For antenna systems capable of independent beam steering in elevation, the vertical pattern may be formed electronically to focus the radiated power more effectively into the ray "take-off" angles that illuminate the surveillance region. This has the potential to improve radar sensitivity when the prevailing ionospheric conditions are such that the surveillance region is illuminated by a relatively narrow band of elevation angles.

1.3.4 Target RCS

The major dimension of man-made targets such as aircraft, missiles, maritime vessels, and terrain vehicles are commensurate with the wavelengths of HF signals, $\lambda = 10$–100 m. Consequently, target RCS behavior falls in the Rayleigh-resonance scattering regime for OTH radar. On the other hand, microwave frequency signals have wavelengths in the order of centimeters, which places the RCS behavior of the same targets in the optical region for conventional radar systems. The large disparity between signal frequencies leads to fundamentally different target RCS characteristics for OTH and microwave radars.

For example, the RCS of the smallest manned targets or unmanned targets of comparable dimension targets such as a 10-m-long missile, will lie in the Rayleigh region when the OTH radar operates near the low end of the HF band. In this region, target RCS falls sharply with decreasing frequency (approximately to the fourth power of frequency) and exhibits relatively little aspect dependence. In contrast, the RCS of such targets will demonstrate resonance-region characteristics when the radar operates near the high end of the HF band. In the resonance region, the target RCS may increase or decrease within a confined range as the frequency changes and also exhibits greater sensitivity to viewing geometry than in the Rayleigh region (but tends to be higher on average).

To provide quantitative indications, the angle-averaged scalar RCS of a fighter-sized aircraft at typical OTH radar operating frequencies may be in the range $\sigma = 10$–20 dBsm. Larger aircraft such as commercial airliners may have mean RCS values of perhaps $\sigma = 20$–30 dBsm, while large steel-hulled ocean-going vessels can have values of $\sigma = 30$–40 dBsm or higher. With the exception of the smallest man-made targets mentioned previously, RCS values for targets with scattering characteristics that fall in the resonance region at OTH radar frequencies may at times be greater than those encountered for the same target in the optical scattering regime at microwave frequencies (Skolnik, 2008b).

1.3.5 Integration Time

Doppler processing is performed over a radar dwell time T, which is also referred to as the coherent processing interval or CPI (equal to the waveform repetition period multiplied by the number of pulses integrated). The CPI of an OTH radar typically varies from $T = 1$–4 seconds for aircraft detection to $T = 10$–40 seconds or more for ship detection. These CPIs are in the order of 100 to 1000 times longer than those typically employed by microwave radar systems, where dwell times are usually measured in tens of milliseconds. With respect to a microwave radar, this provides an OTH radar with 20–30 dB of additional coherent gain against noise. When combined with a 20-dB relative increase in average transmit power, these two factors effectively compensate for the spreading loss deficit of 40 dB incurred by an OTH radar due to the order of magnitude increase in path-length.

It is worth remarking that OTH radars have the possibility of using such long CPIs without encountering target range- or azimuth-cell migration issues due to the coarse resolution of such systems in both dimensions. Long OTH radar dwell times also provide exceedingly fine Doppler frequency resolution. Fine Doppler resolution is needed to separate slow-moving target echoes from surface-clutter returns, particularly in ship-detection applications. For example, the lowest (surface-mode) Doppler frequency resolution of 0.1 Hz provides a radial velocity resolution of 1 m/s (3.6 km/h or 1.9 knots) at a carrier frequency of 15 MHz. The same velocity resolution can be achieved with 20 Hz of Doppler frequency resolution ($T = 50$ ms) at 3 GHz. Fine Doppler-frequency resolution is also important in OTH radar to resolve independent groups of multipath echoes that arise from different targets in the same or similar spatial resolution cell, particularly in aircraft-detection applications.

1.3.6 Total Losses

In Eqn. (1.3), the total losses term $L = L_p L_s$ combines the propagation losses L_p and system losses L_s. Skywave propagation losses result from two main processes known as deviative and non-deviative absorption when attention is restricted to one-hop paths (i.e., no ground reflections). The latter occurs predominantly in the D-region of the ionosphere between heights of about 60–90 km. In the D-region, the energy of an incident HF signal is not reflected or significantly deviated but partially absorbed due to collisions between electrons and neutral air molecules. Across the HF band, the amount of non-deviative absorption is inversely proportional to the square of operating frequency to a good approximation and is normally highest during the middle of the day when the ionization density in the D-region reaches its maximum. Under normal (mid-latitude D-region) conditions, a two-way path loss of perhaps $L_p = 3$–6 dB may be expected at typical OTH radar operating frequencies during the day for one-hop propagation.

At night, the D-region effectively disappears at mid-latitudes and non-deviative absorption often becomes negligible. On the other hand, deviative absorption occurs when the equivalent vertical frequency of a radio wave is close to the critical frequency of a reflecting layer. This type of absorption and the definition of these frequencies will be covered in the next chapter. A lower layer may also obscure the radar signal from reaching higher-altitude layers in the ionosphere and hence limit the signal power density that can propagate to longer ranges in a process known as "blanketing." This phenomenon will also be discussed in Chapter 2.

System losses in OTH radar have similar causes to those encountered in a microwave radar. This includes losses in the analog parts of the receiving system (e.g., the antenna and front ends) as well as those in the digital signal-processing stages (e.g., windows used to control spectral leakage). System losses can easily accumulate to $L_s = 9$–12 dB or higher, particularly when low-sidelobe windows are used for pulse compression, Doppler processing and array beamforming. This system loss estimate may be added to the day and night propagation loss estimates (excluding deviative absorption and blanketing), to yield rough but representative values for L between 9 and 18 dB.

1.3.7 Propagation Factor

The propagation factor F_p lumps together a collection of diverse propagation phenomena which will be discussed more thoroughly in Chapter 2. Briefly, this term incorporates the effects of: (1) polarization mismatch loss due to Faraday rotation (interference between ordinary and extraordinary magneto-ionic components), which changes the polarization-state of the signal in time over a two-way ionospheric path, (2) focussing or de-focussing of the low- and high-angle rays as well as the interference of unresolved multipath components due to propagation modes reflected from different ionospheric layers, and (3) fading caused by scatter from dynamic electron-density irregularities in the E and F regions of the ionosphere between altitudes of 100–600 km. In favorable conditions, the propagation factor can yield a signal power density enhancement of $F_p = 3$–6 dB.

1.3.8 Ambient Noise

In the OTH radar equation, the noise power spectral density N_o is dominated by the contribution of *external* sources. Whereas in microwave radar systems, N_o is typically dominated by internal (thermal) noise generated within the receive system. The power spectral density of external background noise due to natural sources, which is observed in regions of the HF spectrum not occupied by other users, often exceeds the internal receiver noise level of an OTH radar by margins of 10–30 dB. In remote locations, atmospheric sources tend to dominate background noise in the lower HF band, while galactic noise tends to dominate in the upper HF band. A fundamental difference between OTH and microwave radars is that the value of N_o (and hence SNR) varies significantly as function of time of day and season, as well as receiver location and frequency and azimuth/elevation for a directional beam antenna. Variations in azimuth and elevation may also be observed by the highly directional OTH radar receive antenna.

In remote areas, the background noise spectral density in an unoccupied frequency channel may be around −185 dBW/Hz in quiescent conditions during the day near the middle of the HF band. Unintentional man-made radiation from electrical machinery or out-of-band emissions from other users in the HF band inevitably contribute to the perceived background noise level. The proximity of the receive site to populated centers can raise the effective "background noise" spectral density by over 10 dB in rural areas and 20 dB in residential areas with respect to remote areas. This emphasizes the importance of siting the OTH radar receiver well away from cities and industrial areas.

It is important to note that the *external* noise spectral density term N_o in the radar equation is not completely independent of receive antenna gain G_r. Recall that the receive antenna gain be expressed as $G_r = G_e \times G_a$, where G_e is the gain pattern of an individual antenna element (assumed identical for all elements) and G_a is the array factor contribution due to the electronic beam steering process. As the element pattern is relatively broad, it provides effectively equal gain to signals and noise, which means that an increase in G_e (and hence G_r) is associated with a commensurate increase in N_o. On this basis, it has been argued that antenna elements closely matched to the operating frequency on receive do not provide any SNR benefit.

In the case of spatially white external noise, this view is clearly justified as the only component of receive antenna gain that can increase SNR in the radar equation is the array factor, which depends on the number of elements and their relative spacing in practice. However, if the external noise exhibits sufficient spatial structure, and the receive elements are well-matched to the operating frequency, the array factor can in principle be adapted to reduce spatially colored external noise incident from directions other than the useful signal direction closer to the internal noise floor. This subtle point can potentially have important implications for improving SNR via the use of better-matched antenna elements on receive.

1.3.9 Numerical Example

A first-order calculation of the output SNR resulting for a particular OTH radar scenario may be performed by inserting representative values for the various terms in Eqn (1.3). These numerical values are most conveniently substituted into the logarithmic form of the radar equation in Eqn. (1.4), where all terms are expressed in dB. The resulting SNR should not be interpreted as an accurate measure of performance for an actual OTH radar system; rather, this simple exercise merely serves to provide a rough indication of the potential feasibility for target detection in the assumed example.

$$\frac{S}{N}(\text{dB}) = \{P_{ave} + G_t + G_r + T + \lambda^2 + \sigma + F_p\} - \{10\log(4\pi)^3 + L + N_o + R^4\} \qquad (1.4)$$

A detailed SNR performance analysis requires sophisticated models for the radar system, propagation channel, target scattering, and noise environment. Such models need to simultaneously account for the joint statistical properties of the various terms in the OTH radar equation. For example, climatological models based on empirical databases may be used to generate reference ionospheric conditions and noise environmental statistics. This approach has been taken in Root and Headrick (1993) as well as in Headrick, Root, and Thomason (1995) to derive realistic and site-specific evaluations of OTH radar performance.

Consider an OTH radar system with average transmit power $P_{ave} = 200$ kW (53 dBW) operating at a carrier frequency of $f_c = 20$ MHz, where the transmit antenna gain is $G_t = 20$ dB and the receive antenna gain is $G_r = 30$ dB. The corresponding wavelength is $\lambda = 15$ m ($\lambda^2 = 24$ dBm^2). Now assume that a daytime ionosphere allows this frequency to illuminate a region containing a target at a slant-range of $R = 3000$ km ($R^4 = 259$ dBm^4). This corresponds to a ground distance of around 2900 km for a one-hop propagation path with a virtual reflection height of $h_v = 300$ km. Suppose there is a fighter-sized aircraft in the surveillance region that presents a polarization-averaged radar cross section of $\sigma = 15$ dBm^2 to the radar and that its radial velocity can be assumed constant over the CPI. The Doppler shift of the target echo is presumed to be sufficiently high for it to be well-resolved from surface-clutter using a CPI of $T = 1$ seconds ($T = 0$ dBs).

To be clear, the target echo is assumed to compete for detection in a disturbance environment dominated by external background noise after Doppler processing. The two-way propagation path is assumed to yield a focussing gain of $F_p = 3$ dB and a path loss of $L_p = 6$ dB due to absorption. Combining this path loss with a system loss of $L_s = 10$ dB results in a total loss of $L = 16$ dB. For a receiver in a remote location, the background noise power spectral density in a clear channel during the day near a frequency of 20 MHz may be approximated by a nominal value of $N_o = -185$ dBW/Hz. Substituting these numerical values into Eqn. (1.4) yields an output SNR of 22 dB, as computed in Eqn. (1.5). This "back of the envelope" calculation indicates the potential to detect the aircraft target using a threshold 15 dB above the noise-floor to maintain an acceptable false-alarm rate.

$$\frac{S}{N} = \{53 + 20 + 30 + 0 + 24 + 15 + 3\} - \{33 + 16 - 185 + 259\} = 22 \text{ dB} \qquad (1.5)$$

1.4 Nominal System Capabilities

The radar equation does not provide information about the minimum and maximum range of the OTH radar coverage, the dimensions of a simultaneously illuminated DIR, the size of individual resolution cells within a DIR, or the accuracy with which the system can geographically locate detected targets. Quantitative indications regarding these aspects are necessary for an overall appreciation of the nominal capabilities of an OTH radar system. This section lists the main factors which limit the range coverage and DIR footprint dimensions as well as the resolution and accuracy of an OTH radar system.

1.4.1 Minimum and Maximum Range

The ability of the ionosphere to reflect HF radio waves back to Earth depends on a number of factors, the main ones being the signal frequency and the angle at which the signal rays are incident on the ionosphere. Figure 1.14 conceptually illustrates these dependencies for one-hop propagation via a single ionospheric layer. In reality, the process of signal reflection from the ionosphere has many subtleties that are not represented by Figure 1.14, which only attempts to illustrate the basic concepts. A more detailed description of HF signal reflection from the ionosphere is provided in Chapter 2.

Figure 1.14a depicts three rays launched at different elevation angles for a signal of constant frequency. The rays launched at higher elevation angles are reflected over shorter ground ranges from points deeper within the ionosphere, where the electron density is relatively higher. Signal rays striking the ionosphere at angles of incidence greater than a certain (frequency-dependent) critical angle are insufficiently refracted toward the Earth and pass through the ionosphere as *escape* rays on a deflected path. The simple example shown in Figure 1.14a assumes the signal frequency f_1 is above the maximum frequency that the ionosphere can reflect at vertical incidence (i.e., the layer critical frequency).

For operation above the critical frequency, there exists a finite minimum ground range from the transmitter below which no signal rays can be returned to Earth by the regular process of ionospheric reflection. This minimum ground range corresponds to that of the

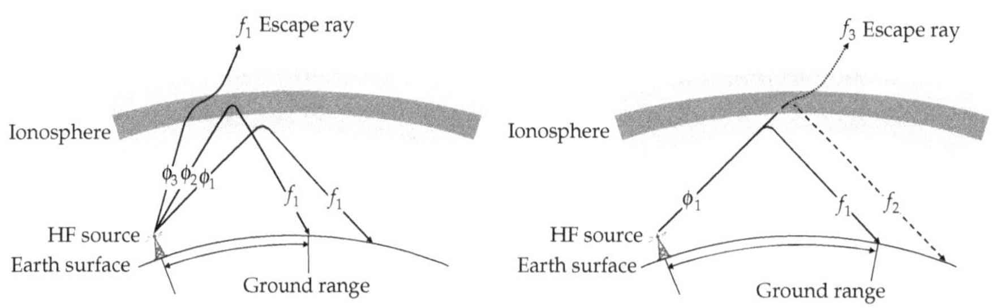

(a) Varying elevation (constant frequency). (b) Varying frequency (constant elevation).

FIGURE 1.14 The electron density of the conceptual ionospheric layer peaks in the middle region (dark gray) and tapers off at higher and lower altitudes (light gray). At a constant signal frequency (solid lines), rays with higher elevation angles are reflected from greater altitudes in the ionosphere, where the electron density is greater, and propagate to a shorter ground range until a critical angle is reached where the ray escapes the ionosphere. For a constant ray elevation angle, a signal of higher frequency is reflected from a greater altitude in the ionosphere, where the electron density is greater, and propagates to a longer ground range until penetration occurs (i.e., when the signal frequency is such that the ray elevation angle equals the critical angle).

ray with elevation angle ψ_2 in Figure 1.14a, known as the *skip-ray*. No illumination can be provided by virtue of skywave propagation at ranges less than this minimum ground range called the *skip-distance*. The area on the Earth's surface encompassed by the locus of points extending out from the transmitter to the skip-distance in all directions is known as the *skip-zone*. A skywave OTH radar cannot perform surveillance inside the skip-zone because the signal power density illuminating this region is too low for effective operation. Note that surface-wave propagation can potentially be used to illuminate a limited interval of ground ranges within the skip-zone.

The skip-distance effectively defines the minimum ground range of the OTH radar coverage. The skip-distance is a function of signal frequency and ionospheric conditions. In principle, it may be shortened by decreasing the signal frequency to provide coverage at closer ranges. However, it is preferable to avoid using frequencies near the low end of the HF band in practice, as a combination of factors can degrade OTH radar performance at these lower operating frequencies.[4] The signal power density returned to ground typically maximizes in a range region that immediately follows the skip-zone. When SNR is regarded as the most important performance metric, it is often highly desirable for an OTH radar to perform surveillance in a range region not far beyond the edge of the skip-zone.

A skip-zone of roughly 1000 km is commonly tolerated in most skywave OTH radar systems to achieve acceptable detection sensitivity and coordinate registration accuracy. This represents the nominal minimum ground range of the system and may be accounted for by setting back the OTH radar site from the closest point of the intended coverage. Such a constraint on site selection may pose serious problems for geographically small countries, unless the coverage of interest is well beyond sovereign borders. While the minimum ground range is related to the nominal lower frequency limit of an OTH radar that operates with continuous waveforms, its value is by no means fixed and may vary by hundreds of kilometers as propagation conditions change. This is in contrast to a conventional microwave radar, where the minimum range is often fixed by the pulse length of the radiated waveform, during which the receiver needs to remain closed.

As illustrated in Figure 1.15(b), an increase in signal frequency causes a ray of constant elevation to be reflected from a greater altitude in the ionosphere (where the electron density is greater) before being returned to Earth over a relatively longer ground range. Figure 1.14(b) also shows that there exists a maximum frequency beyond which a ray of constant elevation angle will not be reflected at oblique incidence, but rather pass through the ionosphere and propagate into space as an escape ray. Hence, the maximum ground range for single-hop skywave propagation is determined by the signal rays launched (with adequate gain) at the lowest take-off angles of the antenna beam and the virtual height in the ionosphere at which these rays are reflected at the maximum frequency before penetration occurs.

For example, a signal ray with a take-off angle of 5 degrees (with respect to the ground-plane of the transmit antenna) reflected from a virtual height of 300 km is returned to ground at a range of about 3000 km after a single (one-hop) reflection from the ionosphere. In practice, the variation in maximum ground range of a one-hop skywave path can be up to 1000 km or more as ionospheric conditions change. Provided that the transmit and receive antennas have adequate gain at low elevation angles of around 5 degrees, the maximum ground range of an OTH radar has a nominal value of around 3000 km.

[4]OTH radar operating frequencies much lower than about 7 MHz are often avoided when possible for several reasons to be explained in Chapter 3.

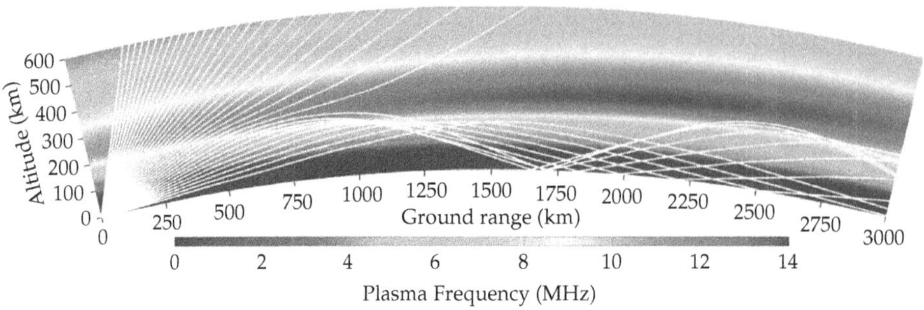

FIGURE 1.15 Ray-tracing through a model ionosphere showing the propagation paths taken by a fan of signal rays radiated at different elevations from the transmitter (generated using the DSTO developed PHaRLAP ray-tracing engine). The escape rays at high-elevation angles are clearly evident in the display, as is the skip-zone that propagates out to a ground range of about 1600 km in this case. The ray with lowest elevation angle is returned to a ground range slightly beyond 3000 km. The second-hop reflections, which propagate the signal well beyond ground ranges of 3000 km, are also illustrated. © Commonwealth of Australia 2011.

The nominal range coverage of a skywave OTH radar is generally considered to be roughly between 1000 and 3000 km. The maximum range limit can in principle be extended to perhaps 4000 km in favorable propagation conditions if the antennas have sufficiently high gain at very low (near grazing) elevation angles. On the other hand, OTH radar designs that allow for the use of relatively lower operating frequencies in the HF band can potentially shorten the minimum range limit to say 500 km. However, attempts to increase the one-hop range coverage of an OTH radar outside of the nominal minimum and maximum limits often comes at the expense of reduced performance.

The nominal OTH radar range coverage of 1000–3000 km corresponds to skywave propagation via a single ionospheric reflection near the path mid-point. The Earth's surface forward-scatters a significant portion of the one-hop energy. This forward scatter can be reflected another time by the ionosphere and be returned to ground at ranges of perhaps 6000 km or more via two-hop propagation. This process can clearly repeat itself to produce multi-hop propagation. Multi-hop propagation and other more exotic ionospheric modes to be described later can enable HF signals to propagate extremely long distances, including around the world. However, with each hop the signal is significantly attenuated by the two-way path spreading loss, the multiple passages through the ionosphere (particularly the sunlit D-region), and intermediate ground reflections, which also absorb signal energy. For these reasons, skywave OTH radars are generally expected to yield satisfactory performance over one-hop ionospheric paths between ground ranges of approximately 1000 and 3000 km.

1.4.2 Dwell Illumination Region

Figure 1.15 shows the skywave propagation paths taken by a fan of HF signal rays emitted at a constant frequency over a band of elevation angles that obliquely impinge upon a reference ionosphere with a modeled electron-density distribution. The propagation paths of the different signal rays were computed by numerical ray tracing using the DSTO-developed software toolbox called PHaRLAP (Cevera, 2010). Immediately following the skip-zone, the signal power density (illumination intensity) rises sharply and typically reaches its maximum value in this range region due to a phenomenon known as skip-focussing. Beyond the range of maximum illumination intensity, the signal power density striking the ground tends to gradually decay (monotonically decrease) as the range increases.

At longer ranges (well away from the edge of the skip-zone), the illumination intensity eventually becomes too weak for effective OTH radar operation; whereas inside the skip-zone, the signal power density returned by the regular process of ionospheric reflection almost vanishes completely. Hence, at a particular frequency, there exists a useful range interval in which the signal power density returned to the Earth's surface is sufficiently high for effective OTH radar operation. This interval defines the *range-depth* of a DIR in the OTH radar coverage. Recall that a DIR is also referred to as a surveillance region or transmitter footprint in the OTH radar nomenclature. Among other factors, the range-depth depends on ionospheric conditions, operating frequency, and the gain of transmit/receive antenna patterns in elevation angle. In practice, the range-depth typically varies between 500 and 1000 km, although the ionosphere may support much larger range-depths at times, particularly for large targets during daytime.

On the other hand, the cross-range dimension of the surveillance region is determined by the main lobe width of the transmit antenna beam in azimuth and clearly increases in proportion with range. In many existing systems, the main lobe of the OTH radar transmit beam is intentionally broadened to search a wider area while at the same time maintaining sufficient gain for target detection in noise. OTH radar transmit beams have nominal widths of about 10 degrees in azimuth depending on aperture size, operating frequency, and steer direction. This yields surveillance regions with a cross-range dimension of roughly 200–500 km over the 1000–3000 km coverage. To summarize, the OTH radar DIR may be approximately 500–1000 km long in ground range and 200–500 km wide in cross-range. The dimensions of the DIR define the area of the surveillance region over which targets can be simultaneously illuminated and potentially detected by an OTH radar using a single frequency.

The position of the surveillance region may be scanned across the OTH radar coverage by electronically steering the transmit beam in azimuth. OTH radars using a single linear antenna aperture for transmit (and receive) typically provide a 90 degree arc of coverage within $\pm 45^\circ$ of the array boresight. At the same time, the carrier frequency of the radar signal may be increased or decreased to shift the surveillance region further or closer, respectively, within the minimum and maximum range limits of the system. Based on the nominal dimensions of a DIR stated above, surveillance of the overall radar coverage in range (1000–3000 km) may require 2–4 transmitter footprints of 500–1000 km range-depth to be stacked with the DIR boundaries abutting in range[5]; whereas to cover a 90-degree azimuthal sector may require ten or more transmit beams steered one main-lobe width apart. Hence, the entire OTH radar coverage may require 40 or more DIRs to tile.

As the radar needs to dwell for the coherent integration time on each surveillance region to detect targets, and also needs to revisit these regions at a rate fast enough for effective tracking of possibly maneuvering targets, the radar can only be scanned over a limited number of different surveillance regions at a time. As a consequence, it is generally not possible for an OTH radar to concurrently perform surveillance over the entire coverage area without degrading the sensitivity of the system. OTH radars therefore schedule one or more tasks using a limited number of surveillance regions that are positioned over strategically selected areas. The active surveillance regions are

[5]In a practical system, the DIR range depth may at times be limited by real-time processing and operator display constraints rather than physical phenomena.

scheduled in an interleaved manner on the radar time-line and visited in turn using a scan policy determined by the region revisit rate requirements and the priority assigned to each task or mission.

1.4.3 Resolution and Accuracy

The OTH radar DIR is in turn composed of many resolution cells that grid the surveillance region in range and azimuth. The (group) range resolution ΔR, defined as half the time-delay resolution multiplied by the speed of light in free space c, is given by the well-known formula $\Delta R = c/2B$, where B is the radar signal bandwidth. In the HF spectrum, the bandwidth that can be gainfully employed for OTH radar operation is primarily limited by user-congestion, which restricts the availability of clear channels to relatively small bandwidths usually not much greater than 5–10 kHz at night and 20–30 kHz during the day. A secondary (physical) limitation is due to frequency dispersion in the ionosphere, which seldom supports coherent skywave propagation across signal bandwidths greater than about 100 kHz. In practice, OTH radars often need to use narrow bandwidths in the order of 5–50 kHz. The range resolution of an OTH radar may therefore vary between 3 and 30 km. This is typically 1000 times poorer than the range resolution of a conventional microwave radar.

Resolution in azimuth $\Delta\theta$ varies inversely with the length of the receive antenna aperture D measured in wavelengths λ. For a linear aperture, the half-power (−3 dB) width of a classical beam steered within ±45 degrees from the broadside direction may be approximated as $\Delta\theta = \lambda/D$ radians. For example, a linear aperture 3 km in length provides a resolution between 0.2 and 2.0 degrees over the HF band. In the middle of the HF band ($f_c = 15$ MHz, $\lambda = 20$ m) and around the center of the OTH radar range coverage ($r = 2000$ km), a 3-km-long aperture yields a cross-range resolution $\Delta\ell = r\Delta\theta$ of about 15 km, which is equivalent to the range resolution $\Delta R = c/2B$ for a typical OTH radar bandwidth of $B = 10$ kHz.

As illustrated in Figure 1.16, this aperture-bandwidth combination produces OTH radar resolution cells that are approximately 15 km long in the range dimension and 15 km wide in cross-range for the previously stated operating frequency and ground range. In this case, each resolution cell has an area of 225 million square meters. This very large backscattering area is one of the main factors contributing to the very high (40–80 dB) clutter-to-useful-signal ratio. More often, the cells are not square-shaped but elongated more in one dimension or the other depending on the relative values of the range and cross-range resolutions. The example OTH radar DIR shown in Figure 1.16 is 900 km long (60 range cells) and 300 km wide (20 azimuth cells), which yields a total of 1200 spatial resolution cells.

Once a target is detected in a particular range-azimuth cell, localization requires the conversion from radar coordinates to geographic position. This transformation is far from straightforward in OTH radar due to the uncertainty of the HF signal path through the ionosphere. Real-time propagation-path information provided by auxiliary sensors is used to compute coordinate registration tables to perform the transformation from radar group-range and angle-of-arrival to ground position in latitude and longitude. More will be said on the topic of coordinate registration in Chapter 4.

Regarding the precision of an OTH radar, the term accuracy refers to how closely a detected target or point scatterer can be located relative to its true position in geographic coordinates. Accuracy may be within 5 km when a target echo is registered relative to

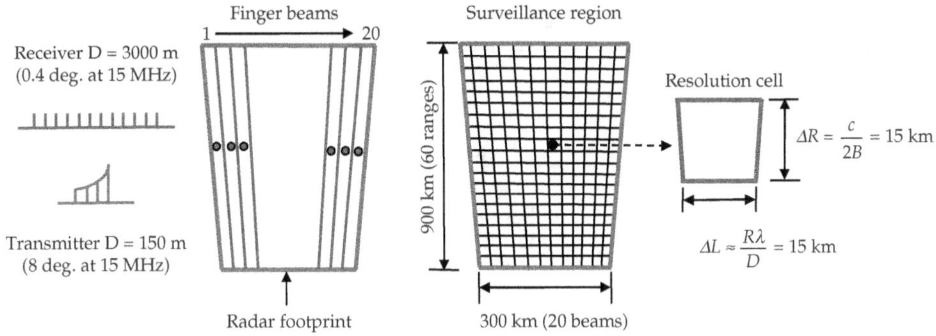

FIGURE 1.16 Example range-depth and cross-range dimensions of an OTH radar surveillance region or DIR. The two panels on the left show the different transmit and receive antenna apertures and the 20 high-resolution finger beams formed on receive within the transmitter footprint. The two panels on the right show the radar resolution cells within the DIR and the representative size of an individual range-azimuth cell for $B = 10$ kHz and $R = 2000$ km.

Characteristic	Nominal Values	General Remark
Operating Frequency	3–30 MHz	Higher during the day (e.g., 15–25 MHz), and lower at night (e.g., 7–12 MHz)
Wavelengths	10–100 m	Comparable to dimensions of man-made targets (e.g., aircraft and ships)
Surface Coverage	6–12 million km^2	Simultaneous coverage area in a single surveillance region (DIR) ~ 200,000 km^2
Ionosphere		
Height Region	100–600 km	Requires real-time and site-specific monitoring using a dedicated network of ionosondes
Number of Modes	1–4	Dominant modes typically resolved by the radar for a single target
Frequency Agility	5:1	Needed to scan the DIR within the range coverage as ionospheric conditions change
Range Coverage		
Minimum Range	500–1000 km	Determined by the size of the skip-zone at the lowest OTH radar design frequency
Maximum Range	3000–4000 km	For one-hop paths and high transmit/receive antenna gains at low elevation angles
DIR Range Depth	500–1000 km	Depends on both ionospheric propagation conditions and operating frequency

TABLE 1.3 Nominal characteristics and capabilities of a representative skywave OTH radar system broadly based on current systems in Australia and the US.

Characteristic	Nominal Values	General Remark
Azimuth Coverage		
Linear Aperture	60–90 deg.	Electronic beam steering is used to scan TX and RX array in azimuth (cone angle)
2D Aperture	180–360 deg.	Depends on element design and array geometry (e.g., L-shaped, Y-shaped)
DIR Cross-Range	200–500 km	Determined by main-lobe width of transmit beam and range of surveillance region
Transmit		
Aperture Size	100–150 m	Broad transmit beam increases simultaneous coverage in a DIR at expense of lower SNR
Azimuth Beamwidth	8–12 deg.	Requires a low number of transmit beams to span the entire arc of coverage
Elevation Beamwidth	~40 deg.	Needed to provide skywave illumination at different ranges within the radar coverage
Antenna Gain	15–25 dB	Traded off against coverage rate or region revisit time for target tracking
Average Power	200–600 kW	Detection and tracking of fast-moving and low-RCS targets against noise
Receive		
Aperture Size	2–3 km	Wide receive apertures improve radar sensitivity against noise and clutter
Azimuth Beamwidth	0.2–2.0 deg.	Computed for 3-km-long linear aperture over the HF band (Rayleigh resolution limit)
Antenna Gain	25–35 dB	Depends on antenna element gain at the operating frequency and array factor
Waveform		
Bandwidth	5–50 kHz	Often limited by user-congestion in the HF band (observed at the radar site)
CIT	1–40 s	Air (1–4 s), surface (10–40 s), limited by coverage rate and frequency-stability of path
PRF	2–60 Hz	Air (20–60 Hz), surface (2–8 Hz), trades off range and Doppler ambiguities
Duty Cycle	100%	Two-site (quasi-monostatic) systems typically based on LFMCW operation

TABLE 1.3 Nominal characteristics and capabilities of a representative skywave OTH radar system broadly based on current systems in Australia and the US. *(Continued)*

Characteristic	Nominal Values	General Remark
Resolution		
Group-Range	3–30 km	Finer for ship-detection (e.g., 5–10 km) than for air-detection (e.g., 15–20 km)
Cross-Range	10–50 km	Depends on receive antenna beamwidth, increases with range of resolution cell
Doppler Frequency	0.025–1 Hz	CIT is often limited by requirements on region revisit rate for target tracking
Relative Velocity	0.25–10 m/s	Rate of change in echo group-range evaluated for an operating frequency of 15 MHz
Ambiguities		
Group-Range	2500–75000 km	Range ambiguous radar echoes (clutter) are usually only an issue for air-mode tasks
Relative Velocity	90–900 km/h	Air-mode 900 km/h (50 Hz PRF), surface-mode 90 km/h (5 Hz PRF) at a frequency of 15 MHz
Bearing	Possibly none	Antenna array design to suppress grating lobes and provide high front-to-back ratio for a ULA
Localization Accuracy		
Absolute	10–20 km	Requires good quality real time and site-specific propagation-path information
Relative	<5 km	For targets near a known a reference point with simultaneous and identifiable radar returns

TABLE 1.3 Nominal characteristics and capabilities of a representative skywave OTH radar system broadly based on current systems in Australia and the US. *(Continued)*

an identifiable return from a known reference point (KRP) in the vicinity of the target that is received simultaneously by the same radar. In the absence of a KRP, the absolute range error may be around 10–20 km when good quality real-time propagation path information is available. Angle errors are typically less than 0.5 degrees, but can exceed 1 degree in adverse conditions, particularly due to the presence of tilts and the passage of traveling wave-like disturbances in the ionosphere.

The early-warning surveillance capability of an OTH radar evidently comes at the expense of lower spatial resolution and poorer target localization accuracy with respect to a conventional microwave radar. While OTH radars are not able to provide precise target position information, their accuracy is generally high enough to effectively cue LOS sensors onto targets of interest. This emphasizes the potential for the coordinated use of complementary sensor types in an integrated surveillance system that is likely to yield more effective results (and more efficient use of resources) than a system relying on the use of one type of sensor alone. The nominal characteristics and capabilities of a hypothetical skywave OTH radar that is representative of currently operational systems are summarized in Table 1.3.

PART I

Fundamental Principles

CHAPTER 2

Skywave Propagation

An understanding of the characteristics of the ionosphere and its properties as a propagation medium for HF radio waves is indispensable for an appreciation of OTH radar system design and operation. The first purpose of this chapter is to provide an introduction to the formation, structure, and variability of the large-scale features of the ionosphere at a fundamental level. The second is to describe the basic principles of skywave propagation for HF radio waves that impinge upon the ionosphere at vertical and oblique incidence when the transmitter and receiver are on the Earth's surface.

Several authoritative texts addressing these subjects in considerable detail are readily available for readers interested in delving further. In light of the vast amount of existing literature, this chapter does not attempt to provide an advanced coverage of the ionosphere and skywave propagation. Rather, it is primarily intended for readers with little or no prior background in these areas. With this purpose in mind, there is a perceived need to describe the underlying concepts with a greater focus on aspects that directly impact OTH radar systems, and this is one of the main motivations for this chapter.

An in-depth discussion of the physical and chemical processes occurring in the ionosphere or the detailed mathematical theory of radio-wave propagation in this complex medium is beyond the scope of this text. A comprehensive coverage of theoretical and experimental results published on these topics can be found in a number of authoritative contributions exclusively dedicated to these subjects. This includes the definitive texts by Davies (1990), Budden (1985a), Ratcliffe (1972), and McNamara (1991), and more recently, the excellent books by Schunk and Nagy (2009) and Kelly (2009). These treatments contain a wealth of information and may be consulted for an exhaustive review of the most important works in these fields.

This chapter is organized into four sections. The first provides a brief historical overview of the study of the ionosphere and early radio-wave propagation experiments that led to its discovery. This is followed by a description of the processes responsible for the formation and structure of multiple layers in different altitude regions of the ionosphere. The main characteristics of these layers and their significance to skywave propagation in OTH radar systems is highlighted. The second section discusses the morphology of the ionosphere in space and time. Here, particular emphasis is placed on large-scale variability and its effect on the design and operation of OTH radar systems. This section includes a description of radio probing techniques at vertical incidence to determine the characteristics of the overhead ionosphere, as well as the origins of disturbances and storms that can strongly influence the performance of many HF systems.

The third section deals with the subject of radio-wave reflection from the ionosphere at oblique incidence, which is of more direct interest to OTH radar. The skywave propagation relationships existing between signal carrier frequency, ray elevation angle, and path ground distance for the prevailing ionospheric structure are described in this section. The final section identifies various types of ionospheric signal modes that may arise in practice, including the magneto-ionic splitting of a radio wave into ordinary and extraordinary waves. This section also describes a number of different physical mechanisms that cause amplitude and phase fading of signals reflected from the ionosphere. Specific models for the observed fading processes are not discussed in this chapter, but will be developed and experimentally validated in Part II of this text.

2.1 The Ionosphere

The ionosphere is considered to be that region of the Earth's upper atmosphere where naturally formed ionized gas (plasma) exists in sufficient concentration to affect the propagation of radio waves. The plasma is created by electromagnetic radiation and energetic particles from the sun that penetrate into the Earth's atmosphere and ionize neutral atoms and molecules to produce free electrons and positive ions. Note that the ionosphere is not only a terrestrial phenomenon, but also exists on other planets which are surrounded by a neutral-gas envelope due to gravitational attraction.

Once the ionization is formed, the plasma is subject to a plethora of different processes, including chemical reactions, diffusion, wave disturbances, plasma instabilities, and transport phenomena. A detailed understanding of these processes, which operate over a wide range of temporal and spatial scales, requires in-depth knowledge of several disciplines, including plasma physics, chemical kinetics, atomic theory, and fluid mechanics. In this section, it is sought to provide a basic description of the ionosphere and its main characteristics that affect HF signal propagation in a relatively straightforward manner without resorting to a demanding level of specialist knowledge.

In the ionosphere, it is the concentration of free electrons, referred to more commonly as electron density, that is effectively responsible for influencing radio-wave propagation. The much heavier positive ions are significantly less mobile and do not respond in the same manner as free electrons to the rapid oscillations of an incident electromagnetic wave. The electron density in the ionosphere can exhibit large and at times sharp variations with height. The electron density height profile can affect radio-wave propagation over a relatively wide spectrum ranging from extremely low frequency (ELF) to very high frequency (VHF). This chapter is concerned with the properties of the ionosphere as a propagation medium for HF signals.

The altitude region in the ionosphere of prime interest to a skywave OTH radar extends roughly between 60 and 600 km above the Earth's surface. It is in this height region that the processes of HF signal absorption, refraction, and reflection are most pronounced. During the day, HF signal refraction and reflection generally occur at heights above 100 km, while the lower height region of the ionosphere below approximately 90 km is mainly responsible for absorption of radio wave energy. Before proceeding to discuss the formation, structure, and morphology of the ionosphere in more detail, some historical perspectives regarding the discovery and uses of the ionosphere leading up to OTH radar are provided as background.

2.1.1 Historical Overview

Probably the first manifestation of the existence of the ionosphere noticed by mankind is the natural phenomenon of the visible aurora in the upper atmosphere. The term aurora borealis (northern lights) was the name given to this phenomenon by Galileo Galilei around 1621 (Eather 1980), while the first formally recorded observation of the aurora in the southern hemisphere (aurora australis) was by James Cook in 1773. The first known suggestion of the presence of an electrically conducting layer in the Earth's atmosphere, now referred to as the ionosphere, came in 1839 when physicist Carl Gauss speculated that the causes of small regular and irregular daily variations observed in the terrestrial magnetic field might be explained by the presence of electric currents in the atmosphere. It was the enigmatic phenomenon of the aurora that led Gauss to speculate the existence of such currents. However, the physical origin of these fascinating displays of colored light, which often appear as luminous arches or other changing forms in the night sky was not correctly deduced until the end of the nineteenth century.

2.1.1.1 Early Radio Experiments

During the period 1864–1873, the Scottish theoretical physicist and mathematician James Clerk Maxwell elegantly formulated the unification of classical electromagnetic theory. This celebrated work is widely recognized as one of the greatest scientific achievements, considered by some to be comparable to those of Isaac Newton and Albert Einstein. The existence of electromagnetic waves, predicted by Maxwell's equations, was subsequently demonstrated experimentally in 1887 by the German physicist Heinrich Hertz, who was the first to use a set of practical apparatus to generate and detect radio waves. About a decade later, a young Italian inventor Guglielmo Marconi adapted and assembled a collection of previously disconnected knowledge and technology of that era to develop and demonstrate the first "wireless" (radio) telegraphy communication system. This system enabled coded messages to be sent across the English channel in 1899.

In his most famous experiment on December 12, 1901, Marconi transmitted radio communication signals across the Atlantic Ocean from Poldhu, Cornwall, England, to St. John's, Newfoundland, Canada, over a ground distance of approximately 3500 km. Using 120 m of wire as a receiving antenna (kept aloft by a high-flying kite), Marconi first heard the clicks of the Morse code letter "S" from the telephone headpiece (Maslin 1987). The ability to establish such long-distance wireless communications was a monumental discovery. From the work of Maxwell and Hertz, it was known that radio waves traveled in straight lines and were deflected only when they encountered a medium with different electrical properties. Not surprisingly, this event sparked tremendous scientific interest to understand the mystery of how radio waves had made their way so far around the Earth to "bridge" the Atlantic.

2.1.1.2 Proposed Physical Explanations

At first, it seemed natural to suppose that the phenomenon allowing such long-distance propagation was one of diffraction. Consequently, several physicists and mathematicians turned their attention to finding the solution satisfying Maxwell's equations for the problem of wave diffraction over a spherical surface of finite conductivity. Initial attempts to explain Marconi's results were based on a propagation mode known as a surface-wave, where a significant portion of radio-wave energy flows around the curvature of the Earth in a height region that extends all the way down to the (sea-air) interface. Among the earliest contributors to the theory of surface-wave propagation were J. Zenneck and A. Sommerfeld. However, calculations showed that the diffraction effect which could be attributed to this propagation mode was inadequate to explain the observed bending of the radio waves. As it turns out, the surface-wave mode represents an important phenomenon for communication and radar applications over shorter but still over-the-horizon distances. This topic will be revisited in Chapter 5.

In 1902, the year following Marconi's trans-Atlantic experiment, a separate thread of investigation led two scientists, Oliver Heaviside in Britain and Arthur Kennelly in the United States, to independently postulate the existence of a conducting layer in upper atmosphere Ratcliffe (1967). According to Kennelly and Heaviside, the conducting layer contained free electrical charges that could reflect radio waves and allow them to be received around the curvature of the Earth. This initial proposal was subsequently further refined by radio engineer W.H. Eccles, who suggested that the number of ions in the atmosphere increased as a function of height (McNamara 1991). Eccles expressed the idea that radio waves passing into regions in the atmosphere with a positive ionization gradient vertically would be progressively bent to follow a path around the curvature of the Earth. However, both the theory of a conducting layer and the notion of an ion concentration increasing with altitude lacked direct experimental evidence. It would do so for more than two decades after the conducting layer theory was first postulated by Heaviside and Kennelly.

In 1902, Marconi carefully documented a shore-to-ship experiment aboard the S.S. Philadelphia where radio signals sent daily from the Poldhu station were recorded as the ship headed west from Great Britain. Marconi was able to receive messages over distances in the order of 1000 km during day but up to 3000 km at night. Marconi was the first to notice that medium- and long-wave radio transmissions could propagate over significantly longer distances at night. Perhaps the earliest physical theory of the conducting layer emerged shortly after these observations (Lodge 1902) and independently by J.E. Taylor (Taylor 1903). It was suggested that the conducting layer was produced by the ionizing action of ultraviolet light from the sun in the upper atmosphere. Importantly, this suggestion first implied solar control of radio wave propagation.

2.1.1.3 Point-to-Point Communication and Broadcasting

As early radio experiments indicated that longer wavelengths provided longer range propagation, the initial trend was to develop systems that operated below the HF band. Originally, wireless telegraphy was predominantly applied for ship-to-shore communications using the medium frequency (MF) band. A dramatic example of the impact of this technology was its vital role in calling for salvage efforts to the Titanic in 1912.

Marconi returned to experimenting with higher frequencies around 1915. His renewed interest in the HF band was mainly driven by the ability to form directional antennas of reduced size that could focus the transmitted energy into narrow beams. Marconi later found that so-called "short waves" could provide reliable communications over thousands of kilometers at much lower power than the hundreds of kilowatts needed at longer wavelengths. Further experimentation revealed that he could maintain contact anywhere in the world at almost any time of day or night, the key to this being the appropriate choice of operating frequency.

These observations reversed the trend from high-power low-frequency communication systems, which were expensive and hence less viable for large networks, to

(a) Marchese Guglielmo Marconi (1874–1937).

(b) Sir Edward V. Appleton (1892–1965).

FIGURE 2.1 Nobel laureates Marchese Guglielmo Marconi for "contributions to the development of wireless telegraphy" and Sir Edward V. Appleton for "contributions to the knowledge of the ionosphere."

low-power HF stations based on the beam system. The beam system was adopted for the Imperial Wireless Chain that linked the United Kingdom, Canada, Australia, South Africa, and India in the mid-1920s. Such systems required highly skilled operators who had considerable understanding of propagation conditions. Avoiding other communications in progress also became important as the number of stations and hence signal traffic grew. This required frequency and time coordination of all stations in the network to maintain satisfactory operation. In 1927, the BBC commenced short-wave broadcasting to the empire. Much of the development was pioneered by Marconi, who was awarded the Nobel Prize in 1909 in recognition of his "contributions to the development of wireless telegraphy." Among the many legacies left by Marconi, pictured in Figure 2.1a, is the *Centro Radioelettrico Sperimentale* located at Torre Chiaruccia, near Santa Marinella, Italy, where Marconi conducted radio location experiments from shore to his ship Elettra.

2.1.1.4 Discovery of the Ionosphere

Between 1912 and 1914, engineers Lee de Forest and L.F. Fuller from the US Federal Telegraph Company conducted propagation experiments between Los Angeles and San Francisco over a ground distance of approximately 500 km using signals at two closely spaced carrier frequencies (a main wave and a compensation wave). The main and compensation waves were received simultaneously with quite different amplitudes and it was proposed that this was due to the interference between the waves reflected from the hypothesized conducting layer and the direct-path signal propagated via the surface-wave mode. Based on the observed fading characteristics, an effective reflection height of 99 km was calculated. This constituted the first estimate of the height of the conducting layer postulated by Heaviside and Kennelly.

The experimental proof of the existence of a reflecting layer in the upper atmosphere came in 1924 with the "frequency changing" experiments of Appleton and Barnett. Ground-based soundings were performed using a BBC transmitter at Bournemouth, where continuous-wave signals were transmitted with a slowly changing carrier frequency such that fading characteristics due to the interference between the surface-wave and the wave reflected by the hypothesized reflecting layer could be measured at a site near Cambridge. By comparing signal intensities received simultaneously on a loop and vertical antenna, Appleton and Barenett demonstrated the existence of a downcoming wave, which we refer to as a skywave. By virtue of the known transmitter-to-receiver separation, a path length difference equation allowed the reflecting layer height to be deduced as 100 km (Appleton and Barnett 1925).

Near sunset, the height of reflection was found to change abruptly, and from this observation, the stratified nature of the ionosphere was first inferred by Appleton in the winter of 1925. The convention of using E, F, and later D, to denote the different layers in the ionosphere was introduced by Appleton. The reason for this nomenclature is described by Sir Edward Appleton in a letter dated March 20, 1943, reproduced in (Silberstein 1959). Appleton named the first layer that he recognized the E-layer for the Electric field of the downcoming wave. Subsequently, when he recognized reflections from a second layer at higher altitude, he proceeded to call it the F-layer. Appleton subsequently conjectured that there may be lower altitude layers, so he decided to name the first two layers he discovered E and F, thus allowing for an alphabetical designation of remaining layers that might come to light as a result of further work.

In 1925, an independent experimental study that also demonstrated the existence of multiple downcoming waves was conducted by Breit and Tuve using a near-vertical pulse-sounding technique. Specifically, Breit and Tuve showed that a pulse of short duration sent upward from a transmitter produced two (and at times more) echoes at the receiver located a few kilometers away. The first echo being due to the direct (surface-wave) path and subsequent echoes coming from waves returned by the reflecting layers. This technique allowed the downcoming waves to be resolved from the direct-path signal in time-delay such that their parameters could be measured directly rather than being deduced based on interference effects. The pulse-sounding technique may be considered as an early form of radar and subsequently came into widespread use as a method for studying the structure and dynamics of the ionosphere.

As the reflecting layers were believed to be composed of electrically charged particles called ions, the region became known as the ionosphere. A detailed account of the naming of the ionosphere is given in Gillmor (1998). Robert Watson-Watt (later developer of radar) first coined the word "ionosphere" in a letter to the United Kingdom Research Board in 1926, but the term did not formally appear in the literature until a few years later, Watson-Watt (1929). In 1927, Sydney Chapman developed a physical theory to describe the structure of layers in the ionosphere. At about this time, ionospheric physics became a field in its own right. Its development was heavily influenced by Appleton, pictured in Figure 2.1b, who was awarded the Nobel Prize in 1947 for his "contributions to the knowledge of the ionosphere" (Clark 1971).

2.1.1.5 The Post-War Years

Up until the second world war, many properties regarding the formation, structure, and dynamics of the ionosphere had been described (Green 1946), but its detailed physics and chemistry were still not well understood. The war provided a strong impetus to develop radio communications technology and radar systems. The great expansion of the ionospheric sounding network during these years was mainly motivated by the need to make routine predictions of radio propagation conditions around the world. This is in contrast to the motivation behind the first ionospheric sounding stations, which were primarily set up for scientific purposes to discover the causes and characteristics of the different layers in the ionosphere. The ionospheric data gathered during the war was almost exclusively used for practical purposes and very few studies were performed to discover their physical significance and meaning.

An important step toward the systematic collection and organization of worldwide ionospheric data in uniform manner has its origin in the International Geophysical Year (IGY) 1957–1958. In addition to data gathered by ground stations, in-situ measurements made by instruments on rockets and satellites provided further information on the density, composition, temperature, electric/magnetic fields, and other properties of the ionosphere. Renewed focus on the scientific analysis of such data led to a more detailed understanding of the physics and chemistry of the ionosphere in the post-war years.

Besides the main traditional uses of the ionosphere for long-distance point-to-point HF communication and broadcasting, direction-finding systems that locate the position of HF emitters using signals received at one or more sites were also developed for civil and military uses. Such systems demanded more detailed information about the ionosphere than HF communication systems, and moreover, such information was required at the time that measurements were made. The level of real-time knowledge demanded of the ionosphere increased even further with the introduction of OTH radar systems used for wide-area surveillance and remote sea-state sensing in the late 1950s.

2.1.2 Formation and Structure

The ionosphere is naturally formed by the action of electromagnetic radiation and corpuscular emissions from the Sun that ionize neutral molecules and atoms in the Earth's upper atmosphere to produce free electrons and positive ions. The resulting electron-ion gas is regarded as a weakly ionized plasma since the maximum concentration of charged particles in the ionosphere is typically less than 1 percent of the air. As free electrons and positive ions are created and destroyed together, the ionosphere is on the whole electrically neutral to a high degree of approximation.

The ability of the ionosphere to reflect HF signals depends upon the physical and chemical processes that drive production, loss, and transport of free electrons. Consequently, the operation of systems (such as OTH radar) that rely critically on skywave propagation is heavily influenced by the characteristics of these processes. The fundamental principles of ionization production, loss, and transport will now be described at a rudimentary level to explain the formation and structure of the ionosphere in the most basic manner.

2.1.2.1 Production Processes

Production refers to the formation of electrons and positive ions from neutral atoms and molecules. With reference to Eqn. (2.1), *photoionization* is the name given to the process by which a photon transfers its energy $h\nu$ to an electron causing it to be released from a neutral atom or molecule P, resulting in a free electron e^- and positive ion P^+. Here, h is Plank's constant and ν is the photon's frequency.

$$P + h\nu \rightarrow P^+ + e^- \tag{2.1}$$

The most important reactions involve molecular nitrogen N_2, molecular oxygen O_2, and nitrous oxide NO, particularly at lower altitudes, while those involving atomic oxygen O and hydrogen H become more significant at higher altitudes. Note that molecular and atomic species of different chemical composition can only be ionized by photons with certain wavelengths. It turns out that the wavelengths involved in photoionization reactions are emitted by the Sun as spectral lines and continuum radiation in the extreme ultra-violet (EUV) and X-ray regions of the electromagnetic spectrum.

The *rate* of production associated with a particular photoionization reaction is proportional to both the intensity of ionizing radiation and the density of the chemical species it ionizes. Hence, the rate of production over all photoionization reactions depends not only on the intensity of the incident solar radiation over a spectrum of wavelengths, but also on the density of neutral particles with different chemical compositions in the atmosphere. As both radiation intensity and particle density vary as a function of altitude, the rate of production changes with height in the atmosphere.

The original Chapman theory describes the rate of ionization production in a horizontal plane-stratified atmosphere for a single component isothermal gas by absorption of a parallel beam of monochromatic solar radiation. These simplifications may be removed to derive the rate of production formula for a multi-component atmosphere ionized by a spectrum of solar radiation, as described in Matsushita and Campbell (1967). However, the equations become more complicated and this detracts from the main intent here, which is to explain the basic principles of the Chapman theory.

Starting from the hydrostatic equation for the vertical pressure distribution of a perfect gas under the influence of gravity, it is possible to show that atmospheric density (number density of neutrals) at a height h above the Earth's surface is given by ρ in Eqn. (2.2), where ρ_0 is the neutral density at the surface ($h = 0$) and $H_p = KT/mg$ is the pressure scale height that depends on the particle mass m, the absolute temperature T, Boltzmann's constant K, and acceleration due to gravity g.

$$\rho = \rho_0 e^{-h/H_p} \tag{2.2}$$

Now denote I_∞ as the incident solar flux density (radiation intensity) in photons per unit area which strikes the top of the atmosphere obliquely at an angle χ relative to the vertical, referred to as the *zenith* angle. The initial radiation intensity I_∞ decreases as the rays penetrate further into the atmosphere due to absorption and therefore becomes a function of height $I(h)$. The change in radiation intensity $\mathrm{d}I(h)$ with height is given by Eqn. (2.3), where σ is the mass absorption coefficient or "cross section" of the gas.

$$\mathrm{d}I(h) = I(h)\sigma\rho \sec\chi \,\mathrm{d}h \tag{2.3}$$

Substituting Eqn. (2.2) into Eqn. (2.3) and integrating over h yields the radiation intensity height profile $I(h)$ in Eqn. (2.4), where $z = (h - h_0)/H_p$ and $h_0 = H_p \ln\{\sigma\rho_0 H_p\}$ are referred to as the reduced and reference heights, respectively.

$$I(h) = I_\infty \exp\left[-\sec\chi e^{-z}\right] \tag{2.4}$$

The rate of free-electron production q at a particular height is proportional to the rate of absorption at that height, such that $q = \eta \cos\chi \,\mathrm{d}I(h)/\mathrm{d}h$, where η is the number of electron-ion pairs produced by one photon. Differentiating Eqn. (2.4) yields the rate of production as the function $q(z, \chi)$ in Eqn. (2.5), where $q_0 = I_\infty\eta/H_p e$ is a constant and e is the natural exponent. This equation is referred to as the Chapman function or Chapman formula. It expresses the rate of ionization production $q(z, \chi)$ as a function of reduced height z, zenith angle χ, and the constant q_0.

$$q(z, \chi) = q_0 \exp [1 - z - \sec \chi e^{-z}] \tag{2.5}$$

Ionization production maximizes at a height where the product of radiation intensity and neutral gas density is highest. Radiation intensity decreases at lower altitudes because of absorption, while the neutral gas density decreases at higher altitudes due to gravity. The net result of these two competing effects is that production peaks at some intermediate height. As we shall see shortly, the peak in electron density production at this intermediate height ultimately gives rise to the formation of a layer in the ionosphere.

By differentiating Eqn. (2.5), the height of maximum ion production is given by h_m in Eqn. (2.6). Note that the reference height h_0 is the height of maximum ion production when the sun is directly overhead ($\chi = 0$). The peak value occurring at height h_m is given by q_m in Eqn. (2.6). Note that q_0 is the peak rate of ion production when the radiation is incident vertically. A change in the incident solar flux density I_∞ alters the peak value but not the height of maximum production; whereas a change in zenith angle alters both the peak value and the height of maximum production.

$$\begin{cases} h_m = h_0 + H_p \ln\{\sec \chi\} \\ q_m = q_0 \cos \chi \end{cases} \tag{2.6}$$

The process of photoionization occurs only during the day when the sun illuminates the atmosphere. However, photoionization is not the only process by which free electrons can be created. A different production mechanism occurs when neutrals in the atmosphere are bombarded by a stream of charged particles that reach the Earth from the Sun. Collisions between these energetic particles, which are carried through the interplanetary medium by the solar wind, and neutrals in the Earth's atmosphere can also lead to the production of free electrons and positive ions. This process is known as *collisional ionization.*

2.1.2.2 Loss Processes

Free electrons may be lost by two main processes known as *recombination* and *attachment*. In contrast to photoionization, which only occurs during the day, both loss processes occur continuously during the day and night. Despite the lack of photoionization at night, it transpires that these loss processes are never completely effective in eroding the ionosphere. Above heights of approximately 210 km, the free-electron density in the ionosphere remains at sufficiently high levels to affect the propagation of radio waves throughout the night until photoionization recommences at dawn. In simple terms, the ionosphere is depleted of ionization during the night but never completely disappears.

With reference to Eqn. (2.7), loss by recombination refers to the process whereby a free electron recombines with a positive ion to form a neutral particle P and (radiated) excess energy $h\nu$.

$$e^- + P^+ \rightarrow P + h\nu \tag{2.7}$$

Importantly, the recombination rate of an electron with a molecular ion is about *five orders of magnitude* greater than that with an atomic ion. As indicated by Eqn. (2.8), atomic ions A^+ are first converted to molecular ions M^+ before recombination takes place in what are known as *charge transfer* reactions. Charge transfer reactions may be interpreted as an intermediate step in the recombination process.

$$A^+ + M \rightarrow M^+ + A + h\nu \tag{2.8}$$

An example of a charge transfer reaction involves atomic oxygen ions $A^+ = O^+$ exchanging charge with molecular oxygen $M = O_2$. There are many different charge transfer reactions, not all of them limited to the form of Eqn. (2.8). Some involve charge exchange between two different molecular species, while others result in the formation of molecular ions with different chemical composition to the reacting species. The more significant charge transfer reactions are described in Davies (1990).

Dissociative recombination is the name given to the specific reaction in Eqn. (2.9), where an electron e^- recombines with a positive molecular ion M^+ to form a neutral molecule M and excess energy radiated as ultraviolet, visible, or infrared airglow. Note that the rate of electron loss by the process of recombination is effectively determined by the rate of dissociative recombination, which is five orders of magnitude greater than the rate of recombination between an electron and atomic ion. The main dissociative recombination reactions involve the molecular species NO^+, O_2^+, and N_2^+.

$$e^- + M^+ \rightarrow M + h\nu \tag{2.9}$$

The rate of recombination varies with altitude and depends on the density of molecular ions at different heights. The molecular ion density is in turn influenced by the density of atomic ions and neutral molecules, which are converted to molecular ions via the charge transfer reactions described previously in the first step of a two-step recombination process. The charge transfer reactions proceed more rapidly in regions of the ionosphere where the molecular density is higher. As a result, heavier molecular ions are generally found in greater densities at lower altitudes, while lighter atomic ions are present in greater proportion at higher altitudes.

Below about 140 km, the charge transfer reactions are very fast due to the relatively high density of molecular species, so the rate of electron loss is governed by the dissociative recombination step in Eqn. (2.9). In this case, the rate of loss L is proportional to the product of the electron and molecular ion densities. Since the density of free electrons N_e equals that of positive ions in an electrically neutral plasma, the rate of loss due to dissociative recombination is given by the square law in Eqn. (2.10), where α is the mean dissociative recombination coefficient for the molecular ions.

$$L = \alpha N_e^2 \tag{2.10}$$

On the other hand, the much slower recombination rate of electrons with positive atomic ions becomes a significant factor at higher altitudes, where atomic ions are present in greater proportion due to the lower abundance of heavier molecular species. In regions of the ionosphere above roughly 210 km, the charge transfer reactions proceed at a relatively slow rate, while the lifetime of a positive atomic ion is relatively long in the absence of such a reaction. The charge transfer step therefore has a greater influence on the rate of electron loss in higher altitude regions where the lighter atomic ions dominate. In this case, the rate of electron loss depends linearly on N_e and is given by Eqn. (2.11), where β is a linear loss coefficient proportional to the neutral molecular density.

$$L = \beta N_e \tag{2.11}$$

The transition between the square law and linear loss formulas that determine the rate of dissociative recombination at lower and higher altitudes, respectively, is not abrupt but occurs gradually as a function of height. At intermediate heights, approximately between 140 and 210 km, the rate of loss may be specified in terms of the mixed relation given by Eqn. (2.12), as described in Davies (1990). Note that Eqn. (2.12) reverts back to the square law in Eqn. (2.10) when $\beta \gg \alpha N_e$ and the linear loss formula in Eqn. (2.11) when $\beta \ll \alpha N_e$.

$$L = \beta\alpha N_e^2/(\beta + \alpha N_e) \tag{2.12}$$

Electrons can also be lost by a different process in Eqn. (2.13), whereby they effectively become attached to neutral molecules to form negatively charged ions. The rate of loss associated with this process, referred to as *attachment*, depends on the neutral density at the altitude considered. It follows that the rate of electron loss due to this process is also given by the linear relationship in Eqn. (2.11). Attachment is more prominent in the lower part of the ionosphere where the density of neutrals is higher. Variations in atmospheric molecular density can therefore have a marked effect on electron density when attachment is the dominant loss process.

$$e + M \rightarrow M^- \tag{2.13}$$

2.1.2.3 Layer Formation

The most important property of the ionosphere from the perspective of OTH radar, and other HF systems that rely on skywave propagation, is the resultant free-electron density. This is determined by the net interaction of production, loss, and transport processes at a particular time, location, and height. Based on the principle of particle conservation, the rate of increase in electron density is given by the continuity equation in Eqn. (2.14).

$$\frac{dN_e}{dt} = \{\text{Production}\} - \{\text{Loss}\} - \{\text{Transport}\} \tag{2.14}$$

Ignoring transport mechanisms for the moment, the difference between production and loss reflects the net rate of increase of electron density. When the rate of electron loss is determined by the square law $L = \alpha N_e^2$, the continuity equation takes the form of Eqn. (2.15).

$$\frac{dN_e}{dt} = q - \alpha N_e^2 \tag{2.15}$$

Quasi-equilibrium is reached when production and loss are almost balanced such that the rate of change of electron density is close to zero. Substituting the Chapman function $q = q(z, \chi)$ and setting $dN_e/dt = 0$ in Eqn. (2.15) yields the steady state electron density height profile $N_e(h, \chi)$ in Eqn. (2.16). This profile is called the α-Chapman layer.

$$N_e(h, \chi) = N_{e0} \exp\left[\frac{1}{2}(1 - z - \sec\chi e^{-z})\right] \tag{2.16}$$

By differentiating the profile $N_e(h, \chi)$ with respect to h and setting the derivative to zero, the maximum electron density is given by $N_m(\chi)$ in Eqn. (2.17). The maximum value of the α-Chapman layer occurs at the height of peak production given by h_m in Eqn. (2.6).

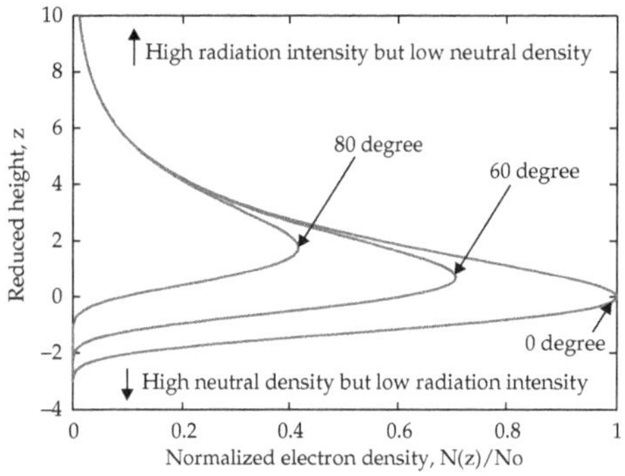

(a) α-Chapman layer (square loss).

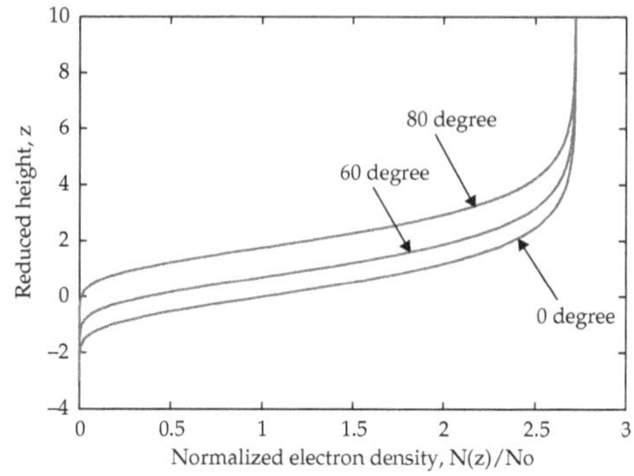

(b) β-Chapman layer (linear loss).

FIGURE 2.2 Layer electron density profiles as a function of reduced height and solar zenith angle χ using the Chapman production formula and loss processes based on the linear and square law.

Normalized electron density profiles of the α-Chapman layer are plotted as a function of reduced height for different zenith angles in Figure 2.2a.

$$N_m(\chi) = N_{e0}\sqrt{\cos\chi} \tag{2.17}$$

The α-Chapman layer is most likely to be a suitable model in the lower ionosphere, below about 140 km, where the square law for the rate of electron loss may be considered appropriate. On the other hand, when the rate of loss follows the linear law $L = \beta N_e$, the continuity equation is given by Eqn. (2.18), where the linear loss coefficient β is proportional to molecular density.

$$\frac{\mathrm{d}N_e}{\mathrm{d}t} = q - \beta N_e \tag{2.18}$$

At equilibrium, the electron density height distribution for this case takes the form of Eqn. (2.19), which is sometimes referred to as the β-Chapman layer. As illustrated in Figure 2.2b, this profile has no maximum except at very great heights where there is an insufficient number of molecules that can be ionized (Davies 1990). In practice the greater influence of transport phenomena at heights where the linear loss law may be considered appropriate (above approximately 210 km) implies that this alternative Chapman layer profile will often not accurately describe the actual electron density distribution at such altitudes.

$$N_e(h,\chi) = N_{e0}\exp\left(1 - \sec\chi e^{-z}\right) \tag{2.19}$$

Although the α- and β-layer height profiles were derived under rather idealized physical conditions, the Chapman theory provides a valuable basis for more sophisticated methods that generalize its underlying assumptions to model the electron density height profile more accurately. A variety of alternative parametric models for the electron density height profile due to one or more ionospheric layers can be found in Davies (1990). In particular, the multi-segment quasi-parabolic (MQP) layer profile has found widespread use in the ionospheric modeling community due to its mathematically tractable form.

Production and loss processes are largely responsible for determining the structure and morphology of the ionization height profile below about 140 km. Movement of ionization becomes increasingly important to account for the formation of peaks in ionization at altitudes greater than about 210 km. Specifically, the three main plasma transport mechanisms are known as neutral winds, diffusion, and electromagnetic drift. These transport processes are briefly identified below. Further descriptions of their physical causes and characteristics are provided in Davies (1990).

Neutral winds are driven by a system of atmospheric pressure gradients generated by temperature differences between the day- and night-side of the Earth due to solar heating.[1] Below about 140 km, the neutral wind motion is transferred onto the ions by collisions and subsequently onto the electrons by Coulomb attraction, so the plasma motion is largely determined by the neutral motion. However, at higher altitudes, where the neutral density is lower, atmospheric winds cannot move the plasma effectively across the magnetic field lines. When the wind blows toward the equator at night, the plasma will in this case follow the magnetic field lines and is lifted to higher altitudes, where the recombination rate is lower. This phenomenon contributes to the continuous presence of significant ionization at higher altitudes throughout the night. During the day, the neutral winds blow poleward and the plasma descends to lower altitudes, where the recombination rate is higher, but at which time photoionization is already replenishing the electron density level.

Diffusion of electrons and positive ions in an ionized gas ("ambipolar diffusion") differs from diffusion in a neutral gas due to the influence of the Earth's magnetic field as well as the electrostatic field created by charge separation. The speed of diffusion increases rapidly at higher altitudes because it varies inversely with the neutral density. It also maximizes along the direction of the magnetic field as free electrons are inhibited from crossing these lines when no external electric field is present. This process is often referred to as field-aligned diffusion. For example, the vertical component of plasma transport by the process of field-aligned diffusion disappears at the magnetic equator, where the direction of geomagnetic field lines is horizontal.

Electromagnetic drift refers to the process by which plasma moves across the Earth's magnetic field lines due to the presence of an external electric field. In the presence of crossed electric and magnetic fields, electrons and ions drift in the same direction perpendicular to the plane containing the field lines. In contrast to vertical diffusion, which vanishes at the magnetic equator, the electromagnetic drift mechanism can cause the plasma to rise rapidly upward to very high altitudes. This phenomenon is particularly relevant during daylight hours when an eastward pointing electric field is set up near the equator by neutral winds.

2.1.3 The D-, E-, and F-Regions

It was mentioned previously that the electron density in the ionosphere arises from a multi-component gas and a spectrum of ionizing radiation, and that ionization may also be produced via collisions between neutrals and energetic particles carried by the solar wind into the atmosphere. At a certain time and geographic location, the net effect is an electron density height profile that typically exhibits a number of local maxima at

[1] Neutral winds are also coupled to the differential rotational velocity of the Earth's surface with latitude.

different altitudes. The number of local maxima, their magnitude(s), and the altitude(s) at which they form varies with time and location, but the electron density peaks that are of primary interest to (terrestrial) HF systems occur in three main height regions known as the D-, E-, and F-regions of the ionosphere.

Figure 2.3 notionally illustrates an example electron density height profile that may be considered representative of a (hypothetical) mid-latitude ionosphere on a summer day. Although the ionosphere is a lightly ionized plasma, the electron density can vary by two or more orders of magnitude as a function of height. The local peaks formed in the D-, E-, and F-regions are often referred to as ionospheric layers. The electron density height profiles of different layers have structures that merge smoothly or superimpose onto one another and are therefore not visible in isolation when multiple layers are present. As illustrated for the D-region in Figure 2.3, the manifestation of a layer may appear more as an inflexion in the electron density height profile rather than a well-defined peak.

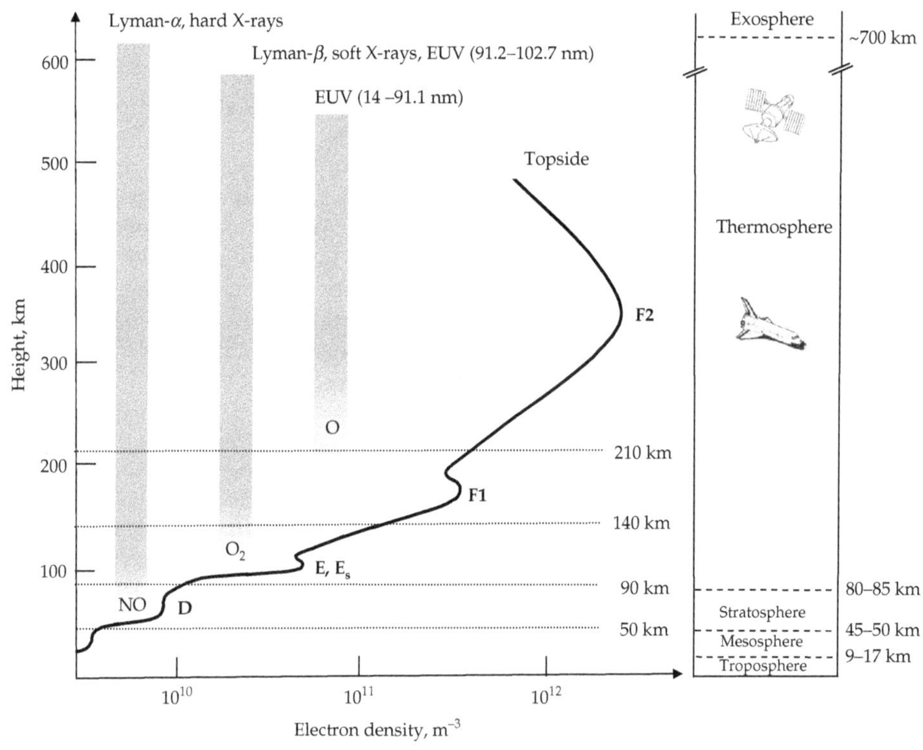

FIGURE 2.3 Notional electron density height profile of a hypothetical mid-latitude ionosphere on a summer day showing the formation of a number of local peaks or layers in different altitude regions. The main components of solar radiation (spectral lines and continuum radiation) that create the ionosphere and some examples of the chemical species ionized are also indicated. As far as radio-wave propagation for terrestrial HF systems is concerned, only the bottom side of the ionosphere below the height of the F2-layer peak electron density is of practical importance. The structure and dynamics of the ionosphere between heights of about 60 and 600 km is typically of primary interest to such systems. © Commonwealth of Australia 2011.

Morphology of the electron density height profile with respect to time and location is not only dependent on changes in the rate of production and loss. Such processes tend to dominate in the D- and E-regions, but in the F-region, the redistribution of ionization due to transport mechanisms plays a key role in establishing the global peaks in electron density. Further discussion regarding the large-scale spatial and temporal variability of the ionosphere is provided in the next section. The general properties of ionospheric layers formed in different height regions will now be described.

2.1.3.1 D-Region

The lowest region of the ionosphere generally considered for HF signal propagation is the D-region. The D-layer is produced between altitudes of 50–90 km mainly due to the ionization of molecular species by hard solar X-rays with wavelengths of 0.1–1.0 nm as well as the ionization of nitric oxide by the Lyman-α spectral line (121.6 nm). Solar radiation at longer wavelengths is mostly absorbed by the atmosphere at higher altitudes.

The electron densities in the D-region are typically two to three orders of magnitude smaller than those encountered in the E- or F-regions. Such ionization levels are not sufficiently high to reflect or significantly refract HF radio waves. From the viewpoint of HF propagation, the primary influence of the D-region is to *absorb* radio wave energy. In other words, the D-region does not effectively deviate the direction of HF signal propagation but it can significantly attenuate the power density of the signal that penetrates through it to reach the upper layers of the ionosphere.

The relatively high air particle density at D-region altitudes (compared to those in the E- and F-regions) increases the collision frequency between electrons and neutrals. The loss of radio wave energy due to these collisions is referred to as *non-deviative* absorption because the direction of signal propagation is essentially unaffected. The amount of absorption depends on the ionization level in the D-region, which exhibits a fairly regular dependence on $\cos \chi$ at mid-latitudes. As a result, absorption is usually greatest around the middle of the day and in summer months with the maximum generally occurring near the sub-solar point.

Across the HF band, non-deviative absorption varies *inversely* with the square of radio frequency to a good approximation. The use of higher frequencies is therefore preferred (when possible) for the operation of HF systems such as OTH radar. A higher frequency minimizes signal attenuation in the D-region and hence increases the signal power density incident on the higher reflecting layers. Several other advantages of using higher frequencies for OTH radar operation will be discussed in Chapter 3.

Enhanced ionization of all particle species in the D-region can occur due to an intense solar flare. Such events can dramatically increase the amount of absorption on paths that traverse the affected area. In addition, absorption occurs each time a radio wave passes up or down through the D-region, which means that single-hop skywave propagation experiences relatively less attenuation than multi-hop skywave propagation over a sunlit path. In unfavorable conditions (i.e., high D-layer ionization and low operating frequency), non-deviative absorption can reduce the signal strength by 10–20 dB per hop over a two-way path, which has significant implications for OTH radar systems limited by local noise sources not propagated by the ionosphere.

The D-layer effectively vanishes at night due to the fast recombination and attachment processes that prevail after sunset. The absence of D-region absorption allows HF signals to propagate effectively over long distances at night via the skywave mode. While this can provide a benefit for OTH operation over long distances using (nighttime) frequencies in the lower HF band, it also reduces isolation in the sense that HF systems become more prone to the reception of noise and interference from very distant (natural and man-made) sources.

2.1.3.2 E-Region

The next height region in the ionosphere is the E-region, which extends roughly between 90 and 140 km. In addition to the normal E-layer, anomalous sporadic-E ionization may also be present, as discussed separately below. The normal E-layer is formed due to radiation from soft X-rays with wavelengths of 0.8–14 nm, the Lyman-β spectral line (102.6 nm), and EUV continuum radiation in the wavelength band 91.2–102.7 nm that ionizes molecular oxygen.

The peak electron density in the E-region is typically two orders of magnitude higher than the D-region and often occurs at a height of about 100 km during daylight hours. Ionization levels in the sunlit E-region are sufficiently high to reflect HF signals. Moreover, the E-layer is renown for providing the most frequency-stable propagation. This is particularly important for OTH radar systems engaged in slow-moving target detection and tracking applications, where useful signals need to be discriminated against clutter returns by Doppler processing.

The height of the E-layer is relatively low compared to layers formed in the F-region. Consequently, oblique reflection from the E-layer limits HF signal propagation to ground ranges not exceeding a natural "horizon" of about 2000 km for a single-hop path. A ground range of 2000 km assumes an E-layer virtual reflection height of 100 km and a signal take-off angle of approximately 1 degree. In practice, one-hop propagation by reflection from the E-layer may be restricted to ground ranges less than 1500 km depending on the gains of the transmit and receive antennas at low elevation.

Dissociative recombination between electrons and molecular ions including NO^+ and O_2^+ almost completely erode ionization in the E-layer after sunset. Transport mechanisms are also responsible for redistributing E-region ionization to higher altitudes in the evening and at night. The normal E-layer effectively disappears after sunset, apart from a small amount of residual ionization that persists throughout the night.

Due to the square-law loss process occurring at E-region heights, the ionization structure of the E-layer may be approximated by the α-Chapman profile. Departures from this idealized profile arise in practice due to complications such as scale-height gradients and the presence of plasma transport mechanisms not accounted for by the model. Although not as significant as in the D-region, a degree of non-deviative absorption also occurs in the lower E-region where the neutral density is relatively high.

A different type of absorption known as *deviative* absorption occurs when the radio wave undergoes reflection, or a significant change in propagation direction, near the height of maximum ionization in a layer. In other words, deviative absoption is most pronounced when the signal is reflected at a point in the ionosphere where the rate of change of electron density with height approaches zero. This phenomenon, which may occur in the E- and F-regions, will be explained later in terms of a parameter known as the layer critical frequency.

2.1.3.3 Sporadic-E

Sporadic-E is an erratically occurring layer that forms at E-region heights but is much less predictable than the normal E-layer. Unlike the normal E-layer, which is only present during the day, sporadic-E ionization may be present during the day or night. As its name suggests, sporadic-E is a transient phenomenon that often persists for short periods of time, typically a few hours or less.

Its cause is not well-understood. In mid-latitude regions, one theory is that it partly arises due to the compression of very fine metallic debris from vaporised meteors that form into narrow highly reflecting sheets due to the interaction of sheared winds and the geomagnetic field (Whitehead 1989). Mid-latitude sporadic-E is a relatively thin layer, typically only a few kilometers thick, which at times is broken up as random clouds or irregularly shaped patches of abnormally intense ionization. The clouds may be tens to hundreds of kilometers in length with gaps in between to create a partially translucent layer that allows HF signals to pass through into the F-region.

Attempts to predict the presence and characteristics of mid-latitude sporadic-E have been relatively unsuccessful. However, the appearance of mid-latitude sporadic-E is observed to be more prominent in the daytime and early evening during summer months. Although sporadic-E is often regarded to be one of the most frequency stable layers for HF signal propagation (providing near "mirror-like" reflection with very little spectral contamination), its high spatial variability when patchy can result in multiple off–great-circle path reflections as well as other problems due to lack of continuity in propagation support.

The production mechanisms responsible for sporadic-E are quite different in the low and high latitudes. At low latitudes, sporadic-E is thought to arise from plasma instability caused by neutral winds that circulate in the atmosphere under pressure gradients and drive a concentration of enhanced currents at E-layer altitudes known as the "equatorial electrojet" (Davies 1990). Equatorial sporadic-E is normally a daytime phenomenon with little seasonal variation. At high latitudes, sporadic-E is thought to arise due to the precipitation of energetic electrons from the magnetosphere. High-latitude sporadic-E also exhibits little seasonal variation, but it occurs most often at night.

Occasionally, the sporadic-E layer can be opaque in nature with an ionization density higher than any other layer in the ionosphere. In this case, its highly reflective properties effectively screen HF signals from reaching the F-region. This phenomenon is known as "blanketing" sporadic-E. On one hand, this may allow an OTH radar to achieve frequency-stable single-mode propagation using frequencies at the upper end of the HF band (or even beyond 30 MHz), which can significantly improve detection and tracking performance for slow-moving surface vessels. On the other hand, obscuration due to blanketing sporadic-E can temporarily deny long-distance propagation via the F-region, which limits the OTH radar range coverage to one-hop ground distances not greater than 2000 km.

2.1.3.4 F-Region

The maximum electron density in the ionosphere usually occurs in the F-region, which is also the highest region of the ionosphere considered for skywave propagation. During the day, the F-region may "split" into two layers, namely, the F1-layer between heights of 140 and 210 km, and the F2-layer typically above 210 km. The bifurcation of the F-region ionization height profile into two distinct (F1- and F2-layer) peaks was previously illustrated in Figure 2.3.

More precisely, ionization densities greater than those of the normal E-layer are formed during the day in the F1-layer height interval. However, the appearance of the F1-layer with a distinct (well-resolved) electron density peak between 140 and 210 km depends to a large extent on the characteristics of the overlying F2-layer. Quite often, the height, thickness, and peak level of the F2-layer is such that it effectively subsumes the relatively weaker F1-layer below it. In other words, the F1-layer is susceptible to being "masked" by the bottom side of the F2-layer. This can give rise to a daytime electron density height profile in which only the F2-layer peak is visible.

The higher altitude of the F1-layer compared to the E-layer typically allows for one-hop skywave propagation over a maximum ground range of 2000 to 3000 km. The F1-layer is less α-Chapman like than the E-layer, but similar to the E-layer, it also effectively disappears at night with the exception of a small amount of residual ionization. For this reason, the normal E-layer and F1-layer are said to be "sun-following" or *solar-controlled* layers.

Unlike the normal E- and F1-layers, ionization in the F2-layer persists throughout the night at sufficiently high levels to effectively operate OTH radar systems. The electron density of the F2-layer may be one or two orders of magnitude higher than that of the normal E-layer. At mid-latitudes, the peak electron density in the F2-layer typically occurs between 210 and 350 km, but in the equatorial region, it may form at significantly greater heights. The F2-layer enables one-hop propagation up to ground ranges of 3000–4000 km.

The F2-layer is by far the most important layer for OTH radar and other HF systems that rely on skywave propagation. There are three main reasons for this. First, the F2-layer usually reflects signals with the highest frequency. This is often associated with a number of significant performance advantages for OTH radar, including lower ambient noise, higher spatial resolution, and increased target RCS. Second, the F2-layer is the highest altitude layer and therefore provides one-hop propagation to the longest ground ranges. Specifically, F2-layer propagation can increase the range coverage of an OTH radar by as much as 2000 km relative to E-layer propagation (i.e., by a factor of two). Third, the F2-layer is always present throughout the day and night, which permits (or does not preclude) continuous operation.

With the exception of sporadic-E, the F2-layer is regarded as the least Chapman-like of all layers in the ionosphere. The formation of the F2-layer peak is the result of transport phenomena rather than recombination (Davies 1990). Consequently, the F2-layer is much less regular than the normal E- and F1-layers. Relative to the solar-controlled layers, the F2-layer exhibits a variability in space and time that is more difficult to predict. Indeed, the F2-layer is said to be *geomagnetically controlled*. This is because certain plasma transport phenomena that contribute to the distribution of ionization in the F2-layer are strongly influenced by the Earth's magnetic field.

One reason for the longevity of the F2-layer is the redistribution of electrons to higher altitudes at night due to a combination of neutral winds, field-aligned diffusion, and electromagnetic drift. Another reason for the continuous presence of the F2-layer is the relatively slower rate of dissociative recombination above 210 km due to the relatively lower density of neutral molecular species at higher altitudes. The greater proportion of lighter atomic ions at higher altitudes (particularly the ionization of atomic oxygen by radiation in the wavelength band 14–91.1 nm), coupled with the very slow recombination rate of electrons with positive atomic ions also contributes to the longevity of the F2-layer. For these reasons, the electron density in the F2-layer survives throughout the night, albeit in reduced form, until the next day when production by photoionization resumes.

Although the gross characteristics and behavior of the F2-layer can be predicted to a certain degree, its detailed structure and dynamics cannot be forecast as accurately as those of solar-controlled layers. More to the point, the properties of the F2-layer cannot

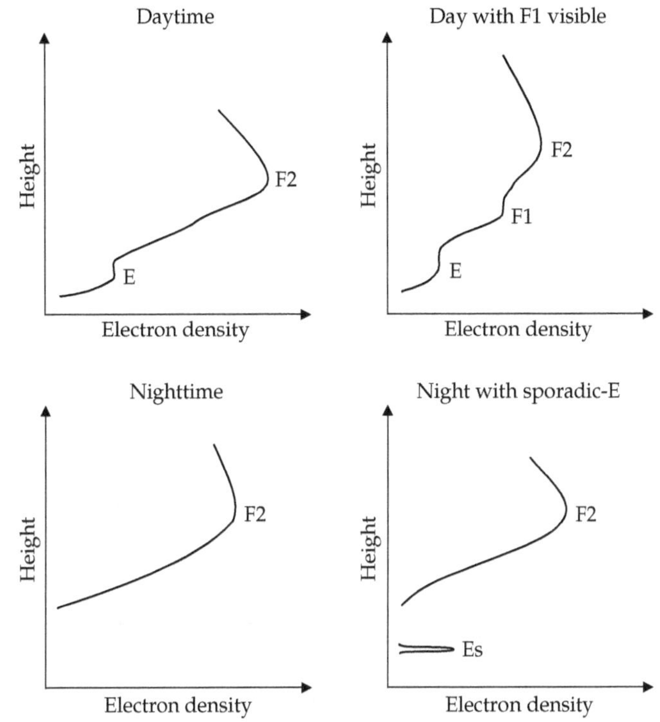

FIGURE 2.4 Four sketches qualitatively illustrating the diurnal variation of the electron density height profile in the ionosphere. The F1-layer may become visible during the day when it is not overshadowed by the F2-layer, while sporadic-E may be present in the day or night.

be modeled in advance to the fidelity required for optimizing the performance of OTH radar systems. An important consequence of this is that dedicated sounding networks need to be employed to more precisely determine the prevailing state of the ionosphere for effective OTH radar operation.

While such a network is of practical value for ascertaining the characteristics of the F2-layer (in a real-time and site-specific manner), a similar argument also applies for the lower altitude layers which are not entirely predictable, especially sporadic-E. Such sensors may be viewed as adjunct subsystems to the main OTH radar and their main functions will be discussed in Chapter 3. Figure 2.4 qualitatively summarizes the day/night variation in the electron density height profile of the ionosphere using simple illustrations.

2.1.3.5 Solar Activity

An important aspect not discussed thus far is that the electron density in the ionosphere is significantly dependent on changes in solar activity, which affect the intensity of solar radiation and corpuscular emissions incident upon the Earth's atmosphere. Changes in solar activity are often measured in terms of the sunspot number. The sunspot number has been recorded since the invention of the telescope by Galileo in 1610. Over time, the

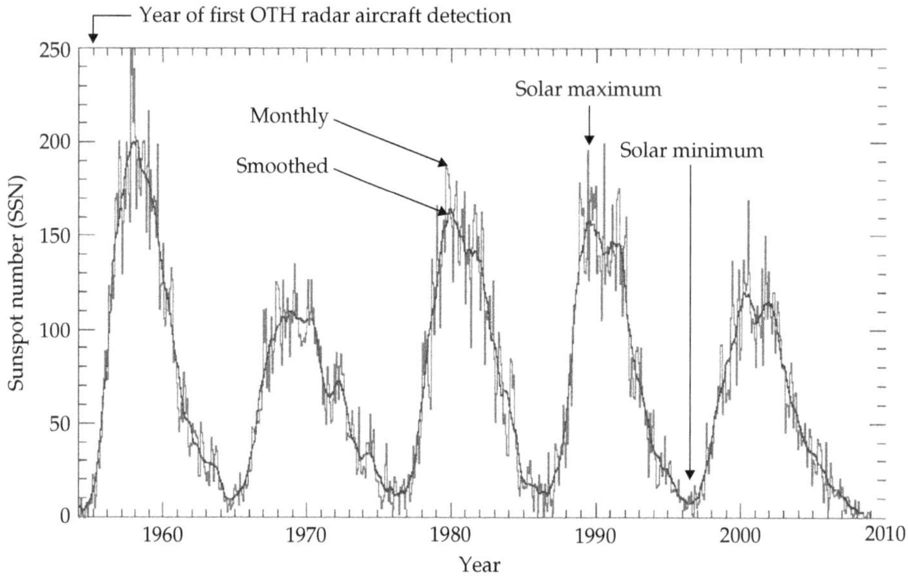

FIGURE 2.5 Observed monthly sunspot number (SSN) and 12-month running mean R_{12} (smoothed SSN) since the mid-1950s. The years corresponding to the peaks and troughs of the solar cycle are referred to as solar maximum and minimum, respectively.

number of sunspots were observed to vary, and Gauss had speculated that a correlation existed between the number of sunspots and the variability of the Earth's magnetic field.

Around 1840, an amateur astronomer Heinrich Schwabe noticed an apparent 10-year periodicity in the sunspot number. This observation led the professional astronomer Rudolf Wolf from the Zurich observatory to examine an extensive set of historical data dating back to 1700. In 1856, Wolf and his team determined that the number of sunspots varied over an 11-year period known as the solar cycle. In 1890, Maunder drew attention to a 70-year period from 1645 to 1715 during which almost no sunspots were observed. This is known as the "Maunder minimum" and is thought to form part of a larger scale (superimposed) variation. As shown in Figure 2.5, the sunspot number is observed to oscillate between zero to a few sunspots at the solar minimum and 100 or more sunspots at the solar maximum in a quasi-periodic manner over the 11-year solar cycle.

An increase in sunspot number is generally associated with higher levels of ionization in the ionosphere. At high sunspot number, the peaks of electron density not only have higher values but also tend to form at greater altitudes, particularly in the F-region. These factors can significantly influence the capabilities of OTH radar systems. For example, an OTH radar can typically use higher operating frequencies near the solar maximum, which often has a beneficial impact on system performance.

Enhanced ionization also leads to a reduction in the dimensions of the skip-zone at a given frequency, which improves the radar coverage by effectively decreasing the minimum range at the lowest design frequency. On the other hand, a raised layer altitude increases the ground-range of single-hop propagation at higher frequencies, which extends the maximum range of the system. In summary, a high sunspot number effectively enhances skywave propagation support and this can help to extend the overall coverage as well as improve the performance of an OTH radar.

However, high solar activity is also associated with an increase in disturbances, such as solar flares, which can at times adversely affect the quality and continuity of skywave propagation. While solar disturbances may occur at any time during the solar cycle, their frequency and severity often increase around the solar maximum. In other words, the benefits of enhanced ionization levels and greater layer altitude during periods of high solar activity often come at the expense of relatively less stable propagation.

With rare exceptions, solar disturbances normally have a detrimental impact on OTH radar performance. The causes and characteristics of different types of solar disturbances as well as their potential impact on HF systems will be discussed in the next section.

2.2 Spatial and Temporal Variability

When discussing the morphology of the ionosphere in space and time, it is convenient to distinguish between variations caused by the sun in its usual or *quiet* state and less typical variations occurring when certain regions on the sun occasionally change to a disturbed or *active* state. Both forms of variability significantly affect HF signal propagation and are therefore important for OTH radar. This section considers the large-scale variability of the ionosphere under the quiet sun.

The first part of this section recalls the basic theory of radio sounding at vertical incidence used to measure several key ionospheric parameters of interest to OTH radar. The spatial and temporal variability of the ionosphere is described in the second part of this section with reference to measurements and models of these key parameters. The effect of ionospheric disturbances and storms caused by different types of solar events are briefly identified and discussed in the third part of this section.

Variability of the ionosphere under the quiet sun occurs over a wide range of temporal and spatial scales. In the context of OTH radar systems, the operator responds to large-scale changes in the ionosphere by modifying the choice of waveform parameters, such as carrier frequency, bandwidth and pulse repetition frequency, or other selectable system parameters such as full- or half-aperture operation to best advantage for the mission(s) in progress. On the other hand, small-scale changes of the ionosphere require system adaptation on a dwell-by-dwell basis at the signal and data processing levels, during which time the operating parameters remain constant.

Large-scale temporal variability typically refers to diurnal changes of the electron density distribution in hour of day (or every few minutes in certain conditions), which is approximately commensurate with the lifetime of an OTH radar task using a fixed frequency and waveform parameters. Large-scale spatial variability may be viewed on the scale of the OTH radar footprint (i.e., from one surveillance region to another), where the system adapts by selecting the operating parameters on a region-by-region basis.

On a coarser level, temporal variability includes seasonal and solar-cycle changes in the ionosphere, which superimpose on the diurnal variation. Whereas the spatial variability of the ionosphere on a global scale may be considered in three magnetic latitude regions, namely, the low-latitude or equatorial region (within about 20 degrees of the dip-equator), the high-latitude or polar region (over the caps and within the auroral ovals[2]), and the

[2]Areas centered on the Earth's magnetic poles at latitudes between about 65–70 degrees.

mid-latitude region. Note that the Earth's magnetic field is well-approximated by an offset dipole field,[3] which means that the magnetic and geographic equators and poles do not coincide.

Small-scale temporal and spatial variability of the ionosphere may be thought of as "random" electron density fluctuations with respect to the underlying large-scale variation (which may be thought of as a gradually evolving "mean" ionization structure). The latter effectively determines the background state of the ionosphere at a particular time and location. This "smoothed" ionospheric structure changes over time but may be regarded quasi-stationary during relatively short intervals, usually not greater than about 10 minutes at mid-latitudes. Near the dawn and dusk terminators, or when ionospheric disturbances or storms are present, the gross characteristics of the ionosphere change more rapidly even in mid-latitude regions.

A vast amount of data pertaining to the large-scale variability of the ionosphere has been acquired by ground-based radio probing instruments or ionosondes. Routine measurements of the overhead ionosphere made by an international network of sounding stations have provided valuable climatological information to statistically characterize the morphology of the ionosphere in space and time.

This section describes the essential principles of radio-wave reflection from the ionosphere at vertical incidence to explain the main features of the ionogram traces produced by these systems. The significance of several key ionospheric layer parameters extracted from such records, and empirical models that describe the large-scale temporal and spatial variability of the ionosphere in terms of such parameters are then discussed.

2.2.1 Radio Sounding at Vertical Incidence

There are a number of fundamentally different experimental techniques for studying the ionosphere. Such techniques may be broadly divided into remote-sensing and in-situ measurements. In-situ measurements of the ionosphere are made with instruments on rockets or satellites that can directly collect data on its physical properties such as particle density, chemical composition, temperature, and electric and magnetic fields. Remote-sensing methods may be further classified as active or passive techniques. The latter infer information about the ionosphere by sensing natural emissions from either ground or space-based platforms, such as the aurora, natural airglow, and the absorption of cosmic radio waves.

On the other hand, active remote-sensing techniques make use of dedicated man-made signals that are transmitted specifically to probe the ionosphere. Examples include total electron content (TEC) measurements using topside signals from GPS satellites and radio probing of the ionosphere based on HF signals transmitted and received by terrestrial sounders. Here, attention is confined to signal reflection from the ionosphere at vertical incidence as this forms the basis of an active ground-based remote-sensing technique known as vertical incidence sounding (Hunsucker 1991).

2.2.1.1 Plasma and Critical Frequency

The reflection of radio waves at vertical incidence occurs at a height in the ionosphere where the electron density is sufficiently high to respond to the rapid oscillations of the incident electromagnetic wave. More specifically, reflection at vertical incidence occurs

[3]Here, the term "offset" refers to the displacement of the Earth's magnetic axis from its rotational axis.

at the point where the radio wave frequency does not exceed the maximum frequency of internal oscillation of the plasma, referred to as the *plasma frequency* (Davies 1990).

When the influence of collisions and the Earth's magnetic field can be neglected, the expression for the plasma frequency is given by f_N in Eqn. (2.20), where N_e is the electron density, q is the electron charge, m_e is the electron mass, and ε_o is the permittivity of free space. The plasma frequency of electrons is much greater than that of positive ions. Indeed, the latter are too massive to respond to the rapid oscillations of an incident radio wave and do not contribute significantly to the reflection process.

$$f_N = \left\{ \frac{N_e q^2}{4\pi^2 \varepsilon_o m_e} \right\}^{\frac{1}{2}} \tag{2.20}$$

When a radio wave is launched upward and encounters the overhead electron density profile, reflection at vertical incidence occurs at the height in the ionosphere where the plasma frequency first equals (or is greater than) the radio wave frequency. It is evident from Eqn. (2.20) that the plasma frequency is proportional to the square root of electron density N_e. By substituting the various constants into Eqn. (2.20), f_N is well-approximated by the simple expression in Eqn. (2.21), where f_N is in Hertz and N_e is in electrons per cubic meter.

$$f_N = 9.0\sqrt{N_e} \tag{2.21}$$

The *critical frequency* of an ionospheric layer is defined as the plasma frequency at the height of *maximum ionization* in the considered layer. For the various ionospheric layers identified previously, this important parameter is commonly denoted by f_oE, f_oEs, f_oF1, and f_oF2. For a given ionospheric layer, the critical frequency is given by Eqn. (2.22), where $\max\{N_e\}$ refers to the peak electron density within that layer. A signal with a frequency that exceeds the critical frequency of a layer will penetrate through that layer at vertical incidence and propagate to higher regions of the ionosphere or into space.

$$f_o = 9.0\sqrt{\max\{N_e\}} \tag{2.22}$$

The electron density in the ionosphere typically reaches its global maximum at the peak of the F2-layer. Hence, the F2-layer critical frequency f_oF2 is often the highest frequency that will be reflected back to Earth at vertical incidence. Signals with frequency higher than the maximum layer critical frequency will penetrate through the ionosphere at vertical incidence and propagate into space.

For an example (daytime) F2-layer with peak electron density $\max\{N_e\} = 10^{12}$, the critical frequency is $f_o\text{F2} = 9$ MHz. In this case, a signal of frequency greater than 9 MHz would not be returned to Earth by the regular process of ionospheric reflection at vertical incidence. Importantly, the ionosphere is able to reflect signals of frequency higher than the maximum layer critical frequency at *oblique* incidence. The topic of reflection at oblique incidence will be covered in Section 2.3.

2.2.1.2 Phase and Group Refractive Index

With reference to a single frequency component of a possibly modulated radio wave signal, the *phase* refractive index of a propagation medium μ is defined in Eqn. (2.23), where c is the speed of light in a vacuum and v_p is the phase velocity of the considered frequency component in the medium. The phase velocity may be interpreted as the rate at which any given point of constant phase on the wave (such as the crest) propagates through space in the medium. Specifically, the phase velocity is given by $v_p = \omega/\kappa$, where $\omega = 2\pi f$ is the angular frequency and $\kappa = 2\pi/\lambda$ is the wavenumber of the considered (monochromatic) component in the medium.

$$\mu = \frac{c}{v_p} \tag{2.23}$$

As described in Davies (1990), the phase refractive index corresponding to a plasma frequency f_N and radio-wave frequency f is given by the simple expression in Eqn. (2.24), under the assumptions of no collisions or superimposed magnetic field. When the plasma frequency equals the radio-wave frequency $f = f_N$, the phase refractive index falls to zero and signal reflection occurs.

$$\mu^2 = 1 - \frac{f_N^2}{f^2} \tag{2.24}$$

The dependence of μ on f and f_N means that the ionosphere is a dispersive and inhomogeneous propagation medium for radio waves, respectively. A more comprehensive expression for μ including the effects of collisions and the Earth's magnetic field is described in Budden (1985a). This more complex formula will be presented later to expose the anisotropic property of the radio refractive index in the ionosphere.

From Eqn. (2.24), it is evident that the phase refractive index is less than unity for $f_N > 0$. This implies that the phase velocity of the radio wave in the ionosphere exceeds the speed of light in a vacuum. Indeed, the phase velocity of a wave may be much greater than the velocities of the individual particles in the wave. However, the phase velocity of a wave is not associated with energy transfer and does not indicate the rate at which the information-carrying "waveform" is moving through the medium.

In a dispersive medium such as the ionosphere, the phase velocity is in general not the same as the group velocity v_g, which is defined in Eqn. (2.25). The group velocity may be interpreted as the rate at which the modulation envelope of the radio wave propagates through the medium.

$$v_g = \frac{\partial \omega}{\partial \kappa} \tag{2.25}$$

In analogous manner to the phase refractive index, the *group* refractive index μ' is defined as the ratio between the speed of light in a vacuum c and the group velocity of the radio wave v_g in the medium. From Davies (1990), the group refractive index can be expressed in terms of the phase refractive index according to Eqn. (2.26), where the substitutions $v_g = \partial\omega/\partial\kappa$, $c = \mu v_p$, and $v_p = f\lambda$ have been made to arrive at the rightmost expression in Eqn. (2.26).

$$\mu' = \frac{c}{v_g} = \frac{\partial}{\partial f}(\mu f) \tag{2.26}$$

When the formula for μ in Eqn. (2.24) is substituted into Eqn. (2.26), it is readily shown that the group refractive index is given by the inverse of the phase refractive index, as in Eqn. (2.27). The simple relation $\mu\mu' = 1$ holds for the special case of no superimposed magnetic field (Davies 1990).

$$\mu' = \frac{1}{\mu} \tag{2.27}$$

As the plasma frequency varies from $f_N = 0$ below the ionosphere to a value of $f_N = f$ in the ionosphere, the group refractive index varies from $\mu' = 1$ to $\mu' = \infty$. In other words, the group velocity of a radio wave slows down from $v_g = c$ prior to entering the ionosphere (in free space, strictly speaking) to $v_g = 0$ at the point of reflection.

2.2.1.3 Real and Virtual Height

The layer critical frequencies and the heights at which signals are reflected from different layers in the ionosphere represent parameters of practical interest to HF systems such as OTH radar. It is important to distinguish between the *real* and *virtual* height of signal reflection in the ionosphere. The real height h_r corresponds to the actual physical height above the Earth's surface at which reflection takes place. On the other hand, the virtual height h_v is defined in Eqn. (2.28) as the time-of-flight τ between the transmission of the signal and the reception of its reflected echo at the same point on the ground (i.e., round-trip delay) multiplied by the speed of light in free space and divided by two. As the group velocity of a radio wave slows down when it enters the ionosphere, the real height h_r is always less than the virtual height h_v.

$$h_v = \frac{1}{2}\int_0^{\tau} c \, \mathrm{d}t \tag{2.28}$$

By rearranging the group refractive index formula $c = \mu' v_g$, and noting that $v_g = \mathrm{d}h/\mathrm{d}t$ for the case of vertical incidence, the term $c \; \mathrm{d}t$ is equivalent to $\mu' \; \mathrm{d}h$. Substituting this equivalence into Eqn. (2.28) and replacing the integration over time-of-flight with integration over real height yields Eqn. (2.29), where the limits of integration from the ground $h = 0$ to the true reflection height h_r, without dividing by two, reflects the reciprocity of the upward and downward signal paths. This basic equation relates the virtual reflection height (time-of-flight) to the overhead group refractive index profile. Under the previously stated assumptions of no collisions or superimposed magnetic field, variation of the group refractive index $\mu' = 1/\mu$ with h depends on the radio frequency f and the electron density height profile through the plasma frequency parameter f_N.

$$h_v = \int_0^{h_r} \mu' \, \mathrm{d}h \tag{2.29}$$

The virtual height may alternatively be written as an integral over plasma frequency f_N in Eqn. (2.30). Here, the integration limits extend from $f_N = 0$ to $f_N = f$ as the signal is reflected at the height where the plasma frequency first equals the radio frequency (for a continuously differentiable electron density profile).

$$h_v = \int_0^{f} \mu' \left\{ \frac{\mathrm{d}h}{\mathrm{d}f_N} \right\} \mathrm{d}f_N \tag{2.30}$$

The expressions in Eqn. (2.29) and Eqn. (2.30) indicate that the structure of the overhead electron density profile can be inferred by radio probing at vertical incidence, i.e., from a trace of h_v against f. In the following section, the form of Eqn. (2.30) will help to provide useful insights for interpreting an important feature of this trace, which is also referred to as a vertical incidence ionogram.

2.2.1.4 Vertical Incidence Ionogram

A ground-based vertical incidence sounder (VIS) or *ionosonde* is an instrument that transmits radio waves vertically upward and measures the time-delay between the radiated signal and the reception of its echo reflected back down from the ionosphere as a function of frequency. The time-delay may be measured using either amplitude modulated pulse waveforms of short duration (pulse sounder), or linear frequency modulated continuous waveforms (chirp sounder), see Poole and Evans (1985). Both methods have advantages and disadvantages (Davies 1990).

The signal frequency is typically swept through the HF band, such that a record of time-of-flight against frequency is obtained. The resulting trace is called a vertical incidence ionogram. VIS systems typically operate over a frequency interval within the 2–20 MHz band. The frequency resolution of traditional systems is around 50 kHz, while the group-range resolution is about 3 km. Modern VIS systems that employ state-of-the-art hardware and signal processing can yield measurements with higher resolution and accuracy; (Reinisch 2009).

Figure 2.6 conceptually shows the relationship between an idealized VIS trace and a representative (daytime) electron density height profile in the ionosphere. Signals of frequency less than a certain minimum frequency, denoted by $f_{\min}$ in Figure 2.6, do not produce detectable echoes due to significant (non-deviative) absorption in the D-region. Extraction of the parameter $f_{\min}$ is considered important as it effectively represents the sole index of absorption provided by ionograms.

As the frequency increases beyond $f_{\min}$, radio wave energy is able to effectively penetrate through the D-region (twice) such that echoes start to appear due to reflections from the E-region. The virtual height of signal reflection increases gradually with frequency at first, but then increases very rapidly as the E-layer critical frequency is approached. This results in the formation of a sharp *cusp* in the ionogram trace at the radio frequency where the signal is reflected from the real height of peak electron density in the E-layer.

To understand the reason for this, we refer back to the expression for the virtual height in Eqn. (2.30). Note that the term $\mathrm{d}h/\mathrm{d}f_N$ tends to infinity as the gradient of plasma frequency with respect to height tends to zero, i.e., $\mathrm{d}f_N/\mathrm{d}h \rightarrow 0$. This occurs at the point of peak electron density in a layer, where the rate of change in electron density with height approaches zero. In addition, when a radio wave is reflected from the local maximum of electron density in the E-layer, for example, we have that $f = f_N = f_o E$, which implies that the group refractive index $\mu' \rightarrow \infty$ in the integrand. This combination of effects causes the integral for the virtual height in Eqn. (2.30) to approach infinity when the radio frequency approaches the layer critical frequency.

The changing slope of the cusp signifies that the virtual height increases progressively more rapidly than the true height as the wave frequency approaches the layer critical frequency. This phenomenon, known as *retardation*, is present over the whole trace but is most accentuated near the layer critical frequency. In practice, the cusps observed on measured ionograms do not extend to an infinite height. This is not only due to the finite resolution of experimental equipment, but also because the reflected signal is heavily attenuated at the layer critical frequency due to the process known as *deviative* absorption (Davies 1990).

Once the signal frequency increases beyond the E-layer critical frequency, the radio wave penetrates through the E-layer and propagates into the F-region. As shown in Figure 2.6, a similar process to that described for the E-layer is repeated for the F1- and F2-layers. These higher altitude layers respectively reflect signals of higher frequency from

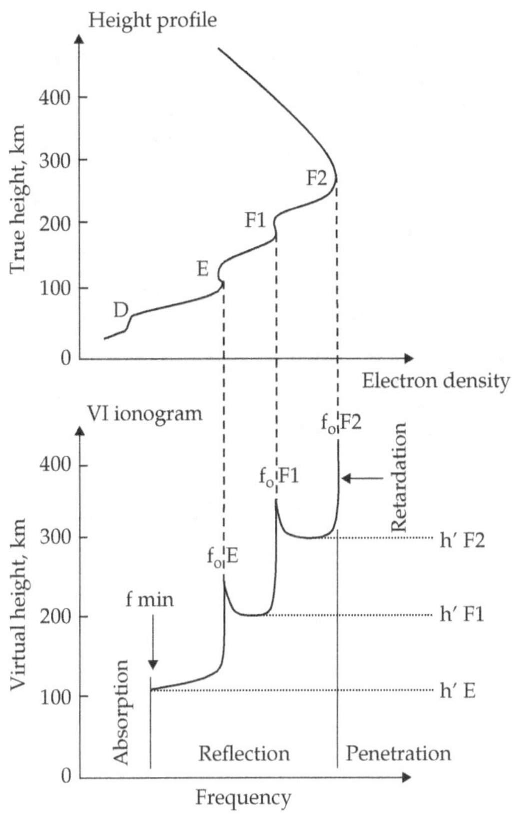

FIGURE 2.6 Conceptual illustration of the relationship between electron density height profile in the ionosphere and the vertical incidence ionogram trace, which is a plot of signal virtual reflection height (time-of-flight) against radio-wave frequency. © Commonwealth of Australia 2011.

greater virtual heights until the critical frequency of the F2-layer is reached. Beyond this point, the transmitted signals are not returned to Earth but penetrate the ionosphere at vertical incidence. Determination of the cusp locations enables the layer critical frequency parameters f_oE, f_oF1, and f_oF2 to be estimated from the ionogram trace. The (minimum) layer virtual height parameters, denoted by h'E, h'F1, and h'F2, are extracted from the ionogram as the minimum heights of the echoes observed for the corresponding layers where the trace is most horizontal.

A host of other ionospheric parameters may be derived by interpreting the ionogram trace (Piggot and Rawer 1972). In particular, mathematical techniques can be applied to "invert" the VIS trace to obtain an estimate of the overhead electron density height profile. In addition to estimating the peak electron densities, the profile also indicates the true heights and thicknesses of layers in the ionosphere. The inversion procedure generally requires the use of sophisticated numerical techniques. In practice, the implementation of automatic inversion methods that yield reliable results using real ionogram traces is far from straightforward. However, such procedures are important in that they allow

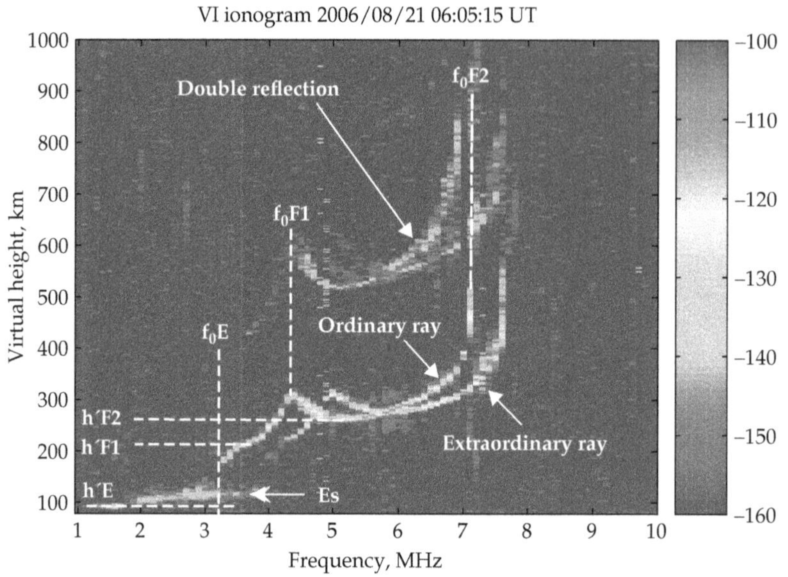

FIGURE 2.7 Real VIS ionogram measured by a sounding station located at Kalkarindji, Northern Australia, geodetic latitude −17.444°, longitude 130.829°. The ionogram was recorded on 21 August 2006 at 06:05 universal time (UT), or 14:47 local time (LT). Estimates of the layer critical frequencies and (minimum) virtual reflection heights made by manual scaling of the ionogram have been overlayed for reference. © Commonwealth of Australia 2011.

electron density height profiles of the ionosphere to be measured at specific locations in near real time. As will be described in more detail later, a real time ionospheric model is valuable for accurate coordinate registration in OTH radar.

Real ionograms are much more complicated than the idealized picture of Figure 2.6 for several reasons. For instance, the presence of the Earth's magnetic field causes a radio wave incident on the bottom of the ionosphere to split into two (ordinary and extraordinary) characteristic waves with different polarization.[4] These two so-called o- and x-waves experience a different refractive index in the ionosphere and are reflected independently, or nearly so, which leads to the production of separate traces in the ionogram. There is also the possibility of multiple ionospheric reflections, due to one or more intermediate ground reflections, or mixed echoes returned by reflection from two or more different layers. In addition, higher layers may not always produce echoes due to blanketing from lower layers.

For these (and other) reasons, recorded ionograms exhibit a wide variety of characteristics with respect to the conceptual diagram sketched in Figure 2.6. To illustrate an example, Figure 2.7 shows a real VIS trace collected on a winter day at a station in Northern Australia. In this record, there is clear evidence of magneto-ionic splitting of the incident signal into ordinary and extraordinary waves, which are indicated by the pair

[4]The *birefringent* property of the ionosphere is discussed in Section 2.4.

of well-resolved traces in Figure 2.7. Doubly reflected echoes involving an intermediate ground-bounce are also evident and labeled in this display. Another feature of the ionogram is the partial blanketing of the normal E-layer by sporadic-E at lower frequencies.

The second edition of the URSI Handbook for Ionogram Interpretation and Reduction (Piggot and Rawer 1972) contains the agreed standard rules and conventions for manual scaling and reduction of vertical incidence ionograms. Considerable latitude is provided in this document for stations to independently adapt and refine the criteria for ionogram interpretation. However, this handbook (also known as UAG-23A) provides a common guide to enable the meaningful comparison of ionospheric data recorded by different sounder networks.

The advent of digital ionosondes has led to the use of automatic scaling procedures to avoid the tedious effort of manual scaling (Reinisch and Huang 1983). Specifically, a scaling algorithm known as ARTIST was developed to automatically extract the trace of the ordinary wave from the VIS ionogram.[5] The electron density height profile is then calculated from this extracted trace using an inversion technique, such as those described in Titheridge (1988) and Paul (1977).

2.2.2 Measurements and Models

Models of the ionosphere fall into two main categories. Firstly, there are models that describe the physical and chemical processes occurring in the ionosphere. Alternatively, there are models that describe the characteristics of radio-wave propagation in this medium. Both types of models may be based on theoretical principles, empirical observations, or a mixture of both. Although models in the former category are clearly related to those in the latter by cause and effect, their formulation and development is motivated from quite different perspectives.

As far as OTH radar system design and operation is concerned, the effect of the ionosphere on HF signal propagation is of more direct interest than the physical and chemical processes that give rise to the ionosphere per se. For this reason, this section mainly deals with models related to the properties of radio-wave propagation in the ionosphere. However, a knowledge of the underlying ionospheric processes that lead to the observed radio-wave propagation properties is important. The description of quantitative models for such processes is beyond the scope of this text, but qualitative explanations of certain important ionospheric phenomena will be provided in this section.

2.2.2.1 Climatological Ionospheric Models

Since the International Geophysical Year (1957–1958), a great deal of information regarding the ionosphere has been generated from experimental measurements collected by a worldwide network of sounders. Data recorded at approximately hourly intervals by a large number of widely spaced systems has enabled a geographic distribution of ionospheric control points to be regularly monitored. The assimilation of data recorded at different stations over several decades has resulted in the compilation of valuable synoptic databases for studying ionospheric structure and variability.

Detailed analysis of these databases has led to the development of empirically derived climatological models that characterize the observed patterns of behavior in the form of

[5]To date, the latest version of this program (ARTIST 5) is reported in Reinisch (2009).

smoothed ionospheric maps of electron density distribution (Rawer and Bilitza 1989). The climatological models used to generate ionospheric profiles at a selected location and time form an integral part of the freely available International Reference Ionosphere (IRI) (Bilitza n.d.). Indeed, the IRI climatological model is widely used in conjunction with numerical ray-tracing techniques[6] to predict the characteristics of skywave propagation for a variety of HF systems.

In certain applications, such as HF communications over a point-to-point ionospheric link, climatological models can assist with system planning, design, and operation through the prediction of propagation conditions for the considered circuit, including maximum and minimum usable frequencies, elevation angles, mode probabilities, delay dispersion, and path loss due to absorption (Ferguson and McNamara 1986), all of which influence system performance. Propagation information may also assist spectrum licensing authorities to determine an appropriate allocation of day/night frequency channels so that systems can maintain reliable service while minimizing interference between stations. Climatological models may also provide valuable insights for planning and design of OTH radar systems in the absence of site-specific measurements.

However, the greater sensitivity of OTH radar performance to disparities between the predicted and actual state of the ionosphere over the ionospheric control points of interest limits the practical use of climatological models from the real-time operation perspective. Climatological models are ultimately based on statistical measures of variability computed from historical data; for example, the median and upper/lower deciles of the layer critical frequencies at a particular time and location.

Even in the (relatively benign) mid-latitude regions, the day-to-day variability of the ionosphere with respect to predictions can be significant for OTH radar applications, particularly the F2-layer and sporadic-E. The inability of climatological models to accurately forecast the state of the ionosphere means that they are of limited use for providing real-time advice to guide OTH radar operation. As small departures from the optimal choice of operating parameters can have a dramatic impact on OTH radar performance, maintaining a site-specific real-time ionospheric model (RTIM) is critical for effective OTH radar operation. The main features and uses of the RTIM maintained by an OTH radar will be described in Chapter 3.

2.2.2.2 Solar-Controlled Layers

The temporal and spatial variability of different altitude regions in the ionosphere may be characterized directly in terms of the morphology of electron density distribution. It may also be described in terms of the variability of a reduced set of radio-wave propagation parameters including, but not limited to, layer critical frequencies and virtual reflection heights, for example. To provide a first-order appreciation of the variations in the lower layers of the ionosphere, the large-scale morphology of the E- and F1-layers is described here in terms of spatial and temporal changes in the layer critical frequency.

The E- and F1-layers are often called "sun-followers" because the large-scale variation of these layers is strongly dependent on changes in the intensity of solar radiation, and hence the rate of production by photoionization. As mentioned previously, the intensity of ionizing radiation is significantly influenced by the level of solar activity, which is often measured in terms of the sunspot number. The active regions on the sun which

[6]Ray-tracing is used to model radio-wave propagation through a specified "snapshot" of the ionosphere.

increase the flux density of ionizing radiation are actually the *plage* regions surrounding the sunspots, but these regions are not as easy to observe, so sunspots serve as indicators of solar activity. This dependence clearly implies long-term changes in E- and F1-layer critical frequencies over the 11-year solar cycle. The flux density of ionizing radiation also depends on the sun's zenith angle χ at the considered time and location. This in turn implies diurnal and seasonal variations of the E- and F1-layer critical frequencies.

A method for predicting the normal E-layer critical frequency parameter f_oE at any time and place is described in Muggleton (1975). A statistical accuracy (median of the rms deviation between predicted and measured values) of about 0.1 MHz in f_oE may be expected using this method, except in auroral zones and where the sun remains below the horizon throughout the day. The method provided by Muggleton has been accepted as the standard (IRI) procedure for predicting f_oE. The findings of Trost (1979) have been incorporated into the IRI model to improve the nighttime variation. Muggleton's method is accurate and relatively simple to calculate but it cannot be expressed succinctly in terms of a single equation for all cases. The reader is referred to the original paper (Muggleton 1975) and the open-source code (Bilitza n.d.) for a complete specification of the IRI standard.

A more succinct, but slightly less accurate, empirical formula that is nevertheless in good agreement with measured values of f_oE appears in Davies (1990). To a first approximation, the *daytime* critical frequency of the E-layer is given by Eqn. (2.31), where f_oE is in MHz, χ is the solar zenith angle, and R_{12} is the 12-month running-mean sunspot number. Changes in R_{12} and χ effectively model the temporal and spatial variability of the daytime E-layer critical frequency.

$$f_o\text{E} = 0.9[(180 + 1.44R_{12})\cos\chi]^{0.25} \qquad (2.31)$$

An important qualification is that the formula in Eqn. (2.31) is not valid at magnetic latitudes greater than about 70 degrees, where particles from the magnetosphere can contribute significantly to E-region ionization. The model in Eqn. (2.31) provides a reasonably good fit to f_oE during the day, usually to within 0.2 MHz of observed values. Additional formulas are reported in Davies (1990) for modeling the sunrise, sunset, and midnight E-layer critical frequencies. Such expressions account for the small amount of residual ionization that remains in the nighttime E-region. The critical frequency of the normal E-layer reaches a minimum of about 0.5 MHz at night.

Figure 2.8 illustrates examples of the temporal and latitudinal variation of f_oE using Eqn. (2.31) for values of f_oE above 1 MHz. In particular, Figure 2.8a shows the diurnal variation of f_oE predicted at the Kalkarindji site in central Australia. Curves have been plotted for August and December of 2006 (a year near the solar minimum) and August of 2001 (a year near the solar maximum). Inspection of the solid and dashed curves in Figure 2.8a illustrates that seasonal variability manifests itself as a change in both the peak value of f_oE as well as the longevity of ionization above a certain level throughout the day. A comparison of the solid and dash-dot curves in Figure 2.8a shows that the solar-cycle variation significantly affects the peak value of f_oE.

The model of f_oE corresponding to the solid curve in Figure 2.8a is directly relevant to the real VI ionogram trace in Figure 2.7. More specifically, the predicted value at around 06:05 UT (14:47 LT) is approximately 3.2 MHz, which agrees closely with the value of f_oE in Figure 2.7. As stated previously, predictions using Eqn. (2.31) are typically within 0.2 MHz of observed values. Figure 2.8b illustrates the geographic variation of the local noon E-layer critical frequency for the same months and years over a geodetic latitude

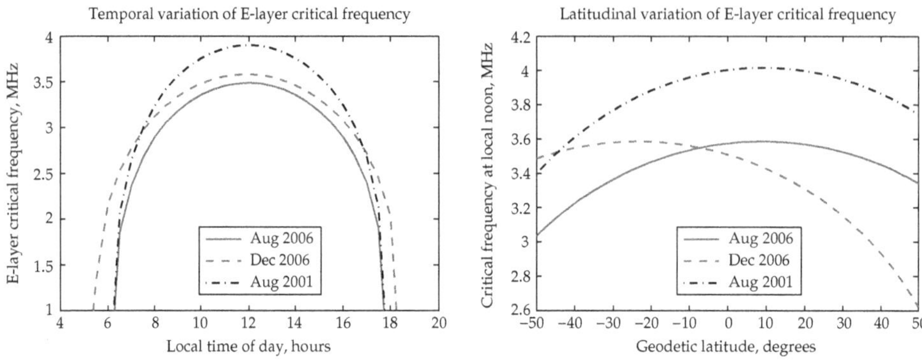

(a) Temporal variation of f_0 E. (b) Latitudinal variation of f_0 E.

FIGURE 2.8 Empirical model describing first-order, large-scale variation of the E-layer critical frequency as a function of time (diurnal, seasonal, solar cycle) and location (geodetic latitude).

band between ±50 degrees. As expected, seasonal variations of f_oE are greatest at higher latitudes (difference between the solid and dashed curves). The sunspot number affects the scale much more than the structure of the latitudinal variation (comparison of the solid and dash-dot curves). The height of peak ionization in the E-layer also varies but is mostly confined to an interval of 100–120 km during daylight hours.

The bifurcation of the F-region into F1- and F2-layers is a phenomenon observed only during the day. The visibility of the F1-layer depends largely on the peak electron density, height, and thickness of the overlying F2-layer. The F1-layer peak is observed more often when f_oF2 is low, such as at low sunspot numbers and during ionospheric storms. In the northern hemisphere, the F1-layer tends to appear more frequently in summer months, but this phenomenon is not prominent in the southern hemisphere. The F1-layer exhibits a more variable height of peak ionization than the E-layer, but the peak often forms at altitudes between 140 and 210 km. Experimental observations suggest that large-scale variability of the F1-layer critical frequency is approximately linearly related to R_{12}. Diurnal, seasonal, and geographic variations are superimposed on those of the solar-cycle due to changes in the sun's zenith angle χ.

A comprehensive method for accurately estimating f_oF1 in terms of magnetic latitude and a quantity known as the ionospheric index defined in Davies (1990) is described by Ducharme, Petrie, and Eyfrig (1971). This method has been accepted as the IRI standard with the only minor modification being the use of the magnetic dip latitude instead of the magnetic dipole latitude (Bilitza, n.d.). The specific formulas used to compute f_oF1 in IRI are also not expressed as a single equation and these formulas will not be repeated here for brevity. Although not as comprehensive as the IRI standard, an alternative expression for f_oF1 given by the single formula in Eqn. (2.32) is empirically found to provide a good fit against measured data (Davies 1990).

$$f_o\text{F1} = (4.3 + 0.01 R_{12}) \cos^{0.2} \chi \tag{2.32}$$

Figure 2.9, in the same format as Figure 2.8, shows the variability of f_oF1 using Eqn. (2.32). The predicted value of f_oF1 for August 2006 at 14:47 LT is approximately 4.4 MHz, which

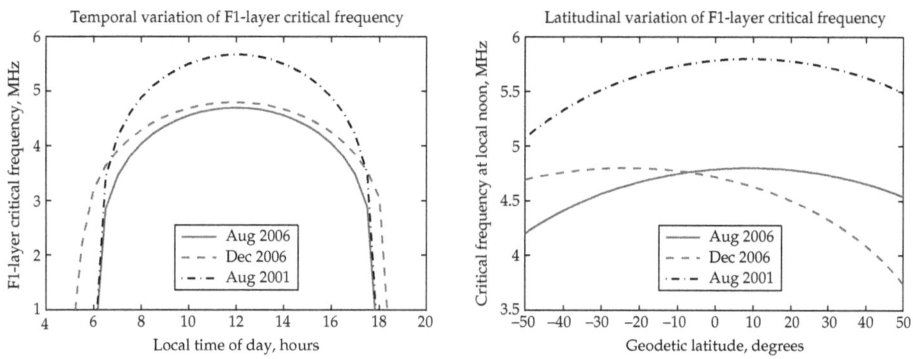

(a) Temporal variation of f_0 F1.

(b) Latitudinal variation of f_0 F1.

FIGURE 2.9 Empirical model describing first-order, large-scale variation of the F1-layer critical frequency as a function of time (diurnal, seasonal, solar cycle) and location (geodetic latitude).

also agrees well with the value observed in the real ionogram of Figure 2.7. With respect to the E-layer, the F1-layer exhibits higher critical frequencies and a more variable height (when resolved from the F2-layer). The linear dependence on R_{12} also affects the solar-cycle variation differently.

To first-order, the remaining structural features of the large-scale F1-layer variation follow a similar general pattern to the E-layer. The E- and F1-layers are relatively stable ionospheric regions for HF propagation. However, they only support daytime operation of HF systems that rely on the skywave mode. In addition, the relatively lower heights of these layers provide for short to medium one-hop path lengths and only reflect low to intermediate frequencies in the HF band compared to the F2-layer.

2.2.2.3 F2-Layer

Whereas simple analytical expressions can be used to yield relatively good predictions of the E- and F1-layer critical frequencies, no formulas exist to accurately forecast the variation of F2-layer critical frequency as a function of time and location. The height of the F2-layer peak also varies more significantly than those of the lower layers. The irregular behavior of the F2-layer is associated with the strong influence exerted on it by the Earth's magnetic field. Plasma transport processes due to the combination of neutral winds, electromagnetic drift, and field-aligned diffusion contribute significantly to changes in the F2-layer electron density distribution.

To illustrate the temporal variability of the F2-layer, Figures 2.10 and 2.11 respectively show the diurnal changes of the layer parameters f_0F2 and h_mF2 based on ionogram data recorded on 2 April 2008 by the VIS system at Kalkarindji. The parameter h_mF2 denotes the real height at which the F2-layer electron density reaches its maximum value. Note that the critical frequency drops to its lowest value not long before dawn but maintains a relatively high value between 3 and 4 MHz during most of the night. Such values are commensurate with the maximum daytime E- and F1-layer critical frequencies.

From Figure 2.10, the parameter f_0F2 reaches a peak value of nearly 11 MHz slightly before noon. However, in the early afternoon, the critical frequency is observed to fall by more than 4 MHz and later recovers to a value greater than 10 MHz in the mid-afternoon.

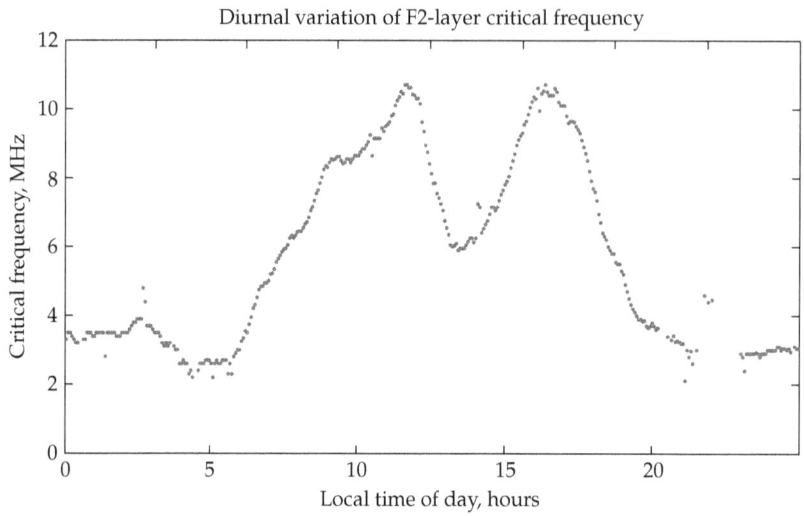

FIGURE 2.10 Diurnal variation in F2-layer critical frequency f_0F2 on 2 April 2008 at Kalkarindji. © Commonwealth of Australia 2011.

This example quantitatively illustrates the significantly higher daytime values of f_0F2 compared to the f_0F1 and f_0E predictions as well as the relatively weaker dependence of f_0F2 on solar zenith angle.

The variation of the real height parameter h_mF2 is noisier than that of f_0F2 in Figure 2.11. This is partly because the estimates of h_mF2 are derived using an inversion process that

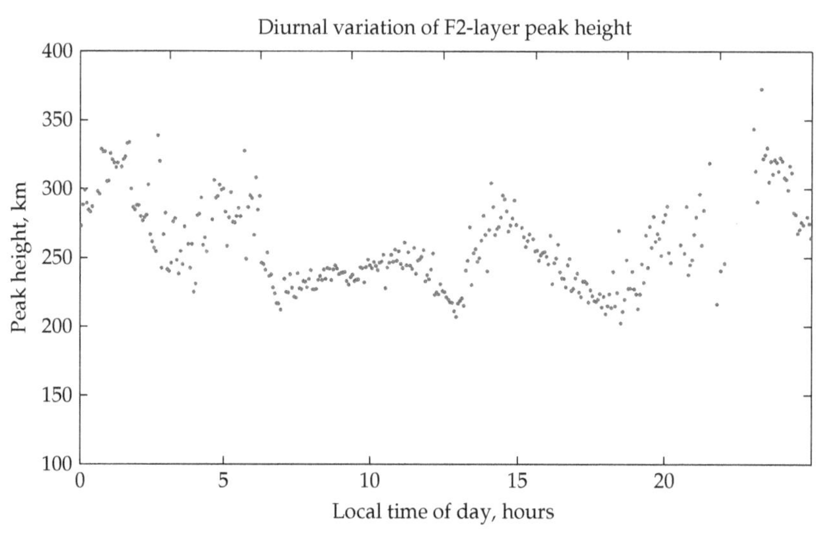

FIGURE 2.11 Diurnal variation in F2-layer peak height h_mF2 on 2 April 2008 at Kalkarindji. © Commonwealth of Australia 2011.

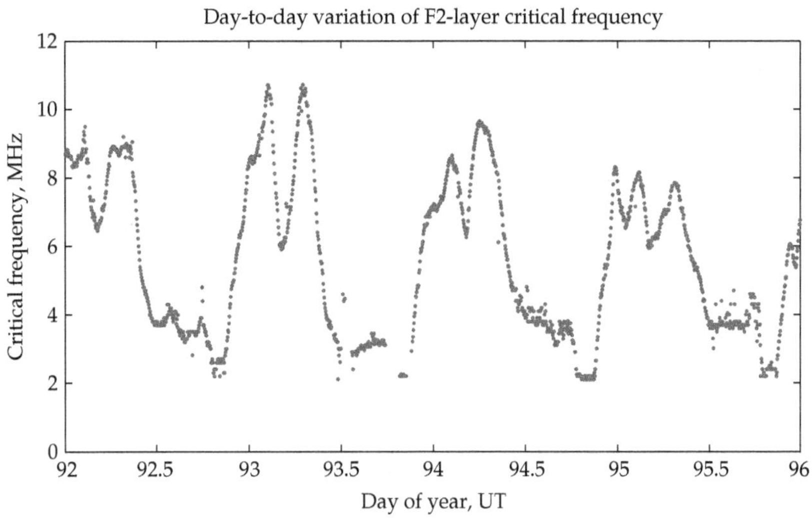

FIGURE 2.12 Day-to-day variation of F2-layer critical frequency f_0F2 over the first four days of April, 2008, at Kalkarindji. © Commonwealth of Australia 2011.

is sensitive to ionogram features. A discussion of inversion procedures can be found in Titheridge (1988) and Paul (1977). In Figure 2.11, the height of the F2-layer peak rises to about 325 km at night and falls to about 225 km during the day. This large difference illustrates the significant variability in the height of the F2-layer peak over a 24-hour period. During the early afternoon, the real height parameter h_mF2 increases by almost 100 km in less than 2 hours. This example illustrates that the F2-layer characteristics can change rapidly well away from the dawn and dusk terminators, even in the relatively benign mid-latitude region.

Figures 2.12 and 2.13 respectively illustrate the variation of f_0F2 and h_mF2 measured at Kalkarindji over four consecutive days starting on 1 April 2008 (day 92). Qualitatively speaking, the diurnal variation exhibits a similar gross behavior from day to day when viewed on a macroscopic level. However, in quantitative terms, the variability of the measured F2-layer parameters at a given time of day is significant, even over two consecutive days.

It is clear from Figure 2.12 that any attempt to fit a monthly median model to the diurnal variation of the F2-layer critical frequency would at certain times lead to significant departures between predicted and actual measurements on any given day. At mid-latitudes, the upper and lower deciles of f_0F2 depart from the median typically by ± 25 percent. In other words, it is not possible to accurately forecast f_0F2 in advance using historical monthly median data. Other F2-layer parameters, such as peak height and thickness, are also not amenable to accurate prediction using monthly median models.

Climatological models are based on synoptic databases and can only make predictions about the present or future state of the ionosphere at a given location in a statistical sense. The variability of actual values with respect to predictions are significant for OTH radar, which represents one of the most demanding types of HF system as far as knowledge about skywave propagation is concerned. Relatively small differences between predicted

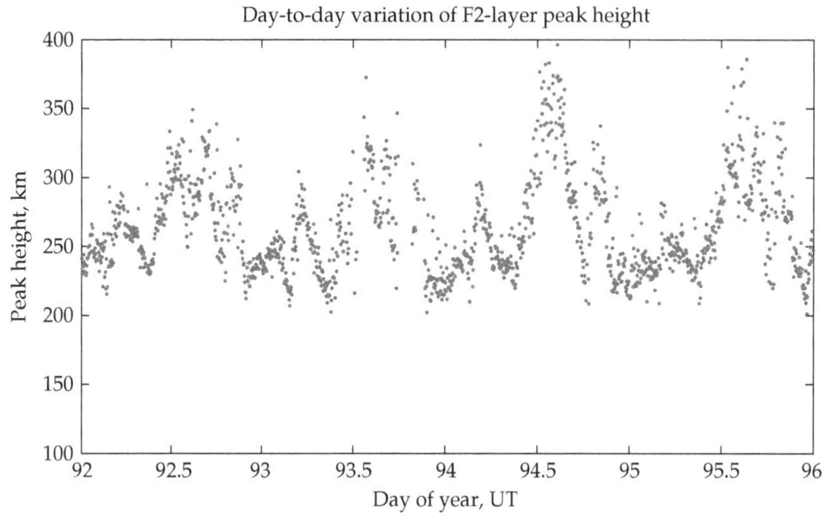

FIGURE 2.13 Day-to-day variation of F2-layer peak height h_mF2 over the first four days of April, 2008, at Kalkarindji. © Commonwealth of Australia 2011.

and actual propagation conditions can lead to substantial degradations in performance. For this reason, OTH radars cannot rely on climatological models for effective day-to-day operation.

To significantly reduce statistical errors, a real-time ionospheric model (RTIM) based on site-specific measurements is required. The RTIM not only guides the selection of operating parameters to enhance OTH radar target detection and tracking performance, but as described in Chapter 3, it also provides propagation-path information needed for accurate coordinate registration and track association.

Despite the less predictable behavior of the F2-layer with respect to solar-controlled layers, some general F2-layer characteristics can be identified. In mid-latitude regions, the critical frequency of the F2-layer is usually lowest shortly before dawn and rises to a maximum around the mid-afternoon. Summer f_0F2 levels are generally greater than those in winter at mid-latitudes. However, in the northern hemisphere, the peak electron density tends to be higher in the middle of the day in winter than in summer. This phenomenon is called the *seasonal anomaly*. Its origins are believed to be a result of seasonal changes in the relative concentrations of atomic and molecular species in the F-region, which alter the balance between production and loss in the summer time. This anomaly is usually absent at mid-latitudes in the southern hemisphere, particularly during periods of low solar activity.

Over the solar cycle, the F2-layer critical frequencies rise dramatically with increasing sunspot number. The higher critical frequencies near the solar maximum are typically accompanied by an increase in the altitude of the F2-layer peak. Worldwide maps of monthly median f_0F2 can be found in Fox and McNamara (1988), while physical models of the F-region ionosphere on a global scale are discussed in Sojka (1989).

As evident from Figure 2.10, the dawn and dusk terminators result in highly dynamic transformations to the ionosphere. The layer critical frequencies and heights can change markedly on a time-scale of minutes during the passage of each terminator.

Significant horizontal gradients or "tilts" in the ionosphere can also be formed around the onset of sunrise or sunset. This can produce significant off–great circle reflections and sharp changes in propagation conditions on a spatial scale of a few hundred kilometers. An OTH radar is required to frequently update its RTIM and operating parameters during such periods in order to adapt effectively to changing propagation conditions.

2.2.2.4 Low and High Latitudes

At low magnetic latitudes, the interaction between the eastward-pointing electric field due to neutral winds in the equatorial E-region and the near horizontal north-south magnetic field lines causes electrons to flow upward in the ionosphere by the transport process of electromagnetic (E × B) drift Kelley (2009). As the electrons rise to higher altitudes, further vertical movement across the magnetic field lines becomes inhibited in the absence of an external electric field. This forces the electrons to spiral back down along the magnetic field lines by the process of field-aligned diffusion.

The combination of neutral winds, electromagnetic drift, and field-aligned diffusion effectively redistributes electrons in the F2-layer from over the magnetic dip-equator to lower altitudes at higher magnetic latitudes of around ±20 degrees. This phenomenon is is known as the *fountain effect*. The enhanced crests of ionization formed north and south of the magnetic equator as a result of the fountain effect are known as the *equatorial anomaly* or the Appleton anomaly (Appleton 1954). These crests are visible in the climatological map of f_oF2 shown in Figure 2.14.

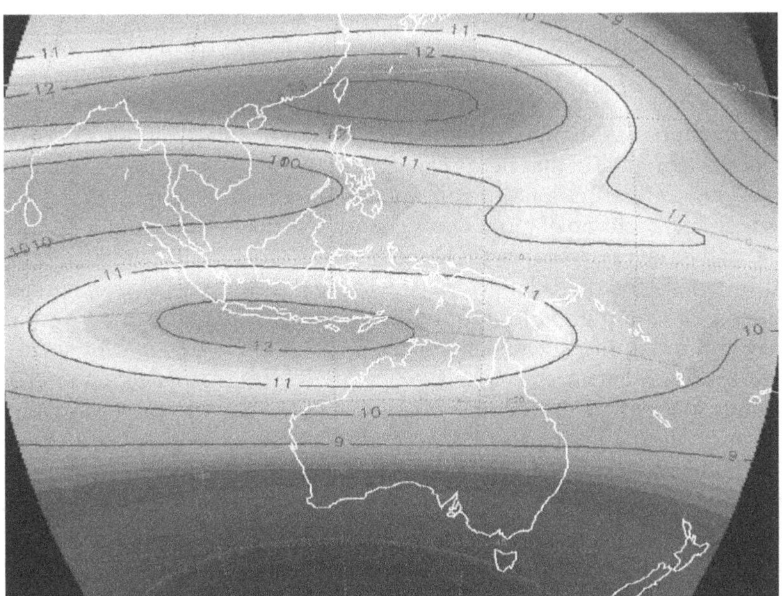

FIGURE 2.14 Climatological map showing iso-contours of f_0F2 on a January day near the solar maximum (06:00:00 UT, 1999). The enhanced ionization of the equatorial anomaly, also known as the Appleton anomaly, are clearly visible on either side of the equator. Note that the crests form around the geomagnetic equator, which is displaced from the geodetic equator.

The equatorial anomaly tends to remain centered at a fixed magnetic dip and sweeps around the Earth from east to west lagging behind the sub-solar point. The ionization is most developed in late afternoon and early evening, particularly near the solar maximum and the equinoxes. The anomaly represents a sharp spatial variation in f_0F2 and h_0F2. The rapid change in critical frequency and large gradients associated with the equatorial anomaly, as well as the rise of the ionosphere at the magnetic equator where the peak of the F2-layer can be as high as 500 km, may under certain conditions allow HF signals to propagate abnormally long-distances on a north-south transequatorial path (in excess of 6000 km) without an intermediate ground reflection.

Signal paths that transit through the equatorial ionosphere are often not frequency-stable due to scattering from dynamic field-aligned electron density irregularities. The reception of backscatter from equatorial regions is generally undesirable for OTH radar as it can lead to unwanted spread-Doppler clutter.

In polar regions and within the auroral ovals, temporal and spatial variations of the ionosphere are heavily influenced by the orientation and convergence of the geomagnetic field lines. The exceedingly high variability of the ionosphere at high latitudes stems largely from the near vertical dip angle of the geomagnetic field lines, which provides a pathway for charged particles carried in the solar wind to penetrate into lower regions of the ionosphere. Charged particles in the solar wind that are intercepted by the Earth spiral around the magnetic field lines and follow them down to relatively low altitudes in the high latitude ionosphere where they increase electron density by the process of collisional ionization, particularly at E- and D-region heights. The polar ionosphere is effectively "unprotected" from solar events, especially near the magnetic poles, where the Earth's magnetic field lines change from closed (connected to their images in the opposite hemisphere) to open, that is, connected to the interplanetary magnetic field (Skolnik 2008b).

Collisional ionization within the auroral ovals produces the aurora borealis (northern lights) and aurora australis (southern lights) at E-region heights of about 100 km. Visible light is emitted when some of the energy carried by the charged particles is radiated upon collision with neutral particles in the ionosphere. Increased ionization in the D-region can lead to higher absorption of radio wave energy, while auroral sporadic-E can lead to screening of higher layers. As far as OTH radar is concerned, signals scattered from the high-latitude regions are often characterized by severe delay and Doppler spread.

Backscatter from the high-latitude ionosphere can seriously degrade performance when received through the antenna sidelobes or range ambiguities. Not surprisingly, most skywave OTH radars are designed to operate with signal reflection points in the mid-latitude regions, where the ionosphere is more frequency stable. Even so, great care must be taken to avoid the inadvertent reception of spread-Doppler clutter from the low- and high-latitude ionospheric regions via waveform selection (to suitably position range ambiguities) and appropriate antenna design (to provide adequate sidelobes and front-to-back ratio).

2.2.3 Disturbances and Storms

Energy does not emanate out from the sun at constant rate, but varies as a function of time and over different regions of the sun's surface. This is because the sun is not a homogeneous plasma; it has internal heating and cooling mechanisms that cause the visible surface of the sun to have constantly evolving regions of lesser or greater activity.

Occasionally, certain regions on the sun's surface may change from their usual "quiet" state to a disturbed or "active" state. Disturbances in solar activity may occur at any time, but their frequency and severity rises and falls over the 11-year solar cycle. Solar maximum is associated with the presence of more frequent and severe disturbances on the sun than solar minimum.

Solar activity is often measured by the number of "dark" sunspots or sunspot groups seen on the sun's surface. Solar radiation flux at a wavelength of 10.7 cm is another common way of measuring the level of solar activity, although sunspot measurements have a longer history of observations. Early in a new solar cycle, few sunspots appear at the solar high latitudes, while as the cycle approaches the solar maximum, an increasing number of sunspots appear closer to the equator.

Solar disturbances have the potential to significantly affect the Earth's magnetosphere, plasmasphere, and ionosphere. The reaction of the Earth's "magneto-plasma" to solar disturbances can in turn have a profound influence on the performance of HF systems that operate by virtue of skywave propagation. The impact of solar disturbances on such systems depends on the time, location, and type of the disturbance on the sun's surface as well as the state of the solar magnetic field. Detailed descriptions of the causes and effects of solar disturbances can be found in Matsushita (1976), Rishbeth (1991), Prolss (1995), Rishbeth and Field (1997), Field and Rishbeth (1997), and Kelly et al. (2004).

Solar disturbances have different levels of severity. In some cases, the characteristics of HF propagation are altered well beyond variations expected under quiet conditions, while in extreme cases, an ionospheric propagation path can disappear completely or "close" for periods of a few hours or even days. For the most part, solar disturbances adversely affect the performance of HF systems such as OTH radar, but this is not always the case, as enhanced ionospheric propagation can also occur (albeit rarely). There are various types of solar disturbances. For example, certain disturbances can have a marked impact on HF propagation at most latitudes, but only during the day, while others may exert an influence in the day or night, with the severity being a function of magnetic latitude.

For certain solar events, the adverse effects on skywave propagation are delayed from the time the disturbances are first observed on the sun. Forecasts and warnings are often available. This is not only of value to system operators for re-prioritizing missions, but also helps diagnose the cause of unusual performance as a natural event rather than equipment failure.

Many nations have a space weather forecast center or agency. For example: the Australian Government Ionospheric Prediction Service (www.ips.gov.au) provides space weather forecasts for the Australia-Oceania and other regions; the US government's official space weather forecast center is the Space Weather Prediction Center (www.swpc.noaa.gov); whereas European Space Agency forecasts can be found at http://esa-spaceweather.net.

2.2.3.1 Solar Flares and Sudden Ionospheric Disturbances

Solar flares result in the emission of an intense burst of X-rays from plage regions that are often located near sunspots in the Sun's chromosphere. This process can last from a few minutes to several hours and occurs most often during periods of high solar activity. If the solar event causing the flare is on the side of the sun facing the Earth, the burst of X-rays is incident on the Earth's atmosphere after approximately 8 minutes. The hard

X-rays penetrate deep into the atmosphere and produce strong ionization in the D-region by the process of photoionization. A flare can temporarily increase the electron density in the D-region by a factor of 10 or more.

The main effect of enhanced D-region ionization on skywave propagation is an increase in non-deviative absorption due to electron-neutral collisions. In severe cases, the increased level of absorption for HF signals can cause practically all of the radio wave energy to be lost, with almost none being transmitted to the higher reflecting layers in the ionosphere. This phenomenon was explained by J.H. Dellinger in 1937 and is sometimes known as the Dellinger effect. More commonly, it is referred to as a shortwave fadeout (SWF), or a daylight fadeout in the HF radio community, since it only occurs on the hemisphere of the Earth that is illuminated by the Sun. Another name often used to describe this type of event is a Sudden Ionospheric Disturbance (SID).

The severity of solar flares may be rated in terms of the X-ray flux density in W/m^2, which can be measured on satellites for wavelengths in the range 10–80 nm. The amount of absorption is highest when the X-ray flux density is large and incident on the D-region at a low solar zenith angle. The most affected HF propagation paths are those on which the signal rays attempt to penetrate through the D-region at the points of highest ionization on the way up or down from the E- and F-region layers. In this case, an SWF may seriously degrade or preclude operation over a sunlit skywave path. Alternatively, an SWF provides welcome isolation when it attenuates signals over ionospheric paths that would otherwise propagate interference and noise from long-distances to an HF system operating on the night-side of the Earth.

In the HF band, non-deviative absorption is inversely proportional to the square of the operating frequency to a good approximation. Consequently, small flares mainly affect the frequencies in the lower part of the HF band. This may leave open the possibility to use higher frequencies with reduced absorption, subject to the E- and F-region layers supporting propagation at such frequencies. On the other hand, large flares can significantly degrade operation over the entire HF band. The disruption due to an SWF is temporary in nature and is typically characterized by an abrupt onset followed by a slower recovery phase.

As illustrated in Figure 2.15, the higher frequencies remain less affected and tend to recover first. Restoration of the ionosphere to its normal state after an SWF may require up to an hour or two. Solar flares significant to the Earth's ionosphere occur on 3 or 4 days per month on average at solar maximum (McNamara 1991). Smaller (less energetic) solar flares occur much more often, perhaps several tens per day at solar maximum, but their effect on the operation of HF systems is proportionally less significant.

2.2.3.2 Coronal Mass Ejections

Coronal mass ejections (CMEs) are caused by magnetic field-line reconnection events on the sun. These events are also responsible for large solar flares, so a CME is often associated with a large flare. During a CME, electrons, protons, and heavy nuclei are accelerated to near the speed of light. The super-heated electrons from CMEs move along the magnetic field lines faster than the solar wind. This results in the formation of a shock front that can impact the Earth's magnetic field causing a geomagnetic storm (to be described later in this section).

The energetic plasma from the CME arrives at the Earth 1–4 days after it has erupted from the sun, causing further geomagnetic storms. A CME affects the Earth's magnetic field via interactions with the perturbed interplanetary magnetic field associated with

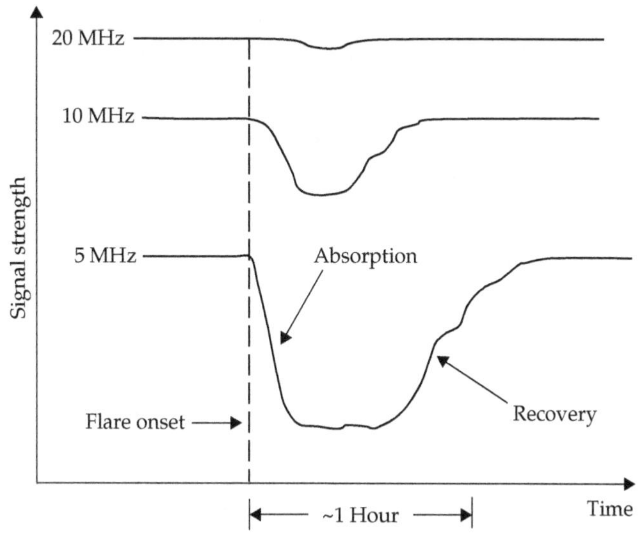

FIGURE 2.15 Conceptual illustration of received signal strength as a function of time at frequencies of 5, 10, and 20 MHz over a daytime skywave path affected by a sudden ionospheric disturbance due to X-rays emitted from a solar flare. Adapted from McNamara (1991).

the CME, causing oscillations and possibly reconnection events in the geomagnetic field. The energetic particles in the CME interact with the Earth the same way as solar proton events (described below). At solar minimum there is about one CME per week. Near solar maximum this increases to an average of 15–20 per week.

2.2.3.3 Solar Proton Events

Protons emitted by the sun may become energized and radiated into the solar wind as a dense stream. This can occur through the action of a large solar flare, or through the bow shock caused by a CME. The dense stream of energetic protons may reach the Earth after a time-delay of a few hours. This is called a "solar proton" event. As the high-energy particle stream enters the Earth's magnetosphere, the particles do not cross the geomagnetic field lines but rather spiral or gyrate around them. This process effectively guides the particles down along the magnetic field lines toward the polar regions, where they penetrate into the ionosphere and increase the electron density in the polar D-region by the process of collisional ionization. This can produce intense absorption on ionospheric paths with polar reflection points in what is known as a polar cap absorption (PCA) event.

The absorption of radio wave energy due to a PCA may be in excess of 20 dB, but remains confined to the polar region, usually within about 20 degrees of the magnetic poles themselves. A PCA can last for up to several days depending on the energy of the proton stream. Several PCA events can be expected each year around the solar maximum. While severe, such events only affect oblique HF propagation paths with mid-points in the polar region. This represents a minority of systems, with the Super Dual Auroral Radar Network (SuperDARN) being an example of an HF system operating at high latitude. The ionosphere in the more densely populated low- and mid-latitude regions is not affected to the same extent by this type of solar event due to the "shielding" effect of the Earth's magnetic field.

2.2.3.4 Geomagnetic and Ionospheric Storms

Coronal holes, large solar flares, and CMEs all give rise to enhanced solar particle flow which may impact on the Earth. Coronal holes are regions of the sun where the magnetic field lines are not closed but open out into space. This phenomenon allows solar particles to stream out unencumbered by the solar magnetic field, which in turn enhances the solar wind from these regions. When an enhanced plasma cloud is incident upon the Earth, it induces electric ring currents in the magnetosphere that disturb the Earth's magnetic field. This results in geomagnetic storms, during which the Earth's magnetic field may vary significantly in both strength and direction. The terrestrial magnetic field is constantly fluctuating, but in the presence of a geomagnetic storm, the variations in strength can be more than an order of magnitude greater than on quiet days.

Geomagnetic storms can produce large changes in the electron density of the F2-layer due to the strong influence it has on the ionization distribution in the upper F-region. Consequently, geomagnetic storms often give rise to ionospheric storms. In general, the ionization structure of the D-, E-, and F1-layers is less affected by geomagnetic storms since the electron density distribution in lower height regions of the ionosphere is not under as tight geomagnetic control. Relative to the lower altitude layers in the ionosphere, the F2-layer is more likely to become unstable during a geomagnetic storm and is therefore more susceptible to significant variations in critical frequency when such events occur.

Ionospheric storms may last for several days, as the coronal hole stream passes over the Earth, or as the CME cloud impacts, or the magnetic field lines redistribute. When the underlying solar disturbances are long-lived, they can result in recurrent storms as the sun rotates every 27 days (at the equator). Moreover, ionospheric storms can affect propagation at middle and low latitudes where the majority of HF systems are located. Specifically, the F2-layer can be severely depleted of electrons at mid-latitudes and its critical frequency may drop by a factor of two or more during major storms. The performance of HF systems that depend on skywave propagation support and stable signal reflection points is therefore prone to degradation as a result of ionospheric storms.

The nature and severity of a storm event depends largely on the energy of the event as well as its location on the sun with respect to the Earth. A strong CME will penetrate deeper into the Earth's magnetic field before the charged particles spiral down along the field lines. This causes an ionospheric storm to extend out from the magnetic poles further toward the magnetic equator. Ionospheric storms arising from weak solar events do not penetrate out as far toward the magnetic equator. The severity of a magnetic storm is measured in terms of the magnetic index (Davies 1990). Geomagnetic disturbance alerts are issued (similar to solar flares) and such warnings also help to diagnose the cause of performance degradations as disturbed propagation conditions rather than system malfunction.

2.2.3.5 Traveling Ionospheric Disturbances

Wavelike disturbances can propagate in the neutral atmosphere due to the oscillation of adiabatically displaced air "parcels" under the restoring force of gravity. The oscillation about the equilibrium height at buoyancy has a characteristic frequency in the atmosphere known as the Brunt-Vaisala frequency. These atmospheric waves, known as acoustic gravity waves (AGW), can transfer their motion to ions via ion-neutral collisions. It has been suggested that this mechanism is responsible for producing wavelike disturbances in the ionosphere known as traveling ionospheric disturbances (TID). Detailed reviews of TID phenomena can be found in Hocke and Schlegel (1996) and Hunsucker (1990).

In the D- and E-regions, the high-collision frequencies cause the plasma motion to follow the neutral motion, but in the F-region, where there are fewer collisions, the ion motion gained from the neutral atmosphere is constrained to move along the geomagnetic field lines. The passage of a TID through the region of signal reflection effectively creates an undulating surface of iso-ionic contours in the ionosphere. A TID may cause large and rapid changes in the radio-wave amplitude, due to focussing and defocussing, as well as Doppler shift, time-delay (group-range), and direction of arrival. The period of the oscillations can vary from 8 to 60 minutes in the F-region. The disturbances can at times be quasi-sinusoidal and may travel horizontally with wavelengths of 50–500 km and speeds ranging from 50 m/s in the E-region to 1000 m/s in the F-region. TIDs of varying scales are generally present all the time. TIDs can cause large target localization and tracking errors in OTH radar when measures are not taken to account for their effects.

2.3 Oblique Propagation

Although radio sounding at vertical incidence provides valuable information about the height structure and dynamics of the overhead ionosphere, oblique propagation is of more direct interest in applications such as point-to-point HF communications, shortwave radio broadcasting, and OTH radar. The process of oblique signal reflection from the ionosphere introduces ray elevation angle and ground distance of the path as additional variables to the electron density height profile and radio wave frequency considered at normal incidence. This section describes the main concepts related to skywave propagation over oblique paths, particularly the relationships existing between signal frequency, ray elevation, and path length (ground distance) for a given electron density distribution in the ionosphere.

The basic equivalence relationships including the secant law, Martyn's theorem, and Breit and Tuve's theorem are described first to explain the fundamental principles of oblique propagation in terms of virtual reflection paths. The various approximations underlying this representation are appropriate for applications requiring propagation models of moderate but not very high accuracy (HF communications being a possible example). For applications requiring higher fidelity models, such as OTH radar, this section also discusses propagation models based on analytical and numerical ray tracing techniques. The latter may be used to accurately model ray paths through an arbitrarily defined electron density snapshot of the ionosphere in three dimensions, taking into account the effects of the Earth's magnetic field.

2.3.1 Equivalence Relationships

To start with a simple example, Figure 2.16 illustrates the scenario where a signal is reflected at oblique incidence from a horizontally uniform ionosphere over a plane Earth. A signal ray with frequency f_o transmitted at point T strikes the bottom of the ionosphere at an angle ϕ_o relative to the normal. As the ray enters the ionosphere, it encounters an

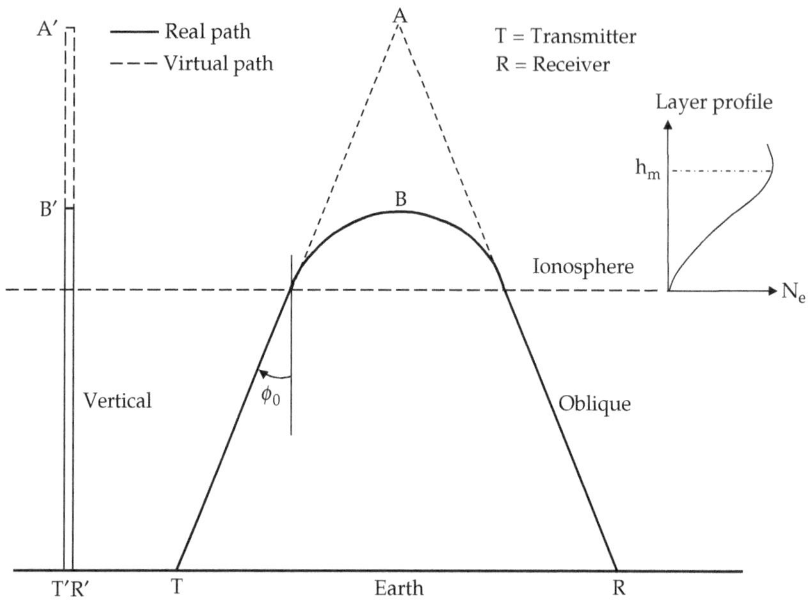

FIGURE 2.16 Equivalent waves reflected obliquely and vertically at the same real height (B and B′) and the same virtual heights (A and A′) assuming a horizontally uniform ionosphere and plane earth. Adapted from Davies (1990).

electron density height profile that increases monotonically below the height of maximum ionization h_m. The change of refractive index with height in the ionosphere causes the signal ray to be continuously deviated in propagation direction such that it follows a curved path through the ionosphere until it reaches point *B*, where reflection occurs.

The reflected ray then propagates along a downward curved path until it emerges from the bottom of the ionosphere, after which it effectively travels in a straight line to point *R* on the ground. As opposed to the real signal path, which traces out the curved trajectory *TBR*, the triangular path *TAR* is referred to as the virtual reflection path. This simple example will serve to illustrate some important principles of oblique propagation for a single horizontally uniform ionospheric layer over a plane Earth. The effect of the curvature of the Earth and the ionosphere will be introduced later on to progressively develop more realistic models of the oblique reflection process.

In addition to illustrating the real and virtual ray paths for the case of oblique reflection, Figure 2.16 also depicts an "equivalent" vertical reflection on the left-hand side. The term equivalent means that a vertically incident signal of frequency f_v is reflected from the same *real* height in the ionosphere as the signal of frequency f_o that strikes the ionosphere obliquely at an angle ϕ_o. The frequency f_v is referred to as the equivalent vertical frequency.

Snell's law may be used to relate the vertical and oblique frequencies reflected from the same true height. Neglecting collisions and the Earth's magnetic field, the relationship between these frequencies is given by Eqn. (2.33), which is known as the *secant law*. The secant law makes an important statement. Specifically, it implies that at a given true height, the ionosphere is able to reflect signals with higher frequency at oblique incidence than it can at normal incidence.

$$f_o = f_v \sec \phi_o \tag{2.33}$$

Breit and Tuve's theorem states that the time-of-flight for the radio wave to travel between points T and R (both located outside of the ionosphere) via point B in the ionosphere is identical to that which would be taken along the virtual reflection path TAR, assuming the radio wave travels at the speed of light in free space over all of the virtual path. In other words, Breit and Tuve's theorem says that the group-path P' defined in Eqn. (2.34) is equal to $TA + AR$. Although the ray path actually follows a curved trajectory through the ionosphere due to refraction, this theorem allows the real ray path with reflection occurring at point B to be replaced by a virtual rectilinear path of triangular geometry as if the ray were "mirror-reflected" at the apex A. In practice, $P' = TA + AR$ is an *approximation* that is often acceptable for applications not requiring high fidelity.

$$P' = \int_{TBR} \frac{ds}{\mu} \tag{2.34}$$

Recall that a signal of equivalent vertical frequency is reflected from the same true height in the ionosphere at vertical incidence as the oblique signal, such that $B = B'$. Martyn's theorem states that the virtual reflection height A for an oblique signal of frequency f_o is equal to the virtual reflection height A' for a vertically incident signal of equivalent vertical frequency f_v. The combination of Martyn's theorem and that of Breit and Tuve enables the equivalent triangular path length P' to be expressed as Eqn. (2.35) for the case of a plane Earth and ionosphere. Here, $h' = A'$ is the virtual height of reflection for a vertically incident signal of equivalent frequency f_v. Since the group velocity of a radio wave slows down when it enters the ionosphere, the real path length TBR is less than the virtual path length P' (also referred to as the *group-path*).

$$P' = 2h' \sec \phi_o \tag{2.35}$$

The assumption of a flat Earth becomes inappropriate for ground distances greater than a few hundred kilometers. An OTH radar typically operates over ground ranges in excess of one thousand kilometers. At such distances, it becomes more appropriate to consider the spherical Earth geometry illustrated in Figure 2.17. Analogously to the previous example, Figure 2.17 illustrates the actual ray path TBR and the equivalent virtual path TAR. Note that the electron density height profile in the ionosphere is assumed to be spherically symmetric (as opposed to horizontally uniform) in this case.

It is readily shown that the group-path P' is given by the formula in Eqn. (2.36), where r_e is the radius of the Earth and D is the great-circle distance (arc-length) between the ground terminals T and R. This formula effectively adjusts the ionospheric height to account for the curvature of the Earth's surface at the path mid-point, which is at a distance of $CF = r_e[1 - \cos(D/2r_e)]$ above the chord TCR at the path mid-point. As expected, Eqn. (2.36) reverts back to the flat Earth formula in Eqn. (2.35) for short distances $D \ll r_e$.

$$P' = 2\left\{h' + r_e[1 - \cos(D/2r_e)]\right\} \sec \phi_o \tag{2.36}$$

The formula in Eqn. (2.36) implicity assumes that f_v is given by the secant law in Eqn. (2.33). This requires the condition $\phi_o = \phi$ in Figure 2.17, which only holds for a flat ionosphere between points P and Q. The curvature of the ionosphere causes the ray angle of incidence ϕ to be greater than the angle of the equivalent triangular path ϕ_o. The curvature of the ionosphere is also not negligible over ground distances commensurate with those of interest to OTH radar.

As explained in Davies (1990), this effect can be accounted for by using the modified secant law in Eqn. (2.37). This expression introduces a correction factor k with values ranging between 1 and 1.2. The multiplier $k \sec \phi_o$ is called the *obliquity factor*. The maximum frequency that can be reflected at oblique incidence from a particular layer is equal to the obliquity factor multiplied by the layer critical frequency.

$$f_o = k f_v \sec \phi_o \tag{2.37}$$

Thus far, no along-path variation of the electron density height profile in the ionosphere between points P and Q has been considered. The influence of the Earth's magnetic field on oblique propagation has also been ignored up to this point. Although these assumptions are clearly unrealistic, it is convenient not to complicate the scenario for the moment and further explain the key relationships existing between signal frequency, ray elevation, and ground distance based on this simplified description of oblique propagation.

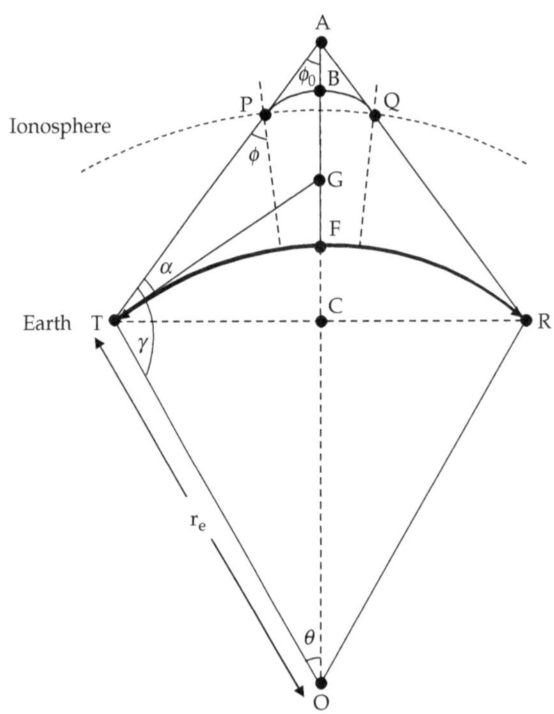

FIGURE 2.17 Conceptual illustration of oblique propagation through a thick curved ionosphere concentric with a spherical Earth. The distance FB is the true reflection height, while FA is the virtual reflection height. The electron density height profile is assumed to be spherically symmetric between points P and Q such that there is no along-path variation of the ionosphere.

Ray tracing procedures used to model oblique propagation in situations where these assumptions are removed will be discussed at the end of this section.

2.3.2 Point-to-Point Links

Skywave propagation over a constant distance D is directly relevant for a point-to-point HF communication link between two fixed ground stations. The study of such propagation is also relevant to OTH radar systems because the surveillance region may be considered as a large number of (two-way) point-to-point circuits. For such systems, an understanding of oblique ionospheric propagation to and from a remote point is required to register the geographic position of target echoes that have been detected localized in the radar coordinates of group-range and angle of arrival.

The principle of propagation path reciprocity is generally applicable for a monostatic OTH radar, and approximately applicable for a quasi-monostatic OTH radar with small bistatic angle, because the outgoing and incoming ray paths are likely to exhibit similar characteristics. In practice, the assumption of reciprocity is often reasonable provided that the elevation patterns of the transmit antenna and backscattering object are not too different. This property can allow the characteristics of a two-way point-to-point propagation path to be inferred from those of the associated one-way path.

Returning to Figure 2.17, it transpires that skywave propagation between the terminals T and R at a signal frequency f_o may occur via one or more different ray paths, or alternatively, it may not occur at all, depending on the electron density height profile. Specifically, the solutions for the virtual ray paths that give rise to oblique propagation between T and R are those which satisfy the previously described equivalence relationships for the assumed (spherically symmetric) electron density height profile.

To identify these ray path solutions, when they exist, the traditional approach is to write $\sec\phi_o$ as function of virtual reflection height h' and the ground distance D for the considered point-to-point link. From the geometry of Figure 2.17, it is possible to write $\sec\phi_o$ in the form of Eqn. (2.38) by using the identity $\sec^2\phi_o = 1 + \tan^2\phi_o$.

$$\sec\phi_o = \sqrt{1 + \left\{\frac{r_e \sin(D/2r_e)}{h' + r_e[1 - \cos(D/2r_e)]}\right\}^2} \tag{2.38}$$

Substitution of Eqn. (2.38) into Eqn. (2.37) yields Eqn. (2.39), where h' and f_v represent the only free parameters for constant values of f_o and D. A plot of virtual height h' against equivalent vertical frequency f_v that satisfies Eqn. (2.39) for particular values of f_o and D is called a *transmission curve*.

$$f_o = kf_v\sqrt{1 + \left\{\frac{r_e \sin(D/2r_e)}{h' + r_e[1 - \cos(D/2r_e)]}\right\}^2} \tag{2.39}$$

By making h' the subject of Eqn. (2.39), the transmission curve may be represented in the form of Eqn. (2.40). This curve represents the locus of all feasible combinations of h' and f_v that can satisfy the equivalence relationships when f_o and D are fixed. The flat-Earth model may be assumed for relatively short ground distances ($D \ll r_e$). In this case, it is readily seen from Eqn. (2.40) that the transmission curve reduces to the much simpler expression $h' = D/\{2\sqrt{(f_o/f_v)^2 - 1}\}$.

$$h' = \frac{r_e \sin(D/2r_e)}{\sqrt{[f_o/(kf_v)]^2 - 1}} - r_e[1 - \cos(D/2r_e)] \tag{2.40}$$

Importantly, the electron density height profile in the ionosphere also constrains h' to be a function of f_v. This function is precisely the vertical incidence ionogram trace, denoted by $h'(f_v)$ in Eqn. (2.41). The integration limit h_r is the height in the ionosphere at which the plasma frequency f_N equals f_v (i.e., where reflection occurs), while μ' is the group refractive index that varies as a function of height in the ionosphere.

$$h'(f_v) = \int_0^{h_r} \mu' \, dh \tag{2.41}$$

It follows that any intersections of the transmission curve in Eqn. (2.40) with the trace $h'(f_v)$ in Eqn. (2.41) yield solutions for the virtual heights of the equivalent triangular paths that obliquely reflect a signal of frequency f_o over a ground distance D. For an arbitrary layer profile $h'(f_v)$, the points of intersection cannot be found analytically in general, but the solutions may be readily computed graphically.[7]

Once the ray virtual refection heights have been identified, other ray parameters such as elevation angle α and group-path P' may be readily determined from the geometry of Figure 2.17. Specifically, the group path is given by substituting Eqn. (2.38) into Eqn. (2.36) to yield Eqn. (2.42), where h' is the virtual reflection height of a particular ray path solution. Note that the expression in Eqn. (2.42) reduces to $P' = 2h'\{1 + (D/2h')^2\}^{1/2}$ for the flat-Earth model ($D \ll r_e$), and as expected, $P' = 2h'$ in the vertical incidence case ($D = 0$).

$$P' = 2\{h' + r_e[1 - \cos(D/2r_e)]\}\sqrt{1 + \left\{\frac{r_e \sin(D/2r_e)}{h' + r_e[1 - \cos(D/2r_e)]}\right\}^2} \tag{2.42}$$

In summary, the equivalence relationships connect several important variables of the virtual ray paths that arise in the case of oblique signal reflection from a spherically symmetric ionosphere. Specifically, the virtual height solution(s) satisfying Eqn. (2.40) and Eqn. (2.41) simulataneously relate the ray angle of incidence ϕ_o, path ground distance D, oblique signal frequency f_o, and electron density height profile $h'(f_v)$ of the ionosphere. Clearly, the ray elevation or "take-off" angle α and group-path P' are uniquely related to (ϕ_0, D) and may be readily determined from the geometry of Figure 2.17.

A significant implication of the described approach, which may be referred to as virtual ray-tracing (VRT), is that oblique propagation path information between two ground terminals separated by *any* ground distance D may be inferred from vertical incidence ionograms measured at the path mid-point in the case of one-hop propagation. This is why the path mid-point is also known as the *control point* in the ionosphere as far as the electron density height profile is concerned.

In practice, the vertical incidence ionogram trace $h'(f_v)$ is due to multiple layers formed in the E- and F-regions of the ionosphere and its form is also influenced by

[7] Analytic solutions are readily found when the layer profile $h'(f_v)$ can be modeled in a mathematically tractable form such as the quasi-parabolic profile discussed in Section 2.3.3.3.

the geomagnetic field. To illustrate the application of the VRT method using a relatively simple example, let's consider the case of a single spherically symmetric layer and no superimposed magnetic field. It is also convenient to generate a representative ionogram trace $h'(f_v)$ using the relatively simple analytical expression of the parabolic layer model in Eqn. (2.43). As described in Davies (1990), h_0 is the height of the bottom of the parabolic layer with critical frequency f_c and semi-thickness y_m.

$$h'(f_v) = h_0 + \frac{1}{2} \ln \left\{ \frac{1 + f_v/f_c}{1 - f_v/f_c} \right\} y_m \qquad (2.43)$$

Later, we shall discuss the use of alternative spherically symmetric parametric layer profile models that permit analytical ray tracing. In contrast to analytical ray tracing, it is emphasized that the VRT method based on Martyn's theorem is quite general in the sense that it may be applied to a spherically symmetric electron density height profile of *arbitrary* form. Ray path calculations using VRT are typically not as accurate as those performed using analytical ray tracing. However, VRT has the redeeming feature that it does not require the layer profile to conform to a mathematically tractable parametric model. In other words, a trace extracted from a measured VI ionogram, arising either from a single layer or the presence of multiple layers, may be used to compute the virtual reflection heights of a path directly with no need to fit an ionospheric model of known parametric form. The significant advantages of analytical and numerical ray-tracing techniques with respect to VRT will be discussed in due course.

Figure 2.18a illustrates an example parabolic layer profile $h'(f_v)$ when the height and semi-thickness parameters are set in the regime of F2-layer values. The layer critical frequency is chosen as $f_c = 7.1$ MHz to reflect the observed value of f_oF2 in the real VI ionogram of Figure 2.7. The height of the bottom of the F2-layer and semi-thickness parameters are set to $h_0 = 210$ km and $y_m = 50$ km, respectively. Inspection of $h'(f_v)$ reveals that the virtual height of signal reflection (at vertical incidence) increases from

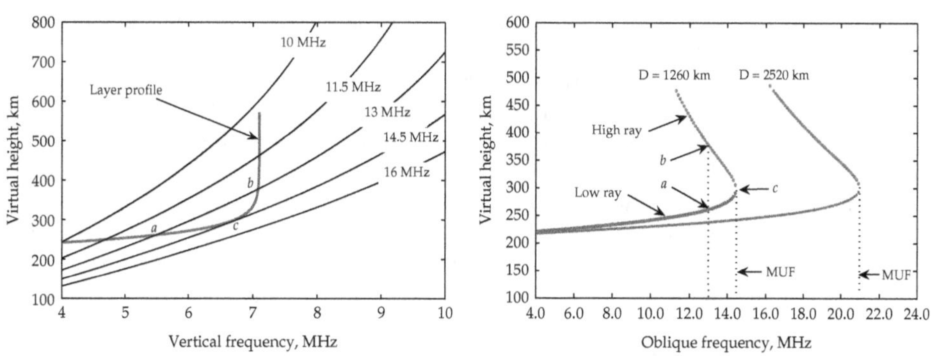

(a) VI ionogram with transmission curves for $D = 1260$ km.

(b) OI ionograms for $D = 1260$ km and $D = 2520$ km.

FIGURE 2.18 Simulated vertical and oblique incidence ionograms for a single parabolic layer with critical frequency $f_c = 7.1$ MHz. The points a, b, and c on the VI ionogram correspond to points with the same label on the OI ionogram for a path ground-distance of $D = 1260$ km. The transmission curves for $D = 2520$ km are not shown in the graphic on the left. © Commonwealth of Australia 2011.

about 250 km to over 500 km as the vertical frequency f_v is swept up toward the layer critical frequency.

While the simulated trace $h'(f_v)$ is not a high fidelity representation of the F2-layer trace in the measured ionogram of Figure 2.7, this relatively simple analytic function is nevertheless indicative of the trace recorded for the ordinary ray of the F2-layer. Note that the presence of the E- and F1-layers effectively masks the F2-layer trace below a frequency of 5 MHz in the real VI ionogram (Figure 2.7). This section of the trace is visible in the simulated layer profile $h'(f_v)$ of Figure 2.18a because the E- and F1-layers are not considered in this example.

Figure 2.18a shows a family of superimposed transmission curves parameterized in oblique frequency f_o for a ground distance of $D = 1260$ km. The virtual reflection heights for this path length at a frequency f_o are given by the points of intersection between the transmission curve corresponding to this frequency and the layer profile $h'(f_v)$. For example, at $f_o = 13$ MHz there are two points of intersection and therefore two virtual reflection heights that satisfy Eqns. (2.40) and (2.43) for the modeled layer profile. It is evident from Figure 2.18a that a signal of frequency $f_o = 13$ MHz propagates over a ground distance $D = 1260$ km by reflection from two virtual heights labeled a and b in this case. The ray reflected with lower virtual height at point a corresponds to a smaller elevation angle and is called the low-angle ray. The ray reflected from point b is called the high-angle ray, or the Pedersen ray. More commonly, they are simply referred to as the *low* and *high* rays.

As the oblique frequency f_o increases, it is apparent from Figure 2.18a that the virtual reflection heights of the low and high rays converge until the transmission curve eventually makes tangential contact with the trace $h'(f_v)$. In this example, tangential contact occurs at point c, where the low and high rays merge to form a single ray known as the "skip-ray." The oblique frequency $f_o = 14.5$ MHz at which tangential contact occurs is known as the maximum useable frequency (MUF) of the point-to-point circuit. Any further increase in frequency beyond the MUF causes propagation between the ground terminals T and R to be lost. Stated another way, there are no solutions compatible with Eqns. (2.40) and (2.43) for $f_o > 14.5$ MHz. For example, note that the transmission curve for $f_o = 16$ MHz does not intersect $h'(f_v)$ in Figure 2.18a. A signal of frequency higher than the MUF is said to skip the ground distance D.

It may at first seem counter-intuitive that the signal at the MUF is not reflected from the height of maximum ionization, where the plasma frequency equals the layer critical frequency. For a fixed ground distance D and signal reflection at the path mid-point, an increase in f_v is associated with a simultaneous decrease in $\sec\phi_o$. The signal at the MUF is reflected at a height where the product $f_v \sec\phi_o$ passes through a maximum. At greater virtual heights, $\sec\phi_o$ decreases more rapidly than f_v is increasing, so the maximum useable frequency is reflected from a height below that of maximum ionization. The MUF should not be confused with the optimum operating frequency, which is usually different for OTH radar. The latter depends on several criteria, such as signal-to-noise ratio, frequency-stability of the path, and multipath-induced Doppler spectrum contamination. The subject of optimum frequency selection will be dealt with in Chapter 3.

The virtual reflection heights of the low and high rays over an oblique path of fixed ground distance D may be plotted as a function of signal frequency f_o. For the parabolic layer profile $h'(f_v)$ in Figure 2.18a, Figure 2.18b shows the resulting curves for the ground distances $D = 1260$ km and $D = 2520$ km. This type of trace is referred to as an oblique incidence (OI) ionogram. The group-path P' may be plotted instead of virtual reflection height on the vertical axis. The points marked a, b, and c on the OI trace for $D = 1260$ km in Figure 2.18b correspond to the points with identical label on the VI trace $h'(f_v)$ in Figure 2.18a. From Figure 2.18b, it can be seen that the low and high rays converge to form a single ray at the MUF, which corresponds to the "nose" of the OI ionogram.

Figure 2.18b compares the OI trace for $D = 1260$ km with that resulting for the same layer profile but a longer ground distance of $D = 2520$ km. The transmission curves for $D = 2520$ km are not shown in Figure 2.18a to avoid cluttering the display. Observe that lengthening the ground distance D increases the MUF and sharpens the nose of the OI ionogram trace. In the presented example, the MUF increases from 14.5 MHz to approximately 21 MHz as D doubles from 1260 km to 2520 km. Clearly, the MUF tends to the layer critical frequency from above as the separation between the ground terminals tends to zero.

Figure 2.19a shows the OI trace for $D = 1260$ km plotted as a function group path P', while Figure 2.19b shows the variation of ray elevation angle α as a function of f_o. Figure 2.20, in the same format as Figure 2.19, illustrates the results for $D = 2520$ km. Note the relatively lower ray take-off angles and the reduced spread of ray elevation angles required for propagation over the longer path. Skywave OTH radar antennas that provide high gain at low elevation angles of around 5 degrees are critical for long-range target detection and tracking.

An actual OI ionogram typically contains contributions from multiple ionospheric layers over a sunlit path. Its form is also influenced by several other factors including the Earth's magnetic field, which produces ordinary and extraordinary waves. Consequently, the structure of a real OI ionogram is much more complex than the simple trace synthesized previously. A practical example of an OI ionogram is shown in Figure 2.21. This ionogram was recorded 5 minutes before the VI ionogram in Figure 2.7. The ground distance of the oblique path was approximately 1260 km, while the mid-point of the path

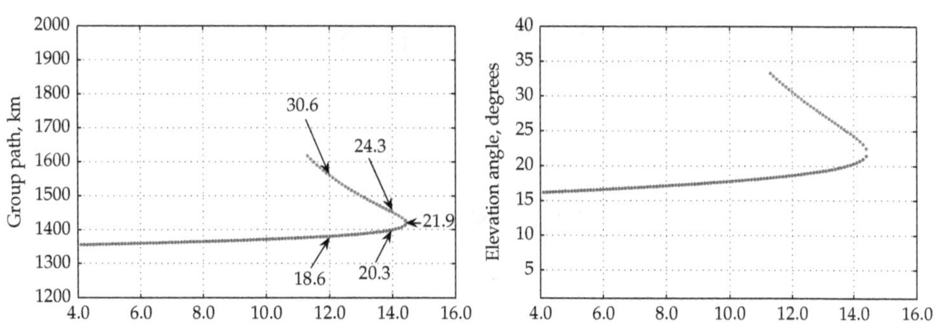

(a) Synthesized oblique incidence ionogram.

(b) Variation of ray elevation angle α.

FIGURE 2.19 Oblique incidence ionogram for $D = 1260$ km plotted with group path P' (as opposed to virtual height h') on the vertical axis. The elevation angles α of the low and high rays are indicated by the numbers next to the trace at certain frequencies in display (a). The variation of α for the low and high rays as a function of frequency is shown in display (b). For the assumed model of $h'(f_v)$ and a path ground distance of $D = 1260$ km, ray take-off angles between 15 and 35 degrees provide skywave propagation between the two terminals at frequencies below the MUF.

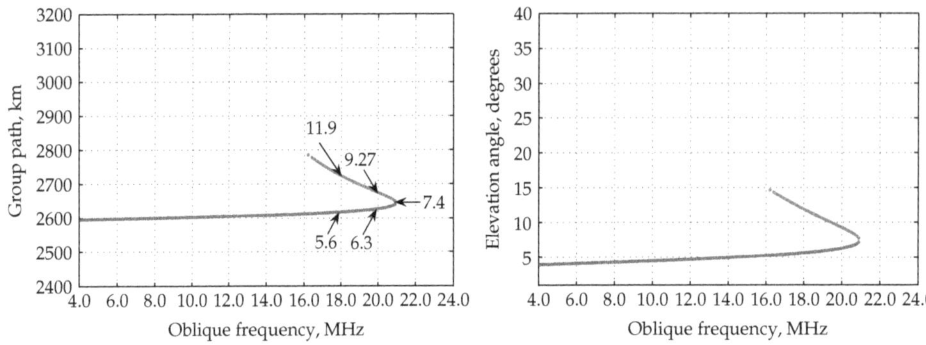

(a) Synthesized oblique incidence ionogram. (b) Variation of elevation angle α.

FIGURE 2.20 Oblique incidence ionogram for $D = 2520$ km plotted as a function group path P'. The elevation angles α of the low and high rays are indicated by the numbers next to the trace at certain frequencies in display (a). The variation of α for the low and high rays as a function of frequency is plotted in display (b). For the assumed model of $h'(f_v)$ and path length of $D = 2520$ km, ray take-off angles between 4 and 15 degrees provide skywave propagation between the two terminals at frequencies below the MUF.

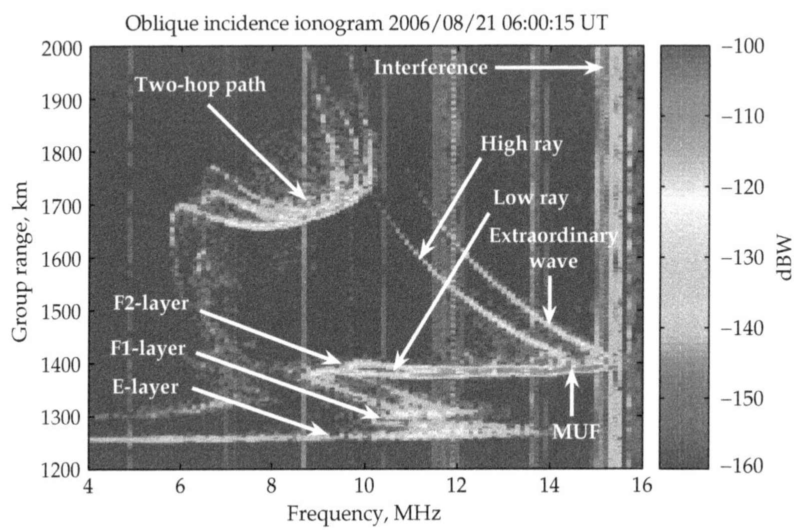

FIGURE 2.21 Oblique incidence ionogram recorded for a 1260 km ground-distance path with the mid-point located near the VI sounder at Kalkarindji, central Australia. Note that the OI ionogram was recorded at a similar time to the VI ionogram in Figure 2.7. The high and low rays of the F2-layer ordinary wave are labeled along with the separate trace corresponding to the F2-layer extraordinary wave. © Commonwealth of Australia 2011.

was less than 200 km from the location of the ionosonde that collected the VI ionogram in Figure 2.7.

The low and high rays associated with the F2-layer ordinary and extraordinary waves are clearly visible in Figure 2.21. These OI traces are associated with the VI traces for the F2-layer ordinary and extraordinary waves that were labeled in Figure 2.7. The resolution of low and high rays in time-delay is generally more difficult for E- and F1-layers, particularly the E-layer which is relatively thin compared to the F2-layer. The high rays of the F1-layer joining into the F2-layer can be discerned in Figure 2.21.

The portion of the real OI ionogram trace corresponding to the low and high rays of the F2-layer ordinary wave in Figure 2.21 may be compared with the simulated trace plotted as a function of P' for $D = 1260$ km in Figure 2.19. Recall that the simulated trace was computed using the equivalence relationships (VRT) based on a parabolic layer model with a height profile that approximated the one measured for the ordinary wave of the F2-layer in the real VI ionogram trace of Figure 2.7. The simulated results in Figure 2.19 predict an MUF of 14.5 MHz for $D = 1260$ km and estimate the group-path of the skip-ray as approximately $P' = 1400$ km. These values agree closely with those observed in the real OI ionogram of Figure 2.21. Despite the rather crude approximations made in this analysis to maintain simplicity, the portion of the F2-layer trace measured for the ordinary wave in the real OI ionogram of Figure 2.21 is represented remarkably well by the simulated trace plotted in Figure 2.19.

2.3.3 Frequency, Elevation, and Ground Range

In the previous description of oblique propagation, the parameter D was held fixed while f_o and ϕ_o were allowed to vary for a given (spherically symmetric) electron density height profile. It is now of interest to consider D as a variable, with f_o or ϕ_o being held fixed in turn. In addition to the VRT method used previously, the topics of analytical and numerical ray tracing are discussed to introduce techniques that provide high fidelity models of radio-wave propagation through a specified ionospheric profile.

In particular, numerical ray tracing permits the assumption of a spherically symmetric ionosphere to be relaxed, thereby allowing for an arbitrarily defined (down-range) variation of the ionization height profile. The effects of the Earth's magnetic field may also be incorporated in numerical ray tracing to model the different ray paths for ordinary and extraordinary waves.

2.3.3.1 Fixed Frequency

Holding the signal frequency f_o constant allows the ground range D of propagation to be studied as a function of ray elevation angle α. This provides an indication of the OTH radar range coverage that can be achieved for a certain antenna elevation pattern at a single frequency. The variation of ground range with ray "take-off" angle at a fixed frequency is conceptually illustrated in terms of equivalent mirror-reflected paths using a representative ionospheric layer[8] in Figure 2.22.

At relatively small (near-grazing) take-off angles, the rays are reflected from low virtual heights in the hypothesized ionospheric layer. As indicated in Figure 2.22, these rays may

[8]Representative for the F2-layer with a spherically symmetric electron density height profile assumed.

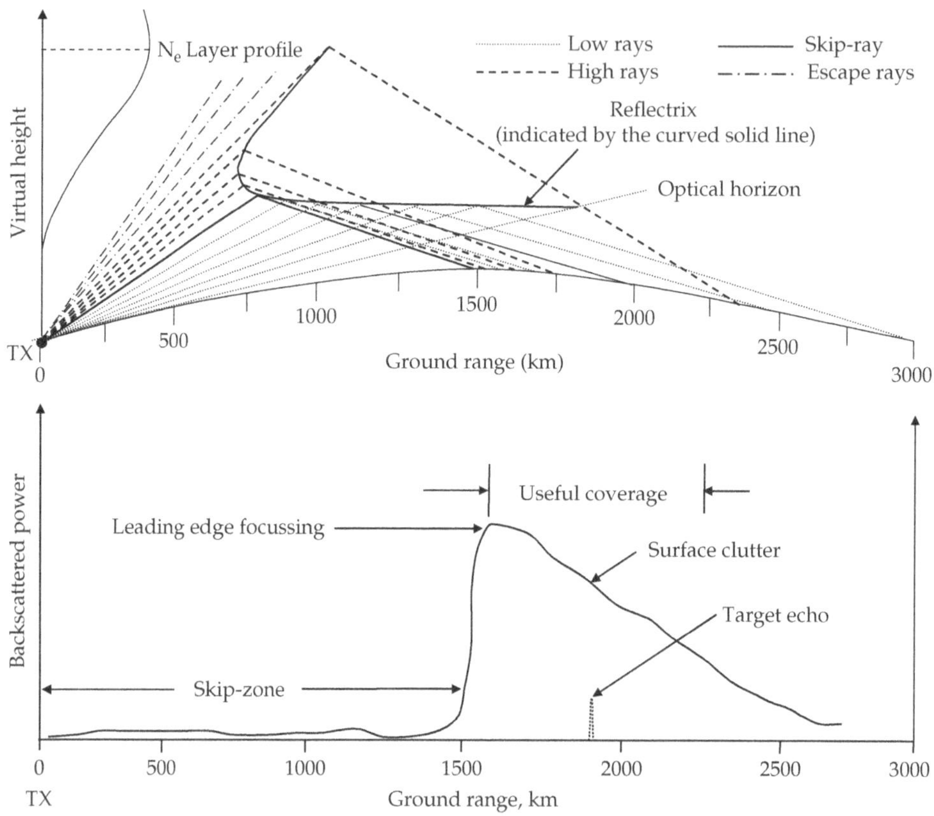

FIGURE 2.22 The signal paths illustrated in the top panel conceptually show the dependence of ground-range on ray elevation angle. Specifically, the sketch shows a fan of rays launched in elevation at a single frequency, while the legend identifies the reflection of high- and low-angle rays from different virtual heights in the ionosphere as well as the skip-ray and escape rays. Note that the locus of the apex of the virtual ray paths traces out a curve known as the reflectrix. The intensity of illumination incident on the ground (and hence the average backscattered power for ground of uniform backscatter coefficient) tends to reach a maximum at a point immediately beyond the skip-zone due to leading edge focussing (bottom panel). For a single layer, the illumination intensity due to one-hop propagation will then gradually decay until a point is reached where the signal-to-noise ratio of target echoes becomes too small for effective OTH radar operation. The ground distance between these two points generally defines the useful range coverage of an OTH radar at a given operating frequency. This represents the range-depth of the OTH radar surveillance region, which is often around 500–1000 km depending on operating frequency, ionospheric conditions, and antenna radiation patterns. © Commonwealth of Australia 2011.

be interpreted as a set of low-angle rays that illuminate points on the Earth's surface at different distances from the transmitter. In other words, each low ray connects a point-to-point circuit with different ground range. As illustrated in Figure 2.22, the rays with lowest elevation angle will typically propagate out to the farthest ground ranges. The maximum ground range of the OTH radar coverage at a particular operating frequency is therefore often limited by the low elevation angle gain of the transmit and receive antennas.

As the elevation angle increases from near-grazing incidence, the rays are reflected from progressively greater virtual heights, where the electron density is higher. Initially, the virtual height does not increase rapidly and the net effect is that the ground range decreases until a minimum is reached. This minimum ground range is called the skip-distance. The area encompassed by the locus of the skip-distance in all directions is referred to as the skip-zone. At the elevation angle corresponding to the minimum ground range, the high and low rays merge into a single ray called the skip-ray (solid line) in Figure 2.22. The focussing of high and low rays immediately beyond the skip-zone produces an enhanced field strength in this region. This effect is known as leading edge focussing.

As no oblique propagation path connects the transmitter to points within the skip-zone, the skip-zone determines the minimum ground range that an OTH radar can illuminate at a fixed frequency by the regular process of ionospheric reflection. When the operating frequency is less than the layer critical frequency, the skip-zone vanishes as signals can be reflected at vertical incidence. The skip-zone grows in size as the operating frequency increases past the layer critical frequency.

Within the skip-zone, illumination is still possible via ground-wave, surface-wave, and space-wave propagation. As far as the detection of surface targets by an OTH radar is concerned, the surface-wave mode potentially allows propagation to a maximum range of about 50 km over good (conductive and even) terrain, and up to about 400 km over sea water. Such propagation will be discussed for the case of HF surface-wave radar systems in Chapter 5.

As the elevation increases beyond that of the skip-ray, propagation is due only to the high rays. Typically, these rays are reflected from virtual heights that increase rapidly with elevation angle. This arises because the gradient of electron density with height on the bottom side of the layer (i.e., below the height of peak electron density) tends to be greater away from the peak than near the peak itself where it approaches zero. Hence, small changes in elevation lead to large increases in the virtual reflection height and ground range of the high rays. Relative to the low rays, this causes the signal power radiated in a small band of elevation angles to be dispersed over a larger ground distance for the high rays. Among other contributing factors, the de-focussing of reflected signal power over the ground due to this effect also causes signals received from the high rays to often have lower power with respect to those received from the low rays at a given point on the ground.

As the elevation continues to approach vertical incidence, there arrives a point at which the equivalent vertical frequency given by the secant law equals the layer critical frequency. The elevation angle at which this occurs is known as the *critical angle*. Naturally, a critical angle less than 90 degrees exists only if the transmitted signal frequency is greater than the layer critical frequency. Above the critical angle, the signal penetrates the ionosphere. This results in so-called "escape rays," which may be deflected on transit through the ionosphere before they propagate into space. The locus of virtual reflection heights traced out as the ray elevation changes from near-grazing incidence to the critical angle gives rise to a curve known as the *reflectrix*.

The bottom panel of Figure 2.22 qualitatively illustrates the typical variation in average backscattered signal power received by an OTH radar as a function of ground range. For a surface with approximately uniform backscattering coefficient, the average backscattered signal power may be considered roughly proportional to the illumination intensity incident on the surface at a particular ground range. Leading edge focussing tends to result in the maximum average power being backscattered from ranges immediately following the skip-zone, where the illumination intensity is typically highest.

As the low and high rays diverge, the received signal power tends to exhibit a gradual decay with increasing ground range. Beyond a certain limiting ground range, the illumination that can be provided at a single operating frequency may be too weak for reliable target detection (i.e., to obtain adequate SNR). This results in a simultaneous useful coverage of finite range-depth at a particular frequency, as illustrated in the bottom panel of Figure 2.22.

2.3.3.2 Fixed Elevation

Fixing the elevation angle ϕ_0 allows the variation in ground range of the signal path to be investigated as the frequency is changed, either for a single ray, or for a fan of rays radiated over a band of elevation angles. As the frequency of the signal increases, a ray incident on the ionosphere at a fixed elevation angle will be reflected from a greater altitude where the electron density is higher. Recall that reflection occurs at the height where the equivalent vertical frequency equals the plasma frequency. For a single thick layer such as the F2-layer, the plasma frequency is assumed to increase monotonically with height until the layer peak.

From Martyn's theorem, the apex of the equivalent signal path is at the virtual reflection height, which also increases with frequency for a ray at fixed elevation under normal ionospheric conditions. Consequently, a ray launched at fixed elevation is typically returned to Earth over a longer ground range as the signal frequency increases. The maximum ground range of a ray with fixed elevation angle is reached when it is reflected from the height of maximum electron density in the ionosphere. Any further increase in frequency will cause a ray at this elevation angle to penetrate the ionosphere.

An OTH radar with fixed but relatively broad transmit and receive antenna patterns in elevation may therefore scan the useful coverage (range-depth) of illumination to different ground ranges by varying the operating frequency. An increase in frequency will move the useful coverage to further ground ranges (up to a certain limit), while a decrease in frequency reduces the size of the skip-zone to enable surveillance at shorter ranges. Importantly, the actual band of elevation angles responsible for illumination at different ranges changes as a function of frequency. Specifically, as the frequency increases, only the rays at lower elevation angles can be reflected from the ionosphere to provide illumination at longer ranges.

The ray-tracing diagrams in Figure 2.23 illustrate the aforementioned points, which are of direct relevance to OTH radar operation. The plots are generated using the IRI model for 15 March 2001, 07:00 UT, assuming the emitter is at geodetic latitude −23.5° and longitude 133.7° (near Alice Springs, in central Australia), with rays transmitted at a bearing of 324.7°T. Geomagnetic field effects are not taken into account. In these diagrams, propagation paths are shown for a fan of rays launched over an identical band of elevation angles at signal frequencies of 20 MHz and 30 MHz.

At the higher frequency (30 MHz), the rays begin to escape the ionosphere at relatively lower elevation angles compared to those which escape at the lower frequency (20 MHz). Note also that the rays at low elevation angles, which do not penetrate the ionosphere at 30 MHz, are reflected from relatively greater heights than the rays with the corresponding elevation angles at 20 MHz. This causes the region of useful coverage,

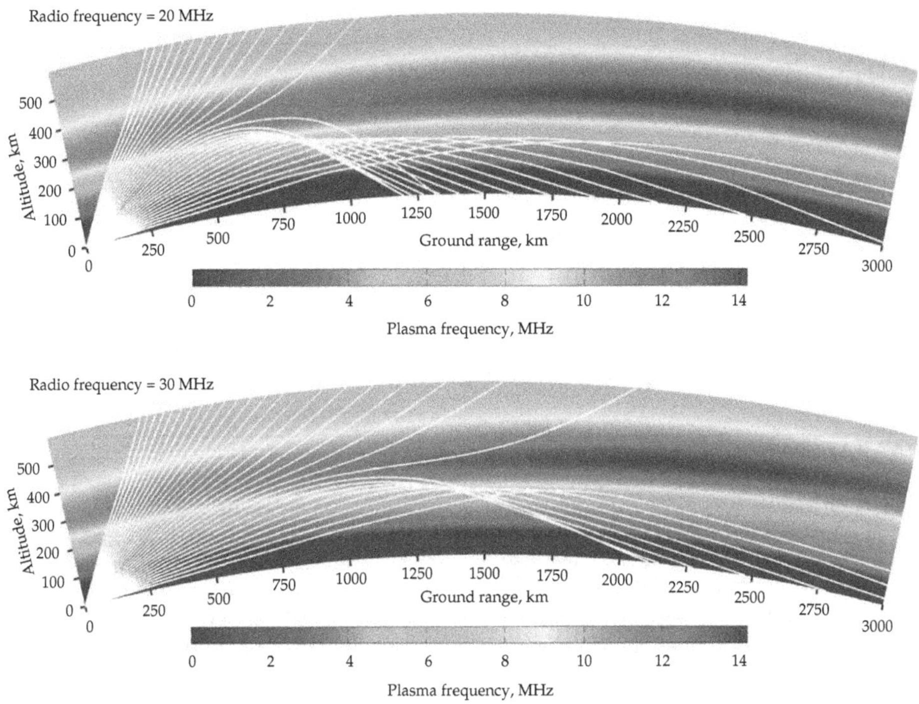

FIGURE 2.23 Ray-tracing diagrams simulating ray paths for an identical band of elevation angles through the same model ionosphere at two different frequencies. Note that rays at a given elevation are reflected from different points in the ionosphere when the frequency is increased from 20 to 30 MHz, and that the region of useful coverage for an OTH radar (starting immediately beyond the skip-zone) shifts further out in ground range at the higher frequency. © Commonwealth of Australia 2011.

commencing immediately beyond the skip-zone, to move further out in ground range by approximately 900 km (starting from about 1200 and 2100 km at 20 and 30 MHz, respectively).

From the perspective of signal power density, Figure 2.24 illustrates that an increase in carrier frequency shifts the edge of the skip-zone to a longer ground range. In this sketch example, raising the frequency from f_1 to f_2 effectively increases the intensity of illumination provided at ranges starting immediately after the skip-zone of f_2 compared to that provided at these ranges when using f_1. The gain in power density at ranges where there is leading edge focussing at f_2 will enhance the signal strength and hence SNR of a target echo at such ranges relative to that observed for f_1. Naturally, this advantage only applies for targets that do not fall inside the skip-zone associated with f_2, as some ranges inside this skip-zone can be effectively illuminated by using the carrier frequency f_1.

In practice, the range-depth of the useful coverage will typically not remain the same at different operating frequencies. Stated another way, the farthest edge of the useful coverage in range will generally not shift out by the same amount as the closest edge (immediately beyond the skip-zone) as the frequency is increased. Quite often, the

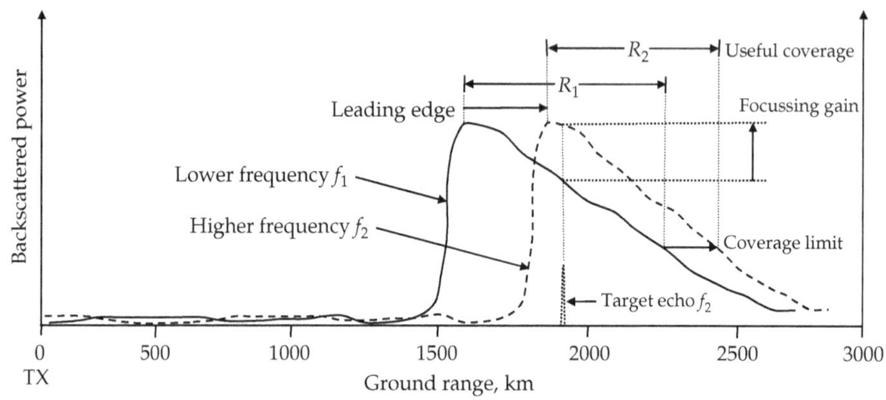

FIGURE 2.24 Notional illustration of backscattered signal power as a function of ground range at two different carrier frequencies f_1 and f_2, where $f_2 > f_1$. At the higher operating frequency, the leading edge moves to a longer ground range, where it provides a focussing gain (increase in signal power density) for targets immediately beyond the skip-zone associated with f_2 relative to that provided at these ranges by the lower frequency (f_1). This results in a stronger target echo when using f_2 compared to that indicated for f_1 in Figure 2.22. On the other hand, the "trailing edge" of the useful coverage, which represents the maximum range limit of the surveillance region for effective OTH radar operation, may not move out in ground range by the same amount as the leading edge when the frequency is changed from f_1 to f_2. Depending on ionospheric conditions, it often occurs that the useful range-depth of the coverage at f_2, denoted by R_2, is less than that for f_1, denoted by R_1. © Commonwealth of Australia 2011.

range-depth of an OTH radar surveillance region is reduced as the skip-zone moves further out, as depicted in Figure 2.24.[9] The described relationships between signal frequency, path length, and ray elevation have important implications for OTH radar system design. In a well-designed OTH radar, the overall range coverage is limited by environmental conditions rather than instrumental factors.

2.3.3.3 Ray Tracing

In applications requiring ray-path calculations of low to moderate accuracy, geometrical optics, or virtual ray-tracing (VRT) based on Martyn's theorem may be used to model radio-wave propagation through a reference or measured ionospheric height profile. The VRT approach assumes the electron density height profile is spherically symmetric and approximates the true ray paths by their equivalent mirror-reflected paths.

The assumption of a spherically symmetric ionosphere does not account for down-range variations of the electron density height profile between the points where the ray paths enter and exit the ionosphere. In addition, the actual (curved) ray paths due to refraction in the ionosphere have group-path characteristics that are only approximated by the "equivalent" virtual (rectilinear) paths. In practice, this further reduces the accuracy of propagation calculations with respect to errors arising from the assumption of a spherically symmetric ionosphere.

[9] An increased range-depth may at times be possible depending on ionospheric conditions.

Provided that the assumed spherically symmetric electron density height profile can be accurately described by a parametric model of appropriate functional form, the canonical equations of the ray paths simplify to a set of mathematically tractable equations that may be solved analytically. By invoking such models, analytical ray-tracing (ART) may be performed to calculate the refracted ray paths through a model ionosphere.

A popular ART approach uses the multi-segment quasi-parabolic (MQP) model of Hill (1979) based on the QP ray-tracing technique described in Croft and Hoogasian (1968). An alternative ART method based on quasi-cubic segment (QCS) models has been proposed in Newton, Dyson, and Bennett (1997). As the ART approach calculates the refracted ray paths exactly, it provides more precise propagation predictions than the VRT method when the assumed ionospheric model accurately represents the true profile near the path midpoint.

The parametric model used for ART may be fitted to empirically derived electron density height profiles, extracted from climatological databases such as IRI, or actual "snapshots" of the ionosphere generated from the application of inversion techniques to real-time measurements recorded by one or more ionosondes. In practice, a site-specific RTIM often combines or blends real-time measurements from a network of sounders with smoothed climatological data to interpolate or extrapolate the model over the region of interest. ART using an MQP model fitted to a site-specific RTIM may be expected to provide good propagation predictions in practical applications.

The accuracy of an ART procedure is limited in practice by the fidelity with which the assumed parametric model of the ionization profile fits the true electron density distribution. From the viewpoint of real-time system operation, a significant advantage of ART is that it provides closed-form expressions for ray parameters such as group range, phase path, and ground distance, which allows the ray parameters of interest to be calculated in a computationally efficient manner. Moreover, ART techniques provide a useful diagnostic tool to validate the implementation of alternative numerical ray-tracing procedures under controlled (simulation) conditions. Specifically, results may be cross-checked for the case of a spherically symmetric ionosphere described by an MPQ or QCS model when there is no out-of-plane propagation due to the presence of the geomagnetic field. (i.e., for propagation parallel or perpendicular to the Earth's magnetic field).

An arbitrary ionization distribution that includes down-range variation of the electron density height profile does not admit to an analytical ray tracing solution. The same can also be said when propagation is not parallel or perpendicular to the Earth's magnetic field. In such situations, ray tracing based purely on numerical methods is necessary. The application of sophisticated numerical ray-tracing (NRT) routines to an arbitrarily defined "snapshot" of the ionosphere, such as the direct output of a site-specific RTIM, has the potential to provide the highest quality propagation predictions, albeit at greater computational expense. Such procedures, which require a representation of the ionosphere gridded in position (latitude/longitude) and height, are considered indispensable for systems such as OTH radar where propagation calculations of high accuracy are critical to successful operation.

When the influence of the geomagnetic field is ignored, the radio refractive index is independent of wave polarization and propagation direction. This allows NRT to be achieved by the successive point-wise application of Snell's law in accordance with the canonical equations of the rays, which correspond to a simplified version of Haselgrove's equations (Haselgrove 1963). Restricting the calculations to 2D also yields a significant speed-up in computations. Figure 2.23 illustrates example outputs of a 2D-NRT engine based on the approach described in Coleman (1998).

A more comprehensive numerical ray-tracing implementation that explicitly represents the delays and losses for both magneto-ionic components produced by the Earth's magnetic field in three dimensions is described in Jones and Stephenson (1975). This widely used technique is based on the integration of the first-order Haselgrove equations and accounts for out-of-plane propagation when the orientation of the Earth's magnetic field does not remain parallel or perpendicular to the wave propagation direction over the ray path.

The DSTO-developed MATLAB ray-tracing toolbox called PHaRLAP (Cervera 2010) provides NRT routines that compute ray paths for both magneto-ionic components in three dimensions based on Jones and Stephenson (1975). The code assumes a cold magneto-plasma with no collisions and that the electron density irregularities in the ionosphere have scale sizes that are large compared to the radio wavelength such that Fermat's principle applies. The output information provided for each ray path includes: (1) geodetic latitude, longitude, and ground range of the ray end-point; (2) group and phase path of the ray; (3) final elevation and bearing of the ray; (4) deviative absorption losses; (5) integrated electron density over the ray path; and (6) polarization state vector along the ray path.

A 3D-NRT example generated using this toolbox is illustrated in Figure 2.25. The effect of the Earth's magnetic field, which causes the incident signal ray to split into ordinary and extraordinary waves (magneto-ionic components), will be discussed further in the following section. In the illustrated example, the two magneto-ionic components not only return to Earth at different ground ranges with respect to each other and the ray

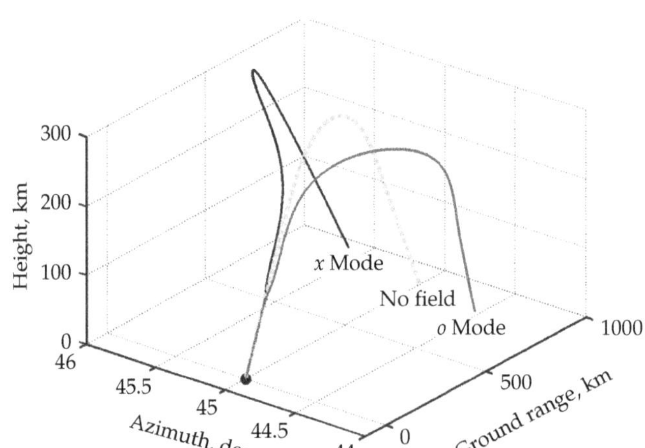

FIGURE 2.25 3D ray-tracing using PHaRLAP toolbox showing the signal paths taken by the ordinary (o) and extraordinary (x) magneto-ionic components compared to the (no-field) ray path when the Earth's magnetic field is neglected. The o and x components split from the incident signal as the ray enters the ionosphere and propagate along separate paths characterized by different ground ranges and bearings (due to out-of-plane propagation) in this example.

associated with the no-magnetic-field case, but also have slightly different azimuths due to the out-of-plane propagation (not captured by a 2D ray-tracing engine).

2.4 Ionospheric Modes

HF signal propagation between a ground-based transmitter and receiver may occur via a number of different skywave paths or ionospheric "modes" not limited to the high- and low-angle rays arising for a single layer under the assumption of no magnetic field. The Earth's magnetic field is superimposed on the electron density distribution of the ionosphere and causes the radio refractive index to depend on wave polarization state and propagation direction. This gives rise to ordinary and extraordinary waves, also known as *magneto-ionic components*, that effectively follow separate paths through the ionosphere.

In addition to the ordinary and extraordinary waves for both the high and low rays, which can arise due to a single ionospheric layer, multipath propagation can also arise due to reflections from two or more physically distinct layers formed in different height regions of the ionosphere. This section provides a brief summary of various ionospheric modes that may be expected for a spherically symmetric ionosphere with one or more reflecting layers. Multipath propagation over two-way paths and some examples of "exotic" propagation modes relevant to HF skywave systems are also discussed.

2.4.1 Ordinary and Extraordinary Waves

Radio-wave propagation in the ionosphere is not only affected by the electron density distribution. The presence of the Earth's magnetic field also influences the radio refractive index in the ionosphere. The celebrated equation for the phase refractive index of a small-amplitude harmonic radio wave of plane polarization in an electrically neutral cold magneto-plasma with homogeneous electron density and a uniform external magnetic field is referred to as the Appleton formula in Davies (1990). This formula may be written as Eqn. (2.44), where $X = N_e q^2/\epsilon_0 m\omega^2$, $Z = \nu/\omega$, $Y_L = qB_L/m\omega$, and $Y_T = qB_T/m\omega$.

$$\mu^2 = 1 - \frac{X}{1 - iZ - \frac{Y_T^2}{2(1-X-iZ)} \pm \left[\frac{Y_T^4}{4(1-X-iZ)^2} + Y_L^2\right]^{1/2}} \tag{2.44}$$

Recall that N_e, q, and m are the electron density, charge, and mass respectively, while ω is the radio-wave angular frequency and ν is the electron-neutral collision frequency. The magnetic field components parallel and perpendicular to the wave propagation direction are denoted by B_L and B_T, respectively. Appleton's formula reverts to the much simpler refractive index expression of Eqn. (2.24) for no collisions ($Z = 0$) and no magnetic field ($Y_L = Y_T = 0$). It is readily shown that making these substitutions in Eqn. (2.44) yields $\mu^2 = 1 - X$ where $X = f_N^2/f^2$ and f_N is the plasma frequency defined in Eqn. (2.20).

While the contribution due to collisions can often be neglected in the E- and F-regions, the effect of the Earth's magnetic field is significant and cannot be ignored. The main point is that a magnetized plasma or "magneto-plasma" is a medium which allows two characteristic waves of different polarization to propagate along essentially independent paths. Specifically, the wave associated with the positive sign before the square bracket in Eqn. (2.44) is called the *ordinary* (o) wave, while the wave associated with the negative sign is called the *extraordinary* (x) wave.

Importantly, the values of Y_L and Y_T not only depend on the magnitudes of the geomagnetic field but also on the direction of wave propagation relative to the orientation of the geomagnetic field lines. Appleton's formula effectively implies that the superposition of the Earth's magnetic field causes the ionosphere to behave as a birefringent and anisotropic propagation medium for radio waves. The derivation of Appleton's formula is beyond the scope of this text, but detailed mathematical descriptions of magneto-ionic theory can be found in Davies (1990), Budden (1985a), and Ratcliffe (1959).

2.4.1.1 Magneto-Ionic Splitting and Polarization

A linearly polarized electromagnetic wave may be thought of as the sum of two counter-rotating circularly polarized waves with the same frequency and amplitude. When such a wave is incident on the bottom of the ionosphere and enters the magnetized plasma, it effectively splits into o and x characteristic waves. In general, these waves have counter-rotating elliptical polarizations as they propagate through the ionosphere. For an observer looking in the direction of the geomagnetic field, the electric field vector of the x-wave appears to rotate in a clockwise sense (i.e., a right-handed wave) irrespective of whether the wave is traveling toward or away from the observer. The o-wave rotates in the opposite (anti-clockwise) sense and is referred to as a left-handed wave with respect to the Earth's magnetic field.

In accordance with Appleton's formula, the o- and x-waves are refracted differently in the ionosphere and propagate along separate paths that can have significantly different times-of-flight or virtual reflection heights. As mentioned previously, a pair of o- and x-waves arises for both the low- and high-angle rays. Magneto-ionic splitting can therefore give rise to four ray paths over a one-way single-hop point-to-point link when reflection from a single spherically symmetric ionospheric layer is considered. The critical frequency and MUF of o- and x-waves can also differ significantly in practice, as illustrated by the VI and OI ionogram traces in Figures 2.7 and 2.21, respectively. Note that the standard convention is to define the MUF in terms of the critical frequency of the ordinary wave.

To illustrate the oblique propagation of o- and x-waves, two special cases may be considered in which the wave normal and geomagnetic field are oriented parallel and perpendicular to each other. In the latter case, referred to as transverse propagation, the longitudinal component Y_L equals zero and $Y_T = Y = qB/m\omega$ in Eqn. (2.44), where B is the magnitude of the magnetic field. The refractive index expressions for the o- and x-waves then simplify to the formulas in Eqns. (2.45) and (2.46), respectively. In this special case, the refractive index of o-wave is identical to the no-magnetic-field value defined in Eqn. (2.24).

$$\text{Transverse propagation } (o\text{-wave}): \quad \mu^2 = 1 - X = 1 - f_N^2/f^2 \qquad (2.45)$$

The refractive index of the x-wave in Eqn. (2.46) has a value less than the no-field value in Eqn. (2.45). Figure 2.26 illustrates the splitting of the incident signal ray into o- and x-waves and the different propagation paths taken by the two magneto-ionic components through the ionosphere in the case of transverse propagation. The o- and x-waves that arise from the incident signal ray striking the ionosphere at a particular elevation are returned to ground at significantly different ground ranges. In practice, the

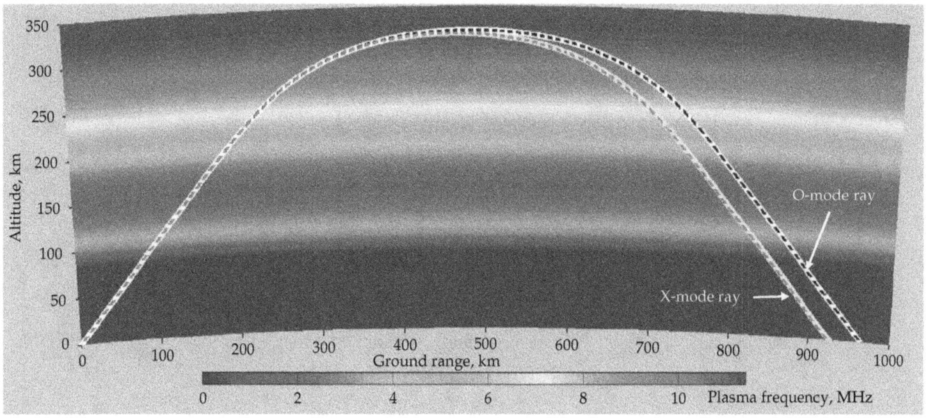

FIGURE 2.26 Numerical ray-tracing of o- and x-waves that split from the same incident signal ray and propagate along different paths through the ionosphere. Results are shown for the case of *transverse* propagation, where the wave normal is perpendicular to the magnetic field. In this special case, the o-wave path is identical to that resulting in the absence of a magnetic field. © Commonwealth of Australia 2011

case of transverse propagation is relevant when the incident ray is launched in the plane of the geomagnetic equator, such that the geomagnetic field lines remain at right angles to the wave normal over the entire trajectory of the o- and x-wave paths.

$$\text{Transverse propagation } (x\text{-wave}): \quad \mu^2 = 1 - \frac{X(1-X)}{1-X-Y^2} \tag{2.46}$$

In the case of parallel propagation, $Y_T = 0$ and $Y_L = Y$. The Appleton formula then reduces to the form of Eqn. (2.47), where the plus sign is associated with the o-wave and the minus sign with the x-wave. In this case, neither refractive index reduces to the no-magnetic-field value of $1 - X$. Note that the refractive index of the o-wave is greater than that of the x-wave, which means that the o-wave path is deviated less than that of the x-wave at the same height in the ionosphere.

$$\text{Parallel propagation}: \mu^2 = 1 - \frac{X}{1 \pm Y^2} \tag{2.47}$$

Figure 2.27 compares the o- and x-wave paths with the no-field path for the case of parallel propagation. In practice, the wave normal of a reflected ray cannot remain parallel to the geomagnetic field lines over the entire path. A reasonable approximation to this conceptual result is when the wave normal is parallel to the geomagnetic field at the apogee of path, where the ray is refracted most in the ionosphere. This occurs when a ray is launched along a magnetic meridian and the path mid-point is at the magnetic equator.

In general, the angle between the wave normal and the geomagnetic field is between 0 and 90 degrees and will also vary along the ray path. This not only causes out-of-plane propagation, as illustrated for both magneto-ionic components in Figure 2.25, but it also changes the polarization state of the wave along the ray path in the ionosphere. Both the

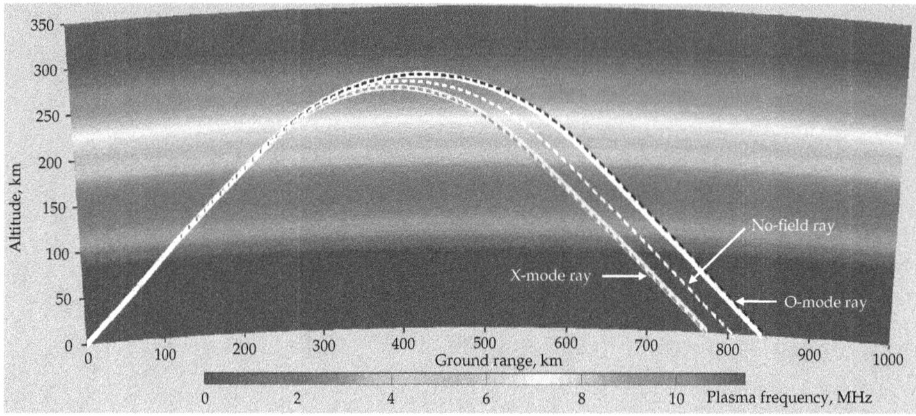

FIGURE 2.27 Numerical ray-tracing of *o*- and *x*-waves that split from the same incident signal ray and propagate along different paths through the ionosphere. Results are shown for the case of parallel propagation, where the wave normal is parallel to the magnetic field. In this case, the *o*- and *x*-wave paths differ and neither of these paths corresponds to the no-magnetic-field path. © Commonwealth of Australia 2011

o- and *x*-waves are in general elliptically polarized, but the eccentricity and inclination of the electric field with respect to the wave normal will continually change along the ray path until the reflected signal emerges from the bottom of the ionosphere.

The *o*- and *x*-waves tend to experience different amounts of absorption or attenuation in the ionosphere. The electric field of the *x*-wave rotates in a direction that increases the speed and hence the radius at which electrons gyrate about the Earth's magnetic field. This effectively leads to an increase in collision frequency with neutrals. The opposite occurs for the *o*-wave, where the sense of rotation of the electric field acts to decrease the speed of electrons and hence the collision frequency with neutrals. Since radio wave energy is lost due to absorption by electron-neutral collisions, the main effect is that the *x*-wave undergoes relatively more attenuation than the *o*-wave (particularly when passing through the D-region). This often results in the appearance of a relatively weaker *x*-wave trace on an oblique incidence ionogram.

2.4.1.2 Faraday Rotation

Of direct interest to OTH radar is the reception of *o*- and *x*-waves on emergence from the ionosphere. At times, it is possible for an OTH radar to resolve the *o*- and *x*-waves in time-delay (group-range) or Doppler-shift. In particular, the magneto-ionic components received as high-angle rays from the F2-layer often exhibit quite different times-of-flight over an oblique path, as illustrated by the well-separated *o* and *x* traces in Figure 2.21. When the magneto-ionic components can be resolved, a vertically or horizontally polarized antenna element can only capture half of the power contained in each wave for the case of a circularly polarized ordinary or extraordinary wave, with the other half being lost due to polarization mismatch.

It is more difficult to resolve the relatively closely spaced *o*- and *x*-waves in the low-angle ray of the F2-layer, which is not as dispersive as the high-angle ray. A similar

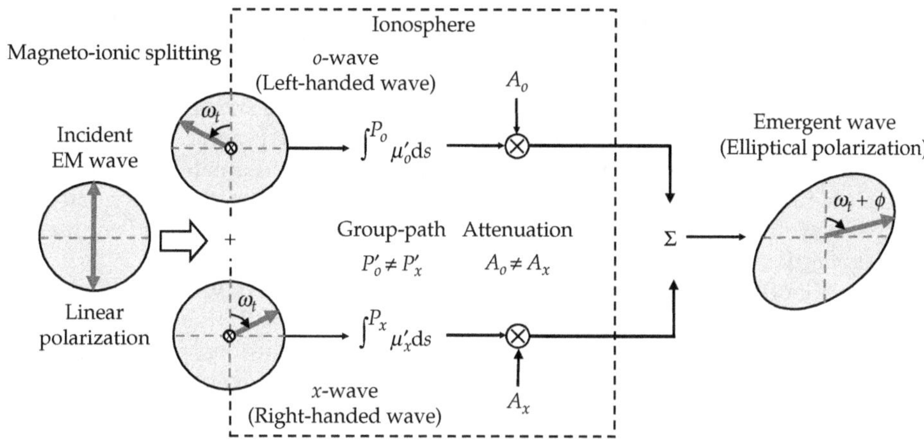

FIGURE 2.28 Conceptual illustration of a linearly polarized incident EM wave being split into *o*- and *x*-waves on entry to the ionosphere. The two characteristic waves generally have elliptical polarization and experience different group-paths and attenuations. On exit from the ionosphere, the reflected characteristic waves produce a resultant wave with elliptical polarization in general. In the special case of circularly polarized *o*- and *x*-waves with equal amplitudes on emergence from the ionosphere, the resultant wave is linearly polarized with the axis of linear polarization being rotated by the angle Ω in Eqn. (2.48) relative to that of the transmitted wave.

situation arises for thinner layers in the ionosphere, such as the E-layer. In this case, the system receives a resultant wave that arises from the interference between unresolved magneto-ionic components. Due to the different phase paths of *o*- and *x*-waves in the ionosphere, the polarization of the resultant wave summed at any point depends on the accumulated phase difference. In the (hypothetical) case where the two emergent waves experience equal attenuation, the outcome is simply a rotation of the axis of linear polarization. This effect is known as *Faraday rotation*.

Faraday rotation is a typical characteristic of birefringent propagation media. The angle of rotation Ω is given by Eqn. (2.48), where μ_o' and μ_x' are the *group* refractive indices over the *o*- and *x*-wave paths, denoted by P_o and P_x respectively, and $\kappa = 2\pi/\lambda$ is the wave number. Recall that the group refractive index is defined by $\mu' = \partial(\omega\mu)/\partial\omega$, where the phase refractive index μ for the *o*- and *x*-waves is given by Appleton's formula.

$$\Omega = \frac{\kappa}{2}\left(\int^{P_o} \mu_o' ds - \int^{P_x} \mu_x' ds\right) \tag{2.48}$$

Note that the relationship $\mu\mu' = 1$ does not hold in the presence of an external magnetic field,[10] so a more general expression for μ' is needed (Budden 1985a). More often, the *o*- and *x*-waves experience different attenuations. The net effect of this is an elliptically polarized wave with the resultant electric field rotated with respect to that of the transmitted wave, as illustrated in Figure 2.28.

[10]The only exception is for the ordinary wave in the case of traverse propagation where $\mu\mu' = 1$.

At a particular time instant, the polarization state of the resultant wave returned to the Earth's surface will vary spatially due to changes in the relative phase between the *o*- and *x*-waves at different points on the ground. When sampled by an array of linearly polarized antenna elements, this produces a spatial interference pattern due to the polarization fringes formed at a given time instant over the ground. The amplitude of the resultant signal received at the same time by the different antennas in the array may fluctuate significantly over a wide aperture. This phenomenon effectively deforms the signal wavefront relative to the plane-wave structure expected in the ideal case of specular reflection for a far-field point source.

At a fixed location, the polarization state of the resultant wave will also change in time due to motions in the ionosphere (e.g., differential Doppler shifts), which alter the relative phase path between the *o*- and *x*-waves. Movement of the ionospheric reflection points over time causes nonstationary Faraday rotation (i.e., movement of the polarization fringes over the ground). For a linearly polarized receiving antenna, such as a vertical monopole or horizontal dipole, this phenomenon manifests itself as polarization fading in time.

Very deep fades in excess of 40 dB may occur when the interfering *o*- and *x*-waves have similar amplitudes. Moreover, a deep fade can last for long periods (e.g., tens of seconds) when the *o* and *x* propagation paths change slowly. The fading may also be very rapid (e.g., multiple cycles within a second) when the *o*- and *x*-waves have significantly different Doppler shifts.

2.4.2 Multipath Propagation

Thus far, attention has been paid to the high- and low-angle rays as well as *o* and *x* characteristic waves in the context of reflection from a single layer in the ionosphere. In practice, a signal is radiated as a fan of rays over a continuum of elevation angles by the transmit antenna. This signal may be reflected one (or more) times by the same or different layers in the ionosphere, such as the normal E-, sporadic-E, and F2-layers, before reaching the receive antenna. We now turn our attention to the description higher-order (multi-hop) and mixed-layer ionospheric modes. To simplify the discussion, the high/low rays and *o*/*x* waves are momentarily ignored such that these alternative ionospheric modes may be described in terms of a single virtual ray path per layer.

The impact of multipath on OTH radar depends on whether the signal represents interference, clutter, or a target echo, and in some cases, whether the different multipath components are resolved in at least one of the canonical radar dimensions (i.e., group-range, direction of arrival, or Doppler frequency). It is instructive to describe the characteristics of different ionospheric modes for the case of one-way propagation first, and to then discuss two-way (point-to-point and wide-area backscatter) paths. As far as OTH radar is concerned, the characteristics of one-way paths are mainly of interest for interference mitigation, whereas two-way point-to-point and wide-area backscatter paths are clearly relevant for the reception of target echoes and clutter returns, respectively.

2.4.2.1 First- and Higher Order Modes

Figure 2.29 illustrates the classical one-hop (first-order) mode for the Es- and F2-layers in the left panel as well as a two-hop (second-order) mode reflected by the Es-layer in the right panel. In the case of two-hop propagation, the effective virtual reflection height

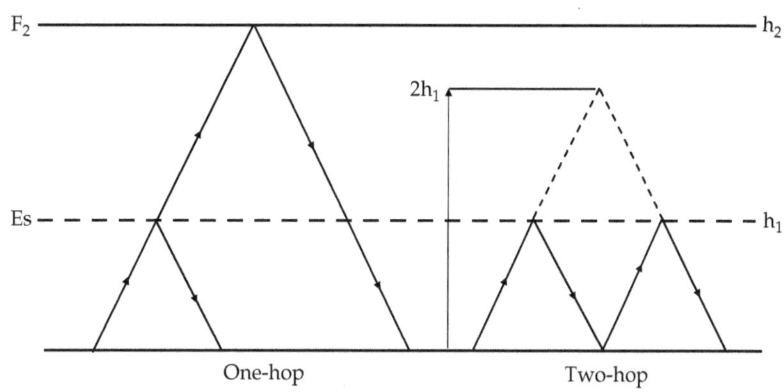

FIGURE 2.29 Conceptual illustration of one-hop modes for two (Es and F2) ionospheric layers over a one-way path on the left, and a two-hop (second-order) mode for the Es-layer on the right.

for the second-order mode is $2h_1$, assuming both ionospheric reflections occur from the same virtual height h_1.

Figure 2.30a plots the ground distance D of the ray path against elevation angle α, with virtual reflection height h as the parameter using Eqn. (2.49). This formula is derived from the geometry of Figure 2.17. The curves for $h_1 = 100$ and $h_2 = 350$ km are indicative of signal path ground distances for one-hop sporadic-E and F2-layer reflections, respectively, while the curve for $2h_1 = 200$ km is indicative of two-hop propagation from the sporadic-E layer.

$$D = 2r_e \left\{ \frac{\pi}{2} - \alpha - \arcsin\left[\frac{r_e \sin(\alpha + \pi/2)}{r_e + h_v} \right] \right\} \tag{2.49}$$

Note that one-hop propagation from the F2-layer provides about 1000 km of additional ground distance relative to the two-hop Es mode at elevation angles below 10 degrees.

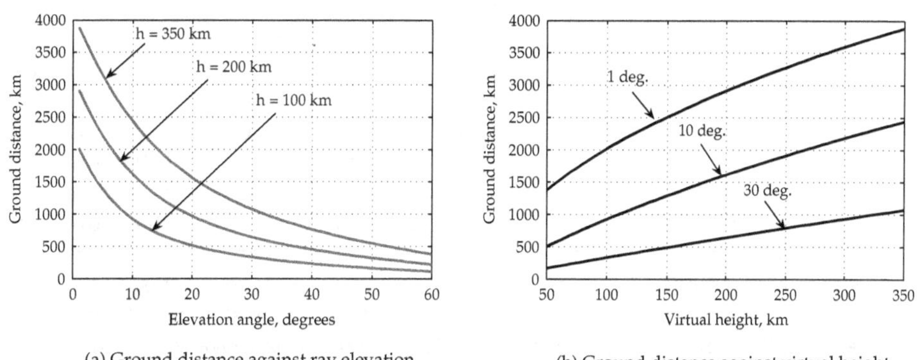

FIGURE 2.30 Ground distance of signal path over a spherical Earth surface as a function of ray elevation angle and virtual reflection height in the ionosphere (at the path mid-point).

Figure 2.30b plots the ground distance D as a function of h with α as the parameter. The importance of low elevation angle rays below 10 degrees for long-distance propagation out to 3000 km is clearly evident in Figure 2.30b. It is also observed from Figure 2.30a that one- and two-hop Es rays with different elevation angles (e.g., approaching 10 and 20 degrees) can be returned to the same ground distance (e.g., 1000 km in this case).

The time-dispersion between one- and two-hop signals received over a fixed ground distance due to the superposition of modes with different group ranges can produce severe frequency-selective fading in an HF communication system, even when a single layer such as Es is isolated for use by frequency selection or elevation control. In such systems, time-dispersion due to multipath can give rise to inter-symbol interference, which is widely recognized as a major source of error for certain types of digital data transmission. When uncompensated, this effect leads to an irreducible error rate that cannot be improved upon by increasing the radiated power.

Higher order modes involving more than two ionospheric reflections from the same layer also occur. However, these multi-hop modes often experience higher attenuation, particularly due to non-deviative absorption in the D-region during the day. The ground reflection component of path-loss depends on the conductivity and dielectric constant of the Earth's surface. For a relatively flat surface, the reflection loss is less for sea than dry ground and typically ranges from 0.5 to 3 dB per reflection for forward scatter. Multi-hop propagation may introduce additional losses due to deviative absorption when the signal reflections occur near the height of peak electron density in a layer.

During the day, the major path-loss contribution for multi-hop modes compared to one-hop modes of comparable path-length arises because the former passes through the D-region on more occasions. Non-deviative absorption depends inversely on the square of operating frequency and may range from 3 to 10 dB or higher per hop depending on the ionization levels in the D-region. D-region absorption significantly attenuates very long-range multi-hop paths during the day. This can effectively reduce mutual interference between widely spaced HF systems operating at the same time on a common frequency.

2.4.2.2 Hybrid or Mixed-Layer Modes

Figure 2.31 illustrates the so-called *M*- and *N*-type propagation modes. For example, the *M* mode can occur when the signal reflected down from the F2-layer does not reach the ground but is reflected from the upper side of the sporadic-E layer (i.e., a topside layer reflection). This M-shaped mode requires an electron density "valley" to be formed between the E- and F-regions.

On the other hand, the *N* mode illustrated in Figure 2.31 involves a reflection from the underside of the Es-layer, followed by a ground reflection and then a reflection from the F2-layer. When more than one reflecting layer is present, multipath propagation over a one-way path can potentially involve first- and higher order modes from the individual layers as well as *M*- and *N*-type modes. Propagation modes that are reflected from more than one layer are sometimes referred to as "hybrid" or mixed-layer modes.

Although the presence of multipath propagation depends on the state of the ionosphere between the considered ground terminals, its characteristics vary with operating frequency as well as the transmit and receive antenna elevation patterns. HF systems may therefore influence the received multipath characteristics by frequency selection and elevation pattern control. Multipath characteristics that may be relevant to the operation of an HF system include mode strengths, time dispersion, Doppler spread, and

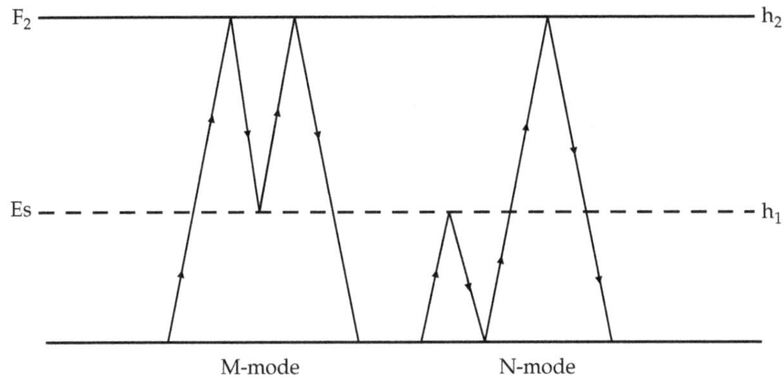

FIGURE 2.31 Conceptual illustration of M- and N-type mixed-layer reflection modes over a one-way path.

angular occupancy. As mentioned previously, the multipath characteristics of one-way skywave propagation are mainly of interest for interference mitigation in OTH radar.

In this case, the angular occupancy of the interference modes is important as this can affect the number of radar beams contaminated by interference. Figure 2.30b shows that for a ground distance of $D = 1000$ km, signals reflected from virtual heights in the ionosphere between 100 and 350 km may be received with elevation angles that range from 10 to 30 degrees. Typically the angular spread of individual propagation modes is confined in elevation such that not all of this interval may be occupied. Nevertheless, multiple ionospheric reflections from different virtual heights can significantly broaden the spatial spectrum of an interference source in elevation.

The characteristics of multipath over an oblique circuit can vary considerably with the choice of operating frequency. Oblique incidence ionograms indicate that multipath from different layers tends to decrease as the MUF of the circuit is approached. An operating frequency near the MUF provides a degree of immunity against multipath by preventing reflections from all layers except the one with highest critical frequency. Operation near the F2-layer MUF also reduces multipath between the high- and low-angle rays, which effectively converge to form a single ray at the MUF.

However, the MUF corresponds to operation on the leading edge, which is very sensitive to small changes in the electron density height profile of the ionosphere at the control point. In simple terms, the signal will practically vanish should the receiver fall into the skip-zone due to a small decrease in the MUF. Consequently, this approach for multipath mitigation is susceptible to ionospheric variations and prone to dramatic variations in signal-strength.

Alternatively, frequency selection may be based on the notion of dominant multipath components by taking the strength of different propagation modes into account. For example, the E- and F1-layers disappear at night and multipath can only be received from the F2-layer (in the absence of sporadic-E). As the frequency is stepped back from the leading edge, the high-angle rays which are reflected from higher in the ionosphere (near the point of maximum electron density) become heavily attenuated with respect to the low-angle rays. In this case, the o- and x-waves in the low-angle ray provide the main contribution, with the o-wave typically being dominant.

When it is impractical to reduce the deleterious impact of multipath propagation on system performance by frequency selection, control of the antenna radiation pattern(s) in elevation may provide an alternative form of immunity by minimizing gain in direction(s) of unwanted signal modes. Importantly, the elevations to be nulled by the transmit and/or receive antenna will depend on the ground distance of the circuit. This may pose problems for systems such as OTH radar, which need to operate over relatively large range-depths simultaneously.

2.4.2.3 Two-Way Paths

The mode structure is further complicated when the signal path involves ionospheric propagation from a transmitter to a remote point, and then by virtue of backscatter from this point, to a receiver located at the same (or similar) position as the transmitter. This type of path is referred to as a two-way point-to-point backscatter path. When a continuum of spatially distributed ground points is considered, the associated ensemble of two-way point-to-point circuits may be reffered to as a wide-area backscatter path.

Provided that the transmitter-to-receiver separation is small compared to the ground distance of the path, such that the bistatic angle is small, and the elevation pattern of the antennas is not too different from that of the backscattering object, the assumption of reciprocal propagation paths is often quite reasonable. In other words, the outgoing signal path from the transmitter to a remote point is an adequate representation of the incoming signal path from the remote point to the receiver.

Restricting attention to two reflecting layers and one-hop propagation Figure 2.32 illustrates that the returned ray may propagate via the same layer as the outgoing ray or via a different layer. This simple example involving only two reflection points illustrates that the transmitted signal may be received via four propagation modes over a two-way path. By distinguishing the outgoing and incoming rays, these paths may be referred to as E-E, E-F, F-E, and F-F modes, where the first letter denotes the forward path and the

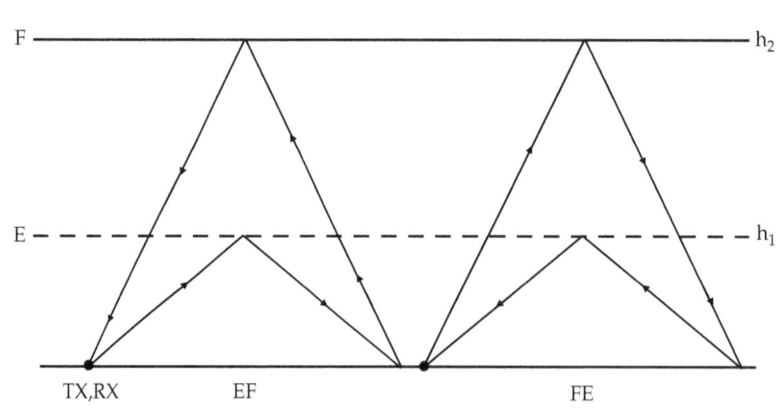

FIGURE 2.32 First-order propagation modes over a two-way point-to-point backscatter path. The arrows indicate the E-F mode (left panel) and F-E mode (right panel), both of which are mixed-layer paths. The E-E and F-F modes involving two reflections from the same layer also exist but are not indicated by the arrows. This gives rise to four propagation modes over a two-way point-to-point path in the illustrated two layer example.

second letter denotes the return path. By the principle of reciprocity, it follows that for M different modes over a one-way path there will be M^2 different modes over a two-way path.

An important consequence of multipath for radar is that a single target can produce multiple echoes with different time-delays, Doppler-shifts, and directions of arrival. When the target multipath can be resolved in one or more radar processing dimensions, the different echoes may trigger independent detections. If such detections are passed onto the tracker, they may give rise to multiple tracks in radar coordinates, despite all the detected echoes originating from one and the same target. This requires methods for associating multiple tracks formed in radar coordinates to a particular target located in geographical coordinates.

On the other hand, unresolved target echoes (propagation modes), such as the E-F and F-E paths with the same group range, do not produce separate detections but can cause deep and rapid fading due to mutual interference. This does not result in multiple tracks, but the fading can at times deny detection which may lead to track loss.

OTH radar surface clutter is returned from a continuum of ground points (land or sea). Multipath propagation often causes surface clutter backscattered from separated points on the ground to be returned with the same group-range, i.e., echoes from physically different surface scatterers returned to the radar via different propagation modes with similar virtual path lengths. This is notionally illustrated in the left panel of Figure 2.33, where backscatter from different ground points $P1$ and $P2$ are respectively returned by the F- and E-layers over the same virtual path lengths (not to scale). This situation may for example cause terrain clutter propagated by one mode and sea clutter propagated by another to superimpose in the same spatial resolution cell.

Furthermore, the superimposed paths correspond to different ionospheric reflection points that may impose different Doppler frequency shifts and spreads on the signal. Hence, multipath propagation can lead to an increase in Doppler spectrum contamination

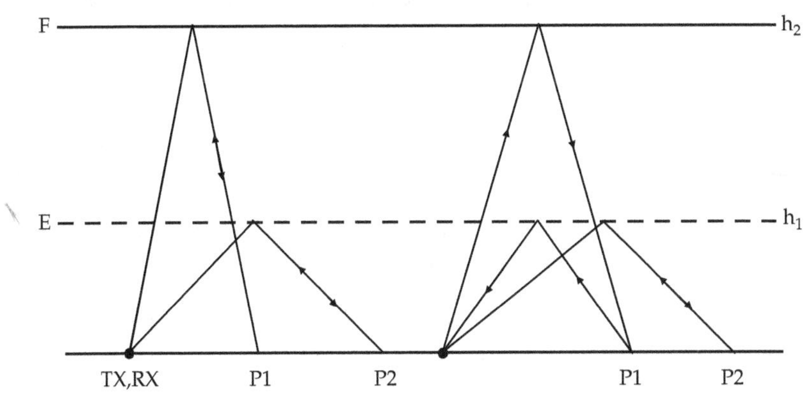

FIGURE 2.33 Multipath can result in clutter backscattered from different ground points $P1$ and $P2$ being returned with the same group-range (i.e., equal virtual path length). The left panel conceptually shows F-F and E-E modes with equal virtual path length received from quite different elevation angles. The right panel shows F-E and E-E modes with equal virtual path length received from closely spaced elevation angles. The points $P1$ and $P2$ are not the same in the left and right panels and the illustrations are not to scale.

when clutter from a relatively disturbed mode is received in the same radar resolution cell as clutter propagated via a frequency-stable mode. This makes it possible for clutter propagated by a disturbed mode (e.g., F-F) to obscure target echoes propagated via a stable mode (e.g., E-E). Such a situation is of concern in OTH radar for slow-moving target detection, where the spectral purity of the clutter Doppler spectrum is of paramount importance.

Elevation pattern control on receive *or* transmit (but not both) is insufficient for countering Doppler spectrum contamination caused by multipath in the two-way propagation scenario. For example, consider a receive antenna that selects the elevation angle of a "stable" E-mode return. The right panel in Figure 2.33 shows that this elevation not only contains the desired E-E mode but also the F-E mode, which has been contaminated due to reflection form the "disturbed" F-layer on the outgoing signal path. For similar virtual path lengths, the E-E and F-E modes will be backscattered from different ground ranges, labeled $P1$ and $P2$ in the right panel. However, the difference in elevation between these two incoming rays may be too small to be resolved by the receive antenna.

By additionally introducing elevation control on transmit, it is in principle possible to illuminate the E-layer (outgoing signal) path and attenuate the transmit antenna gain at elevations corresponding to the F-layer (outgoing signal) path. This effectively leads to the propagation of E-E and E-F modes only, as the F-E and F-F modes are not illuminated by the transmitter. Elevation control on receive can then select the E-layer return by attenuating the F-layer return, which is typically incident from a significantly different elevation angle in the case of signal modes with similar group-range. This leaves the desired E-E path as the dominant mode at the receiver output.

Unfortunately, it is not possible to achieve this type of "mode-selectivity" for all ranges in the OTH radar coverage by forming a single beam pattern on receive and transmit. In practice, the beam patterns need to be adaptive and range-dependent as the elevation angles of the selected and unwanted signal modes are not only unknown a priori, but also change as a function of range from the radar. Modern techniques for circumventing this problem based on the use of multiple transmit waveforms have been discussed in Frazer, Abramovich, and Johnson (2009).

2.4.2.4 Transequatorial Propagation

In addition to the different ionospheric modes already discussed, more "exotic" modes that can allow propagation over longer than usual distances without ground reflections also exist. One of these is a transequatorial propagation (TEP) mode. This mode arises due to the enhancement in electron density approximately 20 degrees north and south of the geomagnetic equator due to the previously described equatorial or Appleton anomaly. This anomaly results in the formation of steep horizontal gradients in the iso-ionic contours of the F2-layer. Specifically, it creates a large-scale upward tilt from the latitudes of the anomaly to the geomagnetic equator, where the critical frequency of the F2-layer typically occurs at a much greater height.

Under certain conditions, the alignment of these tilts is such that HF signal reflection occurs from one crest directly across to the other before the signal is returned to the ground, as illustrated by ray-tracing diagram in Figure 2.34. Because of the two consecutive reflections in the F-region, with only two passages through the absorbing D-region, this mode is often characterized by a high signal strength. The TEP mode can at times propagate HF signals over path lengths of 6000 km or more with remarkably low attenuation and fading rates (McNamara 1973).

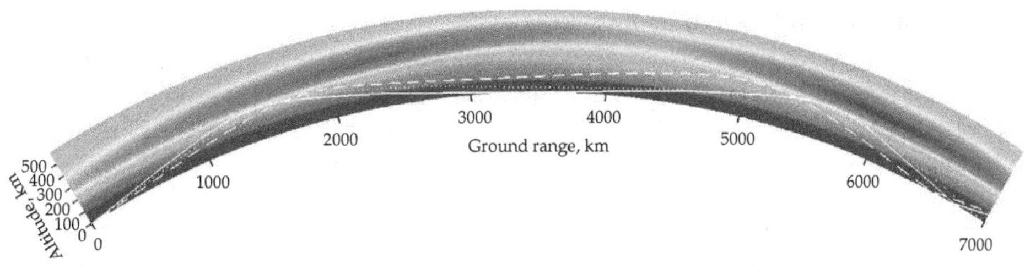

FIGURE 2.34 Ray-tracing diagram showing the transequatorial propagation mode reflected from one crest of the Appleton anomaly to the other across the geomagnetic equator over a ground distance of more than 6000 km without an intermediate ground reflection.

A different type of transequatorial propagation mode that occurs mainly during the evening may be thought of as a "whispering gallery" mode. This mode arises due to the presence of field-aligned irregularities in the F-region that may lead to the formation of "galleries" in which HF signals may be reflected many times at low (near grazing) angles of incidence over long distances. This guided propagation mode favors circuits with ground-points that are connected by the Earth's magnetic field lines (i.e., symmetric about the geomagnetic equator). A different type of "ducted" mode occurs when signal propagation is effectively trapped in an electron density valley formed between E- and F-region layers or the F1- and F2-layers.

As a consequence of the many reflections from field-aligned irregularities in the former type of guided mode, the received signal is likely to exhibit very rapid fading and such propagation is often linked to the phenomenon of spread-F (to be discussed shortly). In OTH radar, propagation modes involving scattering from dynamic field-aligned irregularities can return clutter that is highly spread in Doppler frequency. Minimizing the reception of spread-Doppler clutter due to very long-range backscatter from field-aligned irregularities requires judicious carrier frequency and waveform parameter selection to appropriately position range-ambiguities.

Round-the-world propagation corresponding to a 138 ms time-delay (equivalent to approximately 40,000 km) has also been observed (Rumi 1975). A multi-hop mode would require 40–60 passages through the D-region to reach the antipodal point, which is inconsistent with the remarkably low attenuation and dispersion that has sometimes been observed for such propagation.

The mode thought to allow round-the-world propagation is known as a "chordal mode." It is postulated that the chordal mode allows the signal to propagate by reflection from one part of the ionosphere to another along the inside of a layer with no intermediate ground reflections.[11] On certain circuits, ionospheric conditions are at times suitable for allowing the signal to propagate both in and out of the chordal mode between the transmitter and receiver.

[11] The TEP mode illustrated in Figure 2.34 is also a chordal mode but it returns the signal to ground after reflection from the two electron density enhancements of the equatorial anomaly formed on either side of the magnetic equator.

2.4.3 Amplitude and Phase Fading

Changes in the small-scale or "fine" structure of the ionosphere can affect the amplitude and phase of reflected signals over time-scales much less than the duration of the OTH radar coherent processing interval (in the order of seconds) and over spatial scales much smaller than the length of the receiving antenna aperture (in the order of kilometers). The complex (amplitude and phase) fading of signals reflected by the ionosphere on these temporal and spatial scales are of particular interest to OTH radar from a signal processing perspective. The main mechanisms responsible for the observed fading of skywave signals are briefly identified below. Mathematical models for the complex fading of individual HF signal modes reflected by the ionosphere will described in Part II of this text.

2.4.3.1 Amplitude Fading

The strength of an HF signal received after reflection from the ionosphere is observed to fluctuate in space and time. This variation in signal strength is referred to as amplitude fading and is often described statistically in terms of the fading depth and rate. Fading can be a significant problem for OTH radar as a sufficient signal-to-noise ratio needs to be maintained for reliable target detection. Deep and fast fading is undesirable while slow and shallow fading is often acceptable. In severe cases, fading periods can be as short as fractions of a second, while fading depths may be in excess of 40 dB. Amplitude fading of skywave HF signals can arise due to several physical mechanisms.

For example, amplitude fading may be caused by the focussing or de-focussing of signal rays returned by the ionosphere at a particular point on the ground. A phenomenon known as skip-focussing occurs near the MUF of a circuit in a region immediately beyond the skip-zone. This type of focussing arises due to the convergence of the low- and high-angle ray paths, which can enhance the average power of the returned signal by an order of magnitude relative to that at further ranges where the high and low rays do not converge. The variation in signal amplitude close to the MUF due to ionospheric fluctuations is sometimes referred to as skip-fading. Amplitude fading due to focussing and de-focussing may also occur as a result of traveling ionospheric disturbances. The amplitude of the signal returned from these moving wavelike structures in the ionosphere may be enhanced or diminished as the undulating reflection surface focusses and defocusses the rays at a particular point on the ground over time.

Amplitude fading due to polarization mismatch arises due to a time-varying difference between the polarization state of the incident resultant signal and that of the receiving antenna. The signal polarization can change due to the interference of ordinary and extraordinary waves with a relative phase difference that varies in space and time. The temporal variation in phase-difference $\Omega(t)$ causing nonstationary Faraday rotation is brought about by relative changes in the virtual path lengths of the o- and x-waves through the ionosphere. For example, a differential Doppler shift causes a linear variation of $\Omega(t)$. In the case of circularly polarized o- and x-waves with equal amplitude, a linearly polarized antenna receives a signal amplitude envelope $A(t)$ of rectified cosine shape as the axis of linear polarization of the resultant wave rotates over time.

$$A(t) = A_m|\cos\Omega(t)| \tag{2.50}$$

Figure 2.35 shows a real-data example of polarization fading received on a vertically polarized antenna due to unresolved o- and x-waves in the low-angle ray of the F2-layer.

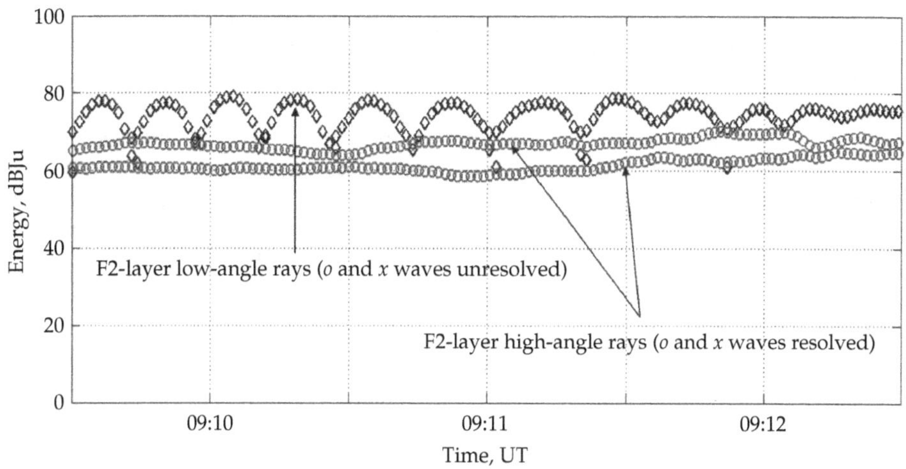

FIGURE 2.35 Real example of polarization fading due to unresolved *o*- and *x*-waves in the F2-layer low-angle ray received by a vertically polarized monopole antenna compared with the simultaneously received *o*- and *x*-waves in the high-angle ray, which are resolved in time-delay. © Commonwealth of Australia 2011.

The quasi-periodic polarization fading pattern resembles a rectified cosine shape, which is indicative of a differential Doppler-shift between the *o*- and *x*-waves. In this case, the fading period is less than 30 seconds and the fading depth is in the order of 20 dB. For OTH radar dwell times of 1 or 2 seconds, this rate is not sufficiently high to cause significant Doppler spread within a CPI. However, the deep fades can submerge echoes from low RCS targets below the noise floor and preclude their detection over several consecutive CPI, which can adversely affect tracking performance. Indeed, deep polarization fading due to Faraday rotation can be the controlling influence on path loss for OTH radar over limited time intervals.

Coupling losses due to polarization mismatch can be avoided by the appropriate combination of signals received on two antennas with orthogonal polarization. In principle, such antenna designs allow a single magneto-ionic component to be selected and isolated for processing. When a single magneto-ionic component can be isolated, either by using polarization diverse antennas or resolving the two characteristics waves in time-delay, the fading is typically slower and shallower than when the *o*- and *x*-waves interfere. The amplitude variation of individual magneto-ionic components is also shown in Figure 2.35. Here, the *o*- and *x*-waves in the high-ray of the F2-layer have been resolved in time-delay.

Fading may also arise due to constructive and destructive wave interference effects not related to those of *o*- and *x*-waves. For example, the interfering components may be the low- and high-angle rays reflected in a particular layer. Alternatively, a number of propagation modes reflected from different ionospheric layers may interfere. Within-mode fading can also arise due to the presence of electron density irregularities in the reflecting surfaces of the ionosphere (Fejer and Kelley 1980). In this case, rapid and deep fading may also occur for a single magneto-ionic component when the signal is diffusely scattered by dynamic electron density irregularities.

If the dimensions of the illuminated irregularity zone are such that the time-dispersion of the rays is much less than the reciprocal of the signal bandwidth, and the irregularities move in a relatively systematic manner, the scattered signal exhibits relatively slow fading, also known as flat-fading. On the other hand, when the time-dispersion of the rays is not small compared to the reciprocal of the signal bandwidth, or if the irregularities move rapidly and in a random manner, fast or frequency-selective fading occurs. The latter type of fading can significantly distort the modulation envelope of the signal and is sometimes referred to as flutter fading. Flutter fading is usually associated with spread-F field-aligned irregularities in the equatorial F-region (Davies 1990).

2.4.3.2 Phase and Frequency

In practice, the electron density height profile in the volume of ionosphere that reflects the incident HF signal is neither spherically symmetric nor static. At times, the plasma concentration may be exceedingly heterogeneous and dynamic due to the presence and movement of electron density irregularities. While the large-scale systematic movement of a layer or group of irregularities upward or downward results in a simple Doppler frequency offset being imposed on the reflected signal, the random motion of individual irregularities relative to one another over time (i.e., unordered movement) imposes an irregular phase modulation on the signal, which manifests itself as Doppler frequency spread. E-layer returns often have almost zero Doppler shift and typically give rise to the most frequency-stable signals. F2-layer returns will rarely have zero-Doppler shift and are more likely to exhibit a higher level of Doppler spectrum contamination. The frequency-stability of skywave signals over the CPI is important in OTH radar because Doppler processing is indispensable for moving-target detection in a powerful surface-clutter environment.

A phenomenon known as spread-F refers to the diffuse scattering of radio waves by irregularities in the F-region that can be magnetic-field aligned in the F2-layer (Huang, Kelley, and Hysell 1993). SDC often arises due to an ionization structure that is more-or-less frozen but drifts in time with respect to the propagation path. Angular scatter received from these structures over a spread of aspect angles relative to the drift velocity gives rise to a spread of Doppler shifts. The term "spread-F" not only refers to the large delay-spread (i.e., time-dispersion) observed on the received echoes, but also the accompanying spread in frequency. In severe cases, fast flutter-fading due to spread-F has the potential to return clutter over a band of Doppler frequencies that occupies most or all of the (aircraft) target velocity search space of an OTH radar.

It is worth distinguishing between Doppler frequency spread, which can apply to a monochromatic signal for example, and frequency dispersion of a signal in the ionosphere, which becomes especially relevant for wideband or spread-spectrum waveforms. As opposed to the coherence time of a signal, which is related to the inverse of its Doppler spread, frequency dispersion is often measured in terms of a coherence bandwidth. The coherence bandwidth of the ionosphere is generally considered to be less than 50–100 kHz for the purposes of effective matched filtering in OTH radar. Modulation envelopes with greater bandwidths than this may experience significant distortion after skywave propagation due to the frequency-dependent refractive index of the ionosphere. A discussion of this affect can be found in Paul (1979).

CHAPTER 3

System Characteristics

The design of an OTH radar system is influenced by a combination of factors related to mission requirements, available technology, environmental conditions, and economic constraints. These factors not only determine the design of the various OTH radar subsystems, but also the practical operation and capabilities of the overall system. The first main objective of this chapter is to describe the general characteristics of the different subsystems in a nominal skywave OTH radar architecture. Naturally, this includes the transmit and receive subsystems, as well as adjunct subsystems that support the operation of the main OTH radar. The second main objective is to explain the underlying factors that drive the key design features of OTH radar subsystems.

The chapter is organized into four sections. The first section describes a number of OTH radar configuration options and site selection considerations. It also reviews the relative merits of different waveform classes for radar applications, and the requirement for OTH radars to control out-of-band spectral emissions. The latter is necessary to minimize interference caused by the radar in adjacent frequency channels occupied by other HF-band users. The second section describes the transmit and receive subsystems and several approaches that have been proposed for OTH radar array calibration. Specifically, the choice of antenna element and array geometry are motivated for the transmit and receive functions. In addition, a number of traditional transmit and receive system designs are contrasted with contemporary alternatives to highlight the benefits provided by recent advances in technology.

The third section describes the frequency management system (FMS), which provides an OTH radar with real-time advice on optimum frequency selection for different mission types and surveillance regions. FMS outputs can also be used to guide the choice of waveform parameters, including the bandwidth, waveform repetition frequency, and CPI, which collectively determine the resolutions and ambiguities of the system in range and Doppler. The considered FMS includes an HF spectrum monitor for clear channel advice, backscatter and VI/OI sounders for propagation advice, as well as a "mini-radar" system for clutter Doppler spectrum assessment. The fourth section provides a brief historical perspective on the evolution of several past and present OTH radar systems, with particular emphasis on research and development in the United States and Australia. Some possible directions for future OTH radar systems are also discussed to conclude the chapter.

To provide a reference point for discussion, the approach taken here is to focus on the characteristics of a "nominal" OTH radar system, which is broadly exemplified by currently operational systems in the United States and Australia. The distinguishing features of several alternative OTH radar designs in other countries are identified throughout the chapter to illustrate examples of the possible diversity in implementation. To contain the scope, the main intent of this chapter is to provide an appreciation of the important concepts and trade-offs underlying OTH radar design without delving into fine detail regarding engineering implementation at the hardware component level. Technical information on these aspects may be found in handbooks specifically dedicated to antenna, transmitter or receiver design, as well as manufacturer specifications of RF devices.

3.1 Preliminary Considerations

The choice of radar configuration and the selection of sites for installing the transmit and receive subsystems significantly affects the operational characteristics and capabilities of an OTH radar system. For this reason, radar configuration and site selection need to be considered in terms of their compatibility with mission types, coverage area, and target classes of interest. In addition to their impact on system performance, such choices can also greatly affect system cost and complexity, which determines that compromises are often required in practice.

The first part of this section describes the three main configuration options that have been proposed for OTH radar systems, along with their relative advantages and disadvantages. The main criteria used to identify suitable sites for the high-power transmitter and highly sensitive receiver are also discussed. The second part of this section considers the waveform characteristics required for successful OTH radar operation, and contrasts the use of continuous versus pulsed waveforms in this application. Methods that may be used to modify a standard linear frequency–modulated continuous waveform (FMCW) in the HF band to satisfy current ITU regulations on out-of-band emissions are briefly described in the third part of this section.

3.1.1 Configuration and Site Selection

OTH radar systems have been operated in truly monostatic, bistatic, and multi-static configurations. In anticipation of the following discussion, it is also useful to define the "quasi-monostatic" configuration, wherein the bistatic angle formed between the lines of bearing from any point in the surveillance region to the two radar sites is small (less than about 5 degrees). This configuration option has been adopted in several operational OTH radar systems for reasons to be explained below. Before proceeding, it is instructive to further clarify the intended meaning of the terms monostatic, bistatic, and multi-static in the OTH radar context.

Monostatic refers to the situation where the radar transmitter and receiver are colocated at a single site. In a monostatic system, the same antenna or array of antennas may be used for transmission and reception by incorporating a high-power duplexor. It is also possible to use all elements in an antenna array for reception, but only a subset of the elements for transmission. Alternatively, an entirely different set of antenna elements may be dedicated for transmission and reception at a single site. Examples of monostatic

OTH radar systems with these characteristics will be described in due course. A defining characteristic of a monostatic (single-site) system is that the signal propagation paths(s) from the radar to and from any point in the surveillance region are reciprocal to a good approximation. In practice, OTH radar operation from a single site requires the use of pulse waveforms to protect the receiver from powerful direct-wave transmissions.

Bistatic refers to the case where the radar transmit and receive systems are located at two well-separated sites. In the context of skywave OTH radar, this configuration often implies that the transmitter and receiver are beyond the line of sight such that the transmitted signal can only be received via skywave or surface-wave propagation. In bistatic systems, the transmitter and receiver have a different viewing geometry of targets and other scatterers in the surveillance region. Moreover, the outgoing and incoming signal propagation paths will pass through different regions of the ionosphere, which may have different characteristics. On the other hand, physical isolation between transmit and receive sites typically allows for operation with continuous (unit duty-cycle) waveforms.

Multi-static configurations deploy a plurality of receivers and/or transmitters and operate these systems simultaneously on the same frequency. Such configurations are less common, but have been developed for surface-wave OTH radar (Anderson, Bates, and Tyler 1999), as well as for HF radar systems that operate by a combination of illumination and echo paths that include skywave, surface-wave, and/or line-of-sight propagation modes. For example, a multi-static HF radar architecture that uses a single transmitter to provide skywave illumination from a remote site and multiple forward-based receivers in the coverage to acquire skin-echoes from targets in the line-of-sight has been described in Frazer (2007). Target detections arising at multiple receive sites may be fed into a centralized tracker to improve target trajectory estimation. Multi-static systems will not be considered further in this section.

3.1.1.1 Single-Site Systems

The skywave OTH radar called "MADRE," which was developed after the second world war by the US Naval Research Laboratory (Headrick and Skolnik 1974), and the French OTH radar known as "Nostradamus," which was developed more recently by ONERA (Bazin et al. 2006), are examples of monostatic (single-site) systems. This configuration provides a number of undeniable advantages with respect to a bistatic (two-site) arrangement. The most important advantages as far as OTH radar systems are concerned will now be identified.

First, a monostatic configuration gives rise to an almost reciprocal signal-propagation path from transmitter to target (outgoing path) and target to receiver (incoming path).[1] This simplifies the process of selecting the most suitable (optimum) operating frequency for the mission at hand. In a bistatic architecture, the outgoing and incoming signal paths are reflected from laterally separated control points in the ionosphere, which may result in quite different propagation characteristics for target-illumination and echo-reception. Besides simplifying the process of optimum frequency selection, a monostatic configuration also simplifies mode identification and coordinate registration with respect to a bistatic radar architecture.

[1] Reciprocity of a two-way skywave propagation path not only depends on the signal passing through the same control point in the ionosphere, but also on the similarity between the elevation patterns of the transmit antenna and backscattering object.

A second advantage of operating with the transmitter and receiver at one location is that it eliminates the potential difficulty of finding and securing an additional radar site, which may cause problems in certain cases due to the significant "real-estate" required to site an OTH radar system. Furthermore, the centralization of all radar functions at a single site eliminates the need to duplicate infrastructure for equipment and operators, which reduces cost. It also obviates the requirement for inter-site communications and synchronization. Nowadays, these may be regarded as economic advantages with respect to a bistatic architecture, but for early OTH radar systems, accurate two-site synchronization posed a significant technical challenge.

Third, monostatic systems typically provide for an expanded set of coverage options with respect to a bistatic arrangement. Recall that the skywave OTH radar footprint (surveillance region) has a finite range-depth, which can be moved to different ranges by changing the operating frequency. Effective two-way skywave propagation generally requires the surveillance region to be at similar range from the transmitter and receiver, such that the operating frequency can be chosen to optimize propagation for both the illumination and echo paths. Clearly, this condition is always satisfied for a monostatic configuration, but for a bistatic OTH radar, it confines the coverage area to regions that are close to being equidistant from the transmitter and receiver. In addition, the antenna elements used for transmission and reception have radiation patterns that typically provide adequate gain over limited sectors in azimuth, the intersection of which defines the feasible coverage area. This factor can significantly reduce the coverage of a bistatic OTH radar relative to a monostatic system.

However, a monostatic OTH radar requires the receiver to be protected (isolated) from the extremely powerful direct-wave signal during transmission. Monostatic OTH radars therefore need to use pulse waveforms to avoid saturating (overloading) the receiver or damaging sensitive equipment. For example, the MADRE OTH radar used simple (amplitude-modulated) pulses of 100-μs duration to provide a group-range resolution of approximately 15 km. For a typical air-mode pulse repetition frequency (PRF) of 50 Hz, this corresponds to a duty-cycle of 0.5 percent for a simple 100-μs pulse.[2] Using pulse waveforms with relatively small duty-cycles attracts a number of penalties with respect to bistatic systems that may operate with continuous waveforms. More will be said on the relative merits of pulsed and continuous waveforms in the next section.

3.1.1.2 Two-Site Systems

Bistatic OTH radar configurations with well-separated transmit and receive sites can operate effectively using continuous waveforms. However, as the inter-site separation increases, the outgoing and incoming signal paths will be reflected by control points spaced further apart in the ionosphere. The likelihood of the illumination and echo paths having significantly different skywave propagation characteristics increases as the locations of the control points diverge. Optimum frequency selection and the mapping of radar coordinates to geographical position therefore becomes more difficult as the length of the transmitter-receiver baseline increases.

The 440-L forward-scatter system described in Willis and Griffiths (2007) and references therein is an example of an unmodulated CW bistatic OTH radar used to detect and localize missile launches by recognizing target Doppler-time signatures and triangulating

[2]FM pulses permit the duty cycle to be extended to perhaps 20 percent for air-mode PRFs.

them over many paths with different viewing geometry. Specifically, transmitters were sited in the Western Pacific (Japan, Guam, Philippines, and Okinawa) with corresponding receivers in Europe (Italy, Germany, and England). The 440-L system was operational between 1970 and 1975 (after which its transmit modules became part of the Australian Jindalee Stage B OTH radar project). Descriptions of other operational bistatic OTH radars (not limited to skywave systems) can be found in Willis and Griffiths (2007).

Widely separated transmit and receive sites that create a large bistatic angle to points within the surveillance region can potentially enhance target RCS and reduce the reception of range-folded clutter due to the unaligned transmit and receive beam steer directions. However, significant disadvantages with regards to frequency selection, coverage area, and coordinate registration, for example, have resulted in this option not being preferred for most skywave OTH radar systems. On the other hand, a small inter-site separation (relative to the target range) allows the radar to operate using continuous waveforms while maintaining a single-site characteristic by virtue of the small bistatic angles to all points in the coverage. As mentioned previously, this is referred to as a quasi-monostatic configuration.

Most if not all continuous-wave skywave OTH radars reported in the literature have been designed with a relatively small inter-site separation. In these quasi-monostatic systems, the transmitter-to-receiver distance represents a compromise between two competing objectives. A lower bound on this distance is determined by the need to provide sufficient isolation to permit operation with continuous waveforms. The upper bound is such that the signal paths to and from a target do not involve significantly different control points in the ionosphere. In other words, the quasi-monostatic configuration enables the benefits of continuous-wave operation to be realized, while retaining the desirable frequency-selection, coverage area, and coordinate registration properties of a single-site system. The main advantages traded for continuous wave operation in a quasi-monostatic configuration, relative to a truly monostatic system, are the need for two radar sites, the duplication of infrastructure, and inter-site communications/synchronization.

The Jindalee OTH radar system operates with an inter-site separation of approximately 100 km. Over dry land, this provides sufficient attenuation of the surface-wave signal on the direct path from the transmitter. A separation of about 100 km also places the receiver in the skip-zone of the transmitter at typical OTH radar operating frequencies. Hence, this architecture effectively isolates the receiver from the strong transmitted signal via both the surface-wave and skywave direct-paths. The bistatic angles subtended for an inter-site separation of 100 km to targets in the coverage area is also relatively small. For example, at a range of 1500 km, an inter-site separation of 100 km corresponds to a bistatic angle of approximately 2 degrees at boresight (along the perpendicular bisector of the transmitter-receiver baseline).

For skywave propagation through a particular ionospheric layer, the signal reflection points for the outgoing and incoming paths in a quasi-monostatic configuration are laterally spaced by tens of kilometers at the path mid-point. For example, an inter-site separation of 100 km results in a control point separation of approximately 50 km at boresight. For a quiet mid-latitude ionosphere, the gross structure of the normal-E and F-region layers are expected to exhibit a high spatial correlation over tens of kilometers (at times not near the dawn or dusk terminators). Sporadic-E is a possible exception as it may exhibit a low spatial correlation over short distances when patchy. It follows that the illumination and echo signal paths will in general experience similar propagation characteristics for radar sites separated by about 100 km.

In a modern two-site OTH radar system, the transmitter is controlled remotely from the receiver site with inter-site synchronization achieved using GPS reference signals. The typically greater cost and complexity of a two-site system needs to be justified in terms of the additional capability it provides with respect to a single-site architecture. A significant benefit that stems from using continuous waveforms is that a higher average power can be radiated with unit duty-cycle waveforms for a fixed peak-power rating of the transmitter to improve target detection performance in a noise-limited environment. Moreover, the ability to control out-of-band spectral emissions without significantly compromising target SNR, or the waveform ambiguity-function characteristics needed for effective OTH radar operation, is more readily achieved using continuous as opposed to pulsed signals (more will be said on this in Section 3.1.3).

A salient benefit of separating the transmit and receive sites is that it enables *independent* half-radar operation simultaneously at different frequencies. The ability to partition the transmit and receive resources for independent OTH radar operation provides significant flexibility to trade detection performance for real-time coverage or coverage rate. In a single site system, out-of-band emissions have the potential to produce significant mutual interference in the half-radar mode of operation. A similar argument applies for an FMS that has its transmitter and receiver located at the same sites as those of the main OTH radar, respectively. Besides offering greater flexibility to operate radar resources and support systems simultaneously at different frequencies, a two-site architecture also provides greater scope to keep the receive site electrically quiet. Even when the transmitter is nominally in the off state, so-called "dark noise" generated by transmitting equipment can produce disturbances in a colocated receiver.

3.1.1.3 Site Selection

Ground with relatively even topography and uniform electrical properties is needed to site the large antenna arrays of a skywave OTH radar system. At decameter wavelengths, the electrical properties of the ground (conductivity and relative permittivity) will exert considerable influence on the gain and radiation pattern characteristics of the HF antennas. For example, the low elevation-angle gain of a vertically polarized ground-based HF antenna is highly dependent on the conductivity of the surface under and near the installation. Ideally, the ground should exhibit relatively homogeneous electrical properties and preferably high conductivity over an extended area, both under and in front of the OTH radar transmit and receive antenna arrays. This helps to maximize low-elevation angle gain for long-distance (one-hop) surveillance, and to minimize differences in the antenna element radiation patterns across the aperture, which are manifested as array calibration errors.

Unlike conventional microwave radar systems, which may be sited on mountain tops or elevated ground to increase range coverage, OTH radars require wide and flat open spaces to site. This generally requires extensive site-preparation works, involving land-clearing and earth-moving, particularly in wooded areas or regions of dense scrub. The ground also needs to be suitable for the installation of large antenna structures, equipment shelters, and other infrastructure for radar operators. Indeed, the original site identified on a map for the Jindalee OTH radar receiver in central Australia was found to be suitably flat but deemed inappropriate upon inspection due to the presence of many creeks and washaways with swampy land on one side (Sinnott 1988).

External HF noise due to natural and man-made sources almost always dominates internally generated (thermal) noise in a well-designed OTH radar receiver. In particular, the man-made noise contribution varies significantly with location. Near urban areas, man-made noise can be strong enough to dominate environmental (atmospheric plus galactic) noise over the entire HF band, and hence limits detection performance against external noise. An electrically quiet location, well away from large residential or industrial areas, is required for the receiver site. In remote locations, the level of man-made noise often falls below the sum of atmospheric and galactic noise across the entire HF band. In practice, it is also appropriate to site the OTH radar transmitter well away from urban centers. This being said, the OTH radar sites should also be easily accessible by road and within a reasonable distance from services. Connection to the mains supply is desirable but not necessary if the site is locally powered by generators, which may also serve as a backup.

The skip-zone phenomenon also needs to be accounted for when selecting OTH radar sites for the intended coverage area. A typical skip-zone limit of about 1000 km clearly requires the OTH radar to be set well back from the nearest edge of the intended coverage. This constraint on site-selection can create a problem for geographically small nations, particularly when areas near sovereign borders need to be monitored. The angle at which the OTH radar views the coverage also needs to be considered carefully, as this may influence the reception of spread-Doppler clutter from disturbed ionospheric regions at low and high (magnetic) latitudes. For example, equatorial spread-Doppler clutter may be received with the full gain of the OTH radar system in the direction of the surveillance region due to range ambiguities, while range-coincident auroral spread-Doppler clutter may be received through the side or back lobes of the transmit and receive antenna patterns.

Continuous-wave OTH radar systems require high attenuation of the self-interference (direct path) signal from the transmitter to ease the already demanding requirements which are set on receiver dynamic range by the high backscattered clutter to target echo power ratio. The OTH radar sites are often selected such that transmit and receive arrays are at extreme steer with respect to each other (i.e., side-on instead of one behind the other). The strength of the direct-path surface-wave signal is therefore further attenuated by the sidelobes of the transmit and receive antennas (beyond that due to physical separation). Isolation from the direct skywave path is mainly provided by the skip-zone phenomenon in the quasi-monostatic configuration. Additionally, the low gain of the transmit and receive antenna elements at high elevation angles also helps to reduce scatter from near vertical incidence in this configuration.

3.1.2 Radar Waveforms

Radar waveforms may be broadly categorized either in terms of signal-implementation characteristics or ambiguity-function properties. The latter basis for signal classification provides a valuable framework to identify the waveform class that is best suited for a particular application, taking into account system objectives with regard to detection and estimation, as well as the expected delay-Doppler distribution of (desired and unwanted) scatterers in the environment. On the other hand, the former basis for classification is appropriate for selecting a specific signal design, within the identified waveform class, that is most compatible with the practical constraints or hardware limitations of the system in relation to waveform generation, transmission, and receiver processing.

Examples of signal-implementation characteristics used to classify waveforms include continuous or pulsed, periodic or nonrecurrent, and the modulation function employed.

Ignoring the possibility to modify signal polarization, modulation can be imposed to change the amplitude, frequency, or phase-state of the carrier signal. Frequency modulation may be linear or nonlinear in time, while phase modulation can involve binary or poly-phase codes. For reasons explained later, radar waveforms are typically synthesized as coherent pulse-trains. In most cases, the modulation is applied during the pulse repetition interval (intra-pulse), although modulation can additionally be applied across successive pulses in the coherent processing interval (inter-pulse). A detailed taxonomy of radar waveforms is beyond the scope of this text, but a comprehensive review can be found in Levanon and Mozeson (2004). The classic texts by Rihaczek (1985), Nathanson (1969), and Cook and Bonfed (1967) are also recommended for an introduction to the theory and application of radar signals.

The matched-filter (correlation-receiver) response to an echo from an ideal point scatterer with delay τ and Doppler shift f is given by the waveform auto-ambiguity function (AF), defined by $\chi(\tau, f)$ in Eqn. (3.1) for the analytic signal $s(t)$, where $*$ denotes complex conjugate and j is the imaginary unit. The mathematical properties of the AF are described in Richards (2005). The AF characteristics of interest to radar systems include resolutions, sidelobe levels, and ambiguities in the delay-Doppler plane.

$$\chi(\tau, f) = \int_{-\infty}^{\infty} s(t)s^*(t-\tau)e^{-j2\pi ft}\,\mathrm{d}t \tag{3.1}$$

Ideally, one wishes to have an appropriate resolution, to search for targets quickly but with some degree of reliability, very low sidelobes, and no ambiguities. Unfortunately, this is not always possible because practical waveforms have finite energy (limited by the system or operational factors) and the AF volume must be conserved. In practice the aforementioned AF characteristics need to be traded off judiciously against one another. The task of the radar signal designer is to find the most appropriate balance by jointly considering operational objectives, system constraints, and environmental conditions.

3.1.2.1 Signal Classes

The bandwidth of an OTH radar waveform is primarily limited by spectrum occupancy (user-congestion) in the HF band, although frequency dispersion in the ionosphere imposes a secondary (physical) restriction when no compensation for this phenomenon is applied. On the other hand, the effective time-on-target or CPI is typically limited by the revisit rate required to establish and maintain tracks over different surveillance regions in the coverage, particularly for maneuvering targets. OTH radar waveforms for aircraft detection typically have a time-bandwidth product (TBP) in the order of 10^4. For example, a bandwidth of 10 kHz and CPI of 1 second. The TBP of OTH radar waveforms used for ship detection are one or two orders of magnitude greater than this due to the larger bandwidths and longer CPIs used. The maximum length of the CPI that can be gainfully employed for ship detection is often limited by the coherence-time of the sky-wave propagation channel. In either case, the combination of operational constraints and physical limitations results in the availability of a finite TBP for OTH radar waveform design.

Although the number of possible radar waveforms is unlimited, there are only a few broadly defined classes of waveforms with distinct resolution properties (Rihaczek 1971). Using the terminology in Rihaczek (1971), signals with a TBP in the order of unity are designated as Class A waveforms and have a ridge-type AF with a resolution cell size of unity. A basic example, which is not of practical interest to OTH radar, is the simple constant-carrier pulse used one at a time with no coherent integration. For a TBP much larger than unity, it is possible to choose between three other waveform classes. These correspond to the thumbtack (Class B1), sheared-ridge (Class B2), and bed-of-nails (Class C) ambiguity function types. Class B1 and B2 waveforms may also be thought of as single-pulse signals, but with a TBP much greater than unity due to the applied modulation. Class C waveforms are coherent pulse-trains (i.e., the periodic repetition of a pulse), where the pulse bandwidth is large compared to the repetition frequency.

The subclassification of single-pulse waveforms into Class B1 and B2 is motivated by Figure 3.1, which illustrates the distinct range-Doppler response characteristics of a bandlimited Gaussian noise signal and a single linear-FM sweep when both waveforms are processed by a *tapered* matched-filter for the purpose of sidelobe suppression. The Class B1 waveform (bandlimited Gaussian noise) exhibits a thumbtack response with a delay-Doppler resolution cell size in the order of 1/TBP. It has no ambiguities, and a sidelobe pedestal that is on average 1/TBP below the main peak in power. On the other hand, the Class B2 waveform (single linear-FM sweep) has a sheared-ridge AF response with a delay-Doppler resolution cells size in the order of unity. This waveform is also characterized by low sidelobes and ambiguity due to range-Doppler coupling.

Class B1 waveforms are desirable from the viewpoint of having high resolution and no ambiguities. However, these advantages come at the expense of a relatively high sidelobe pedestal. The pedestal level is inversely related to the TBP on a power basis. Clutter-to-target echo ratios of 40–80 dB may be encountered in OTH radar applications. A TBP of 10^4 yields a sidelobe pedestal that is 40 dB below the main clutter peak over the *entire* range-Doppler plane. This creates a problem when relatively weak target echoes need to be detected in the presence of much stronger signal reflections. Class B1

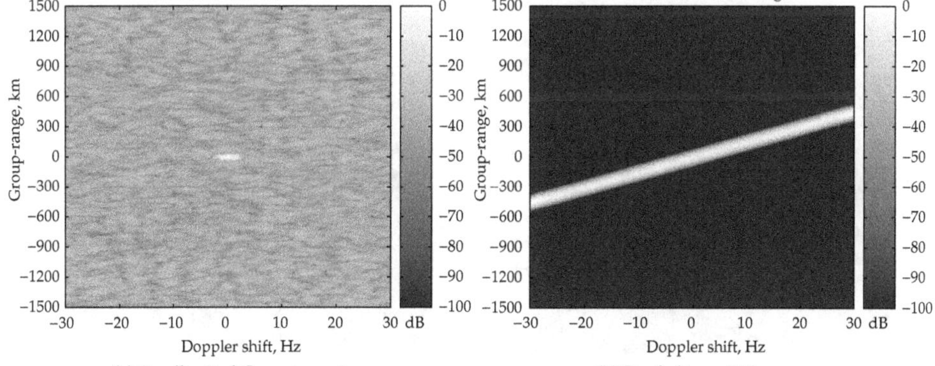

(a) Bandlimited Gaussian noise. (b) Single linear-FM sweep.

FIGURE 3.1 The intensity-modulated displays show the range-Doppler processing magnitude-response of the tapered matched filter to an ideal point scatterer as a function of its displacement in range and Doppler from the matched range-Doppler point. This response is shown for a bandlimited Gaussian noise signal and a single linear-FM sweep. Both waveforms have a nominal bandwidth of 10 kHz and a duration of 1 second (TBP = 10^4). Identical tapering is applied to the matched filter in both cases to lower sidelobe levels in range and Doppler.

waveforms cannot provide sufficient sub-clutter visibility for reliable target detection when using matched-filter processing and the TBP available for waveform design in the OTH radar context. Stated another way, this waveform class is unsuitable for OTH radar due the relatively high level of the sidelobe pedestal when conventional range-Doppler processing techniques are used.

Class B2 waveforms provide for very low sidelobes, but the significant range-Doppler coupling implies that the ability to resolve different point scatterers depends strongly on their relative range and Doppler coordinates. Specifically, the potential for high resolution exists only for scatters with range-Doppler coordinates that do not align with the ridge. The main problem associated with this waveform is that scatters falling along the ridge experience essentially no resolution. In other words, high resolution can be achieved either in range or Doppler, but not in both dimensions independently. This may preclude the separation of a target echo from its multipath components, or the ability to resolve useful signals from clutter echoes at a different range and Doppler that falls on the ridge.

Although OTH radar clutter is often concentrated around zero Doppler frequency, the scatterer distribution in range may extend over thousands of kilometers. Inspection of Figure 3.1b reveals that this can lead to serious interference effects between Doppler-shifted target echoes and 0-Hz clutter returns located at different ranges. Moreover, target tracking relies on the ability to accurately localize echoes in both range and Doppler. This requires waveforms with minimal coupling (ambiguity) between these dimensions. The resolution and ambiguity properties of Class B2 waveforms are therefore incompatible with the OTH radar clutter environment and target localization objective.

The only class of waveform that can simultaneously provide low sidelobe-levels and high-resolution with minimal coupling in range and Doppler are those characterized by a "bed-of-nails" ambiguity function. As mentioned previously, these Class C waveforms are periodic pulse-trains, where the bandwidth of the pulse is large compared to the repetition frequency. Figure 3.2, in the same format as Figure 3.1, shows the conventional (mismatched) range-Doppler processing response for a periodic linear

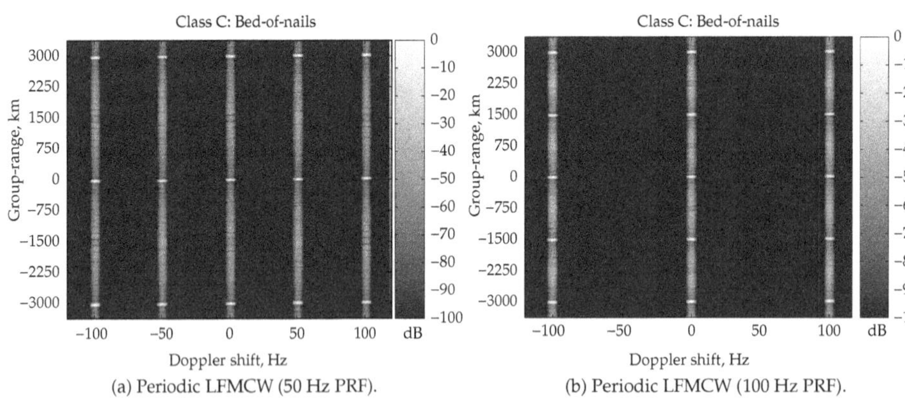

FIGURE 3.2 Tapered matched-filter range-Doppler processing magnitude-responses for a periodic linear frequency–modulated continuous waveform (LFMCW) with 50 Hz and 100 Hz PRF. Both waveforms have a nominal bandwidth of 10 kHz and a duration of 1 second (TBP = 10^4). The PRF determines the locations of concentrated ambiguity "spikes" in the range-Doppler plane. © Commonwealth of Australia 2011.

frequency–modulated continuous waveform with a TBP of 10^4 using two pulse repetition frequencies. Here, the term "mismatched" refers to the use of a tapered matched-filter.

It may be observed in Figure 3.2 that Class C waveforms trade the advantages of low sidelobes and high resolution in exchange for very intense but concentrated ambiguities spaced at regular intervals in range and Doppler. The coordinates of these ambiguities are governed by the choice of PRF, and equations for determining their locations will be described in the next section. For a fixed TBP, Class C (bed-of-nails) waveforms achieve comparable resolution to Class B1 (thumbtack) waveforms and sidelobe levels to Class B2 (sheared-ridge) waveforms, but the energy contained in the sidelobe pedestal for Class B1 waveforms (or the sheared-ridge for Class B2 waveforms) is now concentrated in severe but isolated range and Doppler ambiguities.

In particular, the first- and higher-order range ambiguities can potentially allow clutter echoes returned from very long distances to alias into the unambiguous coverage. The range-folded clutter echoes may arise due to highly dynamic scatterers with significant energy at the Doppler frequencies of target echoes. In such situations, range ambiguities of a Class C waveform can be detrimental to target detection performance in OTH radar. On the other hand, fast-moving targets can produce echoes with Doppler shifts large enough to reach first- or higher-order Doppler ambiguities of a Class C waveform, which can lead to ambiguous target velocity estimates and blind speeds at multiples of the PRF. In OTH radar systems, range and Doppler ambiguities may be of concern for aircraft detection, but are of little or no concern for ship detection.

For Class C waveforms, the locations of ambiguities can be managed by appropriate PRF selection. Specifically, the PRF may be chosen with due regard to the expected characteristics of the range-Doppler distribution of the scatterers in the environment. The key point is that ambiguities associated with Class C waveforms attract practically no penalty when they occur at range-Doppler coordinates where there is negligible scattered energy. In this case, the effective ambiguity function appears to the radar as if it were a "thumbtack" surrounded by very low sidelobes over the entire range-Doppler plane. The ambiguities still exist, but will not cause interference when scatterers are not located at their coordinates with respect to the point interrogated by the matched-filter in the range-Doppler plane.

As illustrated in Figure 3.2, higher PRF waveforms have smaller range ambiguities and larger Doppler ambiguities compared to lower PRF waveforms. For certain radar missions and environmental conditions, it is not always possible to position ambiguities where there is very little or no scatter. A performance penalty is incurred on these occasions if a constant-PRF waveform is used. However, the presence of severe ambiguities at positions controlled by the choice of PRF are considered to be less detrimental to performance than a relatively high sidelobe pedestal level over the entire range-Doppler plane, or effectively no resolution for scatters with certain combinations of range and Doppler.

Class C waveforms are regarded as the most appropriate for OTH radar applications considering the high dynamic-range and delay-Doppler distribution of the scattering environment in combination with the limited TBP values available to operational systems. In the case of a point scatter (e.g., target), use of multiple waveforms with different PRFs may be used to resolve range and Doppler ambiguities. However, in the presence of spatially distributed scatterers (e.g., terrain or sea), use of multiple PRFs may not address the potential problem of target-echo masking due to range-folded spread-Doppler clutter. In this case, inter-pulse modulation can help to manipulate the regions where energy from range-folded returns appears in the range-Doppler plane, such that it does not mask target echoes after conventional processing (Hartnett, Clancy, and Denton 1998).

3.1.2.2 Pulsed and Continuous

The specific choice of OTH radar signal within the domain of Class C waveforms requires identifying the signal implementation that best suits the operational requirements of the system subject to the practical constraints of the available technology. The use of pulsed or continuous waveforms represents a fundamental starting point in the signal-design process. Once this choice is made, other characteristics required for signal generation, such as the modulation type and function employed, may be selected in turn. As already mentioned, monostatic OTH radar configurations are constrained to use waveforms with a duty-cycle less than unity in practice, while two-site OTH radar systems may operate with continuous waveforms.

It is recalled that early OTH radar systems such as MADRE operated with simple (amplitude-modulated) pulse waveforms because this was the only effective option available at the time. Specifically, frequency-modulated continuous wave signals could not be generated with sufficient spectral purity to support effective OTH radar operation in the mid-1950s. Precise synchronization of transmit and receive sites separated by distances in the order of 100 km (to permit operation with unit duty-cycle waveforms) was also challenging when the first skywave systems emerged. As technology advanced, effective two-site OTH radar operation using frequency-modulated continuous waveforms became possible during the 1960s.

The question arises as to the relative merits of using pulsed or continuous waveforms in current single- and two-site OTH radar systems, respectively. An obvious but important attribute of pulsed waveforms is that they allow the previously described operational and economic advantages associated with monostatic OTH radar operation to be realized. In addition, the time-gated nature of pulsed waveforms implies that the *instantaneous* power of the backscattered clutter echo arises from a more limited area of the Earth's surface compared to the continuous-wave case. This eases dynamic range requirements for the waveform generator and allows more efficient amplifiers to be used (Skolnik 2008b).

Pulse waveforms also have a number of significant drawbacks. For a given peak-power rating of the transmitter, pulse waveforms yield a lower average power compared to continuous waveforms that have a constant-modulus envelope. With respect to frequency- or phase-modulated continuous waveforms, this translates to a reduction in target echo SNR, which scales in direct proportion to the waveform duty-cycle. Continuous waveforms offer the opportunity to maximize the average power radiated by the radar system, which can significantly enhance detection performance for fast-moving low-RCS targets, particularly at night when the irreducible level of received external noise may have a spectral density that is 20–30 dB higher than internal noise.

One option to improve noise-limited detection performance using a pulse waveform is to increase the peak-power handling capability of each transmit channel in the antenna array. Another is to increase the number of transmit channels and hence the array gain. While either option or a combination thereof can be pursued, the latter approach is preferred from the viewpoint of relaxing the hardware-related demands placed on the power amplifier and antenna in each transmit channel. On the other hand, increasing the number of transmit channels can result in a reduction of coverage or coverage rate as the antenna gain and beamwidth are inversely related. Not surprisingly, significant compromises are required to obtain the same SNR using pulse waveforms as can be achieved using continuous waveforms with a constant-modulus envelope.

Out-of-band spectrum emissions, which may cause disruption (interference) to other HF band users in adjacent frequency channels, are relatively more difficult to control using pulse waveforms because of their smaller time-bandwidth product (Headrick and Thomason 1998). In practice, out-of-band emissions need to maintained below acceptable levels without significantly degrading the radar waveform's ambiguity function properties, which are needed to effectively fulfill the target detection and parameter estimation objectives of the system. The use of continuous waveforms provides greater scope to satisfy out-of-band emission guidelines determined by local authorities or the International Telecommunications Union (ITU), while maintaining resolution and sidelobe properties appropriate for OTH radar operation. This subject will be pursued in the final part of this section.

3.1.2.3 Linear FMCW

Having identified the need for Class C waveforms, and discussed the significant benefits of continuous-wave operation based on the quasi-monostatic configuration (the design choice preferred by most currently operational skywave OTH radar systems), it remains to decide on the modulation format for each repeated pulse. A modulation envelope with constant modulus is not only desirable from the viewpoint of maximizing the average power for a given peak-power limit, but the generation of constant power is readily compatible with solid-state amplifiers, which are inherently unable to raise their peak-power handling capacity much above their continuous ratings (Stove 1992). A waveform that delivers constant power also places less stringent demands on amplifier linearity. Note that solid-state amplifiers are preferred in OTH radar to enable frequent and rapid switching between possibly widely spaced carrier frequencies during normal operation while maintaining high linearity and spectral purity, as this cannot be achieved using vacuum-tube technology.

A constant-modulus pulse with a TBP much greater than unity may be generated using phase or frequency modulation. Although phase-shift keying and nonlinear frequency modulations have been proposed for continuous-wave radars, the linear-FM pulse has a number of important advantages in practice. First, the target range is proportional to the beat frequency, which allows a simple FFT-based correlator to be used for range processing. Second, analog low-pass filtering can be applied to reduce the bandwidth of the signals after linear-FM mixing or "deramping," which allows the FFT to only process range cells in the useful coverage (i.e., stretch processing). This not only decreases the real-time processing load, but it also reduces the requirements on analog-to-digital converter (ADC) sampling frequency for a given dynamic range, as well as the data rates and volumes flowing through the computing architecture.

These represent important advantages of using linear FM modulation relative to other constant-modulus signals based on phase-coded waveforms or nonlinear frequency-time functions. Such advantages were exploited in early OTH radar systems, which needed to operate with modest ADC technology and limited processing capacity by contemporary standards. Linear-FM modulation or "chirps" with an integer TBP also facilitates the use of computationally efficient algorithms for fully digital pulse compression in modern OTH radar systems (Summers, 1995). Besides reducing real-time processing load, the linear-FM pulse provides excellent range resolution for a given transmission bandwidth

(Turley 2009), which is nominally determined by the frequency excursion of the linear-FM ramp.

When the linear-FM pulse is repeated to form a coherent pulse-train, the large change in instantaneous frequency between the end of one sweep and the commencement of the next causes significant out-of-band emissions. The pure linear-FM pulse may be "shaped" by amplitude-tapering or modifying the frequency-time characteristic of the waveform near the discontinuity to reduce out-of-band emissions without significantly increasing range sidelobe levels. Practical methods to reduce out-of-band emissions without significantly impacting pulse-compression performance will be described in the final part of this section.

For the reasons stated above, the majority of operational two-site OTH radar systems use periodic waveforms based on the repetition of a linear-FM pulse. Figure 3.3 illustrates the frequency-time characteristic of a repetitive linear frequency–modulated continuous waveform (LFMCW) with carrier frequency f_c, bandwidth B, waveform repetition period T_p, and CPI T. The baseband signal is defined by three parameters, namely, the waveform repetition frequency $f_p = 1/T_p$, the frequency excursion of the linear FM-ramp B, and the number of pulses or sweeps in the CPI N. In practice, the LFMCW parameters are chosen in real time depending on the radar mission type and prevailing environmental conditions. In practice, the choice of waveform bandwidth and PRF also needs to be compatible with an integer number of samples per sweep. Criteria used for OTH radar waveform parameter selection will be discussed in Chapter 4.

Table 3.1 relates the resolution and ambiguity properties of the LFMCW in delay and Doppler to the parameters that define this signal. These relationships are also expressed in terms of group range and relative velocity (rate of change of group range) over a two-way monostatic path in Table 3.1. Using typical (air-mode) OTH radar waveform parameters of $B = 10$ kHz, $f_p = 50$ Hz, $T = 1$ second, and $f_c = 15$ MHz, we have that

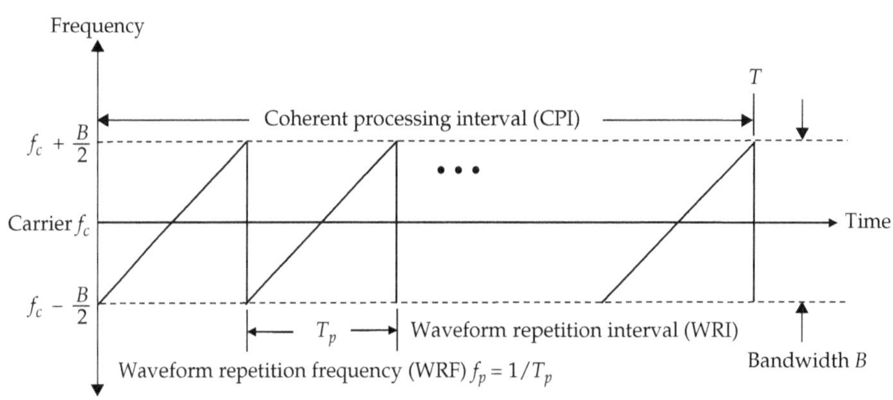

FIGURE 3.3 Instantaneous frequency-time characteristic of a repetitive linear frequency–modulated continuous waveform (LFMCW) with center frequency f_c, nominal bandwidth B, pulse repetition frequency f_p, and coherent processing interval T. The terms pulse repetition frequency (PRF) and waveform repetition frequency (WRF) may be used to distinguish between repetitive waveforms with fractional and unit duty-cycles, respectively, although these terms have often been used interchangeably in the literature. © Commonwealth of Australia 2011.

Repetitive LFMCW	Time delay, s	Group range, m	Doppler shift, Hz	Relative velocity, m/s
Resolution	$\Delta\tau = \frac{1}{B}$	$\Delta R = \frac{c}{2B}$	$\Delta f = \frac{1}{T}$	$\Delta v = \frac{c}{2f_c T}$
Ambiguity	$\tau_{\text{amb}} = T_p$	$R_{\text{amb}} = \frac{c}{2f_p}$	$f_{\text{amb}} = f_p$	$v_{\text{amb}} = \frac{cf_p}{2f_c}$

TABLE 3.1 Expressions for the resolutions and first ambiguities in time delay (group range) and Doppler shift (relative velocity) in terms of the repetitive LFMCW parameters; namely, the pulse repetition frequency $f_p = 1/T_p$, bandwidth B, and coherent processing interval $T = NT_p$. The Doppler frequency resolution and relative velocity ambiguity are sometimes expressed as two-sided intervals symmetric about 0 Hz, i.e., $\Delta f = \pm 1/(2T)$ Hz or $v_{\text{amb}} = \pm cf_p/(4f_c)$ m/s. Second- and higher order ambiguities in range and Doppler occur at integer multiples of first ambiguity. The constant c is the speed of light in free space.

$\Delta R = 15$ km, $R_{\text{amb}} = 3000$ km, $\Delta v = 36$ km/h, and $v_{\text{amb}} = \pm 900$ km/h. For a higher repetition frequency of $f_p = 100$ Hz, $R_{\text{amb}} = 1500$ km, and $v_{\text{amb}} = \pm 1800$ km/h. These first-order range and Doppler ambiguities may be viewed graphically in Figure 3.2.

Jittering the PRF may be used to distinguish between ambiguous and unambiguous echoes, as only the coordinates of the latter remain fixed in range and Doppler when the PRF changes (Richards 2005). Phase coding of successive FMCW sweeps (inter-pulse modulation) enables range-ambiguous echoes to be shifted by a known Doppler offset, as described in Hartnett, Clancy, and Denton Jr. (1998) and Clancy, Bascom, and Hartnett (1999). To improve range resolution when bandwidth is scarce in the congested HF spectrum, it has been proposed to synthesize waveforms occupying two or more disjoint sub-bands of the spectrum close to the desired operating frequency. This requires alternative pulse-compression techniques to maintain low range-sidelobes (Zhang and Liu 2004).

3.1.3 Out-of-Band Emissions

Services operating with allocated frequency channels in the HF band are strictly required to adhere to spectrum management recommendations set by the International Telecommunications Union (ITU) (Int. Telecomm. Union 2006b). In essence, it is necessary to ensure that the spectrum of radiated signals satisfies ITU regulations defined in the form of spectral bounds or "emission masks." The guidelines are designed to minimize mutual interference between users in neighboring frequency channels. For non-allocated services, national spectrum management authorities may determine independent restrictions. In certain countries, the ITU rules are relatively easier to satisfy than the standards imposed by national authorities for non-allocated users.

The ITU also provides resolutions concerning the use of the radio spectrum by primary radar systems (Int. Telecomm. Union 2006a). From January 2003 to the current time, HF radars in Australia are required to conform with ITU regulations in accordance with the spectrum measurement and analysis process (Int. Telecomm. Union 2003). The first part of this section compares the spectrum of a single linear-FM chirp and the repetitive LFMCW, with the emission masks designated for this type of waveform by the ITU. The second part of this section describes the use of amplitude-tapering and finite-flyback techniques for reducing out-of-band emissions caused by the large instantaneous frequency change at the sweep transitions. The impact of these pulse-shaping techniques on radar performance, particularly range-sidelobe levels, is discussed in the final part of this section.

3.1.3.1 LFMCW Spectrum

A detailed analysis of the impact of ITU spectral emission masks on HF radars using a repetitive LFMCW can be found in Turley (2009) and Durbridge (2002). For a frequency f_c and necessary bandwidth B_N, the radiated envelope of the LMFCW spectrum is required to fall by at least 40 dB with respect to the 0 dB reference level (peak of the spectrum) according to the ITU bandwidth formula in Eqn. (3.2). Here, both B and B_{-40} are defined as two-sided bandwidths centered on the operating frequency.

$$B_{-40} = B + 0.0003 f_c \tag{3.2}$$

The instantaneous frequency of a single linear-FM sweep with ramp bandwidth B and chirp duration T_p is given by $f_i(t)$ in Eqn. (3.3), where t denotes time and f_c is the carrier frequency. This frequency-time characteristic was previously illustrated for a coherent train of chirps in Figure 3.3.

$$f_i(t) = \frac{Bt}{T_p} + f_c - \frac{B}{2}, \quad t \in [0, T_p) \tag{3.3}$$

The phase characteristic of the signal can be readily determined as $\phi_p(t)$ in Eqn. (3.4), where the subscript p indicates a single pulse or chirp using the notation in Turley (2009).

$$\phi_p(t) = 2\pi \int_0^t f_i(t')\mathrm{d}t' = \frac{\pi B t^2}{T_p} + 2\pi \left(f_c - \frac{B}{2}\right) t, \quad t \in [0, T_p) \tag{3.4}$$

The analytic linear-FM signal may then be written as $v_p(t)$ in Eqn. (3.5)

$$v_p(t) = \begin{cases} e^{j\phi_p(t)} & t \in [0, T_p) \\ 0 & \text{otherwise} \end{cases} \tag{3.5}$$

The frequency spectrum of this signal can be obtained from its Fourier transform, defined by $V_p(f)$ in Eqn. (3.6).

$$V_p(f) = \int_0^{T_p} v_p(t) e^{-j2\pi f t} \mathrm{d}t \tag{3.6}$$

The normalized power spectral density of the signal used for comparison with emission masks is given by Eqn. (3.7), where $C(x)$ and $S(x)$ are Fresnel integrals (Turley 2009). Following the lead of Turley (2009), we may set $f_c = 0$ without loss of generality, such that the integration limits are $x_1 = \sqrt{2T_p/B}(f + B/2)$ and $x_2 = \sqrt{2T_p/B}(f - B/2)$.

$$|\bar{V}_p(f)|^2 = \frac{B}{T_p}|V_p(f)|^2 = \frac{1}{2}\left\{[C(x_2) - C(x_1)]^2 + [S(x_2) - S(x_1)]^2\right\} \tag{3.7}$$

For a large frequency $|f| \gg B/2$, the upper bounding envelope of the normalized power spectral density $|\bar{V}_p(f)|^2$ can be approximated as in Eqn. (3.8). This is referred to as the large-frequency approximation of the normalized spectral density envelope for a single chirp. This expression is originally due to Regimbal (1965).

$$|\bar{V}_p^E(f)|^2 = \frac{B}{T_p \pi^2} \left[\frac{f}{f^2 - (B/2)^2} \right]^2, \quad |f| > B/2 \tag{3.8}$$

This expression reveals that the out-of-band roll-off rate for a single chirp approaches $1/f^2$ for $|f| \gg B/2$. In other words, the single-chirp roll-off rate is 20 dB/decade in the out-of-band domain. For $|f| < B/2$, the approximate solution for the upper-bounding envelope is given by Eqn. (3.9), which is also due to Regimbal (1965).

$$|\bar{V}_p^E(f)|^2 = \left(1 + \frac{1}{2\pi\sqrt{BT_p}} \frac{B^2}{(B/2)^2 - f^2} \right)^2, \quad |f| < B/2 \tag{3.9}$$

Figure 3.4a compares the theoretical half-spectrum of a single linear-FM pulse using $B = 10$ kHz, $f_p = 250$ Hz with the ITU emission mask for $f_c = 30$ MHz. Beyond the -40 dB frequency, given by the bandwidth formula in Eqn. (3.2), the spectral density is required to roll off at a rate of at least -20 dB/decade in the out-of-band domain until the -60 dB level is reached. The spectrum is then required to stay below this level over the remaining frequency band, where contributions due to spurious signals generally dominate.

It can be observed that the spectrum of a single linear-FM chirp with waveform parameters $BT_p = 250$ is non-compliant with the ITU mask for $f_c = 30$ MHz, and hence for all carrier frequencies in the HF band. The spectrum of an infinite repetitive sequence of chirps can be derived from the spectrum of a single repetition by sampling it at integer

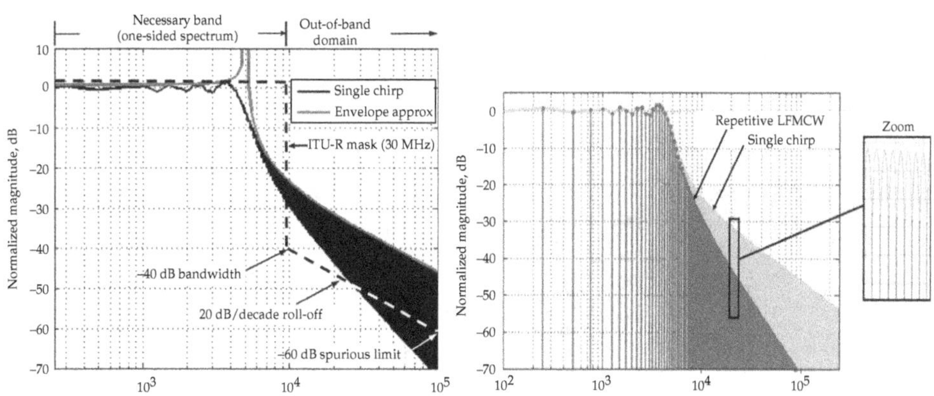

(a) Normalized spectrum of single LFM chirp.

(b) Spectra for a single chirp and repetitive LFMCW.

FIGURE 3.4 The panel on the left shows the theoretical normalized half-spectrum of a single LFM chirp ($B = 10$ kHz, $f_p = 250$ Hz) along with the large-frequency envelope approximation and ITU mask for $f_c = 30$ MHz. The panel on the right shows the theoretical normalized half-spectra for a single LFM chirp ($B = 10$ kHz, $f_p = 250$ Hz) and a repetitive LFMCW containing an infinite sequence of the same LFM chirps. © Commonwealth of Australia 2011.

multiples of the repetition frequency, i.e., at $f = k/T_p$ where k is an integer. This leads to the theoretical (discrete) spectrum of an infinite repetitive LFMCW shown in Figure 3.4b.

Substituting $f = k/T_p$ into the large-frequency approximation for a single chirp and taking the (upper bound) spectrum envelope yields the expression in Eqn. (3.10) for the repetitive LFMCW. It is evident from Eqn. (3.10) that the roll-off rate for a repetitive LFMCW waveform in the out-of-band domain is 40 dB/decade.

$$|\bar{V}^E(f = k/T_p)|^2 = \frac{B^3}{T_p 4\pi^2}\left[f^2 - B^2/4\right]^{-2}, \quad |f| > B/2 \tag{3.10}$$

Figure 3.5a compares the theoretical single-chirp spectrum given in Eqn. (3.7), the theoretical spectrum for the infinite LFMCW sequence with lines at $f = k/T_p$, and the large-frequency approximation for the infinite sequence given in Eqn. (3.10) for $f > B/2$ and in Eqn. (3.9) for $f < B/2$. Note that the approximation follows the LFMCW spectrum very well at high frequencies. Solving for the frequency that yields a normalized spectral value of $-x$ dB in Eqn. (3.10), and multiplying this by 2 for a two-sided spectrum, the x dB bandwidth formula for a infinite repetitive LFMCW is given by:

$$f_{-x\,\mathrm{dB}} = B\left(1 + \frac{2}{\pi\sqrt{BT_p}}10^{x/20}\right)^{1/2} \tag{3.11}$$

This expression, which is derived in Turley (2009), shows that the bandwidth factor $f_{-x\,\mathrm{dB}}/B$ is only a function of the single-chirp TBP $= BT_p$ and the level x. Inserting $x = 40$ dB yields the alternative -40 dB bandwidth formula in Eqn. (3.12). Figure 3.5a shows the mask for the LFMCW waveform using $f_{-40\,\mathrm{dB}}$ given by Eqn. (3.12) and the

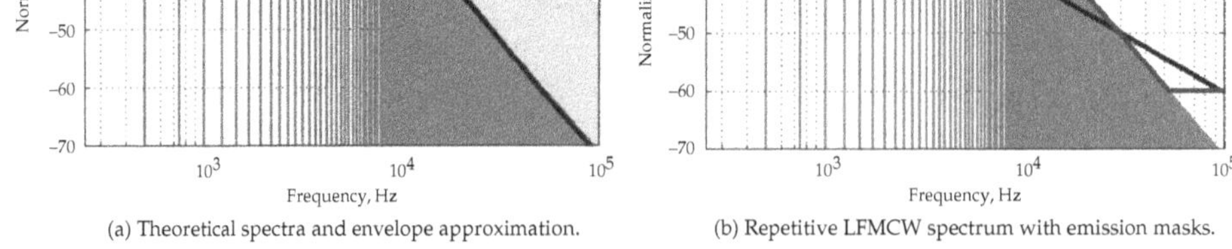

FIGURE 3.5 The left panel compares the theoretical normalized spectra of a single chirp with that of an infinite periodic chirp sequence and the large-frequency envelope approximation for the latter using $B = 10$ kHz and $fp = 250$ Hz. The right panel compares the ITU-R mask with the theoretically derived mask and the infinite chirp sequence spectrum. © Commonwealth of Australia 2011.

40 dB/decade roll-off rate based on Eqn. (3.10). The current ITU mask is also plotted in Figure 3.5a for comparison using $f_c = 30$ MHz.

$$f_{-40\,\mathrm{dB}} = B\left(1 + \frac{200}{\pi\sqrt{BT_p}}\right)^{1/2} \tag{3.12}$$

The alternative (theoretically derived) mask in Eqn. (3.12) matches the −40 dB bandwidth of the repetitive LFMCW spectrum, and also matches the roll-off rate of this spectrum, whereas the current ITU mask is too stringent in the near out-of-band domain and too lenient in the far out-of-band domain. In general, the ITU masks are "tightest" (most difficult to satisfy) for combinations of low single-chirp TBP values and low carrier-frequencies. For aircraft-detection applications, where the TBP values are relatively low, linear FM pulse-shaping techniques are required to satisfy current ITU regulations.

3.1.3.2 Pulse-Shaping Techniques

The main source of out-of-band energy for the repetitive LFMCW is "spectral spreading" due to the frequency-flyback from the end frequency of one sweep to the start frequency of the next. Several pulse-shaping techniques may be applied to improve the spectral performance of such signals by attenuating or modifying the spectral characteristics of the flyback portion. The use of amplitude-modulation to weight down the contribution of the discontinuous frequency step at the sweep transitions is described first. The implementation of a frequency-law with a finite rather than instantaneous flyback time will be discussed second. Both techniques allow out-of-band emissions to be reduced significantly. The choice for a particular system is influenced by practical factors including the ability to realize the signal with high-fidelity in hardware and the impact on target detection performance.

Various amplitude-taper functions may be proposed to control the out-of-band spectral emissions of a repetitive linear FMCW. In this approach, the linear-FM chirp may be multiplied by a window function and then repeated. As in Turley (2009), let us consider amplitude modulation with the raised-cosine taper $a(t)$ given by Eqn. (3.13). This equates to the Hann taper if $\delta = 0.5$, and the Hamming taper if $\delta = 0.54$.

$$a(t) = \delta - (1-\delta)\cos\left(\frac{2\pi t}{T_p}\right), \quad t \in [0, T_p) \tag{3.13}$$

The upper-bound spectrum envelope of the amplitude-modulated repetitive LFMCW (infinite sequence) may be approximated by the expression in Eqn. (3.14) for large frequencies (Turley 2009). The theoretical approximation and actual envelope spectrum for the Hann-tapered repetitive LFMCW is shown in Figure 3.6a. Note that the large-frequency approximation fits the true profile well below the −60 dB level. At very large frequencies, the Hann taper achieves an 80 dB/decade roll-off rate for the infinite periodic sequence and a 60 dB/decade for a single chirp.

$$|\bar{V}^{RC}(f = k/T_p)|^2 = \frac{B^3}{T_p 4\pi^2}\left[\frac{\delta}{f^2 - B^2/4} - \frac{0.5(1-\delta)}{(f-1/T_p)^2 - B^2/4} - \frac{0.5(1-\delta)}{(f+1/T_p)^2 - B^2/4}\right]^2, \quad |f| > B/2 \tag{3.14}$$

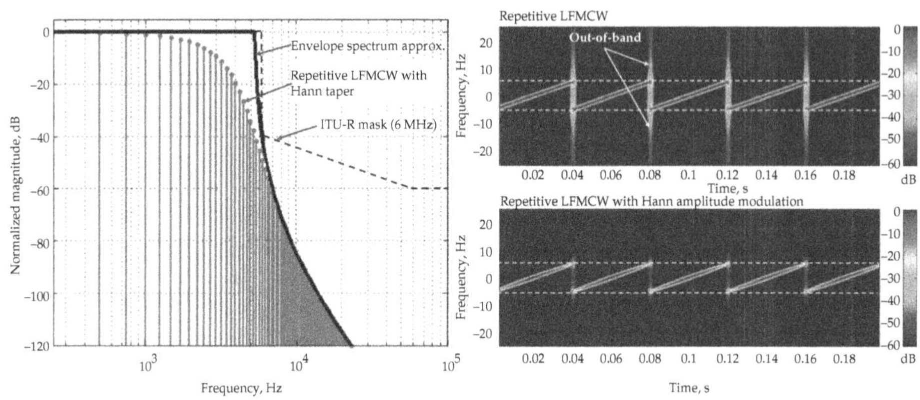

(a) Amplitude-modulated repetitive LFMCW.

(b) Spectrograms before and after pulse-shaping.

FIGURE 3.6 The left panel shows the half-spectrum of the Hann-tapered repetitive LFMCW along with the large-frequency envelope approximation for $B = 10$ kHz, $f_p = 250$ Hz. The right panel shows the reduction of out-of-band energy at the sweep boundaries after the amplitude modulation is applied. © Commonwealth of Australia 2011.

Another option described in Turley (2006) is to apply a cosine-Tukey amplitude-taper over 10 percent of the sweep starting from each extremity. This method is very effective in meeting out-of-band emission requirements with only a small loss in coherent gain and excellent range-sidelobe performance. However, methods based on amplitude modulation rely heavily on the linearity of the high-power amplifier (HPA) to transmit the generated waveform with high fidelity. In Turley (2006), it has not been demonstrated that practical HPAs can achieve the response obtained experimentally at the waveform generator output.

An alternative approach is to modify the frequency-time characteristic rather than the waveform amplitude to satisfy ITU regulations. A simple method to reduce the discontinuity in frequency-step is to introduce a short counter-sloping linear FM segment to join the successive sweeps (i.e., a finite linear-flyback waveform). A detailed analysis of this approach can be found in Durbridge (2002).

Digital waveform generators provide the necessary flexibility to accurately implement such waveforms. The constant-modulus property of a finite-flyback waveform eases demands on amplifier linearity and enables the HPA to deliver constant power. The linear finite-flyback waveform is given by Eqn. (3.15), where the linear-FM ramp rate has been modified to keep the swept bandwidth constant.

$$v_p(t) = \begin{cases} e^{j(2\pi f_c t + \frac{\pi B t^2}{(1-r)T_p})} & t \in [0, (1-r)T_p) \\ e^{j(2\pi f_c t - \frac{\pi B[t-(1-r)T_p]^2}{rT_p} + \phi)} & t \in [(1-r)T_p, T_p) \end{cases} \tag{3.15}$$

Here, f_c is the start frequency of the sweep, B is the swept bandwidth, T_p is the sweep period, r is the flyback ratio, and $\phi = \pi B(1-r)T_p$ ensures phase continuity. The TBP of $v_p(t)$ is reduced by a factor of r with respect to the original waveform ($r = 0$). Typically, only very small values of r not greater than about 5 percent are needed. Figure 3.7a,

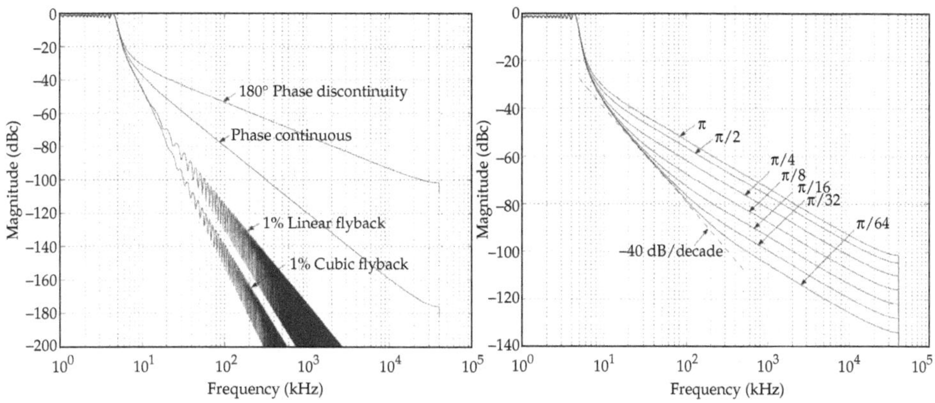

(a) Traditional and flyback LFMCW spectra.

(b) Effect of varying phase discontinuities.

FIGURE 3.7 The left panel compares frequency spectra of traditional LFMCW and flyback LFMCW, while the right panel shows the frequency spectra for traditional LFMCW with different (sweep-to-sweep) phase discontinuities. © Commonwealth of Australia 2011.

reprinted from Durbridge (2002), shows the spectrum of a 1-percent linear flyback waveform with parameters $B = 8.0$ kHz and $f_p = 60.1$ Hz. Note that the 1-percent linear flyback spectrum has a 60 dB/decade roll-off rate and meets current ITU regulations.

Figure 3.7a also shows spectra for a sequence of phase continuous and discontinuous chirps, which respectively correspond to those of a single chirp (20 dB/decade roll-off rate) and the repetitive linear FMCW (40 dB/decade roll-off rate). Figure 3.7a shows the spectra for repetitive LFMCW waveforms with a varying amount of phase discontinuity. As discussed previously, such waveforms may be used to mitigate range-folded clutter (Hartnett et al. 1998). The curves show that the waveform roll-off rate varies from 20 to 40 dB/decade as the phase discontinuity varies from π to zero, respectively.

The linear flyback waveform replaces the discontinuous frequency-step at the sweep boundary with a finite transition that is continuous in phase and frequency. This scheme may be extended to additionally provide continuity in the frequency rate of change between the FM chirp and the flyback. In other words, the instantaneous frequency slopes at either side of the sweep boundaries are equal, along with the phase and frequency. The cubic phase law are given by $\phi(t)$ in Eqn. (3.16), where ϕ_i is the phase at the end of the FM ramp which can be shown to satisfy these three criteria (Durbridge 2002).

$$\phi(t) = \frac{B}{2r^3(1-r)T_p^3}t^4 - \frac{B}{r^2(1-r)T_p^2}t^3 + \frac{B}{2(1-r)T_p^2}t^2 + (f_c + B)t + \phi_i\,, \quad t \in [0, rT_p) \tag{3.16}$$

Figure 3.7a shows the spectrum of a waveform where the 1-percent flyback follows this cubic phase law to provide smooth transitions between the FM ramp and the flyback portion. This spectrum also meets the ITU emission mask and has a large frequency roll-off rate of 80 dB/decade. However, the added complexity of implementing a cubic flyback waveform is not warranted for OTH radar at present since the linear flyback is sufficient to meet current ITU regulations.

3.1.3.3 Range Sidelobes

The effectiveness of a pulse-shaping method should not only be evaluated in terms of its effect on the radiated spectrum. It is also important to consider the influence of the modified waveform on range sidelobe levels, which can impact target detection performance. The analysis in Durbridge (2002) shows that the degradation in range sidelobe levels for the linear flyback waveform increases with the chirp TBP and flyback ratio, but is less sensitive to the individual choices of bandwidth and PRF. The degradation also tends to increase at far (mismatched) ranges in the receiver passband, which can have more of an impact on fully-digital pulse-compression with extended range processing compared to stretch-processing schemes in an analogue deramping receiver.

For the small (1-percent) flyback ratios needed to satisfy ITU emission masks, the increase in range sidelobe levels for the linear flyback waveform are typically less than 3 dB for an aircraft-detection chirp having a TBP of around 100 at ranges that are not highly mismatched. The impact of the finite flyback waveform becomes more significant when the entire spectrum becomes available to a range-correlation receiver. At the largest lags (i.e., most highly mismatched ranges with delays close to half the repetition period), the degradation in range sidelobe levels may be approximately 6 dB for 1-percent flyback and chirp TBP of around 100.

Figure 3.8 illustrates the degradation in range sidelobes for a chirp TBP of 133 using 1- and 5-percent linear flyback. It is reminded that such waveforms are required when operating an OTH radar in aircraft mode, particularly at low carrier frequencies where the −40 dB bandwidth relationship is tightest. Figure 3.9 illustrates the slight effect of changing the PRF on range sidelobes for a constant chirp TBP of 499. The results in these two figures are due to Durbridge (2002).

For a 1-percent linear flyback waveform, the degradation in range sidelobes is much more serious for ship-detection chirps that have a TBP of around 10000. Fortunately, pulse-shaping is not required for typical ship-detection waveform parameter sets, where the standard repetitive LFMCW can meet current ITU emission guidelines. In summary, the range-sidelobe penalty of flyback waveforms is determined primarily by the

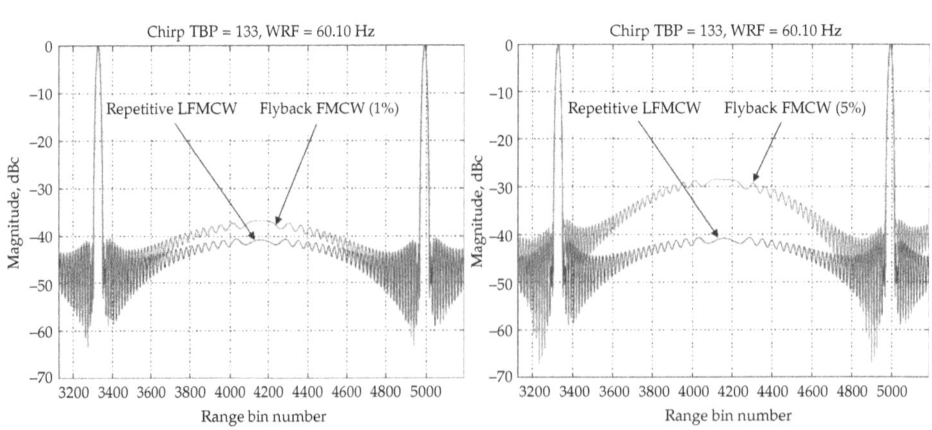

FIGURE 3.8 Effect of 1- and 5-percent flyback ratios on range sidelobe levels for a periodic LFMCW with chirp TBP of 133. © Commonwealth of Australia 2011.

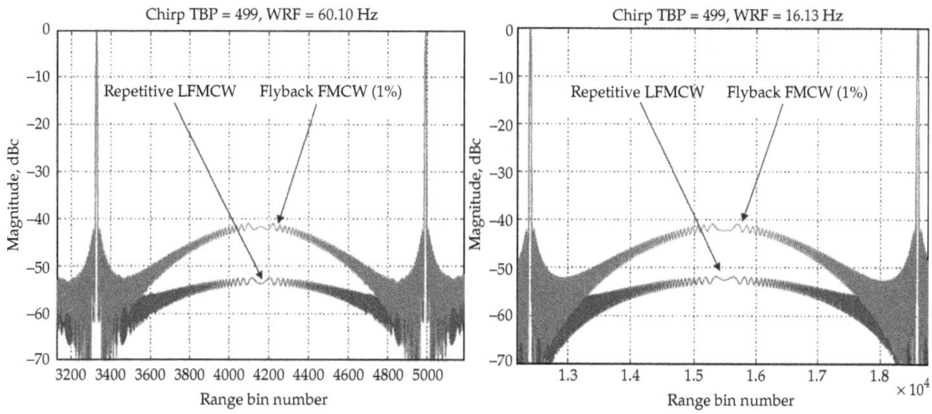

FIGURE 3.9 Effect of changing the WRF on range sidelobe levels for a periodic LFMCW with and without linear flyback using a chirp TBP of 499. © Commonwealth of Australia 2011.

time-bandwidth product and flyback ratio, with only a small contribution from the selection of bandwidth and PRF. A 1-percent linear flyback will meet the current ITU requirements for chirp TBPs greater than 100, while no pulse-shaping is necessary for chirp TBPs greater than 500 (Durbridge 2002).

3.2 Radar Architecture

The architectural characteristics of an OTH radar have traditionally been based on the precept of using a single system used to perform air and surface surveillance tasks. In accordance with this approach, most currently operational skywave OTH radar systems have been designed for aircraft and ship detection as primary and secondary missions, respectively. With reference to a nominal system in this class, the first and second parts of this section describe the general characteristics of the OTH radar transmit and receive subsystems. In each case, particular attention is paid to the factors driving the choice of antenna element and array design. The main issues regarding waveform generation and power amplification are covered for the transmitter, while the essential features of superheterodyne and direct digital techniques are discussed for the receiver. The third part of this section considers the topic of HF array calibration and discusses a number of approaches using internal and external signal sources.

3.2.1 Transmit System

The average transmit power needed for successful OTH radar operation depends on system design and mission requirements. For example, the detection of fast-moving low-RCS targets at night may require an average transmit power approaching 1 MW to enhance SNR, while the detection of slow-moving surface targets during the day may only require an average transmit power of perhaps 10 kW for performance to be clutter-limited. Most currently operational skywave OTH radar systems are typically designed to transmit a maximum average power of hundreds of kilowatts, where this power is radiated by an antenna with a gain of around 15–25 dB.

The OTH radar transmitter generally consists of an array of radiating elements with a separate transmit channel per antenna element. This architecture enables the beam of illumination to be steered electronically over the angular coverage of the OTH radar for wide-area surveillance. The transmit system may be broken down into four essential components: waveform generation, power amplification, antenna elements, and array design. The main characteristics of the transmit antenna are described first to expose the underlying factors which motivate the choice of radiating element and array geometry. This is followed by a description of the waveform generation and power amplification processes within each transmit channel.

3.2.1.1 Antenna Element

The choice of transmit antenna element represents a compromise between competing operational requirements and the need to reduce engineering cost. Important issues to consider for the transmit antenna element include: (1) efficiency, to maximize the radiated power, (2) limiting the voltage standing wave ratio (VSWR) below an acceptable level over the design frequency range, (3) beam pattern characteristics in azimuth and elevation, to ensure that coverage or coverage rate requirements are met without significant illumination in undesired directions, (4) signal fidelity, to radiate waveforms of the required bandwidth with minimal distortion caused by arching (sparking) and phase noise due to mechanical vibration, (5) physical structure and dimensions, particularly the suitability to deploy the element in the desired array configuration with due regard to mutual coupling effects, (6) ground-plane and ground-proximity effects, to improve low elevation-angle gain and mitigate the effects of inhomogeneous ground electrical properties near the aperture, and (7) fully polarimetric or linearly polarized antenna elements, to enable polarization filtering or reduce engineering cost for a given aperture size, respectively. These issues are now discussed in more detail.

OTH radars adapt to changing ionospheric propagation conditions by operating across broad bands of the HF spectrum. Rapid switching between widely spaced frequencies may also be required to illuminate surveillance regions at different ranges in the coverage when multiple tasks are interleaved on the radar time-line. A *broadband* transmit antenna element is therefore required to maintain high radiation efficiency for radar tasks at different frequencies, particularly when target detection is noise-limited. Specifically, such elements need to be well matched over a design frequency range that may span an octave or more without relying on mechanical tuning of the antenna structure. The broadband performance of the transmit antenna element is particularly important for aircraft surveillance missions, where target SNR at the operating frequency is the primary factor limiting detection performance.

At decameter wavelengths, resonant antenna elements are required to be physically large structures. For example, a vertical log-periodic antenna array needs to contain half-wave dipole elements that are 30–40 meters tall to transmit signals efficiently in the lower HF band. For practical reasons described below, OTH radars often employ separate arrays of different transmit antenna elements for operation in two or more designated regions of the HF spectrum. For example, the US OTH-B system used different transmit antennas matched over six sub-bands of the design frequency range (Georges and Harlan 1994). More typically, OTH radars partition the design frequency range into a low and high band such that a well-matched set of transmit antenna elements covers operation in each band.

The factorization theorem states that the radiation pattern for an array of identical antenna elements is given by the product of the element pattern and the array factor. The latter depends on the relative amplitude and phase of the signal that is fed into each antenna element. At a particular operating frequency, the element pattern may be viewed as the modulating envelope of the array factor, the resultant of which gives rise to the transmit beam. It follows that a high-gain transmit beam cannot be electronically steered in directions where the antenna element gain is low. Hence, to effectively illuminate the the OTH radar coverage, the beamwidth of the transmit antenna *element* needs to be commensurately broad in azimuth.

The elevation pattern of the transmit antenna influences the ground ranges that can be effectively illuminated by one-hop skywave propagation. For a nominal OTH radar ground-range coverage of 1000–3000 km, the radiation pattern needs to provide adequate gain at elevation angles between about 5 and 45 degrees. This is necessary because changing the operating frequency to shift the position of the transmitter footprint in ground range also varies the band of elevation angles that illuminate the useful range-depth. An antenna element with a relatively broad radiation pattern in elevation is therefore required to effectively illuminate surveillance regions at different ground ranges within the OTH radar coverage. The band of elevation angles required to illuminate the same range extent will also vary with ionospheric conditions.

Besides ensuring that the radiation pattern characteristics of the antenna element are compatible with the intended coverage, it is also important to minimize illumination in directions that are not of interest. This aspect is particularly important in directions where the gain of the transmit beam cannot be easily controlled by shaping the array factor. For example, element patterns with low gain at high elevation angles are desirable for a linear antenna array to reduce clutter scattered from ionospheric irregularities and meteors at near-vertical incidence on the direct skywave path linking the transmitter to the receiver. Element radiation patterns with a good front-to-back ratio are also desirable for linear arrays to reduce the reception of possibly spread-Doppler clutter from disturbed ionospheric regions behind the OTH radar, and to reduce the potential ambiguity in direction-of-arrival estimation for detected target echoes.

Due to their considerable height, the transmitting elements need to be stabilized by a well-designed supporting structure to reduce phase modulation of the radiated signal caused by mechanical vibration in the presence of wind stress. This source of multiplicative phase-noise imposed on the radar signal is known as Aeolian noise. Spectral contamination of the transmitted signal appears on the backscattered echoes received by the system as Doppler spread. Aeolian noise has the potential to significantly broaden the surface-clutter Doppler spectrum, which can in turn mask target echoes. It is emphasized that Aeolian noise is multiplicative phase-noise imparted on the signal by the physical motion of the transmit (or receive) antenna element. It is not to be confused with Doppler spreading due to signal phase-path variations in the ionosphere or phase-noise generated by nonideal electrical components internal to the system (e.g., in the waveform generator and high-power amplifier). Great care is taken to mechanically dampen high-frequency oscillations of the radiating antenna elements under wind-stress to minimize the impact of Aeolian noise on target detection performance.

Distortion of the radiated signals can also arise from arching or sparking phenomena caused by excessive field-strengths generated in the near-field of the transmit antenna element. The maximum power-handling capacity of the antenna element not only needs to be appropriate for the peak-voltage levels supplied to it, but also needs to account for mutual coupling effects when the element is configured as part of an array. Besides the obvious requirement to minimize VSWR (i.e., the amount of reverse power reflected from the antenna back into the amplifier), a low inductive component of antenna impedance is also desirable over the design frequency range to reduce the amount of forward power that is not effectively radiated but "stored" in the near-field. The frequency response of the antenna also needs to minimize dispersion for the highest anticipated waveform bandwidth such the generated signals are radiated with high fidelity.

Unlike in HF surface-wave radar, where only vertically polarized signals propagate effectively over long-distances above the conductive (saline) sea surface, a skywave OTH radar may operate effectively by transmitting and receiving HF signals with vertical and/or horizontal polarization. Irrespective of the transmit polarization, propagation through the ionosphere will in general transform the signal into an elliptically polarized wave over both the illumination and echo paths. Horizontally polarized antenna elements are typically stacked vertically in two-dimensional curtain arrays that may rise high above the ground to provide independent azimuth and elevation control of the beam pattern. On the other hand, vertically polarized antenna elements may be arranged to form linear or planar arrays on the ground that cover all necessary ray take-off angles due to the relatively broad element pattern in elevation. A reason for the preference to use vertical polarization in a number of currently operational OTH radar systems is related to the lower engineering cost and complexity of the resulting antenna structure. Tall curtain arrays that may be over a hundred meters high are expensive to construct and more difficult to stabilize than ground-based antenna arrays of vertically polarized elements. There is also the complication that the vertical gain pattern of a horizontally polarized element is strong function of its electrical height above the ground (i.e., its height measured in wavelengths).

Examples of radiating elements that have been used in OTH radar transmit antenna arrays include vertically polarized log-periodic dipole arrays (LPDA), horizontal caged-dipoles, biconical antennas, canted dipoles, Yagi antennas, rhombic elements, and tilted monopoles (Skolnik 2008b). A common choice for the transmit antenna element in several existing OTH radar systems is a vertically polarized LPDA. A detailed description of the design, functionality, and performance of this antenna can be found in DuHamel and Isbell (1957), Carrell (1961), and ARRL (1991). Some basic features of the LPDA are now recalled to motivate the choice of this antenna for OTH radar applications.

A log-periodic antenna is an array of dipole elements of varying lengths arranged with a relative spacing from longest to shortest and fed through a common transmission line that alternates 180 degrees in phase from one element to the next. The active section of the antenna, which radiates (or receives) signals most efficiently, changes with frequency as it depends on the location of the "resonant" element that acts most like a half-wave dipole. As the frequency increases, the active section shifts from the longest element at rear of the antenna to the shortest element at the front of antenna. The lower and upper limits of the design frequency range are determined by the lengths of the longest and shortest elements, respectively.

The interaction of the resonant element with neighboring elements in the LDPA provides a front-to-back gain relative to a single dipole. This gain depends on frequency and the LPDA design parameters, which define the element lengths and relative spacings.

The forward gain of an LPDA relative to an isotropic antenna may be around 10 dBi, and the front-to-back ratio may be an order of magnitude higher than this at certain frequencies. In summary, LPDA antennas are suitable for applications such as OTH radar, which require relatively uniform input impedances, low VSWR, and high radiation efficiencies over a broad frequency range. The relatively wide angular coverage provided by the LDPA element (approximately 90 degrees in azimuth and perhaps 5–45 degrees in elevation) combined with the significant forward-gain and high front-to-back ratio are also desirable features for an OTH radar system.

In practice, the radiating dipole elements in the LPDA need to be physically stabilized to reduce Aeolian noise. This is partially achieved by ensuring that the main catenary wire and other wires supporting the radiating dipoles are properly tensioned. Supporting wires also need to be fitted with insulators at appropriate points to minimize their contribution to the radiating structure. In addition, mechanical dampeners are attached to reduce wind-induced vibration of the antenna. It is particularly important to dampen the oscillations excited at higher frequencies since these are the most detrimental from a Doppler-processing perspective. At the highest point of the antenna, on the rear mast or "boom" of the LDPA, a metal rod with a low impedance connection to ground may be used to provide the transmit system with a form of protection against lightning strikes.

The radiation pattern of an HF antenna at low elevation angles is strongly influenced by the electrical properties of the ground, both under and in front of the element, up to distances of a few wavelengths. A highly conductive (metal) mesh screen is often laid over the ground beneath and in front of an LPDA to provide two main benefits. First, the mesh-screen increases antenna gain at low elevation angles to improve sensitivity for long-range propagation paths. Secondly, it stabilizes surface impedance to reduce mismatches between the radiation characteristics of different elements in the array. This minimizes distortions in the radiation pattern (particularly the sidelobe levels) when the elements of the array are phased to form the transmit beam. Mesh ground-screens extending out to 100–200 meters in front of the LPDA have been installed for skywave OTH radar.

An LPDA element combined with a mesh ground-screen provides a broad vertical beam at relatively low cost to cover the band of elevation angles required for one-hop skywave propagation to ground ranges of 1000–3000 km over an azimuth sector up to 90 degrees wide. The LPDA can be made efficient over the entire HF band by making it longer, i.e., adding more dipole elements with appropriate lengths and relative spacings in the array. For reasons described later, it is more cost efficient to use two separate arrays of different LPDA elements for operation in the lower and upper regions of the HF band. Certainly, the LPDA is a rugged and versatile antenna that strikes a good balance between broadband performance, pattern properties, and cost for OTH radar applications.

However, no optimality is claimed for this choice of transmit antenna element, and it is possible that future OTH radars may be based on alternative designs, particularly if one wishes to reap the advantages of two-dimensional transmit arrays. In practice, the physical structure of the LPDA makes this antenna element rather unsuitable for use in two-dimensional array geometries. Only linear arrays of LPDA elements have been implemented for OTH radar applications. Horizontal caged-dipoles and biconical antenna elements have been used in two-dimensional vertical-curtain and ground-based transmit arrays, respectively. More will be said on the topic of two-dimensional arrays below.

3.2.1.2 Array Design

Currently operational skywave OTH radars in the United States and Australia are based on ground-deployed uniform linear arrays (ULA) of vertically polarized (LPDA) transmit antenna elements. A desirable property of the ULA is that it represents a low-cost solution to achieve high azimuth resolution (ignoring the possibility of minimum-redundancy linear arrays). In addition, the ULA geometry simplifies the understanding and management of mutual coupling phenomena, which can be a more complex and challenging issue for two-dimensional arrays. Another practical benefit of the ULA configuration is that it allows classical beams with low sidelobe levels to be electronically formed and steered in a relatively straightforward manner using analog and/or computationally efficient (FFT-based) digital methods. Finally, a ULA is the most natural arrangement for certain antenna elements such as the LPDA, where aligning the curtains side-by-side maintains a known and constant spacing between the resonant dipole elements as a function of operating frequency.

Despite these undeniable advantages, the ULA geometry also has a number of drawbacks. An obvious disadvantage of a single ULA is related to the cone-angle ambiguity (coupling between azimuth and elevation), which effectively precludes the possibility to realize gains in radar sensitivity against clutter and noise that may be provided by a two-dimensional array through vertical selectivity. In addition, the vertical gain of a two-dimensional array can be used to enhance SNR without narrowing the width of the transmit beam in azimuth, which allows for greater coverage or coverage rate with respect to a ULA. The ability to independently control the transmit beam shape in elevation can also help to correctly identify propagation modes over a two-way skywave path for target coordinate registration.

Transmit systems based on a ULA of LDPA elements will be discussed first, as this architecture is directly relevant to a number of currently operational OTH radar systems. Important questions to be addressed for a ULA design include aperture length, element spacing, and options to extend the azimuth coverage. Different implementations of two-dimensional arrays are then discussed with reference to some past and present skywave OTH radar systems.

A transmit ULA with a wide aperture (and many elements distributed at approximately half-wavelength spacing for the highest design frequency) provides a beam with higher gain to improve target detection in a noise-limited environment. However, the higher gain and hence SNR is accompanied by a reduction in the half-power (−3 dB) width of the main lobe, which reduces the azimuth extent of the surveillance region that can be illuminated simultaneously at a particular frequency. In OTH radar, the dwell time needed for Doppler processing is not only determined by SNR considerations, but also the requirement to resolve target echoes from clutter returns (including echoes from other targets). Consequently, the dwell time cannot be reduced below a certain limit, regardless of the SNR improvement provided by a higher transmit antenna gain.

The implication is that the real-time coverage of an OTH radar will require a greater number of surveillance regions to "tile" when the transmit beam is narrow, while the dwell time needed for Doppler processing in each surveillance region cannot be reduced below a certain limit irrespective of the higher transmit antenna gain. When the number of surveillance regions is large, the time taken to return to a particular surveillance region (i.e., the region revisit rate) may become too long for effective tracking, particularly in the case of maneuvering targets. The net result is that using a large transmit aperture will reduce the coverage that can be scheduled on the radar time-line for a given region revisit rate, or the region revisit rate that can be achieved for a given coverage.

For this reason, the choice of transmit aperture size represents a trade-off between sensitivity against noise on the one hand, and coverage or coverage rate on the other. The lower bound on aperture size is determined by the minimum transmit antenna gain that is deemed necessary to achieve sufficient SNR for reliable target detection against noise. The upper bound on the aperture size is determined by the coverage and coverage rate requirements of the system. The latter determines the number of surveillance regions that can be monitored in real-time without significantly compromising target detection and tracking performance. Note that high resolution in azimuth is needed for accurate target localization, but this requirement is decoupled from the transmit array design since it is a function of the *receive* antenna aperture, which may have a much greater length.

Even though it is possible to design an LPDA element with high radiation efficiency over the entire HF band (simply by adding more dipoles and making the LPDA longer), it has often been chosen to use two separate arrays of different LPDA elements designed for operation in low and high bands of the OTH radar frequency range. The motivation for this is mainly connected to the different element spacings required when operating at different frequencies. Specifically, the use of higher frequencies requires a smaller element spacing to avoid grating lobes, but it also allows shorter apertures to achieve the same trade-off between sensitivity (gain) and coverage or coverage rate.

If a single ULA were to be used for transmission over the entire OTH radar design frequency range, this would require longer LPDA elements and the smaller element spacing needed for high-frequency operation. Besides issues related to the construction and cost of longer LPDAs, the smaller element spacing needed for operation at higher frequencies would lead to a significantly oversampled aperture at low frequencies, which is not only unnecessary but also produces undesirable mutual coupling effects that can significantly increase VSWR. This motivates a design based on two separate transmit arrays using LPDA elements of shorter length with ULA spacings tailored specifically to suit low- and high-band operation.

For the ULA configuration, multiple transmit arrays are also required when the total arc of coverage exceeds the azimuth sector that can be illuminated by the antenna element. Skywave OTH radar systems with an azimuth coverage of 60, 90, 180, and 360 degrees have been fielded. For example, the total arc of coverage may be divided into azimuth sectors with a different ULA used to service each sector. The JORN OTH radar at Laverton in Western Australia achieves a coverage of 180 degrees by orienting two ULAs, each with an azimuth coverage of approximately 90 degrees, at right angles (Cameron 1995). Alternatively, the Nostradamus OTH radar (Bazin et al. 2006) provides azimuth coverage over 360 degrees using a two-dimensional array of biconical antenna elements, which have an omni-directional radiation pattern.

Figure 3.10a shows an ariel photograph of the JORN transmit site at Laverton. Note that this system is based on four separate ULAs of LPDA elements. A pair of ULAs is used for low- and high-band operation in each of two 90-degree azimuth sectors. Figure 3.10b shows the JORN OTH radar transmit array at Longreach in Queensland, which uses one pair of colinear ULAs for low- and high-band operation over a 90-degree sector. Figure 3.11 shows an LPDA antenna and the biconical element used to provide 360-degree coverage in the Nostradamus OTH radar. Although mutual coupling is simpler to manage in a ULA, care must be taken when forming beams at extreme steer angles (close to endfire) at frequencies where the element spacing is less than half-wavelength,

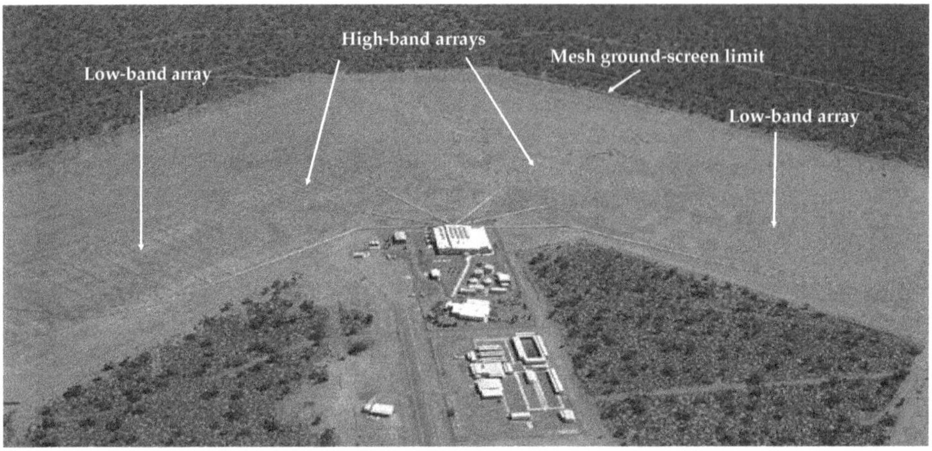

(a) JORN OTH radar transmit site near Laverton (180 degree coverage).

(b) JORN OTH radar transmit site near Longreach (90 degree coverage).

FIGURE 3.10 Aerial photographs of the low- and high-band transmit antenna arrays of the JORN OTH radars near Laverton, Western Australia, and Longreach, Queensland. As indicated, the JORN OTH radar near Laverton makes use of two pairs of low- and high-band ULAs with boresights oriented at right angles to double the azimuth coverage.

as this gives rise to an oversampled aperture. In such cases, the reactive power stored near the array can lead to significant reverse power flowing back into the amplifier chain with the potential to damage equipment.

A side-on view of the JORN OTH radar transmitter array is shown in Figure 3.12. The low-band ULA consists of 14 LPDA elements spaced 12.5 m apart to form an aperture 162.5 m long. Each LPDA element is connected to a pair of (combined) 20-kW amplifiers

(a) Vertical LPDA antenna elements.

(b) Biconical antenna elements.

FIGURE 3.11 Vertical log-periodic dipole array (LPDA) transmit antenna elements and biconical (transmit and receive) antenna elements used in the JORN and Nostradamus OTH radar systems, respectively.

FIGURE 3.12 The JORN OTH radar transmit antenna near Longreach is a uniform linear array of vertically polarized log-periodic dipole array (LPDA) antenna elements. The high-band array is composed of 28 (14 + 14) LPDA elements spaced 5.75 m apart for OTH radar operation between 12 and 32 MHz, while the low-band array is composed of 14 (7 + 7) LPDA elements spaced 12.5 m apart for OTH radar operation between 5 and 12 MHz. Both the high- and low-band transmit arrays can be operated in full- or half-aperture modes. © Commonwealth of Australia 2011.

to yield a total maximum transmit power of 560 kW. The high-band array is composed of 28 LPDA elements spaced 5.75 m apart to form an aperture 155 m long. In this case, each element is connected to a single 20-kW amplifier, resulting in the same total transmit power. For reference, an aperture that is 150 m long yields a transmit beam with a half-power main-lobe width of approximately 9 degrees in azimuth at 15 MHz after tapering is applied to reduce sidelobe levels. This allows a 90-degree arc to be covered by 10 transmit beams in 20 seconds using a 2 second CPI.

To provide a degree of flexibility in selecting the most appropriate compromise between sensitivity (SNR) and coverage or coverage rate, depending on mission requirements and environmental conditions, the transmit array can be operated either in full- or half-aperture mode. In the latter case, the low- and high-band arrays can be split into two independent half apertures containing 7 + 7 or 14 + 14 elements, respectively, to provide twice the coverage or coverage rate. With respect to using the full aperture (for either low- or high-band operation), the SNR loss incurred when using half aperture on both transmit and receive is approximately 9 dB i.e., 6 dB on transmit (half power and antenna gain), and 3 dB on receive.

OTH radar systems designed to achieve sufficient SNR for reliable aircraft detection at night will have considerable excess sensitivity during the day. This is because the background noise spectral density that an OTH radar needs to contend with at night may be 10–20 dB higher than the daytime level. Moreover, the use of higher frequencies in the day is often associated with an increase in RCS for small targets. For aircraft detection, half aperture operation can significantly increase coverage or coverage rate during the day while maintaining adequate SNR. This mode of operation is typically not used for ship detection because high azimuth resolution is required on receive to enhance detection performance in clutter. However, the coverage may often be increased without compromising ship detection performance by broadening or "spoiling" the transmit beam, provided that the transmit antenna gain is still high enough for detection to be clutter-limited.

When ionospheric conditions support propagation over an extended range-depth at a single frequency, a broad transmit beam in elevation can significantly increase the simultaneous coverage. In the case of an LPDA element, which may adequately cover ray take-off angles between 5 and 45 degrees, it more often occurs that the ionosphere limits the range-depth rather than the elevation pattern of the transmit beam. In other words, illumination of the surveillance region is provided by signal rays distributed over a band of elevation angles well within the interval covered by the main lobe of the transmit antenna beam. This situation can lead to significant power being radiated at elevation angles where the rays do not effectively contribute to illuminating the ranges of interest (e.g., escape rays that penetrate the ionosphere at higher elevation angles).

A two-dimensional array provides independent control of the beam pattern in azimuth and elevation. When the band of elevation angles needed to illuminated the range-depth of interest is relatively narrow (e.g., 10–20 degrees wide), a two-dimensional array in principle offers the possibility to better focus the radiated power. This can improve SNR due to the greater power density of the radar signal incident on targets, and SCR by reducing illumination at unwanted elevation angles. As the ionosphere often limits the useful range-depth for a transmit beam with broad coverage in elevation, the increase in radar sensitivity gained from the vertical selectivity of a two-dimensional array will not come at the expense of coverage or coverage rate up to a certain point. However, if fine beams in elevation are used to "spotlight" narrower range-depths than could otherwise

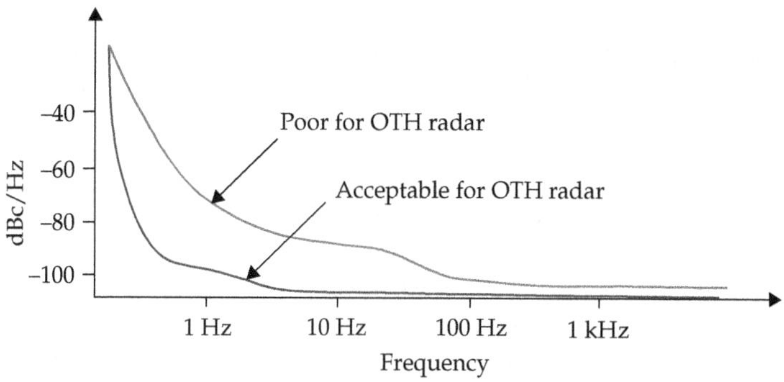

FIGURE 3.13 Notional illustration of phase-noise spectra normalized with respect to the carrier signal level at 0-Hz showing poor and acceptable phase-noise characteristics for OTH radar surveillance applications.

be illuminated via skywave propagation at the operating frequency, the gain in radar sensitivity will attract a penalty of reduced coverage or coverage rate when other areas need to be simultaneously monitored.

The capability to steer and shape the beam pattern characteristics in elevation may also allow different reflecting layers or propagation modes to be preferentially selected for illumination. In ship detection applications, propagation mode selection could be based on minimizing Doppler spectrum contamination, for example. Specifically, the ability to illuminate a surface vessel exclusively via E-layer propagation in the presence of a more disturbed (i.e., less frequency-stable) F2-layer can significantly improve the probability of target detection. However, relative to the ULA geometry, two-dimensional arrays may involve greater costs (depending on the number of elements and configuration), additional constraints regarding the choice of antenna elements for compatibility with the array design, and significant challenges related to mutual coupling phenomena.

For example, the French OTH radar "Nostradamus" is based on a Y-shaped array of biconical antenna elements deployed on the ground. On the other hand, the US OTH radar known as "MADRE" used a planar array of vertically stacked horizontal-dipole elements. A significant advantage of vertical curtain arrays is the higher selectivity they offer for signals at low elevation angles compared to a ground-deployed 2D array. Ground-deployed 2D arrays are more susceptible to undesirable mutual coupling effects when steered at near-grazing elevation, particularly when the signals need to be transmitted (or received) "through" other elements. However, antenna structures rising high above the ground are relatively more complex and expensive to build. Aeolian noise reduction is also generally simpler for ground-deployed arrays due to the lower height of the antenna structure.

3.2.1.3 Waveform Generation

Effective separation of weak target echoes from powerful clutter returns by Doppler processing in OTH radar requires very high spectral purity of the transmitted radar waveform to ensure that system performance remains limited by environmental factors rather than phase and amplitude noise generated by the transmit system. To prevent the system from being undesirably limited by transmitter-site induced noise, it is essential that the phase and amplitude noise present on the radiated signal that is returned to the system by clutter sources remains below the background noise level at the receive beam output. Importantly, an increase in transmit power will not improve target SNR when amplitude and phase noise induced by transmitting equipment on the radiated signal dominates background noise at the receive beam output.

OTH radars with high transmit power and system gain have the potential to achieve high sub-clutter visibility (SCV) levels in frequency-stable ionospheric conditions. This imposes stringent demands on the spectral purity of the radiated signal, and hence the waveform generator, if the system is not to become transmitter-noise limited. The requirements on spurious-free dynamic range to avoid such limitations also need to be met for all waveform parameter combinations set to the same values as employed for radar operations. As depicted in Figure 3.14, an acceptable waveform generator noise characteristic typically needs to fall below −90 dBc/Hz at 1 Hz from the carrier frequency, and below −120 dBc/Hz at 10 Hz frequency offset. An experimental study to determine FMCW waveform generator spectral purity requirements for skywave OTH radar is described in Earl (1998). The results for the system studied indicate that the spurious free dynamic range of the waveform generator is required to be approximately 30 dB above the maximum sub-clutter visibility (SCV).

Traditionally, an analog frequency synthesizer was used to generate the radar waveform, and this reference signal was fed to all amplifier chains and antenna elements in the transmit array via a switchable delay-line distribution network. The transmit beam was electronically steered by means of variable length cables that performed analog (time-delay) beamforming with an amplitude taper (attenuation) applied to reduce sidelobe levels in the radiation pattern. A limitation of this architecture is that phase noise generated in a frequency synthesizer will add *coherently* in the transmit beam. Other drawbacks associated with the analog-beamforming method include the limited flexibility to steer and shape the transmit beam pattern beyond a finite number of hardware-defined

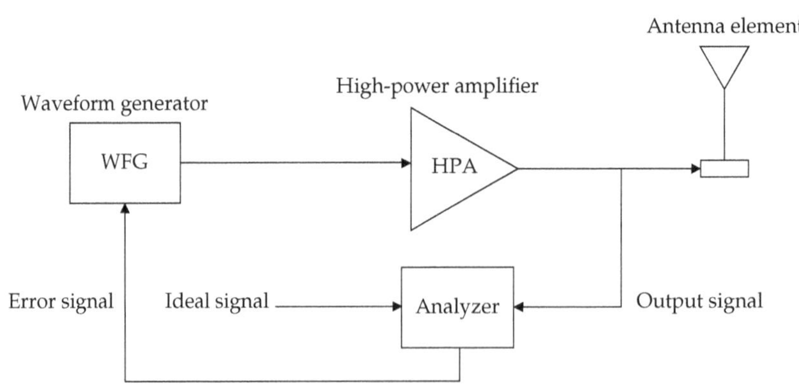

FIGURE 3.14 Conceptual illustration of a feedback loop that can in principle adaptively match the generated radar waveform to the amplifier characteristics for the purpose of improving the spectral purity of the output signal fed to the antenna. The error signal arises due to amplifier imperfections and is used to modify the waveform pulse such that unwanted artifacts at the amplifier output are attenuated.

options, and the reduced set of possibilities for partitioning the transmit aperture (beyond the half-aperture mode) to allow independent radar operation at more than two carrier frequencies.

In modern OTH radars, there is an increasing shift to architectures based on the use of a separate digital waveform generator per antenna element. The ability to define the radar waveform digitally and synthesize the transmit beam in software allows the operator to exercise greater control over the transmit resource. Besides the ability to form the transmit beam digitally, the transmit aperture may be partitioned arbitrarily into subarrays of possibly different length to simultaneously radiate signals in different frequency channels and steer directions. This provides greater versatility to perform different types of OTH radar missions at the same time. Alternatively, a waveform generator per element architecture may be used to radiate a set of different waveforms on the same frequency channel, as in a multiple-input multiple-output (MIMO) system, for reasons to be described in the final paragraph of this subsection.

The phase noise contributed by independent waveform generators in separate (well-isolated) transmit channels may be expected to exhibit little if any correlation from one channel to another. Hence, a waveform generator per element architecture has the advantage that the phase noise from different elements adds incoherently in the radiated signal. For a transmit array with 14–28 elements, this enhances the *radiated* signal-to-noise ratio by an order of magnitude with respect to that in each transmit channel. The attenuation of phase noise in the radiated signal when using a digital waveform generator per element architecture to perform transmit beamforming significantly eases dynamic range requirements on the waveform generator and other equipment in each transmit channel.

An appropriately designed set of "orthogonal" waveforms radiated from different antenna elements or subarrays enables transmit beamforming to be effectively carried out at the receiver after the signals have been radiated, scattered, and acquired. From a radar signal-processing perspective, this allows transmit and receive beams to be formed adaptively and optimized jointly after the data has been recorded. The benefits of performing range-dependent adaptive transmit beamforming at the OTH radar receiver include mode selectivity for spread-Doppler clutter mitigation, as discussed in Frazer, Abramovich, and Johnson (2009). The penalty is a reduction in sensitivity against noise, which is not of concern for ship detection tasks because detection performance is primarily limited by clutter.

3.2.1.4 Power Amplification

As mentioned previously, the average transmit power of an OTH radar can vary from tens or hundreds of kilowatts up to perhaps 1 MW or higher. This significant variation is explained by the diversity in OTH radar system designs, mission types/priorities, performance requirements, and the environmental conditions under which the systems are expected to operate. In most OTH radars, the total power requirement of the system is typically distributed over a number of separate amplifier chains or transmitter channels, each of which feeds an individual antenna element in the array. Dividing the overall power requirement over a large number of transmitting elements not only reduces the maximum power rating required in an individual amplifier chain but also introduces valuable redundancy in case of channel failure within the array.

Some important properties of the power amplifier include: (1) power rating, to satisfy the SNR budget for the pulsed or continuous waveforms used, (2) high efficiency, to reduce operating costs and demands on the air or liquid coolant flow system, (3) high linearity, to comply with out-of-band emission constraints and minimize signal distortion for effective target detection, (4) broadband operation, to allow rapid switching between widely spaced carrier frequencies while maintaining high signal fidelity, and (5) reverse power handling capacity, to prevent signals reflected back into the amplifier by the antenna from degrading performance or damaging equipment. Not surprisingly, trade-offs between competing objectives are invariably required in a practical design.

The final stage of amplification may be achieved using solid-state devices (Hoft and Agi 1986) or conventional vacuum tube technology (Division 1975). Vacuum tubes tend to be more efficient than their solid-state counterparts and were used in several early OTH radar systems. However, solid-state amplifiers have been preferred in modern OTH radars such as JORN and ROTHR. This is mainly because of their ability to rapidly switch between widely spaced operating frequencies as often as every second or so while maintaining high linearity and spectral purity (i.e., broadband operation). Vacuum tubes require retuning for effective operation over wide bandwidths, and are therefore not as suitable for systems requiring large and quasi-instantaneous frequency changes every second or so.

Other potential advantages of solid-state devices relative to vacuum tubes include their compact design and the simplification of power supply. Vacuum tubes are more cumbersome and require a variety of different voltages for the filament, plate circuit, and rest of the transmitter. It is also a held view that solid-state devices have a longer service life provided that ratings are not exceeded, whereas vacuum tubes wear out as their filaments deteriorate from continued operation (ARRL 1991). However, individual transistors cannot withstand the high currents required to generate high power levels, so the currents must be divided over a number of solid-state amplifier devices and combined until the desired final output power is reached. This can be a challenging exercise and may be regarded as a drawback relative to vacuum tubes, which can individually generate higher power levels.

In OTH radar systems, average output power levels of perhaps 10–20 kW per transmit channel may be generated by progressively combining a number of low-power solid-state amplifier modules rated at perhaps 1 kW in tiered stages. This methodology also provides a form of redundancy in the sense that the total output power degrades gracefully when individual units malfunction. At the transmit channel level, the amplifier chains may also be routinely switched from the middle to end elements of array, where they can be driven at lower power levels to extend service lifetime. Note that the radiated waveform properties and transmit beam pattern are influenced primarily by the performance of amplifiers connected to antenna elements near the middle of the aperture, where the power of the output signal is higher. Monitoring the amplifiers continuously during operation allows for rapid response to poor performance or malfunction (e.g., by switching the most reliable amplifier chains into the middle region of the array).

High linearity and power efficiency of the amplifier are often difficult to achieve simultaneously over the entire design frequency range. To minimize nonlinear operation near the point of amplifier saturation, it is desirable to keep the ratio of the peak to average power of waveform near unity to avoid signal distortion or "clipping" effects. Nonlinear operation can also be pronounced near the zero-voltage cross-over point, which is exercised most by the lower amplitude components of the waveform. For this reason, smoothly varying taper functions used to control out-of-band emissions are more robust to distortion in this region of the amplifier characteristic than waveforms with fast-changing low-amplitude ripples.

The amplifier input-output voltage characteristic can also change depending on the previous amplitude properties of the signal. This may be interpreted as a form of memory in the device, which results in waveform-dependent distortion. In practice, the radar waveform may be adaptively modified based on the amplifier characteristics to ensure that the output signal is not significantly contaminated by harmonics or other undesired artifacts related to nonlinear operation and noise. Figure 3.14 illustrates a real-time (analog) feedback loop that may be used to improve the spectral purity of the amplified signal fed into the antennas. In principle, convergence may require a few waveform repetition periods to settle on the pulse shape to be generated for the remainder of the dwell.

3.2.2 Receive System

In two-site OTH radar systems, the receive array is typically based on a quite different antenna element from the one used in the transmit array. Moreover, the receive antenna aperture may be an order of magnitude greater in length than the transmit antenna aperture. The first and second parts of this section describe the main factors driving the choice of antenna element and array design for the receiver, respectively. This is intended to explain the reasons for the significant differences between the receive and transmit antenna arrays in a two-site OTH radar system. The emphasis is again on currently operational skywave OTH radars in the United States and Australia, which serve as archetype systems for the purpose of discussion. Alternative OTH radar designs are also discussed and contrasted with these reference or "nominal" systems.

The third part of this section briefly motivates and describes the general characteristics of the superheterodyne (or simply heterodyne) receiver, which has traditionally been used for OTH radar, and identifies a number of mechanisms that can potentially limit its performance in practice. The fourth part of this section summarizes the main advantages of moving from a traditional heterodyne receiver per antenna subarray architecture to a direct digital receiver (DDRx) per antenna element architecture. In the context of OTH radar systems, DDRx technology currently enables the received signals to be sampled at RF very close to the antenna element.

3.2.2.1 Antenna Element

As opposed to the transmit antenna, which requires broadband elements to radiate the generated power efficiently over the design frequency range, a traditional view is that the choice of receive antenna element is not required to be as well-matched in the lower part of the HF band. This stems from the observation that the external noise spectral density in the HF band will often significantly exceed the internal receiver noise level in unoccupied frequency channels deemed suitable for OTH radar operation, particularly at night when the atmospheric noise level is high at the lower frequencies that need to used. It is argued that any attempt to improve the efficiency of the receiving antenna in an externally noise-limited environment will increase the gain for both useful signals and external noise by the same or similar amount. Based on this reasoning, a receive antenna that is better matched to frequencies in the lower HF band will yield effectively no net gain in signal-to-noise ratio.

Many two-site OTH radar systems have adhered to this design principle and selected a relatively simple receive antenna element for operation over the entire HF band. This approach significantly reduces cost with no "apparent" performance penalty. Since the external noise spectral density that the OTH radar must contend with is typically lower at the higher operating frequencies used during the day, it has been common practice to select a relatively short receive antenna element that is better-matched in the upper HF band, where the margin between external and internal spectral density is often smaller than that at lower operating frequencies used at night. This approach reduces the likelihood of the system becoming internally noise limited during the day. It is also noted that coupling into receive antenna elements with a better VSWR actually has the potential to incur a penalty in performance due to the large-signal environment in adjacent frequency channels and non-linearities in the receiver.

In general, compact antennas with a low height and small footprint are are not only less expensive and more readily configured in a very wide aperture array containing several hundreds of elements, but they are also easier to stabilize for reducing Aeolian noise at the receiver. Despite the use of a simple antenna being preferred in practice, the design choice needs to ensure that the mismatch of the receive element degrades relatively gracefully at lower operating frequencies such that the system remains externally noise limited at night. Using one and the same receive element for both high- and low-band operation is well motivated in OTH radar, particularly when conventional beamforming is used. However, with the increasing use of sophisticated adaptive signal-processing techniques in modern OTH radar systems, this design philosophy (which results in less well-matched antennas at lower frequencies) needs to be exercised with caution, as it may potentially lead to latent limitations in nighttime performance.

Specifically, the nighttime external noise-field is seldom isotropic or spatially white, but often exhibits a degree of spatial structure. A receive antenna that is better-matched in the lower HF band will increase the external-to-internal noise ratio and the signal-to-internal noise ratio by approximately the same amount (assuming the elements have a relatively broad radiation pattern). However, adaptive processing techniques can in principle reject *spatially-structured* external noise more effectively when the external-to-internal noise ratio is higher. This can yield a net benefit in SNR relative to the case where the same techniques are applied using less well-matched antennas. The question of how well the antenna needs to be matched at nighttime frequencies for it not to limit detection performance when adaptive processing is used depends on the relative strength and spatial structure of the background noise, the element geometry of the antenna array and the selectivity of practical receivers to reject large out-of-band signals.

A conventional choice for the OTH radar receive antenna element in the ULA context is based on the twin-whip endfire receive pair (TWERP) concept. This antenna is an appropriately spaced doublet of vertically polarized (monopole) elements whose outputs are combined via a fixed-length cable (delay-line) to provide higher gain in the forward endfire direction (oriented broadside to the ULA) with some attenuation in the reverse direction. For a certain doublet spacing and delay-line length, the front-to-back ratio varies with operating frequency but may be configured to attain its highest value at operating frequencies where most use is anticipated. Note that designs using a fixed analog combiner may optimize front-to-back ratio or gain in the forward direction, depending on which criterion is considered more important.

Figure 3.15a shows the dual-fan antenna elements used by the Jindalee OTH radar near Alice Springs in central Australia. These elements provide slightly better broadband frequency response than a simple whip antenna. The element height is approximately 6 m (quarter wavelength at 12.5 MHz), while the doublet spacing perpendicular to the ULA is 3 m (Earl and Ward 1987). In this architecture, the signal received by the back element passes through a 2.4-m cable that introduces a time-delay designed such that

(a) Dual-fan antenna elements.

(b) Monopole-doublet with elevated feed.

FIGURE 3.15 Dual-fan elements used in the Jindalee OTH radar receive array, near Alice Springs (central Australia), and the elevated-feed monopole-doublets used in the JORN OTH radars at Longreach (Queensland) and Laverton (Western Australia). © Commonwealth of Australia 2011.

the summation of the two element outputs (with the back element output inverted by 180 degrees) yields a cardiod-shaped radiation pattern in azimuth with an adequate front-to-back ratio over the design frequency range.

Figure 3.15b shows the monopole-doublet antenna elements with elevated feed used in the JORN OTH radars. With respect to the fan elements, the use of more rigid metal-tube elements provides better broader performance in a very stable receiving antenna with no need for supporting wires. In addition, the elevated-feed also helps to improve the low elevation-angle gain. The radiation pattern of the receive element is relatively broad and covers elevation angles from about 5 to 45 degrees and azimuths within ±45 degrees of the ULA boresight direction. The azimuth coverage of the element pattern allows high-resolution receive "finger" beams to be electronically steered over a ±45 degree arc by (analog and/or digital) array beamforming. The gain of the receive element pattern at high elevation angles higher than about 60–70 degrees is low to attenuate direct-wave clutter at near-vertical incidence.

Antenna elements based on the TWERP concept are not suitable when the required azimuth coverage is greater than about 90 degrees. Such elements are also not designed for use in two-dimensional arrays. An option to extend the azimuthal coverage beyond 90 degrees is to use separate ULAs of these elements with different orientations, as in the

JORN OTH radar system at Laverton. In the single-site Nostradamus OTH radar system, biconical antennas are used for transmission and reception. This provides broadband performance, omni-directional coverage in azimuth, and independent elevation control using a Y-shaped array (Bazin et al. 2006). Biconical antennas are physically larger (7 m high and 6 m wide) and more complex than the monopole-doublet antenna described previously. Besides the issue of cost for very wide aperture arrays containing hundreds of elements, and the challenge of managing mutual-coupling issues for two-dimensional arrays of well-matched antennas at the operating frequency (especially at low elevation angles), another possible disadvantage of using antennas with a large footprint is that it can limit options for element spacing in certain array configurations.

Alternative receive antenna designs, such those of the MADRE OTH radar in the United States (Headrick and Skolnik 1974) and the "Steel Yard" system built in the former Soviet Union, made use of horizontally polarized dipole antennas. These antennas were stacked vertically to form very tall curtain arrays with back screens fitted to provide adequate front-to-back ratios. Figure 3.16 shows pictures of the Steel Yard OTH radar receive site near Chernobyl. An advantage of such a design relative to a ground-deployed two-dimensional array is that elevation beams with finer resolution can be electronically steered at near-grazing angles. A key point, illustrated here by way of example for the

(a) Vertical curtain arrays of Steel Yard system.

(b) Horizontal cage-dipole antenna elements.

FIGURE 3.16 The previously operational OTH radar system developed by NIIDAR, known in the west as "Steel Yard" or the "Russian Woodpecker," was often heard on short-wave radio bands between 1976 and 1989. The first (eastern) system was built near Chernobyl, while a second (western) system was installed at Komsomolsk-na-Amure in Siberia. The transmit antenna of the western system consisted of two vertical curtain arrays with 13 masts, each containing 10 vertically stacked horizontal cage-dipole elements, while that of the eastern system was originally a vertical log-periodic dipole array. The receive antenna for both systems consisted of two adjacent (high- and low-band) vertical curtain arrays with 30 masts, each containing 10 vertically stacked horizontal cage-dipole elements. The larger receive antenna of the eastern system, shown above, spans a two-dimensional aperture that is about 143 m high and 500 m long. This quasi-monostatic system near Chernobyl operated with a transmit-receive site separation of about 60 km and employed a pseudo-random binary phase-coded pulse waveform with a maximum transmitter power of approximately 1 MW. The system was designed to operate over a frequency range of 4–30 MHz, using a PRF of 10 or 25 Hz, and bandwidth between 10 and 25 kHz. The primary mission of the system was ballistic missile detection, while the secondary mission was aircraft detection.

different OTH radar systems considered, is that the choice of antenna element and array configuration are intimately linked and need to be selected jointly to ensure compatibility with system objectives.

Polarization transformation of an HF signal through the ionosphere is thought to limit the value (benefit versus cost) of using fully-polarimetric antenna elements in OTH radar arrays. The performance benefits of polarization diversity on receive are yet to be convincingly demonstrated for skywave OTH radar applications in order to justify the additional expense and complexity of deploying the associated hardware and signal-processing systems. No currently operational *skywave* OTH radar system designed for surveillance applications is known to exploit polarization diversity on all (or a majority of) the transmit or receive antenna elements.

3.2.2.2 Array Design

A transmit beam that is broad in azimuth is used to simultaneously floodlight a relatively wide surveillance region, whereas a narrow beam is required on receive to provide fine resolution for enhanced target detection and tracking in clutter- or noise-limited conditions. Receive antenna apertures with major dimensions ranging from about 100 m to 3 km with a number of elements between about 20 and 500 have been used in former and current skywave OTH radar systems. The lower limit on receive aperture length is primarily determined by the mission types and priorities, as well as the detection and tracking performance requirements of the system. On the other hand, the upper limit depends on physical limitations, operational factors, and economic constraints. These include loss in spatial coherence due to non-specular signal reflection from the ionosphere (Sweeney 1970), increase in real-time processing load and clarity of operator displays, as well as additional costs associated with the increase in instrumentation and maintenance.

For a given number of available reception channels, a linear array provides classical beams with the highest azimuth resolution. Receive beams with fine azimuth resolution are desirable for several reasons. First, narrow beamwidths reduce the effective clutter backscattering RCS in a radar resolution cell, and Doppler spectrum contamination due to spatial variations in the clutter spectral characteristics. This enhances slow-moving (surface) target detection. Second, a higher gain and smaller spatial passband provides the system with greater immunity against noise and interference signals not incident from the main beam steer direction, respectively. This enhances the detection of fast-moving (aircraft) target echoes. Third, receive beams with high resolution improve target location accuracy and hence tracking performance.

The uniform linear array (ULA) geometry allows receive beams to be digitally formed and steered in different directions in a computationally efficient manner by exploiting the FFT. Moreover, relatively straightforward (amplitude-only) taper functions may be used to lower sidelobe levels in a well-calibrated ULA. Despite the high performance of state-of-the-art computers, efficient spatial processing remains important for current OTH radar systems to enable real-time operation with a receiver-per-element architecture. In receive systems containing close to 500 antenna elements, a significant reduction in computational load at the beamforming stage can be used for processing a greater number of range cells to extend the instantaneous coverage when ionospheric conditions permit, or to accommodate more sophisticated adaptive processing techniques that can enhance target detection and tracking performance.

With respect to two-dimensional ground-based array apertures, the ULA geometry enables mutual coupling effects to be managed more readily. For example, a ULA may be augmented by adding one or more "dummy" antenna elements (not connected to a receiver) at either end of the aperture. This improves the homogeneity of the electromagnetic environment for all receiving elements in all signal DOAs of interest. Furthermore, the low-amplitude tails of the taper functions used for array beamforming help to reduce calibration errors due to mutual coupling end effects in a ULA. For the aforementioned reasons, ULAs of vertically polarized antenna elements based on the TWERP concept have been selected for several currently operational OTH radar systems in the United States and Australia.

In such systems, the relatively broad main lobe of the transmit beam floodlights the surveillance region, while the echoes received by the antenna array are processed in a number of simultaneously formed high-resolution "finger" beams that are electronically steered half a main lobe width apart to cover the surveillance region illuminated by the transmitter. For example, the 2970-m aperture of the JORN receive array has an azimuth resolution of around 0.5 degrees at 15 MHz for angles within 45 degrees from boresight. A total of 20 high-resolution receive beams would be required to cover a transmitter footprint 10 degrees wide. The azimuth sector of this illumination footprint is approximately matched to the JORN transmit ULA aperture of 160 m.

In contrast to the JORN transmit antenna, which employs two separate ULAs with different inter-element spacings for low- and high-band operation, a single ULA is used on receive in this two-site OTH radar design. As the receive antenna element pattern is broad in azimuth, the ULA inter-element spacing needs to be less than half wavelength to avoid potential DOA ambiguities due to grating lobes. The JORN receiver array consists of 480 monopole-doublets with an inter-element spacing of 6.2 m. This provides half-wavelength spatial sampling at a daytime operating frequency of approximately 24 MHz. At lower (nighttime) frequencies, which may be between 7 and 12 MHz for example, the receive array becomes highly oversampled. This opens up a large interval of "invisible" space where beams that do not correspond to physical angles can be steered. Provided the array is well calibrated, and strong near-field sources are not present, such beams may be used to estimate the internal noise level of the receive system.

Figure 3.17 shows the receiving aperture of the Jindalee OTH radar. This array is 2766 m long and consists of 462 dual-fan antenna elements grouped into 32 subarrays with a receive channel per subarray achitecture (Sinnott and Haack 1983). Each subarray combines 28 adjacent antenna elements with 50% overlap between the elements shared by neighboring subarrays. An analog (delay-line) beamformer combines the antenna elements in each subarray to form a subarray beam pattern with a main lobe of similar width in azimuth to the transmit beam. The grating lobes arising due to the large distance between the subarray centers (64 m for the Jindalee OTH radar) are suppressed by the low-gain response of the subarray beam pattern sidelobes. While this avoids DOA ambiguities due to grating lobes, a disadvantage of this scheme relative to the receiver-per-element architecture is that the receive finger beams are constrained to lie within the main lobe of subarray pattern as opposed to the much broader antenna element pattern.

The receiver-per-subarray architecture also has a number of advantages. An obvious practical benefit that originally motivated this design is a large reduction in the required number of receptions channels, which attracted a large proportion of the cost in early OTH radar systems. Moreover, the signal-processing load is also significantly reduced to permit real-time operation using modest computing resources. Note that the reduction includes range processing in each receiver as well as beamforming over different receives. Using the subarray outputs can also reduce demands on receiver dynamic range by

FIGURE 3.17 The Jindalee OTH radar receive array located at Mt Everard, 20 km north-west of Alice Springs, in central Australia (*photograph courtesy of Dr Gordon Frazer*). This system, formerly known as the Jindalee Facility Alice Springs (JFAS), is a quasi-monostatic OTH radar that uses a linear FMCW with a transmit-receive site separation of approximately 100 km. The transmit antenna is a two-band ULA of 8 (low band) and 16 (high band) LPDA elements for operation over a frequency range of 5–28 MHz. The transmit aperture is 137 m long and can be electronically steered to ±45 degree relative to boresight. The maximum transmit power of the Stage B system developed by DSTO Australia between 1978 and 1985 was 160 kW. The receive antenna is a linear array of 462 × 5.5-m high dual-fan elements grouped into 32 overlapped subarrays. The receive aperture is 2766 m long and has a receiver per subarray architecture (i.e., 32 reception channels). The receive beam may be steered to ±45 degree in azimuth by a combination of analog (delay-line) beamforming for subarray pattern steering and digital beamforming to generate the final high-resolution "finger" beams. The primary mission of the JFAS system is aircraft detection, with ship detection being the secondary mission. © Commonwealth of Australia 2011.

attenuating powerful interference that enters the wideband front end through the *sidelobes* of the subarray pattern. These advantages were particularly important for early OTH radar systems.

Clearly, OTH radar antenna designs based on the ULA geometry also have a number of drawbacks. Specifically, the coupling between azimuth and elevation into a single "cone angle," which alone parameterizes the ULA steering vector, effectively precludes independent steering of the receive beam in azimuth and elevation. This ambiguity on receive restricts the ability to effectively select propagation modes by spatial filtering in elevation. The presence of multiple layers reflecting signals from different virtual heights in the E- and F-regions of the ionosphere may lead to clutter, noise, and interference being incident with the same cone angle as useful signals, despite the unwanted signals arriving from possibly quite different DOAs in azimuth and elevation. This can potentially degrade target detection performance with respect to two-dimensional array apertures that enable independent control of the receive beam in azimuth and elevation.

Another drawback is that the measurement of echo great-circle bearing is required for target location but only the cone angle is observable in a ULA. For off-boresight echoes arriving with unknown elevation, the coupling between azimuth and elevation leads to the "coning effect," which represents a form of ambiguity as far as target bearing estimation is concerned. More precisely, when the cone angle is interpreted directly as the great-circle bearing, the apparent azimuth of a signal shifts toward boresight as the elevation angle of incidence increases. Knowledge of the target echo virtual reflection height in the ionosphere as a function of group-range for the prevailing skywave signal modes is required to correct for coning when estimating the target bearing. This propagation-path information needs to provided by a real-time ionospheric model (RTIM) to be discussed later in this chapter.

The limitations of the ULA geometry have led some OTH radar designers to deploy a two-dimensional (nonlinear) receive aperture, such as the Y-shaped array of the Nostradamus system pictured in Figure 3.18. Relative to a single ULA, which requires antenna elements with adequate front-to-back ratio to avoid the potential for DOA ambiguity, a two-dimensional array may use omni-directional antenna elements to provide 360-degree coverage in azimuth. This is because two-dimensional array beamforming eliminates the ambiguity between azimuth and elevation.

3.2.2.3 Heterodyne Receivers

Skywave OTH radar receivers are required to operate over a broad frequency range that may span several octaves as the system responds to changes in ionospheric propagation conditions. Furthermore, such systems often need to switch rapidly between widely spaced carrier frequencies every second or so to acquire signals in the optimum manner for the different missions, tasks, and regions that are concurrently scheduled on the radar time-line. The frequency-agility required for effective OTH radar operation necessitates the use of a receiver with a wideband front-end. A switchable network of *broadband* (preselector) filters are typically inserted in the RF front-end to bandpass the segment of the HF spectrum containing the carrier frequency. These preselector filters are designed with different bandpass characteristics that collectively cover the design frequency range of the system. The band-switching time can be made coincident with other delays inherent to the OTH radar at a frequency change (i.e., within the inter-dwell gap).

In addition, HF radar receivers are required to detect very weak target echoes in the presence of powerful surface clutter and external background noise with a varying power spectral density. This requires receivers with high dynamic range for effective Doppler processing and acceptable sensitivity to ensure that the detection of fast-moving targets is not limited by internal noise. Moreover, this must be accomplished in a crowded HF spectrum that contains a large number of strong man-made signals in spectrally adjacent channels that propagate to the system over a one-way path. This calls for highly selective receivers in a broad sense and places very stringent demands on linearity and spurious free dynamic range.

In the context of a multi-channel antenna array, it is also necessary for the receiver transfer functions to be well-matched over the passband to avoid severe limitations in spatial dynamic range. In general, this includes all stages of the receiver chain along the signal path from the antenna element to the output of the analog-to-digital converter

FIGURE 3.18 The French skywave OTH radar "Nostradamus," located about 80 km west of Paris, operates by means of linear-FM or phase-coded pulse waveforms in a truly monostatic configuration, where a subset of elements in the array are used as reciprocal antennas for transmit and receive. Specifically, the transmit antenna is a three-branch star (Y-shaped) array consisting of 3 × 32 biconical elements, each 7 m high and 6 m wide, that are randomly distributed along 3 arms oriented 120 degrees apart. Each arm of the transmit array has a length of 128 m and a breadth of 80 m. On the other hand, the receive antenna is a Y-shaped array containing 3 × 96 biconical elements randomly distributed along 3 arms of length 384 m and breadth 80 m. Stated in the simplest manner, a third of the array is used for transmitting and the entire array is used for receiving. The receiving elements are grouped into 3×16 subarrays with a digital receiver per subarray to yield a total of 48 reception channels. This design provides for omni-directional (360-degree) coverage and independent electronic beam steering in azimuth and elevation over a frequency range of 6–28 MHz. The maximum transmit power of the system is approximately 50 kW. The Nostradamus OTH radar was developed by ONERA with the primary mission of aircraft detection and secondary missions of ship detection and remote sensing. The first (reported) target detections made by this system were in 1994.

(ADC). A significant reduction in spatial dynamic range not only degrades the performance of conventional beamforming, particularly in terms of raising sidelobe levels, but can also greatly reduce the effectiveness of adaptive beamforming used to mitigate co-channel interference. It follows that great care needs to be taken in HF receiver design to avoid OTH radar performance being limited by instrumental imperfections rather than environmental factors (Pearce 1998).

The analog receiver design most widely used for radar applications is based on the superheterodyne (or heterodyne) architecture described in Skolnik 2008a. This type of receiver has been used in a number of past and present OTH radar systems to fulfil the aforementioned requirements. The basic principle of the heterodyne receiver is to convert the signal frequency to a convenient intermediate frequency (IF) by using a variable-frequency local oscillator (LO) and a mixing stage, such that the desired signal can be amplified and filtered more effectively and conveniently at the selected IF than would be possible at the original radio frequency (RF).

The IF stage is tuned to a fixed frequency that is independent of the radar's operating frequency. This greatly simplifies the optimization of circuits in the IF stage, which provide much of the gain and selectivity in a heterodyne receiver. A fixed IF stage also enables the same gain and selectivity to be maintained over the entire RF tuning range of the receiver. This would be an impossible task for circuits that need to be tuned over a wide range of operating frequencies. In simple terms, a variable-frequency LO plus a fixed IF stage is more convenient to realize in hardware than tuneable filters and amplifiers. A judicious choice of IF also simplifies the design of efficient amplifiers and highly selective filters, while allowing the use of technologies that cannot be tuned "on-the-fly" (e.g., a high-Q crystal filter).

Figure 3.19 shows a simplified block diagram of the heterodyne receiver used to perform stretch processing in the Jindalee Stage B OTH radar system. Recall that this system was implemented with a reception channel per antenna subarray. Analog beamforming of the 28 dual-fan antenna elements in each subarray occurs via a Dolph-Chebychev amplitude-taper weight (attenuation), which is used to control the sidelobe response of the subarray pattern, and a switchable network of different-length cables (delay-lines), which is used to steer the direction of the subarray pattern main lobe, before summing the output signals in a linear combiner to form the subarray beam output. The analog beamforming network of each subarray is situated close to the corresponding set of antenna elements in the array.

Each subarray output is then fed through a low-noise RF power amplifier and a broadband preselector filter not shown in Figure 3.19. The latter provides front-end selectivity in order to attenuate signals that are out-of-band with respect to the OTH radar operating frequency. A broadband preselector filter is typically placed ahead of the first

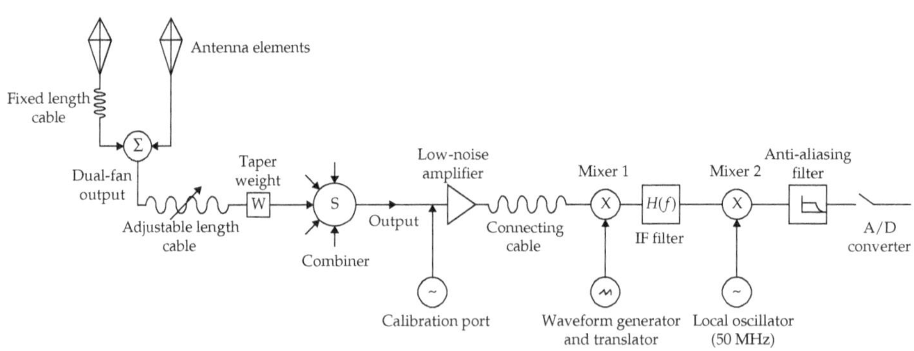

FIGURE 3.19 Simplified block diagram showing the hardware architecture of a reception channel in the Stage B Jindalee OTH radar. This system consisted of (50-percent overlapped) subarrays of dual-fan antenna elements with an analog heterodyne receiver per subarray output that performed stretch processing. © Commonwealth of Australia 2011.

mixing stage in the heterodyne receiver to reduce degradations in sensitivity caused by the reception of interference at the image frequency and distortion products introduced by potentially nonlinear active devices in subsequent stages of the receiver chain. After the subarray outputs pass through the RF amplifier and preselector filter, the signals are fed through a long length of cable that connects each analog-beamformed output to its respective receiver housed in the main building at site.

In the stretch-processing scheme of Figure 3.19, the HF signal entering the receiver is first mixed or "deramped" with a linearly swept repetitive frequency–modulated continuous waveform (FMCW). Synchronization between the radar sites allows this signal to be generated locally at the receiver as a time-delayed and frequency-translated replica of the transmitted waveform. At the mixer output, the beat-frequency signal falls within an intermediate frequency (IF) passband above 50 MHz. Here, the signal is bandpass filtered before being down-converted to near baseband using a 50-MHz fixed-frequency local oscillator. The baseband signal is then input to a low-pass (anti-aliasing) filter before being sampled at or above the Nyquist rate (typically a few kilohertz) by an analog-to-digital converter (ADC). It is now of interest to discuss the individual parts of a heterodyne receiver architecture (with and without stretch processing) in more detail.

The primary function of the RF amplifier is to provide the gain necessary to establish the noise figure of the system, which places an upper bound on radar sensitivity in a noise-limited environment. The receiver noise figure is simply the noise factor expressed in decibels and effectively measures the degradation in output SNR relative to that which would result due to thermal noise in an input termination at standard temperature (usually taken as 290 K). In other words, the noise figure provides a measure of the additional noise that is generated by electrical devices internal to the receiver (a noiseless device has a noise figure of 0 dB). It is important to clarify that receiver noise figure is not related to HF noise from external sources, which is considered to be part of the "signal" in this discussion.

Each individual stage of the receiver will contribute to the overall noise figure. However, if the gain of a stage is high, its noise figure will tend to override or "mask" the noise figure contributions of the stages following it. The use of low-noise active devices with adequate gain in the front-end (as close as possible to the antenna) can therefore help to achieve a low receiver noise figure. However, for reasons described shortly, it is prudent to use the minimum gain necessary in the RF amplifier to establish the receiver noise figure. Any further gain that may be necessary to drive the ADC with appropriate voltage levels can be obtained from other amplifiers in the IF (and possibly baseband) stages of the receiver.

Figure 3.19 indicates that the low-noise amplifier (LNA) is mounted before the long transmission line which connects the subarray output to its associated receiver. The transmission line acts as a lossy amplifier, and if placed as the first stage of a receiving system, it will limit the noise figure to that of the transmission line at best. A low-noise RF amplifier with sufficient gain is placed ahead of the long transmission line to "mask" the noise added by the latter. To achieve a low noise figure and high dynamic range, automatic gain control (AGC) should not be applied to the RF amplifier because the AGC circuit changes the operating characteristic of the amplifier from Class A to a less-linear mode (ARRL 1991).

In a well-designed receiver chain, the noise figure is primarily established in the RF amplifier and/or mixer stages. Hence, the gain of the RF amplifier is often set high enough to override the mixer noise, but not higher. Excessive gain in the RF stage can lead to amplifier instability and self-oscillation under certain conditions. This has the potential to produce unwanted spurious signals and deteriorate the noise figure in a process known as desensitization. Moreover, excessive gain in the RF stage should be avoided to prevent overloading the mixer. Reducing the signal level that enters the nonlinear mixing stage helps to reduce susceptibility to spurious signals generated in subsequent stages of the receiver.

In summary, the RF amplifier gain is adjusted to establish an appropriate receiver noise figure, such that target detection performance ultimately remains limited by external noise (or clutter) rather than internal noise. The RF amplifier is not driven with excessive gain to prevent significant deviations from linear operation and to avoid overloading the non-linear mixing stage. The RF amplifier gain therefore represents a compromise between achieving an appropriate noise figure, while ensuring high linearity and minimum overloading in the mixer. For OTH radar, an "appropriate" noise figure needs to be considered with respect to the lowest anticipated levels of (spatially unstructured) external noise at the receiver site, which varies diurnally, seasonally, and over the 11-year solar cycle.

In general, higher receiver noise figures can be tolerated at night without limiting the system's detection performance. This is because OTH radars often need to use frequencies in the lower HF band at night, where the power spectral density of the received external noise (dominated by atmospheric sources) is relatively high despite the receive antennas being less well-matched than in the upper HF band. On the other hand, the received external noise spectral density may be 20 dB lower in level during the day at the higher operating frequencies typically used by OTH radar systems. This places an upper bound on the noise figure to ensure that the sensitivity of the system is not instrumentally limited.

The purpose of the mixer in a heterodyne receiver is to convert the received RF signal energy to a higher or lower intermediate frequency (IF) where amplification and filtering can be achieved more effectively and conveniently than would be possible at the native signal frequency. The two mixer inputs are the signal received at the output of the RF amplifier and preselector filter and the signal generated by a variable-frequency local oscillator (or a time-delayed and frequency-translated copy of the transmitted waveform in the case of stretch-processing). At the output, the signal of interest is contained in the first-order mixing product.

Specifically, the fundamental component or center-frequency of the local oscillator is tuned to position the desired signal energy in a fixed intermediate-frequency band at the mixer output. In the case of high-side LO injection, the signal of interest in the first-order mixing product is often the difference (as opposed to sum) frequency component. In other words, the IF is equal to the difference between the LO frequency and that of the desired RF signal input to the mixer. In OTH radar applications, the received RF signals are upconverted to an IF that may be between 50 and 100 MHz.

The mixer is one of the most important parts of a high-performance heterodyne receiver because it is at this point that the greatest consideration for dynamic range exists. The spurious-free dynamic range (SFDR) of a receiving system may be defined as the decibel difference or ratio between the largest tolerable input signal level that can be accommodated during reception (without causing distortion products that are discernible at the sensitivity level of the receiver) and the minimum detectable signal level referenced to the input, which may be calculated from a knowledge of the noise figure and receiver bandwidth. Stated another way, the decibel difference between the minimum detectable signal level and the input level which will produce distortion products equal to the minimum detectable signal level referred to the input is the SFDR of the system.

When excessive amounts of signal energy are permitted to reach the mixer, a point is reached where the output no longer increases linearly with the input. This nonlinear phenomenon, known as gain compression, reduces the sensitivity of the mixer to all input signals in the passband. Moreover, when more than one signal is present in the passband going into the mixer, the naturally nonlinear mixing process can produce appreciable spurious responses in the form of intermodulation distortion (IMD) products. As the input signal level to the mixer increases, energy from higher order IMD products may begin to appear above the noise-floor in the receiver passband. Overloading the mixer can seriously degrade SFDR, and hence the ability of the system to detect weak target echoes in the presence of powerful man-made signals received in spectrally near frequency channels that are not rejected by the preselector filter.

Nonlinear mixing can produce three main types of intermodulation distortion (IMD) products with the potential to obscure target echoes. Out-of-band IMD products arise due to nonlinear mixing between two or more powerful interference signals that enter through the wideband front-end of the receiver with the potential to form products overlapping the desired signal bandwidth at IF. Nonlinearity at any stage in the receiver can lead to mixing of the clutter echo with itself. This gives rise to in-band IMD products that can mask target echoes due to the clutter energy at the output occupying a broader Doppler spectrum than that of the clutter at the input. Finally, nonlinear mixing of the received radar echoes with a powerful out-of-band interferer can transfer the interferer modulation onto the in-band signals in a process known as cross-modulation. Willis and Griffiths (2007) state that typical IMD requirements for OTH radar referred to the input are for third-order intercept levels of 35–40 dBm and second-order intercepts of 80–90 dBm. The influence of cross-modulation on the attainable dynamic range of HF radar receivers is discussed in Earl (1987).

In summary, the mixer should have only enough pre-amplification to overcome the mixer noise. It is desirable to have a "strong mixer" that can handle high signal levels without being adversely affected to improve receiver dynamic range. When the receiver dynamic range is insufficient (i.e., poorly matched to the signal environment), strong out-of-band signals that are input to the mixer after passing through the wideband front-end can cause undesired interference to (and distortion of) the desired signals. When external noise due to man-made and natural sources exceeds that of the mixer, it is in principle possible to realize higher dynamic range by not including an RF amplifier. A quantitative analysis of receive system linearity requirements for HF radar can be found in Earl (1991) and Earl, Kerr, and Roberts (1991).

The local oscillator (LO) must operate in a frequency-stable and spectrally pure manner. It should have no significant short- or long-term drifts in fundamental frequency, extremely low phase-noise spectral density relative to the peak value of the fundamental energy, and be reasonably free of spurious responses over a wide spectrum. The SFDR of the LO may be defined as the ratio of the rms value of fundamental signal to the rms value of the strongest spur regardless of where it occurs in the LO output frequency spectrum. Excessive levels of phase noise can have a serious effect on mixer performance, particularly for OTH radar receivers with limited selectivity in the RF front-end leading into the first mixer.

Practical LO signals have limited spectral purity due to phase-noise and spurs. Although the LO noise-floor and spurs are typically at a very low level with respect to the peak value of the fundamental, they may potentially spread over a broad frequency spectrum. This can become a problem because the wideband front-end permits powerful man-made signal in the spectral vicinity of the radar signal to simultaneously enter the mixer. Specifically, unwanted interference due to strong man-made signals in the crowded HF band can appear in the IF filter passband by virtue of reciprocal mixing. Reciprocal mixing refers to the process by which powerful out-of-band signals interact with phase noise and spurs of the LO in the mixer to form unwanted products that overlap the desired signal bandwidth at IF. A tuned circuit with high-Q may be inserted in the LO chain to reduce the level of phase noise and spurs entering the mixer.

Fundamentally, there are two issues associated with the local oscillator that can make reciprocal mixing one of the toughest specifications to meet in a practical OTH radar design. First, there is the broadband phase-noise spectral density, which extends over wide frequency range "far" from the fundamental. This needs to be kept extremely low to prevent energy from out-of-band signals appearing in the desired signal passband at the mixer output. Second, it is important to avoid the smearing of clutter energy in Doppler frequency due to phase-noise in a narrow band of frequencies spectrally "near" to the fundamental. For this reason, the specification of phase-noise is sometimes described in terms of "near" and "far" spectral purity ratios. The specific issue of reciprocal mixing for OTH radar has been considered in Earl (1997).

As mentioned previously, the intermediate frequency is the difference between the LO frequency and the desired signal frequency at the mixer input. The choice of IF needs to be well outside the tuning frequency range of the receiver and be relatively clear of radio interference. Typically, the IF is within an order of magnitude of the receiver's RF tuning range and is selected from an established set of standard IFs that are kept clear of channel allocations. The specific choice of IF represents a compromise between several competing objectives. On one hand, a low IF enables the high selectivity and gain required of this fixed stage to be implemented more effectively and conveniently in a practical sense. On the other hand, a higher IF improves the image-rejection capability of heterodyne receivers for reasons that are now described.

The main function of the IF filter is to help reject unwanted mixer products that fall outside of the receiver passband. In the case of high-side LO injection, the image frequency is equal to the desired signal frequency at the mixer input plus twice the IF. As any signals at the image frequency will fall inside the IF filter passband at the mixer output, it follows that interference at the image frequency needs to be rejected prior to the mixer. In HF radar applications, it is possible to use a high enough first IF so that a fixed-tuned preselector filter can reject image signals in the receiver front-end without sacrificing frequency-agile operation over the entire HF band. Image rejection requirements for OTH radar have been considered in Earl (1995a).

In a broad sense, selectivity refers to the rejection of unwanted (out-of-band) signal energy in any part of the receiver, not limited to the IF bandpass filter. This includes selectivity in the receiver front-end to reject interference at the image frequency and powerful out-of-band signals that can cause overloading in subsequent stages of the receiver. It also includes selection circuits or filters used to reject energy in the LO signal at frequencies other than the fundamental. While the selectivity of the heterodyne receiver is effectively set by 3-dB bandwidth of the IF filter network, selectivity embraces all parts of the receiver in this context. In the 30–100 MHz frequency range, where it is difficult to build high-Q inductors, and because coaxial cavities are very large, an IF filter based on the helical resonator is an excellent choice (ARRL 1991). High-Q crystal filters have also been used to provide bandpass selectivity at IF.

The IF filter is usually followed by one or more IF amplifier stages (not shown explicitly in Figure 3.19), which provide the additional gain required to drive the ADC. Such amplifiers also compensate for IF filter insertion loss. As a rough guide in order of magnitude, the input signals may need to be amplified from microvolts at the antenna to volts at the input to the ADC. This requires the receiver chain to provide a voltage gain in the order of 60 dB. This gain needs to be distributed optimally across the RF, IF, and possibly baseband stages of the receiver to achieve a low noise figure and high dynamic range. The IF-strip typically provides most of the necessary gain.

In the stretch-processing receiver of Figure 3.19, a fixed-frequency local oscillator (offset slightly from the fixed-IF) is used to downconvert the bandpass filtered and amplified signals to near baseband. The bandwidth of the low-pass (anti-aliasing) filter is set according to the ADC sampling frequency and determines the maximum range-depth that can be processed by this type of receiver for a given radar signal bandwidth. Signal distortion arising due to timing jitter and quantization errors in the ADC need to be minimized to preserve the spectral characteristics of the input signals over a dynamic range consistent with the attainable sub-clutter visibility (SCV) for effective Doppler processing. ADC dynamic range requirements for the described OTH radar system are discussed in Earl and Whitington (1999).

Analog heterodyne receivers based on the stretch-processing principle were first used in the WARF OTH radar system operated by SRI in the mid-1960s. This architecture was later adopted in various forms for the Jindalee OTH radar. The stretch-processing receiver is of significant importance in the story of OTH radar and has some undeniable advantages in FMCW radar applications not limited to HF systems. For this reason, more will be said about this type of receiver in the following chapter. Nowadays, analog heterodyne receivers can sample the entire radar signal bandwidth, which in principle allows all unambiguous range cells to be processed using fully-digital pulse compression. With respect to the stretch-processing receiver, such designs do not limit the instantaneous range-depth of an OTH radar system and provide a level of spur suppression commensurate with the chirp time-bandwidth product.

Depending on the ADC technology available, a heterodyne receiver may sample the radar signal directly at a second IF which may be lower than the first by an order or magnitude. In a direct-IF sampling heterodyne receiver, the variable-frequency LO input to the first mixer of Figure 3.19 is a CW signal instead of a time-synchronized replica of the transmitted waveform, while the fixed-frequency LO input to the second mixer in Figure 3.19 translates the desired signal to a lower IF rather than baseband. This architecture enables receiver gain, selectivity, dynamic range, and noise figure to be managed in a flexible manner. The second stage IF output signal is then directly sampled by an ADC with sufficient resolution (effective number of bits) and SFDR.

More recently, it is possible for state-of-the-art ADC devices to sample the received RF signals directly at a point behind the low-noise amplifier and possibly a preselctor filter using a simplified front-end with no analog mixers. The potential advantages of using a direct digital receiver (DDRx), which can in principle allow fully-digital downconversion and filtering of multiple signals from RF to baseband, will be briefly discussed in the next section. Our discussion here continues to consider the analog heterodyne receiver in either the stretch-processing or direct IF-digital configuration.

The transfer functions (frequency responses) of the different reception channels in the antenna array need to be tightly matched across the receiver passband for effective conventional beamforming or adaptive processing. As indicated in Figure 3.19, the Jindalee Stage B system made provisions for a calibration signal to be injected at a port located immediately behind each subarray output. This signal is designed to probe the channel transfer functions such that compensation for differences in the gain and phase responses of each reception channel accumulated along the signal paths following the injection point can be estimated and applied.

Internal calibration signals can probe a combination of broadband components, such as the LNA, preselector filter, and connecting cables in the receiver front-end, as well as narrow components, such as the IF filter and anti-aliasing filter. The gain and phase responses of each channel are measured at an appropriate frequency resolution over the receiver passband and digital corrections are applied in the form of a calibration table to correct for differences in the channel transfer functions. Drifts in the reception channel characteristics over time require the calibration table to be updated routinely, even when the OTH radar operating frequency is not changed. Obviously, the radiation patterns of the subarrays cannot be measured using a calibration scheme based on internal signals.

3.2.2.4 Direct Digital Receivers

With respect to the proven effectiveness of analog heterodyne receivers in OTH radar applications, such designs also have a number of drawbacks. Besides the complexity of generating and distributing a number of variable- and fixed-frequency LO signals for mixing and translation, there is also the complexity of overall system design, which needs to properly balance interactions between various subsystems (LO, mixers, IF-strip, distribution system). Second, analog devices are prone to temporal drifts in operating characteristics due to temperature changes (short-term) and component deterioration (long-term), which necessitates the routine application of sophisticated calibration procedures. Third, analog heterodyne receivers can only process a single radar channel at a time, which means that the full aperture of the antenna array cannot be exploited to perform multiple simultaneous radar tasks or support functions without installing multiple receivers in parallel behind each antenna-element or subarray. Finally, high-performance heterodyne receivers are relatively expensive, so the cost of a receiver-per-element configuration can be prohibitive for large OTH radar arrays that contain hundreds of elements.

Direct digital receiver (DDRx) technology based on state-of-the art ADC devices currently enables the entire HF band to be sampled directly near the output of each antenna element with a bit-resolution and SFDR appropriate for OTH radar applications. A high-performance anlog section is placed ahead of the digital section in a DDRx to provide the required amount of pre-amplification and selectivity. A well-designed digital section should not limit the performance of a DDRx, which should faithfully preserve the spectral content of signals across a dynamic range that is sufficiently wide to capture target echoes, clutter returns, and other man-made signals in the HF band. In other words, additive and multiplicative distortions introduced by imperfections in the ADC need to remain below the sensitivity level (noise-floor) of the receiver *after* radar signal processing is applied.

The gain settings in different sections of a DDRx need to be distributed optimally to achieve high SFDR and low noise figure such that sensitivity remains externally noise limited. The conversion bandwidth of present-day 16-bit ADC devices is over 100 MHz, while OTH radar signals have bandwidths of a few tens of kHz. In this type of receiver, where the radar channel bandwidth is a small fraction of the conversion bandwidth, the likelihood of the largest spur being contained in the radar channel is significantly reduced. In addition, CW spurs from the ADC are suppressed by the chirp time-bandwidth product after digital pulse compression, which is typically greater than 20 dB for aircraft detection and up to 40 dB for ship detection. Moreover, CW spurs from different ADC devices will be incoherent across the different reception channels of the array and may therefore be suppressed further after digital beamforming.

A DDRx that can sample the entire HF band allows modern OTH radars to perform direct downcoversion (digital mixing, selective filtering, and data decimation) to acquire different signals received simultaneously in multiple frequency channels by passing the ADC output through a number of on-board direct digital downconverters (DDCs). The output of each DDC may potentially be input to a channelizer if multiple signals need to be extracted from the DDC output, as well as an arbitrary resampler if a specific (non-integer) downsampling ratio is required. The output data rate passed onto the signal-processing stage should be decimated according to the desired signal bandwidth to lower the rate of data flow and computational complexity. As discussed in the next chapter, a resampler that can implement a fractional exact time-delay is useful for array beamforming in a wide aperture phase-array system. The DDRx approach avoids some of the aforementioned limitations associated with the use of analog devices in a heterodyne receiver. In particular, array calibration becomes more straightforward as all narrowband filters are digital. Issues such as reciprocal mixing are still present, although in a slightly modified form (Skolnik 2008b). Specifically, the reciprocal mixing phenomenon manifests itself due to the noise spectral density of the ADC clock.

A DDRx must operate in a large-signal environment that includes strong interference from AM radio stations and other HF transmitters. Spurious responses generated due to large signals within the ADC may superimpose on target echoes and limit detection performance. The high power of such signals therefore drives the dynamic range requirement of the ADC. To ease demands on dynamic range without diminishing the flexibility of DDRx, a bandpass preselector filter may be installed in the front-end to attenuate strong broadcast signals below the lowest design frequency in the HF band and to prevent aliasing beyond the maximum sampling frequency of the ADC. It has been demonstrated that DDRx designs based on the insertion of a well-designed bandpass preselector filter in the front-end do not attract significant performance degradations with respect to state-of-the-art analog receivers (Skolnik 2008b).

The cost of a DDRx is currently less than a quarter of the price of a high-performance heterodyne receiver. This makes a DDRx-per-element architecture economically feasible for wide-aperture antenna arrays with approximately 500 elements. In this configuration, the high-resolution receive finger beams are no longer constrained to lie in the main lobe of the antenna subarray pattern, but may be steered anywhere within the much broader azimuth coverage of the antenna element pattern. Moreover, a DDRx allows multiple frequency channels within the HF band to be simultaneously processed in parallel. Relative to an analog heterodyne receiver-per-subarray architecture, which is constrained to select a single narrwoband frequency channel at a time in the main lobe of the subarray pattern, a DDRx with multiple downconverters opens up new possibilities for exploiting the full aperture of the receive array. The capability to digitize the entire HF band and process multiple frequency channels in parallel, coupled with the expanded azimuth coverage of a DDRx-per-element array, may be exploited to perform a number

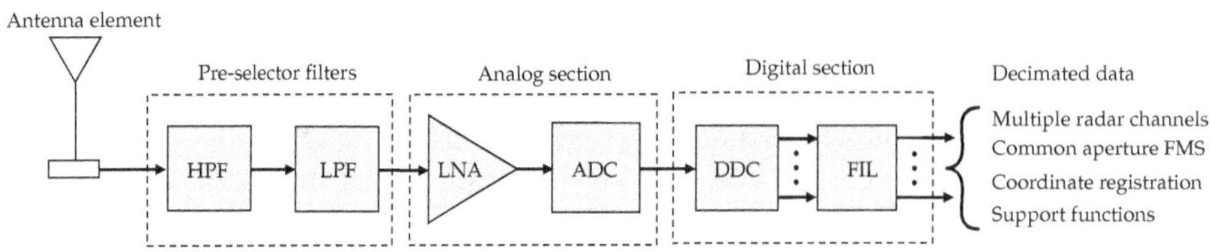

FIGURE 3.20 Simplified block diagram of a direct digital receiver that samples the entire HF band. A high-pass filter (HPF) may be used to attenuate strong broadcast signals below the HF band. To prevent aliasing, this is followed by a low-pass filter (LPF) with a cutoff frequency that may exceed 30 MHz. A low-noise amplifier (LNA) with variable gain is used to control the signal level input to the ADC. The digital outputs are in-phase and quadrature samples that are subsequently fed into multiple digital downcoverters (DDC). The digital section of the receiver can be configured to simultaneously downconvert and filter signals in multiple frequency channels. The output data rates are typically decimated according to the channel bandwidths with the down-sampled data being passed on to the processing stage. Possible applications for the data streams acquired in parallel are identified on the right hand side.

of different OTH radar surveillance missions, common aperture frequency management tasks, and other radar support functions, all at the same time.

Figure 3.20 shows a simplified block diagram of a DDRx with an analog preselector filter. When the feeder cables between the antenna element and equipment shelter are long, a low-noise RF amplifier may also be inserted at the base of the antenna (not shown). The preselector filter may be implemented as the cascade of a high-pass filter (HPF), which provides attenuation for signals in the broadcast bands (below about 2 MHz), and an anti-aliasing low-pass filter (LPF) with a cut-off frequency determined by the ADC sampling rate. The preselector filter output is amplified and digitized to produce in-phase and quadrature components (i.e., complex samples). The samples at the ADC output can be fed into multiple DDC channels, each of which extracts a selected portion of the HF spectrum using a numerically controlled oscillator, low-pass filter, and integer downsampler. The multiple simultaneous frequency channels can then be passed onto a number of processing stages at the output of the DDRx.

A DDRx with multiple downconverters allows the system to simultaneously acquire and process signals received at different frequencies at the spatial resolution of the full receiving array. Note that the transmit resource still needs to be partitioned for simultaneous OTH radar operation at different carrier frequencies, but the full gain and azimuth coverage of the receive array is available for servicing the radar task on each frequency. For aircraft detection in particular, this can provide greater flexibility to trade sensitivity for coverage. For example, when using the half-aperture mode of operation on transmit, the DDRx approach would not incur the additional 3-dB SNR penalty on receive since the full aperture may be used. Designs based on single analog heterodyne receiver per element can only work in this mode if the receive array is also operated as independent halves. This leads to a 3-dB loss in sensitivity against white-noise, not to mention a commensurate reduction in beamwidth, which degrades the rejection of spatially structured disturbance signals and target localization accuracy.

The ability to receive on different frequencies without compromising the spatial resolution of the full receive aperture can potentially allow aircraft and ship detection tasks to be scheduled concurrently on the radar time-line. As slow-moving target detection is primarily clutter-limited, the transmit resource could be mainly devoted to aircraft detection with only a few radiating elements used for ship detection (e.g., less than half-aperture provided the transmitter has a digital waveform generator per element architecture). The DDRx could simultaneously acquire radar signals at different frequencies optimized for the air and ship detection tasks and processes them in parallel using the full receive antenna aperture to improve both SNR and SCR. With a heterodyne receiver, ship detection tasks using the full receive aperture would need to interleaved in sequence with air tasks on the radar time-line. This approach is not very effective in practice due to the long CPIs needed for ship detection and high region revisit rates required for aircraft tracking.

Besides opening up the possibility for multiple radar channels and different mission types using the full receive aperture, a DDRx-per-element receiver architecture could also support frequency management functions on a common aperture. This includes the possibility to perform high-fidelity propagation-path assessment by combining coherent backscatter sounding and passive frequency channel evaluation data to rank channels based on direction-sensitive SNR measurements. Clutter Doppler spectrum assessment could be performed at the spatial resolution of the main OTH radar without consuming resources needed for its surveillance duties, while OI/VI sounding would yield a detailed appraisal of mode structure for the interrogated ionospheric paths. A frequency management system based on common aperture measurements is likely to provide higher fidelity real-time advice to guide OTH radar operation compared to using a co-located but scaled-down FMS array with an aperture that is an order of magnitude narrower. In addition, downconverters may potentially be allocated to remote-sensing tasks, CR-related functions (e.g., land-sea mapping), and passive radar using emitters of opportunity, for example.

The larger number of spatial degrees of freedom and wider instantaneous azimuth coverage of a DDRx-per-element architecture also significantly improve the scope for adaptive processing to mitigate unwanted disturbances including interference and clutter. Relative to a heterodyne receiver, where selectivity is achieved by analog filters at IF and baseband, the narrowband section of a DDRx is fully digital and exhibits relatively less susceptibility to degradations in spatial dynamic range caused by nonidentical reception channels in the array. Having access to the decimated A/D samples containing the full radar signal bandwidth and perhaps adjacent frequency channels also permits a wider range of signal-processing techniques to be used with respect to an analog receiver that implements stretch processing. In particular, an extended set of range cells can be formed, which increases the instantaneous coverage when ionospheric propagation conditions are favorable.

3.2.3 Array Calibration

The receive antenna pattern resulting after conventional array beamforming is required to have low sidelobe levels, particularly in the presence of strong and directional interference or clutter. In practice, array calibration errors may arise due to instrumental uncertainties in a multi-channel antenna system, as well as environmental factors local to the receiver site. Array calibration errors ultimately lead to discrepancies between the presumed and actual measurement characteristics of the receive antenna. Such discrepancies have the potential to significantly raise sidelobe levels and distort the main lobe characteristics of the resulting beampattern. An ability to estimate and apply compensation for different types of array calibration errors is therefore essential to successful OTH radar operation, especially since target detection performance depends heavily on the ability to effectively suppress off-azimuth disturbance signals.

At the antenna-sensor level, array calibration errors may arise due to nonidentical element radiation patterns (gain and phase response), heterogeneity of ground electrical properties in the vicinity of the aperture, mutual coupling between the elements of the array, and uncertainties in relative sensor positions, for example. Following the antenna elements, internal array calibration errors may arise due to gain and phase mismatches in the feeder cables, nonidentical wideband receiver front-ends, and differences between receiver transfer functions over the radar signal bandwidth (i.e., differences between passband frequency responses from one channel to another). The latter can be a significant issue in the narrowband section of an analog (heterodyne) receiver.

The first part of this section broadly classifies different types of array calibration errors and briefly discusses their potential impact on useful signal detection and estimation. This motivates the second and third parts of this section, which respectively describe the use of internal and external signal sources to estimate compensation for the different types of array calibration errors. To simplify the discussion, attention is focused on the analog heterodyne receiver, as the advantages of using direct digital receivers to reduce potential sources of internal array calibration errors have already been mentioned. In addition, the intent of this section is to describe and contrast the relative merits of different array calibration approaches, rather than to delve into specific algorithms. The reader is referred to Farina (1992) and Lewis, Kretschmer, and Shelton (1986) for detailed descriptions of array calibration principles and techniques.

3.2.3.1 Impact of Calibration Errors

Array calibration errors can be broadly classified into two categories; those which occur due to the nonideal characteristics of the antenna element and the wideband front-end of each reception channel (i.e. the "broadband" components), and those which occur after demodulation in the selective part of the heterodyne receiver up to the A/D converter output (i.e., the "narrowband" components). The antenna sensors, feeder cables, and low-noise amplifiers in the front-end are so-called broadband components because they are designed for operation over wide bands of the HF spectrum, while the analog IF bandpass and anti-aliasing filters in the signal paths leading into the A/D converter are narrowband components as they are designed to pass only signals within a frequency band that is small compared to the carrier frequency. Calibration errors involving broadband components may vary with carrier frequency and signal DOA, but are effectively independent of passband frequency. On the other hand, calibration errors involving narrowband components typically vary over the passband, but are often independent of carrier frequency and signal DOA in the heterodyne receiver architecture.

Antenna array imperfections that can limit OTH radar performance mainly arise due to nonidentical element radiation patterns (gain and phase), mutual coupling effects, ground electrical properties, and small errors in relative sensor positions. Differences in the characteristics of amplifiers and cables from one reception channel to another in the wideband front-end can also limit performance as these errors superimpose on the above-mentioned antenna array imperfections. Unlike the latter, calibration errors due to broadband components behind the antennas are independent of signal DOA.

Passband-independent calibration errors are often referred to as array manifold errors (Ng, Er, and Kot 1994) or array steering vector mismatches (Compton 1982). Such errors effectively cause the perceived signal wavefront to differ from the incident one by a

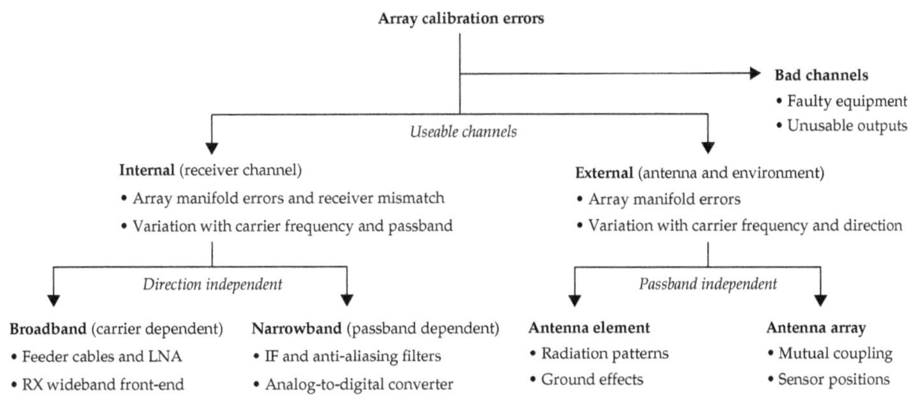

FIGURE 3.21 Potential sources and characteristics of array calibration errors. Internal sources occur due to hardware imperfections in the reception channels behind the antenna elements, while external sources arise from the uncertain properties of the antenna elements and the interaction of the array with the local environment. The term receiver mismatch refers to differences in the channel transfer functions across the radar signal passband, while array manifold errors are considered effectively constant over the receiver passband. Internal calibration errors may be carrier-frequency and passband-frequency dependent due to the presence of both broadband and narrowband components in each reception channel, respectively, but these errors are effectively independent of signal DOA. On the other hand, array manifold errors caused by antenna element and local environment uncertainties vary with carrier frequency and signal DOA, but are effectively independent of passband frequency due to their broadband characteristics. © Commonwealth of Australia 2011.

multiplicative complex scalar at each reception channel in a narrowband system. Array manifold errors contributed by the feeder cables, low-noise amplifier, and wide-band front-end of the receiver are independent of signal DOA and may be corrected by applying a single complex (gain and phase) digital correction per reception channel that varies with carrier frequency. On the other hand, array manifold errors contributed by the antenna elements will additionally depend on signal DOA, in general. The signal-processing area concerned with estimating compensation for array manifold errors is commonly known as "array calibration."

When mutual coupling dominates the direction-sensitive component of array manifold errors in a ULA, the nature of this contribution can usually be approximated in the form of a banded matrix due to the electromagnetic coupling of any given antenna element with a limited number of its neighboring elements. In the case of homogeneous mutual coupling, this banded matrix exhibits a Toeplitz structure, and its effect can be minimized by properly handling the elements near the ends of the ULA. This may involve deploying dummy elements to augment the ULA and/or applying a low-sidelobe array taper with a low-amplitude weighting for the end elements. The problem of estimating compensation for non-homogeneous mutual coupling is a difficult problem in practice. The reader is referred to (Solomon, 1998) for a detailed discussion of this topic in relation to OTH radar.

Array manifold errors can degrade OTH radar performance in several respects. This includes: (1) loss of coherent gain against spatially uncorrelated noise for target echoes, which adds on to SNR losses caused by straddling and tapering, (2) main lobe distortion

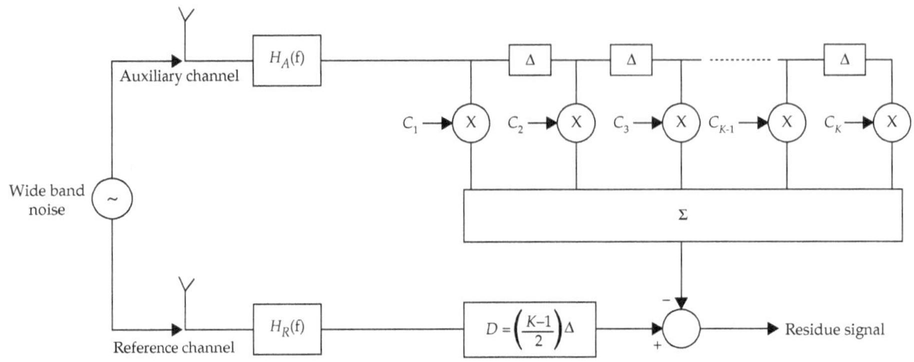

FIGURE 3.22 A two-channel tap-delay line compensator (adaptive processor) configured in self-cancelation mode and excited by a broadband noise input. The tap-delay line weights that best compensate for receiver mismatch minimize the output power of the residue signal.

and beam pointing errors, which can reduce the DOA estimation accuracy of peak interpolation procedures, and perhaps most importantly, (3) higher sidelobe levels, which increases the vulnerability of the system to useful-signal masking by off-azimuth interference and clutter. These issues, which affect signal detection and estimation, provide strong motivation to compensate for array manifold errors. Figure 3.21 summarizes the potential sources and characteristics of array calibration errors that may be compensated using internal and external signal sources.

Aside from conventional beamforming, array manifold errors can lead to useful signal cancelation when adaptive beamforming is applied, particularly in the case of unsupervised training. This occurs because array manifold errors represent mismatches in the mathematically defined array steering vector model, which is used to protect the useful signal from being canceled (Kelly 1989). However, array manifold errors do not effect the interference cancelation performance of adaptive beamformers or sidelobe cancelers because the adaptive algorithm automatically adjusts to fixed amplitude and phase distortions of the interference wavefront (Farina 1992). In supervised training applications, where useful signals are assumed to be absent from the data snapshots used to adapt the weight vector, relatively small array manifold errors are not expected to significantly degrade useful signal reception.

Now let us turn our attention to imperfections arising in the narrowband section of the heterodyne receivers. The transfer functions of the analog filters in each receiver can deviate or "mistune" from their nominal (desired) frequency response due to temperature drifts and the non-ideal behavior of hardware components. Unlike array manifold errors, the departures in gain and phase relative to the nominal frequency response are independent of carrier frequency in the narrowband section of a heterodyne receiver, but usually vary over the passband of interest. More importantly, the variation of these errors across the passband is generally not identical from one receiver to another. The resulting differences in frequency response over the passband of different reception channels is often referred to as receiver mismatch.

The main effect of receiver mismatch is to reduce the spatial coherence of signals acquired by the array. Even small frequency response mismatches can significantly reduce the correlation coefficient between signals at the output of a pair of channels that are excited by the same input. In a multi-channel receiver system, this phenomenon acts to reduce the spatial dynamic range of the array by increasing the spatial rank of the output signal beyond that of the input signal. In other words, a broadband signal incident on the array with a perfectly coherent (rank-one) wavefront will have a spatial covariance matrix of rank greater than unity at the narrowband receiver output when frequency response mismatch is present.

In practice, receiver mismatch can severely limit the interference nulling capability of adaptive *spatial* processors. This occurs because a reduction in the correlation coefficient between signals at the outputs of different receivers will decrease the maximum achievable cancelation ratio when a single complex weight per (spatial) reception channel is used to linearly combine the signals at the processor output. Although useful signal reception is also effected, the main consequence of receiver mismatch is that it causes an instrumental limitation on the achievable interference cancelation ratio (Abramovich, Kachur, and Struchev 1984).

Typically, receiver mismatch is compensated by inserting an adaptively tuned digital transversal filter (tap delay-line), or its frequency domain equivalent, at the output of each reception channel and probing the receiver passbands with a common reference signal (Monzingo and Miller 1980). However, calibration techniques based on internal signals do not correct for uncertainties at the antenna-sensor level, or those in sections of the receiver chain prior to the location where the signal is injected. Ideally, compensation is required for array manifold errors as well as receiver mismatch in order to equalize the global transfer functions of all reception channels. This can be achieved by using a mixture of internal and external signal sources for array calibration.

Thus far, we have discussed the various types of array calibration errors, their potential impact on performance, and the form of compensation required, based on the assumption that none of the antenna elements or receive channels in the array are faulty to a point that is beyond use. In large OTH radar receiving arrays, it can occur that a number of elements or reception channels fail, resulting in outputs that cannot be used for spatial processing. For this reason, procedures are required to identify "bad" channels and to devise an appropriate remedy for array beamforming. One approach is to modify the array taper weights based on the remaining "good" antenna elements until the faulty equipment can be repaired. Alternatively, techniques based on linear prediction (discussed in Chapter 4) may be applied to generate synthetic but statistically representative data for the missing elements. The deleterious impact of a faulty receiver connected to an antenna element near the middle of the array can be reduced by switching in a good receiver (previously connected to an end element of the array) in its place.

3.2.3.2 Internal Calibration Sources

Internal signals are typically injected to calibrate the reception channels of the array on an individual basis with respect to a desired frequency response characteristic. The standard procedure is to inject calibration signals behind the antenna elements at one or more points along the receiver chain using a well-calibrated signal distribution network, with one of the calibration ports being situated as close as practicable to the antenna feed point (Pearce 1997). The wideband section of each reception channel may be probed using a single tone at the carrier frequency required for OTH radar operation, while the narrowband sections need to be probed by signals that excite the full receiver bandwidth.

The open-loop receiver outputs are analyzed to estimate digital corrections as a function of receiver number and passband frequency bin in the form of a two-dimensional calibration table. Calibration tables may be computed and applied at regular intervals to mitigate temporal variations in the reception channel characteristics. New calibration tables also need to be calculated at the start of a new mission or task when the operating frequency is changed, particularly if a recent calibration table computed for a nearby frequency is not available. Internal signals can in principle enable accurate calibration of all components in the signal path downstream from the injection point (calibration port), but such schemes cannot correct for imperfections in the antenna elements.

Internal signals used to probe the receiver passband may be deterministic (e.g., a frequency shifted set of CW tones or a linear FMCW signal) or stochastic (e.g., bandlimited white noise). The latter has been used to estimate digital compensation in an HF multi-channel receiver system, and to assess the performance of a calibration solution for the array in terms of both plane-wave and receiver-rejection tests (Frazer 2001). Passband-dependent errors due to receiver mismatch are traditionally compensated using tap-delay lines. The basic aspects of this scheme are now briefly recalled.

A two-channel model of a tap-delay line compensated adaptive processor is shown in Figure 3.22. The reference and auxiliary channel transfer functions are denoted by $H_R(f)$ and $H_A(f)$, respectively. A powerful source of broadband noise is injected to probe the narrowband channel transfer functions and the complex-valued transversal filter coefficients are adjusted so that the power of the difference signal at the output is minimized.

Closed-form solutions for the optimal filter coefficients may be derived based on the unconstrained least-squares criterion; see Monzingo and Miller (1980), Lewis et al. (1986), and Farina (1992). The effectiveness of the compensation is measured in terms of the cancelation ratio, which is the ratio of the input wideband noise power to the output residue power. A good compensator yields a high cancelation ratio. The achievable cancelation ratio depends on the characteristics of the frequency mismatch, the dimension of the compensator (i.e., the number of taps K), and the product between the tap-delay (Δ) and analog-filter bandwidth.

A pole-zero error model of the channel transfer functions was used by Lewis et al. (1986) to derive analytical expressions for the maximum achievable cancelation ratio as a function of the adaptive canceler system parameters. The author used these results to propose a procedure for "optimizing" the design of an adaptive canceler with respect to system parameters and constraints. The relationship between cancelation ratio and adaptive canceler system parameters was also analyzed for sinusoidal amplitude and phase mismatches in Monzingo and Miller (1980) and for triangular amplitude and linear phase mismatch by Farina (1992).

An experimental adaptive array system was used by Teitelbaum (1991) to study the effectiveness of a 31-tap transversal digital filter for channel equalization. For a two-channel system, measurements of the transfer functions over a 600-kHz bandwidth revealed amplitude and phase mismatches within plus or minus 2 dB and 30 degrees, respectively. Such mismatches were found to limit the cancelation ratio to approximately 20 dB. This ratio was restored to approximately 65 dB after equalization. These results demonstrate that the receiver mismatch does not need to be very large for the cancelation ratio to drop by 45 dB. Such a large degradation due to calibration errors can have a significant impact on system performance when powerful interference is present.

Compensation for receiver mismatch can also be applied in the frequency domain by performing an FFT on data segments at the output of the reference and auxiliary reception channels (Abramovich et al. 1984). In this case, compensation is applied via the direct weighting of FFT outputs in the auxiliary channel(s) over the frequency range of interest. The optimum weighting for a particular point in the passband is determined by the ratio of the reference to auxiliary channel response at that frequency. The statistical variability of the estimated frequency domain corrections due to finite sample effects was quantified in Abramovich and Kachur (1987). Identical compensation can be provided by an equivalent tap-delay line which has the same number of taps as the number of time samples in the FFT and a tap spacing equal to the delay between the samples input to the FFT. Under these conditions, the equivalent tap-delay line filter coefficients are given by the inverse FFT of the frequency domain compensation weights (Compton 1988b).

Figure 3.23 compares the real-data frequency responses measured for a reference and auxiliary OTH radar reception channel using internal calibration signals. For each channel, the experimental measurements were performed by injecting deterministic signals into a calibration port that contains a low-noise amplifier, feeder cable, and heterodyne receiver in the signal path.The gain mismatch between these two channels varies by up to 15% over the passband in Figure 3.23a, while the peak-to-peak phase mismatch (plotted against the right vertical axis of Figure 3.23b) is about 8 degrees.

This provides a quantitative indication of the gain and phase mismatches that may be encountered in practice for a heterodyne HF receiver. The frequency bins that sample the passband may be interpreted as range cells in the stretch-processing receiver based on linear-FMCW deramping. These passband-dependent mismatches were found to degrade interference rejection by more than 10 dB at the output of adaptive spatial processing in Fabrizio, Gray, and Turley (2001).

Figure 3.24 compares experimentally measured frequency responses for the *reference* channel at two carrier frequencies spaced approximately 1 MHz apart. These two measurements were made 12 minutes apart. Note that the gain and phase mismatches between the responses measured at two widely spaced carrier frequencies, indicated by the

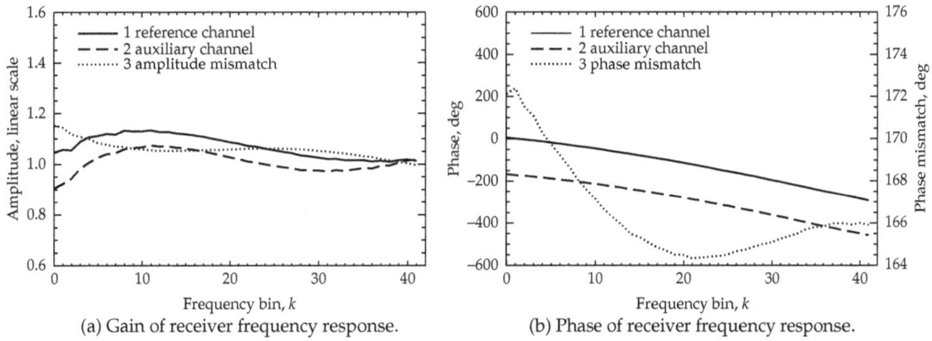

FIGURE 3.23 Gain and phase response of reference channel transfer function (curve 1) and auxiliary channel transfer function (curve 2). Curve 3 shows the gain and phase mismatch as the ratio of the reference to auxiliary channel response over the passband of interest. The phase mismatch is plotted against the right vertical axis of (b). © Commonwealth of Australia 2011.

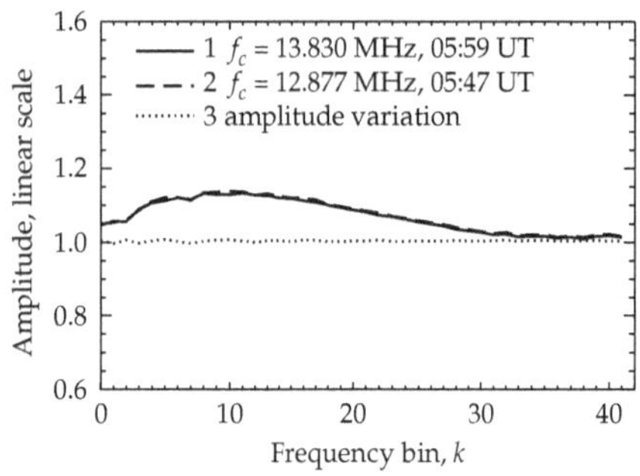

(a) Amplitude of receiver frequency response.

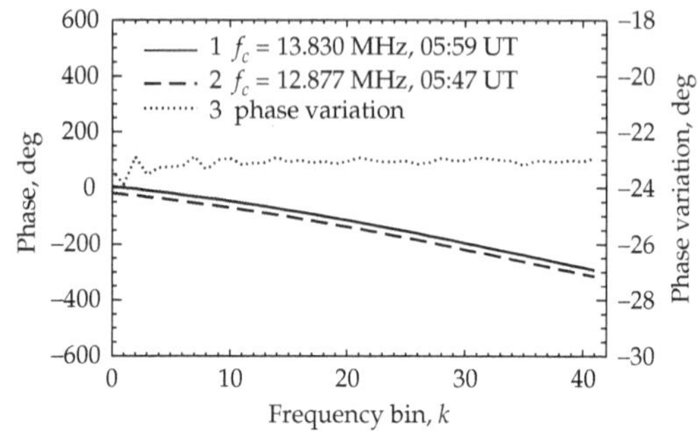

(b) Phase of receiver frequency response.

FIGURE 3.24 Curves 1 and 2 in (a) and (b) show the (amplitude and phase) transfer functions of the reference channel measured at two different carrier frequencies 12 minutes apart. Curve 3 in both plots shows the variation in amplitude and phase over the passband of interest with the latter plotted against the right vertical axis of (b). © Commonwealth of Australia 2011.

dotted lines, are effectively related by a single complex constant with negligible variation over the passband. This is expected because passband-dependent calibration errors occur in the fixed narrowband section of a heterodyne receiver where the analog filter characteristics are independent of carrier frequency. Specifically, the constant phase-offset is mainly due to the change in electrical length of the feeder cable in the signal path at the two different carrier frequencies. The difference in gain of the wide-band front-end is observed to be negligible over 1 MHz. These results show that the wideband section of the reception channel does not significantly contribute to receiver mismatch in the analog heterodyne architecture.

3.2.3.3 External Calibration Sources

The main drawback of using only internal signals for calibration is that array manifold errors due to the antenna elements cannot be observed. External sources are required to have a number of characteristics in order to be appropriate for the calibration of a very wide OTH radar receive aperture. Ideally, external sources should be far enough away to be in the far-field of the antenna array, possess high SNR that is in excess of the peak-to-sidelobe level demanded of the calibration solution, be incident on the array with sufficiently well-known spatial structures (plane wavefronts), and probe an appropriate number of azimuth and elevation angles in the OTH radar coverage. Moreover, external sources should permit the array to be calibrated at different frequencies, both during the day and night. This extensive set of competitive requirements makes the problem of OTH radar array calibration very difficult in practice.

With reference to Figure 3.25, external signals for array calibration may be grouped into near-field or far-field sources, and then be further subdivided as active or passive sources (the latter requiring illumination from the radar to produce a suitable calibration signal). The relative advantages and shortcomings of using these alternative calibration sources will be discussed below. It has been argued that external sources of opportunity for array calibration can make relocatable HF radars more cost effective and rapidly deployable with a reduced need for site preparation (Solomon et al.

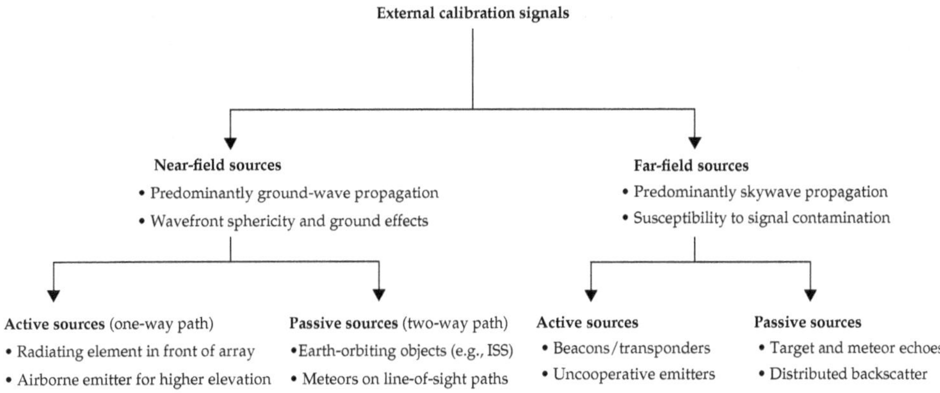

FIGURE 3.25 A (non-exhaustive) taxonomy of active and passive signal sources that can potentially be used to calibrate the receiving antenna array of an HF radar. Note that ground-wave propagation refers to the sum of space-wave and surface-wave propagation, while the acronym ISS stands for International Space Station. © Commonwealth of Australia 2011.

1998). Another motivation for using external sources of opportunity is to provide calibration measurements independent of those made using standard equipment and procedures, which become particularly valuable at times when the latter malfunction (Bourdillon and Delloue 1994).

An external calibration source may conveniently be placed a few hundred meters in front of the antenna array. However, a ground-based emitter in the near-field has limited utility for array calibration in the OTH radar context. First, the calibration signals are effectively received via ground-wave (space-wave and surface-wave) propagation as opposed to skywave propagation. In addition, surface-wave propagation is strongly influenced by inhomogeneities in the electrical properties of the ground in comparison to skywave signals received from higher elevation angles. The main issue with using such sources for array calibration is that manifold errors experienced by skywave signals incident from higher elevations may be quite different to those measured at grazing incidence.

Antenna radiation patterns at higher elevations may be measured and calibrated using airborne signal sources in the line of sight. In this case, the calibration signal is mainly received via space-wave propagation with a possible surface-wave contribution. For example, the signals may be radiated (or relayed using a transponder) from a hovering helicopter, where the platform moves to a different position each time a measurement is taken. A simple antenna with a well-known radiation pattern, such as a whip, may be located near the array to provide a reference. The signals received by all of the elements to be calibrated may be compared directly against those received simultaneously by the reference antenna. *Relative* antenna pattern measurements are more robust to unknown changes in propagation characteristics and other effects. However, this approach is only suitable for a one-off radiation pattern measurement of a receive or transmit antenna array rather than routine calibration.

Array calibration may also be performed using an HF radar echoes reflected from large earth-orbiting objects in the line of sight. For example, the International Space Station (ISS) was used as a (passive) point source to calibrate an OTH radar receive array in

McMillan (2011). However, such procedures are best performed at night using signal frequencies well above the maximum layer critical frequency to minimize ionospheric effects on the line-of-sight propagation path. Experimental results in McMillan (2011) demonstrate the feasibility of this approach. However, the usable frequencies are necessarily higher than those required for OTH radar operation at night because the signal needs to pass through the ionosphere, while calibration of the array at the daytime OTH radar frequencies must be performed during the night for the same reason.

Signals received from ground-based sources in the far-field of the array via one-way skywave propagation, such as those radiated by beacons, transponders, or emitters of opportunity, have also been used for array calibration. For example, a technique that uses AM radio broadcasts of opportunity to calibrate for receiver mismatch was proposed and tested in Fabrizio et al. (2001). Figure 3.26 shows the reduction of passband-dependent mismatch after applying compensation estimated from an AM source of opportunity to the auxiliary channel in the experiment discussed previously in this subsection.

Far-field point scatterers of opportunity, such as target echoes and meteor returns, were utilized for HF array calibration in Fernandez, Vesecky, and Teague (2003) and Solomon et al. (1998), respectively. In the latter study, multiple echoes were analyzed jointly to estimate compensation for array manifold errors; see also Solomon et al. (1999). This scheme relies on a sufficient number of identifiable discrete echoes, which may not be available for an array with hundreds of antenna elements. Besides this requirement, another potential problem of relying on two-way skywave backscatter for array calibration is that the echo SNR will not be high enough to yield calibration solutions with sufficiently low sidelobe levels.

Array manifold calibration techniques based on far-field sources propagated via single-hop skywave paths appear to be the most promising option as far as the use external signals is concerned. However, such techniques are based heavily on the premise that such signals impinge on the array as planar wavefronts after ionospheric reflection. This assumption may be reasonably accurate at times, but care must be taken since the ionosphere can impart significant wavefront distortions (Fabrizio 2000). The main point

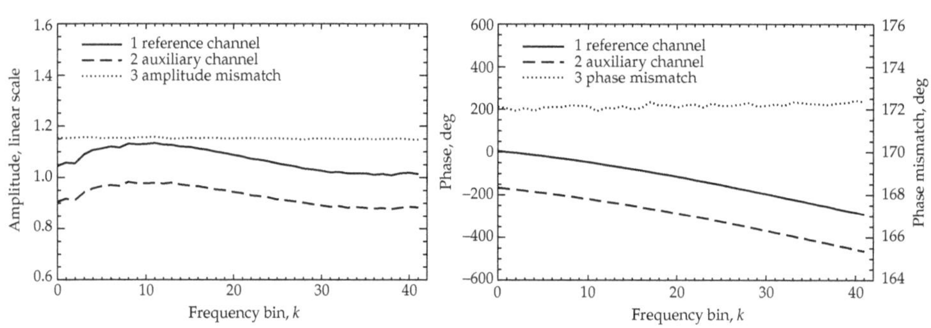

(a) Amplitude of receiver frequency response.

(b) Phase of receiver frequency response.

FIGURE 3.26 Curve 1 shows the reference channel transfer function, while curve 2 shows the auxiliary channel transfer function after compensation for channel mismatch is applied. Curve 3 shows the mismatch between the reference and auxiliary channels over the passband of interest. The phase mismatch is plotted against the right vertical axis of (b). © Commonwealth of Australia 2011.

is that calibration of array manifold errors at the antenna-sensor level using external sources is very challenging for the large antenna arrays of skywave OTH radar systems.

3.3 Frequency Management

The performance of skywave OTH radar systems is strongly influenced by the choice of operating frequency. The optimum choice of carrier frequency for a particular mission type and surveillance region location depends strongly on the prevailing HF signal-plus-noise environment and ionospheric propagation conditions. Specifically, deviations from the optimum carrier frequency in the order of hundreds of kilohertz can significantly affect the quality of skywave propagation (SNR and/or SCR). On the other hand, differences of a few kilohertz can substantially change the level of interference received from other narrowband users of the HF spectrum. System performance is not only sensitive to relatively small departures from the optimum carrier frequency, but also to waveform design and parameter selection. It follows that a poor choice of operating frequency and waveform can lead to serious degradations in target detection and tracking performance.

Statistical descriptions of the ionosphere based on extensive analysis of historical data gathered at different sites around the world can be used to predict monthly median propagation conditions for a particular skywave circuit and time. The question arises as to the utility of these empirically derived climatological models for guiding OTH radar operation. In practice, it is found that the random variability of actual ionospheric conditions with respect to median predictions is significant enough to cause poor performance of OTH radar systems. Statistical forecasts based on climatological models need to be supported by real-time and site-specific HF propagation information for more effective OTH radar operation.

To optimize the selection of carrier frequency and waveform parameters, it is considered imperative for an OTH radar system to incorporate a suite of auxiliary sensors that can evaluate ionospheric propagation conditions and monitor spectral occupancy at the time of operation from the radar site(s). In other words, the requirement is for a real-time and site-specific frequency management system (FMS) that provides OTH radar operators with high-fidelity advice to select the most suitable carrier frequency and waveform parameters for the mission at hand. The amount of operator intervention needed in this process may be reduced by incorporating a control system to interpret the sensor data and automatically configure radar operations based solely on mission requirements.

The techniques used for optimum frequency selection in OTH radar differ markedly from those used in point-to-point HF communication systems. This is because the metrics used for OTH radar frequency selection, such as signal-to-noise ratio, Doppler spectrum contamination, and multi-mode content, have relative priorities that change depending on the target class of interest. Another factor that influences frequency selection for OTH radar is need to simultaneously illuminate a large surveillance region rather than a localized area. As the value of each performance metric changes as a function of frequency, time, and location, optimum frequency selection for OTH radar would be time-consuming for operators in the absence of an automated FMS.

High-fildelity propagation-path information is also required to accurately transform radar coordinates to geographic position to produce ground-registered radar tracks. Besides the real-time uses of FMS data products in an operational OTH radar system, off-line analysis of synoptic environmental data archives that routinely accumulate calibrated measurements (even when the radar is not operating), may be used to derive quantitative estimates of radar performance which are valuable for guiding future system design.[3] This is recognized as another important role of the FMS.

This section briefly overviews the main characteristics of the FMS originally conceived for the Jindalee OTH radar in Earl and Ward (1986). The first part of this section describes the backscatter sounder and mini-radar systems, which provide information regarding signal power and Doppler spectrum characteristics for propagation-path assessment, respectively. The second part describes the spectral surveillance and background noise monitors, which identify clear frequency channels in the HF band and measure the environmental noise spectral density in different directions, respectively. Vertical/oblique incidence sounders and channel scattering function equipment, which provide detailed information on multimode-propagation structure, are discussed in the third part of this section. The reader is referred to Earl and Ward (1986) or Earl and Ward (1987) for a detailed description of the Jindalee OTH radar FMS.

3.3.1 Propagation-Path Assessment

Due to the vast coverage of OTH radar (millions of square kilometers), it is impractical to perform ionospheric propagation-path assessments over the entire search area by relying on point-to-point (vertical and oblique) ionospheric sounders alone. The large number of ionospheric control points that need to be monitored would require a dense ionosonde network, which may be prohibitive from a cost perspective. Moreover, the remote locations of many control points in the coverage often makes it infeasible or inconvenient to install and maintain such systems, particularly beyond sovereign coastlines or borders.

The technique used to measure the signal power density that is propagated by the ionosphere to a remote location and then scattered from the Earth's surface back along a similar path to a receiver located at or close to the transmit site is that of backscatter sounding (BSS) (Croft 1972). In the two-site (quasi-monostatic) configuration, the BSS transmit and receive subsystems are colocated with those of the main OTH radar facility to provide site-specific measurements of the backscattered signal power density in the form of an ionogram. The BSS ionogram is essentially a measure of the signal power density backscattered from land and sea surfaces (as well as other scatterers) in the coverage area as a function of beam direction, group-range, and operating frequency. This in turn provides a quantitative indication of the signal power density that illuminates targets in a surveillance region of the main radar at different operating frequencies. The power density of the illumination is not directly observable in this process due to the unknown backscattering coefficient of the Earth's surface.

The backscatter sounder provides an assessment of the propagation conditions in terms of the returned signal power. Combining this information with background noise spectral density measurements made at the same site allows frequency channels to be ranked in terms of SNR at a particular time and for a particular surveillance region. SNR is the primary performance metric for the detection of fast-moving targets (e.g. aircraft). However, the BSS ionogram does not provide any information about the spectral characteristics of the backscattered echoes. Propagation-path assessment in terms of clutter Doppler spectrum contamination is of primary importance for the detection of slow-moving targets (e.g., ships). For this reason the FMS includes a low-power frequency-agile miniradar

[3] OTH radar performance modeling based on environmental data is discussed in Skolnik (2008b).

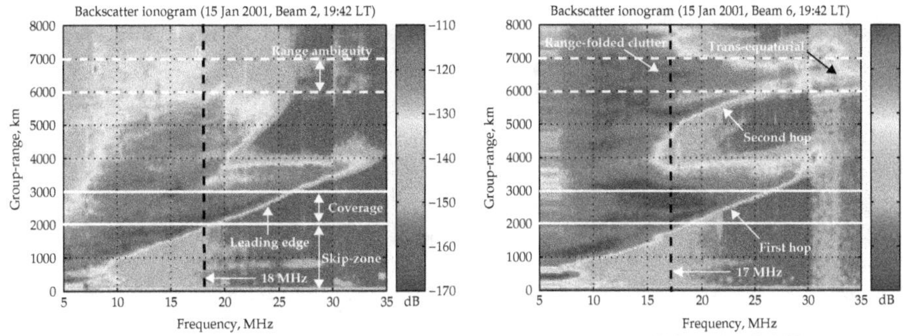

(a) Beam 2 at 19:42 LT.

(b) Beam 6 at 19:42 LT.

FIGURE 3.27 Backscatter ionograms measured simultaneously by the Jindalee FMS system in the evening (19:42 LT) of a summer day (15 January 2001) on two beams spaced approximately 45 degrees apart. © Commonwealth of Australia 2011.

system to measure the spectral characteristics of the received signal. The key features of the backscatter sounder and miniradar system used for skywave propagation-path assessment are described below.

3.3.1.1 Backscatter Sounder

Backscatter sounding represents an efficient technique for gaining an overall appraisal of propagation conditions over a wide area from the OTH radar site(s). In the Jindalee OTH radar system, a *single* vertical LDPA antenna element located at the transmitter site (Harts Range, approximately 100 km to the northeast of Alice Springs, central Australia) is dedicated to the FMS and used to floodlight the entire OTH radar coverage area to the northwest of Australia. At the receiver site (Mt. Everard, approximately 20 km to the northwest of Alice Springs), a ULA of 28 dual-fan elements are used to feed a beamformer, resulting in the generation of eight receiver beams contained within the (approximately 90-degree) arc of the FMS transmit beam.

The transmitted linear FMCW sweeps up at a rate of 100 kHz/s, such that it takes 4 minutes to record an ionogram between the frequency limits of 6 and 30 MHz for all of the eight FMS beams. Specifically, the received signal power is measured as a function of propagation delay at 200-kHz frequency increments with a group-range resolution of approximately 50 km and a group-range coverage of up to 10000 km. Note that each 200 kHz increment is used to obtain multiple estimates of the clutter power-delay profile on all 8 beams using an effective signal bandwidth of 3 kHz.[4] The long (multi-hop) distance coverage of the backscatter ionogram is useful for identifying potential sources of clutter that may fold into the OTH radar surveillance region through first- and higher-order range ambiguities. This assists the radar operator to select an appropriate waveform PRF.

Figures 3.27a and 3.27b show two backscatter ionograms recorded simultaneously by the Jindalee FMS at 19:42 LT on two beams spaced approximately 45 degrees apart. The differences between the received signal power distributions in group-range and

[4]Estimates corrupted by strong RFI are rejected, while the remaining estimates are consolidated into a single power-delay profile for each 200 kHz segment of the spectrum. See Earl (1991) for BSS RFI clean-up.

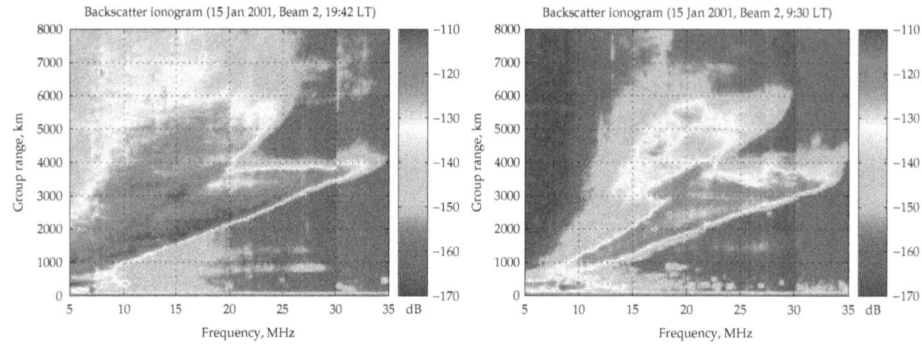

(a) Beam 2 evening ionogram at 19:42 LT.

(b) Beam 2 morning ionogram at 09:30 LT.

FIGURE 3.28 Backscatter ionograms measured by the Jindalee FMS system in the evening and morning of the same day on the same FMS beam.© Commonwealth of Australia 2011.

frequency, particularly at long distances in this example, clearly demonstrates the azimuthal dependence of propagation conditions. Figure 3.28b shows a morning ionogram recorded in the same beam as the evening ionogram in Figure 3.28a. The differences between these two ionograms illustrates the significant time-of-day dependence of skywave propagation. Some more detailed features of the illustrated BSS data examples will now be discussed.

The increase in minimum group-range of received signal backscatter with operating frequency gives rise to a so-called "leading edge," which is labeled in Figure 3.27a. This feature is typical of F-region propagation. In this nighttime example, only the F2-layer affords skywave propagation above 10 MHz and effectively no backscatter is received at group-ranges prior to the leading edge at a particular frequency. In this case, the leading edge of the F2-layer in the BBS ionogram represents the skip-distance measured in group-range as opposed to ground-range along the FMS beam direction. For a given group-range, enhanced illumination and hence backscatter due to ionospheric focussing occurs at the frequency which places the group-range immediately beyond the skip-zone. As illustrated in Figure 3.27a for a hypothetical surveillance region with a group-range extent of 2000–3000 km, the strongest backscatter is received at a frequency near 18 MHz in the direction of beam 2. In Figure 3.27b, this frequency is about 17 MHz for a surveillance region with identical group-range extents but shifted approximately 45 degrees in azimuth.

Figures 3.27a and 3.27b also show that significant clutter power is returned from long-distances via multi-hop skywave paths and possibly trans-equatorial propagation modes. Typical nighttime PRFs used for aircraft detection may lie between 30 and 50 Hz, which have first-order group-range ambiguities of 5000 and 3000 km, respectively. If the selected PRF has a group-range ambiguity that coincides with strong backscatter, potentially Doppler-spread clutter can fold into the surveillance region and obscure target echoes. It is evident from Figures 3.27a and 3.27b that the same choice of PRF can lead to strong range-folded clutter in the direction of beam 6 but not in that of beam 2. In this case, both the carrier frequency and waveform PRF need to be adjusted depending on the direction of the surveillance region.

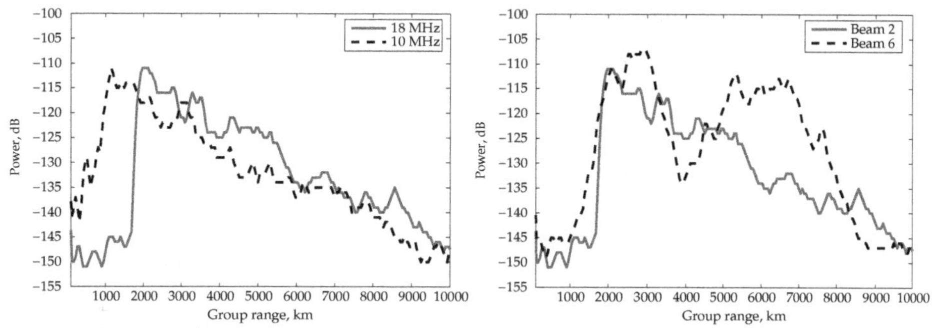

(a) Illustration of frequency dependence in beam 2.

(b) Illustration of beam dependence at 18 MHz.

FIGURE 3.29 Variation of backscattered signal power received by the Jindalee FMS as a function of group-range at two operating frequencies and in two beam steer directions. The data were recorded at 19:42 LT on 15 January 2001.© Commonwealth of Australia 2011.

Figure 3.29 shows the backscattered signal power received as a function of group-range for two different operating frequencies and beam directions at a particular time. The example in Figure 3.29a shows that the range region of strong backscatter power, which starts immediately beyond the skip-zone, moves out in group-range from roughly 1000 km to 2000 km as the operating frequency is increased from 10 MHz to 18 MHz. This clearly illustrates that the extents of the surveillance region or radar footprint with strongest backscattered signal power may be shifted in range by varying the operating frequency. On the other hand, Figure 3.29b shows that quite different backscattered signal power profiles may result for the same frequency when the direction of propagation is changed.

Lack of knowledge regarding the normalized backscatter coefficient of the Earth's surface can lead to inaccurate interpretation of the true power density that illuminates a target. The normalized backscatter coefficient varies with topography and moisture over land and can change by as much as 20 dB or more over the ocean depending on sea-state. For example, a very clam or flat sea surface results in little backscatter as more signal energy is "specularly reflected" forward. In this case, relatively weak backscatter from the region of interest does not necessarily indicate that ionospheric propagation is poor. In practice, the backscatter ionogram provides a good relative indication of the signal power density that illuminates different regions of the OTH radar coverage as a function of frequency. Importantly, the ability to perform a relative assessment allows the frequencies to be meaningfully ranked for a particular surveillance region on the basis of signal power.

A BSS system generally requires higher transmit power to generate ionograms compared to VI/OI sounders which operate over a one-way point-to-point circuit. This is to compensate for the much greater two-way path spreading loss, and the greater number of passages through the absorptive D-region during the day. Losses due to surface reflection combined with low backscatter coefficients (particularly for relatively flat surfaces at near grazing incidence) also contributes to a reduction in received echo strength. A BBS system may need to operate with a transmit power of 10–20 kW and a receive-transmit antenna gain product of perhaps 10–20 dB. It is also worth noting that backscatter sounding techniques are not limited to OTH radar. They are also used in HF communications for broadcasting effectively to sites that are required to maintain radio silence.

3.3.1.2 Mini-Radar System

The BSS provides a wide-band assessment of propagation conditions over the entire radar coverage in terms of backscattered signal power, but it does not provide information regarding the Doppler spectrum characteristics of the received echoes. To assess the spectral characteristics of the returned signals, the FMS incorporates a low-power frequency-agile "miniradar" capable of measuring the Doppler profile of backscattered energy. This system interrogates a number of narrowband channels at well-spaced carrier frequencies that sample a selected portion of the HF spectrum. For each frequency channel probed, the mini-radar measures the distribution of backscattered signal power in group-range, Doppler frequency, and beam direction. In essence, the miniradar output is a conventional range-Doppler map of the received signals, which is evaluated for each interrogated frequency in each of the eight FMS beams.

The spectral characteristics of the received signal depend on whether the echoes are backscattered from land or sea, and on the properties of skywave propagation over a two-way path. In the absence of spectral contamination from the ionosphere, the Doppler spectrum of echoes returned from land would be characterized by a single peak at zero hertz (broadened by the point spread function of the system), whereas that returned from the sea would appear as two dominant peaks (possibly asymmetrically spaced about 0-Hz due to surface currents) in addition to a second-order continuum. Each propagation path through the ionosphere modifies these ideal Doppler profiles by introducing a frequency offset (Doppler shift) due to the regular component of signal phase-path variation over the CPI, and a spreading (broadening) in Doppler due to random variations of the signal phase-path over the CPI.

In practice, echoes from slow-moving surface targets often need to be detected against a disturbance background dominated by clutter after Doppler processing. The Doppler-spectrum characteristics of clutter received in the same range-azimuth resolution cell as a target echo vary markedly with operating frequency. The choice of operating frequency can dramatically effect detection performance in a clutter-limited environment. Doppler spectrum contamination therefore becomes a primary performance metric for OTH radar frequency selection in ship-detection applications. Stated another way, the spectral purity of the returned signals is a more appropriate indicator of channel quality than SNR for such missions. Importantly, the most frequency-stable channels are usually not the same as those providing maximum SNR for a given ionospheric circuit.

The miniradar in the Jindalee FMS uses the same transmitting and receiving facilities as the BSS. In order to provide Doppler information for a particular frequency channel, a narrowband repetitive linear FMCW is transmitted and received by the system over a designated dwell time. The waveform repetition frequency, swept bandwidth, and dwell time are variable over a range of values, as is the number and spacing of the carrier frequencies interrogated within the HF band. To guide frequency selection for ship detection, the mini-radar waveform parameters might be set to 5-Hz WRF, 20-kHz bandwidth, and 25.6-second dwell time (128 sweeps), for example, while the carrier frequency spacing may be around 1 MHz subject to clear channel availability. The backscattered echoes are received on eight FMS beams and subjected to range-Doppler processing after downconversion and filtering.

The angular resolution of the Jindalee mini-radar is 16 times less than that of the main OTH radar. Its broader beam therefore samples backscatter from a comparatively larger volume of irregularities in the ionosphere. This tends to increase variability among the illuminated irregularities, which may lead to a pessimistic estimate of the spectral

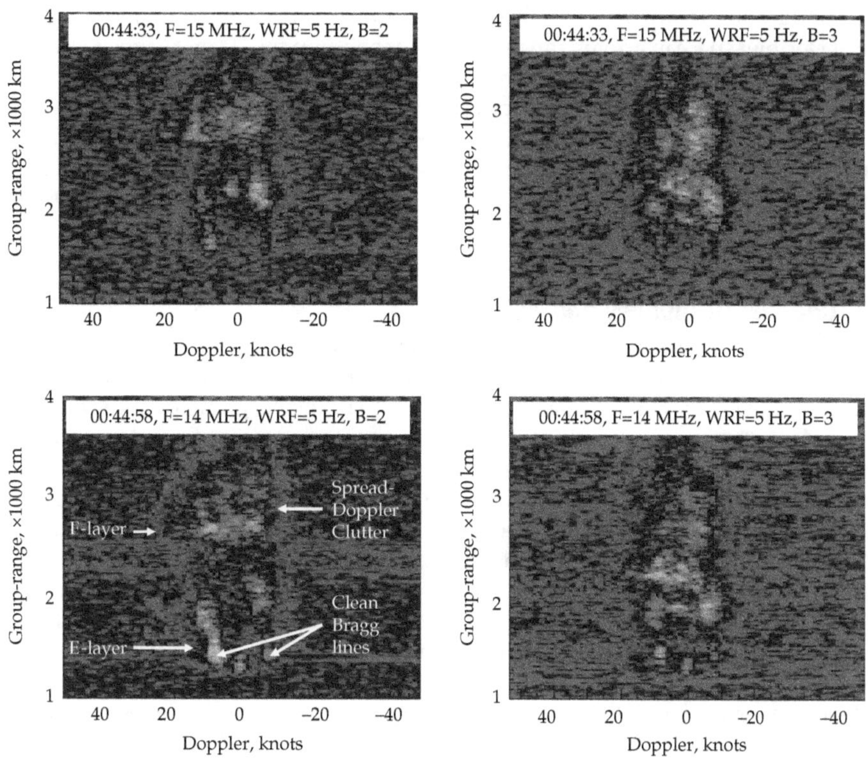

FIGURE 3.30 Examples of Jindalee mini-radar data showing range-Doppler maps of backscattered echoes at a frequencies of 14 MHz and 15 MHz in FMS beams 2 and 3 (spaced approximately 10 degrees apart). © Commonwealth of Australia 2011.

contamination observed on the main OTH radar. Nevertheless, the mini-radar is a valuable tool for guiding frequency selection in the main radar because it can effectively determine the relative suitability of different frequencies for ship-detection tasks. An automated frequency advice system for OTH radar ship detection based mainly on miniradar measurements is described in Barnes (1996).

Figure 3.30 shows mini-radar displays recorded at two frequencies spaced 1 MHz apart and in two adjacent FMS beams spaced approximately 10 degrees apart. The clutter Doppler spectrum clearly varies with range and beam as the properties of the skywave propagation paths change. The ensemble of mini-radar range-Doppler maps acquired across eight FMS beams for all interrogated frequency channels enables an operator to assess the spectral characteristics of the propagation path relevant to the main OTH radar surveillance region. The ability to rank different frequency channels based on Doppler spectrum contamination using real-time site-specific mini-radar measurements greatly assists operators to optimize frequency selection in ship-detection applications.

In a multipath environment, clutter echoes backscattered from a different patches of the Earth's surface may be received with the same group-range and cone angle. The superposition of clutter echoes received via multiple propagation paths with different

Doppler shifts and spreads will effectively contribute to broadening the clutter-occupied region of Doppler space in a given range-azimuth resolution cell. Target echoes associated with a dominant propagation mode can potentially be masked by clutter returns from a different propagation mode that is Doppler shifted relative to the dominant mode. For ship detection, the vulnerability of a channel to multimode-induced spectral contamination should also be considered for frequency selection. This type of contamination can degrade detection performance even when the individual modes have small Doppler spreads.

3.3.2 Channel Occupancy and Noise

In contrast to broadcasting, communication, and a number of other services in the HF band, OTH radars are not allocated frequencies but are permitted to use broad bands of the HF spectrum on a secondary basis. Permission to radiate is granted on the condition that the radar causes no discernable interference to other users of the spectrum, and that operation on protected frequencies (e.g., emergency channels) is forbidden at all times. In other words, OTH radars operate with strict adherence to a policy of noninterference by finding clear frequency channels of appropriate bandwidth in between other users of the HF band. Channel occupancy in the HF band not only depends on the spatial and spectral distribution of HF emitters, but also on the prevailing ionospheric conditions. As all of these factors are dynamic and unpredictable, an OTH radar requires a colocated spectrum monitor to identify clear channels in real time across the system's frequency range.

In clear frequency channels, HF background noise from external sources usually limits detection performance for fast-moving targets (e.g., aircraft). Exceptions arise when spread-Doppler clutter is present, or if the target speed and heading give rise to a low relative velocity, where clutter may limit detection performance. In remote locations, background noise is dominated by atmospheric and galactic sources, and its power spectral density changes significantly with time, location, direction, and frequency in the HF band. Knowledge of the prevailing background noise spectral density in the surveillance region directions at the receiver site is necessary for optimum OTH radar frequency selection based on the SNR criterion.

The spectral surveillance subsystem of the Jindalee FMS includes an omni-directional spectrum monitor and a directional antenna to measure the background noise spectral density. Its function is twofold: (1) to routinely evaluate HF spectrum occupancy, such that only clear channels (not in the forbidden frequency table) are identified and potentially selected for radar operation; and (2) to measure the directional HF background noise spectral density over the radar's angular coverage, such that clear channels eligible for use can be meaningfully ranked based on the SNR performance metric. The latter is achieved by combining background-noise data and backscatter-sounder data received in the same FMS beam.

3.3.2.1 Spectrum Monitor

Despite the shift of many services to satellite, microwave, and fibre-optic links, the HF band remains densely occupied. In addition to AM radio broadcast stations and long-range point-to-point communication systems, which continue to occupy large segments of the HF spectrum, there is an increasing proliferation of HF radars that require relatively wider signal bandwidths for surveillance and remote-sensing applications. To minimize mutual interference among users of the HF band, licenses to transmit are issued on a regional basis by national and international administrators that manage frequency allocations and schedules.

As described in McNamara (1991), a frequency allocation is nominally for a bandwidth of 3 kHz, which means that there are only about 10^4 channels available in the HF band to satisfy all users. In addition, most systems require access to more than a single frequency channel in order to operate effectively over diurnal, seasonal, and solar-cycle variations, which further increases demands on the finite spectrum resource. Not surprisingly, all HF band channels have been allocated many times over worldwide, at times within the same country (McNamara, 1991). To use the spectrum more efficiently, it is in principle possible for certain systems to operate independently on the same allocated frequency channel by using cooperatively designed waveforms that are "non-interfering" after processing in the receiver. For example, this principle may be adopted by cooperative networks of oceanographic HF radars.

An omnidirectional antenna located at the OTH radar receiver site is used to monitor channel occupancy in the Jindalee FMS. Using an omnidirectional (whip) antenna for this purpose ensures that man-made transmissions are not missed due to being masked in beam pattern nulls. Specifically, the HF spectrum analyzer measures the interference-plus-noise power spectral density over a frequency range of 5–45 MHz at a granularity of 2 kHz, with clear frequency channel updates provided at 2- to 10-minute intervals. For evaluation of channel occupancy, the received signal power level is evaluated in all 2-kHz channels of the spectrum. These site-specific estimates are then averaged over 10 passes and an algorithm is used to classify channels as either clear or occupied. For channels that are declared to be clear, a reliability index is ascribed based on the previous history of the channel's occupancy (Earl and Ward, 1986).

Forbidden channels are stored in a look-up table with an in-built control feature that automatically prevents them from being inadvertently selected for radar use, irrespective of whether such channels would otherwise be declared as clear. Figure 3.31 shows an example of spectrum monitor data recorded between 17 and 18 MHz during the day by the Jindalee FMS. Occupied channels are indicated along with the background noise estimate and a clear channel of 100-kHz bandwidth, as measured at the receiver site.

3.3.2.2 Background Noise

While spectrum occupancy is measured using an omnidirectional antenna, the background noise spectral density is measured on eight directional FMS beams that span the OTH radar arc of coverage. The frequency channels declared to be clear by the spectrum monitor are collectively used to estimate the background noise spectral density, with averaged values computed at increments of approximately 1 MHz across the spectrum. As the receiver site is located in a remote area, the background noise spectral density estimates in clear frequency channels are mainly due to atmospheric and galactic noise, which dominate in the lower and upper regions of the HF band, respectively.

Figure 3.32a shows the diurnal variation and frequency dependence of the estimated background noise spectral density for a particular FMS beam at the Jindalee OTH radar receive site. The background noise spectral density is observed to exhibit a relatively uniform profile in frequency around the middle of the day. However, the nighttime background noise spectral density is significantly higher at lower frequencies. In this example, the nighttime background noise spectral density in the lower HF band increases by more than 20 dB compared to daytime levels.

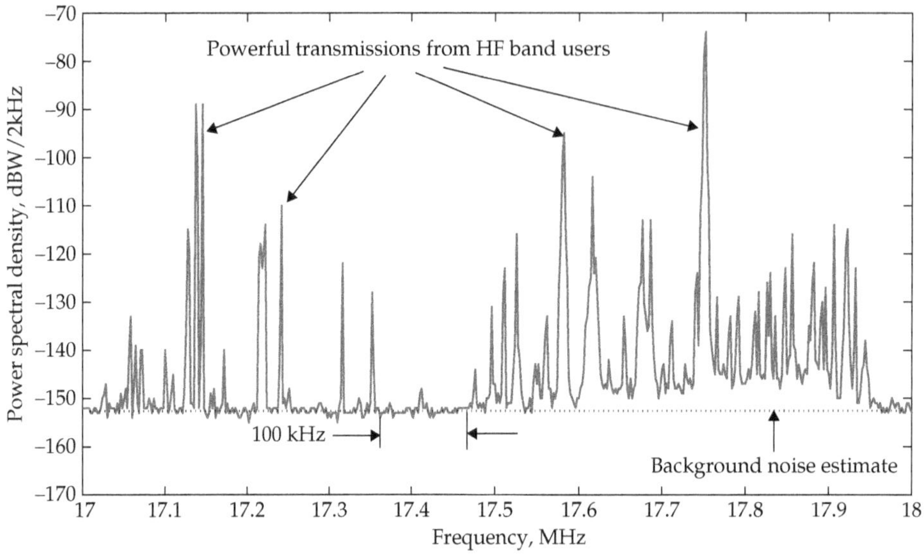

FIGURE 3.31 Practical example of power spectral density containing man-made transmissions and background noise in a 1-MHz band of the HF spectrum. The data was recorded during the day (12:31 LT, March 2002) by the spectrum monitor of the Jindalee FMS located in a remote region of central Australia, 23.532 degree (S) and 133.678 degree (E). The spectrum illustrates the presence of powerful narrowband transmissions originating from different users in the HF band and an unoccupied frequency channel with a bandwidth of about 100 kHz. The power spectral density within this channel is near the estimated background noise level. If this channel is declared to be clear after multiple passes of the spectrum, it would become eligible for OTH radar use provided it is not a forbidden frequency. © Commonwealth of Australia 2011.

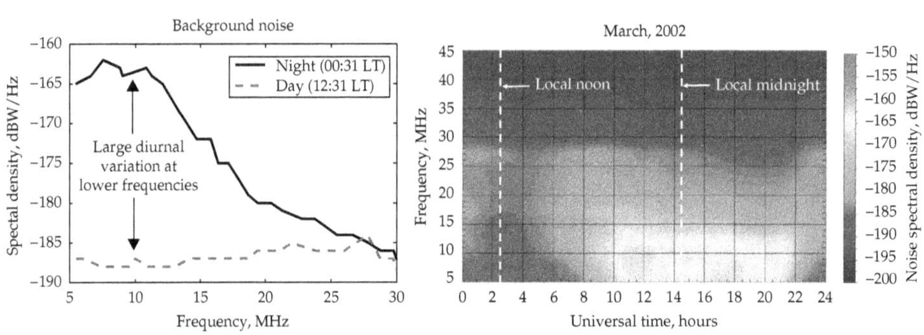

(a) Spectral densities in the day and night.

(b) Monthly median values (March 2002).

FIGURE 3.32 Real-data examples showing the diurnal variation of background noise spectral density on a March day in 2002 and the diurnal variation of monthly median background noise levels measured by the Jindalee FMS system in March 2002. © Commonwealth of Australia 2011.

Figure 3.32b illustrates the monthly median background noise spectral density as a function of frequency and time of day. The significant variation as a function of time and frequency arises due the absence of D-region absorption at night and the lower critical frequency of the F2-layer at night. This enables atmospheric noise below the F2-layer critical frequency to propagate effectively over very long distances at night. Skywave OTH radars are therefore confronted with higher background noise levels at night, when lower operating frequencies need to be used, compared to the daytime, when the ionosphere supports propagation at higher frequencies.

The combination of real-time BSS data and background noise measurements made in the same FMS beam enables clear frequency channels to be ranked on the basis of SNR for a given surveillance region. The optimum frequency yielding maximum SNR changes with time as well as the azimuth and range extents of the surveillance region in the OTH radar coverage. Note that the range of the surveillance region only changes the numerator of the SNR expression (i.e., signal power) as a function of frequency, whereas a change in surveillance region direction will change both the signal power and background noise power as a function of frequency.

In summary, the combination of backscatter sounder and background noise data is required to find the optimum aircraft-detection frequency for a particular surveillance region based on maximizing SNR. Whereas the ensemble of mini-radar clutter range-Doppler maps may be used to select the optimum ship-detection frequency for a particular surveillance region in terms of minimizing Doppler spectrum contamination. In either case, the key to providing automated frequency advice to OTH radar operators is the availability of real-time and site-specific measurements that are calibrated to allow SNR- and SCR-related performance metrics to be evaluated for all directions and ranges within the coverage.

3.3.3 Ionospheric Mode Structure

Backscatter sounding and the mini-radar represent efficient techniques for measuring the power level and spectral characteristics of the returned signal echoes, respectively, but neither system provides propagation-path information in a form that is directly suitable for interpreting the ionospheric mode structure. Specifically, backscatter ionograms and miniradar range-Doppler displays are poorly suited to the resolution and identification of propagation modes, such as the two-way E, F1, and F2 modes. For F-region layers in particular, this includes high- and low-angle rays and magneto-ionic splitting of each ray into ordinary and extraordinary waves. Moreover, "mixed" modes also arise over a two-way path, as described in Chapter 2.

A detailed knowledge of ionospheric mode structure is relevant to the interaction of the radar with a point target. A target located at a fixed ground range and great circle bearing can give rise to multiple echoes detected at different group-ranges and cone angles when the signal bandwidth or antenna aperture is sufficiently large to resolve the propagation delays or angles of arrival of the ionospheric modes present. A sounder capable of identifying the ionospheric modes propagating over a particular skywave circuit, and determining the virtual height(s) of the reflecting layer(s), is not only required to reliably associate radar detections with individual targets, but also to accurately convert group-range to ground-range and cone-angle to great-circle bearing (even in the case of a single propagation mode).

An oblique incidence (OI) sounder provides a direct observation of received signal energy as a function of propagation delay (group-range) and operating frequency over a one-way point-to-point circuit. The mode-structure for the oblique path may be readily interpreted from the resulting two-dimensional image known as an OI ionogram. The frequency-dependence and group-ranges of so-called "mixed-modes" that would occur for the same oblique path over a two-way circuit may be inferred from one-way measurements by convolving the ionogram with itself, i.e., by invoking the principle of reciprocity. A vertical incidence (VI) sounder is based on the same principle as an OI sounder but observes the overhead ionosphere at near vertical incidence using appropriate transmit and receive antennas separated by a small distance.

Although OI/VI ionograms can assist with frequency selection for a particular mission, the main role of the deployed OI/VI sounders is to routinely probe strategically chosen ionospheric control points within the OTH radar coverage to facilitate the generation of a real-time ionospheric model (RTIM). A site-specific RTIM in the region of interest is valuable for associating multipath tracks to individual targets as well as for coordinate registration (CR), i.e., the process of converting tracks from radar coordinates to geographic positions. This section discusses OI and VI sounders and their relationship to the RTIM. The channel scattering function (CSF), which provides delay and Doppler information for the ionospheric modes present in a single narrowband frequency channel at a time, will be described at the end of this section.

3.3.3.1 Vertical and Oblique Sounders

The Jindalee OI/VI sounder systems are based on a linear FMCW that is swept through the HF band at a rate of 100 kHz/s. In the original Jindalee FMS, the VI sounder used a delta antenna at the transmit site to direct signals vertically upward such that the antenna at the receiver site acquired the echoes returned by the ionosphere after a near-overhead reflection. For the oblique path, a remote low-power broadband transmitter feeding a bowtie antenna was used. In both cases, the received signal is downconverted, filtered, and analyzed in 610-ms time frames, which corresponds to an effective bandwidth of 61 kHz and a one-way group-range resolution of 4.9 km. OI/VI ionogram traces covering the full HF band may be updated in under 5 minutes. "Clean-up" algorithms are typically used to remove narrowband RFI in order to display the ionogram traces more clearly.

Figure 3.33 shows example OI ionograms recorded by the Jindalee FMS system for the Darwin to Alice Springs HF link (over a ground distance of approximately 1250 km) during the day and night. The variation in mode content as a function of frequency across the HF band and the change in mode content with time of day at a particular frequency is clearly evident in these examples. Inspection of the ionogram trace allows the ionospheric modes that propagate the signal at a particular frequency to be identified, as described in Chapter 2. The virtual reflection height corresponding to each ionospheric layer affording propagation may also be estimated from the power-delay profile due to the known positions of the transmit and receive sites. Automatic recognition of specific trace features enables ionospheric parameters such as the layer critical frequencies and heights to be extracted using software tools from the recorded ionograms without the need for manual interpretation.

OI sounders may be used to determine the strength and time dispersion of ionospheric modes to optimize frequency selection in long-range point-to-point HF communication systems. Link quality for certain systems can be improved by selecting carrier frequencies where skywave propagation is effectively due to a single dominant mode of adequate

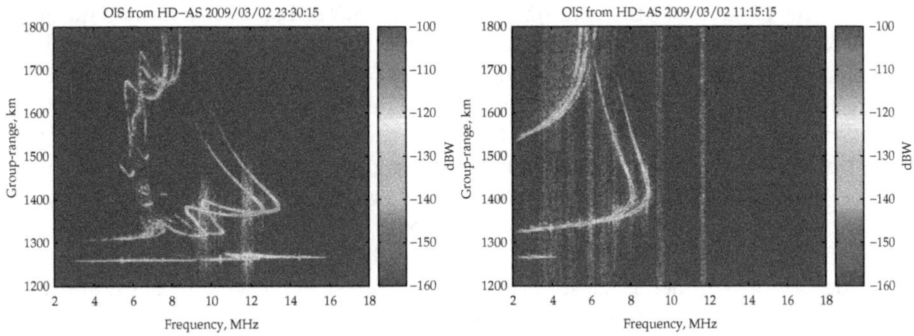

(a) Daytime ionogram (14:30 LT). (b) Nighttime ionogram (02:15 LT).

FIGURE 3.33 OI ionograms recorded by the Jindalee FMS for an ionospheric circuit linking a transmitter near Darwin (Northern Australia) and a receiver near Alice springs (central Australia) over a ground distance of approximately 1250 km. © Commonwealth of Australia 2011.

SNR. In Figure 3.33b, the best range of operating frequencies for almost single-mode propagation occurs between 6 and 7 MHz. Over this range of frequencies, propagation is mainly due to the low-angle ray of the F2-layer. Both magnetoionic components are present, but cannot be resolved in group delay. In OTH radar applications such as ship-detection and remote sea-state sensing, an OI sounder located near or within the surveillance region can at times help to select frequencies that reduce multimode-induced Doppler spectrum contamination.

However, the main benefit of OI/VI sounder data to OTH radar is not so much for frequency selection as it is to generate and update the RTIM. This requires a coordinated network of OI/VI sounders that routinely interrogate a strategically chosen set of ionospheric control points within the OTH radar coverage. When the end point of a desired oblique path is not accessible for deploying an OI sounder station (e.g., over the ocean), a VI sounder located at the mid-point of the desired path can be used to monitor the ionospheric control point and predict the mode content of all oblique circuits with the same midpoint.

An OI ionogram provides accurate information on ionospheric mode structure for skywave paths with terminals close to those of the measured circuit. Of particular interest is the density of the network required to establish a high-fidelity RTIM over the OTH radar coverage. This depends in part on the extent to which the mode structure identified on a particular path is relevant to adjacent paths. As the number of sounders that can be deployed is far less than the number of control points that need to be monitored for OTH radar, RTIMs often need to extrapolate experimental measurements to predict the state of the ionosphere at locations where no real data is available.

Typically, VIS data may be used to predict F2-layer behavior to within 300–400 km for a quiet mid-latitude ionosphere at a time not near the terminators. Traveling ionospheric disturbances can significantly reduce the spatial scales over which the measurements may be extrapolated with adequate accuracy. The normal E-layer and F1-layer are relatively more predictable, and quite accurate empirical formulas for both layer critical frequencies based on climatological models available. Sporadic-E is more difficult to characterize in detail, and predictions tend to break down more rapidly with distance and/or time compared to the F2-layer. As the confidence of extrapolations based on a limited number of measured data points decreases, statistically smoothed ionospheric models based on historical (synoptic) databases may be progressively blended into the RTIM.

A high-fidelity wide-area RTIM is a fundamental prerequisite for mode identification and coordinate registration in OTH radar. Inversion procedures may be applied to OI/VI sounder traces of frequency versus virtual path-length (delay) to model the variation of electron density with true height in the ionosphere. Alternatively, a multi-segment quasi-parabolic (MQP) electron density height-profile model can be fitted directly to the ionogram trace. Figure 3.34 shows a sample ionospheric model of plasma frequency against true height parameterized in terms of three physical QP layers (E, F1, and F2), smoothly joined by inverse segments to form a monotonically increasing profile (i.e., without valleys). The parameters of this model vary as the RTIM assimilates ionospheric data measured at different times and locations by the sounder network. The RTIM may

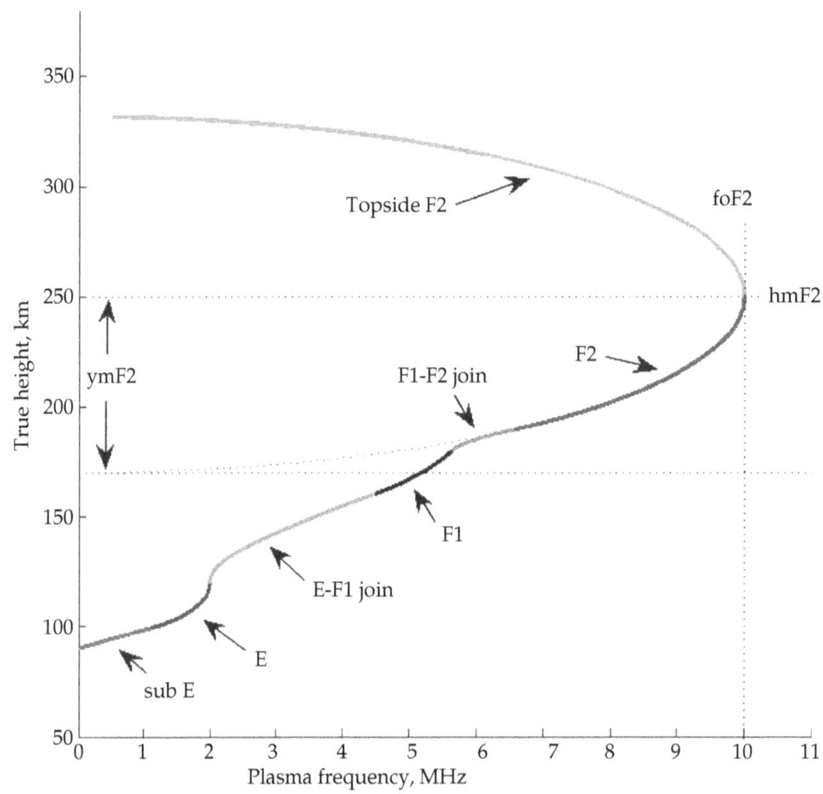

FIGURE 3.34 Example of a multi-segment quasi-parabolic (MQP) ionospheric profile model used to represent the true height variation of plasma frequency for three physical layers (E, F1, and F2). As illustrated for the F2-layer, the QP profile is defined by three physical parameters referred to as the critical frequency foF2, true height hmF2, and semi-thickness ymF2. In the MQP model, the individual layer profiles are smoothly joined by inverse segments to form a resultant electron density profile that monotonically increases with height (with the possible exception of sporadic-E). © Commonwealth of Australia 2011.

also change in response to other types of inputs such as known reference points (KRPs) and ionospheric data from climatological models.

Ray-tracing techniques may be applied to the RTIM to predict signal-propagation path characteristics over very wide areas. In particular, the MQP profile permits computationally efficient analytical ray-tracing procedures to be used (Dyson and Bennett 1988), subject to the assumption of a locally spherically-symmetric electron-density height-profile and the use of a first-order approximation for magneto-ionic (o/x) ray splitting; see Bennett, Chen, and Dyson (1991), and Chen, Bennett, and Dyson (1992). Accurate propagation-path information is required for mode-linking and coordinate registration of the OTH radar tracks.

3.3.3.2 Channel Scattering Function

The channel scattering function (CSF) provides an assessment of ionospheric propagation in a narrowband frequency channel over a one-way path by measuring the distribution of received signal power as a function of path delay, Doppler-shift, and possibly angle-of-arrival if a receiver array is used. As opposed to the OI sounder, which integrates the total received power at a particular group-range over all Doppler shifts, the two-dimensional CSF spectrum measures the delay-Doppler coupling of the received power.

Figure 3.35 shows an experimental CSF spectrum recorded by a single receiver. This example illustrates the offsets and spreads of the three dominant propagation modes

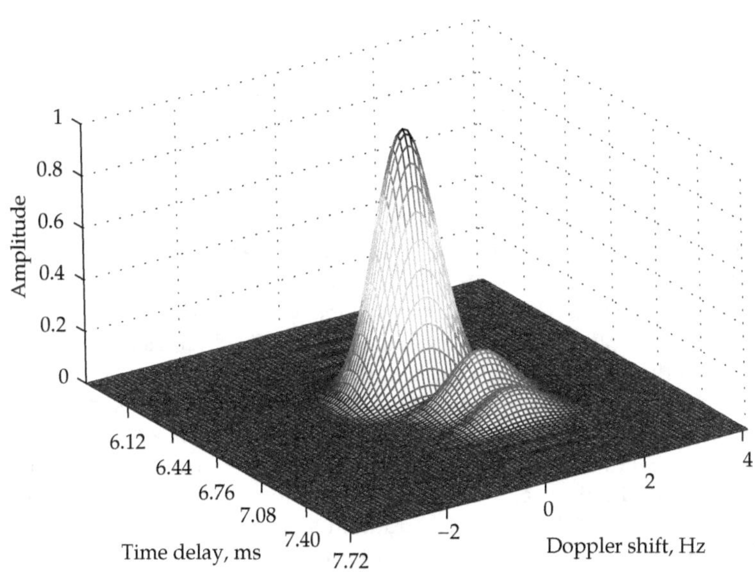

FIGURE 3.35 Channel scattering function recorded on a single receiver for an ionospheric path that linked sites near Longreach and Darwin, Australia, over a ground distance of approximately 1850 km (April 2003). The peak with highest amplitude and smallest delay is contributed by the low ray of the F2-layer, which contains the unresolved *o*- and *x*-waves. The other two peaks with progressively smaller amplitudes and larger delays are due to the resolved *o*- and *x*-waves in the high ray of the F2-layer, respectively. © Commonwealth of Australia 2011.

in the delay-Doppler plane. In other words, the CSF provides a detailed evaluation of multimode time and frequency dispersion (offsets and spreads) for a particular HF path and narrowband channel.

In essence, the CSF is configured as a one-way oblique-path radar using low-power transmissions and range-Doppler processing based on a repetitive linear FMCW swept over a narrow bandwidth. The purpose of the CSF is not to detect targets, but to provide a detailed assessment of the limits imposed on narrowband HF signal detection and estimation by the ionospheric propagation channel. The transmitted signal may be acquired on an array of antennas connected to a multi-channel digital receiver. This extends the CSF measurement capability to include offset and spread in angle of arrival, jointly with delay and Doppler (i.e., a three-dimensional spectrum of the received signal).

A detailed analysis of CSF data recorded on the very wide antenna aperture of the Jindalee OTH radar receiving array will be presented in the second part of this text to reveal the fine structure of skywave propagation and to model the angle-delay-Doppler characteristics of a quiet mid-latitude ionospheric channel. The CSF measured on an array is perhaps the most fundamental indicator of the characteristics of a narrowband skywave HF channel for an OTH radar using linearly polarized receive antennas.

3.4 Historical Perspective

This section provides a brief historical perspective on the evolution of skywave OTH radar systems in the United States (US) and Australia. The operational significance and major highlights of the most important OTH radar systems in the United States and Australia will be discussed separately to provide a chronological thread of national development. For brevity, OTH radar research and development outside of the United States and Australia will not be covered in detail. However, early HF radar work in the United Kingdom (UK) during World War 2 (WW2) is briefly recalled at the start of this section to provide background leading up to the emergence of OTH radar. On the other hand, future prospects for next-generation OTH radar systems are mentioned at the conclusion of this section to indicate the possible way ahead.

There are numerous other countries outside of the United States and Australia with previous or currently active HF radar programs, although only a few of these programs have led to the development of operational skywave OTH systems. For example, substantial efforts have been invested to develop skywave OTH radar systems in Russia, Ukraine, China, and France. A description of OTH radar developments and achievements in Russia and Ukraine can be found in Evstratov et al. (1994), while a detailed discussion of the French OTH radar "Nostradamus" is available in Bazin et al. (2006). General descriptions of OTH radar and ionospheric studies conducted in China appear in Zhou and Jiao (1994), Li (1998), and Guo, Ni, and Liu (2003).

3.4.1 Past and Present Systems

The development of HF radar during WW2 evolved almost exclusively for line-of-sight (LOS) surveillance applications. Much less attention was paid by comparison at the time to the possibility of target detection and tracking at OTH ranges. The Chain Home

(CH) system, developed on the east coast of the United Kingdom after the success of the Daventry experiment in 1935, was an operational network of integrated HF LOS radars. The real-time air picture provided by the CH system allowed the Royal Air Force (RAF) to deploy limited fighter resources effectively at the required time and location in the Battle of Britain. This decisive capability first demonstrated the operational value of radar for air-defense during wartime.

The CH system operated with a high peak-power of 350 kW (subsequently 750 kW) at frequencies between 20 and 30 MHz. The radar signal was a simple AM-pulse of $20\mu s$ duration with a repetition frequency of 25 or 12.5 Hz. The transmitted waveform was synchronized to the frequency of the mains to provide a common time-reference such that different CH stations did not interfere with each other. The CH transmit and receive systems were separated for isolation, but the radar configuration maintained a quasi-monostatic characteristic with respect to the coverage. The broad transmit beam provided "floodlight" illumination over the entire surveillance volume, while the receive antenna array of orthogonal half-wave dipoles (stacked on wooden towers 215-ft tall) was capable of azimuth and elevation angle estimation. A detailed description of the CH system can be found in Neale (1985a).

Although the CH system was not intended to detect targets beyond the LOS, skywave signals backscattered from the Earth's surface at OTH ranges were occasionally received. These long-range returns represented a source of nuisance or clutter for CH radar operators when they appeared. Later during WW2, technology became available to generate sufficient power for radar systems to operate at frequencies above the HF band. Using higher frequencies allowed the development of more compact and competitive radar systems for the LOS surveillance application relative to those implemented at HF. This motivated the transition to LOS surveillance radars that operated in the UHF and microwave frequency region.

The first operational bistatic radars capable of aircraft detection and localization using noncooperative transmissions were developed in Germany and installed at various sites across the English Channel during WW2 to exploit the floodlight illumination of the CH system. A technical description of the *Klein Heidelberg* passive radar system and the Electronic Warfare (EW) context in which it evolved is provided in Griffiths and Willis (2010). The companion paper Griffiths (2013) discusses the operational significance and performance of two other German WW2 HF radars, ELEFANT and SEE-ELEFANT. These systems were used to perform the first *active* bistatic radar experiments with the potential to detect and *localize* target echoes.

It is reported in Trenkle (1979) that an Arctic convoy of ships near Jan Mayen Island was detected using the ELEFANT radar located at Castricum, Netherlands, at a range of 2200 km, under favourable ionospheric conditions in August 1944. This report is confirmed by two other references in Griffiths (2013). It is unclear how frequently or reliably such detections could be made, or if the radar had been cued onto the targets. In any case, such a report may indicate the first ship detection by an HF radar system at OTH ranges using the skywave propagation mode.

3.4.1.1 MADRE OTH Radar

Following WW2, the Naval Research Laboratory (NRL) in the United States considered the use of HF radar with the specific intention of exploiting skywave propagation to detect targets at ranges an order of magnitude greater than was previously possible using conventional LOS radar systems. By the late 1940s, experimental investigations with this newly envisioned goal in mind were ongoing by a number of groups in several different countries (Headrick and Thomason 1998). However, a general feeling at the time was that HF radar had little use except for in the application of oblique backscatter sounding to provide a propagation-path indicator for frequency selection in long-range HF communication systems.

Due to the "look-down" geometry of the skywave propagation path, echoes backscattered from air or surface targets at OTH ranges cannot be discerned from much more powerful ground clutter returns in a single radar pulse. Besides the very large clutter-to-signal ratio, the low signal-to-noise ratio in a single radar pulse posed another major obstacle to the success of OTH radar. It was quickly recognized that OTH radar would only succeed if echoes from moving targets could be discriminated against the much stronger but "stationary" surface-clutter returns by matched-filter Doppler processing. This technique required Doppler integration of the received echo energy over a number of phase-coherent radar pulses. In principle, such processing would not only resolve the clutter issue, but also provide the valuable SNR gain needed for reliable target detection against noise.

Effective matched-filter Doppler processing required HF receivers with sufficiently high dynamic range to acquire faint target echoes and powerful clutter returns without distorting the spectral characteristics of the signals, as well as the technology to perform coherent integration of the received echoes over a processing interval. Moreover, for the Doppler processing technique to work effectively, the ionosphere must not induce significant spectral contamination on the HF radar signals reflected from one or more layers over a two-way path. For this reason, it was necessary to measure the frequency stability of the ionosphere for the propagation of narrowband HF signals over long-distance skywave paths. This represented an essential first step toward establishing the feasibility of a full-scale OTH radar in the region of interest.

In the 1950's, an experimental system capable of storing signal samples on a magnetic drum to allow playback with time compression and Doppler processing was developed at NRL based on several emerging signal-processing technologies. This system was initially designed for frequencies in the lower VHF and upper HF bands as means to increase SNR in radar systems. It had been developed at a time of general interest decline for skywave OTH radar and was not originally intended for such application (Headrick and Thomason 1998). However, when this new technology was used to examine high-resolution Doppler spectra of ground backscatter received via skywave propagation, the Earth echo was found to be well localized in Doppler frequency despite its very large amplitude. This important finding led to a resurgence of interest in skywave OTH radar and the newly developed equipment was later put to use in that direction.

In 1956, NRL completed a definitive set of experiments confirming that the necessary prerequisites to allow aircraft detection with OTH radar could be satisfied by a full-scale system (Thomason 2003). The ionosphere was found to be sufficiently frequency stable for aircraft target detection, while the ability to perform matched-filter Doppler processing of radar echoes using "magnetic drum recording equipment" (MADRE) was successfully demonstrated in 1958. These developments attracted the sponsorship required to pursue the construction of the first skywave OTH radar at Chesapeake Bay in the United States. This radar was named MADRE, which perhaps reflects the observation made in Headrick and Skolnik (1974) that "the signal processor was the key element to the success achieved with OTH radar."

MADRE began operations in the fall of 1961, and demonstrated its ability to detect and track aircraft targets on North Atlantic air routes at ranges an order of magnitude greater than conventional microwave radars. With further improvements to the dynamic range of the data processor, surface vessels were detected and manually tracked in 1967. A detailed and authoritative description of the MADRE system appears in Headrick and Skolnik (1974). Thomason (2003) provides an excellent historical account of OTH radar development in the United States, while the characteristics and applications of several HF radars, including the MADRE system, can be found in Headrick and Thomason (1998).

MADRE was a single-site (truly monostatic) OTH radar system that made use of the same antenna for transmission and reception. The transmitted waveform was based on a coherent train of simple AM-pulses with 100 ms duration and 25-kW average power. A high-power duplexer enabled signal transmission and echo reception through the reciprocal antenna. Figure 3.36 shows the two primary antennas of the MADRE OTH radar installed at the NRL field site in Chesapeake Bay. The large fixed antenna array is shown on the left of picture, while the smaller rotary antenna appears at the top of the tower on the right.

The high-gain primary antenna was a fixed (98-m-wide and 43-m-high) phased array of horizontal dipoles in corner reflectors stacked two high and ten wide. The antenna beam could be electronically steered in azimuth up to 30 degrees from boresight by adjusting the lengths of feed cables. The phasing applied to the two horizontal rows provided

FIGURE 3.36 The two primary antennas of the MADRE OTH radar system installed on the Naval Research Laboratory (NRL) field site at Chesapeake Bay (*photograph courtesy of NRL and Dr G. Frazer*).

some degree of elevation steering. The high-gain primary antenna was used for target detection and tracking at long ranges via the skywave mode. The low-gain primary antenna consisted of two in-phase horizontal dipoles in a corner reflector mounted on a 200-ft high tower. This smaller antenna was steered by mechanical rotation and routinely used to monitor US ballistic missile launches. A vertically polarized folded triangular monopole with a back screen was used to detect targets via the surface-wave mode over the sea.

MADRE is widely recognized as the first radar system to detect and track aircraft at ranges an order of magnitude greater than conventional microwave radars. This system had modest capabilities and was not intended to represent an operational prototype for wide-area surveillance. Its historical importance lies in the success of exploiting matched-filter Doppler processing for frequency resolution and coherent gain. The MADRE OTH radar represents a ground-breaking system for demonstrating the ability to detect and track targets at ranges of thousands of kilometres via the skywave propagation mode. Indeed, this system was the first to discover most of HF OTH radar's target detection capabilities.

3.4.1.2 Wide-Aperture Research Facility (WARF)

In the 1960s, students and technical staff at Stanford University designed and built an alternative skywave OTH radar system known as the Wide Aperture Research Facility (WARF) at two major field sites in the central valley of California. This revolutionary system was sponsored by the Defense Advanced Research Projects Agency (DARPA) and subsequently became a Standford Research Institute (SRI) International facility in 1967. Experimental measurements of the spatial coherence of skywave signals received by the WARF first showed that antenna apertures more than 2.5-km long could be used to form coherent beams with high gain and resolution for OTH radar systems (Sweeney 1970).

The WARF design philosophy departed significantly from that of the MADRE OTH radar and was paradigm-shifting in a number of respects. Besides employing completely different antenna elements and array designs on transmit and receive (described below), the WARF transmit and receive sites were separated by about 100 nautical miles. Indeed, WARF pioneered the use of linear frequency–modulated continuous waveforms (FMCW) in conjunction with the two-site OTH radar architecture known as the quasi-monostatic configuration. Use of a constant-modulus waveform with unit duty-cycle substantially improved radar sensitivity against noise (SNR), while the larger time-bandwidth product of such waveforms compared to AM-pulses of equivalent range-resolution enabled the out-of-band emissions to be suppressed more effectively, which made HF radar an acceptable occupant of the HF spectrum.

Unlike MADRE, which used the same vertical curtain array of horizontal-dipole antenna elements for transmission and reception, the WARF system employed a dual-band uniform linear array (ULA) of vertical LPDA elements matched to the lower and upper HF band. This dual-band feature combined with the inherent broadband characteristic of the LPDA element allowed WARF to operate over a wider frequency range than MADRE. The WARF receive system was based on 2.55-km long ULA consisting of 256 twin-whip end-fire receiving pair (TWERP) antenna elements spaced 10 meters apart (Barnum 1986). The receive elements were matched to the upper HF band, where the power spectral density of external noise is typically lower. The wide aperture of the WARF receiver array provided fine spatial resolution, which increased radar sensitivity against clutter (SCR) and spatially structured interference. It also improved target localization accuracy.

The relatively broad beam of the WARF transmit antenna was used to floodlight a surveillance region within the OTH radar coverage, while a number of high-resolution receive beams were electronically steered across the illuminated region to simultaneously acquire the backscattered echoes. The WARF system first demonstrated this operational concept, which permitted the instantaneous coverage of the OTH radar to be decoupled from the spatial resolution of the receive antenna. Besides providing an effective wide-area surveillance capability for aircraft targets, the WARF architecture significantly enhanced ship detection performance in a clutter-limited environment with respect to MADRE. WARF is of fundamental importance in the history of OTH radar as it represents a legacy system upon which currently operational designs in both the United States and Australia are based. A detailed description of WARF can be found in Barnum (1993).

3.4.1.3 AN/FPS-95 (Cobra Mist)

Following the success of MADRE, the AN/FPS-95 skywave OTH radar system, also known by its code-name "Cobra Mist," was developed jointly by the US Air Force and the Royal Air Force at Orford Ness in Suffolk, England. The missions of the AN/FPS-95 included the detection and tracking of aircraft as well as the detection of missile and satellite vehicle launchings in Eastern Europe and over western areas of the former Soviet Union. Cobra Mist was also intended to provide a research and development test-bed to determine effective OTH radar techniques for other operational missions. Site works for the project began in mid-1967 and construction was completed by the contractor, Radio Corporation of America (RCA), in July 1971.

Cobra Mist was a single-site OTH radar system that used AM-pulse waveforms, similar to MADRE, but the peak power that could be generated was much greater (up to 3.5 MW). As in MADRE, the same antenna elements were used for transmission and reception. However, the Cobra Mist system was based on 18 LPDA elements or "strings" that were 2200 ft (671 m) long and contained both horizontal and vertical dipoles. The LPDA elements were oriented to form a fan shape that radiated out from a central "hub" point to provide an approximately constant beamwidth independent of frequency. The strings were separated by 7 degrees and occupied a sector of a circle that spanned an angle of 119 degrees. Six adjacent LPDA elements were combined by means of a cable switching network to form a beam. The AN/FPS-95 was the largest, most powerful, and most sophisticated OTH backscatter radar of its time. A detailed description of the system can be found in (Fowle et al. 1979).

The HF radar community as a whole had expected that Cobra Mist would set new standards in performance and capability for the OTH radar art. In practice, quite the opposite transpired. The original plan to commence operation in July of 1972 was delayed because Cobra Mist was plagued by a serious "clutter-related" noise problem of unknown origin that severely limited sub-clutter visibility to values not much greater than 60 dB. As the source of this noise could not be conclusively identified after an extensive series of site tests, the project was terminated abruptly in June 1973. The system was subsequently dismantled and the components removed from site. Malfunction of radar equipment was exonerated from being the cause of the problem, although not with complete certainty. The order of magnitude increase in transmit power with respect to MADRE, coupled with the relatively coarse spatial resolution of the AN/FPS-95 antenna, would have required receivers with high linearity and spurious-free dynamic range to avoid self-inflicted limitations in system performance.

The hypothesis that jamming caused the clutter-related noise could not be confirmed or denied by site tests. Natural phenomena producing spread-Doppler clutter in the high-latitude ionosphere could also not be excluded as the possible cause, particularly as the antenna beam pattern was not designed to suppress such signals effectively. The declassified report (Fowle et al. 1979), submitted by MITRE engineers called upon to determine why the AN/FPS-95 did not meet performance expectations, provides a detailed account of the attempts made to locate and correct the source of the noise, which remains an enigma to this day.

3.4.1.4 AN/FPS-118 (OTH-B)

In the late 1970s, the US Air Force developed a two-site OTH radar system known as the AN/FPS-118 in partnership with General Electric. The initial phase of this project was an experimental system located in Maine that served as a technology demonstrator. The AN/FPS-118 system operated in a quasi-monostatic configuration with an inter-site separation of about 160 km. This system operated using the linear FMCW technique pioneered on the WARF, but the transmit power was an order of magnitude greater. As shown in Figure 3.37, the AN/FPS-118 adopted canted-dipole antenna designs instead of the LPDA elements used in WARF. It is argued that such antennas provided robust low-elevation performance in the presence of reduced ground conductivity, particularly when snow and ice covered the area surrounding the array.

In the early 1980s, the AN/FPS-118 transitioned from an experimental program to a full-scale operational development (OTH-B) under contract with General Electric. The US Air Force OTH-B systems were installed as three-sector radars on the east and west coasts of the United States in Maine and Oregon, respectively. The primary mission of the six OTH-B systems deployed (i.e., three east and west coast radars) was to detect aircraft and cruise missiles across two 180-degree azimuth sectors spanning a one-hop coverage area of more than 30 million square kilometers over the Atlantic and Pacific oceans.

At each OTH-B site, the transmit antenna consisted of 3 contiguous 6-band linear arrays of canted-dipoles with 12 elements per band and 41-m-high backscreens. Each 6-band linear array provided 60 degree coverage in azimuth using antenna apertures of 304, 224, 167, 123, 92, 68 m for operation over a frequency range of 5–28 MHz. The three contiguous transmit arrays were oriented 60 degrees apart to provide 180 degree

FIGURE 3.37 Canted-dipole elements used in the transmit antenna arrays of the US Air Force AN/FPS-118 (OTH-B) radar systems developed by General Electric in Maine and Oregon.

coverage in azimuth at ranges between 900 and 3000 km. The OTH-B radars operated with a maximum transmit power of approximately 1 MW using linear FMCW with PRFs of 10–60 Hz, bandwidths of 5–40 kHz, and CPIs of 1–20 seconds (Skolnik 2008b).

The receive antenna of the OTH-B radar was implemented as three contiguous linear arrays of 246 × 5.4-m vertical monopoles with 20-m backscreens. The receive apertures were 1519 m long and contained 82 reception channels. Each of the arrays provided 60 degree coverage in azimuth with the three boresights oriented 60 degree apart. A detailed description of OTH-B can be found in Georges and Harlan (1994). Hughes (1988) reported on a series of tests validating the capability of OTH-B to detect cruise missiles. A distinguishing feature of this system is that it was the first operational skywave OTH radar in the United States that possessed a genuine wide-area surveillance capability for continental air-defense (Headrick and Thomason 1998). In 1991, the US Air Force put the OTH-B system in "warm storage" in response to the perception of a reduced air threat (Thomason 2003).

3.4.1.5 AN/TPS-71 (ROTHR)

The design and procurement of the US Navy AN/TPS-71 system, also known as the relocatable OTH radar or "ROTHR," was driven by the threat of long-range bombers and missile carriers to the battle group at sea (Thomason 2003). Under the sponsorship of the Chief of Naval Operations, the Space and Naval Warfare Command awarded a contract to the Raytheon Corporation in 1984 for the full-scale development phase of AN/TPS-71. The concept was that ROTHR could be air transported in containers and assembled quickly for operation at preprepared sites to counter the identified threat.

A prototype system was constructed and tested in Virginia, after which it was relocated and tested again. Based on the successful evaluation of the ROTHR system, a total of three production units were built. The first of these systems is currently installed in Virginia, the second in Texas, and the third in Puerto Rico. These current sites were selected to monitor air and surface drug-trafficking routes into the United States, which became of concern after the open-ocean air threat to the battle group diminished. Aerial photographs of the ROTHR transmit and receive sites in Virginia are shown in Figure 3.38.

The ROTHR transmit system in Virginia is situated on a 20-hectare site in the Chesapeake area. The transmit antenna is a two-band linear array of 2 × 16 LPDA elements (75–125 ft or 23–38 m tall) designed to operate at frequencies between 5 and 28 MHz. The maximum average power is around 200 kW, and similar to WARF, the quasi-monostatic configuration enables ROTHR to exploit linear FMCW operation. The instrumented range of selectable PRF, bandwidth, and CPI parameters caters for both aircraft and ship detection missions. The transmit beam may be steered in azimuth over a ±30 degree arc from boresight. The ROTHR system currently provides wide-area surveillance of drug-trafficking routes into the United States at ranges of up to 2000 nautical miles (1 nautical mile is equal to 1.852 km).

The receiving aperture is a uniform linear array of 372 vertical twin-monopole antenna elements steered at endfire to provide an appropriate front-to-back (directivity) ratio. The twin-monopole antenna elements are 5.8 m tall and uniformly spaced approximately 6.95 m apart to yield a receive array with a 2580-m-long aperture. The array has a digital reception channel per element and the receive beam may be steered in azimuth to ±45 degrees from the boresight direction. A detailed description of the US Navy's ROTHR systems, which are heavily based on the WARF design, appears in Headrick and Thomason (1996).

FIGURE 3.38 The top and bottom photographs show the transmit and receive sites of the US Navy AN/TPS-71 Relocatable OTH Radar (ROTHR) located at Chesapeake and New Kent, in Virginia, respectively. *(Pictures courtesy of the ROTHR Program).*

3.4.2 OTH Radar in Australia

Australian HF radar programs commenced in the 1950s. These included experimental investigations into skywave-path stability with an eye to OTH radar (Sinnott 1988), but the initial focus was on line-of-sight observations of RCS enhancements as rockets left and reentered the Earth's atmosphere from a test range near Woomera, in central Australia. RCS enhancements due to ionization in the engine exhaust plume were observed, but this phenomenon was significant only after the missile had reached altitudes of about

100 km with the engine still burning. Such results received considerable interest in the United States because of their potential relevance to ballistic missile detection during the boost phase.

In addition to the scientific impact of this early research, a considerable amount of indigenous skill and experience in performing field trials had already been developed in Australia by the end of the 1960s. This expertise subsequently proved to be of considerable influence and importance in helping to form a collaborative research agreement on OTH backscatter radar between Australia and the United States in 1969. Exposure to US knowledge and technology was a key element that enabled OTH radar development to proceed in Australia.

The current JORN system was preceded by two main phases of the Jindalee project, referred to as Jindalee Stages A and B, as well as an earlier project to quantify ionospheric path stability in the Australian region, known as GEEBUNG. An excellent historical account of the evolution of HF radar in Australia between the 1950s up to the mid-1980s is provided in Sinnott (1988), while Colegrove (2000) summarizes skywave OTH radar developments in Australia from the mid-1980s to the year 2000. This section briefly overviews the various stages of OTH radar research and development in Australia culminating with the JORN system.

3.4.2.1 GEEBUNG

Although it had been revealed that skywave OTH radar had been tested in the United States and that the detection of aircraft targets at ranges of thousands of kilometers had been achieved, it was necessary to confirm that ionospheric conditions in Australia would not preclude effective OTH radar operation. The first experimental studies to ascertain the feasibility of OTH radar in the Australian region commenced in 1970 with a project known as GEEBUNG. This project utilized a frequency-modulated continuous wave transmitter and receiver, both of US design, in addition to communications equipment and antennas of local design.

This project involved a long measurement campaign to monitor the characteristics of ionospheric propagation between two stations separated by a ground distance of 1850 km. The transmit site was located at Mirikata, near Woomera, while the receive site was situated close to Broome, on the north-west coast of Australia. Over an 11-month period, one-way path-loss and Doppler stability measurements of the skywave propagation channel were made by Dr Malcolm Golley. Meanwhile, backscatter experiments to assess the properties of two-way propagation over land and sea surfaces were performed by Dr Fred Earl.

The main recommendation of the data analysis, which appeared as the last sentence of an overview report prepared at the conclusion of GEEBUNG, was that there was no evidence to suggest that the "operational value of an OTH radar in Australia would be reduced to an unacceptable level by inherent physical limitations imposed by the ionosphere." This finding supported the proposal for a pilot radar as the next phase of the project. The GEEBUNG experimental sites were decommissioned in 1972, but the conclusions of the data analysis would be far-reaching. Indeed, the concept of a pilot radar led to the birth of the Australian OTH radar project "Jindalee."

3.4.2.2 Jindalee Stage A

Stage A of the Jindalee project officially commenced in the beginning of 1974. It involved the development of a non-scanning OTH backscatter radar with an architecture based on the design of the WARF system. Two sites were identified in a remote region of central Australia, not far outside of Alice Springs. Specifically, the transmit site was located at Harts Range (100 km north-east of Alice Springs), while the location ultimately chosen for the receive site was at Mt Everard (40 km north-west of Alice Springs). The two sites and the radar boresight were selected to observe targets of opportunity (mainly commercial jets) on an international air route in a north-westerly direction. In particular, the fixed "staring" beam of the Stage A radar provided a limited sector of coverage (a few degrees wide) aligned with the A76 air-traffic corridor.

The Stage A transmit antenna was a ULA of 16 LPDA elements borrowed from the United States, while the receive antenna was a locally designed 640-m-long ULA consisting of 128-whip antennas arranged as doublets. The waveform generator and signal processor were also locally manufactured and represented advancements in the state-of-the-art. The latter was adapted from the Programmable Signal Processor used in the sonar program that led to the Barra system, while the former involved the development of a linear sweep generator that met extremely stringent demands on spectral purity. Establishing an OTH radar technology base in Australia was widely regarded to be an essential part of the project.

The Jindalee Stage A radar commenced operations in 1976, and after years of preparation, it was fitting that an aircraft was detected during the initial checkout phase (Sinnott 1988). Routine logging of target detections over a 2-year period and off-line analysis of recorded data demonstrated a credible aircraft detection capability. A large amount of environmental data was also recorded to guide future OTH radar development. Although not planned, or expected for the capabilities of the Stage A system, ship detection trials performed in December 1977 demonstrated that the detection of surface vessels off the northwest coast of Australia was feasible.

The approval for Stage B of Project Jindalee was conditional on the performance of Stage A. The provision was made for a "hold" period after the completion of Stages A, during which the results of the system would be critically reviewed and assessed before agreement was given for Stage B to proceed. To support the case for a Stage B radar, a live demonstration of the system's capability was organized. A cooperative aircraft was commissioned for a special trail to be witnessed by a high-level delegation that had been invited to visit the receiver site in April 1977. The importance of the success of this trial in winning over the support of senior officers that were present on the day of the demonstration is described in Sinnott (1988).

3.4.2.3 Jindalee Stage B

Based on the success of Stage A, a second much more ambitious phase of the Jindalee project was approved in 1978. Stage B involved the development and demonstration of an OTH radar with operational capabilities. Radar sensitivity was increased by using a larger number of more powerful transmitters in addition to a much larger receive array. The Stage A transmit antenna was retained, but a major outcome of the performance-cost tradeoff study was the design of a receive array with a 2.8-km-long aperture consisting of 462 uniformly spaced dual-fan antenna elements. A 3 km by 100 m strip of land was cleared for the receiving array, and its quite remarkable that a man and his wife installed all of the 924 (single-fan) antenna elements in 32 days while camping at the receiver site (Sinnott 1988).

The requirement for track-while-scan over a sector of least 60 degrees called for much more powerful computing than was used in the Stage A system. As a processor suitable for the task was not commercially available at the time, a custom-designed processor known as the ARithmetically Oriented (ARO) processor was specifically developed for the Stage B radar. The design and implementation of this signal processor, including the programming language, software operating system, and digital hardware, was necessary to ensure that the Stage B radar could fulfill its real-time surveillance requirement without depending on the future availability of commercial off-the-shelf technology that was fit for the purpose.

To reduce system cost and processing load, the 462 antenna elements of the receive array were grouped into 32 overlapped subarrays, with each subarray steered to the direction of the transmitter footprint via analog beamforming. A reception channel per subarray architecture significantly reduced the number of receivers (which represented a significant proportion of the system cost) and eased demands on the ARO signal processor, while allowing high-resolution "finger-beams" to be processed within the surveillance region by digital beamforming (Sinnott and Haack 1983). In favorable environmental conditions, both the transmit and receive apertures could be operated independently as left and right halves in the Stage B design to improve the real-time coverage or coverage rate.

The Stage B radar was software controlled to a much greater extent than the Stage A system, which relied more so on manual configuration of the radar through switch settings and cable connections (Sinnott 1988). The rudimentary display capability of Stage A was replaced by more user-friendly interfaces in Stage B to improve system control for radar operators. The Frequency Management System was also significantly expanded in accordance with the increased radar capability. The wide arc of coverage also meant that more beacons were deployed to support Stage B operation. The Derby installation was upgraded with more capable equipment, while a main beacon site was established in the Darwin region.

Stage B ship detections were achieved for uncued targets in 1983, while automatic aircraft tracking was implemented in 1984. By mid-1984, the Jindalee Stage B radar had substantially achieved its major objective of demonstrating the capabilities required for Australian defence surveillance (Sinnott 1988). The transition from Stage A to Stage B took roughly 6 years to complete. After a series of successful service evaluation trials between 1984 and 1987, the operational system was handed over to Defence Forces and became known as the Jindalee Facility Alice Springs (JFAS). Following a number of system upgrades described in (Colegrove 2000), JFAS was officially operated and managed by the Royal Australian Air Force (RAAF) No. 1 Radar Surveillance Unit (1RSU) on 1 January 1993.

3.4.2.4 Jindalee Operational Radar Network (JORN)

A high-level review to define the way forward after Jindalee Stage B led to a decision to develop a network of new OTH radars known as the Jindalee Operational Radar Network (JORN). After this decision was announced by the Minister in 1986, the refurbishment or upgrade of JFAS assumed an important but secondary role. In the proposed plan, JFAS was also to provide a testbed for developmental purposes that would facilitate the implementation, testing, and transition of future enhancements into JORN.

It is beyond the scope of this text to recount the specific details of the transition from JFAS to JORN, but an authoritative description can be found in Colegrove (2000). For a general description of the JORN architecture and its surveillance capabilities, the reader may consult Cameron (1995). From a Defence procurement perspective, an insightful case study of the acquisition and management of JORN can be found in Markowski, Hall, and Wylie (2010). The main characteristics of the JORN radars developed by Telecom Australia, GEC-Marconi UK, and RLM USA-Australia are now briefly summarized.

The JORN OTH radar systems near Longreach (Queensland) and Laverton (Western Australia) are similar in many respects with the exception that the Laverton radar utilizes two sets of ULA apertures oriented at right angles to achieve 180-degree coverage in azimuth. Both systems operate in a quasi-monostatic configuration and employ linear FMCW signals that are modified to reduce out-of-band spectral emissions. The waveform repetition frequency, bandwidth, and CPI may vary over typical values of 4 to 60 Hz, 4 to 40 kHz, and 2 to 40 seconds, respectively, depending on whether air or ship detection tasks are being performed.

Taking the Laverton OTH radar as an example, the separation between the transmit and receive sites is approximately 80 km. The transmit antenna at Laverton is based on two adjacent 2-band ULAs of LPDA elements. The transmit array contains 14 (low-band) and 28 (high-band) LPDA elements to form apertures that are approximately 160 m long. As described in Colegrove (2000), the maximum average power is approximately 560 kW (i.e., 28 transmitters with 20-kW solid-state power amplifiers). The transmit antenna gain varies from about 20 to 30 dB over the design frequency range of 5 to 32 MHz. The transmit beam of each array may be steered in azimuth to ±45 degrees from boresight, which provides a combined coverage of 180 degrees.

To service the two transmit ULAs with orthogonal boresight directions, the JORN receive antenna at Laverton also consists of two ULA arms oriented at right angles. Each arm of the receiving aperture is 2970 m long and is comprised of 480 twin-monopole elements with an elevated-feed to improve gain at low elevation angles. A photograph of the JORN receive ULA viewed at endfire is shown in Figure 3.39. The receiver per (twin-monopole) element architecture allows fully digital beam steering over a ±45 degree arc relative to boresight. The primary mission of JORN is aircraft detection, while ship detection is the secondary mission. The Royal Australian Air Force accepted JORN into service in April 2003.

3.4.3 Future Prospects

The JORN may well be viewed as the culmination of more than 30 years of OTH radar research and development in Australia. However, it is also important to recognize that JORN was in many respects designed as a low-risk solution based heavily on the proven performance of the predecessor JFAS system. Moreover, the JFAS system involved significant compromises in its design. These were made to satisfy tight budget constraints in the various stages of its development. It is therefore not surprising that fundamentally new threads of research with the potential to deliver not incremental but substantial improvements in OTH radar capability would eventually emerge. This section briefly identifies some possible areas in which next-generation OTH radar systems may move beyond traditional precepts to improve performance and expand functionality.

While the LPDA antenna has many advantages as a transmitting element for OTH radar, the long array of dipoles occupies a large footprint. In addition, the radiation pattern of the LPDA element limits the coverage to about 90 degrees in azimuth. It is also the case that the resonant region of the antenna changes position significantly as a function of operating frequency. When viewed collectively, these characteristics make the LPDA element rather unsuitable for use in two-dimensional (nonlinear) array geometries. In principle, two-dimensional transmit arrays offer greater flexibility to manage the spatial

FIGURE 3.39 Each receiving antenna of the JORN Laverton OTH radar is a uniform linear array of 480 elevated-feed twin-monopole elements spaced 6.2 m apart (aperture length of 2970 m) with a reception channel per element. © Commonwealth of Australia 2011.

distribution of illumination by allowing the transmit beam to be steered and shaped independently in azimuth and elevation. Two-dimensional arrays of omni-directional elements may also be used to provide 360-degree coverage in azimuth using a single antenna aperture.

A possible alternative to the ULA of LPDA elements is to distribute the available power across a larger number of more compact and less expensive antennas that are individually connected to transmit modules with lower power. It is well known that antenna elements comprising cages of wires or metal tubes can be used to achieve electrical shortening in height and broadband performance. A more compact transmit antenna with an omni-directional pattern can be arranged in a two-dimensional array to enable beam steering in elevation and provide 360-degree coverage in azimuth.[5] Moreover, a short rigid antenna provides greater immunity against Aeolian noise, which improves the spectral purity of radiated signals. The need for an extensive ground screen to stabilize surface impedance can also be reduced by driving the active sections against a raised counterpoise insulated from the ground at the base.

A more compact antenna is unlikely to match the broadband performance of the LPDA element, and may only perform effectively over an octave. However, an array based on a large number of such antennas with a digital waveform generator and power amplifier

[5]This class of antenna is used in the Nostradamus OTH radar described previously.

per element would offer the possibility to operate the transmit resource in a more flexible manner, as well as reduce the internally generated phase noise in the transmit beam by averaging independent contributions over a larger number of elements. These may be considered as additional advantages with respect to the aforementioned benefits of a two-dimensional transmit array based on more compact and rigid antenna elements. The issue of mutual coupling requires careful attention in such designs, as problems associated with this phenomenon are more complex to manage in two-dimensional arrays.

Two-dimensional transmit antenna arrays have been described for mode-selective OTH radar systems in Frazer et al. (2010), where the ability to form "non-causal" adaptive range-dependent transmit beams using an appropriately chosen set of "orthogonal" waveforms in a multiple-input multiple-output (MIMO) context is proposed to improve maritime domain awareness in disturbed (multimode) clutter conditions. Further descriptions of mode-selective OTH radars and experimental results appear in Frazer et al. (2010), Frazer, Abramovich, and Johnson (2009), and Abramovich, Frazer, and Johnson (2011).

Similar to the transmit antenna, the design of the JORN receive antenna was heavily based on the proven performance of the JFAS array (which was in turn modeled on the WARF system). The use of receive antenna elements that are appropriately matched in the operating region of the HF band combined with 2D array apertures and adaptive processing on receive has the potential to enhance target detection performance by allowing clutter and noise that exhibits significant structure in space and/or time to be mitigated more effectively than would be possible with a linear array. Furthermore, wideband direct digital receivers capable of downconverting multiple simultaneous frequency channels could potentially be incorporated in future systems such that the full aperture of the receive array may be used to perform multiple radar tasks, common aperture frequency management, coordinate registration, and other support functions, all at the same time.

It is also expected that advances in distributed computing will enable future systems to host more sophisticated adaptive signal-processing algorithms for clutter and interference mitigation, as well as allowing the performance of different signal-processing chains and trackers to be evaluated in real time such that the best output can be displayed to radar operators. At the system level, it may also be possible to further develop radar resource management approaches. Such approaches help to automatically configure radar operations based solely on mission requirements and a knowledge of the prevailing environmental conditions with minimal operator intervention. These alternative concepts have the potential to create pathways for new and exciting capabilities in future operational OTH radar systems.

CHAPTER 4

Conventional Processing

Signal and data processing have long been recognized as key elements to the success of OTH radar (Headrick and Skolnik, 1974). Such systems are inherently required to operate in severe clutter, interference, and noise environments, where the power of a received target echo is typically several orders of magnitude lower than that of the disturbance signal. Here, signal processing refers to a sequence of steps that transform the in-phase and quadrature baseband data samples acquired by the analog-to-digital (A/D) converters in each receiver of a multi-channel system to complex-valued outputs in the canonical radar dimensions of group-range, beam direction, and Doppler frequency. The two most important objectives of these steps, which include pulse compression, array beamforming, and Doppler processing, are to enhance the signal-to-disturbance ratio (SDR) and to resolve the received radar echoes in one or more of the canonical radar dimensions. This improves target detection performance and parameter estimation accuracy in the subsequent data-processing stages.

The amplitude envelope of the signal-processing output is subsequently passed on to a sequence of data-processing steps, which typically include constant false-alarm rate (CFAR) processing, peak detection and parameter estimation, as well as tracking and coordinate registration. The final output of this data processing is displayed to operators as a set of confirmed radar tracks registered in geographic coordinates. The purpose of this chapter is to describe the essential steps in the OTH radar signal- and data-processing chains. This commences from the (decimated) baseband A/D samples data, where we left off in the preceding chapter, and finishes with the geographic track display, which represents the end product of an OTH radar system. The focus in this chapter is on conventional processing techniques for OTH radar and a selection of ameliorative processing steps that are traditionally considered part of the standard sequence. The use of adaptive filtering techniques to enhance system performance in challenging clutter and interference conditions is considered in Part III.

The first section of this chapter provides a general description of the HF signal environment. Besides identifying the main sources of clutter, interference, and noise, this section also discusses the physical characteristics of the various signals received by OTH radars. The second section describes the three rudimentary OTH radar signal-processing steps; namely, pulse compression, array beamforming, and Doppler processing, which are common to many other types of radar systems. The fundamental concepts associated with these core routines are reviewed in this section and specific implementation details relevant to OTH radar systems are also highlighted. The third section discusses a number of operational considerations for air- and surface-target detection in OTH radar.

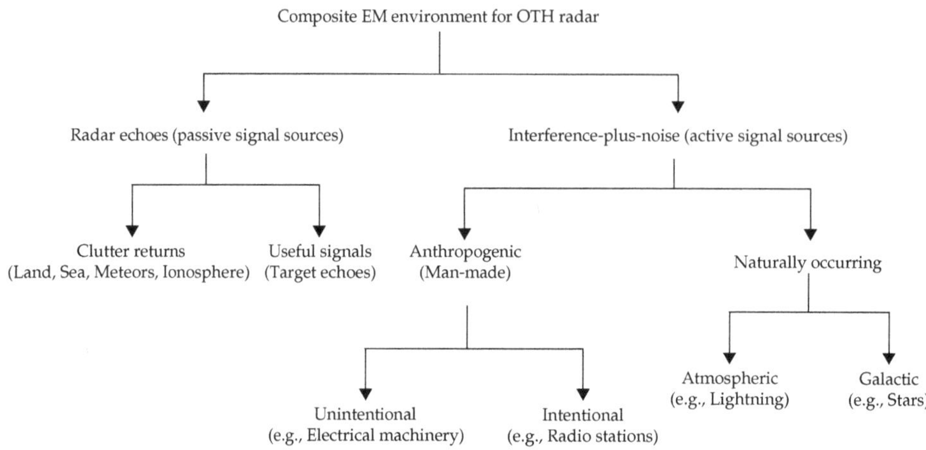

FIGURE 4.1 For an OTH radar, the HF signal environment consists of clutter returns and potentially target echoes, in addition to interference and noise received from natural and man-made sources that emit signals with frequency components in the radar bandwidth. © Commonwealth of Australia 2011.

This includes the critical importance of selecting appropriate operating frequencies and waveform parameters for different radar missions, as well as the significant benefits of applying several ameliorative signal-processing steps. The fourth section of this chapter discusses the subjects of CFAR processing, peak detection and estimation, as well as tracking and coordinate registration. Example results using real OTH radar data are provided where possible to illustrate the practical value of the various processing steps.

4.1 Signal Environment

An OTH radar receives a mixture of different signal types in the HF environment. The diagram in Figure 4.1 broadly categorizes the received signals into two main classes. On the left, the signals are due to echoes of the transmitted radar waveform, while on the right, the signals come from interference and noise sources independent of the radar. The former signals are received from objects in the radar scene that scatter the incident illumination, but do not produce their own independent radiation. For this reason, these objects are sometimes referred to as *passive* signal sources. The latter are received from man-made or natural emissions independent of the radar waveform, but with frequency components that partially or fully overlap the radar bandwidth. These emitters are often referred to as *active* signal sources.[1]

Radar echoes may be further subdivided into target echoes and clutter returns. The former represent skin-echoes from objects of interest to the radar, such as an aircraft or ship in the surveillance region. Target echoes are also commonly referred to as useful signals or desired signals in the radar nomenclature. Unwanted echoes from other objects in the

[1]A transponder (repeater) is a special case of an active source that retransmits a delayed and possibly Doppler-shifted version of the intercepted radar waveform and therefore does not give rise to an independent signal.

radar scene are collectively referred to as clutter. In OTH radar, clutter mainly arises due to backscatter from large areas of the Earth's surface (terrain or sea) that are simultaneously illuminated by the radar. It also arises as a result of direct or indirect volume scatter from the ionosphere. This includes scattering from electron density irregularities, which are naturally formed in the E- and F-regions, as well as transient echoes from ionized meteor trails in the upper atmosphere. Target-like signals with parameters that are significantly mismatched to those which the current radar search is tuned to may also be considered as clutter.

On the other hand, external interference and noise may originate from natural or man-made sources. In the HF band, the main sources of natural noise include *atmospheric* noise due to lightning discharges, and *galactic* noise from the sun and other stars. Man-made signals may be further subdivided as either originating from intentional or unintentional radiators. Examples of the former include short-wave radio broadcasts, point-to-point communication links, HF radar systems, ionospheric sounding stations, and potentially jamming. When received by an OTH radar, such signals are collectively referred to as radio-frequency interference (RFI). Unintentional radiation may arise from industrial machinery and other electrical equipment in use, including vehicle ignition systems and overhead power lines. Such signals are often referred to as man-made noise. The sum of clutter, interference, and noise (man-made and natural) represents the composite disturbance signal the radar must contend with. Internal noise of thermal origin is typically dominated by external noise at the OTH radar receiver output. Internal receiver noise is not explicitly indicted in Figure 4.1.

Restricting attention to a single OTH radar reception channel connected to the antenna element designated as the phase reference of the receiver array, the acquired baseband signal at time t may be expressed by the complex scalar $x(t)$ in Eqn. (4.1). In general, this signal is the superposition of surface and volume clutter $c(t)$, external interference-plus-noise $i(t)$, internal receiver noise $n(t)$, and possibly a target echo $\nu s(t)$, where the states $\nu = 0$ and $\nu = 1$ indicate target presence and absence, respectively. The scope of this section is to provide a description of these signal components and to briefly overview their different physical characteristics. This not only motivates the conventional signal- and data-processing steps discussed in the following sections, but also serves as a prelude to the development of more detailed signal models in Part II.

$$x(t) = \nu s(t) + c(t) + i(t) + n(t) \tag{4.1}$$

4.1.1 Target Echoes

Most targets of interest to a surveillance radar have physical dimensions ranging from a few meters to possibly hundreds of meters, while the dimensions of an OTH radar spatial resolution cell typically vary from 5 to 50 km in range and 15 to 75 km in cross-range. Unlike high-resolution line-of-sight radars, which are capable of resolving individual scattering centers on an aircraft frame or surface vessel by using large signal bandwidths or electrically wide apertures, a target echo is effectively confined to a single range-azimuth cell in OTH radar (ignoring spectral leakage effects and resolved target echoes due to multipath propagation). In simple terms, an aircraft or ship effectively acts as a point scatterer in the OTH radar context.

In addition, the surveillance region of an OTH radar is generally located in the far-field of the receiving array. Specifically, the slant range between radar and target, denoted by

R, normally satisfies the Rayleigh far-field condition in Eqn. (4.2), where D is the major dimension of the receiving aperture, and λ is the radio wavelength. For a uniform linear array (ULA) of length $D = 3$ km, and a mid-band operating frequency of $f_c = 15$ MHz ($\lambda = 20$ m), this condition requires that $R > 900$ km, which is satisfied by the nominal OTH radar skip-distance of approximately 1000 km. OTH radar targets are therefore often modeled as far-field point scatterers of the transmitted waveform for most intents and purposes.

$$R > 2D^2/\lambda \tag{4.2}$$

4.1.1.1 Point Scatterer Model

Suppose the radar transmits a coherent burst of N radio frequency (RF) pulses with duration T_p and pulse repetition interval T in a coherent processing interval (CPI) of NT seconds. The analytic representation of the RF signal transmitted by the system may be written as $r(t)$ in Eqn. (4.3), noting that the waveform actually emitted by the radar is the real part of the analytic signal $\Re\{r(t)\}$. In Eqn. (4.3), $A_r > 0$ is an amplitude factor related to the transmit power and antenna gain, f_c is the carrier frequency, and $\vartheta \in [0, 2\pi)$ is an arbitrary initial phase.

$$r(t) = A_r m(t) e^{j(2\pi f_c t + \vartheta)}, \quad t \in [0, NT) \tag{4.3}$$

The complex envelope of $r(t)$ is a baseband-modulation signal $m(t)$, also referred to as the radar waveform. It is assumed that $m(t)$ takes the form of Eqn. (4.4), where $p(t)$ is the pulse waveform of bandwidth B and $t \in [0, T_p)$. The pulse waveform $p(t)$ is assumed to be narrowband ($B \ll f_c$), with a duration T_p that does not exceed the pulse repetition interval T. The two cases $T_p < T$ and $T_p = T$ correspond to pulse waveform (PW) and continuous waveform (CW) radar operation, respectively.

$$m(t) = \sum_{n=0}^{N-1} p(t - nT) \tag{4.4}$$

The average power is given by $P_{ave} = E_p/T$, where E_p is the pulse energy defined in Eqn. (4.5). Practical transmitters are limited in the peak value of $|p(t)|$. For a simple (amplitude-modulated) pulse, the requirement for high average power competes with the desire for fine range resolution. The former requires a long pulse with high duty-cycle T_p/T, while the latter requires a short pulse with high bandwidth $B = 1/T_p$. A significant advantage of phase or frequency-modulated waveforms is that they provide effectively independent control of the pulse duty-cycle (average power) and pulse bandwidth (range resolution).

$$E_p = \int_0^{T_p} |p(t)|^2 \, dt \tag{4.5}$$

The repetitive linear frequency-modulated continuous waveform (FMCW) is commonly used in two-site OTH radar systems. A unit-amplitude linear FM pulse of nominal bandwidth B is given by the constant modulus (CM) waveform $p(t)$ in Eqn. (4.6), which is also known as a "chirp." The N pulses in the CPI are sometimes referred to as "sweeps." A two-site OTH radar configuration maximizes P_{ave} for a fixed peak value of $|p(t)|$ by transmitting a CM pulse with unit duty-cycle ($T_p = T$). A monostatic OTH radar is required to interrupt the linear FM waveform ($T_p < T$), which trades off average power for the benefits of single-site operation.

$$p(t) = e^{j\pi Bt^2/T_p}, \quad t \in [0, T_p) \tag{4.6}$$

Substituting Eqn. (4.6) into Eqn. (4.3), the transmitted RF chirp may be expressed as $r(t) = A_r e^{j\psi(t)}$ where $\psi(t) = 2\pi(f_c t + Bt^2/2T_p) + \vartheta$ for $t \in [0, T_p)$. The instantaneous frequency of the transmitted signal during a particular linear FM sweep is given by $f(t)$ in Eqn. (4.7). Note that $f(t)$ ramps up linearly from f_c to $f_c + B$ over the pulse repetition interval (PRI). Although f_c is usually defined as the center frequency of the radar signal (as opposed to the initial frequency of each sweep), the latter interpretation is adopted here without loss of generality.

$$f(t) = \frac{1}{2\pi}\frac{\partial\psi(t)}{\partial t} = f_c + Bt/T_p \tag{4.7}$$

Suppose that the radar illuminates a far-field point scatterer (target) with a constant radar cross section (RCS) and relative velocity v over the CPI. Specifically, for a target with group-range $R(t)$, at time t, the relative velocity v is defined in terms of the rate of change of group-range, as in Eqn. (4.8). The negative sign convention implies that a target approaching the radar will have a positive relative velocity.

$$v = -\frac{\partial R(t)}{\partial t} \tag{4.8}$$

The group-range $R(t)$ of the signal path that links the radar to the target is given by Eqn. (4.9), where $\tau(t)$ is the round-trip time delay of the echo measured over the transmitter-target-receiver path by the reference receiver of the antenna array at time t, and c is the speed of light in free space. The definition of a reference receiver will be needed later to extend the target echo model in space for antenna arrays.

$$R(t) = \frac{c\tau(t)}{2} \tag{4.9}$$

Ignoring relativistic effects, we may write $\tau(t) = \tau - 2vt/c$, where $\tau = 2R(0)/c$ is the delay corresponding to the target range at the start of the CPI ($t = 0$). The RF echo returned by the target to the reference receiver, whose location is defined to be at the origin of the spatial coordinate system, can be expressed as $e(t)$ in Eqn. (4.10), where α is the target echo complex amplitude and $r(t)$ is the RF signal transmitted by the radar.

$$e(t) = \alpha r(t - \tau(t)) = \alpha r(t - \tau + 2vt/c) \tag{4.10}$$

Using Eqn. (4.3), we may write $e(t)$ in the form of Eqn. (4.11), where $f_d = 2vf_c/c$ is the Doppler frequency shift of the echo scattered from the target. In this expression, the magnitude and phase of α account for all propagation and scattering effects between the radar and target (including a range-dependent phase term $e^{-j2\pi f_c \tau}$), as well as the receive system characteristics.

$$e(t) = \alpha m(t[1 + 2v/c] - \tau)e^{j2\pi(f_c+f_d)t} \tag{4.11}$$

It is evident from Eqn. (4.11) that the movement of the target acts to compress or stretch the time scale of the radar waveform $m(t)$. Since the change in target echo round-trip delay over the CPI (given by $\tau(NT) - \tau = 2vNT/c$) is expected to be much smaller than the reciprocal of the pulse bandwidth B (i.e., no range migration), the approximation in Eqn. (4.12) is justified for $t \in [0, NT)$. This represents a key assumption in the formulation of the target echo model.

$$m(t[1 + 2v/c] - \tau) \approx m(t - \tau) \tag{4.12}$$

Second, for typical radar waveform parameters and realistic target velocities, the Doppler shift f_d is much smaller than the reciprocal of the pulse duration, such that the condition in Eqn. (4.13) usually holds. In this case, the Doppler shift due to target motion imposes an effectively constant phase modulation over a single PRI that changes by an amount $e^{j2\pi f_d n T_p}$ across the different pulses $n = 0, \ldots, N-1$ of the CPI.

$$f_d T_p \ll 1 \tag{4.13}$$

After substituting the approximation of Eqn. (4.12) into Eqn. (4.11), and multiplying $e(t)$ by $\exp(-j2\pi f_c t)$ to account for complex (I/Q) downconversion in the radar receiver, the baseband target echo is given by $s(t)$ in Eqn. (4.14). In other words, an ideal point-target echo $s(t)$ at baseband is a complex-scaled, time-delayed, and Doppler-shifted version of the modulating radar waveform $m(t)$.

$$s(t) = \alpha m(t - \tau)e^{j2\pi f_d t} = \alpha \sum_{n=0}^{N-1} p(t - nT - \tau)e^{j2\pi f_d t} \tag{4.14}$$

The characteristics of the target echo amplitude α influence the signal-to-disturbance ratio (SDR), and hence radar performance measured in terms of target detection and false-alarm probabilities. The characteristics of α also influence the prospects for target classification. On the other hand, the target echo parameters, including round-trip time-delay τ, Doppler frequency shift f_d, and angle of arrival (to be incorporated later), provide a means to resolve echoes from different scatterers, as well as multipath echoes from the same scatterer. The determination of these parameters also enables the position and relative velocity of targets to be estimated for localization and tracking purposes.

4.1.1.2 Radar Cross Section

From the radar equation discussed in Chapter 1, the square of the received target-echo amplitude $|\alpha|^2$ is directly proportional to the target RCS σ. The RCS of a scatterer is traditionally defined as the effective area of a hypothetical object that intercepts the power density of an incident plane wave and reradiates the total intercepted power isotropically to produce the same power density at the receiver as the scatterer being considered.

Mathematically, the RCS of an object is defined in Eqn. (4.15), where E_i is electric field strength of the plane wave incident on the object, and E_s is the electric field strength of the scattered wave measured at a distance R from the object. The limit $R \to \infty$ in the RCS definition may be interpreted as a requirement for the far-field (plane-wave) condition to apply in a practical setting.

$$\sigma = \lim_{R \to \infty} 4\pi R^2 \frac{|E_s|^2}{|E_i|^2} \tag{4.15}$$

The RCS of a radar target is in general a function of illumination frequency, transmit and receive polarizations, viewing angle for a monostatic configuration (i.e., target aspect), or scattering geometry in a bistatic system. The RCS also depends on the target's physical structure, component dimensions, and material composition. Although RCS is measured in units of square meters, it may be orders of magnitude larger or smaller than the physical cross-sectional area presented by the target to the radar along the viewing direction.

Moreover, the effective target RCS may need to be modified from its free-space value in practice to account for mutual coupling effects due to the presence of the Earth's surface, as well as propagation effects such as (unresolved) multipath. Some of the major factors influencing target RCS in the skywave OTH radar context will be discussed below. The reader interested in delving further on the general subject of RCS is referred to Knott, Shaeffer, and Tuley (1993) for a comprehensive coverage of this topic.

For a fixed illumination frequency and scattering geometry, a target RCS value may be defined for each combination of transmit and receive polarization. The basis selected for describing the polarization of a radio wave is arbitrary, but a decomposition of the plane containing the electric field into horizontal (H) and vertical (V) components is often chosen. This (monochromatic) description using a linear (H,V) basis leads to the definition of a two-by-two polarization scattering matrix in Eqn. (4.16).

$$\begin{bmatrix} E_s^{[V]} \\ E_s^{[H]} \end{bmatrix} = \begin{bmatrix} S_{VV} & S_{VH} \\ S_{HV} & S_{HH} \end{bmatrix} \begin{bmatrix} E_i^{[V]} \\ E_i^{[H]} \end{bmatrix} \tag{4.16}$$

The first subscript letter of each matrix element signifies the E-vector polarization of the scattered echo, while the second letter signifies the E-vector polarization of the incident illumination. The target RCS for each of the four different transmit and receive polarization combinations is given by the modulus squared of the associated complex scalars S_{VV}, S_{VH}, S_{HV}, and S_{HH}. Note that for a monostatic radar, the cross-polar terms S_{VH} and S_{HV} are equal in a reciprocal medium.

A fully polarimetric target RCS description may be useful in certain radar systems to model complex scattering processes for classification purposes. When either the target illumination path or the echo return path involve line-of-sight or surface-wave propagation, specific transmit-receive polarization combinations may be of interest (Willis and Griffiths 2007), particularly as only the polarization component matched to that of the receive antenna can be detected. However, Faraday rotation in the ionosphere modifies the polarization state of the HF signal which illuminates the target from that leaving the transmitter, as well as the polarization state of the signal scattered from the target to that arriving at the receiver.

The change in polarization imposed on the signal as it propagates through the ionosphere is arbitrary and time-varying. In skywave OTH radar systems, the polarization transformation occurring over both the transmitter-to-target and target-to-receiver paths is unknown, and it has been argued that a polarization scattering matrix description of the target RCS is unnecessary (Skolnik, 2008b). For this reason, such a description is often dispensed with in preference to a (polarization-averaged) scalar RCS representation.

For a surveillance radar operating in the HF band, the target RCS characteristics fall in the Rayleigh-resonance scattering regime as opposed to the optical region at microwave frequencies. In the wavelength range $\lambda = 10$–100 m, ships and manned aircraft have major dimensions that put them in the resonance scattering regime. Here, the target RCS can be larger than in the optical region, but may fluctuate significantly with frequency.

The target RCS in the Rayleigh region can also be sensitive to target aspect or scattering geometry.

In many cases, the HF RCS of a target depends more on the gross dimensions of the dominant scatterers and their relative locations rather than the detailed shapes of the reflecting surfaces, particularly if the shaping is on a scale much less than one wavelength. In particular, the dimensions of large conductive segments on a target, such as the aircraft wings or fuselage, play a prominent role in determining the HF RCS. As pointed out in Willis and Griffiths (2007), such targets may exhibit wing spans or fuselage lengths commensurate with half of the radio wavelength λ, resulting in RCS values that can approach λ^2. Most military targets are characterized by free-space RCS values that may range from tens to hundreds of square meters for (manned) aircraft, up to perhaps tens of thousands of square meters for very large surface vessels.

At the other extreme, the RCS characteristics of very small aircraft (e.g., unmanned aerial vehicles) and surface vessels (e.g., "go-fast" boats), as well as certain types of missiles (e.g., cruise missiles), will lie in the Rayleigh region across the lower part of the HF band. In this region, the RCS exhibits reduced sensitivity to target aspect and a strong dependence on the target's gross dimensions. To a good approximation, the RCS is proportional to the fourth power of frequency in the Rayleigh region. For small targets (in particular), the change in RCS characteristics can be significant when viewed over the entire HF band.

For example, the RCS of a missile of length 10 m is in the Rayleigh region at a frequency of 3 MHz ($\lambda = 100$ m). The same target lies in the resonance scattering region at a frequency of 30 MHz ($\lambda = 10$ m). Figure 4.2 summarizes the main factors influencing target RCS for OTH radar systems, and provides an order-of-magnitude indication of nominal RCS values for different (broadly defined) target classes. The range of values quoted for the nominal RCS should be interpreted as a rough indication only, since the instantaneous target RCS can vary significantly (e.g., by up to ten or more decibels) from its nominal or mean value averaged over signal polarization and target aspect.

The free-space RCS of aircraft in the HF band may be estimated from experimental anechoic-chamber measurements made on scale models at appropriately scaled frequencies, or by carrying out full-scale measurements in the field using reference scatterers or transponders for calibration. Alternatively, target RCS may be estimated by computer simulation using the method-of-moments modeling approach, as in the numerical electromagnetic code (NEC), for example. For targets with highly conducting surfaces, RCS estimates within a few decibels of experimentally measured values have been derived by numerical modeling (Skolnik 2008b).

Method-of-moments modeling results for the HF RCS of a military aircraft have been reported in Robinson (1989) as well as in Lyon and French (1999) for different incident and scattered signal polarization combinations. The monostatic (backscatter) RCS of the F-18 fighter was computed as a function of aspect angle in Skolnik (2008b) using NEC2 and a wire-grid representation of the aircraft (based on a plastic kit model) at frequencies of 12, 18, and 30 MHz for horizontal (HH) polarization and a declination (look-down) angle of 5 degrees. The computed RCS has a mean value of approximately 20 dBsm, but can drop to values below 10 dBsm as a function of aspect angle and frequency.

Full-scale target RCS measurements at HF have been made in the field for small and large surface vessels. The high conductivity of the sea at HF significantly modifies the effective RCS of surface targets relative to their free-space values. Modeled RCS values for small and large surface vessels were confirmed against experimental measurements in

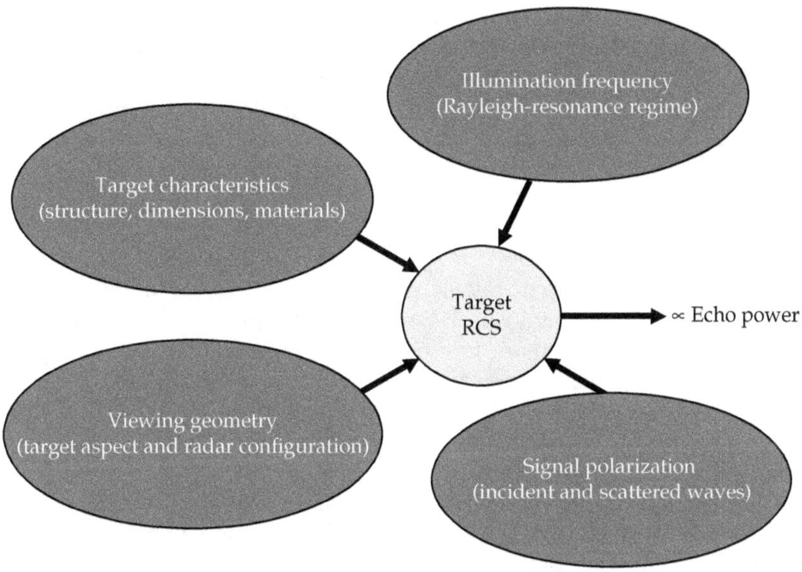

Nominal RCS	Large	Medium	Small
Aircraft targets	Commercial airliner 20–30 dBsm	Fighter-sized 10–20 dBsm	Missile 0–10 dBsm
Surface targets	Cargo freighter 30–50 dBsm	Patrol boat 20–30 dBsm	Go-fast boats 0–10 dBsm

FIGURE 4.2 Summary of main factors influencing target RCS, and nominal RCS values for different (broadly defined) target classes in the HF band. The table provides a rough indication of the expected range of nominal or mean RCS values. For a particular viewing geometry, incident and scattered polarization, and operating frequency, the instantaneous RCS of the target may differ significantly from its nominal or mean value. © Commonwealth of Australia 2011.

Leong and Wilson (2006). Comparisons between modeled and experimentally measured RCS values of two "go-fast" boats at different frequencies were reported in Dinger et al. (1999). Calculation of RCS based on standard computational methods may not be as reliable for targets composed largely of materials such as fiberglass or wood that cannot be modeled as PEC.

4.1.1.3 Complex Amplitude Variations

In practice, the complex amplitude α of a target echo is not constant, but fluctuates over time within the CPI and from one CPI to another. Fading of the target echo in magnitude and phase can arise due to a number of physical mechanisms. For example, the target may not present a steady RCS, or maintain a constant relative velocity, due to maneuvers. Target dynamics can alter both the magnitude and phase of the received echo complex amplitude. Furthermore, the radar and target are not in free space, which means that the target echo complex amplitude can also fluctuate as a result of variations in propagation conditions. This includes Faraday rotation (polarization fading) and unresolved multipath (wave interference) over a two-way skywave path. Target echo fading may be represented by incorporating a time-varying echo complex amplitude model $\alpha(t) = \mu(t)e^{j\gamma(t)}$ with fluctuating magnitude $\mu(t)$ and phase $\gamma(t)$, as in Eqn. (4.17).

$$s(t) = \alpha(t)m(t-\tau)e^{j2\pi f_d t} = \mu(t)m(t-\tau)e^{j(2\pi f_d t+\gamma(t))} \tag{4.17}$$

Target echoes propagated over a two-way skywave path via well-separated (E and F region) layers in the ionosphere (e.g., E-E, F-F modes) are typically resolved by OTH radars due to their significant difference in group-range. However, mixed layer paths (e.g., E-F, F-E) and magneto-ionic components (i.e., the o- and x-waves) are often more difficult to resolve in time delay, Doppler frequency, or direction of arrival (DOA). Mutual interference between unresolved ionospheric modes of similar strength can cause deep and rapid fading of the received target echo.

In particular, the interference of o- and x-waves produces a spatial pattern of polarization fringes (Barnum 1973), such that an aircraft target flies through a field of changing polarization, typically with a full rotation taking place in 10–50 km of travel (Willis and Griffiths 2007). For a receiving antenna with linear (V or H) polarization, fading due to time-varying Faraday rotation also arises because only one component of the echo polarization can be sensed. Even when a single magneto-ionic component of the target echo can be resolved by the radar, fading may arise due to electron density irregularities in the ionosphere that diffusely scatter (rather than specularly reflect) the signal. Traveling ionospheric disturbances can also cause the target echo to fluctuate significantly as a result of signal focussing and de-focussing effects.

For an airborne target, complex fading of the received echo can potentially arise as a consequence of unresolved multipath reflections from the Earth's surface. Figure 4.3 illustrates the direct and indirect (surface-reflected) ray paths between the radar and an

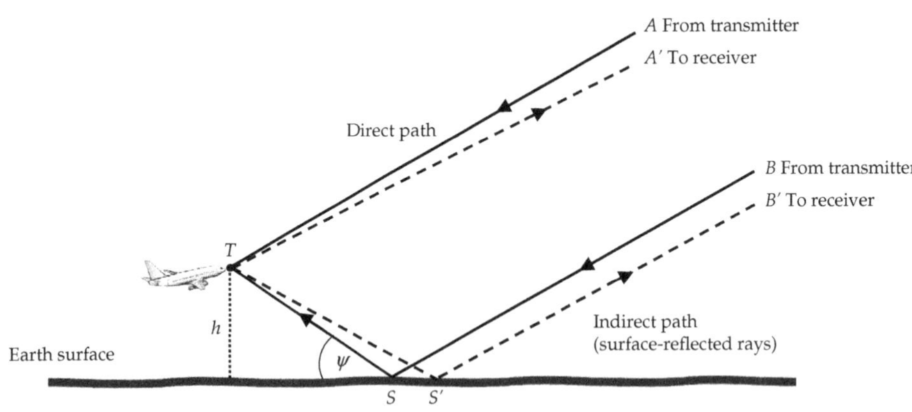

FIGURE 4.3 Direct and indirect (surface-reflected) ray paths shown for a single ionospheric propagation mode and an airborne target at altitude h. The outgoing (transmitter-to-target) and incoming (target-to-receiver) signal paths coincide for a monostatic system, but are slightly different for a bistatic OTH radar configuration. © Commonwealth of Australia 2011.

aircaft at altitude h. Even when attention is restricted to a single ionospheric mode and specular reflection from a smooth sea surface, there are four possible (altitude-dependent) multipath returns from the target, each having a slightly different group-range, Doppler shift, and DOA.

The two mixed-paths involve a single reflection from the Earth's surface at grazing angle ψ, i.e., paths $AT-S'B'$ and $BS-TA'$. Note that for a monostatic radar configuration, the points A, B, and S coincide with the points A', B', and S', respectively. For an OTH radar system with well separated transmit and receive sites, the transmitter-to-target and target-to-receiver paths will be slightly different, as indicated in Figure 4.3. It is instructive to discuss altitude-dependent multipath slightly further, as this topic is not always considered explicitly in the OTH radar context.

The difference in coordinates between the direct and surface-reflected multipath returns in group-range, Doppler shift, and DOA depend on the target altitude, ionospheric state, and receiver/transmitter locations with respect to the target. In practice, the separations are far too small to resolve using a single CPI, typical OTH radar parameters, and conventional signal processing. Consequently, the main discernible effect of altitude-dependent multipath is both intra-dwell and dwell-to-dwell fading of the target echo due to the superposition of unresolved returns with time-varying (relative) amplitudes and phases. This type of multipath is mainly considered for targets over the sea, where the losses due to surface reflection are relatively small and fading effects most pronounced.

For a monostatic (or quasi-monostatic) OTH radar, the difference in one-way group-range between the direct path AT and indirect path BST may be approximated as δR in Eqn. (4.18). This simple approximation is valid for an ionospheric reflection point that is far from the target and surface-reflection point, with the assumption of a locally flat Earth model between the target and surface-reflection point. The use of high-resolution techniques to resolve the direct and indirect paths in range or Doppler have been exploited for target altitude estimation in OTH radar, as described in Praschifka, Durbridge, and Lane (2009), and references therein.

$$\delta R = 2h \sin \psi \tag{4.18}$$

For example, consider a target with altitude $h = 13$ km at a ground distance of 1200 km from a monostatic OTH radar. In the case of E-layer propagation with a virtual reflection height of 100 km, this corresponds to a grazing angle of about $\psi = 7$ degrees. In this case, $\delta R \approx 3$ km, which agrees well with experimental measurements made over a one-way path for a similar scenario in Praschifka, Durbridge, and Lane (2009). The challenge for altitude estimation is not only to resolve the direct and surface-reflected paths for a single ionospheric mode, but also to associate these paths correctly when multiple ionospheric modes (including closely spaced magneto-ionic components) are present.

From a signal-processing perspective, the question arises as to statistical characteristics of the target echo fluctuations in magnitude $\mu(t)$ and phase $\gamma(t)$. In the radar literature, target echo fluctuation models for radar were first introduced by Swerling in the 1950s, and were subsequently republished in Swerling (1960) and Swerling (1997). Recall that the Swerling I model assumes $\mu(t)$ is Rayleigh distributed and varies independently from scan to scan (dwell to dwell), but is invariant over the different pulses of a particular scan. The Swerling II model is similar, but assumes faster fluctuations, such that $\mu(t)$ additionally varies independently from pulse to pulse. The Rayleigh density for $\mu(t)$ is consistent with the assumption of zero-mean IID Gaussian distributed real and imaginary parts of $\alpha(t)$, where the phase $\gamma(t)$ is uniformly distributed over $[0, 2\pi)$. Notionally, this is representative of a target with a large number of independent scatterers, none of which dominate.

The Swerling III and IV models are analogous to the Swerling I and II models in terms of representing slow (scan to scan) and fast (pulse to pulse) fluctuations, respectively, but the density function of $\mu(t)$ is assumed to be Ricean instead of Rayleigh in the Swerling III and IV models. This is consistent with a complex-Gaussian description of $\alpha(t)$ with a nonzero mean. The Swerling III and IV models are notionally representative of a target composed of a single dominant (non-fluctuating) scatterer along with an assembly of smaller independent scatterers. The traditional Swerling models have since been generalized to incorporate other distributions not limited to the Rayleigh and Ricean densities, as described in Shnidman (2003). The ideal point-scatterer model described previously with a non-fluctuating (deterministic) complex amplitude α is sometimes known as the Swerling zero or Swerling V model.

In practice, an actual target echo typically exhibits complex amplitude fluctuations that are partially correlated from pulse to pulse, i.e., neither coherent, nor independent. This observed behavior lies somewhere between the Swerling I (or III) model, and the Swerling II (or IV) model, which represent the two extreme cases of coherent and incoherent scattering, respectively. Target echoes received by OTH radar are often best approximated by the Swerling I (or III) models as opposed to the Swerling II (or IV) models. One reason for this is that the target echo complex amplitude usually exhibits a relatively high inter-pulse correlation coefficient (near unity), and is hence far from being statistically independent from pulse to pulse. Another reason is that the relatively long region revisit rates, which may be two or three orders of magnitude greater than the pulse repetition interval, result in fluctuations with a relatively low scan-to-scan correlation coefficient such that $\alpha(t)$ may be considered approximately independent from scan-to-scan.

Target echo fluctuations with a high inter-pulse correlation coefficient do not significantly affect coherent integration gain. Indeed, the coherent integration gain is quite robust to small gain and phase variations in the target echo complex amplitude over the different pulses of the CPI. However, the same cannot be said for the sidelobe levels after Doppler processing. While Doppler sidelobe levels are of concern for powerful clutter signals, those of a (non-maneuvering) target echo usually fall below the background noise floor when a low sidelobe level taper is used to control spectral leakage. For these reasons, the expression derived for an ideal point-target echo $s(t)$ in Eqn. (4.14) is generally regarded as an appropriate first-order model for non-maneuvering targets provided that the OTH radar CPI is limited to a few seconds and the range resolution is greater than a few kilometers (i.e., no range migration over the CPI). This model will therefore be adopted in the following as a basis for describing the conventional signal and data-processing steps used for target detection and parameter estimation.

4.1.2 Clutter Returns

OTH radars receive powerful clutter echoes due to backscatter from large areas of the Earth's surface. The surface clutter power received in a single range-azimuth resolution cell may be 40–80 dB greater than that of an aircraft target echo. This very low signal-to-clutter ratio (SCR) arises because of the inherent "look-down" geometry of a skywave OTH radar and the coarse spatial resolution of such systems in both range and cross-range. For fighter-sized and commercial aircraft, SCR values ranging from perhaps −40 to −60 dB are more typical, while the SCR of echoes from large surface vessels can potentially exceed −40 dB. The large difference between clutter and target echo power levels places stringent demands on waveform spectral purity and receiver dynamic range for effective Doppler processing.

The target detection performance of an OTH radar system is directly impacted by the power and spectral characteristics of surface and volume clutter received across the different spatial resolution cells of the surveillance region. The clutter characteristics are influenced by the properties of the surface scatterers as well as the ionospheric paths that propagate the echoes to the receiver. This section briefly describes the main characteristics of surface and volume clutter received by an OTH radar. The first two parts of this section discuss factors affecting the power, Doppler spectrum characteristics, and spatial variability of land and sea clutter, respectively. The third part of this section considers ionospheric clutter received due to meteors and ionospheric irregularities.

4.1.2.1 Terrain Clutter

The surface clutter power that competes for detection with a target echo is proportional to the effective backscatter RCS σ_c of the scattering patch that returns clutter in the same spatial resolution cell as the target. To maintain simplicity, consider a single ionospheric mode and a one-hop path to the surveillance region. For this mode, assume that the radar system gain and propagation channel effects are similar for both the clutter and target echoes received in a particular spatial resolution cell. In this case, a first-order indication of the received SCR (prior to Doppler processing) may be approximated in the simplest manner by Eqn. (4.19), where σ is the target RCS. This expression provides a useful starting point for discussion.

$$\text{SCR} = \frac{\sigma}{\sigma_c} \tag{4.19}$$

The effective clutter RCS σ_c may be written as in Eqn. (4.20), where σ_o is the normalized backscatter coefficient characterizing the reflective properties of the surface in terms of an effective RCS per unit area, and A is the area of the scattering patch that contributes to the clutter echo received in the considered spatial resolution cell. The dimensionless quantity σ_o (dB) depends on the electrical properties and topography of the scattering surface, in addition to the signal frequency, grazing angle-of-incidence, and polarization. To illustrate the key points, the surface properties are assumed to be approximately homogeneous within a single (range-azimuth) resolution cell.

$$\sigma_c = \sigma_o \times A \tag{4.20}$$

The surface area A (m^2) is essentially determined by the ground range and cross-range resolutions of the radar system, denoted by ΔR and ΔL, respectively, as well as the geometry of the skywave propagation path. The ground range resolution may be approximated by $\Delta R = (c/2B)\cos\psi$, where c is the speed of light in free space, B is the radar signal bandwidth, and ψ is the grazing angle-of-incidence of the skywave path to the scattering surface. The cross-range resolution may be written as $\Delta L = R\Delta\theta$, where $\Delta\theta$ is the azimuth resolution of the receive antenna beam, and R is the range of the resolution cell from the radar. The azimuth resolution may be approximated by the Rayleigh resolution limit $\Delta\theta = \lambda/D$, where D is the receive aperture length, and $\lambda = c/f_c$ is the carrier wavelength. Specifically, $\Delta\theta$ is the angular separation between the peak and first null of a conventionally formed antenna beampattern for a ULA steered at broadside.

This yields the approximate expression in Eqn. (4.21) for the surface area of the scattering patch that contributes clutter to a single OTH radar resolution cell.

$$A = \Delta R \times \Delta L = \frac{c}{2B}\cos\psi \times \frac{\lambda}{D}R \tag{4.21}$$

The relatively simple formulas in Eqn. (4.19), (4.20), and (4.21) indicate that the SCR in a given resolution cell is proportional to the product between the bandwidth and (electrical) aperture of the radar system. For an OTH radar signal bandwidth $B = 10$ kHz, and aperture length $D = 3$ km, the resolution cell area A varies approximately between 80 and 90 dBm^2, depending on operating frequency f_c and range R. Assuming a normalized backscatter coefficient of $\sigma_0 = -20$ dB yields $\sigma_c = 60$–70 dBsm. A fighter-sized aircraft target may have a nominal RCS of $\sigma = 10$–20 dBsm. Hence, a rough estimate of the SCR for this target class may be −40 to −60 dB, which is 60–80 dB too low for reliable detection prior to Doppler processing (i.e., in a single radar pulse). The focus here is not so much on the accuracy of this estimate as on illustrating that suppression of the clutter echo by increasing the range and azimuth resolutions to enable detection (without Doppler processing) is impractical for OTH radar. Moreover, multipath propagation tends to increase the average clutter power due to the the sum of independent clutter contributions from different ionospheric modes received in the same resolution cell. Multipath effects are clearly not taken into account by Eqn. (4.19), (4.20), and (4.21).

The normalized backscatter coefficient σ_0 of land surfaces is influenced by several factors. The presence of irregular topographical features, such as rough or mountainous terrain, will tend to enhance the backscatter coefficient relative to flat land with similar electrical properties. In residential or industrial areas, the presence of buildings and other large man-made structures will also tend to enhance the backscatter coefficient. On the other hand, spatial and/or temporal changes in surface conductivity, due to rain for example, can significantly modify the backscatter coefficient of land with identical topography. Land surface conductivity depends largely on the concentration of salts in the soil as well as the presence of moisture. Moisture disassociates electrolytes and allows the ions greater mobility in an aqueous solution. This increases conductivity because electric charges are conducted through the movement of ions in the solvent when an external electric field is present. Not surprisingly, dry desert plains and flat regions covered by snow or ice tend to exhibit relatively low values of σ_o, while large cities and mountainous areas (particularly in the tropics) tend to exhibit relatively high values of σ_0.

The normalized backscatter coefficient of land surfaces may range very roughly from about −40 dB to perhaps −10 dB or higher at typical OTH radar viewing angles. The presence of cities can result in sudden changes of received clutter power due to an abrupt spatial variation (increase) in the value of σ_o above the surrounding background. Identification of local "discontinuities" in received clutter power in a particular resolution cell and its correct association to a geographically known reference point (KRP) may be exploited for coordinate registration in OTH radar. A target detected simultaneously by the same radar near a KRP that marks the radar return with a distinguishable signature may be localized with higher precision (i.e., relative position errors can be significantly smaller than absolute errors). The topic of coordinate registration will be discussed at the end of this chapter.

Due to the exceedingly low SCR values encountered in OTH radar applications, an important property of the clutter echo relates to its Doppler spectrum characteristics. In frequency-stable ionospheric conditions, most of the clutter power returned from land is typically concentrated in a relatively small band of Doppler frequencies centered near 0 Hz. The land clutter spectrum may not maximize at 0 Hz because signal reflection from the ionosphere can impose a positive or negative Doppler offset that is typically range, azimuth, and propagation mode dependent. This frequency-offset effectively shifts the centroid or peak of clutter Doppler spectrum associated with a particular propagation mode by a certain amount (often less than a 1–2 Hz at a carrier frequency in the middle of the HF band) to one side of zero Doppler. On the other hand, random changes of the signal phase path during the CPI causes the land clutter energy to spread in Doppler frequency. This may be interpreted as multiplicative noise that effectively broadens the width of the clutter Doppler spectrum. The amount of spread is also often range, azimuth, and propagation mode dependent, particularly for F-region reflections.

Considering a single ionospheric propagation mode, the clutter Doppler frequency shift and spread characteristics received at a particular resolution cell and time can be represented in terms of a statistically expected power spectral density function, denoted by $s(f)$, where f is the Doppler frequency. A simple parametric model to approximate $s(f)$ for the case of land clutter is the power function in Eqn. (4.22). This profile is determined by a frequency offset parameter f_m, which determines the mean Doppler shift or centroid frequency of the distribution, and a frequency spread parameter f_w, which models the spectral width of the distribution. The amplitude α_c determines the maximum value of the spectral density at $f = f_m$, and hence controls the received clutter power. The parameter n shapes the fall-off characteristics of $s(f)$ with frequency about the peak value (a higher value of n corresponds to a more rapid fall-off with frequency).

$$s(f) = \frac{\alpha_c}{1 + [(f - f_m)/f_w]^n} \tag{4.22}$$

At a frequency near the middle of the HF band (e.g., $f_c = 15$ MHz), the mean Doppler frequency shift imposed on a skywave signal due to F-layer propagation over a quiet daytime mid-latitude ionospheric path (not near the terminators) typically has a magnitude of less than 2 Hz. Moreover, for an air-mode CPI between 1 and 4 seconds long, the power of the land clutter Doppler spectrum may fall off by 60–80 dB with respect to its maximum value at a Doppler frequency within 5 Hz from the peak in frequency-stable ionospheric conditions. On the other hand, an aircraft traveling at a constant velocity gives rise to an echo with a well-defined Doppler shift that is typically in the range $f_d = 5$–50 Hz. In other words, Doppler processing can increase the SCR of an aircraft target echo by margins of 60–80 dB relative to that in a single pulse within a few Hertz from the land clutter peak. Indeed, for fast-moving targets with a high relative velocity, SCR ceases to be the factor that limits detection performance as the target echo will often lie in a Doppler-frequency region where external noise dominates clutter.

Random process realizations that closely approximate the pulse-to-pulse correlation properties of the received land clutter echo may be generated using auto-regressive (AR) models of relatively low order (Marple 1987). The simplest (first-order) AR model described by the recursive relation in Eqn. (4.23) is often used to represent the baseband complex envelope of the land clutter signal $c(t)$. Here, T is the pulse repetition interval, $\rho(T) \in [0, 1]$ is the inter-pulse complex correlation coefficient, and $\xi(t)$ is innovative (circularly-symmetric) zero-mean white Gaussian noise with variance equal to the clutter power. For PRIs less than $T = 0.01$ seconds, the value of $\rho(T)$ is typically above 0.99 in frequency-stable ionospheric conditions. As described in Marple (1987), the statistically expected power spectral density $s(f)$ of a sampled first-order AR process has a (periodic) Lorentzian form, which is related to a value of $n = 2$ in the power function of Eqn. (4.22).

$$c(t) = \rho(T)c(t - T) + \sqrt{1 - |\rho(T)|^2}\xi(t) \tag{4.23}$$

In practice, the presence of different ionospheric layers (and magneto-ionic splitting) leads to multipath skywave propagation. As a result of the different virtual heights of signal reflection, the clutter received in a particular OTH radar spatial resolution cell may be the sum of multiple contributions scattered from separate patches of the Earth's surface. This scenario is sketched in Figure 4.4, where scattering patches 1 and 2 give rise to clutter echoes in the same radar resolution cell (i.e., same group-range and cone angle of arrival) from possibly quite different types of surfaces (e.g., one land and the other sea). Moreover, clutter echoes scattered from different patches may be subjected to different Doppler frequency shifts and spreads due to propagation via independent ionospheric paths. The superposition of clutter echoes propagated by different ionospheric modes therefore tends to broaden the clutter-occupied region of Doppler space in a spatial resolution cell with respect to the Doppler contamination present on individual path contributions.

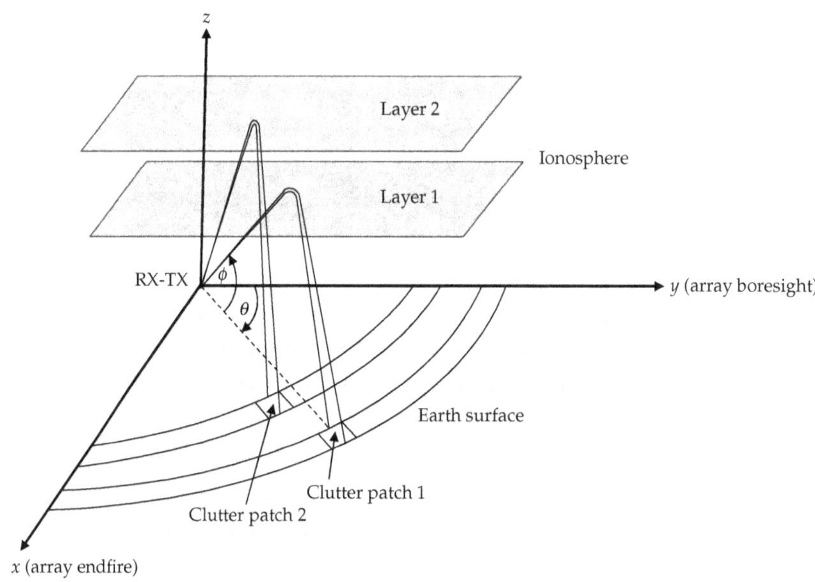

FIGURE 4.4 Notional illustration of two clutter-scattering patches illuminated by signal reflection from ionospheric layers at different heights for an OTH radar with a receiving ULA oriented along the x-axis. The dimensions of the clutter patches are determined by the range and cross-range resolutions of the system. The clutter echoes scattered from patches 1 and 2 have similar group-ranges and cone angles-of-arrival, and are therefore received by the radar in the same spatial resolution cell. However, the scattering patches have different ground ranges and great-circle bearings relative to the radar. In other words, clutter backscattered from different geographic areas of the Earth's surface may be received in the same spatial resolution cell of the radar when multipath is present.

4.1.2.2 Sea Clutter

Understanding the characteristics of the HF sea-echo is of fundamental importance to ship detection in OTH radar. Experimental measurements made by the Naval Research Laboratory have demonstrated that the clutter power returned from the Atlantic ocean surface is typically an order of magnitude greater on average than that returned from regions in the central United States (Skolnik 2008b). This difference occurs because salt water has a relatively high conductivity, and specific Fourier components of ocean wave-height spectrum give rise to a Bragg-resonant scattering mechanism that will be described below.

The power and Doppler spectrum characteristics of the clutter echo backscattered from the ocean surface depend heavily on ionospheric conditions and sea state, the latter being a function of the prevailing surface-wind speed and direction. The normalized backscatter coefficient of a sea surface that is fully developed by local winds at the radar operating frequency is theoretically expected have an average value of about $\sigma_0 = -20$ dB when the wind blows in a direction approximately parallel to the radar beam. This value may decrease by several decibels as the orientation of the surface-wind changes from being parallel to perpendicular with the azimuth of the radar beam.

Although the normalized backscatter coefficient of a fully developed sea may be more than 10 dB higher than that of dry and flat land, it is perhaps 10 dB or more lower than that of rough/mountainous terrain in tropical areas. On the other hand, the value of σ_0 for the ocean can at times be smaller than -40 dB when the sea surface is exceptionally calm. This is because a skywave signal incident on a "mirror-smooth" sea-surface at shallow grazing angle is almost specularly reflected in the forward direction, with very little energy being backscattered toward the radar. The value of σ_0 can change abruptly within an OTH radar surveillance region over land, due to the presence of cities for example, but its spatial variation over open ocean areas is relatively smooth by comparison. This is due to the near-uniform conductivity of the ocean surface and the relatively gradual changes in sea state.

In addition to different clutter-power properties, there are significant differences in the Doppler spectrum characteristics of land and sea clutter. These differences become particularly evident when the clutter spectrum is observed at high Doppler-frequency resolution (as in OTH radar ship-detection tasks, where a CPI of 20 seconds or more may be used). Figure 4.5a shows an intensity-modulated range-Doppler map at the output of an OTH radar beam where the range coverage straddles a number of land-sea boundaries. Note that the structure of the clutter Doppler spectrum alternates from a profile that has a single dominant peak to one with two well-resolved peaks as the range-gate moves over land and sea surfaces, respectively.

The Doppler spectra shown in Figure 4.5b are line plots extracted from Figure 4.5a at range resolution cells containing backscatter from land and sea surfaces, as indicated by the two dashed horizontal lines in Figure 4.5a. In this example, the land clutter is stronger than sea clutter at a comparable range. As expected, land clutter is concentrated in a narrow band of Doppler frequencies centred near 0 Hz. On the other hand, the sea clutter exhibits two well-resolved spectral components on either side of zero Doppler frequency. This feature of the clutter Doppler spectrum, which clearly distinguishes echoes from land and sea surfaces, provides a useful discriminant for identifying the locations of land-sea boundaries in radar coordinates. These boundaries may be correlated with coastline maps to provide an alternative (spatially distributed) KRP source for coordinate registration in OTH radar.

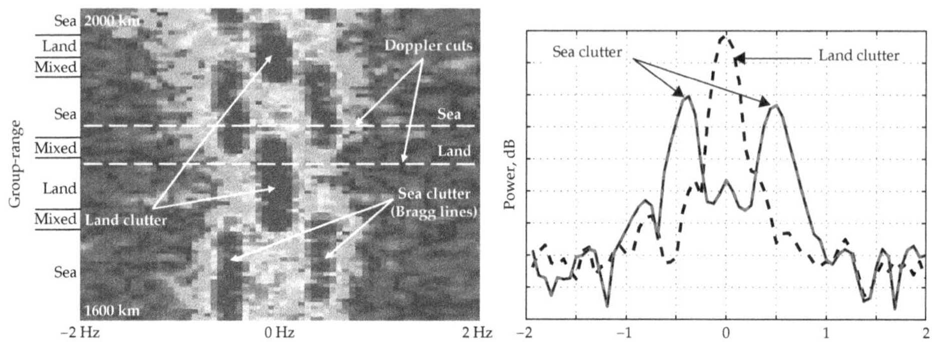

(a) OTH radar range-Doppler map in littoral area.

(b) Land- and sea-clutter Doppler spectra.

FIGURE 4.5 Variation of clutter power and Doppler spectrum characteristics received by a skywave OTH radar as a function of group-range in a particular beam. The data were recorded at a carrier frequency of 19.440 MHz. The group-range limits of the display are approximately from 1600 to 2000 km, while the Doppler frequency interval of ±2 Hz corresponds to unambiguous relative velocities between ±55 km/h. © Commonwealth of Australia 2011.

The question arises as to the mechanism responsible for the observed pair of spectral components in the ocean-clutter Doppler spectrum. The answer resides in the physics that describes the motion of water waves on the ocean surface and the interaction of these waves with the HF radar signal in the scattering process. The basic phenomenology involved is explained below, with a more detailed description provided in Chapter 5.

To a good approximation, the ocean surface can be decomposed into a spectrum of surface gravity waves that satisfy the dispersion relation in Eqn. (4.24), where ω and κ are the angular frequency and wavenumber of the water wave, respectively, g is acceleration due to gravity, and d is the water depth.

$$\omega^2 = g\kappa \tanh(\kappa d) \tag{4.24}$$

In other words, the ocean surface can be represented as a superposition surface gravity waves, or sinusoidal "Fourier surfaces," which are parameterized by different wavenumbers and propagation directions and weighted by a continuous directional wave-height spectrum. In deep water, the dispersion relation reduces to the simpler form of Eqn. (4.25) for water waves of length less than $2d$. For the water waves that give rise to resonant scatter at HF (5–50 m in length), modeling the ocean surface height in terms of a directional spectrum of surface gravity waves satisfying the deep water dispersion relation in Eqn. (4.25) is applicable in open ocean areas not too close to coastlines.

$$\omega^2 = g\kappa \tag{4.25}$$

The interaction between a radio wave and the sea surface was investigated in Barrick (1972a, b) to derive an expression for the Doppler power spectral density structure of the scattered field. The derivation assumes the wave heights are much smaller than the radio wavelength, such that the ocean surface may be considered a slightly rough reflector of the radar signal. This condition is often met in the HF band when the sea is not too rough. Under such conditions, the Doppler spectrum of the scattered radio wave may be expressed in terms of the ocean directional wave-height spectrum in the form of a perturbation series expansion (Barrick 1972a).

Barrick's model of the sea-echo Doppler spectrum will be discussed in more detail in Chapter 5. This section briefly describes the origin of the dominant (first-order) components of the expansion, which give rise to the observed pair of spectral peaks in the sea-clutter Doppler spectrum of Figure 4.5. To explain this feature, it is observed from Eqn. (4.25) that the phase velocity v_p of a water wave is given by Eqn. (4.26), where L is the wavelength.

$$v_p = \frac{\omega}{\kappa} = \sqrt{\frac{gL}{2\pi}} \tag{4.26}$$

Specifically, the most powerful clutter contributions returned to the radar are associated with resonant echoes scattered from two specific components of the directional wave-height spectrum. These two specific components of the ocean wave-field are referred to as the advancing and receding Bragg wave-trains. In the backscatter case, the Bragg wave-trains move directly toward and away from the radar, respectively.

Importantly, the Bragg wave-trains have a wavelength L that satisfies the backscatter (monostatic) resonance condition in Eqn. (4.27), where λ is the radio wavelength and ψ is the grazing angle at which the signal is incident on the sea surface. Figure 4.6 illustrates the reflection of an HF skywave signal from the advancing Bragg wave-train. The condition

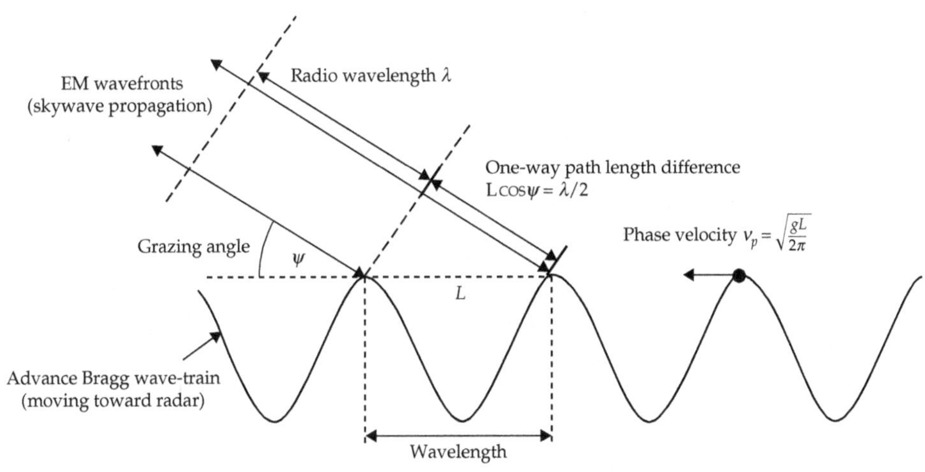

FIGURE 4.6 Resonant (first-order) sea clutter is produced by the advancing and receding Bragg wave-trains, which give rise to two discrete spectral components in the echo Doppler spectrum (neglecting the effect of ionospheric broadening). The condition for resonant scattering is illustrated for skywave propagation and the advancing Bragg wave-train, which moves directly toward the radar. The two-way path length difference between reflections of the signal from points in phase on the water wave equals an integer multiple of the radio wavelength λ. The same condition applies for the receding Bragg wave-train, which has the same wavelength, but moves in the opposite direction (i.e., directly away from the radar).

in Eqn. (4.27) results in fully constructive interference of the echoes returned from points in phase on this water wave because the *two-way* path-length difference is precisely equal to an integer multiple of λ.

$$L \cos \psi = \lambda/2 \tag{4.27}$$

This phenomenon is analogous to the Bragg wave-trains acting as diffraction gratings for the incident radio wave. Second- and higher-order terms of the perturbation series expansion, which respectively involve scattering from two or more different surface gravity waves in the directional wave-height spectrum (before the signal is returned to the radar), contribute a continuum of spectral components to the sea-clutter Doppler spectrum. These terms will be described in the next chapter. By substituting the expression for L satisfying Eqn. (4.27) into Eqn. (4.26), it is readily shown that (in deep water) the Bragg wave-trains move with a phase velocity v_p given by Eqn. (4.28). More precisely, this is the phase velocity of the Bragg wave-trains relative to the underlying body of water, which may itself be moving due to the presence of surface currents.

$$v_p = \pm \frac{1}{2} \sqrt{\frac{g\lambda}{\pi \cos \psi}} \tag{4.28}$$

Neglecting the influence of surface currents, the Bragg wave-train velocity imposes a Doppler frequency shift of $f_b = 2v_p \cos \psi/\lambda$ on the backscattered first-order clutter. Using the expression for v_p in Eqn. (4.28), the Doppler shift f_b imposed on the first-order clutter echoes is given by Eqn. (4.29). For the case of near-grazing incidence $\psi \rightarrow 0$, an approximate expression for the Bragg wave-train Doppler shifts is $f_b = \pm 0.102\sqrt{f_c}$, where f_c is the carrier frequency in MHz and f_b is in units of Hertz. In summary, scattering from the Bragg wave-trains produces two so-called "Bragg lines" in the sea-clutter Doppler spectrum with a frequency separation of $2|f_b|$.

$$f_b = \pm \sqrt{\frac{g \cos \psi}{\pi \lambda}} \tag{4.29}$$

The experimental data in Figure 4.5 was recorded at a frequency of $f_c = 19.440$ MHz. In this case, skywave propagation was supported by the sporadic-E layer, which has a natural one-hop (ground distance) horizon of around 2000 km. Sporadic-E propagation over longer ground ranges may occur via multi-hop paths. According to Eqn. (4.29), the Bragg lines should have a Doppler shift of approximately $f_b = \pm 0.45$ Hz as $\psi \rightarrow 0$. The separation between the Bragg lines in Figure 4.5b agrees well with the expected value of $2|f_b| = 0.9$. However, the Bragg lines are not spaced symmetrically about 0 Hz. The two dominant frequency components are shifted slightly toward positive Doppler frequency and centered at approximately $\delta f = 0.05$ Hz.

As the land clutter is centered very close to zero Doppler frequency, this excludes the possibility of an ionospheric Doppler shift, since both sea and land clutter are returned by one and the same (sporadic-E) layer. An ionospheric Doppler shift close to 0 Hz is typical for skywave propagation via the sporadic-E or normal-E layers. The observed displacement of the Bragg lines by a Doppler shift δf may be attributed to surface currents, which modify the radial velocities of the advancing and receding Bragg wave-trains by an equal amount v_s. This in turn displaces the Bragg line Doppler shifts, as in Eqn. (4.30), where the displacement frequency is given by $\delta f = 2v_s \cos \psi/\lambda$. For $\delta f = 0.05$ Hz and $\psi \rightarrow 0$, the effective radial component of surface-current velocity is estimated as $v_s = 0.38$ m/s (0.74 kt) in this example.

$$f_b = \pm\sqrt{\frac{g\cos\psi}{\pi\lambda}} + \delta f \tag{4.30}$$

Inspection of Eqn. (4.29) reveals that the Bragg-line Doppler frequencies f_b vary with the *square root* of operating frequency. On the other hand, the Doppler shift of a target echo moving with constant relative velocity varies in proportion to the operating frequency. The difference between these two Doppler-shift relationships raises the possibility for changing the operating frequency to unmask slow-moving surface targets that are obscured by powerful Bragg lines at a particular frequency. However, a 20% or greater change in carrier frequency may be needed to unmask a target echo under a Bragg line when using a 20-second CPI and classical (FFT-based) Doppler processing. In skywave OTH radar systems, this procedure is therefore only effective when the ionosphere supports frequency-stable HF signal propagation to the surveillance region (i.e., with minimal Doppler spectrum contamination) at frequencies spaced a few MHz apart. Although this is infrequently the case for F-region propagation, the approach may be feasible when an intense and frequency-stable sporadic-E layer with critical frequency well above that of the F2-layer is present at the path mid-point. The subject of dual-frequency operation to unmask target echoes under the Bragg lines will be discussed in the context of HF surface-wave radar in the next chapter.

The power of the first-order spectral components with positive and negative Doppler frequencies are proportional to the square of the amplitudes (heights) of the advancing and receding Bragg wave-trains, respectively. Consequently, a Bragg line reaches its maximum amplitude in the Doppler spectrum when surface winds blowing in a direction parallel to the radar beam excite and fully develop the Bragg wave-train component of the ocean wave-field that gives rise to it. In practice, the dominant Bragg line is often sufficiently strong to mask echoes from the largest surface-vessels.

Although typically 20–30 dB weaker than the first-order clutter, the second-order clutter is distributed over a relatively broad continuum of Doppler frequencies, which in general extends between and beyond the Bragg lines. Second-order clutter contributions therefore have the potential to obscure useful signals over a larger portion of the target velocity search space compared to first-order clutter. The power of second-order clutter tends to increase with both operating frequency and sea-state. High sea-states may raise the second-order clutter continuum by 10–20 dB relative to calm seas, which can have a significant (detrimental) impact on the detection of small- and medium-sized surface vessels.

The pulse-to-pulse correlation properties of skywave OTH radar clutter returned from the sea via a single ionospheric mode in a particular range-azimuth cell can (for signal-processing purposes) be modeled in the simplest manner using a second-order AR process, with realizations generated according to Eqn. (4.31). Here, the AR coefficients b_i for $i = 1, 2$ determine the location of the poles inside the unit circle and hence the clutter Doppler spectrum structure (including the effect of ionospherically induced Doppler broadening), while the value of σ_ξ^2 determines the clutter power. The AR model parameters are expected to vary from one range-azimuth cell to another, while multipath propagation may be modeled using AR processes of higher order. For a detailed description of AR processes, the reader is referred to Marple (1987).

$$c(t) = -\sum_{i=1}^{2} b_i c(t - iT) + \sigma_\xi^2 \xi(t) \tag{4.31}$$

4.1.2.3 Ionospheric Clutter

An OTH radar may receive volume clutter backscattered from dynamic electron density irregularities in the ionosphere and ionized meteor trails via direct (i.e., line-of-sight) paths or indirect paths. The latter may involve one or more oblique ionospheric reflections, possibly in addition to oblique Earth surface reflections. Although the two aforementioned sources of volume clutter are governed by entirely different physical mechanisms, the echoes received from these scatterers are often collectively referred to as ionospheric clutter.

Ionospheric clutter has the potential to severely limit OTH radar performance because the energy of the received echoes can at times spread over significant portions (or all) of the OTH radar's Doppler search space. Indeed, the upper atmosphere is home to HF-signal scatterers with velocities that are orders of magnitude higher than natural scatterers on the Earth's surface. As a result, ionospheric clutter can potentially degrade the detection of echoes from both slow- and fast-moving targets when its level rises above the background noise-floor. The main characteristics of ionospheric clutter received due to meteors and electron density irregularities are now discussed in turn.

Echoes arising due to meteors that enter the Earth's atmosphere are an omnipresent source of ionospheric clutter to an OTH radar. Detailed descriptions and mathematical modeling of the radar response to meteor scatter at high and medium radio frequencies can be found in Thomas, Whitham, and Elford (1988), Cervera and Elford (2004), and Cervera et al. (2004), while the physical characteristics of meteors and their ionized trails have been documented in the early work of McKinley (1961). The reader is referred to these authoritative contributions for a comprehensive coverage of these topics.

Meteoroids orbit the Sun on individual paths (sporadic meteoroids) or as members of streams with a common orbit. The latter gives rise to meteor showers when the Earth intersects the meteoriod stream. The initial mass m_∞ of a meteoroid before it enters the Earth's atmosphere may range from 10^{-10} g to perhaps tens of grams (although there is no upper limit for lower frequency events), while the particle velocity V may vary from 10–70 km/s. Meteoroids with lower velocities than about 10 km/s enter into a geocentric orbit (space debris), while meteors with higher velocities than about 70 km/s escape the sun's gravity (interstellar material). With the exception of very small meteroids, the particles experience very little deceleration as they disintegrate in the atmosphere. Beyond a limiting mass, the cumulative flux of meteors is approximately proportional to $1/m_\infty$, and at the smaller end of the mass spectrum, sporadic meteors greatly outnumber shower meteors.

A meteoroid entering the Earth's atmosphere heats up, and atoms begin to evaporate off the surface of the particle in a process known as ablation. Energetic atoms are ionized upon collision with a neutral constituent and the ions thermalize after approximately 10 collisions. This process leaves a trail of enhanced ionization that often forms between altitudes of 90–120 km. A meteor trail typically extends 10–15 km in height, while radius of the plasma column may vary from perhaps 15 m to about 1 m in the vicinity of the meteoroid. The characteristics of the meteor trail depend largely on m_∞, V, and the zenith angle of the meteoroid path χ.

For radar purposes, a trail is characterized by its maximum ionization line density q_M in electrons per meter. The quantity q_M may be modeled as in Eqn. (4.32), where q_z is the maximum zenithal line density, i.e., the maximum value of the electron line density that would be produced if the meteoroid was incident vertically, while $a = 0.965(V/40)^{-0.028}$ and $b = 0.84 + 0.02(V/40)^{-3.5}$ are empirically determined constants that depend on the particle's velocity V expressed in units of km/s (Cervera and Elford 2004). For example, a zenithal maximum ionization line density of $q_z = 10^{10}$ m^{-1} corresponds to a meteoroid mass of about 10^{-6}g and a velocity of approximately 35 km/s.

$$q_M = q_z(\cos\chi)^b, \quad q_z \propto (m_\infty)^a \tag{4.32}$$

As stated in Thomas, Whitham, and Elford (1988), an artificial but convenient boundary between under-dense trails ($q_M < 10^{-14}$ m^{-1}) and over-dense trails ($q_M > 10^{-14}$ m^{-1}) may be defined with respect to the radio-wave scattering process. The transition from under-dense to over-dense trails is actually a gradual one taking place over a range of electron line densities between 10^{13} and 10^{15} m^{-1}. In under-dense trails, each electron may be regarded as an independent scatterer, while an over-dense trail scatters more like a macroscopic conducting surface (similar to the ionosphere). As far as OTH radar is concerned, the number of echoes from under-dense trails greatly exceeds those from over-dense trails. The plasma density in the vicinity of the meteoroid can be high enough to give rise to over-dense scattering.

Meteor echoes are known to have two main components, known as the head and trail echo. The head echo is thought to arise from a localized region of ionization (hot plasma) in the vicinity of the meteoroid, which is continually regenerated at lower altitudes as the particle descends.[2] This echo is short-lived on the time scale of an OTH radar CPI, as the meteoroid path only intercepts the radar field of view for a brief instant. The latter is scattered from the ionized trail (cooled plasma column) formed along the meteoroid trajectory. Meteor trails are capable of producing relatively longer lasting echoes that may persist from a few tenths of a second to more than a second, depending on the trail characteristics and radar frequency. The ionized trail dissipates by ambipolar diffusion, turbulence, and chemical processes. Diffusion dominates for a few seconds, while turbulence is the chief mechanism for trails lasting 30 seconds or more (Thomas, Whitham, and Elford 1988).

An individual meteor echo is generally confined to a small number of range cells (typically under three for typical OTH radar bandwidths), but contamination may arise over large range extents because it is possible for many meteor trails to be illuminated by the radar beam over all direct and indirect paths. Meteor echoes mostly appear spread in Doppler due to their transient nature on the scale of an OTH radar CPI. The echoes may also appear Doppler shifted, as the trail ionization can drift under the influence of neutral winds in the upper atmosphere.

After the trail is formed, diffusion causes the echo power to decay with time. This decay is modeled according to the exponential law in Eqn. (4.33), where the time constant τ_c depends on trail characteristics and radio frequency, while the initial echo power P_0

[2]Head echo is not from the meteoroid, as the particle is far too small to effectively scatter HF signals.

additionally depends on radar system parameters such a transmit power and antenna gains. The echo has an effective Doppler bandwidth of approximately $1/\tau_c$ Hz after Fourier transformation. The value of τ_c decreases with increasing frequency. Meteor echoes with Doppler bandwidths ranging from 10 to 30 Hz are often observed by HF radars with strengths that may exceed the received background noise level by 10–40 dB. Although clutter returned from the Earth's surface is much stronger than ionospheric clutter from meteors, the former occupies a significantly narrower Doppler width (about 1–2 Hz over typical OTH radar CPI in frequency-stable ionospheric conditions).

$$P(t) = P_0 \exp\left(-\frac{t}{\tau_c}\right) \tag{4.33}$$

The impact of meteor clutter in OTH radar systems depends on the rate and spatial distribution of received echoes that are sufficiently powerful to potentially deny the detection of useful signals. Of particular interest to OTH radar operation is the variation of this rate with time of day, day of year, propagation mode, operating frequency, slant range, and beam direction. Computer models for predicting these rates were developed and compared with experimental HF radar observations in Thomas, Whitham, and Elford (1988).

Sporadic meteor echoes occur more or less randomly throughout the year, with rates that may vary between 0.1 and 5 detectable events per second in a particular OTH radar surveillance region. However, the meteor flux, and consequently the rate of detected echoes, can increase dramatically during meteor showers, which occur at predictable times of the year. Meteor showers such as the Leonids and Eta Aquarids can last from a few days to tens of days. At such times, the prevailing meteor flux can lead to echo rates that are an order of magnitude higher than the underlying sporadic meteor background.

The rate of meteor echoes also exhibits significant diurnal variation. Both sporadic and meteor shower echoes are received more frequently in the morning hours. Depending on the pointing direction of the OTH radar beam (azimuth and elevation), the peak rate in the case of meteor shower echoes typically occurs around 06:00 local time (LT) in the direction of the apex of the Earth's travel. The minimum rate for such echoes is observed around 18:00 LT in the opposite (anti-apex) direction. Other significant components of the non-isotropic sporadic meteor distribution, including the helion, anti-helion, and torroidal source radiants, are described in Cervera et al. (2004).

As the OTH radar surveillance region and operating frequency are constrained by mission requirements, the mitigation of meteor echoes received by the system is traditionally attempted in the signal-processing domain. In particular, spatial processing on receive may be used to attenuate echoes not incident from the radar look direction. The use of conventional beamforming can be quite effective if the antenna array is well calibrated, particularly when the receiver allows for independent control of the beam pattern in azimuth and elevation.

For meteor echoes in the main beam that cannot be effectively suppressed by spatial processing, the only option is rejection in the time domain by exploiting the transient nature of such echoes relative to target echoes. An ameliorative time-domain signal-processing step that excises meteor echoes received during the radar dwell will be described later in this chapter. As the demand for OTH radars with higher sensitivity grows, echoes from smaller and smaller meteors may be expected to rise above the background noise floor. Although echoes from these meteors have relatively lower power, a potential issue is that such meteors are much more numerous and the associated clutter may impact target detection over a greater proportion of range, beam, and Doppler cells.

The term *spread-Doppler clutter* (SDC) is often used to describe all undesired echoes that have an energy distribution not concentrated around a well-defined Doppler shift. In the case of SDC backscattered from ionospheric (electron density) irregularities with different scale sizes and velocities, the spread of echo energy in Doppler frequency arises due to the rapid and random fluctuations of the signal phase-path over the CPI. In contrast to (short lived) meteor echoes, the SDC signal scattered from ionospheric irregularities is present over the entire dwell time.

Furthermore, unlike the (spatially localized) contamination from an individual meteor echo, SDC echoes from ionospheric irregularities are usually received over a continuous interval of ranges and beams. Besides this important difference, meteor clutter will often not obscure the detection of echoes from the same target over a significant number of consecutive dwells. On the other hand, SDC from ionospheric irregularities can degrade sub-clutter visibility across large portions of the radar search space over many consecutive dwells, and is therefore more likely to adversely effect tracking.

An introduction to the physical formation and synoptic characteristics of ionospheric irregularities can be found in Davies (1990) and references therein. So-called "fast" SDC can at times be severe enough to obscure regions of the Doppler frequency search space that contain echoes from high-speed aircraft targets. Ionospheric phenomena that gives rise to SDC tend to be most prevalent at nighttime, particulary at low and high magnetic latitudes, but may also occur (albeit less frequently) in mid-latitude regions. For OTH radars with primary reflection points in the mid-latitude ionosphere, significant SDC may be received from regions outside the radar coverage through first- and higher order range ambiguities of the radar waveform. Alternatively, SDC echoes with group-ranges coincident with those of the OTH radar coverage may be received through the side or back lobes of the antenna pattern when the surveillance region is not in the same direction from the OTH radar as the irregularities.

The mitigation of SDC from dynamic ionospheric irregularities by means of signal processing has proven to be a difficult challenge in practical OTH radar systems. Alternative approaches to reduce the impact of such phenomena include appropriate waveform PRF selection to manage range ambiguities, and non-recurrent waveform design to manipulate the regions of Doppler space affected by range-ambiguous echoes (Clancy, Bascom, and Hartnett 1999). Judicious waveform design and parameter selection is important for OTH radars with a coverage that points toward the equator, as range-ambiguous SDC can at times be received due to the equatorial spread-F phenomenon.

On the other hand, an OTH radar located at higher mid-latitudes may receive SDC from ionospheric irregularities in the auroral region. In this case, the SDC can be received at the same group-ranges as useful signals (i.e., range-coincident), but the Doppler spread echoes may be incident from directions different to those of useful signals. Obviously, such SDC cannot be mitigated by controlling range ambiguities. In this case, careful transmit and receive antenna design is needed to attenuate the radar response to SDC that may enter through the side or back lobes of the combined (co-array) radiation pattern of the system.

4.1.3 Noise and Interference

Electromagnetic noise in the HF band establishes a lower limit on the strength of a target echo that can be reliably detected by an OTH radar. Environmental noise (i.e., originating from natural sources external to the radar) almost always dominates internal receiver noise of thermal origin at the operating frequencies of interest to OTH radar. The median environmental noise power spectral density will generally exceed −175 dBW/Hz around the middle of the HF band, while the internal noise level of an HF receiver may be −195 dBW/Hz. In other words, the median environmental noise power is often 20 dB or more higher than the internal receiver noise at frequencies of interest to OTH radar (Barnum and Simpson 1997).

This section discusses the two main environmental sources of HF noise, which are broadly classified as being of atmospheric or galactic origin, as well as man-made noise and interference that comes from unintentional and intentional radiators, respectively. The relative importance of different noise sources to OTH radar operation not only changes as a function of time of day, season, and geographic location, but it also varies with operating frequency. Most aircraft targets compete for detection in a Doppler frequency region where external noise dominates clutter. It is therefore important to understand the morphology of the power spectral density and spatial (i.e., direction-of-arrival) properties of different HF noise and interference sources, particularly in terms of their potential impact on SNR, and hence OTH radar detection performance. This is particularly relevant for the detection of fast-moving targets with potentially low RCS.

4.1.3.1 Atmospheric and Galactic Noise

A major component of environmental noise in the HF band is atmospheric noise due to lightning. Lightning discharges generate profuse amounts of radio-wave energy over a broad frequency spectrum. The power spectral density level of radio emissions from a lightning strike does not maximize in the HF band, but tends to increase with decreasing frequency into the medium-frequency (MF) 300–3000 kHz and low-frequency (LF) 30–300 kHz bands. However, components of the signal radiated in the HF band may be received over very long distances by virtue of skywave propagation.

Atmospheric noise generated by electrical storms around the world often constitutes the dominant source of environmental noise received by an OTH radar system. More precisely, atmospheric noise is often the main contributor to the background (ambient) noise power spectral density in the lower half of the HF band, particularly at nighttime, when D-region absorption disappears and the ionosphere propagates HF signals at such frequencies more effectively over very long distances.

Atmospheric noise is typically observed as a sequence of short-duration bursts of possibly high amplitude that are superimposed on a practically time-continuous background of relatively lower intensity. The stronger (impulsive) component of atmospheric noise usually arises from thunderstorm activity that is near or local to the receiver site. Sources "near" to the receiver are assumed to be propagated by one (or two)-hop skywave paths with relatively low losses, while those "local" to the receiver may be propagated effectively via either surface-wave or line-of-sight modes. On the other hand, the time-continuous background component arises from a large number of more distant sources that typically propagate to the receiver via long-range (multi-hop) skywave paths with relatively higher attenuation due to spreading and absorption.

On a worldwide scale, it is estimated that there are about eight million lightning discharges per day, which roughly equates to about 100 "flashes" per second. Atmospheric noise received from the sum of all lightning strikes results in a power spectral density that varies with frequency, time of day, season of year, and geographic location, due to changing meteorological and ionospheric conditions.

The sum of all atmospheric noise contributions from remote sources (distant from the receiver) gives rise to an underlying power spectral density envelope that varies relatively smoothly (slowly) as a function of frequency and time. Specifically, on the scale of the OTH radar signal bandwidth and CPI, this background component of atmospheric noise exhibits a locally flat power spectral density of constant level, and is observed as temporally white Gaussian noise of constant power when sampled at the Nyquist rate.

In contrast, strong impulsive noise received from thunderstorm centers near or local to the receiver site do not deliver energy at a constant rate. A single lightning stroke has a duration of about 30 μs, with cloud-to-ground discharges typically radiating more energy than cloud-to-cloud discharges. Most lightning strikes are composed of multiple (3–4 or more) individual strokes separated by a relatively longer interval of perhaps 40–50 ms. Successive strokes can give rise to a strobing effect since they may share the same discharge channel as the leader, but are usually of weaker intensity as the source of the charge is progressively depleted. A detailed description of the different physical processes that give rise to the lightning discharge, and the wide variety of complex structures that may be manifested, can be found in Uman (1987).

The non-Gaussian component of atmospheric noise is in general received as a series of "pulses" or bursts that may exhibit a large dynamic range in amplitude. In addition, clustering of the pulses is often observed, as discussed in Hall (1966) and Shivaprasad (1971). The duration of an individual lightning strike is about 100–400 ms, which is relatively short compared to a typical OTH radar CPI. Over tropical land masses, where thunderstorm activity tends to be most concentrated, the average flash rate density can reach $0.25/\text{km}^2/\text{day}$ or higher. For an OTH radar surveillance region of one million square kilometers, this is equivalent to about 3 flashes per second on average. During less and more active storms, the number of flashes may vary from 1 to 5 per second (Barnum and Simpson 1997).

However, it should be kept in mind that the rate of very strong bursts (received with an instantaneous power level of say 20 dB or more above the background continuum) may be significantly less than the overall flash rate. With the exception of certain localized regions prone to storm activity, the frequency of lightning generally diminishes at higher latitudes and over open ocean areas. Figure 4.7 shows an experimental power versus time record of atmospheric noise sampled with a time granularity of 250 ms over a 1 minute interval by an HF receiver connected to a monopole antenna near Darwin, Australia, on a November day. The three strongest impulses received 20–40 dB above the background level have a burst duration of 1–2 sample bins (i.e., under 500 ms).

Excluding lightning events local to the receiver site, the worst impulsive noise scenario for an OTH radar occurs when an active electrical storm is present in the surveillance region. Thunderstorms in the same direction (azimuth extents) as the surveillance region, but located at longer ranges in the two-hop skywave coverage, may also have a significant impact on OTH radar performance. Contamination from a single impulsive-noise event can reduce the SNR of a target echo by 20 dB or more at the signal-processing output. An OTH radar requires mitigation strategies for impulsive noise received via one- or two-hop skywave propagation paths, as well as local impulsive noise received via the surface wave and line-of-sight modes. Signal-processing techniques to address this problem will be discussed in the third section of this chapter.

From the 1960s to the 1980s, a worldwide effort was made to measure the characteristics and variations of atmospheric noise. The results have been documented in a series of International Radio Consultative Committee (CCIR) publications collectively known as CCIR Report 322 (CCIR 1964, 1983, and 1988). These reports quantify the distribution of atmospheric noise as a function of frequency and time throughout the world in

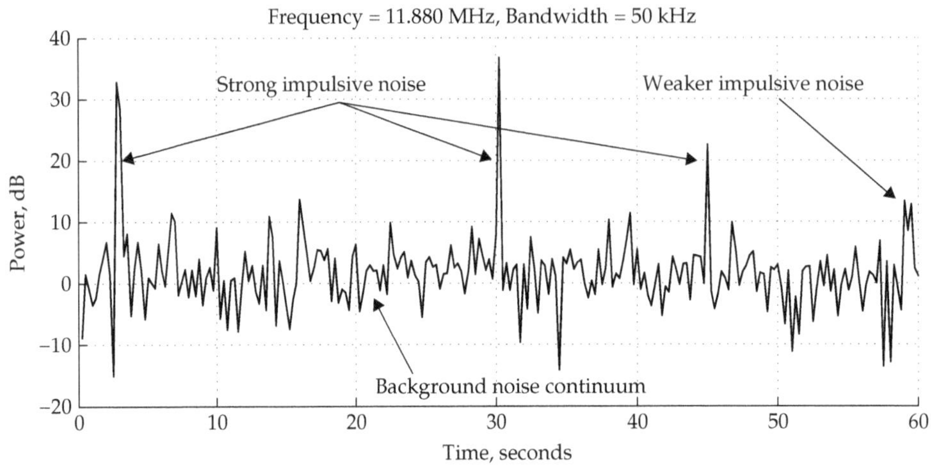

FIGURE 4.7 Example of the amplitude envelope of strong and weaker impulsive noise bursts superimposed on a continuous environmental noise background of lower level. The data were recorded at 11.880 MHz by an HF receiver of 50-kHz bandwidth connected to a monopole antenna in the Darwin region (north Australia) at 15:17 LT, November 1998.

terms of an environmental noise figure (expressed in decibels) for a lossless short vertical antenna over a perfectly conducting ground plane. Empirically derived models are provided as seasonal world maps showing the median and decile values of the atmospheric noise distribution in frequency during four-hour time blocks of the day.[3] CCIR Report 322 was eventually superseded by the International Telecommunications Union (ITU) standard HF noise model (Int. Telecomm. Union 1999a), and its subsequent updates published in later recommendations, although the CCIR reports remain a common reference.

It is useful to recall the relationship between noise figure and power spectral density as it applies to external sources. The noise power received from external sources may be expressed in terms of a noise factor, f_a, defined by $f_a = p_n/kT_0B$, where p_n is the noise power available from an equivalent lossless antenna (W), $k = 1.38 \times 10^{-23}$ is Boltzmann's constant (J/K), T_0 is a reference temperature taken as 288 K (15°C), and B is the effective receiver bandwidth (Hz). The environmental noise figure is given by $F_a = 10 \log f_a$, and may be expressed as in Eqn. (4.34), where $S = 10 \log (p_n/B)$ is the noise power spectral density available from an equivalent lossless antenna, and $10 \log (kT_0) = -204$ dBJ is a constant. This formula shows that the noise power spectral density in dBW/Hz, i.e., the power in a 1-Hz band relative to 1 W, is given by $S = F_a - 204$. The CCIR reports publish their noise data in terms of F_a in decibels for a lossless short vertical antenna over a perfectly conducting ground plane.

$$F_a = 10 \log (p_n/B) - 10 \log (kT_0) = S + 204 \tag{4.34}$$

[3]The data analysis excludes local thunderstorm contributions at the individual collection sites.

There have been varying levels of agreement between documented HF noise models and experimental data measured at sites independent of those contributing to the models. Discrepancies identified in Northey and Whitham (2000) were attributed to the CCIR model including data from a limited number of sites, particularly in the Southern Hemisphere (only two Southern Hemisphere sites were used in the first CCIR report compared with 25 in the Northern Hemisphere), and the model not including a solar cycle dependency. The occurrence of such discrepancies highlights the importance of performing site-specific HF noise measurements to inform the planning and design of OTH radar systems.

A different atmospheric noise modeling approach was taken in Kotaki and Katoh (1984), where estimates of the atmospheric noise level were calculated using global maps of thunderstorm activity combined with HF signal-propagation models. Similar to the CCIR and ITU models, a possible drawback of this model for OTH radar is that it does not account for the directional dependence of atmospheric noise. The spatial spectrum of background noise is unlikely to be uniform in azimuth and elevation. Background noise data free of strong impulsive components has the potential to exhibit a degree of spatial structure. This is not surprising given that major sources of atmospheric noise emanate from concentrated regions of thunderstorm activity. Global maps of lightning occurrence were combined with HF ray-tracing propagation calculations to form a direction-sensitive model of atmospheric noise in Coleman (2000). Such models may be used to guide OTH radar receive antenna design.

Traditional signal-processing models of atmospheric noise, which incorporate the statistics of burst duration, time between bursts, and probability densities in amplitude and phase have been proposed by Hall (1970) as well as Spaulding and Washburn (1985), for example. At a more detailed level, models that simulate the burst structure of impulsive noise to yield expressions for the received signal waveform are described in Lemmon (2001) and references therein. Whereas the aforementioned synoptic models of background atmospheric noise may be used as an input to the radar equation to analyze the (SNR) performance of a proposed OTH radar system design, models describing the statistical characteristics of stronger impulsive components may be used to guide the development of signal-processing strategies for mitigating such disturbance signals.

The second main source of naturally occurring HF noise is extraterrestrial or galactic noise. The power spectral density of galactic noise reaching terrestrial receivers from the sun and other stars in the Milky way becomes significant for radar systems that operate at frequencies extending from about 15 MHz to 100 GHz. At frequencies below 15 MHz, the level of galactic noise is limited by the presence of the ionosphere (for reasons to be described in a moment), while at frequencies above 100 GHz, it is mainly limited by atmospheric absorption. Internal receiver noise tends to dominate galactic noise above frequencies of about 250 MHz and therefore becomes limiting for radars operating at UHF and microwave frequencies.

As far as OTH radar is concerned, galactic noise tends to dominate atmospheric noise at the upper end of the HF band. This is partly because galactic noise can only impinge on Earth's surface at frequencies that are high enough to penetrate the ionosphere (i.e., above the maximum layer critical frequency). Galactic noise at lower frequencies is either reflected from the topside of the ionosphere or absorbed. For an OTH radar receiver located in an electrically quiet (i.e., remote) area, galactic noise tends to be the limiting type of external noise source in the upper half of the HF band.

Figures 4.8 and 4.9 show experimental measurements of the power spectral density received by an omni-directional (whip) antenna across 2-kHz wide channels of the HF

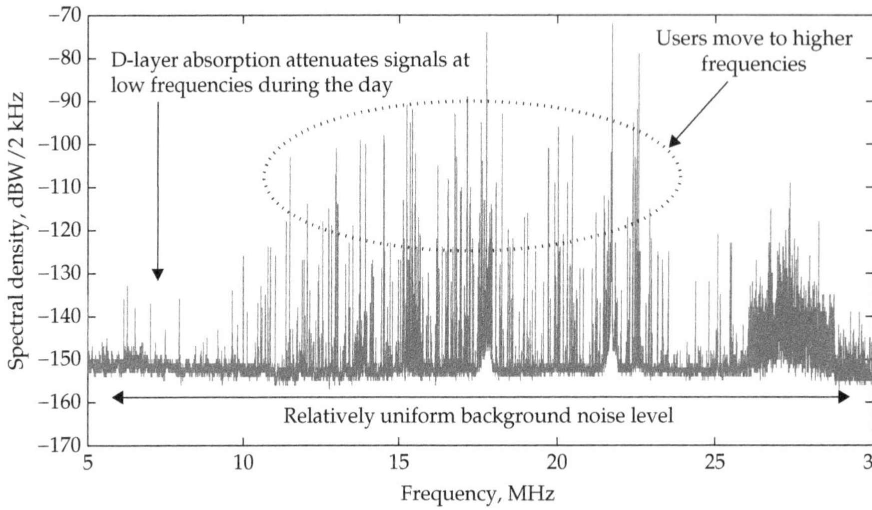

FIGURE 4.8 Example of a daytime power spectral density measurement in the HF band. The spectrum contains a large number of strong narrowband man-made signals superimposed on an underlying environmental noise background. Atmospheric noise is significantly attenuated during the day (particularly at lower frequencies) due to D-layer absorption, especially for sources propagated over long-distance (multi-hop) skywave paths. On the other hand, users take advantage of increased ionospheric support for skywave propagation at higher frequencies during the day, which explains the congestion in the middle and higher regions of the HF spectrum. The data were recorded by the spectrum monitor of the Jindalee frequency management system, located at latitude 23.532 degree (S) and longitude 133.678 degree (E), at 12:31 LT in March 2002. © Commonwealth of Australia 2011.

band at a remote location in the day and night, respectively. The high amplitude "spikes" in these figures are due to narrowband man-made signals (intentional transmissions), which will be described in the following section. Background noise usually refers to the sum of atmospheric, galactic, and man-made noise, with the latter coming from unintentional radiators. The underlying (environmental noise) envelope of the power spectral density in the two figures provides a rough indication of the background noise level received near noon and midnight local time. The time and frequency dependence of this profile is important because it provides an estimate of the potentially irreducible background noise level that an OTH radar receives during operation on a clear frequency channel.

It is evident from Figure 4.9 that the background noise power spectral density varies by a margin of about 20 dB between the lower and upper ends of the HF band. A comparison of Figure 4.9 and Figure 4.8 also reveals that the background noise level varies by a similar margin between local noon and midnight at frequencies near the low end of the HF band. The significant diurnal variation in power spectral density is partially explained by the very lossy multi-hop skywave paths existing during the day (due to D-layer absorption), which attenuates atmospheric noise contributions from long ranges, particularly at lower frequencies, and by there being no D-layer absorption at night with few or no skywave paths to terrestrial sources at the higher frequencies. These variations have significant

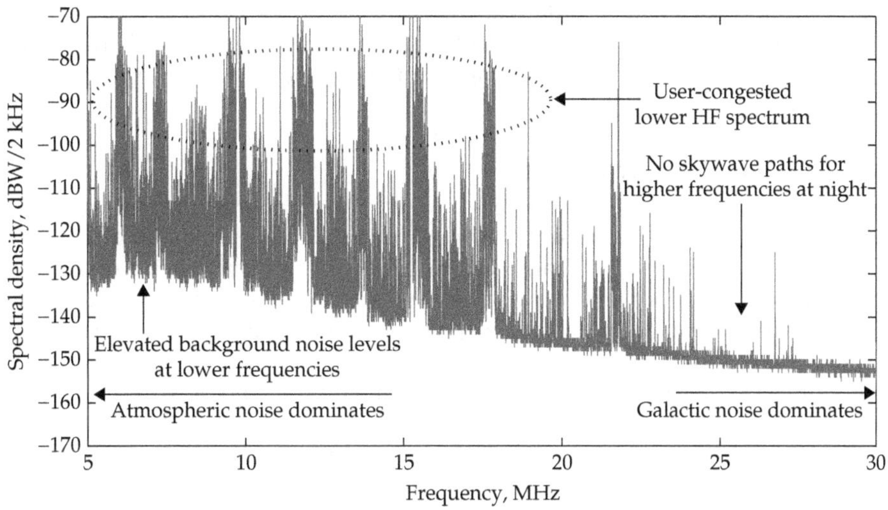

FIGURE 4.9 Example of nighttime power spectral density showing greater user congestion and higher background noise levels at lower frequencies in the HF band than for daytime. Users are forced to move to lower frequencies at night because of reduced ionospheric support for skywave propagation at higher frequencies. On the other hand, the disappearance of the D-layer at night enables the ionosphere to propagate atmospheric noise sources (and man-made signals) at lower frequencies with much less attenuation over very long distances. The data were recorded by the spectrum monitor of the Jindalee frequency management system on the same day as Figure 4.8 at 00:31 LT. © Commonwealth of Australia 2011.

implications for OTH radar performance as a function of time of day and operating frequency when SNR limits target detection performance.

HF receivers may be designed with a low noise figure to achieve internal noise spectral densities of perhaps −195 dBW/Hz (−162 dBW/2kHz) or better. With reference to the specific examples in Figures 4.8 and 4.9, this internal receiver noise level is some 30 dB lower than the background noise spectral density in the lower HF band at night, and perhaps 5 dB lower than the daytime background noise spectral density in the middle and upper regions of the HF band. Recall that an OTH radar is often required to operate in the lower HF band at night due to the reduced critical frequency of the ionosphere, but takes advantage of the increased support for skywave propagation at higher frequencies during the day to operate in the middle and upper regions of the HF band. The detection of small fast-moving targets is therefore more challenging for an OTH radar at night because the system needs to contend with higher background noise levels and potentially smaller RCS values in the lower HF band.

If the noise-field exhibits significant spatial structure at lower frequencies at night, adaptive array processing can potentially be used to reduce the external noise power below that in the conventional beamformer output. If the potential reduction in nighttime background noise by adaptive processing is significant relative to the internal noise floor of the system, there would be a valid argument for an OTH radar design that uses better matched antennas at lower frequencies. However, if the external noise does not possess sufficient spatial structure to achieve a sufficiently high cancelation ratio at the adaptive processor output, there would be little or no SNR benefit to using better matched antennas at lower frequencies. The answer to this question clearly depends on noise characteristics at the receiver site, the adaptive processing techniques used, and the properties of the antenna array.

4.1.3.2 Man-Made Noise and Interference

Man-made emissions may be broadly categorized as originating from intentional or unintentional radiators. To distinguish between the terms "noise" and "interference" in relation to man-made sources, signals received by an OTH radar from intentional (but uncooperative) radiators are referred to as radio-frequency interference (RFI), or simply interference, while emissions received from unintentional radiators are referred to as man-made noise.

Information-carrying signals from intentional radiators mainly arise from licensed users of the HF spectrum. Applications include fixed point-to-point communications, mobile communications (between land, maritime, and aeronautical platforms), as well as short-wave radio broadcasting. It is common for certain systems to rely primarily on alternative means of communication (e.g., microwave links, fiber optics, or via satellite), and to maintain an HF capability for secondary (back-up) purposes. The HF band resource is shared among users by allocating separate portions of the spectrum to different services. The efficiency of its use depends on the judicious coordination of frequencies and transmission times to minimize mutual interference. The schedules also need to be designed such that systems can maintain an acceptable grade of service as propagation conditions change. The HF band resource is regulated on a regional basis with primary and secondary services often being identified in shared regions of spectrum.

Powerful broadcast stations in the HF band may have 200–500-kW transmitters and antenna gains of perhaps 10–20 dB. Signals received from these emitters via skywave propagation can have power spectral densities 60–80 dBW/Hz above the background noise level. With the notable exception of spread-spectrum signals, transmissions from individual man-made sources occupy very narrow bands of the HF spectrum, typically between 3 and 15 kHz. As a result, the composite interference-plus-noise spectral density appears as a set of powerful "spikes" at different frequencies (due to HF band users), superimposed on a background noise envelope of lower intensity that varies relatively smoothly as a function of frequency. HF radars are permitted to use wide bands of the spectrum on a secondary basis, provided that emergency channels are barred from selection and no discernible interference is caused to other users. In other words, HF radars operate on the principle of noninterference by selecting only "clear channels" (i.e., unoccupied frequencies) that are available between other users of the spectrum, with protected channels being automatically excluded on a permanent basis.

Occupied frequency channels in the HF band can change rapidly as users respond to diurnal variations in ionospheric conditions, especially near the terminators. It is evident from Figure 4.8 that many users exploit the enhanced support for skywave propagation in the middle and higher regions of the HF spectrum during the day when the ionosphere is well established. At night, the ionosphere provides reduced or no support for skywave propagation at higher frequencies, so users are effectively forced to operate in lower regions of the HF spectrum, as evident in Figure 4.9. The crowding that occurs in the lower HF band, coupled with the disappearance of the D-layer, results in high user-congestion at night. This can at times make it extremely difficult to find clear frequency channels of appropriate bandwidth on which to operate an OTH radar.

Unintentional man-made noise chiefly arises from electrical equipment in use, such as electric motors, power lines, vehicle ignition systems, arc-welders, and neon signs, to name a few. The spectral density of man-made noise decreases with increasing frequency but varies significantly with location. Close to densely populated residential or industrial areas, the level of man-made noise increases over the entire HF band relative to that in rural or remote areas. Indeed, man-made noise can dominate atmospheric and galactic noise over much or all of the HF band when the receiver is located in the proximity of large city centers. In recent years, concerns have risen with respect to noise generated by "broadband over power lines" (BPL) internet access, which may operate at frequencies in the HF band. The BPL signal can leak from the power lines and cause disruption to HF systems. Certain providers have installed filters to attenuate certain frequencies in the HF spectrum. However, the uncanceled emissions are believed to interfere with HF communication and OTH radar systems located near BPL deployments.

Models based on CCIR reports that describe the power spectral density of man-made noise for receiver sites located in different (broadly defined) types of areas are given by $N_0(f)$ in Eqn. (4.35), where the parameter β depends on the type of area considered (residential, rural, or remote), f_c is in units of MHz, and "ln" denotes natural logarithm. Figure 4.10 plots the power spectral densities of man-made noise according to this model (Lucas and Harper 1965). The SNR advantage of siting sensitive receivers far away from highly populated areas to reduce the man-made noise contribution may be 10 dB or more. Receiver locations separated by about 100 km or more over dry land from the nearest urban areas generally provide sufficient isolation for man-made noise not to be the limiting noise contribution. In remote locations, man-made noise is typically low enough such that atmospheric and galactic noise often dominate in the lower and upper regions of the HF band, respectively.

$$N_0(f_c) = -\beta - 12.6 \times \ln(f_c/3), \text{ where } \begin{cases} \beta = 136 \text{ residential} \\ \beta = 148 \text{ rural} \\ \beta = 164 \text{ remote} \end{cases} \tag{4.35}$$

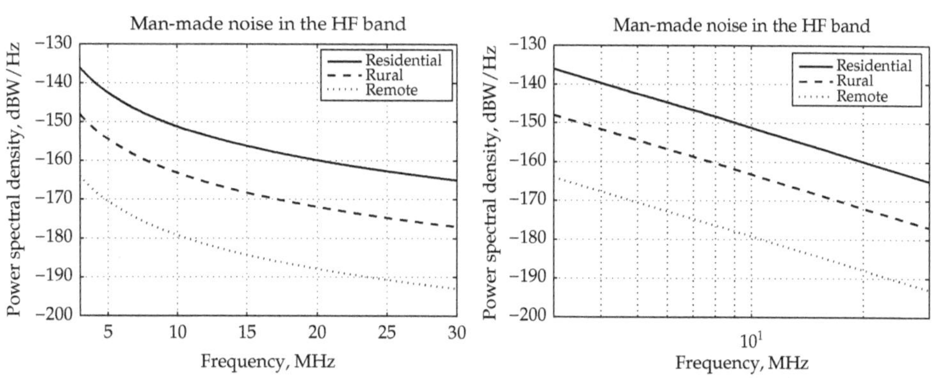

(a) Linear frequency scale.

(b) Log frequency scale.

FIGURE 4.10 Power spectral density models for man-made noise in the HF band assuming the receiver is located in a residential, rural, and remote area. The curves are plotted using a linear- and log-frequency scale for comparison.

Although use of the HF band is regulated, incidental interference between different users may occasionally occur, either because regulations have not been followed, or as a result of unforeseen propagation conditions. Interference received from an individual man-made source is often highly directional. This property allows receive antennas with radiation patterns characterized by a narrow mainlobe width and low sidelobe levels to reduce the angular passband of the system, such that any off-azimuth interference signals are highly attenuated. This feature of OTH radar systems provides significant immunity against incidental man-man interference that does not arrive from the same direction as the target echo. Alternatively, adaptive processing in the spatial and/or temporal domain(s) may be performed to mitigate powerful RFI entering through the conventional antenna pattern sidelobes and/or main beam. The subject of adaptive processing for interference mitigation will be considered in Part III.

Deliberate interference intended to disturb a system such as radar is referred to as *jamming*. Analogous to microwave radar, HF jamming sources may be located on the target platform itself (self-screening), with the signal being received in the mainlobe of the antenna beam, or at a separate (stand-off) location, with the signal often being received through the sidelobes of the antenna beam. The jamming signal may be incoherent with the radar waveform (e.g., noise-like) and operate in a "spot" or "barrage" mode by spreading its energy over a narrow or broad range of frequencies, respectively. Such signals attempt to raise the noise floor over the entire target search space in range and Doppler to impair the detection performance of an OTH radar system. Alternatively, the jamming signal may be coherent with the radar waveform, as in the case of deception jamming. This type of jamming may be employed to generate multiple false targets (decoys) with realistic range and Doppler frequency signatures. Such signals attempt to impair the tracking performance of an OTH radar system.

Self-screening sources located within the surveillance region can potentially deny the detection of the host platform and other targets in the jammer (azimuth) strobe at possibly different ranges to the host platform. Such sources can expect near-optimum propagation to the radar receiver (i.e., with high average power), as the choice of operating frequency for the surveillance region is often selected based on this criterion. In the case of stand-off sources located arbitrarily with respect to the surveillance region (e.g., a ground-based emitter), propagation conditions will generally be suboptimum. However, such sources may have greater power and antenna gain at their disposal, allowing the signals to reach the radar receiver with significant strength. For a comprehensive review of radar ECM and ECCM techniques, the reader is referred to Skolnik (2008c).

4.2 Standard Routines

Conventional processing for OTH radar consists of a sequence of predetermined signal- and data-processing steps. These steps transform the multi-channel input data samples acquired over a number of CPI by the OTH radar receiver array into ground-registered target tracks, which represent the final product of an OTH radar system. The block diagram in Figure 4.11 illustrates the sequential flow of rudimentary signal- and data-processing steps that represent the core of the OTH radar conventional processing chain.

Specifically, the signal-processing chain aims to enhance the useful signal-to-disturbance ratio (SDR) and to resolve the received radar echoes in the dimensions of

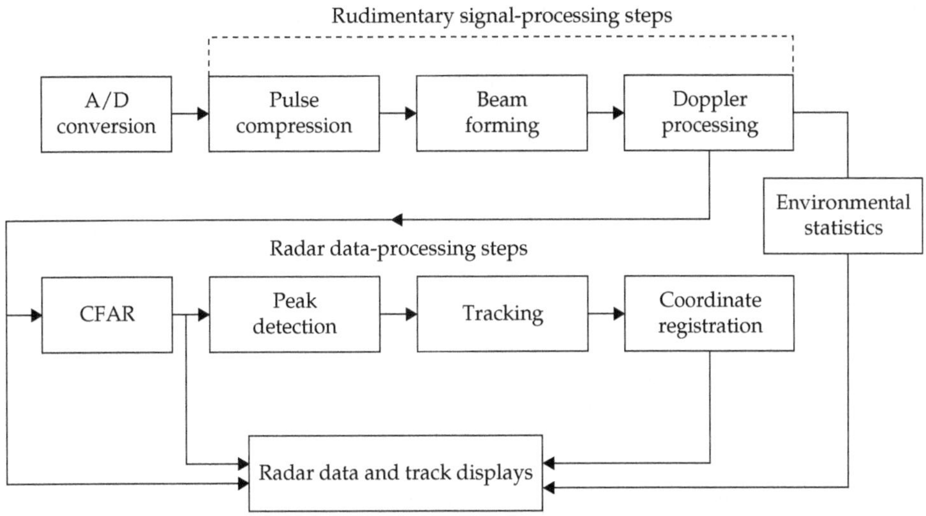

FIGURE 4.11 Rudimentary OTH radar signal and data-processing steps from A/D conversion to the operator displays. The signal-processing steps convert the input data cube (acquired in the dimensions of samples, pulses, and receivers), to an output data cube in the canonical radar dimensions of group-range, beam direction, and Doppler frequency. A number of ameliorative processing steps, including transient disturbance mitigation, data extrapolation, and ionospheric Doppler correction (to be explained in due course), are typically inserted at various points in this "core" signal-processing chain. Environmental statistics related to clutter and noise are often computed as a by-product of conventional processing and displayed to operators in real time. As indicated in the diagram, the operators may also view the azimuth-range-Doppler (ARD) data images before and after CFAR processing, as well as the ground-registered track display. © Commonwealth of Australia 2011.

group-range (pulse compression), relative velocity (Doppler processing), and direction of arrival (array beamforming), while the data-processing steps (to be described in the next section) are primarily responsible for target detection and tracking. This section recalls fundamental concepts associated with the three core signal-processing steps, which are common to many other types of radar systems. Further reading on the subject of radar signal processing can be found in several excellent texts, including Skolnik (2008a), Nathanson et al. (1999), Richards and Scheer (2010), and Melvin and Scheer (2013).

The order in which three rudimentary signal-processing steps are performed is in theory interchangeable. However, in many practical systems, the standard procedure is to optimize computational efficiency by giving precedence to steps for which there exists most scope for data reduction. This policy reduces unnecessary processing in subsequent steps along the chain. Typically, the pulse compression and array beamforming steps only retain a subset of useful range and beam cells (from the overall number processed), which can provide a substantial reduction in data samples with respect to the total number of samples-per-pulse and receivers, respectively. On the other hand, Doppler processing typically results in no data reduction for OTH radar, as the number of frequency bins retained is usually not lower than the number of pulses integrated over the CPI. In the context of conventional processing, it is therefore more computationally efficient to beamform only the range cells retained, as opposed to all

A/D samples in the PRI, and to Doppler process only the pulse-trains in the beams that span the surveillance region, as opposed to those in all receivers of the array.

In addition to the rudimentary radar signal-processing steps, a suite of ameliorative processing techniques is also usually applied to address specific environmental phenomena that can potentially degrade system performance. An important consideration for OTH radar is the mitigation of impulsive noise bursts and transient meteor echoes, for example. This section additionally describes a number of ameliorative processing techniques that are traditionally considered part of the conventional processing chain. The signal-processing output is a complex-valued data cube of much lower volume (dimensions) compared to the input data-cube, with resolution cells in the canonical radar dimensions of group-range, beam direction, and Doppler frequency. At the end of the signal-processing chain, the phase information is often discarded, and the envelope of the azimuth-range-Doppler (ARD) data cube is passed on for data processing.

As indicated in Figure 4.11, the rudimentary data-processing steps include constant false-alarm rate (CFAR) processing, peak detection and parameter estimation, as well as tracking, and coordinate registration. Data processing also involves significant data reduction, with the number of peaks generated per dwell being around five orders of magnitude less than the number of input (A/D) data samples acquired by the system during a CPI. The tracker output represents the ultimate "distillation" of multiple CPI of data acquired by the radar over the mission lifetime. This enormous data reduction not only permits the information to be transferred efficiently between sites, but also allows the radar outputs to be displayed in an effective manner for interpretation by human operators. In addition to the geographic track display, OTH radar operators may also view intermediate outputs at different stages of the signal- and data-processing sequence, as indicated in Figure 4.11. During normal operations, environmental radar performance statistics, such as sub-clutter visibility, minimum detectable radial speed, and noise level, may also be computed and stored in synoptic databases to guide future mission planning and system design.

4.2.1 Pulse Compression

The pulse compression step, also referred to as *range processing* (it is chosen here not to distinguish between the two terms), operates on the decimated A/D samples data acquired by each receiver of the array to produce a number of group-range resolution cells for each transmitted pulse of the radar waveform. Conventional pulse compression is based on the use of a matched-filter bank that processes the input data samples to produce a set of group-range resolution cells. The first main objective of pulse compression is to enhance the SNR of radar echoes received with a round-trip time delay that matches each of the group-range resolution cells processed. The second main objective is to resolve the received radar echoes on the basis of their differences in round-trip time delay, such that contributions from independent scatterers received over different virtual path lengths can be effectively separated into different group-range resolution cells.

The SNR gain provided by the matched-filter operation enhances target detection performance against noise and interference, while the ability to resolve a target echo from other radar echoes (possibly arising from the same target due to multipath) in group-range enhances detection performance against clutter. Importantly, pulse compression not only improves detection performance, but also enables the group-range of a detected target echo to be estimated for localization purposes. This section considers two alternative pulse compression (range processing) methods for OTH radar. The first is known as chirp deramping or stretch processing, while the second involves the direct application of a fully-digital matched filter to the decimated A/D data samples. The benefits and shortcomings of each method are contrasted.

Before proceeding, it is useful to recall the matched filter concept as it applies to the analytic representation of the baseband point-target echo $s(t)$ derived in Section 4.1.1. With reference to Figure 4.12, the received baseband signal $x(t)$ is input to a matched filter with impulse response $h(t)$ given by Eqn. (4.36). In words, the impulse response of the matched filter (used for range processing) is the time-reversed and conjugated version of the pulse waveform $p(t)$.

$$h(t) = p(-t)^* \tag{4.36}$$

For the moment, we ignore the additive disturbance components (clutter and noise) in $x(t)$, and consider only the matched filter response to the ideal target echo $s(t)$ derived in Section 4.1.1. In this case, the matched filter output $y(t)$ is given by the convolution of $s(t)$ with $p(-t)^*$ in Eqn. (4.37). The digital samples of the continuous-time signal $y(t)$ acquired at the Nyquist rate (or at higher frequency) are often referred to as range gates. This description is consistent with a purely analog matched-filter implementation, wherein the digital samples at the A/D converter output yield the group-range resolution cells directly.

$$y(t) = s(t) * p(-t)^* = \int_{-\infty}^{\infty} s(u)p^*(u-t)\,du \tag{4.37}$$

Substituting the target echo expression $s(t) = \alpha m(t-\tau)e^{j2\pi f_d t}$ into Eqn. (4.37) leads to Eqn. (4.38), where it is recalled that $m(t)$ is the baseband radar waveform, τ is the round-trip delay, and f_d is the Doppler shift. To avoid defining a number of different scale factors, the complex amplitude α is assumed to absorb any appropriate modification of the original constant in the following development.

$$y(t) = \alpha \int_{-\infty}^{\infty} m(u-\tau)p^*(u-t)e^{j2\pi f_d u}\,du \tag{4.38}$$

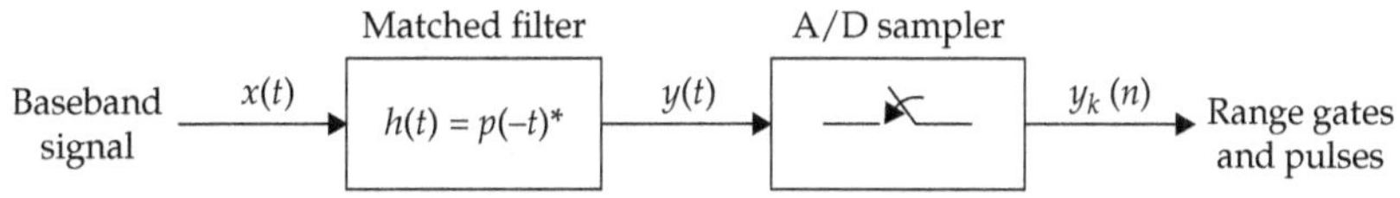

FIGURE 4.12 Representation of the analog matched filter implementation followed by A/D conversion to sample the range-gates $k = 1, \ldots, K$ at pulse n. The impulse response $h(t)$ of the matched filter is given by the time-reverse and conjugate of the pulse waveform $p(t)$. A realizable matched filter has the causal impulse response $h(t) = p(T_p - t)^*$, where the time delay T_p equals the pulse duration, such that $h(t) = 0$ for $t < 0$. For the purpose of illustrating the main concepts, we may adopt the non-causal impulse response $h(t) = p(-t)^*$ to simplify notation, and focus on the matched-filter response to the baseband target echo signal $s(t)$, as opposed to the received signal $x(t)$, which additionally contains clutter, interference, and noise.

The baseband radar waveform $m(t)$ is in turn comprised of N identical and uniformly spaced pulses over the CPI denoted by $p(t)$. Substituting the expression for $m(t)$ given by Eqn. (4.4) into Eqn. (4.38) yields Eqn. (4.39).

$$y(t) = \alpha \int_{-\infty}^{\infty} \sum_{n=0}^{N-1} p(u - nT - \tau) p^*(u - t) e^{j2\pi f_d u} \, du \tag{4.39}$$

Now, perform the change of variable $u' = u - nT - \tau$, such that the output $y(t)$ may be written in the more convenient form of Eqn. (4.40). Note that the constant α has absorbed the delay dependent phase term $e^{j2\pi f_d \tau}$ in Eqn. (4.40).

$$y(t) = \alpha \sum_{n=0}^{N-1} e^{j2\pi f_d nT} \int_{-\infty}^{\infty} p(u') p^*(u' - [t - nT - \tau]) e^{j2\pi f_d u'} \, du' \tag{4.40}$$

The term in the integral of Eqn. (4.40) may be recognized as the ambiguity function of the pulse waveform $p(t)$, denoted by $\chi_p(t - nT - \tau, f_d)$ when this function is evaluated at time delay $t - nT - \tau$ and Doppler shift f_d. Using this notation, the matched filtered output may be written as in Eqn. (4.41), where $\nu_d = f_d T$ is defined as the normalized Doppler frequency.

$$y(t) = \alpha \sum_{n=0}^{N-1} \chi_p(t - nT - \tau, f_d) e^{j2\pi \nu_d n} \tag{4.41}$$

To generate a number K of range gates in each pulse repetition interval, where the range-gate indices $k = 0, \ldots, K - 1$ correspond to the round-trip delays τ_k, the output $y(t)$ is sampled at time instants $t_{kn} = \tau_k + nT$ for $n = 0, \ldots, N - 1$. This results in a series of complex samples $y_k(n) = y(t_{kn})$ indexed by pulse number n and range cell k in Eqn. (4.42). Increments in range k are sometimes referred to as *fast-time* samples, while those in pulse n are referred to as *slow-time* samples. For target echo Doppler shifts much smaller than the reciprocal of the pulse duration, the range profile may be approximated as $\chi_p(\tau_k - \tau, 0)$. This response reaches its maximum value of $\chi_p(0, 0) = E_p$ at the range cell that matches the target echo group-range exactly, i.e., where $\tau_k = \tau$.

$$y_k(n) = \alpha \chi_p(\tau_k - \tau, f_d) e^{j2\pi \nu_d n} \tag{4.42}$$

In contrast to the purely analog matched filter implementation, the stretch processing and cross-correlation receivers described in the following sections represent partially- and fully-digital pulse compression methods, respectively. Since the target Doppler shift is typically sufficiently small to be ignored for pulse compression in practice, such schemes may be applied separately (i.e., repeated) for each pulse of the radar waveform in what may be referred to as pulse-by-pulse processing.

4.2.1.1 Stretch Processing

In early OTH radar systems, A/D converters were unable to sample the full bandwidth of the transmitted signal at the Nyquist rate with a sufficiently high dynamic range for effective Doppler processing. In addition, the data rates generated needed to be kept low for the signal processor to operate in real time due to the limited capacity of the computing resources available at the time. This motivated a hybrid (analog/digital) pulse

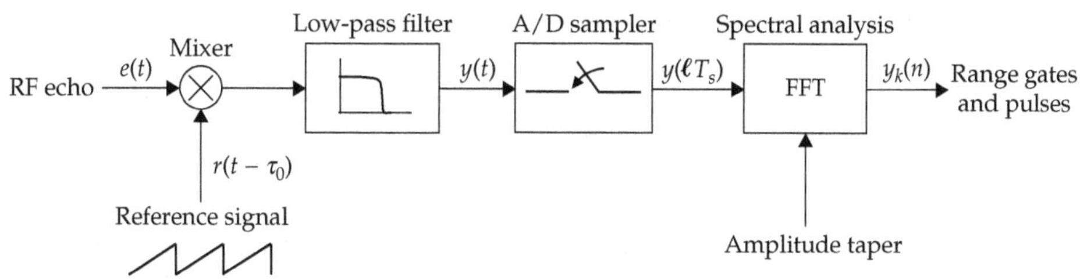

FIGURE 4.13 Representation of the chirp deramping (stretch processing) pulse compression method using an analog mixer fed with the received RF signal (only the target echo component $e(t)$ is considered) and a delayed local copy of the transmitted waveform $r(t - \tau_0)$. The mixer output $y(t)$ is input to an analog low-pass filter with cutoff frequency f_b. The filter output is sampled at the Nyquist rate (or higher) and digital spectral analysis is applied to the time series $y(\ell T_s)$ for $\ell = 0, \ldots, L-1$ acquired in each pulse repetition interval, typically using an FFT with amplitude taper. The frequency bins $k = 1, \ldots, K$ of the FFT output $y_k(n)$ correspond to the different range gates processed in each radar pulse n. © Commonwealth of Australia 2011.

compression scheme known as chirp deramping, or stretch processing, which avoids the need to directly sample the full radar signal bandwidth B while (approximately) retaining the nominal group-range resolution of $c/2B$. The different elements of this range processing scheme are shown in Figure 4.13.

Rather than performing down conversion of the received signal using a CW local oscillator slaved to the carrier frequency, the reference signal fed into the analog mixer is a copy of the RF radar waveform $r(t)$ shifted by a known delay τ_0 relative to the transmitted signal. This delay determines the nearest group-range of interest $R_0 = c\tau_0/2$, which effectively corresponds to the nearest edge of the OTH radar surveillance region in group-range. The output of an ideal in-phase and quadrature mixer fed by the reference signal $r(t - \tau_0)$ and a conjugated version of the RF target echo $e(t)$ is given by $y(t)$ in Eqn. (4.43).

$$y(t) = e^*(t)r(t - \tau_0) \tag{4.43}$$

Confining attention to the first linear FM pulse of $r(t - \tau_0)$ and the previously described analytic representation of the target echo $e(t)$ over the time interval $t \in [\tau_0 + \tau, \tau_0 + T]$, the mixer output $y(t)$ takes the form of Eqn. (4.44), where the Doppler shift of the echo is ignored due to the relatively short pulse duration, and α is a complex amplitude that absorbs an immaterial phase term. This mixing operation is notionally illustrated for the first pulse of the CPI in Figure 4.14. In practice, analog downconversion and filtering are often carried out via one or more intermediate frequency stages, but we shall not be concerned with these practical implementation details to describe the basic concepts.

$$y(t) = \alpha e^{j2\pi B(\tau - \tau_0)t/T} = \alpha e^{j2\pi f_e t} \tag{4.44}$$

For a single target echo, the analytic (I/Q) chirp deramping process leads to a complex exponential signal $y(t)$ at the mixer output with a frequency proportional to the relative delay $\tau - \tau_0$ between the target echo and the reference waveform fed into the mixer. More precisely, the difference frequency between the target echo and reference signal is given by f_e in Eqn. (4.45). When multiple echoes are received over a continuum of time delays, each echo contributes its own spectral component with a frequency that depends on the relative delay $\tau - \tau_0$ of the echo.

$$f_e = B(\tau - \tau_0)/T \tag{4.45}$$

This relationship implies that the frequency of each echo contribution is linearly related to the group-range of the scatterer R, as in Eqn. (4.46). This equation, which links the group-range R, relative delay $\tau - \tau_0$, and difference frequency f_e applies over the interval $t \in [\tau_0 + \tau, \tau_0 + T]$. The relative delay is assumed to be positive and small compared with the pulse duration (i.e., $\tau_0 \leq \tau \ll T$), such that the subsequent low-pass filtering stage in Figure 4.13 need only pass a relatively small bandwidth of (nonnegative) frequencies.

$$R = R_0 + \frac{c(\tau - \tau_0)}{2} = R_0 + f_e \frac{Tc}{2B} \tag{4.46}$$

In practice, the analog mixing process may involve only the real (in-phase) components of the received and reference signals. This produces sum and difference frequencies at the mixer output. The sum frequency is removed by the low-pass (anti-aliasing) filter prior to A/D conversion in Figure 4.12. In addition, the low-pass filter cutoff frequency f_b removes the large (negative) difference frequencies that occur during a brief mismatch period $t \in [nT + \tau_0, nT + \tau_0 + \tau]$ in each pulse following the first ($n > 0$). The low-pass filter also removes large (positive) difference frequencies arising from echoes with group-ranges falling outside of the surveillance region $R \in [R_0, R_0 + R_m]$. In other words, the low-pass filter cutoff frequency f_b determines the maximum difference frequency f_e that is passed, and hence range-depth R_m of the surveillance region, as given by Eqn. (4.47).

$$R_m = \frac{f_b T c}{2B} \tag{4.47}$$

The key point here is that the low-pass filter cut off frequency f_b may be an order of magnitude lower than the radar waveform bandwidth B without significantly degrading range resolution or compromising the range depth of the surveillance region when the latter is limited by ionospheric propagation. For example, using $B = 10$ kHz and $T = 0.02$ s ($f_p = 50$ Hz), a value of $f_b = 2$ kHz provides a range depth of $R_m = 600$ km from Eqn. (4.47). The stretch processing method eases demands on A/D sampling rate for a required wordlength (number of bits) and spurious free dynamic range, as well as reducing data rates and real-time signal-processing load. For a point scatterer with a group-range $R \in [R_0, R_0 + R_m]$, the output of the anti-aliasing filter sampled at (or above) the Nyquist rate $f_s = 1/T_s$ yields the time-series $y(\ell T_s)$ in Eqn. (4.48) for $\ell = 0, \ldots, L - 1$ in a single pulse or FMCW sweep.

$$y(\ell T_s) = \alpha e^{j2\pi f_e \ell T_s} \tag{4.48}$$

As the echo group-range R is linearly related to the downconverted signal frequency f_e, range processing is performed by digital spectral analysis of the samples $\{y(\ell T_s)\}_{\ell=0}^{L-1}$. The range-processed output y_k is written in terms of a weighted discrete Fourier transform (DFT) in Eqn. (4.49). Fast Fourier transform (FFT) algorithms are used in practice to

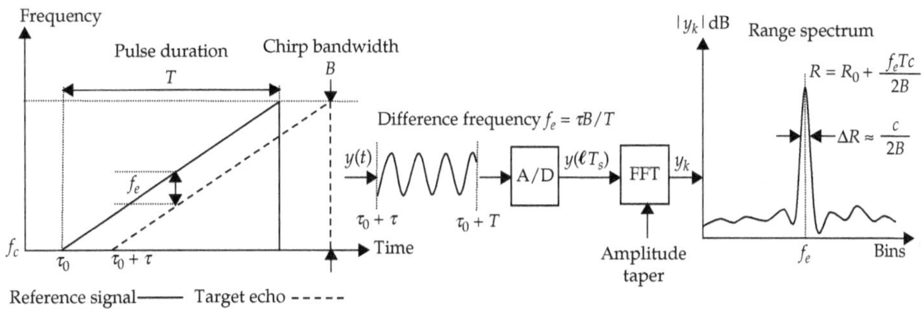

FIGURE 4.14 Summary of stretch processing pulse compression scheme implemented as a chirp-deramping (mixing) process followed by low-pass filtering (not shown), A/D conversion, and digital spectral analysis to form the range spectrum. In the diagram, the transmit start time is at the origin, while the reference signal delay is set to τ_0. The target echo has a delay $\tau_0 + \tau$ with respect to the transmitted waveform. The relationship between difference frequency f_e at the mixer output and group-range R of the echo relative to $R_0 = c\tau_0/2$ is indicated along with the group-range resolution in the illustrated range spectrum. © Commonwealth of Australia 2011.

compute the DFT output more efficiently. An amplitude-only taper function $w(\ell)$ is applied to control spectral leakage (i.e., range sidelobes) in Eqn. (4.49).

$$y_k = \sum_{\ell=0}^{L-1} w(\ell)\ y(\ell T_s)\ e^{-j2\pi k\ell/L} \tag{4.49}$$

The ranges cells y_k correspond to the FFT frequency bins $f_k = k/LT_s$ for $k = 0, \ldots, K-1$, where the maximum frequency f_K is less than or equal to the low-pass filter bandwidth f_b. As illustrated in Figure 4.14, this leads to a range spectrum where the contribution of a point-target echo exhibits a maximum amplitude response at the frequency bin that corresponds to the echo difference frequency f_e. The relationship between FFT frequency bin f_k and group-range cell R_k in the range spectrum y_k is given by Eqn. (4.50).

$$R_k = R_0 + \frac{f_k T c}{2B} \tag{4.50}$$

For $\tau \ll T$, the frequency resolution corresponding to the DFT integration time of $T - \tau$ is approximately $\Delta f \simeq 1/T$ for a uniform (rectangular) window. Note the group-range resolution ΔR is related to Δf by Eqn. (4.51). Inserting $\Delta f \simeq 1/T$ implies that the group-range resolution is $\Delta R \simeq c/2B$. Although resolution is important when echoes from multiple scatterers with different group-ranges are present, we shall continue to consider the case of a single point scatterer to discuss the concepts of "scallop loss" due to straddling and sidelobe levels in the range spectrum.

$$\Delta R = \Delta f \frac{Tc}{2B} \tag{4.51}$$

If the L-point FFT is performed with uniform shading weights $w(\ell) = 1$, the frequency-domain output for a single echo with difference frequency f_e is a scaled version of the periodic sinc function, also known as the asinc function, denoted by $S(f)$ in Eqn. (4.52), where $f_s = 1/T_s$ is the sampling frequency. The coordinate of the maximum of $S(f)$ occurs at the echo difference frequency $f = f_e$, where the value is given by $S(f_e) = \alpha L$.

$$S(f) = \alpha e^{-j[\pi(L-1)(f-f_e)T_s]} \frac{\sin[\pi(f - f_e)LT_s]}{\sin[\pi(f - f_e)T_s]}, \quad f \in [0, f_s) \tag{4.52}$$

The range cells effectively sample this response at the FFT frequency bins, i.e., $y_k = S(f_k)$. Repeating this procedure for each of the $n = 1, \ldots, N$ pulses of the CPI yields the output $y_k(n)$ in Eqn. (4.53). The peak amplitude of y_k occurs at the FFT frequency bin f_k that most closely matches f_e. The value of f_e will in general not coincide exactly with one of the FFT frequency bins. A scallop loss results due to this straddling because none of the range cells sample the maximum of $S(f)$. Scallop losses relative to the peak of $S(f)$ may be as high as 3.9 dB in power (amplitude squared) for a rectangular window. This is equivalent to the loss in SNR for square-law detection. Such losses may be mitigated at the expense of higher processing load, e.g., increasing the FFT length by zero padding.

$$y_k(n) = S(f_k)e^{j2\pi \nu_d n} = y_k e^{j2\pi \nu_d n} \tag{4.53}$$

For a rectangular window, range sidelobes appear when f_e is not perfectly matched to any f_k. High range sidelobes cause problems when multiple echoes with different delays and varying strengths are received in the same beam and Doppler cell. Specifically, the presence of faint target echoes may be obscured by the range sidelobes of stronger clutter returns (e.g., meteor echoes). The sinc function has a peak sidelobe level (SLL) of −13 dB relative to the maximum value. Meteor echoes may be 30 dB or higher above the noise floor, so it is important to lower the range sidelobes below those of the uniform window by using an appropriate taper function $w(\ell)$ in Eqn. (4.49).

The choice of taper involves a compromise between several competing factors. On one hand, very low sidelobe tapers increase SNR loss due to mismatch (to be covered later). They also reduce range resolution by increasing the −3.0-dB mainlobe width (i.e., the frequency separation between half-power points in the mainlobe response). On the other hand, low sidelobe levels decrease spectral leakage for radar echoes that are spaced more than the mainlobe width apart in difference frequency. In addition, a greater mainlobe width provides robustness against scallop loss due to straddling effects. An authoritative description on the use of window functions for digital spectral analysis, including the definition of various windows and their relative figures of merit, can be found in Harris (1978).

4.2.1.2 Fully Digital Implementation

State-of-the art direct digital receiver (DDRx) technology permits all of the HF spectrum to be sampled close to the antenna element with a spurious free dynamic range (and other performance characteristics) appropriate for OTH radar. The received OTH radar echoes may therefore be sampled directly at RF or at IF (after a pre-selection filter) and digitally downconverted to baseband. Digital filtering is then performed to extract the signals in the radar channel and remove (attenuate) signals at other frequencies. After digital downcoversion and filtering, the data rate is decimated in accordance with the radar signal bandwidth. Modern computing platforms enable digital signals acquired at rates commensurate with the OTH radar waveform bandwidth to be distributed across a network of CPUs and processed in real time. Such advances allow for the implementation of a fully-digital pulse compression scheme.

Relative to the stretch-processing method described previously, a significant advantage of fully-digital pulse compression is that it permits all range cells in the repetition period to be processed as opposed to a subset determined by the reference delay offset and low-pass filter bandwidth. Extended processing in range may be used to increase the OTH radar coverage at times when ionospheric propagation supports target detection over greater range depths. Alternatively, it may be used to process range cells inside the skip-zone for sampling interference and noise in the absence of powerful clutter contamination during the radar dwell. Such range cells may be used as training data for interference and noise mitigation by adaptive processing in OTH radar.

The discrete-time impulse response function of the digital matched filter is obtained by time-reversing and conjugating the transmitted pulse waveform samples $p(\ell T_s)$, where ℓ is the sample number and T_s is the sampling period after decimation. The causal digital matched-filter impulse response h_ℓ is given by Eqn. (4.54), where the constraint $T_p \leq T$ implies that $h_\ell = 0$ for $\ell < 0$.

$$h_\ell = p^*(T - \ell T_s) \tag{4.54}$$

Restricting attention to the first pulse in the CPI, the baseband samples received due to a single target echo are given by s_ℓ in Eqn. (4.55), where τ is the round-trip time delay. As explained previously, the Doppler shift of the echo is typically small enough to be neglected over the relatively short PRI for the purpose of range processing on a pulse-by-pulse basis.

$$s_\ell = \alpha p(\ell T_s - \tau) \tag{4.55}$$

The digital matched-filter output y_ℓ is the convolution of h_ℓ and s_ℓ in Eqn. (4.56), where the integer $L = T/T_s$ is defined as the impulse response duration expressed in sample bins. Recall that the impulse response duration is equal the pulse repetition interval T in a continuous wave system where $T_p = T$. The limits $m \in [\ell - L, \ldots, \ell]$ apply in Eqn. (4.56) because the impulse response h_ℓ is only nonzero for $\ell \in [0, L]$.

$$y_\ell = \sum_{m=-\infty}^{\infty} s_m h_{\ell-m} = \sum_{m=\ell-L}^{\ell} s_m h_{\ell-m} \tag{4.56}$$

Substituting $s_m = p(mT_s - \tau)$ from Eqn. (4.55), and $h_{\ell-m} = p^*(T - [\ell - m]T_s)$ from Eqn. (4.54), into the discrete convolution sum of Eqn. (4.56) yields Eqn. (4.57). In the hypothetical case of an echo with zero time delay $\tau = 0$, the amplitude of the matched-filter output y_ℓ maximizes when the term in the square brackets $\ell T_s - T$ equals zero, i.e., at time sample $\ell = L$. Hence, the first (zero-delay) range sample is the matched-filter output evaluated at time $\ell = L$, which corresponds to the impulse response duration.

$$y_\ell = \alpha \sum_{m=\ell-L}^{\ell} p(mT_s - \tau) p^*(mT_s - [\ell T_s - T]) \tag{4.57}$$

In general, for an echo with delay $\tau \geq 0$, the matched-filter output y_ℓ will reach its greatest amplitude at time sample $\ell = L + k$, where the delay kT_s relative to the transmitted waveform is best matched to τ. Stated simply, the causal range samples are the matched-filter outputs y_ℓ evaluated at integer multiples of the sampling period $\ell = L + k$ for

$k = 0, \ldots, K - 1$. To write an expression for the range samples directly, we make the change of variables $k = \ell - L$ in Eqn. (4.57) to yield Eqn. (4.58).

$$y_k = \alpha \sum_{m=k}^{k+L} p(mT_s - \tau) p^*(mT_s - kT_s) \tag{4.58}$$

The range sample denoted by y_k is matched to an echo delay $\tau_k = kT_s$ or group-range $R_k = c\tau_k/2$. It is evident from the expression in Eqn. (4.59) that the matched-filtering operation is equivalent to a cross-correlation receiver. In other words, pulse compression is performed by cross-correlating a delayed and conjugated version of the transmitted pulse waveform samples with the acquired signal samples. Note that the latter is assumed to arise from a single point-scatterer echo in Eqn. (4.59).

$$y_k = \alpha \sum_{m=k}^{k+L} p(mT_s - \tau) p^*(mT_s - \tau_k) \tag{4.59}$$

The summation in Eqn. (4.59) may be denoted by $\chi'_p(\tau_k - \tau, 0)$, where the first and second arguments are the relative delay and Doppler shift between the two cross-correlated pulse waveforms. The range-processed output $y_k(n)$ that results for a single target echo over the different pulses $n = 0, \ldots, N - 1$ is given by Eqn. (4.60). Evaluating the digital cross-correlation at the delay of the echo $\tau_k = \tau$ yields the maximum amplitude response $\chi'_p(0, 0) = \sum_{m=0}^{L} |p(mT_s)|^2$ at the pulse-compressor output.

$$y_k(n) = y_k e^{j2\pi \nu_d n} = \alpha \chi'_p(\tau_k - \tau, 0) e^{j2\pi \nu_d n} \tag{4.60}$$

In practice, pulse compression is performed on the received signal samples $x(\ell T_s)$, which contain clutter echoes, interference-plus-noise, and possibly multiple target echoes. Moreover, the terms in the correlation sum are typically shaded by a set of amplitude taper weights $\{w(m)\}_{m=0}^{L}$, which are designed to trade a slight loss in SNR and range resolution for lower range sidelobes and greater robustness to scallop losses caused by range straddling. Accounting for the received data samples and window function weights, the fully digital pulse compression scheme transforms the acquired samples $x(\ell T_s)$ to the range-gated pulse outputs $y_k(n)$ according to Eqn. (4.61).

$$y_k(n) = \sum_{m=k}^{k+L} w(m - k) x(mT_s + nT) p^*([m - k]T_s), \quad \begin{cases} k = 0, \ldots, K - 1 \\ n = 0, \ldots, N - 1 \end{cases} \tag{4.61}$$

Computationally fast algorithms for range processing by digital pulse compression are available in the case of linear FMCW signals provided that the product of the pulse repetition interval and the signal bandwidth is an integer (Summers 1995). The group-range associated with the first retained range cell depends on the location of the nearest edge of the surveillance region (if skip-zone range cells are not required), while the number of retained range cells depends on the range depth of the surveillance region (which is ultimately limited by ionospheric propagation conditions or system constraints such as processing load and operator displays).

4.2.2 Array Beamforming

Array beamforming may be described as the process of linearly combining (summing) amplitude-weighted and phase-shifted signals received by the individual sensors of a multi-channel (narrowband) system to produce a resultant signal known as the beam output. The complex weights (ampltiude and phase) effectively modify the radiation pattern of the antenna array and hence the relative contributions of signals received from different directions in the beam output. In OTH radar, conventional beamforming is based on the application of a spatial matched filter to the "snapshot" of data samples acquired by the different receivers of the array at a particular range gate and pulse number. The formation of multiple beams steered in different directions by this process is also known as electronic or digital beam steering.

Conventional beamforming enhances the signal-to-disturbance ratio (SDR) relative to that in a single antenna element by providing coherent gain to signals arriving from the radar steer or "look" direction, while attenuating interference and clutter incident from other directions by means of beam patterns with low sidelobes. Besides enhancing SDR and resolving received signals based on differences in their direction of arrival, the other important role of conventional beamforming is to estimate the angular coordinates of detected target echoes for localization and tracking. Although this section describes conventional beamforming performed at the OTH radar receiver, the main concepts are reciprocal in the sense that they are also applicable to the transmitter.

4.2.2.1 Antenna Element and Subarray Patterns

A receiver-per-element architecture was not feasible in certain early OTH radar systems, such as the Jindalee Facility Alice Springs (JFAS) Stage B OTH radar, which was built more than three decades ago in central Australia. High-quality HF receivers were expensive at the time, and only limited computer-processing capacity was available to process the outputs. To reduce system cost and processing load, individual antenna elements in the receiving aperture were grouped into a manageable number of subarrays with a single reception channel per subarray architecture. For example, the Jindalee OTH radar receive system is based on a ULA composed of 462 dual-fan antenna elements. The full aperture is divided into 32 uniformly spaced subarrays, each consisting of 28 dual-fan elements, with adjacent subarrays overlapped by 50 percent (i.e., with half the elements shared by neighboring subarrays on either side).

The subarray patterns are steered by linearly combining the analog signals received in the different antenna elements using a network of switchable length cables that implement true time-delay-and-sum (i.e., broadband) beamforming at RF. Constructive reinforcement of useful signals is achieved at the linear combiner output by ensuring that the relative time delay inserted between any two elements via the different length cables is equal to the difference in path length traveled by a signal incident as a plane wave from the radar steer direction divided by the speed of light in free space. Signals impinging from other directions do not perfectly cohere at the linear combiner output with these inserted time delays and thus experience varying degrees of destructive interference or attenuation relative to the useful signals. Typically, the analog beamforming process also includes amplitude tapering to lower the subarray beampattern sidelobes.

The analog beamforming yields a subarray radiation pattern $G_s(\theta, \phi)$ in azimuth θ and elevation ϕ with higher directivity than the element radiation pattern $G_e(\theta, \phi)$. The subarray outputs are individually downconverted, filtered, sampled, and subsequently combined in a digital beamforming operation to be described in the next section. This operation contributes an array pattern factor $G_a(\theta, \phi)$ defined as the radiation pattern that would result when a number of identical *isotropic* antenna elements are combined in the digital beamformer. For identical but non-isotropic subarray antenna patterns, the factorization theorem states that the overall receive antenna pattern $G(\theta, \phi)$ is given by the product of the subarray pattern $G_s(\theta, \phi)$ and the array pattern factor $G_a(\theta, \phi)$, as in Eqn. (4.62).

$$G(\theta, \phi) = G_a(\theta, \phi) \times G_s(\theta, \phi) \tag{4.62}$$

The subarray antenna pattern behaves as a modulating envelope, while the array pattern factor $G_a(\theta, \phi)$ is controlled by the choice of complex weights used to linearly combine the receiver subarray outputs in the digital beamformer. In the Jindalee OTH radar design, the receiver subarray pattern $G_s(\theta, \phi)$ has a mainlobe width commensurate with that of the transmit antenna pattern, while $G(\theta, \phi)$ may be interpreted as the pattern of a (high resolution) finger beam formed on receive. In modern OTH radars with a receiver-per-element architecture, the finger-beam pattern $G(\theta, \phi)$ is given by Eqn. (4.63), where $G_A(\theta, \phi)$ is the array pattern factor when all of the antenna elements in the aperture are combined by digital beamforming. Since the element pattern $G_e(\theta, \phi)$ is relatively broad, finger beams may be digitally steered over a much wider angular sector than that spanned by the mainlobe of the analog subarray pattern.

$$G(\theta, \phi) = G_A(\theta, \phi) \times G_e(\theta, \phi) \tag{4.63}$$

The digital beamforming concepts discussed below are applicable to both the digital receiver-per-subarray and receiver-per-element architectures. The versatility of using a digital receiver-per-element comes at the expense of a significantly higher processing load for the range-processing and array-beamforming steps, not to mention data-transfer challenges in a distributed computing system. However, the additional spatial degrees of freedom, and the ability to steer simultaneous beams over a wide arc, provides a number of significant operational advantages, particulary when adaptive beamforming is used to suppress interference and clutter signals not incident from the radar look direction.

4.2.2.2 Digital Beamforming

Consider a monochromatic plane-wave signal incident on the antenna array from a far-field point source with azimuth $\theta \in [-\pi, \pi)$ and elevation $\phi \in [0, \pi/2]$. Using the standard right-hand coordinate system in Figure 4.15, the three components of the signal wavevector $\mathbf{k} = [k_x, k_y, k_z]^T$ are given by Eqn. (4.64), where the superscript T denotes transpose, λ is the radio wavelength, and $\mathbf{u}(\theta, \phi)$ is the unit vector in direction (θ, ϕ). The (θ, ϕ) dependence of $\mathbf{k}$ has been dropped for notational convenience.

$$\mathbf{k} = -\frac{2\pi}{\lambda}\mathbf{u}(\theta, \phi) = -\frac{2\pi}{\lambda}[\cos\phi \sin\theta, \cos\phi\cos\theta, \sin\phi]^T \tag{4.64}$$

Assume the array consists of M identical sensors disposed at arbitrary locations with respect to the origin of the coordinate system, as determined by the sensor position vectors $\mathbf{r}_m$ for $m = 0, \ldots, M-1$ in Eqn. (4.65). The uniform linear array (ULA) geometry depicted in Figure 4.15 will be the main focus in this section, although we do not confine ourselves to this special case for now. Without loss of generality, we may assign the first

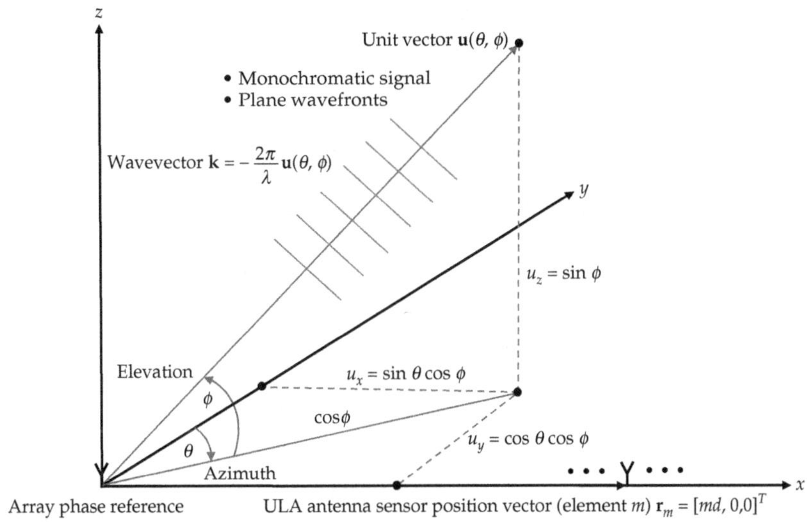

FIGURE 4.15 Illustration of a three-dimensional coordinate system with the first antenna sensor located at the origin (array phase reference) and the other antenna sensors of a ULA disposed along the x-axis with position vectors $\mathbf{r}_m = [md, 0, 0]^T$ for $m = 0, \ldots, M-1$. A monochromatic plane-wave signal with wavevector $\mathbf{k}$ impinges on the ULA from an elevation angle $\phi \in [0, \pi/2]$ and azimuth angle $\theta \in [-\pi/2, \pi/2)$. The convention used for positive azimuth is clockwise from the y to x axis, as indicated in the diagram, while the definition of elevation is consistent with a ground-deployed array in the xy-plane.

sensor of the array ($m = 0$) as the phase reference and assume this sensor is located at the origin, i.e., $\mathbf{r}_0 = \mathbf{0}$.

$$\mathbf{r}_m = [x_m, y_m, z_m]^T \tag{4.65}$$

The scalar function that describes the monochromatic plane-wave signal field at time t and position $\mathbf{r}$ may be expressed as $s(t, \mathbf{r})$ in Eqn. (4.66), where $\omega = 2\pi f_c$ is the angular frequency, A is the amplitude, and ψ is the phase at the origin at time $t = 0$. The term $\mathbf{k} \cdot \mathbf{r}$ is the scalar (inner) product between the signal wavevector and position vector.

$$s(t, \mathbf{r}) = A\exp\left(\omega t - \mathbf{k} \cdot \mathbf{r} + \psi\right) \tag{4.66}$$

Denote the signal received by the first sensor at the origin as $s_0(t) = A\exp\left(j2\pi f_c t + \psi\right)$. In practice, the frequency $f_c = c/\lambda$ may be interpreted as the carrier of the (narrowband) OTH radar signal, while (θ, ϕ) may be interpreted as the direction of arrival of a target echo propagated by a single skywave signal mode. From Eqn. (4.66), the signal received at sensor m is given by $s_m(t)$ in Eqn. (4.67), which is related to the signal received at the origin by a phase shift $\exp\left(-j\mathbf{k} \cdot \mathbf{r}_m\right)$ that is dependent on sensor location $\mathbf{r}_m$.

$$s_m(t) = s_0(t)e^{-j\mathbf{k} \cdot \mathbf{r}_m} \tag{4.67}$$

It is convenient to define the M-dimensional vector of signals received by the array as $\mathbf{s}(t)$ in Eqn. (4.68). This vector is sometimes referred to as the array "snapshot" vector. We shall name $\mathbf{s}(t)$ the useful-signal snapshot, as the term snapshot vector typically refers to the overall received signal, which additionally includes clutter, interference, and noise.

$$\mathbf{s}(t) = [s_0(t), \ldots, s_{M-1}(t)]^T \tag{4.68}$$

It is also useful to define the M-dimensional vector $\mathbf{v}(\theta, \phi)$ in Eqn. (4.69), which is known as the array steering vector for a signal direction of arrival (θ, ϕ). The first element of $\mathbf{v}(\theta, \phi)$ is unity, as $\mathbf{r}_0 = \mathbf{0}$ by definition. Note that the frequency dependence of $\mathbf{v}(\theta, \phi)$ is implicit and embedded in the definition of $\mathbf{k}$.

$$\mathbf{v}(\theta, \phi) = [e^{-j\mathbf{k}\cdot\mathbf{r}_0}, \ldots, e^{-j\mathbf{k}\cdot\mathbf{r}_{M-1}}]^T \tag{4.69}$$

The useful-signal snapshot in Eqn. (4.68) may be expressed in the compact form of Eqn. (4.70) . This expression for the useful signal snapshot is general in the sense that it applies for an arbitrary array element geometry, wherein the individual antenna sensors may be disposed in one, two, or three dimensions.

$$\mathbf{s}(t) = s_0(t)\mathbf{v}(\theta, \phi) \tag{4.70}$$

For a ground-deployed ULA oriented along the x-axis, the position vectors are given by $\mathbf{r}_m = [md, 0, 0]^T$, where d is the spacing between adjacent sensors. In this case, the inner-product terms in the steering vector expression are given by Eqn. (4.71).

$$-\mathbf{k}\cdot\mathbf{r}_m = 2\pi md \cos\phi \sin\theta/\lambda \tag{4.71}$$

At zero elevation, $\cos\phi = 1$ and the ULA steering vector is parameterized by the azimuth angle θ alone. Specifically, the steering vector reduces to the familiar expression for $\mathbf{v}(\theta)$ in Eqn. (4.72). In the context of OTH radar, this steering vector model is only appropriate for surface-wave propagation.

$$\mathbf{v}(\theta) = [1, e^{j2\pi d \sin\theta/\lambda}, \ldots, e^{j2\pi(M-1)d \sin\theta/\lambda}]^T \tag{4.72}$$

The effect of signal elevation cannot be ignored in skywave OTH radar systems. For this reason, it is convenient to work in terms of a "cone" angle φ, which parameterizes the ULA steering vector for all signal azimuth and elevation angles. The cone angle is defined according to the relationship in Eqn. (4.73).

$$\sin\varphi = \cos\phi \sin\theta \tag{4.73}$$

Using the definition of cone angle, the ULA steering vector may be written as in Eqn. (4.74), where $z(\varphi)$ is known as the phase factor. An ambiguity between azimuth and elevation exists for a ULA because different combinations of ϕ and θ can lead to the same value of φ and hence steering vector in Eqn. (4.74).

$$\mathbf{v}(\varphi) = [1, z(\varphi), \ldots, z^{M-1}(\varphi)]^T, \quad z(\varphi) = e^{j2\pi d \sin\varphi/\lambda} \tag{4.74}$$

For an off-boresight signal ($\theta \neq 0$), an increase in elevation ϕ causes a reduction in φ. Hence, when φ is interpreted directly as the "apparent" azimuth, without appropriate correction for coning, the perceived bearing of the signal shifts toward boresight as the elevation angle increases. In order to obtain great-circle bearing estimates for skywave OTH radar, real-time propagation-path advice is required to provide an accurate estimate of the mode elevation angle.

In addition, the OTH radar signal is not monochromatic in practice since it has a baseband modulation envelope $m(t)$ of finite bandwidth B. A signal is often said to be narrowband provided the bandwidth B is much smaller than the carrier frequency f_c. For typical OTH radar waveforms, the fractional bandwidth B/f_c is in the order of 10^{-3}. When the modulation envelope is incorporated, the useful signal received at the origin $s_0(t)$ is given by Eqn. (4.75).

$$s_0(t) = Am(t)e^{j2\pi f_c t+\psi} \tag{4.75}$$

An alternative definition of narrowband relevant to array beamforming is based on time-bandwidth product. Specifically, the time taken for the signal wavefront to propagate across the array is assumed to be small compared to the reciprocal of the signal bandwidth B. The delay of the signal at receiver m relative to that at the origin is given by τ_m in Eqn. (4.76). The minus sign indicates that the signal wavefront reaches the origin later than the receivers on the positive x-axis when θ is positive.

$$\tau_m = -\frac{\mathbf{u}(\theta, \phi) \cdot \mathbf{r}_m}{c} = -md \sin\varphi/c \tag{4.76}$$

In this case, the signal received at element m is related to that at the origin by Eqn. (4.77). For a modulation envelope $m(t)$ with uniform spectral density over a bandwidth B, the correlation coefficient is given by $\rho(\tau_m) = \text{sinc}(B\tau_m)$, where $\text{sinc}(x) = \sin(\pi x)/(\pi x)$ is the sinc function. For the modulation envelope to remain highly correlated over the time delay τ_m, the narrowband condition $B|\tau_m| \ll 1$ is required.

$$s_m(t) = s_0(t - \tau_m) = Am(t - \tau_m)e^{j2\pi f_c(t-\tau_m)+\psi} \tag{4.77}$$

The signal $s_0(t)$ may be considered narrowband from an array beamforming perspective provided that $B|\tau_M| \ll 1$ for all cone angles of interest. In the worst case, $|\tau_M| = (M-1)d/c$ for a ULA when $|\sin\varphi| = 1$. Here, $|\tau_M|$ is the time taken for a signal incident at endfire to propagate across the aperture. These considerations imply that the signal may be considered narrowband for all directions of arrival if the condition in Eqn. (4.78) is satisfied, where $D = (M-1)d$ is the length of the ULA aperture.

$$BD/c \ll 1 \tag{4.78}$$

Stated simply, this definition of narrowband says that the modulation envelope of the signal changes slowly compared to the time taken for the signal wavefront to traverse the array. In this case, the approximation $m(t - \tau_m) \simeq m(t)$ is justified as the correlation coefficient $\rho(\tau_m) \to 1$ for all receivers m. Substituting $m(t)$ for $m(t - \tau_m)$ in Eqn. (4.77) yields Eqn. (4.79):

$$s_m(t) = s_0(t)e^{-j2\pi f_c \tau_m} \tag{4.79}$$

Using Eqn. (4.76), it is readily shown that $2\pi f_c \tau_m = \mathbf{k} \cdot \mathbf{r}_m$. Hence, the expression in Eqn. (4.77) may be approximated as Eqn. (4.80) for a ULA and a narrowband signal. However, OTH radar receive antenna apertures may be 3 km long, so the narrowband assumption can start to break down between endpoints of the array, particulary at higher signal bandwidths and for steer angles closer to endfire.

$$s_m(t) = s_0(t)e^{-j\mathbf{k}\cdot\mathbf{r}_m} = s_0(t)e^{j2\pi md \sin\varphi/\lambda} \tag{4.80}$$

In practice, this issue can be resolved by employing fractional sampling in each receiver to digitally impose a true time delay that effectively "focuses" the aperture to the centroid of the surveillance region. The finger beams may be steered around this centroid direction without significant loss in spatial coherence. To a good approximation, the useful-signal snapshot received by a ULA may therefore be written as Eqn. (4.81).

$$\mathbf{s}(t) = s_0(t)\mathbf{v}(\varphi) \tag{4.81}$$

The radar operations of down conversion, filtering, and range processing are applied in identical manner to the signals in all receivers of the array. Denote the transformation of the received signal $s(t)$ to the digital samples $y_k(n)$ processed at range-gate k of pulse n by the operation $\mathcal{T}\{\cdot\}$ in Eqn. (4.82)

$$y_k(n) = \mathcal{T}\{s(t)\} \tag{4.82}$$

This operation is applied in each reception channel of the array to yield the range gates and pulses in receiver m. Let the corresponding output be denoted by $y_k(n, m) = \mathcal{T}\{s_m(t)\}$. Using this notation, define the M-dimensional vector $\mathbf{y}_k(n)$ in Eqn. (4.83) as the spatial snapshot of digital samples processed by the array at range k and pulse n.

$$\mathbf{y}_k(n) = [y_k(n, 0), \ldots, y_k(n, M-1)]^T \tag{4.83}$$

Using Eqn. (4.81), the snapshot $\mathbf{y}_k(n)$ may be written in the form of Eqn. (4.84), where the indexes k and n have been momentarily dropped to simplify notation, i.e., $\mathbf{y} = \mathbf{y}_k(n)$, and $y_0 = y_k(n, 0)$. To summarize, the complex-valued snapshot vector $\mathbf{y}$ corresponds to the array output at a specific but unnamed range-gate and pulse number when only a point-target echo (useful signal) is considered.

$$\mathbf{y} = \mathcal{T}\{\mathbf{s}(t)\} = y_0\mathbf{v}(\varphi) \tag{4.84}$$

To determine the array pattern response when the system is steered to a cone angle φ_0, we evaluate the conventional beamformer output $y(\varphi_0, \varphi)$ in Eqn. (4.85). Here, the spatial matched filter $\mathbf{v}(\varphi_0)$ corresponding to a cone angle φ_0 is applied to the snapshot vector $\mathbf{y} = y_0\mathbf{v}(\varphi)$ containing a signal with possibly different cone angle φ. The symbol $\dagger$ denotes the Hermitian (conjugate transpose) operator.

$$y(\varphi_0, \varphi) = \mathbf{v}^\dagger(\varphi_0)\mathbf{y} = y_0\left\{\mathbf{v}^\dagger(\varphi_0)\mathbf{v}(\varphi)\right\} \tag{4.85}$$

Using Eqn. (4.74), the beam output $y(\varphi_0, \varphi)$ can be written as in Eqn. (4.86), where the spatial phase factor $z(\varphi) = \exp(j2\pi d \sin\varphi/\lambda)$ and $z_0 = z(\varphi_0)$. By summing the terms of the geometric progression, the beam output is given by the expression on the right-hand side of Eqn. (4.86), where $z = z(\varphi)z_0^{-1}$.

$$y(\varphi_0, \varphi) = y_0 \sum_{m=0}^{M-1} \left(z(\varphi)z_0^{-1}\right)^m = y_0\left(\frac{1 - z^M}{1 - z}\right) \tag{4.86}$$

The conventional beamformer output $y(\varphi_0, \varphi)$ may be written in the alternative form of Eqn. (4.87), where the complex scalar $z = \exp[j2\pi d(\sin\varphi - \sin\varphi_0)/\lambda]$.

$$y(\varphi_0, \varphi) = y_0 z^{(M-1)/2}\left(\frac{z^{M/2} - z^{-M/2}}{z^{1/2} - z^{-1/2}}\right) \tag{4.87}$$

The numerator inside the parentheses of Eqn. (4.87) may be simplified as in Eqn. (4.88), where the identity $(e^{j\chi})^n = \cos(n\chi) + j\sin(n\chi)$ has been used. A simplified expression for the denominator is obtained by substituting $M = 1$ in Eqn. (4.88).

$$z^{M/2} - z^{-M/2} = 2j \sin[M(\pi d/\lambda)(\sin\varphi - \sin\varphi_0)] \tag{4.88}$$

By inserting these expressions into Eqn. (4.87), the amplitude and phase response of the array pattern $\mathcal{G}_A(\varphi_0, \varphi) = y(\varphi_0, \varphi)/y_0$ may be written in the form of Eqn. (4.89). Similar to the gain pattern (which relates to power), the amplitude and phase response of the ULA pattern $\mathcal{G}(\varphi_0, \varphi)$ is equal to $\mathcal{G}_A(\varphi_0, \varphi)$ multiplied by the amplitude and phase response of the antenna sensor pattern.

$$\mathcal{G}_A(\varphi_0, \varphi) = e^{j(M-1)(\pi d/\lambda)[\sin\varphi - \sin\varphi_0]} \left\{ \frac{\sin[M(\pi d/\lambda)(\sin\varphi - \sin\varphi_0)]}{\sin[(\pi d/\lambda)(\sin\varphi - \sin\varphi_0)]} \right\} \tag{4.89}$$

The normalized array beampattern $P(\varphi_0, \varphi)$ defined in Eqn. (4.90) represents the *power* gain contributed by the array pattern factor as a function of φ_0 and φ *relative* to the maximum value of $\mathcal{G}_A^2(\varphi_0, \varphi_0) = M^2$, which occurs when the steer and signal cone angles align. Important characteristics of the beampattern include the 3.0-dB mainlobe width (angular resolution), grating lobes (angular ambiguities), and the sidelobe structure (peak levels and nulls), to be discussed in the following subsection.

$$P(\varphi_0, \varphi) = \left\{ \frac{|\mathcal{G}_A(\varphi_0, \varphi)|}{\mathcal{G}_A(\varphi_0, \varphi_0)} \right\}^2 = \frac{1}{M^2} \left\{ \frac{\sin[M(\pi d/\lambda)(\sin\varphi - \sin\varphi_0)]}{\sin[(\pi d/\lambda)(\sin\varphi - \sin\varphi_0)]} \right\}^2 \tag{4.90}$$

The conventional beamformer output in Eqn. (4.85) can for signal-processing purposes be written as Eqn. (4.91), where $y(m) = y_0 e^{j2\pi f(\varphi)m}$ are the samples of the snapshot vector $\mathbf{y}$, and $f(\varphi) = d\sin\varphi/\lambda$ is the normalized spatial frequency corresponding to a cone angle φ. This expression shows that conventional (matched-filter) beamforming for a ULA is equivalent to spectral analysis of the array data in the spatial dimension (i.e., a DFT applied to the receiver outputs in a particular range cell and pulse number).

$$y(\varphi_0, \varphi) = \sum_{m=0}^{M-1} y(m) e^{-j2\pi f(\varphi_0)m} \tag{4.91}$$

In practice, the OTH radar receiving array is digitally steered to a number of cone angles denoted by φ_b for $b = 1, \ldots, B$ such that the beams cover or "tile" the surveillance region. This operation may be performed by evaluating the DFT samples $y_k^{[b]}(n)$ in Eqn. (4.92) at the normalized spatial frequencies $f(\varphi_b) = d\sin\varphi_b/\lambda \in [0, 1)$, where the cone angles $\varphi_1, \ldots, \varphi_B$ correspond to the desired finger-beam steer directions. The subscripts k and n used previously to index range-gate and pulse number, respectively, have been reintroduced to complete the notation in Eqn. (4.92). A spatial taper function $w(m)$ to control the beampattern sidelobe response has also been included in Eqn. (4.92).

$$y_k^{[b]}(n) = \sum_{m=0}^{M-1} w(m) y_k(n, m) e^{-j2\pi[d\sin\varphi_b/\lambda]m} \tag{4.92}$$

As $f(\varphi_b)$ is proportional to $\sin\varphi_b$ and not φ_b, the FFT will not yield beam steer directions that are equally spaced in cone angle. However, the FFT may be used provided that system steer angles φ_b associated with the FFT frequency bins $f(\varphi_b) = b/M$ are acceptable for

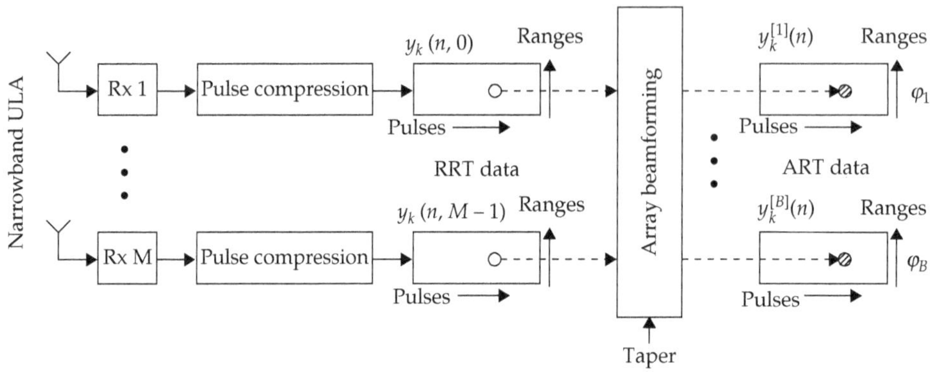

FIGURE 4.16 Array beamforming applied to each range (k) and pulse (n) processed by the M receivers of the array. In this case, the input data to the beamformer is receiver-range-time (RRT) data. The output of the beamformer is a set of finger beams steered to different cone angles $\varphi_1, \ldots, \varphi_B$ and the collection of all finger beams is referred to as an azimuth-range-time (ART) data. © Commonwealth of Australia 2011.

the retained set of finger beams $b \in [0, M-1]$. Note that the properties of the receive antenna pattern are identical irrespective of where in the signal-processing chain the array beamforming step is applied. Figure 4.16 illustrates the application of array beamforming after pulse compression.

4.2.2.3 Array Pattern Characteristics

The conventional beampattern characteristics of interest include the 3.0-dB mainlobe width, sidelobe levels, grating lobes, and null locations. For a ULA, these characteristics vary with aperture length, inter-sensor spacing, signal frequency, as well as steer angle and taper function. To simplify the analysis, attention is restricted to a ULA steered in the boresight direction with a uniform taper (rectangular window). In this case, the normalized conventional beampattern is given by Eqn. (4.93).

$$P(\varphi) = \frac{1}{M^2}\left\{\frac{\sin\left[\frac{M\pi d}{\lambda}\sin\varphi\right]}{\sin\left[\frac{\pi d}{\lambda}\sin\varphi\right]}\right\}^2 \tag{4.93}$$

The 3.0-dB mainlobe width, also referred to as the beamwidth, is defined as the angular separation between half-power points on the pattern. Provided that the array aperture is much larger than the radio wavelength ($Md \gg \lambda$), the conventional beamwidth for a ULA steered at broadside with a uniform taper is approximated by Eqn. (4.94). Relative to the boresight beamwidth in Eqn. (4.94), the beamwidth of a ULA increases as the steer angle approaches endfire ($\varphi \to \pm\pi/2$). This is because $\sin\varphi$ changes more slowly than φ for angles off boresight. The Rayleigh limit $\lambda/(Md)$, defined as the separation between the peak and first null of the conventional pattern steered at boresight, is often adopted as a nominal beamwidth value for a ULA.

$$\Delta\varphi \simeq \frac{0.89\lambda}{Md} \text{ radians}, \quad \Delta\varphi \simeq \frac{50\lambda}{Md} \text{ degrees} \tag{4.94}$$

Angular resolution is related to the inverse of the beamwidth, and therefore increases with the electrical length of the aperture, or in proportion to aperture length and operating frequency. Fine angular resolution is desirable in general, but a very small beamwidth requires the formation of a greater number of beams to capture radar echoes effectively over a surveillance region with a given angular coverage. Forming a large number of beams per dwell may pose problems in subsequent parts of the signal- and data-processing chains as it increases real-time processing load. A large number of beams also increases the difficulty of designing an appropriate data display for radar operators.

Moreover, the spatial coherence of skywave signal wavefronts tends to decrease as a function of distance due to ionospheric propagation effects. Increasing the aperture size beyond a certain point may therefore lead to diminishing performance benefits. Besides these limitations, very large receive apertures are also expensive to build and maintain. In practice, there are operational, phenomenological, and economic reasons to limit the array aperture size.

For a ULA steered at broadside, *grating lobes* in the beampattern $P(\varphi)$ occur whenever the phase factor satisfies the condition $z(\varphi) = e^{j2\pi d \sin\varphi/\lambda} = 1$. This arises at angles φ_m given by Eqn. (4.95) for integers m that satisfy $|m\lambda/d| \leq 1$. For example, if $d/\lambda = 1$ then $m = 0, \pm 1$. In this case, the value $m = 0$ corresponds to the mainlobe at $\varphi = 0$, while the values $m = \pm 1$ correspond to grating lobes formed at $\varphi = \pm 90°$ using Eqn. (4.95). The distance between adjacent subarray centers in the Jindalee OTH radar is approximately $d = 84$ m, which leads to an under-sampled aperture. In this system, the grating lobes are significantly attenuated by the modulating envelope of the subarray antenna pattern at all frequencies and steer angles of interest.

$$\varphi_m = \arcsin(m\lambda/d) \tag{4.95}$$

For a large number of receivers $M \gg 1$, the numerator of the beampattern $P(\varphi)$ changes much more rapidly than the denominator. In this case, the peak sidelobe levels occur approximately when the numerator $\sin(Md \sin\varphi/\lambda)$ reaches a maximum at angles given by $\varphi = \varphi_m$, where φ_m satisfies Eqn. (4.96).

$$Md \sin\varphi_m/\lambda = (2m-1)\frac{\pi}{2} \quad \text{for} \quad m = \pm 2, \pm 3, \ldots \tag{4.96}$$

Substituting $\sin\varphi_m = (2m-1)\pi\lambda/(2Md)$ into the beampattern expression $P(\varphi)$ provides approximate values for the sidelobe levels, which are given by $P(\varphi_m)$ in Eqn. (4.97). For $M = 10$ receivers, this formula approximates the value of the first sidelobe ($m = \pm 2$) as $P(\varphi_m) = -13.2$ dB.

$$P(\varphi_m) \approx \{M \sin(\pi[2m-1]/2M)\}^{-2} \tag{4.97}$$

On the other hand, beampattern nulls $P(\varphi_m) = 0$ occur when the numerator is equal to zero and the denominator is greater than zero. This implies that nulls are formed when the condition in Eqn. (4.98) is satisfied. In other words, the nulls at $\varphi_m = \arcsin(m\lambda/Nd)$ occur when the phase difference across the ends of the array coincides with an integer number of cycles.

$$Md \sin\varphi_m/\lambda = m \quad \text{for} \quad m = \pm 1, \pm 2, \ldots \tag{4.98}$$

Although grating lobes arise for an under-sampled aperture, beams in nonphysical or "invisible" space can be formed in an over-sampled aperture. The normalized spatial frequency for a ULA is given by $f(\varphi) = d\sin\varphi/\lambda = \rho\sin\varphi$. This frequency has an unambiguous interval of $f(\varphi) \in [-0.5, 0.5)$. However, for the physically visible cone angles $\varphi \in [-\pi/2, \pi/2)$, we have that $\sin\varphi \in [-1, 1]$, so the normalized spatial frequencies $f(\varphi) \in [-\rho, \rho]$ span all of visible space. In an over-sampled aperture, where $\rho = d/\lambda < 0.5$, this leaves an interval of normalized spatial frequencies in Eqn. (4.99) that corresponds to so-called "invisible space."

$$\rho < |f(\varphi)| < \frac{1}{2} \tag{4.99}$$

Conventional beams matched to normalized spatial frequencies in invisible space cannot contain energy from sources external to the array (neglecting sidelobe leakage). This condition strictly holds for a perfectly calibrated array and sources in the far field of the aperture. An advantage of forming "invisible beams" is that they provide an estimate of the system internal noise level during operation. However, array calibration errors and near-field sources can bias this estimate, as some energy from external signals will in practice leak into the invisible beams.

Ideally, the beampattern of the antenna consists of a narrow mainlobe with high peak gain and no sidelobes. In practice, the peak gain and beamwidth of the mainlobe must be traded off against the sidelobe levels by using illumination functions or aperture tapers. In OTH radar, beampattern sidelobe levels need to be kept low to prevent off-azimuth disturbance signals (clutter, interference, and noise) from masking weaker target echoes received through the mainlobe. When particularly strong man-made RFI is present, it is possible that the sidelobe levels of a conventional (tapered) pattern may not be low enough to prevent the disturbance level from rising above the white-noise floor in a significant number of beams. This motivates the use of adaptive beamforming techniques discussed in Part III.

4.2.3 Doppler Processing

In OTH radar, the much stronger power of clutter echoes relative to target echoes in a given range-azimuth cell implies that Doppler processing is essential for target detection. Even in the hypothetical absence of clutter, Doppler processing would still be required in OTH radar to reliably detect weaker target echoes against background noise due to the relatively low SNR in a single radar pulse.[4] In addition to enhancing target detection performance against clutter and noise, Doppler processing also provides target relative velocity information, which is valuable for tracking.

Doppler processing broadly refers to temporal filtering or spectral analysis of the slow-time data samples received in a particular range-azimuth cell over the set of pulses comprising the CPI. In OTH radar systems, Doppler processing operates on the sequence of N slow-time samples $y_k^{[b]}(n)$ received over $n = 0, \ldots, N-1$ pulse repetition intervals for each range gate (k) and azimuth cell (b). This section discusses two forms of Doppler processing. The first may be described as temporal (slow-time) filtering to attenuate the clutter in a moving target indicator (MTI) context, while the second is coherent Doppler processing based on digital spectral analysis of the slow-time data samples.

[4]Some microwave radar systems can at times afford not to perform coherent integration for target detection due to their much finer spatial resolution, different viewing geometries, and lower noise levels compared to OTH radar.

Doppler processing for maneuvering targets is also briefly discussed. This section concludes with an illustration of the practical application of the pulse compression, array beamforming, and Doppler processing steps to real OTH radar data.

4.2.3.1 Moving Target Indicator

A computationally efficient method to attenuate clutter in a given range-azimuth cell is to apply a digital filter in the slow-time domain known as a moving target indicator (MTI). As OTH radar clutter is often concentrated near zero Doppler frequency, a high-pass filter is required to suppress the clutter without attenuating target echoes at higher Doppler frequencies. MTI filters are often implemented as finite impulse response (FIR) tapped-delay lines, although recursive or infinite impulse response (IIR) filters may also be employed. The simplest one-pulse MTI canceler produces an output $z_k^{[b]}(n)$ given by Eqn. (4.100). The Doppler frequency response of this filter has a single null at zero Hertz, which removes/attenuates clutter components with zero/small Doppler shifts, respectively.

$$z_k^{[b]}(n) = y_k^{[b]}(n) - y_k^{[b]}(n-1) \tag{4.100}$$

This simple MTI filter is unlikely to provide the high level of clutter cancelation required for OTH radar applications. Assume that the Doppler frequency response of the clutter $\mathbf{H}_c(f)$ received in particular range-azimuth cell can be accurately represented by an autoregressive (AR) model of relatively low order $\kappa \ll N$, as in Eqn. (4.101). Here, the real constant α_c is scaled to match the clutter power, $\{b_i\}_{i=0}^{\kappa}$ are the AR process coefficients, and $\{p_i\}_{i=1}^{\kappa}$ are the poles of the frequency response with a modulus less than or equal to unity. Note that $b_0 = 1$ and $z = \exp(j2\pi f T)$, where f is the Doppler frequency and T is the pulse repetition interval.

$$\mathbf{H}_c(f) = \frac{\alpha_c}{1 + \sum_{i=1}^{\kappa} b_i z^{-i}} = \frac{\alpha_c}{\prod_{i=1}^{\kappa}(1 - p_i z^{-i})} \tag{4.101}$$

The FIR filter with frequency response $\mathbf{H}_w(f) = 1/\mathbf{H}_c(f)$ that "whitens" the clutter is implemented as the κ-pulse moving-average (MA) canceler in Eqn. (4.102). It is possible to precompute a number of filters with various frequency response characteristics to account for different clutter Doppler spectrum properties, and to use the output of the one that yields the signal $z_k^{[b]}(n)$ with lowest power. Although such a method is computationally efficient, it is likely that none of the precomputed filter candidates is well-matched to the prevailing clutter Doppler spectra, which are often highly variable in practice. This scheme can therefore lead to an intolerable degradation in clutter cancelation performance.

$$z_k^{[b]}(n) = y_k^{[b]}(n) + \sum_{i=1}^{\kappa} b_i y_k^{[b]}(n-i) \tag{4.102}$$

Alternatively, the AR parameter vector $\mathbf{b} = [1, b_1, \ldots, b_\kappa]^T$ may be estimated adaptively from the data using the techniques described in Marple (1987). For example, the AR parameter vector $\mathbf{b}$ may be estimated as the solution of the Yule-Walker equations in Eqn. (4.103). Here, $\mathbf{u}_1 = [1, 0, \ldots, 0]^T$ is the $\kappa + 1$-dimensional unit vector with the first element equal to unity, $\mathbf{y}_k^{[b]}(n) = [y_k^{[b]}(n), \ldots, y_k^{[b]}(n+\kappa)]^T$ for $n = 0, \ldots, N-\kappa-1$ are the training data vectors, and $\hat{\mathbf{R}}_\kappa$ is a sample estimate of the clutter temporal (pulse-to-pulse) covariance matrix in range k and beam b. In a locally homogeneous clutter environment, the training data are taken from range-azimuth cells neighboring the one filtered for reasons described in Chapter 10.

$$\mathbf{b} = \frac{\hat{\mathbf{R}}_{\kappa}^{-1}\mathbf{u}_1}{\mathbf{u}_1^T\hat{\mathbf{R}}_{\kappa}^{-1}\mathbf{u}_1}, \quad \hat{\mathbf{R}}_{\kappa} = \sum_{n=0}^{N-\kappa}\mathbf{y}_k^{[b]}(n)\mathbf{y}_k^{[b]\dagger}(n) \tag{4.103}$$

The MTI output $z_k^{[b]}(n)$ in Eqn. (4.102) is a slow-time domain signal that contains residual clutter, noise, and potentially one or more Doppler-shifted target echoes. In this case, target detection is based on the amplitude envelope of this output, with target presence being declared if $|z_k^{[b]}(n)|$ exceeds an appropriate detection threshold. The traditional MTI approach has computational advantages, but it also has some significant drawbacks.

First, it does not estimate the Doppler shift (relative velocity) of any detected targets, which would provide valuable information for tracking. Second, it provides no insight regarding the number of targets present when a detection is declared. This does not allow for multiple target echoes in the same range-azimuth cell to be distinguished. Third, it provides a lower SNR gain for target echoes in a white-noise background relative to coherent integration over all pulses in the CPI.

For these three main reasons, the MTI approach is not used for Doppler processing in OTH radar. However, the adaptive clutter filtering concept is very useful in OTH radar. As described later, it can be applied to efficiently detect the pulses in the CPI contaminated by transient disturbances (e.g., impulsive noise, meteor echoes), as well as to interpolate or extrapolate slow-time clutter samples, which is needed for two important ameliorative processing steps to be described later in the chapter.

4.2.3.2 Spectral Analysis

Relative to the MTI approach, Doppler processing by spectral analysis of all N slow-time samples in the CPI provides better detection performance and additional information at the expense of greater computational complexity. Specifically, this scheme provides (1) higher SNR gain to enhance the detection of target echoes in Doppler frequency regions dominated by noise, (2) the ability to resolve multiple target echoes with different Doppler shifts in the same range-azimuth cell to identify the number of echoes present and to prevent fading due to mutual interference between unresolved echoes, and (3) an estimate of the relative velocity of each detected echo in both sign and magnitude to enhance tracking performance.

Digital spectral analysis is performed by computing the discrete Fourier transform (DFT) of the slow-time data according to the well-known expression in Eqn. (4.104), where $\{w(n)\}_{n=0}^{N-1}$ is a window function used to control Doppler spectrum sidelobes. The index n in $z_k^{[b]}(n)$ is the frequency bin index corresponding to a Doppler shift of $n/(NT)$ Hertz. A wide variety of FFT algorithms have been developed to optimize the number of floating point operations (additions/multiplications) and memory usage in order to minimize computational time for different data lengths and processor characteristics (Burrus and Parks 1985).

$$z_k^{[b]}(n) = \sum_{m=0}^{N-1} w(m)y_k^{[b]}(m)e^{-j2\pi mn/N} \tag{4.104}$$

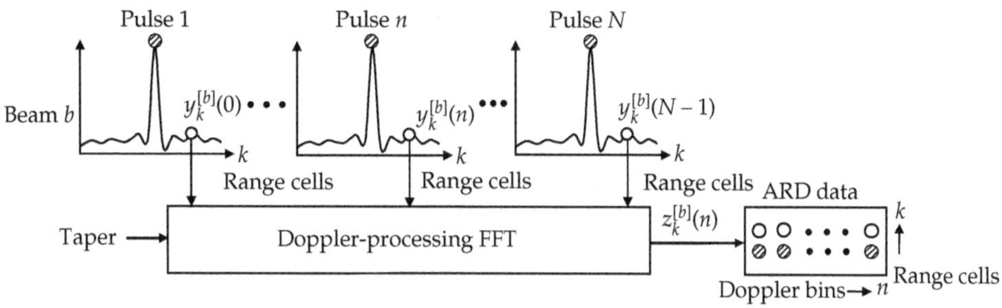

FIGURE 4.17 Coherent Doppler processing by digital spectral analysis of the slow-time samples after pulse compression and array beamforming over all N pulses (linear FM sweeps) in the CPI. The Doppler processing FFT of length N is performed on each range-azimuth cell in turn to produce azimuth-range-Doppler (ARD) data.

The slow-time samples in each range-azimuth cell are Doppler processed in turn via an N-point tapered FFT to produce an azimuth-range-Doppler (ARD) map, as shown in Figure 4.17. Spectral samples in the vicinity of zero Doppler frequency are typically dominated by powerful surface-clutter returns, while frequency bins associated with higher Doppler shifts are often dominated by external noise. The different frequency bins of the Doppler spectrum are searched individually for target echoes and when a sample that represents a local peak is found to lie above the detection threshold, it not only indicates the presence of a target echo in the associated range-azimuth cell, but it also provides an estimate of the target's (approaching or receding) relative velocity.

A single (ideal) point-target echo with constant RCS and Doppler shift f_d over the CPI has a slow-time signature given by $y_k^{[b]}(n)$ in Eqn. (4.105), where α is the complex amplitude of the target echo in range cell k and azimuth cell b.

$$y_k^{[b]}(n) = \alpha e^{j2\pi f_d nT} \tag{4.105}$$

For a rectangular window $w(m) = 1$, the continuous frequency domain output of this sampled signal is the periodic sinc function $S(f)$ in Eqn. (4.106). Here, $f_p = 1/T$ is the pulse repetition frequency. The DFT in Eqn. (4.104) samples the function $S(f)$ at the Doppler frequencies $f_n = n/(NT)$, where Doppler bins in the two regions $n \in [0, N/2)$ and $n \in [N/2, N)$ correspond to positive and negative target relative velocities, respectively.

$$S(f) = \alpha e^{-j[\pi(N-1)(f-f_d)T]} \frac{\sin[\pi(f - f_d)NT]}{\sin[\pi(f - f_d)T]}, \quad f \in [0, f_p) \tag{4.106}$$

The main problem of a rectangular window is the relatively high sidelobe levels, which allow strong clutter echoes near 0 Hz to potentially mask much weaker target echoes that are well displaced from the clutter in Doppler frequency. The highest (first) sidelobe of $S(f)$ is only about 13 dB down from the peak when using a rectangular window, while the Doppler resolution (or half-power mainlobe width) is approximately equal to $0.89/NT$. A loss in Doppler resolution, which degrades detection performance in one or two additional frequency bins close to the dominant clutter spectral components, is deemed acceptable for reducing the clutter sidelobe levels, which enhances target detection performance over the rest of the Doppler spectrum in a powerful clutter environment. A comprehensive summary of the performance of different (amplitude-only) windows can be found in Harris (1978).

When a window is applied to control spectral leakage, lower sidelobe levels are not only traded off against frequency resolution, but also come at the expense of SNR-processing loss. Relative to a uniform window, an amplitude taper reduces both the coherent gain on useful signals as well as the variance of white noise at the Doppler processing output, but not in the same proportion. The processing loss due to windowing is defined as the output signal-to-white-noise ratio that results when the window is applied (SNR_w) divided by that arising for a uniform window (SNR_u), with both SNRs evaluated at the maximum of the spectrum ($f_n = f_d$). The SNR-processing loss L_p for a window with real-valued weights $w(m)$ is given by the simple expression in Eqn. (4.107). Table 4.1 lists the 3.0-dB mainlobe width, SNR-processing loss, peak sidelobe level, and maximum scallop loss for a variety of commonly used windows.

$$L_p = \frac{\text{SNR}_w}{\text{SNR}_u} = \frac{|\sum_{m=1}^{M} w(m)|^2}{M\sum_{m=1}^{M} |w(m)|^2} \tag{4.107}$$

Scallop loss varies depending on the mismatch between the actual signal Doppler frequency and the most well-matched FFT frequency bin. The worst case arises when the signal Doppler shift lies exactly between two adjacent frequency bins. For example, the maximum scallop loss for a rectangular window is 3.9 dB, whereas the wider mainlobe of Blackmann-Harris (minimum four sample) window has a maximum scallop loss of only 0.83 dB. For systems that evaluate the Doppler spectrum at the FFT bins, robustness to scallop loss represents a form of compensation for the SNR-processing loss of a low-sidelobe window function (e.g., 3.0 dB for the Blackmann-Harris window).

Parametric spectrum estimation techniques, such as those discussed in Marple (1987), may be applied for Doppler processing. However, the tapered FFT is often used in OTH

Window Function	3.0-dB BW, Bins	SNR Loss, dB	Peak SLL, dB	Scallop Loss, dB
Rectangular	0.89	0	−13	3.92
Hann	1.44	−1.76	−32	1.42
Hamming	1.30	−1.34	−43	1.78
Dolph-Chebyshev (50 dB)	1.33	−1.43	−50	1.70
Dolph-Chebyshev (70 dB)	1.55	−2.10	−70	1.25
Blackman-Harris (92 dB)	1.90	−3.00	−92	0.83

TABLE 4.1 Figures of merit for some commonly used window functions. Windows that provide for lower sidelobes are typically associated with higher SNR-processing loss and greater mainlobe width (decreased resolution), but are often more robust to scallop loss when the signal frequency does not perfectly match that of any processed Doppler bin.

radar due to its robustness when the signals are well resolved. In ship-detection applications, high resolution and low sidelobes in Doppler are ideally needed simultaneously. Alternative Doppler processing techniques have been applied for the detection of slow-moving targets in OTH radar using short CPI lengths; see Barnum (1986), Root (1998), and Fabrizio et al. (2004b), for example.

4.2.3.3 Maneuvering Targets

Conventional (FFT-based) Doppler processing is based on the assumption that the Doppler shift of all target echoes is constant or approximately so over the radar CPI. In practice, certain targets may have rapidly varying relative velocities (maneuvering or accelerating targets), which causes the echo energy to spread significantly in the Doppler spectrum. Smearing of the echo energy over multiple Doppler frequency bins can lead to significant loss in coherent integration gain and hence SNR. The SNR loss may exceed 10 dB for an accelerating target when conventional Doppler processing is used. Besides degrading target detection performance, smearing in Doppler complicates the task of relative velocity estimation. Valuable information about the target is often contained in the echo's Doppler-time signature. The capability to estimate the time-varying Doppler frequency content of the target echo (i.e., Doppler law) can therefore provide important insights for target classification.

In such situations, joint time-frequency analysis provides an alternative method for effective maneuvring target detection and instantaneous Doppler frequency estimation, including any multicomponent Doppler features of the echo. A wide variety of time-frequency analysis techniques have been proposed in the literature. The reader is referred to Cohen (1989), Cohen (1995), and Qian (2002) for a detailed description of various time-frequency transforms and their properties. Each transform or distribution has its own particular strengths and weaknesses with respect to resolution, concentration, computational complexity, and cross-term interference. Some techniques provide high resolution but are computationally demanding, while others provide lower resolution but are relatively faster to compute. Certain methods strike a good compromise between speed and resolution, but exhibit susceptibility to cross-term interference.

Time-frequency analysis of the slow-time sample sequence received in a particular range-azimuth cell may find applications in OTH radar when there are aircraft or ship target maneuvres, as well as in the observation of rockets during the boost phase and during mid-course flight (Zhang, Amin, and Frazer 2003). A significant challenge in the OTH radar environment is that the time-frequency distribution is required to operate in the presence of clutter that is typically 40–80 dB stronger than the target echo. The study in Thayaparan and Kennedy (2004) succinctly describes and compares the performance of 12 different time-frequency transforms applied to resolve the Doppler-time signature of several accelerating aircraft target echoes against sea clutter using experimental HF radar data.

Application of the Wigner-Ville distribution to HF line-of-sight radar measurements of a theatre ballistic missile during powered flight were presented in Frazer (2001). In this practical application, the Doppler law of the missile was estimated in the presence of transient meteor scatterers at a temporal resolution of approximately 20 ms. Robust techniques for estimating time-varying target echo Doppler signatures in a low SCR environment have also been reported for OTH radar applications in Zhang, Amin, and Frazer (2003) and in Wang et al. (2003).

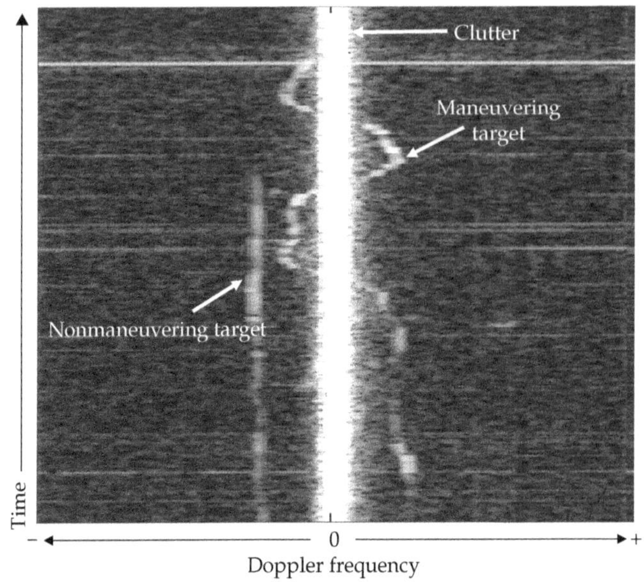

FIGURE 4.18 Waterfall display showing a vertically stacked sequence of real OTH radar Doppler spectra obtained for a single range-azimuth cell over a number of consecutive CPI. This display shows echo traces produced by maneuvring and nonmaneuvering aircraft targets. Such displays can assist operators to associate detected echoes that move across the 0 Hz Doppler frequency line (blind speed) containing strong surface clutter. © Commonwealth of Australia 2011.

To observe the Doppler-time signature of a maneuvering target echo effectively, an OTH radar may dwell continuously on a single surveillance region without inter-dwell gaps. The uninterrupted stream of coherent radar waveform pulses may be processed not as independent CPIs but as a series of partially overlapped CPIs, such that the time between Doppler spectrum updates is a small fraction of the coherent integration time. In addition, attention may be focused on a particular area of interest corresponding to one or more range gates in a certain beam, for example. The Doppler spectrum updates in a single range-azimuth cell may be stacked vertically one CPI after another to provide a display of the target's relative velocity against time. Figure 4.18 shows an example of the Doppler frequency versus time profile in a range-azimuth cell that contains an echo from a maneuvering target. This type of display may be referred to as a waterfall display.

4.2.3.4 Practical Examples

Figure 4.19 illustrates the step-wise transformation of the input OTH radar data cube acquired in the dimensions of receivers, fast-time, and slow-time in a single CPI to the azimuth-range-Doppler (ARD) output data resulting after the application of pulse compression, Doppler processing, and array beamforming. The order of the latter two steps has been interchanged relative to the previously described sequence to more clearly illustrate the impact of array beamforming on the target echo. The first transformation in this practical example shows the effectiveness of Doppler processing, which resolves the surface clutter and a (faint) aircraft target echo into different frequency bins across

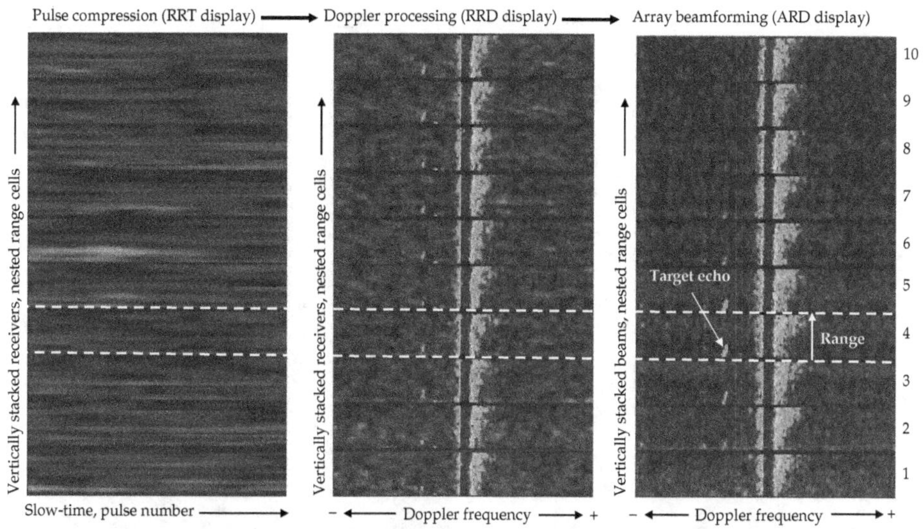

FIGURE 4.19 Step-wise transformation of the input OTH radar data cube after the pulse compression, Doppler processing, and array beamforming steps are applied to a single CPI of data. The leftmost panel shows a receiver-range-time (RRT) display, where the receiver (subarray antenna) outputs are stacked vertically with nested range cells and pulses (slow-time samples) arranged horizontally. The middle panel shows a receiver-range-Doppler (RRD) display, where the horizontal axis has been transformed from pulse number to frequency bin after Doppler processing. The clutter is concentrated near zero Doppler frequency, and an aircraft target echo is very faintly visible in the receiver outputs (only a subset of receivers in the array are shown). The rightmost panel shows an azimuth-range-Doppler (ARD) display, where the receiver outputs are transformed to 10 vertically stacked finger beams after array beamforming (only a subset of finger beams are shown). The aircraft target echo is easily distinguished in the final ARD display.

all receivers (i.e., from the leftmost to middle panel of Figure 4.19). Array beamforming provides further coherent gain to enhance the SNR of the target echo as well as localizing it in azimuth (i.e., from the middle to rightmost panel of Figure 4.19).

The importance of using a low-sidelobe window for Doppler processing is illustrated in Figure 4.20. A uniform window results in the real aircraft target echo being submerged by high clutter sidelobes, which spread over all Doppler bins (left panel), while the application of an appropriate window to lower clutter sidelobes allows the target echo to be clearly distinguished (right panel). This is quite a striking example showing the benefit of using windows to trade resolution for sidelobe levels in a real-world application.

The top panel of Figure 4.21 shows a zoomed-in view of the range-Doppler display for the beam containing the target echo in Figure 4.20. This display reveals multiple echoes received from a single target due to multipath propagation in the ionosphere. Note that different echoes can be resolved by the OTH radar in range and Doppler. The three clearly resolved target echoes are associated with the two-way propagation modes labeled in the left margin of Figure 4.21. The number 1 preceding each letter denotes one-hop skywave propagation, while the first and second letters separated by a hyphen denote the layer

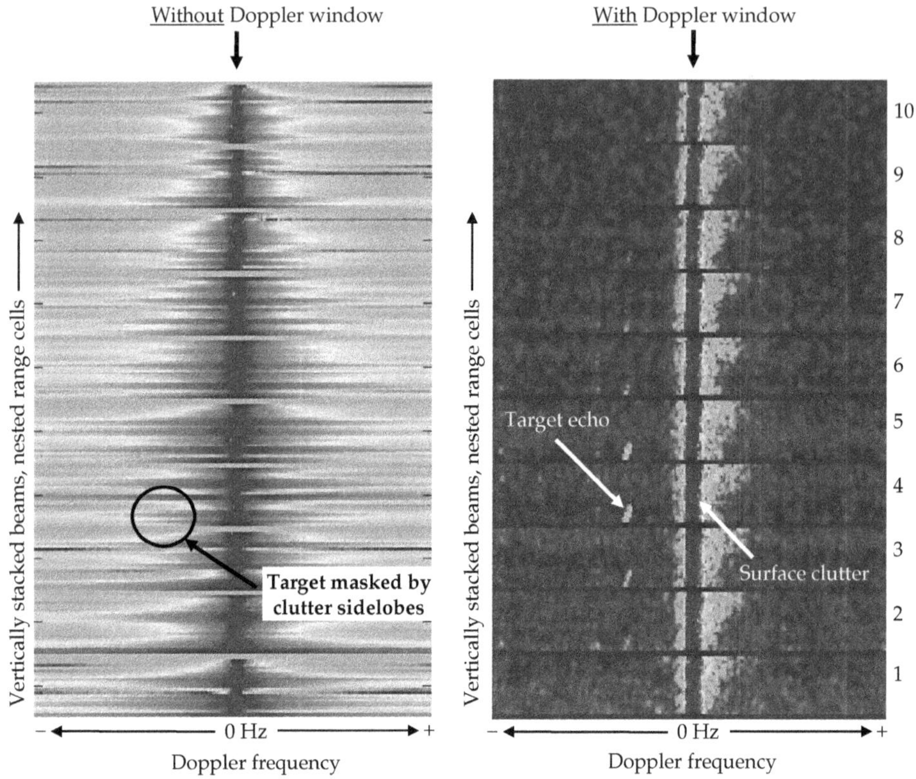

FIGURE 4.20 Practical example illustrating the importance of the Doppler processing window function to control (lower) clutter sidelobe levels. A uniform window results in the target echo being masked by high clutter sidelobes in the ARD display (left panel), while the use of a low-sidelobe window allows the target echo to be clearly distinguished (right panel). © Commonwealth of Australia 2011.

in the ionosphere reflecting the signal on the radar-to-target and target-to-radar paths, respectively.

In Figure 4.21, the group-range extends from approximately 1200 to 1575 km, while the Doppler frequency limits of ±33 Hz correspond to an unambiguous relative velocity interval of ±1115 km/h at the carrier frequency of $f_c = 15.979$ MHz. The radar signal bandwidth was 8 kHz, which yields a range resolution of about 19 km. The number of range cells retained is $K = 20$. The PRF is $f_p = 66$ Hz and the number of pulses integrated is $N = 256$, which corresponds to a CPI of approximately 4 seconds. These parameters will be used below to explain the basic characteristics of the multipath target echo structure.

The bottom panel of Figure 4.21 shows the beam spectra of the target echoes for each of the three resolved propagation modes. Each response is plotted as the magnitude squared of the beam outputs at the range-Doppler cell containing the target echo. As the $1E-1F$ and $1F-1E$ modes are unresolved in both range and Doppler, the beam spectra

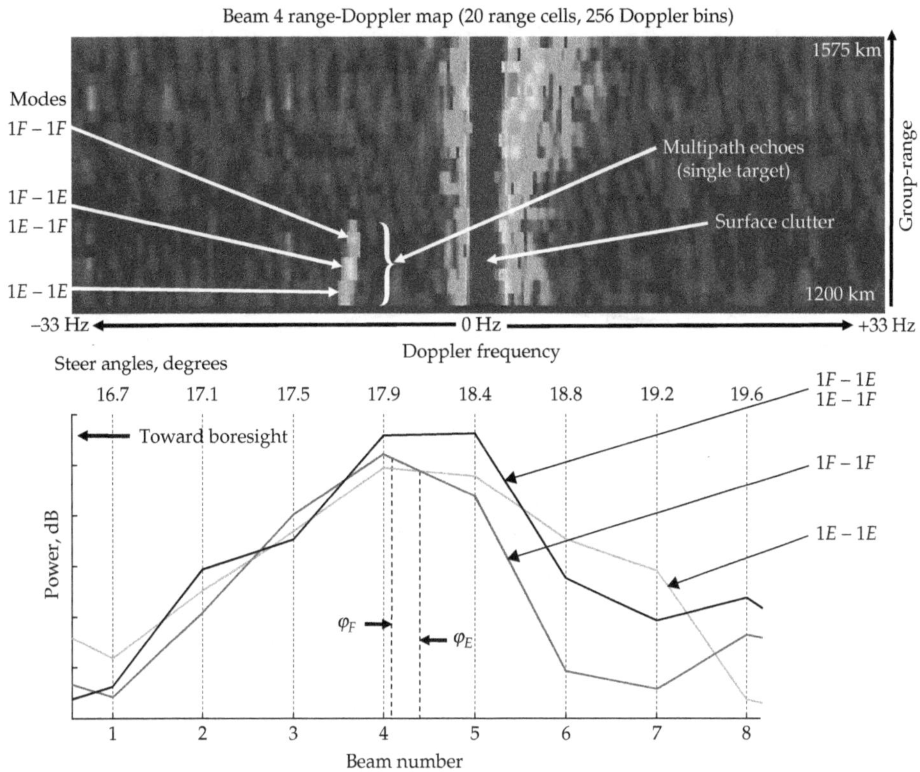

FIGURE 4.21 The OTH radar beam output in the top panel shows an example of a real aircraft target producing multiple echoes that are resolved in range and Doppler. The $1F-1F$ mode has a longer group-range and smaller Doppler-shift than the $1E-1E$ mode, while the $1E-1F$ and $1F-1E$ cannot be resolved due to their similar group-ranges and Doppler shifts. The bottom panel shows the beam response of each resolved multipath target echo component (the cone angle in degrees relative to boresight is indicated as a function of beam number). The interpolated peak of the $1F-1F$ mode is relatively closer to boresight than that of the $1E-1E$ mode due to the coning effect, which acts to shift the apparent azimuth of a mode toward boresight of a ULA as the elevation angle of the echo increases.

for these two modes cannot be observed separately, unlike the $1E-1E$ and $1F-1F$ modes, which are clearly resolved. Interpolated estimates of the $1E-1E$ and $1F-1F$ mode cone angles are indicated by φ_E and φ_F, respectively, in Figure 4.21. The F-layer mode is incident from a higher elevation angle than the E-layer mode, which explains why the F-layer mode is received with an "apparent" azimuth (cone angle) closer to the ULA boresight.

Figure 4.22 illustrates a scenario that assists to explain the range and Doppler characteristics of the multipath target echoes observed in the real-data display of Figure 4.21. Using typical values for the E- and F-layer virtual reflection heights, the group-ranges corresponding to the various propagation modes for a (low-altitude) aircraft target at a ground range of 1200 km are listed in Figure 4.22. The elevation angles of the incoming E

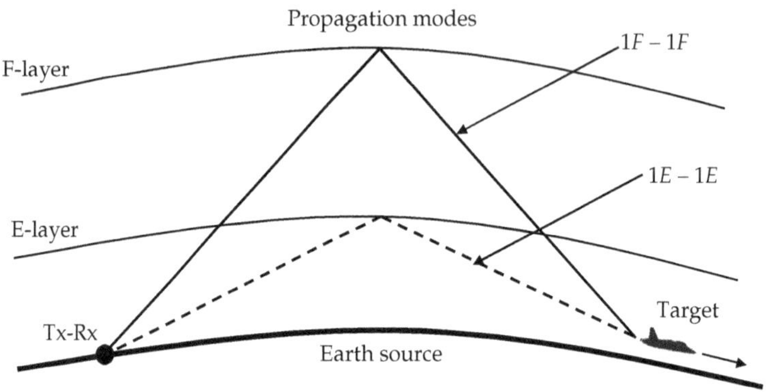

Example scenario

- E-layer virtual height = 100 km
- F-layer virtual height = 250 km
- Path ground distance = 1200 km
- Target radial velocity = 400 km/h
- Operating frequency = 16 MHz

Propagation mode	Doppler shift, Hz	Group-range, km
$1E - 1E$	–11.8	1225 km
$1F - 1E$ $1E - 1F$	–11.5	1273 km
$1F - 1F$	–11.2	1321 km

FIGURE 4.22 Sketch example of multipath (E- and F-layer) propagation between the radar and a receding aircraft target. Using the parameters listed, the predicted group-ranges and Doppler shifts for the various target echo propagation modes are tabulated. This assumes a spherical Earth geometry with concentric ionosphere and a target flying at low and constant altitude above the Earth's surface. © Commonwealth of Australia 2011.

and F-layer rays are calculated as approximately 6.7 and 19.5 degrees using spherical Earth geometry. For a target flying horizontally with a radial velocity component of 400 km/h, the Doppler shifts projected along the different (two-way) mode elevation angles have been calculated for a frequency of 16 MHz in Figure 4.22.

In this sketch example, the difference in Doppler shift between adjacent modes in descending order of group-range is about 0.3 Hz, which corresponds to a little over one Doppler bin for a CPI of 4 seconds. Although no truth is available regarding the scenario for the real-data example, this Doppler shift separation between the different modes of the target echo is consistent with the separation observed in Figure 4.21. Note that the mode with the lowest group-range has the highest Doppler shift. This relationship can help to distinguish a velocity ambiguous target, since an echo that wraps once around Doppler frequency will exhibit the opposite relationship between mode group-range and Doppler shift.

The maximum difference in group-range between the modes is about 96 km in the sketch example. This corresponds to about 5 group-range resolution cells, or a quarter of the processed range-depth, which is also consistent with the range spread observed for the echoes in the real-data display of Figure 4.21. With reference to the values listed in the table of Figure 4.22, it is also apparent from Figure 4.21 that the real target modes are clustered around a Doppler shift of about −11 Hz over a group-range extent roughly between 1220 and 1320 km. This sketch example is not intended to confirm the real-data measurements (as no truth information is available), but rather to illustrate how the characteristics of the multipath target echo structure in range and Doppler are related to the geometry of ionospheric propagation.

4.3 Operational Considerations

The success of air and surface target detection tasks depends on several factors related to the design and operation of an OTH radar system. Indeed, the effectiveness of air and surface target detection tasks depends on such different criteria that to optimize performance for one task or the other at the system level would lead to two quite different OTH radar designs. In practice, aircraft, ships, and other target classes often need to be detected and tracked using a single system. In this case, the OTH radar architecture represents a compromise between the competing objectives associated with these two fundamentally different types of tasks. The weighting given to different aspects of the OTH radar design reflects the relative priorities accorded to air and surface missions. Once the system has been configured based on these priorities, the remaining flexibility lies in the selection of operating parameters.

Bearing in mind the inevitable limitations of signal and data processing, the success of a particular OTH radar mission is largely determined by the properties of the useful signals, clutter, and noise in the received data. Not surprisingly, these properties depend critically on the choice of carrier frequency and radar waveform parameters. The carrier frequency and waveform parameters need to be chosen to best suit the prevailing environmental conditions and target class of interest. Section 4.3.1 discusses the main operational considerations affecting the choice of OTH radar carrier frequency, waveform bandwidth, pulse repetition frequency, coherent integration time, and array aperture usage, for air and surface tasks. System design aspects were covered previously in Chapter 3.

Once the data has been acquired, it also transpires that total reliance on the previously described rudimentary signal-processing steps can at times lead to significant degradations in output SDR over large portions of the ARD data cube. For this reason, the rudimentary routines are often supported by a number of ameliorative signal-processing steps. Some of these steps are included to improve the immunity of the radar to adverse clutter and noise conditions, such as severe impulsive noise or transient meteor echoes. Other steps are inserted into the signal-processing chain to make better use of the radar resource in terms of improving coverage or coverage rate. Sections 4.3.2 and 4.3.3 discuss these two types of ameliorative signal-processing steps, which are often considered indispensable for effective OTH radar operation and hence regarded as part of standard processing.

4.3.1 Air and Surface Tasks

Surveillance radar targets may be classified using a host of different criteria. For example, targets may be very broadly categorized in terms of their nominal RCS values at the radar operating frequency (small or large) and their expected domain of velocities (slow or fast). It is tempting to say that surface-vessels fall into the slow-moving and large-RCS target class, while aircraft fall into the fast-moving and small-RCS target class. As this is mostly the case, it has become common practice to refer to OTH radar tasks as air- or surface-mode tasks. This natural division of target classes may inadvertently lead to the notion that OTH radar operating parameters are either optimized to detect aircraft or surface vessels, irrespective of their RCS and velocity characteristics.

In reality, the carrier frequency and waveform parameters are not so much optimized with respect to a physical target entity (aircraft or surface vessel), as to whether the main factor limiting target detection is likely to be noise or clutter. Within the instrumented domain of OTH radar operating parameters, the carrier frequency and waveform design are optimized for target detection in two broadly defined scenarios, (1) fast-moving targets with potentially low RCS that are likely to compete for detection in a background dominated by noise, or (2) relatively slow-moving targets with possibly high RCS that are typically embedded in a clutter-dominated disturbance.

To mention an example, the detection of small fast boats may be limited by noise rather than clutter in frequency-stable ionospheric conditions. In such situations, the choice of carrier frequency may be based on criteria typically used for air-mode as opposed to surface-mode tasks. In other words, the target RCS and velocity have a greater influence on parameter choice than whether the target is an aircraft or surface vessel. In the following, we shall adhere to the usual convention and refer to different OTH radar tasks as simply air and surface tasks for convenience of discussion. However, it should be kept in mind that, from a radar perspective, the carrier frequency and waveform parameters are actually optimized for target detection in noise or clutter.

4.3.1.1 Carrier Frequency

For air tasks, the choice of carrier frequency is primarily based on maximizing SNR. This entails selecting a frequency that on an average provides the highest SNR for the air target(s) in the surveillance region. In practice, maximizing the ratio between the average power of the signal backscattered from the surveillance region, and the background noise power in clear frequency channels of appropriate bandwidth, represents the standard method for optimum frequency selection based on the SNR criterion.

Subtleties arise because the SNR depends on the background noise power at the output of (high-resolution) receive antenna finger beams steered across the surveillance region rather than that measured by an omnidirectional "whip" antenna or subarray. In addition, the target RCS characteristics and mean power density of the *incident* illumination (as opposed to that backscattered from the surveillance region) are usually not considered directly in this process, which measures CNR rather than target SNR. However, these factors often turn out to be second-order effects and are hence not the main drivers for frequency selection.

The requirement to maximize SNR often implies that the carrier frequency is selected slightly below the maximum useable frequency (MUF) associated with the closest range of the surveillance region. Such a choice effectively places the nearest range of coverage slightly beyond the edge of the skip-zone. This seeks to exploit the natural focussing gain provided by ionospheric reflection in the region immediately following the skip-zone. Care must be taken to avoid operating too close to the leading edge, as a reduction in the MUF due to changes in ionospheric conditions can cause the nearest ranges of the coverage to fall inside the skip-zone. Fortunately, the average power density of the HF signal returned to ground by the ionosphere does not decay rapidly as the frequency decreases below the MUF defined previously. This creates a frequency interval of perhaps 1 MHz wide over which radar signal propagation to the geographic area of interest is close to optimum in terms of the average signal power density illuminating the target. Besides the signal power, the Doppler coherence of the clutter (and target) echoes will also affect detection performance.

In accordance with OTH radar policy of noninterference, the choice of carrier frequency is also subject to operating on a clear frequency channel deemed to be free of other HF band users. Ideally, one or more clear frequency channels of appropriate bandwidth can be identified within the optimum frequency interval for radar signal propagation. In this case, one or more of these candidate frequencies may be used in turn when the radar revisits the surveillance region. When user-congestion is high, it may be necessary to use a suboptimum frequency yielding a lower SNR. It is also conceivable that secondary factors such as system gain and spatial resolution may enter into consideration in a more explicit way than described, but we shall not be concerned with this level of detail here.

In practice, the OTH radar operator receives frequency advice specific to a particular surveillance region from a real-time frequency management system (FMS) that automatically ranks different frequency channels based on the SNR performance metric. In this system, backscatter sounder (BSS) measurements made in the direction of the surveillance region provide an indication of the variation in average signal power over different ranges in the surveillance region as a function of frequency. This data is combined with clear channel advice from a spectrum surveillance system and background noise spectral density estimates from a directional antenna using a beam that covers the surveillance region. FMS data from these different subsystems is automatically processed to provide the operator with a ranking of unoccupied frequency channels (of at least the required bandwidth) that maximize the average SNR over the surveillance region.

The highest average clutter power backscattered from the surveillance region, which is typically received at a frequency slightly below the leading edge of the BSS ionogram, often comes at the expense of higher signal-fading rates and depths. The signals returned from ranges immediately beyond the skip-zone tend to exhibit higher Doppler spectrum contamination with respect to signals returned from further ranges, where the influence of high rays is typically less significant. However, such effects are often of secondary concern to the SNR performance metric for aircraft detection due to the high relative velocities and hence Doppler shifts of the targets. Empirical evidence suggests that the enhancement in mean signal power outweighs the effects of reduced signal coherence (i.e., broadening of the Doppler spectrum characteristics of the received clutter and target echoes) when SNR is the primary performance metric for frequency selection.

As already mentioned, the frequency that places the first range of the surveillance region immediately beyond the edge of the skip-zone is also the highest frequency that can be used to provide illumination over the entire surveillance region by the regular process of ionospheric reflection. A clear frequency channel slightly below this MUF is chosen in practice to optimize SNR and hence target detection performance in a noise-limited environment. With respect to the alternative of using lower frequencies (well below the leading edge), a frequency slightly below the MUF also tends to be favorable from a target RCS perspective. This is especially the case for targets with small major dimensions, since the average RCS of a small target (which may be on the verge of falling into the Rayleigh scattering regime at lower frequencies), tends to increase as the scattering regime shifts further towards the resonance region at higher operating frequencies.

Besides the aforementioned advantages of ionospheric focussing gain and potentially higher target RCS, the power spectral density of environmental noise is typically lower at higher operating frequencies. Although the background noise spectral density may exhibit a relatively uniform profile as a function of frequency during the day, the reduction in noise level with increasing frequency is significant at night for reasons described previously. In this case, operation close to the MUF for the nearest range of the surveillance region also tends to coincide with the lowest environmental noise spectral density. The combination of all these factors points to the use of an operating frequency not far below that of the leading edge of the BSS ionogram at the nearest group-range of the surveillance region, subject to clear channel availability. This summarizes the general principles behind optimum carrier frequency selection for air-mode tasks, where SNR represents the primary target detection performance metric.

The detection of slow-moving targets is more likely to be limited by clutter rather than noise. In the case of ship detection, this includes first- and second-order sea-clutter components, both of which can mask target echoes. Carrier frequency selection for surface-mode tasks is therefore primarily based on the criterion of minimizing Doppler spectrum contamination of the backscattered echoes in order to maximize the signal-to-clutter ratio (SCR). The radar signal power density illuminating the radar footprint and noise power spectral density are not to be completely ignored for surface-mode tasks, but these become secondary factors to the temporal coherence of the signal phase-path over the CPI when it comes to the detection of slow-moving target in a clutter-limited environment. Figure 4.23 notionally depicts the dependence of target visibility on echo Doppler-shift and target RCS (echo strength) against a real ocean-clutter Doppler spectrum.

Skywave propagation with reduced multipath contamination and signal-fading typically arises at frequencies well below the leading edge when F-layer propagation is considered. Such frequencies are lower than those normally used for air-mode tasks at a

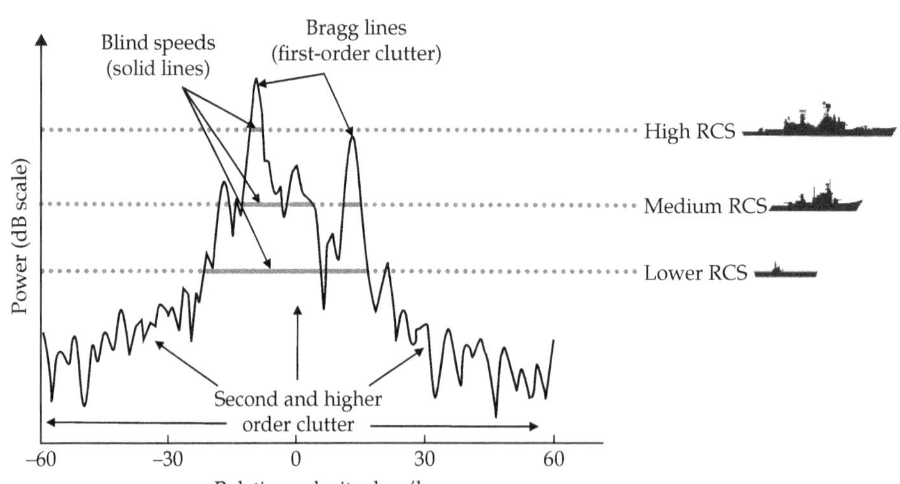

FIGURE 4.23 Real-data example of an ocean-clutter Doppler spectrum recorded by an OTH radar and illustration of the target blind speed intervals as a function of echo strength. Very-high RCS targets (40–50 dBsm) can only be masked by the Bragg lines and have a narrow blind-speed interval, while lower RCS targets with echo amplitudes comparable to the second-order clutter continuum may be masked over a much broader band of Doppler frequencies (relative velocities), especially at high sea states. © Commonwealth of Australia 2011.

given surveillance region and time. At these lower frequencies, the high rays experience greater attenuation and the main contribution is due to the ordinary and extraordinary waves in the low ray, which tends to provide more frequency-stable propagation in the context of F-layer reflections. For large surface vessels, the use of lower frequencies will not cause the target RCS to fall into the Rayleigh scattering regime, while the higher background noise spectral density encountered at lower frequencies at night will not affect detection performance in a clutter-limited environment.

The reception of multipath echoes with comparable strength tends to be most pronounced as the signal frequency is reduced slightly below the MUF of a point-to-point circuit. This is due to the low- and high-angle rays separating, as well as the *o*- and *x*-waves in the high ray of the F-layer separating. Multipath components received with different ionospheric Doppler shifts and spreads in the same radar resolution cell can significantly contribute to clutter Doppler spectrum broadening, and hence the obscuration of slow-moving target echoes. As the frequency is reduced well below the leading edge, the high rays become more highly attenuated and their multipath contributions are less significant. However, at lower frequencies, the likelihood of simultaneous propagation via different ionospheric layers in the E- and F-regions is higher, and this may also give rise to multipath-induced spectral contamination of the clutter Doppler spectrum.

Figure 4.24 shows the impact of multipath propagation and ionospheric phase-path variations (with range-dependent Doppler shifts and spreads) on the received clutter Doppler spectrum. A major problem associated with multipath is that disturbed

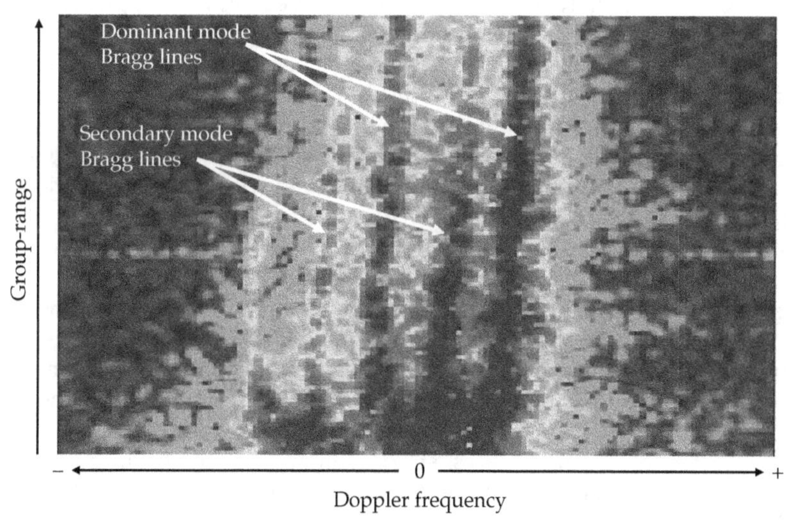

FIGURE 4.24 Range-Doppler map showing the deleterious effect of multipath on the clutter Doppler spectrum. More disturbed or weaker modes may not support target detection but the clutter from such modes may obscure regions of Doppler space where target echoes propagated by the most dominant and frequency-stable mode are present. Apart from the different Doppler shifts imposed by the ionosphere on the dominant and secondary modes illustrated in this example, significant range dependence of the Bragg-line Doppler shifts and spreads is observed for each of the two propagation modes. © Commonwealth of Australia 2011.

(or weak) propagation modes may not allow for target detection but can nevertheless superimpose clutter in regions of Doppler space where target echoes corresponding to sufficiently strong and frequency-stable propagation modes may be present. This situation arises frequently in practice because the ionosphere often imposes nonidentical Doppler shifts and spreads on different propagation modes, as evident for the dominant and secondary modes indicated in Figure 4.24. The other feature is that the ionosphere imposes range-dependent Doppler shifts and spreads on individual propagation modes. This phenomenon causes the Bragg lines to drift in Doppler frequency and change their width as a function of range for each mode labeled in Figure 4.24.

Sporadic-E often provides frequency-stable propagation and this layer can sometimes have a higher critical frequency than the F2-layer, even though it may not be blanketing. In such situations, the surveillance region may be illuminated effectively via sporadic-E propagation with relatively low spectral contamination for (one-hop) ground distances less than 2000 km. Intense sporadic-E ionization can at times allow for extended coverage over greater ground distances via multi-hop propagation. However, this propagation mode is not entirely predictable and may only last for relatively short periods of time, typically a few hours or less. The normal E-layer is also typically frequency-stable and provides an alternative for ground distances less than 2000 km, but only during the day, and F-layer propagation may also be present (beyond a limiting range) due to its usually higher critical frequency.

In the absence of intense sporadic-E ionization with a higher critical frequency than the F2-layer, the choice of optimum frequency for surface-mode tasks can be difficult to determine from BSS ionograms alone. For this reason, operators often consult mini radar soundings that routinely measure the backscattered-clutter Doppler spectrum as a function of beam direction and group-range at a number of suitably spaced carrier frequencies. Despite having lower sensitivity than the main OTH radar, the mini radar is generally regarded as the most reliable source of information when it comes to selecting the frequency for a surface task. Oblique incidence ionograms indicate the one-way mode structure of a point-to-point ionospheric circuit, and may also be used to guide frequency selection with respect to minimizing multimode-induced spectral contamination.

Frequency selection based on the exclusive optimization of SNR or SCR represent the two extreme cases for air- and surface-mode tasks, respectively. In certain scenarios of practical interest, targets need to be detected in a disturbance background that is not strongly dominated by either noise or clutter. In general, both criteria need to be considered for frequency selection rather than optimizing only one with no consideration for the other. For example, it may occur that the SCR is adequate for detecting a small surface vessel (due to the excellent purity of the clutter Doppler spectrum at the selected frequency), but that focusing solely on this criterion has resulted in an SNR that is too low for target detection. In many cases, striking an appropriate balance between the two optimization criteria is required for successful OTH radar operation. SNR and SCR should therefore not be considered as exclusive performance indicators, but rather as primary and secondary performance metrics, depending on the mission at hand.

4.3.1.2 Bandwidth

The ionosphere is a frequency-dispersive propagation medium and this property can at times be appreciable on the scale of the transmitted signal bandwidth. Specifically, the different frequency components of an HF signal can experience dispersion in group path, Doppler shift, and polarization state during transit through the ionosphere. Such phenomena place an upper bound on the bandwidth of an HF signal that can be gainfully employed to enhance range resolution in OTH radar systems (without applying some form of compensation). This limit is sometimes defined in terms of a coherence bandwidth that varies with ionospheric conditions. For OTH radar purposes, the maximum usable bandwidth that can be gainfully employed to enhance range resolution using standard processing is typically considered to be about 100 kHz or less for F-layer propagation. The question arises as to the factors driving the choice of OTH radar signal bandwidth for air and surface tasks within the bounds imposed by this physical limit.

A higher radar signal bandwidth provides finer group-range resolution. This is often desirable from the viewpoint of reducing the spatial resolution cell size of the system, and hence the effective RCS of backscattered clutter that competes for detection with (slow-moving) target echoes. In the case of fast-moving targets, where the echoes need to be detected against noise, fine range resolution provides greater scope to resolve returns from the same target (due to multipath propagation), or target echoes from independent scatterers such as meteors. When spread-Doppler clutter is present over the radar's velocity search-space, fine range resolution can also reduce the clutter power competing for detection with fast-moving target echoes.

In summary, greater signal bandwidths improve target detection against clutter by enhancing the SCR in the individual OTH radar resolution cells. Importantly, a wider signal bandwidth can also improve the localization accuracy of detected target echoes in radar coordinates not limited to group-range. This is because the peak interpolation procedures in range, azimuth and Doppler assume the presence of a single well-resolved target echo in at least one of the three canonical radar dimensions for accurate parameter estimation. Higher range resolution helps to separate target echoes, which not only alleviates rapid fading due to unresolved multipath, but also allows for more accurate parameter estimation.

On the other hand, the desire to use a higher bandwidth may render the task of finding clear frequency channels appropriate for radar operation more difficult in the crowded HF band, especially at nighttime when the lower HF spectrum becomes highly congested. As the requirement for bandwidth increases, the number of clear frequency channels deemed suitable for radar operation can diminish very rapidly. Reduced clear channel availability provides less scope for optimizing the characteristics of OTH radar signal propagation to the surveillance region, or to lower the background noise spectral density, both of which influence SNR. Moreover, opening up the radar bandwidth increases the potential to receive incidental man-made interference. The main point is that using a large signal bandwidth may significantly compromise the maximum achievable SNR relative to that possible with a lower bandwidth.

The net outcome is that air-mode tasks tend to employ relatively narrow bandwidths, most often between about 5 and 15 kHz. This primarily reflects the importance of maximizing SNR for the detection of fast-moving targets with potentially low RCS. The 5–15 kHz bandwidth interval is also considered to provide adequate group-range resolution (10–30 km) for target echo detection and localization in radar coordinates. Range-folded spread-Doppler clutter is often of greater concern at night for air-mode tasks, but the SNR penalty for increasing the bandwidth at this time can be severe due to the dense occupancy of the lower HF band. For this reason, even narrower bandwidths are typically used at night (5–8 kHz) compared to the daytime (8–15 kHz).

In contrast, surface tasks tend to use higher bandwidths that typically vary between 15 and 30 kHz. This primarily reflects the importance of minimizing clutter power in each radar resolution cell. Apart from lowering the effective clutter RCS, range cells of small size provide another important benefit for slow-moving target detection. Specifically, the clutter Doppler spectrum characteristics may change significantly with group-range when the Doppler resolution is very fine (e.g., 0.05 Hz for a 20 seconds CPI). This may arise due to the presence of land-sea boundaries in the radar coverage, or ionospheric Doppler shifts and spreads that vary with the group-range of the signal path. A smaller range cell confines the "collecting interval" in delay and hence reduces the degree of variation in the clutter Doppler spectrum that is captured in a particular spatial resolution cell. Figure 4.25 illustrates a real OTH radar data example of ships with different relative velocities competing for detection against a mixture of sea and land clutter where the received clutter Doppler spectrum characteristics are clearly range and beam dependent.

Similar concepts apply for the angular resolution of the system. A larger receive antenna aperture reduces the cross-range dimension of the radar resolution cell. This

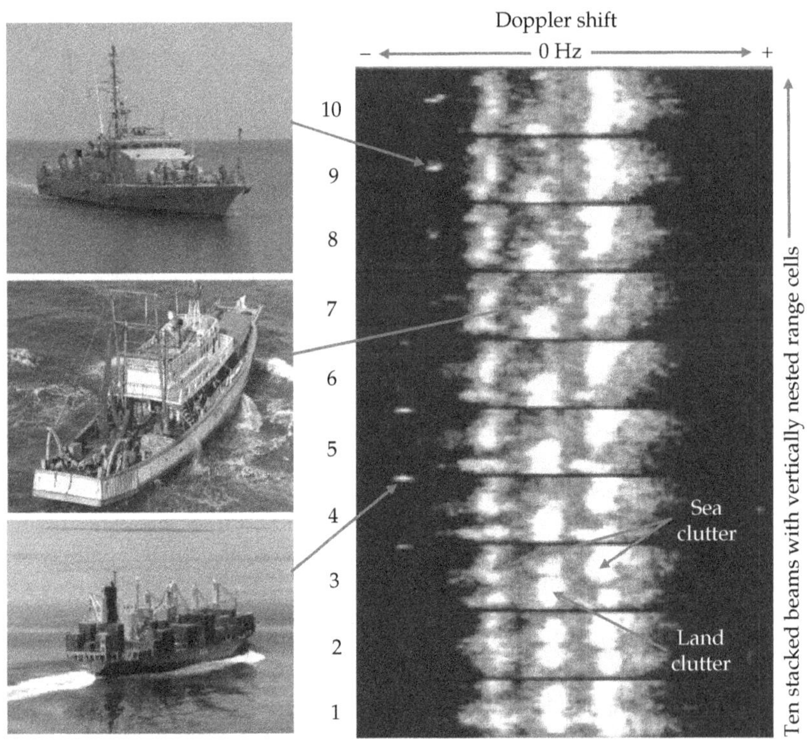

FIGURE 4.25 Real OTH radar data showing the detection of a patrol boat, fishing vessel, and container ship. Doppler spreading of the first-order clutter returns is mainly caused by irregular phase-path variations imposed on the radar signal by two-way ionospheric (skywave) propagation over the CPI. This phenomenon, combined with the presence of land clutter, effectively narrows the Doppler spectrum region between the Bragg lines where the clutter level remains sufficiently low to enable target detection. Both of these effects contribute to the difficulty of detecting the fishing vessel between the Bragg lines in beam 7.
© Commonwealth of Australia 2011.

enhances SCR for slow-moving target detection and may decrease the amount of clutter Doppler spectrum broadening due to angular variations in the clutter Doppler spectrum characteristics. The full aperture of the OTH radar receive antenna is therefore often used for ship detection. However, the transmit beam may be broadened (using appropriate tapers or half aperture operation) to provide a wider arc of coverage without compromising detection performance in a clutter-limited environment. It is cautioned that the transmit antenna gain can in some cases impact SCR. Independent half-aperture OTH radar operation (on transmit and receive) is generally more suitable for air-mode tasks to increase coverage or coverage rate, at the expense of lower SNR. This mode of operation is more appropriate during the day when the achievable SNR is typically higher and spread-Doppler clutter is often less problematic.

4.3.1.3 Pulse Repetition Frequency

Use of a single (periodic) radar waveform with a constant pulse repetition frequency (PRF) entails a compromise between group-range and Doppler frequency (relative velocity) ambiguities. For a monostatic radar, the first group-range and relative velocity ambiguities, denoted by R_{amb} and v_{amb}, respectively, are defined in Eqn. (4.108) where f_p is the waveform PRF and f_c is the carrier frequency.[5] Higher order ambiguities occur at integer multiples of R_{amb} and v_{amb}. For the purpose of discussion, we identify the low-PRF ($f_p < 10$ Hz), medium-PRF ($f_p = 10$–40 Hz), and high-PRF ($f_p > 40$ Hz) operating regimes for OTH radar.

$$R_{amb} = \frac{c}{2f_p}, \quad v_{amb} = \pm\frac{cf_p}{4f_c} \tag{4.108}$$

Range ambiguities can cause problems for OTH radar because clutter may be returned over very long distances via skywave propagation. This may include single- or multi-hop paths, transequatorial propagation, and chordal modes, for example. Although range-ambiguous clutter echoes received from long distances are generally weaker than those backscattered from the surveillance region, they can be significantly spread in Doppler frequency, particularly when the propagation path involves scattering from dynamic (field-aligned) electron density irregularities. For an OTH radar pointing equatorward, this commonly occurs at night when the backscattered radar signal passes through disturbed ionospheric regions at low magnetic latitudes (e.g., spread-F). Importantly, the received Doppler-spread clutter may be sufficiently strong to mask target echoes when it folds into the radar surveillance region through a range ambiguity.

While the possibility of detecting range-aliased target echoes also exists (particularly at very close ranges), this issue is considered to be less problematic in practice than the reception of spread-Doppler clutter. Target echoes propagated by very-long-range (multi-hop) skywave paths are typically submerged below the noise floor, particularly when D-layer absorption is high. On the other hand, Doppler ambiguities can cause problems with respect to target echoes in the surveillance region. Fast-moving targets can potentially wrap around the unambiguous Doppler space and fold back onto the surface clutter occupied region near 0 Hz. The first so-called "blind speed" corresponds to a relative velocity $v = 2v_{amb}$, where the target echo has a Doppler shift equal

[5] Note that the nearest range-ambiguous region can be located at a distance that is further or closer to the radar than the intended surveillance region.

to the PRF and therefore falls on 0 Hz after wrapping around the Doppler spectrum once.

As far as ship detection in OTH radar is concerned, a combination of range and velocity ambiguities that does not involve a compromise in practice can be found using a constant PRF. This is because the values of f_p needed to ensure velocity-unambiguous operation for most surface vessels fall in the low-PRF regime. For example, a PRF of $f_p = 5$ Hz yields an unambiguous relative velocity interval of $v_{amb} = \pm 90$ km/h at $f_c = 15$ MHz. The first group-range ambiguity for this PRF is $R_{amb} = 30{,}000$ km. With the possible exception of round-the-world propagation (the circumference of the Earth is about 40,000 km), these range ambiguities occur at distances that are far enough away to be of little concern for surface-mode tasks.

However, compromises may at times be required for aircraft detection when a single waveform is used. The high-PRF regime is normally most suitable for air-mode tasks in the day, when long-range (multi-hop) backscatter that can potentially return spread-Doppler clutter from beyond the surveillance region experiences relatively high attenuation due to passing more times through the D-layer. In addition, equatorial spread-Doppler clutter resulting from the formation of field-aligned irregularities is typically a late afternoon and nighttime phenomenon. For example, a PRF of $f_p = 50$ Hz yields an unambiguous relative velocity interval of $v_{amb} = \pm 900$ km/h at $f_c = 15$ MHz, which is sufficient for many (but not all) aircraft targets. On the other hand, the first group-range ambiguity for this PRF (relative to the first group-range of the surveillance region) is $R_{amb} = 3{,}000$ km. For a given number of pulses (sweeps), a higher PRF reduces the CPI, which increases the coverage or coverage rate.

Range-folded clutter is much less of a problem when it is either highly attenuated, or not significantly shifted or spread in Doppler frequency, relative to that backscattered from the surveillance region. Such situations often permit OTH radar operation in the high-PRF (velocity unambiguous) regime for air-mode tasks during the day. However, at night, spread-Doppler clutter is more prevalent, and D-layer absorption disappears, which increases the susceptibility of the radar to receiving strong range-folded spread-Doppler clutter. Hence, the medium-PRF regime may be more appropriate for air-mode tasks at nighttime. In this case, the possibility of velocity ambiguity may be accepted for faster aircraft targets in exchange for greater immunity to range-ambiguous spread-Doppler clutter contamination, which can degrade the detection of aircraft with lower relative velocities. A lower PRF also increases SNR for a given number of sweeps.

In principle, multiple waveforms with different PRFs can be used to resolve range and Doppler ambiguities. One method is to change the PRF within the CPI to resolve the ambiguities during a single dwell. However, this approach can significantly complicate Doppler processing. An alternative is to maintain a constant PRF in a particular CPI, and to jitter the PRF from scan to scan. This makes Doppler processing more straightforward, but resolving ambiguities for multiple time around echoes requires more than a single dwell. Jittering the PRF enables Doppler and/or range ambiguous target echoes to be identified. However, such a scheme may not represent a solution to the range-ambiguous spread-Doppler clutter problem. This is because the clutter in question typically occupies a considerable range extent, so jittering the PRF simply slides the contaminated region across the range-Doppler plane, but does not uncover the useful signals. Other types of (nonrecurrent) waveforms may be used to manipulate the regions of range-Doppler search space occupied by range-ambiguous spread-Doppler clutter, such that this disturbance do not obscure useful signals (Clancy, Bascom, and Hartnett 1999).

4.3.1.4 Coherent Processing Interval

A long coherent processing interval (radar dwell time) enhances the signal-to-white-noise ratio and Doppler-frequency resolution of the system. For conventional Doppler processing, these two quantities are directly proportional to the OTH radar dwell time in the case of a non-fluctuating target with uniform velocity and skywave propagation via a frequency-stable ionosphere. The main drawback of long dwell times is that they significantly consume the radar resource. For a given coverage area to be searched in real time by the OTH radar, longer dwells lead to reduced region revisit rates, which can degrade tracking performance. Alternatively, for given region revisit rates, longer dwells decrease the total coverage area that can be simultaneously scheduled on the radar time-line.

For air tasks, the lower bound on the dwell time is mainly determined by the need to provide adequate SNR for the detection of potentially low RCS targets, and sufficient Doppler resolution to separate target echoes from clutter (as well as other target echoes). The latter is required to identify the number of targets and to accurately estimate their relative velocities. The upper bound on dwell time for air tasks is mainly limited by coverage and coverage rate requirements. Under normal conditions, the air-mode CPI is less frequently limited by issues such as (ionospheric) channel coherence time, and target acceleration over the CPI. Air-mode dwell times ranging from 1 to 4 seconds are considered to represent a reasonable compromise between the competing objectives of SNR gain and Doppler resolution on the one hand, and coverage or coverage rate on the other.

Quite different considerations apply for surface-mode tasks. The lower bound on dwell time is primarily driven by the requirement for very fine Doppler resolution to separate slow-moving target echoes from land and sea clutter. In the case of ship detection, the resonant water-wave components on the sea surface can have velocities that are very close to those of targets, which implies that high Doppler resolution is required. On the other hand, the upper bound on dwell time may be limited by the coherence time of the ionospheric propagation channel or the need to concurrently schedule (interleaved) air-mode tasks. Unless the target echo energy integrates more effectively than the clutter in the Doppler frequency bin containing the target echo, extending the dwell time past a certain point will not provide additional benefit in terms of improving SCR. Moreover, the impact of very low region revisit rates on ship-tracking performance should not be ignored.

Even though such vessels often move relatively slowly, the target echoes experience superimposed changes in group-range and Doppler shift due to ionospheric variations. For surface tasks with long region revisit times, the contribution of ionospheric path variations to the range and Doppler parameters of the target echo can be more significant than those imparted by the motion of the vessel itself. The target echo path may be subjected to irregular changes in group-range and Doppler shift from one dwell revisit to the next as the signal reflection points move in the ionosphere. The ionospheric contribution is effectively perceived by the radar as a superimposed but fictitious target "maneuver" (even if the target motion is steady). Consequently, very low region revisit rates can have a detrimental effect on ship-tracking performance.

OTH radar dwell times between about 20 and 60 seconds are typically considered appropriate for surface-mode tasks. Table 4.2 lists some representative OTH radar operating parameters for air- and surface-mode tasks, including their associated resolutions and ambiguities. It is emphasized that there is not a standard set of operating parameters for each type of task. The actual parameters used in practice may clearly vary with respect

Task	B, kHz	ΔR, km	D, m	ΔL, km	f_p, Hz	R_{amb}, km	v_{amb}, km/h	T_{CIT}, s	Δv, m/s
Air	10	15	1000	30	50	3000	± 900	1	10
Ship	30	5	3000	10	5	30000	± 90	20	0.5

TABLE 4.2 Nominal values of OTH radar operating parameters for air and surface tasks. From left to right, the table lists typical values for OTH radar signal bandwidth, range resolution, aperture length, cross-range resolution, PRF, first range ambiguity, first velocity ambiguity, CPI, and velocity resolution. Quantities that depend on the carrier frequency and ground range have been computed for a mid-band frequency of 15 MHz and a ground range of 1500 km. These are representative parameter values for a hypothetical (archetype) OTH radar system.

to these nominal values depending on the specific nature of the radar mission and the prevailing environmental conditions.

4.3.2 Transient Disturbance Mitigation

Strong motivation exists to improve the immunity of OTH radar systems to powerful disturbance signals that are short-lived on the time scale of the CPI. HF disturbance signals in this category include impulsive noise bursts due to lightning discharges, transient meteor echoes, and man-made interference signals that overlap the radar bandwidth for brief instants (e.g., a frequency-swept ionospheric sounder signal). The reception of such disturbances cannot be avoided by frequency selection or the adjustment of radar waveform parameters. This has resulted in the development of ameliorative signal-processing techniques for transient disturbance mitigation. The application of such techniques to impulsive noise excision (INE) historically represents one of the first data-dependent processing steps to be widely implemented in operational OTH radar systems.

The term "impulsive noise" generally refers to disturbances received from lightning strikes with an amplitude well above the underlying background noise level. Although the phenomenology associated with impulsive (atmospheric) noise is entirely different to that which gives rise to meteor echoes and short-duration man-made interference, the transient nature of the disturbance signal remains a common factor. This shared property has meant that ameliorative signal-processing techniques used to mitigate these transient disturbances are very similar at a fundamental level. Ameliorative processing for transient disturbance mitigation is often associated with the INE application because such techniques were originally introduced for this purpose. We shall therefore use impulsive noise as a vehicle for describing the main principles of such processing, with the understanding that similar concepts apply for transient meteor echoes and short duration man-made interference. Exceptions where significant differences occur will be pointed out in the following discussion.

One or more lightning bursts received in a CPI can significantly degrade SNR over the entire range-Doppler search space when conventional processing is applied. The impulsive noise signal is incoherent with the radar waveform, so its energy is spread over all range cells after pulse compression. On the other hand, spreading in Doppler occurs because spectral analysis of the slow-time data samples by Fourier transformation has the effect of smearing impulsive energy in the frequency domain. Similar arguments apply for short duration man-made interference (independent of the radar waveform).

Transient meteor echoes are spread in Doppler for the same reason mentioned above, but such echoes are often well-localized in range because the scattered signal is coherent with the radar waveform.

Impulsive noise due to lightning has the potential to degrade radar sensitivity by tens of decibels when no compensation is applied. Specifically, the SNR in a particular beam of a badly contaminated radar dwell may be reduced by up to 30 dB or more relative to quiescent conditions, while average SNR degradations of around 10–20 dB are not uncommon when active storms are located in the direction of the surveillance region at ranges in the one- or two-hop skywave coverage. Such a dramatic reduction in radar sensitivity can effectively preclude the detection of many aircraft targets.

As mentioned previously, an individual lightning strike has a duration of perhaps 100–400 ms, while the impulse rates may vary from one to five flashes per second during less and more active storms. Although not all flashes significantly increase the received noise level, degradations in SNR may at times persist over a number of successive radar dwells. The probability of impulsive noise corruption over several consecutive air mode CPI in a particular surveillance region and beam can at times be high enough to impair or disrupt target tracking when compensation in the form of ameliorative signal processing is not applied.

An individual impulsive noise burst received from a far-field event via a one-hop propagation mode is typically well localized in direction of arrival. A receive antenna pattern with low sidelobe levels can often suppress off-azimuth impulsive noise signals effectively. However, spatial processing cannot cancel impulsive noise entering through the mainlobe of the receive antenna pattern without significantly compromising useful signal reception. For a single far-field impulsive noise source propagated over a one-hop ionospheric path, the number of affected conventional beams is typically between one and three, when attention is restricted to a single ionospheric mode.

More than three finger beams can be affected when multiple bursts from spatially separate sources are received during the same CPI, or if multipath propagation with significant coning is present for ULA beams steered away from boresight, or if burst is strong enough to raise the noise-floor in the antenna sidelobe region. Indeed, the reception of strong impulsive noise from a near-field source can potentially deny detection in all conventionally formed beams, range cells, and Doppler bins of the affected CPI. Ameliorative processing techniques are therefore required to enhance detection performance against transient disturbances such as impulsive noise, meteor echoes, and short-duration man-made interference.

4.3.2.1 Principles and Techniques

In the signal-processing chain of Figure 4.11, the transient disturbance mitigation step is generally performed after pulse compression and array beamforming but prior to Doppler processing (i.e., on the slow-time samples in each range-azimuth cell). The rational behind this is that such processing is applied to suppress transient disturbances that have been inadequately canceled by conventional beamforming. In other words, such routines operate on the slow-time data samples received in the finger beams and ranges of the surveillance region and are mainly intended to remove transient disturbances received through the mainlobe of the antenna beampattern.

The specific implementation details vary, but the key elements of many techniques are quite similar. The first step often involves the identification of the affected samples in the slow-time domain. The set of corrupted slow-time samples are then set to zero or "excised" from the slow-time data. Finally, the missing data are reconstructed using interpolation methods based on linear prediction to restore clutter coherence over the data segments where contamination was detected. Typically, this procedure is repeated on a range-azimuth cell-by-cell basis to capture spatially localized contamination more effectively, such as range-formed meteor echoes.

Numerous investigations on the mitigation of transient disturbances received by OTH radar have been conducted and reported in the literature. The reader is referred to Barnum and Simpson (1997), Lu, Kirlin, and Wang (2003), Kramer and Williams (1994), Guo, Sun, and Yeo (2005), Turley (2003), and references therein for examples not limited to "excise-and-interpolate" procedures. An alternative approach for impulsive noise mitigation was proposed in Guo, Sun, and Yeo (2005). This approach assumes a parametric model for clutter signals and impulsive noise and then estimates the model parameters from the received data. Model components having parameters consistent with impulsive noise characteristics are then identified and subtracted from the data.

With respect to the traditional excise-and-interpolate approach, alternative model-based methods may be more suitable in situations where transient interference corrupts a large fraction of the CPI. However, model-based techniques are vulnerable to mismatches between the assumed analytical representation of the signals and the data characteristics that are actually received. There is also the difficulty of estimating model parameters accurately from finite data records. The impulsive noise rejection technique described in Turley (2003) is briefly discussed below as an exemplar of the "excise-and-interpolate" approach.

The initial step of impulsive noise detection is often not directly possible in the slow-time domain because the contaminating signals are typically masked by powerful surface clutter echoes. To identify data samples corrupted by impulsive noise, a common method is apply an (high-pass) FIR notch-filter to attenuate the dominant surface clutter returns. The coefficients of this filter may be derived adaptively, as discussed in the previous section, based on a κth-order AR model of the slow-time clutter samples. A value of $\kappa = 5$ is proposed in Turley (2003). An alternative method is to pre-Doppler process the data, remove the dominant clutter components by zeroing the Doppler bins over a narrow band frequencies centered around zero Hertz, and then inverse-Fourier transform the modified spectrum to yield the slow-time data for impulsive noise detection (Barnum and Simpson 1997). The former technique is often adopted, not only because it is efficient and effective, but also because estimates of the clutter AR coefficients are subsequently needed for the interpolation step.

The availability of skip-zone range cells may be exploited to directly identify the contaminated slow-time samples without clutter filtering. The skip-zone range cells allow impulsive noise to be observed on a pulse-by-pulse basis against much lower clutter levels and background noise. This approach is suitable for detecting contamination due to lightning or man-made interference but not meteor echoes, as the transient backscatter to be removed is contained in the operational range cells (where it is often submerged by clutter).

Once a slow-time sample sequence with minimal residual surface-clutter is obtained for each range-azimuth cell, the detection of corrupted samples is based on the simple property that the transient disturbance has a short duration but high amplitude relative to the mean or median of the sample sequence. Appropriate thresholds can therefore be set to identify and excise the slow-time samples contaminated by transient disturbances. Note that the excision step is performed on the original slow-time data rather than the data used to detect the locations of corrupted samples.

Assuming the uncontaminated data can be accurately described by an AR model of relatively low order, the excised data samples may be replaced by linear prediction using a weighted combination of the adjacent unaffected samples. In Turley (2003), the AR model parameters (linear prediction coefficients) b_i for $i = 1, \ldots, \kappa$ are estimated from the "good" (uncontaminated) data samples using Nuttall's method in conjunction with Andersen's implementation of Burg's maximum-entropy spectral analysis technique, which ensures that the filter coefficients are stable.

Replacement samples for a particular segment of excised slow-time data are derived by using neighboring uncontaminated data windows as inputs to the regression process. Specifically, the missing or "bad" data samples are replaced by using the estimated regression coefficients to predict forward and backward into the corrupt data segment, as in Eqn. (4.109), where $y_k^{[b]}(n)$ is the acquired data sample at beam b, range k, and pulse n.

$$\text{Forward}: \ \hat{y}_k^{[b]}(n) = \sum_{i=1}^{\kappa} b_i y_k^{[b]}(n-i)\,, \quad \text{Backward}: \ \check{y}_k^{[b]}(n) = \sum_{i=1}^{\kappa} b_i^* y_k^{[b]}(n+i) \quad (4.109)$$

Clearly, the predictions $\hat{y}_k^{[b]}(n)$ and $\check{y}_k^{[b]}(n)$ are subsequently denoted by $y_k^{[b]}(n)$ in the two recursive relations of Eqn. (4.109) when making subsequent predictions forward or backward, respectively. The forward and backward predictions are then combined using a weighted average to form the interpolated data sample $\tilde{y}_k^{[b]}(n)$, as in Eqn. (4.110). The relative weighting $w(n)$ is applied because estimation errors tend to grow as the predicted number of samples increases in either direction. For interpolated data segments that are more than a few samples long, it is therefore advisable to favor the prediction in the direction with shorter length to the sample being replaced. A weighting function $w(n)$ given by a raised cosine taper is proposed in Turley (2003) so as to provide a smooth transition into the contaminated data segment.

$$\tilde{y}_k^{[b]}(n) = w(n)\hat{y}_k^{[b]}(n) + [1 - w(n)]\check{y}_k^{[b]}(n), \quad w(n) \in [0, 1] \quad (4.110)$$

4.3.2.2 Real-Data Examples

The top panel in Figure 4.26a shows an intensity-modulated range-Doppler display in a conventionally processed finger beam for a radar dwell that has been contaminated by impulsive noise. The received contamination due to lightning raises the mean noise level over the entire range-Doppler map and significantly lowers the SNR of a real aircraft target echo in this beam. The bottom panel in Figure 4.26a illustrates the reduction in mean noise level and improvement in target echo SNR when transient disturbance mitigation is applied immediately prior to Doppler processing. Incorporating this ameliorative processing step into the OTH radar signal-processing chain enables the real aircraft target echo to be readily detected in this example. Figure 4.26b compares the Doppler spectra at the target azimuth-range cell with and without transient disturbance mitigation. The practical benefit of such processing is clearly evident.

The top panel of Figure 4.27a shows the range-Doppler map resulting for an OTH radar beam contaminated by a transient meteor echo. The meteor head echo is short-lived on the scale of the radar CPI, while the ionized trail may give rise to a longer lasting and possibly coherent echo. A long-lasting trail echo on the scale of the radar CPI may be Doppler-shifted as the enhanced ionization drifts under the influence of winds in the upper atmosphere. When a coherent trail echo persists over most or all of the CPI, it produces a target-like return that is well-localized in Doppler frequency.

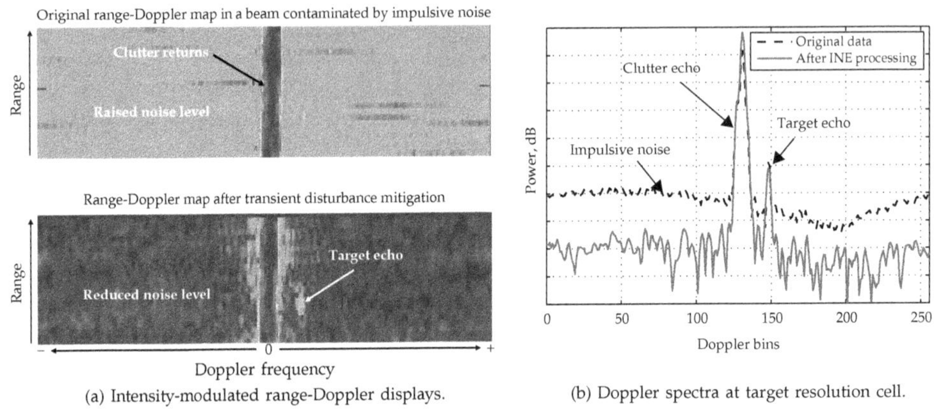

FIGURE 4.26 The intensity-modulated displays on the left show a conventional range-Doppler map for an air-mode dwell with impulsive noise contamination (top panel), and the output after transient disturbance suppression is applied (bottom panel) in an OTH radar finger beam containing a real aircraft target echo. The Doppler spectra on the right confirms the improvement in SNR at the target range-azimuth cell after the transient disturbance suppression step is applied. © Commonwealth of Australia 2011.

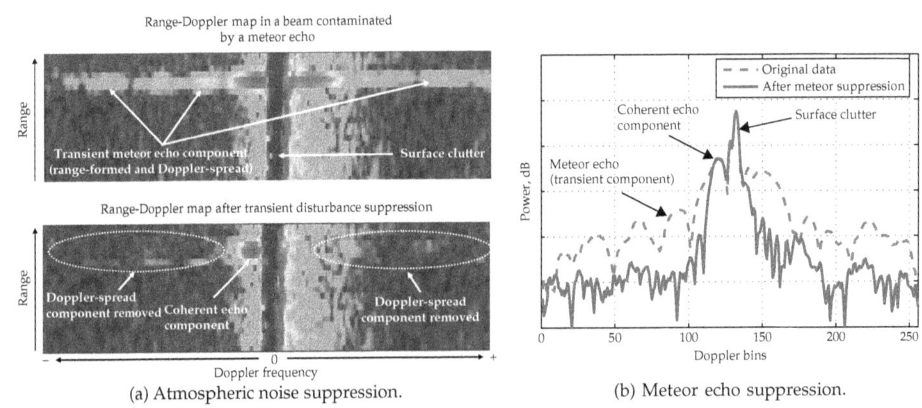

FIGURE 4.27 The range-Doppler displays on the left show the output of conventional processing when meteor echo contamination is present (top panel), and the output after the transient disturbance suppression is applied (bottom panel). The ameliorative processing effectively removes the Doppler-spread meteor echo component but leaves behind the Doppler-shifted (coherent echo) component, which appears target-like. The Doppler spectra on the right illustrate the improvement in sub-clutter visibility after the transient disturbance suppression step is applied. © Commonwealth of Australia 2011.

The bottom panel in Figure 4.27a shows that the Doppler spread component of the meteor echo has been effectively removed by the transient disturbance suppression step. In this example, a coherent (target-like) meteor echo component is also present with a Doppler shift corresponding to a relative velocity of about 50 km/h, as indicated in the bottom panel of Figure 4.27a. Figure 4.27b compares the Doppler spectra in the range-azimuth cell containing the meteor echo with and without ameliorative processing. It may be observed that the coherent (target-like) component is not attenuated by such processing, while the Doppler-spread component has been reduced across the entire velocity search space.

Transient disturbance suppression techniques are equally applicable to short duration man-made interference, but they are ineffective for persistent interference that overlaps the receiver bandwidth for significant portions or all of the CPI. Figure 4.28 illustrates four range-Doppler maps corresponding to four adjacent finger beams steered approximately one mainlobe-width apart. This example shows continuous-wave (CW) man-made radio frequency interference (RFI) that is present over the entire dwell time. This (unmodulated) RFI manifests itself as a vertical "stripe" localized in azimuth and Doppler but spread over all range cells. On this occasion, the RFI does not obscure a pair of much weaker (multipath) target echoes due to the difference in Doppler frequency. Modulated man-made signals produce so-called broadband RFI that additionally spreads over all Doppler cells. Adaptive processing techniques that can remove CW and broadband RFI will be discussed in Part III.

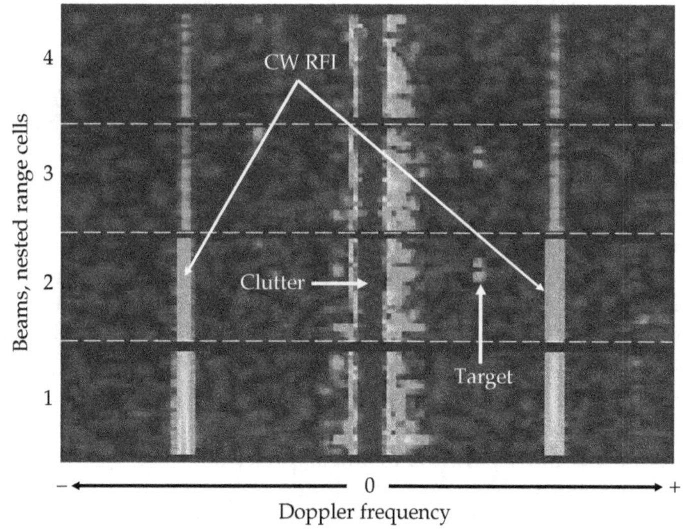

FIGURE 4.28 Range-Doppler maps processed by an OTH radar in four adjacent finger beams steered approximately one mainlobe-width apart showing continuous-wave (CW) radio frequency interference (RFI) localized in azimuth and Doppler but spread over all range cells. The interference does not obscure the target returns in this case because of the different Doppler frequencies of CW RFI and the two (multipath) target echoes. © Commonwealth of Australia 2011.

4.3.3 Data Extrapolation and Signal Conditioning

The desire to increase the coverage or coverage rate of an OTH radar system without significantly compromising performance has led to the development of data extrapolation (DATEX) techniques. When applied in the slow-time domain, such techniques attempt to retain the coherent integration gain and Doppler frequency resolution properties of a normal dwell using a CPI that is a only fraction of the length (typically half a normal dwell). The key elements of the DATEX approach will be discussed in this section with respect to the Doppler-processing application, which is intended to improve the coverage or coverage rate of an OTH radar.

The DATEX methodology may also be applied for aperture extrapolation (APEX) or bandwidth extrapolation (BANDEX). The former application improves aperture usage, either by increasing the SNR gain and angular resolution of the receiving system using the full aperture, or to achieve similar performance as the full aperture using a half aperture, which frees up radar resources for other tasks. On the other hand, the latter application improves the efficiency of spectrum usage in the congested HF band by providing similar pulse-compression gain and range resolution using one half of the bandwidth normally required. The DATEX approach was originally described in Swingler and Walker (1988).

An additional step that is commonly inserted into the radar signal-processing chain is known as ionospheric distortion correction (IDC). For both air and surface tasks, IDC aims to compensate for range and beam dependent Doppler shifts (and possibly spreads depending on the type of algorithm employed) that are imposed on radar echoes by the ionosphere. IDC improves the performance of the subsequent constant false-alarm rate (CFAR) processing step, as it effectively "aligns" the clutter spectra received in different ranges and beams with the CFAR kernel in the Doppler dimension. IDC is beneficial for false-alarm reduction, useful signal detection, and target relative velocity estimation.

4.3.3.1 Data Extrapolation (DATEX)

The ability to reduce the CPI by a factor of two with little or no penalty in target SNR or Doppler resolution has a significant operational impact for OTH radar. Specifically, such a capability frees up the radar resource to double the coverage for a given coverage rate. This has motivated the development of alternative spectral analysis methods with respect to conventional Doppler processing. High-resolution spectrum estimation techniques based on parametric (AR) models have previously been considered for OTH radar Doppler processing (Barnum 1986). However, the performance of such techniques is sensitive to AR-model order selection. Moreover, significant performance degradations may be encountered when attempting to capture signal components with very low SNR in the AR model. This issue can be a problem for relatively weak target echoes, which may not be represented in the resulting AR Doppler spectrum.

An alternative is to apply the estimated AR model parameters for data extrapolation beyond the ends of the CPI by means of linear prediction, using a similar approach to that described for the interpolation of corrupted data samples within the CPI. Data extrapolation may be used to extend the received slow-time sample sequence by half of its original length at either end of the CPI. In other words, this procedure introduces synthetic data with similar spectral characteristics to the acquired data as if the former had been collected before and after the actual CPI. This effectively doubles the number of slow-time samples integrated by Doppler processing without consuming additional radar resources. The question arises as to the performance of this technique relative to a radar dwell of equivalent length containing only real data.

Consider an N-point slow-time sequence of real-data samples $\{y_k^{[b]}(n)\}$ in range k and beam b that are indexed by $n = N/2+1, \ldots, 3N/2$, where N is assumed to be even. The DATEX method extrapolates the CPI using the estimated linear prediction coefficients $\{b_i\}_{i=1}^{\kappa}$, as in Eqn. (4.111), to yield a slow-time data sequence with $2N$ samples. Here, we have defined the intervals for the future and past sample as $\mathcal{F} = [3N/2+1, 2N]$ and $\mathcal{P} = [1, N/2]$, respectively. DATEX may be interpreted as a pre-Doppler processing step that extends the received (real data) CPI into the past and future using an AR model of relatively low order $\kappa \ll N$. Extrapolation factors less than or greater than two are clearly possible, but experimental results suggest that a factor of two provides a good compromise between resolution and variance for OTH radar Doppler spectrum estimation.

$$y_k^{[b]}(n) = \sum_{i=1}^{\kappa} b_i y_k^{[b]}(n-i) \text{ for } n \in \mathcal{F}, \quad y_k^{[b]}(n) = \sum_{i=1}^{\kappa} b_i^* y_k^{[b]}(n+i) \text{ for } n \in \mathcal{P} \tag{4.111}$$

There are two main advantages to extrapolating the data by $N/2$ samples in both forward and backward directions rather than by N samples in one direction only. First, the data is extended by minimum length of $N/2$ either side of the real data in the CPI to achieve a total sample length of $2N$. Second, the real-data samples are located in the central section of the extended CPI, such that the window used for Doppler processing gives relatively higher importance to the acquired data and weights down the extrapolated synthetic data. This ensures that the extrapolated data at either end of the CPI contributes relatively less energy to the resulting Doppler spectrum. At the same time, this contribution is critical for controlling the Doppler sidelobes of the most dominant clutter spectral components.

The DATEX technique is robust to model-order selection because only the Doppler spectrum characteristics of the dominant clutter components need to be well represented in the estimated AR model. This is sufficient to preserve the spectral properties of the dominant clutter components as if they had been collected over an interval twice the actual length. The inability of the estimated AR model parameters to capture the relatively weak target echoes, and hence represent these components in the extrapolated data, is not a problem because the peak of the target echo Doppler spectrum is controlled mainly by the real data in the central part of the extended CPI. This not only makes the technique robust to the absence of target echo components in the extrapolated data, but also improves the coherent gain on the target echo relative to that for a dwell of length N. This improvement arises because of the relatively higher window amplitude in the middle portion of the extended dwell. Although the sidelobes of the target echo Doppler spectrum may be raised, this is not of practical concern, as sidelobe levels of −30 dB or so are typically low enough for the target Doppler sidelobes to fall below the noise floor.

The main benefits of applying DATEX prior to conventional Doppler processing may be summarized as follows. First, it provides approximately 2–3 dB of additional coherent gain for target echoes against white noise relative to the original CPI. Second, it doubles the Doppler frequency resolution to separate target echoes from the dominant clutter components without increasing the radar dwell time. From an operational viewpoint, this allows the coverage or coverage rate of the OTH radar to be doubled, without a significant compromise in performance. Relative to standard AR spectrum estimation techniques, DATEX is more robust to model-order selection and the issue of capturing low SNR (target echo) components in the estimated AR parameters. Clutter Doppler spectra that are characterized by high dynamic range and band-limitedness are ideally suited to robust performance enhancement by DATEX. Similar conclusions hold when this methodology is applied for APEX and BANDEX in the antenna element and fast-time sample domains, respectively.

DATEX methods were first introduced to skywave OTH radar by Turley and Voight (1992). A comparison of DATEX with a related technique described by Frazer (2001) was reported in Gadwal and Krolik (2003). To improve the performance of DATEX in practice, care needs to be taken to minimize bias in the AR parameter estimates due to corruption from impulsive noise. The influence of outliers can be reduced using Nuttall's method (Nuttall 1976). Another refinement involves dealing with the possibility of non-stationary AR parameters over the radar dwell, which may be of concern for relatively longer surface-mode CPIs. Time variation of the AR model coefficients may be mitigated by performing the future and past predictions (extrapolations) based on estimates computed from the last and first halves of the (real data) collection interval, respectively.

Poor quality extrapolation arising from contaminated real-data samples located near the extremities of the acquired CPI may be avoided using the minimum filter error technique discussed in Turley and Voight (1992). In addition, the estimated AR model parameters needed to apply DATEX may also used to filter out clutter and interpolate data samples corrupted by transient disturbances within the CPI. This provides a computationally efficient framework for performing DATEX and transient disturbance suppression. An alternative (non-parametric) processing technique that may also be employed for both data extrapolation and interpolation was proposed in Sacchi, Ulrych, and Walker (1998). This iterative technique involves solving a series of weighted least-squares problems, and is found to perform comparably to DATEX in practical examples (Turley 2008).

4.3.3.2 Ionospheric Distortion Correction (IDC)

Skywave radar echoes may be subjected to regular and random phase-path variations over the CPI. These variations can significantly change the clutter Doppler spectrum characteristics received in different spatial resolution cells. Specifically, the clutter Doppler spectrum in a particular range-azimuth cell may be subjected to a mean Doppler shift (frequency translation), due to the regular component of phase-path variation over the CPI, whereas any random component of phase-path variation results in smearing (frequency broadening) of the clutter Doppler spectrum. The Doppler shifts and spreads imposed on OTH radar echoes received via a particular (two-way) ionospheric mode are range- and azimuth-dependent in general. The spatial variability of the ionospheric Doppler shifts and spreads is also propagation-mode dependent, and typically more pronounced for propagation modes involving F-region reflections.

In OTH radar, ionospheric Doppler shifts and spreads can have a detrimental impact on useful signal detection, false-alarm rate, and target relative velocity estimation. Severe Doppler spread induced on HF signals by the ionosphere may pose problems for fast-moving (aircraft) target detection against clutter (e.g., equatorial spread-F). However, most ionospheric Doppler broadening mechanisms encountered in the relatively quiet mid-latitude region are more of an issue for surface-mode tasks. The estimation and correction of ionospheric Doppler shifts and spreads in each range-azimuth resolution cell leads to more effective Doppler processing and CFAR processing (whitening). The aim of ionospheric distortion correction (IDC) is to estimate and remove ionospheric Doppler shifts in each spatial resolution cell, and if possible, to apply compensation for reducing the impact of ionospheric Doppler spread.

Several IDC techniques have been proposed to compensate for the variation in signal Doppler frequency over the CPI prior to coherent processing. These include the clutter "de-smearing" methods of Parent and Bourdillon (1987) and Abramovich, Anderson, and Solomon (1996), for example. Such techniques are known to be most effective in situations where the spectral contamination is caused by a slowly varying phase-path smearing function, which is imparted on the signal by a single fluctuating ionospheric layer. Estimation and correction of the smearing function received in different resolution cells can significantly reduce clutter Doppler spectrum broadening, and hence improve slow-moving target detection.

Such techniques are often considered less appropriate in conditions where significant multipath propagation contributes to the clutter received in the spatial resolution cell under test. This is because each ionospheric propagation mode will have a different smearing function, all of which cannot be compensated for simultaneously by applying a single complex (gain and phase) correction per pulse. Unfortunately, IDC techniques for correcting ionospheric Doppler spread have not, as yet, found widespread use in operational OTH radar systems, perhaps due to the limited conditions under which they can operate effectively.

Another objective of IDC is to remove the mean ionospheric Doppler shift, such that the dominant land and sea clutter spectral components in different spatial resolution cells are aligned with respect to a common 0 Hz reference. The ionospheric Doppler shift correction step analyzes the peaks in the clutter Doppler spectrum at each range-azimuth cell to classify whether land and/or sea clutter is present. Smoothed estimates of the ionospheric Doppler shift are computed over a local neighborhood of range and azimuth cells using a model with a parameter introduced to account for ocean surface currents. This parameter may be estimated using land clutter as a stationary Doppler reference when the coverage is over littoral areas.

The mean Doppler shift measurements resulting over different neighborhoods are used to align the Doppler spectra of all resolution cells. This not only improves the effectiveness of CFAR processing, but also provides corrected Doppler frequency estimates for the detections passed onto the tracker. Besides enhancing detection and tracking performance, ionospheric Doppler shift removal also improves the interpretability of the ARD data displayed to operators. The reader is referred to the above-mentioned citations on IDC processing for examples of performance on real data. The remainder of this section illustrates the practical benefits of DATEX using real OTH radar data.

4.3.3.3 Real-Data Examples

The range-Doppler displays in Figure 4.29 illustrate the two main advantages of DATEX in an aircraft detection application. The top panel shows the range-Doppler map at the output of a conventional beam for a processed CPI composed of 64 (real data) sweeps. These 64 sweeps were extracted from the central section of the original OTH radar CPI, which actually contained 128 sweeps of real data.

The strong multipath echoes corresponding to target A have low Doppler shifts and are close to the powerful surface-clutter peak near 0 Hz. The relatively short (64-sweep) CPI of approximately 1 second duration is not sufficiently long to properly resolve the multipath echoes of target A from the clutter when a low-sidelobe taper needs to be used for coherent processing to detect the weaker target echo at higher Doppler shift (target B). This target is obviously well resolved from the clutter, but it has lower SNR and is on the verge of not being detected against the background noise.

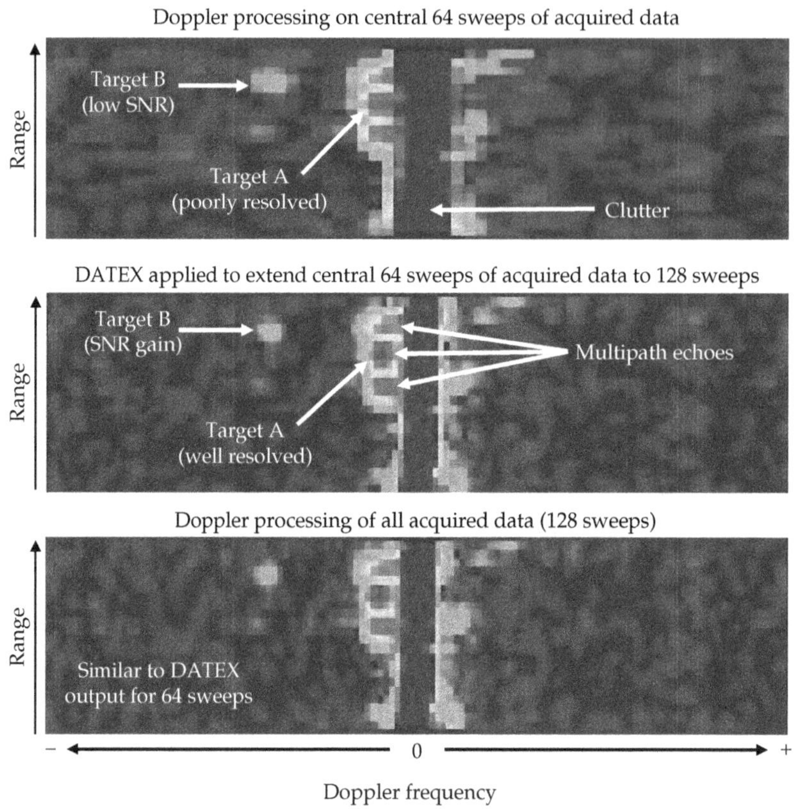

FIGURE 4.29 The top panel shows a range-Doppler map formed using 64 sweeps of real data in a beam containing echoes from two targets. Multipath echoes from a real aircraft target with low relative velocity are poorly resolved from the more powerful clutter returns near zero Doppler frequency (target A). An echo from a different target with higher relative velocity is well resolved from the clutter, but has a much lower SNR (target B). The middle panel shows the range-Doppler map resulting when the same 64 sweeps of real data are extrapolated to 128 sweeps prior to Doppler processing. The enhanced Doppler resolution and SNR gain enables targets A and B to be more clearly distinguished from noise and clutter, respectively. The bottom panel shows the range-Doppler map that results when 128 sweeps of real data are coherently processed. The almost identical middle and bottom panels illustrates that DATEX can be used to halve the radar dwell time without a noticeable effect on performance. © Commonwealth of Australia 2011.

The middle panel in Figure 4.29 shows the range-Doppler output when DATEX is applied to extrapolate the same 64 sweeps of real data by 32 sweeps either side of the CPI to a yield an effective dwell time of approximately 2 seconds (128 sweeps). It is emphasized that the extrapolated data is generated in software and that only 1 second of actual radar time is needed to collect the data input to DATEX. The middle panel illustrates that DATEX provides enhanced Doppler resolution, which helps to separate the multipath echoes of target A from the strong surface clutter near zero Doppler frequency, as well as increased SNR, which helps to detect the weaker echo from target B against background noise at higher Doppler frequency.

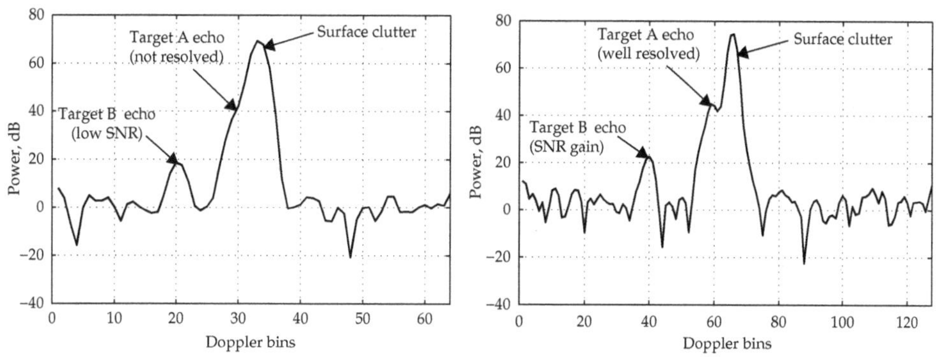

(a) 64 real-data sweeps without DATEX.

(b) Same 64 sweeps with DATEX to 128 sweeps.

FIGURE 4.30 Conventional Doppler spectra using 64 sweeps of acquired data without and with data extrapolation to 128 sweeps. The enhanced Doppler resolution permits the stronger aircraft target echo to be resolved from the more powerful surface clutter (target A). The SNR gain provided by DATEX also increases the visibility of the weaker aircraft target echo against noise (target B). © Commonwealth of Australia 2011.

The bottom panel in Figure 4.29 shows the range-Doppler map that results when all 128 sweeps of real data acquired during the CPI are used for Doppler processing. This CPI of data was collected using a radar dwell time of 2 seconds, but is almost indistinguishable from the DATEX output, which only requires the radar to collect 1 second of data. The Doppler spectra in Figure 4.30, extracted from the range-Doppler maps in Figure 4.29, clearly shows the two main benefits of DATEX for detecting the echoes of targets A and B in the same range-azimuth cell.

4.4 Detection and Tracking

At the signal-processing output, the phase information in the ARD data is typically discarded and the magnitude envelope of the samples are input to the data-processing sequence as an ARD image. The main steps in this sequence are constant false-alarm rate (CFAR) processing, peak detection, and parameter estimation, followed by tracking and coordinate registration. The first part of this section overviews the main concepts related to CFAR processing for OTH radar, including the choice of disturbance probability density function, reference data kernel design, as well as cell averaging and ordered statistics methods for threshold estimation. The second part of this section describes the peak detection and parameter estimation steps, which generate the input data to the tracker. The final part of this section summarizes a number of approaches that have been proposed for tracking and coordinate registration in OTH radar.

4.4.1 Constant False-Alarm Rate Processing

The main objective of classical CFAR processing is to transform the amplitude $y = |x|$ or square-law measurement $y = |x|^2$ of the complex sample x in an unnamed resolution cell of the ARD data to generate a test statistic $z = T(y)$ that belongs to a known reference distribution such that a fixed threshold can be used to provide a constant and predictable false-alarm rate over the entire data cube while maintaining an acceptable probability of detection for target-like signals. An additional benefit of CFAR processing is that it enables target echoes to stand out as the main features of the resulting "whitened" ARD image z, which is often viewed by radar operators in association with the track display. A comprehensive treatment on the subject of CFAR for "automatic detection" can be found in Minkler and Minkler (1990). An in-depth coverage of CFAR principles and techniques can also be found in the excellent texts of Levanon (1988), Nathanson, Reilly, and Cohen (1999), and Richards (2005).

The three main elements of classical CFAR processing are represented in the flow chart of Figure 4.31. In no particular order, the first element involves the identification of a suitable density function with the flexibility to capture variations in the statistical properties of the disturbance using one or more free parameters. Distribution-free CFAR techniques also exist, but they tend to require large data volumes when relatively low constant false-alarm rates are required (Sarma and Tufts 2001). The second element is kernel design, which refers to the selection of a CFAR window that may extend in one or more ARD data-cube dimensions. The shape and size of the CFAR window determines the reference cells in the neighborhood of the sample under test used to estimate the unknown disturbance PDF parameters for that same sample. The third element is the method used to compute the parameters of the whitening transformation from the

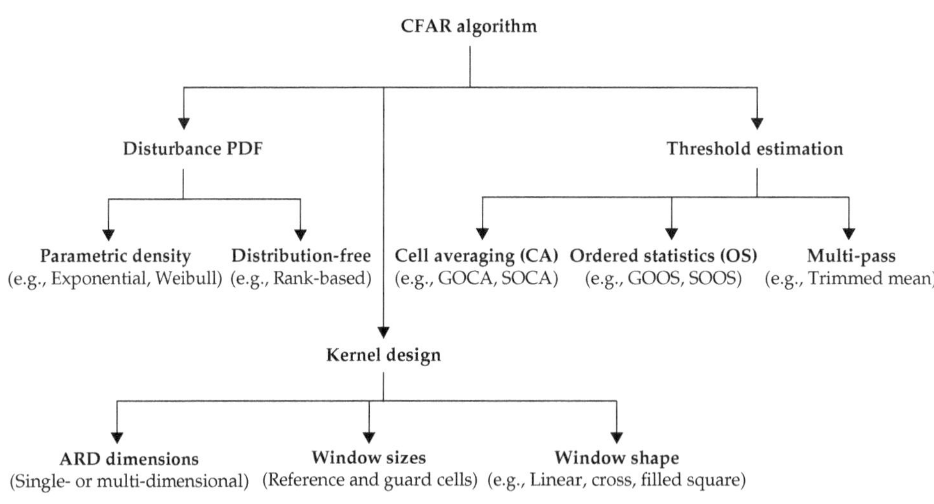

FIGURE 4.31 The main elements of a classical CFAR algorithm include: (1) the identification of a suitable probability density function with one or more free parameters to capture the possibly variable statistical characteristics of the disturbance, (2) the selection of an appropriate kernel design to estimate the unknown disturbance PDF parameters using reference cells near the sample under test, and (3) the specification of a robust technique to estimate the unknown disturbance PDF parameters and data-dependent threshold from the reference data. These elements may be adapted based on the location of the test sample within the ARD data cube. © Commonwealth of Australia 2011.

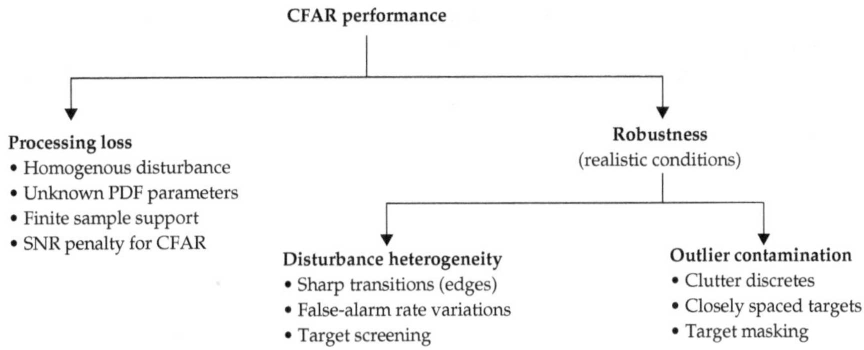

FIGURE 4.32 The main performance metrics for evaluating CFAR performance include: (1) processing loss in a homogeneous disturbance environment (i.e., effective SNR penalty incurred for CFAR processing) due to the unknown disturbance PDF parameters, (2) robustness against potentially sharp changes in disturbance characteristics (e.g., disturbance power fluctuations over the data cube) measured in terms of false-alarm rate variations and useful signal screening in the vicinity of transitions, and (3) resistance to outlier contamination caused by the presence of clutter discretes or target echoes in the reference cells, measured in terms of immunity to useful signal masking. © Commonwealth of Australia 2011.

reference data. In practice, the whitening transformation often involves data-dependent normalization of the test sample. The choices of disturbance PDF, kernel design, and threshold estimation method may be invariant or adapted as a function of test-sample location within the ARD data cube.

The main criteria used to assess the performance of a CFAR technique are shown in Figure 4.32. In a homogeneous disturbance environment with a known PDF but unknown parameters, CFAR detection is associated with a processing loss due to the lack of a priori knowledge regarding the disturbance PDF parameters. This loss is expressed in terms of an effective SNR penalty with respect to the case where the disturbance PDF parameters are perfectly known. Some degradation in the probability of detection is therefore traded for the benefit of a constant and predictable false-alarm rate (in homogenous conditions).

Other CFAR performance metrics are important for robustness in realistic disturbance environments. Practical CFAR techniques require a degree of localization to counter heterogeneity of the disturbance statistics throughout the data cube. For example, good edge performance is required to handle sharp transitions in disturbance properties near clutter ridges. Robustness to disturbance heterogeneity is measured in terms of variations in false-alarm rate and susceptibility to target screening in the vicinity of sharp disturbance power transitions. Another important issue is the presence of outliers in the reference cells, due to closely spaced targets for example, which can significantly bias the disturbance PDF estimate. Immunity to outliers is typically measured in terms of resistance to useful signal masking.

4.4.1.1 Detection and False-Alarm Probabilities

Let the complex scalar x denote the signal-processing output in an unnamed resolution cell of the ARD data cube. A decision needs to be made about whether this test sample contains disturbance only $x = d$, or if it additionally contains a useful signal, i.e., $x = s + d$. The square-law detector operates on the nonnegative scalar $y = |x|^2 \in [0, \infty)$ to decide whether a target is present or absent based on the binary hypothesis test in Eqn. (4.112). In the null hypothesis H_0, the value of y is attributed to the sum of clutter, interference, and noise alone. In the alternative hypothesis H_1, the value of y is attributed to the presence of a useful signal in addition to the disturbance.

$$\begin{cases} H_0 : y = |d|^2 \\ H_1 : y = |s + d|^2 \end{cases} \tag{4.112}$$

The standard procedure for deciding between the two hypothesis is threshold detection in Eqn. (4.113). A target is declared present if y exceeds a selected threshold T (i.e., H_1 is accepted as true). If $y \leq T$, the target is deemed to be absent and H_0 is accepted as true. Detection performance is assessed in terms of false-alarm and detection probabilities, denoted by P_{FA} and P_D, respectively. The former is the probability that a target is declared present when it is in fact absent (i.e., H_1 is accepted when H_0 is actually true or "in force"). The latter is the probability that a target is declared present when it is in fact present (i.e., H_1 is accepted when H_1 is in force).

$$y \underset{H_0}{\overset{H_1}{\gtrless}} T \tag{4.113}$$

For a given threshold T, P_{FA} and P_D may be computed from the probability density function of the measurement y conditioned on the null and alternative hypothesis, respectively. The former density is denoted by $p_y(y|H_0)$, while the latter is denoted by $p_y(y|H_1)$. For a one-sided acceptance test, these probabilities are defined by Eqn. (4.114). Clearly, P_D and P_{FA} need to be traded off against each other, since they rise and fall together as T changes. The plot of P_D versus P_{FA} as T varies is called a receiver operating characteristic (ROC) curve. For a particular ROC curve, the selection of threshold T sets the operating point, and hence the performance metrics (P_{FA} and P_D) of the test.

$$P_D = \int_T^{\infty} p_y(y|H_1)\mathrm{d}y, \quad P_{FA} = \int_T^{\infty} p_y(y|H_0)\mathrm{d}y \tag{4.114}$$

To motivate CFAR processing, assume the complex-valued disturbance process d has independent and identically distributed (IID) zero-mean Gaussian real and imaginary parts with unknown variance $\sigma^2/2$. To be clear, we assume that d is described by a circular-symmetric complex normal distribution with the correlation properties in Eqn. (4.115). In this case, $|d|$ has a Rayleigh distribution, and $|d|^2$ has a chi-square distribution with two degrees of freedom. The latter is also referred to as an exponential distribution.

$$\mathrm{E}\{dd^*\} = \sigma^2, \quad \mathrm{E}\{d^2\} = 0 \tag{4.115}$$

For the square-law measurement $y = |x|^2$, the disturbance PDF $p_y(y|H_0)$, and associated cumulative density $f_y(y|H_0)$, are given by Eqn. (4.116), where the only free parameter is the disturbance power or variance σ^2. Empirical analysis of collected OTH radar data suggests that ARD samples of lower power found well outside the clutter-occupied region of Doppler space tend to follow this distribution quite well. These samples are mainly dominated by external background noise.

$$p_y(y|H_0) = \frac{1}{\sigma^2}e^{-y/\sigma^2}, \quad f_y(y|H_0) = 1 - e^{-y/\sigma^2}, \quad y \in [0, \infty] \tag{4.116}$$

The value of P_{FA} for a threshold T is given by Eqn. (4.117). The exponential dependence of P_{FA} on σ^2 shows that the false-alarm rate is sensitive to changes in disturbance power, which occur in practice over different regions of the ARD data cube. For example, surface clutter has much higher power than background noise, but its region of influence is typically limited to a certain number of Doppler bins. Directional RFI may raise the disturbance level significantly above the background noise, but perhaps only in a few beams. Transient meteor echoes can also raise the disturbance power, but usually over a localized set of ARD resolution cells.

$$P_{FA} = 1 - f_y(T|H_0) = e^{-T/\sigma^2} \tag{4.117}$$

A direct comparison of the raw measurements y with a fixed threshold T would therefore lead to large variations in P_{FA} over the ARD data cube and from dwell to dwell. Moreover, threshold levels low enough to detect target echoes in raw measurements y would result in very large numbers of unwanted detections triggered by clutter and interference in other regions of the data cube. If the disturbance power σ^2 in each test sample y were known a priori, a clairvoyant threshold $T = \alpha\sigma^2$ could be set to provide a constant and predictable false-alarm rate over all ARD samples given by Eqn. (4.118). In this (hypothetical) case, the threshold multiplier $\alpha = -\ln P_{FA}$ can then be set according to the desired value of P_{FA}.

$$P_{FA} = e^{-\alpha} \tag{4.118}$$

Equivalently, it would be possible to perform the simple whitening transformation (normalization) in Eqn. (4.119), such that a fixed detection threshold $T = \alpha$ applied to z maintains the constant false-alarm rate of Eqn. (4.118), despite variations in σ^2 from one test sample to another. In practice, the disturbance power is unknown a priori and must be estimated from the data on a sample by sample basis. CFAR techniques for adaptively estimating suitable thresholds or equivalent whitening transformations will be discussed in the following parts of this section.

$$z = T(y) = \frac{y}{\sigma^2} \tag{4.119}$$

Ensuring the false-alarm rate remains within an acceptable range of values is vital for an operational radar system. A false-alarm rate that is too large can lead to a high false track rate and unmanageable computational demands on subsequent data-processing steps, while a very low false-alarm rate may unnecessarily degrade the probability of target echo detection. In addition, practical tracking filters are often designed and tuned based on approximate knowledge of the expected false-alarm and detection probabilities. As far as the latter is concerned, it can be shown that P_D is given by Eqn. (4.120) under the stated assumptions, where $\chi = |s|^2/\sigma^2$ is the SDR in a single resolution cell at the signal-processing output, see Richards (2005).

$$P_D = e^{-\alpha/(1+\chi)} \tag{4.120}$$

4.4.1.2 Cell Averaging CFAR

As a knowledge of σ^2 is required to set appropriate thresholds, this parameter needs to be estimated for each test sample in the ARD output using "training" data from a set of N reference cells $\{y_i\}_{i=1}^N$ in the neighborhood of the test sample y. The CFAR window defines a kernel of reference cells that are assumed to contain independent disturbance samples with the same statistical characteristics as the disturbance in the test sample. The maximum-likelihood estimate of σ^2 is in this case given by the average $\hat{\sigma}^2$ of the reference cell (square-law) measurements, as in Eqn. (4.121).

$$\hat{\sigma}^2 = \frac{1}{N}\sum_{i=1}^{N} y_i \tag{4.121}$$

One or more "guard" cells immediately adjacent to the test sample in each CFAR window dimension are typically excluded from the reference kernel. These cells are excluded to reduce the possibility of target signal components entering into the reference cells due to spectral leakage effects, as this can bias the disturbance estimate. The number of guard cells depends on the target echo point spread function and should also account for straddling effects. Most tapers used for conventional processing cause a spread of about three cells, so one guard cell either side of the test sample in each dimension of the CFAR window usually suffices. The cell-averaging detection threshold may then be estimated as $\hat{T}$ in Eqn. (4.127).

$$\hat{T} = \alpha\hat{\sigma}^2 \tag{4.122}$$

As $\hat{\sigma}^2$ and $\hat{T}$ are random variables, so too will be the probability of false alarm $\hat{P}_{FA} = e^{-\hat{T}/\sigma^2}$. The detector is considered to be CFAR with respect to the unknown disturbance PDF parameters (i.e., the disturbance power σ^2 in this case) if the *expected value* of the false-alarm rate is independent of σ^2. It can be shown that the statistically expected false-alarm rate is given by $\bar{P}_{FA}$ in Eqn. (4.123) (Richards 2005). The multiplier α can be selected using Eqn. (4.123) to yield a predictable and constant (statistically expected) false-alarm rate. This method is known as the cell-averaging (CA) CFAR scheme. In the limit of $N \to \infty$, $\bar{P}_{FA}$ tends to its asymptotically expected value given by Eqn. (4.118).

$$\bar{P}_{FA} = \mathrm{E}\{\hat{P}_{FA}\} = \left(1 + \frac{\alpha}{N}\right)^{-N} \tag{4.123}$$

Similarly, the probability of detection is also a random variable $\hat{P}_D = e^{-\hat{T}/(1+\chi)}$. The expected value of $\hat{P}_D$ is given by $\bar{P}_D$ in Eqn. (4.124) for the assumed Gaussian disturbance model and a single-cell SDR χ. The quantity $\bar{P}_D$ also tends to its asymptotically expected value in Eqn. (4.120) as $N \to \infty$.

$$\bar{P}_D = \mathrm{E}\{\hat{P}_D\} = \left(1 + \frac{\alpha}{N(1+\chi)}\right)^{-N} \tag{4.124}$$

Combining Eqn. (4.123) with Eqn. (4.124), eliminating α, and making χ the subject, yields the expression for χ_N in Eqn. (4.125). The variable χ_N is the SDR required to obtain a certain $\bar{P}_D$ and $\bar{P}_{FA}$ as a function of the number of independent samples averaged N.

Using Eqn. (4.118) and Eqn. (4.120), it is relatively straightforward to derive the SDR required in the asymptotic case as $\chi_\infty = \ln(P_{FA}/P_D)/\ln P_D$.

$$\chi_N = \frac{(\bar{P}_D/\bar{P}_{FA})^{1/N} - 1}{1 - \bar{P}_D^{1/N}} \tag{4.125}$$

The processing loss for the CA-CFAR scheme under ideal homogeneous disturbance conditions is given by the ratio ρ_N in Eqn. (4.126). For a constant $\bar{P}_D$, the processing loss decreases with N and the value of $\bar{P}_{FA}$. As a guide, for the interval $N = 10$–20 independent samples and values of $\bar{P}_{FA} = 10^{-4}$–10^{-6}, the CA-CFAR-processing loss varies between about $\rho_N = 1$ dB and $\rho_N = 3$ dB when $\bar{P}_D = 0.9$.

$$\rho_N = \frac{\chi_N}{\chi_\infty} \tag{4.126}$$

In practice, the aforementioned ideal conditions are seldom encountered. The kernel will often not contain statistically independent and homogeneous disturbance samples. The assumption of statistical independence is violated even for white noise due to tapers used in signal processing, while the homogeneous disturbance assumption typically does not hold due to changes in the disturbance PDF or its parameters over the reference cells. In addition, some reference cells may be contaminated by outliers (e.g., other target-like signals) that are unrelated to the disturbance in the test sample. CFAR performance is sensitive to kernel design, which needs to be carefully tailored to the characteristics of the disturbance environment. The kernel may also be adapted to match the expected disturbance phenomenology by tuning both window shape and size parameters depending on the location of the test cell within the ARD data cube.

CFAR windows with large dimensions are preferable in that they contain a higher number of reference cells for estimating the disturbance PDF parameters. This reduces statistical errors due to finite sample support in a homogeneous environment. CFAR windows of larger size also provide greater immunity to biasing of the estimates when the kernel contains outliers. However, the assumption of a homogeneous disturbance environment is often not realistic in practice. As the window size increases, the statistical properties of the disturbance can change significantly within the kernel. This may cause the false-alarm rate to vary because the sample set in the kernel can have statistical characteristics that are not representative of the disturbance in the sample under test. Kernels containing smaller sample sets preserve local information better and are therefore less vulnerable to disturbance heterogeneity. However, a CFAR window containing a small number of samples is more susceptible to statistical errors and biasing effects caused by outliers.

To illustrate these effects, consider a one-dimensional CFAR window with $N/2$ cells leading and lagging the test sample and one guard cell either side. Figure 4.33 shows the performance of the CA-CFAR technique for $N = 20$ and a threshold multiplier yielding $\bar{P}_{FA} = 10^{-3}$. Four targets are injected at sample numbers 50, 115, 200, and 205. A detection arises when the CA-CFAR threshold (dashed curve) is below the target peak. A disturbance with sharp changes in power from $\sigma^2 = 10$ dB to $\sigma^2 = 40$ dB has been simulated, as indicated in Figure 4.33.

Processing loss occurs because the CA-CFAR threshold will on average exceed the clairvoyant threshold (dotted line) to maintain a given false-alarm rate in homogenous conditions. A detected and screened (undetected) target are shown at samples 50 and 115, respectively, both with 20 dB SDR. Screening occurs due to the change in disturbance

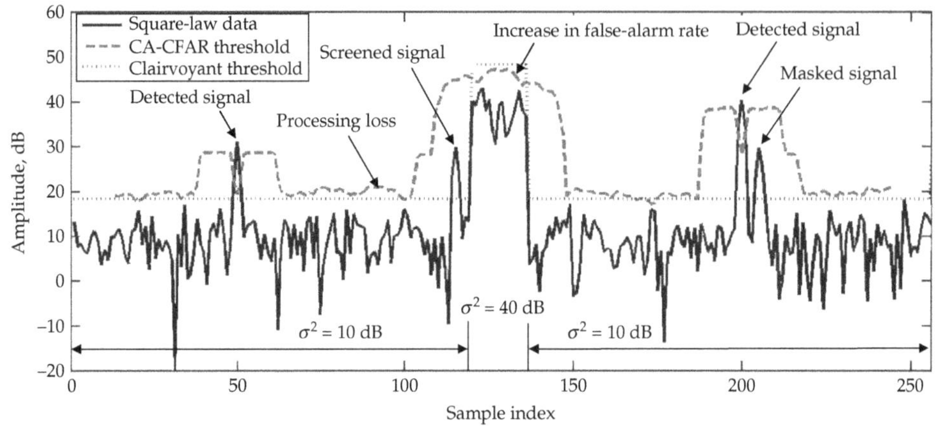

FIGURE 4.33 Simulated data example of CA-CFAR–processing performance. The kernel has 20 reference cells divided equally between the leading and lagging windows, and one guard cell either side of the test sample. The disturbance power changes instantaneously from $\sigma^2 = 10$ dB to $\sigma^2 = 40$ dB and then back to $\sigma^2 = 10$ dB to simulate sharp transitions in disturbance (e.g., clutter) characteristics. Four signals are injected into the lower power disturbance region, but only two are detected above the CA-CFAR threshold. The other two are not detected due to screening and masking effects. © Commonwealth of Australia 2011.

power in the leading and lagging windows of the kernel. A weaker target (at sample 205) masked by a closely spaced stronger target (at sample 200) illustrates the detrimental effect of "outliers" in the kernel. The CA-CFAR threshold is lower than the clairvoyant value in the high power disturbance region. This increases the false-alarm rate, although no false detections occur in this example.

The smallest-of-cell-averaging (SOCA) CFAR method aims to reduce signal screening and masking effects by estimating the disturbance power independently in the lagging and leading windows and then calculating the threshold $\hat{T}_{so}$ based on the smallest-of value denoted by $\hat{\sigma}^2_{min}$ in Eqn. (4.127). On the other hand, the greatest-of-cell-averaging (GOCA) CFAR method aims to reduce false alarms due to sudden disturbance power variations by estimating the threshold $\hat{T}_{go}$ based on the greatest-of value $\hat{\sigma}^2_{max}$. The threshold multipliers α_{so} and α_{go} to achieve the same $\bar{P}_{FA}$ as the CA-CFAR scheme are different for the SOCA and GOCA schemes. Expressions for computing appropriate multipliers to yield a desired false-alarm rate are given in Weiss (1982).

$$\hat{T}_{so} = \alpha_{so}\hat{\sigma}^2_{min}\,, \quad \hat{T}_{go} = \alpha_{go}\hat{\sigma}^2_{max} \tag{4.127}$$

Figure 4.34, in the same format as Figure 4.33, shows that SOCA-CFAR processing trades off robustness to false alarms in a heterogeneous disturbance environment for greater immunity to signal screening and masking effects. Although the SOCA-CFAR method is less vulnerable to sharp disturbance power transitions and the presence of outliers from a signal-detection perspective, as illustrated by the detection of all four targets in Figure 4.34, such a technique is not recommended when robustness to false alarms in a heterogeneous disturbance environment is of primary concern. In this example, SOCA-CFAR gives rise to numerous false alarms in the region of higher disturbance power.

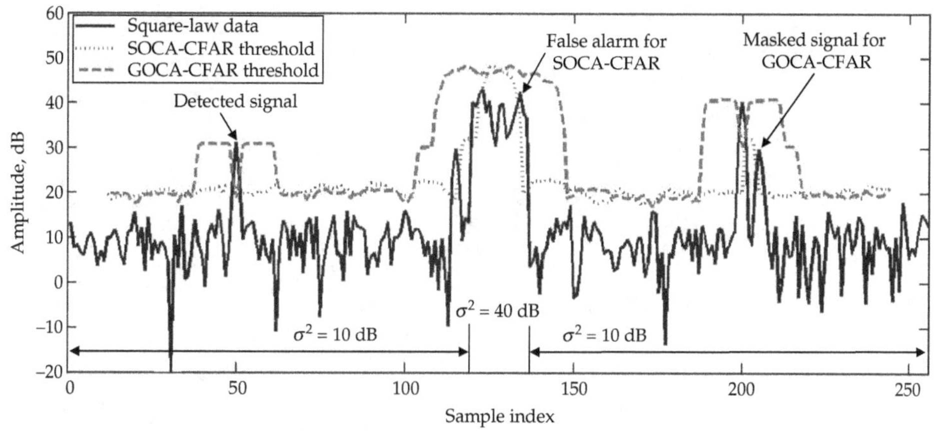

FIGURE 4.34 Simulated data example of SOCA- and GOCA-CFAR–processing performance. The kernel design, interference characteristics, and injected signals are identical to those described for the CA-CFAR scheme in Figure 4.33. SOCA-CFAR detects all four signals at the expense of numerous false alarms in the high-power disturbance region. GOCA-CFAR is robust to false alarms caused by sharp transitions in disturbance power, but is susceptible to signal screening and masking effects. © Commonwealth of Australia 2011.

GOCA-CFAR is more robust to false alarms caused by abrupt changes in disturbance power, but is susceptible to signal screening and masking effects, as also illustrated in Figure 4.34.

Figure 4.35 conceptually illustrates a cross-shaped CFAR window with reference cells in range and Doppler and the center at the (current) test sample location. With reference to this two-dimensional CFAR window, the disturbance statistics can vary markedly in vertical (range) and horizontal (Doppler) windows depending on the location of the test sample in the range-Doppler map. For example, the surface-clutter properties vary sharply in Doppler frequency but are relatively more homogeneous in range. The statistics of surface clutter in a particular test sample are therefore better captured using reference cells in the vertical window of the CFAR-processing kernel. On the other hand, transient meteor echoes tend to be confined in range and smeared across Doppler, so the statistics of this disturbance type are better captured by the samples in the horizontal window of the kernel.

The GOCA-CFAR method has been found to be effective in OTH radar when a two-dimensional (cross-shaped) range-Doppler kernel is employed and disturbance power estimates are calculated independently in the vertical and horizontal windows. In this case, the whitening (i.e., normalizing) transformation for the test sample is based on the greatest-of value, as defined in Eqn. (4.128). Here, r and d are range and Doppler cell indices of the current sample, while K and L are the number of reference cells in the vertical and horizontal windows, respectively.[6] The conditions $|k| > 1$ and $|\ell| > 1$ exclude the test sample plus a guard cell either side from the calculation of α_r and α_d, respectively.

[6]The previously used variables r, d, K, and L are redefined in Eqn. (4.128) to simplify the notation.

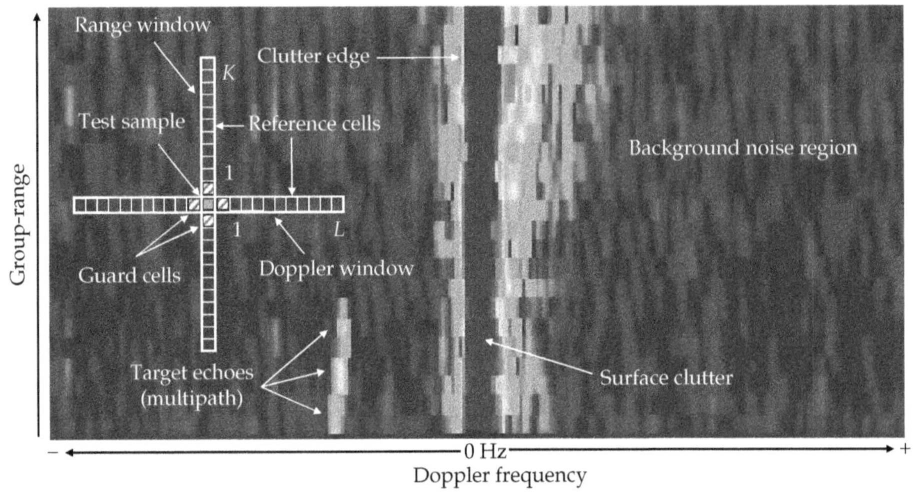

FIGURE 4.35 Illustration of a two-dimensional (cross-shaped) CFAR window extending in range and Doppler. The sample under test is surrounded by one guard cell in each dimension. The kernel is translated in range and Doppler according to the location of the test sample, but clearly needs to be modified near the edges of the range-Doppler map where there is no data available to fill the (leading or lagging) range and Doppler windows of length K and L, respectively.

The quantity $\max(\alpha_r, \alpha_d)$ represents the greatest-of estimate of the *local* disturbance level used to normalize each sample in turn. This operation yields the whitened ARD output $z(d, r)$.

$$z(d,r) = \frac{y(d,r)}{\max(\alpha_r, \alpha_d)}, \quad \alpha_r = \sum_{k=-K}^{K} \frac{y(d, r+k)}{2(K-1)}, \quad \alpha_d = \sum_{\ell=-L}^{L} \frac{y(d+\ell, r)}{2(L-1)}, \quad \begin{cases} |k| > 1 \\ |\ell| > 1 \end{cases} \tag{4.128}$$

Relative to CA-CFAR, the two-dimensional GOCA-CFAR method provides a degree of adaptivity in the kernel direction with test cell location. This provides robustness to false alarms when the dominant disturbance energy is aligned in either range or Doppler. Adopting the greatest-of philosophy reflects the importance of minimizing false alarms. This comes at the expense of greater processing loss compared to the CA-CFAR scheme in homogeneous regions of the range-Doppler map, and less immunity to masking caused by outliers (Turley 1997).

Window lengths greater than about 20 samples are recommended for each kernel dimension in practice. In OTH radar, the CFAR window dimensions can be made larger in the noise-dominated region of the range-Doppler map to reduce processing loss, as this region is relatively more homogenous than the region containing surface clutter. Greater immunity to signal masking can be achieved by conditionally excluding any detected outliers using a multi-pass trimmed mean technique, as described in Ritcey (1986 and 1989), for example.

4.4.1.3 Ordered Statistics

In very heterogeneous disturbance environments, estimates need to be calculated from highly localized data using small CFAR windows. In such situations, sample support can become a problem for CA-CFAR and its extensions to SOCA- or GOCA-CFAR. More specifically, GOCA-CFAR techniques become very susceptible to outliers when small sample sets are deemed necessary, as their presence can significantly bias the disturbance estimate under such conditions. Methods based on ordered statistics (OS) may be used to provide more robust estimates for single-mode distributions not limited to the exponential PDF (Rohling 1983). OS-CFAR ranks the reference cell samples $\{y_i\}_{i=1}^{N}$ in order of ascending amplitude to form a sequence $\{y_{(1)}, \ldots, y_{(N)}\}$. Element k of the sequence $y_{(k)}$ is known as the kth ordered statistic. In OS-CFAR, the kth ordered statistic is selected as an estimate of the disturbance, and the threshold is set as a multiple of this value in Eqn. (4.129).

$$\hat{T}_{os} = \alpha_{os}\, y_{(k)} \tag{4.129}$$

An expression for the threshold multiplier that achieves a certain false-alarm rate for a particular value of k is provided in Richards (2005). Selection of the order or rank parameter k provides scope for empirical or adaptive tuning. For a given PDF, the rank may be selected to estimate the mean value and thus provide a robust version of the CA-CFAR test statistic. For example, the median may be selected as it exhibits good edge detection performance while being robust to outliers.

The greatest-of philosophy may be combined with the OS approach and the cross-shaped (range-Doppler) kernel to develop a greatest-of ordered statistics (GOOS) 2D-CFAR scheme. For example, the data whitening transformation may involve amplitude normalization by the highest of the medians computed from the range and Doppler windows, denoted by β_r and β_d, respectively. The GOOS 2D-CFAR whitener in Eqn. (4.130) may be expected to perform well in OTH radar applications.

$$z(d,r) = \frac{y(d,r)}{\max(\beta_r, \beta_d)} \tag{4.130}$$

Thus far, we have assumed a complex-Gaussian disturbance process with a Rayleigh distributed amplitude envelope. This description is often appropriate for the noise-dominated regions of the range-Doppler map. On the other hand, the amplitudes of clutter-dominated samples tend to follow a Weibull distribution more closely. The Weibull PDF is given by Eqn. (4.131), where $y = |x|$ is now the magnitude (i.e., linear measurement) of the complex ARD data-cube sample x, B is a scale parameter, and C is a shape parameter. The Rayleigh distribution corresponds to the special case of $C = 2$ in Eqn. (4.131). Values of $C = 0.33$ have been observed for clutter-dominated disturbance samples (Turley 2008). This two-parameter distribution provides additional flexibility relative to the Rayleigh distribution for CFAR processing in OTH radar systems.

$$p(y) = \frac{C}{B}\left(\frac{y}{B}\right)^{C-1} \exp\left[-\left(\frac{y}{B}\right)^{C}\right], \quad x \geq 0 \tag{4.131}$$

A method for estimating the Weibull scale and shape parameters is described in Turley (2008), with provisions made to adaptively increase sample size in homogeneous (noise-dominated) regions to reduce processing loss, while maintaining good edge performance in heterogeneous (clutter-dominated) zones. Once the PDF parameters are estimated, CFAR processing involves transforming each data sample $y(d,r)$ to a test statistic $z(d,r)$ that is ideally described by a unit-variance Rayleigh PDF. For the Weibull distribution, the transformation $z = T(y)$ is given by Eqn. (4.132), where $B(r,d)$ and $C(r,d)$ denote the local scale and shape parameter estimates, respectively. Estimates based on the OS

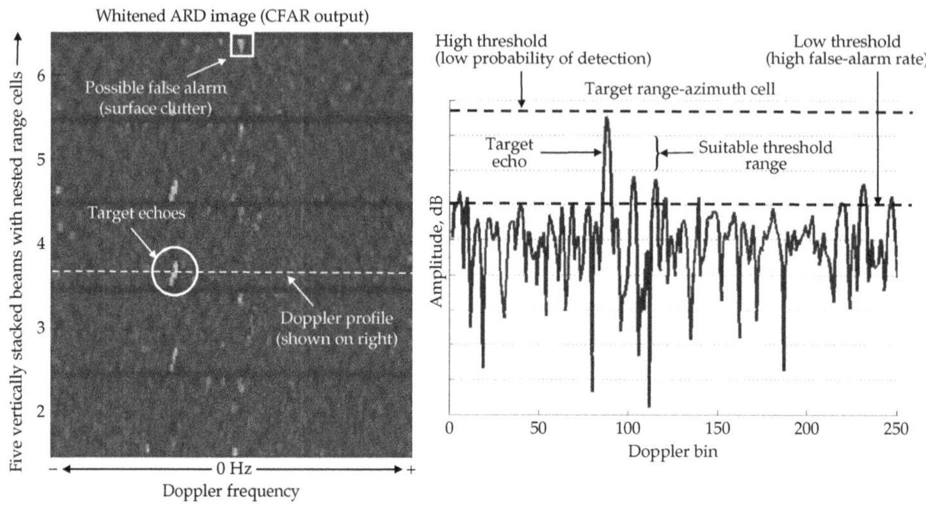

FIGURE 4.36 Practical example showing the application of the GOOS-CFAR technique to real OTH radar ARD data using a 2D (cross-shaped) kernel in range and Doppler. The left panel shows a whitened ARD image containing a real aircraft target echo, and some false alarms due to clutter. The right panel shows a whitened Doppler profile at a range-azimuth cell containing a target echo. A suitable threshold setting allows this echo to be detected with no false alarms in the Doppler profile. © Commonwealth of Australia 2011.

approach may be used in conjunction with two-parameter PDFs such as the Weibull distribution (Rifkin 1994).

$$z(d,r) = \left(\frac{y(d,r)}{B(d,r)}\right)^{C(d,r)/2} \tag{4.132}$$

The considerations in this section point to the use of a two-parameter PDF in conjunction with a two-dimensional kernel matched to the expected clutter, interference, and noise profiles, where the threshold estimates are based on cell averaging or ordered statistics using the greatest-of approach. This summarizes the main elements of conventional CFAR processing for OTH radar. The left panel of Figure 4.36 illustrates the application of the GOOS whitening approach to real OTH radar data shown previously in Figure 4.19.

The right panel of Figure 4.36 shows the Doppler profile in a range-beam cell containing an aircraft target echo. A suitable threshold may be set to detect the target echo with no false alarms in this range-azimuth cell. Although the surface clutter has clearly been normalized to the background level, the whitened ARD image exhibits a small number of false alarms near zero Doppler frequency. Importantly, the target echo is clearly visible above a relatively uniform background, which allows for approximate CFAR detection using a constant threshold setting over the entire ARD image.

4.4.2 Threshold Detection and Peak Estimation

In practice, the whitened ARD image $z(d,r)$ is not passed directly onto the threshold detection circuit. One reason for this is that multiple detections can result for a single (well-resolved) target echo if one or more samples neighboring the peak sample in range, Doppler, or beam also exceed the detection threshold. This commonly occurs when the target echo straddles two resolution cells, particularly as the use of tapers increases the mainlobe width of the point spread function (PSF).

To address this issue, and reduce the amount of further data processing, only the whitened ARD samples that qualify as local peaks in all three data-cube dimensions are considered for threshold detection (i.e., only data samples with an amplitude greater than its immediate neighbors in range, Doppler, and beam). Considering only the local peaks of the CFAR output not only ensures a single detection for a single (well-resolved) target echo, but also reduces false alarms and computational load in subsequent data-processing steps, without altering the probability of detection. Combining the requirement for the sample to be a peak with the CFAR test statistic, we obtain the hybrid detector in Eqn. (4.133), where only the peaks $z_p(d,r)$ are tested for signal presence.

$$z_p(d,r) \underset{H_0}{\overset{H_1}{\gtrless}} \gamma\,, \quad z_p(d,r) = \begin{cases} z(d,r) & \text{local peak} \\ 0 & \text{otherwise} \end{cases} \tag{4.133}$$

When a CFAR output sample is identified as a local peak and its value exceeds the threshold, further processing is carried out to generate the peak information input to the tracker. The main parameters of interest are the peak location in ARD coordinates and the peak amplitude. Target echoes typically do not fall exactly on a resolution cell of the ARD data. If the integer bin coordinates of the detected peak sample are interpreted directly as the target parameters, the estimation error may be as much as half a bin in *each* radar-processing dimension. Since the conventional ARD resolution cells can be relatively coarsely spaced, the accumulation of estimation errors over three radar dimensions may be significant.

It is possible to interrogate a finer-spaced grid of ARD resolution cells (e.g., in Doppler by zero-padding the acquired data and computing an FFT of larger size). However, this method is inefficient because it interpolates the entire spectrum, whereas greater estimation accuracy is only needed at the locations of the detected peaks. A relatively simple and computationally efficient technique involves fitting a parabola through the peak sample z_p and its two nearest neighbors, denoted by z_{p-1} and z_{p+1}, in each data-cube dimension separately. Specifically, the parabola is fitted to these three points with the sample amplitudes expressed in decibel scale. This analytical interpolation approach is tantamount to locally approximating the mainlobe of the point spread function (PSF) by a parabola when expressed in decibel scale.

There are a few points of caution when using this approach. First, the approximation of the PSF by a parabola is appropriate provided that the three points used to fit the curve do in fact lie in the mainlobe of the PSF. This condition is satisfied as long as the input data is not under-sampled, the sample separation at the output of spectral analysis is not greater than the FFT bin spacing, and window functions that lower sidelobe levels at the expense of broadening the mainlobe are employed. Another important point is that this procedure assumes a single well-resolved target echo and the mainlobe structure of a PSF associated with conventional processing. The presence of unresolved target echoes, or the use of adaptive processing with possibly distorted mainlobe characteristics, may also degrade the effectiveness of this approach.

The coefficients of the interpolating parabola $z(f) = a_0 + a_1 f + a_2 f^2$ in the continuous variable f (normalized by the bin width) are given by the unique solution of the linear equations in Eqn. (4.134). Here, the vector $\mathbf{z} = [z_{p-1}, z_p, z_{p+1}]^T$ contains the peak sample z_p, assumed to be centered at the origin $f = 0$, and the two samples immediately either side of the peak, located at unit bin spacing $f = \pm 1$. The vector $\mathbf{a} = [a_0, a_1, a_2]^T$ contains the parabola coefficients, while $\mathbf{M}$ is the coordinate matrix for the three sample points used.

$$\mathbf{z} = \mathbf{M}\mathbf{a}, \quad \mathbf{M} = \begin{bmatrix} 1 & f-1 & (f-1)^2 \\ 1 & f & f^2 \\ 1 & f+1 & (f+1)^2 \end{bmatrix} \tag{4.134}$$

The coordinate of the maximum of the interpolated peak (in non-integer bins) relative to the location of the highest amplitude sample is given by $\hat{f}_p$ in Eqn. (4.135). In other words, $\hat{f}_p$ is the displacement between the bin containing the peak sample and a more precise estimate of the true peak location. The amplitude of the interpolated peak $\hat{z}_p$ is found by evaluating the parabola at the stationary point, i.e., $\hat{z}_p = z(\hat{f}_p)$. It is advisable to perform the interpolation using the amplitude envelope of the *unwhitened* (i.e., pre-CFAR) ARD samples at the location of the peak detected in the whitened ARD. This is because CFAR processing can distort the expected structure of the target echo PSF.

$$\hat{f}_p = \frac{-a_1}{2a_2} = \frac{-[z_{p+1} - z_{p-1}]}{2(z_{p-1} - 2z_p + z_{p+1})} \tag{4.135}$$

Figure 4.37 illustrates the interpolation procedure applied to the beam spectrum of a real target echo corresponding to the $1E - 1E$ mode in Figure 4.21. When this procedure is applied to the three data cube dimensions separately, the output is an estimate of the interpolated peak amplitude and more precise (non-integer bin) estimates of the peak ARD coordinates. This information is assembled into a data structure and passed onto the tracker for each peak that is detected in the whitened ARD image.

An additional peak parameter that may be utilized for tracking is peak curvature defined by Eqn. (4.136). The PDF of this parameter is found to differ significantly for target and clutter echoes received by OTH radar, particularly in the Doppler domain. This difference may be exploited to reduce the false track rate in highly cluttered environments (Colegrove and Cheung 2002).

$$a_2 = \frac{z_{p-1} - 2z_p + z_{p+1}}{2} \tag{4.136}$$

4.4.3 Tracking and Coordinate Registration

OTH radars are required to establish and maintain tracks on possibly a large number of targets while continuing to search the surveillance region for new targets in what may be referred to as the "track-while-scan" mode. The penultimate stage in the data-processing sequence is tracking applied in radar coordinates to the peak data extracted from each radar scan. Although specific algorithm implementation details can vary significantly, the great majority of multiple target tracking systems essentially consist of three core components, namely, the tracking filter, data association, and track management. The first part of this section briefly overviews the fundamental roles of each component at an elementary level and identifies some generic approaches that may be proposed for tracking in OTH radar.

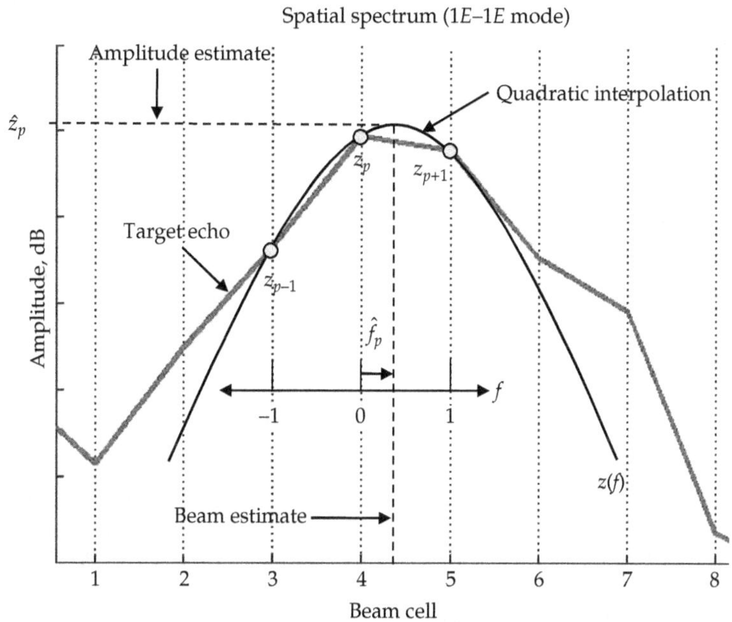

FIGURE 4.37 Illustration of target echo parameter estimation using quadratic interpolation between the peak sample and its immediate neighbors in the beam spectrum. When applied across all three ARD data dimensions, this yields more accurate (non-integer) estimates of the target peak location in range, Doppler, and beam, as well as peak amplitude or SNR, for the tracker. © Commonwealth of Australia 2011.

Detailed descriptions of the methodologies which may be used in a variety of multiple target tracking systems can be found in several excellent texts, including Bar-Shalom and Fortmann (1988), Stone, Barlow, and Corwin (1999), Blackman and Popoli (1999), and Ristic, Arulampalam, and Gordon (2004). When tracking is performed in radar coordinates, further processing is required to convert the tracks to geographic coordinates based on available propagation-path information. In general, it is also required to associate tracks formed on resolved multipath echoes to individual targets. The second part of this section briefly discusses the topics of coordinate registration and track association in OTH radar.

4.4.3.1 Tracking

The multiple model unified probabilistic data association filter or MM-UPDAF algorithm described in Colegrove and Davey (2003) represents a benchmark tracking approach for OTH radar systems. This approach effectively unifies the Probabilistic Data Association Filter (PDAF) developed by Bar-Shalom and Tse (1975) with a number of additional features based on practical experience and extensive data analysis.

Specifically, the MM-UPDAF algorithm incorporates a certain number of $m = 1, \ldots, M$ target motion models that describe the evolution of the target state vector $\mathbf{x}_m(k)$ over time or dwell number k with in an interactive multiple model (IMM) framework. The target state vector $\mathbf{x}_m(k)$ typically includes the group-range r, group-range rate $\dot{r}$, azimuth θ, azimuth rate $\dot{\theta}$, and amplitude A, as in Eqn. (4.137).

$$\mathbf{x}_m(k) = [r, \dot{r}, \theta, \dot{\theta}, A]^T \tag{4.137}$$

The state vector is updated according to the target dynamics model in Eqn. (4.138), where $\mathbf{F}_m(k)$ is a known state transition matrix that embodies the equations of motion for each model m, while model errors including target maneuvres are represented by the Gaussian distributed process noise vector $\mathbf{u}_m(k)$ with zero-mean and known covariance. To a good approximation, a linear model for state vector update may be assumed in OTH radar due to the long range of the surveillance region from the radar. Although constant acceleration and other target dynamic models are possible, the ones discussed here are considered to be constant velocity models.

$$\mathbf{x}_m(k+1) = \mathbf{F}_m(k)\mathbf{x}_m(k) + \mathbf{u}_m(k) \tag{4.138}$$

Within the IMM framework, multiple models may be introduced for different purposes. For example, they may be used for target track initiation to account for the possibility of velocity ambiguities. They may also be used to enhance tracking of maneuvering targets by defining multiple models with different process noise covariance. In the latter case, the flexibility of the multiple model approach can in principle allow estimation errors to be kept relatively low during a maneuver, while at the same maintaining high accuracy during nonmaneuvering portions of the track.

Further generalizations indicated by the term "unified" in the acronym MM-UPDAF involve the incorporation of multi-component nonuniform clutter distributions, target visibility, and a nearest neighbor methodology for each target model. Details of the MM-UPDAF algorithm can be found in Colegrove and Davey (2003). The measurement vector $\mathbf{y}(k)$, resulting for a particular (unnamed) peak detected in dwell k, typically consists of the interpolated peak coordinates in azimuth $\theta(k)$ and group-range $r(k)$, as well as the interpolated estimates of Doppler shift $f_d(k)$ and peak amplitude $A(k)$.

$$\mathbf{y}(k) = [\theta(k), r(k), f_d(k), A(k)]^T \tag{4.139}$$

The choice of tracking filter used to perform the recursive target state estimation depends in part on the coordinate system selected for tracking. In OTH radar, tracking is often performed in radar coordinates with the established tracks subsequently registered to geographic position. This procedure tends to simplify the measurement model, which relates the sensor measurement vector to the target state vector. To improve mathematical tractability, and reduce processing load, the measurement model is approximated by the linear equation in Eqn. (4.140), where $\mathbf{y}_m(k)$ is the measurement vector at time k, $\mathbf{H}_m(k)$ is a known measurement matrix for model m, and $\mathbf{v}_m(k)$ is a zero-mean Gaussian distributed measurement noise vector with a known but possibly changing covariance.

$$\mathbf{y}_m(k) = \mathbf{H}_m(k)\mathbf{x}_m(k) + \mathbf{v}_m(k) \tag{4.140}$$

The most commonly used tracking filter to perform the recursive target state estimation is the Kalman filter (KF). This elegant and computationally efficient algorithm is known to be optimum in the Bayesian setting under the assumption of a linear model in which both the process and measurement noise are Gaussian. The filtering operation is carried out using a bank of Kalman filters, one for each model $m = 1, \ldots, M$.

A nonlinear measurement model can arise when tracking is performed in Cartesian coordinates.[7] In this case, an extended Kalman filter (EKF) implementation is appropriate. The EKF performs analytical linearization of the nonlinear process and measurement functions by using a first-order Taylor series expansion about the current state estimate. Nonlinear filters such as the EKF, unscented Kalman filter (UKF), and particle filters fulfill important roles in situations where the KF is not optimal, but such filters are not considered further here.

Data association refers to the process of associating a discrete set of measurements at time k to either a set of existing tracks, to a newly initiated track, or to no tracks. In an operational system, the probability of detection for a true target is less than unity, while the false alarm rate is inevitably greater than zero. The former implies that a peak detection originating from a true target, which has given rise to an existing track, may not be available at each time k. Moreover, detections may arise due to the presence of several other targets in the surveillance region, or false alarms due to clutter, interference, and noise. Hence, the determination of which measurement corresponds to which target, if any, is not straightforward in practice. In particular, difficulties can arise when independent targets have similar radar coordinates, or false alarms are triggered close to the target locations.

This raises the problem of data association, which in general involves the definition and evaluation of a distance measure between a track and a measurement. Popular methods for data association include the Global Nearest Neighbor (GNN) approach, Joint Probabilistic Data Association (JPDA), and the Probabilistic Multiple Hypothesis Tracker (PMHT). The GNN method first performs a gating of measurements to screen out unlikely candidates for track update. In other words, measurements which do not fall within a specified error tolerance relative to the filter prediction are excluded as candidates for track update. It then selects the "nearest" gated (validated) detection according to the defined distance measure and uses this measurement to perform the track update.

The GNN method ignores any uncertainty in the selection of a particular measurement when there are multiple valid candidates. The JPDA approach differs in that it estimates the track update based on the relative significance of all validated measurements located within a specified neighborhood. The tracking filter is then fed a centroid measurement formed by taking a weighted average of all the validated measurements. The filter covariance is also adjusted by an amount dependent on the scatter of the validated measurements. In OTH radar applications, the JPDA approach is found to be more robust than the GNN data association method when the tracker operates in a cluttered environment. A comparison of this approach with the PMHT method may be found in Colegrove, Davey, and Cheung (2003), where the metrics used to assess performance include: (1) track establishment, (2) track maintenance, (3) track error, and (4) false tracks.

Track status is typically defined in terms of three life stages, commonly referred to as tentative, confirmed, or deleted. A tentative track may be initiated on a measurement that has not been associated to any of the existing tracks. Track confirmation is based on the number of updates performed on a tentative or "pending" track over a period of time or number of scans. Track deletion is based on the length of time or number of scans without a track update, after which the track confidence falls to an unacceptably low level. The

[7] This may be attempted in HFSW radar, where skywave propagation path knowledge is not required.

notion of track confidence or quality is typically introduced for track management as a criterion for switching between the different track life stages.

The geographic display Figure 4.38 shows an example of real OTH radar tracks formed in radar coordinates on resolved multipath returns from a single aircraft target. This particular example illustrates significant mode-dependent range and bearing fluctuations due to the suspected presence of a wave-like traveling ionospheric disturbance (TID). The presence of multiple tracks formed on a single target motivates the need for track-to-target association. Coordinate registration is required to convert multipath tracks in radar coordinates to a single track that estimates the geographic positions of the target to radar operators. The coordinate registration task is quite challenging when propagation-path fluctuations due to (large- and small-scale) TIDs need to be corrected.

So far, the input to the tracker is assumed to be a set of point measurements (candidate peaks), which have been threshold detected from the whitened ARD image resulting at each scan. The tracking algorithm effectively links the measurements over time to estimate the target parameters of interest. In other words, the standard tracking approach is to perform the detection and estimation functions sequentially. This approach

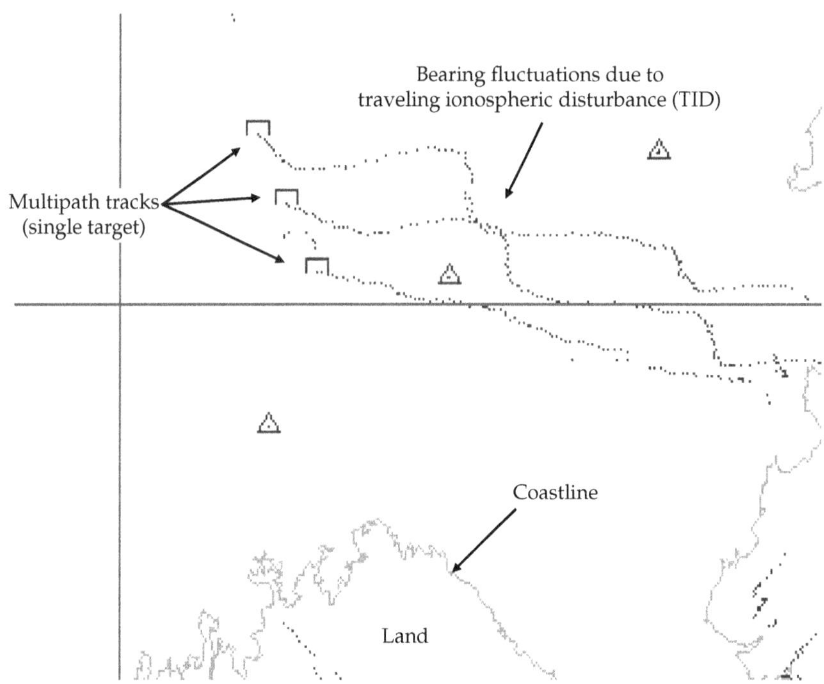

FIGURE 4.38 Geographic track display showing three tracks formed on multipath echoes from a single target with significant mode-dependent range and bearing fluctuations due to the suspected presence of a TID. The relatively straight track nearest to the coastline (with lowest group-range) corresponds to the $1E - 1E$ propagation mode, which is unaffected by the TID. The middle track arises due to mixed-modes ($1E - 1F$ and $1F - 1E$), while the track with highest group-range is due to the $1F - 1F$ mode. These tracks involve at least one F-region reflection, and are hence affected by a TID passing through the F-region. © Commonwealth of Australia 2011.

is appropriate for targets with relatively high SNR. However, a sufficiently high probability of detection for low SNR targets can only be achieved by reducing the detection threshold. This gives rise to a higher rate of false detections, which causes the tracker to generate false tracks. An alternative approach is to eliminate the hard binary decisions on a scan-by-scan basis and to perform tracking directly on the ARD image. This field of research is known as track-before-detect (TkBD).

In the TkBD framework, detection decisions are not made on each radar dwell, but are rather deferred to the output of the tracker. In principle, this allows the algorithm to exploit target PSF features in the ARD image and temporal correlation to improve both decision performance and track-estimation accuracy. The TkBD methodology is expected to provide an enhanced capability to track low SNR targets with respect to the standard tracking paradigm. The discussion of such techniques, which may also utilize phase information in the unwhitened ARD, is beyond the scope of this text. The reader may consult Davey (2009), and references therein, for a detailed description of TkBD techniques using complex-valued and amplitude-only data measurements.

4.4.3.2 Coordinate Registration and Track Association

Coordinate registration (CR) uses the available skywave propagation-path information to effectively determine a mapping between radar coordinates in group-range and cone angle, also known as slant range and apparent azimuth, to geographic position on the Earth's surface in latitude and longitude, or equivalently in ground distance and great-circle bearing from the radar. A mapping needs to be determined for each ionospheric mode that links the radar system to a point in the surveillance region over a two-way path. A variety of different approaches may be proposed to tackle the CR problem. Broadly speaking, these approaches fall into two main categories; (1) those which rely on knowledge in the form of an ionospheric model that allows signal propagation-path information to be deduced from the application of ray-tracing procedures (Dyson and Bennett 1992), and (2) those which exploit opportunistic scatterers or natural topographical features on the Earth's surface that enable distinguishable radar returns to be referenced against known geographic positions. A combination thereof is often favorable in practice.

In the first category, there is the option to use a smoothed ionospheric model based on historical synoptic data that statistically accounts for seasonal and daily variations of the ionosphere in the region of interest. However, due to the inherently unpredictable behavior of the ionosphere, such models are unable to forecast the detailed character of skywave propagation to the high level of fidelity required for accurate CR. More robust CR systems are based on the construction of a site-specific real-time ionospheric model (RTIM) based on experimental measurements made in the region of interest. This avoids a *complete* reliance on a smoothed ionospheric model, and hence reduces the degree of uncertainty about the state of the ionosphere for CR purposes. For economic and logistic reasons, smoothed ionospheric models often need to be blended into the RTIM as the spatial (and temporal) separation from points where real data is available increases.

RTIM parameters based on a multi-segment quasi-parabolic (MQP) representation of the ionospheric height profile may be estimated by interpreting data collected from a network of sounders, which are specifically deployed to routinely probe different signal reflection points within the operating region. For example, this may include a set of vertical and oblique incidence sounders, as well as beacons or transponders, that periodically gather ionospheric data at a number of different locations (Cameron 1995). The accuracy of the RTIM depends on several factors, including the geographical disposition and density of the network, the rate at which the RTIM is updated relative to the changes occurring in the ionosphere, the quality of the sensor data, the effectiveness of data interpretation techniques, and the appropriateness of the parametric model used to represent the ionospheric structure (height profiles). A significant practical issue is that a sounder network of limited density can only sample a very small portion of the coverage area interrogated by an OTH radar.

Another option is to use the main OTH radar itself to gather information for CR using opportunistic sources. Depending on the specific methodology, this may or may not come at the expense of precluding the system from performing its main surveillance duties. For example, the identification of echo signatures arising from geographically known reference points (KRPs) may be incorporated into the CR process as a by-product of normal operation. This may include the detection of certain targets, such as commercial airliners or container ships, whose positions are known from ADS-B or AIS self-reporting systems, respectively. Hub points where aircraft target tracks commence and terminate may be associated with the known positions of airports. Sharp spatial variations in backscattered clutter power due to RCS enhancements may also be associated with scatterers in known positions, such as cities for example (Barnum and Simpson 1998).

In contrast to these localized KRPs, changes in the backscattering properties of the Earth's surface over the surveillance region provides a spatially distributed referencing source, which may also be exploited for CR. For example, changes in the surface clutter Doppler spectrum characteristics as the radar coverage traverses land-sea boundaries may be aligned with coastline maps to derive CR transformations. A CR system can in principle attempt to merge information from both dedicated sensors and opportunistic KRP sources in the coverage to derive the RTIM and associated CR table of estimated radar-to-ground coordinate corrections for each outbound and inbound propagation mode combination. An integrated CR system architecture with the flexibility to incorporate information from a diversity of sources as they become available in real time is expected to provide the best performance (i.e., the most robust conversions from radar to ground coordinates). More information on CR can be found in Chapter 3.

As illustrated previously, the existence of multipath signal propagation over a two-way skywave path causes multiple radar echoes to be returned from a single target. These echoes are typically received with different coordinates in group-range, Doppler shift, and cone angle. In many cases, the multipath target echoes can be resolved in at least one of the three data-cube dimensions. The presence of resolved multipath often leads to independent peak detections at different radar coordinates for a particular target. The tracker may in turn initiate separate tracks by filtering the peaks detected from different modes over time.

Since the true number of targets present is unknown, various plausible track-to-target associations may exist. In practice, it is desirable to display only one correctly ground-registered track per target. This gives rise to the challenging problem of associating multipath tracks to individual targets. The process of resolving the track-to-target ambiguities is known as track association, while that of combining the associated tracks is known as "track fusion." One approach attempts to address this problem by performing target tracking directly in ground coordinates (Pulford and Evans 1998). More commonly, track fusion algorithms are based on associating target tracks already formed in radar coordinates, either with or without the use of CR information.

For example, the technique described in Percival and White (1997 and 1998) recursively constructs all track-to-target association hypotheses and evaluates the probability of each hypothesis in order to decide on the track-to-target assignment. In this approach, each track update is treated independently of past updates and this results in an "instantaneous" track fusion estimate based purely on spatial relationships. Track history can additionally be incorporated to account for both temporal and spatial relationships in the association process, as described in Sarunic and Rutten (2001). The problem of target mode-linking with uncertain coordinate registration (i.e., requiring joint ionospheric and target state estimation) has been addressed in Rutten and Percival (2001), Pulford (2004), and Anderson and Krolik (2002).

CHAPTER 5
Surface-Wave Radar

Whereas skywave over-the-horizon (OTH) radar represents the only ground-based sensor capable of target detection and tracking at ranges of thousands of kilometers (without the aid of a satellite), two-way surface-wave propagation in the HF band represents an alternative mode that may be exploited to provide wide-area surveillance over shorter but nevertheless OTH distances in the order of perhaps hundreds of kilometers. This classification of OTH radar is often referred to as high-frequency surface-wave (HFSW) radar or groundwave radar. HFSW radar systems use vertically polarized signals and typically operate in the lower half of the HF band, where the relatively high conductivity of saline water allows the surface-wave mode to propagate effectively over the sea to illuminate targets well beyond the line of sight. Besides surveillance applications, HFSW radars are also frequently used for remote sea-state sensing and wind-field mapping.

The focus of this chapter is on HFSW radars used for surveillance, where ships and aircraft represent the primary and secondary targets to be detected and tracked, respectively. In general, HFSW radar systems operate by space-wave propagation, similar to conventional microwave radar systems, and surface-wave propagation, which dominates in the diffraction region where the line-of-sight path is occluded by the Earth's curvature. As the OTH capability of HFSW radar is of particular interest here, the emphasis is on target detection and tracking via the surface-wave mode. Although skywave and surface-wave OTH radar systems share several characteristics in common, there are a number of important areas where the differences are significant. In view of the preceding discussion on skywave systems, the main purpose of the current chapter is to provide an overall appreciation of HFSW radar principles and techniques, as well as to identify and describe the key features that distinguish these two kinds of OTH radar systems.

The chapter is organized into five sections. Section 5.1 describes the concept of operation for HFSW radar, the general characteristics and nominal capabilities of such systems, as well as their potential roles in military and civil applications. Section 5.2 explores the properties of the surface-wave propagation mode in more detail, particularly with regard to attenuation characteristics over short and long distances, as well as the influence of atmospheric refraction, sea-surface roughness, and the presence of mixed land-sea paths. Section 5.3 discusses various environmental factors that can potentially limit the performance of practical HFSW radar systems, including clutter scattered by the sea surface and from the ionosphere, as well as HF interference and noise. Section 5.4 discusses HFSW radar system design and practical implementation aspects, including radar configuration, site selection, transmit and receive systems, as well as signal and data-processing architectures. Frequency selection criteria for effective HFSW radar operation are described in Section 5.5, where the benefits of dual or multi-frequency operation are also explained. The chapter concludes with a description of several example HFSW radar systems developed in Australia, Canada, and the United Kingdom.

5.1 General Characteristics

An authoritative description of HFSW radar systems is provided in the excellent review articles of Sevgi, Ponsford, and Chan (2001) and Ponsford, Sevgi, and Chan (2001). The reader is also referred to the more recent articles of Ponsford and Wang (2010) and Apaydin and Sevgi (2010) for a summary of HFSW radar performance and propagation modeling, respectively. The fundamental aspects of HFSW radar are recalled in this section to introduce the basic operational principles, system characteristics, nominal capabilities, and practical applications of such systems.

5.1.1 Principle of Operation

To describe the principle of operation of an HFSW radar, it is first necessary to identify the two main components that contribute to ground-wave propagation over the Earth, namely, the space-wave and surface-wave modes. These components of ground-wave propagation are briefly described in the first part of this section. In light of the well-known characteristics of space-wave propagation, the second part of this section describes the general properties of surface-wave propagation. This motivates the operational concept of HFSW radar, which is described in the third part of this section.

5.1.1.1 Ground-Wave Propagation

Suppose a transmitting antenna is located at point T above a perfectly conducting ground plane, as shown in Figure 5.1. The voltage V induced in a receiving antenna located at position R is expressed as the vector sum of direct and surface-reflected waves, as in

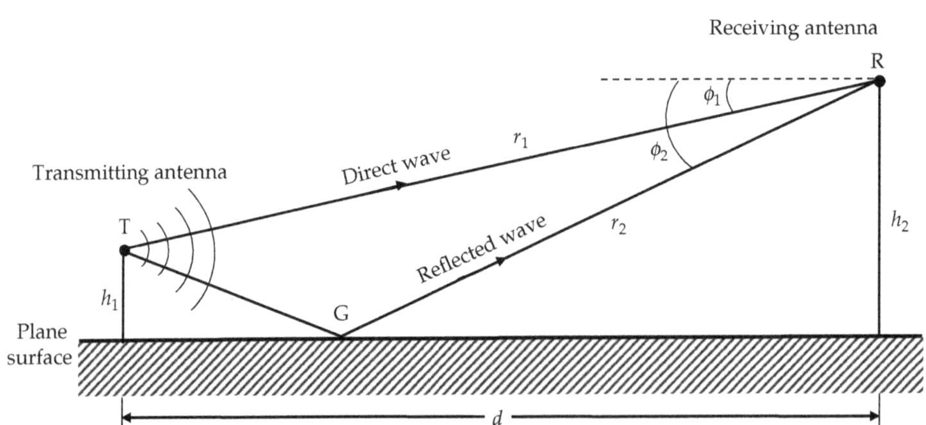

FIGURE 5.1 Illustration of direct and reflected signal paths between a transmitting antenna (T) and a receiving antenna (R) located above a plane surface.

Eqn. (5.1). Here, r_1 and r_2 are the direct and indirect (total) path-lengths, respectively, $k_0 = 2\pi/\lambda$ is the radio wavenumber, I is the current in the transmitting antenna, K is a constant, U_1 and U_2 account for antenna radiation patterns, and R is the Fresnel plane-wave reflection coefficient (Barclay 2003). The sum of direct and surface-reflected waves in Eqn. (5.1) is referred to as space-wave propagation.

$$V = KI\left\{U_1\frac{\exp(-jk_0r_1)}{r_1} + U_2R\frac{\exp(-jk_0r_2)}{r_2}\right\} \tag{5.1}$$

A realistic Earth surface has finite conductivity. In this case, an accurate description of the field at R requires a third contribution to be added, as in Eqn. (5.2), where the reflection coefficient R depends on wave polarization state, transmitted radio frequency, surface electrical properties, and the angle of reflection at point G. The final term in Eqn. (5.2), which is needed to satisfy Maxwell's equations with the applied boundary conditions, represents the surface-wave mode. The attenuation factor F will be discussed in due course.

$$V = KI\left\{U_1\frac{\exp(-jk_0r_1)}{r_1} + U_2R\frac{\exp(-jk_0r_2)}{r_2} + (1-R)F\frac{\exp(-jk_0r_2)}{r_2}\right\} \tag{5.2}$$

For a surface of finite conductivity, ground-wave propagation consists of both space-wave and surface-wave propagation. It transpires that the inclusion of the surface-wave term in Eqn. (5.2) is required for a proper representation of the field at R in the HF band and at lower frequencies, especially when one or both terminals are within a few wavelengths from the ground. For radio frequencies above the HF band, the traditional representation in Eqn. (5.1) is considered to be appropriate for many applications.

When the terminals are close to the surface, the signals received via the direct and reflected (space-wave) paths tend to cancel one another as $R \to -1$, such that propagation between T and R effectively arises only due to the surface-wave mode. For this reason, this mode is of great importance to radio systems operating at HF and in lower frequency bands, as it is the only physical mechanism by which ground-wave signals can propagate between T and R in the absence of a space-wave contribution. For example, surface-wave propagation is exploited for MF/LF broadcasting and LF/VLF communications. Importantly, it also supports the operation of OTH radar systems in the HF band.

5.1.1.2 Surface-Wave Characteristics

The first main property of surface-wave propagation is that only vertically polarized signals have the potential to produce appreciable field-strengths in the diffraction zone, or at shorter ranges for terminals near the surface, where the direct and reflected space-wave signals tend to cancel. In the case of horizontal polarization, the boundary condition at the surface effectively precludes radio waves from propagating effectively over land or sea via the surface-wave propagation mechanism. Horizontally polarized signals experience very high surface-wave attenuation and are therefore of limited or no practical use for radar or communication systems, unless space-wave or skywave propagation modes can be used to fulfill operational objectives.

Second, the surface-wave mode is not to be confused with "anomalous propagation" due to atmospheric refraction, also known as ducting, which is commonly considered in the microwave radar literature. The surface wave mode owes its presence to currents that flow in ground of finite conductivity. As a result, this propagation mode is not reliant on the presence of an atmosphere for its existence. Indeed, surface waves would be able to propagate beyond the line of sight in the absence of an atmosphere with due regard to diffraction effects. In practice, the variation of radio refractive index with altitude in the troposphere needs to be accounted for to accurately model the field arising at R due to ground-wave propagation. More will be said on this in Section 5.2.

Third, signals propagated by the surface-wave mode are characterized by relatively slow fading-rates compared to skywave signals, which has significant advantages for (slow-moving) surface-target detection in HFSW radar. Surface-wave propagation generally gives rise to very stable signals over terrain, while signals propagated over the sea surface may be subject to slow fading due to tidal effects and the dynamics of water waves. Fourth, signals propagated by the surface-wave mode experience negligible dispersion over relatively large bandwidths that may exceed 100 kHz. With respect to skywave propagation, the larger coherence bandwidth of the surface-wave mode potentially enables wideband operation without the need to compensate for dispersion on receive.

An understanding of the path-loss descriptor for surface-wave propagation is not only a subject of great theoretical interest to the mathematical physicist, but is also of significant practical value to the system engineer. The rate at which the surface-wave power density decays varies significantly with distance and becomes progressively more severe than the inverse-square law (describing free-space loss due to spreading) as the range increases. For vertically polarized signals, surface-wave attenuation with distance is governed by many factors including: (1) radio frequency, (2) ground conductivity, (3) terminal heights, (4) surface roughness, (5) surface inhomogeneities, and (6) atmospheric refraction. The behavior of the path-loss descriptor for surface-wave propagation will be discussed in Section 5.2.

Broadly speaking, the rate at which the surface-wave field-strength decays increases with distance for terminals at constant heights above a surface with uniform conductivity. At frequencies in the lower HF band, the rate of change in attenuation becomes very rapid beyond ground ranges of about 200 km for paths over the sea. At these long ranges, the surface-wave effectively enters a "dB per kilometer" power-loss regime. This attenuation regime engages at shorter ranges as the radio frequency increases, or as the ground conductivity decreases. For an HFSW radar, this implies that only small increases in detection range can be achieved by large increases in transmit power at long ranges. For example, an order of magnitude increase in transmit power may increase the maximum detection range by about 10 km in the dB per kilometer regime.

5.1.1.3 HFSW Radar Concept

The British Chain Home air-defence HF radars used during World War II exploited horizontal polarization to produce strong reflections from aircraft wings, and to assist in the formation of elevation beams to provide altitude information for line-of-sight detections. As described previously, skywave OTH radars have been implemented with vertical or horizontal polarization, noting that the ionospheric propagation medium will in general transform the polarization state of an HF signal to elliptical regardless of that transmitted by the system. On the other hand, an HFSW radar is restricted to using vertical polarization if it is to exploit surface-wave propagation to detect and track targets over the sea at ranges of hundreds of kilometers.

Figure 5.2 notionally illustrates the operational concept for a single-site (monostatic) HFSW radar that transmits and receives vertically polarized signals over a two-way

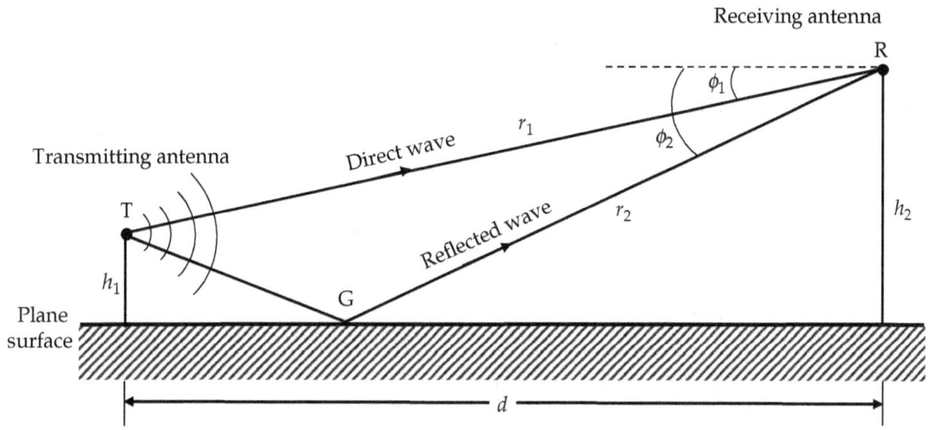

FIGURE 5.2 Notional illustration of the concept of operation for HFSW radar. Space-wave propagation may be exploited for the detection of high-altitude aircraft targets within the interference region, whereas surface-wave propagation enables the detection of surface-vessels and low-flying aircraft in the diffraction zone. © Commonwealth of Australia 2011.

(backscatter) path. As illustrated in Figure 5.2, the system operates by a combination of space-wave and surface-wave propagation to illuminate and receive echoes from targets at different ranges and altitudes. Two-site (bistatic) HFSW radar configurations are also common, with the potential for different propagation mechanisms to occur on the transmitter-to-target and target-to-receiver paths. The operating frequency of an HFSW radar is often between 3 and 15 MHz, although higher (or lower) frequencies may be used depending on target class, range of interest, and prevailing environmental conditions. In practice, HFSW radar systems are located on or close to the shore line to maximize coupling between the radar signals and the surface-wave propagation mode over the conductive sea surface.

Similar to conventional radar systems, an HFSW radar can exploit space-wave propagation to detect targets in the line of sight (i.e., within the interference region). For such targets, an HFSW radar may utilize frequencies in the upper HF band to enhance the detection and localization of echoes returned from small (low-RCS) aircraft. However, for a radar system with antennas near sea-level, space-wave propagation is not as effective for the detection of low-altitude aircraft or surface vessels because the direct and reflected signals tend to cancel in the region near the surface. Moreover, space-wave propagation is effectively absent at OTH ranges under normal atmospheric conditions. Motivated by the latter situation in particular, we now turn our attention to describing the use of surface-wave propagation in HFSW radar, which distinguishes the operation of this type of radar from its skywave (and microwave) counterparts.

The surface-wave mode provides a mechanism to effectively detect surface-vessels and low-altitude aircraft located in the diffraction zone. In the HF band, only the use of lower frequencies enable the surface-wave mode to propagate effectively over OTH ranges above the conductive (saline) sea surface. Specifically, this "guided" mode allows the radar signal to propagate in a region above the sea-air interface and to follow this interface along a path around the curvature of the Earth such that targets can be illuminated at ranges of hundreds of kilometers. The vertically polarized component of the echo backscattered by the target is able to propagate to the HFSW radar receiver by virtue of the same physical mechanism along an almost reciprocal path in a monostatic system. Target echoes acquired by the receiver, along with clutter returns, man-made interference, and environmental noise, are then processed to enable detection and tracking.

As in skywave OTH radar systems, Doppler processing is indispensable for HFSW radar to increase both the signal-to-clutter ratio (SCR) and signal-to-noise ratio (SNR). Specifically, target echoes need to be resolved from powerful surface-clutter returns backscattered by the sea at grazing incidence, and in some cases, volume-clutter received at near vertical incidence from the ionosphere. Coherent integration of the useful signal energy against potentially strong interference and noise in the lower HF band (received via surface-wave and skywave modes) is also essential for successful operation.

The detection of large surface-vessels at ranges of 200–400 km often requires the HFSW radar to operate in the 3–7 MHz frequency band to reduce surface-wave path-loss. Higher frequencies are generally more appropriate for the detection of small targets located on or not far above the sea surface at shorter ranges. Clearly, the chief advantage of HFSW radar as a coastal surveillance system with respect to a ground-based microwave radar resides in its capability to detect surface-vessels and low-flying aircraft at ranges well beyond the radio horizon.

5.1.2 Architecture and Capabilities

The first part of this section describes the main architectural characteristics of HFSW radar systems to provide a preliminary indication of practical implementation aspects for the transmitter and receiver, as well as radar configurations and site selection. This is followed by a discussion of the HFSW radar equation in the second part of this section, where attention is paid to subtle issues regarding the definition of antenna gains and target radar cross section (RCS) in the presence of the ground. The third part of this section is concerned with HFSW radar performance, including nominal detection and tracking capabilities for different target classes, potential coverage in range and azimuth, as well as the resolutions and accuracy of a typical system. Many of these topics will be revisited and discussed in more detail later in the chapter.

5.1.2.1 Basic System Description

HFSW radar systems may be broadly divided into two classes; low-power systems used primarily for oceanographic research, including remote sea-state sensing and wind-field mapping, and high-power systems designed specifically for surveillance applications (detection and tracking of man-made targets). In the former category, a large number of commercial systems are currently in use, while in the latter category, relatively few operational systems have been developed, and less still are currently in an operational role. Low-power HFSW radars designed for remote-sensing as the primary mission have occasionally been used as dual-purpose systems due to their limited hard-target detection capabilities.

To introduce the general characteristics and capabilities of HFSW radar systems used in surveillance applications, it is convenient to restrict our attention to an archetype system that more or less typifies the essential features of real-world HFSW radars used for this purpose. For such intents, a practical implementation broadly exemplified by the Iluka and SECAR systems developed in Australia, or the SWR-503 system developed in Canada, serves as a suitable starting point for an initial description. Specific differences between a number of (past and present) HFSW radars, in terms of system architectures and operational capabilities, will be discussed in the final section of this chapter. Before proceeding, it is prudent to note that HFSW radar systems used primarily for remote sea-state sensing can have significantly different implementation characteristics (besides transmit power) with respect to the surveillance systems described below.

HFSW radar systems designed for surface and air target detection as their primary and secondary missions, respectively, have been deployed in single-site (e.g., SWR-503) and two-site (e.g., Iluka and SECAR) configurations. Traditionally, these monostatic and bistatic (or quasi-monostatic) systems have been based on the use of phase-coded pulse waveforms and frequency modulated continuous waveforms (FMCW), respectively. In the example systems cited above, a vertical log-periodic dipole array (LPDA) is used as the transmit antenna element, while a uniform linear array (ULA) based on one or more combined-monopole elements is used as the receive antenna. Specifically, antenna elements that are endfire-steered monopole-doublets or quadlets have been preferred over single monopoles to provide front-to-back directivity on receive. Figure 5.3 shows a picture of the LPDA transmit antenna and the receiving ULA of the Iluka HFSW radar when it was deployed near Darwin in far north Australia.

In a two-site system, the transmitter and receiver are typically separated by 50–100 km to provide the sufficient isolation for continuous wave operation. Ideally, both sites are located at a height close to sea level and in the vicinity of the shoreline, preferably near the high tide water mark, for effective coupling of the system to and from the surface wave mode on transmit and receive, respectively. The transmit site typically requires a 200-m-by-200-m area for installation of the large LPDA element, which needs to be well-matched at frequencies in the lower HF band. On the other hand, the receive site requires a relatively straight stretch of coastline up to 500 m in length (and perhaps 50 m wide) to accommodate the relatively long aperture of the receive array.

(a) Transmit antenna: vertical log-periodic dipole array.

(b) Receive antenna: uniform linear array of 32 monopole elements.

FIGURE 5.3 Transmit and receive antennas of the Iluka HFSW radar deployed near Darwin in far north Australia. The transmitter antenna is a single vertical LPDA element, while the receiving ULA consists of 32 monopole elements spaced approximately 16-m apart to form an aperture 500-m long. © Commonwealth of Australia 2011.

The LDPA transmit antenna has a relatively broad beam in azimuth, with the potential to provide instantaneous coverage over a 90–120 degree sector. In the aforementioned systems, the transmit beam effectively floodlights the entire HFSW radar surveillance region simultaneously using a single frequency. Unlike in skywave OTH radar systems, no step-scanning of the transmitter footprint in range or azimuth is required in this case. On the other hand, the ULA antenna is based on a multi-channel receiver with a digital receiver per element. A set of higher resolution finger beams are electronically steered to receive echoes incident from different directions within the broad illumination pattern of the transmit beam. In this case, the revisit time is equal to the CPI, which may be up to hundreds of seconds long owing to the high stability of surface-wave signals.

The relatively predictable characteristics of surface-wave propagation (relative to skywave propagation) has in most systems not justified the use of adjunct sensors for frequency management in HFSW radar—with the notable exception of a real-time spectrum monitor to ensure interference-free operation. This is consistent with the view of HFSW radar being a relatively simple and low-cost sensor that may be deployed at relatively short notice to provide OTH surveillance of surface-vessels and low-flying aircraft in coastal regions. Additional sensors capable of providing real-time knowledge of ionospheric conditions have the potential to assist frequency selection in HFSW radar, particularly to reduce the incidence of clutter returned via skywave paths.

Adaptive signal processing for mitigating clutter, interference, and noise is considered vital for the effective operation of HFSW radar systems in adverse environmental conditions. Environmental factors and their impact will be described in Section 5.3. Dual-frequency or multi-frequency operation is also used as a strategy to improve HFSW radar performance, noting that such operation is often better suited for target detection via the surface-wave mode compared to the skywave mode. Further descriptions of adaptive signal processing and multi-frequency operation, including frequency selection criteria for HFSW radar, will be discussed in Sections 5.3 and 5.4.

5.1.2.2 HFSW Radar Equation

The noise-limited version of the monostatic HFSW radar equation expresses the SNR at the detection circuit by S/N in Eqn. (5.3), where P_t is the average transmitted power (W), G_t and G_r are the transmit and receive antenna gains in the target direction, respectively, σ is the target backscatter radar cross section (m^2), T is the effective integration time (s), L_b is the basic one-way transmission loss for the propagation path that links the radar to the target, L_s is the total radar system loss, λ is the radio wavelength (m), and N_o is the overall (external plus internal) noise spectral density (W/Hz) in the radar bandwidth. Clearly, S and N represent the received target echo power (P_r) and noise power (P_n), which are respectively defined by the numerator and denominator of the intermediate expression in Eqn. (5.3). This expression is valid in the diffraction region, as it assumes the surface-wave alone is active. Alternative HFSW radar equations for clutter-limited detection and bistatic systems will be described later in this section.

$$\frac{S}{N} = \frac{P_t G_t G_r \sigma / \left[L_s L_b^2 \lambda^2 / (4\pi) \right]}{N_o / T} = \frac{P_t G_t G_r \sigma T 4\pi}{L_s L_b^2 \lambda^2 N_o} \tag{5.3}$$

The basic transmission loss L_b is defined in Eqn. (5.4). Note that this definition combines the surface-wave attenuation factor, denoted by $A < 1$, with the free-space loss due to spreading for a target at range d over a one-way path (Sevgi, Ponsford, and Chan 2001). A distinguishing feature of the HFSW radar equation with respect to the conventional radar equation is the inclusion of the vertically polarized surface-wave (field-strength) attenuation factor A, which depends on radio frequency, ground range, antenna/target heights, as well as the electrical properties, roughness, and homogeneity of the surface along the propagation path. The additional attenuation of the surface-wave relative to free space propagation, which is represented by the term A, is a key element of the HFSW equation.

$$L_b = \left(\frac{4\pi d}{\lambda A}\right)^2 \tag{5.4}$$

A detailed discussion of the HFSW radar equation can be found in Milsom (1997), where care is taken to resolve several sources of potential confusion regarding the definitions of antenna gains and target RCS in Eqn. (5.3). In typical HFSW radar scenarios, the antenna and target heights above the ground are less than a wavelength, and the distance separating the radar from the target is large compared to the terminal heights. In this case, the specification of appropriate antenna gains and target RCS requires two ground-related effects to be taken into account. As explained in Milsom (1997), an inconsistent set of parameter definitions in Eqn. (5.3) can inadvertently bias the predicted SNR by margins in excess of 10 dB.

The first effect is related to a doubling of the field strength of a horizontally propagating radio wave when a vertically polarized antenna is radiating above a reflecting surface. This so-called "ground-plane" effect results in a quadrupling of the power flux density, or a 6-dB increase relative to the free space antenna gain. As this effect is present on both the transmit and receive signal paths, the gain of each antenna needs to be adjusted with respect to its free space value to account for this phenomenon. As far as the radar equation is concerned, the ground-plane effect over a two-way path may either be built into the definitions of the transmit and receive antenna gains, or incorporated as a propagation factor. The former option is taken here.

For ground of finite conductivity, the doubling of the electric field strength is related to the Fresnel plane-wave reflection coefficient for vertical polarization, which tends to a value of -1 near grazing incidence for a flat reflecting surface. This causes the direct and surface-reflected (space-wave) signal paths to cancel when the terminals are located immediately above the surface, but it acts to double the field-strength of the surface-wave contribution in Eqn. (5.2). For a perfectly conducting ground-plane, the reflection coefficient has a value of 1 and the surface wave contribution disappears. In this case, doubling of the field strength is related to the constructive interference between the direct and ground-reflected paths, as described in Milsom (1997).

The second phenomenon relates to mutual coupling between the antenna and its image in the ground plane, which results in an increase of radiation resistance from its free-space value. This "ground-proximity" effect also applies to the target, but only becomes a significant factor when the antenna or target height is less than one wavelength. In the lower HF band, the wavelengths are long (up to 100 m), which often means that mutual coupling with the ground cannot be ignored. The influence of the ground-plane and ground-proximity effects on the definitions of antenna gains G_t and G_r in Eqn. (5.3) is discussed first. The modification of the free-space target RCS to account for the ground-proximity effect in the definition of σ is then discussed.

For vertically polarized antennas located above a ground-plane of finite conductivity, the effective antenna gains are given by Eqn. (5.5), where g_t and g_r are respectively defined as the transmit and receive antenna gains relative to an isotropic radiator at the same position, also known as RISP gains.[1] The factor of 4 included in both the transmit and receive antenna gains accounts for the 6-dB increase in power flux density due the ground-plane effect on the outgoing and incoming signal paths, respectively, while the terms Δ_t and Δ_r represent the losses in effective antenna gains due to the ground-proximity effect.

$$G_t = g_t \frac{4}{(1+\Delta_t)}, \quad G_r = g_r \frac{4}{(1+\Delta_r)} \tag{5.5}$$

For a vertical Hertzian-dipole above a perfect electric conductor (PEC) ground plane, the value of Δ is given by Eqn. (5.6), where $k_0 = 2\pi/\lambda$ is the radio wavenumber and h is the dipole height above the ground (Milsom 1997). The value of Δ tends to zero for a height greater than about one wavelength, whereas $\Delta = 1$ when the (transmit or receive) antenna is located immediately above the PEC surface. For a finitely conducting surface, additional loss terms enter into G_t and G_r through modified values of Δ_t and Δ_r.

$$\Delta = \left[3/(2k_0h)^2\right]\left[\sin(2k_0h)/(2k_0h) - \cos(2k_0h)\right] \tag{5.6}$$

Using the definition of (Shearman 1983), the effective RCS of a target above a reflecting surface may be related to its free space value σ' according to Eqn. (5.7), where $\Delta_x \to 0$ for target heights greater than about one wavelength, and $\Delta_x = 1$ when the target is located immediately above a PEC ground-plane. In the latter case, a 6-dB reduction in target RCS occurs due the mutual coupling between the target and its image in a PEC ground plane. The denominator $(1 + \Delta_x)$ is squared in Eqn. (5.7) because the ground proximity effect acts on the power collecting *and* reradiating properties of the target. As described for the antennas, the effect of a finitely conducting surface modifies the effective target RCS through the term Δ_x.

$$\sigma = \frac{\sigma'}{(1+\Delta_x)^2} \tag{5.7}$$

The free-space target RCS σ' is expressed in Eqn. (5.8) as the product of the RISP gain g_x of an equivalent (matched) antenna that represents the target's echoing properties in the backscatter direction, and the scattering cross-section of a (matched) isotropic element in free-space, which is equal to $\lambda^2/(4\pi)$. When using the HFSW radar equation in Eqn. (5.3), the definition of σ in Eqn. (5.7) is consistent with the antenna gain definitions in Eqn. (5.5) provided that L_b is given by Eqn. (5.4). The value of A is less then one for lossy ground and this term is also known as the Norton surface-wave attenuation function. Recall that

[1]For example, the RISP gain is 1.5 for a vertical Hertzian dipole on a PEC plane, and 1.64 for a vertical $\lambda/4$ monopole on a PEC plane. Further discussion on the use of RISP gains in preference to dBi values can be found in Milsom (1997).

this form of the HFSW radar equation applies for a monostatic system and noise-limited detection in the diffraction zone, where space-wave (and skywave) propagation to the target are considered to be absent.

$$\sigma' = g_x^2 \left(\frac{\lambda^2}{4\pi}\right) \tag{5.8}$$

As stated in (Milsom 1997), the internationally recognized definition of basic transmission loss (BTL) is the ratio of the radiated power to the power available from a matched receiving antenna, when the actual antennas are replaced by isotropic antennas at the same locations as the actual antennas. For a one-way path above a PEC ground-plane, the BTL for terminals separated by a horizontal distance d is given by ℓ_b in Eqn. (5.9) when this definition is used. Here, Δ_T and Δ_R are given by Eqn. (5.6) depending on the heights of the transmit and receive antennas, respectively. Note that when both antennas are immediately above the PEC ground-plane, we have that $\Delta_T = \Delta_R = 1$, and ℓ_b is equivalent to the free-space BTL given by $(4\pi d/\lambda)^2 = (2k_0 d)^2$.

$$\ell_b = \left(\frac{2\pi d}{\lambda}\right)^2 (1 + \Delta_T)(1 + \Delta_R) \tag{5.9}$$

Now consider the BTL on each leg of a two-way radar signal path. For the transmitter-to-target path, we have that $\Delta_T = \Delta_t$ and $\Delta_R = \Delta_x$ in Eqn. (5.9), while for the target-to-receiver path, we have that $\Delta_T = \Delta_x$ and $\Delta_R = \Delta_r$. Assuming the transmit and receive antennas experience the same loss in effective gain due to the ground-proximity effect (i.e., mutual coupling with the ground) it follows that $\Delta_t = \Delta_r$, which implies that ℓ_b given by Eqn. (5.10) is the identical for both signal paths.

$$\ell_b = \left(\frac{2\pi d}{\lambda}\right)^2 (1 + \Delta_a)(1 + \Delta_x), \quad \text{where } \Delta_a = \Delta_t = \Delta_r \tag{5.10}$$

Using the BTL formula in Eqn. (5.10), the HFSW radar equation in Eqn. (5.3) may be written in the equivalent form of Eqn. (5.11), where $l_b = \ell_b/A^2$ is the BTL for a finitely conducting ground with the term Δ_a adjusted accordingly. The term A^2 describes the surface-wave *power* loss over a one-way path. Assuming one is prepared to work in terms of RISP antenna gains g_t and g_r, it is argued in (Milsom 1997) that Eqn. (5.11) represents a more convenient form of the radar equation than Eqn. (5.3) because it employs the ITU-R definition of BTL ℓ_b in (Int. Telecomm. Union 1994), and the free-space radar cross section σ', which is independent of target height. Moreover, the term ℓ_b can be readily computed for arbitrary path geometries using the GRWAVE program of Rotheram (1981b), which is recommended by the ITU-R and automatically handles all the mutual coupling terms associated with Eqn. (5.9).

$$\frac{S}{N} = \frac{P_t g_t g_r \sigma' T 4\pi}{l_b^2 L_s \lambda^2 N_o} \tag{5.11}$$

The noise spectral density is assumed to be uniform over the radar bandwidth and given by $N_o = kT_e F_a$, where k is Boltzmann's constant (1.38×10^{-23} J/K), T_e is the effective system temperature (290 K), and F_a is the noise factor dominated by external sources. A point often overlooked is the effect of receive antenna gain on SNR in externally noise-limited environments. Most HFSW radars employ a receiving array of antenna elements with a gain pattern that is broad in azimuth. After spatial processing, the receive beam has a gain $G_r = G_e G_a$, where G_e and G_a are the antenna element and array factor gains in the target direction, respectively.

If the external noise is weakly correlated across antenna elements of the array, and its spectral density in the receiver is significantly higher than the internal (thermal) noise level, an increase in antenna element gain G_e will have little effect on the received SNR as it increases both the useful signal (target echo) and external noise by an equal amount to a good approximation. In this case, only the array factor component G_a of the receive antenna gain G_r will effectively contribute to increasing the SNR in the HFSW radar equation of Eqn. (5.3). As external noise-limited conditions often arise in practice (particularly at night in the lower HF band), caution should be exercised when attempting to increase SNR by using better-matched *receive* antenna elements.

As a sketch example, consider an HFSW radar operating at 5 MHz with an average power of 1 kW (30 dBW). Assume the product of the transmit and receive antenna gains (including ground effects) is 25 dB, and that a ship with an effective RCS of 30 dBsm is moving with a relative velocity of 15 knots at a range of 200 km from the radar. It is assumed that the target echo Doppler shift is high enough for detection to be limited by noise for a relatively calm sea surface in the absence of ionospheric clutter. Using surface-wave attenuation curves for a smooth sea surface provided in the next section, the quantity $L_b^2(\lambda^2/4\pi)$ in Eqn. (5.3) is approximately equal to 230 dB. Since reciprocity applies for a single-site system, the two-way path loss has been calculated by doubling the one-way path-loss in decibels. The received power P_r is then given by Eqn. (5.12) with all quantities expressed in decibels.

$$P_r = P_t + (G_t + G_r) + \sigma - 230 = 30 + 25 + 30 - 230 = -145 \text{ dBW} \tag{5.12}$$

Using a nighttime noise spectral density of $N_o = -155$ dBW/Hz for a coastal radar site not located near a residential area, and assuming a 100 second (20 dBs) CPI yields the received noise power of $P_n = -155 - 20 = -175$ dBW. Assuming total system losses are limited to $L_s = 12$ dB, the predicted SNR for this example is given by Eqn. (5.13).

$$S/N = P_r - P_n - L_s = -145 + 175 - 12 = 18 \text{ dB} \tag{5.13}$$

Beyond about 200 km, the noise-limited version of the HFSW radar equation may not be appropriate for operating frequencies below the F2-layer critical frequency due to the presence of ionospheric clutter. Ionospheric clutter can also be received at shorter ranges of around 100 km via E-region propagation during the day, and at night via sporadic-E, with the potential to spread over the entire target velocity search-space. In addition, high sea-states can raise the surface clutter level and increase its Doppler spectrum extent. The surface-wave path-loss is also significantly influenced by surface-roughness, particularly at higher frequencies and longer ranges. In many cases, a clutter-limited version of the radar equation is more appropriate to predict detection performance, especially for surface targets.

It is also noted that system losses at short ranges may be high in a monostatic HFSW radar due to eclipsing associated with pulsed waveform operation. As the receiver needs to remain closed for the pulse duration, the minimum target detection range may be limited to a few tens of kilometers in a monostatic system (typically 20–50 km depending on the waveform) due to the high eclipsing loss incurred at shorter ranges. Eclipsing also affects long ranges due to blanking at the end of the PRI due to the next pulse. In some cases, this may be more important because the signals are so weak, whereas at short range there is often excess signal strength for detection. Such losses are accounted for by the term L_s in a monostatic system. A bistatic configuration avoids the eclipsing issue and may exploit continuous waveforms to increase the average power for a fixed transmitter peak-power rating.

The received target-echo power P_r for a bistatic radar system is given by Eqn. (5.14), where σ is now the bistatic target RCS, $L_t = [4\pi d_t/(\lambda A_t)]^2$ is the BTL for the transmitter-to-target path of length d_t, and $L_r = [4\pi d_r/(\lambda A_r)]^2$ is the BTL for the target-to-receiver path of length d_r. The form of Eqn. (5.14) implicitly assumes surface-wave propagation on both the outgoing and incoming signal paths. The expression for the received noise power P_n is unchanged from that described for the monostatic configuration. For a given total path length $d_t + d_r$, a bistatic configuration is typically characterized by enhanced sensitivity (SNR) in regions near the transmitter or receiver.

$$P_r = \frac{P_t G_t G_r \sigma (4\pi/\lambda^2)}{L_s L_t L_r} \tag{5.14}$$

The surface clutter-limited version of the radar equation is often appropriate for targets with relative velocities less than about 15 knots. For moderate to high sea-states and ranges less than around 100 km, second-order sea clutter typically dominates external noise and ionospheric clutter during the day and night. In each transmitted radar pulse, the received echo power from the target and a colocated scattering patch of the sea surface will be amplified by roughly the same system gain, so the SCR may be expressed as Eqn. (5.15) to a first-order approximation, where σ_c is the effective RCS of the sea surface that scatters clutter into the range-azimuth and Doppler resolution cell of the target echo. The target is assumed to have a steady RCS σ and constant relative velocity, such that the useful signal integrates perfectly over the CPI.

$$\frac{S}{C} = \frac{\sigma}{\sigma_c} \tag{5.15}$$

For a monostatic radar, the effective RCS of sea clutter can be written in the form of Eqn. (5.16), where $\Delta R = c/(2B)$ is the range resolution for a radar signal bandwidth B, and $\Delta L = R\Delta\theta$ is the cross-range resolution for a receive antenna beamwidth $\Delta\theta$ at the target range R. The term $\sigma_{vv}^0(\omega_d)$ is defined as the effective cross-section per unit area of the sea surface integrated over the analysis bandwidth, which is commensurate with the Doppler frequency resolution of the radar. This term is evaluated at the target echo Doppler radian frequency ω_d.

$$\sigma_c = \sigma_{vv}^0(\omega_d)\Delta R \Delta L \tag{5.16}$$

Specifically, the quantity $\sigma_{vv}^0(\omega_d)$ is given by the convolution integral in Eqn. (5.17), where $\sigma_{vv}(\omega)$ is the normalized Doppler spectral cross section per unit area per unit rad/s bandwidth of the sea surface for vertically polarized signals that are backscattered at grazing incidence, as defined in (Barrick 1972b), and $W(\omega)$ is the power spectral density of the Doppler processing window function normalized such that the total clutter power is conserved (Maresca and Barnum, 1982). Ignoring the relatively small contribution of higher-order terms, the quantity $\sigma_{vv}(\omega) = \sigma_1(\omega) + \sigma_2(\omega)$ is the sum of a first-order clutter contribution $\sigma_1(\omega)$, consisting of two discrete spectral components known as Bragg lines, and a second-order clutter continuum $\sigma_2(\omega)$ that spreads over a band of Doppler frequencies. Theoretical expressions for $\sigma_1(\omega)$ and $\sigma_2(\omega)$ as a function of the sea-surface directional wave-height spectrum are derived in (Barrick 1972b), and these will be recalled in Section 5.3.

$$\sigma_{vv}^0(\omega_d) = \frac{1}{2}\int_{-\infty}^{\infty} \sigma_{vv}(\omega)W(\omega - \omega_d)\mathrm{d}\omega \tag{5.17}$$

Substituting the previously defined parameters into Eqn. (5.15), and the beamwidth approximation $\Delta\theta = \lambda/D$, where D is the length of the receive antenna aperture, yields the SCR expression Eqn. (5.18). The form of Eqn. (5.18) indicates that the SCR is proportional to the bandwidth-aperture product of the system, and inversely proportional to the target range. Provided the target-echo Doppler-shift does not fall exactly one of the Bragg-line frequencies, a longer CPI can in principle be used to resolve the target echo from the first-order echo, and reduce the effective scattering cross-section per unit area of the second-order sea echo by narrowing the analysis bandwidth of the Doppler bin containing the useful signal (Maresca and Barnum 1982). For a perfectly coherent target, the dependence of SCR on CPI is embedded in the term $\sigma_{vv}^0(\omega_d)$.

$$\frac{S}{C} = \frac{\sigma}{\sigma_{vv}^0(\omega_d)}\frac{2B}{c}\frac{D}{R\lambda} \tag{5.18}$$

To illustrate a simple example, consider an HFSW radar operating at 5 MHz ($\lambda = 60$ m) with a bandwidth $B = 20$ kHz, using a receive aperture of $D = 500$ m and a CPI of $T = 100$ s. Assume that a surface vessel with an effective RCS of σ dBsm is located at a range of $R = 100$ km from the radar. Inserting these parameters into Eqn. (5.18) yields a spatial resolution cell with a surface area of approximately 80 dBsm. For a scattering patch of this size, the SCR is given by Eqn. (5.19) with all quantities expressed in decibels.[2]

$$\frac{S}{C} = \sigma - \left(\sigma_{vv}^0(\omega_d) + 80\right) \tag{5.19}$$

In a surface clutter-limited environment, target blind speeds occur when the SCR is less than the detection threshold. Simulation results showing blind speeds for $S/C < 10$ dB and target RCS values of $\sigma = 30$ dBsm and $\sigma = 50$ dBsm against first- and second-order clutter using theoretical expressions of $\sigma_1(\omega)$ and $\sigma_2(\omega)$ can be found in (Maresca and Barnum 1982). As pointed out in (Maresca and Barnum 1982), a longer CPI serves both to confine the first-order echo to a narrow range of frequencies, and to reduce the level of the second-order sea echo competing for detection in the target Doppler bin.

5.1.2.3 Nominal System Performance

HFSW radar can fill an important surveillance gap for maritime nations that lies between the maximum range coverage of a coastal microwave radar and the 200 nautical mile limit of the Exclusive Economic Zone (EEZ). With respect to a land-based microwave radar, HFSW radar represents an alternative system to detect and track

[2]Using the antenna gain definitions in Eqn. (5.3), $\sigma_{vv}^0(\omega_d)$ has a value of approximately -32 dB for a fully developed sea and a semi-isotropic angular spreading function when ω_d falls on a Bragg line Doppler frequency. Depending the form of $\sigma_2(\omega)$, the value of $\sigma_{vv}^0(\omega_d)$ may be 20–40 dB (or more) below the first-order contribution at Doppler frequencies that place the target echo in the second-order clutter continuum for a CPI of $T = 100$ s.

targets at altitudes down to sea level up to a ground range of a few hundred kilometers. For low-flying aircraft and large surface-vessels, its potential range coverage can be an order of magnitude greater than that of a conventional line-of-sight radar in normal atmospheric conditions. Monostatic HFSW radars typically have a minimum detection range of tens of kilometers associated with pulse-waveform operation. This minimum range limit is complementary to the coverage of a microwave radar, which makes a compelling argument for the use of colocated (integrated) HFSW and microwave radar systems.

With respect to skywave OTH radar systems, HFSW radars are considerably smaller in size, significantly less expensive, and not as complex to operate. Moreover, the skywave OTH radar coverage is typically limited to a minimum range of about 1000 km due to the skip-zone phenomenon discussed in Chapter 2. This constraint can be a problem for geographically small nations when it is not possible to find a suitable radar site which is set back this required minimum distance from the surveillance area of interest. Another significant practical advantage with respect to skywave systems stems from the fact that HFSW radars are not reliant on the ionosphere as a propagation medium. Disturbed ionospheric conditions can at times preclude ship detection at all ranges in a skywave OTH radar system.

The relatively benign (frequency-stable) properties of surface-wave propagation are favorable for detecting slow-moving surface vessels against sea clutter. Relative to skywave systems, HFSW radars may therefore provide more reliable performance in coastal surveillance applications within the EEZ. In addition, HFSW radars are capable of more accurate target localization and tracking due to the greater certainty associated with the surface-wave propagation path. Multipath is also much less of an issue in HFSW radar, particularly at OTH ranges where the space-wave contribution is absent. This effectively eliminates the multipath track to target assignment problem, which is an issue is skywave systems. However, similar to skywave systems, HFSW radars also lack target height information. An estimation method for HFSW radar has been proposed in Howland and Clutterbuck (1997).

Having identified some of the niche characteristics of HFSW radar with regard to both line-of-sight and skywave OTH radar systems, the remainder of this section describes the nominal surveillance capabilities of an HFSW radar. The coverage and resolution of a typical system in range, azimuth, and Doppler is described first. This is followed by a summary of the nominal detection capabilities, which are often specified in terms of maximum detection range for different target classes as a function of sea-state and day or nighttime operation. Tracking performance is then discussed, particularly target location accuracy in terms of steady-state rms errors in range and azimuth.

The "instrumented" coverage area of an HFSW radar defines the region in which target detection and tracking can be attempted by the system. The maximum range is typically between 400 and 500 km. The minimum range for a monostatic system is typically between 20 and 50 km, while ranges less than that of the direct-wave path may be processed in a bistatic system. The azimuth extent typically varies from 90 to 120 degrees, depending on the transmit/receive antenna designs and the radar configuration. The azimuth coverage may be greater or smaller depending on the geography of the surveillance area. These range and azimuth boundaries yield an overall coverage area in the order of one hundred thousand square kilometers.

Resolution is often defined as the −3 dB (half-power) width of the main lobe response when the matched filter is applied without the use of a taper. The range resolution of an HFSW radar is typically not larger than 15 km ($B = 10$ kHz) and not smaller than 3 km ($B = 50$ kHz), although channel occupancy in the HF spectrum often prevents resolutions less than 5 km from being achieved in practice. The Doppler resolution may have limits of 0.001–1 Hz, corresponding to CPI lengths between 1 and 1000 s. The receive beamwidth in azimuth for a 500 m long ULA typically varies between 6 and 10 degrees, and is a function of both operating frequency and steer direction.

The maximum detection range not only depends on system characteristics but also on target type, sea-state, ionospheric conditions, and external noise. The latter two factors exhibit significant day/night dependence, while high sea-states are caused by strong winds that have been blowing with sufficient fetch and duration. Surface vessels with small relative velocities produce echoes within the second-order clutter continuum in the Doppler spectrum. Reliable detection of such targets at an acceptable false alarm rate requires the SCR to be greater than a threshold of about 10 dB. Figure 5.4 provides an indication of what might be expected as the maximum detection range for several target classes, presented in an alternative manner (Ponsford, D'Souza, Kirubarajan 2009). The specified ranges represent nominal values in favorable environmental conditions. Note that performance may be significantly worse in adverse environmental conditions.

During the daytime, detection of very large ocean-going vessels may be expected throughout most of the EEZ. Lower frequencies (3–7 MHz) are often used to reduce surface wave path-loss at longer ranges without changing the RCS scattering regime of such targets. The detection of large surface-vessels with displacements greater than about 3000 tons against second-order clutter is found to exhibit little dependence on sea-state. This is because only the first-order sea clutter components are powerful enough to mask echoes from large surface-targets. At nighttime, environmental noise has a higher level in the lower HF band and ionospheric clutter is typically more difficult to avoid. The combination of these effects can limit the maximum detection range of very large vessels at night compared to daytime values.

Frequency	3–7 MHz			3–7 MHz			3–7 MHz			7–15 MHz			7–15 MHz		
Night	280	260	240	250	210	140	190	90	-	30	-	-	120	110	90
Day	400	380	360	330	230	140	210	110	-	60	-	-	220	190	160
Sea-State	1–2	3–4	5–6	1–2	3–4	5–6	1–2	3–4	5–6	1–2	3–4	5–6	1–2	3–4	5–6
Target Class	Large cargo vessel (40–50 dBsm)			Offshore trawler (30–40 dBsm)			Mid-shore trawler (20–30 dBsm)			Go-fast boat Radial vel. > 20 kts (0–10 dBsm)			Low-flying aircraft (10–20 dBsm)		
Disturbance	Second-order sea clutter and ionospheric clutter									External noise and interference					

FIGURE 5.4 Indicative values of what may be expected of a state-of-the art HFSW radar in terms of maximum detection ranges in kilometers as a function of target class, sea-state, and day/night operation at the optimal frequency in favorable environmental conditions. Typical operating frequency intervals and dominant disturbance types are also indicated, noting that ionospheric clutter may be present beyond about 90 km and 210 km during the day and night, respectively, with the potential to influence aircraft detection when significantly spread in Doppler. The specified ranges are for practical systems similar to SECAR or SWR-503. The SECAR system is designed with an average power of 5 kW, and an effective receive aperture of about 8 wavelengths at 5 MHz.

On the other hand, maximum detection range for large and medium-sized fishing vessels against second-order clutter is significantly influenced by sea-state. The second-order clutter cross section per unit area increases with sea-state and operating frequency. In addition, the radar resolution cell area (and hence clutter RCS) is proportional to the target range. For smaller vessels with wooden or fiberglass hulls and lengths of around 20 m, it may be attempted to increase the RCS and target echo Doppler shift by increasing the operating frequency. However, these benefits are often offset by an increase in the second-order clutter level and Doppler spectrum extent. The maximum detection range for these target classes also decreases at night for the same reasons mentioned above.

Echoes from fast-moving targets such as aircraft and go-fast boats will generally have a Doppler shift high enough for detection to be against the greater of environmental noise or ionospheric clutter when the latter is significantly spread in Doppler. Go-fast boats 5–7 m in length may be detected at relatively short ranges using higher frequencies, which may be between 15 and 20 MHz or higher. This prevents the RCS from falling dramatically in the Rayleigh region at the expense of increased surface-wave path loss, particularly at ranges beyond about 60 km. Such vessels do not usually travel with sufficient speed in high sea-states to place the echoes outside of the second-order clutter continuum, which typically precludes their detection. The detection of low-flying aircraft with higher RCS at longer ranges is often limited by SNR, although high sea states can significantly increase path loss due to surface roughness, particularly at higher frequencies.

Modern HFSW radars can automatically initiate, maintain, and terminate tracks in real time. The tracker may need to process up to 1000 detections per CPI, with the capability to maintain and update 500 or more tracks simultaneously before the completion of the next CPI. For a target with approximately constant velocity, the maximum steady-state root-mean-square (rms) tracking error is typically less than 1 km in range and 1 degree in azimuth. The range rms error is typically smaller (500 m) during the day when lower spectrum congestion allows greater signal bandwidths to be used, whereas the azimuth rms error is typically greater for steer angles near to the endfire direction of the ULA, where it may approach 2 degrees.

Radar tracks are typically updated at intervals between 1 and 10 s for an aircraft target, and 50 to perhaps 250 s for surface vessels. These update rates are equal to the CPIs typically used for aircraft and ship detection, respectively. The update rates can be quicker, if overlapped or parallel CPIs are used. For an HFSW radar that floodlights the entire surveillance region with a broad transmit antenna beam, the update rates apply to all tracks established over the coverage area. Targets are declared to be present on the basis of confirmed tracks that have reached a sufficiently high confidence level. The false track rate for confirmed tracks can in certain systems be limited to less than 5 per day.

Importantly, the coverage of an HFSW radar is not occluded by the presence of small islands along the propagation path or raised topographical features (including other vessels), which can produce strong shadow regions for a ground-based microwave radar. Clearly, HFSW radar performance is also not affected by rain and hail, which can produce strong weather clutter in conventional radar systems. It has been claimed that HFSW can offer highly autonomous 24-h/day, 7-day/week operation in all weather conditions with one of the lowest operating costs per unit coverage area of all radar types (Thayaparan and MacDougall 2005).

Drawbacks of this technology relative to conventional microwave radars include the relatively poorer spatial resolution, the need for larger infrastructure in coastal areas, and a higher degree of susceptibility to adverse sea-state and ionospheric conditions.

The latter can degrade performance by increasing the level of spread-Doppler clutter, external noise, and man-made interference received by the radar. Despite the lower spatial resolution and target localization accuracy of HFSW radars relative to coastal microwave radars, their OTH surveillance properties continue to make them an attractive proposition for monitoring the EEZ. However, only a few HFSW radar installations with surveillance as the primary mission are reported to be operational worldwide, and commercial products of this kind have not yet realistically entered the marketplace.

5.1.3 Practical Applications

The United Nations Convention on the Law of the Sea (UNCLOS) grants all maritime nations sovereign rights over an area of ocean known as the Exclusive Economic Zone (EEZ). This zone extends out to 200 nautical miles from shore (approximately 370 km). In return, maritime nations are required to establish and maintain administration, law enforcement, and environmental protection over this area. The responsibility to exercise jurisdiction within the EEZ requires that vessel activity be persistently monitored in real time. No single sensor can fulfil all the requirements related to target detection and tracking, positive identification, and ascertaining the purpose of activity of both cooperative and uncooperative surface vessels continuously in real time. It is therefore of interest to consider how the unique capabilities of HFSW radar may be integrated as part of a system of sensors to assist a maritime nation's security forces to protect its sovereign rights within the EEZ.

The main focus of this section is to describe the potential role of HSWR radar as an element of an integrated maritime surveillance (IMS) system based on the fusion of data from several different sensors to provide a recognized maritime picture (RMP) of activity within the EEZ. The first part of this section describes the potential contribution of HFSW radar in establishing the RMP, while several military and civil applications of the resulting RMP are discussed in the second part of this section. The reader is referred to Ponsford, D'Souza, Kirubarajan 2009 and references therein for authoritative descriptions of the architecture and processes required to generate an effective RMP by virtue of a "layered surveillance" approach. The alternative application of HFSW radar for remote sea-state sensing is briefly discussed in the third part of this section.

5.1.3.1 Maritime Surveillance

To establish a comprehensive RMP in the EEZ, persistent monitoring of surface-vessels is needed to provide real-time information regarding target location, speed, heading, and track history, as well as positive identification and knowledge of activities for both cooperative and noncooperative vessels where possible. As every type of sensor has strengths and weaknesses, building up the RMP may be achieved through the acquisition and distribution of real-time information gathered by an integrated network of active and passive sensors, each of which offers a unique range of capabilities. In Ponsford, D'Souza, Kirubarajan 2009, integrated maritime surveillance is defined as the systematic collection of data from a range of sensors and sources and the assemblage of a picture of the position and movement of both cooperative and noncooperative vessels. The capabilities (and limitations) of different sensors that can be brought to bear for this purpose are now discussed.

In average atmospheric conditions, the microwave radar horizon is typically limited to a ground range of about 20–30 km for a platform a few tens of meters above sea level.

The range coverage may be extended to perhaps 40–60 km by elevating the radar antenna to higher ground when this is possible in coastal regions. Anomalous propagation, also known as super-refraction or "ducting," can extend the radio horizon significantly in practice, but this phenomenon is not controllable and may not be reliable. Airborne radars can provide greater coverage, but such systems are expensive to procure, operate, and maintain for continuous surveillance of the EEZ, particularly for countries with long coastlines. Maritime traffic can be monitored by coast guard sea or air patrols, albeit intermittently, and at great cost per surface area. Skywave OTH radars provide vast coverage but require very large installations, and moreover, the detection of surface vessels is susceptible to disturbed ionospheric conditions. On the other hand, satellite technology may have neither the spatial nor temporal resolution to satisfy the EEZ surveillance requirement when considered on their own.

Active sensors have the advantage of being able to provide the location and track of a target without requiring cooperation from vessel. Although a degree of target classification may also be possible in certain cases, a significant drawback of active sensors such as radar is that they do not provide vessel identification. Positive identification may be obtained through self-reporting from a cooperative vessel, from direct observation by sea or air patrols, or from passive sensors, which rely on vessels broadcasting their location either intentionally or unintentionally. Intentional systems include automated identification systems (AIS), which broadcast the vessels identification, speed, heading, rate of turn, and other information at VHF with a nominal range of about 40 km. As mentioned in Ponsford, D'Souza, Kirubarajan 2009, unintentional methods involve locating the whereabouts of a vessel based on the interception of RF emissions.

HFSW radar can provide inexpensive surveillance of the EEZ with the capability to detect and track surface targets continuously in all types of weather. This quite unique capability for persistent surveillance over broad areas is an important attribute that gives HFSW radar a prominent role in an IMS system. Specifically, an HFSW radar can provide a background layer of current and past activity in terms of track data, onto which data or information from other active or passive sensors can be associated and mapped. For example, HFSW radar tracks may be associated with positively identified vessels based on AIS data (Ponsford, D'Souza, Kirubarajan 2009). Where inconsistencies occur, such as a radar track that does not have associated AIS data, or vice versa, reconnaissance assets can be deployed to investigate.

Track data generated by an HFSW radar is sufficiently accurate to vector-in higher precision assets (with relatively more circumscribed coverage) to positively identify contacts made in areas where suspicious activity has been detected. HFSW radar can also be used to broadly classify detected targets in terms of size (radar cross section), track history, heading, and speed. By providing current locations and tracks of unidentified contacts, HFSW radar can improve the usage and effectiveness of reconnaissance resources, such as patrol boats and aircraft. The significant contribution that HFSW radar can make to an RMP in the aforementioned respects is currently of great interest to maritime nations that are required to provide law enforcement and environmental protection within the EEZ.

5.1.3.2 Military and Civil Uses

In light of the previous discussion on maritime surveillance, alternative perspectives for airborne target detection in a military context are given here. The first HFSW radar programs with military applications in mind were sponsored by DARPA and the US

Navy in 1967 for missile defence (Barrick 2003). Missile detection represents a very challenging problem for HFSW radar because sea-skimming missiles have very low RCS at the operating frequencies needed for effective surface-wave propagation at OTH ranges. Frequencies approaching 20 MHz may be required, and missile detection ranges that are perhaps only 40 percent greater at best than ground-based microwave radars have ever been projected under optimal conditions (Barrick 2003).

HFSW radar is perceived to provide longer range aircraft detection performance than microwave radar for low-flying targets, but this early-warning advantage has to date not been considered sufficient or important enough to warrant significant investment in HFSW radar technology for this application. Although ground-based microwave radars have a relatively shorter range coverage for low-flying aircraft, they can provide more information and better accuracy with faster updates. The last HFSW radar program in the United States was terminated in about 1989 without any system entering into operational use. Raytheon in the then transferred their HFSW interests to their Canadian counterpart.

Detection and tracking of ships at OTH ranges up to the EEZ limit is the surveillance application most suited to HFSW radar. This is because the RCS of ocean-going vessels is larger than for other military targets. In addition, the target motion is often relatively steady, which allows for effective Doppler processing using long CPI. Although HFSW radars have detected and tracked large vessels to ranges of more than 400 km, the expense of such systems and the need for large areas of coastal real estate appears to have been problematic enough to keep them out of mainstream use. Whether HFSW comes into vigorous operational use against military or other hard-target applications ultimately depends on its perceived cost versus benefit ratio (Barrick 2003).

In addition to military uses, HFSW radars contribute to a number of civil applications. This includes the protection of offshore assets such oil and gas production platforms, where threats can also arise from natural hazards such as icebergs (Sevgi, Ponsford, and Chan 2001). Other applications include search and rescue of vessels in distress, and to detect/deter unauthorized activities such as people smuggling, drug trafficking, illegal fishing, and the dumping of pollutants (Anderson et al. 2003). An emerging application relates to the protection of passenger ships and merchant vessels from pirates on high-risk shipping lanes (Guinvarch et al. 2006).

5.1.3.3 Remote Sensing

In remote-sensing applications, HF radar signal echoes scattered from the ocean surface may be interpreted to estimate the directional wave-height spectrum of the sea, as well as to map surface-currents and winds in both speed and direction. In essence, surface-currents may be estimated using Doppler-shift measurements of the first-order echoes backscattered from the ocean surface (Essen, Gurgel, and Schlick 2000), while a measurement of the second-order continuum is needed to estimate the directional wave-height spectrum by inversion (Lipa 1977). Wind measurement requires a combination of first- and second-order information and the use of model-fitting techniques based on the assumption of locally wind-driven waves (Wyatt et al. 2006).

The accuracy of estimated ocean surface and wind-field parameters depends on many factors, including the radar system resolution in range, azimuth, and Doppler, the spatial and temporal variability of the measured parameters in the interrogated resolution cell, the useful signal-to-interference plus noise ratio, the presence of unwanted hard-target echoes due to ships (Gurgel and Schlick 2005), and the accuracy of the underlying model assumptions, which can start to break down for particular combinations of operating frequency and wave-height spectra or in areas of shallow water (Lipa and Barrick 1986). A network of HFSW radars with overlapped coverage is typically used to improve estimation accuracy and resolve ambiguities for surface-current and wind-field parameters.

A large number of experimental and commercial HFSW radar systems have been deployed world wide for remote-sensing purposes. Examples include the coastal ocean dynamics application radar (CODAR) or Seasonde (Lipa and Barrick 1983) and the Wellen Radar (WERA) HF radar (Gurgel et al. 1999). The application of HF-radar techniques to oceanography is beyond the scope of this text, but the reader is referred to Skolnik (2008b) and references therein for a more detailed discussion of this subject. It is noteworthy that there has been an increasing trend to exploit the proliferation of oceanographic HFSW radars for dual-use applications, i.e., remote sensing and ship detection.

5.2 Propagation Mechanism

In 1907, the German mathematical physicist Johann Zenneck analyzed the solution of Maxwell's equations for a vertically-polarized plane wave in the presence of a planar boundary separating free space and a half space of finite conductivity (Zenneck 1907). In addition to the familiar space-wave propagation modes (direct and reflected waves), it was found that a propagation mode with the characteristics of a surface-wave is also present in the solution of Maxwell's equations when boundary conditions for a surface of finite conductivity are applied. In practice, the plane-wave scenario analyzed by Zenneck would require a hypothetical source of infinite extent to generate near the surface. Nevertheless, such results were of significant theoretical interest and stimulated further development of the theory on surface waves for more realistic radio-wave sources.

As the presence of the ionosphere was not known (or proven) at the time of Marconi's famous wireless telegraphy experiment in December 1901, scientists first attempted to explain the reception of radio waves across the Atlantic in terms of the surface-wave mode. Even though this phenomenon did not correctly explain Marconi's observations, Zenneck's solution was indeed compatible with the existence of a mode that propagates in a region that follows the space above a plane (or curved) surface of finite conductivity to allow the reception of radio signals beyond the line of sight. The understanding of this propagation mechanisms is not only of interest to the mathematical physicist, but is also of great practical value to radio systems operating in the HF, MF, LF, and VLF bands for applications such as broadcasting, communications, and radar.

The production of significant surface wave field strengths in the vicinity of a finite-conductivity surface at operationally useful ranges applies only for vertical polarization. The boundary condition effectively precludes horizontally polarized radio waves from propagating by this physical mechanism in a region near the surface due to the much higher attenuation experienced by such signals. In reality, the energy of the surface-wave is not confined to flow in a well-localized region immediately above the surface. Despite the nomenclature suggesting otherwise, a large portion of the surface-wave energy is contained at significant heights above the air-ground interface. The action of a finitely conducting surface is such that at some distance from the radiating source the distribution of field strength with height extends all the way down to the space immediately above the interface.

The mathematics describing surface-wave propagation is complex in the general case, and not easy to understand without delving into cumbersome equations. The purposes of this section is to describe the fundamental principles and essential features of the surface-wave mode. One of the the most interesting characteristics of surface-wave propagation is the path-loss descriptor. The behavior of the surface-wave attenuation function has been studied for a variety of different problems. Figure 5.5 illustrates some of the main aspects that need to be considered in the problem formulation. This includes defining the source of radio waves, the media electrical properties, relative terminal locations, as well as the interface and path characteristics.

Depending on the specific scenario considered and approximations made, several mathematical models of the surface-wave field strength have been derived. The first formula for a flat Earth is attributed to Sommerfeld (1909), but the first manageable analytical solution to the problem was developed by Norton (1937), who provided an engineering approximation to Sommerfeld's work. Path-loss formulas for a spherical Earth with homogenous surface conductivity were later contributed by van der Pol and Bremmer

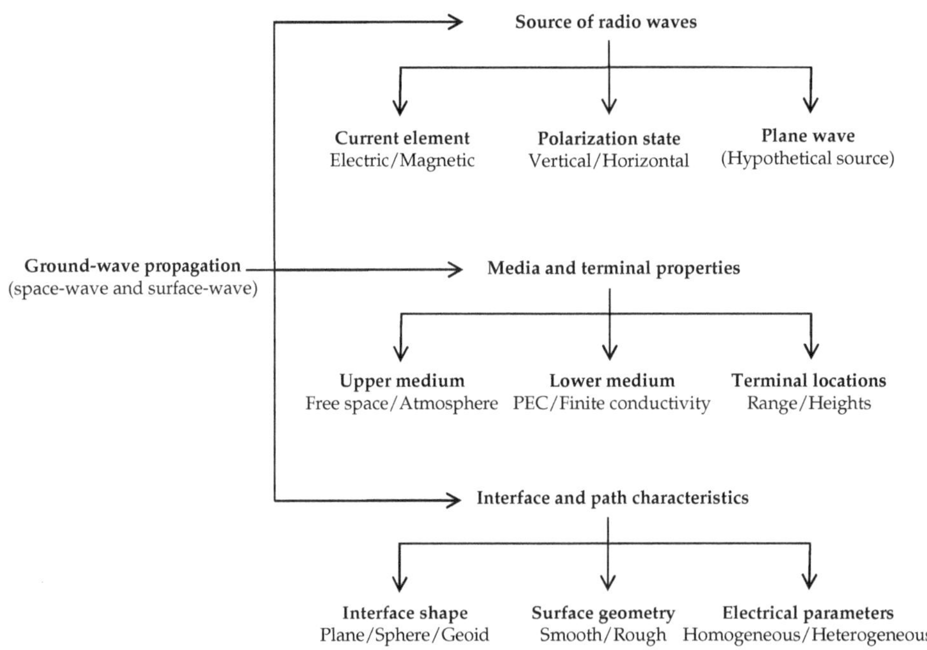

FIGURE 5.5 The theory of ground-wave propagation has been developed for a wide variety of different problems broadly defined in terms of the radio wave source, media and terminal properties, as well as interface and path characteristics. For each considered scenario, the analysis often additionally depends on a number of physical assumptions and mathematical approximations. The categories listed under each branch serve to identify some of the elements that may be used to distinguish between different problem formulations. This simple breakdown is not intended to capture the entire taxonomy of scenarios investigated in the literature, or the subtle details of the problems addressed, and should therefore not be interpreted as an exhaustive indication of the conditions under which results have been derived. © Commonwealth of Australia 2011.

(1937a), Norton (1941), and Fock (1945). Berry and Chrisman (1966) provided a Fortran implementation of Norton's equations for a homogeneous spherical Earth. Later models incorporated sea-surface roughness (Barrick 1972a), tropospheric refraction (Rotheram 1981b), and a piecewise implementation of Norton's equations for propagation over irregular terrain with inhomogeneous conductivity (Ott 1992).

This section is divided into the three parts. The first provides a basic introduction to surface wave propagation for vertically polarized signals assuming the terminals are located in free space above a smooth homogeneous surface of finite conductivity. The plane and spherical interfaces are considered to represent propagation over the Earth at short and long distances, respectively. The second part describes the effect of tropospheric refraction on surface wave propagation in the HF band and discusses the key features of the ground-wave propagation model known as GRWAVE (Rotheram 1981b), which is currently the ITU-R standard. The final part of this section explains the influence of surface roughness and heterogeneity on path-loss when propagation occurs in high sea-states or over mixed land-sea paths, respectively.

5.2.1 Short and Long Distances

The initial work of Zenneck on ground-wave propagation was subsequently extended to allow for more practical sources of radio waves by considering the fields arising due a short vertical current element, which may be regarded as a point source. The characteristics of the electric and magnetic fields set up by the spherical wave expanding out from a vertical current element situated in free space above a plane homogeneous surface of finite conductivity will be described first for the case of ground-wave propagation to a receiving point located above the boundary separating the two media. The assumption of a plane interface may be considered appropriate for propagation over the Earth's surface provided the transmit and receive terminals are separated by a relatively short distance, which will be quantified in due course.

At longer ranges, when the heights of the transmitter and receiver are such that two terminals are not in the line of sight, it becomes necessary to account for diffraction over a curved Earth to properly model the fields at the receiving point. In this case, the fields produced by ground-wave propagation exist purely due to the surface-wave mode. A significant step toward the development of propagation models appropriate for longer ranges involved the replacement of the plane interface with a smooth homogeneous spherical surface of finite conductivity. Attenuation functions for the fields set up by a vertical current element above a finitely conducting Earth surface with realistic ground constants will be described following the plane-interface analysis. Further refinements to the ground-wave propagation model, which include the effects of tropospheric refraction, surface roughness, and heterogeneous ground properties, will be incorporated successively in the following sections.

5.2.1.1 Plane Surface

With reference to Figure 5.6, Sommerfeld (1909) first derived expressions for the electric and magnetic field-strength components arising at point "R" due to a short vertical current element (dipole moment) radiating from point "T," with both terminals situated in free space above a plane homogeneous surface of finite conductivity. In general, the resultant field at point R is the sum of a space-wave contribution, due to the interference between the direct and reflected signals, plus a surface-wave contribution that results when Maxwell's equations are required to satisfy boundary conditions for

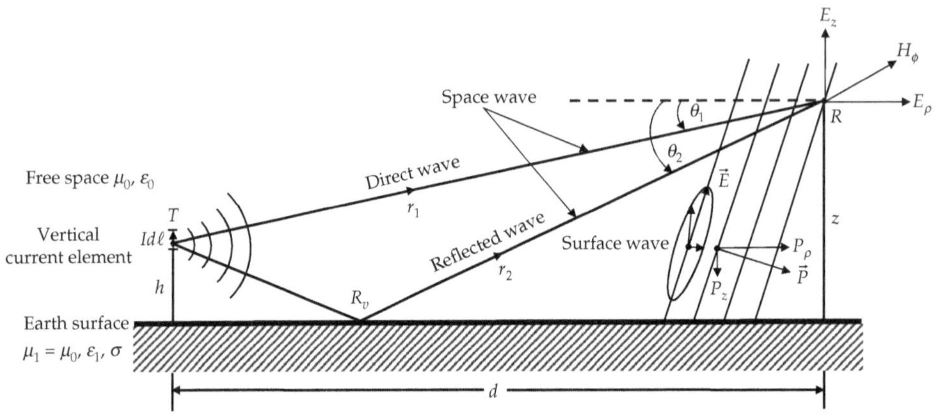

FIGURE 5.6 Notional illustration of ground-wave propagation above a flat homogeneous surface that separates two media with different electrical properties. The electric and magnetic field-strength components at point "R" result from a short vertical current element radiating at point "T." The diagram shows a side-on view of the space-wave contribution (direct and reflected paths) in addition to the surface-wave mode. Both terminals are assumed to be situated in free space with electrical constants μ_0, ε_0, and $\sigma_0 = 0$. The media in the lower half-space has electrical parameters $\mu_1 = \mu_0$ and ε_1 with finite conductivity σ. A cylindrical coordinate system (z, ρ, ϕ) is used, where z is the displacement in the vertical direction (height), ρ is radial distance from the origin in the plane of the interface between the two media, and ϕ is the angle in this plane with respect to an arbitrary reference direction. Defining the origin as the point on the surface directly below the transmitter, the points T and R have cylindrical coordinates $(h, 0, 0)$ and (z, d, ϕ), respectively, where d is the "ground" distance separating the terminals in the radial direction ϕ.

a lossy (i.e., imperfectly conducting) ground. The approach used by Sommerfeld was to represent the spherical wave generated by the dipole moment $Id\ell$ (i.e., the product of source current and length), as a spectrum of cylindrical waves. A detailed account of Sommerfeld's original formulation, as well as the alternative formulation of Weyl for a homogenous flat ground of finite conductivity, can be found in Maclean and Wu 1993.

An approximate expression for the fields, compatible with the formulations of Sommerfeld and Weyl, led in turn to the engineering solution of Norton (1936, 1937). These approximate expressions for the vertical and radial electric field-strength components, denoted by E_z and E_ρ, respectively, and the horizontal component of magnetic field-strength H_ϕ, are stated without proof here. The derivation of these formulas, including a description of the approximations made, can be found in (Maclean and Wu 1993). Specifically, the vertically polarized electric field component E_z at point R is given by Eqn. (5.20). Space-wave propagation corresponds to the first two terms in the square brackets, which model the direct and reflected paths, respectively. The third term represents the surface-wave mode.

$$E_z = \frac{jk_0 Id\ell Z_0}{4\pi}\left[\cos^2\theta_1 \frac{\exp(-jk_0 r_1)}{r_1} + R_v \cos^2\theta_2 \frac{\exp(-jk_0 r_2)}{r_2} + (1 - R_v)F(w)\cos^2\theta_2 \frac{\exp(-jk_0 r_2)}{r_2}\right] \tag{5.20}$$

The quantities k_0, Z_0, R_v, $F(w)$, and w will be defined after the formulas for the other two field components at point "R" are given. The radial component of electric field strength E_ρ is given by Eqn. (5.21). Under the stated assumptions, a vertical current element does not produce a horizontal (into the page) component of electric field, i.e., $E_\phi = 0$. Any horizontal polarization component of the electric field that may be radiated by a practical (vertically disposed) antenna at "T" will have a surface-wave contribution that is heavily attenuated at "R" for realistic surfaces and ground distances of interest to an HFSW radar.

$$E_\rho = \frac{-jk_0 I d\ell Z_0}{4\pi}\left[\cos\theta_1 \sin\theta_1 \frac{\exp(-jk_0 r_1)}{r_1} + R_v \cos\theta_2 \sin\theta_2 \frac{\exp(-jk_0 r_2)}{r_2} - (1 - R_v)\Delta_0 F(w)\cos\theta_2 \frac{\exp(-jk_0 r_2)}{r_2}\right] \tag{5.21}$$

The horizontal magnetic-field component H_ϕ is given by Eqn. (5.22). Although receivers can incorporate ferrite rods that are sensitive to the magnetic field of a radio wave, the planning and design of most radio systems is normally based on predictions of the electric field. For the surface-wave mode and "ground" of relatively high conductivity, such as the sea, H_ϕ is approximately related to the major electric field component E_z by $H_\phi = -E_z/Z_0$, where Z_0 is the impedance of free space. As stated in Barclay 2003, we shall not consider H_ϕ further as it suffices to design in terms of the electric field.

$$H_\phi = \frac{-jk_0 I d\ell}{4\pi}\left[\cos\theta_1 \frac{\exp(-jk_0 r_1)}{r_1} + R_v \cos\theta_2 \frac{\exp(-jk_0 r_2)}{r_2} - (1 - R_v) F(w)\cos\theta_2 \frac{\exp(-jk_0 r_2)}{r_2}\right] \tag{5.22}$$

From the geometry of Figure 5.6, it is straightforward to show that the lengths of the direct and reflected ray paths from T to R are given by $r_1 = [d^2 + (z - h)^2]^{1/2}$, and $r_2 = [d^2 + (z + h)^2]^{1/2}$, respectively. It is also apparent that $\cos\theta_1 = d/r_1$, $\cos\theta_2 = d/r_2$, $\sin\theta_1 = (z - h)/r_1$, and $\sin\theta_2 = (z + h)/r_2$. The term k_0 is the free-space wavenumber, given in Eqn. (5.23), where λ and f are the wavelength and frequency of the radio wave, respectively. Here, $\omega = 2\pi f$ is the radian frequency, and $c = (\mu_0\varepsilon_0)^{-1/2}$ is the speed of light in a vacuum. The permeability and permittivity of free space are $\mu_0 = 4\pi \times 10^{-7}$ T$\cdot$m /A and $\varepsilon_0 = 8.85 \times 10^{-12}$ C^2/N$\cdot$m^2.

$$k_0 = \frac{2\pi}{\lambda} = \frac{\omega}{c} = 2\pi f(\mu_0\varepsilon_0)^{1/2} \tag{5.23}$$

To define the remaining terms, we recall that a surface of finite conductivity has a complex permittivity given by ε_1 in Eqn. (5.24), where ε_r is the relative permittivity of the medium in the lower half-space, and σ is the conductivity measured in S/m. The minus sign preceding the imaginary part of ε_1 is associated with the convention $e^{j\omega t}$ to represent the harmonic variation of the radio wave. Use of $e^{-j\omega t}$ is consistent with a positive sign preceding the imaginary part of ε_1 in Eqn. (5.24), as described in (Barrick 1971). The former convention corresponding to Eqn. (5.24) will be adopted here.

$$\varepsilon_1 = \varepsilon_r\varepsilon_0 - j\sigma/\omega \tag{5.24}$$

In the following, it is useful to define the term $v = k_0/k_1$, where $k_1 = \omega(\mu_1\varepsilon_1)^{1/2}$ is of analogous form to k_0 using the complex permittivity ε_1 given by Eqn. (5.24). As $\mu_1 = \mu_0$ is assumed for the problem considered, it is readily shown that $v = (\varepsilon_0/\varepsilon_1)^{1/2}$. From Eqn. (5.24), v can be written in the form of Eqn. (5.25), where $\chi = \sigma/(\omega\varepsilon_0)$.

$$v = \left(\frac{\varepsilon_0}{\varepsilon_1}\right)^{1/2} = \frac{1}{(\varepsilon_r - j\chi)^{1/2}} \tag{5.25}$$

The impedance of free space is given by $Z_0 = (\mu_0/\varepsilon_0)^{1/2} = 120\pi$ Ohms, while the normalized surface impedance is defined as $\Delta_0 = Z_s/Z_0$, where Z_s is the impedance of the surface in Ohms. For a smooth interface, the normalized surface impedance can be expressed in the form of Eqn. (5.26), as described in (Wait 1964) and (Maclean and Wu 1993).

$$\Delta_0 = \frac{1}{Z_0}\left(\frac{\mu_1}{\varepsilon_1}\right)^{1/2}\left(1 - \frac{\varepsilon_0\mu_0}{\varepsilon_1\mu_1}\cos^2\theta_2\right)^{1/2} \tag{5.26}$$

In the considered scenario, where $\mu_1 = \mu_0$, it is straightforward to show that Δ_0 can be written as in Eqn. (5.27). For surfaces of relatively high conductivity, such as the sea, the magnitude of v is small compared to unity. In such situations, the normalized surface impedance is well approximated by $\Delta_0 \simeq v$. We shall make use of this expression to describe certain characteristics of surface wave propagation later in this section.

$$\Delta_0 = v[1 - v^2\cos^2\theta_2]^{1/2} \tag{5.27}$$

Referring back to the electric field formulas in Eqn. (5.20) and Eqn. (5.21), the term R_v is defined as the plane-wave complex Fresnel reflection coefficient for vertical polarization and is given by Eqn. (5.28). The expression on the right hand side of Eqn. (5.28) is obtained by using Eqn. (5.27) and Eqn. (5.25) to write Δ_0 in terms of the electrical constants of the surface. The form of this expression is identical to that specified for the case of parallel propagation in Chapter 6 of Barclay (2003).

$$R_v = \frac{\sin\theta_2 - \Delta_0}{\sin\theta_2 + \Delta_0} = \frac{(\varepsilon_r - j\chi)\sin\theta_2 - [(\varepsilon_r - j\chi) - \cos^2\theta_2]^{1/2}}{(\varepsilon_r - j\chi)\sin\theta_2 + [(\varepsilon_r - j\chi) - \cos^2\theta_2]^{1/2}} \tag{5.28}$$

It transpires from the derivation reported in (Maclean and Wu 1993) that the argument w of the attenuation function $F(w)$ is given by Eqn. (5.29). On the right hand side of Eqn. (5.29), the complex scalar w is written in terms of R_v and v by first using Eqn. (5.28) to make the substitution $(\sin\theta_2 + \Delta_0)^2 = 4\Delta_0^2/(1 - R_v)^2$, and then substituting the expression for Δ_0 given in Eqn. (5.27).

$$w = \frac{-jk_0r_2}{2\cos^2\theta_2}(\sin\theta_2 + \Delta_0)^2 = \frac{-j2k_0r_2v^2}{\cos^2\theta_2(1 - R_v)^2}(1 - v^2\cos^2\theta_2) \tag{5.29}$$

Finally, the surface-wave attenuation function $F(w)$ is given by the expression in Eqn. (5.30), where erfc($\cdot$) is the complementary error function, see Abramowitz and Stegan (1964). This completely defines the approximate formulas for the vertical and radial electric field components in Eqn. (5.20) and Eqn. (5.21), respectively. We now turn our attention to describing the influence of the space-wave and surface-wave contributions in scenarios of practical interest to HFSW radar.

$$F(w) = 1 - j\sqrt{\pi w}\exp(-w)\text{erfc}(j\sqrt{w}) \tag{5.30}$$

In the special case of both terminals located immediately above the plane interface, we have that $\theta_1 = \theta_2 = 0$, and from Eqn. (5.28), $R_v = -1$ for a finite-conductivity surface. It follows that the direct and surface-reflected waves in Eqn. (5.20) act to cancel one another, such that the space-wave contribution becomes negligible at "R." This leaves the surface-wave component as the dominant contribution to the resultant field. This scenario is relevant to a coastal HFSW radar and the ship detection application. In this case, the fields E_z and E_ρ can be approximated by the relatively simpler expressions in Eqn. (5.31), which are solely due to the surface-wave.

$$E_z = j60k_0 I d\ell F(w)\frac{\exp(-jk_0 r_2)}{r_2}, \quad E_\rho = j60k_0 I d\ell \Delta_0 F(w)\frac{\exp(-jk_0 r_2)}{r_2} \tag{5.31}$$

As the magnetic field component of the surface-wave is horizontal, the vertical and radial electric field components of the surface-wave are associated with forward and downward Poynting vector components, respectively, as illustrated in Figure 5.6. For ground-based transmit and receive terminals, the argument w of the attenuation function $F(w)$ can also be simplified. Specifically, it is observed from Eqn. (5.29) that w reduces to Eqn. (5.32) for $R_v = -1$ and $\theta_2 = 0$.

$$w = \frac{-jk_0 r_2 v^2}{2}(1 - v^2) \tag{5.32}$$

Using Eqn. (5.31) and Eqn. (5.27), the ratio of radial to vertical surface-wave electric field components can be approximated by Eqn. (5.33) for a surface of high conductivity since $\Delta_0 \simeq v$ when the magnitude of v is much smaller than unity. As v is complex in general (with nonzero real and imaginary parts) for a finite-conductivity surface, the resultant surface-wave electric field is elliptically polarized due to the relative phase shift between E_z and E_ρ. The relative magnitudes and phases of E_z and E_ρ are such that the major axis of the ellipse is tilted in the direction of propagation, as depicted in Figure 5.6.

$$\frac{E_\rho}{E_z} \simeq v = \frac{1}{[\varepsilon_r - j\sigma/(\omega\varepsilon_0)]^{1/2}} \tag{5.33}$$

Some representative values of $\varepsilon_r - j\chi$ and $|v|$ encountered for various surface types at radio frequencies of 3 and 10 MHz are listed in Table 5.1. Specific data on the electrical properties of the Earth's surface in different geographical regions can be found in (Int. Telecomm. Union 1999b). It is clear from Eqn. (5.33) that the ellipticity of the surface-wave polarization and forward tilt angle of the propagating wavefront depend on the electrical properties of the lower medium (ε_r, σ) and the radio frequency ω. For a highly conducting surface, such as seawater, $|v|$ is very small in the HF band, so the resultant electric field is almost linearly polarized and the wavefront is tilted only slightly in the forward direction.

Note that the forward component of the Poynting vector is associated with the vertical component of the electric field and energy flow from T to R along the surface. The downward component of the Poynting vector is associated with the radial component of the electric field and energy flow into the surface. The downward component of the

Surface Type	$\varepsilon_r - j\chi$, 3 MHz	$\lvert v \rvert$, 3 MHz	$\varepsilon_r - j\chi$, 10 MHz	$\lvert v \rvert$, 10 MHz
Atlantic ocean ($\sigma = 4.0$, $\varepsilon_r = 80$)	$80 - j24000$	0.0065	$80 - j7200$	0.012
Fresh-water lake ($\sigma = 10^{-1}$, $\varepsilon_r = 80$)	$80 - j600$	0.04	$80 - j180$	0.07
Good moist soil ($\sigma = 10^{-2}$, $\varepsilon_r = 10$)	$10 - j60$	0.13	$10 - j18$	0.22
Poor dry ground ($\sigma = 10^{-3}$, $\varepsilon_r = 5$)	$5 - j6$	0.36	$5 - j1.8$	0.43

TABLE 5.1 Indicative values of complex dielectric permittivity for different surface types at two frequencies in the lower HF band. The conductivity σ is expressed in S/m, while ε_r is the relative permittivity. The numbers listed should be interpreted as representative values. For system planning purposes, data on specific ground types for particular paths of interest may be found using the geographic conductivity maps published in the ITU-R Recommendation P.832-2 (1999).

Poynting vector owes its presence to currents induced in the lower medium and arises in practice because the Earth's surface is not a perfect reflector. The downward component of the Poynting vector causes a portion of radio wave energy to be dissipated in the ground as the surface wave progresses. It is this absorption of radio wave energy in ground of finite conductivity that causes the power of the surface-wave to decay more rapidly with distance than the inverse-square law describing free-space propagation.

For a fixed radio frequency, a surface of higher conductivity results in a relatively smaller radial component of electric field and more energy returned to the propagating vertical field, which allows energy to flow along the surface. On the other hand, for a surface of low conductivity, such as poor dry soil, the relatively larger radial component of electric field causes the ellipticity of polarization and wavefront tilt angle to increase, which leads to greater surface-wave attenuation. It is evident from Eqn. (5.33) that similar arguments apply when the radio frequency is changed for a given (finite-conductivity) surface. The additional attenuation of the surface-wave vertical electric fieldstrength component with respect to free-space propagation, where the field strength decays inversely with distance, is represented by the term $F(w)$ for terminals separated by more than a few wavelengths in the Sommerfeld-Norton theory.

For practical purposes, it is useful to plot curves of the magnitude of the vertical electric field strength $|E_z|$ in terms of a reference radiator with well-known properties. This allows the absolute value of E_z to be readily scaled for a particular system design with a different antenna gain and transmit power relative to the reference radiator. According to the formula for E_z in Eqn. (5.31), the vertical field strength depends on the current-length product of the Hertzian vertical dipole. The ITU-R standard is to select $Id\ell = 5\lambda/(2\pi)$, such that characteristics of the Hertzian vertical dipole are identical to those of a short vertical monopole radiating 1 kW. Over a perfectly conducting plane, such a monopole will generate a field strength of 300 mV/m at a distance of 1 km along the plane. Hence, for this reference radiator, the surface-wave field strength $|E_z|$ (mV/m) established at a distance d (km) along the plane for a radiated power of P (kW) is given by Eqn. (5.34), where $\mathcal{F} = |F(w)|$.

$$|E_z| = \frac{300}{d}\sqrt{P}\,\mathcal{F} \tag{5.34}$$

If P is defined as the average power in kW, then $|E_z|$ is the root-mean-square (rms) value of the electric field strength in mV/m. At ranges within a few wavelengths from the source, $|F(w)| \simeq 1$, noting that the attenuation function $F(w) = 1$ at $w = 0$. At these close ranges, the field strength decays approximately as $1/d$ (inversely with distance), similarly to the free-space field. Beyond a few wavelengths from the source, it can be shown that $F(w)$ follows $-1/(2w)$. It is evident from Eqn. (5.32), that $|F(w)| \propto 1/d$ at these longer ranges. In other words, the field strength makes a transition into a regime where it becomes inversely proportional to the *square* of distance at ranges greater than a few wavelengths, where $\mathcal{F} \propto 1/d$ in Eqn. (5.34).

For propagation over the Earth, this attenuation regime remains valid for relatively short distances and small terminal heights, where the Sommerfeld-Norton theory can be applied. More specifically, the theory is restricted to ranges less than approximately $10\lambda^{1/3}$ km and terminal heights below about $35\lambda^{2/3}$ m, where λ is the radio wavelength in meters (Barclay 2003). For example, between frequencies of 3 and 10 MHz ($\lambda = 30 - 100$ m), the maximum range at which the theory remains valid varies between 30 and 45 km, while the maximum terminal height varies between 340 and 750 m.

Although the fall off in surface-wave field strength in the regime where $\mathcal{F} \propto 1/d$ is substantially greater than that for space-wave propagation, vertically polarized signals in the HF and lower frequency bands can nevertheless produce appreciable field strengths at useful ranges to support the effective operation of broadcasting, communications, and radar systems based on the surface-wave mode. A convenient engineering approximation for the plane-earth attenuation factor $\mathcal{F}$ is given by the empirical formula in Eqn. (5.35) due to Norton (1935), where p is called the numerical distance and b is the phase constant.

$$\mathcal{F} = \frac{2 + 0.3p}{2 + p + 0.6p^2} - \left(\frac{p}{2}\right)^{1/2} \exp(-5p/8) \sin b \tag{5.35}$$

For vertical polarization, the numerical distance and phase constant are given by the approximate expressions in Eqn. (5.36), where $x = 60\sigma\lambda$. Recall that λ is the wavelength (m), σ is the ground conductivity (S/m), and ε_r is the relative permittivity of the ground. It is stated in (Norton 1935) that use of Eqn. (5.35) is restricted to $b < 30°$, i.e., for propagation over ground of average conductivity at frequencies less than about 10 MHz. Such conditions apply for a shore-based HFSW radar with coverage over the sea. Note that $\mathcal{F} = 1$ for $p = 0$, and $\mathcal{F} \to 1/(2p)$ as p tends to infinity.

$$p = \frac{\pi d}{\lambda x} \cos b, \quad b = \tan^{-1}\left(\frac{\varepsilon_r + 1}{x}\right) \tag{5.36}$$

The point of transition from inverse-distance attenuation near to the source to the regime where the field strength becomes inversely proportional to the *square* of distance for large p depends on terminal separation d and the electrical properties of ground. It will be shown that this transition occurs at further ranges from the source for a highly conductive ground. According to Norton (1941), the formula in Eqn. (5.35) may be applied for the case of a spherical earth with small error out to a distance of about $80/f^{1/3}$ km, where f is the frequency in MHz. This expression is approximately equivalent to the distance $10\lambda^{1/3}$ km, where λ is in meters.

When both the transmitting and receiving antennas are elevated by more than $35\lambda^{2/3}$ m above a plane surface, the influence of the surface wave contribution becomes minimal and may often be neglected such that the resultant field strength at R is mainly due to space-wave propagation, i.e., the sum of the direct and ground reflected waves. In such situations, the use of classical ray theory or geometric optics to calculate the resultant field is known to provide acceptable results. In the HF band, a possible example of where this may be appropriate is the case of high-altitude aircraft-to-aircraft links, where both terminals are located many wavelengths above the ground. For systems operating at VHF and higher frequencies, the influence of the surface-wave mode can typically be ignored for most practical applications.

5.2.1.2 Spherical Surface

To derive more realistic models of surface-wave propagation over the Earth's surface at longer ranges for two terminals that are not in the line of sight, the plane interface needs to be replaced by a spherical surface, as illustrated in Figure 5.7. Formulas for the electric and magnetic field intensities set up at point P due to a vertical current element $Id\ell$, with both terminals situated in free space above a *perfectly conducting* sphere, were originally derived by Watson (1919). To account for more realistic Earth surface characteristics, van

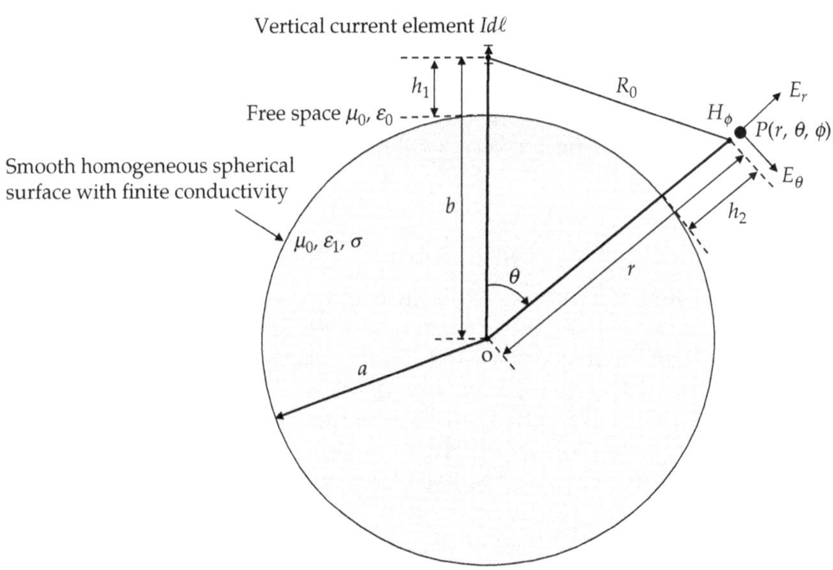

FIGURE 5.7 Geometry for ground-wave propagation above a smooth homogeneous spherical surface of finite conductivity. A short vertical current element $Id\ell$ is situated in free space along the $\theta = 0$ axis, at a height h_1 (radius b) above a sphere of radius a. Defining the origin at the center of the sphere (point O), the receiving point P is situated at a height h_2 (radius r) above the surface with spherical coordinates (r, θ, ϕ). The length of the chord between the vertical current element and the receiving point is R_0. © Commonwealth of Australia 2011.

der Pol and Bremmer calculated the fields resulting for a finitely conducting sphere in a collection of highly mathematical works [(van der Pol and Bremmer 1937a), (van der Pol and Bremmer 1937b), (van der Pol and Bremmer 1938), (van der Pol and Bremmer 1939)]. A summary of the theory used to address what was regarded as one of the most challenging problems in mathematical physics at the time is provided in Bremmer (1949).

A graphical method for computing the ground-wave field intensity was subsequently developed by Norton (1941), such that the complex theoretical results derived for a curved Earth could be readily applied in practice by the system engineer. Later contributions that furthered the understanding of radio wave propagation over a smooth finitely conducting spherical surface were made by Fock (1945) in the former Soviet Union and Wait (1964) in the United States. In his second approach, Fock solved a second-order differential equation of the parabolic type for the attenuation factor relative to the free-space field that satisfied the boundary conditions on the sphere and at the current element. The solution in Leontovich and Fock (1946) was found to be consistent with Fock's earlier analysis and algebraically identical to that obtained by van der Pol and Bremmer. This section simply states and describes the celebrated residue series formula for the surface-wave attenuation factor, which is based on Fock's second approach. The reader is referred to Maclean and Wu (1993) for a comprehensive description of its derivation.

Define the complex function W as the additional attenuation factor experienced by the radial electric field strength component E_r at point P due to the presence of the Earth's surface relative to the field that would have arisen at this point in free space. In other words, the reference to be multiplied by W is the radial electric field that would have arisen at point P in the absence of the Earth. Note that the radial electric field component in spherical coordinates E_r is oriented vertically with respect to the surface. This is not to be confused with the radial electric field component E_ρ defined previously for the plane-interface analysis using cylindrical coordinates, which was oriented in a direction parallel to the surface. It is shown in Maclean and Wu (1993) that W can be written in the form of Eqn. (5.37). This formula is known as the Residue series.

$$W = \exp\left[j(k_0(R_0 - d) - \pi/4)\right] 2\sqrt{\pi x} \sum_{s=1}^{\infty} \frac{\exp(-jxt_s)}{t_s - q^2} \frac{w_1(t_s - y_1)}{w_1(t_s)} \frac{w_1(t_s - y_2)}{w_1(t_s)} \tag{5.37}$$

With reference to Figure 5.7, the various terms in this formula are given by $A = (k_0 a/2)^{1/3}$, $x = A\theta, \theta = d/a, q = -jA\Delta_0, y_1 = k_0 h_1/A, y_2 = k_0 h_2/A$, and $R_0 = (b^2 + r^2 - 2br\cos\theta)^{1/2}$, while t_s are the roots of the complex differential equation in Eqn. (5.38).

$$w_1'(t) - q w_1(t) = 0 \tag{5.38}$$

The function $w_1(t)$ can be expressed in terms of the Airy function, denoted by Ai[·]. The reader is referred to Abramowitz and Stegan (1964) for a description of the Airy function extended to the complex plane.

$$w_1(t) = 2\sqrt{\pi}\exp(-j\pi/6)\mathrm{Ai}[t\exp(-j2\pi/3)] \tag{5.39}$$

The roots of Eqn. (5.38) depend on the value of q, which is complex in general. It is useful to define t_s^0 and t_s^∞ as the roots corresponding to the real values of $q = 0$ and $q = \infty$, respectively. In other words, t_s^0 are the roots of $w_1'(t) = 0$, whereas t_s^∞ are the roots of $w_1(t) = 0$. From Maclean and Wu (1993), the values of t_s^0 and t_s^∞ are listed in Table 5.2 for increasing s.

s	t_s^0	t_s^∞
1	$1.01879 \exp(-j\pi/3)$	$2.338107 \exp(-j\pi/3)$
2	$3.24820 \exp(-j\pi/3)$	$4.087949 \exp(-j\pi/3)$
3	$4.82010 \exp(-j\pi/3)$	$5.520560 \exp(-j\pi/3)$
4	$6.16331 \exp(-j\pi/3)$	$6.786708 \exp(-j\pi/3)$
5	$7.37218 \exp(-j\pi/3)$	$7.944134 \exp(-j\pi/3)$
≥ 6	$[1.5(s-3/4)\pi]^{2/3} \exp(-j\pi/3)$	$[1.5(s-1/4)\pi]^{2/3} \exp(-j\pi/3)$

TABLE 5.2 The roots of the differential equation $w_1'(t) - qw_1(t) = 0$ for real values of $q = 0$ and $q = \infty$ are denoted by t_s^0 and t_s^∞, respectively.

For small values of q that satisfy $|q| < |t_s^0|$, the roots $t_s(q)$ of Eqn. (5.38) may be evaluated using a Taylor series expansion around $q = 0$. The first five terms of this expansion appear in Eqn. (5.40). This expression is suitable for low radio frequencies and surfaces of high conductivity, such as seawater in the lower HF band. In this case, Eqn. (5.40) provides a sufficiently accurate calculation of the roots for most practical applications.

$$t_s(q) = t_s^0 + q/t_s^0 - q^2/2(t_s^0)^3 + q^3\{1/[3(t_s^0)^2] + 1/[2(t_s^0)^5]\}$$
$$-q^4\{7/[12(t_s^0)^4] + 5/[8(t_s^0)^7]\} + \cdots \tag{5.40}$$

For large values of q, the roots $t_s(q)$ may be evaluated using a Taylor series expansion around $q^{-1} = 0$. The first six terms of this expansion appear in Eqn. (5.41). This formula is most appropriate for ground of low conductivity and at high radio frequencies. As such, it is not of direct interest to HFSW radar as far as surface-wave propagation over the sea is concerned. However, Eqn. (5.41) may be useful for calculating the surface-wave attenuation between the transmitter and receiver of a bistatic radar system over ground of low conductivity.

$$t_s(q) = t_s^\infty + 1/q + t_s^\infty/(3q^3) + 1/(4q^4) + (t_s^\infty)^2/(5q^5) + 7t_s^\infty/(18q^6) + \cdots \tag{5.41}$$

The residue series formula for calculating the additional surface-wave attenuation W is appropriate for the considered scenario provided that: (1) the receiving point P is located in the geometrical shadow region of the radiating current element, and (2) the ground distance d is greater than $10\lambda^{1/3}$ km, where λ is the free-space wavelength in meters. If either of the conditions in Eqn. (5.42) are not satisfied, the convergence of the residue series may be poor and the results unreliable. In this case, alternative formulas are provided in (Maclean and Wu 1993).

$$\text{Shadow region: } d > (2a)^{1/2}(\sqrt{h_1} + \sqrt{h_2}), \quad \text{Long range: } d > 10\lambda^{1/3} \tag{5.42}$$

The main result is that the Earth's curvature introduces significant diffraction effects in the HF band and at lower frequencies. The representation of the electric field strength received at long distances and in the shadow region reveals a third surface-wave attenuation regime, where the residue series formula predicts that the field strength decays as an exponential function of distance. In the special case of both terminals located on the surface of the sphere (i.e., $r = b = a$), it is shown in Maclean and Wu (1993) that the radial electric field intensity E_r at the receiver can be approximated by Eqn. (5.43), where W is given by Eqn. (5.37) for $y_1 = y_2 = 0$. For terminals on the surface, and ground distances less than a few hundred kilometers, the chord length R_0 is well approximated by the arc distance d to describe the inverse distance component of the field strength attenuation.

$$E_r = \frac{jk_0 Z_0 I d\ell}{4\pi} \frac{\exp - jk_0 R_0}{d} W \tag{5.43}$$

By substituting $Z_0 = 120\pi$ and the reference radiator dipole moment $Id\ell = 5\lambda/(2\pi)$ into Eqn. (5.43), the magnitude of the radial electric field strength $|E_r|$ for ground-based terminals can be written as in Eqn. (5.44), where $|E_r|$ is in units of mV/m, d is the distance in km, and P is the radiated power in kW. The field strength at the receiving terminal may be readily adjusted to account for the actual radiated power and antenna gain of transmitting system relative to the 1-kW reference radiator.

$$|E_r| = \frac{300}{d}\sqrt{P}\left|\sqrt{\pi x}\sum_{s=1}^{\infty}\frac{\exp(-jxt_s)}{t_s - q^2}\right| = \frac{300}{d}\sqrt{P}\mathcal{W} \tag{5.44}$$

This equation is analogous to Eqn. (5.34), except that the Sommerfeld-Norton attenuation factor $\mathcal{F} = |F(w)|$ for a plane Earth is replaced by the residue series attenuation formula $\mathcal{W} = |W|$ in Eqn. (5.45). It is shown later that there is a significant region of overlap near ranges of approximately $10\lambda^{1/3}$ km where both methods are considered to be valid and the predictions $\mathcal{F}$ and $\mathcal{W}$ are in good agreement with each other. This means that no extrapolation is required between the two theoretical results. Under the stated conditions, the surface-wave field strength attenuation may be approximated by $\mathcal{F}$ at ranges less than $10\lambda^{1/3}$ km and $\mathcal{W}$ at longer ranges.

$$\mathcal{W} = \left|\sqrt{\pi x}\sum_{s=1}^{\infty}\frac{\exp(-jxt_s)}{t_s - q^2}\right| \tag{5.45}$$

For transmit and receive terminals immediately above the surface of a sphere, the terms $\mathcal{F}^2$ or $\mathcal{W}^2$ represent the additional surface-wave power attenuation beyond the free-space loss due to spreading (inverse square law). The basic transmission loss (BTL) L_b used in the radar equation is given by the free-space BTL $L_{fs} = (2k_0 d)^2$ combined with the additional power loss A^2 due to the presence of the finitely conducting sphere, as in Eqn. (5.46), where A is the surface wave field strength attenuation. At ranges where the Sommerfeld-Norton flat-earth theory applies, the term A is replaced by $\mathcal{F}$ in Eqn. (5.46), while at longer ranges, A is replaced by $\mathcal{W}$ using the residue series formula.

$$L_b = \frac{L_{fs}}{A^2} = \left(\frac{2k_0 d}{A}\right)^2 \tag{5.46}$$

The BTL in decibels is given by $L_b(\text{dB}) = L_{fs}(\text{dB}) - 20\log A$. The free-space BTL in decibels can be written as $L_{fs}(\text{dB}) = 32.44 + 20\log f + 20\log d$, where f is the frequency in MHz and d is the range in km, see Barclay (2003). From Eqn. (5.44), we have that $20\log d - 20\log A = 109.54 - 20\log E$ for the reference radiator, where E is the electric field strength in μV/m, and A equals $\mathcal{F}$ or $\mathcal{W}$ depending on the range. It follows that

L_b in decibels can be related to the vertical component of electric field strength E in μV/m using the simple formula in Eqn. (5.47). This formula enables reference curves, which are often plotted in terms of field strength for broadcasting and communication system design, to be related to the definition of BTL, which is often used in range or SNR equations for radar performance prediction.

$$L_b(\text{dB}) = 142.0 + 20 \log f(\text{MHz}) - 20 \log E(\mu\text{V/m}) \tag{5.47}$$

Figure 5.8 shows BTL curves plotted against range in linear and log scale at a radio frequency of 10 MHz for terminals immediately above a smooth spherical sea surface of conductivity $\sigma = 4$ S/m and relative permittivity $\varepsilon_r = 80$. The first 16 roots corresponding to the Taylor series expansion about $q = 0$ were used to compute the residue series (dashed line). Note that the free-space BTL curve (dotted line) has a slope of 2 when the BTL and range are expressed in decibels (inverse-square law). The solid and dashed lines correspond to L_b when A is given by $\mathcal{F}$ and $\mathcal{W}$, respectively. At ranges greater than $10\lambda^{1/3} \simeq 30$ km, the Sommerfeld-Norton flat-earth theory (solid line) becomes inappropriate and diffraction effects are taken into account by the residue series formula (dashed line). Below a range of about 30 km, the convergence of the residue series is poor and the Sommerfeld-Norton flat-earth formula may be used, as explained perviously.

It is observed that the Sommerfeld-Norton curve has a slope of 2 near zero range and transitions to a slope of 4 beyond a limiting range that varies with frequency and the electrical properties of the surface, as illustrated below. On the other hand, the residue series curve is associated with an exponential path loss and therefore has a slope that increases in proportion with range. This is sometimes known as the "dB per kilometer" path-loss regime. Figure 5.9, in the same format as Figure 5.8, shows the curves resulting for terminals located immediately above a smooth spherical surface of conductivity $\sigma =$ 0.01 S/m and relative permittivity $\varepsilon_r = 10$, which is representative of good soil. A comparison of the solid lines in Figures 5.8b and 5.9b reveals that slope of the BTL transitions from the free-space value of 2 (inverse-distance field strength variation) to 4 (inverse-square distance field strength variation) at a shorter range for the surface of lower conductivity.

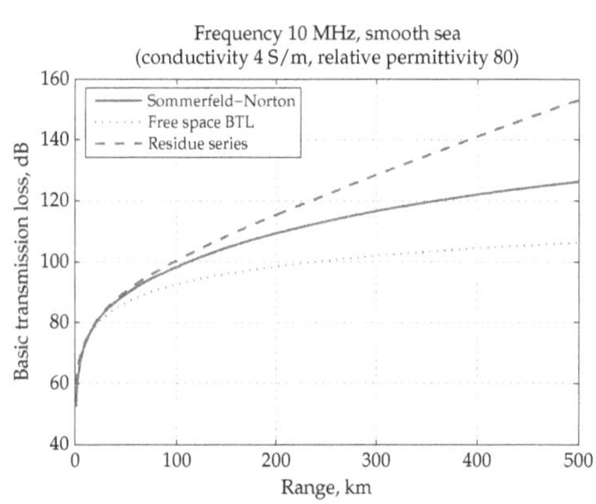

(a) Linear range scale.

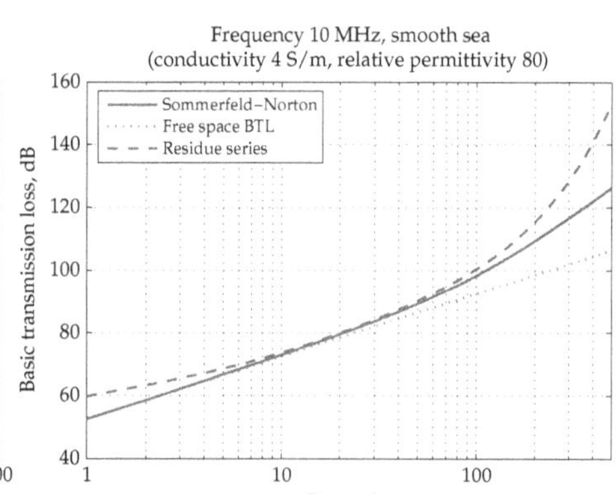

(b) Log range scale.

FIGURE 5.8 Basic transmission loss (BTL) curves at a radio frequency of 10 MHz for terminals above a smooth spherical sea surface with $\sigma = 4$ S/m and $\varepsilon_r = 80$. © Commonwealth of Australia 2011.

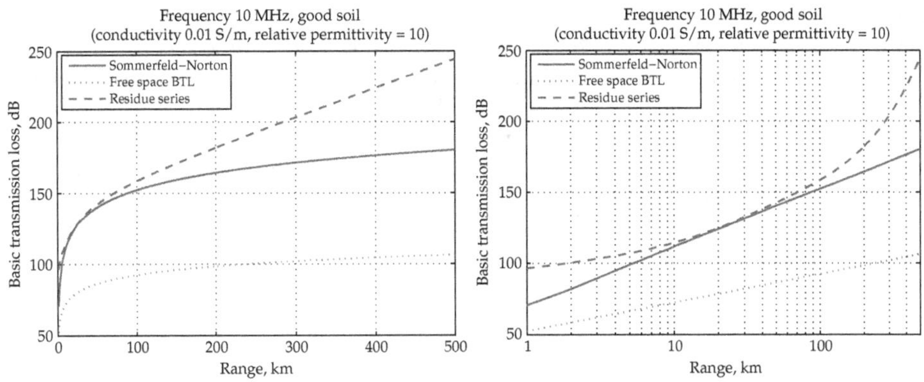

(a) Linear range scale.

(b) Log range scale.

FIGURE 5.9 Basic transmission loss (BTL) curves at a radio frequency of 10 MHz for terminals above a smooth spherical terrain surface with $\sigma = 0.01$ S/m and $\varepsilon_r = 10$. © Commonwealth of Australia 2011.

Focusing on surface-wave propagation in the diffraction zone, where the residue series formula applies, Figure 5.10a shows how the BTL varies with range for different radio frequencies when the transmit and receive terminals are located immediately above a smooth sea surface. These curves assume an effective Earth radius factor of 4/3 to take tropospheric refraction effects into account, as discussed in the following section. At a given range, the BTL expressed in decibels needs to be doubled for two way propagation between a shore-based (monostatic) HFSW radar system and surface target. The significant advantage of operating at lower frequencies in terms of reducing the surface-wave path loss is clearly evident from Figure 5.10a. However, this benefit needs to be traded off

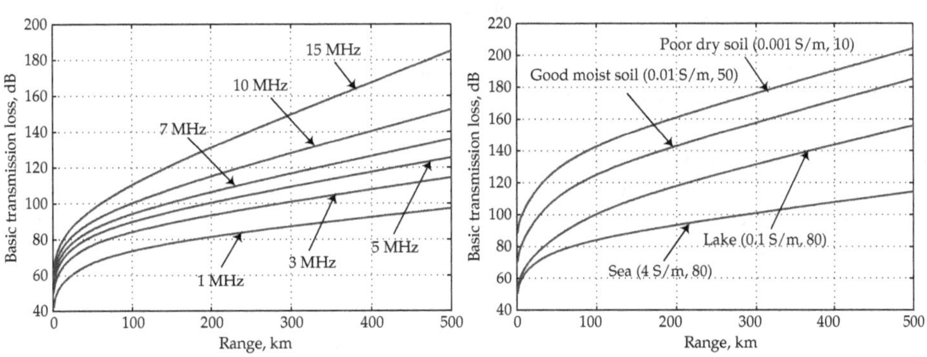

(a) BTL dependence on radio frequency (sea surface).

(b) BTL dependence on ground-type (3 MHz).

FIGURE 5.10 The left panel shows basic transmission loss curves at different radio frequencies for terminals located immediately above a smooth spherical (sea) surface with conductivity $\sigma = 4$ S/m and relative permittivity $\varepsilon_r = 80$. The right panel compares basic transmission loss curves for smooth spherical surfaces with different electrical parameters at a given radio frequency of 3 MHz. © Commonwealth of Australia 2011.

against the requirement for larger antennas on transmit, a typically higher noise spectral density at night, and the potential reduction of RCS for smaller targets.

Figure 5.10a also shows the rapidly accelerating fall-off in signal strength at ranges beyond 200 km. At these farthest ranges of an HFSW radar coverage, an order of magnitude increase in transmit power may only provide an additional 10 km of detection range in a noise-limited environment. Figure 5.10b shows the relatively higher BTL values for surface-wave propagation over smooth spherical surfaces with lower conductivity than the sea. Recalling that the BTL in decibels needs to be doubled for two-way propagation, the representative curves for poor/good soil and a "fresh-water" lake illustrate why HFSW radar is only effective at long ranges when the surface wave propagates over the sea.

The conductivity of an inland sea containing a mixture of fresh and saline water depends on the type and concentration of electrolytes in aqueous solution. Conductivity is approximately proportional to salt concentration in the case of sodium chloride, but an increase in conductivity with salt concentration does not hold for all types of electrolytes. The high BTL over poor dry ground is considered beneficial when the radar transmit and receive sites need to be isolated for effective continuous wave operation. The curves for ground types with relatively low conductivity were calculated using the first 16 roots corresponding to a Taylor series expansion about $q = \infty$ in the residue series formula. For reasons described in the next section, the effective Earth radius factor also needs to be modified slightly in the lower HF band for ground of relatively lower conductivity than the sea (Rotheram 1981b).

5.2.2 Tropospheric Refraction

The early work of Sommerfeld, Norton, Van der Pol, Bremmer, and Fock assumed that ground-wave propagation occurred in free space. This effectively ignored the presence of the atmosphere. The radio refractive index in the Earth's atmosphere is affected by air pressure, temperature, and humidity that vary with height, time, and location. Under normal (globally averaged) atmospheric conditions, the radio refractive index decreases with height and radio waves are refracted toward the Earth. The incorporation of tropospheric refraction effects represents a further refinement to the ground-wave propagation model. In this case, it is no longer assumed that the radio waves in the upper medium travel in straight lines.

The variation of radio refractive index with height follows an exponential profile on average to a good approximation. When both terminals are located near to the Earth's surface, the assumption of a (locally) linear variation of refractive index with height allows the affect of the atmosphere on signal propagation to be taken into account by a surprisingly simple modification of the free-space theory discussed previously. Specifically, tropospheric refraction effects may in this case be incorporated by substituting the true Earth radius with an effective radius that is greater than the actual value by a frequency-dependent multiplicative factor. For signal frequencies above the HF band, a correction factor of approximately 4/3 is often used to compensate for the bending of the ray path due to refraction in the troposphere at altitudes where the linear model for the height variation of refractive index is appropriate.

For signals in the HF and lower frequency bands, this factor monotonically decreases with frequency under normal atmospheric conditions until refraction may be considered negligible below about 10 kHz. In normal atmospheric conditions, the effective Earth radius factor nominally varies between 1 and 4/3 between the VLF and VHF bands (Rotheram, 1981b) . In the presence of anomalous propagation, the correction factor may need to be adjusted outside of this normal range of values (between 1 and 4/3) to account for either super-refraction or sub-refraction phenomena, as the case may be.

The Sommerfeld-Norton flat-earth theory was later extended by Bremmer (1958) and (Wait 1956) to account for linear variation of the tropospheric refractive index with height. The resulting formula is in the form a series where the first term is the original flat-Earth expression, and subsequent terms are proportional to the inverse powers of the effective earth radius. Importantly, modification of the free-space theory by an effective Earth radius applies when the terminals are close to ground and the radio frequency is above 10 MHz. The problem of applying this simple modification in the case of elevated terminals or for signal frequencies below 10 MHz will now be discussed.

In normal atmospheric conditions, the height profile of the radio refractive index $n(h)$ may be approximated by the exponential model in Eqn. (5.48), where the two (globally averaged) parameters are the surface refractivity $n_s = 1.000315$ and the scale height $h_s = 7.35$ km. It follows that when either or both terminals are not located near to the Earth's surface, the signal propagation path traverses a refractivity profile that is nonlinear. More precisely, it transpires that the assumption of a linearly varying refractive index profile becomes less accurate and no longer valid at all radio frequencies when either or both terminals are elevated by about 1 km or more above the Earth's surface.

$$n(h) = 1 + (n_s - 1)\exp(-h/h_s) \tag{5.48}$$

In this case, radio wave propagation cannot be modeled using free-space theory and an effective Earth radius to account for tropospheric refraction effects. At radio frequencies below about 10 MHz, calculations based on this traditional method may also be unreliable when both terminals are located near to the surface, particularly over ground of low conductivity (Rotheram 1981b). To improve the accuracy of ground-wave field strength predictions, an alternative propagation model known as GRWAVE was developed and implemented as a FORTRAN computer program by Rotheram (1981b). GRWAVE not only accounts for a refractivity height profile of exponential form, but also combines the contributions of space-wave and surface-wave propagation for arbitrary terminal ranges and heights above a smooth homogeneous spherical surface of finite conductivity.

Moreover, the GRWAVE model described in Rotheram (1981b) is not limited to the HF band and handles a very wide frequency spectrum. In the case of surface-wave propagation, it incorporates the previously discussed inverse-distance, inverse-square-distance, and exponential field-strength attenuation regimes as the range between the terminals increases. In essence, GRWAVE may be interpreted as a seamless interpolation between a number of methods that cater for space-wave propagation and the three surface-wave field-strength attenuation regimes. The assumptions of an exponential refractivity height profile and smooth homogeneous spherical surface of finite conductivity are reminded. The nominal range and height boundaries at which the different methods apply are now briefly summarized.

- *Geometrical optics (ray theory):* This method applies in the direct radiation zone or interference region where the transmitter and receiver are in the line of sight. In this region, the resultant signal is only due to space-wave propagation provided that both terminals are located sufficiently high above the ground for the surface-wave contribution to be negligible. This is nominally for terminal heights greater than about $35\lambda^{2/3}$ m at ranges less than $10\lambda^{1/3}$ km, where λ is radio wavelength in meters. At longer ranges, the calculation of field strength by geometrical optics is applicable for elevated terminals with heights greater than about $35\lambda^{2/3}$ m when the transmitter and receiver are within the line of sight. In such cases, the field strengths vary inversely with distance and the resultant signal may be calculated by summing the complex amplitudes of the signals arriving via the direct and surface-reflected paths at the receiver location. The ray path calculations are performed with due regard to the complex reflection coefficient and tropospheric refraction effects using an exponential refractivity profile.

- *Extended Sommerfeld-Norton (flat-Earth theory):* This method is applicable for terminals separated by distances less than about $10\lambda^{1/3}$ km when the transmitter and/or receiver is located at a height not greater than approximately $35\lambda^{2/3}$ meters above a finitely conducting surface. In this case, the contribution of the surface wave becomes significant and must be accounted for in addition to space-wave propagation. Use of the extended flat-Earth formula with terms up to order three can provide accurate results out to distances of approximately $15\lambda^{1/3}$ km, which provides a considerable region of overlap with the residue series convergence region (discussed next). For terminals located immediately above a surface of finite conductivity, the direct and reflected paths tend to cancel and surface-wave propagation becomes the dominant ground-wave contribution. In this case, the field strength varies inversely with the square of distance up to the aforementioned range limit of the flat-Earth theory, where diffraction effects around a curved Earth may be ignored.
- *Residue series (mode summation):* This final method is applicable at the farthest ranges when the terminals are beyond the radio horizon, also referred to as the diffraction zone or shadow region. For both terminals situated at small heights above a surface of finite conductivity, the residue series formula using up to nine terms converges for ground range greater than approximately $10\lambda^{1/3}$ km. In the more general case of an elevated transmitter and/or receiver, the method applies for ranges greater than approximately $10\lambda^{1/3}$ km provided the terminals are separated by a sufficiently long distance such that no line-of-sight path exists between the transmitter and receiver. In this third and furthest (over-the-horizon) range regime, the residue series formula predicts that the surface-wave field strength decays exponentially with distance.

Fortunately, there is a considerable region of range and height domain overlap between the validity and convergence of the various methods used to compute the field strength to cater for almost all terminal geometries. The results based on different methods are found to be in good agreement in these regions of overlap. Perhaps the only exception is for elevated terminals that are just within the radio horizon at a range greater than $15\lambda^{1/3}$ km. This scenario cannot be covered accurately by the extended Sommerfeld-Norton theory. In this case, the agreement between geometrical optics and the residue series formula is not as close as for other regions of overlap, particularly when one of the terminals (receiver or transmitter) is close to the ground (Rotheram 1981b).

For the case of ground-based terminals and propagation over the highly conductive sea surface, it transpires that modifying the free-space theory by using an effective earth radius factor of 4/3 provides a quite accurate prediction of the field in the diffraction zone

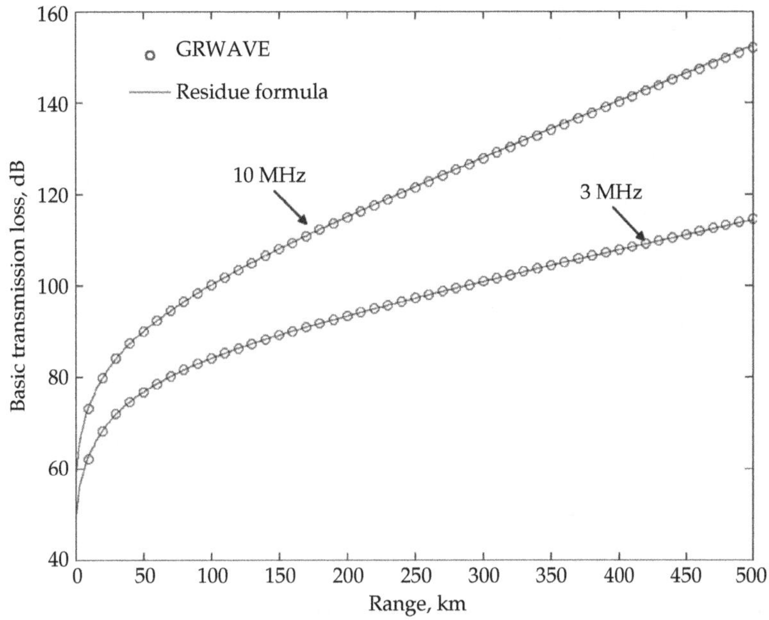

FIGURE 5.11 Comparison of GRWAVE basic transmission loss (BTL) curve with the results computed using the previously defined residue series formula at radio frequencies of 3 MHz and 10 MHz. Recall that BTL represents the *one-way* path loss. The surface is assumed to be smooth with a homogeneous conductivity of 4 S/m and dielectric constant (relative permittivity) of 80 to model losses over a calm sea. Both terminals are assumed to be located immediately above a spherical surface using an effective Earth radius multiplier of 4/3. *(Courtesy of Dr. R. Rydolls, Defence R & D Canada.)*

at frequencies between 3 and 10 MHz compared to those resulting for an exponential refractivity profile. However, this simple modification is not appropriate for a surface of low conductivity, as shown in Rotheram (1981b). Use of a 4/3 effective Earth radius is also not appropriate for all (terrain and sea) surfaces at frequencies below the HF band.

As illustrated in Figure 5.11, using the residue series formula in Eqn. (5.45) and an effective Earth radius of 4/3 yields results that are in good agreement with GRWAVE for ranges beyond about 40 km at 3 MHz when both terminals are located immediately above a smooth surface with electrical parameters representative of the Atlantic ocean. GRWAVE is currently adopted as the ITU-R standard method to predict ground-wave propagation based on the exponential refractive index height profile since these results have the potential to differ significantly from the linear model under certain conditions. An exhaustive set of reference curves plotted using GRWAVE under different conditions can be found in Rotheram (1981b).

The BTL calculations using Eqn. (5.45) and an effective Earth radius of 4/3 may also be compared with the two-way path loss prediction results reported in Skolnik (2008b). These results, reproduced from Skolnik (2008b) in the left panel of Figure 5.12, are based on the propagation code of Berry and Chrisman (1966). The right hand panel shows the calculations based on the residue series formula for three frequencies in the lower

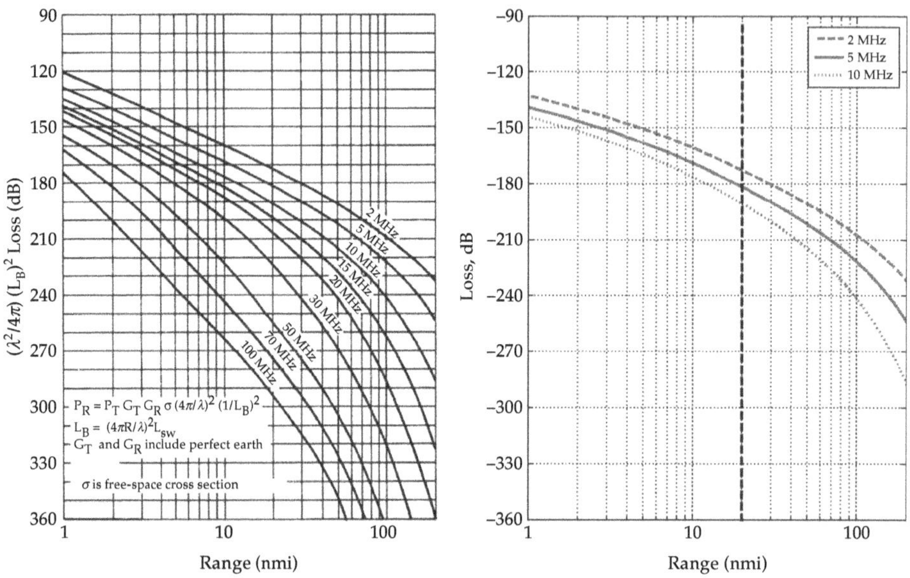

FIGURE 5.12 Comparison of two-way path loss predictions using the residue series formula with a 4/3 Earth radius factor and the Berri-Chrisman propagation code used for HFSW radar performance analysis. The surface is assumed to be smooth, target and antenna heights are 2 m, the ocean conductivity is taken as 5 S/m, and the dielectric constant is 80. As expected, the agreement is very close beyond about 20 nmi, where the residue series formula is applicable.

HF band. Beyond a range of about 20 nmi (37 km), where the residue series converges, the agreement between the curves is better than 1 dB up to EEZ limit of 200 nmi (370 km). These results not only provide a confirmation of the numerical calculations, but also indicate that a relatively simple procedure may be used to predict BTL (or two-way path loss) for HFSW radar performance modeling at OTH ranges when the terminals are close to a calm sea surface.

A practical demonstration of the agreement between GRWAVE model predictions and experimental measurements of signal strength as a function of range is illustrated for two frequencies in Figure 5.13, also reproduced from Skolnik (2008b). The experimental data was recorded by a coastal HFSW radar system over a one-way path as the transmitter was moved by a small boat out to a range of approximately 110 km. To eliminate the problem of normalizing the experimental measurements to account for antenna gains and system losses, the GRWAVE curves have been scaled such that they align with the measurements at an arbitrarily chosen range of 40 km. Once this normalization is made, it is evident that the model predictions agree reasonably well with the experimental measurements over the interrogated range extent. As pointed out in Skolnik (2008b), GRWAVE seems to underestimate attenuation slightly at 7.72 MHz, but overestimate it at 12.42 MHz. The roughness of the ocean surface was low in this case (sea-state 1 to 2).

5.2.3 Surface Roughness and Heterogeneity

In practice, the Earth's surface may not be smooth along the propagation path due to irregular land topography and ocean waves, both of which produce a rough air-ground

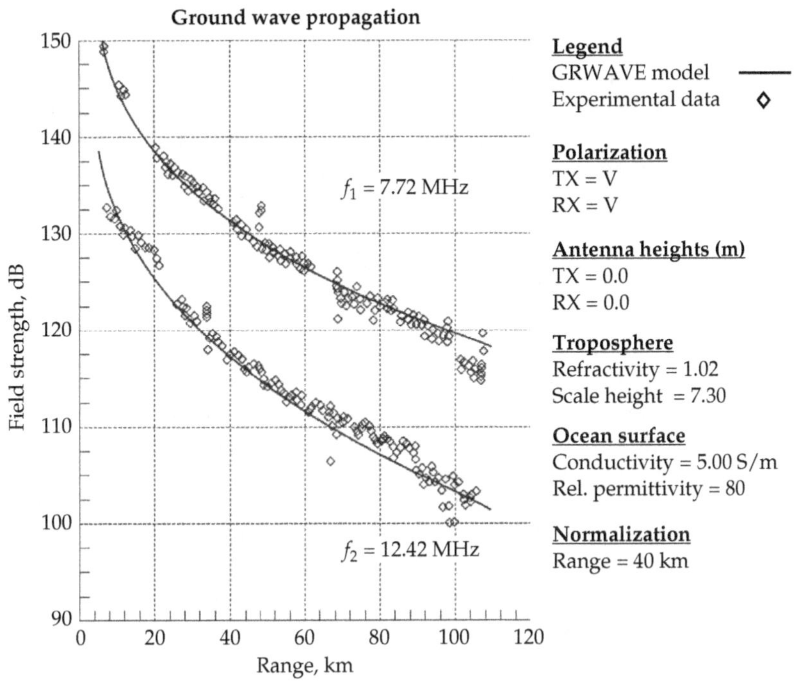

FIGURE 5.13 Experimental measurements of one-way surface-wave field-strength attenuation made at two frequencies by a shore-based HFSW radar receiver using a transmitter on a surface vessel. The experimental data is compared with GRWAVE model predictions using the parameters listed on the right hand side.

interface. Of particular interest to HFSW radar is the effect of sea-state on surface-wave field-strength attenuation (path loss) as a function of range. This topic was investigated in Barrick (1971) for vertically polarized signals at grazing incidence. The main findings are briefly summarized below.

Predictions of the excess path loss arising due to a non-smooth surface were derived by considering a slightly rough interface between air and a finitely conducting medium assuming that: (1) the surface height variations above the mean level are small relative to the radio wavelength, (2) the surface slopes are sufficiently small for nonlinear effects to be neglected, and (3) the medium below the interface is highly conducting such that the Leontovich boundary condition applies. These assumptions are satisfied by the ocean surface for all sea-states of practical interest up to the mid-VHF region.

The approach taken by Barrick was to derive an effective surface impedance in terms of the surface-height spatial spectrum, and then to conceptually replace the slightly rough surface by a smooth flat surface with this effective impedance for basic transmission loss calculations using standard methods. Barrick derived the effective surface impedance for two empirical spatial height spectrum models, namely, the directional Neumann-Pierson and semi-isotropic Phillips wind-wave spectrum models. The effective surface impedance is composed of two terms. The first is due to the impedance of the lower medium when the surface is perfectly smooth, while the second accounts for the effect of roughness, which depends on the amplitudes of the surface-height spectral components present.

For a given radio frequency and sea-state, the excess loss in decibels attributable to surface roughness with respect to a smooth sea is defined as the basic transmission loss (BTL) predicted for a rough sea minus the nominal value calculated for a smooth sea. Families of curves showing excess loss versus terminal separation (with both transmitter and receiver located immediately above the surface) are provided by Barrick (1971), with radio frequencies and sea-state as parameters, over a range interval between 10 and 1000 km. It is found that normal sea-state variations do not result in significant losses below about 2 MHz. However, the excess loss increases significantly with sea-state and distance between the terminals at higher frequencies. It is found that the excess loss due to sea-surface roughness tends to peak between 10 and 15 MHz (Barrick 1971), which is typically near the upper frequency limit of most HFSW radar systems.

Figure 5.14 shows the excess loss accumulated over a *two-way* reciprocal path due to surface roughness with respect to a smooth sea as a function of wind speed for a radio frequency of 10 MHz. The calculations assume the terminals are located immediately above the ocean surface with conductivity $\sigma = 4$ S/m and relative permittivity $\varepsilon_r = 80$. The Earth radius is taken as 6370 km, with an effective Earth radius factor of 4/3. The effective normalized surface impedance used for each wind speed corresponds to the Phillips semi-isotropic wind-wave spectrum model, see Barrick (1971). The excess loss is calculated according to the residue series formula using the effective normalized surface impedance for a non-smooth sea and its value for a smooth sea surface with identical electrical parameters.

Figure 5.14b shows that the excess loss over a two-way path is predicted to be small even for long ranges beyond the EEZ limit (370 km) at sea-state 2 (10-knot wind). The excess loss cannot be neglected, however, for higher sea-states at a radio frequency of

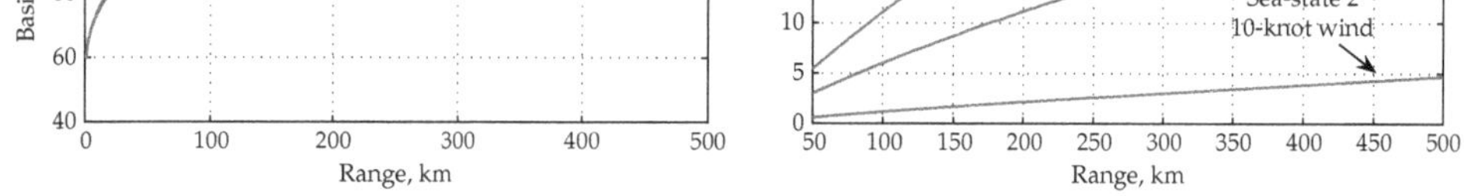

(a) Basic transmission loss (one-way path).

(b) Two-way path excess loss for a rough sea.

FIGURE 5.14 Predicted basic transmission loss (left panel) and *two-way* path excess loss (right panel) accounting for ocean surface roughness at a radio frequency of 10 MHz for wind speeds of 10, 20, and 30 knots (i.e., sea-states 2, 4, and 6, respectively) assuming Phillips' wind-wave spectrum model and terminals located immediately above a sea surface with conductivity $\sigma = 4$ S/m and relative permittivity $\varepsilon_r = 80$. An Earth radius of 6370 km, and earth radius factor of 4/3, were assumed to compute the curves using the residue series formula.

10 MHz. Specifically, the two-way excess loss at 10 MHz is predicted to be greater than 17 dB at the EEZ limit for sea states 4 and above. The results in Figure 5.14b show the excess loss at a single frequency for a variable sea state. We now consider the excess loss for a given sea-state at different frequencies.

Figure 5.15 shows the predicted two-way excess loss for sea state 4 (20-knot wind) with radio frequency as the parameter. These curves indicate that excess losses due to surface roughness are predicted to be relatively small for frequencies below 5 MHz at sea state 4. However, increasing the frequency to 10 and 15 MHz results in significant excess losses of approximately 12 and 20 dB, respectively, at a range of 200 km. The main conclusion is that excess loss due to surface roughness is a potentially important effect for HFSW radar, particularly at radio frequencies above 5 MHz and wind speeds greater than 10 knots. The excess loss increases with rougher seas, higher frequencies, and longer ranges, as a general rule of thumb.

There are a couple of interesting exceptions to the last statement. First, a marginal negative excess loss (i.e., signal enhancement due to roughness) is predicted for high sea-states near the low frequency limit of the HF band (3 MHz) in Barrick (1971). This enhancement is believed to be caused by an increase in impedance that is purely reactive when the ocean waves present have small lengths compared to the radio wavelength. However, this predicted increase in signal strength is very small (less than 1 dB) and unlikely to be ever detected. Second, the sea-state loss is predicted to decrease at radio frequencies greater than about 15 MHz. This occurs because the increase in surface impedance of a smooth sea at frequencies beyond about 15 MHz becomes a more significant proportion of the total effective impedance and therefore reduces the relative contribution due to roughness.

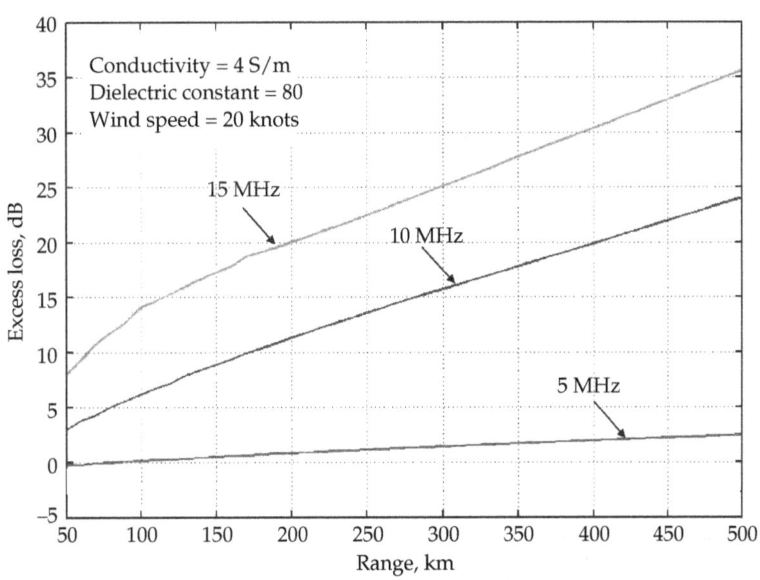

FIGURE 5.15 Predicted two-way path excess loss at different frequencies due to ocean surface roughness for a wind speed of 20 knots (sea-state 4) assuming Phillips' semi-isotropic ocean-wave spectrum. © Crown 2010. Government of Canada. (*Courtesy of Dr. H. Leong, Defence R & D Canada.*)

The variation in excess loss was also calculated as function of receiver height above a rough ocean surface at a given range in Barrick (1971). At a frequency of 10 MHz, and for a transmitter located immediately above a smooth sea surface, Figure 5.16a shows the predicted "height-gain" in decibels at different ranges (i.e., decrease in BTL relative to that resulting for a ground-based receiver). Since reciprocity applies, these curves are also valid when the transmitter and receiver positions are interchanged. The height-gain effect is observed as the terminal height moves upward at a given range. Over the first 500 m in altitude, the BTL increases by a maximum of 2–3 dB relative to that at the surface for all ranges considered. This illustrates that surface-wave energy is not confined to a small height region near the surface, but extends to a considerable altitude, and can therefore effectively illuminate low-flying aircraft well beyond the radio horizon.

As the altitude approaches 1 km, there is a monotonic increase in the signal as the receiving point begins to move out of the shadow region and into the direct radiation zone, or interference region. Figure 5.16b shows the excess loss as a function of receiver height at different ranges when the smooth sea is replaced by a rough sea that is fully developed by a 25-knot wind. The Phillips wind-wave spectrum model is assumed. The excess losses relative to a smooth sea increases by a maximum amount of about 2–3 dB in the region where the surface wave dominates. At greater altitudes, the excess loss decreases monotonically to zero as the receiver moves toward the region lit by space-wave propagation. In this example, sea-surface roughness is predicted to cause a relatively small (2–3 dB) additional loss in signal with respect to that immediately above the rough surface at heights of up to a few hundred meters.

In practice, ground-wave propagation may not occur above a homogeneous surface with uniform conductivity over the entire path. For example, propagation over the ocean surface may in some cases be interrupted by islands or land masses that lie on the path linking the two terminals. Propagation over ground with changing electrical properties is often referred to as *mixed-path* propagation. The problem of predicting the field strength

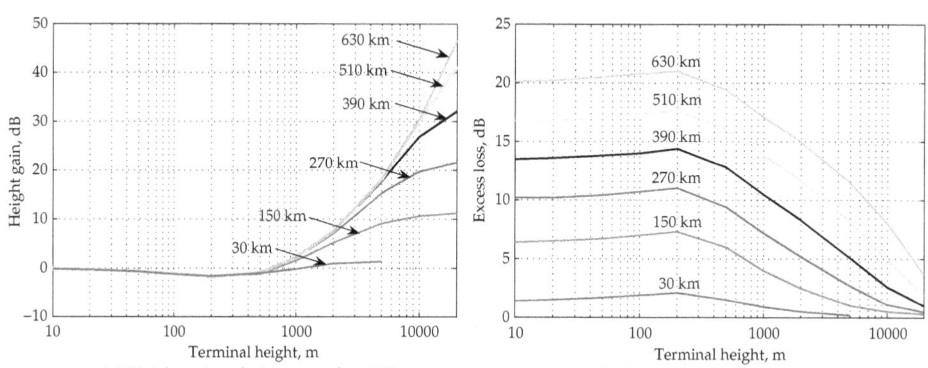

(a) Height gain relative to surface BTL.

(b) Excess loss for a rough sea.

FIGURE 5.16 Height-gain effect for a transmitter immediately above the sea surface ($\sigma = 4$ S/m, $\varepsilon_r = 80$) and a receiver at different ranges and heights above the surface. The left panel shows the height-gain effect (in decibels) relative to the basic transmission loss for a receiver located immediately above a smooth sea surface with the curves parameterized in range. The right panel shows the excess loss predicted for each range as a function of receiver height at sea state 5 (25-knot wind) assuming the Phillips' ocean-wave spectrum model. Source of numerical data from Barrick (1971a).

resulting when a surface wave propagates over two or more segments of ground with different conductivity was considered in Eckersley (1930).

The method proposed by Eckersley is based on joining the attenuation curves computed over the various homogenous segments of the ground at each transition along the path. While this approach may seem reasonable, the predictions do not agree well with empirical measurements made over actual mixed paths. Moreover, this method does not satisfy the reciprocity principle, which requires the attenuation to be identical when the transmitter and receiver positions are exchanged.

To improve the estimate of surface-wave attenuation over electrically heterogeneous ground segments, Millington proposed to apply Eckersley's method twice, once in the forward direction, and once in the reverse direction, with the two different predictions averaged to ensure that reciprocity is satisfied (Millington 1949). This rather intuitive technique was carefully tested in a controlled experiment and found to accurately predict the field strength over a 210-km path with a land-to-sea boundary transition occurring approximately 80 km from the transmitter (Millington and Isted 1950).

To account for more complex paths that involve multiple surface-type transitions, Millington's procedure may be repeated for each homogeneous section to compute an estimate over the total path. Figure 5.17 conceptually illustrates the application of

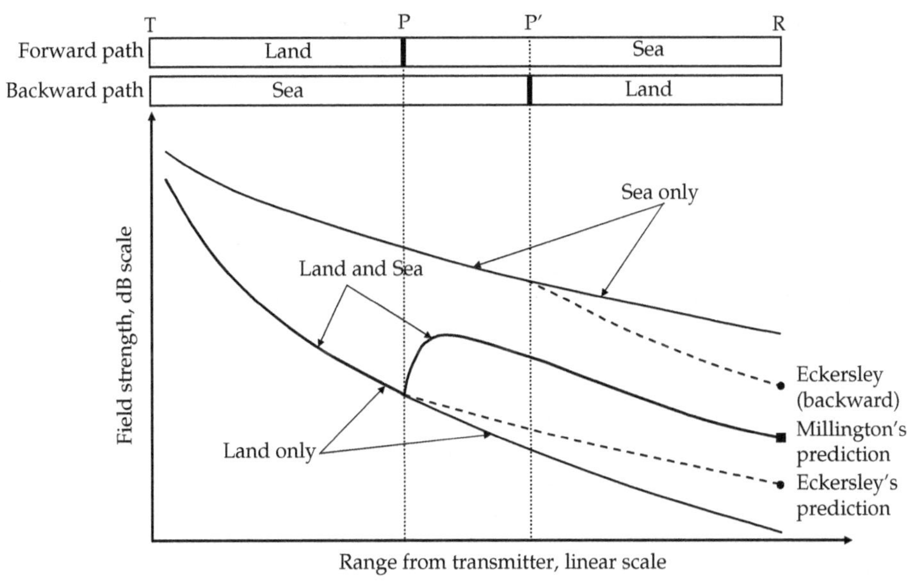

FIGURE 5.17 Conceptual illustration of Millington's method for the prediction of surface-wave field strength attenuation over a path between a ground-based transmitter (T) and receiver (R) with a transition in surface conductivity at point P. It is assumed that point P is well separated from both T and R, and that both ground types have homogeneous electrical properties. The curve depicted by a thick solid line illustrates the field strength predicted using Millington's method for the forward path, while the two thin solid lines represent the curves associated with propagation over a single homogeneous land/sea surface. The dashed lines illustrate Eckersley's prediction for the forward path, and the application of Eckersley's approach to the reverse path (i.e., with the positions of the transmitter and receiver exchanged). In the latter case, the ground-type transition occurs at point P′ with respect to the transmitter. © Commonwealth of Australia 2011.

Millington's method to predict the surface-wave attenuation over a path containing a single land-to-sea transition. The relationship of Millington's estimate to Eckersley's forward and backward predictions is also shown. Millintgon's average of the two results (expressed in decibels) can be interpreted as a constraint that forces reciprocity to hold.

As may be expected, passage of a radio wave over a terrestrial surface that changes from high to low conductivity along the path results in a sudden decrease of the field strength close to the surface immediately after the boundary. This is not surprising since a poorly conducting ground causes an increase in surface-wave attenuation. It is perhaps less expected that a transition from a region of low to high conductivity causes a sharp *increase* in field strength near the surface immediately after the boundary. This phenomenon is referred to as Millington's effect. For example, it is observed that the surface-wave field strength near the interface drops suddenly as the radio wave propagates over a sea-to-land boundary, and continues to decay rapidly as the signal propagates over land, but then exhibits an abrupt increase or "recovery" at the land-to-sea boundary.

Millington explains the recovery in terms of the redistribution of energy that occurs in the vertical plane (elevated portions of the wavefront) around the surface conductivity transitions. More energy is distributed high above the surface over land and becomes redistributed closer to the surface as the propagation path encounters the sea. However, the recovery is by no means complete. The presence of land results in a net additional attenuation with respect to a path of the same length over the ocean only. The additional loss may be small for a low number of small and sparsely separated islands that are not too close to the receiver or transmitter. On the other hand, the presence of wide and long islands along the path can effectively occlude signals propagated by the surface wave.

Millington's approach for predicting surface-wave attenuation over mixed paths with changes in ground conductivity is widely used in practice and has been recommended by the ITU-R. Millington's method was proposed without mathematical analysis or proof. The method is based on a combination of existing surface-wave propagation theory, scientific intuition, and the physical requirement to satisfy the reciprocity principle. The latter being achieved by applying Eckersley's method in both the forward and backward direction and averaging the two results expressed in decibels. An appealing feature of Millington's method is its inherent simplicity and suitability for practical applications.

Alternative models that provide a piecewise implementation of Norton's equations for surface-wave propagation over irregular terrain with inhomogeneous conductivity can be found in Ott (1992). For readers interested in delving further, a formal mathematical theory of propagation over inhomogeneous spherical ground involving two- and three-section mixed paths appears in Maclean and Wu (1993). Sevgi (2003) has developed a computer program to model surface wave propagation over inhomogeneous ground, with attention paid to hybrid paths over the ocean involving multiple islands.

5.3 Environmental Factors

Disturbance signals that compete for detection with target echoes may be classified as: (1) clutter, such as that scattered directly from ocean and land surfaces at near-grazing incidence, or volume scatter received from the ionosphere and meteors in the upper atmosphere at near vertical incidence, as well as unwanted radar echoes returned by a combination of surface and volume scatterers, and (2) interference and noise originating from external (natural and man-made) sources that emit signals with a spectral density that partially or fully overlaps the radar bandwidth.

Detection performance is limited by the dominant disturbance type that the HFSW radar must contend with at the coordinates of the target echo in beam, range, and Doppler. This not only depends on radar operating parameters, but also upon the prevailing environmental conditions, including sea-state, ionospheric structure, and the interference-plus-noise field. An understanding of the physical phenomena responsible for the various disturbance signal characteristics provides a basis for reducing their impact on the performance of HFSW radar systems.

This section discusses a number of environmental factors that can potentially limit the performance of HFSW radar systems. The first part of this section discusses sea clutter, which often poses the greatest impediment to the detection of slow-moving (small and medium-sized) surface vessels out to ranges of about 150–200 km in the absence of ionospheric clutter. The second part of this section discusses ionospheric clutter, which has the potential to obscure both slow− = and fast-moving target echoes at ranges beyond about 100 km. The third part of this section discusses external interference and noise, which is typically the limiting disturbance type at all ranges for aircraft and fast-moving surface vessels with relative velocities greater than about 20 knots (provided spread-Doppler ionospheric clutter is not limiting in these resolution cells). The raised background noise levels at night also limits the detection of large ships (not masked by the Bragg lines) at ranges beyond about 150–200 km when ionospheric clutter is not dominant.

5.3.1 Sea Clutter

Surface clutter received due to the scattering of vertically polarized radar signals from wind-driven water waves in the ocean has been the subject of extensive theoretical and experimental research. Even though much of the literature describing the backscattered surface-wave sea echo is directly related to the use of HFSW radars for remote sea-state sensing, such results are also relevant to surveillance applications when interpreted from a target detection perspective. Knowledge of the dependence between the physical properties of ocean water waves and the (statistically expected) sea clutter characteristics may be used to guide HFSW radar system design and operation. Seminal works in this area are due to Barrick and others. For example, authoritative descriptions of physics-based models of the sea echo can be found in (Barrick 1972a) and (Barrick 1972b).

This section is divided into the three parts. The first part provides a brief review of fundamental oceanographic principles regarding the physical properties of wind-driven water waves. An appreciation of these principles is indispensable to understanding the structure of the sea echo. For readers interested in more information on this subject, a detailed coverage of the physical properties of wind-driven water waves can be found in (Kinsman 1965). The second part of this section summarizes the main results of Barrick and others by recalling the mathematical models derived for the first- and second-order sea echo Doppler spectral cross sections. The third part of this section discusses the impact of the described HF sea clutter characteristics on the detection of slow-moving targets in HFSW radar systems.

5.3.1.1 Basic Ocean Wave Principles

In order to quantitatively interpret the essential characteristics of the sea echo received by an HFSW radar, it is necessary to formulate a model of the sea surface and to recall certain physical properties of wind-driven water waves. To this end, let the scalar $z(\mathbf{r}, t)$ be a random process that represents a realization of the ocean surface height at position $\mathbf{r}$ and time t relative to the horizontal xy-plane, which represents the mean ocean surface height in a confined area of interest S.

$$\{z(\mathbf{r}, t), \ \mathbf{r} = [x, y]; \ \mathbf{r} \in S\} \tag{5.49}$$

For convenience of discussion, assume that the second-order statistics of this process are temporally stationary over a limited observation interval $t \in [t_0, t_0 + T_o]$ and spatially homogeneous over the surface area S. The surface S may be defined as the area of a radar spatial resolution cell, where the receive antenna is assumed to have a narrow beam. Standard models characterize the statistical properties of the ocean surface by its directional wave-height spectrum or power spectral density $S(\boldsymbol{\kappa})$ given by Eqn. (5.50), where $\boldsymbol{\kappa}$ is the directional ocean wavevector, ω is the angular frequency of the associated water wave, and $< \cdot >$ denotes ensemble average.

$$S(\boldsymbol{\kappa}) = \int\int \mathrm{d}\mathbf{s} \int < z(\mathbf{r}, t) z(\mathbf{r} + \mathbf{s}, t + \tau) > e^{j(\boldsymbol{\kappa} \cdot \mathbf{s} - \omega \tau)} \mathrm{d}\tau \tag{5.50}$$

The wind-generated ocean waves responsible for producing the HF sea clutter echo are known as surface gravity waves because the principal restoring force on the perturbed water mass is gravity. Wind-driven waves with a length L greater than 1.73 cm are gravity waves (Phillips 1966). Here, we are only concerned with this type of wave, as opposed to capillary waves (of shorter length), where surface tension acts as the dominant restoring force. Ocean waves may be generated by other mechanisms, including earthquakes and planetary forces, but only wind waves will be considered here.

Specifically, for a water wave of length L and period T between successive crests, the angular frequency $\omega = 2\pi/T$ is not a free variable but depends on the wavenumber $\kappa = |\boldsymbol{\kappa}| = 2\pi/L$. This dependence is described by the dispersion relation in Eqn. (5.51), where g is acceleration due to gravity and d is the water depth. This explains why the spectrum $S(\boldsymbol{\kappa})$ in Eqn. (5.50) is solely a function of $\boldsymbol{\kappa}$.

$$\omega^2(\kappa) = g\kappa \tanh(\kappa d) \tag{5.51}$$

In deep water, where $d > L/2$, the term $\tanh(\kappa d)$ tends to unity and the dispersion relation reduces to the simple formula in Eqn. (5.52), where the κ-dependence of ω is dropped for notational convenience. Strictly speaking, the deep-water dispersion relation in Eqn. (5.52) is most appropriate for small amplitude water waves that propagate in open ocean areas well away from coastlines (Barrick 1972b).

$$\omega^2 = g\kappa \tag{5.52}$$

For reasons to be described in the following section, the water waves that contribute most to the HF sea echo have a length $L = \lambda/2$ equal to half the radio wavelength λ. For radio wavelengths in the HF band, the water waves mainly responsible for the HF sea echo range from 5 to 50 m in length. Hence, the deep-water dispersion relation may be considered valid in water of depth $d > 25$ m for the longest of these waves, where the influence of the sea floor may be neglected. Restricting attention to cases where the deep-water dispersion relation in Eqn. (5.52) may be considered valid, the phase velocity v of a water wave of length L is given by Eqn. (5.53).

$$v = \frac{\omega}{\kappa} = \sqrt{\frac{gL}{2\pi}} \tag{5.53}$$

Stated simply, the phase velocity of a water wave varies with the square root of its length and is unique to waves of a certain length. This physical property of ocean waves plays an important role in determining the Doppler spectrum characteristics of the sea echo. Having described the basic properties of a single surface gravity wave in deep water, we now turn our attention to the formation and characteristics of the directional wave-height spectrum $S(\boldsymbol{\kappa})$ defined in Eqn. (5.50).

As the surface wind begins to blow on a mirror-smooth sea, it first produces small waves with lengths of a few centimeters (Miles 1957). These short waves become larger in amplitude due to wind stress until a limiting spectral density is reached. Beyond this limit, to be quantified in due course, energy is transferred by a process of nonlinear wave interaction from these short waves to produce slightly longer waves (Hasselmann 1961). In a growing sea, these longer waves also become larger in amplitude and eventually reach the limiting spectral density associated with waves of this length. The wave interaction process continues to produce slightly longer waves until waves with a phase velocity approximately equal to the wind speed have been generated. When the characteristic limiting spectral density is reached by all waves under a certain length, this set of waves are considered to be completely aroused by the wind.

For a constant velocity wind blowing with sufficient fetch and duration, ocean waves of a given length therefore do not continue to grow indefinitely but reach a point of equilibrium where the energy added by the wind to the waves is equal to that lost due to wave breaking and other dissipation phenomena. During this process, nonlinear wave interactions continue to transfer energy from waves which have reached equilibrium to waves of longer length. This process continues until the wind has completely aroused all ocean waves down to a cutoff frequency, where the maximum wave phase velocity is approximately equal to the wind speed. When the wind has completely aroused all the ocean waves down to this cutoff frequency, the directional wave-height spectrum is saturated and the sea is said to be fully developed for a given wind speed.

At equilibrium, the wave system reaches a steady state and $S(\boldsymbol{\kappa})$ may be represented in the form of Eqn. (5.54), where $F(\kappa)$ is a nondirectional wave spectrum, and $G(\phi, \kappa)$ is an angular spreading function that describes the azimuth distribution of wave energy relative to a reference direction. It is convenient to define ϕ as the angle of wave propagation with respect to the radar beam direction, which is taken as the positive x-axis ($\phi = 0$). Note that $\phi \in [-\pi, \pi)$ denotes the direction the wave is traveling to, not the direction it is coming from. The directional wave spectrum $S(\boldsymbol{\kappa})$ is also often written as $S(\kappa, \phi)$, where $\kappa = |\boldsymbol{\kappa}|$ is the wavenumber, or alternatively as $S(\kappa_x, \kappa_y)$, where $\kappa_x = \kappa \cos\phi$ and $\kappa_y = \kappa \sin\phi$.

$$S(\boldsymbol{\kappa}) = F(\kappa)G(\kappa, \phi) \tag{5.54}$$

The normalization is such that the mean square surface-height $h^2 =< z(\mathbf{r}, t)^2 >$ is given by Eqn. (5.55). The term $\kappa \equiv (\kappa_x^2 + \kappa_y^2)^{1/2}$ in the integral on the right hand side of Eqn. (5.55) arises as a result of the change of variables from cartesian to polar form. Specifically, κ is the Jacobian matrix determinant associated with this transformation.

$$h^2 = \int_{-\infty}^{\infty}\int_{-\infty}^{\infty} S(\kappa_x, \kappa_y)\, \mathrm{d}\kappa_x\, \mathrm{d}\kappa_y = \int_{0}^{\infty}\int_{-\pi}^{\pi} S(\kappa, \phi)\, \kappa\, \mathrm{d}\kappa\, \mathrm{d}\phi \tag{5.55}$$

Before proceeding to discuss the forms of $F(\kappa)$ and $G(\kappa, \phi)$ in more detail, it is recalled that the nondirectional wave spectrum $F(\kappa)$ is defined as the directional wave spectrum $S(\kappa, \phi)$ integrated over all angles ϕ. This definition implies that the angular spreading function $G(\kappa, \phi)$ is normalized such its value integrated over all angles is equal to unity, as in Eqn. (5.56).

$$F(\kappa) = \int_{-\pi}^{\pi} S(\kappa, \phi)\, \mathrm{d}\phi \quad \Rightarrow \quad \int_{-\pi}^{\pi} G(\kappa, \phi)\, \mathrm{d}\phi = 1 \tag{5.56}$$

Various nondirectional wave-height spectrum models have been proposed based on theoretical analysis and empirical observations (Kinsman 1965). Two popular nondirectional wind-wave spectrum models for a fully developed sea are the Phillips model (Phillips 1966) and the Pierson-Moskowitz model (Pierson and Moskowitz 1964). These two models are selected for discussion here as they convey the basic concepts in a simple manner.

The Phillips model is given by $F(\kappa)$ in Eqn. (5.57), where u is the wind speed and B is a dimensionless quantity known as the Phillips constant in the spatial wavenumber spectrum. In the equivalent temporal wavenumber spectrum (derived below), the Phillips constant is sometimes defined as $\alpha = 2\mathrm{B}$ (Maresca and Barnum 1982). In any case, an approximate value of $\mathrm{B} = 0.005$ has been widely adopted.

$$F(\kappa) = \begin{cases} \alpha/(2\kappa^4) = B/\kappa^4 & \text{for } \kappa \geq g/u^2 \\ 0 & \text{for } \kappa < g/u^2 \end{cases} \tag{5.57}$$

The Pierson-Moskowitz model may be written as $F(\kappa)$ in Eqn. (5.58), where $\alpha = 0.0081$ and $\beta = 0.74$ are dimensionless constants, $k_c = g/u^2$ is the spatial wavenumber cutoff, and u is the wind speed at a height of 19.4 ft above the sea surface. It is clear from Eqn. (5.57) and Eqn. (5.58) that these two models differ mainly in the postulated form of the lower end cutoff. The exponential term in Eqn. (5.58) represents the rapid decay in spectral density for waves with phase velocities greater than the wind speed.

$$F(\kappa) = \frac{\alpha}{2\kappa^4} \exp\left[-\beta \left(\frac{\kappa_c}{\kappa}\right)^2\right] \tag{5.58}$$

From Eqn. (5.55), it is also possible to write h^2 according to Eqn. (5.59), where $\mathcal{S}(\omega, \phi) = \mathcal{F}(\omega)\mathcal{G}(\omega, \phi)$ is the directional wave spectrum expressed as a function of radian frequency (temporal wavenumber) ω. The term $\kappa\ \partial\kappa/\partial\omega$ is equal to the Jacobian matrix determinant associated with the transformation of the directional wave spectrum from spatial to temporal wavenumber.

$$h^2 = \int_{0}^{\infty}\int_{-\pi}^{\pi} S(\kappa, \phi)\, \kappa\, \frac{\partial\kappa}{\partial\omega}\, \mathrm{d}\omega\, \mathrm{d}\phi = \int_{0}^{\infty}\int_{-\pi}^{\pi} \mathcal{F}(\omega)\mathcal{G}(\omega, \phi)\, \mathrm{d}\omega\, \mathrm{d}\phi \tag{5.59}$$

Using the deep-water dispersion relationship $\kappa = \omega^2/g$, we have that $\kappa\ \partial\kappa/\partial\omega = 2\omega^3/g^2$, while from Eqn. (5.54), $S(\kappa, \phi) = F(\omega^2/g)G(\omega^2/g, \phi)$. Substituting these expressions into Eqn. (5.59), it is readily shown that $\mathcal{F}(\omega) = F(\omega^2/g)2\omega^3/g^2$ and $\mathcal{G}(\omega, \phi) = G(\omega^2/g, \phi)$.

From Eqn. (5.58), it follows that the Pierson-Moskowitz wave spectrum model has the equivalent form $\mathcal{F}(\omega)$ of Eqn. (5.60) in the temporal wavenumber domain, while the Phillips model can be written as $\mathcal{F}(\omega) = \alpha g^2/\omega^5$ for $\omega \geq g/u$ and $\mathcal{F}(\omega) = 0$ for $\omega < g/u$.

$$\mathcal{F}(\omega) = \frac{\alpha g^2}{\omega^5} \exp\left[-\beta \left(\frac{g}{\omega u}\right)^4\right] \tag{5.60}$$

The $1/\omega^5$ envelope of $\mathcal{F}(\omega)$ in the equilibrium region reflects the characteristic limiting spectral density of ocean waves as a function of frequency. More sophisticated multi-parameter models may be used to represent the wave spectrum produced by fetch-limited and finite-duration winds (Hasselmann et al. 1976), but such models will not be considered here.

Figure 5.18a shows examples of the Pierson-Moskowitz spectral form for different wind speeds along with the $1/\omega^5$ Phillips equilibrium limit (spectral density asymptote). On the other hand, Figure 5.18b plots the ocean wave phase velocity (dashed line) and length (solid line) as a function of ocean wave frequency according to the deep water dispersion relation. The lengths of resonant waves ($L = \lambda/2$) at the radio frequencies of 3, 4, 6, and 15 MHz are labeled on the solid line in Figure 5.18b. The linear scale on the vertical axis of Figure 5.18b is common to both curves and shows the wave phase velocity in knots and length in meters.

Figure 5.18a illustrates that an increase in wind speed does not increase the spectral density of waves in the equilibrium region. An increase in wind speed contributes to building up the spectral density of longer and faster waves at lower frequencies until the equilibrium limit for these waves is reached. It is also evident that moderate wind speeds less than about 15 knots are sufficient to fully develop the resonant waves at radio

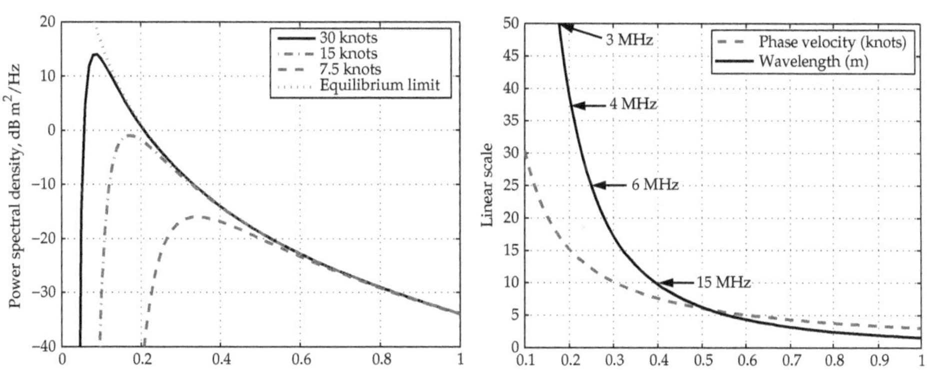

(a) Pierson-Moskowitz spectra and Phillips equilibrium limit.

(b) Phase velocity and wave length against frequency.

FIGURE 5.18 Nondirectional ocean wave-height spectral density functions based on the Pierson-Moskowitz model for different wind speeds. This model assumes a nearly constant velocity wind blowing with sufficient duration and fetch to fully develop the sea, as well as the presence of a single wave-system of wind-driven waves generated by local winds (no swell). The phase velocity and length of ocean waves as a function of frequency are plotted on the right-hand side using a common vertical axis. The lengths of resonant waves are also indicated for different radio frequencies on the solid line.

frequencies greater than 4 MHz. However, as described previously, these resonant ocean waves will not be completely aroused unless an approximately constant wind blows for a sufficiently long time and over a sufficiently long fetch (defined as the horizontal distance over which a nearly constant wind has been blowing). A lookup table that shows the duration and fetch required to fully develop the sea for a given wind speed appears in Barrick (1972b).

Waves generated by local winds give rise to a rough ocean surface that subsequently interacts with the wind and modifies the characteristics of airflow adjacent to the surface. Even if the the wind is not already turbulent due to atmospheric instability and wind shear effects, this interaction causes the airflow above the sea to become turbulent as it passes over the rough surface. For each wave frequency above the low-end cutoff, the eddying wind produces waves with a spread directions around a dominant direction corresponding to that of the mean wind. The polar diagram of a directional wind-wave spectrum model is described by the angular spreading function $G(\kappa, \phi)$.

At this point, it is convenient to make a change of variables by defining $G(\kappa, \phi) = G'(\kappa, \phi')$, where $\phi' = \phi - \phi_w$ is the angle of wave propagation relative to the mean wind direction ϕ_w. A relatively simple model for $G'(\kappa, \phi')$ is the semi-isotropic distribution in Eqn. (5.61). In this model, all angles within the half-plane (180-degree sector) centered symmetrically about ϕ_w are equally favored in amplitude by the waves.

$$G'(\kappa, \phi') = \begin{cases} 1/\pi & \phi' \in [-\pi/2, \pi/2) \\ 0 & \text{otherwise} \end{cases} \tag{5.61}$$

A more commonly adopted angular spreading function is based on the cardioid-shaped profile in Eqn. (5.62), where the κ-dependence is made implicit for notational convenience. The spreading parameter s is a function of the wind speed and wavelength (κ). This parameter takes even integer values $\{s = 2n, n \in Z^+\}$, where n may range from 1 to 15 or so (Shearman 1983). Smaller values of s are typically associated with shorter waves, while the maximum value typically occurs around the long-wave cutoff.

$$G'(\phi') = A \cos^s(\phi'/2)\,, \quad \phi' \in [-\pi, \pi) \tag{5.62}$$

The normalization A is a constant satisfying the condition on the right hand side of Eqn. (5.56). For $s = 2, 4$, and 8 it can be shown that $A = 1/\pi, 4/(3\pi)$, and $64/(35\pi)$, respectively. As explained in (Maresca and Barnum 1982), other values of A may be calculated using the expression in Eqn. (5.63).

$$A = \frac{1}{\pi} 2^{s-1} \frac{\Gamma^2(s/2+1)}{\Gamma(s+1)} \tag{5.63}$$

HF radars have sufficient sensitivity to measure sea echo components from water waves propagating against the wind. Such waves can arise due to nonlinear wave-wave interactions, reflection processes, wave-current interactions, and swell (Skolnik 2008b). The cardioid-shaped polar diagram may be modified to account for upwind-propagating waves. For example, the angular spreading function model of Eqn. (5.64) was described in (Tyler et al. 1974).

$$G'(\phi') = A'[\varepsilon + (1-\varepsilon)\cos^s(\phi'/2)]\,, \quad \phi' \in [-\pi, \pi) \tag{5.64}$$

The positive scalar ε represents the small fraction of wave energy traveling in the direction opposite to the wind. The value of ε may be in the order of 1 percent to model upwind-propagating waves. This modification accounts for the fact that both Bragg lines are visible in the HFSW radar Doppler spectrum–even when the radar beam direction is parallel to the mean wind direction. The associated normalization factor A' is given by Eqn. (5.65). Figure 5.19 illustrates the form of $G'(\phi')$ for $\varepsilon = 0.01$ and different values of the spreading parameter $s = 2, 8, 30$. An alternative model for $G'(\phi')$ is described in (Long and Trizna 1973).

$$A' = (2\pi\varepsilon + (1-\varepsilon)/A)^{-1} \tag{5.65}$$

In summary, these directional wave spectrum models describe a system of deep-water surface gravity waves driven by (nearly constant) local winds with sufficient duration and fetch to fully develope the sea down to a minimum wave cutoff frequency determined by the mean wind speed (i.e., not growing or decaying seas). Specifically, these wind waves are a system of ocean waves in equilibrium that are set up by winds blowing above the considered area of ocean. The resulting wave system gives rise to a rather disorganized or "random-appearing" ocean surface-height profile (Barrick 1972b).

When waves move out of the area in which they were originally aroused by the winds, these waves change their shape and settle down to what is known as swell. Swell appears less random and more nearly sinusoidal due to the different speeds at which waves of different frequencies propagate out from the area in which they were generated. Swell can arise from storms thousands of miles away and enter into a region of local wind waves to produce an additional wave system. In the presence multiple wave systems, the ocean directional wave-height spectrum cannot be represented by the simple models described in this section. In this case, partitioning of the wave systems is required such that the sea surface can be modeled as the sum of different directional wave-height spectra.

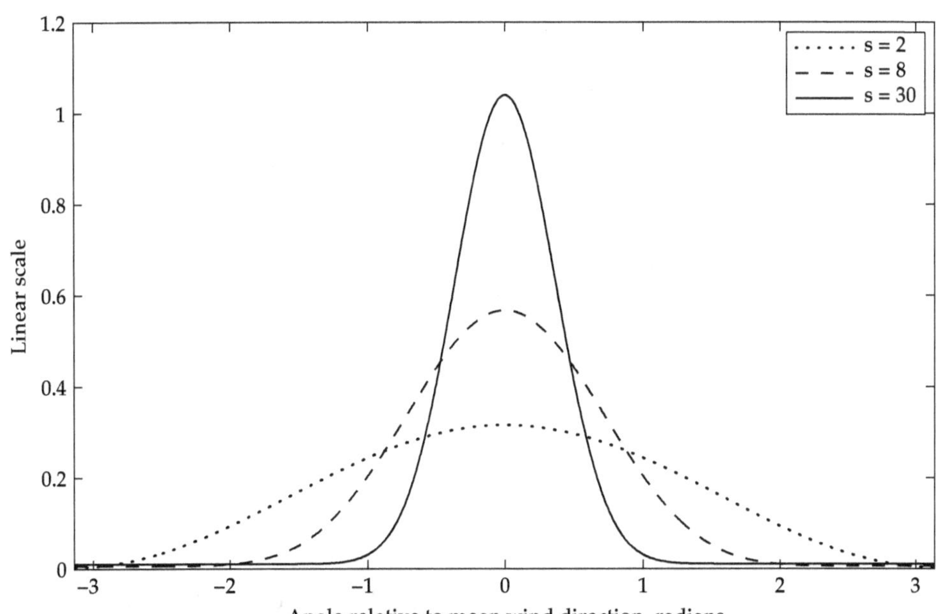

FIGURE 5.19 Examples of wave energy angular spreading function $G'(\phi')$ according to Eqn. (5.64) using $\varepsilon = 0.01$ and values of $s = 2, 8, 30$.

5.3.1.2 First- and Second-Order Echoes

The derivation of a mathematical representation of the HFSW radar sea echo requires a physics-based model of the interaction between the dynamic ocean surface and the incident radio wave for the considered scattering geometry. Electromagnetic scattering from rough surfaces has been modeled using the techniques of classical physical and geometrical optics. However, such approaches are limited to surfaces whose principal radii of curvature are much greater than the radio wavelength. For radio waves with decameter wavelengths, this generally does not hold for the ocean surface. Consequently, such methods are considered unsuitable for modeling HF signals scattered from the sea-air interface.

Alternatively, an approximate expression for the Doppler spectrum of HF radio waves scattered by the sea surface may be derived by extending the boundary perturbation approach used by Rice for static rough surfaces (Rice 1951) to the case of a dynamic ocean surface. Based on this approach, the first- and second-order spectral cross sections per unit area of the ocean surface were derived in (Barrick 1972a) and (Barrick 1972b) under three conditions: (1) the height of the surface is small relative to the radio wavelength, (2) surface slopes are small compared to unity, and (3) the impedance of the surface is small in terms of the free-space wave impedance. Providing the sea is not too rough, such assumptions may be considered valid over the HF band.

Restricting attention to the HFSW radar application, we shall consider the case of vertically polarized radio waves transmitted and received at grazing incidence to sea surface. The total clutter cross section per unit area σ^0 ($\mathrm{m}^2/\mathrm{m}^2$) integrated over all Doppler frequencies is defined by Eqn. (5.66), where ω is the Doppler frequency in radians per second, and $\sigma^0(\omega)$ is the total clutter cross section per unit area per unit (rad/s) bandwidth for vertical polarization at grazing incidence, as defined in Barrick (1972a). To be clear, $\omega = \omega_s - \omega_i$, where ω_i and ω_s are the radian frequencies of the incident (transmitted) and scattered (received) radio waves, respectively. This is not to be confused with the water wave angular frequency in the previous subsection.

$$\sigma^0 = \frac{1}{2}\int_{-\infty}^{\infty} \sigma^0(\omega)\mathrm{d}\omega \tag{5.66}$$

The spectral cross section per unit area $\sigma^0(\omega)$ may be decomposed as the sum of first-order $\sigma_1^0(\omega)$, second-order $\sigma_2^0(\omega)$, and higher-order $\sigma_h^0(\omega)$ scattering terms, as in Eqn. (5.67). Substitution of Eqn. (5.67) into Eqn. (5.66) gives the corresponding first, second, and higher order (Doppler-integrated) cross sections per unit area, denoted by σ_1^0, σ_2^0, and σ_h^0, respectively. Higher order terms provide relatively less energetic contributions to the sea echo and will not be considered further. The first- and second-order terms will now be described in more detail.

$$\sigma^0(\omega) = \sigma_1^0(\omega) + \sigma_2^0(\omega) + \sigma_h^0(\omega) \tag{5.67}$$

Historically, it was Crombie (1955) who first correctly deduced that water waves on the ocean surface effectively behave as diffraction gratings to the incident radio waves and that the dominant return is due to the Bragg scatter mechanism analogous to that of light rays scattered by a diffraction grating. The expression for the first-order spectral cross section $\sigma_1^0(\omega)$ of the sea in Eqn. (5.68) was derived in (Barrick 1972a) and (Barrick 1972b) under the aforementioned conditions.

$$\sigma^0{}_1(\omega) = 2^6\pi k_0^4 \sum_{m'=\pm 1} S(m'[\mathbf{k}_s - \mathbf{k}_i])\delta(\omega - m'\omega_B) \tag{5.68}$$

In Eqn. (5.68), $\mathbf{k}_i$ and $\mathbf{k}_s$ represent the incident and scattered radio wavevectors, respectively, $k_0 = |\mathbf{k}_i| = 2\pi/\lambda$ is radio wavenumber, $\delta(\cdot)$ is the Dirac delta function, and $m' = \pm 1$. The term $\omega_B = \sqrt{g|\mathbf{k}_s - \mathbf{k}_i|}$ is known as the Bragg angular frequency. In Barrick's derivation, the three-dimensional (spatial-temporal) symmetrical surface roughness spectrum $W(\kappa_x, \kappa_y, \omega)$ defined by Rice (1951) is replaced by the directional wave height spectrum *measured* by oceanographers, denoted by $S(\boldsymbol{\kappa})$ in Eqn.(5.68).

In a monostatic system, sea clutter is received due to backscatter such that $\mathbf{k}_s = -\mathbf{k}_i$. Examination of Eqn. (5.68) reveals that first-order scatter arises from the presence of only two ocean waves within the entire spectrum $S(\boldsymbol{\kappa})$. For the backscatter case, the associated wavevectors are given by Eqn. (5.69). These two ocean waves have the same wavenumber $\kappa = 2k_0$ and spatial length equal to half the radio wavelength $L = \lambda/2$. The two solutions $m' = \pm 1$ correspond to water waves of this length moving directly toward ($m' = 1$) and away ($m' = -1$) from the radar along the antenna beam direction.

$$\boldsymbol{\kappa} = m'(\mathbf{k}_s - \mathbf{k}_i) = -2m'\mathbf{k}_i \tag{5.69}$$

These specific components of the directional wave spectrum are known as Bragg wave trains or "resonant waves." The phase velocities of the Bragg wave trains impose Doppler shifts on the first-order clutter echoes given by $m'\omega_B = \omega_s - \omega_i$. Substituting $\mathbf{k}_s = -\mathbf{k}_i$ into the expression $\omega_B = \sqrt{g|\mathbf{k}_s - \mathbf{k}_i|}$ yields the Bragg angular frequency Doppler shift for the backscatter case in Eqn. (5.70).

$$\omega_B = \sqrt{2gk_0} \tag{5.70}$$

Using Eqn. (5.69) and Eqn. (5.70), it is possible to write the backscatter first-order spectral cross section as in Eqn. (5.71). Recall that this expression applies for vertical polarization and scattering at grazing incidence in a monostatic radar configuration. Normal convention is to assume that the radar beam is pointed in the positive x-axis direction, so the advancing wavevector $\boldsymbol{\kappa} = -2\mathbf{k}_i$ is equivalent to $(\kappa_x, \kappa_y) = (-2k_0, 0)$, or $\kappa = 2k_0$ and $\phi = \pi$. Also note that $\phi' = \phi - \phi_w = \pi - \phi_w$ in this case. Analogous arguments apply for the receding wavevector $\boldsymbol{\kappa} = 2\mathbf{k}_i$.

$$\sigma_1^0(\omega) = 2^6\pi k_0^4[S(-2\mathbf{k}_i)\delta(\omega - \omega_B) + S(2\mathbf{k}_i)\delta(\omega + \omega_B)] \tag{5.71}$$

The previously described product model $S(\boldsymbol{\kappa}) = F(\kappa)G(\kappa, \phi)$ allows us to write $S(2\mathbf{k}_i) = F(2k_0)G(2k_0, 0)$ and $S(-2\mathbf{k}_i) = F(2k_0)G(2k_0, \pi)$. It is also reminded that $G(2k_0, 0)$ is equivalent to $G'(2k_0, -\phi_w)$ or $G'(2k_0, \phi_w)$ for a cardiod-shaped angular spreading function described in Eqn. (5.64), and similarly, we have that $G(2k_0, \pi) = G'(2k_0, \phi_w + \pi)$. Substituting the latter expressions into Eqn. (5.71) yields Eqn. (5.72).

$$\sigma_1^0(\omega) = 2^6\pi k_0^4 F(2k_0)[G'(2k_0, \phi_w)\delta(\omega - \omega_B) + G'(2k_0, \phi_w + \pi)\delta(\omega + \omega_B)] \tag{5.72}$$

Assuming the Bragg wave trains are fully developed, and that the nondirectional spectrum is described by the Phillips model in Eqn. (5.57), the expression for $\sigma_1^0(\omega)$ in Eqn.

(5.72) simplifies to Eqn. (5.73). Note that the κ^{-4} fall off in spectral density at the equilibrium limit is canceled by the k_0^4 multiplier of the first-order scattering coefficient $\sigma_1^0(\omega)$ in Eqn. (5.72).

$$\sigma_1^0(\omega) = 4\pi B[G'(2k_0, \phi_w)\delta(\omega - \omega_B) + G'(2k_0, \phi_w + \pi)\delta(\omega + \omega_B)] \tag{5.73}$$

This suggests that the first-order cross section per unit area is approximately invariant over a frequency range where the resonant Bragg wave trains can be assumed to be fully developed or nearly so. Specifically, the integrated first-order cross section per unit area is given by Eqn. (5.74).

$$\sigma_1^0 = \frac{1}{2}\int_{-\infty}^{\infty} \sigma_1^0(\omega)d\omega = 2\pi B[G'(2k_0, \phi_w) + G'(2k_0, \phi_w + \pi)] \tag{5.74}$$

For the semi-isotropic model and a wind direction in the positive-x half-plane i.e., $\phi_w \in [-\pi/2, \pi/2)$ we have that $G'(2k_0, \phi_w) = 1/\pi$ and $G'(2k_0, \phi_w + \pi) = 0$. Clearly, the opposite applies for a wind direction in the negative-x half-plane. By making these substitutions in Eqn. (5.75), it is readily shown that the first-order backscatter cross section per unit area under these conditions is equal to $2B$, or the Phillips constant α defined in (Maresca and Barnum 1982), as in Eqn. (5.74). Note that Barrick's derivation of $\sigma^0 = -17$ dB in his original paper (Barrick 1972a) corresponds to a value of $B = 0.01$. In practice, the value of B may vary between 0.004 and 0.04 depending on fetch, with $B = 0.005$ being a good approximation. In this case, the value in Eqn (5.75) is -20 dB, and this value shall be adopted as a reference for the semi-isotropic model hereafter.

$$\sigma_1^0 = 2B = \alpha \tag{5.75}$$

This analysis is based on Barrick's definition of the incident electric field strength, which assumes free-space antenna gains. An alternative definition of σ_1^0 is based on effective antenna gains over a conductive surface. The latter definition incorporates the doubling of the surface-wave electric field strength into the antenna gains rather than σ_0 for both the illumination and echo signal paths. This leads to a value of σ_1^0 that is 16 times less than the Phillips constant α, Teague, Tyler, and Stewart (1975) and Shearman (1983). Both values are equally valid providing consistent definitions are adopted in the radar equation, as described in Milsom (1997).

The actual cross section can be much less than that theoretically expected when the resonant waves are not fully developed or when the directional distribution of the wave system at the Bragg wavenumber does not maximize along the antenna beam direction. The latter effect is due to the proportionality in Eqn. (5.76). When care is taken, the roughly constant cross section per unit area property of the first-order sea echo provides a natural reference scatterer in the field, which is valuable for estimating path loss and target RCS (Barrick 1977).

$$\sigma_1^0 \propto [G'(2k_0, \phi_w) + G'(2k_0, \phi_w + \pi)] \tag{5.76}$$

In summary, the first-order sea echo manifests itself as two discrete spectral components commonly referred to as Bragg lines. The energy in each component is proportional to the wave-height directional spectrum evaluated at the corresponding Bragg wave-vector. From Eqn. (5.73), the ratio between the Bragg line peaks is given by Eqn. (5.77).

By adopting a suitable model for $G'(2k_0, \phi')$ and finding the value of ϕ that best fits the observed Bragg-line ratio, it is possible to infer the mean wind direction.

$$\frac{\sigma_1^0(\omega_B)}{\sigma_1^0(-\omega_B)} = \frac{G'(2k_0, \phi')}{G'(2k_0, \phi' + \pi)} \tag{5.77}$$

The dependence of Bragg-line Doppler shift on radio frequency is now examined in more detail. In deep water, the phase velocity of a wave of length $L = \lambda/2$ is given by the dispersion relationship in Eqn. (5.53) as $v_B = \sqrt{g\lambda/(4\pi)}$. The Doppler frequency shift imposed on echoes backscattered from ocean waves moving radially toward and away from the radar with this phase velocity is equal to $m' f_B$ where $f_B = \pm 2v_B/\lambda$. Substituting $v_B = \sqrt{g\lambda/(4\pi)}$ into this expression yields the magnitude of the Doppler frequency shift f_B in Eqn. (5.78). Note that $\omega_B = \sqrt{2gk_0} = 2\pi f_B$ for $L = \lambda/2$, as expected.

$$f_B = \sqrt{\frac{g}{\pi\lambda}} \tag{5.78}$$

A simple expression for f_B appears in Eqn. (5.79), where f_B is in Hz and $f_o = c/\lambda$ is in MHz. Unlike the Doppler shift on a target echo, which is linearly related to carrier frequency, the Doppler shifts of the first-order clutter returns vary with the square-root of carrier frequency. As illustrated later, the departure between the linear and square-root relationships of the Doppler shift to the carrier frequency can be exploited in practice to enhance the target detection and tracking performance of HFSW radar systems against first-order sea clutter.

$$f_B \simeq 0.102\sqrt{f_o(\text{MHz})} \tag{5.79}$$

In the presence of a surface current with uniform relative velocity v_s, the motion of the Bragg wave trains is superimposed on that of the current. In this case, the Bragg line Doppler shifts are not symmetrical about the carrier frequency but displaced by an equal amount $f_s = 2v_s/\lambda$ about zero Doppler frequency, as in Eqn. (5.80). Estimation of this frequency displacement from Doppler spectrum records across different range-azimuth cells and over time forms the basis of surface-current mapping.

$$f_B = f_s \pm \sqrt{\frac{g}{\pi\lambda}} \tag{5.80}$$

In practice, the first-order clutter echoes are generally not perfectly coherent over the relatively long HFSW radar CPI. Doppler broadening of the Bragg lines may occur due to surface current turbulence in the radar resolution cell, nonlinear wave-current interactions, and bathymetric affects near coastlines, including the presence of shallow water, where the sea floor influences the dynamics of resonant waves. At this point, we turn our attention to the description of second-order clutter.

Barrick's equation for the second-order sea echo scattering cross section per unit area per unit (rad/s) bandwidth is given by $\sigma_2^0(\omega)$ in Eqn. (5.81). This term arises from a combination of electromagnetic and hydrodynamic effects. Graphical representations of the physical process generating $\sigma_2^0(\omega)$ can be found in (Lipa and Barrick 1986). The two main mechanisms responsible for the production of the second-order sea echo are conceptually illustrated in Figure 5.20 and are now briefly described.

$$\sigma_2^0(\omega) = 2^6 \pi k_o^4 \sum_{m_1, m_2 = \pm 1} \int\int |\Gamma(m_1 \boldsymbol{\kappa}_1, m_2 \boldsymbol{\kappa}_2)|^2 S(m_1 \boldsymbol{\kappa}_1) S(m_2 \boldsymbol{\kappa}_2)$$
$$\times\, \delta(\omega - m_1 \sqrt{g\kappa_1} - m_2 \sqrt{g\kappa_2}) \mathrm{d}\boldsymbol{\kappa}_1 \mathrm{d}\boldsymbol{\kappa}_2 \qquad (5.81)$$

In the case of electromagnetic effects, second-order clutter is due to signals that are Bragg scattered first from one ocean wave train and then another in such a manner that the twice reflected signal is returned to the radar. Specifically, if the effective spacing of the crests or troughs of an ocean wave train as viewed along the radar beam direction is equal to one-half the radio wavelength, then diffractive scattering will occur along the surface of the specular reflection angle. If these scattered radar waves encounter a second sea wave of suitable wavelength and direction, then the signal will be returned to the radar by a similar process. This is the basis of the electromagnetic second-order scattering mechanism illustrated for the case of two incoming waves in Figure 5.20.

Unlike the first-order echoes, which arise from only two specific components of the directional wave-height spectrum, second-order scatter is due to a sum of contributions resulting from different pairs of ocean wave trains. Importantly, the entire directional wave-height spectrum $S(\boldsymbol{\kappa})$ is involved in this double scattering process. This is because each contribution to the second-order sea echo arises from a different pair

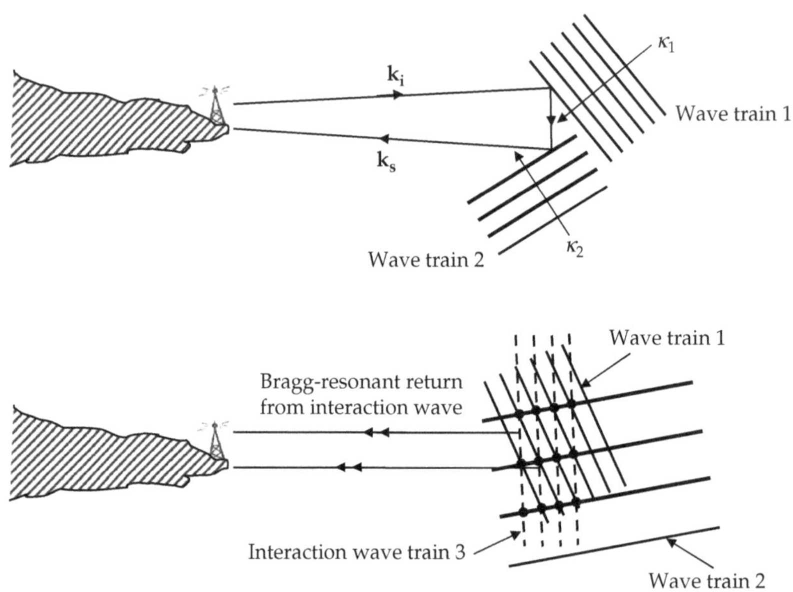

FIGURE 5.20 Notional illustration of the main electromagnetic (top panel) and hydrodynamic (bottom panel) second-order scattering mechanisms. The top panel shows "corner-reflector" scattering from two wave trains traveling at right angles. The bottom panel shows Bragg-resonant scattering from wave train 3, which is the result of an interaction product between wave trains 1 and 2.

of ocean wave-trains that are collectively parameterized by a continuum of wavelengths and propagation directions.

However, only pairs of ocean wave trains in $S(\kappa)$ with certain lengths and propagation directions can combine to produce an intermediate and twice scattered radio wave with the latter being directed toward the radar. Specifically, the two wavevectors must be compatible with returning the incident radio wave to the radar by diffractive scattering along the angle of specular reflection from each wave train in this double bounce process. Stated mathematically, a second-order contribution can arise from two ocean wavevectors κ_1 and κ_2 that satisfy the backscatter condition in Eqn. (5.82).

$$\mathbf{k}_s - \mathbf{k}_i = -2\mathbf{k}_i = \kappa_1 + \kappa_2 \tag{5.82}$$

In particular, the electromagnetic second-order backscatter mechanism referred to as "corner-reflector" diffractive reasonant scatter arises from two wave trains crossing at right angles to each other, which requires the additional condition in Eqn. (5.83) to be satisfied. The wavenumber relationships pertinent to the corner-reflector model may be derived from the law of conservation of momentum for the scatter applied to the intermediate and twice scattered wave, as described in Trizna (1982).

$$\kappa_1 \cdot \kappa_2 = 0 \tag{5.83}$$

With reference to Figure 5.20, the Doppler shift of the echo returned to the radar depends on the orientation of the "corner" with respect to the look direction. As the corner is rotated by choosing appropriate pairs of waves satisfying Eqn. (5.82) and Eqn. (5.83) from the directional wave-height spectrum, the Doppler shift accumulated over the first and second diffractions varies as $\omega_1 + \omega_2$ in Eqn. (5.84) for incoming wave trains. This passes through a turning point (maximum value) when the radar beam direction bisects the corner $\varphi = \pi/4$, and often causes a peak to appear in the echo Doppler spectrum at $2^{3/4}$ times the first-order Bragg frequency ω_B. Here, φ is defined as the angle between the backscattered radio wave and the ocean wave vector κ_1.

$$\omega_1 + \omega_2 = \omega_B(\sqrt{\cos\varphi} + \sqrt{\sin\varphi}) \tag{5.84}$$

Hydrodynamic effects arise due to the interaction between two crossing sea-waves κ_1 and κ_2, which produce second-order waves $\kappa_1 \pm \kappa_2$. The coupling between the circular particle motions in the two crossing waves gives rise to these (evanescent) interaction waves. The interaction waves do not propagate freely like gravity waves but can nevertheless contribute to second-order scatter via the Bragg mechanism (Barrick 1972b). In this case, κ_1 and κ_2 must also satisfy Eqn. (5.82), but without the constraint in Eqn. (5.83). If a third wave train, produced by the interaction of two crossing sea waves, is perpendicular to the radar look direction and satisfies the $L = \lambda/2$ condition, Bragg resonant backscatter occurs, as illustrated in Figure 5.20.

Harmonic second-order scatter is a special case of the hydrodynamic mechanism where the interacting waves travel in the same direction directly toward or away from radar. As the particle motion is circular and not transverse to the direction of a water wave, the shape of the wave is not sinusoidal but has a trochoidal profile characterized by a sharp crest and shallow trough. This profile may be decomposed into a fundamental sinusoidal component and a number of harmonics. For a particular radio wavelength λ, resonant backscatter may occur from the fundamental and harmonic components with lengths $L = n\lambda/2$ where $n \in Z^+$. These sea waves have phase velocities and Doppler shifts relative to the first-order term ($L = \lambda/2$) of $v_B\sqrt{n}$ and $\omega_B\sqrt{n}$, respectively. The discrete spectral line contribution for the harmonic $n = 2$ is sometimes observed in the sea echo Doppler spectrum.

The relationship between the radio and ocean wavevectors for both electromagnetic and hydrodynamic second-order scattering mechanisms is graphically illustrated by the vector diagram in Figure 5.21. Here, the spatial wavenumber component p is defined to lie along the radar beam with q in the perpendicular direction. The definition of κ_1 and κ_2 in Figure 5.21 ensures that the second-order backscatter condition is satisfied for any point (p, q), which represents the intersection of κ_1, κ_2, and the wavevector of the intermediate radio wave $\mathbf{k}_m$. As the point (p, q) moves, different pairs of ocean waves with lengths and directions constrained by Eqn. (5.82) become involved and contribute to the second-order scattering process (Lipa and Barrick 1986).

The integral in Eqn. (5.81) reflects the fact that there are infinitely many pairs of waves κ_1 and κ_2 that can satisfy this vector triad. The two coordinates p and q are sometimes used as integration variables in place of κ_1 and κ_2 in Eqn. (5.81) to emphasize that κ_1 and κ_2 are constrained by Eqn. (5.82). As the interacting pair of ocean waves may have a radial velocity component toward or away from the radar, there are four possible combinations of interacting wave trains in the directional wave-height spectrum. These are denoted by $m_1\kappa_1$ for $m_1 = \pm 1$, and $m_2\kappa_1$ for $m_2 = \pm 1$. A plus sign indicates a wave-train with an approaching radial phase velocity component, while a minus sign indicates a wave train with a receding radial phase velocity component.

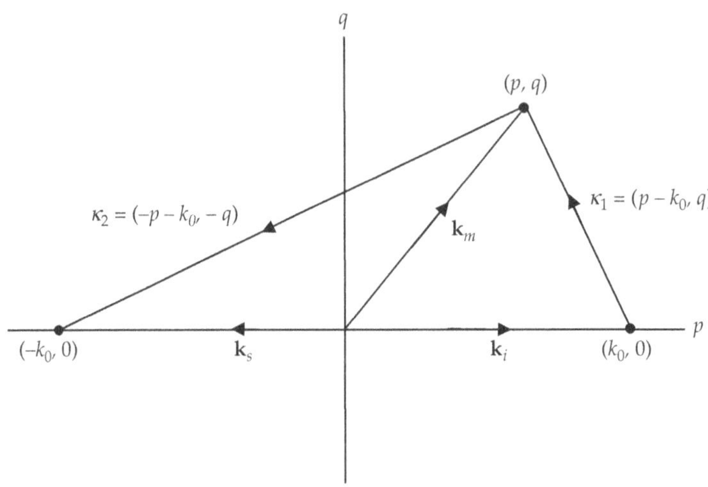

FIGURE 5.21 In the second-order scattering process, the incident radio wave ($\mathbf{k}_i$) interacts with the first ocean wave train κ_1 to produce an intermediate scattered radio wave ($\mathbf{k}_m$). This intermediate wave interacts with a second ocean wave train κ_2 to produce a radio wave scattered back toward the radar ($\mathbf{k}_s$). The pair of ocean wavevectors involved in this process are constrained to satisfy the vector triad $\kappa_1 + \kappa_2 = \mathbf{k}_s - \mathbf{k}_i = -2\mathbf{k}_i = (-2k_0, 0)$. The pair of ocean waves that produce second-order scatter may propagate either toward or away from the radar ($m_{1,2} = \pm 1$). At the integration point (p, q), this gives rise to four possible combinations of interacting wave trains in the directional wave-height spectrum.

The angular Doppler frequency ω_D imposed on the second-order contribution due to ocean waves $m_1\boldsymbol{\kappa}_1$ and $m_2\boldsymbol{\kappa}_2$ is given by Eqn. (5.85), where ω_s and ω_i are the scattered and incident radio wave angular frequencies, while $\kappa_1 = |\boldsymbol{\kappa}_1|$ and $\kappa_2 = |\boldsymbol{\kappa}_2|$ are the wavenumbers of the two wave trains. In other words, the Doppler shift resulting on the first diffraction is equal in magnitude to the frequency $\omega_1 = \sqrt{g\kappa_1}$ and is equal to $\pm\omega_1$ depending on the direction of wave train 1. Similarly, a shift of $\pm\omega_2$ results from the second diffraction, where $\omega_2 = \sqrt{g\kappa_2}$, such that the total accumulated Doppler shift is given by Eqn. (5.85).

$$\omega_D = \omega_s - \omega_i = m_1\sqrt{g\kappa_1} + m_2\sqrt{g\kappa_2} \tag{5.85}$$

As pointed out in Lipa and Barrick (1986), the case $m_1 = m_2$ defines the region of second-order scatter corresponding to Doppler frequencies outside the Bragg lines, $\omega_D^2 > 2gk_0$, while the case $m_1 \neq m_2$ defines the region of second-order scatter corresponding to Doppler frequencies between the Bragg lines, $\omega_D^2 < 2gk_0$. Second-order backscatter may be interpreted as a two-wave extension of first-order (single-wave) backscatter, which satisfies the condition $\boldsymbol{\kappa} = -2\mathbf{k}_i$. In this case, the directly approaching and receding resonant waves $m'\boldsymbol{\kappa}$ give rise to radian Doppler shifts $\omega_D = m'\omega_B$, where $m' = \pm 1$ and $\omega_B = \sqrt{g\kappa} = \sqrt{2gk_0}$.

The second-order scattering kernel or coupling coefficient $\Gamma(m_1\boldsymbol{\kappa}_1, m_2\boldsymbol{\kappa}_2)$ may be decomposed as the sum of an electromagnetic $\Gamma_{EM}(m_1\boldsymbol{\kappa}_1, m_2\boldsymbol{\kappa}_2)$ and hydrodynamic $\Gamma_H(m_1\boldsymbol{\kappa}_1, m_2\boldsymbol{\kappa}_2)$ term, as in Eqn. (5.86). The dependence on polarization enters through $\Gamma(m_1\boldsymbol{\kappa}_1, m_2\boldsymbol{\kappa}_2)$, and expressions for both contributions to the coupling coefficient can be found in (Barrick 1972b). At near-grazing incidence, the scattering coefficient of a highly conducting surface such as the sea is much larger for vertical than for horizontal polarization. The cross-polar scattering coefficients typically assume intermediate values (Skolnik 2008b).

$$\Gamma(m_1\boldsymbol{\kappa}_1, m_2\boldsymbol{\kappa}_2) = \Gamma_{EM}(m_1\boldsymbol{\kappa}_1, m_2\boldsymbol{\kappa}_2) + \Gamma_H(m_1\boldsymbol{\kappa}_1, m_2\boldsymbol{\kappa}_2) \tag{5.86}$$

At this point, it is worth lending some concreteness to the discussion by considering a practical example of the sea echo Doppler spectrum. Figure 5.22 illustrates an actual sea echo Doppler spectrum recorded by a monostatic HFSW radar in a single range-azimuth resolution cell at a frequency of 4.1 MHz. The first-order spectral lines due to the advancing and receding Bragg wave trains are indicated along with the second-order continuum. The latter occupies a Doppler region in between and outside of the two Bragg lines. In this case, the second-order clutter discrete component with spectral line at $2^{1/2}$ times the Bragg frequency f_B can be discerned and is labeled in Figure 5.22.

Although the power in the second-order clutter continuum is usually 20–30 dB weaker than that of the first-order echoes, the former can limit detection performance in the Doppler frequency region where echoes from slow-moving surface vessels are sought. Importantly, the contribution $\sigma_2^0(\omega)$ has the potential to mask target echoes over a wider distribution of Doppler frequencies (radial velocities) relative to the more powerful but discrete first-order echoes $\sigma_1^0(\omega)$. To model detection performance in terms of blind speeds, theoretical sea echo Doppler spectra may be computed by evaluation of $\sigma^0(\omega)$ for an assumed directional wave-height spectrum $S(\boldsymbol{\kappa})$, as in Maresca and Barnum (1982). Examples of sea echo Doppler spectra for different wind speeds and directions are also illustrated in Ponsford, Sevgi, and Chan (2001), whereas example spectra for different transmit and receive polarizations can be found in Skolnik (2008b).

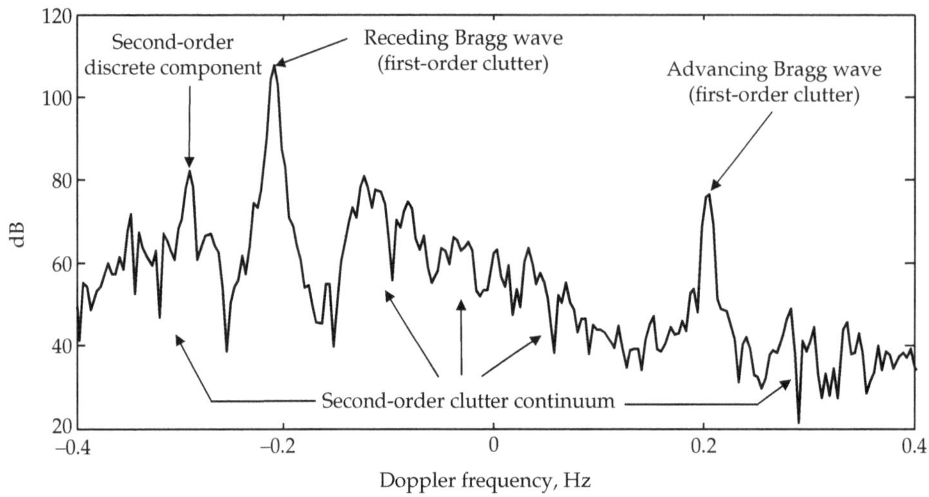

FIGURE 5.22 Example Doppler spectrum received by a monostatic HFSW radar system at a frequency of 4.1 MHz. The magnitudes of the first-order echoes are proportional to the ocean wave-height spectrum $S(\boldsymbol{\kappa})$ evaluated at the respective Bragg wavevectors $\boldsymbol{\kappa} = \pm 2\mathbf{k}_i$. The magnitude and structure of the second-order continuum is determined by the ocean wave-height spectrum including all wave frequencies and directions. Data is © Crown 2010. Government of Canada. (*Courtesy of Dr. H. Leong, Defence R & D Canada.*)

The described relationship between the sea-surface directional wave-height spectrum $S(\boldsymbol{\kappa})$ and the sea echo Doppler spectrum in the case of vertical polarization provides an opportunity for HFSW radar to remotely estimate sea-state parameters via an inversion procedure. Such procedures are by no means straightforward and will not be covered here, but the problem has been addressed in Hisaki (1996), for example. The dependence of $S(\boldsymbol{\kappa})$ on wind speed and direction may be exploited to infer the surface wind-field under the assumption of local wind-driven waves. Key features of the sea echo Doppler spectrum and the information that may potentially be extracted is summarized in Figure 5.23. The description of higher-order sea clutter and alternatives to Barrick's first- and second-order theory are beyond the scope of this text.

5.3.1.3 Impact on HFSW Radar

In surveillance applications, ship echoes received by an HFSW radar can potentially be masked by sea clutter over considerable range and Doppler frequency intervals. Target detection may be limited by powerful first-order spectral components (Bragg lines) over a very limited portion of the Doppler spectrum, or the relatively lower level second-order continuum over a wider band of Doppler frequencies. Broadly speaking, surface vessels with relative velocities less than about 20 knots have the potential to be obscured by the sea echo. External noise eventually dominates sea clutter at long ranges (typically beyond about 150–250 km for second-order clutter) due to the heavy attenuation of the surface-wave with distance. The HFSW radar range-Doppler map in Figure 5.24 shows an example of sea clutter characteristics and a ship target echo located adjacent to a Bragg line.

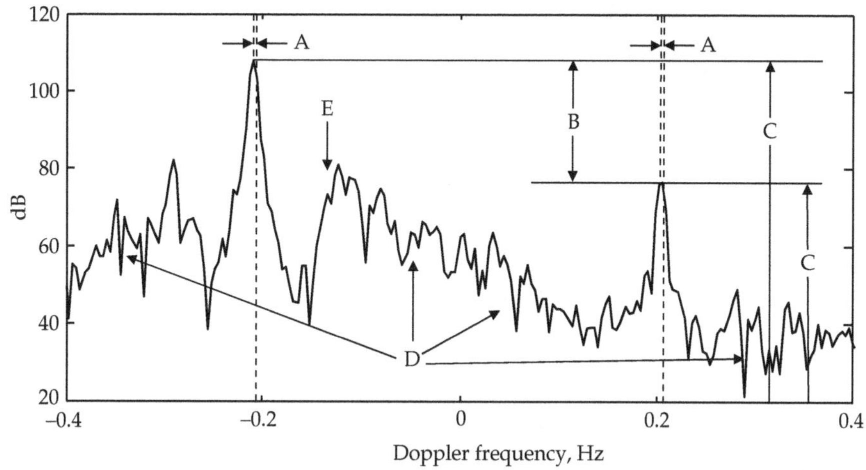

FIGURE 5.23 Relationship of HF radar sea echo Doppler spectrum characteristics to sea-state and surface-wind conditions. (A) Doppler shift of the first-order Bragg lines from their expected values: radial component of surface current. (B) Ratio of the advance to recede first-order Bragg line amplitudes: mean direction of wind-driven waves and hence surface wind. (C) Magnitude of first-order Bragg lines: ocean wave-height spectral values for the resonant wave frequency in two reciprocal directions along the radar beam. (D) Magnitude and shape of second-order continuum: ocean wave-height spectrum for all wave frequencies and directions. (E) Separation of inner edge of second-order structure from first-order Bragg lines: low-frequency cutoff of ocean wave-height spectrum. Data is © Crown 2010. Government of Canada. (*Courtesy of Dr. H. Leong, Defence R & D Canada.*)

As the wind speed increases beyond that necessary to fully develop the resonant waves, the first-order echoes are not predicted to increase in strength after saturation is reached at a given radio frequency. However, the second-order continuum contains contributions from the entire wave spectrum, including ocean waves of length greater than the resonant waves. Strong winds can excite and increase the spectral density of waves that are longer and faster than the resonant waves, which provide additional contributions to the second-order continuum. Typically, the power and Doppler frequency extent of the second-order clutter continuum will increase at a given radio frequency as the sea-state grows. For example, the troughs often appearing next to the Bragg lines at low sea-states may become filled by second-order clutter at higher sea-states.

The detection of very large ships with a mean (angle-averaged) RCS of 40–50 dBsm is limited primarily by the strong first-order echoes in a very confined region of the Doppler spectrum. An HFSW radar can usually detect very large ships against second-order clutter, so detection performance for this target class is only a weak function of sea-state. As a result, the detection of large vessels (typically greater than 1000 tons) is often limited by external noise at long ranges if not masked by Bragg lines. Increasing radiated power or system gain can therefore extend the detection range of such targets.

On the other hand, sea-state can strongly influence the detection of low-speed small and medium-sized surface vessels (typically less than 1000 tons) with a mean RCS of 30 dBsm, or less. The detection of such vessels is a strong function of the continuum

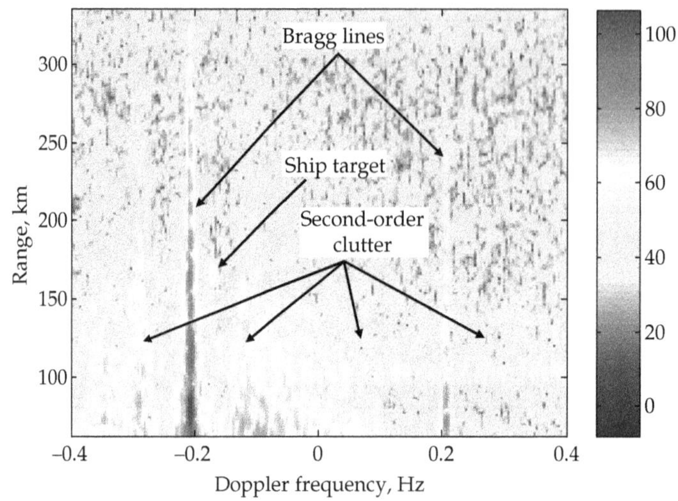

FIGURE 5.24 Range-Doppler characteristics of sea clutter recorded by an HFSW radar on the east coast of Canada. An example Doppler spectrum from this data appears as a line plot in Figure 5.22. A ship target echo is visible at a range of about 160 km in the second-order clutter trough next to the receding Bragg line. At higher sea-states, this trough would be filled with a higher level of second-order clutter. © Crown 2010. Government of Canada. (*Courtesy of Dr. H. Leong, Defence R & D Canada.*)

level at the radial velocity of the target, which is determined by the radar frequency as well as the wind speed and direction relative to the radar beam. Detection of small and medium-sized (low-speed) vessels is therefore heavily influenced by wind speed and direction, as well as the choice of operating frequency.

The detection of smaller surface vessels traveling against the wind is typically possible for radial velocities greater than about 20 knots. However, the detection of small surface vessels becomes more difficult when the target moves in the same direction as the prevailing wind and when the wind blows parallel to the radar beam. In this case, detection may be possible in low sea-states (sea-state 2 or less), but becomes unlikely in high states unless the target is moving with a high radial speed in excess of 20 knots. A quantitative analysis of target blind (radial) speeds as a function of target RCS, sea-state (mean wind speed and direction), operating frequency, and CPI length can be found in Maresca and Barnum (1982). Although this analysis was conducted for skywave OTH radar systems, the general findings for the lower (nighttime) frequencies are also relevant to HFSW radars.

For a given sea-state, the power contained in the second-order clutter continuum relative to that in the Bragg lines tends to rise as the radio frequency increases beyond that at which the resonant waves have reached their equilibrium limit. In other words, the cross section per unit area of the second-order continuum tends to grow with operating frequency for a given sea-state, while that of the first-order echo is predicted to remain approximately constant with frequency in the region where the resonant waves are fully developed.

The first- and second-order scattering cross sections per unit area of the sea may therefore be reduced by using lower radio frequencies, where the sea surface appears smoother to the electromagnetic wave and the resonant waves are less likely to be fully developed. At higher frequencies, the resonant Bragg waves have smaller wavelengths and may be saturated by relatively light winds, while the Doppler spectrum of second-order continuum is typically higher in level (relative to the Bragg lines) and has a form that is more sensitive to wind speed and direction than at lower radio frequencies.

If the radar is operated at a low frequency, where the sea is not fully developed, the energy contained in the Bragg lines and second-order continuum will be significantly reduced. Under such conditions, it often possible to detect low-speed small and medium-sized vessels to significantly greater ranges than at high frequencies. An exception may apply for very small vessels if the RCS falls into the Rayleigh region at the lower frequency.

Despite lower operating frequencies often being preferred for ship detection in HFSW radar, the use of higher frequencies may under certain circumstances improve target SCR. The underlying basis for this is that a higher carrier frequency can shift the Doppler frequency of a target echo from a location near or within the Bragg lines to a region well outside of the Bragg lines. The second-order clutter level outside the Bragg lines also increases at a higher carrier frequency, but the clutter spectral density at the Doppler frequency of the target echo may be less than that competing with the echo from the same target at a lower carrier frequency.

A higher frequency also reduces the scattering patch area due to the finer angular resolution of classical beamforming. It may also improve target RCS, particularly for smaller surface vessels. The benefits of using higher frequencies for clutter-limited detection tend to be more pronounced for small surface vessels at short ranges, where the advantages of higher Doppler shift, reduced resolution cell size, and potentially higher RCS may outweigh the increase in cross section per unit area of the second-order clutter integrated over all Doppler frequencies. This assumes that detection at the target range remains limited by second-order clutter despite the additional surface-wave attenuation at higher frequencies.

In a clutter-limited environment, it is well known that the detection of radar echoes from point targets can be improved by reducing the size of the spatial resolution cell. Unfortunately, the signal bandwidth cannot be increased beyond a few tens of kilohertz due to heavy user congestion in the lower HF band. In addition, it may be not be feasible or convenient to increase the aperture of the receive antenna beyond a maximum length of a few hundred meters, either because of economic or operational reasons, including site constraints due to land topography in coastline regions. Such factors limit the ability to reduce the spatial resolution cell area of an HFSW radar beyond a certain size.

Provided the target echo Doppler shift does not coincide with that of a Bragg line, it may be possible to improve the target SCR by increasing the CPI. Second-order clutter is a finite-bandwidth signal with relatively shorter temporal coherence than a target echo, assuming the target moves with a nearly constant velocity over the CPI. Longer integration times can therefore reduce the second-order clutter energy competing with a perfectly coherent target echo in a particular Doppler frequency bin by narrowing the analysis bandwidth.

The benefit of using longer CPIs for target detection against sea clutter is not only predicted in Maresca and Barnum (1982), but has also been observed in practice (Menelle, Auffray, and Jangal 2008). To address the issue of optimizing the CPI for different types of targets, the received pulse-trains can be partitioned within the signal processing system and processed using a number of different CPI. However, the radar CPI is often optimized according to the class of target to be detected, and other issues such as range-migration and maneuvering targets (which results in Doppler spread), so it cannot be increased indefinitely without incurring penalties in detection and tracking performance. Consequently, the option of using longer CPIs is also restricted. Attempts to enhance SCR by increasing the CPI eventually become counterproductive when overall system performance is considered.

As mentioned previously, the second-order clutter power relative to that in the Bragg lines tends to increase with sea-state at a given radio frequency. An experimental study of the relationship between average wave height (an indicator of sea-state) and mean second-order clutter power between the Bragg lines was conducted in Leong (2002) at a radio frequency of 3.1 MHz. An approximate linear relationship was found when both quantities were expressed on a decibels scale. Specifically, doubling of the average wave-height increased the mean power of the second-order clutter between the Bragg lines by about 13 dB. This provides an approximate guide for scaling target detection performance against second-order clutter on the basis of sea-state.

The effect of sea-state on the detection of large and small vessels competing against second-order ocean clutter was investigated by Leong and Ponsford (2008) using frequencies of 3.1 and 4.1 MHz. Two classes of ships were considered, large cargo freighters with gross registered tonnage (GRT) in the order of several tens of thousands of tons, and smaller vessels with a GRT of about 1000 tons. The difference in angle-averaged RCS for these two types of targets was estimated to be approximately 10 dBsm. It was concluded that very large vessels could be detected against the second-order continuum regardless of sea-state due to their high RCS (in the order of 40 dBsm). The detection of large vessels may be precluded by Bragg line masking at any range (Leong, Helleur, and Rey 2002) or the low signal-to-noise ratio at ranges beyond about 150 km at night.

Second-order sea clutter was found to have a significant impact on the detection of smaller vessels, but at high sea-states only. It was concluded that small ships were difficult to detect in very rough seas with significant wave heights between 5.5 and 6.5 m beyond ranges of about 100 km. The same targets also could also not be detected in rough seas with significant wave heights between 3.6 and 4.2 m at ranges beyond about 150 km. The effective clutter RCS increases with range due to the greater cross-range dimension of the resolution cell, while the RCS of a point target with constant aspect and speed clearly remains the same. Sea clutter scattered by rough to very rough seas can therefore strongly impact the detection of small ships at longer ranges where its level is above the noise.

The results in Leong and Ponsford (2008) show that the radar tends to perform better at the lower frequency of 3.1 MHz compared to 4.1 MHz for the detection of the considered targets against sea clutter. In practice, the optimum choice of operating frequency in a sea clutter–limited environment not only depends on the ocean directional wave-height spectrum, but also on the target parameters, including RCS behavior, radial velocity, range, and azimuth relative to the mean wind direction. The signal frequency represents the main parameter that the radar can control to optimize the detection of smaller surface vessels against second-order clutter. However, for a realistic ocean directional wave-height spectrum, no single frequency can maximize the statistically expected SDR for all target locations and radial speeds.

5.3.2 Ionospheric Clutter

Although HFSW radars do not rely on skywave propagation to operate beyond the line of sight, it is obviously not possible to simply "turn off" the ionosphere for such systems in practice. Unfortunately, not all of the radio wave energy emitted by an HFSW radar is coupled to the surface-wave mode. Part of the radiated signal inevitably propagates upward and impinges on the ionosphere. Under certain conditions, some of the incident skywave energy will be reflected or scattered by the ionosphere and returned to the receiver. Reflections from the ionosphere represent a source of disturbance to an HFSW radar. Unwanted echoes received by an HFSW radar via the ionosphere are collectively referred to as *ionospheric clutter*.

Ionospheric clutter is detrimental to HFSW radar because it has the potential to mask target echoes at operational ranges from about 90 km up to the EEZ limit and beyond. When scattering in the ionosphere occurs from dynamic electron density irregularities, the returned echoes may be significantly spread in Doppler and contaminate the entire velocity search space. Ionospheric clutter that is spread in range and Doppler can seriously impair the target detection performance or remote sensing capabilities of an HFSW radar.

Ionospheric clutter is considered by many practitioners as one the greatest impediments to achieving consistent HFSW radar performance at ranges beyond about 90 km. For this reason, significant attention has been paid to understanding the properties of ionospheric clutter and developing techniques to reduce its impact on operational systems. This section discusses ionospheric clutter path types, range-Doppler characteristics, frequency dependence, and spatio-polarimetric properties to motivate a number of mitigation approaches.

5.3.2.1 Path Typologies

Ionospheric clutter may arise as a result of the radar signal being propagated in several different ways. An investigation into the most effective ionospheric clutter paths (i.e., those with the greatest potential to cause disturbance in an HFSW radar) was conducted by Sevgi, Ponsford, and Chan (2001) using the ICEPAC simulation software package. Three primary self-interference path categories may be identified. The first corresponds to a direct skywave path, where the transmitted signal is reflected by one or more layers in the ionosphere toward the receiver from a virtual height that may range from 90 to 400+ km, i.e., the one-way transmitter-ionosphere-receiver skywave path.

This type of path is associated with ionospheric reflections at elevation angles near vertical incidence (NVI) for a single-site HFSW radar that operates below the maximum critical frequency of the ionosphere. Clearly, bistatic HFSW radar systems with a relatively small inter-site separation are also subject to the reception of ionospheric clutter via the NVI path below the maximum useable frequency of the oblique circuit. In practice, direct backscatter from the ionosphere can at times be received from directions different to that of "specular" reflection. In a single-site system, for example, ionospheric clutter may be received from directions other than the NVI path due to large-scale ionization gradients and electron density irregularities, as well as transient echoes backscattered from meteor trails in the upper atmosphere.

The second type of signal path corresponds to two-way skywave propagation involving intermediate scattering from the Earth's surface. In this case, ionospheric clutter may be received at lower elevation angles due to the signal being backscattered by land or sea surfaces at potentially long distances from the radar. An example of the Earth surface backscatter (ESB) path is transmitter-ionosphere-sea-ionosphere-receiver. Operation above the maximum layer critical frequency can help to mitigate radar echoes returned via the direct NVI path, but it will not necessarily eliminate ionospheric clutter received over oblique ESB paths from longer ranges due to surface scatterers that are beyond the edge of the skip-zone.

The third so-called "mixed-path" category arises due to a combination of skywave and surface-wave propagation between transmitter and receiver. This is not to be confused with mixed-path surface-wave propagation over segments of land and sea, which was discussed previously. The simplest examples in this third category include the signal path: transmitter-ionosphere-sea-receiver, with the final leg being via the surface-wave mode, and vice-versa, i.e., the transmitter-sea-ionosphere-receiver path, with the first leg being via the surface-wave mode. The former of these examples has the potential to be particularly insidious as the received disturbance has the same polarization as the target echo and also arrives at near-grazing elevation angle.

5.3.2.2 Range Occupancy

During the day, radar echoes from the normal E-layer or sporadic-E, as well from the upper D-region or mesosphere, may potentially affect HFSW radar performance at operational ranges between about 90 and 130 km. This range band applies for a single-site system when ionospheric clutter is received from the D- and E-regions by virtue of the direct NVI path. Radar resolution cells at longer ranges may be corrupted by ionospheric clutter propagated via ESB paths. Although D-region ionization and the normal E-layer effectively disappear after sunset, sporadic-E may be present at various times throughout the day or night.

Ionization levels that are sufficiently high to reflect signals in the lower HF band are always present in the F-region. Ionospheric clutter received from the F-region may contaminate longer ranges in the interval between about 200 and 400 km or greater via the NVI and mixed-paths. Ionospheric clutter received from the F-region via ESB paths will be at ranges outside of the operational HFSW radar coverage, but such returns can potentially mask target echoes when they are folded in through range-ambiguities.

In any given radar dwell, ionospheric clutter may not occupy the entire range extents quoted in the previous paragraphs for the E- and F-regions and the NVI path. For example, ionospheric clutter from a relatively thin sporadic-E layer may only contaminate a 10–15 km range band contained within the 90–130-km interval. In addition, not all ionospheric regions may simultaneously contribute ionospheric clutter as the disturbance type that limits detection performance at a particular operating frequency. For example, mainly E-region returns will be received during the day when operation is below the E-layer critical frequency, despite the presence of F-region layers above it.

In addition, the range occupancy of ionospheric clutter varies over time, particularly when returned from the F-region. Figure 5.25 shows the variation in nighttime F-region ionospheric clutter power received as a function of range by a single-site HFSW radar on the east coast of Canada. Note that the range at which the nearest (NVI) echoes are received varies from below 250 to about 400 km in about 5 hours between 00:00 and 05:00 UT. Ionospheric clutter possibly returned via mixed-paths is also evident in the display.

In unfavorable conditions, the range occupancy of ionospheric clutter can be broad and potentially contaminate the entire coverage from 90 km up to the 200 nautical mile EEZ limit (370 km). In particular, ESB and mixed-path propagation described previously can significantly increase the range occupancy of ionospheric clutter beyond that of the NVI path. It is also important to distinguish between range-coincident and range-ambiguous ionospheric clutter. The former is received from group-ranges or virtual paths lengths less than the first range ambiguity of the (periodic) radar waveform, while the latter can fold into the radar coverage from much longer range due to first- and higher order range ambiguities. The latter is also known as "range-wrapped" or "range-aliased" clutter.

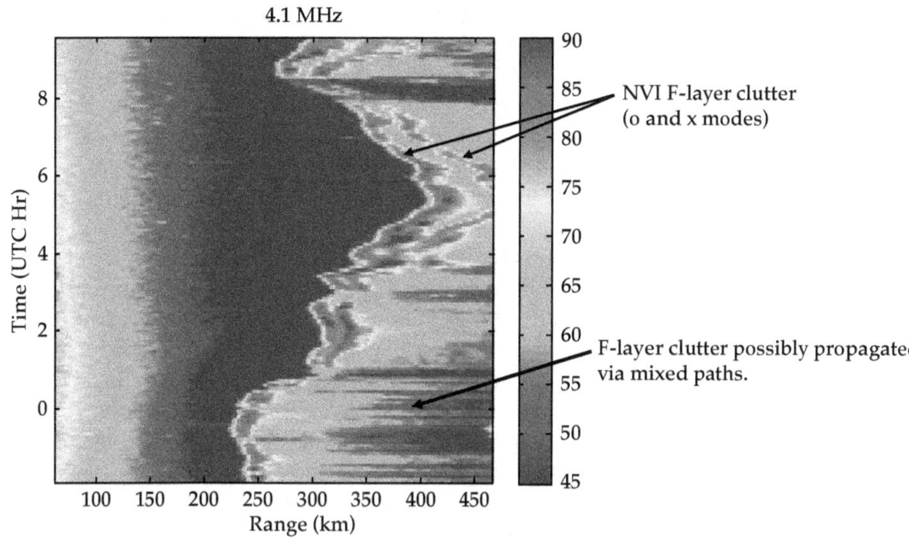

FIGURE 5.25 Example of nighttime variation of ionospheric clutter power from the F-region as a function of range as received by a single-site HFSW radar. The intensity modulated display shows the average clutter power in decibels (uncalibrated) received between the Bragg lines along a fixed beam at 4.1 MHz. The range of the nearest (NVI) echoes is observed to vary by more than 150 km over a 5 hour period. Magneto-ionic splitting of the skywave signal into ordinary (o) and extraordinary (x) waves, which produces multiple NVI returns at different ranges, is indicated in the display. The relatively weaker ionospheric clutter that is distributed continuously over longer ranges trailing the strong NVI echoes is most probably received via mixed paths. © Crown 2010. Government of Canada. (*Courtesy of Dr. H. Leong, Defence R & D Canada.*)

In the following, attention is mainly restricted to range-coincident ionospheric clutter received via the NVI path.

5.3.2.3 Doppler Characteristics

Experimental investigations confirm that the Doppler characteristics of ionospheric clutter can vary markedly in both frequency spread and centroid. The Doppler spectrum profile of ionospheric clutter may be qualitatively classified as being due to either *specular* reflection or *diffuse* scattering. The former corresponds to reflection of the radar signal from a frequency-stable ionospheric layer, which may impose a Doppler shift on the echo, but very small Doppler spread compared to the waveform PRF. In this case, the received signal is almost discrete in Doppler frequency with similar spectral characteristics to a target echo. Although specular returns can be very powerful, most of the energy is concentrated in a very narrow band of Doppler frequencies after coherent integration. These ionospheric clutter echoes are often not a problem as far as target masking is concerned, but they may cause false alarms in surveillance systems, or confuse the interpretation of the sea echo spectrum in remote sensing applications.

On the other hand, diffuse ionospheric clutter may occupy a broad band of Doppler frequencies that may approach or exceed the waveform PRF when the radar signal is scattered from dynamic electron density irregularities in the ionosphere. Contamination by so-called "fast" spread-Doppler clutter can at times mask aircraft target echoes with radial speeds of perhaps 100–200 m/s. The presence of diffuse ionospheric clutter can significantly increase the potential for target echo obscuration in the affected ranges. Figure 5.26 illustrates these two forms of ionospheric clutter, both of which are received as time-continuous signals that persist over the entire CPI. Clutter that is spread in Doppler due to the short duration of the echo relative to the CPI (e.g., transient meteor echoes) was discussed in Chapter 4.

Various physical mechanisms responsible for producing diffuse ionospheric clutter may be identified based on a combination of experimental analysis and simulations. One class of spread Doppler clutter is thought to arise due to scattering from dynamic small-scale geomagnetic field-aligned electron density irregularities produced by Kelvin-Helmholtz instabilities (Abramovich et al. 2004). Spread Doppler clutter may also arise due to large-scale atmospheric gravity waves or traveling ionospheric disturbances, which temporally modulate the signal phase-path. Another possible spread Doppler clutter mechanism is the interference of unresolved ionospheric propagation modes including "micro-multipath" rays scattered from a layer that has a spherically inhomogeneous electron density distribution.

The physical mechanisms responsible for diffuse ionospheric clutter in the E- and F-regions of the equatorial, auroral, and mid-latitude ionosphere differ significantly, but tend to be more prevalent during high sunspot years of the solar cycle when flare activity and hence the potential for geomagnetic storms is greatest. Although E-region echoes are typically more frequency-stable than F-region echoes, diffuse ionospheric clutter with significant spread in Doppler frequency may also be received from electron density irregularities in the normal E and sporadic-E layers (Thayaparan and MacDougall 2005). The reader may consult Davies (1990) for an in-depth review of the physical mechanisms responsible for spread Doppler clutter in different height regions of the ionosphere, as well as their synoptic dependence on time-of-day, season, year in the solar cycle (solar activity), and magnetic latitude.

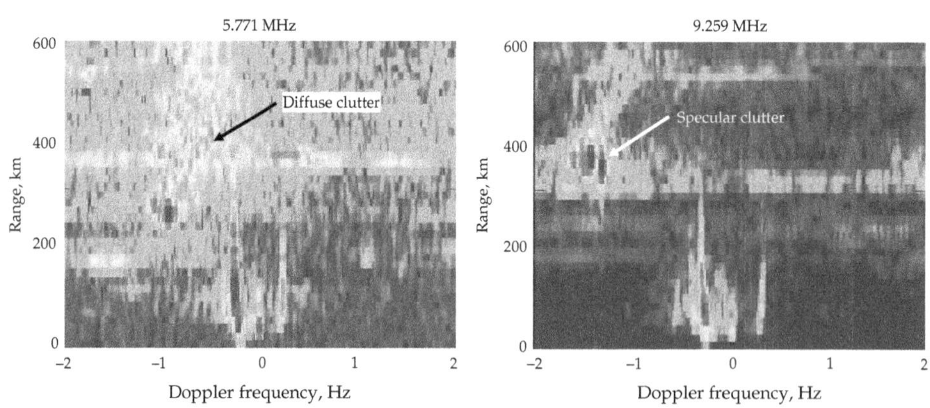

FIGURE 5.26 Range-Doppler maps recorded by the Australian Iluka HFSW radar in a fixed beam steer direction at operating frequencies of 5.771 and 9.259 MHz. The frequency dependence of the received ionospheric clutter characteristics (range occupancy and Doppler spread) is evident in these displays, recorded approximately 20 minutes apart. © Commonwealth of Australia 2011.

In practice, an HFSW radar may receive a mixture of specular and diffuse ionospheric clutter. Note that purely specular (i.e., perfectly coherent) ionospheric clutter does not exist in practice, strictly speaking. In good (frequency-stable) ionospheric conditions, Doppler spread clutter components caused by imperfect temporal coherence may be received several tens of decibels below the strength of the main specular reflection peak in the Doppler spectrum. However, due to the extremely high compression gain of HFSW radar systems, which may use coherent integration times in the hundreds of seconds for slow-moving target detection, such components can still be high enough above the noise floor to limit detection performance.

As stated in (Abramovich et al. 2004), many existing ionospheric sounders cannot capture ionospheric properties over such a wide dynamic range. Current ionospheric models incorporated into software packages such as ICEPAC also find it challenging to accurately portray the characteristics of ionospheric clutter over dynamic ranges commensurate with those of modern HFSW radar systems.

Experimentally validated models capable of representing ionospheric clutter to the fine level of detail required in surveillance applications are unfortunately not available at present. This is particularly the case for systems that receive clutter from tropical or arctic regions of the ionosphere, where the physical mechanisms responsible for generating Doppler spread echoes are more prevalent (and complex) than at mid-latitudes. Such models would be of significant value to guide mitigation techniques based on adaptive processing.

5.3.2.4 Frequency Dependence

The reception of ionospheric clutter and its range-Doppler characteristics depend strongly on the choice of operating frequency. During the day, ionization in the E- and F-regions can potentially give rise to ionospheric clutter, while ionization in the D-region is mainly responsible for attenuation of skywave signals. Ionospheric clutter from the E- and F-regions can only be received via (at least) two passages through the D-layer in the day. At operating frequencies lower than the D-layer critical frequency, skywave propagation is effectively not supported due to high absorption in the D-region. Operating frequencies lower than about 3 MHz tend to be highly attenuated around the middle of the day. Operation below the D-layer critical frequency, or at frequencies near the low end of the HF band, can therefore greatly reduce ionospheric clutter received from the E- or F-regions during daylight hours.

At night, when the D-layer disappears due to electron-ion recombination, an HFSW radar may be operated above the maximum critical frequency of the ionosphere at the expense of greater surface-wave path-loss. This frequency typically corresponds to the F2-layer critical frequency, although sporadic-E can at times be the most dense layer. In principle, radio waves with frequencies above the maximum layer critical frequency will penetrate through the ionosphere at vertical incidence. In practice, ionospheric clutter may be received on NVI paths even when operating above the nominal F2-layer critical frequency due to the presence of electron density irregularities (Abramovich et al. 2004).

Moreover, oblique reflection may generate ionospheric clutter via ESB and mixed paths at frequencies above the nominal F2-layer critical frequency. The operating frequency may be raised above the maximum useable frequency (MUF) for oblique paths up to certain range. This creates a skip-zone around the transmitter, such that no backscatter can be received from the Earth's surface within the skip-zone by the regular process of

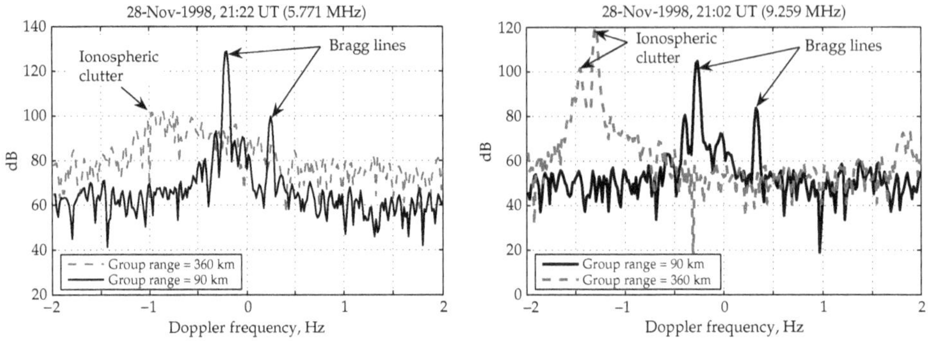

(a) Diffuse ionospheric clutter (5.711 MHz). (b) Specular ionospheric clutter (9.259 MHz).

FIGURE 5.27 Doppler spectra showing examples of diffuse and specular ionospheric clutter at a group range of 360 km (dashed lines). Doppler spectra at a range of 90 km containing mainly sea surface clutter with essentially no ionospheric clutter contamination are shown for comparison (solid lines). The data are from the range-Doppler display in Figure 5.26.

ionospheric reflection. Unfortunately, this often requires using frequencies that are too high to support effective surface-wave propagation at long distances up to the EEZ limit.

During the day, it may be possible to choose a frequency that selectively minimizes the reception of ionospheric clutter from either the E- or F-regions, but not from both layers simultaneously. For example, an operating frequency close to but not higher than the E-layer critical frequency will mainly produce E-region echoes, since the incident signals are reflected from the E-layer and do not reach the higher-altitude F-region. On the other hand, a signal frequency higher than the E-layer critical frequency (but not above the F2-layer critical frequency) will effectively penetrate through the E-region and reflect from F-region layers only.

Ionospheric clutter is typically a greater problem at night due to the disappearance of the absorptive D-layer and reflective E-layer, which may be exploited as a screen to reduce contamination from the F-region during the day. With the possible exception of sporadic-E, only the F2-layer is present at night, and the nighttime F2-layer is prone to generating diffuse (i.e., spread-Doppler) ionospheric clutter.

The range occupancy and Doppler spectrum characteristics of F2-layer ionospheric clutter may vary significantly with operating frequency. A comparison of F2-layer ionospheric clutter Doppler spectra received by an HFSW radar in the same range-azimuth cell at two well-spaced carrier frequencies is illustrated in Figure 5.27. Alternative methods to frequency selection for combating the ionospheric clutter problem, including antenna design, adaptive processing, and multi-frequency operation, are clearly needed and will be discussed later in this chapter.

5.3.2.5 Spatial and Polarimetric Properties

When NVI skywave returns are considered to be the dominant source of ionospheric clutter, the use of transmit and receive antenna elements with a deep and broad null at high elevation angles represents a primary mitigation strategy. However, the design of the antenna element alone is unlikely to completely remove all vestiges of NVI ionospheric clutter. Adaptive beamforming provides additional scope to discriminate between ionospheric clutter and target echoes based on differences in signal direction of arrival (DOA) or wavefront structure. Provided the ionospheric clutter has a spatial covariance matrix of relatively low rank, and the disturbance does not enter through the main lobe of the antenna beam pattern, considerable potential exists for improving the signal-to-clutter ratio by adaptive beamforming.

It has been argued that antenna arrays with elements deployed in two dimensions on the ground (e.g., L- or T-shaped arrays), provide enhanced vertical selectivity at high elevation angles to better cancel ionospheric clutter received via NVI paths. However, such array geometries may not be compatible with site constraints in coastline regions. Moreover, if there is a significant component of ionospheric clutter arriving at the receiver via the surface wave mode at near-grazing angle (due to mixed-path propagation), the added cost and complexity of a two-dimensional receive array design may not deliver the expected performance benefits. MIMO radar techniques that exploit 2D (nonlinear) arrays on receive and transmit along with a carefully chosen set of "orthogonal" waveforms may provide a form of remedy for the ionospheric clutter problem. Such systems have been discussed by Frazer, Abramovich, and Johnson (2007) in the context of skywave OTH radar.

Experimental analysis of ionospheric clutter has revealed that its spatial properties are extremely variable from one case to another (Xianrong, Feng, and Hengyu 2006a). At times, specular or diffuse ionospheric clutter returns are well localized in DOA, while at other times, the disturbance is broadly spread in azimuth (cone angle for a ULA) regardless of its Doppler spectrum characteristics. A commonly observed characteristic is that the spatial properties of ionospheric clutter are heterogeneous in time-delay and therefore need to be estimated on a range-by-range basis. Moreover, in a single range cell, the spatial properties of ionospheric clutter may additionally vary with pulse number or Doppler frequency. These characteristics significantly complicate the implementation of effective adaptive processing techniques for ionospheric clutter mitigation, particularly in terms of identifying sufficient and suitable training data.

The polarization of the received ionospheric clutter signal depends on the propagation path. Echoes received via NVI and ESB paths may be expected to be elliptically polarized, in general, as such paths involve skywave propagation directly to the receiver. On the other hand, mixed-path ionospheric clutter received via the surface-wave propagation mode in the final leg will be vertically polarized. In situations where the dominant ionospheric clutter contribution is elliptically polarized, an auxiliary set of horizontal antenna elements may provide further scope to improve SCR by means of adaptive spatio-polarization filtering.

In principle, such filtering enables the mitigation of elliptically polarized ionosphere clutter that enters through the main lobe of the receive beam pattern, where the latter is formed by using the vertically polarized antennas in the array. In practice, effective suppression of ionospheric clutter in the polarization domain requires very high correlation between the horizontal and vertical polarization components of the disturbance signal. If this correlation is not sufficiently high, or if the surface-wave contribution of ionospheric clutter on the mixed-path is significant, the potential to exploit the polarimetric properties of ionospheric clutter for mitigation is diminished.

In Abramovich et al. (2004), auxiliary horizontal loop antennas were used in conjunction with a ULA of 32 twin-monopole (doublet) elements to test this mitigation strategy. Ideally, the loop antennas do not pick up the pure ground wave signals. In practice, mutual coupling and other imperfections-limited surface-wave isolation in the loop antennas to several tens of decibels relative to the doublets. As stated in (Abramovich, Anderson, Gorokhov, and Spencer, 2004), experimental trials demonstrate that inter-polarization correlation for Doppler spread ionospheric clutter is unfortunately too low to provide any substantial mitigation capability. However, the intra-polarization spatial correlation properties were found to be quite high for both vertical and horizontal polarization components, which motivates the use of adaptive beamforming for sidelobe ionospheric clutter suppression.

5.3.3 Interference and Noise

The power spectral density of the external noise in the radar bandwidth ultimately places an upper limit on the maximum detection range of large vessels as well as smaller (air and surface) targets with radial speeds above approximately 20 knots. The main components of background noise were identified in Chapter 4, i.e., irreducible galactic noise, thunderstorm spherics, and man-made noise from unintentional radiators. In the lower HF band, atmospheric noise dominates at night in rural/remote locations and is higher in level near the equator than at the poles. Depending on the site location, incidental man-made noise may be limiting during the day in the lower HF band.

HFSW radar operation near the lower end of the HF spectrum (3–7 MHz) is often required in practice to reduce path-loss for the surface-wave propagation mode, particularly for the detection of ships at long ranges in excess of 200 km. A potential drawback of operating an HFSW radar in this frequency region is that user-congestion and the background noise level are typically greater than at higher frequencies. The reports (CCIR 1964, 1983, and 1988) and (CCIR 1970) provide HF noise spectral density estimates as a function of receiver location for different times of the day and seasons, taking into account the contribution of man-made sources separately. Operation in the lower HF band is particularly challenging at night due to the increased user-congestion and higher background-noise spectral density compared to daytime.

The significant diurnal dependence of clear channel availability and noise spectral density in the lower HF band is strongly linked to changes in ionospheric propagation conditions. During daylight hours, multi-hop skywave paths between 3 and 7 MHz are very lossy due to D-layer absorption. At such frequencies, HF interference and noise from long-distance sources are therefore highly attenuated over multi-hop paths during the day. In addition, many HF services that rely on the skywave mode tend to operate at higher frequencies during the day to exploit the increased ionospheric propagation support. The combination of very lossy long-range (multi-hop) skywave paths at low frequencies during the day, and the migration of many users to higher frequencies at this time, has the effect of reducing channel occupancy and external noise spectral density in the lower HF band compared to nighttime levels.

The D-layer effectively disappears at night and this typically allows HF signals in the 3–7 MHz band to propagate over long distances with reduced attenuation. Moreover, many HF systems are forced to move to lower frequencies at night due to the diminished support for skywave propagation at higher frequencies. In other words, the reduced ionization density in the nighttime F-region causes many users to crowd into the lower HF band, where operation via the skywave mode remains possible.

The combination of these factors results in a much greater number of interference and noise sources becoming "visible" to an HFSW radar at night. The higher user-congestion in the lower HF band at night can make it very difficult to find a clear channel on which to operate the radar, while the higher external noise spectral density can limit the detection of both fast- and slow-moving targets at relatively shorter ranges. The difference between day and night external noise spectral density levels at a fixed frequency in the lower HF band may be up to 20 dB or more.

Figure 5.28 shows the noise figure variation in 4-hour time blocks with season at a frequency of 4.1 MHz for an electrically quiet site on the east coast of Canada based on data from the final report of (CCIR 1964, 1983, and 1988). The maximum noise levels are observed to occur between the hours of 20:00 and 04:00 local time (i.e., columns 3 and 4 on the bar chart). The minimum noise levels are observed to occur between the hours of 08:00 and 16:00 local time (i.e., columns 1 and 6).

When D-layer absorption disappears shortly after sunset, the nighttime external noise level rises by about 15 dB with respect to the daytime level. As much of this noise is unstructured in space and time, its level cannot be reduced significantly by signal processing. HFSW radar performance therefore degrades at night. This is not only due to the degradation in SNR at all ranges, but also because ionospheric clutter from the F-region at longer ranges tends to be diffuse and is more difficult to mitigate for the reasons mentioned previously.

Incidental interference that overlaps the radar bandwidth from other users of the HF spectrum is often highly directional and sometimes structured in time. A variety of signal processing techniques that may operate in the temporal, spatial, and polarization domains (separately or jointly) have been developed to mitigate incidental man-made interference. For example, short-lived interference from transmitters such as

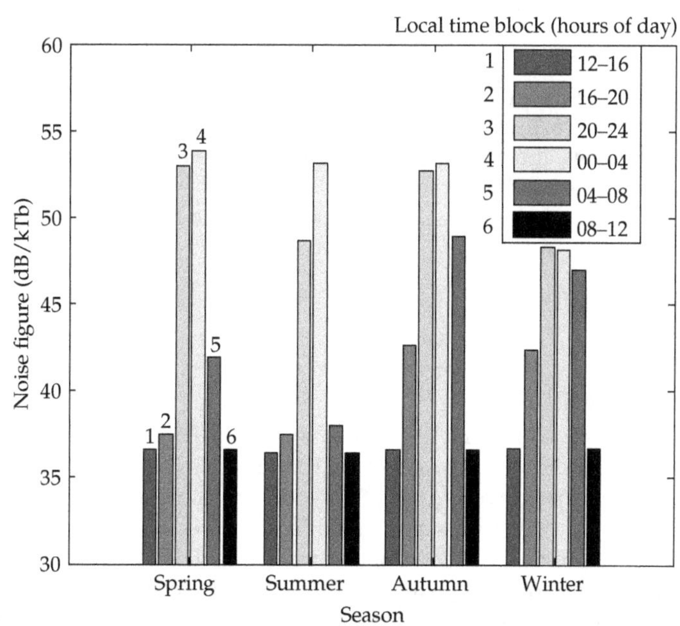

FIGURE 5.28 Noise figure at 4 MHz for a quiet site on the east coast of Canada. © Crown 2010. Government of Canada. (*Courtesy of Dr. H. Leong, Defence R & D Canada.*)

frequency-swept ionospheric sounders may be canceled using time-domain processing techniques similar to those used for suppressing atmospheric noise (lightning discharges) provided the impulse or burst duration is short compared to the CPI, as described in Chapter 4. The theoretical and practical performance of several such methods has been analyzed and compared using experimental HFSW radar data in Lu et al. 2010.

For time-continuous or long-lived interference on the time scale of the HFSW radar CPI, adaptive beamforming represents an alternative approach for rejecting persistent signals that overlap the radar bandwidth, but are not incident from directions in the main lobe of receive beam pattern (i.e., sidelobe interference). When such interference propagates to an HFSW radar via the skywave mode, the spatial non-stationarity of the disturbance signals over the relatively long CPI needs to be taken into account for effective mitigation. A robust adaptive beamforming technique suitable for practical implementation in modern HSWR radar systems was developed and experimentally tested in Fabrizio, Gershman, and Turley (2004).

Narrowband interference has been effectively canceled using a mismatched signal processing method in Ponsford, Dizaji, and McKerracher (2005). In this procedure, orthogonal phase codes are used to create a matched channel that is sensitive to radar echoes, external interference and noise, and a mismatched channel that is only sensitive to external interference and noise. The outputs of the two channels are cross-correlated to estimate the interference and noise in the matched channel. The narrowband interference is then removed by subtracting this estimate from the matched channel. Results for this technique have been reported in Ponsford, Dizaji, and McKerracher (2003).

Skywave interference will in general exhibit an elliptical polarization state that varies in space and time due to Faraday rotation. However, the surface-wave mode can only propagate HF signals with vertical polarization effectively over the sea. By incorporating auxiliary horizontally polarized receive antennas in addition to the vertically polarized main antennas, an HFSW radar can in principle employ adaptive spatio-polarimetric filtering to cancel skywave interference entering through the main lobe of the receive beam (formed using the vertically polarized antennas). In theory, this technique provides a means to reject skywave interference received from a similar direction to target echoes (i.e., within the main beam). This potential capability represents a distinguishing feature between surface-wave and skywave OTH radar systems to be discussed further in the next section.

5.4 Practical Implementation

The practical implementation of an HFSW radar system is driven by a number of factors that influence the choice of radar configuration, the selection of radar site(s), the antenna element and array designs, as well as the receive and transmit subsystem architectures. The latter includes radar waveforms and signal/data processing techniques, which are of paramount importance to the success of an operational HFSW radar system. In view of the preceding discussion on skywave OTH radar, this section focuses on practical implementation aspects that are specifically relevant to HFSW radar. The first part of this section is concerned with HFSW radar configuration and site selection, the second considers the transmit and receive subsystems, while the third discusses signal and data processing.

5.4.1 Configuration and Siting

Radar configuration refers to the relative positions of the transmitter and receiver. For example, in a monostatic configuration, the transmitter and receiver are located at a single site, while in a bistatic configuration, the transmitter and receiver are located at two well-separated sites. A multi-static HFSW radar, wherein a single transmitter services multiple receivers, may also be considered to improve track accuracy, but such configurations will not be considered here. The first part of this section briefly reviews the benefits and shortcomings of monostatic and bistatic architectures for HFSW radar, noting that systems designed primarily for surveillance within the EEZ have been implemented in both configurations. The second part of this section discusses the selection and preparation of sites for (land-based) HFSW radar systems, as both of these aspects are prerequisites for effective operation.

5.4.1.1 Radar Configuration

In the context of HFSW radar, a monostatic configuration often does not imply the use of a reciprocal antenna for receive and transmit.[3] Rather, this term often refers to the use of a single-site, which typically incorporates physically distinct but effectively colocated transmit and receive systems. A significant advantage of the monostatic HFSW radar configuration relative to bistatic systems is the expanded set of geographical coverage options. The field of view of a bistatic system is nominally limited by the overlap of the angular coverage of the transmit and receive antenna patterns. As these radiation patterns have a confined beamwidth in azimuth, the area/volume of overlap clearly maximizes for a monostatic configuration. Moreover, monostatic configurations typically lead to operational systems that are more economical to field, simpler to maintain, and easier to deploy at short notice. In addition, coastal real estate is only required at one site.

Despite these undeniable advantages, two-site systems have been preferred in several cases. A bistatic configuration permits operation using continuous waveforms, which provides higher average power than the pulse waveforms used in monostatic systems (for a given transmit peak-power rating). The inter-site separation required for bistatic systems depends on several factors, including the dynamic range of the receiver, the radiated signal power, the conductivity of the ground along the direct-wave path, and the transmit/receive antenna patterns in the direction linking the two sites. Ideally, the direct path is mainly over dry land of low conductivity in a direction well outside the azimuth sector of the radar coverage as viewed from the receiver. In practice, distances that provide sufficient isolation between the HFSW radar transmitter and receiver may vary between 40 and 100 km. Synchronization between two distant sites is usually achieved via a GPS time reference.

In principle, the transmitter and receiver do not need to be well separated when pulse waveforms are used in a single-site system. However, it has been found that separations of tens to hundreds of meters may be necessary to prevent so-called "dark noise" generated by the transmitter in the off-state from entering the receiver during echo reception. Perhaps counter-intuitively, this requirement can at times mean that the coastal length and surface area needed to accommodate a single-site system may actually exceed the combined costal lengths and surface areas of two separate radar sites in a bistatic

[3]Reciprocal antennas have been used in an HFSW radar system to be described in Section 5.5.

configuration. A comparative analysis of monostatic and bistatic HFSW radar configurations, which considers waveform characteristics, coverage area, detection performance, and site selection, is reported in Marrone and Edwards (2008).

In contrast to a monostatic system, radar echoes from targets and the sea surface are not received via backscatter in a bistatic system, but rather by a process of side-scatter. The measured group-range of a side-scattered radar echo from a (point) surface-target in the surveillance region is the sum of the path-lengths from transmitter-to-target and target-to-receiver. The measured group-range of the echo therefore locates the target on an ellipse whose two foci are at the positions of the transmitter and receiver. The intersection of the line emanating out from the receiver at the estimated azimuth of the target echo with the locus of the ellipse (corresponding to the measured group-range of the echo) determines the location of the target. This concept is illustrated in the plan-view sketch on the left-hand side of Figure 5.29.

While assessing the relative merits of monostatic and bistatic HFSW radar configurations, one needs to consider the impact of different viewing geometries on the expected Doppler spectrum characteristics of the target echo and sea clutter for the intended HFSW radar coverage and mission priorities. This is because viewing geometry can significantly affect target detection and tracking performance for a certain mission type and coverage area (Wang, Dizaji, and Ponsford 2004). With reference to Figure 5.29, the bistatic Doppler shift of a target echo (defined as the group-range rate) is given by f_D in Eqn. (5.87), where λ is the radio wavelength and v is the target speed. On the other hand, the Doppler shifts of the Bragg lines is given by f_B in Eqn. (5.87), where $\beta = \phi_T + \phi_R$ is referred to as the bistatic angle. For a given baseline L, Eqn. (5.87) implies that the Bragg lines are spaced closer together in Doppler frequency as the bistatic angle increases. This explains why the Bragg line separation in Doppler frequency narrows for a bistatic system as the group-range decreases. Indeed the Bragg-lines are observed to merge (at zero Doppler

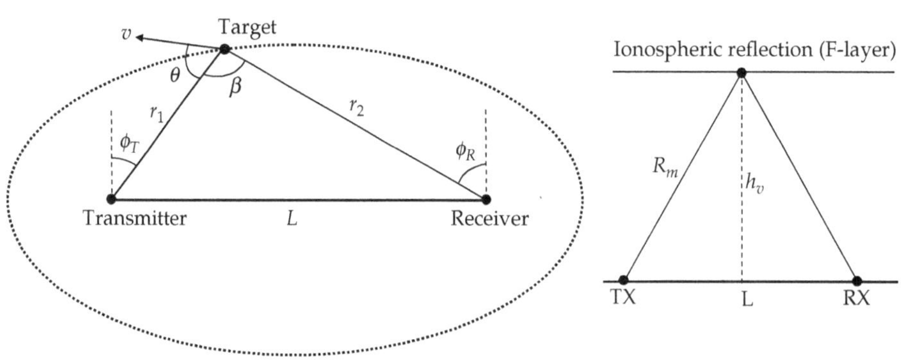

FIGURE 5.29 The diagram on the left shows a plan view of the bistatic radar scattering geometry for a surface target. The transmitter and receiver are separated by a distance L, the bistatic angle at the target location is β, while the total path-length of the target echo is $r_1 + r_2$. The measured group-range of the echo $(r_1 + r_2)/2$ locates the target on the ellipse, while the echo angle of arrival ϕ_R determines the position of the target on the ellipse. The diagram on the right shows a side-on view of a near vertical incidence ionospheric clutter path for the bistatic radar configuration. A single F-layer reflection with virtual h_v and total (direct-wave) path length R_m is illustrated.

frequency) in a bistatic system as the group-range tends to that of the direct wave, i.e., when $\beta = \pi$.

$$f_D = \frac{2v}{\lambda}\cos(\theta + \beta/2)\cos(\beta/2), \qquad f_B = \pm\sqrt{\frac{g}{\pi\lambda}\cos(\beta/2)} \tag{5.87}$$

A key point is that optimizing the radar configuration for a particular deployment needs to take into account the anticipated speeds and headings of the targets of interest with due regard to the characteristics of land and ocean surfaces in the surveillance area, including the synoptic behavior of surface-currents and the directional wave-height spectrum. Evaluating the probability of detecting target echoes against first- and second-order sea clutter using such information represents an important input for optimizing HFSW radar design and viewing geometry. A poor choice of radar configuration (and sites) can significantly degrade the capabilities of an HFSW radar system (Anderson 2007).

The group-range and angle-of-arrival distributions of ionospheric clutter also differ for monostatic and bistatic systems. The sketch on the right in Figure 5.29 illustrates that the group-range $R_m/2$ of the NVI ionospheric clutter path is approximately related to the inter-site separation L and virtual height of signal reflection h_v by Pythagoras' theorem, assuming a relatively short baseline length L (under 200 km). The influence of site separation on the coverage of an HFSW radar contaminated by NVI ionospheric clutter from the nighttime F-layer was analyzed in Leong (2006). It was concluded that a bistatic configuration can potentially allow the detection of ships at greater ranges than a monostatic system at night when ionospheric clutter is limiting. It was also argued in Leong (2006) that the elliptical constant group-range contours of a bistatic HFSW radar are more suitable for the surveillance of areas near water entries along coastlines.

5.4.1.2 Site Selection

A critical requirement for HFSW radar is to effectively couple vertically polarized signals (transmitted and received by the system) to the surface-wave propagation mode. A major factor influencing the coupling of radio wave energy to the surface-wave mode over the sea is the position of the antenna element(s) relative to the highly conducting saline water surface. Specifically, the height of the antennas above sea level and the distance of the antennas from the shoreline represent important criteria for HFSW radar site selection. This is because transmit and receive sites located on or very close to the coastline are necessary to effectively couple the ground-plane of the antenna to the ocean surface.

Ideally, the antennas should be placed as near as possible to the high-tide water mark to improve coupling of transmitted and received signals to the surface-wave mode. Locating the antennas even one or two wavelengths above sea level can introduce appreciable site losses (Berry and Chrisman 1966). The deleterious effect of antenna height on path loss has been validated in experiments. It has been found that raising the transmit antenna height by 35 m above sea level (with the receiver at sea level) results in an additional loss of approximately 6 dB over a one-way 140-km path. Elevating the antenna by a further 35 m incurred practically no additional penalty (Anderson et al. 2003).

An HFSW radar receive site typically requires relatively straight stretches of coastline at least 200–500 m in length to accommodate the major dimension of the antenna array. A receive site approximately 50-m wide is required to account for the antenna doublet separation, the length of ground radials and an isolation corridor. This translates to about 10–40 thousand square meters of coastal real estate. In addition, a site with relatively even topography and uniform ground electrical properties is preferable, particularly to minimize sources of array calibration errors. The combination of the above-mentioned factors often implies that HFSW radar site selection is heavily constrained in practice. Remote sea-state sensing systems, such as CODAR, utilize compact antennas that occupy a much smaller footprint, which significantly eases site constraints.

The HFSW radar receiver site should be well isolated (i.e., sufficiently distant) from residential and industrial areas to reduce the spectral density level of man-made noise below that of atmospheric noise in the lower HF band. If possible, it is also preferable for an HFSW radar system not to be located in regions of concentrated thunderstorm activity, where background noise levels tend to be higher. The impact of interference and noise at the site location additionally depends on the angular coverage of the receive beams steered for surveillance with respect to the directions of dominant sources, as well as the ability of these sources to reach the radar via ground-wave or skywave propagation. The large variability in ionospheric conditions with magnetic latitude and time of day also needs to be considered for site selction, as the occurrence and severity of Doppler-spread ionospheric clutter depends strongly on both of these factors.

HFSW radars operate effectively over open ocean areas, particularly in regions where water salinity is high and where high sea-states occur less frequently. Surface roughness can significantly attenuate the surface-wave signal, especially at frequencies near the middle of the HF band. High sea-states also increase the received surface-clutter power in Doppler intervals where slow-moving target echoes need to be detected. The impact of local bathymetry on the received clutter properties, including dominant ocean currents and tides in the coverage, should be considered when selecting the radar site(s). Careful attention also needs to be paid to the presence of land masses or islands, particularly those which are wide or long and close to the transmitter or receiver, as surface-wave attenuation can be significantly increased by propagation over significant stretches of terrain. The relatively lower conductivity of fresh-water lakes, or inland seas where ocean water mixes with fresh water, may preclude the effective operation of HFSW radar systems. For this reason, sites looking out to open ocean but adjacent to significant outflows of fresh water may be sub-optimal.

HFSW radars are almost always land-based systems. However, there has been considerable interest in evaluating the feasibility of mobile shipborne HFSW radar installations. Because of the long radio wavelengths in the lower HF band, the entire ship will radiate as the antenna, which can degrade the resulting antenna gain and radiation pattern properties. Mutual electromagnetic interference (EMI) with other shipboard systems is also a significant issue. An electromagnetic compatibility (EMC) study was conducted in Li et al. (1995) for the case of HF antennas mounted on a Navy ship. An experimental analysis of HFSW radar sea clutter received on a moving platform appears in Xie, Yuan, and Liu (2001). The possibility of fielding multistatic systems involving shipborne sites has been discussed in Baixiao et al. (2006).

5.4.2 Radar Subsystems

An HFSW radar is composed of three main subsystems; the transmit system, the receive system, and an HF spectrum occupancy monitor for clear channel advice. The latter, which may be regarded as being part of the receive signal processing system, is not described here as it was previously discussed in for skywave OTH radar as an element of the frequency management system. As there is also considerable overlap between the main subsystem design concepts for skywave OTH and HFSW radar, topics such as transmit/receive antenna/array design, and signal processing will be covered briefly here.

5.4.2.1 Transmit System

Skywave OTH radars have been implemented using antennas with vertical or horizontal polarization, as ionospheric propagation is supported for both, but the polarization of an HFSW radar transmit antenna must be vertical to effectively support surface-wave propagation at OTH ranges. Regarding the choice of transmit antenna element, the main aspects of concern are antenna efficiency over the design frequency range, azimuth coverage to floodlight the surveillance area, high gain at low elevation angles for effective coupling to the surface-wave mode, low gain at high elevation angles to reduce illumination of skywave clutter sources, and adequate front-to-back ratio for systems that only look for targets in a forward direction.

Other important considerations include the susceptibility of the antenna structure to Aeolian noise caused by mechanical vibration induced by winds, which may be strong in coastal regions, the capability to radiate HF signals at the required power levels without significant distortion due to arcing and other nonlinear phenomena, as well as cost and ease of deployment in the field. A ground screen, radial wires, or counter poise is typically used to improve the low elevation angle gain of the antenna.

A number of past and present systems, including the high-power site of the Iluka HFSW radar, have selected the vertical log-periodic dipole array (LPDA) as the transmit antenna element. For maximum efficiency in the 5–15 MHz frequency range, resonant antenna elements need to be large structures. The radiating elements of an LDPA antenna may be 10–40 m high and require an area of about 200-m by 200-m to install, including the supporting structure, ground screen, and equipment compound. An HFSW radar using a single LPDA transmit antenna floodlights the entire coverage area with a relatively broad beam that may be 90–120 degrees wide in azimuth (at low elevation angles). Unlike skywave OTH radars, this permits the entire surveillance area to be illuminated simultaneously at the expense of reduced sensitivity in a noise-limited environment. HFSW radar systems based on multiple transmit antenna elements and a digital waveform generator per element architecture that facilitates electronic beam steering have also been developed and will be described in the final section of the chapter.

The average transmit power of HFSW radar systems designed for surveillance up to the EEZ limit is typically between 1 and 10 kW. Solid-state devices offer an attractive alternative to vacuum tube amplifiers in HFSW radar applications that demand operation over a wide frequency band with rapid frequency changes. For example, this capability is desirable to enable dual-frequency operation by generating interleaved radar pulses centered at two widely spaced operating frequencies (Leong and Ponsford 2004). By comparison, HFSW radars used for remote sea-state sensing typically transmit an order of magnitude lower power. For example, the low-power site of the Iluka HFSW radar was based on a simple dart antenna with a 100-W transmitter, while other low-power systems such as OSMAR have used a three-element YagiUda antenna driven by a 200-W power amplifier.

When an HFSW radar is surrounded by ocean (e.g., located on a narrow peninsula or mounted on a ship), transmit antennas with a good front-to-back ratio are important to reduce scattering which occurs from behind and then in front of the transmit location. This mechanism, referred to as the second part of second-order clutter (Ponsford 1993), can significantly raise the "wings" of the clutter Doppler spectrum in the frequency region outside of the Bragg lines. This adversely affects the detection of slow-moving targets by an amount that depends on sea-state and viewing geometry. The experiments described in Ponsford (1993) show that a figure-of-eight transmit antenna pattern increased the clutter spectral density at Doppler frequencies outside of the Bragg lines by 9 dB compared to a cardioid pattern for an HFSW radar largely surrounded by sea.

Two-site HFSW radar systems have predominantly used signals based on the repetitive linear FM continuous waveform. A constant-modulus waveform with unit-duty cycle not only maximizes the average power for a given transmitter peak-power rating, but also places less stringent demands on the linearity of power amplifiers. The repetitive linear FM continuous waveform can also maintain excellent ambiguity function properties after temporal or spectral tapering is applied to reduce out-of-band spectral emissions. Moreover, computationally efficient matched filtering techniques may be applied to such waveforms, as described in Chapter 4.

On the other hand, single-site systems have exploited sequences of phase-coded pulse waveforms, such as binary or Frank quadrature phase codes (Ponsford, Dizaji, and McKerracher 2003). The phase codes used in successive pulses of the CPI can be changed to mitigate range-ambiguous (second-time-around) radar echoes scattered by long-range (skywave) clutter sources. In any case, effective Doppler processing in HFSW radar requires high spectral purity of the radiated waveform. Phase noise characteristics better than 80–100 dBc/Hz at 1 Hz from the carrier are often necessary to avoid instrumental (as opposed to physical) limitations to system performance.

5.4.2.2 Receive System

The majority of HFSW radars used for surveillance have adopted a receive antenna based on a uniform linear array (ULA) of vertically polarized singlet, doublet, or quadruplet monopole elements. With respect to 2D arrays, the widespread use of the ULA geometry in HFSW radar systems is motivated by the relatively simpler array calibration and data processing, as well as the ease of deployment on straight stretches of coastline, where the array may be oriented along the shore with boresight perpendicular to the water's edge. For example, the Australian Iluka system was based on a ULA of 32 monopole antennas with an aperture of 500 m and a digital receiver per element architecture. Such a ULA provides a receive beam with a main lobe about 5 degrees wide in azimuth at an operating frequency of 6 MHz. Two dummy elements (not connected to a receiver) are often placed at either end of the ULA to improve the homogeneity of mutual coupling, particularly for the first and last receiving elements.

The receive antenna elements do not need to be as well-matched over the design frequency range as the transmit element. The main justification for this is that the radar operates in an externally noise-limited environment (presumably with little or no spatial structure). In this case, a well-matched receive antenna element increase the gain for radar signals and external noise by the same amount, which yields no improvement in SNR. The use of (relatively short) monopole elements on receive significantly reduces the cost and footprint of the antenna array, besides being more robust to Aeolian noise. However, it has been argued that more efficient (broadband) antenna elements on receive may improve SNR when the external noise field exhibits significant spatial structure and adaptive beamforming is used.

Ideally, the receive antenna element needs to provide high gain at low elevation angles over a wide azimuth sector (equal to that of the coverage), an appropriate front-to-back ratio for a forward-looking ULA, and a broad null for signals arriving at near-vertical incidence. An ideal vertical monopole antenna on a perfectly conducting ground plane has an omnidirectional pattern in azimuth and a maximum gain at grazing incidence. The gain decreases with increasing elevation and approaches zero near vertical incidence. A ULA based on vertical monopole antenna elements arranged as doublets (separated by about 15 m to provide adequate front-to-back ratio in the lower HF band) over a ground mesh-screen, which stabilizes the input impedance of vertical radiators and improves coupling of the antenna ground-plane to the surface-wave mode, represents a cost-effective solution that is consistent with the aforementioned objectives.

The ground screen is typically placed both under and in front of the receive array. In an attempt to increase antenna gain at low elevation angles, certain HFSW radar systems have used ground screens that extend all the way into the sea. On the other hand, low gain at high elevation angles is required to attenuate ionospheric clutter received via the NVI path. In practice, antenna design is not sufficient to eliminate the intense overhead and near-overhead echo, or the reception of skywave interference and noise, as such signals may be incident over a range of elevation angles. The combination of antenna design, frequency management, and signal processing is required to reduce performance limitations imposed by skywave disturbance signals in real-world systems.

In addition to the main (vertically polarized) antenna elements in the receive array, a number of auxiliary antennas with horizontal polarization may be incorporated to augment the receive array (possibly in two dimensions). Since ionospheric clutter and interference signals arriving via the skywave mode are elliptically polarized, the auxiliary antennas can, in principle, be used to cancel ionospherically propagated disturbances received by the main array using polarization filtering. This topic will be discussed in the next section with application to interference rejection. Although two-dimensional arrays, such as L-shaped and T-shaped geometries, have the potential to yield superior performance against clutter and interference received via the skywave mode, no currently operational HFSW radar system designed with surveillance as its primary mission has been implemented with this feature to date.

In state-of-the-art HFSWR radar systems, the classic superheterodyne receiver has been superseded by direct digital receivers (DDRx) that can sample the entire HF band at the antenna element after a bandpass preselect or filter. Multiple narrowband frequency channels can be digitally down converted and low-pass filtered to enable simultaneous multi-frequency operation as well as other radar support functions such as common aperture environmental noise monitoring. As far as adaptive processing is concerned, the DDRx architecture is relatively less susceptible to degradations in spatial dynamic range of the antenna array caused by nonidentical reception channel transfer functions (frequency responses), which often occurs in the selective (analog IF) sections of a classical superheterodyne receiver.

5.4.3 Signal and Data Processing

The rudimentary conventional signal- and data-processing steps described for skywave OTH radar in the previous chapter also apply to HFSW radar. The pulse compression, array beamforming, and Doppler processing steps follow identical principles, so these will not be repeated here. The main differences in relation to system resolution and accuracy are briefly summarized. Two signal-processing applications specific to HFSW radar that have not been discussed in detail so far will be described; namely, the mitigation of

NVI ionospheric clutter mitigation by adaptive processing, and the rejection of skywave interference by polarization filtering. CFAR and tracking techniques that show promise for HFSW radar are also identified.

The surface-wave mode is more frequency-stable than skywave propagation, which allows much longer CPIs to be gainfully employed in HFSW radar. This is particularly important for the detection of large surface vessels, where the target velocity also tends to remain steady for long periods of time. In many HFSW radar systems, the transmitter floodlights the whole coverage simultaneously such that the CPI equals the revisit rate. Relative to a step-scanning transmitter, which steers the beam to different directions in the coverage, this allows for greater time-on-target and Doppler-frequency resolution for a given revisit rate, which may improve detection performance in a clutter-limited environment. Ship-detection CPIs may extend into the hundreds of seconds in HFSW radar, while aircraft-detection CPIs typically range from 2 to 10 seconds.

The surface-wave mode is also less frequency-dispersive than the skywave mode. In principle, this allows radar bandwidths of up to 100 kHz or more to be employed for fine range resolution (1.5 km). The physical limitation imposed by the coherence bandwidth of the propagation channel is much less restrictive for an HFSW radar than for a skywave OTH radar. However, high user-congestion in the lower HF band often constrains the maximum group range resolution to about 5 km (30 kHz) in practice. While the range and Doppler resolutions of an HFSW radar are typically higher than a skywave OTH radar, the angular resolution is comparatively lower due to the relatively smaller receive apertures and typically lower operating frequencies. The receive beams may have half-power main lobe widths of about 5–10 degrees. This provides a cross-range resolution of roughly 4–8 km at a range of 50 km, and about 25–50 km at a range of 300 km.

It is also important to distinguish between the spatial resolution cell size and target localization accuracy. Provided the target echo is well resolved from other radar echoes in at least one of the three radar processing dimensions (azimuth, range, or Doppler), the localization accuracy of a target echo with high signal-to-noise ratio may be 5–10 times greater than the resolution in each processing dimension when the peak coordinates are estimated by interpolation between the highest amplitude sample and its two immediate neighbors. A salient point is that the high Doppler resolution of an HFSW radar can indirectly enhance the accuracy of target echo location estimates in range and azimuth by resolving a high SNR target echo from clutter returns and other target echoes.

Several CFAR techniques tailored specifically to the HFSW radar signal environment have been proposed and tested, using real data where an individual CPI may have more than one million pixels. The reader is referred to Wang, Wang, and Ponsford (2011) and Lu et al. (2004), as well as Dzvonkovskaya and Rohling (2006) and Dzvonkovskaya and Rohling (2007), and references therein, for specific details of CFAR implementations and their application to real data. After CFAR processing, a plot extractor detects and estimates the parameters of all peaks in range, azimuth, and Doppler that exceed a predetermined threshold and forwards the extracted "hits" in each CPI to a tracker where successive detections are associated to form tracks. The tracker needs to filter out many false alarms over time to keep the false track rate low while maintaining an appropriate probability of detection.

The multiple-hypothesis tracker (MHT) has been successfully implemented in some operational HFSW radar systems. As explained in Ponsford, Sevgi, and Chan (2001), this tracker incorporates a deferred decision approach by maintaining multiple track options over a number of updates until enough confidence is built up to establish a track and remove the other competing alternatives. The minimum number of associated detections required to confirm a track is normally limited by the requirement to maintain a low false track rate.

The output of the track validation procedure is the declaration of a set of confirmed tracks that have satisfied the track promotion logic (e.g., two associated detections for tentative, five associated detections for confirmed, and seven misses to delete). Most false tracks arise due to ionospheric clutter and ocean clutter peaks. False track rates better than 0.25 per hour are quoted in Ponsford, Dizaji, and McKerracher (2003) in an experiment where the system successfully tracked all reported targets.

With respect to skywave OTH radar, coordinate registration is a simpler problem in HFSW radar due to the greater certainty regarding the surface-wave propagation path. For an established track, track accuracy is typically better than 0.5 km in range and 0.25 degrees in azimuth (Ponsford, Dizaji, and McKerracher 2003). Track position errors are often dominated by system biases that can be removed by calibration. In addition, a target normally produces a single surface-wave echo as opposed to a number of well-resolved echoes, which commonly occurs due to multipath in the skywave propagation channel. This effectively eliminates the multiple track to target assignment problem present in skywave OTH radar systems.

5.4.3.1 Ionospheric Clutter Mitigation

Adaptive processing may be combined with multi-frequency operation as well as receive and transmit antenna design as part of an overall ionospheric clutter mitigation strategy. In the absence of experimentally validated signal-processing models for ionospheric clutter, a number of empirical observations of this phenomenon are useful for guiding adaptive processor design. The NVI ionospheric clutter signals that cause significant problems for HFSW radar beyond ranges of about 90 km are typically distributed in range-Doppler and are time-continuous over the CPI. For this reason, spatial and space-time adaptive processing have been identified as the main classes of signal processing techniques to mitigate this disturbance type.

Although it may be anticipated that ionospheric clutter received in a particular range cell will exhibit a degree of directionality, it is perhaps less expected that the spatial structure of ionospheric clutter can change significantly from one range cell to next. In other words, the spatial characteristics of ionospheric clutter in the coverage are often found to be highly heterogeneous in range. This has a number of implications for adaptive filter design. First, the adaptive filter needs to be updated on a range-by-range cell basis with training data taken only from the operational range cell to be processed. This means that no "target-free" (supervised) secondary data is available for training the adaptive filter when all the pulses in the CPI or Doppler cells are used to estimate the disturbance covariance matrix.

The combination of unsupervised training data and finite sample support slows convergence rate and leaves the target highly susceptible to self-cancelation, particularly in the presence of array calibration errors. If the presumed useful signal response vector does not accurately match the actual (received) one, the target is interpreted as a disturbance by the adaptive processor and consequently canceled. Second, the training data may also contain powerful first-order ocean clutter that does not need to be rejected by the adaptive filter since these returns are effectively dealt with by Doppler processing. The presence of powerful sea clutter in the training data can seduce the adaptive filter into canceling surface returns in preference to the ionospheric clutter, which consumes adaptive degrees of freedom without providing significant benefits.

For this reason, post-Doppler techniques have been adopted such that only Doppler cells free of dominant sea clutter returns are used for training (Abramovich, Anderson, Lyudviga, Spencer, Turcaj, and Hibble 2004). However, a potential problem stems from the fact that the most energetic ionospheric clutter components often occupy the "low speed" Doppler cells, where second-order sea clutter and target echoes also reside. Moreover, the heterogeneity in angular distribution of ionospheric clutter may not be confined to the range alone, but may also be present across Doppler cells. This can lead to performance losses when the Doppler cells used for training at a given range are different to the ones processed by the adapted filter. The effectiveness of post-Doppler techniques also degrades when the ionospheric clutter in a particular range is associated with a non-stationary angular distribution over the relatively long HFSW radar CPI. An alternative pre-Doppler technique for ionospheric clutter mitigation has been proposed in (Fabrizio and Holdsworth 2008).

5.4.3.2 Polarization Filtering

Skywave interference is elliptically polarized, in general, and can therefore be received by antennas with vertical and horizontal polarization. On the other hand, only vertically polarized antenna elements are expected to receive the radar signal propagated by the surface-wave mode. Adaptive polarization filtering to suppress skywave interference was applied in a practical HFSW radar by Madden (1987) using a single auxiliary antenna with horizontal polarization. Another experimental investigation incorporating up to four horizontal dipoles configured as two separate crosses behind a main array of six vertically polarized elements was conducted in Leong (1997).

In practice, the radar signal is unfortunately not completely absent from the outputs of the horizontally polarized antennas due to imperfections or misalignment of these elements. For effective adaptive polarization filtering, the presence of surface clutter and target echoes in the auxiliary antennas needs to be minimized. The horizontal antennas therefore need to be carefully designed and installed to minimize the reception of the useful signal. The sample matrix inverse (SMI) method was used in Leong (1997) to estimate the adaptive processor weights in each waveform repetition interval by taking training samples from the farthest range bins (i.e., near the end of the pulse repetition interval), which contained skywave interference but minimal surface-wave clutter.

It was found that system performance improved with the use of an increasing number of horizontal antennas. Interference suppression levels greater than 13 dB were achieved using four horizontal antennas in Leong (1997). For systems based on two auxiliary antennas, configurations using orthogonally oriented horizontal dipoles performed the best. In this case, the location of the dipoles appeared to have little effect on performance (i.e., whether the orthogonal antennas were separated or colocated). Use of only one horizontal dipole antenna achieved about 4–6 dB interference suppression, depending on the orientation of the horizontal element.

Performance when using vertical and horizontal antenna elements as the auxiliaries was also compared in Leong (2000). When the target echo and interference are incident from a similar direction, the target echo is canceled along with the interference when only vertical monopoles were used as auxiliary elements. This is because the null formed on the interference also cancels the target. However, the target echo could be protected (i.e., not canceled) in the main beam interference scenario when the horizontal antennas were used as auxiliary elements. This result, which represents an important benefit sought from spatio-polarization filtering, was achieved despite the Bragg lines being visible at the outputs of the horizontal antennas (6–8 dB lower than those observed in the vertical antennas).

The study concluded that the interference was not suppressed as effectively when using horizontal (as opposed to vertical) antennas as the auxiliaries. This is because the skywave interference received by the main array of vertical antennas is better correlated with the interference received by the auxiliary vertical antennas than that received by the horizontal antennas. A similar observation was reported in an independent experimental study that used 16 elevated-feed vertical monopoles and 16 horizontal-loop antennas in a ULA that was approximately 300-m long (Abramovich, Spencer, Tarnavskii, and Anderson 2000).

5.5 Operational Considerations

For optimum performance, HFSW radar operation needs to be tailored to the target class of interest, the range-Doppler region where the target echo is expected, and the prevailing environmental conditions, where the latter includes sea-state and ionospheric structure, as well as HF spectrum occupancy and background noise spectral density. Similar to skywave OTH radar, the appropriate choice of carrier frequency is of fundamental importance to the successful operation of HFSW radar. The carrier frequency effectively controls the tradeoff between several competing factors that strongly influence target detection and parameter estimation performance. Specifically, carrier frequency selection impacts surface-wave path loss (as a function of range), target RCS characteristics, ocean and ionospheric clutter properties, background noise spectral density, as well as system gain and azimuth resolution.

The first part of this section reviews a number of basic RCS concepts relevant to HFSW, where the dominant component of polarization is constrained to be vertical, and coupling of the target with the sea surface cannot be ignored for surface vessels and low-altitude aircraft. For a comprehensive treatment on the topic of RCS, the reader is referred to Knott, Shaeffer, and Tuley (1993) and Ruck, Barrick, Stuart, and Krichbaum (1970). The second part of this section describes the main criteria used for frequency selection in HFSW radar and explains the significant advantages of dual or multi-frequency operation, which have been exploited in a number practical systems. The third part of this section contrasts the designs and capabilities of three representative HFSW radar systems with different architectural characteristics.

5.5.1 Radar Cross Section

The RCS of a target, such as a ship or aircraft, depends on many factors, including target dimensions and structure, electrical properties of construction materials, illumination frequency, viewing geometry, and the polarization(s) used on transmit and receive. The target RCS appears in both the noise and clutter-limited versions of the radar equation. Realistic estimates of target RCS are therefore valuable for calculating the energy budget required to satisfy performance requirements at the radar design stage, or for predicting the detection performance of a given system against a particular target class.

A detailed knowledge of RCS characteristics may also assist to determine the most favorable operating mode to detect a certain type of target. Moreover, an understanding of RCS signatures in the frequency and/or polarization domains may be exploited as a discriminant for target classification (Strausberger, Garber, Chamberlain, and Walton 1992). Information regarding the expected target RCS fluctuations may also help to determine the potential value of using echo amplitude as an additional parameter for associating detections in a tracking system. Considerable effort has therefore been invested to model and measure the RCS characteristics of surveillance radar targets.

In skywave OTH radar, targets are illuminated by a signal with time-varying polarization due to (non-stationary) Faraday rotation in the ionosphere. In this case, various structures on the target, such as the wings or fuselage of an aircraft, can at different times contribute in a dominant way to the overall RCS as the rotating polarization becomes more or less matched to different canonical structures on the target. The situation is not as general for HFSW radar because surface-wave propagation is only effectively supported for vertical polarization, both to and from the target. Consequently, the received echo is primarily a result of scattering from vertical (conductive) elements of the target structure. The RCS of maritime vessels is therefore particularly sensitive to the target's height above the sea surface.

The presence of metal masts or frames that effectively increase the vertical extent of a maritime vessel have the potential to significantly increase RCS when the surface-wave mode alone is active. For example, it is shown in Trizna (1982) that the RCS of a small surface vessel (fishing boat) is dominated by the quarter wavelength monopole contribution of a 16.6-meter-long metal mast with a peak RCS near 30 dBsm at the resonant frequency of 4.5 MHz. Owing to the characteristics of this element, the RCS drops rapidly to values below 10 dBsm at frequencies spaced 1 MHz either side of the resonant frequency. The vertical polarization of the surface-wave means that the illumination frequency and viewing geometry play prominent roles in defining the target RCS feature space in HFSW radar (Wu and Deng 2006).

5.5.1.1 Rayleigh-Resonance Region

The RCS characteristics of a target are highly dependent on the major dimension D of its conductive segments relative to the radar wavelength λ. Broadly speaking, there are three scattering regimes where both qualitative and quantitative differences in RCS behavior occur. The low-frequency regime, where the target dimensions are much less than the radar wavelength, is known as the *Rayleigh* scattering region ($D \ll \lambda$). The intermediate frequency regime, where the target dimensions are of the same order as the radar wavelength, is referred to as the *resonance* or *Mie* scattering region ($D \approx \lambda$). The high-frequency regime, where the target dimensions become much larger than the radar wavelength, is called the *optical* scattering region ($D \gg \lambda$). Although the resonance region may be approximately regarded as the frequency interval where $0.1 \leq \lambda/D \leq 10$, there are no distinct boundaries between the different scattering regimes. From a physical perspective, the RCS characteristics of a target transition gradually from those of one regime to another as the frequency changes.

HFSW radars typically operate in the lower HF band (3–15 MHz), which corresponds to a wavelength range between 20 and 100 m. At such frequencies, most aircraft and surface vessels have a major dimension comparable to the radar wavelength, which places their RCS in the resonance scattering region. More precisely, the RCS characteristics of large (ocean-going) ships lie in the resonance region over the entire HF band. However, the smallest aircraft, cruise missiles, and go-fast boats have RCS characteristics that fall within the Rayleigh region near the low end of the HF band.

In the Rayleigh region, the RCS is strongly influenced by the target's gross dimensions as opposed to its detailed structure or shape. It also exhibits reduced sensitivity to viewing geometry. The RCS of a target illuminated at low frequency has characteristics that become more like an isotropic scatterer. Perhaps most importantly, the target RCS decreases rapidly with the fourth power of frequency (to a good approximation) in the Rayleigh region.

In the resonance region, a number of different elements (conductive segments) on the target may contribute in a significant way to the resultant RCS. Depending on the radar viewing geometry and the relative positions of the dominant scatterers on the target, the resultant RCS may fluctuate significantly with aspect angle at a fixed operating frequency. On the other hand, the vector sum of scattered contributions from different elements of the target may interfere in a constructive manner at one frequency and destructively at another for a given radar viewing geometry and target aspect angle. This causes the target RCS to exhibit peaks and troughs (strengthen or weaken) within a confined range as the operating frequency is changed. The backscatter target RCS, and hence echo power received by a monostatic HFSW radar, can vary significantly (by more than 10 dB) as a function of target aspect and illumination frequency in the resonance region.

Estimates of the free-space target RCS may be suitable in certain radar applications, particularly for systems operating at UHF and microwave frequencies. In the lower HF band, coupling between the highly conductive sea surface and a surface-vessel cannot be ignored as far as the impact on effective target RCS is concerned. In addition to the approximate image field resulting for a surface-vessel on a calm sea surface, ocean roughness and surface-vessel motion can have an appreciable influence on the effective target RCS observed by an HFSW radar. Target RCS signatures in both aspect angle and illumination frequency have been simulated for the case of a stationary surface vessel over a calm sea, as described in the next section. Modeling of the target RCS becomes more difficult when vessel motion (including pitching and rolling) is introduced and different sea-state conditions are considered.

The radar cross section of large vessels (>1000 tons) at HF exhibits quite complex behavior, but the angle-average value may be roughly approximated by the empirical formula $\sigma = 52 f D^{3/2}$, where σ is the vessel's free-space RCS in square meters, f is the radar frequency in MHz, and D is ship size in thousands of metric tons. The RCS of small and medium-sized vessels (< 1000 tons) is dominated by their vertical metallic superstructure. If this superstructure is grounded to the ocean surface, or isolated from the ground, then the RCS of the vessel can be approximated by that of a grounded monopole antenna, or dipole antenna, respectively.

A resonant dipole has an RCS that approaches λ^2, while the RCS of an equivalent resonant monopole is 6 dB lower. The RCS below resonance may be assumed to exhibit an f^4 roll-off. Due to the large (decameter) wavelengths at HF, small boats with a low vertical extent will have very small RCS. Such targets may only be detected against background noise, which requires their radial speed to be high enough for the echo to be resolved in Doppler frequency from the ocean clutter.

5.5.1.2 Modeling Approaches

Analytical techniques based on geometrical optics, physical optics, or the geometrical theory of diffraction may be used to provide rough estimates of RCS values provided the target can be modeled reasonably well in terms of a few simple canonical structures.

The RCS characteristics of simple structures including a perfectly conducting cylinder, monopole element, and hemisphere, are illustrated in Skolnik (2008b).

For more precise modeling of target RCS, it is often necessary to resort to numerical methods. In this case, finite-difference time-domain (FDTD) or method-of-moments (MOM) techniques may be employed to compute RCS estimates. These numerical techniques employ time and frequency domain formulations, respectively, to provide full-wave solutions of Maxwell's equations in three dimensions.

A popular program based on the MOM formulation is the numerical electromagnetic code (NEC). There are several versions of NEC, with NEC-2 (1981) being the latest version of the code that is openly available. To imitate complex metal structures, actual design drawings of an aircraft or ship are often used to construct a wire-grid model of the target. Meshing is typically assumed to be composed of perfect electric conductor (PEC) material and specific criteria are followed regarding the choice of grid spacing and wire radius for a given segment length in the target model. An example comparing experimentally measured and simulation results for the RCS of a large surface vessel using NEC-2 will be illustrated below.

Numerical RCS modeling should not only account for the complex target structure, but also its interaction with the ocean surface. Wire mesh models of a surface-vessel are typically attached to an infinite PEC ground plane to simulate a flat and perfectly conductive ocean surface. This attempts to capture the scattering influences from the target's image as well as any potential coupling. An alternative is to compute the target free-space RCS with the wire grid model connected to its reflection in the horizontal plane. Incorporating ocean roughness and dealing with dynamic target interactions with the water surface is more difficult, as is the modeling of targets that do not have highly conductive surfaces. To validate simulated RCS estimates, it is common to compare results obtained using different numerical techniques, or to reconstruct complex models of targets with already available (experimentally measured) RCS values.

5.5.1.3 RCS Interactions

The spatial resolution cell size of an HFSW radar is very large compared to the dimensions of a target, so it often occurs that a single range-azimuth resolution cell contains more than one target. In this case, mutual RCS interactions between unresolved targets may be significant from a detection or classification viewpoint. In practice, the effect of RCS interactions becomes most noticeable in an HFSW radar when echoes from different targets occupy the same range-azimuth and Doppler cell, as in the case of nearby stationary targets, for example.

In Sevgi (2001), strong signal-strength fluctuations have been observed on the echo from an off-shore oil platform on the Canadian East Coast due to the presence of shuttle tankers that stop close to the platform for loading. The oil rigs are over 100 m high and have an RCS in the order of 50 dBsm, while the tankers in the proximity of the oil rig are some 400 m in length and have an angle-averaged RCS of approximately 40 dBsm. The usually steady echo from the oil rig can vary by more than 10 dB due to the presence of the tanker. Echo power fluctuations caused by mutual RCS interactions effects typically exhibit different fading characteristics with respect to those expected for a single maneuvering) target. This may potentially be used to infer the presence of multiple targets in the same radar resolution cell.

The concept of exploiting RCS coupling for the detection of a target behind a second target via forward scattering from the latter has been considered in Guinvarch et al.

(2006). The authors speculate that a collection of large vessels queued in the vicinity of each other around a congested port or in a narrow strait can form an "electromagnetic barrier" to conventional X-band coastal radars, which may preclude the detection of smaller targets located in the shadow regions. The shadow regions are less pronounced in the HF band, provided the considered location is not immediately behind the large vessel.

While the backscatter RCS tends to dominate at low frequencies, it has been observed that the forward scatter RCS begins to dominate as the frequency is increased. A study in Solomon, Leong, and Antar 2008 demonstrated that out of all the bistatic angles, forward scattering yielded the most potential for matching or exceeding the monostatic (backscatter) RCS values above a certain frequency regardless of the target aspect angle. For the large surface vessels considered, the relative increase in RCS reached as high as 30 dB with the broadside target aspect being the only orientation that maintained equivalence between the forward/backscatter RCS and did not show an increase at higher frequencies. Although this trait existed over the entire HF band for very large vessels, this feature engaged at different frequency points for smaller vessels, after which the forward scattering advantage was consistently maintained.

5.5.1.4 Experimental Measurements

While the RCS of ships and aircraft have been extensively studied in the optical region, only a limited number of investigations have been conducted in the HF band. Empirical expressions derived for the free-space RCS of vessels at microwave frequencies may not translate accurately to those observed by an HFSW radar. For example, RCS expressions based solely on vessel gross-tonnage and radar frequency do not explicitly consider the significance of vessel height or the influence of the conductive sea surface on RCS for HFSW radar.

Field experiments to measure RCS in the HF band may be conducted by using natural reference scatterers with known RCS, such as Bragg lines in a fully developed sea (Leong 2006), or reference signals from well-calibrated transponders with an identifiable Doppler offset located in the same area as the target (Dinger et al. 1999). When attempting to measure the dependence of RCS on target aspect, the range of the vessel needs to be sufficiently small such that the echo can be reliably detected above the background disturbance level even when the RCS passes through potentially deep nulls.

The HF-RCS of large and small ships illuminated by the surface-wave mode were studied in Leong and Wilson (2006). Specifically, RCS modeling results obtained by numerical simulation were compared with real measurements made by HFSW radar at frequencies of 3.1 and 4.1 MHz for two known vessels: a 2405-ton Canadian Coast Guard ship (Teleost) and a 36-kiloton cargo-freighter vessel (Bonn Express), both shown in Figure 5.30. The measurements were calibrated based on the assumption of a fully developed sea at the operating frequency, where the first-order scattering coefficient (echoing cross section per unit area) and the surface area of the radar resolution cell containing the test vessel are known.

Due to their different Doppler frequency shifts, the echo from the vessel may be separated and compared against the strength of the Bragg lines, which serve as a natural reference for making calibrated RCS measurements. This experimental method for estimating target RCS is appropriate when persistent winds with a sufficient speed and fetch fully develop the sea at the radar operating frequency. It is also convenient if the beam direction is close to being parallel with the mean wind direction when knowledge of the angular spreading function at the Bragg wave train frequency is not available. The

Bonn Express | Teleost

FIGURE 5.30 The left panel shows the Bonn Express, a 36 kiloton cargo-container with a length of 236 m and an estimated bridge height of 30 m. The right panel shows the Teleost, a medium-sized (2405 ton) coast guard ship with length 63 m, breadth 14.2 m and maximum mast height of 24 m. © Crown 2010. Government of Canada. (*Courtesy of Dr. H. Leong, Defence R & D Canada.*)

lower the operating frequency, the less likely the sea will be fully developed due to the increasing wavelength and higher speed of the ocean waves responsible for producing the Bragg lines in the Doppler spectrum.

It was found that both vessels had a comparable angle-averaged RCS of approximately 40 dBsm at 3.1 and 4.1 MHz, despite the Teleost having a gross tonnage 15 times less than the container ship. This result was attributed to the large contribution of the A-frame on the Teleost, which at a height of 24 m is near the resonant quarter-wavelength height at 3.1 MHz. Figure 5.31 shows that NEC accurately estimated the measured RCS values for the Bonn Express at 4.1 MHz, including the dependence of RCS on aspect angle at points where experimental data was obtained. Close agreement between predicted and measured RCS values as a function of target aspect have also been reported for the Teleost in Podilchak, Leong, Solomon, and Antar (2009).

The close agreement between experimental and simulated results in the bottom panel of Figure 5.31 provides convincing evidence that the backscatter RCS passes through peaks and troughs as a function of target aspect angle at a frequency of 4.1 MHz. A comparison of the modeling results in the top and bottom panels of Figure 5.31 predicts that the locations of these peaks and troughs differ significantly at carrier frequencies spaced 1 MHz apart at the lower extreme of the HF band.

For certain target aspects, the difference in RCS at these two frequencies can potentially enhance or reduce the received echo power by margins in excess of 10 dB. The dependence of target RCS over this frequency range may be exploited to improve detection performance by means of dual-frequency operation. As pointed out in Leong and Wilson (2006), the enhanced target RCS arising for a large vessel near broadside explains why vessels traveling tangentially have occasionally been detected at very long ranges by HFSW radars.

At one extreme, large ferry ships have measured RCS values approaching 50 dBsm (Menelle et al. 2008), while at the other, small boats with lengths ranging from 5 to 8 m typically have RCS values of 0–10 dBsm (or less) in the lower HF band. For operating frequencies between 3 and 5 MHz, the RCS of medium-sized vessels with a gross registered

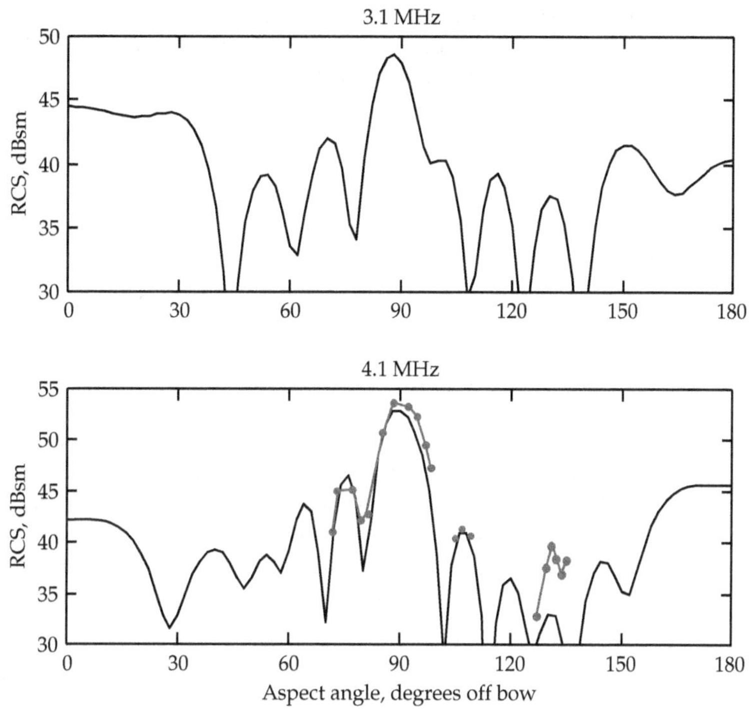

FIGURE 5.31 Radar cross section as a function of aspect angle for the Bonn Express cargo freighter at 3.1 MHz and 4.1 MHz. Numerical modeling results are shown with a solid line, while measured data are shown by the dots joined with a solid line. A dual-frequency HFSW radar would have a better chance of maintaining track at aspect angles near 62, 70, 100, 105 degrees. © Crown 2010. Government of Canada. (*Courtesy of Dr. H. Leong, Defence R & D Canada.*)

tonnage of about 1000 tons was estimated to be about 25–30 dBsm at aspects less than 25 degrees from the stern-on or bow-on directions (Leong 2007). In practice, mean RCS values should be interpreted with caution as the actual target RCS can vary by an order of magnitude or more relative to the mean in the resonance region depending on the combination of operating frequency and target aspect.

The detection and tracking of small surface-craft, also known as go-fast boats (GFB) (Dinger et al. 1999), or rigid inflatable boats (RIB) (Blake 2000), is of significant interest for border protection as their high speed, good endurance, and low vertical profile allows them to cover long distances very rapidly and covertly. Besides their low height (often less than 1 m above the sea surface), these boats are generally constructed with minimal metallic content. Depending on design and configuration, such targets may only have one or two significant conducting segments. Other materials, such as fibreglass and wood, are often used for the floor or hull, while plastic inflatable tubes may be used to provide buoyancy. This combination of attributes makes the detection of such vessels challenging for any type of radar, particularly when the sea is not calm.

A detailed experimental and simulation study of the RCS behavior of this target class appears in Dinger et al. (1999). The RCS of two go-fast boats was measured at bow-on and stern-on aspects with respect to the HFSW radar receiver as a function of operating frequency when the vessel speed was approximately 20 knots. For a 25-ft GFB, the RCS increased gradually from a value of about 0 dBsm at 6 MHz to 9 dBsm at 15 MHz. A 21-ft GFB exhibited a similar increase with frequency, but the RCS was 7 to 10 dB lower.

For both boats, no peaks or resonant structures were observed in the variation of RCS with frequency, which is consistent with Rayleigh scattering region behavior. In (Blake 2000), the RCS of a 5.4-m-long RIB was estimated as 5 dBsm at 20 MHz assuming that only a metallic A-frame at the rear of the RIB reflects the radar signal. Although the RCS is low, an important consideration for detection is that such targets can travel at high speeds of perhaps 20–40 knots in low sea-states.

The highly conductive sea water displaced by a GFB traveling at high speed is also thought to contribute to raising the RCS observed in practice with respect to values predicted by numerical modeling for the static scenario. In addition, unlike slow-moving vessels that need to be detected against sea clutter, echoes from GFB/RIB targets often have Doppler-shifts that fall outside of the second-order clutter continuum. In this region of the Doppler spectrum, the disturbance level is generally dominated by a *range-independent* background noise floor. OTH detection and tracking of small surface targets using HFSW radar may therefore be feasible at relatively short ranges of a few tens of kilometers in the Doppler frequency region beyond the second-order clutter continuum when sufficiently high frequencies are used, as demonstrated in Dinger et al. (1999) and Blake (2000).

For this target class, frequencies ranging from 10 to 15 MHz or higher may be used to increase RCS and reduce the background noise spectral density. At relatively short ranges of a few tens of kilometers, the benefits of higher target-RCS and a lower noise-floor at higher frequencies typically outweigh the loss due to greater surface-wave attenuation when detection performance is limited by SNR (Emery 2003), particularly for low sea-states where the excess path loss due to surface-roughness is negligible.

Long CPIs are preferable to increase coherent gain on the target echo for detection against noise. In practice, CPIs of about 30–60 seconds represent a compromise for such targets as excessive Doppler smearing can occur due to target acceleration. As pointed out in Blake (2000), a typical marine radar has a maximum detection range of 15–20 km under normal atmospheric conditions, while the detection of an RIB-type target could well fail at shorter ranges. HFSW radar may therefore provide an alternative shore-based surveillance system for the detection of small fast boats at short ranges up to and beyond the conventional radar horizon.

5.5.2 Multi-Frequency Operation

Frequency selection for HFSW radar involves some different considerations with respect to those discussed previously for skywave OTH radar. Clearly, a common requirement is for the carrier frequency to be selected with regard to the prevailing HF spectrum occupancy to maintain interference-free operation. As described for skywave OTH radar, a spectral surveillance system that identifies clear frequency channels in real time at the receiver site is a prerequisite for successful HFSW radar operation.

The first part of this section describes the main factors driving frequency selection in HFSW radar for the detection of two broadly defined target classes, namely, fast-moving low-RCS targets and slow-moving high-RCS targets. The second part of this section explains the various benefits of multi-frequency operation, particularly to improve the performance of an HFSW radar system against target RCS fluctuations, Bragg line masking, and ionospheric clutter contamination.

5.5.2.1 Frequency Selection

The main factors driving HFSW radar frequency selection for target detection over the horizon include: (1) surface-wave attenuation or path-loss, (2) background noise spectral density, (3) angle-averaged target RCS, (4) target Doppler frequency, (5) azimuth resolution on receive, and (6) ionospheric clutter contamination. All of these factors vary with operating frequency.

For targets beyond the line of sight, the path loss experienced by surface-wave propagation over the ocean increases dramatically with radio frequency. In other words, the power density of the signal that illuminates a target in the diffraction zone is significantly greater at lower frequencies. This is the single major reason that motivates HFSW radar operation at lower frequencies.

Operation at higher frequencies is typically associated with lower background noise spectral density levels and increased clear channel availability, particularly at night. Use of higher frequencies also reduces susceptibility to ionospheric clutter contamination, particularly via the direct NVI skywave path. The receive antenna beam has narrower main lobe at higher frequencies, which improves azimuth resolution and accuracy. Fine azimuth resolution reduces the effective sea clutter RCS in a radar resolution cell and increases system immunity against spatially structured interference.

For small targets in particular, the angle-averaged RCS is often greater at higher frequencies. In addition, the higher target echo Doppler shift associated with an increase in carrier frequency may improve the detection of slow-moving targets against sea clutter by placing the useful signal in a frequency region containing a lower disturbance level. Figure 5.32 indicates that frequency selection for HFSW radar involves finding an appropriate compromise between reducing surface-wave path-loss (favored at lower frequencies) and all other aforementioned competing factors, which are typically favored at higher frequencies.

For the purpose of simplifying the discussion, we may broadly consider HFSW radar operation in "low" and "high" frequency regimes denoted by 3–7 and 7–15 MHz, respectively. HFSW radars are usually not operated near the upper end of the HF band unless space-wave propagation to a line-of-sight target exists. We may also consider two nominal HFSW radar missions, namely, the detection of slow-moving high-RCS targets (i.e., large ships) at long ranges between 200 and 400 km, and the detection of fast-moving

Frequency Choice	Signal Attenuation	Azimuth Resolution	Target RCS	Target Doppler	External Noise	Ionospheric Clutter
Lower (3–7 MHz)	Higher field strengths OTH	Less accurate localization	Tends to Rayleigh	Lower frequency shift	Hard to find clear channels	Higher susceptibility
Higher (7–15 MHz)	Attenuation rises rapidly	Higher directivity	Tends to resonance	Higher frequency shift	Often lower external noise	Increased immunity

FIGURE 5.32 HFSW radar frequency selection requires striking a balance between reducing surface-wave attenuation (favored at lower frequencies) and all other competing factors listed above (favored at higher frequencies).

low-RCS targets (e.g., small aircraft) at short to intermediate ranges, defined as 50–100 and 100–200 km, respectively. We may also define a close range band of less than 50 km, noting that these are only indicative values.

The detection of fast-moving targets is often limited by background noise rather than clutter after Doppler processing. A possible exception occurs when spread-Doppler ionospheric clutter contaminates the range-azimuth cell containing the target. Ignoring this possibility for the moment, frequency selection for this target class essentially reduces to a tradeoff between three competing objectives: (1) surface-wave path-loss, (2) target RCS, and (3) background noise spectral density. For small (fighter-sized) aircraft, the target RCS tends to increase with frequency, but so too does the path-loss. At short to intermediate ranges, these two effects tend to roughly cancel each other out such that their product does not change dramatically on average in the "low" and "high" frequency regimes. However, the irreducible (spatially uncorrelated) component of the background noise spectral density tends to decrease at higher frequencies, particularly at night.

Moreover, it is usually less difficult to find clear frequency channels of the appropriate bandwidth at higher frequencies, particularly at night. Indeed, it is typically found that there are almost twice as many channels of a fixed bandwidth in 7–15 MHz as there are in 3–7 MHz. Ionospheric clutter is also less of a problem at higher frequencies on the NVI path. These considerations mean that it is often best to operate in the 7–15 MHz frequency regime to detect fast-moving low-RCS targets at short to intermediate ranges. Use of frequencies in the upper HF band leads to a significant increase in path loss at ranges beyond 50 km. At such ranges, the additional path loss is less likely to be canceled by an increase in target RCS. In summary, the "high" frequency regime is often considered to be more appropriate for the detection of fast-moving small-sized targets at short to intermediate ranges.

At close ranges, the surface-wave path-loss increases less rapidly with increasing frequency than it does at further ranges. At ranges less than a few tens of kilometers, increasing the frequency beyond 15 MHz is more likely to improve the net product between target RCS and path-loss, particularly for GFB targets that rise only a meter or so above the sea surface. Frequencies beyond 15 MHz may therefore be considered appropriate for the detection of small (low-flying) aircraft or GFB targets at close ranges.

On the other hand, operation in the "low" frequency regime (3–7 MHz) is necessary to keep path loss to within manageable limits at long ranges. Unfortunately, small targets fall well within the Rayleigh scattering region at such frequencies and their RCS drops precipitously. Many HFSW radar systems therefore find it difficult to reliably detect fast-moving small-sized targets at long ranges.

The RCS of large ships remains in the resonance region over the entire HF band so operation in the "high" frequency regime is unlikely to significantly increase the RCS for this target class. To optimize the detection of large ships at long ranges, the main option for increasing the received target echo power is to decrease the surface-wave path loss by operating in the "low" frequency regime (3–7 MHz). At low frequencies and long ranges, slow-moving targets are often detected against second-order sea clutter and potentially ionospheric clutter. Depending on sea-state, the power and Doppler spectrum extent of the second-order sea clutter tends to diminish as the operating frequency is lowered. On the other hand, in the "high" frequency regime (7–15 MHz), the surface-wave attenuation becomes prohibitively large at long ranges, especially for high sea-states. In this case, the path loss at long ranges can be large enough to submerge echoes from large ships and second-order sea-clutter below the background noise floor.

For these reasons, the "low" frequency regime is often considered more appropriate for the detection of large ships at long ranges. However, a major concern with operation below the maximum critical frequency of the E- or F-region is ionospheric clutter. This may at times justify the use of higher operating frequencies for the detection of large surface vessels at long ranges, particularly at night when spread-Doppler clutter may be received below the F2-layer critical frequency. Raising the operating frequency to reduce spread-Doppler clutter returned via the NVI path from the F2-layer comes at the expense of higher surface-wave attenuation and hence reduced detection range. During the day, when E-region ionization is well developed, dual-frequency operation may be used to combat the ionospheric clutter problem, as described in the next section.

An HFSW radar may at times be tasked to search for a specific type of target with known RCS characteristics. For example, the target of interest may be a certain type of fishing vessel such as the one considered in Trizna (1982). In this specific case, the target RCS for vertical polarization is dominated by a metal mast with a resonant frequency of 4.5 MHz. As shown in Trizna (1982), the target RCS may vary by as much as 10–15 dB across the "low" frequency regime 3 – 7 MHz. When a priori knowledge about the target RCS characteristics is available, the operating frequency may be tailored accordingly to maximize the power of the received echo. Clearly, the broad categorization of "low" and "high" frequency regimes is too coarse for frequency optimization in these specific cases.

5.5.2.2 Dual-Frequency Operation

Dual- or multi-frequency operation in HFSW radar differs from traditional frequency agility applied to interleaved dwells in the sense that the radar may contemporaneously rather than sequentially transmit and receive signals on different carrier frequencies. The data stream acquired in each frequency channel is processed independently (i.e., pulse compression, transient excision, Doppler processing, beamforming, CFAR, and peak detection). The waveform parameters set for each carrier frequency may be optimized for the detection and tracking of either ship or aircraft targets, but not both simultaneously. This is mainly due to the two types of target having very different velocities and detection update rate requirements for effective tracking.

When the waveform parameters on different frequencies are configured to detect the same target class (i.e., ships or aircraft), the peak detections resulting from the multiple processed data streams may be fed into a common tracker which associates them. The simultaneous detection of ships at one frequency and aircraft at another offers significant flexibility, but it does not yield the benefits of combining observations for the same targets illuminated at different frequencies. For simplicity, we restrict our attention to the use of only two frequencies (i.e., dual-frequency operation) for the detection and tracking of a single target class. The detection and tracking of ships is considered, as surface vessels often represent the primary target in HFSW radar.

The simultaneous use of two carrier frequencies can provide several advantages for HFSW radar (Leong and Wilson 2006). The first major advantage provided by dual-frequency operation is an improved capability to track ships through first- and second-order sea clutter. Due to the square-root relationship between the Bragg-line Doppler frequency and the radar carrier frequency, the relative separation between the target echo and nearest Bragg line in the Doppler spectrum is a function of carrier frequency. Provided the HFSW radar operates at two well-spaced carrier frequencies, a target echo masked by a powerful Bragg line at one carrier frequency (i.e., a blind speed) will be separated in Doppler frequency from the Bragg line at another carrier frequency.

Consider a practical example of a surface-vessel with a relative velocity of 9 m/s (17.5 knots). At a carrier frequency of 3 MHz, the target Doppler shift is 0.18 Hz for a monostatic radar configuration. The Doppler frequency of the nearest Bragg line is separated from that of the target echo by only 0.004 Hz in this case, which cannot be resolved with a CPI under 200 s. At a carrier frequency of 4 MHz, the target echo Doppler shift is 0.24 Hz, while the advancing Bragg line Doppler frequency is approximately 0.20 Hz. This increases the Doppler frequency separation by an order of magnitude to 0.04 Hz. A CPI of 100 s provides a classical Doppler frequency resolution of 0.01 Hz. For this CPI, the separation between the target echo and advancing Bragg line is equal to four FFT frequency bins at 4 MHz, but less than half a frequency bin at 3 MHz. Experimentally recorded HFSW radar data that illustrates this practical benefit of dual-frequency operation is shown in Figure 5.33.

To summarize, echoes from a particular target received simultaneously by the same radar at two different carrier frequencies cannot both fall on a Bragg line in the Doppler spectrum. In an HFSW radar, the problem of Bragg-line masking can be addressed by dual-frequency operation provided the carrier frequency separation and CPI length are chosen appropriately. Specifically, the frequency separation and CPI length need to be such that the square-root versus linear Doppler frequency relationship can be resolved for the target relative velocities of interest.

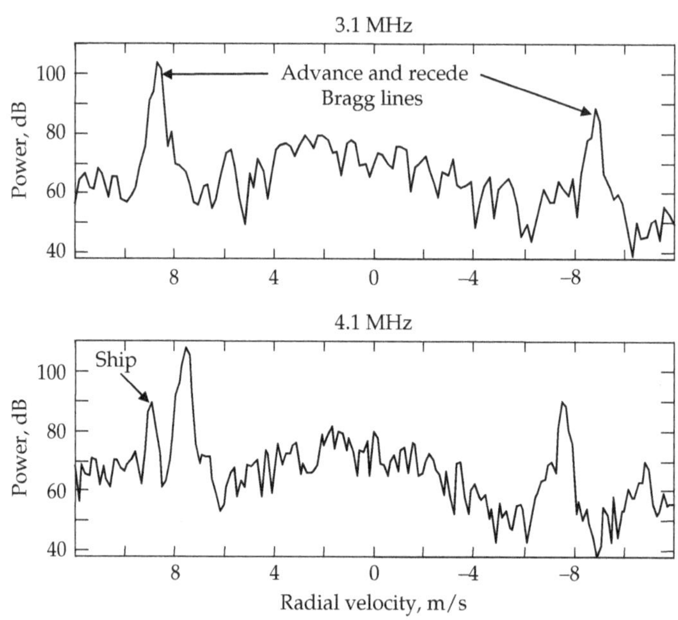

FIGURE 5.33 Target echoes from ships may not be detected when the Doppler frequency of the echo is close to that of a Bragg line. The top panel shows a situation where an echo from a ship target is masked by the advancing Bragg line at 3.1 MHz. Dual-frequency operation allow the target echo to be clearly distinguished from the advance Bragg line at 4.1 MHz (bottom panel). © Crown 2010. Government of Canada. (*Courtesy of Dr. H. Leong, Defence R & D Canada.*)

In addition, the two frequencies should not be so far apart from the optimum frequency such that the target echo can be reliably detected in both frequency channels when not masked by a Bragg line. This enables an HFSW radar to detect ship echoes against second-order clutter plus noise regardless of the target relative velocity. A similar argument applies for target echoes masked by certain discrete second-order spectral components in the sea-clutter Doppler spectrum, which also follow a square-root relationship, similar to the Bragg lines.

A second advantage of dual-frequency operation is that it provides an alternative method for mitigating ionospheric clutter received on the NVI path. During the day, the approach is to select one operating frequency below the E-layer critical frequency and the other above it. The following discussion focuses on NVI ionospheric clutter because this path typically gives rise to the strongest returns and is often most problematic in terms of obscuring target echoes in the HFSW radar coverage.

Ionospheric clutter is returned by the E-layer at the lower operating frequency (which is typically set below 4 MHz at noon local time) and this reflection contaminates a band of range starting at around 100 km. The first affected range depends on the virtual reflection height of the NVI path, as well as the transmitter-receiver separation. Importantly, the E-layer prevents reflections from the higher altitude F-layer when operating below the E-layer critical frequency.

On the other hand, the signal at the higher frequency penetrates through the E-layer at near vertical incidence. This signal may be reflected back by the F-layer, resulting in clutter being observed beyond a distance of about 220 km. In principle, the lower and higher frequencies are affected by ionospheric clutter returned from the E-layer or F-layer, respectively, but not from both layers at the same time. Hence, a target echo at a given range will in principle be free of ionospheric clutter contamination in at least one of the frequency channels. Figure 5.34 shows a practical example of daytime ionospheric clutter received simultaneously by an HFSW radar at two different frequencies.

The mitigation of ionospheric clutter at night is more challenging because the D- and E-layers effectively disappear after sunset, but sufficient ionization to reflect HF signals remains in the F-region. At night, ionospheric clutter contamination from the F2-layer at ranges beyond about 200 km cannot be avoided by dual-frequency operation, unless one of the operating frequencies is higher than the F2-layer critical frequency. However, operation above the relatively high F2-layer critical frequency may not be a feasible option for HFSW radar due to the greater surface wave path-loss incurred. In other words, the use of higher frequencies attracts a considerable performance penalty for the detection of large ships at long ranges. Figure 5.35 illustrates the longer-range ionospheric clutter received from the F2-layer at night.

As shown previously, large surface vessels are found to have RCS values that exhibit significant aspect and frequency dependence in the lower HF band. At a given operating frequency, the target RCS will pass through (imperfect) nulls and enhancements as the aspect angle changes. Large ships may not be detected by an HFSW radar at aspect angles that fall in a deep RCS null. Importantly, RCS nulls may occur at quite different aspect angles when the operating frequency is changed by about 1 MHz near the low end of the HF band. This is evident in Figure 5.31, which shows the RCS variation of a large cargo freighter with aspect angle at 3.1 and 4.1 MHz based on NEC modeling (solid line) and the corroborating experimental measurements at 4.1 MHz (dotted line).

For target aspect angles near 28, 45, 62, 70, 75, 100, 105, 110, 130, and 150 degrees, it is observed from the modeling that the RCS passes through a relatively low value at one

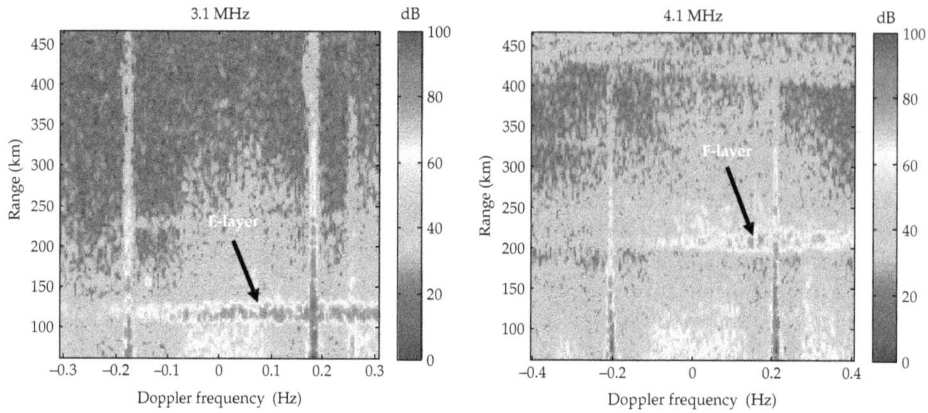

FIGURE 5.34 Range-Doppler maps showing *daytime* ionospheric clutter received simultaneously at two frequencies on Feb 20, 2002. The critical frequency of the E-layer at the time of the experiment was between 3.1 and 4.1 MHz. The display on the left (3.1 MHz) shows spread-Doppler ionospheric clutter returned from the E-layer at ranges between 110 and 130 km. Ionospheric clutter from the F-region is hardly noticeable in this display. The display on the right (4.1 MHz) shows spread-Doppler ionospheric clutter returned from the F-region, with the strongest returns occupying ranges between about 200 and 250 km. No trace of ionospheric clutter from the E-layer is visible when operating at the higher frequency. © Crown 2010. Government of Canada. (*Courtesy of Dr. H. Leong, Defence R & D Canada.*)

operating frequency and an enhancement at the other. Specifically, the predicted difference in RCS at operating frequencies of 3.1 and 4.1 MHz can be higher than 10 dB for certain aspect angles. Note that the deepest RCS nulls rarely line up at the same aspect angle at both frequencies. Simultaneous radar operation on two well spaced frequencies may therefore reduce the likelihood of missed target detections due to RCS nulls. Tracking ship targets through RCS nulls is considered to represent the third major advantage of dual-frequency operation. Another potential benefit is that relatively short-lived narrowband interference will only impact one frequency channel.

5.5.3 Example Systems

World-wide, perhaps a dozen or more countries have previous or currently active HFSW radar programs. This includes Canada, Unites States, Australia, Great Britain, Germany, France, China, Russia, Japan, and Turkey, for example. This section describes a selection of representative HFSW radar systems that have been designed and operated with surveillance applications as the primary mission. The performance of these systems against different target classes, as claimed in the literature, are also quoted to provide an indication of nominal capabilities. The HFSW radars discussed in more detail are the two-site Iluka and SECAR systems, developed in Australia, the single-site SWR-503 system, developed in Canada, and the truly monostatic BAE surface-wave radar, developed in the United Kingdom.

For the reader interested in delving further, information on several other former and present HFSW radar systems may be found in a number of excellent articles. For example,

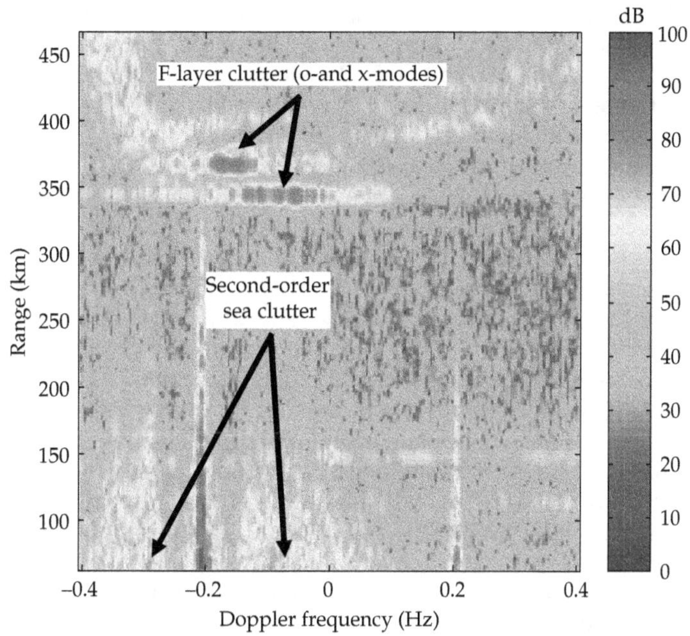

FIGURE 5.35 Range-Doppler map showing *nighttime* ionospheric clutter returned from the F2-layer at a frequency of 4.1 MHz. The presence of resolved ordinary (o) and extraordinary (x) magneto-ionic components, with the former being received at lower range, is indicated in the display. © Crown 2010. Government of Canada. *(Courtesy of Dr. H. Leong, Defence R & D Canada.)*

the HFSW radar demonstrator in Biscay Bay, France, developed by ONERA and the French MOD (DGA) is described in Menelle et al. (2008), the Wellen Radar (WERA) system originally developed by the University of Hamburg, Germany, is described in Dzvonkovskaya et al. (2008) and references therein, while progress in HFSW radar activities in China have been reported in Liu, Xu, and Zhang (2003) and (Liu 1996).

5.5.3.1 Iluka and SECAR Systems

Research and development in HFSW radar by the Australian government started in 1993 when an experimental system was designed and deployed by the Defence Science and Technology Organisation (DSTO) at a site north of Adelaide, South Australia (Anderson, Bates, and Tyler, 1999). Following the successful trials of this experimental system, a much larger and more capable radar named Iluka was designed, built, and installed near Darwin in northern Australia. Several different organizations were involved in the Iluka project, including DSTO, Telstra Applied Technologies (TAT), and the Centre for Sensor Signal and Information Processing (CSSIP).

A series of scientific experiments were conducted using the Iluka system in 1998. TAT later sublicensed its HFSW radar interests to Daronmont Technologies (DarT), which developed a production version of the radar named Surface-wave Extended Coastal Area

Radar (SECAR). In conjunction with scientists and engineers from DSTO and CSSIP, DarT conducted extensive trials with SECAR in 2000 from a receive site on Bathurst Island, north of Darwin. The Iluka and SECAR systems are not currently deployed in an operational role, but both systems represent examples of bistatic HFSW radars designed for the surveillance of air and surface targets within the EEZ.

The Iluka HFSW radar was designed to operate in the 5–15 MHz frequency range, although capable of operating at higher frequencies if needed. This system consisted of a primary high-power transmit site, located at Stingray Head (65 km south-west of Darwin), and a secondary low-power transmit site at Lee Point (10 km north-east of Darwin). The receive site, located at Gunn Point (30 km north-east of Darwin) was separated by about 95 km from the high-power site, and approximately 18 km from the low-power site. The secondary low-power site was mainly used to investigate the dependence of sea clutter characteristics on illumination geometry.

The receiver site was sufficiently isolated from both transmitters to allow operation with unit duty-cycle frequency modulated continuous waveforms (FMCW). The Iluka receiving system was based on a 500 m long ULA of 32 vertical monopoles, each connected to a well-calibrated HF receiver with boresight oriented in a westerly direction (at right-angles to the shoreline). At the low-power site, a 250-W amplifier drove a pair of monopoles spaced 20-m apart and fed in anti-phase. A single log-periodic dipole array (LDPA) antenna connected to 10-kW transmitter was used at the high-power site. This radar routinely detected 42 m patrol boats at ranges beyond 150 km, and small low-flying aircraft 20-m above sea surface at ranges in excess of 200 km. Due to the accompanying space-wave propagation, commercial aircraft at high altitudes could be detected at much greater distances.

The architectural characteristics of the SECAR radar are summarized in Table 5.3, while Figure 5.36 shows pictures of the SECAR transmit antenna and receiver array. A comprehensive description of the SECAR system can be found in (Anderson et al. 2003). The SECAR system has demonstrated detection performance of large surface vessels, such as Frigates and offshore trawlers out to the EEZ limit (370 km). Smaller fishing boats and GFBs have been detected up to ranges of 120 km and 70 km, respectively, while low-flying fighter-sized aircraft may be detected to ranges in excess of 200 km. A salient advantage of the FMCW bistatic HFSW radar configuration is that targets such as GFBs can be detected effectively at close (and long) range (a few tens of kilometers). This can be problematic for single-site or truly monostatic HFSW radar systems that use pulse waveforms due to the significant eclipsing loss experienced at close and long ranges as a result of the finite pulse duration.

5.5.3.2 SWR-503

Raytheon Canada System Limited (RCSL) and the Canadian Department of National Defence (DND) developed two single-site HFSW radars which commenced operation off the east coast of Newfoundland at Cape Race and Cape Bonavista in 1999. These two radars, designated SWR-503, are separated by 226 km approximately on the longitudinal line 53° W to provide overlapped coverage of an area in the Grand Banks. The operating frequency range is nominally between 3 and 5 MHz, which is best suited to the detection of medium to large surface vessels at intermediate to long distances as the primary mission. The detection and tracking of small private aircraft and GFBs is more challenging at these low frequencies due to the significantly lower RCS of such targets plus the need to use short CPIs for effective Doppler processing and rapid detection updates.

System	SECAR
Manufacturer	Daronmont Technologies, Australia
Configuration	Bistatic HFSW radar
Inter-site-separation	50–150 km
Average power	5 kW
Transmit antenna	Single vertical log-periodic with ground screen
Receive antenna	16 or 32 twin-monopole elements with ground screen
ULA aperture	200–500 m
Beamwidth	3–9 degrees (500-m aperture)
Frequency range	4–16 MHz
Waveform	Repetitive linear FMCW
Bandwidth	10–50 kHz
WRF	4–50 Hz
CPI	1–120 seconds (equal to revisit rate)
Coverage arc	120 degrees
Instantaneous range-depth	100–500 km
Primary mission	Ship detection
Secondary missions	Aircraft detection, remote-sensing

TABLE 5.3 Nominal architecture and operating parameters of the SECAR HFSW radar system.

Detailed descriptions of SWR-503 can be found in Ponsford, Sevgi, and Chan (2001), as well as in Leong (2002) and Leong, Helleur, and Rey (2002). The key operational parameters of the SWR-503, as quoted in these references, are listed in Table 5.4. The HFSW radar at Cape Race employs a log-periodic transmit antenna and a ULA receive antenna composed of 16 doublet elements. The receiving doublets, each consisting of two kite-shaped monopoles, are phased end-fire and the separation between adjacent doublets is 33.33 m. The doublet elements are spaced half-wavelength apart at the radar design frequency of 4.5 MHz. The electronically steered receive beams have a main lobe width of approximately 9 degrees, but azimuth estimates typically have an error of less than 1 degree for well-resolved radar echoes. The coverage arc is ±60 degrees from boresight. The HFSW radar at Cape Bonavista is very similar to the Cape Race system, except a that quadlet is used as the receive antenna element instead of a doublet.

The transmitter generated an average power of 1.6 kW, and sea clutter was typically found to be the limiting disturbance to ranges greater than 350 km during the day (using

(a) Log-periodic transmit antenna.

(b) Monopole-doublet receiver array.

FIGURE 5.36 The SECAR HFSW radar transmit antenna is single vertical log-periodic dipole array (LDPA), while the receive antenna is a uniform linear array of twin-monopole elements configured to provide front-to-back directivity. © Commonwealth of Australia 2011.

SWR-503	Cape Race, Newfoundland
Radar configuration	Monostatic
Transmit power	16 kW peak, 1.6 kW average
Transmit antenna	7-element log-periodic monopole (Gain 8 dBi)
Receive antenna	ULA of 16 kite-shaped doublets
Doublet separation	33.33 m
Beam width	9 degrees
Frequency range	3–5 MHz
Waveform	Sequence of phase-coded pulses
PRF	250 Hz
Pulse bandwidth	20 kHz
Sampling rate	100 kHz
CPI lengths	10–1000 s
Maximum range	500 km (large surface targets)
Coverage arc	120 degrees

TABLE 5.4 Nominal design parameters of the HFSW radar at Cape Race (46:39:07 N, 53:05:24 W), Newfoundland, Canada. The boresight of the receiving array is 121 degrees clockwise from true north.

CPIs of hundreds of seconds). However, at night the radar is usually externally noise limited beyond approximately 150 km. The system could employ an 8-bit complementary phase-coded sequence on transmit, which permits range sidelobe reduction and high pulse repetition frequency (PRF) operation through the suppression of range-wrapped ionospheric clutter. For example, a phased-coded pulse may have a length of 400 microseconds with a bit duration of 50 microseconds (Leong 2002). The radar could also use "mismatched" phase-code sequences to enable the suppression of strong narrowband co-channel interference signals (Ponsford, Dizaji, and McKerracher 2003).

The nominal operating bandwidth of 20 kHz provides a group-range resolution of 7.5 km, while oversampling with a period of 10 microseconds can yield a measurement accuracy of about 0.3 km for well-resolved radar echoes. The cross-range resolution is roughly equal to 50 km at the 200 nmi (370 km) EEZ limit. CPIs of 10 seconds (air mode), 100 seconds (ship mode), and 1000 seconds (stationary targets) have been used to provide relative velocity resolutions of approximately 5, 0.5, and 0.05 m/s, respectively, at 3 MHz. Extensive trials have been conducted to demonstrate beyond the horizon detection of aircraft to 300 km, surface targets to 500 km, and icebergs to 300 km. Medium sized ships such as the Ville de Quebec (length 436 ft, height 140 ft) were tracked to ranges of up to 235 km, while smaller vessels such as the Anne S. Pierce (length 117 ft, height 55 ft) and Artic Pride (length 64 ft, height 45 ft) were tracked to approximately 110 km (Leong, Helleur, and Rey 2002).

5.5.3.3 BAE SWR System

The HFSW radar system developed by BAE systems in the United Kingdom is described in the two-part paper of Dickel, Emery, and Money (2007). Its predecessor version was described by Emery, Money, and Matthewson (2004). The first distinguishing feature of this system is that it uses a single uniform linear array (ULA) of antenna elements for transmission and reception. This truly monostatic HFSW radar has the advantage of a relatively small real-estate footprint by dispensing with the need for a separate transmit antenna or a second radar site. A truly monostatic or single-site system is not subject to variations in performance caused by the changing bistatic angle to different points in the coverage. A picture of the BAE HFSW radar array is shown in Figure 5.37.

The ULA is typically composed of 16 identical antenna elements, each configured as a doublet pair of tetrahedral dipoles phased in the endfire direction with approximately 15 dB front-to-back ratio (Boswell, Emery, and Bedford 2006). All antennas in the array are used for reception, while the middle six antennas are connected to a bank of solid-state high-power duplexers, which enable transmission and reception using the same (i.e., reciprocal) elements. The smaller transmit aperture provides a relatively broad beam in azimuth, similar to that provided by a single LPDA antenna. Besides requiring less real estate than an LPDA antenna, using an array of rigid transmit antennas tends to reduce Aeolian noise (phase noise) caused by movement of radiating elements under wind stress. It also allows the transmit beam to be digitally steered and shaped. However, these benefits may come at the expense of reduced frequency range with respect to an LPDA, as compact elements typically exhibit broadband performance over an octave.

The six tetrahedral dipole antenna elements in the middle region of array are driven by individual digital waveform generators connected to 1-kW solid-state power amplifiers. The waveform generators are fully programmable to simultaneously transmit different waveforms in a multiple-input multiple-output (MIMO) architecture. For example, the

FIGURE 5.37 Uniform linear array of 16 (doublet pair) tetrahedral dipole antenna elements used in the BAE surface-wave radar. The middle six elements are fitted with high-power duplexers and used as reciprocal antennas for transmission and reception.

system can be programmed to synthesize a simple pulse, linear or nonlinear chirps, and stepped frequency signals up to a maximum bandwidth of 30 kHz.

The ability of the middle six antenna elements to independently transmit different radar waveforms at different carrier frequencies represents another distinguishing feature of this system. A digital waveform generator per element allows the transmit resource to be divided among multiple frequencies without degrading performance in a clutter-limited environment (i.e., low-power floodlight illumination using a subset of the six transmit elements at each frequency). A "spotlight" mode that involves electronic steering of narrow transmit beams with higher gain using all six radiating elements may be used in noise-limited conditions.

As described in Dickel, Emery, and Money (2007), a doublet at one end of the array is configured with an inverse pattern to serve as a Rear Lobe Blanking (RLB) antenna, while a doublet at the other end of the array is used for environmental monitoring as well as beamforming. The use of wide-band receivers that sample the signal at the RF element level allows for digital down conversion and processing of up to four simultaneous frequency channels per element. Signal and data processing up to the plot extraction stage are run in parallel for each of the four down converted channels. Different processing steps/parameters may be used in each channel to contemporaneously perform air and surface target surveillance, for example. The peak data resulting from each of the four channels is then fused to produce target tracks in the final processing stage.

The simultaneous use of up to four distinct carrier frequencies, combined with plot level fusion of the processed outputs from different frequency channels provides a higher level of immunity to incidental man-made interference, target RCS fades, and Bragg-line masking. Multi-frequency operation also provides greater scope to optimize frequency selection for different target classes, ranges, and speeds under the prevailing environmental conditions. The BAE HFSW radar has nominal a frequency range of 8–16 MHz and can provide coverage across a 110 degree sector. Surface vessels may be tracked to ranges of 220 km over an area of approximately 30,000 square kilometers, while aircraft may be tracked simultaneously out to 350 km (large commercial airliners) over a coverage twice the size.

Aircraft target localization accuracy is approximately 1 km in range and 0.5 degrees in azimuth relative to air traffic control (ATC) radar acquisitions Dickel, Emery, and Money (2007). In Emery, Money, and Matthewson (2004), the rms position error for a ferry ship tracked by HFSW radar were measured at 1.3 km in range and 0.7 degrees in azimuth relative to known GPS ground-truth information. This accuracy is considered good for a long-range sensor with inherently "coarse" spatial resolution. Target echoes become partially eclipsed by the transmit pulse duration at close ranges, and degradation in detection performance is experienced. SNR degradations due to eclipsing loss become significant below about 20 km (Emery, Money, and Matthewson 2004), which represents the nominal minimum range of the system.

PART II

Signal Description

CHAPTER 6

Wave-Interference Model

A wide variety of high-frequency (HF) systems rely on the ionosphere as a radio wave propagation medium for operation over long distances well beyond the line-of-sight. Examples include HF communication links for transferring data or speech between remote sites, direction-finding networks for locating the position of distant HF emitters, and OTH radars used for early-warning wide-area surveillance. In practice, the performance of HF systems utilizing the ionosphere not only depends on the choice of operating parameters, but also on the characteristics of the propagation medium. For this reason, there is great interest in analyzing and modeling the properties of HF signals reflected by the ionosphere, particularly those which can potentially limit system performance.

A popular model for skywave HF signals received by antenna arrays consists of a superposition of multiple plane waves that emanate from the different ionospheric reflection points. Specifically, the traditional signal-processing model assumes that a single specularly reflected plane wave component is received for each distinct propagation mode. This model is rather simple, far too simple in fact to satisfactorily describe the actual signal structure received by practical HF arrays. Typically, HF signals undergo diffuse scattering from a number of spatially extended but often localized regions in the ionosphere. A question that frequently arises is whether the complex fading of the received signals can be decomposed as the sum of a relatively small number of plane wave components with slightly different directions-of-arrival and Doppler shifts, such that the fading pattern observed on the ground can be represented accurately by the interference of these waves over a limited period of time.

The main objective of this chapter is to quantitatively assess the virtues and limitations of the wave interference model for describing the observed signal fading phenomena in space and time. Section 6.1 discusses the applications of the wave interference model in different contexts and defines the scope of this model for the purpose of this chapter. Section 6.2 describes a multichannel scattering function experiment to demonstrate the existence and time-varying characteristics of wavefront distortions for resolved ionospheric modes received on a very wide aperture antenna array. Section 6.3 mathematically describes the wave-interference model and a super-resolution parameter estimation technique for the identification of closely spaced component rays. A fitting accuracy measure is also described in this section to assess the performance of the wave-interference model using real data. Section 6.4 presents an extended set of experimental results to illustrate the domain of applicability and shortcomings of the wave interference model for a relatively quiet mid-latitude ionospheric channel.

6.1 Deterministic Description

Compensation for performance degradations incurred as a result of propagation effects, which can significantly distort the received signal structure relative to that expected under ideal or benign conditions, is often sought in the form of ameliorative signal processing. Modern signal-processing algorithms are usually designed and optimized on the basis of mathematical models that are presumed to reflect the characteristics of actual signals received in practice, either deterministically or in a statistical sense. Not surprisingly, the effectiveness of model-based algorithms in real-world systems is largely determined by the fidelity with which the assumed model is able to represent the properties of the acquired data. Importantly, modern systems capable of sensing fine scale features of the received signal structure require models of commensurate sophistication and accuracy to appropriately guide signal processor design.

The models which have traditionally been used as bases for the development of many conventional signal-processing techniques have been rather simple, particularly for state-of-the-art HF systems such as OTH radar, where signals are sampled by very well calibrated wide aperture antenna arrays using receivers with high spectral purity and dynamic range. In a number of important signal-processing applications, traditional models do not provide a sufficiently realistic description of the received signals. At the moment, an important frontier in OTH radar development lies in the area of improving the received signal models and developing more advanced signal-processing techniques based on these improved models to enhance system performance.

HF propagation models derived from consideration of detailed ionospheric physics provide valuable phenomenological insights and can potentially yield high quality representations of the received signal structure. However, the physical parameters these models depend upon are generally not known with a high degree of certainty, and in many cases, the mathematical formulations are quite detailed and not always tractable for the purpose of designing and optimizing real-time signal-processing algorithms. These factors can restrict their utility in a number of practical applications, particularly when the physics-based model does not lend itself readily to the generation of synthetic data sample realizations for signal-processing analysis.

An alternative approach is to represent the observed characteristics of the received data from a pure signal-processing perspective, without detailed consideration of the physics involved beyond basic principles. Although such models are often convenient for analysis, they can at times be over-simplistic and unrepresentative of the real data. This can lead to significant discrepancies between expected system performance and that encountered in practice. Between these two extremes, there is scope to construct improved signal-processing models which represent the empirically observed data characteristics more accurately than their traditional counterparts, while retaining a mathematical formulation that is amenable for guiding the development of advanced signal-processing algorithms.

6.1.1 Background and Scope

The ionosphere is typically composed of multiple reflecting regions or layers, so propagation in this medium is usually by multipath components or signal modes. At a more detailed level, the process of signal reflection from each region or layer is by no means "mirror-like" and can at times induce appreciable distortion on the individual signal modes. Both multipath propagation, due to reflections from distinct ionospheric regions, and within-mode distortion, due to diffuse scattering processes local to each reflection point, are phenomena that have the potential to degrade or impair system performance.

The simplest received signal model results when each propagation mode is considered to be specularly reflected from an ideal ionospheric layer with a spherically symmetric refractive index profile that varies smoothly as a function of height. In this case, the process of diffuse scattering is neglected and the signal received from a far field source is modeled as a superposition of plane waves that correspond to the different propagation modes. Each plane wave component is parameterized by a complex amplitude, direction of arrival (azimuth/elevation), Doppler shift, and polarization state. Such a model represents what may be referred to as the *gross structure* of a skywave HF signal. In many respects, this deterministic model has played a central role in the development of many signal detection and parameter estimation techniques for HF systems not limited to OTH radar.

In practice, the radio refractive index of the ionosphere is known to be heterogeneous, dynamic, dispersive, and anisotropic as far as HF signals are concerned. To develop more sophisticated signal models and processing techniques for HF systems, it is therefore necessary to move away from the ideal notion of an ionospheric layer behaving as a smooth "copper sheet" reflector of radio waves. More precisely, the ionospheric reflection process diffusely scatters an incident HF signal such that a cluster of rays is returned to ground. The complex morphology of the ionosphere and geomagnetic field that permeates it will introduce distortions that can influence the amplitude, phase, and polarization state of a reflected signal mode in both space and time. It is the difference between the actual signal characteristics and those expected in the ideal case of specular reflection which gives each propagating mode its *fine structure.*

A variety of models have been proposed to explain the fine structure of individual ionospheric modes. One popular model describes fine structure as a superposition of a small number of submodes or rays that have similar Doppler frequencies and are closely spaced in direction of arrival (DOA). The different rays in this model are presumed to result from relatively few specular reflection points in an irregular ionospheric layer that move as the electron density distribution changes in time. Interference between the different rays produces a complex (amplitude and phase) fading pattern over the ground to represent the fine structure of a signal mode. The accuracy with which this model can represent the fine structure of real ionospheric modes as a function of the number of rays assumed, and the length of the observation interval, is of considerable interest and represents the main subject of this chapter.

The combination of experimental and mathematical procedures used to estimate the parameters of the constituent rays are collectively referred to as wavefront analysis (WFA) techniques. In practice, WFA has traditionally been applied to decompose the multipath gross structure of the received signal by assuming that each propagation mode undergoes specular reflection. The lack of analysis on isolated ionospheric modes has led to much debate regarding the capability of the wave interference model to additionally represent fine structure. A summary of reported experimental results on the analysis of gross and fine structure is included in the next section to provide a brief synopsis of the research performed in this area and the main conclusions drawn.

The focus of this chapter is to quantify the ability of the wave interference model to represent the *fine structure* of signal modes. A number of qualifications are required to clearly define the scope of such a model with this objective in mind. First, the analysis considers oblique ionospheric channels that propagate narrowband HF signals from a far field source to a receiving antenna array. Second, we are concerned with the case of single-hop skywave propagation with no intermediate ground reflection(s). Third, the skywave signal is received in one (linear) component of polarization at each antenna element (vertical in this case). The study of narrowband signals received over point-to-point ionospheric links using a very wide aperture multichannel receiver array is not only important to HF communication and direction-finding systems, but is also of great relevance to the problem of HF interference rejection in OTH radar.

Moreover, a detailed understanding of one-way propagation between a point source and a sensor array provides a useful basis from which to infer the characteristics of two-way channels, which are relevant to modeling radar signal echoes from targets, for example. For the monostatic transmit-receive radar configuration, the reciprocity principle often enables the outgoing and incoming signal transfer functions to be modeled in identical manner, while certain approximations can be made to account for the quasi-monostatic geometry often employed by two-site OTH radar systems. In addition, two-way channel models may be combined with surface scattering models to describe clutter returns from terrain or sea surfaces.

The natural starting point is therefore to characterize the one-way ionospheric channel for the simplest case of single-hop propagation. The current chapter is concerned with the propagation of narrowband signals with bandwidths in the order of tens of kilohertz. Although wideband HF systems exist, it is worth noting that most OTH radar systems and HF arrays operate using narrowband signals. More specifically, the chapter is devoted to models that describe the space-time characteristics of signals received by vertically polarized antenna elements. While the potential benefits of polarization diversity have been investigated at HF, most current operational skywave OTH radar systems receive a single (vertical or horizontal) polarization component.

From an experimental data viewpoint, the emphasis is on relatively quiet mid-latitude paths as opposed to the typically more disturbed polar and equatorial ionosphere. The low- and high-latitude ionosphere is significantly more complex, and this may partly explain why most current skywave OTH radars operate with ionospheric control points located in the mid-latitude region. In summary, the objective of this chapter is to evaluate the accuracy with which an estimated wave interference model can represent the complex valued samples of individual narrowband HF signal modes received by a vertically polarized wide aperture antenna array over a one-way single-hop mid-latitude ionospheric path.

6.1.2 Gross Structure of Composite Wavefields

The ability to resolve and estimate the parameters of multiple propagation modes of a signal reflected from the ionosphere is of fundamental importance to the successful operation of many HF systems. For example, HF direction-finding systems based on interferometry are known to produce fluctuating and often inaccurate estimates of the signal angle of arrival and hence emitter position when unresolved multipath is present. In HF communications systems, multipath components with similar strengths but different time delays and Doppler shifts can produce deep and rapid frequency-selective fading of the received signal envelope. This type of fading can significantly increase the data transmission error rate. In OTH radar applications, multipath not only affects target detection and tracking, but can also have a major influence on the effectiveness of adaptive processing for interference rejection.

Historically, HF array systems commonly used classical beamforming techniques to resolve the gross structure of multipath propagation by estimating the DOA of each signal mode. A major obstacle encountered by such systems was that many array apertures were not always large enough to resolve the different propagation modes. As a result, the initial emphasis was often more on avoiding multipath rather than to isolate the different propagation modes by array processing. Wavefront testing methods were developed in Treharne (1967) and more recently in Warrington, Thomas, and Jones (1990) so that estimates of the emitter DOA in HF direction-finding systems were only taken at times when the received wavefront closely resembled a plane wave. These techniques, which often rely heavily on the relative fading between modes to provide times of quasi-unimodal propagation, severely restrict the times and circumstances under which suitable data can be acquired, and are therefore considered to be of limited utility Hayden (1961).

In the early 1980s, super-resolution algorithms were developed in the field of array signal processing to enhance the resolution capabilities of sensor arrays. For example, the MUltiple SIgnal Classification (MUSIC) algorithm described in Schmidt (1979) sparked tremendous interest in subspace methods and led to the development and analysis of various super-resolution techniques. In contrast to the vast quantity of published theoretical analysis and computer simulation results in this area, there has been comparatively little reported on the practical application of such algorithms to resolve the DOAs of propagation modes in the HF environment. The experimental studies described below illustrate the various difficulties encountered when super-resolution techniques have been applied in real HF multipath environments.

A 16-element linear antenna array with a 120-m aperture was used in Creekmore, Bronez, and Keizer (1993) to estimate the propagation mode DOAs from known AM radio broadcasts of opportunity. MUSIC and three other super-resolution techniques were applied to resolve the assumed number of ionospheric modes for each source. The authors concluded that the propagation modes appeared to be "spatially extended" due to temporal variations in the ionosphere. In other words, the discrete plane wave model assumed by the adopted techniques did not appear to be strictly valid in practice. The observed fine structure significantly complicated the process of identifying the correct number of modes and associating a single bearing per mode, as both quantities appeared to fluctuate over time.

An irregular two-dimensional array consisting of 8 elements with an effective aperture of about 8 wavelengths was used in Tarran (1997) to determine the azimuth and elevation angles of signal modes propagated from a known transmitter over a 1235-km mid-latitude path. The MUSIC algorithm was used to estimate the direction of arrival of two dominant modes at a rate of 30 azimuth/elevation angle measurements per second. Despite the fast rate of the measurements, the resulting azimuth/elevation scatter plot demonstrated spreads in the order of a few degrees for each of the modes in azimuth and elevation. This led the author to conclude that ionospheric reflection can cause very rapid and significant fluctuations in the mode DOAs. Such investigations provided further evidence that a plane wave model with fixed direction of arrival is not representative of an individual signal mode even over very short time intervals.

A uniformly spaced 6-element circular array of 50-m diameter was used in Moyle and Warrington (1997) to estimate the DOAs of modes propagated over a controlled 778-km mid-latitude ionospheric path. The ionospheric path was controlled in the sense that oblique sounding records were used to identify the mode structure prevailing during the course of the experiment. Although three distinct propagation modes were resolved in range by the ionogram at the system operating frequency, the application of MUSIC, and a number of other super-resolution algorithms were unable to resolve the directions-of-arrival of all three propagation modes. The authors concluded that the inability to resolve all three modes may have been due to the poorly calibrated reception channels in the array and the relatively small aperture employed.

A 7-element V-shaped array with a major dimension of 350 m was used by Zatman and Strangeways (1994) to resolve multiple ionospheric modes received from HF emitters of opportunity. MUSIC and the Direction-of-Arrival by Signal Elimination (DOSE) algorithms described in Zatman and Strangeways (1994) were applied to estimate the azimuth and elevation angles of the various propagation modes. While neither algorithm performed consistently well, it was concluded that the inability of MUSIC to resolve the propagation modes was possibly due to the high correlation existing between different paths, the additive noise present in the data, and the effects of mutual coupling between antenna elements.

A well-calibrated uniform linear array (ULA) with a 1.4-km aperture was used in Fabrizio et al. (1998) to receive two powerful radio broadcasts of opportunity. One of these sources was received over a controlled mid-latitude ionospheric path, while the other source propagated via the equatorial ionosphere. To resolve highly correlated modes, the 16 digital receivers were divided into sub-apertures of 12 receivers to form spatially smoothed MUSIC spectra. Spatial smoothing significantly improved the capability of MUSIC to resolve the propagation modes that were identified on oblique sounding records for the path. The time evolution of MUSIC spectra indicated a gradual variation of the mode cone angles for both sources of opportunity during a typical OTH radar coherent processing interval. Variations over a few seconds were found to be within fractions of a degree for the modes propagated on the mid-latitude path, and one degree for the modes propagated via the equatorial region.

Some important conclusions arise from the reported experimental results. The first is that conventional beamforming will in general struggle to resolve multipath gross structure and estimate the DOA's of different propagation modes using aperture sizes less than about 100 m. MUSIC and other subspace methods can in principle provide higher resolution with respect to conventional beamforming, but the main drawback of such techniques is that they are sensitive to the plane wave assumption. Consequently, the realization of super-resolution in practice is often limited by propagation effects (mode fine structure) and instrumental factors (array calibration errors). The former not only distorts the spatial structure of the signal relative to the plane wavefront, but can give rise to distortions that exhibit a time-varying or non-stationary behavior.

Broadly speaking, resolving the gross structure of a multipath signal reflected by the ionosphere requires well-calibrated antenna arrays with a wide electrical aperture, relatively low noise levels, and in many cases the use of super-resolution techniques that operate over short time intervals and are insensitive to inter-mode correlations. Modern HF systems stand to benefit from a more detailed understanding of mode fine structure, particularly since this phenomenon can at times be the performance limiting factor in state-of-the-art HF systems, where great care is taken to reduce the influence of array imperfections, site errors, and noise.

6.1.3 Fine Structure of Individual Modes

There are two main ways of observing and analyzing the received signal wavefronts on a mode-separated basis. A received signal can be reduced to its component modes by exploiting differences in either time-of-arrival, to resolve modes as a function of group range, or the regular component of ionospheric phase path variation, to resolve the modes in Doppler frequency. The former requires modulated waveforms and has the advantage of being capable of performing separation over very short time intervals (corresponding to the pulse duration). However, large bandwidths are required to resolve modes with small differences in group range. This may lead to problems with co-channel interference in the crowded HF band and frequency dispersion of the signal in the ionosphere. The latter option can be performed with continuous wave (CW) signals, which are less susceptible to interference and clearly not subject to dispersion, but may require long CPIs to resolve modes with similar Doppler shifts. This increases the time interval between successive mode wavefront observations, which is not suitable for observing variations over short time scales (within the CPI).

One of the first detailed experimental investigations of fine structure using a very wide aperture array was conducted in Sweeney (1970). A 2.5-km uniform linear array (ULA) was used to sample the amplitude and phase of ionospheric modes propagated over a 2550-km ground distance mid-latitude path. The ULA was composed of 8 non-overlapping subarrays, each consisting of 32 vertical whip antennas spaced 10 m apart. The 32 vertical whips in each subarray were connected to an analog beamformer to form a subarray output. The 8 subarray outputs were downconverted and sampled by individual well-calibrated receivers. A linear FMCW waveform was used in one experiment to separate the different propagation modes in delay, while in another, the modes were separated in Doppler using CW signals. The principal aim of the analysis was to examine the discreteness (spectral purity) of the received modes in azimuth, range, and Doppler.

Particular attention was paid to the mode wavefront structure in order to determine the extent to which fine structure degrades the azimuthal pattern properties of a very wide aperture array. While single-hop modes appeared discrete in azimuth, range, and Doppler to the resolution of the array, it was noticed that double-hop modes exhibited considerable spread in all three dimensions. Based on these results, it was concluded that for single-hop modes the presence of mode fine structure did not significantly affect the "performance" of a very wide aperture antenna array. The author postulated that the observed spreading on double-hop modes was being caused by intermediate ground reflection from very rough (mountainous) terrain near the path midpoint.

The performance of a very wide aperture antenna array takes on a different meaning depending on whether the signal represents a useful echo received in the mainlobe of the antenna pattern or interference received in the sidelobes. When the signal represents interference to be removed by adaptive beamforming, the impact of fine structure is experienced in the vicinity of the relatively steep "nulls" of the adapted antenna pattern as opposed to the much broader mainlobe. In this case, performance is not measured in terms of the loss in coherent gain due to wavefront distortions, but rather in terms of the degradation in output signal-to-interference plus noise ratio (SINR), which is highly sensitive to mismatches between the antenna pattern null and the fine structure of an interference signal wavefront.

The impact of mode fine structure on the interference cancelation performance of adaptive beamforming in a very wide aperture antenna array was reported in Fabrizio et al. (1998). The performance of the beamforming system was found to depend heavily on the rate at which the adaptive weights were updated. The rate yielding best performance represented a balance between slower updates to increase the amount of training data and faster updates to counter variations in the interference wavefronts caused by mode fine structure. The authors concluded that variations in the wavefronts of ionospheric modes over time intervals commensurate with OTH radar CPI have the potential to severely degrade the interference cancelation performance of a very wide aperture adaptive antenna array.

Another experimental investigation of fine structure was carried out in Rice (1973), where a 32-element ULA with a 1.2-km aperture was used to measure the phase-fronts of ionospheric modes propagated over a 911-km mid-latitude path. Different modes were resolved in time-of-arrival using an FMCW waveform and identified based on oblique sounder data. The unwrapped phase-fronts received from six propagation modes exhibited varying degrees of phase nonlinearity. In particular, the phase-front of a mode reflected from the F2-layer was much more linear than those observed for modes reflected simultaneously by lower ionospheric layers. It was concluded that the mechanism leading to distorted phase-fronts is associated with phenomena near the height of reflection, rather than diffraction effects arising from the passage of rays through lower height regions of the ionosphere from where other modes are reflected. The author attributed the observed phenomena to within-mode wave interference effects. This interpretation considers each mode to be composed of a number of submodes which have nearly the same transit time (unresolved in range) but slightly different angles of arrival and Doppler shifts to account for the nonlinear phase-front variations observed at 1-minute intervals in Rice (1976).

The wave interference interpretation of fine structure is supported by measurements made at vertical incidence by Felgate and Golley (1971). The authors employed an array of 89 elements to fill a circular area with a diameter of approximately 1 km. A waveform with a 70-microsecond pulse duration and 50-Hz repetition frequency was used to separate and identify the propagation modes in group range. The amplitude pattern produced by each mode over the ground was sampled in time and presented as an intensity-modulated photographic display. Periodic fringe patterns consisting of alternate bright and dark bands were frequently observed for individual modes. The authors suggested that the regularity of these fringe patterns was produced by the interference between a small number of discrete rays returned to ground from different specular reflection points in the ionospheric layer. The motion of fringes over the ground with respect to time was attributed to changes in either the horizontal or vertical position of these specular reflection points, which alters the phase relationship between the different rays.

The assumption that a mode consists of a small number of specularly reflected rays with similar Doppler shifts and closely spaced angles of arrival was also assumed in Clark and Tibble (1978). An 8-element vertical antenna array with a height of 74 m was used to measure the elevation angles of arrival of CW ionospheric modes separated on the basis of Doppler shift. It was found that the elevation angles of certain modes fluctuated in a sinusoidal fashion by more than 5 degrees during a 90-second interval. The authors commented that such results seemed unrealistic and that the most probable explanation for the large excursions in elevation angle was their inability to resolve the rays that comprised the fine structure of the analyzed modes.

It is evident that the wave interference model of mode fine structure is supported by several independent experimental investigations. This deterministic description has the advantage of being mathematically tractable from a signal-processing perspective, particularly when the number of rays required to represent the observed fading pattern is relatively small. Mode fine structure can also be interpreted more readily from a physical perspective when the number of rays is small. However, an experimental analysis that attempts to estimate the spatial and temporal parameters of the interfering rays and then directly compares the simulated fine structure with actual measurements recorded by a wide aperture antenna array is required to ascertain the validity and accuracy of the wave interference model in a more convincing manner.

6.2 Channel Scattering Function

The *multi-sensor* channel scattering function (CSF) experiment described in this section allows mode fine structure to be analyzed jointly in space and time over typical OTH radar CPI lengths. The first objective of the analysis is to demonstrate the existence and characteristics of this phenomenon, while the second is to assess the capability of the wave interference model to represent the observed mode characteristics in a deterministic manner. A distinguishing aspect of this experiment is that amplitude and phase measurements were made on a mode separated basis by a very wide aperture array at fine temporal resolution, as described in more detail below.

Experimental data were collected by the Jindalee OTH radar receiver located near Alice Springs in Central Australia. This facility is based on a 2.8-km long uniform linear array with 32 reception channels. Its main architectural features were described in Chapter 3. To probe the ionosphere, a test transmitter was positioned in the far field of the array and cooperatively radiated a known signal from a vertical whip antenna. The transmitter was located near Darwin, approximately 1265 km to the north of the receiver site and at a great circle bearing close to 22 degrees from boresight. The trial also made use of an oblique incidence sounder, which routinely records ionograms to determine the mode content for this particular ionospheric circuit as part of the Jindalee Frequency Management System (FMS).

The array was tuned to receive a narrowband FMCW signal from the test transmitter, which was linearly swept over a 20-kHz bandwidth at a rate of 60 sweeps per second using a fixed carrier frequency of 16.110 MHz. A communication link makes it possible to synchronize the emitted FMCW signal with a local copy of this signal at the receiver site, such that the modes can be separated on the basis of time-delay. The absolute group range can be estimated for each propagation mode. Clear channel advice from the Jindalee FMS system confirmed that the 20-kHz channel with a center frequency of 16.110 MHz was free of co-channel interference from other users at the time of the experiment.

The data were acquired by each receiver as a sequence of coherent processing intervals (CPI) or "dwells." Each dwell of data is recorded in approximately 4.2 seconds, during which a total of 256 phase coherent FMCW sweeps are emitted and received by the system. Adjacent dwells were separated by an inter-dwell gap of about 0.5 seconds as part of the normal operating procedure. A total of 47 dwells were recorded during the experiment between 06:17 and 06:21 UT on 1 April 1998. The oblique incidence ionogram was collected on the same day at 06:23 UT, shortly after the CSF data.

A 20-kHz bandwidth yields a group range resolution of 15 km for a one-way path. This resolution was traded off to about 20 km in order to control range sidelobes using a Hamming window. As the various propagation modes appear over a finite interval in group range, only a portion of the range spectrum needs to be retained for further processing. In this case, a total of 42 range samples covering a range depth of 615 km between 1055 and 1670 km were retained for further processing. The effective bandwidth of the ionogram trace is close to 60 kHz, which translates to a group range resolution of approximately 5 km for a one-way path.

6.2.1 Ionospheric Mode Identification

Figure 6.1 shows the oblique incidence ionogram recorded for the Darwin-to-Alice Springs skywave link. The ionogram trace may be interpreted to determine the mode content of the ionospheric circuit. This involves identifying the number of propagation modes and the ionospheric regions that reflected them. For a narrowband signal, the mode content is estimated as the point(s) of intersection between the ionogram trace and a line drawn vertically at the operating frequency. It is clear from Figure 6.1 that the mode content changes as a function of operating frequency. The variation in the number of propagation modes and their group ranges with frequency illustrates the dispersive nature of the skywave HF channel.

At the CSF operating frequency of 16.110 MHz, the ionogram resolves five distinct propagation modes at group ranges of 1290, 1300, 1430, 1475, and 1540 km. To identify the ionospheric layers responsible for propagation, the virtual height of reflection in the ionosphere is calculated for each mode. By assuming a spherical earth model and reflection from a concentric ionospheric layer, the virtual ionospheric height of reflection h_v is given by Eqn. (6.1), where $r_e = 6370$ km is the Earth's radius, $d = 1265$ km is the ground distance of the path, and g_r (km) is the group range of the mode measured

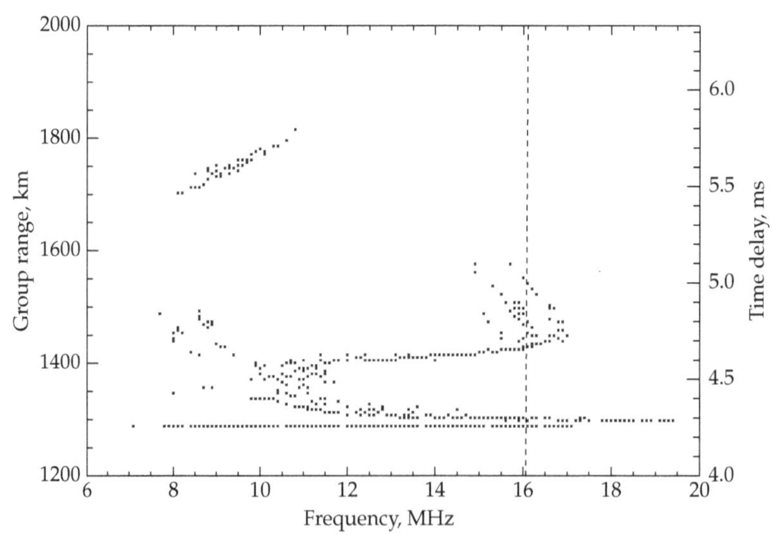

FIGURE 6.1 Oblique incidence ionogram indicating the mode content for the Darwin-to-Alice Springs ionospheric circuit as a function of carrier frequency on 1 April 1998 at 06:23 UT. © Commonwealth of Australia 2011.

by the ionogram. Using Eqn. (6.1), the virtual heights corresponding to the different propagation modes are calculated as 99, 122, 303, 349, and 408 km, respectively.

$$h_v = \frac{g_r \sin[\pi - d/2r_e - \arcsin(2r_e \sin(d/2r_e)/g_r)]}{2\sin(d/2r_e)} - r_e \quad (6.1)$$

The lowest reflecting layer with a virtual height of 99 km is identified as mid-latitude sporadic-E. This layer normally forms at altitudes between 90 and 110 km and is often characterized by a relatively flat trace with respect to frequency in the ionogram. The mode corresponding to a virtual reflection height of 122 km also exhibits a flat ionogram trace with respect to frequency and is identified as a reflection from possibly the same mid-latitude sporadic-E layer. It is reasonable to ask how two reflections from the same ionospheric layer can arrive with different time-delays or group ranges. A possible explanation is that a signal can be reflected from a point in the ionosphere that does not lie on the great circle plane and therefore travels a further distance compared to the signal resolved at a lower group range. This situation may arise in the sporadic-E layer due to a reflection from a different cloud of ionization that is off the great circle plane.

The propagation mode with a group range of 1430 km and a virtual height of 303 km is composed of the ordinary and extraordinary magneto-ionic components in the low-angle ray of the F2-layer, which are too close to be resolved by the oblique incidence sounder. The propagation mode with a group range of 1475 km and virtual height of 349 km corresponds to the ordinary magneto-ionic component of the high-angle ray in the F2-layer. Finally, the mode with the largest group range of 1540 km and virtual height of 408 km is due to the extraordinary magneto-ionic component of the high-angle ray in the F2-layer. The one-hop sporadic-E reflection is denoted by $1E_s$ while the F2-layer reflection corresponding to the low-angle ray is denoted by $1F_2$. The resolved ordinary and extraordinary magneto-ionic components in the high-angle ray are referred to as $1F_2(o)$ and $1F_2(x)$, respectively.

6.2.2 Nominal Mode Parameters

An individual propagation mode reflected from a localized region in the ionosphere over a one-way path normally appears discrete in group range when the resolution cell size is in the order of 20 km. Quite often, this group range resolution is sufficient for resolving different propagation modes reflected by physically distinct ionospheric regions into separate range bins. Figure 6.2 shows the mean power-delay profile of the received signal averaged over all receivers in the array and the period of data collection. The noise level evident in the flat portion of the spectrum prior to range cell 12 indicates the signal-to-noise ratio of each resolved mode over the period of data collection.

The four peaks in the power-delay profile of Figure 6.2 appear at group ranges of 1290, 1435, 1480, and 1540 km. The locations of these peaks agree well with the group ranges of the modes resolved by the ionogram. Recall that the group range resolution of the power-delay profile is in the order of 20 km, which is not high enough to resolve the two sporadic-E modes in the ionogram at group ranges of 1290 and 1300 km. The mean SNR of the mode(s) represented by each peak of the power-delay profile is estimated as the ratio between the magnitude of the peak and the background noise level estimated from the range cells below 1200 km. Clearly, no modes can be present in these range cells for a 1265-km ground-distance path. In ascending order of group range, the four resolved propagation modes have SNRs of approximately 34.5, 40.5, 34.0, and 20.0 dB.

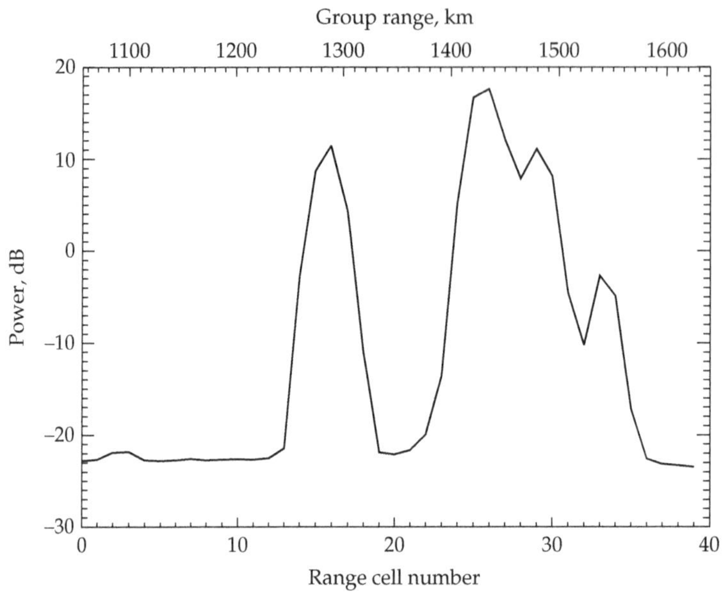

FIGURE 6.2 Average power-delay profile (group range power spectrum) recorded for the Darwin-to-Alice Springs ionospheric link by the Jindalee array at $f_c = 16.110$ MHz on 1 April 1998 between 06:17 and 06:21 UT. © Commonwealth of Australia 2011.

The capability of the array to isolate the contributions of individual propagation modes into different range bins enables the space-time characteristics of each mode to be studied independently (i.e., separately from the other modes). Specifically, the sequence of "slow-time" samples from one pulse to another in a particular range cell provides information regarding the Doppler characteristics of the propagation mode contained in a particular range cell. Whereas the "array snapshot" samples recorded across different receivers in a particular pulse provides information on the spatial characteristics of the propagation mode in the interrogated range cell. Before proceeding to the fine structure analysis, it is of interest to estimate the nominal gross structure parameters of the ionospheric circuit under study. Apart from estimating the nominal signal-to-noise ratio and time-delay, this also involves estimating the mean direction-of-arrival and Doppler shift of each resolved propagation mode over the period of data collection.

The mean cone angle-of-arrival and Doppler frequency shift of a propagation mode can be estimated by evaluating the conventional angular and Doppler power spectrum using the data in the range cell containing the mode of interest with averaging performed over the data collection interval. Figures 6.3 to 6.6 show the angular and Doppler power spectra resulting for range cells $k = 16, 26, 29, 33$ which contain the four resolved signal modes. The mean cone angle-of-arrival and Doppler shift of the mode(s) received in each range cell are estimated as the locations of the maxima in the angular and Doppler power spectra, respectively. These estimates are listed in Table 6.1, which summarizes the multipath gross structure parameters of the HF link.

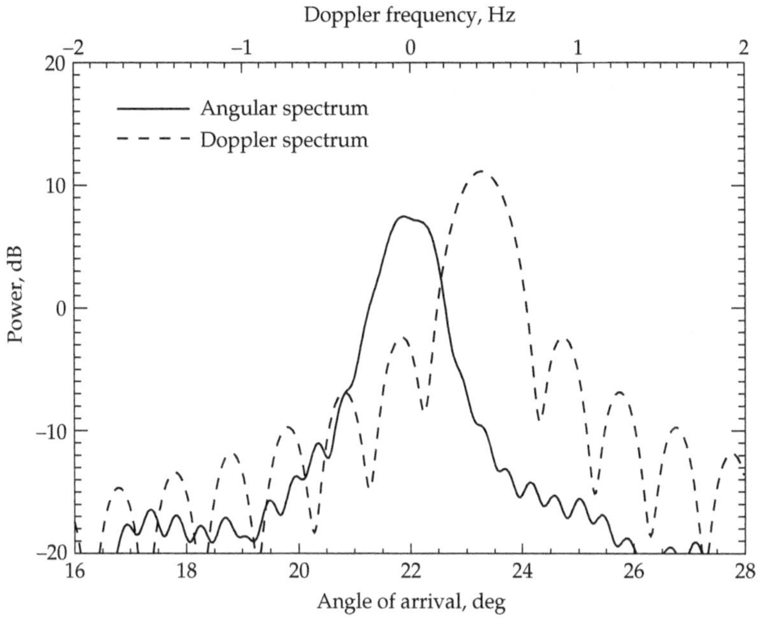

FIGURE 6.3 Conventional Doppler and angle-of-arrival spectrum for the $1E_s$ mode.

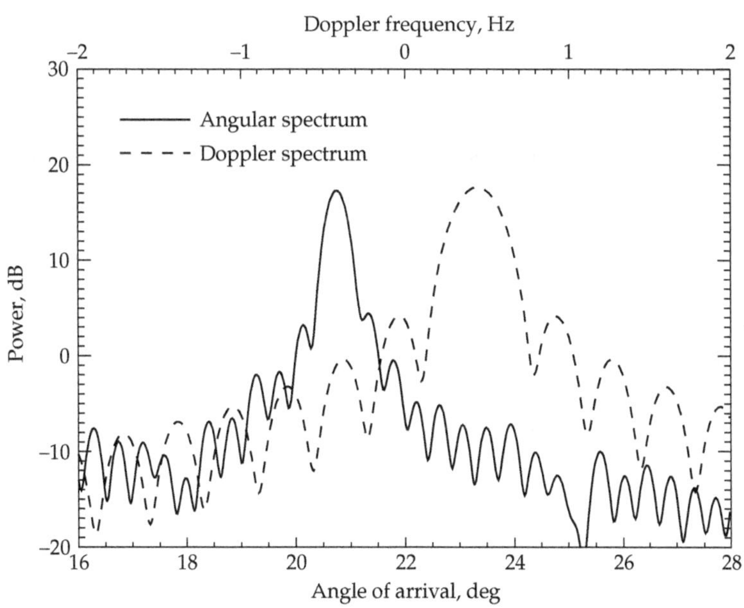

FIGURE 6.4 Conventional Doppler and angle-of-arrival spectrum for the $1F_2$ mode.

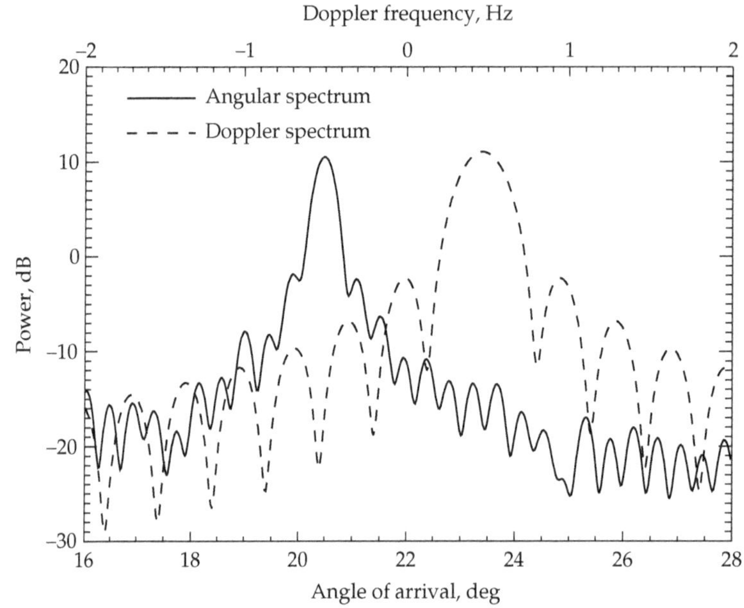

FIGURE 6.5 Conventional Doppler and angle-of-arrival spectrum for the $1F_2(o)$ mode. © Commonwealth of Australia 2011.

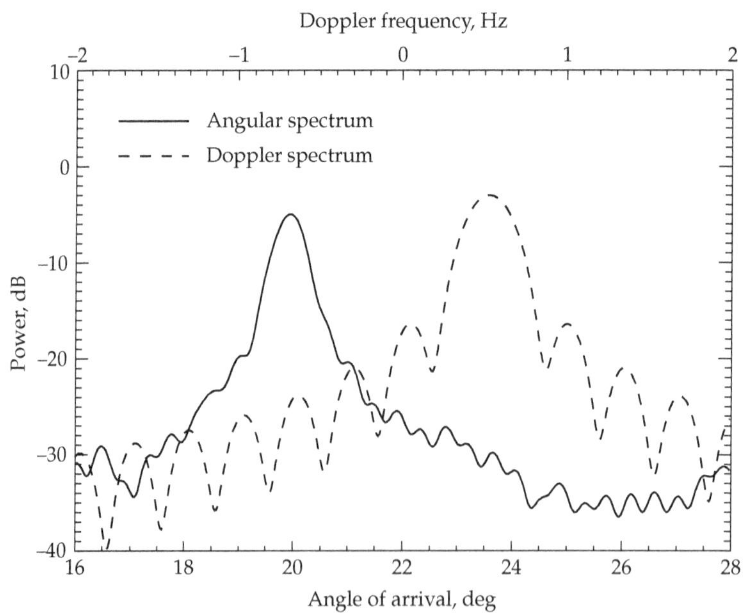

FIGURE 6.6 Conventional Doppler and angle-of-arrival spectrum for the $1F_2(x)$ mode. © Commonwealth of Australia 2011.

Mode	g_r, km	h_v, km	SNR, dB	Δf, Hz	θ, deg
$1E_s$	1290	99	34.5	0.42	21.9
$1F_2$	1430	303	40.5	0.44	20.8
$1F_2(o)$	1475	349	34.0	0.46	20.5
$1F_2(x)$	1540	408	20.0	0.53	19.9

TABLE 6.1 Parameters describing the gross structure of multipath propagation for the Darwin-to-Alice Springs skywave HF link on 1 April 1998 between 06:17 and 06:21 UT.

6.2.3 Fine Structure Observations

Examples of the mode wavefields sampled in space and time over a single CPI may be visualized as intensity-modulated displays in Figures 6.7 to 6.10. These four displays show the real component of the complex valued wavefields sampled over receivers and pulses at range cells $k = 16, 26, 29, 33$ which coincide with the four peak locations in Figure 6.2. A specularly reflected (plane wave) signal mode would appear as a two-dimensional (space-time) sinusoid in these displays. The spatial frequency of the sinusoid is related to the cone angle-of-arrival, while the temporal frequency is related to the Doppler shift. In the ideal case of specular reflection, the wavefront crests are expected to have equal magnitude and form straight line ridges of constant slope in these displays (i.e., perfectly linear phase-fronts).

Highly non-planar wavefronts are expected in Figure 6.7 because the wavefield in this range cell results from a superposition of unresolved sporadic-E modes that most likely have different directions-of-arrival. Wavefronts that are more planar in structure

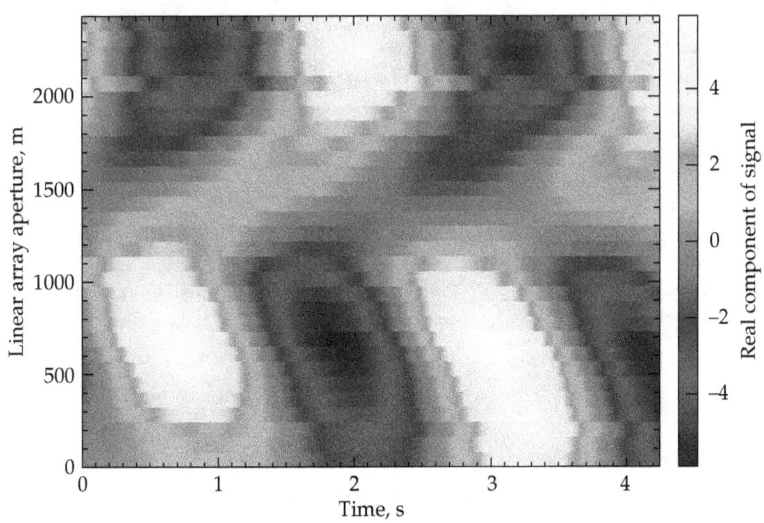

FIGURE 6.7 Real component of the $1E_s$ mode signal field sampled in space and time by the main array (range cell $k = 16$). © Commonwealth of Australia 2011.

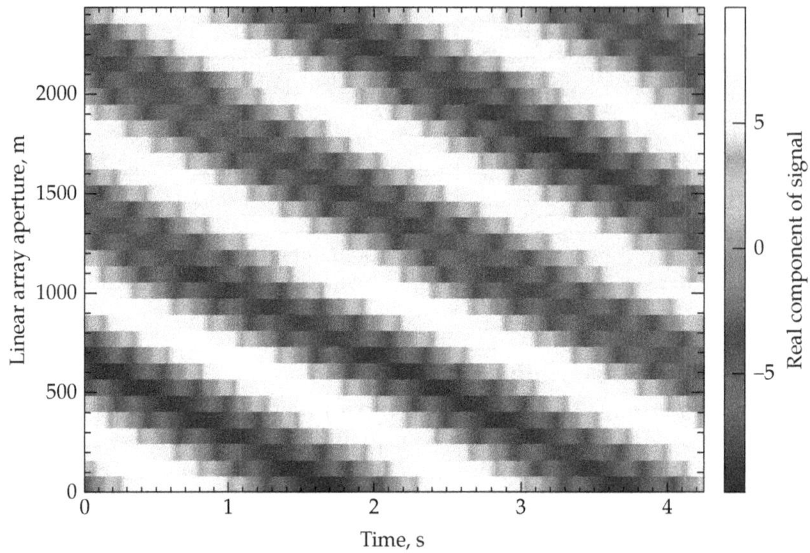

FIGURE 6.8 Real component of the $1F_2$ mode signal field sampled in space and time by the main array (range cell $k = 26$). © Commonwealth of Australia 2011.

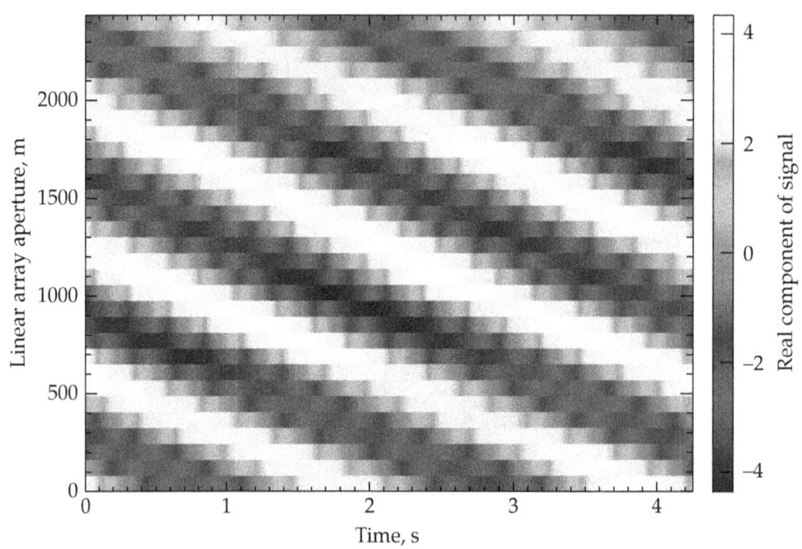

FIGURE 6.9 Real component of the $1F_2(o)$ mode signal field sampled in space and time by the main array (range cell $k = 29$). © Commonwealth of Australia 2011.

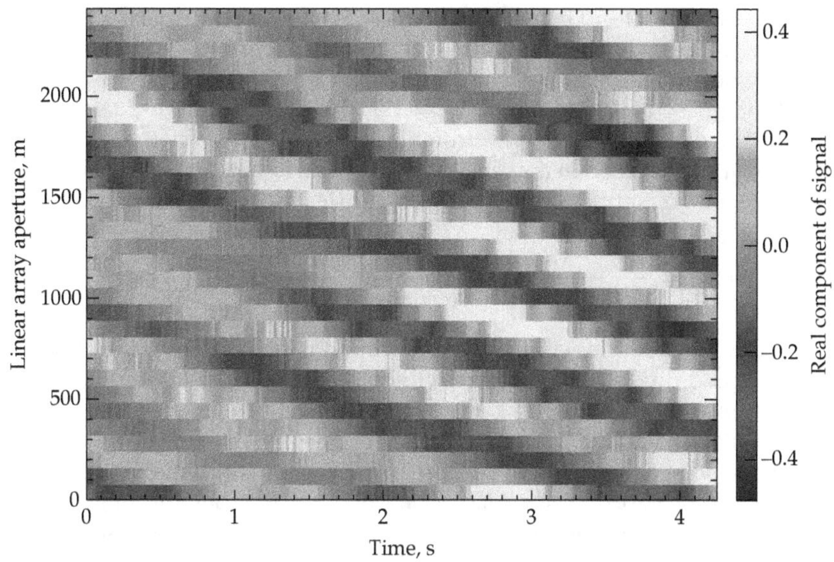

FIGURE 6.10 Real component of the $1F_2(x)$ mode signal field sampled in space and time by the main array (range cell $k = 33$). © Commonwealth of Australia 2011.

are observed for the $1F_2$ mode in Figure 6.8, which closely resembles a two-dimensional sinusoid. Figures 6.9 and 6.10 display the results for the $1F_2(o)$ and $1F_2(x)$ magneto-ionic components, respectively. The $1F_2(o)$ mode also exhibits relatively planar wavefronts, but it is apparent that the ionospheric reflection process has significantly distorted the $1F_2(x)$ mode. Such distortions are attributable to the presence of ionospheric irregularities because a single magneto-ionic component (i.e., the extraordinary wave of the F_2-layer high-angle ray) has been effectively isolated and is not contaminated by other components that could be theoretically expected for a smooth ionospheric layer.

Experimental observations in Rice (1976) and Sweeney (1970) made on a very wide aperture array suggest that a mode wavefront can often be regarded as having a more or less planar large-scale structure with some degree of amplitude and phase corrugations superimposed. At a particular time instant, these corrugations may be viewed as the spatial modulation imparted by the ionosphere on an underlying plane wavefront corresponding to the ideal case of specular reflection. Wavefront tests can be devised to detect the existence of mode fine structure over short time intervals. It is of interest to quantify: (1) the degree of departure of a mode wavefront relative to the plane wave model of best fit at a particular time, and (2) the dynamics of mode wavefront variations as a function of time due to changes in the relative gain and phase of the signal between receiver outputs.

Define $\mathbf{x}_k(t) \in \mathcal{C}^N$ as the complex N-dimensional array snapshot vector recorded in range cell k and PRI t. A measure of the similarity between the *spatial structure* of two array snapshots recorded at the same range cell k but in different pulses t and $t + \Delta t$ is given by the magnitude squared coherence (MSC) function in Eqn. (6.2). The MSC is unity when the snapshots $\mathbf{x}_k(t)$ and $\mathbf{x}_k(t + \Delta t)$ are related by a complex scalar (i.e., have the same spatial structure) and equals zero when the two snapshots become orthogonal. A value between these two extremes indicates the degree of spatial structure variation over Δt PRI, which translates to a time interval of $\tau = \Delta t / f_p$ seconds ($f_p = 60$ Hz). Unlike the RMS phase deviation measure used in Rice (1976), the MSC takes both the amplitude and phase of the mode wavefronts into account. Moreover, the condition of unit MSC is invariant to constant array calibration errors, local scattering effects, and the temporal phase rotation introduced by the ionospherically-induced Doppler shift.

$$\xi_k(\Delta t) = \frac{|\mathbf{x}_k^\dagger(t)\mathbf{x}_k(t+\Delta t)|^2}{\mathbf{x}_k^\dagger(t)\mathbf{x}_k(t)\ \mathbf{x}_k^\dagger(t+\Delta t)\mathbf{x}_k(t+\Delta t)}, \quad 0 \le \xi_k(\Delta t) \le 1 \tag{6.2}$$

Figures 6.11 to 6.14 show the cumulative distributions of MSC values evaluated for different modes and time intervals τ. These distributions were obtained by evaluating MSC for all pairs of array snapshots separated by $\tau = \Delta t / f_p$ seconds in 47 dwells of data with each dwell containing $P = 256$ snapshots. The minimum temporal separation of $\tau = 1/60$ seconds corresponds to curve 1 in all figures and may be regarded as a quasi-instantaneous measure of the MSC. The MSC distributions for longer temporal separations of 1, 2, and 3 seconds are represented by curves 2, 3, and 4, respectively.

Due to the presence of additive noise, the MSC values will not be exactly equal to unity even when the mode wavefront shape remains perfectly rigid. The effect of additive noise on the MSC distribution is conservatively illustrated by curve 1 in all figures. This assumes that the change in wavefront shape due to physical phenomena is negligible for

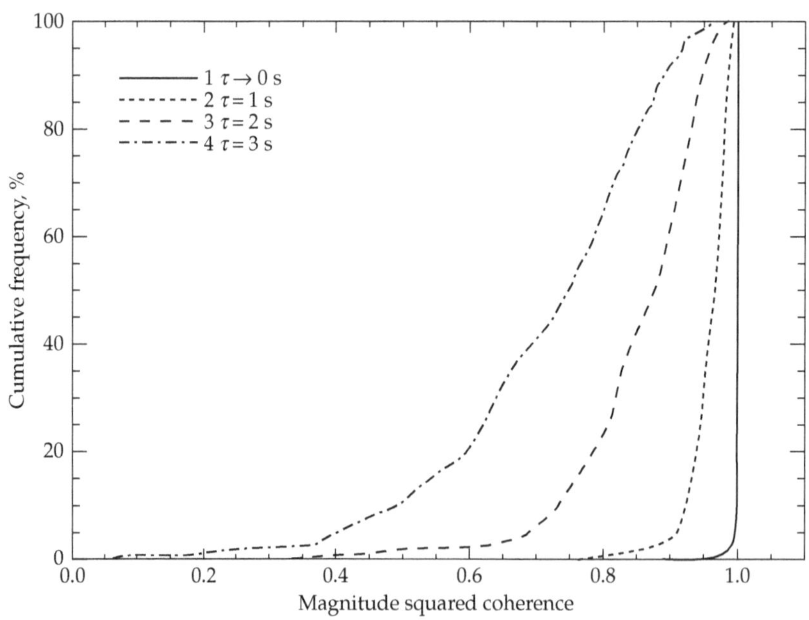

FIGURE 6.11 Cumulative distribution of the MSC for the $1E_s$ mode.

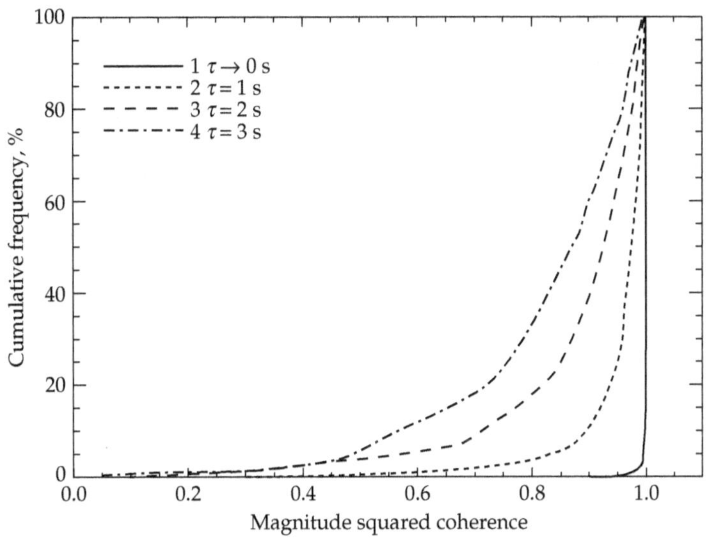

FIGURE 6.12 Cumulative distribution of the MSC for the $1F_2$ mode.

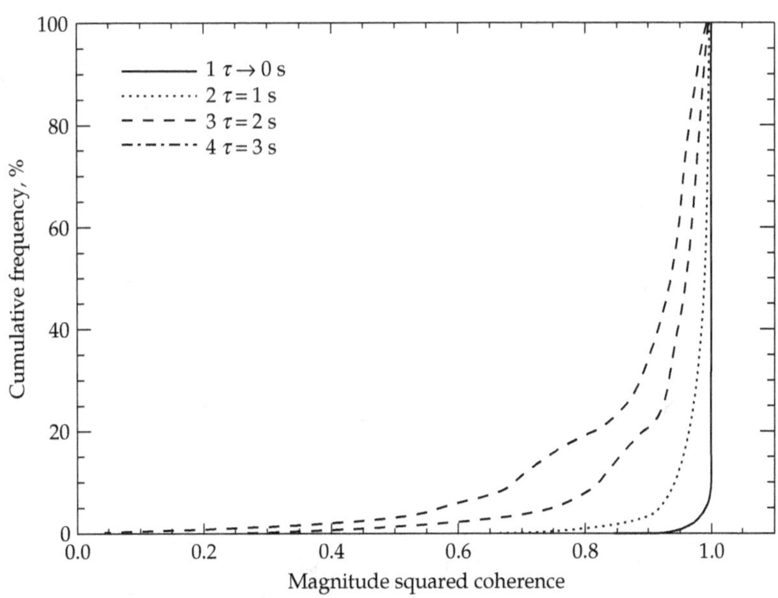

FIGURE 6.13 Cumulative distribution of the MSC for the $1F_2(o)$ mode.

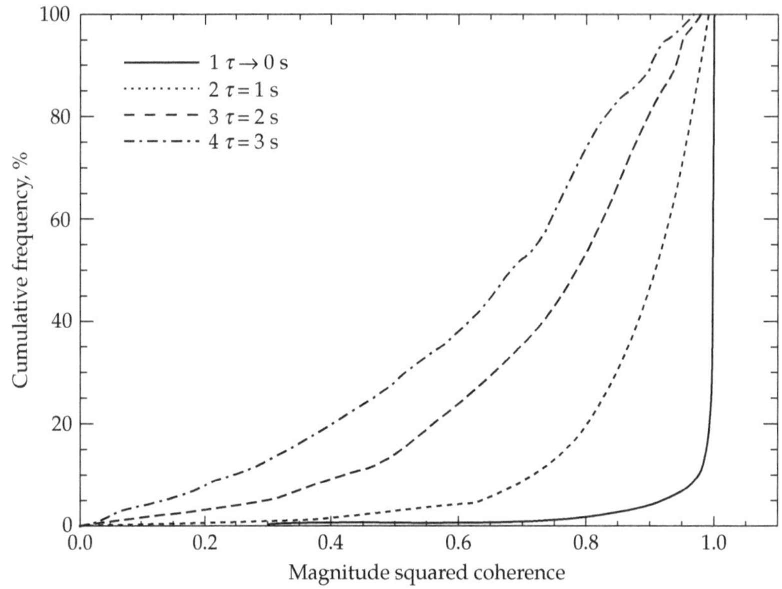

FIGURE 6.14 Cumulative distribution of the MSC for the $1F_2(x)$ mode.

$\tau = 1/60$ seconds and that departures in the MSC from unity are due to additive noise. If the mode wavefronts retained the same shape over a longer time interval τ, the MSC distributions is expected to resemble curve 1 in all figures because the MSC distribution would be invariant to temporal separation. The results clearly demonstrate that the MSC is highly dependent on temporal separation. This not only confirms the presence of time-varying wavefront distortions due to mode fine structure, but also shows that the mode wavefronts evolve in a correlated manner over time and become progressively different as the temporal separation increases.

The significant reduction in MSC from curve 1 to curve 2 illustrates that changes in spatial structure are appreciable over time intervals as short as 1 second. Curve 1 in Figure 6.13 indicates that 99 percent of the MSC values for the $1F_2(o)$ mode are above 0.95 for $\tau = 1/60$ seconds, while approximately 70 percent of the MSC values drop below 0.95 when the temporal separation is increased to 3 seconds (curve 4 in Figure 6.13). On the basis of this result, it may be claimed with 99 percent confidence that 70 percent of the array snapshots separated by a time interval of $\tau = 3$ seconds do not exhibit the same level of similarity in spatial structure as when the same data snapshots are separated by $\tau = 1/60$ seconds.

The main point is that for very short temporal separations, in the order of an OTH radar PRI, the mode wavefront structure remains essentially fixed, but as the temporal separation increases from a fraction of a second to a few seconds, the dissimilarity between the mode wavefronts gradually increases and becomes quite significant. Although these results quantitatively measure changes in mode wavefront structure as a function of temporal separation, they provide no information regarding the nature of such variations. For instance, it is useful to understand whether such changes are caused by shifts in the nominal cone angle-of-arrival of the underlying plane wavefront, or whether they primarily arise due to fluctuations in the wavefront amplitude and phase corrugations or "crinkles."

An alternative MSC function that provides this information is formulated in Eqn. (6.3), where $\mathbf{s}(\theta)$ is the plane wave array steering vector for a cone angle-of-arrival θ. The value of θ that maximizes $\rho_k(t, \theta)$ at time t is denoted by θ_{max} and represents the cone angle-of-arrival of the plane wave that best fits the array snapshot $\mathbf{x}_k(t)$ in a least squares sense. The maximum value of the MSC itself $\rho_k(t, \theta_{max})$ is a measure of the goodness of fit between the array snapshot $\mathbf{x}_k(t)$ and the *best fitting* plane wave model. In simple terms, $\rho_k(t, \theta_{max})$ indicates the size of amplitude and phase crinkle on the wavefront. A value of unity is only attained when $\mathbf{x}_k(t)$ is a plane wave, while lower values indicate the degree of departure of $\mathbf{x}_k(t)$ from the nearest point on the plane wave array manifold.

$$\rho_k(t, \theta) = \frac{|\mathbf{s}^\dagger(\theta)\mathbf{x}_k(t)|^2}{\mathbf{s}^\dagger(\theta)\mathbf{s}(\theta)\ \mathbf{x}_k^\dagger(t)\mathbf{x}_k(t)}, \quad 0 \leq \rho_k(t, \theta) \leq 1 \tag{6.3}$$

The quantities θ_{max} and $\rho_k(t, \theta_{max})$ were evaluated for each mode over four successive CPI and plotted in Figures 6.15 to 6.18 as a function of slow time t. Note the inter-dwell gap of approximately 0.5 seconds where no data is recorded. These curves illustrate the temporal behavior of the maximizing argument θ_{max}, which defines the plane wave model of best fit, and the MSC function $\rho_k(t, \theta_{max})$, which measures the goodness of fit, over time frames commensurate with typical OTH radar CPI. Different vertical axis scales have been used in order to clearly show the variations for different modes.

With the exception of Figure 6.15, which corresponds to the unresolved sporadic-E modes, the value of $\rho_k(t, \theta_{max})$ is reasonably close to unity. This supports the view of essentially planar wavefronts. Moreover, the plane waves of best fit to the mode wavefronts

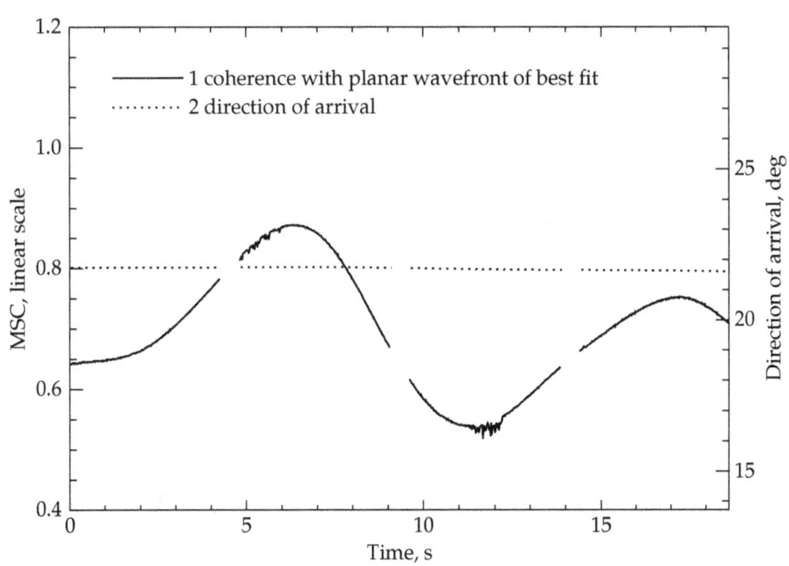

FIGURE 6.15 Analysis of planarity for the array snapshot $\mathbf{x}_k(t)$ containing the $1E_s$ mode ($k = 16$). © Commonwealth of Australia 2011.

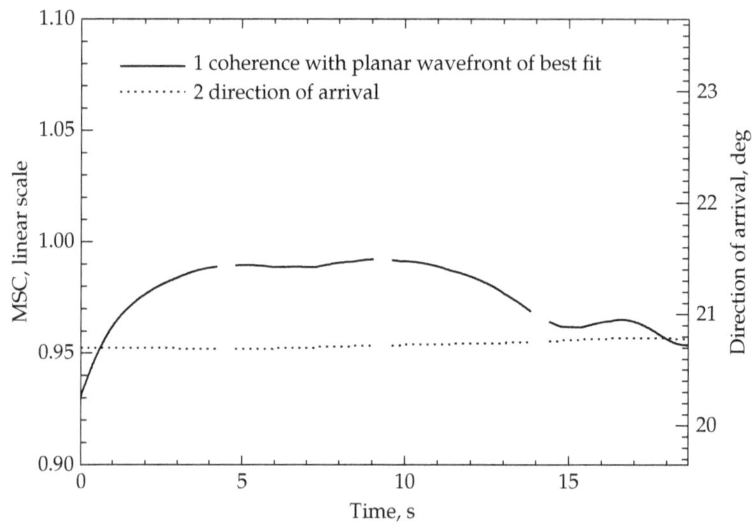

FIGURE 6.16 Analysis of planarity for the array snapshot $\mathbf{x}_k(t)$ containing the $1F_2$ mode ($k = 26$). © Commonwealth of Australia 2011.

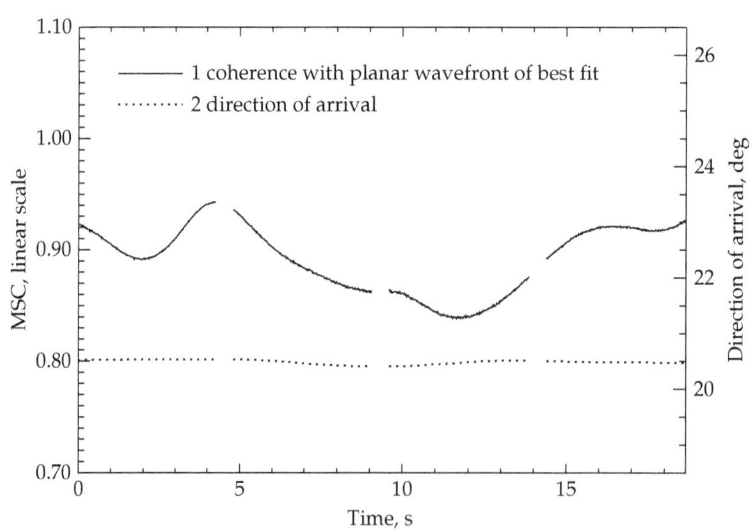

FIGURE 6.17 Analysis of planarity for the array snapshot $\mathbf{x}_k(t)$ containing the $1F_2(o)$ mode ($k = 29$). © Commonwealth of Australia 2011.

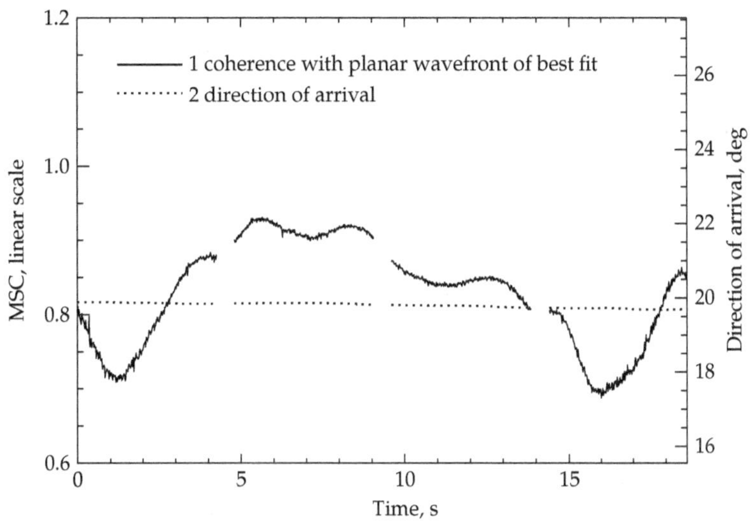

FIGURE 6.18 Analysis of planarity for the array snapshot $\mathbf{x}_k(t)$ containing the $1F_2(x)$ mode ($k = 33$). © Commonwealth of Australia 2011.

remain relatively constant in angle of arrival over a few seconds. This indicates that it is mainly the wavefront distortions or "crinkles" which are changing over time. It is also evident that the degree of fit to the best matched plane wave changes in a rather smooth fashion when sampled at the pulse repetition interval. This indicates that the amplitude and phase modulations imparted on the plane wave of best fit vary in a correlated manner when observed from one PRI to another at a temporal resolution of 1/60 seconds.

The value of $\rho_k(t, \theta_{max})$ is also subject to additive noise and the presence of array manifold errors. Variations due to additive noise are random from one PRI to another and superimpose on the smooth variations caused by the physical processes evolving in the ionosphere. The effect of noise is most apparent in the Figure 6.18, which corresponds to the mode with lowest signal-to-noise ratio, but is hardly noticeable in the other three figures. Array manifold errors are assumed to be fixed over an observation interval in the order of seconds so their presence would contribute a bias in $\rho_k(t, \theta_{max})$ but not a variation.

The observed variations in $\rho_k(t, \theta_{max})$ may therefore be attributed to spatial distortions induced by the ionospheric reflection process on the different propagation modes. An important point is that the precise form of these variations differs substantially from one mode to another. This strongly suggests that spatial distortions induced on a mode reflected by a particular ionospheric region are independent of those induced on another mode reflected by a physically distinct ionospheric region. The practical significance of this observation will be exploited in the final chapter of this text.

The collection of experimental results presented in this section confirm the view that a mode wavefront may be pictured as having an essentially planar spatial structure with some degree of amplitude and phase corrugations superimposed. The analysis has also yielded additional information regarding the time-evolution of individual mode wavefronts over time intervals commensurate with typical OTH radar CPI. While the underlying mean plane wavefront does not vary significantly over a few seconds, the size and shape of the wavefront distortions change gradually in a correlated manner. This leads to the interpretation that the ionospheric reflection process induces changing amplitude and phase distortions about a mean plane wavefront that become progressively de-correlated over time.

6.3 Resolving Fine Structure

To resolve mode fine structure, it is necessary to describe a wave interference model of the HF channel that deterministically represents the signal samples recorded by the antenna array in space and time under standard assumptions and approximations. A mathematical model of the *gross structure* of the received signal is described first based on the notion of specular reflection from each ionospheric region. This traditional model is then extended by incorporating multiple interfering rays for each propagation mode to parametrically represent fine structure. Robust super-resolution techniques appropriate for estimating the fine structure parameters are then identified and described.

6.3.1 Signal Representation

The complex envelope of the baseband space-time digital samples received by the system due to propagation mode $m = 1, 2, \ldots, M$ are denoted by $g_m(\ell, t, n)$ in Eqn. (6.4). The fast-time samples acquired during a pulse repetition interval (PRI) are indexed by $\ell = 0, 1, \ldots, L-1$, where $L = 320$ raw A/D samples per sweep in this experiment. The index $t = 0, 1, \ldots, P-1$ refers to slow-time samples within each CPI, where $P = 256$ PRI. The indices ℓ and t reference samples collected in time, while the index $n = 0, 1, \ldots, N-1$ references spatial samples collected by the $N = 32$ receivers of the antenna array. The terms T_s and f_p respectively denote the fast-time sampling period in seconds and the pulse repetition frequency (PRF) in Hertz. For the collected data, $T_s = 52$ microseconds and $f_p = 60$ Hz.

$$g_m(l, t, n) = A_m \exp\left(j2\pi\{\Delta u_m \ell T_s + \Delta f_m t/f_p\} + j\{\mathbf{k}_m \cdot \mathbf{r}_n + \gamma(\tau_m)\}\right) \tag{6.4}$$

Within a given PRI, the frequency of a deramped FMCW signal mode is proportional to its time-delay τ_m relative to the reference signal when stretch processing is applied. For a single-hop mode propagated over the ground distance of 1265 km, τ_m typically ranges between 4 and 6 milliseconds. The difference or beat frequency is defined by Δu_m in Eqn. (6.5), where $f_b = 20$ kHz is the FMCW sweep bandwidth and $LT_s = 1/f_p$ is the duration of the PRI in seconds. This group-range dependent frequency manifests itself as a regular phase progression $e^{j2\pi \Delta u_m \ell T_s}$ over fast-time samples within each sweep in Eqn. (6.4). The change in group range of a signal mode due to ionospheric motion over typical OTH radar CPI is extremely small compared to the range cell size. Since range cell migration is not an issue in this case, the difference frequency Δu_m after deramping may be assumed fixed over a CPI of 4.2 seconds.

$$\Delta u_m = \frac{f_b \tau_m}{LT_s} \tag{6.5}$$

An ionospheric layer may exhibit a (large-scale) uniform component of motion that imposes a regular Doppler shift on the reflected signal mode. The Doppler shift imposed by the ionosphere on mode m is defined by Δf_m in Eqn. (6.6), where $f_c = 16.110$ MHz is the carrier frequency, $c = 3.0 \times 10^8$ m/s is the speed of light in free space, and v_m is the relative velocity (group range rate) of the mode. A positive Doppler shift indicates that the effective reflection point is moving "downwards," shortening the phase path of the mode with respect to time, while the reverse applies when the Doppler shift is negative.

$$\Delta f_m = \frac{2v_m f_c}{c} \tag{6.6}$$

The relative velocity v_m of the reflection point is typically less than 10 m/s for quiet mid-latitude paths. This corresponds to a Doppler shift of 1 Hz for a single reflection at a carrier frequency of 15 MHz. Over a 4.2-second CPI, the effective displacement of the reflection point at this velocity is 42 m, which results in a differential time-delay of $\delta\tau_m = 1.4$ ns between endpoints of the CPI. As the time-bandwidth product condition in Eqn. (6.7) is satisfied, the Doppler effect manifests itself as a regular phase progression or frequency shift $e^{j2\pi\Delta f_m t/f_p}$ in Eqn. (6.4). For $v_m = 10$ m/s, it would take over 15 minutes before the change in virtual reflection height causes the mode to migrate by one range cell.

$$\delta\tau_m f_b \ll 1 \tag{6.7}$$

The initial phase of mode m at the start of the CPI relative to that of the transmitted waveform is given by $\gamma(\tau_m)$ in Eqn. (6.8) for a linear FMCW signal. This term depends on the mode time-delay τ_m at the beginning of the CPI in the reference (first) receiver and determines the starting phase relationship among the modes. The scalar A_m is an attenuation factor that accounts for all losses in the mode amplitude between the transmitter and receiver.

$$\gamma(\tau_m) = e^{j2\pi\left\{f_c\tau_m + \frac{f_b\tau_m^2}{LT_s}\right\}} \tag{6.8}$$

With reference to Figure 6.19, the ULA is aligned along the x-axis with the reference antenna at the origin. The N subarray centers of the Jindalee array are spaced $\Delta d = 84$ m

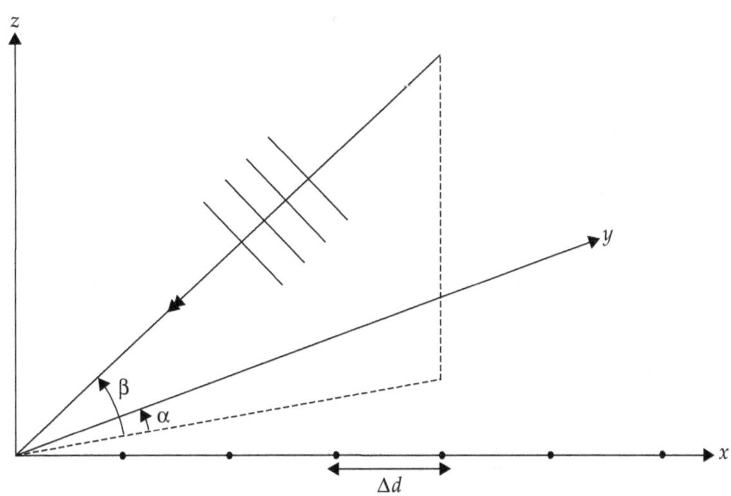

FIGURE 6.19 Three-dimensional coordinate system showing a plane wave incident from azimuth α and elevation β on a ULA aligned along the x-axis.

apart. Consider a plane wave signal mode incident upon the ULA at azimuth α_m relative to boresight and elevation β_m after specular reflection from the ionosphere. In this experiment, the great circle bearing of the test transmitter is 22 degrees. The elevation β depends on the virtual reflection height and typically varies between 5 and 35 degrees for single-hop ionospheric modes over a 1265-km mid-latitude path. By defining $\boldsymbol{k}_m = \frac{2\pi}{\lambda}\boldsymbol{u}(\alpha_m, \beta_m)$ as the signal wavevector, where $\boldsymbol{u}(\alpha_m, \beta_m) = [\cos\beta_m \sin\alpha_m, \cos\beta_m \cos\alpha_m, \sin\beta_m]^T$ is a unit vector in the wave propagation direction and $\lambda = c/f_c$ is the carrier wavelength, the relative phase of the carrier at receiver n in the array with respect to the reference receiver at the origin is calculated as the inner product $\boldsymbol{k}_m \cdot \boldsymbol{r}_n$ in Eqn. (6.9), where $\boldsymbol{r}_n = [n\Delta d, 0, 0]^T$ is the position vector of receiver n.

$$\boldsymbol{k}_m \cdot \boldsymbol{r}_n = \frac{2\pi}{\lambda} n\Delta d \cos\beta_m \sin\alpha_m = \frac{2\pi}{\lambda} n\Delta d \sin\theta_m \tag{6.9}$$

The differential time-delay between opposite ends of the array is given by Eqn. (6.10) which is in the order of 2–3 microseconds for $\alpha_m = 22$ degrees, $\beta_m = 5$–35 degrees, and $(N-1)\Delta d = 2.8$ km. Such delays correspond to a time-bandwidth product much less than unity for a radar signal bandwidth $f_b = 20$ kHz, so the phase relationship over different receivers may be written as $e^{j\boldsymbol{k}_m \cdot \boldsymbol{r}_n}$ in accordance with the narrowband assumption. The spatial phase difference may be interpreted in terms of the *cone* angle-of-arrival θ_m subtended by the mode wavevector and the x-axis. When the cone angle is interpreted as a bearing for a source off boresight, the apparent azimuth of the source approaches boresight as the mode elevation angle increases. This "coning effect" often allows different propagation modes from a single source to be resolved in cone angle by a ULA.

$$\boldsymbol{u}(\alpha_m, \beta_m) \cdot \boldsymbol{r}_{N-1}/c = (N-1)\Delta d \cos\beta_m \sin\alpha_m / c \tag{6.10}$$

Range processing is performed by taking the weighted fast Fourier transform (FFT) of the fast-time samples in each PRI. The range bins retained for further processing are indexed by $k = 0, 1, \ldots, K-1$, where $K = 42$ corresponds to a range depth of 630 km in this case. The range processing FFT of Eqn. (6.4) over all pulses and receivers yields the output $s_m(k, t, n)$ in Eqn. (6.11), where $c_m = A_m e^{j\gamma(\tau_m)}$ is the mode complex amplitude and $\psi_m(k) = f(\frac{k}{LT_s} - \Delta u_m)$ is the range spectrum point spread function defined by the Fourier transform of the Hanning window used for range processing. This function is normalized such that $f(0) = 1$.

$$s_m(k, t, n) = c_m \psi_m(k) \exp\left(j2\pi\{\Delta f_m t/f_p + n\Delta d \sin\theta_m/\lambda\}\right) \tag{6.11}$$

The received data $x(k, t, n)$ is the sum of M modes $s_m(k, t, n)$ for $m = 1, \ldots, M$ and uncorrelated noise. The array snapshots $\mathbf{x}_k(t) = [x(k, t, 0), \ldots, x(k, t, N-1)]^T$ received by the antenna array at range bin k and PRI t may be written as in Eqn. (6.12), where $f_m = \Delta f_m/f_p$ is the mode Doppler shift normalized by the PRF and $\mathbf{s}(\theta_m) = [1, z_m, \ldots, z_m^{N-1}]^T$ is the plane wave array steering vector defined in terms of $z_m = e^{j2\pi\Delta d \sin(\theta_m)/\lambda}$.

$$\mathbf{x}_k(t) = \sum_{m=1}^{M} c_m \psi_m(k) \mathbf{s}(\theta_m) e^{j2\pi f_m t} + \mathbf{n}_k(t) \tag{6.12}$$

The additive noise is often modeled as temporally and spatially white with second-order statistics given by Eqn. (6.13). Here, $\mathbf{n}_k^{[n]}(t)$ denotes the nth element of $\mathbf{n}_k(t)$, $E\{\cdot\}$ is the statistical expectation operator, and $\delta(\cdot)$ is the Kronecker delta function. Residual noise correlation in range due to the FFT window function has been ignored here.

$$E\{\mathbf{n}_{k_1}^{[n_1]}(t_1)\mathbf{n}_{k_2}^{[n_2]*}(t_2)\} = \sigma_n^2\, \delta(k_1 - k_2)\, \delta(t_1 - t_2)\, \delta(n_1 - n_2) \tag{6.13}$$

Let k_m be the range cell most closely matched to mode m. Provided the different modes are well resolved in group range, the M mode waveforms can be effectively separated into different range cells $\{k_m\}_{m=1}^{M}$ with negligible contamination from neighboring modes. In other words, contributions due to the mismatched modes may be neglected at cell k_m when the spectral leakage in range falls below the noise level. In this case, we can make the approximation $\psi_m(k) = \delta(k - k_m)$, such that $\mathbf{x}_{k_m}(t)$ is effectively given by Eqn. (6.14).

$$\mathbf{x}_{k_m}(t) = c_m \mathbf{s}(\theta_m) e^{j2\pi f_m t} + \mathbf{n}_{k_m}(t) \tag{6.14}$$

It has been conjectured that individual propagation modes may be composed of a small number of (presumably specular) components or "submodes" unresolved in time-delay but with different Doppler shifts and closely spaced directions-of-arrival. For a smooth layer, the unresolved submodes may correspond to the four theoretically expected rays (high/low, ordinary/extraordinary) or a subset of them, whereas for a single resolved magneto-ionic component, these submodes can only originate from electron density irregularities in the ionosphere. The wave interference model of fine structure represents the mode in range k_m as the sum of R_m submodes or rays in Eqn. (6.15).

$$\mathbf{x}_{k_m}(t) = \sum_{r=1}^{R_m} c_{mr}\, \mathbf{s}(\theta_{mr}) e^{j2\pi f_{mr} t} + \mathbf{n}_{k_m}(t) \tag{6.15}$$

The distinction between the gross and fine structure of an *individual* propagation mode resolved in range is that the latter contains more than one term (i.e., $R_m > 1$) in the wave interference model. The wave interference model in Eqn. (6.15) is deterministic, whereas the Doppler shift, direction of arrival, and other parameters of real ionospheric modes are expected to change over time. The scope of such a model is therefore to represent mode fine structure over a limited time commensurate with typical OTH radar CPI (i.e., in the order of a few seconds).

6.3.2 Parameter Estimation

Define the NP-dimensional stacked vector $\mathbf{z}_{k_m} = [\mathbf{x}_{k_m}^T(0), \mathbf{x}_{k_m}^T(1), \ldots, \mathbf{x}_{k_m}^T(P)]^T$ as the space-time data sampled over all receivers and pulses in range cell k_m during the CPI. By restricting attention to a single range cell for the time being, the data vector corresponding to the analyzed mode may be simply referred to as $\mathbf{z}$. The signal component in $\mathbf{z}$ is assumed to result from the interference of R rays that model the fine structure of the signal mode. In accordance with Eqn. (6.15), the data vector may be written in the form of Eqn. (6.16), where $\mathsf{A}(\varphi)$ is a $NP \times R$ mixing matrix to be defined shortly, the vector $\mathbf{c} = [c_1, \ldots, c_R]$ contains the ray complex amplitudes (i.e., magnitudes and initial phases) as its elements, and $\mathbf{n}$ is the stacked vector of additive noise defined similarly to $\mathbf{z}$.

$$\mathbf{z} = \mathsf{A}(\varphi)\mathbf{c} + \mathbf{n} \tag{6.16}$$

The R columns of the mixing matrix $\mathsf{A}(\varphi)$ are composed of ray space-time steering vectors $\mathbf{a}(\theta_r, f_r) \in \mathcal{C}^{NP}$ for $r = 1, \ldots, R$. The matrix $\mathsf{A}(\varphi)$ is therefore defined by the functional form of the space-time manifold $\mathbf{a}(\theta, f)$ and the fine structure parameter vector $\varphi = [\theta_1, \theta_2, \ldots, \theta_R, f_1, f_2, \ldots, f_R]^T$, which contains the cone angles-of-arrival and normalized Doppler frequency shifts of the R rays.

$$\mathsf{A}(\varphi) = [\mathbf{a}(\theta_1, f_1), \mathbf{a}(\theta_2, f_2), \ldots, \mathbf{a}(\theta_R, f_R)] \tag{6.17}$$

The NP-dimensional space-time manifold $\mathbf{a}(\theta, f)$ may be expressed as the Kronecker product ($\otimes$) between the N-dimensional spatial steering vector $\mathbf{s}(\theta)$ defined previously and the P-dimensional temporal steering vector $\mathbf{v}(f) = [1, e^{j2\pi f}, \ldots, e^{j2\pi(P-1)f}]$. The manifold $\mathbf{a}(\theta, f)$ has a Vandermonde structure, which implies that any given steering vector cannot be expressed as a linear combination of other steering vectors on the manifold.

$$\mathbf{a}(\theta, f) = \mathbf{s}(\theta) \otimes \mathbf{v}(f) \tag{6.18}$$

Ideally, the maximum likelihood estimate of the ray parameters is given by Eqn. (6.19). However, it is quite common for the parameter vector and ray complex amplitudes to be estimated separately rather than jointly. Provided the parameter estimation technique is chosen and implemented judiciously, as described in the next section, such approaches can yield comparable estimates to the ML technique in a more computationally efficient manner.

$$\{\hat{\mathbf{c}}, \hat{\varphi}\}_{\mathrm{ML}} = \arg\min_{\mathbf{c},\varphi} \|\mathbf{z} - \mathsf{A}(\varphi)\mathbf{c}\|^2 \tag{6.19}$$

For the assumed number of rays R, the first task is to estimate the parameter vector φ from the data $\mathbf{z}$. Based on this estimate, denoted by $\hat{\varphi}$, it is possible to reconstruct an approximation of the mixing matrix as $\mathsf{A}(\hat{\varphi})$. The ray complex amplitudes may then be estimated as the vector $\hat{\mathbf{c}}$ that provides the best least squares fit to the measured data $\mathbf{z}$ in accordance with Eqn. (6.20), where $\|\cdot\|^2$ represents the L2-norm (also known as the Frobenius norm $\|\cdot\|_F$) and $\mathsf{A}^+(\hat{\varphi}) = [\mathsf{A}^\dagger(\hat{\varphi})\mathsf{A}(\hat{\varphi})]^{-1}\mathsf{A}^\dagger(\hat{\varphi})$ is the Moore-Penrose pseudo-inverse of $\mathsf{A}(\hat{\varphi})$. Stated another way, the ray complex amplitudes $\hat{\mathbf{c}}$ are estimated so as to minimize the power of the residual error between the model $\hat{\mathbf{z}} = \mathsf{A}(\hat{\varphi})\hat{\mathbf{c}}$ and measured data $\mathbf{z}$.

$$\hat{\mathbf{c}} = \arg\min_{\mathbf{c}} \|\mathbf{z} - \mathsf{A}(\hat{\varphi})\mathbf{c}\|^2 = \mathsf{A}^+(\hat{\varphi})\mathbf{z} \tag{6.20}$$

Once the model order is selected and the ray parameters are estimated, the second task is to evaluate the accuracy with which the wave interference model represents the fine structure of real ionospheric modes sampled in space and time. A quantitative measure of the match between the simulated space-time data and the experimentally received signal samples is required to evaluate the performance of the wave-interference model as well as to establish criteria for accepting or rejecting its validity. A performance metric that is intuitively appealing and relatively simple to calculate is the ratio of the energy in the residual error $\|\mathbf{z} - \hat{\mathbf{z}}\|^2$ to that of the experimental data $\|\mathbf{z}\|^2$. The model-fitting accuracy (MFA) metric defined in Eqn. (6.21) expresses the fit of the model to the data based on this criterion as a percentage.

$$\mathrm{MFA}(\%) = \left\{1 - \frac{\|\mathbf{z} - \mathsf{A}(\hat{\varphi})\hat{\mathbf{c}}\|^2}{\|\mathbf{z}\|^2}\right\} \times 100 \tag{6.21}$$

The model $\hat{\mathbf{z}} = \mathsf{A}(\hat{\varphi})\hat{\mathbf{c}}$ is considered to be a satisfactory representation of the received data $\mathbf{z}$ when the MFA is sufficiently close to its upper limit of 100 percent. The meaning of "sufficiently close" should be defined with respect to the signal-to-noise ratio as the MFA will be less than 100 percent even if the model exactly replicates the signal component of the received data due to the presence of additive noise. From Eqn. (6.21), it is relatively simple to show that the expected value of the MFA for perfect signal modeling and uncorrelated additive noise is given by SNR/(SNR+1), where SNR is the signal-to-noise ratio. For the high SNR modes in the analyzed data, the MFA will most likely be limited by modeling errors rather than additive noise.

As far as resolving the fine structure of a propagation mode is concerned, the problem of estimating the ray parameter vector and complex amplitudes is a challenging one in practice. This is because the constituent rays will typically have closely spaced Doppler shifts and cone angles-of-arrival. Two rays with the same Doppler shift but different angles of arrival will produce a non-planar wavefront at the receivers that does not change over time. On the other hand, two rays with the same angle of arrival but different Doppler shifts will produce a plane wave that exhibits fading in time. In general, different rays have slightly different angles of arrival and Doppler shifts, which results in time-varying and non-planar wavefronts. This qualitative interpretation of a mode wavefront is consistent with the experimental observations made in Section 6.2.3.

Classical spectrum estimation techniques based on the FFT algorithm (periodogram) are computationally efficient.[1] A disadvantage of such techniques is that the mainlobe width prevents the resolution of two or more frequency components that are mutually spaced closer than the reciprocal of the data length. Another disadvantage is that strong signals can leak through the sidelobes and potentially mask a relatively weaker signal in the main lobe. The single dominant peaks in Figures 6.3 to 6.6 clearly show the inability of classical angle and Doppler spectrum estimation to resolve multiple rays within a mode due to the aforementioned drawbacks. Such techniques are therefore unsuitable for resolving the fine structure of individual propagation modes in the HF environment.

Other methods based on linear prediction Marple (1987) or the minimum variance distortionless response (MVDR) estimator Capon (1969) may be considered, but the best frequency resolution and estimation performance has been attributed to techniques based on the *eigenstructure* of the sample covariance matrix. The basis for improved performance, especially at lower signal-to-noise ratios, stems from the division of the information contained in the sample covariance matrix into signal and noise vector subspaces. The tremendous interest in subspace methods mainly originates from the initial development of the MUltiple SIgnal Classification (MUSIC) algorithm, Schmidt (1981). Since the introduction of MUSIC, a large variety of different super-resolution algorithms have emerged. An excellent summary of these algorithms and their relative merits can be found in Krim and Viberg (1996). A description of MUSIC is not repeated here for brevity, but the reader is referred to Marple (1987) for a detailed exposition of the essential concepts.

The maximum likelihood (ML) estimator mentioned earlier is a parametric method that can outperform MUSIC, particularly when two or more rays are coherent or highly correlated over the observation interval, see Krim and Viberg (1996). However, the ML estimator is computationally intensive relative to MUSIC, and perhaps

[1] The term spectrum estimation is used rather than frequency estimation because the periodogram also estimates the power of the sinusoids in the data.

more importantly, its convergence to the global optimum of the objective function in Eqn. (6.19) is not always guaranteed. In addition, sub-aperture smoothing techniques described in Pillai (1989) can be applied to improve the performance of MUSIC in the presence of correlated arrivals. MUSIC is identified as a suitable compromise between resolution power, computational complexity, and the capability to perform joint space-time frequency estimation in this application.

While MUSIC inherently estimates the parameters of different rays separately, it is possible to estimate the spatial and temporal frequencies of a particular ray jointly. Application of space-time MUSIC allows the different rays to be discriminated simultaneously in two dimensions rather than only in one. For example, two rays with almost identical Doppler shifts may not be resolved in temporal frequency, but if their cone angles-of-arrival are sufficiently different, it is theoretically possible to resolve the two rays as different peaks in the space-time domain. Another advantage of joint space-time processing is that pairing of the ray angle-of-arrival and Doppler frequency estimates occurs automatically since each ray is resolved on a two-dimensional (space-time) search grid. Disadvantages include the extra computational effort due to the use of covariance matrices with higher dimensionality, the larger amount of data required to estimate such matrices accurately, and the two-dimensional search instead of two one-dimensional searches.

6.3.3 Space-Time MUSIC

To describe the space-time MUSIC technique, let $\mathbf{x}_k(s,t)$ be an N_s-dimensional array snapshot vector recorded by N_s receivers of the antenna array starting at receiver number s, where $s \in [0, N-N_s]$ and $N_s < N$. Unlike the full-aperture array snapshot $\mathbf{x}_k(t) \in \mathcal{C}^N$, which contains all receiver outputs, $\mathbf{x}_k(s,t) \in \mathcal{C}^{N_s}$ represents a sub-aperture of the ULA that contains N_s consecutive receiver outputs with the first element corresponding to receiver index s. The space-time data vector $\mathbf{z}_k(s,t,\Delta t) \in \mathcal{C}^{N_s N_t}$ in Eqn. (6.22) is formed by stacking the N_s-dimensional array snapshots $\mathbf{x}_k(s,t)$ recorded over N_t slow-time samples spaced Δt PRI apart such that $t \in [0, P - N_t\Delta t]$ and $N_t\Delta t < P$. It is assumed that the values of N_s and N_t are chosen such that $N_s N_t > R$, and that the R rays have distinct parameter tuples $\{\theta, f\}$.

$$\mathbf{z}_k(s,t,\Delta t) = [\mathbf{x}_k^T(s,t), \mathbf{x}_k^T(s,t+\Delta t), \ldots \mathbf{x}_k^T(s,t+(N_t-1)\Delta t)]^T \tag{6.22}$$

The space-time data vector $\mathbf{z}_k(s,t,\Delta t)$ can be written as Eqn. (6.23), where $\mathsf{A}(\varphi)$ is the reduced dimension $N_s N_t \times R$ mixing matrix. To avoid complicated notation, the sub-aperture ray mixing matrix in Eqn. (6.23) is not distinguished from the full-aperture version defined previously in Eqn. (6.17).

$$\mathbf{z}_k(s,t,\Delta t) = \mathsf{A}(\varphi)\mathbf{s}_k(t) + \mathbf{n}_k(s,t,\Delta t) \tag{6.23}$$

The M-dimensional signal vector $\mathbf{s}_k(t)$ in Eqn. (6.24) contains the complex waveforms $s_m(k,t,s)$ recorded for each ray at the starting receiver (s). The space-time vector of uncorrelated white noise $\mathbf{n}_k(s,t,\Delta t)$ is constructed in analogous fashion to Eqn. (6.22) and represents the additive noise component of the space-time vector $\mathbf{z}_k(s,t,\Delta t)$.

$$\mathbf{s}_k(t) = [s_1(k,t,s),\ s_2(k,t,s) \cdots s_M(k,t,s)]^T \tag{6.24}$$

Traditionally, the space-time sample covariance matrix used to compute the MUSIC spectrum is estimated using the full array aperture. When two or more rays are coherent (i.e., have the same Doppler shift), the standard MUSIC spectrum fails to provide consistent estimates of all the ray angles-of-arrival and Doppler shifts. This occurs due to a rank deficiency in the dimension of the signal subspace, as discussed in Pillai (1989). To rectify this situation, the idea of spatial smoothing described in Shan, Wax, and Kailath (1985) and Pillai and Kwon (1989) may be employed to "de-correlate" the rays while preserving the angular and Doppler information. Specifically, the forward-backward spatial smoothing technique is used to form the sample space-time matrix $\hat{\mathsf{R}}_z(k)$ in Eqn. (6.25), where $P' = P - N_t \Delta t$, $N' = N - N_s$ and J is the square $N_s N_t \times N_s N_t$-dimensional exchange matrix with ones on the anti-diagonal and zeros elsewhere.

$$\hat{\mathsf{R}}_z(k) = \sum_{t=0}^{P'} \sum_{s=0}^{N'} \mathbf{z}_k(s,t,\Delta t)\mathbf{z}_k^{\dagger}(s,t,\Delta t) + \mathsf{J}\mathbf{z}_k^{*}(s,t,\Delta t)\mathbf{z}_k^{T}(s,t,\Delta t)\mathsf{J} \tag{6.25}$$

The statistically expected space-time covariance matrix $\mathsf{R}_z(k)$ takes the form of Eqn. (6.26), where $\mathsf{S}_k \in \mathcal{C}^{R \times R}$ is referred to as the spatially smoothed source covariance matrix. Although the true source covariance matrix defined as $E\{\mathbf{s}_k(t)\mathbf{s}_k^{\dagger}(t)\}$ is not full rank when two or more signals are coherent, it can be shown that the rank of the spatially smoothed source covariance matrix increases by one with probability one for each spatial average (Pillai 1989). This property allows MUSIC to yield consistent estimates of the ray parameters when two or more rays are coherent or highly correlated. The cost of sub-aperture smoothing is a reduction in resolution and the maximum number of rays that can be resolved.

$$\mathsf{R}_z(k) = \mathsf{A}(\varphi)\mathsf{S}_k\mathsf{A}^{\dagger}(\varphi) + \sigma_n^2 \mathsf{I} \tag{6.26}$$

Assuming S_k has full rank R, the eigen-decomposition of the positive definite Hermitian matrix $\mathsf{R}_z(k)$ may be expressed as in Eqn. (6.27), where the R columns of Q_s contain the eigenvectors spanning the signal subspace. A critical point is that the span of the signal eigenvectors is identical to the range space of $\mathsf{A}(\varphi)$. On the other hand, the $N_s N_t - R$ columns of Q_n are noise eigenvectors spanning the orthogonal subspace to Q_s and hence $\mathsf{A}(\varphi)$. The diagonal matrix $\boldsymbol{\Lambda} = \text{diag}[\lambda_1, \ldots, \lambda_R]$ contains the largest eigenvalues associated with the R principal eigenvectors, while the smaller (noise) eigenvalues are all equal to σ_n^2.

$$\mathsf{R}_z(k) = \mathsf{Q}_s \boldsymbol{\Lambda} \mathsf{Q}_s^{\dagger} + \sigma_n^2 \mathsf{Q}_n \mathsf{Q}_n^{\dagger} \tag{6.27}$$

As the eigenvectors in the signal subspace are orthogonal to those in the noise subspace, the ray vectors $\mathbf{a}(\theta_r, f_r)$ are also mutually orthogonal to the noise subspace eigenvectors. In other words, $\mathsf{Q}_n^{\dagger}\mathbf{a}(\theta, f) = 0$ when the parameters $\theta = \theta_r$ and $f = f_r$ for $r = 1, 2, \ldots, R$. Moreover, the parameter tuples $\{\theta_r, f_r\}$ corresponding to the R rays are the only possible cone angle and Doppler shift combinations that can satisfy this orthogonality condition because any collection of distinct space-time signal vectors $\mathbf{a}(\theta, f)$ forms a linearly independent set. The spatially smoothed MUSIC spectrum $p_{mu}(\theta, f)$ in Eqn. (6.28) may therefore be computed by projecting a spectrum of signal vectors $\mathbf{a}(\theta, f)$ against the *estimated* noise subspace $\hat{Q}_n$ calculated from the *sample* space-time covariance matrix $\hat{R}_z(k)$ in Eqn. (6.25).

$$p_{mu}(\theta, f) = \frac{\mathbf{a}^\dagger(\theta, f)\mathbf{a}(\theta, f)}{\mathbf{a}^\dagger(\theta, f)\hat{Q}_n\hat{Q}_n^\dagger\mathbf{a}(\theta, f)} \tag{6.28}$$

The MUSIC spectrum exhibits sharp peaks in the vicinity of the true ray angles-of-arrival and Doppler shifts. The coordinates of these peaks yields the parameter vector estimate $\hat{\varphi}$. The magnitude of the peaks in the MUSIC spectrum should not be interpreted as the amplitude of the rays. The ray complex amplitudes are estimated separately using the least squares procedure in Eqn. (6.20).

6.4 Experimental Results

The described space-time MUSIC technique requires the model order or number of rays to be specified. A method based on information theory that uses the eigenvalues of the sample covariance matrix to estimate the number of rays was developed by Wax and Kailath (1985). Although such a method is applicable here, it should be kept in mind that the objective is to determine how well a small number of rays can represent mode fine structure. A model involving a high number of rays can always be constructed in principle to represent mode fine structure very accurately. However, such a model is of limited utility for a number of reasons. First, the physical significance of the model is more readily interpreted when the data characteristics can be represented by a small number of rays; and second, models of very high order are less mathematically tractable and require many parameters to specify.

At the other extreme, a one-ray model is too simplistic and does not satisfactorily represent the observed HF channel characteristics. Models of intermediate complexity are of most practical benefit because they are challenging enough to allow the performance of different algorithms to be meaningfully compared and ranked, but not so challenging as to defeat them all. The question arises as to what constitutes the meaning of the term "relatively few rays" or "a small number of rays." Two- and three-ray wavefields were synthesized by Gething (1991) in order to compare the performance of different wavefront analysis techniques. The same model orders were adopted by Rice (1982) to analytically derive probability density functions for a defined measure of wavefront planarity. In this section, the ability of the model to characterize mode fine structure using up to four rays is analyzed. The major justification for going to four rays is that it coincides with the number of rays that are theoretically expected after reflection from a single smooth ionospheric layer.

6.4.1 Preliminary Data Analysis

In addition to the model order R, the sub-aperture dimensions also need to be specified in terms of the number of receivers N_s, the number of temporal taps N_t, and the spacing between adjacent taps Δt. While higher values of N_s, N_t, and Δt improve resolution, the larger sample covariance matrix dimension that results leads to an increase in the variance of the MUSIC spectrum estimates when using a finite set of space-time data. Use of data from multiple CPI to stabilize these estimates is not recommended because there is a greater chance of the ray parameters changing as the observation interval increases.

There are no definitive guidelines for selecting the "optimum" array sub-aperture dimensions when the signal environment is unknown a priori. In the following analysis, the space-time MUSIC spectra were estimated using $N_s = 16$ receivers, $N_t = 16$ temporal taps, and $\Delta t = 12$ PRI (i.e., a 0.2-second tap spacing). Based on the premise of limiting the duration of the analysis to a single CPI of data at a time, these values were empirically chosen to trade off the resolution and variance of the parameter estimation technique so as to yield satisfactory performance.

To compare the experimental application of one- and two-dimensional MUSIC for the problem of resolving mode fine structure, the smoothed spatial-only MUSIC spectrum is obtained by setting $N_t = 1$ and $N_s = 16$, while the temporal-only version is computed by setting $N_s = 1$ and $N_t = 16$ with $\Delta t = 12$. The first example considers the $1E_s$ mode received in a particular CPI. The real component of the space-time wavefield received for this mode was previously illustrated in Figure 6.7.

Curves 2 and 3 in Figure 6.20 illustrate the spatial-only MUSIC spectra assuming a model order of four and five rays respectively. Note that the four peaks resolved for $M = 4$ become broader when $M = 5$ is assumed but a fifth peak does not appear. This suggests the presence of four dominant rays that can be resolved in angle-of-arrival. Curve 1 in Figure 6.20 shows the standard MUSIC spectrum *without* spatial smoothing for the same data (i.e., $N_s = N = 30$). Curve 4 in Figure 6.20 relates to the right vertical axis and represents the MVDR spectrum using the same spatially smoothed sample spatial covariance matrix as that used to compute the MUSIC spectra in curves 2 and 3.

A comparison of curves 2 and 4 in Figure 6.20 illustrates that the spatial-only MUSIC algorithm can resolve four closely spaced rays while the MVDR only resolves three. The presence of the fourth ray is indicated by a slight bulge in the MVDR spectrum at the angle of the unresolved ray. This illustrates the superior resolution of the subspace

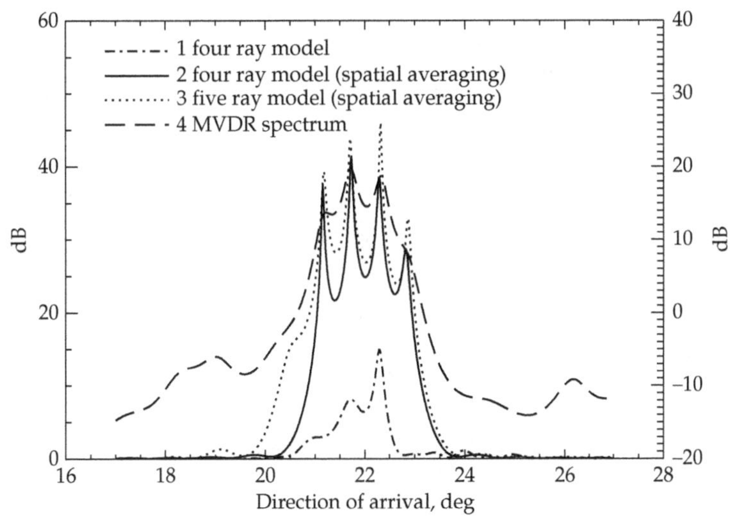

FIGURE 6.20 Spatial MUSIC spectra for the $1E_s$ mode ($k = 16$). The angles of arrival corresponding to the four peaks in Curve 2 are 21.2, 21.7, 22.3, and 22.8 degrees. The scale on the right vertical axis corresponds to the MVDR spectrum (curve 4) only.

method, as expected. A comparison of curves 1 and 2 in Figure 6.20 illustrates the benefit of spatial smoothing that allows four rays to be resolved. The poor performance of the standard (unsmoothed) MUSIC algorithm in this example suggests the presence of highly correlated arrivals during the observation interval. This suggests the presence of rays with very similar Doppler shifts.

Using the same data, curves 1, 2, and 3 in Figure 6.21 illustrate the temporal-only MUSIC spectra for an assumed model order of one, two, and three rays, respectively. All three MUSIC spectra relate to the left vertical axis and were evaluated for the sample temporal covariance matrix using $N_t = 16$ and $\Delta t = 12$. Curve 4 relates to the right vertical axis and shows the MVDR spectrum computed using the same sample covariance matrix as MUSIC. Despite four rays being resolved in the angular MUSIC spectrum, only two rays can be resolved in Doppler frequency by MUSIC. The inability of temporal-only MUSIC to resolve more than two rays is consistent with the previous suggestion that some of the rays have very similar Doppler shifts.

Although the MVDR spectrum is often referred to as a high-resolution estimator, it cannot resolve the two rays resolved by MUSIC in Figure 6.21. This is a further illustration of the superior performance of the MUSIC super-resolution algorithm. Nevertheless, the MVDR spectrum is important because it indicates that most of the received power is distributed in an interval of Doppler frequencies between 0.4 and 0.6 Hz. If four dominant rays are present, then the Doppler frequencies of these rays are expected to be in this interval.

The discrepancy between the number of rays resolved in angle-of-arrival and Doppler frequency creates a problem when it comes to analyzing the wavefield since it is not possible to unambiguously assign a spatial and temporal frequency to each ray. This real-data

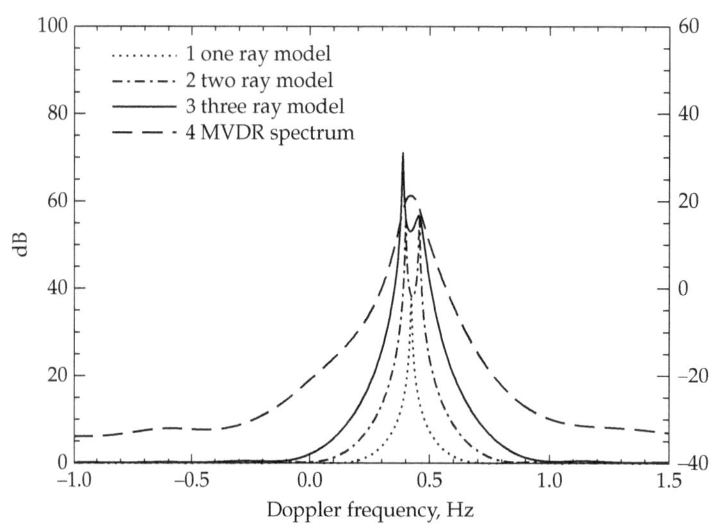

FIGURE 6.21 Temporal MUSIC spectra for the 1 E_s mode ($k = 16$). The Doppler frequencies corresponding to the two peaks in Curve 2 are 0.40 and 0.46 Hz. The scale on the right vertical axis corresponds to the MVDR spectrum (curve 4) only.
© Commonwealth of Australia 2011.

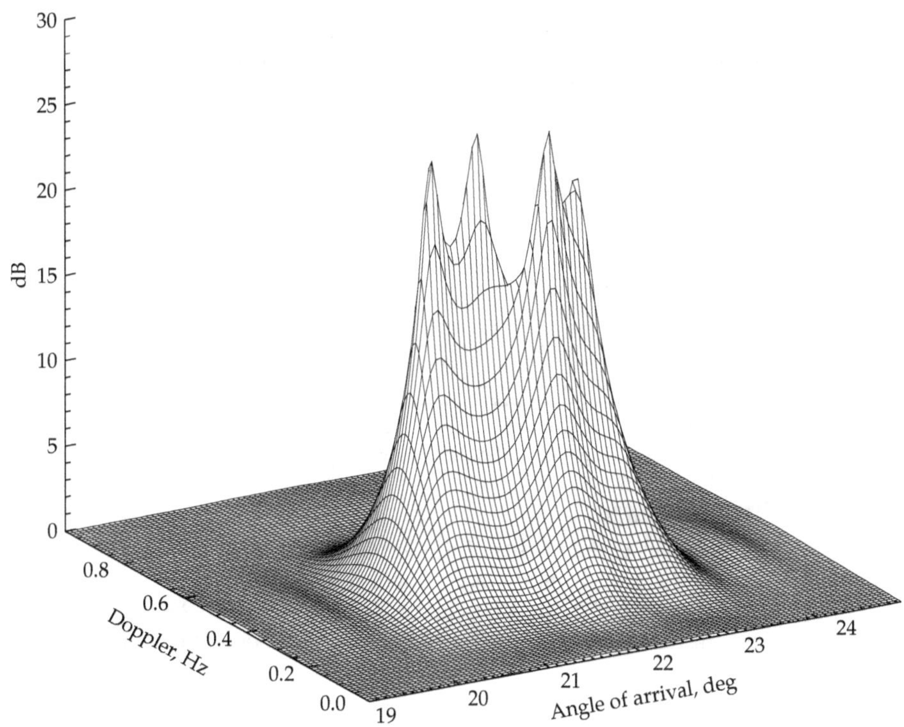

FIGURE 6.22 Space-time MUSIC spectrum for the $1E_s$ mode assuming four rays. © Commonwealth of Australia 2011.

example serves to illustrate the principal motivation for using space-time MUSIC rather than estimating the parameters via two one-dimensional MUSIC spectra. The space-time MUSIC spectrum derived from the same data using $N_s = 16$, $N_t = 16$, $\Delta t = 12$ and assuming four rays is shown as a three-dimensional surface plot in Figure 6.22.

The parameters of the four rays resolved for the 1Es mode by the space-time MUSIC spectrum in Figure 6.22 are listed in Table 6.2. This includes the amplitude and initial phase of each ray that were estimated using the least squares procedure in Eqn. (6.20). Note that the location of each peak resolved by the temporal-only and spatial-only MUSIC spectra is consistent with the coordinates of one of the four peaks resolved in the space-time MUSIC spectrum. The space-time MUSIC spectrum also confirms that three of the four rays have almost identical Doppler shifts. These three rays were not resolved by temporal-only MUSIC, but were resolved by the smoothed spatial-only and space-time MUSIC techniques due to their different angles of arrival.

Examples of the space-time MUSIC spectra resulting for the $1F_2$, $1F_2(o)$, and $1F_2(x)$ modes are shown in Figures 6.23 to 6.25. For each mode, the model order was chosen to represent the number of rays that were frequently required to yield a fitting accuracy greater than 90 percent over the entire data set containing 47 CPI. The ray parameters estimated from these space-time MUSIC spectra are also listed in Table 6.2.

Mode	Angle of arrival, deg	Doppler, Hz	Magnitude, linear	Phase, deg
$1E_s$ ray 1	21.2	0.43	1.10	77.3
$1E_s$ ray 2	21.7	0.44	2.57	−174.8
$1E_s$ ray 3	22.4	0.40	2.16	−8.8
$1E_s$ ray 4	22.8	0.43	0.65	−127.4
$1F_2$ ray 1	20.1	0.50	0.54	141.4
$1F_2$ ray 2	20.7	0.48	2.76	54.5
$1F_2$ ray 3	21.3	0.37	0.64	104.7
$1F_2(o)$ ray 1	20.4	0.48	3.93	−50.0
$1F_2(x)$ ray 1	19.7	0.51	0.67	61.4
$1F_2(x)$ ray 2	20.1	0.52	0.88	160.8

TABLE 6.2 Ray parameters estimated from space-time MUSIC spectra computed for the $1E_s$, $1F_2$, $1F_2(o)$, and $1F_2(x)$ modes using a single CPI of data. © Commonwealth of Australia 2011.

6.4.2 Model-Fitting Accuracy

Figures 6.26 to 6.29 summarize the results of this experimental analysis by showing the fitting accuracy achieved for each mode as a function of the number of rays and the particular CPI of data analyzed. Note that certain CPI in Figures 6.26 to 6.29 do not contain a measurement for all model orders because the number of peaks resolved by

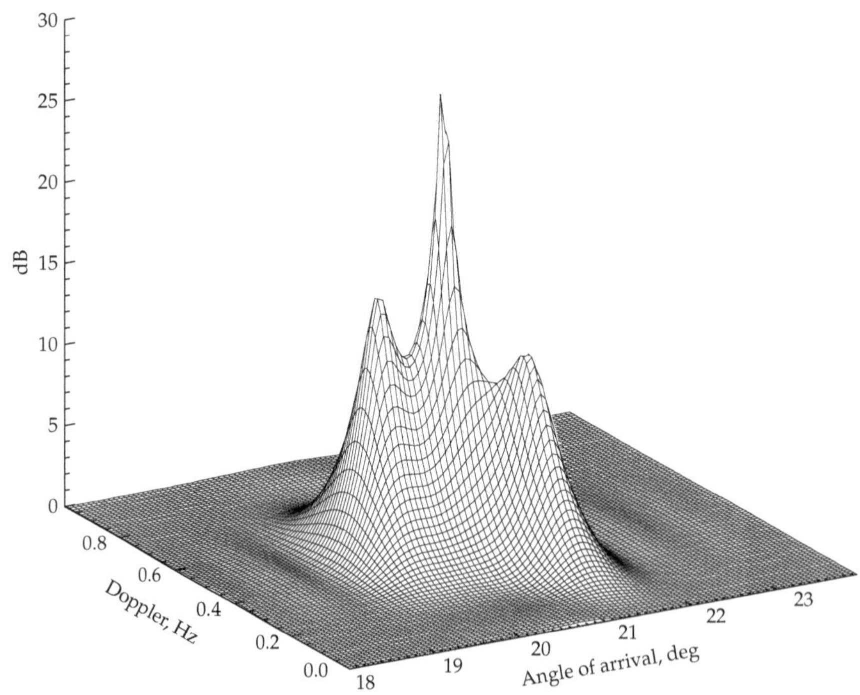

FIGURE 6.23 Space-time MUSIC spectrum for the $1F_2$ mode assuming three rays. © Commonwealth of Australia 2011.

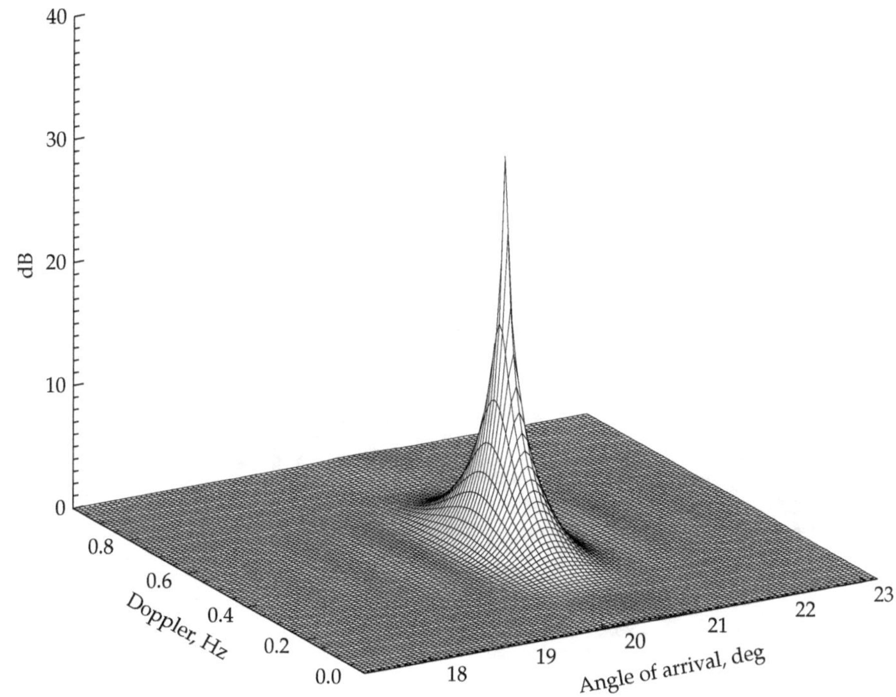

FIGURE 6.24 Space-time MUSIC spectrum for the $1F_2(o)$ mode assuming one ray. © Commonwealth of Australia 2011.

the space-time MUSIC spectrum does not always coincide with the assumed number of rays. This situation may arise when the model order is overestimated or if the rays are present but cannot be resolved. The fitting accuracy is not displayed when the number of peaks does not agree with the assumed model order.

It is important to note that the failure of MUSIC to resolve the expected number of peaks does not imply that a wave interference model having less than the assumed number of rays cannot adequately represent the data. For example, the assumption of two and three rays for the $1F_2$ mode in dwell 22 gives rise to one and two peaks, respectively, so the symbols corresponding to these model orders do not appear for dwell number 22 in Figure 6.27. However, the fitting accuracy achieved by the two resolved rays when three rays were assumed was 93 percent.

Figure 6.30 shows the real component of the model wavefield using the ray parameters estimated for the $1E_s$ mode in Table 6.2. Upon visual inspection, this simulated wavefield is very similar to the experimentally recorded one shown in Figure 6.7. A quantitative comparison between the real and simulated complex wavefields is made in terms of the MFA metric defined in Section 6.3.2. In this example, the estimated four ray interference model yields a fitting accuracy of 96 percent. From a physical perspective, the two rays with cone angles less than the great circle azimuth of the test transmitter (22 degrees) and the two rays with cone angles larger than 22 degrees are possibly reflected from different clouds of sporadic-E ionization.

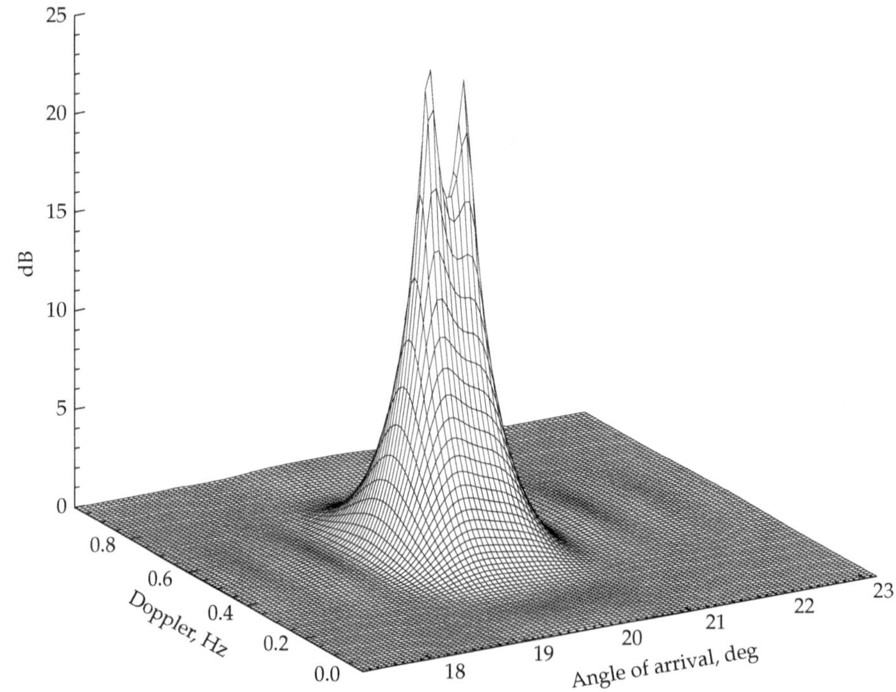

FIGURE 6.25 Space-time MUSIC spectrum for the $1F_2(x)$ mode assuming two rays. © Commonwealth of Australia 2011.

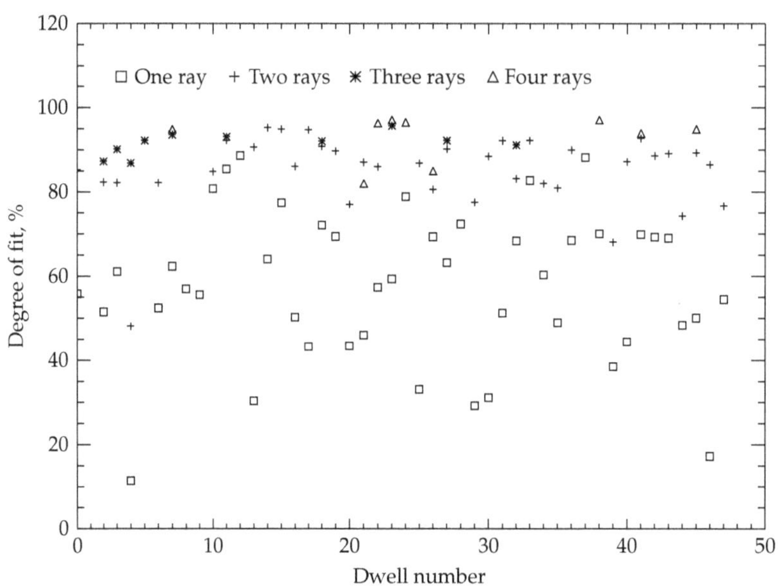

FIGURE 6.26 Fitting accuracy for the $1E_s$ mode. © Commonwealth of Australia 2011.

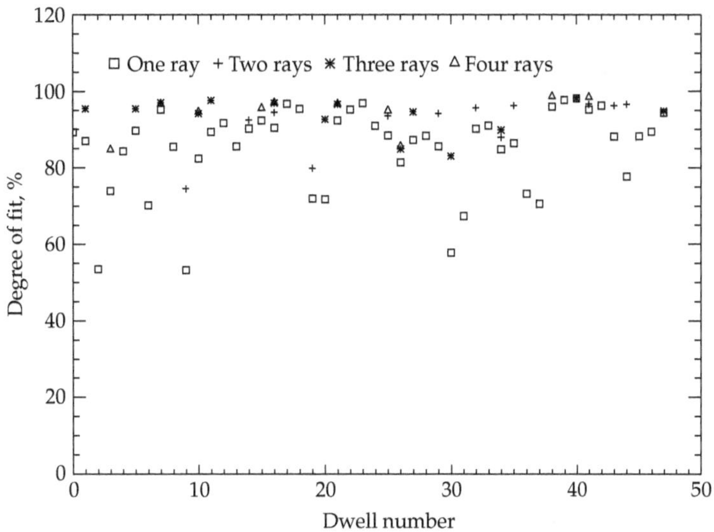

FIGURE 6.27 Fitting accuracy for the $1F_2$ mode. © Commonwealth of Australia 2011.

At least two rays are expected for the $1F_2$ mode as this mode is theoretically composed of both the ordinary and extraordinary rays in the low-angle path. Inspection of Figure 6.27 reveals that three-rays are resolved more often than two-rays for this mode. Furthermore, a three-ray model at times yields a significantly better fitting accuracy compared with the two-ray model. For example, the three-ray model depicted by the

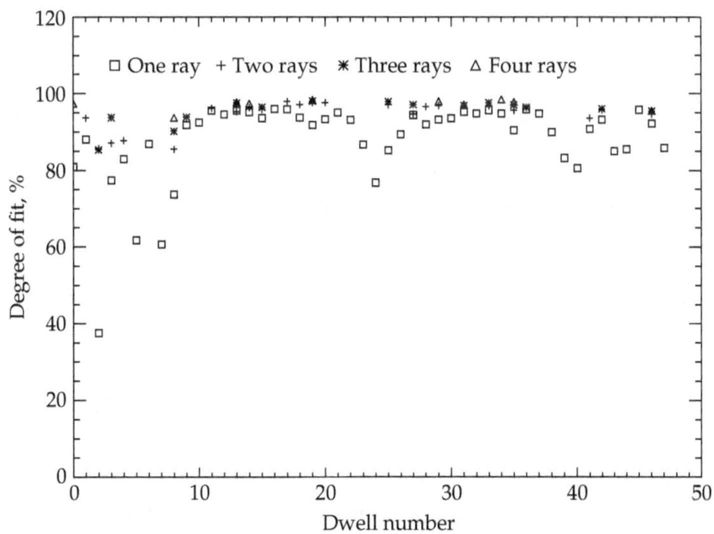

FIGURE 6.28 Fitting accuracy for the $1F_2(o)$ mode. © Commonwealth of Australia 2011.

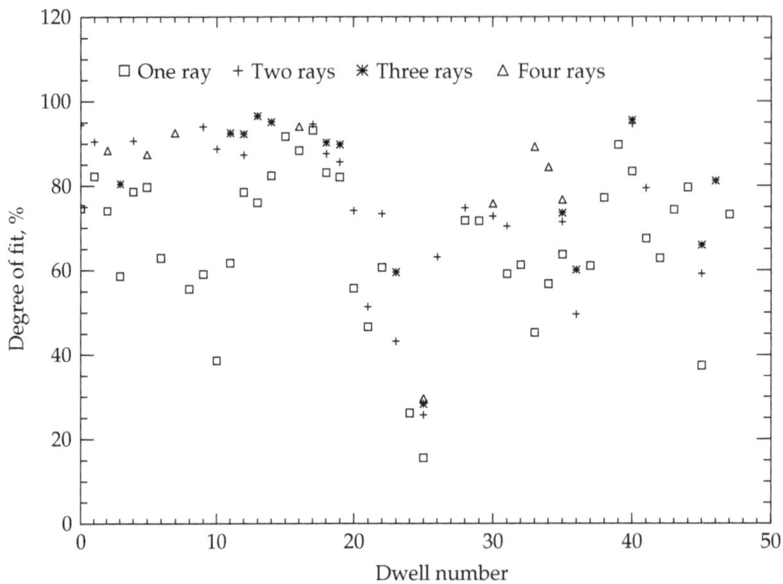

FIGURE 6.29 Fitting accuracy for the $1F_2(x)$ mode. © Commonwealth of Australia 2011.

space-time MUSIC spectrum in Figure 6.23 led to a fitting accuracy of 95 percent as compared to 87 percent for a two-ray model. When a third ray is resolved, it is thought to account for ionospheric irregularities in F-region.

In the absence of ionospheric irregularities, a one-ray model is expected to provide an accurate description of the wavefield produced by a single magneto-ionic component. As shown in Figure 6.28, this was found to be the case for the $1F_2(o)$ mode in most of the CPI analyzed. A high fitting accuracy using a single ray also indicates that the ULA has a steering vector that is accurately matched at the carrier frequency to the one expected in the absence of calibration errors. In the example of Figure 6.24, a single ray provided a fitting accuracy of 95 percent for the $1F_2(o)$ mode.

Figure 6.29 shows that a single ray is not able to represent the $1F_2(x)$ mode wavefields as well as those of the $1F_2(o)$ mode. The extraordinary magneto-ionic component is reflected from a different region in the F_2-layer, which on this occasion appears to be relatively more disturbed than the region that reflected the ordinary magneto-ionic component. For the $1F_2(x)$ mode, it was found that two or more rays were often needed to yield fitting a accuracy above 90 percent. Two rays were resolved more often than three or four rays for this particular mode and the fitting accuracy achieved using a two-ray model was usually considerably higher than that for one ray. For example, the two-ray model corresponding to the spectrum in Figure 6.25 resulted in a fitting accuracy of 94 percent, whereas a one-ray model only fitted the data to an accuracy of 82 percent.

To illustrate the time-variation of ray parameters from one dwell to another, consider a two-ray representation of the $1E_s$ mode. Figure 6.31 shows the variation in the estimated cone angles-of-arrival and Doppler shifts of the two rays for dwells in which the fitting accuracy was greater than 85 percent. The time-variation of the ray angles-of-arrival is

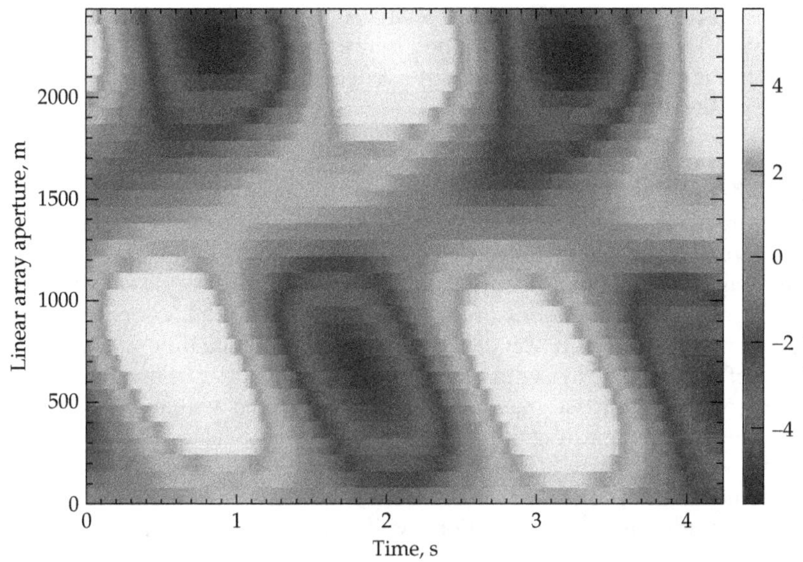

FIGURE 6.30 Real component of the space-time wavefield simulated for the $1E_s$ mode using the estimated four-ray model. © Commonwealth of Australia 2011.

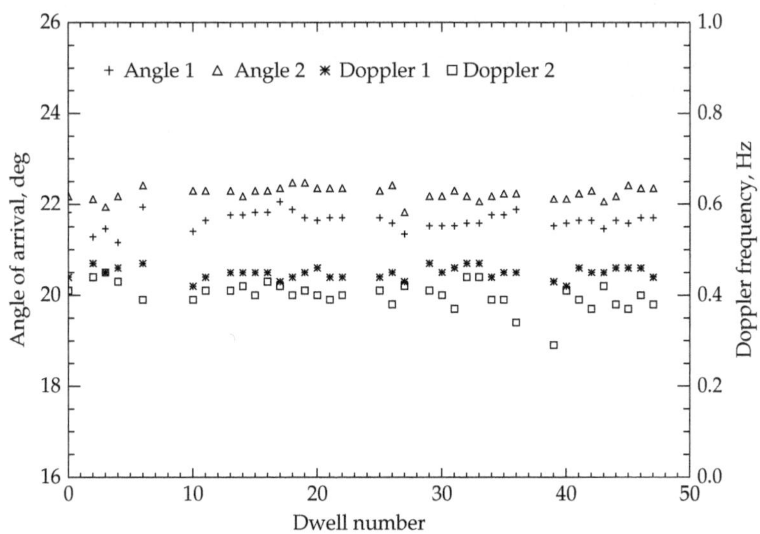

FIGURE 6.31 Dwell-to-dwell variation of the estimated angles of arrival and Doppler shifts of a two-ray model for the $1E_s$ mode. © Commonwealth of Australia 2011.

in the order of 0.5 degrees over the 47 CPI analyzed, while the Doppler shifts vary by approximately 0.1 Hz. The observed variation in ray parameters from dwell to dwell is expected due to the movement of the effective reflection points in the ionosphere over time intervals in the order of a few minutes.

6.4.3 Summary and Discussion

The fine structure of HF signal modes reflected from different layers in the ionosphere is considered by some investigators to result from a superposition of relatively few specularly reflected rays having similar Doppler shifts (to account for the observed temporal fading) and closely spaced directions-of-arrival (to account for the observed wavefront distortions). However, the resolution of a complex-valued space-time mode wavefield into a number of component rays with a very wide aperture antenna array has received insufficient attention in practice. Such analysis is considered valuable as it provides quantitative information regarding the capability of such a model to represent the mode wavefields received by a very wide aperture array over typical OTH radar CPI.

After a preliminary analysis of mode fine structure that demonstrated the existence and characteristics of mode wavefront distortions within typical OTH radar CPI, the MUSIC space-time parameter estimation technique was used to resolve the fine structure of individual ionospheric modes recorded by the Jindalee receiving array on a controlled mid-latitude propagation path. A performance measure was developed to assess the accuracy with which the model was able to fit the experimentally recorded wavefields, and the physical significance of the resolved rays was also tentatively interpreted on the basis of an oblique incidence ionogram recorded for the path.

The experimental results indicate that over time intervals in the order of a few seconds, the ionospheric modes received by the Jindalee antenna array could be represented to fitting accuracies above 90 percent using four rays or less over a large percentage of the analyzed data set. However, time-variations of the ray parameters were observed from one CPI to another as a result of motions in the ionosphere. These variations imply that as the CPI increases, the number of rays required to maintain a given modeling accuracy will also increase. Consequently, the wave interference model may be regarded as appropriate for representing a quiet mid-latitude HF channel in a *deterministic* manner over an interval of a few seconds, but is not suitable for representing the characteristics of a longer data set received over an interval of a few minutes.

CHAPTER 7

Statistical Signal Model

In applications such as OTH radar, it is useful to have an accurate model for the space-time complex (amplitude and phase) fading of HF signals reflected from different ionospheric regions or layers over time intervals that *exceed* a few seconds. For reasons mentioned in the previous chapter, the deterministic wave-interference characterization of mode fine structure becomes inconvenient and more difficult to interpret intuitively when the observation interval is lengthened from a few seconds to a few minutes. If the complex fading process can be treated as a stationary random process over a time interval of a few minutes in relatively stable ionospheric conditions, it is then possible to conveniently model the space-time characteristics of the received signals in a statistical sense.

This chapter is concerned with the development and experimental validation of space-time statistical models for resolved HF signal modes received over an interval of a few minutes by a very wide aperture antenna array after a single-hop oblique reflection from the ionosphere. In an OTH radar system, such time intervals are often commensurate with the typical lifetime of the optimum operating frequency selected for a particular OTH radar task. The objective is to develop models that are mathematically tractable for analysis and capable of accurately describing the experimentally observed second-order statistics of different propagation modes using relatively few parameters whose physical significance can be readily interpreted.

Section 7.1 provides background information and a brief review of the statistical models used to characterize HF signals returned to the ground by the ionosphere. A variety of empirical and physics-based models that describe HF signal reflection from an irregular and dynamic ionosphere are considered, but special attention is paid to statistical models that have been experimentally validated. Section 7.2 describes statistical models to represent the complex fading of an HF signal mode reflected from a randomly fluctuating ionosphere. The relationship between the deterministic wave-interference model and these (stationary) statistical models is also discussed.

It is emphasized that verification of the ionospheric physics responsible for the observed signal fading characteristics is beyond the scope of this text. This chapter is primarily devoted to comparing the statistical properties of the proposed models against those of the received signals. The hypothesis tests used to accept or reject the validity of the proposed statistical models from sample realizations of actual fading process are then described and applied to the experimental data. Section 7.3 validates a model for the temporal second-order statistics of resolved signal modes across all receivers in the array, while Section 7.4 validates models for the spatial and space-time correlation properties of resolved signal modes.

7.1 Stationary Processes

The first part of this section provides background on the use of statistical models to describe HF signals reflected by the ionosphere and also defines the scope of the stationary statistical models considered in this chapter. The second part of this section contains a review of specific statistical HF signal models reported in the literature. Of particular interest are models that may be used to describe the statistical properties of signals sampled in time by a single receiver or those sampled in space and time by antenna arrays. This review provides motivation for a more detailed understanding of the spatial and *space-time* statistical properties of HF signal modes reflected from localized regions or layers in the ionosphere. The third part of this section discusses a number of extensions that allow these properties to be incorporated into realistic multi-sensor HF channel models that can be used for both signal-processor design and analysis.

7.1.1 Background and Scope

An HF signal mode returned to the ground by a localized ionospheric region has an amplitude and phase that is known to fluctuate in space and time relative to the ideal signal structure expected in the hypothetical case of specular reflection. This observation leads to the conclusion that an individual region or layer in the ionosphere cannot be considered static or spherically symmetric but must have properties that are time varying and spatially heterogeneous. In simple terms, what is usually thought of as a single down-coming plane wave must in reality be a collection of waves scattered by the presence and movement of electron density irregularities or perturbations in the ionosphere.

In the previous chapter, a superposition of relatively few plane waves with different angles-of-arrival and Doppler shifts was used to model the space-time complex fading of HF signal modes reflected by different ionospheric layers over time intervals in the order of a few seconds. Although the wave-interference model was often able to accurately represent the experimentally observed fading processes with less than four rays, it was found that the parameters of this model changed significantly as a function time for observation intervals longer than a few seconds.

Consequently, over time intervals of a few minutes, an accurate representation of these fading processes can only be obtained by regularly updating the model parameters or by increasing the model order significantly. Both of these alternatives lead to a specification of the fading process that becomes cumbersome and difficult to interpret intuitively. For this reason, the wave interference model may be inappropriate in applications that require a complete data set received over several minutes to be characterized in a concise manner using relatively few physically meaningful parameters.

It is evident from the preceding experimental data analysis that signal fading produced on the ground by the ionospheric reflection process is better treated as a random process when the observation interval increases beyond a few seconds. Unlike the deterministic wave-interference model, the fading due to a random process is not entirely predictable. Such processes are usually specified in terms of their statistical properties, one of the most important being the auto-correlation function in the dimensions of space, time, or space-time. These second-order statistics completely define a Gaussian distributed random process, for example.

If the statistical properties of the fading are governed by a random process with a joint probability density function that is invariant over the observation interval, then the fading process is said to be *stationary*. If the joint density of the random process is the same when measured at different locations over a region of interest, then the process is said to be *spatially homogeneous*. A random process is referred to as *spatially stationary* when the joint density of the observations made at two locations separated by a fixed distance in a given direction is independent of the absolute location of the two points. When this density function depends only on the distance between two points and not the relative orientation of one from the other, the process is referred to as *isotropic*.

The fading of a single propagation mode is known to be dispersive in frequency and nonstationary in time and space, but if attention is restricted to narrowband signals with bandwidths less than approximately 20 kHz, time intervals in the order of a few minutes, and apertures not larger than a few kilometers, then it is more likely that signal samples collected by the antenna array over a relatively quiet mid-latitude ionospheric path can be adequately described by a stationary complex-valued space-time random process. The statistical characterization of narrowband signals over such temporal and spatial scales is of great relevance to the operation of OTH radar systems. In particular, models that can accurately represent the statistical properties of the data samples acquired by actual systems are useful for guiding the design and optimization of adaptive array signal processing algorithms in HF systems such as OTH radar.

Signal models can be broadly categorized as empirical models, which are formulated mainly on the basis of experimental data analysis, or physical models, which are derived from consideration of the natural phenomena assumed to generate the observed data characteristics. Experimentally validated empirical models have been developed to describe the *temporal-only* characteristics of resolved HF signal modes in single-receiver systems; see (Watterson, Juroshek, and Bensema 1970). Specifically, the amplitude and phase fading of individual propagation modes has been shown to be adequately described by stationary complex Gaussian random processes with temporal auto-correlation functions or power spectral densities of Gaussian form.

The advantage of such a description is that it concisely provides a user with a direct indication of the signal mode Doppler shifts and spreads, both of which have the potential to influence system performance. For HF systems using antenna arrays, a space-time statistical model of HF signal modes reflected from localized ionospheric regions represents the natural extension of the temporal-only model experimentally validated in Watterson et al. (1970). To redress the lack of experimental analysis needed for such an extension, the current chapter is devoted to the mathematical description and experimental validation of space-time statistical models for resolved HF signal modes propagated over a single-hop mid-latitude oblique ionospheric path.

7.1.2 Measurements on HF Signals

A large quantity of experimental measurements have been published on the statistical properties of HF signals reflected from the ionosphere. It is therefore necessary to establish a crude sorting scheme that can be used to classify such measurements. For example, one may ask if the measurements have been made at one antenna or at multiple spaced antennas, and if the amplitude and/or phase of the received signal has been recorded. Other important factors that influence the data characteristics are the temporal resolution and aperture length over which the measurements have been made, and whether the data arises due to a single magneto-ionic component (ordinary or extraordinary wave), a single mode composed of both o and x characteristic waves, or two or more superimposed modes. In general, the characteristics of the measurements will also depend on whether observations are made on single-hop or multi-hop ionospheric paths, whether layers in the E-, Es-, or F-regions were involved in signal reflection, and if the geographic location of the reflection points is in the mid-latitude, auroral, or equatorial region.

A large list of references to experimental and theoretical results on the subject of amplitude fading at a single antenna are contained in the CCIR (1970) report. In many studies, the amplitude of the received signal is treated as a random variable and assigned a probability density function. Experimental evidence suggests that amplitude fading over time intervals in the order of a few minutes can be approximated fairly closely by the Rayleigh distribution or the Nakagami-Rice distribution, see Barnes (1992) for example. The Rayleigh distribution is actually a special case of the Nakagami-Rice distribution when the received signal contains no steady or deterministic amplitude component. Both of these parametric distributions have been widely accepted as being representative of amplitude fading under a variety of conditions.

The probability density function of the signal amplitude is necessary but not sufficient to construct a satisfactory representation of the fading process. Specification of the correlation between amplitude samples received over different time separations is also required to describe the fading rate of the signal. In CCIR (1970), the auto-correlation function of the amplitude fluctuations is assumed to have a Gaussian form and the fading period is defined as the standard deviation of the Gaussian auto-correlation function that best fits the observed measurements. Under multimode conditions, values in the range of 0.5–2.5 seconds are typical, whereas for a single magneto-ionic component, fading periods in the order of minutes were found in Balser and Smith 1962. This clearly illustrates the need to distinguish between different kinds of measurements.

Experimental measurements of the *spatial* correlation coefficient between the signal amplitudes received simultaneously at spaced antennas were also made by Balser and Smith (1962) over a 1600-km mid-latitude path. A pulsed waveform was used to separate the different propagation modes, and the amplitudes of the resolved modes were sampled by a uniform linear array of 6-whip antennas spaced 610 m apart. For single modes, it was found that the diversity distance, defined as the spatial separation at which the correlation coefficient falls to 0.5, was in the order of 40 wavelengths for single-hop paths, and 10 wavelengths for multi-hop paths. As pointed out in Sweeney (1970), these values correspond to an angular spread of about 0.2 degrees and 1 degree, respectively. These measurements, and earlier ones referenced in Booker, Ratcliffe, and Shinn (1950), have led researchers to picture a single ionospheric mode as a single ray coherently reflected by a smooth ionosphere surrounded by a cone of incoherent rays produced by the roughness of the ionosphere. The former is often referred as the specular component by analogy with specular reflection from a mirror, while the latter are known as diffracted components (i.e., the diffusely scattered "micro-multipath" rays). The coherence ratio of a particular mode is defined as the ratio of the power in the specular component to that in the diffracted components and is a measure of the size of the amplitude and phase "crinkles" in the reflected signal wavefronts.

Several investigators have estimated the coherence ratio of ionospheric modes using amplitude measurements at a single antenna, or phase difference measurements between a pair of spaced antennas. When a specular component is present, the amplitude is assumed to be distributed according to the Nakagami-Rice density, which is parameterized by the coherence ratio. A more sensitive technique for estimating the coherence ratio is based on the theory of phase measurements at spaced receivers

(Whale and Gardiner 1966). Starting from the Nakagami-Rice distribution, the authors derived curves for the standard deviation of the phase difference measured at two spaced antennas in terms of the coherence ratio and the spatial auto-correlation function of the diffracted components.

A comprehensive summary of experimental measurements of coherence ratio can be found in Gething (1991). For example, Bramley (1951) quoted the square root of the coherence ratio $b = 2.5$ as being fairly typical, while Hughes and Morris (1963) calculated values ranging from $b = 0.4$ to $b = 1.9$. Continuous wave signals were used in the latter case and the data may have contained more than one propagation mode. The analysis of Warrington, Thomas, and Jones (1990) concluded that the coherence ratio was between 0 and 7 under "approximately single-moded propagation," but values greater than 40 were estimated when the same data were interpreted as specular component with a *variable* direction-of-arrival plus a cone of diffracted components. The estimates of coherence ratio vary markedly from one analysis to another. This prompted Gething (1991) to comment that "even allowing for the expected variability of the ionosphere from day to day, results obtained by various investigators do not seem entirely consistent."

In the investigation of Boys (1968), it was inferred that there is no specular component but only a group of rays coming from very similar directions with no common phase history. In other words, Boys (1968) regarded the true coherence ratio as zero, which implies that the reflected signal is considered as a completely scattered wave whose spatial correlation function depends on the angular power spectrum of the diffracted components scattered by the roughness of the ionosphere. This view was supported by experimental measurements of a Radio Australia broadcast (17.870 MHz) that was propagated by a one-hop reflection from the F2-layer over a 2700-km mid-latitude path. Phase measurements were made at 10 vertical whip antennas spaced approximately 50 m apart with one of the central antennas providing the phase reference. Over a 5-minute interval, Boys (1968) noted that the plane wave of best fit to the data did not wander in direction-of-arrival. The author concluded that over such time intervals the characteristics of the received wavefield are best expressed in terms of the angular power spectrum of the wavefront distortions. This power spectrum was assumed to have a Gaussian form and the standard deviation was found to be between 0.1 and 0.3 degrees.

A purely statistical representation of the reflected wave is consistent with the theory developed in Booker et al. (1950). The authors thought of the ionosphere as an irregular phase screen or diffraction grating which randomly varies the electric field distribution re-radiated beyond the horizontal surface of the screen. It was mathematically shown that the generalized spatial auto-correlation function of the electric field received on a parallel plane a finite distance away from the screen (say on the ground) is the same as that of the electric field distribution which gives rise to it just beyond the horizontal surface of the screen. In other words, the angular power spectrum of the signal received on the ground is given by the Fourier transform of the spatial auto-correlation function of the electric field irregularities immediately beyond the rough diffracting screen.

A phase-changing screen is the usual condition as far as the ionosphere is concerned (Whale and Gardiner 1966). As also explained in Boys (1968), a variation in the electron density causes a change in the real part of the refractive index and this causes a change in phase. From a physical viewpoint, a phase-changing screen seems realistic as most of the irregularities occur in the E- and F-regions, where the collision frequency and hence the attenuation is small (Boys 1968). Alternatively, in Bowhill (1961), the irregularities are viewed as forming a "buckled" specular reflector that acts as a phase screen for both normally and obliquely incident waves. At oblique incidence, the phase change imposed on the wave is reduced by a factor of $\cos\theta_i$, where θ_i is the angle relative to normal incidence. This is related to the phenomenon that an optical surface becomes more nearly a specular reflector at grazing incidence ($\theta_i \rightarrow 90$ degrees). Bowhill (1961) notes that the same effect, described in terms of the Rayleigh roughness criterion, is also observed for radio waves.

Many investigators have concentrated on ionospheric models based on phase-changing screens, but it should be noted that the theory of diffraction from an irregular ionosphere in Booker et al. (1950) is based on a *single* ionospheric pass or reflection. This theory is appropriate for a quiescent ionosphere that one would normally consider to be smooth, but not a highly disturbed region that contains field-aligned irregularities associated with spread-F, for example. Moreover, signals from very distant sources usually involve multiple ionospheric reflections and intermediate scattering from terrain or sea surfaces. Such signals do not satisfy the requirements of this theory and are therefore not expected to obey such models. It is often assumed that the two-dimensional spatial auto-correlation function of the phase screen has a two-dimensional Gaussian form, as in Bramley (1955) and Bowhill (1961) for example. However, these spatial models have not been validated against experimental data.

Using the theory in Booker et al. (1950), it is possible to define a *characteristic* size of irregularities as the distance between two points on the ground for which the generalized auto-correlation of the signal falls to a value of 0.5, as suggested in Briggs and Phillips (1950). In the study of Briggs and Phillips (1950), three receivers placed at the corners of an isoceles right angled triangle with side 130 m were used to observe the amplitude fading of "single echoes" reflected by the ionosphere at vertical incidence using a pulse waveform transmitter. By assuming that fading was predominantly due to a diffraction pattern of constant form that drifted over the ground with a constant velocity, the temporal auto-correlation function at the receivers was translated into a spatial auto-correlation function, from which irregularity scale sizes of 100–200 m were estimated for the F-region.

The first experimental confirmation of a stationary statistical model for the temporal amplitude and phase characteristics of resolved narrowband HF signal skywave modes propagated over a period of a few minutes over a mid-latitude ionospheric path was reported in Watterson et al. (1970). The authors used formal hypothesis tests to show that the baseband signal sampled at a single receiver was adequately described by a zero-mean complex Gaussian process that produces Rayleigh amplitude fading. This model is consistent with a completely scattered wave interpretation. The temporal modulation sequences imposed by the ionosphere on individual signal modes were also found to be statistically independent. The observed complex (amplitude and phase) fluctuations were characterized by their Doppler spectrum, which in general was shown to be the sum of two Gaussian functions of frequency, one for each magneto-ionic component.

7.1.3 Extension to Antenna Arrays

An experimentally confirmed *space-time* statistical model that can accurately describe the complex fading of narrowband HF signals reflected by a number of distinct but localized regions in the ionosphere would be of significant value for guiding the development and practical implementation of robust spatial and space-time adaptive processing algorithms in operational HF systems such as OTH radar. The experimentally validated model for single-receiver systems in Watterson et al. (1970) confirmed three main assumptions regarding the statistical properties of complex fading for resolved narrowband HF signal modes reflected by the ionosphere over a period of a few minutes. The three main validated assumptions may be summarized as follows:

- The amplitude and phase fluctuations of an individual propagation mode were assumed to be described by a stationary (circular-symmetric) complex-Gaussian random process with independent and identically distributed real and imaginary parts that produce Rayleigh fading.
- The complex Gaussian random processes that temporally modulate different propagation modes reflected from physically distinct but spatially localized regions (layers) of the ionosphere were shown to be mutually independent.
- The power spectral density of the random modulation process imparted on a single propagation mode were confirmed to be those of a Gaussian function of frequency parameterized by a Doppler shift and spread parameter.

For some time, it has been unclear how to extend the Watterson model to describe the statistical properties of narrowband HF signal modes received by *antenna arrays*. This is partly due to the lack of experimental measurements analyzed for this specific purpose. The aim of this chapter is to extend the work of Watterson et al. (1970) by analyzing the *spatial* and *space-time* statistical characteristics of a quiet mid-latitude HF channel. The Gaussian scattering assumption verified in Watterson et al. (1970) is used as a foundation to construct hypothesis tests for accepting or rejecting assumptions postulated for the spatial and space-time second-order statistics of resolved propagation modes with a known level of confidence. Specifically, the main objective is to determine the validity of the following hypothesized extensions on a *mode-separated* basis:

- *Spatial homogeneity*: The structure of the temporal auto-correlation function of a signal mode is assumed to be the same across all receivers in a very wide aperture array, with the possible exception of a scale change (power difference) from one receiver to another.
- *Spatial stationarity*: The spatial statistics of the amplitude and phase wavefront fluctuations are assumed to depend only on receiver separation and not their absolute location in the array.
- *Mean plane wavefront*: The mean wavefront of a resolved signal mode is assumed to have a plane wave structure parameterized by a fixed direction-of-arrival over an observation interval of a few minutes.
- *Spatial correlation coefficient*: The correlation coefficient between a pair of receivers in the array is assumed to be an exponentially decaying function of receiver separation.
- *Space-time separability*: The space-time second-order statistics of a signal mode are assumed to be accurately described as a Kronecker product of the temporal-only and spatial-only auto-correlation functions.

The question may be posed as to the connection between the wave interference model used to characterize short data sets in the order of a few seconds and a stationary statistical model used to characterize longer data sets collected over a few minutes. To provide a possible answer it is instructive to compare the physical interpretation of both models. A common interpretation for the different specular components in the wave-interference model is that they originate from spatially separate reflection points in a particular ionospheric layer, each of which is assumed to be effectively smooth over at least one Fresnel zone. The common physical interpretation of the statistical model is that signal reflection effectively occurs over a continuous localized region of an ionospheric layer that may be regarded as a rough reflecting surface. These two interpretations are based on different notional representations of the ionospheric reflection process.

At a particular time instant, the rough reflecting surface in the ionosphere may be replaced by an equivalent phase-changing screen as far as representing the effect on signal reflection is concerned. This phase screen may in turn be resolved into its spatial Fourier components. Each (two-dimensional) spatial frequency component present in the spectral decomposition of the surface gives rise to a specular signal component reflected in a direction determined by the two-dimensional spatial frequency of the field at the screen. If the corrugations or "buckles" of the irregular ionospheric surface do not vary rapidly with distance in any direction, then only a few spatial frequency components are required to accurately approximate the character of the reflecting surface and hence the equivalent phase screen. The dominant spatial frequency components that approximate this structure at a particular time instant will each radiate their own specular component. These superimpose with one another to produce a wave-interference fading pattern on the ground.

As the surface changes shape with time, its character will be represented by different spatial frequency components. However, the structure of such a surface evolves slowly on the time-scale of a few seconds as signals received from isolated modes over such intervals are almost coherent when propagated by a quiet mid-latitude ionosphere. It is therefore plausible that the spatial frequency components which dominate the structure of the surface at a particular time instant do not change significantly over a period of a few seconds, such that the time-evolution of the surface is described by a superposition of these components with changing relative phases. The relative phases are determined by the differential Doppler shifts associated with the dominant spatial frequency components. This is a possible interpretation of the space-time wave interference model in terms of reflection from a slowly evolving irregular surface.

Over longer periods of time, in the order of a few minutes, the surface will be described by a large number of different spatial frequency components that radiate energy over a distribution or "continuum" of directions. If such a signal can be treated as a statistically stationary signal over the observation interval, then the probability density function of the radiated components in direction-of-arrival and Doppler shift forms the space-time power density of the signal. This is the essence behind the statistical signal model description. Apart from this qualitative explanation of the connection between the wave-interference and statistical signal models, it is noted that the available data cannot be used to make definite conclusions regarding the different *physical interpretations* of either model. Consequently, the verification of the physical origin of the wave-interference model and its hypothesized connection to the statistical model is beyond the scope of this text.

7.2 Diffuse Scattering

Consider a ground-based HF transmitter that illuminates a volume of ionosphere in which the electron density distribution is not static or horizontally uniform (spherically symmetric) due to the presence of dynamic plasma concentration perturbations or irregularities. Assume that the incident signal is returned to ground by diffuse scattering from a spatially extended but localized region in a single layer of the ionosphere over a one-hop oblique path. The returned signal mode is then sampled in amplitude and phase by a ground-based array of antennas located over-the-horizon from the transmitter such that the transmitter is in the far-field of the receiving aperture.

By representing the effect of a random ionosphere on the received signal in terms of an equivalent irregular reflection surface or phase-changing screen, the first part of this section formulates a rudimentary mathematical expression for the signal received on the ground at a particular time instant. The second part of this section incorporates ionospheric variability to write a basic expression for the fading of the received signal in space and time. Under certain simplifying assumptions, this expression is subsequently used in the third part of this section to derive the generalized space-time auto-correlation function of the received signal in terms of the statistical properties of the conceptual irregular reflection surface, or equivalent phase-changing screen. To avoid complicating this description, which is only intended to convey elementary concepts in a straightforward manner, the effect of the Earth's magnetic field on radio wave propagation in the ionosphere is ignored.

It is also emphasized that the mathematical expressions used to describe the received signals in this section are not derived from physical models of the ionosphere or radio wave scattering in this medium. The basic principles introduced to derive these expressions may be regarded as analogous replacements of the true physics occurring in the ionosphere as far as the statistical characteristics of the received signals are concerned. This approach is frequently used for the purpose of deriving representative mathematical models of the received signals in a relatively simple manner. The detailed theory can be found in Budden (1985) for readers interested in delving further. In other words, this section does not attempt to describe the physics which is actually occurring to produce the observed complex fading phenomena, but rather to use some well-known idealized principles to derive rudimentary expressions for the received signals and their space-time second-order statistics.

While the ionosphere is in reality a thick medium through which radio waves are progressively refracted, we shall consider the diffuse scattering process in terms of a conceptual irregular reflection surface, or equivalent a horizontal phase-screen mirror located at the virtual reflection height of the signal mode (i.e., at the apex of the virtual propagation path) and centered at the path mid-point or control point in the ionosphere. As discussed in Chapter 2, the ionosphere will in general support more than one propagation mode between two ground points due to the presence of different layers. Each layer may in turn support low- and high-angle rays as well as *o* and *x* magneto-ionic components. In this case, each of the modes needs to be represented by a thin phase screen model as a minimum requirement. Ray tracing procedures, used in conjunction with a reference ionosphere can provide an indication of the propagation modes present over a particular path at a given time. The time-varying irregular reflection surface used to conceptually represent signal scattering by the ionosphere into the receiver is assumed to have a finite spatial extent. Experimental results suggest that the dimensions of its sides are typically in the order of a few kilometers for relatively stable ionospheric conditions Gething (1991). Regular large-scale vertical movement of the mean horizontal plane of the irregular surface with respect to time may be interpreted as imposing a mean Doppler shift on the signal, while changes in the spatial structure or "buckles" of the irregular surface about this horizontal plane causes the signal received on the ground to fade.

To relate the statistical properties of the received signal to those of the conceptual irregular reflecting surface, it is first necessary to develop an equation that expresses the field on the ground for a given realization of the surface. As mentioned previously, the effect of time-varying irregularities is assumed to alter the phases of the scattered ray paths such that the "rough" scattering surface may be replaced by an equivalent phase-changing screen that radiates the downcoming signal from the mean horizontal plane.

As described in Booker et al. (1950), diffraction patterns arising from irregular screens may be applied for describing the complex fading of radio waves transmitted through or returned to ground by the ionosphere. Although this method may provide a valid statistical representation of the received signal, the phase-screen approach should not be regarded as a physical model of the diffuse scattering process. Hence, the statistical characteristics of the phase-screen cannot be used to deduce detailed information about the actual ionospheric fluctuations.

7.2.1 Mathematical Representation

Consider a monochromatic transverse electromagnetic plane wave with wave vector $\vec{k}_i$ incident at an angle θ_i with respect to the normal of a flat reflecting surface in the xy-plane. This horizontal plane is assumed to be located at the virtual reflection height of the signal and centered at the control point of the skywave path. The electric field $\vec{E}_i(\vec{r}_s)$ incident on the reflecting surface at position $\vec{r}_s$ is then given by,

$$\vec{E}_i(\vec{r}_s) = \vec{E} \exp(-j\vec{k}_i \cdot \vec{r}_s) \tag{7.1}$$

where $\vec{E}$ represents the electric field strength phasor after the path loss from the transmitter has been taken into account. In the ideal case of perfect specular reflection from a flat surface, the *reflected* field distribution $\vec{E}_r(\vec{r}_s)$ *at the surface* is given by Eqn. (7.2), where $\vec{k}_r$ is the wave-vector of the reflected wave such that the angle-of-incidence equals the angle-of-reflection $\theta_i = \theta_r$.

$$\vec{E}_r(\vec{r}_s) = \vec{E} \exp(j\vec{k}_r \cdot \vec{r}_s) \tag{7.2}$$

Assume that the properties of the received signal reflected from an irregular ionosphere can be modeled in terms of a conceptual "buckled" reflection surface defined by the vertical displacement function zs (x, y) about the mean horizontal plane. This surface is conceptual in the sense that it is effectively used to represent the signal characteristics as opposed to the ionospheric reflection process. Nevertheless, it is instructive to relate the signal characteristics to the properties of this effective reflection surface.

An alternative approach is to translate the effective surface displacements $z_s(x, y)$ occurring at a given time instant into equivalent phase shifts that are imposed on the field reflected from the mean horizontal plane. In other words, the concept is to replace the irregular reflection surface by an equivalent flat "phase screen" fixed in the xy-plane that imposes a phase modulation on the reflected field. The phase modulation imposed at each point on the screen has a value that is determined by the prevailing displacement of the effective irregular reflection surface at that point.

For propagation in the direction of $\vec{k}_r$, a radiating point on the surface with a displacement $z_s(x, y)$ normal to the xy-plane can be replaced by an identical radiating point which is on the xy-plane at position $\vec{r}_{xy} = [x, y, 0]^T$ and phase shifted by $\exp\{-j\frac{2\pi}{\lambda} z_s(x, y) \cos\theta_i\}$. This transformation is only strictly valid for a receiving location that is far from the radiating surface and in the direction of specular reflection, since

$z_s(x, y)\cos\theta_i$ is the difference in path length between the actual and substituted radiating points to such a location.

After this transformation, the reflected signal can be thought of as arising from a radiating phase screen in the xy-plane with a field distribution given by,

$$\vec{E}_r(\vec{r}_{xy}) = \vec{E}e^{-j\frac{2\pi}{\lambda}f(\vec{r}_{xy},\theta_i)}\exp(j\vec{k}_r\cdot\vec{r}_{xy}) \tag{7.3}$$

where the position vector $\vec{r}_{xy}$ spans the phase screen which is of finite spatial extent in xy-plane and $f(\vec{r}_{xy},\theta_i) = z_s(x, y)\cos\theta_i$. As described by Booker et al. (1950), the angular spectrum $P(\vec{k})$ of the reflected signal is given by the spatial Fourier transform of the electric field distribution $\vec{E}_r(\vec{r}_{xy})$ at the phase screen. When evanescent waves are neglected, the angular spectrum of waves $P(\vec{k})$ that propagate with wave-vector $\vec{k}$ can be calculated according to the spatial Fourier transform in Eqn. (7.4).

$$P(\vec{k}) = \int_{\text{screen}} \vec{E}_r(\vec{r}_{xy})e^{-j\vec{k}\cdot\vec{r}_{xy}}d\vec{r}_{xy} = \vec{E}\int_{\text{screen}} e^{-j\frac{2\pi}{\lambda}f(\vec{r}_{xy},\theta_i)}e^{-j(\vec{k}-\vec{k}_r)\cdot\vec{r}_{xy}}d\vec{r}_{xy} \tag{7.4}$$

In the case of a perfectly flat reflecting surface, the function $f(\vec{r}_{xy},\theta_i) = 0$. As the dimensions of the screen are much larger than the wavelength λ, it can be deduced from Eqn. (7.4) that the angular spectrum of the reflected wave tends to the delta function $P(\vec{k}) \to \delta(\vec{k}_r)$ when $f(\vec{r}_{xy},\theta_i) = 0$. This corresponds to a specular reflection of the incident wave, as expected.

It is also evident from Eqn. (7.4) that the effect of a rough reflecting surface on the angular spectrum depends on the angle of incidence θ_i, as for a given surface displacement $z_s(x, y)$, the amount of phase modulation is determined by θ_i through the relationship $f(\vec{r}_{xy},\theta_i) = z_s(x, y)\cos\theta_i$. As the ray tends toward grazing incidence ($\theta_i \to 90$ degrees), the phase modulation imparted by the equivalent screen is reduced by a factor of $\cos\theta_i$. This makes the reflecting surface appear relatively less "rough" to the radio wave, such that it becomes more nearly specularly reflected when closer to grazing incidence.

Once the angular spectrum of the returned signal is known for a given surface displacement function, it is possible to represent the signal received by an array in the plane $z = z_g$ on the ground by a superposition of plane waves.

$$\vec{E}_r(\vec{r}_g) = \int P(\vec{k})e^{j\vec{k}\cdot\vec{r}_g}d\vec{k}, \quad \vec{r}_g = [x, y, z_g] \tag{7.5}$$

For over-the-horizon propagation, the only field measured by the array is the reflected field $\vec{E}_r(\vec{r}_g)$ as there is no line-of-sight between the transmitter and receiver. For a wavevector with components $\vec{k} = [k_x\ k_y\ k_z]^T$, it is possible to substitute $\vec{k}\cdot\vec{r}_g = \vec{k}\cdot\vec{r}_{xy} + k_z z_g$ into Eqn. (7.5) to yield,

$$\vec{E}(\vec{r}_g) = \int Q(\vec{k}, z_g)e^{j\vec{k}\cdot\vec{r}_{xy}}d\vec{k}, \quad \vec{r}_{xy} = [x, y, 0] \tag{7.6}$$

where $Q(\vec{k}, z_g) = P(\vec{k})e^{jk_z z_g}$. It is evident from Eqn. (7.6) that $Q(\vec{k}, z_g)$ represents the spatial Fourier transform of the field existing in the plane $z = z_g$, which contains the antenna array.

As $z_g \to \infty$, the electric field intensity observed on such a plane is referred to as the Fraunhoffer pattern in diffraction studies. The Fresnel pattern applies for finite z_g, and this pattern corresponds more nearly to that formed by an ionospherically reflected radio wave on the ground. The relationship between the spatial Fourier transform of the field at the phase screen and that on a parallel plane a finite distance away from this screen will be exploited later in this section to derive models for the generalized space-time auto-correlation function of the signal received by the array.

7.2.2 Varying Ionospheric Structure

This section extends on the preceding analysis to account for an irregular conceptual reflecting surface with a time-varying structure. This allows the space-time statistical properties of the signal received on the ground to be written in terms of the statistical properties of the time-varying surface or equivalent phase screen. It is shown that the generalized *space-time* auto-correlation of the Fresnel diffraction pattern is, under certain conditions, the same as that of the field distribution at the screen that gives rise to it. The analysis proceeds similarly to that undertaken in Booker et al. (1950) for the spatial-only auto-correlation function of the electric field produced on a plane a finite distance away from a two-dimensional random diffracting screen.

7.2.2.1 Second-Order Statistics

To take temporal changes of the irregular reflection surface into account, the quantity $P(\vec{k}, t)$ is introduced and defined as the angular spectrum of the field that results from the time-varying surface displacements $z_s(x, y, t)$ and phase screen function $f(\vec{r}_{xy}, \theta_i, t) = z_s(x, y, t)\cos\theta_i$. This angular spectrum is calculated according to Eqn. (7.4) by substituting $P(\vec{k}, t)$ for $P(\vec{k})$ and $f(\vec{r}_{xy}, \theta_i, t)$ for $f(\vec{r}_{xy}, \theta_i)$. The relationship $Q(\vec{k}, t) = P(\vec{k}, t)e^{jk_z z_g}$ follows from a similar argument to that described in the preceding section, with the implicit dependence of $Q(\vec{k}, t)$ on z_g being dropped for notational convenience.

Although the ionosphere changes in a continuous manner, it does so very slowly compared with the propagation time between the surface and the ground (typically a few microseconds), or within the OTH radar pulse repetition interval for that matter (typically less than 0.1 second). Hence, the reflected field generated at a particular instant t will propagate to the plane containing the array and remain more or less constant for a finite period of time. This allows the array to measure the field produced by an effectively "frozen" ionospheric structure before the field changes by an amount that is sufficiently large for the measuring instrument to detect.

The space-time Fourier transform $P(\vec{k}, f_v)$ that describes the spatial and temporal variations of the field disturbance *at the phase screen* is given by Eqn. (7.7), where f_v represents the different frequency components in the temporal fluctuations of the electric field disturbance at the screen.

$$P(\vec{k}, f_v) = \int_{-\infty}^{\infty} P(\vec{k}, t)e^{-j2\pi f_v t}dt \tag{7.7}$$

Similarly, the space-time Fourier transform of the field in the plane $z = z_g$ is given by $Q(\vec{k}, f_v) = P(\vec{k}, f_v)e^{jk_z z_g}$. The Wiener-Khintchine theorem states that the auto-correlation function is given by the inverse Fourier transform of its power spectrum. In this case, we have an angle-frequency power spectrum $|P(\vec{k}, f_v)|^2$ for the field at the screen, so the generalized auto-correlation function of the field at the screen $r(\vec{\Delta}d, \Delta t)$ is given by,

$$r(\vec{\Delta}d, \Delta t) = \frac{\iint |P(\vec{k}, f_v)|^2 e^{j\vec{k}\cdot\vec{\Delta}d} e^{j2\pi f_v \Delta t} d\vec{k} df_v}{\iint |P(\vec{k}, f_v)|^2 d\vec{k} df_v} \tag{7.8}$$

where Δt is the time interval and $\vec{\Delta d} = [\Delta x, \Delta y, 0]^T$ is the vector displacement over the screen. In the plane $z = z_g$, the auto-correlation function is given by the same theorem,

$$r_z(\vec{\Delta d}, \Delta t) = \frac{\iint |Q(\vec{k}, f_v)|^2 e^{j\vec{k}\cdot\vec{\Delta d}} e^{j2\pi f_v \Delta t} d\vec{k} d f_v}{\iint |Q(\vec{k}, f_v)|^2 d\vec{k} d f_v} \tag{7.9}$$

but $|P(\vec{k}, f_v)|^2 = |Q(\vec{k}, f_v)|^2$ which implies that $r_z(\vec{\Delta d}, \Delta t) = r(\vec{\Delta d}, \Delta t)$. In other words the generalized space-time auto-correlation of the field distribution measured in the plane $z = z_g$ (i.e., that of the Fresnel diffraction pattern) is the same as that of the field distribution at the screen that gives rise to it. This important result is a space-time generalization of the one-dimensional spatial-only result reported in Booker et al. (1950). Strictly, the equivalence is only for the generalized (normalized) auto-correlation function as the amplitude of the field strength diminishes as the reflected signal propagates from the screen to the plane containing the array.

To calculate the generalized space-time auto-correlation function of the field received by the antenna array, the time-varying field at the screen $\vec{E}_r(\vec{r}_{xy}, t)$ is introduced. This field is given by Eqn. (7.3) after replacing the "frozen" surface displacement function $f(\vec{r}_{xy}, \theta_i)$ by its time-varying counterpart $f(\vec{r}_{xy}, \theta_i, t)$. Once this substitution is made, the space-time auto-correlation function of the field distribution at the screen is given by,

$$r(\vec{\Delta d}, \Delta t) = \frac{\iint \vec{E}_r(\vec{r}_{xy}, t)\vec{E}_r^*(\vec{r}_{xy} + \vec{\Delta d}, t + \Delta t) d\vec{r}_{xy} dt}{\iint |\vec{E}_r(\vec{r}_{xy}, t)|^2 d\vec{r}_{xy} dt} \tag{7.10}$$

and from the equivalence stated above, $r(\vec{\Delta d}, \Delta t) = r_z(\vec{\Delta d}, \Delta t)$. By expanding Eqn. (7.10), the space-time auto-correlation function of the field measured by the array can be written as Eqn. (7.11), where the carrier-frequency dependent term $e^{-j\omega\Delta t}$ is neglected and the dependence of the function $f(\vec{r}_{xy}, t)$ on the angle of incidence θ_i has been dropped for notational convenience.

$$r_z(\vec{\Delta d}, \Delta t) = \exp(-j\vec{k}_r \cdot \vec{\Delta d}) \iint e^{j\frac{2\pi}{\lambda}[f(\vec{r}_{xy}+\vec{\Delta d}, t+\Delta t) - f(\vec{r}_{xy}, t)]} d\vec{r}_{xy} dt \tag{7.11}$$

The ionosphere is known to be a dispersive, heterogeneous, and nonstationary propagation medium for radio waves. However, if attention is restricted to narrowband signals (with bandwidths less than 20 kHz) and time intervals in the order of a few minutes, the statistical properties in a localized region of ionosphere are more likely to be dispersionless, homogeneous, and stationary to a good approximation. Homogeneity implies that the spatial variation of the surface displacement function has statistical properties that are independent of the spatial reference point on the surface, while stationarity implies that the ensemble statistics of these variations are independent of the time origin considered.

It is then possible to associate a probability density function (PDF) $p[\delta(\vec{\Delta d}, \Delta t)]$ for the *differential* surface displacement function $\delta(\vec{\Delta d}, \Delta t) = f(\vec{r}_{xy} + \vec{\Delta d}, t + \Delta t) - f(\vec{r}_{xy}, t)$, which is independent of the absolute position $\vec{r}_{xy}$ and time t, but is dependent on spatial separation $\vec{\Delta d}$ and time interval Δt. Once this description is accepted, the ensemble statistics governed by the PDF in Eqn. (7.12) are the same as those evaluated for a particular space-time realization of the surface in Eqn. (7.10).

$$r_z(\vec{\Delta d}, \Delta t) = \exp(-j\vec{k}_r \cdot \vec{\Delta d}) \int e^{j\frac{2\pi}{\lambda}\delta(\vec{\Delta d}, \Delta t)} p[\delta(\vec{\Delta d}, \Delta t)] d\delta(\vec{\Delta d}, \Delta t) \tag{7.12}$$

Eqn. (7.12) shows that for a statistically stationary and spatially homogeneous (conceptual) irregular reflecting surface, the generalized space-time auto-correlation function of the field measured by an array on the ground is the *characteristic function* of the probability density that describes the space-time statistical properties of the differential displacements in the reflecting surface. To derive models for the auto-correlation functions, it is then necessary to assume a model for the relative surface displacements in terms of the joint (space-time) probability density function $p[\delta(\vec{\Delta d}, \Delta t)]$.

7.2.3 Auto-Correlation Functions

There are many models that could be proposed for $p[\delta(\vec{\Delta d}, \Delta t)]$. For example, if there is a steady drift of irregularities moving at a velocity $\vec{v}_p$ across the reflection region, then the surface may be modeled as one of constant form that shifts in horizontal position over time. Such a model would result in an unchanging diffraction pattern that moves at a velocity $\vec{v}_p$ over the ground. In this case, the temporal auto-correlation function for a time interval Δt is the same as the spatial auto-correlation function for a separation $\vec{\Delta d} = \vec{v}_p \Delta t$.

Although such models have been considered in several studies, there will in general be a random component that changes the structure of the surface over time. These random fluctuations may be present in addition to the steady drift component. If there is no steady drift and only random fluctuations, then such fluctuations may be more correlated in one direction than in another, implying that the spatial auto-correlation function measured by the array depends on its orientation in the $z = z_g$ plane. This can have significant implications for interference rejection on two-dimensional and linear antenna arrays.

The simplest models occur when the spatial correlation of the surface fluctuations only depends on distance $|\vec{\Delta d}|$ (i.e., an isotropic surface) and the space-time PDF is separable, which implies that $p[\delta(\vec{\Delta d}, \Delta t)] = p_s[\delta(|\vec{\Delta d}|)]p_t[\delta(\Delta t)]$, where $p_s[\delta(|\vec{\Delta d}|)]$ is the spatial PDF and $p_t[\delta(\Delta t)]$ is the temporal PDF. In practice, this type of model may be appropriate in the absence of large-scale structures such as traveling ionospheric disturbances (TIDs), which are expected to impose a coupling between the spatial and temporal PDFs.

A separable model may not be appropriate for a number of reasons in practice, but the evaluation of the one-dimensional correlation functions $r_s(|\vec{\Delta d}|)$ and $r_t(\Delta t)$ from the corresponding PDFs serves as a useful starting point. In accordance with Eqn. (7.12), a separable PDF model implies that the space-time auto-correlation function is given by the product of the spatial and temporal auto-correlation functions $r_z(\vec{\Delta d}, \Delta t) = r_s(|\vec{\Delta d}|)r_t(\Delta t)$.

Let us consider two limiting cases for which these correlation functions can be related analytically to the PDF of the relative surface displacements. The first case corresponds to situations where the temporal interval Δt is *greater* than the time interval over which the velocities of the points on the surface normal to the xy-plane remain constant. In this case it is assumed that the surface displacement probability $p_t[\delta(\Delta t)]$ is that of a random walk possibly around a regular mean fluid motion v_z that is normal to the xy-plane.

$$p_t[\delta(\Delta t)] = \frac{1}{\sigma_t(\Delta t)\sqrt{2\pi}} \exp - \left\{ \frac{(\delta(\Delta t) - v_z \Delta t \cos\theta_i)^2}{2\sigma_t^2(\Delta t)} \right\} \tag{7.13}$$

The variance of the distribution $\sigma_t^2(\Delta t) = D_t|\Delta t|\cos\theta_i$ is assumed to vary linearly with temporal separation. The positive constant D_t is a measure of how quickly the surface changes. The dependence on $\cos\theta_i$ comes from the definition of relative displacement

$\delta(\Delta t) = f(\vec{r}_{xy}, \theta_i, t+\Delta t) - f(\vec{r}_{xy}, \theta_i, t)$ and the definition of $f(\vec{r}_{xy}, \theta_i, t) = z_s(x, y, t)\cos\theta_i$. The temporal auto-correlation function resulting for this probability density is given by Eqn. (7.14), where $\Delta f_d = v_z \cos\theta_i/\lambda$ is the regular component of Doppler shift and $|\vec{k}_s|^2 = \left(\frac{2\pi}{\lambda}\right)^2$.

$$r_t(\Delta t) = \int_{-\infty}^{\infty} e^{j\frac{2\pi}{\lambda}\delta(\Delta t)} p_t[\delta(\Delta t)]\, d\delta(\Delta t) = e^{j2\pi\Delta f_d \Delta t} e^{-|\vec{k}_s|^2 D_t \cos\theta_i |\Delta t|} \tag{7.14}$$

It can be seen from Eqn. (7.14) that the magnitude of $r_t(\Delta t)$ is a decaying exponential with a time constant that depends on D_t and $\cos\theta_i$. The phase of $r_t(\Delta t)$ is linear and depends on the regular component of Doppler shift Δf_d.

In the other limit, it is assumed the time interval Δt is shorter than the time interval over which the velocities of points on the surface normal to the xy-plane is effectively constant. The fluid velocity probability distribution $p(v)$ is assumed to be Gaussian with mean velocity v_z and root mean square velocity σ_v.

$$p(v) = \frac{1}{\sigma_v\sqrt{2\pi}} \exp - \left\{ \frac{(v - v_z)^2}{2\sigma_v^2} \right\} \tag{7.15}$$

In this case, the relative displacement probability distribution is related to the relative velocity probability distribution through the relation $p_t[\delta(\Delta t)] = p(v\Delta t \cos\theta_i)$. Substitution of this relation into the integral of Eqn. (7.16) yields the following temporal auto-correlation function.

$$r_t(\Delta t) = \int_{-\infty}^{\infty} e^{j\frac{2\pi}{\lambda}\delta(\Delta t)} p_t[\delta(\Delta t)] d\delta(\Delta t) = e^{j2\pi\Delta f_d \Delta t} e^{-|\vec{k}_s|^2 \sigma_v^2 \cos\theta_i{}^2 \Delta t^2} \tag{7.16}$$

It can be seen from Eqn. (7.16) that the magnitude of $r_t(\Delta t)$ is described by a Gaussian amplitude envelope with a variance that depends on σ_v^2 and $\cos\theta_i{}^2$, while the phase is linear and depends on the mean component of Doppler shift Δf_d, as before.

The spatial auto-correlation function $r_s(|\vec{\Delta d}|)$ can be determined in a similar manner by defining the spatial probability distribution $p_s(|\vec{\Delta d}|)$ as that of a random walk about the mean horizontal plane at time t.

$$p_s[\delta(|\vec{\Delta d}|)] = \frac{1}{\sigma_s(|\vec{\Delta d}|)\sqrt{2\pi}} \exp - \left\{ \frac{\delta(\vec{\Delta d})^2}{2\sigma_s^2(|\vec{\Delta d}|)} \right\} \tag{7.17}$$

The variance of the plasma displacements is assumed to be a linear function of distance $\sigma_s^2(|\vec{\Delta d}|) = D_s|\vec{\Delta d}|$, where the constant D_s is a measure of the roughness of the surface. In this case the spatial auto-correlation function is given by

$$r_s(|\vec{\Delta d}|) = e^{-j\vec{k}_r\cdot\vec{\Delta d}} \int_{-\infty}^{\infty} e^{j\frac{2\pi}{\lambda}\delta(|\Delta d|)} p_s[\delta(|\vec{\Delta d}|)] d\delta(|\vec{\Delta d}|) = e^{-j\vec{k}_r\cdot\vec{\Delta d}} e^{-|\vec{k}_s|^2 D_s |\vec{\Delta d}|} \tag{7.18}$$

It has an amplitude which is an exponentially decaying function of distance $|\vec{\Delta d}|$, while its phase is linear for a uniform linear array with adjacent sensors displaced by $\vec{\Delta d}$. If the variance $\sigma_s^2(|\vec{\Delta d}|) = D_s'|\vec{\Delta d}|^2$ varies quadratically with separation, then the amplitude of the spatial auto-correlation function will take on a Gaussian form.

$$r_s(|\vec{\Delta d}|) = e^{-j\vec{k}_r\cdot\vec{\Delta d}} \int_{-\infty}^{\infty} e^{j\frac{2\pi}{\lambda}\delta(|\Delta d|)} p_s[\delta(|\vec{\Delta d}|)] d\delta(|\vec{\Delta d}|) = e^{-j\vec{k}_r\cdot\vec{\Delta d}} e^{-|\vec{k}_s|^2 D_s' |\vec{\Delta d}|^2} \tag{7.19}$$

Recall that the power spectrum corresponding to a decaying exponential has a Lorentzian profile, whereas an auto-correlation function with a Gaussian amplitude envelope gives rise to a power spectrum with a Gaussian profile (Kreyszig 1988). Correlation functions with amplitude envelopes that fall off rapidly with time or distance give rise to broad power spectrums and vice versa. The linear phase shift in the auto-correlation function simply shifts the centre of the power spectrum (whose shape is determined solely by the amplitude envelope of the auto-correlation function in this case) to a position that corresponds to the mean Doppler shift or direction-of-arrival of the signal.

7.3 Temporal Statistics

To validate a proposed model for the second-order statistics of signal fading, it is necessary to analyze the significance of any deviations between the hypothesized auto-correlation function and that estimated from the experimental data. If the probability distribution of these differences is known under the assumption that the model is correct, it is possible to derive appropriate confidence bounds and hence determine whether the observed differences can be reasonably attributed to statistical errors arising due to finite sample effects, or if there is good reason to reject the hypothesized auto-correlation function model.

In Watterson et al. (1970), it was experimentally validated that the temporal fading of a skywave signal mode can be represented by a random process with a Gaussian auto-correlation function. In practice, the parameters of the Gaussian auto-correlation function are not known a priori for the received signal mode and must therefore be estimated from the available data. The estimated parameters may be used to define a hypothesized model for the statistically expected auto-correlation function. The validity of this model may then be assessed with respect to the sample auto-correlation function estimated from the data. The first part of this section is concerned with estimating the parameters of the hypothesized temporal auto-correlation function model that best fits the sample auto-correlation sequence (ACS) estimated from the data.

The second part of this section theoretically derives the probability density functions for the magnitude and phase of the sample ACS under the assumption that the model is correct. After confirming these theoretical results by simulation, hypothesis tests that can be used to accept or reject the validity of the proposed temporal ACS model are formulated and applied. Besides checking the validity of the Gaussian temporal ASC model proposed by Watterson on the analyzed data set, these hypothesis tests also determine whether there is good reason to believe that the parameters of this model remain invariant across the different receivers of a very wide aperture antenna array. This statistical characteristic of a signal mode, referred to as the spatial homogeneity of the temporal ACS, represents an important aspect of the model for antenna arrays that could not be validated in the previously described (single-receiver) studies.

7.3.1 Parameter Estimation Method

The auto-correlation function of a time-sampled random process is often evaluated at a number of discrete points separated by an integer multiple of the sampling period, which equals the PRI in this case. The term auto-correlation sequence (ACS) is used to describe the samples of the auto-correlation function (ACF) evaluated at a number of lag spacings. The temporal ACS of the complex random process $x_k^{[n]}(t)$ sampled by receiver n at PRI $t = 1, 2, \ldots, P$ in range cell k may be estimated as the *unbiased* sample ACS $\hat{r}_n(\tau)$ in Eqn. (7.20). The k-dependence of $\hat{r}_n(\tau)$ has been dropped as the modes resolved in different range cells are considered separately.

$$\hat{r}_n(\tau) = \frac{1}{P-\tau} \sum_{t=1}^{P-\tau} x_k^{[n]}(t) x_k^{[n]*}(t+\tau) \quad \begin{cases} \tau = 0, 1, \cdots, Q-1 \leq P \\ n = 0, 1, \cdots, N-1 \end{cases} \tag{7.20}$$

A parametric model of the *statistically expected* ACS $r_n(\tau)$ can be written in the form of Eqn. (7.21), where the scalar parameters a, b, and c_n are the coefficients of a quadratic polynomial $P_n(\tau)$ in lag index τ. As described in a moment, constraints may be imposed on the polynomial coefficients so as to represent Gaussian or exponentially decaying autocorrelation functions as two special cases. Clearly, higher-order polynomial functions may be proposed, but these will not be considered here.

$$r_n(\tau) = e^{P_n(\tau)} = e^{a\tau^2 + b\tau + c_n} \tag{7.21}$$

The power of the signal in receiver n is given by the value of the ACS at zero lag $r_n(0) = e^{c_n}$, where c_n is constrained to being a *real* scalar that may vary with receiver number n. If the Doppler power spectrum model is assumed to be a Gaussian function of frequency centered about a mean Doppler shift, then the ACS $r_n(\tau)$ has an amplitude envelope that is a Gaussian function of τ, while the phase of $r_n(\tau)$ is linearly related τ with a slope determined by the mean Doppler shift. In Eqn. (7.21), this rather popular model corresponds to a real value of a and an imaginary value of b.

On the other hand, if the Doppler power spectrum model is assumed to be a Lorentzian function of frequency centered about a mean Doppler shift, the ACS $r_n(\tau)$ is required to have an amplitude envelope that is a decaying exponential function of τ, while the phase of $r_n(\tau)$ is again linear with a slope determined by the mean Doppler shift. In Eqn. (7.21), this corresponds to $a = 0$ and a complex value of b. The constraints imposed on the coefficients of the polynomial $P_n(\tau)$ for a Gaussian or Lorentzian power spectrum with mean Doppler shift f (normalized by the PRF) appear in Eqn. (7.22). The operators $\Re\{\cdot\}$ and $\Im\{\cdot\}$ return the real and imaginary parts of the argument, respectively. We also require $\Im\{c_n\} = 0$ in all cases, as the ACS at zero lag equals the signal power.

$$r_n(\tau) = e^{a\tau^2} e^{b\tau} e^{c_n} \quad \begin{cases} a < 0, \ \Re\{b\} = 0, \ \Im\{b\} = 2\pi f & \text{Gaussian} \\ a = 0, \ \Re\{b\} < 0, \ \Im\{b\} = 2\pi f & \text{Lorentzian} \end{cases} \tag{7.22}$$

The two above-mentioned models have previously been used to represent the *temporal* ACS of HF signals obliquely *backscattered* by ionospheric irregularities; see Villain et al. (1996) and Hanuise et al. (1993). The authors concluded that it was preferable to interpret the statistical properties of the data in terms of the ACS rather than the power spectrum because the different features of the two aforementioned models can be more easily distinguished visually in the lag domain. Model interpretation and analysis based on the ACS has a further advantage in that it is more mathematically tractable from both a parameter estimation and statistical validation perspective.

The *sample* ACS $\hat{r}_n(\tau)$ may be expressed in the form of Eqn. (7.23), where the complex scalar function $\hat{P}_n(\tau) = \ln\{\hat{r}_n(\tau)\}$ will in general not exactly match the quadratic polynomial model $P_n(\tau)$ assumed in Eqn. (7.21). The order of the polynomial model for the

ACS may be extended beyond two, and the parameter constraints in Eqn. (7.22) may be relaxed to allow for a wider variety of ACS models that differ from the Gaussian or decaying exponential forms. Despite the greater flexibility this would provide, alternative models of higher complexity will not be considered here. Models of higher order that could fit the sample ACS more accurately could easily be proposed, but the amount of statistical uncertainty and variability often associated with estimating the ACS from experimental data will in general not warrant the search for extremely accurate models. In addition, more sophisticated models typically become more difficult to understand intuitively and often lead to more complex implementations when computer simulations are required to generate sample data realizations.

$$\hat{r}_n(\tau) = e^{\hat{P}_n(\tau)} \tag{7.23}$$

Before proceeding to describe the parameter estimation method for the two candidate ACS models, the dependence of the power term e^{c_n} on receiver number n is explained. The power has been allowed to vary from one receiver to another for two reasons. First, the potential existence of instrumental errors that effect the amplitude of measurements, such as differences between the antenna sensor, gains, can be absorbed by the term e^{c_n} such that the remaining parameters of the model can be adjusted to suit the shape or structure of the sample ACS. Second, it is of primary interest to determine whether the ACS or power spectrum *structure* complies with a given parametric model, and whether the parameters of this model may be assumed to be identical for all receivers. The power level or amplitude scaling, that may be different from one receiver to another is often of secondary importance as far as *temporal* processing is concerned. However, differences in scale may be important in *spatial* and *space-time* processing applications, so this aspect of the model will be revisited in Section 7.4.

As the sample ACS is unbiased, it seems reasonable to estimate the parameters of the temporal ACS model according to the least squares criterion in Eqn. (7.24), where $\mathbf{v} = [a\ b\ c_1 \cdots c_N]^T$ is the model parameter vector, $\|\cdot\|_F$ denotes the $L2$ or Frobenius norm, and the minimization is performed subject to the constraints on the elements of $\mathbf{v}$ in Eqn. (7.22), which depends on the particular model being considered. The model parameters a and b are effectively free to estimate the best fit to the sample ACS *structure* measured over *all* receivers, while the possibly different *scales* of the ACS measured in different receivers are absorbed by the parameters c_n for $n = 0, \ldots, N-1$.

$$\hat{\mathbf{v}} = \arg\min_{\mathbf{v}} \sum_{n=0}^{N-1} \sum_{\tau=0}^{Q-1} \|\hat{r}_n(\tau) - r_n(\tau)\|_F \tag{7.24}$$

Substituting the assumed model $r_n(\tau) = e^{P_n(\tau)}$ and sample ACS $\hat{r}_n(\tau) = e^{\hat{P}_n(\tau)}$ into Eqn. (7.24) yields Eqn. (7.25). This is a nonlinear optimization problem which may be solved using iterative methods. However, if the model and sample ACS are quite similar at the *minimum*, the value of $P_n(\tau) - \hat{P}_n(\tau)$ will be close to zero and $e^{P_n(\tau)-\hat{P}_n(\tau)}$ will have a value near unity.

$$\hat{\mathbf{v}} = \arg\min_{\mathbf{v}} \sum_{n=0}^{N-1} \sum_{\tau=0}^{Q-1} \|\hat{r}_n(\tau)[1 - e^{P_n(\tau)-\hat{P}_n(\tau)}]\|_F \tag{7.25}$$

In this case, the term $e^{P_n(\tau)-\hat{P}_n(\tau)}$ can be accurately approximated by a first-order Taylor series expansion $1 + P_n(\tau) - \hat{P}_n(\tau)$. Substituting the first-order truncated Taylor series expansion into Eqn. (7.25) yields the following modified cost function that corresponds to an approximate least squares criterion.

$$\tilde{\mathbf{v}} = \arg\min_{\mathbf{v}} \sum_{n=0}^{N-1} \sum_{\tau=0}^{Q-1} \|\hat{r}_n(\tau)\|_F \, \|P_n(\tau) - \hat{P}_n(\tau)\|_F \tag{7.26}$$

By substituting $P_n(\tau) = a\tau^2 + b\tau + c_n$ and $\hat{P}_n(\tau) = x_n(\tau) + jy_n(\tau)$ into the objective function of Eqn. (7.26), where $x_n(\tau) = \Re\{\hat{P}_n(\tau)\}$ and $y_n(\tau) = \Im\{\hat{P}_n(\tau)\}$ are the real and imaginary parts of $\hat{P}_n(\tau)$, respectively, the parameter estimation problem can be rewritten in the form of Eqn. (7.27).

$$\tilde{\mathbf{v}} = \arg\min_{\mathbf{v}} \sum_{n=0}^{N-1} \sum_{\tau=0}^{Q-1} \|\hat{r}_n(\tau)\|_F \, \|\{a\tau^2 + b\tau + c_n\} - \{x_n(\tau) + jy_n(\tau)\}\|_F \tag{7.27}$$

This unconstrained optimization problem can be expressed in the more compact form of Eqn. (7.28), where $\mathbf{p} = \left[\mathbf{q}_1^T, \ldots, \mathbf{q}_N^T\right]^T$ is an NQ-dimensional stacked vector of the Q measurements made in each receiver $\mathbf{q}_n = [\hat{P}_n(0), \ldots, \hat{P}_n(Q-1)]^T$, while the $NQ \times NQ$ diagonal matrix $\mathbf{W} = \text{diag}\left[\mathbf{w}_1^T, \ldots, \mathbf{w}_N^T\right]$ is composed of the weighting elements, which are given by the modulus squared of the sample ACS measurements in each receiver $\mathbf{w}_n = [|\hat{r}_n(0)|^2, \ldots, |\hat{r}_n(Q-1)|^2]^T$ for $n = 0, \ldots, N-1$.

$$\tilde{\mathbf{v}} = \arg\min_{\mathbf{v}} \; (\mathbf{p} - \mathbf{M}\mathbf{v})^\dagger \mathbf{W} (\mathbf{p} - \mathbf{M}\mathbf{v}) \tag{7.28}$$

The matrix $\mathbf{M}$ in Eqn. (7.28) is constructed according to Eqn. (7.29). Each NQ-dimensional column of $\mathbf{M}$ is composed of N stacked Q-dimensional column vectors, denoted by $\mathbf{a}, \mathbf{b}, \mathbf{1}, \mathbf{0}$, whose elements are given in Eqn. (7.29) for $\tau = 0, 1, \ldots, Q-1$. In shorthand notation, $\mathbf{M} = [\mathbf{1} \otimes \mathbf{a}, \mathbf{1} \otimes \mathbf{b}, \mathbf{I}_N \otimes \mathbf{1}]$, where the symbol $\otimes$ denotes Kronecker product and $\mathbf{I}_N$ is the N-variate identity matrix.

$$\mathbf{M} = \begin{bmatrix} \mathbf{a} & \mathbf{b} & \mathbf{1} & \mathbf{0} & \cdots & \mathbf{0} \\ \mathbf{a} & \mathbf{b} & \mathbf{0} & \mathbf{1} & \cdots & \mathbf{0} \\ \vdots & \vdots & \vdots & \vdots & \vdots & \vdots \\ \mathbf{a} & \mathbf{b} & \mathbf{0} & \mathbf{0} & \cdots & \mathbf{1} \end{bmatrix}, \quad \begin{cases} \mathbf{a}^{[\tau]} = \tau^2 \\ \mathbf{b}^{[\tau]} = \tau \\ \mathbf{1}^{[\tau]} = 1 \\ \mathbf{0}^{[\tau]} = 0 \end{cases} \tag{7.29}$$

The solution of the *unconstrained* weighted quadratic least-squares optimization problem in Eqn. (7.28) is given by the minimizing argument $\tilde{\mathbf{v}}$ expressed in Eqn. (7.30). This unconstrained solution refers to the situation where no restrictions are imposed on the real and imaginary values of the complex parameters in the vector $\mathbf{v}$.

$$\tilde{\mathbf{v}} = (\mathbf{M}^\dagger \mathbf{W} \mathbf{M})^{-1} \mathbf{M}^\dagger \mathbf{W} \mathbf{p} \tag{7.30}$$

In the present case, it is of interest to constrain the values of certain parameters in $\mathbf{v}$ to yield a model of the ACS that is consistent with the hypothesized Gaussian or decaying exponential representation. When one or more parameters in $\mathbf{v}$ are required to be either purely real or imaginary, as they are for both models of interest, it is better to treat the real and imaginary parts of the least squares optimization problem separately. To do this, we may define $\mathbf{p}_r = \Re\{\mathbf{p}\}$ and $\mathbf{p}_i = \Im\{\mathbf{p}\}$ as the real and imaginary parts of the measurement vector $\mathbf{p}$. This allows parameter estimation for either model to be partitioned into two separate (real-valued) *unconstrained* least squares problems in Eqn. (7.31).

$$\begin{cases} \tilde{\mathbf{v}}_r = \arg\min_{\mathbf{v}_r} (\mathbf{p}_r - \mathbf{M}_r\mathbf{v}_r)^\dagger \mathbf{W}(\mathbf{p}_r - \mathbf{M}_r\mathbf{v}_r) \\ \tilde{\mathbf{v}}_i = \arg\min_{\mathbf{v}_i} (\mathbf{p}_i - \mathbf{M}_i\mathbf{v}_i)^\dagger \mathbf{W}(\mathbf{p}_i - \mathbf{M}_i\mathbf{v}_i) \end{cases} \tag{7.31}$$

These problems can be solved by tailoring the general solution in Eqn. (7.30) to the specific case arising for the real and imaginary parts of the considered model. For the Gaussian case, $\mathbf{M}_r = [\mathbf{1} \otimes \mathbf{a}, \mathbf{I}_N \otimes \mathbf{1}]$ and $\mathbf{v}_r = [a, c_1, \ldots, c_N]^T$, while $\mathbf{M}_i = [\mathbf{1} \otimes \mathbf{b}]$ and $\mathbf{v}_i = [b_i]^T$. On the other hand, for the decaying exponential case, $\mathbf{M}_r = [\mathbf{1} \otimes \mathbf{b}_r, \mathbf{I}_N \otimes \mathbf{1}]$ and $\mathbf{v}_r = [b_r, c_1, \ldots, c_N]^T$, while $\mathbf{M}_i = [\mathbf{1} \otimes \mathbf{b}_i]$ and $\mathbf{v}_i = [b_i]^T$, where b_r and b_i are the real and imaginary parts of the polynomial coefficient b. In expanded form, these solutions correspond to the least-squares problems in Eqn. (7.32), which collectively estimate the parameter vectors of the Gaussian model $\tilde{\mathbf{v}}_G = [a, b_i, c_1, \ldots, c_N]^T$ and the Lorentzian model $\tilde{\mathbf{v}}_L = [0, b, c_1, \ldots, c_N]^T$ from the sample ACS.

$$\begin{aligned} \tilde{\mathbf{v}}_G &= \arg\min_{\mathbf{v}} \textstyle\sum_{n=0}^{N-1} \sum_{\tau=0}^{Q-1} \|\hat{r}_n(\tau)\|_F [(x_n(\tau) - a\tau^2 - c_n)^2 + (y_n(\tau) - b_i\tau)^2] \\ \tilde{\mathbf{v}}_L &= \arg\min_{\mathbf{v}} \textstyle\sum_{n=0}^{N-1} \sum_{\tau=0}^{Q-1} \|\hat{r}_n(\tau)\|_F [(x_n(\tau) - b_r\tau - c_n)^2 + (y_n(\tau) - b_i\tau)^2] \end{aligned} \tag{7.32}$$

Note that *linearly constrained* complex-valued least squares problems can be dealt with analytically using the theory described in Kay (1987), but it is not possible to constrain the real and imaginary parts of a complex variable separately using this approach. Consequently, these standard results cannot be used to estimate the model parameters required in this particular application.

7.3.2 Hypothesis Acceptance Test

Once the parameters are estimated, it is possible to generate a hypothesized model for the expected ACS and to assess the significance of any departures between the model ACS and the sample ACS. If these differences are relatively small (i.e., small enough to be reasonably attributed to estimation errors), then the temporal second-order statistics of within-mode fading may be declared *spatially homogeneous*, as no strong reason exists to reject the hypothesized model of the temporal ACS, which is described by identical parameters in all receivers of the array except for possibly a scale change.

It is important to statistically quantify what is meant by "significant" departure and to state the degree of confidence associated with any conclusion drawn from a comparison between the second-order statistics of the measured data and the hypothesized ACS model. The marginal asymptotic distributions of the magnitude and phase of the sample ACS are derived in Appendix A for a stationary complex Gaussian random process that produces Rayleigh fading. The distribution of the sample ACS depend on the number of samples P used for averaging in Eqn. (7.20), as well as the statistically expected ACS of the random process. Using the large-sample distributions derived in Appendix A, it is possible to derive magnitude and phase error-bounds with respect to the hypothesized ACS model that are expected to contain (say) 90 percent of the estimates.

If either the magnitude or phase of a particular lag in the sample ACS lies outside the established error bounds, then such a departure is deemed to be significant and provides a strong reason to reject the hypothesized ACS model. If the entire ACS lies within these bounds, then there is no strong reason to reject the proposed ACS model, which may be accepted as valid with significance level of (say) 90 percent. This criteria is used to accept or reject the validity of the assumed Gaussian or decaying exponential spatially homogeneous temporal ACS model. For each propagation mode analyzed, the sample ACS distributions and corresponding error bounds are calculated on a case-by-case basis as they depend on both the assumed model type and associated parameters.

Before presenting the experimental results, it is possible to confirm the validity of the sample ACS distributions derived in Appendix A by simulating a large number of realizations of a complex-Gaussian random process with known second-order statistics. The sample ACS is computed for each realization, so that for each lag component of the ACS, it is possible to evaluate the statistical distribution of the estimate. For a large number of realizations, the numerically computed distributions for the magnitude and phase (or real and imaginary parts) may be compared directly with the theoretically derived probability density functions. For this purpose, a stationary first-order autoregressive (AR) Gaussian process $z(t)$ was simulated according to the recursive relation in Eqn. (7.33).

$$z(t) = \alpha z(t-1) + \sqrt{1-|\alpha|^2}\, n(t), \quad t = 0, 1, \ldots, P-1 \tag{7.33}$$

The complex scalar α is the AR(1) random process parameter, chosen such that $|\alpha| < 1$ to ensure stability (i.e., pole inside the unit circle), while the innovations $n(t)$ are described by a complex (circularly symmetric) Gaussian driving white-noise process with the following second-order statistics.

$$E\{n(t)n^*(t+\tau)\} = \delta(\tau), \quad E\{n(t)n(t+\tau)\} = 0 \tag{7.34}$$

The ACS of the AR(1) random process $z(t)$ coincides with the exponentially decaying model described in the previous section. From Eqn. (7.33) and (7.34), it can be shown that the statistically expected auto-correlation function of the process $z(t)$ is given by Eqn. (7.35).

$$r_z(\tau) = E\{z(t)z^*(t+\tau)\} = \alpha^\tau \tag{7.35}$$

In the simulation study, a total of 10,000 realizations were generated using values of $P = 10,000$ and $\alpha = 0.99$. For each realization, the unbiased sample ACS $\hat{r}_z(\tau)$ was evaluated at a number of lag indices $\tau = 0, 1, \ldots, Q-1$ with $Q = 30$. A value of $\tau = 20$ was arbitrarily chosen as an example lag component of the sample ACS to compare the cumulative density functions derived from theory with those computed via simulations.

At a lag interval of $\tau = 20$, the expected value of the ACS is $r_z(20) = \alpha^{20} = 0.818$, and as the sample ACS is unbiased, this is equal to the mean value of $\hat{r}_z(20)$ averaged over an infinite number of realizations. Due to finite sample effects, the interrogated lag component of the sample ACS will be complex-valued in general, so the cumulative densities may be presented either in terms of the real and imaginary parts or in terms of magnitude and phase.

Figures 7.1 and 7.2 show the theoretical and simulated cumulative densities of the real and imaginary parts of the sample ACS at lag component $\tau = 20$. The theoretical curves correspond to Gaussian probability density functions. The mean values used for the theoretical curves in Figures 7.1 and 7.2 are the statistically expected values of the real and imaginary parts of the ACS at this lag component. The variances associated with the theoretical curves are determined by the expressions derived in Appendix A. It is evident that the theoretical curves accurately match those obtained by simulation. Figures 7.3 and 7.4 show analogous results for the magnitude and phase cumulative distributions.

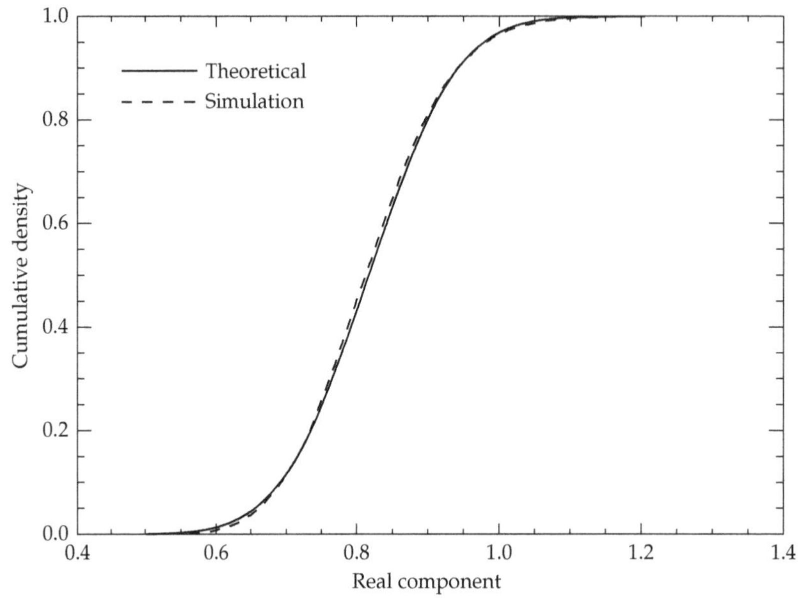

FIGURE 7.1 Cumulative densities for the real part of the sample ACS $\hat{r}_z(\tau)$ for an AR(1) process with $\alpha = 0.99$, $\tau = 20$, and $P = 10,000$. © Commonwealth of Australia 2011.

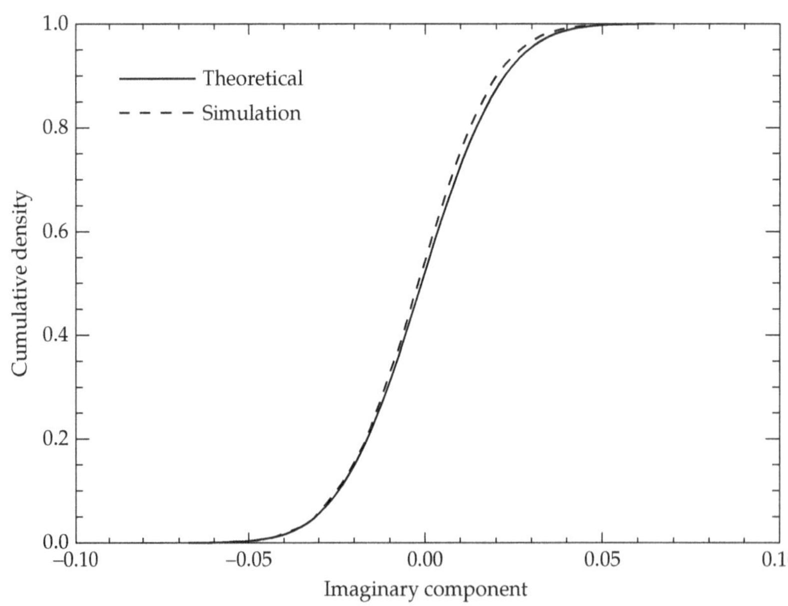

FIGURE 7.2 Cumulative densities for the imaginary part of the sample ACS $\hat{r}_z(\tau)$ for an AR(1) process with $\alpha = 0.99$, $\tau = 20$, and $P = 10,000$. © Commonwealth of Australia 2011.

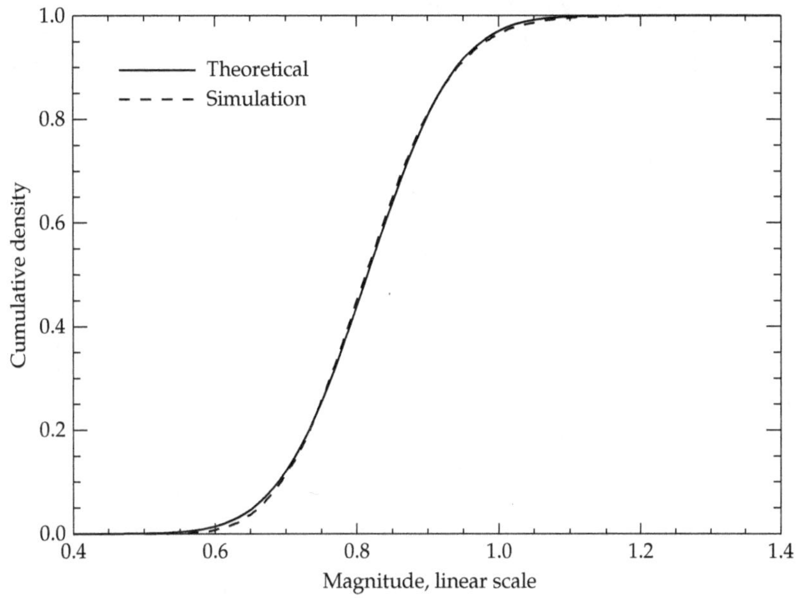

FIGURE 7.3 Cumulative densities for the magnitude of the sample ACS $\hat{r}_z(\tau)$ for an AR(1) process with $\alpha = 0.99$, $\tau = 20$, and $P = 10,000$. © Commonwealth of Australia 2011.

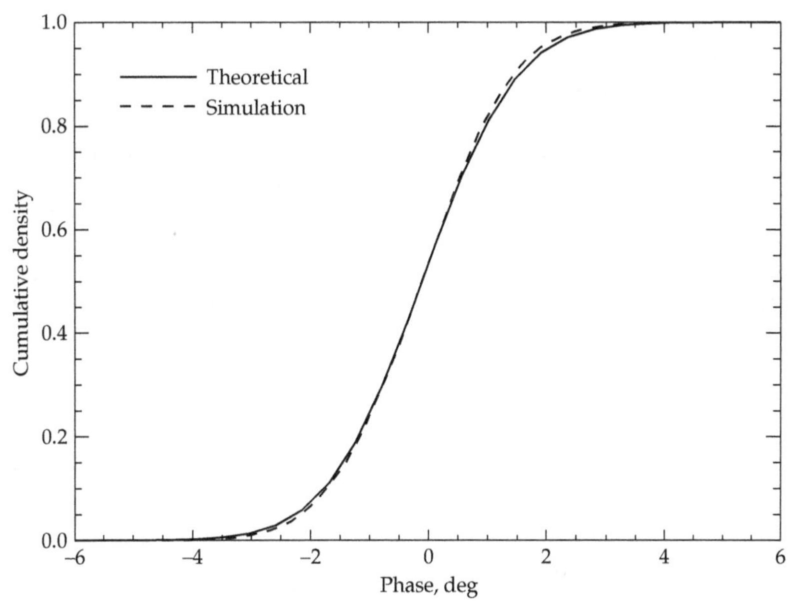

FIGURE 7.4 Cumulative densities for the phase of the sample ACS $\hat{r}_z(\tau)$ for an AR(1) process with $\alpha = 0.99$, $\tau = 20$, and $P = 10,000$. © Commonwealth of Australia 2011.

7.3.3 Spatial Homogeneity Assumption

The Gaussian model for the temporal ACS of a skywave signal mode was experimentally confirmed by Watterson et al. (1970), but it was not possible for the authors of that work to determine whether the parameters of such a model could be assumed constant over all receivers in a very wide aperture array. In this experiment, the temporal ACS of each resolved propagation mode was computed separately in $N = 30$ receivers using 47 CPI of data collected over a period of about 4 minutes. The total number of samples available for estimating the temporal ACS in each receiver was $47 \times 256 = 12{,}032$. A set of $Q = 30$ uniformly spaced sample lags were calculated by averaging a sum of lagged products. The ACS starts at zero lag with consecutive lags equally spaced at multiples of 6 PRI (0.1 seconds) apart.

Using the ACS estimates computed separately in each of $N = 30$ receivers, the Gaussian model parameters were estimated for each resolved propagation mode using the method described in Section 7.3. The estimated temporal ACS parameters are shown in Table 7.1. Under the spatial homogeneity assumption, the Gaussian temporal ACS model estimated for a particular mode is assumed to have the same form in all $N = 30$ receivers up to a scaling factor c_n. The ACS scale estimates c_n resulting for the different receivers $n = 1, \ldots, N$ are not shown in Table 7.1, but will be analyzed in the next section dealing with spatial statistics.

Table 7.1 also lists the two parameters describing the Doppler power spectrum that corresponds to the estimated temporal ACS model. Specifically, the parameters f and B_t denote the mean Doppler shift (frequency displacement) and the Doppler bandwidth (frequency spread) of the Gaussian power spectrum, respectively. The mean Doppler shift is calculated as $f = b/j2\pi\Delta t$ for a lag interval $\Delta t = 0.1$ seconds, since the model parameter b represents the change in phase per unit a lag interval. The mean Doppler shifts in Table 7.1 are practically the same as those estimated for the same modes using conventional processing in the previous chapter. The Doppler bandwidth is calculated as the inverse of the lag time for which the magnitude of the temporal auto-correlation function model falls by a factor of $1/e$. For the Gaussian model, $B_t = \{\Delta t\sqrt{1/a}\}^{-1}$. The Doppler bandwidths in Table 7.1 are of the same order as those measured in Shepherd and Lomax (1967) and Watterson et al. (1970) on mid-latitude ionospheric paths.

Based on the hypothesized model for the statistically expected temporal ACS structure in each receiver, it is possible to derive confidence bounds for the experimentally measured sample ACS at each lag point using the theoretical results in Appendix A. At each lag point, these upper and lower bounds specify the interval expected to contain a certain percentage of sample ACS values when the random process is statistically described

Mode	a	b	B_t, Hz	f, Hz
$1E_s$	-2.09×10^{-4}	$j \times 0.268$	0.145	0.43
$1F_2$	-2.24×10^{-4}	$j \times 0.277$	0.149	0.44
$1F_2(o)$	-9.06×10^{-5}	$j \times 0.294$	0.095	0.47
$1F_2(x)$	-1.97×10^{-4}	$j \times 0.332$	0.140	0.53

TABLE 7.1 Gaussian temporal ACS and Doppler spectrum model parameters estimated for the different propagation modes of the considered HF link.

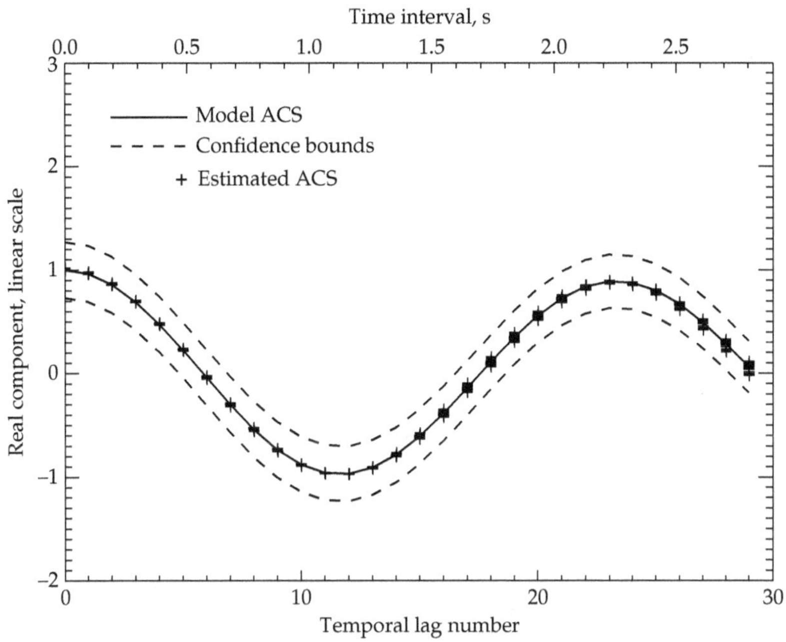

FIGURE 7.5 Temporal auto-correlation function for real component of $1E_s$ mode.

by the hypothesized model. If one or more lag components in the sample ACS are not contained within (say) the 90 percent confidence bounds, then there is strong reason to believe the hypothesized model is not suitable for describing the data. Conversely, when the entire sample ACS lies within the confidence bounds, there is no strong reason to reject the hypothesized model, which may then be accepted.

The solid line in Figure 7.5 shows the *real* component of the Gaussian ACS model for the $1E_s$ mode based on the parameters estimated for this mode in Table 7.1. The dashed lines show the upper and lower deciles of the sample ACS distributions expected at each lag value under the assumption that this model is valid when $47 \times 256 = 12,032$ data points are used to estimate the sample ACS. The experimental sample ACS measured in each receiver is first normalized by the estimated power scaling factor e^{c_n} for $n = 1, 2, \ldots, N$. The resulting $N = 30$ sample ACS values computed using the experimental data at each temporal lag are then over-plotted using a + symbol in Figure 7.5. Figure 7.6, in the same format as Figure 7.5, shows the results for the *imaginary* component of the $1E_s$ mode temporal ACS.

The oscillatory behavior of the real and imaginary parts with respect to lag number is due to the mean Doppler shift. The higher the mean Doppler shift, the higher the frequency of these oscillations. The rate of decay of the amplitude envelope in both the real and imaginary parts is due to Doppler spread. The higher the Doppler spread, the faster the magnitude of the ACS falls with respect to lag number. Figures 7.7–7.12 show the results for the other modes in the same format. In all cases, the Gaussian temporal ACS model estimated for each mode in Table 7.1 agrees well with the experimentally observed sample ACS values measured in *all* $N = 30$ receivers.

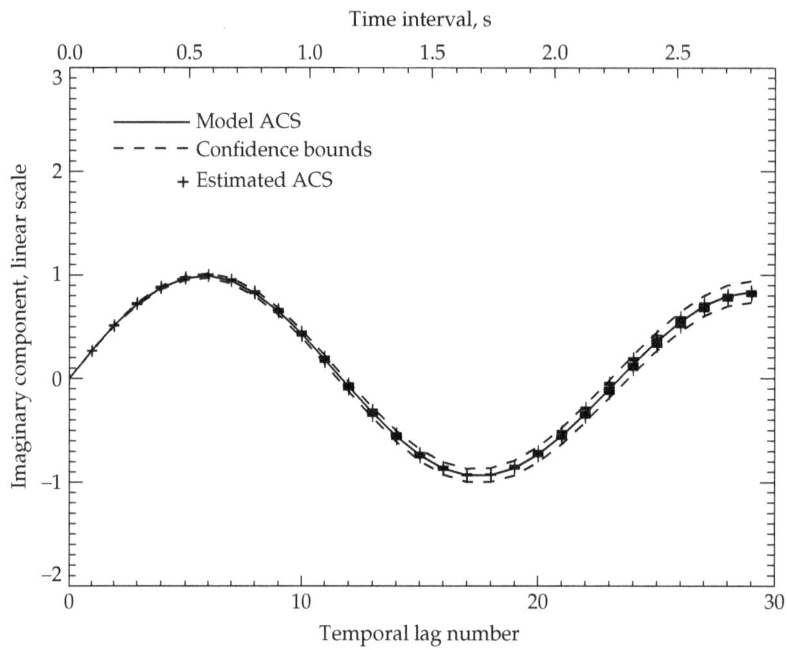

FIGURE 7.6 Temporal auto-correlation function for imaginary component of $1E_s$ mode. © Commonwealth of Australia 2011.

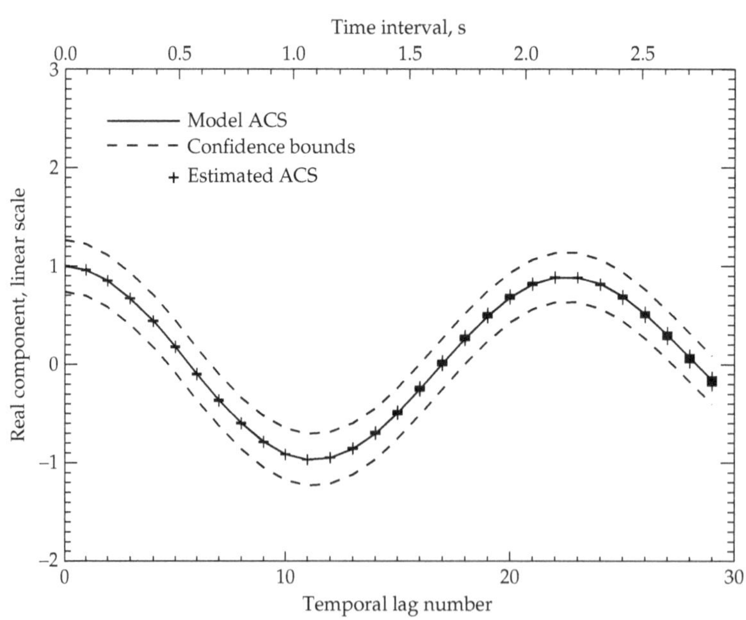

FIGURE 7.7 Temporal auto-correlation function for real component of $1F_2$ mode. © Commonwealth of Australia 2011.

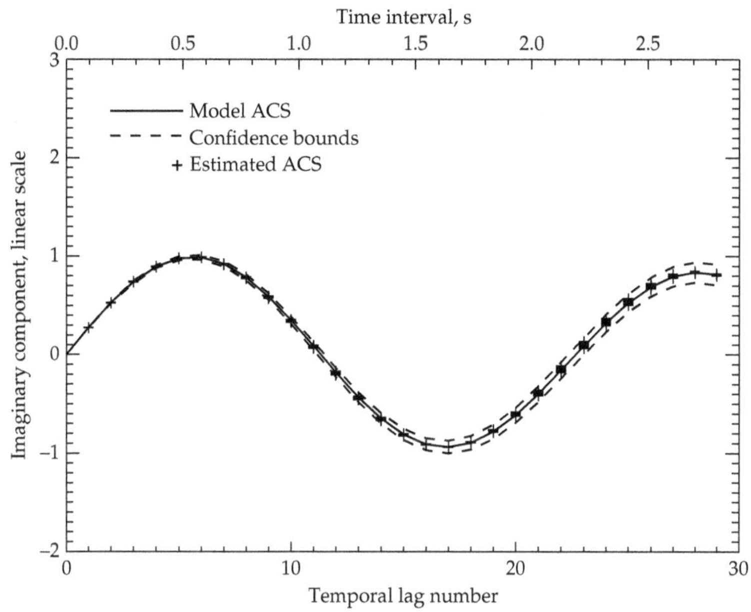

FIGURE 7.8 Temporal auto-correlation function for imaginary component of $1F_2$ mode. © Commonwealth of Australia 2011.

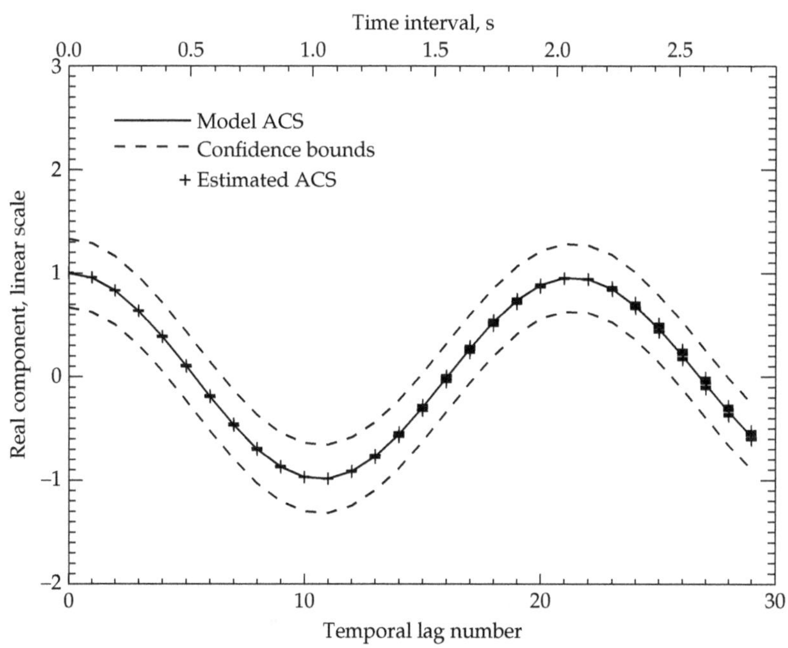

FIGURE 7.9 Temporal auto-correlation function for real component of $1F_2(o)$ mode. © Commonwealth of Australia 2011.

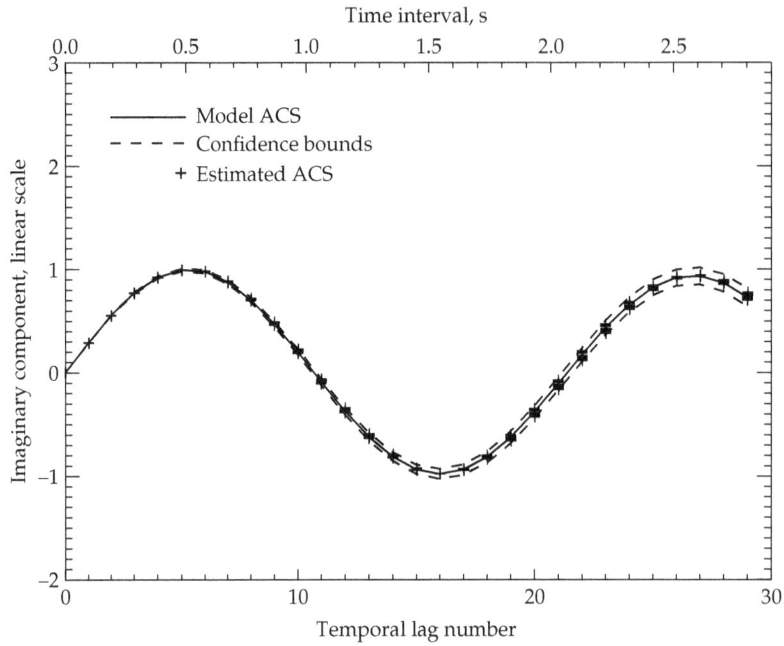

FIGURE 7.10 Temporal auto-correlation function for imaginary component of $1F_2(o)$ mode. © Commonwealth of Australia 2011.

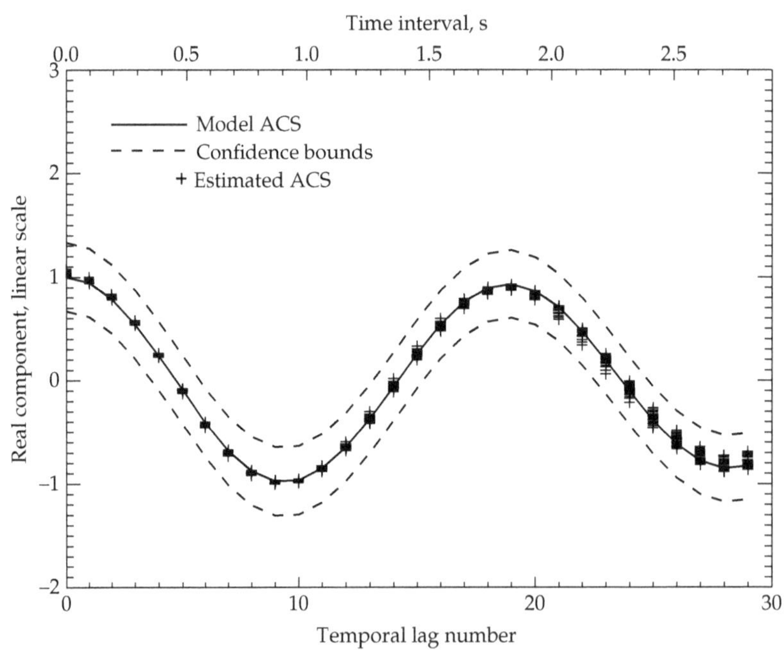

FIGURE 7.11 Temporal auto-correlation function for real component of $1F_2(x)$ mode. © Commonwealth of Australia 2011.

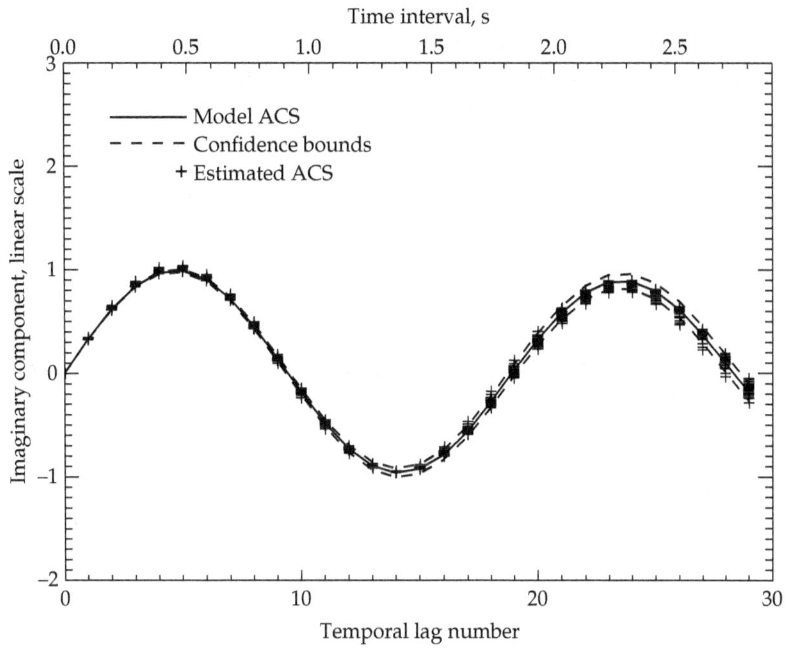

FIGURE 7.12 Temporal auto-correlation function for imaginary component of $1F_2(x)$ mode. © Commonwealth of Australia 2011.

Based on these results, there is no strong reason to reject a Gaussian-shaped temporal ACS model for HF signal modes reflected by a relatively quiet mid-latitude ionosphere. This reconfirms the findings of Watterson et al. (1970) based on single-receiver data. In addition, the variability of the normalized temporal sample ACS across different receivers is almost always contained within the upper/lower decile confidence bounds. This observation, which could not be made on a single receiver by Watterson et al. (1970), leads to the conclusion that there is no strong reason to reject the assumption of spatial homogeneity for signal modes propagated by the analyzed mid-latitude HF channel. In other words, for a very wide aperture antenna array that spans a length of nearly 3 km, it has been demonstrated that the temporal ACS of these modes received at spatially separated points along the array can be described by a Gaussian model with the *same* mean Doppler shift and Doppler spread parameters. Apart from a possibly different power level or scaling of the ACS in different receivers, these results suggest that the temporal mode fading statistics can be assumed spatially homogeneous over very wide apertures for a relatively quiet mid-latitude HF channel.

7.4 Spatial and Space-Time Statistics

Mode wavefronts received due to signal reflection from a localized ionospheric region appear to have a mean structure that is planar with time-varying amplitude and phase modulations superimposed. The time-varying wavefront "crinkles" may be interpreted as a random spatial fading process that causes the angular spectrum of the received signal mode to spread around the mean direction-of-arrival (much like temporal fading causes the Doppler spectrum to spread about the mean Doppler shift). Assuming the spatial fading process is stationary over an observation interval of a few minutes, it is of interest to determine the validity of the spatial covariance matrix model in Eqn. (7.36). Such a model is consistent with the interpretation of the spatial fading process described above, and has been previously adopted to describe spatially spread signals in Paulraj and Kailath (1988) and Ringelstein, Gershman, and Bohme (1999).

$$\mathbf{R}_x = \mathrm{E}\{\mathbf{x}_k(t)\mathbf{x}_k^H(t)\} = \sigma_s^2[\mathbf{s}(\theta)\mathbf{s}^H(\theta)] \odot \mathbf{B} \tag{7.36}$$

In Eqn. (7.36), the symbol $\odot$ denotes Shur-Hadamard or element-wise product, σ_s^2 is the mean square value or power of the signal mode (assumed to be the same in all receivers), $\mathbf{s}(\theta)$ is the steering vector of a ULA for cone angle θ, and $\mathbf{B}$ is an "angular spreading" matrix with its (i, j)th element equal to the magnitude of the spatial correlation coefficient between receivers i and j. This coefficient is defined as ρ_{ij} in Eqn. (7.37), where $r_x^{[i,j]}$ for $i, j = 0, \ldots, N-1$ denotes the (i, j)th element of the statistically expected spatial covariance matrix $\mathbf{R}_x$.

$$\mathbf{B}^{[i,j]} = \rho_{ij} = \frac{|r_x^{[i,j]}|}{\sqrt{r_x^{[i,i]} r_x^{[j,j]}}} \tag{7.37}$$

In this model, the *mean* plane wavefront is assumed to have a fixed direction-of-arrival that uniquely determines the *phase* of all spatial covariance matrix elements $r_x^{[i,j]}$, just as the phase of the temporal ACS was solely determined by the mean Doppler shift. This is tantamount to defining the mean wavefront of a signal mode as the one that makes the real and imaginary components of the random modulation process *statistically independent*. In this case, the spatial modulations contribute only to the *magnitude* of the second-order statistics in $\mathbf{R}_x$. For an arbitrarily defined mean wavefront represented by the N-dimensional vector $\mathbf{s}$, the elements of the spatial covariance matrix based on such a model are given by Eqn. (7.38), where $\mathbf{s}^{[n]}$ denotes the nth element of $\mathbf{s}$.

$$r_x^{[i,j]} = \sigma_s^2 \rho_{ij} \mathbf{s}^{[i]} \mathbf{s}^{[j]*} \tag{7.38}$$

When the mean wavefront is assumed to be a plane wave, the spatial covariance matrix $\mathbf{R}_x$ can be expressed in terms of the phase-only matrix $\mathbf{s}(\theta)\mathbf{s}^H(\theta)$ that determines the phase of all elements $r_x^{[i,j]}$ through the mean DOA parameter θ, and the real-valued matrix $\mathbf{B}$ that determines the magnitude of all elements $r_x^{[i,j]}$ through a parametric correlation coefficient model to be described shortly. Defining the phase $\measuredangle \mathbf{s}^{[n]}(\theta) = \phi_n$, the proposed model may be expressed in terms of the magnitude and phase of the spatial covariance matrix elements in Eqn. (7.39).

$$\left|r_x^{[i,j]}\right| = \sigma_s^2 \rho_{ij}, \quad \measuredangle r_x^{[i,j]} = \phi_i - \phi_j \tag{7.39}$$

A model for the spatial correlation coefficients ρ_{ij} in a very wide aperture antenna array has not been experimentally verified for resolved HF signal modes by formal hypothesis testing. Similarly, the plane wave model often assumed for the *mean wavefront* of an HF signal mode reflected from a localized ionospheric region has not been experimentally confirmed. The phase structure of the wavefront is important in applications where the signal is regarded as a *desired* signal to be received by the system. The correlation properties of the wavefront are of interest when the signal represents *interference*, as the correlation coefficients strongly influence the achievable cancelation ratio in adaptive spatial processing systems.

The synthesis of mean wavefront and correlation coefficient models yields a spatial covariance matrix model of the array data. For a stationary Gaussian distributed process, such a model completely defines the spatial statistics of a signal mode. This section develops a hypothesis testing framework that can be used to accept or reject formulated models for the mean wavefront and spatial correlation coefficients with a known level of confidence. The accuracy with which the experimentally measured two-dimensional *space-time* ACS of a propagation mode can be deduced from the product of its spatial-only and temporal-only ACS models is also quantified in this section (i.e., the assumption of space-time statistical separability).

7.4.1 Correlation Coefficients

The multiplicative amplitude and phase modulations are assumed to be described by a zero-mean complex Gaussian process that produces Rayleigh fading in the spatial dimension. This statistical description follows from the Gaussian scattering assumption validated by Watterson et al. (1970). Before proceeding to the validation stage, it is first necessary to describe a model for the correlation coefficients ρ_{ij} in the matrix $\mathbf{B}$. As for the temporal ACS, the Gaussian and decaying exponential amplitude envelope models are considered for the correlation coefficients.

$$\rho_{ij} = e^{P(|i-j|)} \quad \begin{cases} P(d) = e^{ad^2} & \text{Gaussian} \\ P(d) = e^{bd} & \text{Decaying exponential} \end{cases} \tag{7.40}$$

In accordance with the assumption of a stationary process, such models assume that the second-order statistics of the spatial fading process depend only on the spacing between receivers $d = |i - j|$ and not on their location (i, j) in the array. In practice, the real-valued model parameters $a < 0$ or $b < 0$ in Eqn. (7.40) that pertain to a particular propagation mode are not known a priori. These parameters may be estimated from the *unbiased* spatial sample covariance matrix (SCM) $\hat{\mathbf{R}}_x$ calculated using real data in Eqn. (7.41).

$$\hat{\mathbf{R}}_x = \frac{1}{P}\sum_{t=0}^{P-1} \mathbf{x}_k(t)\mathbf{x}_k^H(t) = \left[\hat{r}_x^{[i,j]}\right]_{i,j=0}^{N-1} \tag{7.41}$$

For multivariate Gaussian snapshots $\mathbf{x}_k(t)$, the spatial SCM represents the maximum likelihood estimate of $\mathbf{R}_x$. Using the *invariance property* of the maximum likelihood (ML) estimator Kay (1987), it follows that the ML estimate of the magnitude of the correlation coefficient is given by $\hat{\rho}_{i,j} = \left|\hat{r}_x^{[i,j]}\right| / \sqrt{\hat{r}_x^{[i,i]}\hat{r}_x^{[j,j]}}$. As $\mathbf{R}_x$ is assumed to be a Toeplitz matrix, further averaging of the estimates may be performed over different receiver pairs spaced apart by an equal distance $d = |i - j|$. This yields the correlation coefficient function $\hat{\rho}(d)$ in Eqn. (7.42), which is equivalent to the spatial ACS envelope normalized such that $\hat{\rho}(0) = 1$.

$$\hat{\rho}(d) = \frac{1}{N-d-1}\sum_{i=0}^{N-d-1} \hat{\rho}_{i,i+d} \tag{7.42}$$

The values of a or b that provide the best least-squares fit between the real-data measurements $\hat{\rho}(d) = e^{\hat{P}(d)}$ and the correlation coefficient model $\rho_{ij} = e^{P(d)}$ in

Eqn. (7.40) can be computed using the same parameter estimation method described for the temporal ACS in Eqn. (7.43). It is then a matter of assessing the significance of the departure between the assumed model and the sample estimates so that a decision can be made to accept or reject the hypothesized model with a known confidence level.

$$\tilde{b} = \arg\min_{b} \sum_{d=0}^{N-1} \|\hat{\rho}(d)\|_F [\hat{P}(d) - bd]^2 \tag{7.43}$$

For a multivariate complex Gaussian distributed process, the sampling properties of the maximum likelihood estimate of the magnitude squared coherence $\hat{\rho}^2$ were found by Goodman (1963). If n statistically independent samples are used to estimate the magnitude squared coherence ρ^2 between a pair of receivers, it can be shown that the distribution of the sample estimate $\hat{\rho}^2$ is given by Eqn. (7.44), where ρ^2 is the statistically expected value.

$$p(\hat{\rho}^2) = (n-1)(1-\rho^2)^n(1-\hat{\rho}^2)^{n-2}F(n, n; 1; \rho^2\hat{\rho}^2) \tag{7.44}$$

The term $F(x, y; w; z)$ is the classical hypergeometric function defined in Eqn. (7.45), where we have also defined $(x)_k = x(x+1)\ldots(x+k-1)$. This series is guaranteed to converge if the argument of the function z has a modulus less than unity. In this case, we have that $z = \rho^2\hat{\rho}^2 < 1$, so the function always converges. The density function $p(\hat{\rho}^2)$ was used in Carter (1971) for frequency domain coherence estimation. The same distribution may be used to derive confidence intervals that may be used to accept or reject the parametric model assumed for the spatial correlation coefficients.

$$F(x, y; w; z) = \sum_{k=0}^{\infty} \frac{(x)_k(y)_k}{(w)_k}\frac{z^k}{k!} \tag{7.45}$$

As consecutive array snapshots are not statistically independent due to the significant temporal correlation, it is necessary to determine the equivalent number of statistically independent observations n to be used in Eqn. (7.44). For a large number of samples P, Priestly (1981) describes a method to determine the equivalent number of statistically independent samples for a correlated Gaussian random process. The first step is to compute the variance of the sample variance. For a complex random process, this is given by σ_r^2 in Eqn. (7.46), where $r(k)$ is the expected correlation between data points spaced k samples apart and $\hat{r}(0)$ is the sample variance of the random process, as described in Thierren (1992).

$$\sigma_r^2 = \mathrm{E}\{[\hat{r}(0) - r(0)]^2\} = \frac{1}{P}\sum_{k=-(P-1)}^{P-1}\left(1 - \frac{|k|}{P}\right)|r(k)|^2 \tag{7.46}$$

Using Eqn. (7.46), the variance of the sample variance is $r^2(0)/P$ for P *statistically independent* samples as $P \to \infty$. Hence, the effective number of independent observations for a correlated process is given by $P_e = r^2(0)/\sigma_r^2 \leq P$. The calculated value of P_e to the nearest integer may then be substituted for the number of independent samples n in Eqn. (7.44) to derive appropriate confidence intervals. If the estimates $\hat{\rho}_{ij}$ fall within the specified bounds for all receiver pairs (i, j), then the model for the spatial correlation coefficients is accepted, otherwise it is rejected.

The sample spatial covariance matrix $\hat{\mathbf{R}}_x$ used to compute the various correlation coefficients was estimated from $N = 30$ receivers using Eqn. (7.41), where $P = 12{,}032$ array snapshots (recorded in 47 dwells). For a separation of d receiver spacings, equivalent to $d \times 84.0$ m in this experiment, there are $N - d$ correlation coefficients that can be formed using $P = 12{,}032$ array snapshots. These $N - d$ correlation coefficients were further averaged according to Eqn. (7.42) in order to form the envelope $\hat{\rho}(d)$ used for model parameter estimation. It is postulated that an exponentially decaying function of receiver separation describes the behavior of the statistically expected spatial correlation coefficients. As the array snapshots used to form the sample spatial covariance matrix are correlated and therefore *not* statistically independent, an effective number of independent samples was also calculated using the procedure described in the previous paragraph.

Once the model parameter b is estimated from the SCM, and the confidence bounds are specified for the sample correlation coefficients based on an effective number of statistically independent samples, it is of interest to analyze the behavior of the estimated mode spatial correlation coefficients as a function of both receiver spacing and absolute position in the array. For a spatially stationary process, the correlation coefficient between two receiver outputs only depends on receiver separation and not the absolute receiver positions in the array. Hence, along with the mean wavefront analysis in the next section, this analysis will indicate the validity of both the spatial stationarity assumption as well as the exponentially decaying model proposed for the spatial ACS envelope.

Figures 7.13–7.16 show the hypothesized models for the propagation modes as solid lines and the associated confidence bounds (upper and lower deciles) as dashed lines. The

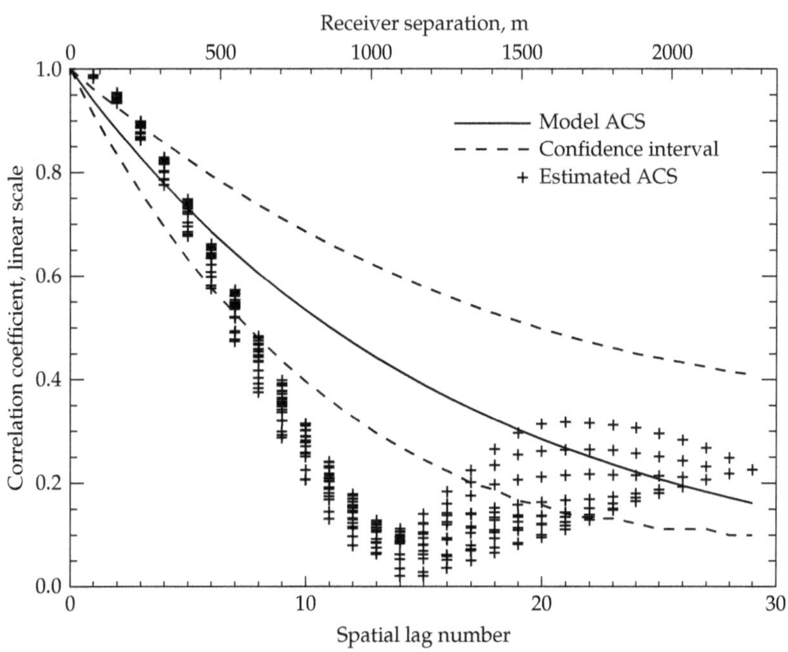

FIGURE 7.13 Spatial correlation coefficients for $1E_s$ mode. © Commonwealth of Australia 2011.

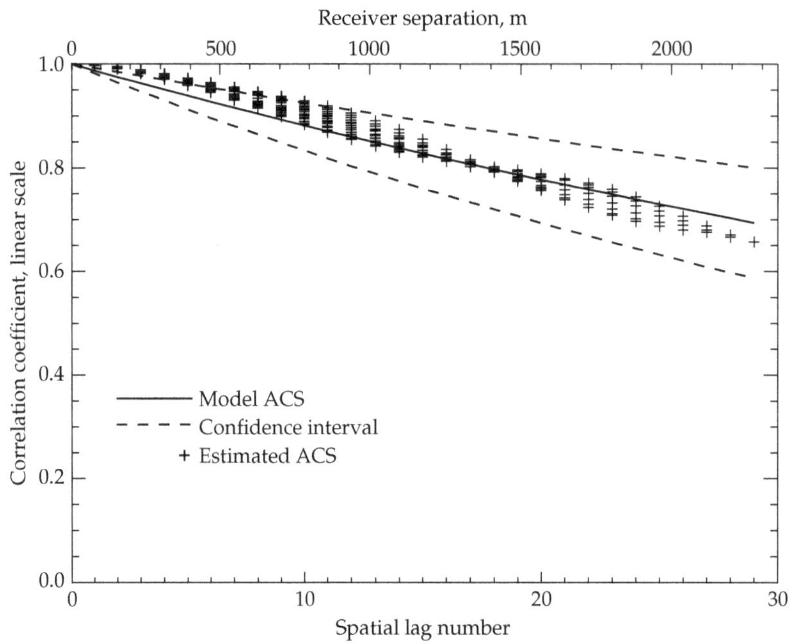

FIGURE 7.14 Spatial correlation coefficients for $1F_2$ mode. © Commonwealth of Australia 2011.

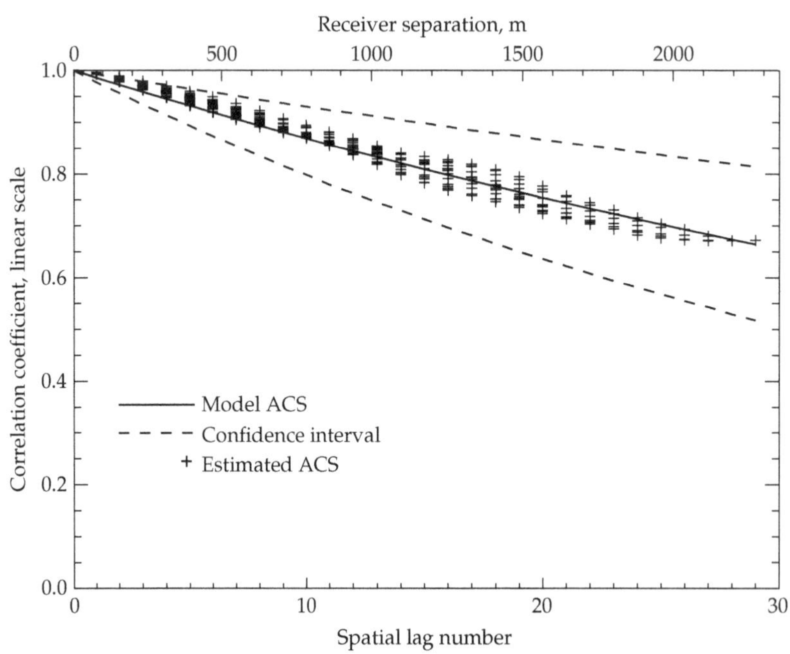

FIGURE 7.15 Spatial correlation coefficients for $1F_2(o)$ mode. © Commonwealth of Australia 2011.

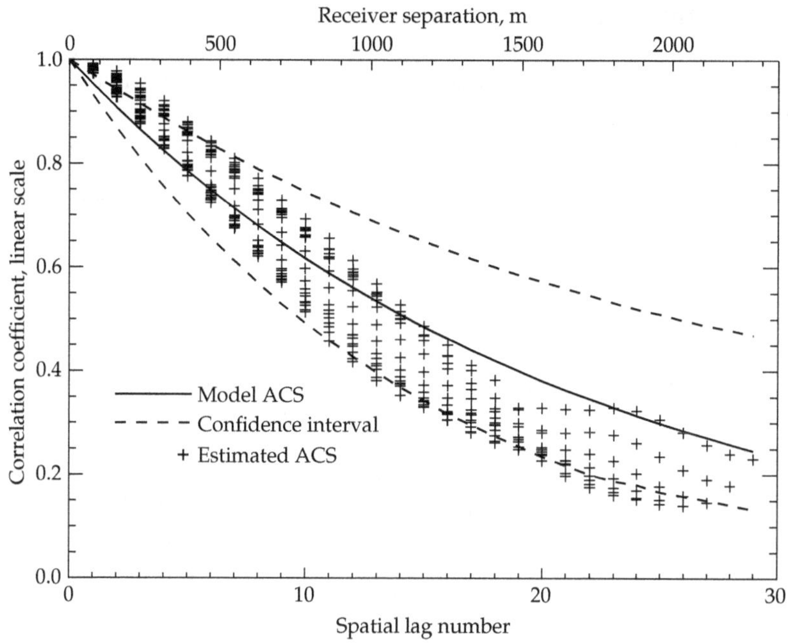

FIGURE 7.16 Spatial correlation coefficients for $1F_2(x)$ mode. © Commonwealth of Australia 2011.

actual estimates of the spatial correlation coefficients calculated from the experimental data have been overplotted using + symbols in each figure. There are more points marked with a + at smaller receiver separations because there are more pairs of receivers with such separations in the array. Clearly, only one estimate is available for the maximum spatial separation that occurs between the first and last receivers of the array.

It is evident from Figure 7.13 that correlation coefficients estimated for the $1E_s$ mode are not well described by the spatially stationary exponentially decaying model. This is not surprising because it was shown in the preceding chapter that this mode is likely to be composed of two sporadic-E reflections coming from distinct clouds of ionization that are unresolved in range but have significantly different angles-of-arrival. The superposition of two different spatial frequencies is expected to produce regularly spaced nulls (beating) in the spatial ACS. The first (imperfect) null is evident at a receiver separation of approximately 1150 m, while it appears that the correlation coefficient is falling toward a second null slightly past 2500 m. In contrast to the $1E_s$ mode, the remaining signal modes are in very good agreement with the postulated statistical model. The great majority of the measurements lie inside the confidence bounds for these three other modes. Hence, for well-resolved signal modes reflected from a localized region in the ionosphere, there is no strong reason to reject the spatially stationary exponentially decaying model for the correlation coefficient over a relatively quiet mid-latitude path.

7.4.2 Mean Plane Wavefront

The plane wave $\mathbf{s}(\theta)$ is the hypothesized model for the mean wavefront of a single propagation mode. In the presence of array manifold errors, the mean wavefront received by the system may not be planar. Even in the absence of such errors, it is conceivable that ionospheric phenomena may contribute to departures between the mean wavefront and the plane wavefront. To verify the mean plane wavefront hypothesis, it is first necessary to estimate the cone angle-of-arrival θ and amplitude A of the plane wave that best fits the mean wavefront $\hat{\mathbf{s}}$ (extracted from the sample spatial covariance matrix) according to the least squares criterion in Eqn. (7.47).

$$\hat{A}, \hat{\theta} = \arg\min_{A,\theta} \|\hat{\mathbf{s}} - A\mathbf{s}(\theta)\|_F \tag{7.47}$$

From Eqn. (7.38), it follows that the magnitude of the mean wavefront $|\mathbf{s}^{[n]}|$ at element n can be estimated as $|\hat{\mathbf{s}}^{[n]}|$ in Eqn. (7.48), since $\rho_{nn} = 1$ by definition and the power scaling term σ_s^2 is taken into account by the amplitude parameter A in Eqn. (7.47). The phases $\measuredangle \mathbf{s}^{[n]} = \phi_n$ for $n = 1, 2, \ldots, N$ can be found by solving a set of linear equations satisfying the phase difference relationship in Eqn. (7.48) using the elements (i, j) in the upper or lower triangle of the spatial sample covariance matrix. When all such elements are used, the equations become overdetermined for $N > 3$, and a least squares technique may be employed to estimate the phase structure of the mean wavefront.

$$|\hat{\mathbf{s}}^{[n]}| = \sqrt{\hat{r}_x^{[n,n]}}, \quad \measuredangle \hat{r}_x^{[i,j]} = \phi_i - \phi_j \tag{7.48}$$

Once the mean wavefront has been extracted from the sample spatial covariance matrix, the plane wave parameters of best fit may be estimated by performing a one-dimensional search for the minimizing cone angle-of-arrival $\hat{\theta}$ and amplitude $\hat{A}$. To assess the significance of the differences between the sample variances $\hat{r}_x^{[n,n]}$ over the N receivers and the value $\hat{A}^2$ expected for the assumed model, the distribution derived in Appendix A may be used to calculate confidence intervals for testing this feature of the model. For the mean plane wavefront model to be accepted, all of the sample variances are required to lie within the upper and lower deciles of the distribution. The reception channels are assumed to be well matched, and no allowances are made for potential differences between the gains of different reception channels.

Figures 7.17–7.20 show the mean power $\hat{A}^2$ of best fit (solid line), confidence bounds (dashed lines), and the actual power estimated in the different receivers as "+" symbols for each propagation mode. The $1E_s$ and $1F_2(x)$ reflections have a significant number of measurements that lie outside the confidence bounds. This suggests the plane wavefront model is not representative of the mean wavefront for these propagation modes. In the case of the $1E_s$ mode, this is understandable due to the two unresolved reflections that are present, but for the $1F_2(x)$ mode, this result is somewhat unexpected and indicates that the spatial structure of the ionosphere that propagated this single magneto-ionic component was relatively more disturbed than the regions which propagated the $1F_2$ and $1F_2(o)$ modes in the analyzed data. This is supported by the fact that the correlation coefficient measurements for the $1F_2(x)$ mode in Figure 7.16 are comparatively more spread than those of the $1F_2$ and $1F_2(o)$ modes. The $1F_2$ and $1F_2(o)$ modes satisfy the constant power condition of the mean plane wavefront model.

To fully validate the proposed spatial covariance matrix model, it is also necessary to investigate the linearity of the phase across the aperture. The distribution for the phase

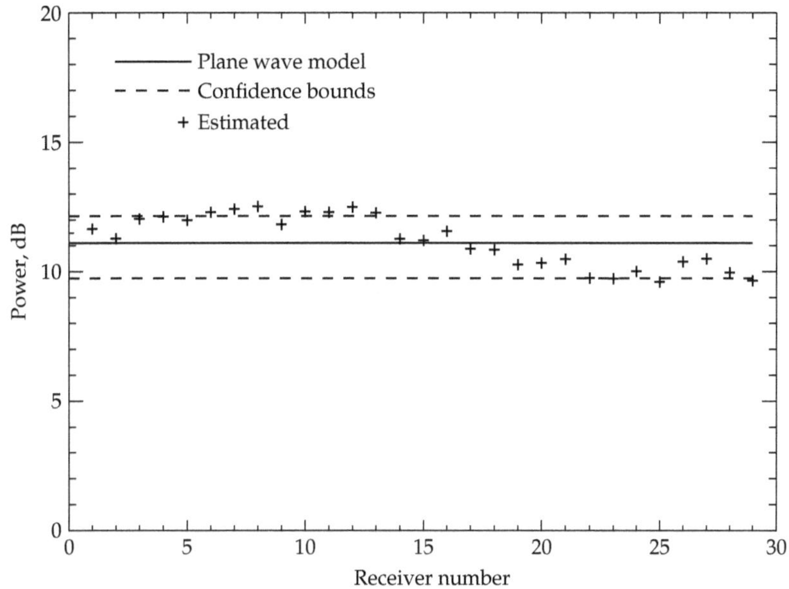

FIGURE 7.17 Power across aperture for $1E_s$ mode. © Commonwealth of Australia 2011.

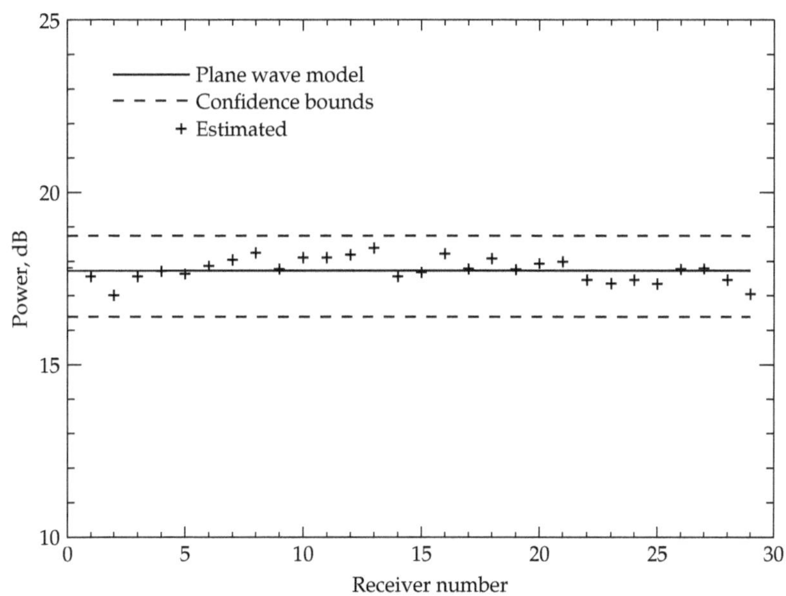

FIGURE 7.18 Power across aperture for $1F_2$ mode. © Commonwealth of Australia 2011.

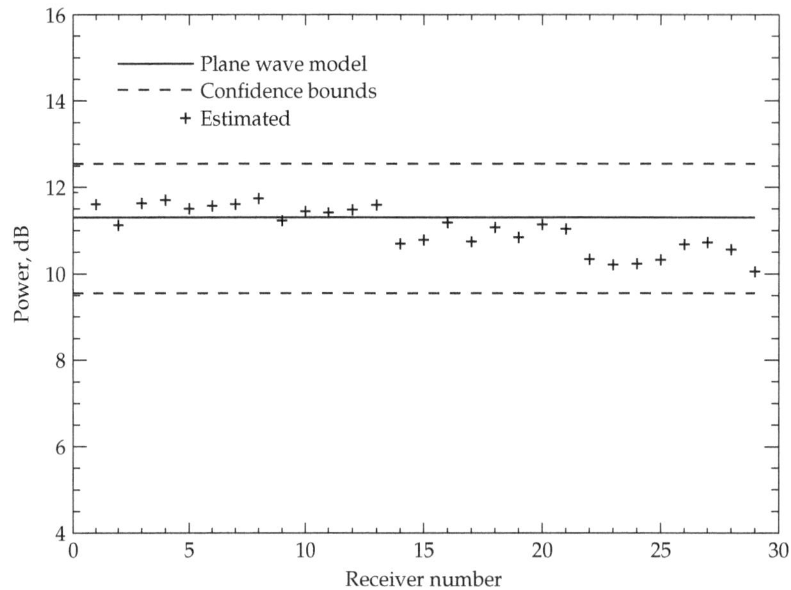

FIGURE 7.19 Power across aperture for $1F_2(o)$ mode. © Commonwealth of Australia 2011.

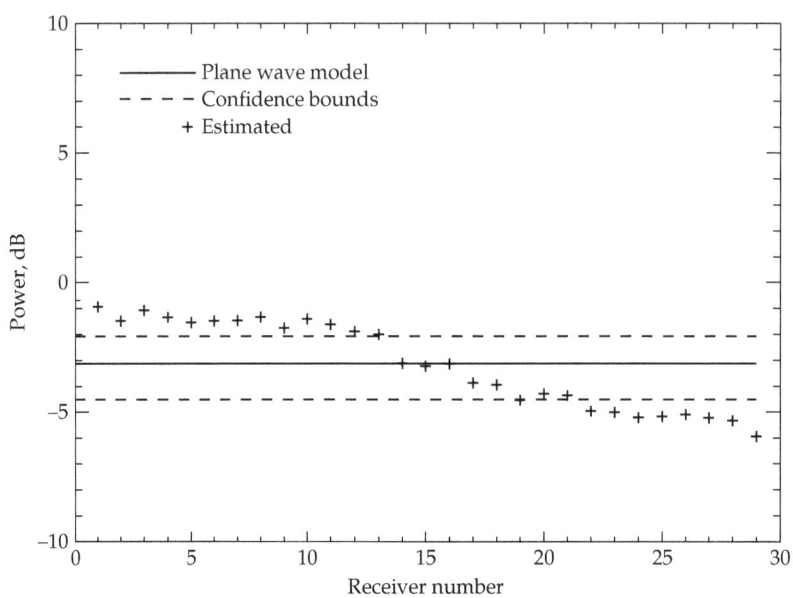

FIGURE 7.20 Power across aperture for $1F_2(x)$ mode. © Commonwealth of Australia 2011.

of the temporal ACS derived in Appendix A may also be used for the phase of the spatial ACS. The latter is expected to be a linear function of receiver number with a slope that depends on the cone angle-of-arrival of best fit $\hat{\theta}$. Note that the distribution of the phase of a sample spatial lag about its expected value is identical to that of a sample temporal lag when the magnitudes of the expected correlations in the spatial and temporal domains are the same and the same number of samples are used to form them. Based on this equivalence, confidence intervals for the phase of the sample spatial lags may be derived using the formulas in Appendix A.

The expected linear phase progression of the spatial ACS is calculated for each mode using the corresponding estimate of the cone angle-of-arrival of best fit $\hat{\theta}$. Since the spatial sample covariance matrix is Hermitian, all of the information is contained in the elements of either the upper or lower triangle of the matrix (including the main diagonal). It is therefore required to check whether the phase progression of the elements in each column of either the upper or lower triangle of the spatial SCM falls entirely within the established confidence bounds. For the plane wavefront model to be accepted, all of the sample spatial lag phases are required to lie within the upper and lower decile confidence bounds relative to the linear phase front of best fit.

Figures 7.21–7.24 show the mean linear phase-front (solid line), the upper and lower confidence bounds (dashed lines), and the estimated phases corresponding to different spatial lags and reference receivers ("+" symbols) for each of the propagation modes. More than one symbol appears for all but the largest lag number in Figures 7.21–7.24 because each receiver in the array may be designated as the spatial phase reference. In all figures, the right vertical axis shows the cone angle-of-arrival, which depends on the slope of the linear phase-front. The cone angle of best fit for a particular propagation

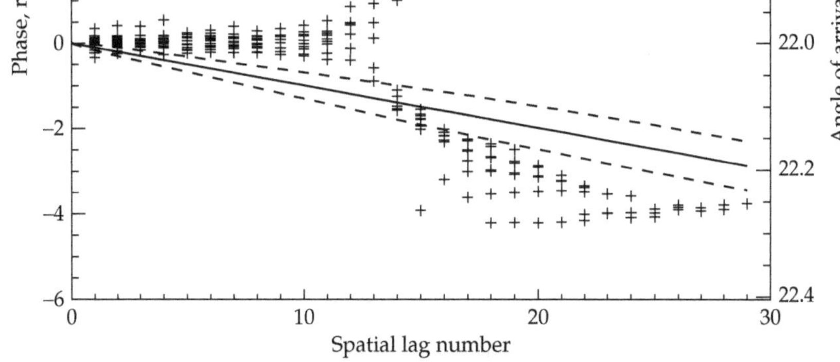

FIGURE 7.21 Phase measurements across the aperture for $1E_s$ mode with different receivers in the array used as the spatial reference. © Commonwealth of Australia 2011.

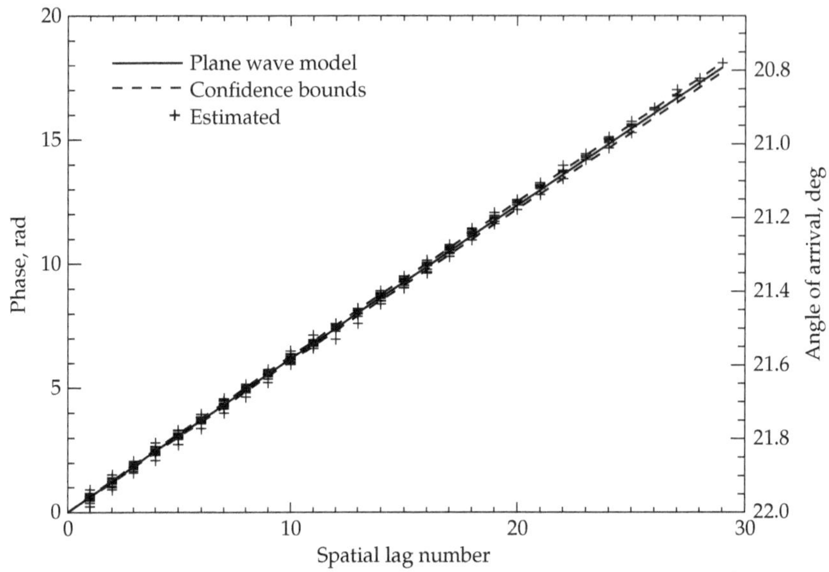

FIGURE 7.22 Phase measurements across the aperture for $1F_2$ mode with different receivers in the array used as the spatial reference. © Commonwealth of Australia 2011.

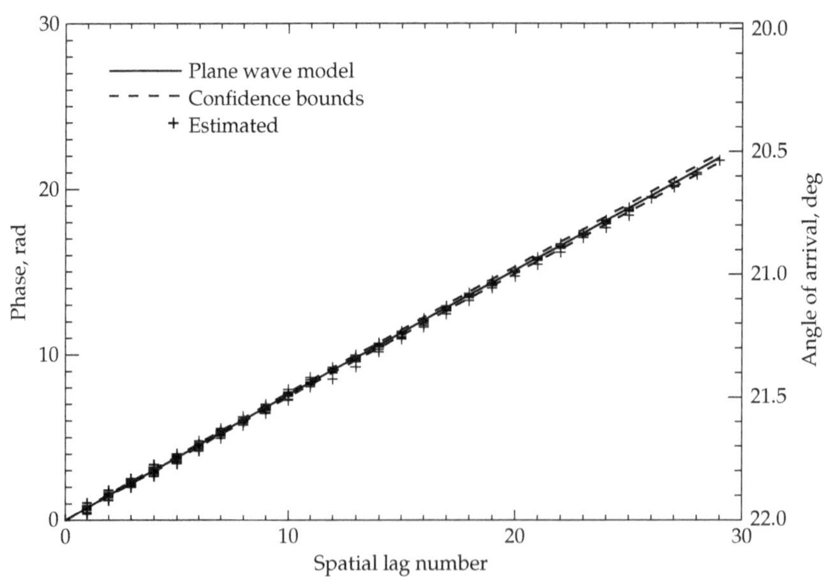

FIGURE 7.23 Phase measurements across the aperture for $1F_2(o)$ mode with different receivers in the array used as the spatial reference. © Commonwealth of Australia 2011.

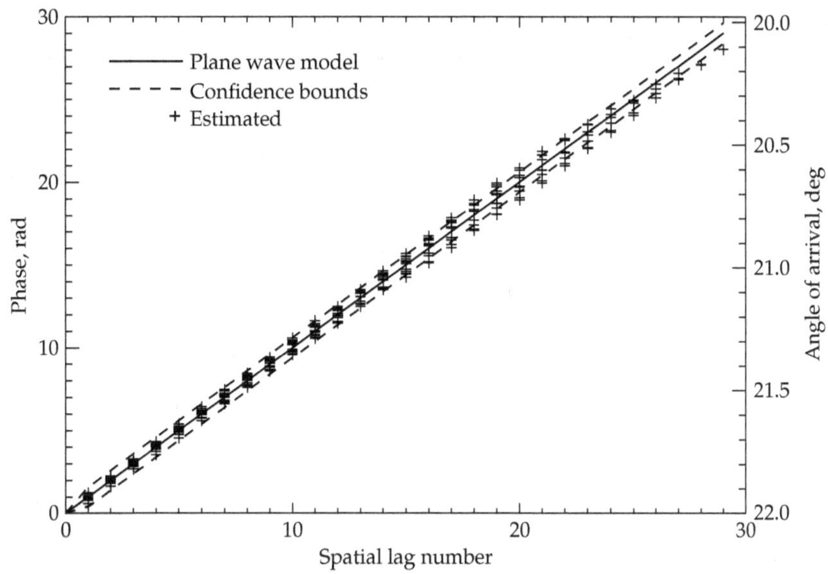

FIGURE 7.24 Phase measurements across the aperture for $1F_2(x)$ mode with different receivers in the array used as the spatial reference. © Commonwealth of Australia 2011.

mode is indicated by the intersection of a straight line from the origin with the right vertical axis.

As for the power measurements, it should be kept in mind that phase measurements are also affected by instrumental errors such as phase mismatches between the responses of different reception channels. Despite the potential existence of these mismatches, it was assumed that the receivers were identical and no allowance was made to account for the potential influence of calibration errors by widening the confidence bounds. If the model is valid and such errors are present, more than the expected percentage of measurements are expected to lie outside the confidence bounds calculated under the assumption of no errors. Hence, the presence of gain and phase mismatches between the responses of the reception channels will tend to favor the rejection of plane wavefront model.

It is clear from Figure 7.21 that the $1E_s$ mode is far from having a planar phase-front. This is expected due to there being two reflections with significantly different cone angles-of-arrival. Figures 7.22 and 7.23 indicate that the mean wavefront of the $1F_2$ and $1F_2(o)$ modes have much more linear phase-fronts. The majority of the phase measurements made on these two modes fall within the confidence bounds. As the constant power assumption has already been validated for the $1F_2$ and $1F_2(o)$ modes, it follows that there is no strong reason to reject the mean plane wavefront model for these two modes. A higher spread of the phase estimates is observed for the $1F_2(x)$ mode in Figure 7.24, but the large majority of the measurements lie within the confidence bounds. This aspect of the plane wave model appears to be valid for the $1F_2(x)$ mode despite the fact that the power of this mode varied beyond the confidence bounds across the aperture.

Mode	b	B_s, deg	θ, deg	FA (%)
$1E_s$	−0.0628	1.58	22.20	24.0
$1F_2$	−0.0126	0.32	20.75	97.8
$1F_2(o)$	−0.0141	0.36	20.48	98.9
$1F_2(x)$	−0.0483	1.22	19.98	94.0

TABLE 7.2 Decaying exponential spatial ACS parameters estimated for the different propagation modes. The fitting accuracy achieved by the separable space-time ACS model is also listed.

The spatial ACS model parameters resulting for all modes are listed in Table 7.2. The parameter b in Table 7.2 is the coefficient of the estimated exponentially decaying function in Eqn. (7.40) for a receiver spacing of 84 m. The cone angle of best fit estimated according to Eqn. (7.47) is listed as θ in this table. The spatial bandwidth B_s of the signal mode is calculated by first determining the distance at which the magnitude of the spatial ACS falls by a factor of $1/e$. The inverse of this distance is then computed to yield a spatial frequency in units of m^{-1}. This spatial frequency may be expressed in terms of a cone angle for the ULA at the operating wavelength. This cone angle, which is a one-sided measurement of the width of the angular spectrum, is then doubled to provide a measure of spatial bandwidth or angular spread given by $B_s = 2 \times \arcsin(b\lambda/\Delta d)$ degrees. Spatial bandwidths of around 1 degree at a frequency of 16.110 MHz agrees well with the previous experimental results described in the opening section of this chapter.

7.4.3 Space-Time Separability

It is of interest to determine whether the sample *space-time* ACS $\hat{r}(\Delta d, \Delta t)$ measured from experimental data can be represented with reasonable accuracy in terms of the previously validated temporal-only ACS $r_t(\Delta t)$ and spatial-only ACS $r_s(\Delta d)$ models. The expected second-order statistics of a stationary space-time random process are said to be *separable* when the space-time ACS $r(\Delta d, \Delta t)$ can be expressed as a product of the temporal-only and spatial-only ACS, as in Eqn. (7.49).

$$r(\Delta d, \Delta t) = r_s(\Delta d) \times r_t(\Delta t) \tag{7.49}$$

Hypothesis tests may be derived to accept or reject space-time separability of the model by using the results in Appendix A. A simpler method for measuring whether a separable model is accurate is to quantify the fitting accuracy between the separable model of the space-time ACS $r(\Delta d, \Delta t)$ in Eqn. (7.49) and the sample space-time ACS $\hat{r}(\Delta d, \Delta t)$ estimated directly from the experimental data. A separable model may be considered appropriate when the fitting accuracy defined in Eqn. (7.50) exceeds a relatively high value of say 95 percent. The fitting accuracy is computed by synthesizing a space-time ACS model according to Eqn. (7.49) using the previously estimated Gaussian temporal ACS and exponentially decaying spatial ACS models for each propagation mode. The fitting accuracies resulting for the different propagation modes over 900 space-time lag points (i.e., 30 spatial lags times 30 temporal lags) are also listed in Table 7.2.

$$\text{Fitting Accuracy} = 1 - \frac{\sum_{\Delta d=0}^{N-1} \sum_{\Delta t=0}^{Q-1} |\hat{r}(\Delta d, \Delta t) - r(\Delta d, \Delta t)|^2}{\sum_{\Delta d=0}^{N-1} \sum_{\Delta t=0}^{Q-1} |\hat{r}(\Delta d, \Delta t)|^2} \tag{7.50}$$

As expected, the space-time ACS of the $1E_s$ mode is not well represented by this separable model because the spatial ACS model was previously rejected. The $1F_2$ and $1F_2(o)$ modes are very well represented by the separable space-time ACS model indicating that for these modes the sample space-time ACS can be deduced with high accuracy from a knowledge of the experimentally validated temporal-only and spatial-only ACS models. The $1F_2(x)$ mode is quite well represented by the separable model but not as well as the other two F-region modes. This may be due to the relatively more disturbed characteristics of the $1F_2(x)$ mode in the analyzed data set, which caused the mean plane wavefront model to be rejected for this single magneto-ionic component.

Figures 7.25–7.32 illustrate the match between the experimental sample space-time ACS and the parametric model of a separable space-time ACS in terms of the real and imaginary components for each of the different propagation modes. Although 900 space-time lags were used to compute the fitting accuracy, only 100 space-time lags (10 spatial lags by 10 temporal lags) are displayed in these figures for clarity. The 10 lags chosen in

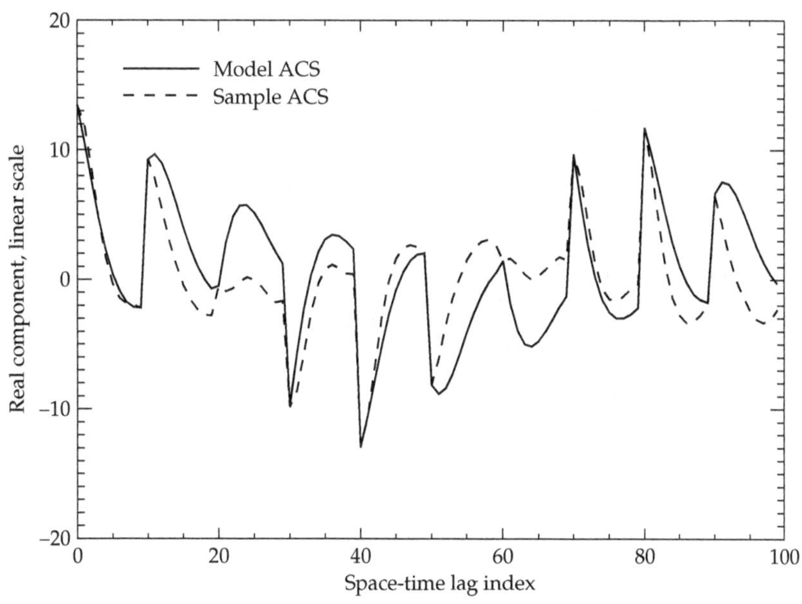

FIGURE 7.25 Real component of space-time ACS for $1E_s$ mode.

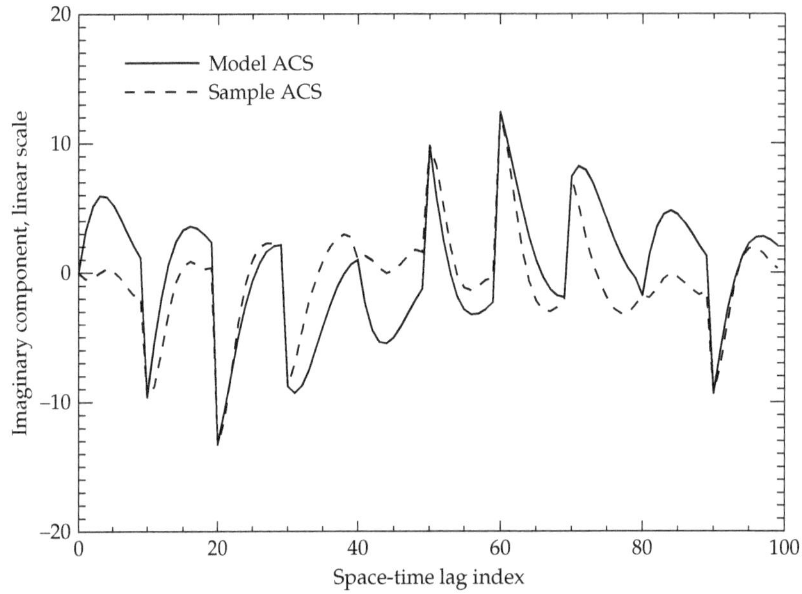

FIGURE 7.26 Imaginary component of space-time ACS for $1E_s$ mode.

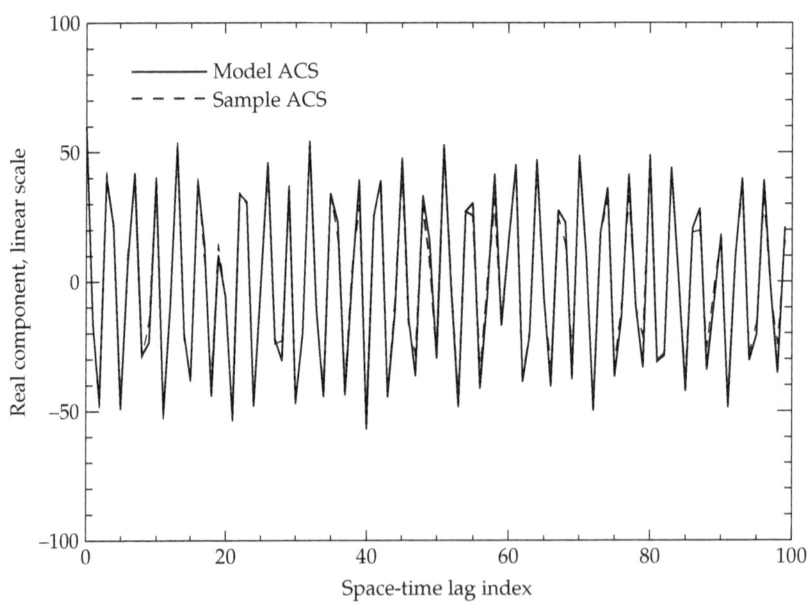

FIGURE 7.27 Real component of space-time ACS for $1F_2$ mode.

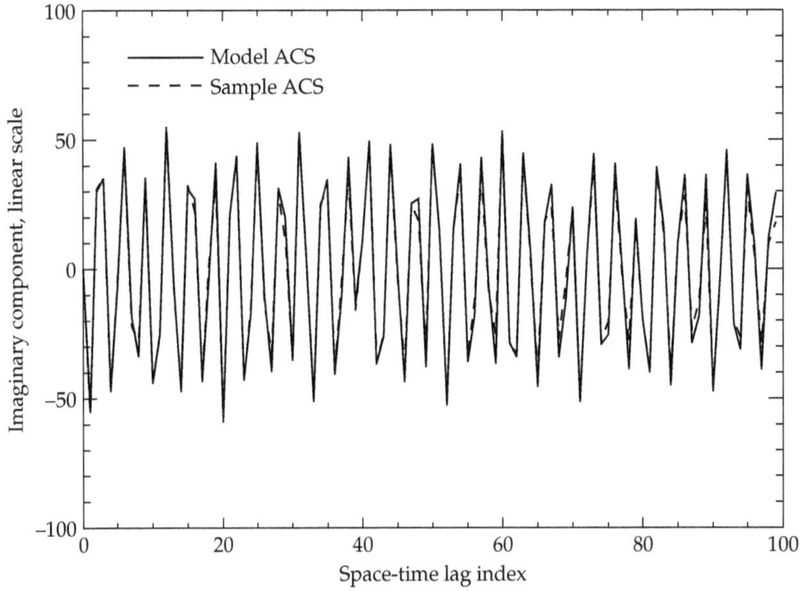

FIGURE 7.28 Imaginary component of space-time ACS for $1F_2$ mode.

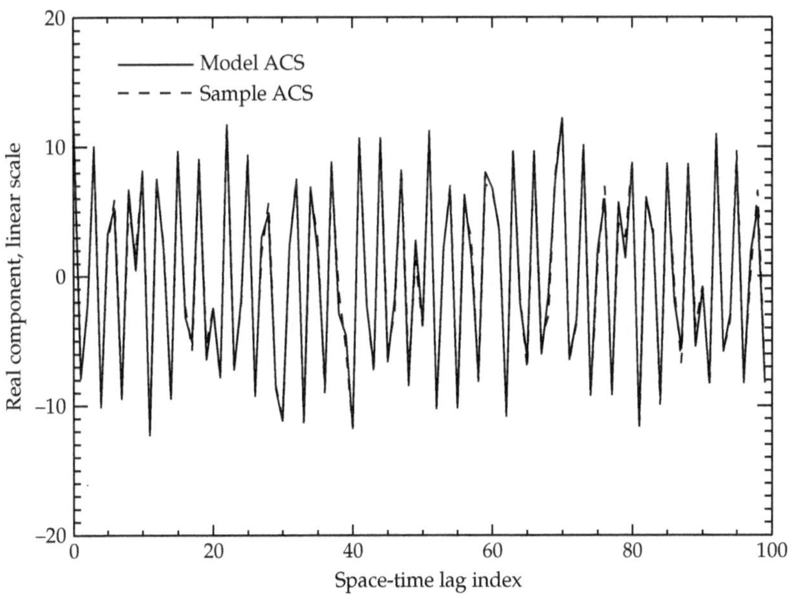

FIGURE 7.29 Real component of space-time ACS for $1F_2(o)$ mode.

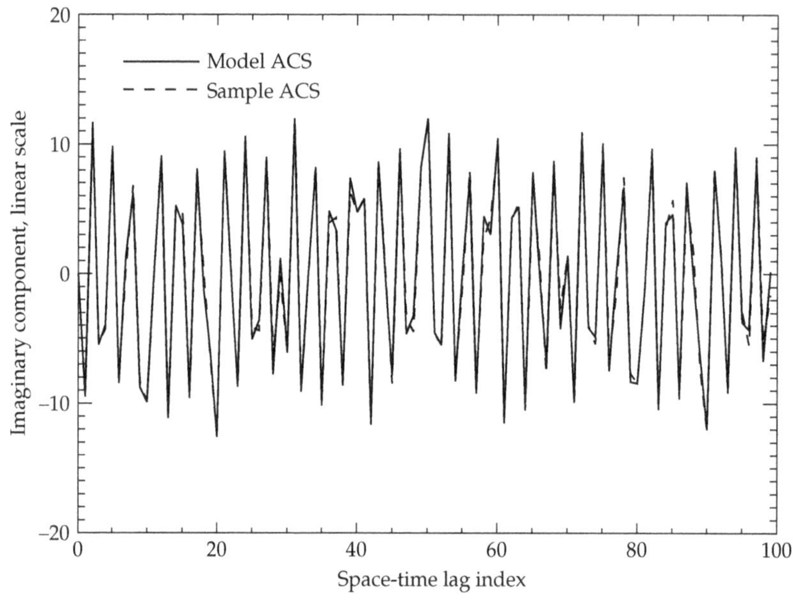

FIGURE 7.30 Imaginary component of space-time ACS for $1F_2(o)$ mode.

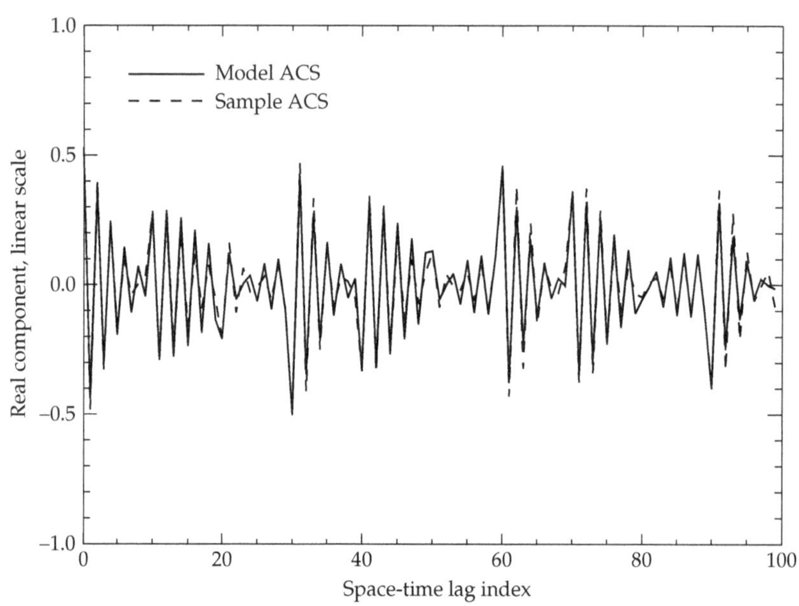

FIGURE 7.31 Real component of space-time ACS for $1F_2(x)$ mode.

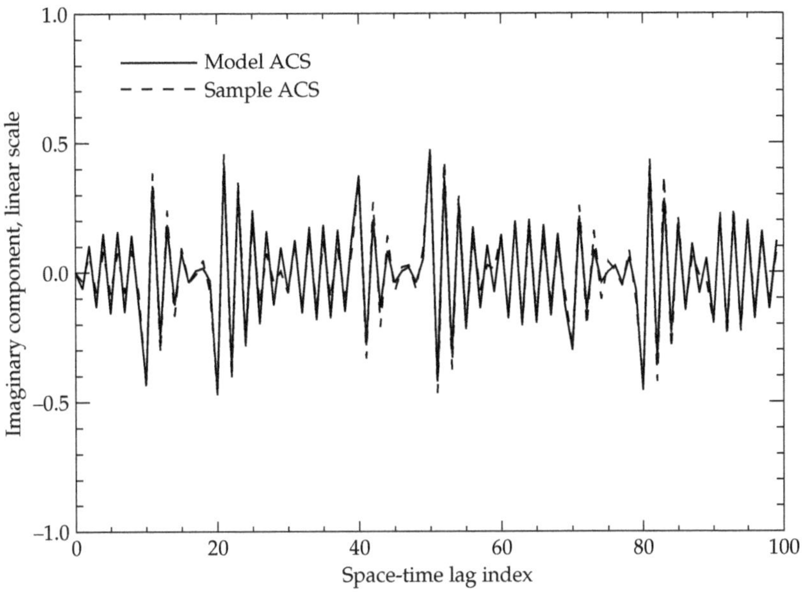

FIGURE 7.32 Imaginary component of space-time ACS for $1F_2(x)$ mode.

each domain correspond to every third lag such that the plotted components are given by $\tilde{r}(i, j) = \hat{r}(3i\Delta t, 3j\Delta d)$ for $i, j = 0, 1, ..., 9$. These data points have been stacked into a one-dimensional vector referenced by the space-time lag index $\ell = i \times 10 + j$. The ability of the parametric and separable space-time ACS model to almost exactly replicate the experimental sample space-time ACS over 900 lag points in Figures 7.27 and 7.28 indicates that the second-order statistics of single modes reflected by the mid-latitude ionosphere can be well represented in a statistical sense by the postulated stationary and separable space-time ACS model over time intervals of a few minutes.

CHAPTER 8

HF Channel Simulator

The deterministic wave-interference model analyzed previously may be interpreted as a traditional array signal-processing model based on the superposition of plane-wave signals. While this representation has many attractive features that facilitate model identification and numerical simulation, it may not always be the most appropriate type of model for representing HF signals reflected by the ionosphere. In certain adaptive signal-processing applications of considerable importance to OTH radar, use of such a model can lead to misleading results that are not consistent with those encountered in practice. This chapter describes an alternative *array signal-processing* model that represents the space-time characteristics of HF signal modes reflected from the ionosphere in a *statistical* sense. In addition, parameter-estimation techniques for this model are proposed and validated using real data.

The described model simulates the fine structure of a signal mode as a realization of a wide-sense stationary multi-channel Gaussian random process defined by second-order statistics consistent with those observed for real ionospheric modes in the preceding chapter. Besides the capability of this model to parametrically represent a wide variety of statistical characteristics, additional attributes include: (1) mathematical tractability to simplify model identification based on limited data sizes, (2) concise description in terms of a small set of physically meaningful parameters that can be used to guide signal-processing algorithm selection, and (3) straightforward software implementation for convenient generation of computer simulations to assess the performance of different signal-processing techniques. The mathematical formulation reverts to the traditional array signal-processing model as a special case for certain parameter choices.

The chapter is divided into four sections. Section 8.1 provides a brief review of models based on point and extended signal sources, as well as existing methods to estimate the parameters of coherently and incoherently distributed signals. Section 8.2 describes the generalized Watterson model used to simulate realizations of narrowband HF signals reflected by the ionosphere. Section 8.3 presents two alternative parameter-estimation techniques applicable to the generalized Watterson model. The effectiveness of these techniques is experimentally demonstrated using real data containing a mixture of unresolved space-time distributed signal modes in Section 8.4.

8.1 Point and Extended Sources

In many array-processing systems, it is commonly assumed that the received signals originate from point sources in the far-field and that the propagation channel gives rise to perfectly planar wavefronts incident on the array from different directions. However, in wide variety of real-world applications, not limited to radar, the array often receives sources that have been diffusely scattered by one or more spatially extended regions in the environment. In this case, the propagation channel effectively transforms a single point source into one or more *extended* signals at the receiver. This section briefly contrasts the traditional array-processing model with extended or distributed signal descriptions in terms of physical significance, data modeling, and parameter estimation.

8.1.1 Traditional Array-Processing Models

For more than two decades, array signal-processing research was strongly focused on the problem of estimating the directions of arrival of plane-wave signals in the presence of additive noise (Krim and Viberg 1996). This problem is based on the assumption of point sources in the far field of a well-calibrated sensor array, where the spatial data snapshots $\mathbf{x}(t) \in \mathcal{C}^N$ may be represented according to the traditional model in Eqn. (8.1).

$$\mathbf{x}(t) = \mathbf{A}(\underline{\theta})\mathbf{s}(t) + \mathbf{n}(t) \tag{8.1}$$

Here, $\mathbf{A}(\underline{\theta}) = [\mathbf{a}(\theta_1), \ldots, \mathbf{a}(\theta_M)]$ is the steering matrix for a signal DOA parameter vector $\underline{\theta} = [\theta_1, \ldots, \theta_M]$, $\mathbf{s}(t) = [s_1(t), \ldots, s_M(t)]$ is the signal vector, and $\mathbf{n}(t)$ is additive noise. Typically, the main objective is to estimate the signal DOA parameter vector $\underline{\theta}$ from a noisy data set $\mathbf{x}(t)$ using a finite number of snapshots. The source waveforms $s_m(t)$ for $m = 1, \ldots, M$ are also assumed to be unknown. The DOA estimates may subsequently be used in a classic signal-copy procedure to estimate the $M < N$ source waveforms (Schmidt 1981).

Maximum likelihood (ML) DOA estimators are often considered optimal, but finding the solution of the ML criterion function generally requires computationally intensive iterative procedures that are not always guaranteed to converge or reach the global optimum (Ottersten, Viberg, Stoica, and Nehorai 1993). Spectral techniques, such as MUSIC (Schmidt 1979) for example, have lower complexity and yield comparable performance to the ML estimator in the asymptotic case for uncorrelated sources (Stoica and Nehorai 1989). However, such techniques solve for the signal parameters separately, rather than jointly, which may degrade performance for a finite number of samples, particularly for highly correlated sources or when the SNR is low.

For ULAs, the Iterative Quadratic Maximum Likelihood (IQML) technique (Bresler and Macovski 1986) achieves similar performance to the ML method by iteratively solving a series of quadratic optimization problems. A modification to IQML based on the weighted subspace fitting (WSF) concept (Viberg and Ottersten 1991) was introduced in Stoica and Sharman (1990b). The resulting Method of Direction Estimation (MODE) algorithm is essentially in closed form (Krim and Viberg 1996) as the estimates have the asymptotic accuracy of the true optimum after the second IQML pass. Simulation results have shown that MODE performs better than MUSIC and closely to the ML method (Stoica and Sharman 1990b). From a computational complexity and performance point of view, MODE is regarded one of the best DOA estimation algorithms for ULAs.

In a multipath environment, the plane-wave assumption corresponds to the ideal case of specular reflection, a condition that is seldom accurately met for HF signals propagated by the ionosphere. As far as DOA estimation is concerned, the more realistic case of "crinkled" wavefronts amounts to decomposing mode fine structure into a sum of vectors on the traditional array manifold. This is the premise behind the wave-interference model considered in Chapter 6, where each mode is regarded as the sum of a relatively small number of Doppler-shifted plane-wave components with constant DOAs and deterministic complex amplitudes. When this traditional model can provide an accurate description of the received signals using a relatively small number M of component rays that can be resolved for parameter estimation, such a model has the distinct advantage of being able to replicate the received signal in a deterministic manner (neglecting the noise contribution).

However, such a model is also recognized to have some drawbacks. The main problem for HF signals is that each mode is potentially composed of a large number of closely spaced rays with time-varying parameters. In Chapter 6, the ray parameters were found to vary rapidly, and it was not always possible to obtain accurate wave-interference model fits to isolated signal modes using small values of M.

The model identification task becomes more difficult when the various signal modes cannot be resolved on the basis of time-delay or Doppler shift for individual analysis. Estimating the DOAs of possibly a large number of rays summed over all signal modes can represent a formidable task, if not an infeasible one given the limited number of array sensors and spatial resolution available. Even if it were possible to accurately estimate all parameters of all modes, such a model becomes cumbersome as it requires many variables to specify. This greatly detracts from its original intent of providing a concise description of the received signal using relatively few plane-wave components.

8.1.2 Coherent and Incoherent Ray Distributions

For applications requiring high fidelity, a distributed signal model is more appropriate than Eqn. (8.1) to describe the temporal and spatial properties of HF signal modes that exhibit non-planar and time-varying wavefronts. In terms of the DOA spectrum, a distributed signal can be thought of as possessing an angular extent over a continuum of directions. The instantaneous wavefront may be expressed as the superposition of infinitely many rays with different DOAs and complex amplitudes statistically described by the angular distribution of the signal.

Quite often, the angular distribution is modeled by a symmetric spatial power density function, which is defined by a mean DOA parameter that corresponds to the center of the distribution and an angular spread parameter that determines the spatial extent of the source about its mean DOA. The scope of such a model is to accurately portray the characteristics of HF signals in a *statistical* sense, rather than to describe particular realizations of the signal in a deterministic manner.

The space-time properties of a distributed signal received by a sensor array are strongly dependent on the degree of correlation that exists among the complex amplitudes of the rays incident from different directions. Depending on the bandwidth of the source and the time-dispersion of the rays scattered by the medium, as well as the dynamics of the medium over the observation interval, the incoming rays of a distributed signal may exhibit varying degrees of correlation. This can range from fully correlated (coherent rays) to completely uncorrelated (incoherent rays).

The former is known as a coherently distributed (CD) signal and arises when: (1) the maximum time dispersion of the rays is significantly less than the reciprocal of the signal bandwidth, and (2) the source and scattering region may to a good approximation be considered static over the observation interval such that the ray complex amplitudes are effectively time-invariant. The latter is termed an incoherently distributed (ID) source and arises when the ray complex amplitudes change in a random manner over the observation interval.

In general, a CD signal is received as a random but perfectly coherent wavefront that contributes a rank one component to the spatial covariance matrix of the received data. This time-invariant wavefront may bear little resemblance to a plane wave on the array steering vector manifold. For HF signal modes reflected by the ionosphere, a CD signal representation is appropriate over a relatively short time frame, such as the PRI for example, during which the ray parameters may be regarded as essentially "frozen."

An ID source produces time-varying signal wavefronts over the observation interval that may be viewed as independent realizations of a stationary random process defined by its spatial power density function. In the HF environment, this situation may arise over a relatively longer time interval, such as from one CPI to another for example, during which the ray parameters become almost totally decorrelated. Unlike a CD signal, an ID signal is associated with a spatial covariance matrix of full rank N.

An intermediate situation is typically observed over the OTH radar CPI. Specifically, a signal mode is received as a set of wavefronts that change in a relatively smooth or correlated manner from one PRI to another over the CPI. This type of signal is referred as a partially correlated distributed (PCD) signal. The energy of a PCD signal is essentially confined to a multi-rank subspace of the spatial covariance matrix with an effective dimension to between 1 and N.

8.1.3 Parametric Localization of Distributed Signals

When there is a one-to-one mapping between the spatial spectrum model parameters and a CD signal wavefront, the collection of all distinct CD signal "spatial signatures" traces out an unambiguous generalized array manifold defined in terms of the spatial spectrum model parameters. Providing the parametric spatial spectrum model is known for all CD signals, and the number of CD signals is less than the number of sensors, the spatial covariance matrix estimated from the array data may be partitioned into signal and noise subspaces, which allows a MUSIC-like technique to be applied for parametric CD signal localization. This type of approach was proposed in Valaee, Champagne, and Kabal (1995), for example, where the number of search dimensions equals the number of spatial spectrum model parameters.

However, this technique assumes knowledge of a parametric model for the spatial *spectrum* of the CD signal as opposed to the power spectral density. In practice, this is a rather strong assumption because the spatial spectrum of an actual CD signal is in general complex valued (magnitude and phase) and typically represents a realization of a random process with second-order statistics defined by the power spectral density. A polynomial rooting approach for the localization of CD signals that avoids the assumption of a parametrically defined spatial spectrum model was proposed in Goldberg and Messer (1998). This method uses a truncated Fourier series to *approximate* the spatial signature vectors on the CD source "manifold."

Gaussian-shaped spatial power density functions were assumed in Trump and Ottersten (1996) for modeling ID signals in mobile communications. The ML estimator was developed for such signals, but the adopted Newton-type search procedure required accurate initialization and imposed a high computational load. An exhaustive search for the covariance matrix model parameters that provide the best least squares fit to the sample covariance matrix was also proposed, but the number of search dimensions for this algorithm grows as twice the number of sources which can become prohibitively large. A related spatial covariance matrix fitting approach was proposed in Gersham, Mecklenbrauker, and Bohme (1997), where the "coherence loss" or angular spread was assumed to have an identical form for all signals regardless of their mean DOA. In the HF environment, this is an unrealistic assumption as the coherence loss is known to depend on the ionospheric layer that affords propagation. This approach also required computationally expensive genetic algorithms to solve the highly nonlinear parameter-estimation problem.

MUSIC-like estimators for ID source localization were investigated by Meng, Stoica, and Wong (1996) and Valaee, Champagne, and Kabal (1995). Both of the proposed methods are based on the spatial covariance matrix, which does not strictly admit to a noise-only subspace for ID sources. The critical assumption is that the angular spread of all sources is small and that the majority of signal energy is concentrated into a small number of eigenvalues of the spatial covariance matrix. Terming this as the "effective" dimension of the signal subspace, a "quasi-noise" subspace can be identified. The idea is to form a set of model spatial covariance matrices over a grid of mean and spread ID source model parameters. The quasi-noise subspace of the model matrices is then projected against the quasi-signal subspace of the sample covariance matrix and the scalar result inverted so that the ID source parameters can be localized at the points corresponding to the most prominent local peaks of the spectrum. A point of concern for such methods is the predisposition to use the spatial covariance matrix, which strictly has a degenerate noise subspace for ID signals.

As described in the next section, the generalized Watterson model of Abramovich, Gorokhov, and Demeure (1996) was proposed for the HF environment. This is the preferred distributed signal model in this application for a number of reasons. First, the parameters of this model can be chosen to select between CD and ID signals or the traditional point source model. Second, the rational (pole-zero) channel transfer function model adopted for distributed signals is sufficiently flexible to approximate a wide variety of power spectral density functions using relatively few parameters. A third major advantage is the ease with which the model can be used to generate realizations of distributed signals on a computer and its mathematical tractability for parameter-estimation purposes.

A potential disadvantage of this model is that it does not take into account any information that may be contained in the higher order statistics of the data. The scalar-type multi-channel random process used in the generalized Watterson model is consistent with separable space-time statistics, but is not designed for signals with a power spectral density that exhibits angle-Doppler coupling, such as signals reflected from a moving platform. A parameter-estimation technique was not supplied for the generalized Watterson model in Abramovich, Gorokhov, and Demeure (1996). Moreover, the previously described CD/ID localization methods are not based on this model. For this reason, a significant portion of this chapter is devoted to describing this model and addressing the important topic of parameter estimation using real data.

8.2 Generalized Watterson Model

A mathematical model that can accurately simulate, in a statistical sense, the complex fading of HF signals sampled in space and time by a very wide aperture antenna array after reflection from the ionosphere provides a valuable tool for the development and testing of OTH radar signal-processing algorithms. The original skywave HF channel simulator validated for narrowband *single* receiver systems by Watterson, Juroshek, and Bensema (1970) was subsequently generalized in Abramovich, Gorokhov, and Demeure (1996) to the case of narrowband antenna arrays. This so-called generalized Watterson model (GWM) is sufficiently flexible to represent the statistical characteristics of the HF channel analyzed in the preceding chapter. It also has the advantage that digital realizations of the HF channel and received signal modes can be readily generated for computer simulations.

The first part of this section describes the GWM and explains the physical significance of different parameters to provide some insights for modeling various types of signals received by OTH radar systems. The simulation of temporal and spatial fluctuations imposed by the skywave HF channel on each propagation mode is described in the second part of this section. These fluctuations represent the "fine structure" of HF signal modes. The third and final part of this section derives the expected second-order statistics of the GWM and expresses the space-time auto-correlation sequence (ACS) of the data as a function of the model parameters.

To preface the following discussion, it is emphasized that the GWM is based mainly on empirical observations and does not intend or attempt to model specific geophysical phenomena occurring in the ionosphere. The main purpose of this model is to represent the broadly observed statistical characteristics of experimental data received by practical HF antenna arrays. Its purpose is not to describe the scenario-dependent detail, but rather to represent the observed characteristics of collected data more realistically than the traditional array signal-processing model, while retaining a certain degree of mathematical tractability. The GWM therefore provides an alternative to the standard plane-wave model, as the latter is often unsuitable for assessing the performance of signal-processing techniques in OTH radar and possibly other HF systems.

8.2.1 Mathematical Formulation and Interpretation

Consider an N-element antenna array that receives a narrowband signal from a far-field HF source after a single-hop ionsopheric reflection. The point source is assumed to be diffusely scattered by a number of spatially separate but extended regions in the ionosphere such that multiple distributed signal modes appear at the receiver. In this case, the antenna array snapshot $\mathbf{x}_k(t) \in \mathcal{C}^N$ recorded at fast-time sample k and slow-time sample t may be expressed as the superposition of M distributed signal modes, denoted by $\{\mathbf{s}_{km}(t)\}_{m=1}^{M}$, that propagate from source to receiver along different paths and additive noise $\mathbf{n}_k(t)$.

$$\mathbf{x}_k(t) = \sum_{m=1}^{M} \mathbf{s}_{km}(t) + \mathbf{n}_k(t) \qquad \begin{cases} t = 1, 2, \ldots, N_p \\ k = 1, 2, \ldots, N_k \end{cases} \tag{8.2}$$

Each *distributed* signal mode is assumed to consist of possibly a large number or continuum of rays scattered from a localized region of the ionosphere, such that the maximum time-dispersion of the rays is significantly less than the reciprocal of the signal bandwidth. In simple terms, the different rays may be thought of as micro-multipath components that are relatively closely clustered about the nominal path of a propagation mode. The number of distinct signal modes M in the model therefore reflects the number of localized scattering regions in the ionosphere. Restricting attention to the dominant multipath components, the number of distributed signal modes propagated along distinct nominal paths is often relatively small. For example, $M = 5$ was experimentally observed in the analysis of Chapter 6. The GWM mathematically describes each distributed signal mode in the following form:

$$\mathbf{s}_{km}(t) = A_m \mathbf{S}(\theta_m)\mathbf{c}_m(t) g_k(t - \tau_m) e^{j2\pi f_m t} \tag{8.3}$$

Here, the complex scalar function $g_k(t)$ represents the baseband source waveform after down conversion, τ_m is the nominal mode time-delay, f_m is the nominal mode Doppler shift normalized by the pulse repetition frequency f_p, and A_m is the root mean square (RMS) amplitude of the mode. The elements of the $N \times N$ diagonal matrix $\mathbf{S}(\theta_m)$ are defined by the array steering vector associated with the nominal DOA θ_m of mode m. For a narrowband ULA of identical sensors and cone angle of arrival θ_m, these elements are given by Eqn. (8.4), where d is the inter-sensor spacing and λ is the carrier wavelength. The mean DOA, mean Doppler shift, and RMS amplitude parameters of the M modes represent the gross structure of the multipath HF channel.

$$\mathbf{S}(\theta_m) = \text{diag}[1, e^{j2\pi d \sin\theta_m/\lambda}, \ldots, e^{j2\pi(N-1)d\sin\theta_m/\lambda}] \tag{8.4}$$

The (statistically stationary) Gaussian-distributed channel response vector $\mathbf{c}_m(t) \in \mathcal{C}^N$ in Eqn. (8.3) models the random temporal and spatial fluctuations imposed by the ionosphere on mode m. These fluctuations represent the Doppler and angular spread on the signal modes to account for the fine structure of the HF channel. This fine structure is not represented in the traditional array signal-processing model. In general, the channel response vector may be expressed in the form of Eqn. (8.5), where $f_m(\theta, t)$ is the time-varying angular spectrum that defines the complex amplitude of a ray with cone angle θ at time t.

$$\mathbf{c}_m(t) = \int f_m(\theta, t)\, \mathbf{s}(\theta)\, d\theta \tag{8.5}$$

The CD signal description corresponds to a time-invariant angular spectrum over the observation interval and a fixed channel response vector. This assumption is appropriate over fast-time samples k in a particular pulse repetition interval t, as the duration of a typical OTH radar pulse is small relative to the time scale of ionospheric fluctuations. An ID description corresponds to statistically independent angular spectra and temporally uncorrelated channel response vectors, which may be suitable in practice from one OTH radar coherent processing interval (CPI) to another. The PCD model corresponds to gradually changing wavefront distortions that are correlated from one pulse to another in the form of a slow-time dependent channel response vector $\mathbf{c}_m(t)$. The PCD model best matches the observed behavior of the mode wavefront variations from pulse-to-pulse over the CPI.

As far as the power spectral density (PSD) of $\mathbf{c}_m(t)$ is concerned, the two most common parametric PSD models for distributed signals are the Lorenzian- and Gaussian-shaped functions. For a spatial or temporal frequency ω, the symmetric PSD profiles given by

$p_m(\omega)$ in Eqn. (8.6) may be specified for each distributed signal mode m in terms of a mean parameter ω_m, that sets the center of the distribution, and spread parameter a_m, that controls the width of the distribution. The GWM model therefore requires a total of five parameters per propagation mode to specify in its simplest form. That is, a mean and spread parameter in both space and time plus an RMS amplitude.

$$p_m(\omega) = \begin{cases} a_m/\{a_m + (\omega - \omega_m)^2\} & \text{Lorenzian} \\ \exp\{-a_m(\omega - \omega_m)^2\} & \text{Gaussian} \end{cases} \tag{8.6}$$

To model the received signal, the source waveform $g_k(t)$ also needs to be defined. The source waveform $g_k(t)$ is normalized to unit variance since the mode power is equal to A_m^2 by definition. Note that $g_k(t)$ is the radiated waveform independent of propagation effects, which are taken into account by other terms in the GWM. Broadly speaking, two alternative categories may be considered. The first relates to a waveform that is independent of the radar signal, such as that emitted by a source of "interference." The second relates to a waveform that is coherent with the radar signal, such as that reflected by a point scatterer or target.

For an incoherent interference source with a uniform spectral density over the receiver bandwidth, the waveform $g_k(t)$ may be statistically modeled as white noise in fast-time and slow-time, with correlation properties given by Eqn. (8.7), where $\delta_{ii'}$ is shorthand for the Kronecker delta function $\delta(i - i')$. Residual correlation in fast-time k due to the pulse-compression window has been ignored here. Use of the function $\text{sinc}(x) = \frac{\sin(\pi x)}{\pi x}$ to describe the inter-mode correlation coefficients implicitly assumes that the interference PSD is flat over the receiver bandwidth f_b.

$$E\{g_k(t - \tau_m)g_{k'}^*(t' - \tau_{m'})\} = \text{sinc}(f_b[\tau_m - \tau_{m'}])\delta_{kk'}\delta_{tt'} \tag{8.7}$$

When the source waveform is coherent with the radar signal, $g_k(t)$ is a deterministic function of range k and independent of pulse number t. Specifically, $g_k(t - \tau_m)$ may be defined as in Eqn. (8.8), where $W(f)$ is the range processing point spread function normalized to unit power, f_b is the now radar signal bandwidth, and f_p is the radar pulse repetition frequency. If the signal is sampled at the Nyquist rate and the mode delay τ_m falls exactly on a fast-time (range) bin k_m, we have that $g_k(t - \tau_m) = \delta(k_m)$ when spectral leakage due to the range window function is neglected.

$$g_k(t - \tau_m) = W(f_p[k - \tau_m f_b]) \tag{8.8}$$

The additive noise $\mathbf{n}_k(t)$ is assumed to be uncorrelated with the source waveform $g_k(t)$. In general, it arises from a combination of internal receiver noise and external background noise, with the latter typically dominating. On a clear frequency channel, without strong impulsive noise due to lightning, the additive noise may be assumed complex Gaussian distributed with independent and identically distributed real and imaginary parts and second-order statistics given by Eqn. (8.9). Here, $\mathbf{n}_k^{[n]}(t)$ denotes element n of the vector $\mathbf{n}_k(t)$ and $*$ denotes complex conjugate.

$$E\{\mathbf{n}_k^{[n]}(t)\mathbf{n}_{k'}^{[n']*}(t')\} = \sigma_n^2\delta_{kk'}\delta_{tt'}\delta_{nn'} \tag{8.9}$$

More specific assumptions can be made regarding the source waveform and additive noise to model particular scenarios. In other words, the GWM can be readily extended beyond the simple case of additive white noise and the coherent or incoherent signal types described before. These assumptions merely serve to simplify the explanation of the model. More generally, such a model should be viewed as a signal-processing framework with a number of "place holders" that may be modified to accommodate scenario-specific assumptions, as required.

8.2.2 Temporal and Spatial Fluctuations

HF channel fluctuations are statistically described by the random processes $\mathbf{c}_m(t) \in \mathcal{C}^N$ for $m = 1, 2, \ldots, M$ which model complex fading of the various signal modes. Variation of an individual element of $\mathbf{c}_m(t)$ over time t represents the temporal amplitude and phase modulation induced by the ionosphere on mode m. Such modulation results in Doppler spread of the signal in each receiver of the array. On the other hand, changes in the amplitude and phase relationship between the different elements of $\mathbf{c}_m(t)$ over time t models the time-varying wavefront distortions imparted by the ionosphere on mode m. These distortions give rise to angular spread on mode m. In Abramovich, Gorokhov, and Demeure (1996), it is proposed to model the two-dimensional (space-time) random vector $\mathbf{c}_m(t)$ by using a multi-variate scalar type auto-regressive (AR) process of order L, as in Eqn. (8.10).

$$\mathbf{c}_m(t) = \sum_{\ell=1}^{L} \alpha_{m\ell}(f_p)\mathbf{c}_m(t-\ell) + \mu_m \underline{\varepsilon}_m(t) \tag{8.10}$$

The slow-time dependent random vector $\mathbf{c}_m(t)$ may be assumed invariant over fast-time k provided the pulse repetition frequency f_p is significantly greater than the bandwidth of the Doppler spread $B_t(m)$ induced by the skywave channel on each mode m. The *structure* of the Doppler spectrum of mode m, sampled at the pulse repetition interval $1/f_p$, is defined by the scalar AR coefficients $\alpha_{m\ell}(f_p)$ for $\ell = 1, \ldots, L$. On the other hand, the amplitude μ_m of the innovative (temporally white) noise vector $\underline{\varepsilon}_m(t) \in \mathcal{C}^N$ determines the *scale* of the Doppler spectrum. The Doppler bandwidth $B_t(m)$ may be defined as the inverse of the time delay for the temporal ACS of mode m to fall by a factor of $1/e$. In the case of a first-order AR model ($L = 1$), this definition implies that the AR parameter $\alpha_m(f_p)$ is given by Eqn. (8.11).

$$\alpha_m(f_p) = e^{-B_t(m)/f_p} \tag{8.11}$$

The first-order AR model is typically justified when $1/f_p < 0.1$ seconds for a relatively quiet mid-latitude ionosphere. In this case, it is observed in Abramovich, Gorokhov, and Demeure (1996) that $\alpha_{m1}(f_p) \to 1$ and $\alpha_{m\ell} \to 0$ for $\ell = 2, \ldots, L$. Assuming a first-order AR process with parameter $\alpha_m(f_p)$, it is readily shown that the power of the individual elements in the channel vector $\mathbf{c}_m^{[n]}(t)$ is given by Eqn. (8.12) for $n = 1, \ldots, N$.

$$\mathrm{E}\{\mathbf{c}_m^{[n]}(t)\mathbf{c}_m^{[n]*}(t)\} = \frac{\mu_m^2 \mathrm{E}\{\underline{\varepsilon}_m^{[n]}(t)\underline{\varepsilon}_m^{[n]*}(t)\}}{1 - |\alpha_m(f_p)|^2} \tag{8.12}$$

All N elements of the driving white noise sequence $\underline{\varepsilon}_m^{[n]}(t)$ are assumed to have unit power, and the processes for different modes m are assumed to be statistically independent, such that $\mathrm{E}\{\underline{\varepsilon}_m^{[n]}(t)\underline{\varepsilon}_{m'}^{[n]*}(t')\} = \delta_{mm'}\delta_{tt'}$. In this case, and with reference to Eqn. (8.12), the scaling term $\mu_m = \sqrt{1 - |\alpha_m(f_p)|^2}$ ensures that the elements of the channel vector $\mathbf{c}_m$ also have unit variance. This normalization is consistent with the previous definition of A_m as RMS power of mode m. Hence, the first-order AR model for the channel response vector may be written as

$$\mathbf{c}_m(t) = \alpha_m(f_p)\mathbf{c}_m(t-1) + \sqrt{1 - |\alpha_m(f_p)|^2}\, \underline{\varepsilon}_m(t) \tag{8.13}$$

In analogous fashion, the spatial channel fluctuations that give rise to angular spread may be described using the simplest (first-order) AR process in Eqn. (8.14). Here, $\beta_m(d)$ is the spatial AR parameter that depends on the inter-sensor spacing d of the array elements. The complex scalar $\gamma_{mn}(t)$ is a driving white noise process of unit variance. Hence, the elements of the driving white noise vector $\underline{\varepsilon}_m^{[n]}(t)$ for $n = 1, \ldots, N$ are linked by a spatial first-order AR process. The scaling $\nu_m = \sqrt{1 - |\beta_m(d)|^2}$ applied to $\gamma_{mn}(t)$ ensures that all elements of the innovative noise vector $\underline{\epsilon}_m(t)$ have unit variance.

$$\underline{\varepsilon}_m^{[n]}(t) = \beta_m(d)\underline{\varepsilon}_m^{[n-1]}(t) + \sqrt{1 - |\beta_m(d)|^2}\, \gamma_{mn}(t) \tag{8.14}$$

The angular bandwidth of the channel $B_s(m)$ is defined as the inverse of the distance over which the spatial ACS of mode m drops by a factor of $1/e$ when the mean angle of arrival is at boresight. In Eqn. (8.15), the spatial correlation coefficient $\beta_m(d)$ not only depends on antenna separation d and angular bandwidth $B_s(m)$, but also on the mean cone angle of arrival θ_m of mode m. In other words, $B_s(m)$ is defined similarly to $B_t(m)$ for $\theta_m = 0$. As the mean cone angle of arrival tends to endfire $\theta_m \to \pi/2$, the perceived angular spread is reduced by a factor of $|1 - \sin\theta_m| \to 0$. This is because only temporal fluctuations of the wavefront can be observed in the endfire direction.

$$\beta_m(d) = e^{-B_s(m)|1-\sin\theta_m|d} \tag{8.15}$$

The scalar driving white noise sequence $\gamma_{mn}(t)$ is defined as a zero-mean complex Gaussian process with independent and identically distributed (i.i.d.) real and imaginary parts having the correlation properties in Eqn. (8.16). This description is consistent with the assumptions of Rayleigh fading as well as mutually independent modes, as experimentally verified by Watterson et al. (1970).

$$\mathrm{E}\{\gamma_{mn}(t)\gamma^*_{m'n'}(t')\} = \delta_{mm'}\delta_{nn'}\delta_{tt'} \tag{8.16}$$

The first-order AR channel model involves one mean and spread parameter per mode in both space and time plus an RMS amplitude (i.e., five real valued parameters). As far as temporal fluctuations are concerned, the mean Doppler shift f_m and spread parameter $B_t(m)$ may be combined into a complex valued parameter defined as $z_m = \alpha_m(f_p)e^{j2\pi f_m}$ in Eqn. (8.17).

$$z_m = e^{-B_t(m)/f_p + j2\pi f_m} \tag{8.17}$$

This may be interpreted as the temporal pole of mode m in the complex plane, all of which are confined to lie on or inside the unit circle. The modulus of z_m is determined by the dampening factor $e^{-B_t(m)/f_p} \leq 1$, while its argument $2\pi f_m$ is determined by the mean Doppler shift of the mode normalized by f_p. As $B_t(m)$ increases, the pole z_m moves toward the origin of the complex plane and the width of the Doppler spectrum increases. On the other hand, the parameter f_m shifts the argument of the pole z_m and therefore controls the centroid of the Doppler spectrum.

Similarly, the mean angle of arrival θ_m and spatial spread parameter $B_s(m)$ may be combined into a complex valued parameter $w_m = \beta_m(d)e^{j2\pi d \sin\theta_m/\lambda}$ given by Eqn. (8.17).

This may be considered as the spatial pole of mode m in the complex plane, which is also confined to the unit disc. The dampening factor $e^{-B_s(m)|1-\sin\theta_m|d}$ and phase $2\pi d \sin\theta_m/\lambda$ of w_m respectively define the width and centroid of the spatial spectrum of mode m. An attractive feature of the simplest first-order AR representation is that the channel parameters can be specified by a pair of complex poles per mode and an RMS amplitude. This is a rather compact description. Moreover, the parameter values are readily interpretable as physical significance can be attached to the modulus and argument of each pole in terms of the width and offset of the mode spectral density, respectively.

$$w_m = e^{-B_s(m)|1-\sin\theta_m|d + j2\pi d \sin\theta_m/\lambda} \tag{8.18}$$

The main difference between the first-order AR description of the temporal fluctuations and the Watterson model is that the Doppler power spectrum has a Lorenzian rather than Gaussian form. When the detailed shape of the Doppler power spectrum is considered important, an AR model of higher order may be used to approximate a Gaussian function of frequency to the required level of accuracy. However, for many practical purposes, the mean and spread parameter values are of greater significance than the choice between Gaussian or Lorenzian profiles. Regarding the spatial fluctuations, the AR(1) process parametrically models the ACS envelope by an exponentially decaying function of distance, as experimentally verified for resolved ionospheric modes in Chapter 7.

8.2.3 Expected Second-Order Statistics

For the first-order AR model $L = 1$, the mean Doppler shift and DOA of each mode can be absorbed into the channel vector $\mathbf{c}_m(t)$ by substituting the coefficients $\alpha_m(f_p)$ and $\beta_m(d)$ in the temporal and spatial AR(1) model recursions with the complex valued parameters z_m and w_m defined earlier. This leads to the more concise GWM description in Eqn. (8.19).

$$\mathbf{x}_k(t) = \sum_{m=1}^{M} A_m \mathbf{c}_m(t) g_k(t-\tau_m) + \mathbf{n}_k(t) \tag{8.19}$$

The statistically expected space-time auto-correlation sequence (ACS) of $\mathbf{x}_k(t)$ is defined in Eqn. (8.20). Specifically, $r_k(i, j)$ is the correlation for range bin k between data samples separated by a time lag of i/f_p seconds and a distance of jd meters apart. Note that j is used to index receiver separation in Eqn. (8.20).

$$r_k(i, j) = \mathrm{E}\{\mathbf{x}_k^{[n]}(t)\mathbf{x}_k^{[n-j]*}(t-i)\} \quad \begin{cases} i = 0, \ldots, L_t - 1 \\ j = 0, \ldots, L_s - 1 \end{cases} \tag{8.20}$$

Since the different modes are propagated by statistically independent channels, and the additive noise is uncorrelated with the signal waveform, the space-time ACS of the received data is given by the sum of the individual mode correlations and that of the additive white noise in Eqn. (8.21). The independence of the signal waveform $g_k(t)$ and the random space-time channel vectors $\mathbf{c}_m(t)$ for all modes m allows the statistical expectation in Eqn. (8.20) to be separated as two terms in Eqn. (8.21).

$$r_k(i, j) = \mathrm{E}\{g_k(t)g_k^*(t-i)\} \sum_{m=1}^{M} A_m^2 \mathrm{E}\{\mathbf{c}_m^{[n]}(t)\mathbf{c}_m^{[n-j]*}(t-i)\} + \sigma_n^2 \delta_{ij} \tag{8.21}$$

Consider the term outside the summation of Eqn. (8.21) in the two cases where the source waveform is coherent and incoherent with the radar waveform. Recall that a coherent signal may be due to an echo from a point scatterer, while an incoherent signal can arise from an independent interference source. In the former case, the waveform is perfectly correlated from pulse to pulse, excluding channel effects, while in the latter case it may be completely uncorrelated. These two extreme cases are reflected in Eqn. (8.22).

$$\mathrm{E}\{g_k(t)g_k^*(t-i)\} = \begin{cases} 1 & \text{Coherent} \\ \delta(i) & \text{Incoherent} \end{cases} \tag{8.22}$$

It is shown in Appendix B that the space-time second-order statistics of mode m can be expressed as the product of its temporal and spatial ACS. The space-time separability property of the GWM is consistent with experimental measurements made on individual propagation modes in Chapter 7, and allows the term in the summation of Eqn. (8.21) to be written as Eqn. (8.23).

$$\mathrm{E}\{\mathbf{c}_m^{[n]}(t)\mathbf{c}_m^{[n-j]*}(t-i)\} = \mathrm{E}\{\mathbf{c}_m^{[n]}(t)\mathbf{c}_m^{[n]*}(t-i)\} \times \mathrm{E}\{\mathbf{c}_m^{[n]}(t)\mathbf{c}_m^{[n-j]*}(t)\} \tag{8.23}$$

Space-time separability is an inherent property of the GWM. This property is related to the use of a *scalar-type* AR process, and holds for all AR model orders. Use of a scalar-type multi-channel AR process also implies that the temporal ACS of a signal mode is identical in all antenna elements $n = 1, \dots, N$. Such a model conforms with the spatial homogeneity assumption experimentally validated for isolated ionospheric modes in Chapter 7.

For the adopted first-order AR model, it can be readily shown that the ACS structure of each signal mode is completely defined by the complex-valued AR parameter tuple $\{w_m, z_m\}$, and scaled by the power of the mode A_m^2. Specifically, the temporal ACS structure of mode m is given by $\mathrm{E}\{\mathbf{c}_m^{[n]}(t)\mathbf{c}_m^{[n]*}(t-i)\} = z_m^i$, while the spatial ACS structure is given by $\mathrm{E}\{\mathbf{c}_m^{[n]}(t)\mathbf{c}_m^{[n-j]*}(t)\} = w_m^j$. Substituting these functions into Eqn. (8.23) yields the space-time ACS structure of the channel for mode m in Eqn. (8.24).

$$\mathrm{E}\{\mathbf{c}_m^{[n]}(t)\mathbf{c}_m^{[n-j]*}(t-i)\} = z_m^i w_m^j \tag{8.24}$$

Substituting Eqn. (8.24) and Eqn. (8.22) into Eqn. (8.21), the space-time ACS of the received data can be expressed in the form of Eqn. (8.25), where $h_m = A_m^2$ is defined as the power of mode m. An incoherent source only allows the spatial statistics of the HF channel to be observed as the space-time ACS has an expected value of zero for all temporal lags $i > 0$. Moreover, such sources do not allow individual modes to be resolved in range k, so their properties must be studied collectively rather than individually.

$$r_k(i, j) = \sigma_n^2\delta_{ij} + \begin{cases} \sum_{m=1}^{M} h_m z_m^i w_m^j & \text{Coherent} \\ \delta(i)\sum_{m=1}^{M} h_m z_m^i w_m^j & \text{Incoherent} \end{cases} \tag{8.25}$$

An important aspect of the model is that the instantaneous spatial covariance matrix of a signal mode m has unit rank when averaged over a single PRI at slow-time index t. The asymptotic form of this matrix, denoted by $\mathbf{R}_m(t)$, is given by Eqn. (8.26). This occurs because each distributed signal mode appears as a CD signal with a rigid wavefront

$\mathbf{c}_m(t)$ over the relatively short PRI. This wavefront is not planar, in general, and gives to a spatial covariance matrix $\mathbf{R}_m(t)$ of *unit* rank over the PRI.

$$\mathbf{R}_m(t) = \lim_{K\to\infty} \frac{1}{K} \sum_{k=1}^{K} \mathbf{s}_{km}(t)\mathbf{s}_{km}^{\dagger}(t) = h_m \mathbf{c}_m(t)\mathbf{c}_m^{\dagger}(t) \tag{8.26}$$

On the other hand, when the slow-time varying channel vector $\mathbf{c}_m(t)$ is viewed over a relatively long time interval, such as the CPI for example, the spatial covariance matrix $\mathbf{R}_m$ of mode m takes on its statistically expected Teoplitz form in Eqn. (8.26), which is defined by the spatial ACS dimension of Eqn. (8.25).

$$\mathbf{R}_m = \lim_{P\to\infty} \frac{1}{P} \sum_{t=1}^{P} \mathbf{R}_m(t) = h_m \text{Toep}[1, w_m, w_m^2, \ldots, w_m^{N-1}] \tag{8.27}$$

The statistically expected matrix $\mathbf{R}_m$ may be written as in Eqn. (8.28), where $\mathbf{s}(\theta_m)$ is the steering vector for the mean DOA θ_m of mode m, $\odot$ denotes Hadamard (element-wise) product, and $\mathbf{B}_m$ is an $N \times N$ spatial "spreading" matrix with elements defined by $\mathbf{B}_m^{[i,j]} = \beta_m(d)^{|i-j|}$ for $i, j = 1, \ldots, N$.

$$\mathbf{R}_m = \mathbf{s}(\theta_m)\mathbf{s}^{\dagger}(\theta_m) \odot \mathbf{B}_m \tag{8.28}$$

In an analogous manner, the statistically expected temporal covariance matrix of mode m takes the form $\mathbf{Q}_m$ in Eqn. (8.29), where $\mathbf{v}(f_m) = [1, e^{j2\pi f_m}, \ldots, e^{j2\pi f_m(P-1)}]$ is a complex sinusoid at the mean Doppler frequency f_m, and $\mathbf{A}_m$ is the $P \times P$ temporal "spreading" matrix with elements defined by $\mathbf{A}_m^{[i,j]} = \alpha_m(f_p)^{|i-j|}$ for $i, j = 1, \ldots, P$.

$$\mathbf{Q}_m = \mathbf{v}(f_m)\mathbf{v}^{\dagger}(f_m) \odot \mathbf{A}_m \tag{8.29}$$

From the separability property of the GWM, the statistically expected space-time covariance matrix $\mathbf{S}_m$ of mode m is given by the Kroneker product $\otimes$ of the spatial-only and temporal-only covariance matrices in Eqn. (8.30), where $\bar{\mathbf{c}}_m$ is defined as the space-time stacked vector of $\{\mathbf{c}_m(1), \ldots, \mathbf{c}_m(P)\}$. Clearly, the GWM reverts to a monochromatic plane-wave model for the special parameter choices of $\beta_m(d) = 1$ and $\alpha_m(f_p) = 1$.

$$\mathbf{S}_m = \text{E}\{\bar{\mathbf{c}}_m \bar{\mathbf{c}}_m^{\dagger}\} = \mathbf{R}_m \otimes \mathbf{Q}_m \tag{8.30}$$

The number of terms M in modal decomposition of the space-time ACS is assumed to correspond to the number of distributed signal modes reflected by distinct ionospheric regions over a relatively quiet single-hop skywave path. At low and high geomagnetic latitudes, the ionosphere is typically more disturbed and the power spectral densities of the received modes may differ significantly from the Lorentzian profile. In this case, more than one term in the modal decomposition of the ACS may be used to model a particular propagation mode. From a signal-processing perspective, M is then considered as the number of terms in the ACS model without associating a physical interpretation.

While the mathematical form of the GWM does provide additional flexibility to model distributed signals with more sophisticated (possibly asymmetric) spectral density functions in space and time, the ability to associate model parameters to propagation modes is lost in this case when the individual modes cannot be resolved in range. Furthermore, the GWM is not designed to model angle-Doppler coupling of the mode power spectral densities, which may be present due to significant plasma convection phenomena often arising at low and high geomagnetic latitudes. Modeling strong angle-Doppler coupling requires multi-variate rather than scalar-type AR process. Such processes will not be considered here.

8.3 Parameter-Estimation Techniques

This section presents spectral and closed-form methods for estimating the modal pairs (z_m, w_m) and the associated residues h_m of all modes $m = 1, \ldots, M$ from the sample space-time ACS $\hat{r}_k(i, j)$ when the M modes are superimposed and cannot be resolved in range. The modal structure of the statistically expected ACS derived in the previous section is exploited to derive a novel subspace parameter-estimation technique referred to as matched-field (MF) MUSIC, as well as a computationally attractive closed-form parameter estimation based on a least squares optimization criterion. Prior to introducing these alternative techniques, existing methods for estimating the parameters of a mixture of first-order AR signals from the sample space-time ACS will be discussed.

8.3.1 Standard Identification Procedures

Rational (pole-zero) system models are often used to parameterize the power spectral density of random processes. The most general is referred to as the auto-regressive moving-average ARMA(p, q) model that contains p poles and q zeros (Marple 1987). In appendix C, it is shown that the superposition of M independent first-order AR processes gives rise to a signal that is statistically equivalent to an ARMA$(M, M-1)$ process. For the GWM described in this chapter, parameter estimation may therefore be couched as an ARMA$(M, M-1)$ model identification problem.

Maximum likelihood parameter-estimation techniques for ARMA model identification exist (Kay 1987), but even for one-dimensional data sequences, these iterative procedures impose high computational burdens and their convergence is not guaranteed. An ML estimator for the ARMA$(M, M-1)$ model is known (Kumaresan, Scharf, and Shaw 1986), but this procedure relies on a measurement of the *impulse response* of the process for identification rather than a statistical realization of the output data for an arbitrary input signal. In practice, it may only be possible to observe the latter. Under such circumstances, it is quite common to estimate the model parameters using the sample auto-correlation sequence (ACS) computed from the received data.

Although a sample ACS vector of *finite length* is not a sufficient statistic for ARMA model identification (Arato 1961), sub-optimum approaches based on second-order statistics are popular due to the relatively simple system of equations that need to be solved for parameter estimation. It was found by Bruzzone and Kaveh (1984) that the statistical information preserved in the sample ACS depends on the lags included and the characteristics of the ARMA process. The sample ACS is best suited to the estimation of narrowband ARMA processes. The use of more lags compared to the order of the process is also recommended up to a certain limit, as the increased statistical variability in extending the estimation equations can be outweighed by the information gain accompanying the inclusion of later lags.

The extended set of Yule-Walker equations are the backbone of many ARMA parameter estimation techniques (Cadzow 1982). The over-determined Yule-Walker technique proposed in Porat and Friedlander (1985) makes use of the unbiased sample ACS to estimate the AR parameters of the ARMA model first. The MA parameters can be estimated after filtering the original data by the inverse of the estimated AR transfer function (Marple 1987). This non-iterative least squares procedure for ARMA model identification solves for the AR and MA parameters *separately* rather than *jointly*, and is therefore sub-optimal even with respect to the sample ACS upon which it is based.

To best utilize the statistical information preserved in the sample ACS, the ARMA model parameters that provide the best least-squares fit to the unbiased sample lags need to be estimated jointly. As found by Beex and Scharf (1981), this approach leads to nonlinear equations. The least-squares criterion may be modified to estimate the model parameters analytically using a technique similar to Prony's method (de Prony 1795), albeit with inferior performance to the true least squares estimator. A computationally attractive closed-form procedure that jointly solves for the ARMA model parameters yielding the *exact* least squares fit to the unbiased sample ACS has not been proposed to date for the general case of either one- or two-dimensional data.

However, an explicit connection exists between the problem of jointly estimating the parameters of an ARMA(M, $M-1$) process that yield the best least squares fit to an unbiased sample ACS and the problem of estimating the parameters of superimposed exponentially damped complex sinusoids in additive zero-mean Gaussian noise. Although these problems are mathematically related, an important distinction is that parameter estimation is performed on the data samples in the latter case, while estimation is performed on the sample covariances in the former case. Statistical noise due to finite sample effects that corrupt the ACS measurements are often considered to be zero-mean Gaussian distributed. This assumption holds asymptotically for the sample lags by the central limit theorem.

An exact least squares procedure exists for estimating superimposed exponential signals in noise (Bresler and Macovski 1986). Perhaps less known is that this mathematical technique is also applicable to the problem of estimating the parameters of an ARMA(M, $M-1$) model using the sample ACS. However, for the case of two-dimensional (space-time) data, there is no direct generalization of this approach due to the lack of a fundamental theorem of algebra for polynomials in more than one variable. The two-dimensional modal analysis problem was studied by Clark and Scharf (1994), but problems were encountered in pairing the modes. Other methods that decompose the inherently 2-D problem into two 1-D problems have been developed, as in Sacchini, Steedly, and Moses (1993) and Hua (1992), but none of these estimators are minimizers of the *two-dimensional* least squares criterion function.

In this section, the analogy between the two problems is traced further to arrive at a matrix structure that confines distributed signals to a *single* subspace dimension. Under relatively mild conditions, a noise-only subspace can be shown to exist. This in turn permits the application of a generalized "matched-field" (MF)-MUSIC procedure for parameter estimation. Importantly, the MF-MUSIC algorithm can be extended to two dimensions to solve the space-time parameter estimation problem in a way that allows the modes to be paired automatically. In this case, the spectrum manifold is parameterized by the mean and spread parameters of the propagation channel in both the angle and Doppler dimensions, as described in Fabrizio, Gray, and Turley (2000).

A four-dimensional manifold is not highly desirable from a computational standpoint. An alternative closed-form method that separates the 2-D least squares problem into two 1-D least squares problems and then uses a statistically consistent method for pairing the modes across the two dimensions is also presented. This method can be used to jointly estimate the 2-D ARMA(M, $M-1$) model parameters in optimal least squares fashion using the sample space-time ACS. Alternatively, it may be applied directly to data samples for the purpose of two-dimensional modal analysis based on maximum likelihood.

8.3.2 Matched-Field MUSIC Algorithm

A potential problem with using subspace techniques based on the data covariance matrix for ID source parameter estimation is that such sources often occupy the full rank of the covariance matrix. For the case of a single ID source with small angle or Doppler spread, an effective signal subspace of low rank may be defined to allow the use of MUSIC-like estimation methods. However, this approximation breaks down when multiple ID sources or signal modes with possibly large spreads are present, as in this case, the noise subspace of the covariance matrix becomes degenerate. In this section, alternative matrix structures are developed such that an ID signal mode described by the proposed first-order AR model occupies a single subspace dimension. This property allows subspace-based parameter-estimation techniques to be applied when multiple ID signal modes with possibly large angle or Doppler spreads are present.

The sample ACS $\bar{r}(i, j)$ used for parameter estimation is calculated by first averaging the sum of lagged products in Eqn. (8.31) over a CPI to form $\hat{r}(i, j)$, where $N_t = P - i + 1$ and $N_s = N - j + 1$. Averaging of these estimates is then performed across the different CPI to form the mean sample ACS $\bar{r}(i, j)$ over the period of data collection. The subscript k has been dropped as attention is restricted to a single range cell at a time.

$$\hat{r}_k(i, j) = \frac{1}{N_t N_s} \sum_{t=1}^{N_t} \sum_{d=1}^{N_s} \mathbf{x}_k^{[d]}(t) \mathbf{x}_k^{[d+j]*}(t+i) \quad \begin{cases} i = 0, \ldots, L_t - 1 \\ j = 0, \ldots, L_s - 1 \end{cases} \tag{8.31}$$

To introduce the alternative two-dimensional (space-time) parameter-estimation technique, we first define the $(L_s - P_s + 1) \times P_s$ matrix $\mathbf{C}(i)$ and the related $(L_t - P_t + 1)(L_s - P_s + 1) \times P_t P_s$ block matrix $\mathbf{D}$ in Eqn. (8.32), where $M < P_t < L_t$ and $M < P_s < L_s$.

$$\mathbf{C}(i) = \begin{bmatrix} \bar{r}(i, P_s - 1) & \cdots & \bar{r}(i, 1) & \bar{r}(i, 0) \\ \bar{r}(i, P_s) & \cdots & \bar{r}(i, 2) & \bar{r}(i, 1) \\ \vdots & & & \\ \bar{r}(i, L_s - 1) & & & \end{bmatrix} \quad \mathbf{D} = \begin{bmatrix} \mathbf{C}(P_t - 1) & \cdots & \mathbf{C}(1) & \mathbf{C}(0) \\ \mathbf{C}(P_t) & \cdots & \mathbf{C}(2) & \mathbf{C}(1) \\ \vdots & & & \\ \mathbf{C}(L_t - 1) & & & \end{bmatrix} \tag{8.32}$$

Assuming no model errors and neglecting additive noise for the moment, the mean sample ACS tends to its statistically expected form $\bar{r}(i, j) = \sum_{m=1}^{M} h_m z_m^i w_m^j$ for a large number averages. In this case, the matrix $\mathbf{D}$ may be factorized as $\mathbf{D} = \mathbf{FG}$, where the $(L_t - P_t + 1)(L_s - P_s + 1) \times M$ matrix $\mathbf{F}$ is defined by Eqn. (8.33),

$$\mathbf{F} = [h_1 \mathbf{z}_1^{'} \otimes \mathbf{w}_1^{'} \cdots h_m \mathbf{z}_M^{'} \otimes \mathbf{w}_M^{'}] \quad \begin{cases} \mathbf{z}_m^{'} = [z_m^{P_t - 1} \, z_m^{P_t} \cdots z_m^{L_t - 1}]^T \\ \mathbf{w}_m^{'} = [w_m^{P_s - 1} \, w_m^{P_s} \cdots w_m^{L_s - 1}]^T \end{cases} \tag{8.33}$$

and the $M \times P_t P_s$ matrix $\mathbf{G}$ is defined by Eqn. (8.34). The symbol $\otimes$ denotes Kronecker product in Eqn. (8.33) and Eqn. (8.34).

$$\mathbf{G} = [\mathbf{z}_1'' \otimes \mathbf{w}_1'' \cdots \mathbf{z}_M'' \otimes \mathbf{w}_M'']^{\dagger} \quad \begin{cases} \mathbf{z}_m'' = [z_m^0 \; z_m^{-1} \cdots z_m^{-(P_t-1)}]^{\dagger} \\ \mathbf{w}_m'' = [w_m^0 \; w_m^{-1} \cdots w_m^{-(P_s-1)}]^{\dagger} \end{cases} \tag{8.34}$$

Since the vectors $\mathbf{z}_m' \otimes \mathbf{w}_m'$ and $\mathbf{z}_m'' \otimes \mathbf{w}_m''$ have a Vandermonde structure and the mode parameter pairs (z_m, w_m) for $m = 1, 2, \ldots, M$ are assumed to be distinct, it follows that both matrices $\mathbf{F}$ and $\mathbf{G}$ have full rank M. As a result, the $M \times M$ Hermitian matrix $\mathbf{F}^{\dagger}\mathbf{F}\mathbf{G}\mathbf{G}^{\dagger}$ is positive definite with M eigenvalues λ_m and eigenvectors $\mathbf{u}_m$ that satisfy Eqn. (8.35).

$$(\mathbf{F}^{\dagger}\mathbf{F}\mathbf{G}\mathbf{G}^{\dagger})\mathbf{u}_m = \lambda_m \mathbf{u}_m, \quad m = 0, 1, \ldots, M \tag{8.35}$$

Pre-multiplying Eqn. (8.35) by $\mathbf{G}^{\dagger}$ leads to Eqn. (8.36), where $\mathbf{D}^{\dagger}\mathbf{D} = \mathbf{G}^{\dagger}\mathbf{F}^{\dagger}\mathbf{F}\mathbf{G}$ and $\mathbf{q}_m = \mathbf{G}^{\dagger}\mathbf{u}_m$ is an eigenvector of $\mathbf{D}^{\dagger}\mathbf{D}$ with corresponding eigenvalue λ_m.

$$(\mathbf{G}^{\dagger}\mathbf{F}^{\dagger}\mathbf{F}\mathbf{G})\mathbf{G}^{\dagger}\mathbf{u}_m = (\mathbf{D}^{\dagger}\mathbf{D})\mathbf{G}^{\dagger}\mathbf{u}_m = (\mathbf{D}^{\dagger}\mathbf{D})\mathbf{q}_m = \lambda_m \mathbf{q}_m \tag{8.36}$$

In other words, the $(P_t P_s) \times (P_t P_s)$ Hermitian matrix $\mathbf{D}^{\dagger}\mathbf{D}$ is positive semi-definite with M positive eigenvalues λ_m for $m = 1, 2, \ldots, M$ and the remaining $(P_t \times P_s) - M$ eigenvalues equal to zero. Moreover, the M principal eigenvectors $\mathbf{q}_m$ of the matrix $\mathbf{D}^{\dagger}\mathbf{D}$ are formed as a linear combination of the columns in $\mathbf{G}^{\dagger}$ which are composed of the M space-time signal vectors $\mathbf{z}_m'' \otimes \mathbf{w}_m''$. In the presence of lag estimation errors and additive noise, these properties do not hold exactly but they tend to be approximately true. To estimate the parameter pairs (z_m, w_m), the sample lags $\bar{r}(i, j)$ are used to form the matrix $\mathbf{D}$ in Eqn. (8.32). The eigen-decomposition of the Hermitian matrix $\mathbf{D}^{\dagger}\mathbf{D}$ is then expressed in terms of signal and noise subspaces in Eqn. (8.37).

$$\mathbf{D}^{\dagger}\mathbf{D} = \mathbf{Q}_s \mathbf{\Lambda}_s \mathbf{Q}_s^{\dagger} + \mathbf{Q}_n \mathbf{\Lambda}_n \mathbf{Q}_n^{\dagger} \tag{8.37}$$

In Eqn. (8.37), the $P_t P_s \times M$ matrix $\mathbf{Q}_s$ contains the M signal-subspace eigenvectors as its columns and the $M \times M$ diagonal matrix $\mathbf{\Lambda}_s$ contains the M corresponding eigenvalues. The $P_t P_s \times (P_t P_s - M)$ matrix $\mathbf{Q}_n$ and $(P_t P_s - M) \times (P_t P_s - M)$ diagonal matrix $\mathbf{\Lambda}_n$ are defined in analogous fashion and contain the noise subspace eigenvectors and eigenvalues, respectively. The approximate orthogonality between the signal vectors $\mathbf{v}(\phi) = \mathbf{z}''(\alpha, f) \otimes \mathbf{w}''(\beta, \theta)$ and the noise subspace spanned by the columns of $\mathbf{Q}_n$ is exploited to form a MUSIC-like cost function $p(\phi)$ in Eqn. (8.38), where the parameter vector $\phi = [\alpha, \beta, \theta, f]^T$. Since the manifold $\mathbf{v}(\phi)$ is parameterized by the statistical properties of the channel model $[\alpha, \beta]$ as well as the regular (deterministic) components $[\theta, f]$, this technique may be regarded as a "matched-field" (MF) version of MUSIC, and referred to as MF-MUSIC.

$$p(\phi) = \{\mathbf{v}^{\dagger}(\phi)\mathbf{Q}_n\mathbf{Q}_n^{\dagger}\mathbf{v}(\phi)\}^{-1} \tag{8.38}$$

A search for the peaks of the MF-MUSIC cost function $p(\phi)$ over a four-dimensional manifold defined by the parameter vector ϕ yields estimates of both the complex pole locations (z_m, w_m) and the pairing at the same time. Once these parameters have been estimated, the corresponding residues h_m can be estimated by a least squares fit to the sample ACS. The residues h_m for $m = 1, 2, \ldots, M$ are constrained to be real when the

ARMA(M, $M-1$) process is the sum of M independent first-order AR processes. This is because the power of the mth AR process is given by $h_m = A_m^2$. To estimate the residues, a stacked ($L_s L_t$)-dimensional vector $\tilde{\mathbf{r}}$ is defined to contain the space-time sample ACS.

$$\tilde{\mathbf{r}} = [r(0,0), \ldots, r(0, L_s-1), r(1,0), \ldots, r(1, L_s-1), \cdots, r(L_t-1, 0), \ldots, r(L_t-1, L_s-1)]^T \tag{8.39}$$

Similarly, the $L_s L_t \times M$ matrix $\tilde{\mathbf{V}}$ is defined in Eqn. (8.40) to contain the space-time signal vectors that have been estimated by the MF-MUSIC technique for mode m.

$$\tilde{\mathbf{V}} = [\mathbf{z}_1 \otimes \mathbf{w}_1 \cdots \mathbf{z}_M \otimes \mathbf{w}_M] \quad \begin{cases} \mathbf{z}_m &= [z_m^0 \, z_m^1 \cdots z_m^{(L_t-1)}]^T \\ \mathbf{w}_m &= [w_m^0 \, w_m^1 \cdots w_m^{(L_s-1)}]^T \end{cases} \tag{8.40}$$

The residue vector $\mathbf{h} = [h_1 \, h_2 \cdots h_M]^T$ is estimated as the argument that minimizes the difference between the model and sample ACS according to the least squares criterion in Eqn. (8.41). Here, the real valued vector $\mathbf{r}$ and matrix $\mathbf{V}$ are constructed from the real $\Re\{\cdot\}$ and imaginary $\Im\{\cdot\}$ parts of $\tilde{\mathbf{r}}$ and $\tilde{\mathbf{V}}$, respectively.

$$\hat{\mathbf{h}} = \arg\min_{\mathbf{h}} \|\mathbf{r} - \mathbf{V}\mathbf{h}\|^2 , \quad \mathbf{r} = \begin{bmatrix} \Re\{\tilde{\mathbf{r}}\} \\ \Im\{\tilde{\mathbf{r}}\} \end{bmatrix} , \quad \mathbf{V} = \begin{bmatrix} \Re\{\tilde{\mathbf{V}}\} \\ \Im\{\tilde{\mathbf{V}}\} \end{bmatrix} \tag{8.41}$$

The minimizing argument $\hat{\mathbf{h}}$ is given by the well-known formula in Eqn. (8.42), where $\mathbf{V}^+ = [\mathbf{V}^\dagger \mathbf{V}]^{-1} \mathbf{V}^\dagger$ is the Moore-Penrose pseudo inverse of $\mathbf{V}$.

$$\hat{\mathbf{h}} = [\mathbf{V}^\dagger \mathbf{V}]^{-1} \mathbf{V}^\dagger \mathbf{r} = \mathbf{V}^+ \mathbf{r} \tag{8.42}$$

Once all the model parameters have been estimated, the accuracy with which the model ACS fits the sample ACS computed from the data may be quantified. The modeling performance may be assessed in terms of the fitting accuracy (FA) measure defined in Eqn. (8.43).

$$\mathrm{FA} = 1 - \frac{\|\tilde{\mathbf{r}} - \tilde{\mathbf{V}}\hat{\mathbf{h}}\|^2}{\|\tilde{\mathbf{r}}\|^2} \tag{8.43}$$

The estimated ACS model may be extended to an infinite number of space-time lags using the AR recursive relation and Fourier transformed to estimate the power spectral density in the angle-Doppler domain. The resulting two-dimensional spectra provides a measure of the angle-Doppler spread of the resolved signal modes.

MF-MUSIC can also be applied to estimate the parameters of 2-D exponentially damped sinusoidal signals in noise. An important distinction between these two problems is that in the current application the matrix $\mathbf{D}$ contains the estimated data covariances or lags, while in the latter case $\mathbf{D}$ would contain the actual data samples where the signals are corrupted by additive noise. The computational complexity of evaluating the cost function over a four-dimensional manifold may be prohibitive. This motivates the search for an alternative closed-form parameter-estimation technique that does not require a search over a four-dimensional manifold.

8.3.3 Polynomial Rooting Method

To introduce the 2-D least squares method based on polynomial rooting, we define the $L_t \times L_s$ matrix $\mathbf{T}$ such that its $(i, j)^{\text{th}}$ entry equals $r(i, j)$.

$$\mathbf{T} = \begin{bmatrix} r(0,0) & r(0,1) & \cdots & r(0, L_s - 1) \\ r(1,0) & r(1,1) & \cdots & r(1, L_s - 1) \\ \vdots & & \ddots & \vdots \\ r(L_t - 1, 0) & \cdots & \cdots & r(L_t - 1, L_s - 1) \end{bmatrix} \tag{8.44}$$

Based on the modal decomposition of $r(i, j) = \sum_{m=1}^{M} h_m z_m^i w_m^j$, it is possible to factorize the matrix $\mathbf{T}$ as in Eqn. (8.45), where $\mathbf{z}_m = [z_m^0, z_m^1, \ldots, z_m^{L_t-1}]^T$, $\mathbf{w}_m = [w_m^0, w_m^1, \ldots, w_m^{L_s-1}]^\dagger$, the $L_t \times M$ matrix $\mathbf{Z} = [\mathbf{z}_1, \mathbf{z}_2, \ldots, \mathbf{z}_M]$, the $L_s \times M$ matrix $\mathbf{W} = [\mathbf{w}_1, \mathbf{w}_2, \ldots, \mathbf{w}_M]$, and the $M \times M$ diagonal matrix $\mathbf{H} = \text{diag}[h_1, h_2, \ldots, h_M]$.

$$\mathbf{T} = \sum_{m=1}^{M} h_m \mathrm{z}_m \mathrm{w}_m^\dagger = \mathbf{Z}\mathbf{H}\mathbf{W}^\dagger \tag{8.45}$$

In practice, we have an estimate $\hat{\mathbf{T}}$ containing the sample space-time lags. Even in the absence of model mismatch and additive noise, the matrix $\hat{\mathbf{T}}$ will generally have full rank despite $L_t > M$ and $L_s > M$ due to the presence of estimation errors in the sample lags. In analogous manner to Cadzow, Baseghi, and Hsu (1983), we can attempt to reduce the influence of statistical errors on the parameter estimates by taking a reduced-rank approximation of $\hat{\mathbf{T}}$ using a truncated singular value decomposition (SVD). The number of modes M is assumed to be known, but could be estimated as part of the SVD-based rank-reduction transformation in Eqn. (8.46).

$$\tilde{\mathbf{T}} = \sum_{m=1}^{M} \sigma_m \mathrm{u}_m \mathrm{v}_m^\dagger = \mathbf{U}_s \mathbf{\Sigma}_s \mathbf{V}_s^\dagger \tag{8.46}$$

Here, σ_m are defined as the principal singular values of $\hat{\mathbf{T}}$, $\mathbf{u}_m$ are the corresponding L_t-dimensional left singular vectors, and $\mathbf{v}_m$ are the L_s-dimensional right singular vectors for $m = 1, \ldots, M$. The matrices $\mathbf{U}_s = [\mathbf{u}_1, \ldots, \mathbf{u}_M]$ and $\mathbf{V}_s = [\mathbf{v}_1, \ldots, \mathbf{v}_M]$ contain the left and right singular vectors, respectively, while the $M \times M$ diagonal matrix $\mathbf{\Sigma}_s = \text{diag}[\sigma_1, \ldots, \sigma_M]$ contains the singular values. The proposed parameter-estimation method is based on finding the model parameters (h_m, z_m, w_m) $\forall m$ that provide the best least-squares fit to $\tilde{\mathbf{T}}$ by minimizing the Frobenius norm $\|\cdot\|_F$ in the criterion function of Eqn. (8.47).

$$f_{cr}(\mathbf{Z}, \mathbf{H}, \mathbf{W}) = \|\tilde{\mathbf{T}} - \mathbf{Z}\mathbf{H}\mathbf{W}^\dagger\|_F \tag{8.47}$$

This multidimensional optimization problem is separable. For a given matrix $\mathbf{Z}$, it can be shown that the arguments $\mathbf{H}$ and $\mathbf{W}$ that minimize the criterion function satisfy Eqn. (8.48), where $\mathbf{Z}^+ = (\mathbf{Z}^\dagger\mathbf{Z})^{-1}\mathbf{Z}^\dagger$ is the Moore-Penrose pseudo inverse of $\mathbf{Z}$.

$$\mathbf{H}\mathbf{W}^\dagger = (\mathbf{Z}^\dagger\mathbf{Z})^{-1}\mathbf{Z}^\dagger\tilde{\mathbf{T}} = \mathbf{Z}^+\tilde{\mathbf{T}} \tag{8.48}$$

Substitution of Eqn. (8.48) into Eqn. (8.47) yields the criterion function of Eqn. (8.49), which only depends on the matrix $\mathbf{Z}$. Here, $\mathbf{P}_z = (\mathbf{I} - \mathbf{Z}\mathbf{Z}^+)$ is the orthogonal projector onto the null space of $\mathbf{Z}^\dagger$.

$$\hat{\mathbf{Z}} = \arg\min_{\mathbf{Z}} \|(\mathbf{I} - \mathbf{Z}\mathbf{Z}^+)\tilde{\mathbf{T}}\|_F = \arg\min_{\mathbf{Z}} \|\mathbf{P}_z\tilde{\mathbf{T}}\|_F \tag{8.49}$$

Using the relation $\|Q\|_F = \mathrm{Tr}\{QQ^\dagger\}$, where $\mathrm{Tr}\{\cdot\}$ denotes the trace operator, it is possible to rewrite Eqn. (8.49) as Eqn. (8.50), since a cyclic rotation of the elements in the $\mathrm{Tr}\{\cdot\}$ operator does not affect the value of the trace.

$$\hat{\mathbf{Z}} = \arg\min_{\mathbf{Z}} \mathrm{Tr}\{\mathbf{P}_z\tilde{\mathbf{T}}\tilde{\mathbf{T}}^\dagger\mathbf{P}_z^\dagger\} = \arg\min_{\mathbf{Z}} \mathrm{Tr}\{\mathbf{P}_z^\dagger\mathbf{P}_z\tilde{\mathbf{T}}\tilde{\mathbf{T}}^\dagger\} \tag{8.50}$$

The projection matrix is symmetric $\mathbf{P}_z = \mathbf{P}_z^\dagger$ and idempotent $\mathbf{P}_z^2 = \mathbf{P}_z$, so Eqn. (8.50) may be simplified to yield the following compact expression:

$$\hat{\mathbf{Z}} = \arg\min_{\mathbf{Z}} \mathrm{Tr}\{\mathbf{P}_z\tilde{\mathbf{T}}\tilde{\mathbf{T}}^\dagger\} \tag{8.51}$$

Assuming the number of modes M is known or can be estimated by a suitable method using the singular values of $\hat{\mathbf{T}}$, solving for $\mathbf{Z}$ is more tractable (Stoica and Sharman 1990b) when the projection matrix is reparameterized in terms of an $L_t \times (L_t - M)$ Toeplitz matrix $\mathbf{A}$ defined by,

$$\mathbf{A} = \begin{bmatrix} a_M & a_{M-1} & \cdots & a_0 & & \cdots & 0 \\ & \ddots & \ddots & & & \ddots & \\ 0 & & & a_M & a_{M-1} & \cdots & a_0 \end{bmatrix}^\dagger \tag{8.52}$$

with elements $a_0, a_1, \ldots, a_M$ that are the coefficients of the characteristic polynomial $p(z)$.

$$p(z) = a_0 z^M + a_1 z^{M-1} + \cdots + a_M = \prod_{m=1}^{M}(z - z_m) \tag{8.53}$$

Since the matrix $\mathbf{A}$ has full rank $(L_t - M)$, and by construction it is apparent that $\mathbf{A}^\dagger\mathbf{Z} = \mathbf{0}$, it follows that the columns of $\mathbf{A}$ do in fact form a basis to span the null space of $\mathbf{Z}^\dagger$. As a result, the projection matrix can be reparameterized as $\mathbf{P}_z = \mathbf{A}(\mathbf{A}^\dagger\mathbf{A})^{-1}\mathbf{A}^\dagger$ which leads to a criterion function in terms of the polynomial coefficient vector $\mathbf{a} = [a_0, a_1, \ldots, a_M]^T$.

$$\hat{\mathbf{a}} = \arg\min_{\mathbf{a}} \mathrm{Tr}\{\mathbf{A}(\mathbf{A}^\dagger\mathbf{A})^{-1}\mathbf{A}^\dagger\tilde{\mathbf{T}}\tilde{\mathbf{T}}^\dagger\} \tag{8.54}$$

As pointed out in Krim and Viberg (1996), the coefficient vector $\mathbf{a} = [a_0, a_1, \ldots, a_M]^T$ that minimizes Eqn. (8.54) may be estimated analytically by sequentially solving the two quadratic optimization problems in Eqn. (8.55) subject to the linear constraint $\hat{\mathbf{a}}^\dagger\mathbf{e} = 1$, where $\mathbf{e} = [1, 0, \cdots, 0]^T$ ensures a non-trivial solution. The least squares estimates of the temporal poles $\hat{z}_m$ for $m = 1, 2, \ldots, M$ are then given by the roots of the estimated characteristic polynomial $\hat{p}(z)$. The polynomial roots can be found using the algorithm of Aurand (1987), which is guaranteed to converge.

$$\begin{aligned} &(1)\ \hat{\mathbf{a}} = \arg\min_{\mathbf{a}} \mathrm{Tr}\{\mathbf{A}\mathbf{A}^\dagger\tilde{\mathbf{T}}\tilde{\mathbf{T}}^\dagger\} \\ &(2)\ \hat{\mathbf{a}} = \arg\min_{\mathbf{a}} \mathrm{Tr}\{\mathbf{A}(\hat{\mathbf{A}}^\dagger\hat{\mathbf{A}})^{-1}\mathbf{A}^\dagger\tilde{\mathbf{T}}\tilde{\mathbf{T}}^\dagger\} \end{aligned} \tag{8.55}$$

The elements of $\hat{\mathbf{a}}$ that solve the first quadratic problem in Eqn. (8.55) are used to form the matrix $\hat{\mathbf{A}}$ in the second quadratic problem according to Eqn. (8.52). In other words, $(\hat{\mathbf{A}}^{\dagger}\hat{\mathbf{A}})^{-1}$ is a *data-dependent* weighting matrix in the second quadratic problem. By using the truncated singular value decomposition of $\hat{\mathbf{T}}$, it can be shown that more than two iterations are not needed. While this has been demonstrated in simulations for the DOA estimation problem (Stoica and Sharman 1990b), this desirable feature of the two-step method described on previous page will be demonstrated using experimental data later in the chapter.

A more complex set of constraints can be applied in certain situations to ensure that the roots corresponding to the estimated polynomial coefficient vector $\hat{\mathbf{a}}$ lie on the unit circle. These constraints are useful for the direction-of-arrival estimation problem considered in Stoica and Sharman (1990b), where the received signals are assumed to be plane waves. Such constraints should *not* be applied in this instance because the received signals are assumed to be distributed in angle and Doppler. The width of the Doppler distribution for mode m is determined by the distance between the pole $\hat{z}_m$ and the unit circle. In other words, the modulus of $\hat{z}_m$ provides an estimate of the mode Doppler spread, while the argument of $\hat{z}_m$ estimates the mean Doppler shift.

As shown in Bresler and Macovski (1986), both quadratic problems in Eqn. (8.55) may be solved using the same technique as that for deriving linear prediction coefficients by the covariance method. The solution of the first problem in Eqn. (8.55) is given by

$$\hat{\mathbf{a}} = \frac{[\sum_{j=0}^{L_s-1}\mathbf{Y}^{\dagger}(j)\mathbf{Y}(j)]^{-1}\mathbf{e}}{\mathbf{e}^T[\sum_{j=0}^{L_s-1}\mathbf{Y}^{\dagger}(j)\mathbf{Y}(j)]^{-1}\mathbf{e}}, \quad \mathbf{Y}(j) = \begin{bmatrix} \hat{r}(M, j) & \cdots & \hat{r}(1, j) & \hat{r}(0, j) \\ \hat{r}(M+1, j) & \cdots & \hat{r}(2, j) & \hat{r}(1, j) \\ \vdots & & & \\ \hat{r}(L_t - 1, j) & & & \end{bmatrix} \tag{8.56}$$

The estimated coefficients $\hat{\mathbf{a}} = [\hat{a}_0 \cdots \hat{a}_M]^T$ from this first step are then used to form the matrix $\hat{\mathbf{A}}$ according to Eqn. (8.52). The second quadratic problem in Eqn. (8.55) can then be solved in similar fashion to the first:

$$\hat{\mathbf{a}} = \frac{[\sum_{j=0}^{L_s-1}\mathbf{Y}^{\dagger}(j)(\hat{\mathbf{A}}^{\dagger}\hat{\mathbf{A}})^{-1}\mathbf{Y}(j)]^{-1}\mathbf{e}}{\mathbf{e}^T[\sum_{j=0}^{L_s-1}\mathbf{Y}^{\dagger}(j)(\hat{\mathbf{A}}^{\dagger}\hat{\mathbf{A}})^{-1}\mathbf{Y}(j)]^{-1}\mathbf{e}} \tag{8.57}$$

In analogous manner, the argument $\mathbf{W}$ minimizing the criterion function is estimated as Eqn. (8.58), where $\mathbf{P}_w = (\mathbf{I} - \mathbf{W}\mathbf{W}^{+})$ is the orthogonal projector onto the null space of $\mathbf{W}^{\dagger}$ and $\mathbf{W}^{+} = (\mathbf{W}^{\dagger}\mathbf{W})^{-1}\mathbf{W}^{\dagger}$ is the Moore-Penrose pseudo inverse of $\mathbf{W}$.

$$\hat{\mathbf{W}} = \arg\min_{\mathbf{W}} \operatorname{Tr}\{\mathbf{P}_w \tilde{\mathbf{T}}^{\dagger}\tilde{\mathbf{T}}\} \tag{8.58}$$

In this case, the projection matrix is reparameterized as $\mathbf{P}_w = \mathbf{B}(\mathbf{B}^{\dagger}\mathbf{B})^{-1}\mathbf{B}^{\dagger}$ where,

$$\mathbf{B} = \begin{bmatrix} b_M & b_{M-1} & \cdots & b_0 & & \cdots & 0 \\ & \ddots & \ddots & & & \ddots & \\ 0 & & & b_M & b_{M-1} & \cdots & b_0 \end{bmatrix}^{\dagger} \tag{8.59}$$

is constructed from the coefficients of the *spatial* characteristic polynomial $q(w)$.

$$q(w) = b_0 w^M + b_1 w^{M-1} + \cdots + b_M = \prod_{m=1}^{M} (w - w_m) \tag{8.60}$$

Similarly, the polynomial coefficient vector $\mathbf{b} = [b_0, b_1, \ldots, b_M]^T$ is estimated by solving the following two quadratic optimization problems subject to the linear constraint $\hat{\mathbf{b}}^\dagger \mathbf{e} = 1$.

$$\begin{aligned} &(1)\ \hat{\mathbf{b}} = \arg\min_{\mathbf{b}}\ \mathrm{Tr}\{\mathbf{B}\mathbf{B}^\dagger \tilde{\mathbf{T}}^\dagger \tilde{\mathbf{T}}\} \\ &(2)\ \hat{\mathbf{b}} = \arg\min_{\mathbf{b}}\ \mathrm{Tr}\{\mathbf{B}(\hat{\mathbf{B}}^\dagger \hat{\mathbf{B}})^{-1}\mathbf{B}^\dagger \tilde{\mathbf{T}}^\dagger \tilde{\mathbf{T}}\} \end{aligned} \tag{8.61}$$

The vector $\hat{\mathbf{b}}$ that solves the problem in line (1) of Eqn. (8.61) is given by

$$\hat{\mathbf{b}} = \frac{[\sum_{i=0}^{L_t-1} \mathbf{X}^\dagger(i)\mathbf{X}(i)]^{-1}\mathbf{e}}{\mathbf{e}^\dagger[\sum_{i=0}^{L_t-1} \mathbf{X}^\dagger(i)\mathbf{X}(i)]^{-1}\mathbf{e}}, \quad \mathbf{X}(i) = \begin{bmatrix} \hat{r}(i, M) & \cdots & \hat{r}(i,1)\hat{r}(i,0) \\ \hat{r}(i, M+1) & \cdots & \hat{r}(i,2)\hat{r}(i,1) \\ \vdots & & \\ \hat{r}(i, L_s - 1) & & \end{bmatrix} \tag{8.62}$$

The polynomial coefficients in $\hat{\mathbf{b}}$ estimated in this first step are substituted for the true values to form the matrix $\hat{\mathbf{B}}$ according to Eqn. (8.59). This matrix is needed for the problem in line (2) of Eqn. (8.61). The polynomial coefficient vector $\hat{\mathbf{b}}$ that solves this problem is calculated as

$$\hat{\mathbf{b}} = \frac{[\sum_{i=0}^{L_t-1} \mathbf{X}^\dagger(i)(\hat{\mathbf{B}}^\dagger\hat{\mathbf{B}})^{-1}\mathbf{X}(i)]^{-1}\mathbf{e}}{\mathbf{e}^\dagger[\sum_{i=0}^{L_t-1} \mathbf{X}^\dagger(i)(\hat{\mathbf{B}}^\dagger\hat{\mathbf{B}})^{-1}\mathbf{X}(i)]^{-1}\mathbf{e}} \tag{8.63}$$

The spatial poles $\hat{w}_m$ are derived by finding the roots of the estimated polynomial $\hat{q}(w)$. Note that aside from the initial SVD and final polynomial rooting, this algorithm estimates the temporal poles z_m and spatial poles w_m of the M modes from the sample space-time lags in closed form based on the least squares criterion.

Once the M complex roots are estimated in each data dimensions, it is necessary to correctly pair the temporal poles z_m and spatial poles w_m associated with each mode prior to estimating the residues h_m. A brute force pairing method is to try all possible combinations and choose the one which leads to the smallest error according to the criterion function in Eqn. (8.47). For M modes, the number of combinations to be tried is given by $M!/(2(M-2)!)$, which becomes quite large for M greater than five.

An alternative way of pairing the poles exploits the approximate orthogonality between the signal vectors $\mathbf{z}''_m \otimes \mathbf{w}''_m$ and the noise subspace spanned by the columns of $\mathbf{Q}_n$ described in Section 8.3.2. In other words, the spatial pole w_m for $m = 1, \ldots, M$ paired to a particular reference temporal pole z_{m_r} is the one that minimizes the cost $c(m)$ in Eqn. (8.64).

$$c(m) = (\mathbf{z}''_{m_r} \otimes \mathbf{w}''_m)^\dagger \mathbf{Q}_n \mathbf{Q}_n^\dagger (\mathbf{z}''_{m_r} \otimes \mathbf{w}''_m) \tag{8.64}$$

8.4 Real-Data Application

The estimation of space-time AR model parameters when only one propagation mode is present in a particular range cell is relatively straightforward and does not require the techniques presented in the previous section. However, the CSF data analyzed in the Chapters 6 and 7 contains a practical example where two sporadic-E modes are known to be present from an oblique incidence ionogram, but could not be resolved in delay due to the lower range resolution of the CSF data. As these modes are likely to have overlapped angular and Doppler power spectral densities, estimation of the individual mode parameters from the received signal mixture is no longer straightforward.

In this case, alternative parameter-estimation techniques are required to identify the mode spectral densities. This section presents a comparison of MF-MUSIC and the closed-form least squares technique when both methods are applied to real data containing two unresolved sporadic-E modes. Performance is measured in terms of the accuracy with which the estimated model ACS can represent the actual sample ACS. This comparison provides useful information regarding both the validity of the assumed space-time ACS model, as well as the robustness of the proposed parameter-estimation techniques.

8.4.1 Closed-Form Least Squares

The sample ACS $\hat{r}(i, j)$ in Eqn. (8.65) is formed by averaging the sum of lagged products within a CPI spaced apart by $i\Delta_t/f_p$ seconds and $j\Delta_d$ meters, where Δ_t and Δ_d denote an integer number of PRI and receivers, respectively. The number of temporal and spatial averages in a CPI containing P pulses and N receivers is given by $N_t = P - i\Delta_t + 1$ and $N_s = N - j\Delta_d + 1$, respectively. These sample lags $\hat{r}(i, j)$ are then averaged over all CPI in the data set to form the mean ACS $\bar{r}(i, j)$ over the period of data collection. The mean ACS is used for parameter estimation.

$$\hat{r}_k(i, j) = \frac{1}{N_t N_s} \sum_{t=1}^{N_t} \sum_{d=1}^{N_s} \mathbf{x}_k^{[d]}(t) \mathbf{x}_k^{[d+j\Delta d]*}(t + i\Delta t) \quad \begin{cases} i = 0, \ldots, L_t - 1 \\ j = 0, \ldots, L_s - 1 \end{cases} \tag{8.65}$$

The temporal lags were computed at an interval of 0.1 seconds (i.e., Δ_t = 6 PRI for f_p = 60 Hz), while the distance between spatial lags was 84 m (i.e., Δd = 1 sub-array spacing for the Jindalee array). Using all of the $P = 256$ PRI and $N = 30$ receivers available per CPI, a total number of $L_t = L_s = 30$ ACS lags were estimated in each dimension. The lags were estimated in range cell $k = 16$, which contains the two unresolved sporadic-E modes. The estimates $\hat{r}(i, j)$ for each dwell were then averaged over 47 CPI to form the mean ACS $\bar{r}_k(i, j)$ used in further processing.

Table 8.1 lists the parameters estimated using the closed-form least squares technique assuming $M = 2$ modes. The pairing of temporal and spatial parameters was performed

Mode	Δf_m, Hz	$\alpha(\Delta t)$	θ_m, deg	$\beta(\Delta d)$	h_m	$m = 1$, dB	$m = 2$, dB
$m_r = 1$	0.46	0.998	22.2	0.920	9.46	−56.1	−39.3
$m_r = 2$	0.39	0.997	21.7	0.963	5.23	−40.3	−62.7

TABLE 8.1 Distributed signal mode parameters estimated from the mean sample ACS $\bar{r}_k(i\Delta t, j\Delta d)$ using real data known to contain a superposition of two sporadic-E modes.

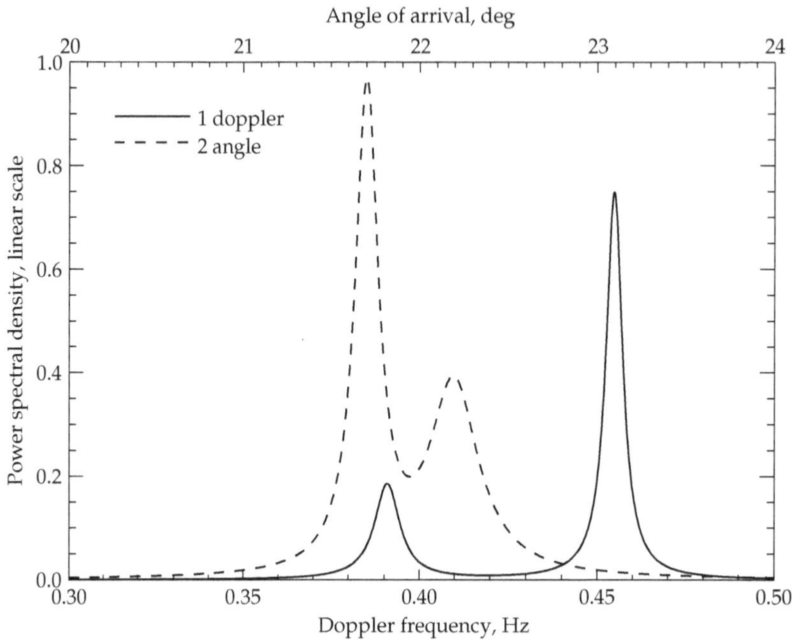

FIGURE 8.1 The spatial and temporal power densities of the two modes in range cell $k = 16$. These densities were calculated by extending the model ACS and performing spectral analysis. The model parameters were estimated from a finite length experimental data sample ACS. The linear scale on the vertical axis is uncalibrated as the absolute signal power is unknown. The angle spectra (dashed line) is associated with the top horizontal axis, while the Doppler spectra (solid line) is associated with the bottom horizontal axis.

by evaluating the function $c(m)$ in Eqn. (8.64) for each temporal pole z_{m_r} and spatial pole w_m combination, where $\{m_r, m\} = 1, 2$. This step is required because the roots of the estimated polynomials $\hat{p}(z)$ and $\hat{q}(w)$ have no ordering. The values of $c(m)$ in Table 8.1 were calculated for $P_s = P_t = M + 1$ to guarantee the existence of a noise-only subspace in the absence of modeling errors. The poles are associated as the pair that yields the lowest values of this function. In other words, the first temporal pole $m_r = 1$ is paired to the first spatial pole $m = 1$. As expected, the combination $m_r = 2$ and $m = 2$ also minimizes $c(m)$ to form the other pair.

The mean cone angle of arrival θ_m and Doppler shift Δf_m estimated for the two modes in Table 8.1 are very similar to the values previously estimated by the (standard) space-time MUSIC algorithm in Figure 6.31. However, the least squares technique assumes a statistical signal model and additionally estimates the temporal and spatial spread parameters $\{\alpha_m(\Delta t), \beta_m(\Delta d)\}_{m=1,2}$ of the two modes. These parameters are used to estimate the power spectral densities of the two modes in Doppler and angle. Using Eqn. (C.7) from Appendix C and the model parameters in Table 8.1, the power spectral densities estimated for the two sporadic-E modes are plotted in Figure 8.1.

Curve 1 shows that the Doppler spread on both modes is in the order of 0.1 Hz. Previous results for the Doppler spread of sporadic-E modes on oblique mid-latitude

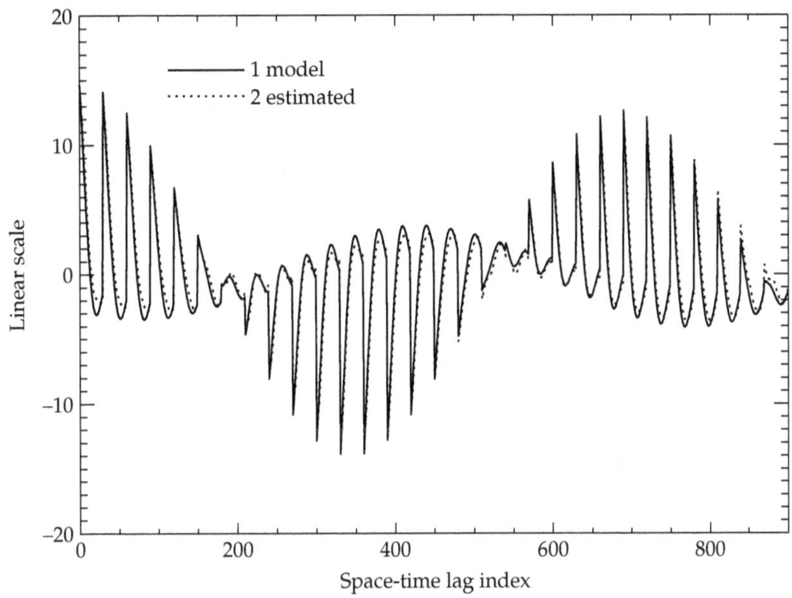

FIGURE 8.2 Real component of the space-time ACS predicted by the model compared with the real component of the space-time ACS estimated from experimental data.

paths are difficult to find, although these results are not dissimilar to the Doppler spread measurements of 0.18, 0.07, and 0.16 Hz published in Shepherd and Lomax (1967) for clean single modes propagated by the F-region over a 4100-km mid-latitude path. The angular spread is approximately 0.2 degrees, which agrees well with measurements of around 0.4 degrees made for single-hop E- and F-region modes in Balser and Smith (1962) for a 1566-km mid-latitude path using a very wide aperture array (2000 ft).

The estimated model ACS provides a fitting accuracy of 97 percent to the experimental sample ACS. Figures 8.2 and 8.3 respectively compare the real and imaginary parts of the modeled space-time ACS with the mean sample ACS. In these figures, the $L_t L_s = 900$ space-time lags have been stacked in the vector $\tilde{\mathbf{r}}$ of Eqn. (8.39). As expected for such a high-fitting accuracy, the model ACS closely resembles the ACS computed from real data across all 900 points using only $M = 2$ modes with five parameters per mode. This remarkable result serves to illustrate the accuracy of the model assumed to describe the space-time ACS of signal modes reflected by the ionosphere over a relatively quiet mid-latitude path, and the effectiveness of the closed-form least-squares parameter-estimation technique for model identification when the ACS arises from a mixture of unresolved (space-time) distributed signal modes.

The 97 percent model fit was achieved using the *two-step* quadratic minimization procedure in Eqns. (8.55) and (8.61). It is possible to repeat this procedure more than twice using an updated estimate of the polynomial coefficients in the weighting matrix at each iteration. Figure 8.4 shows the value of the criterion function to be minimized at different iterations for the spatial and temporal parameter-estimation problems. It is

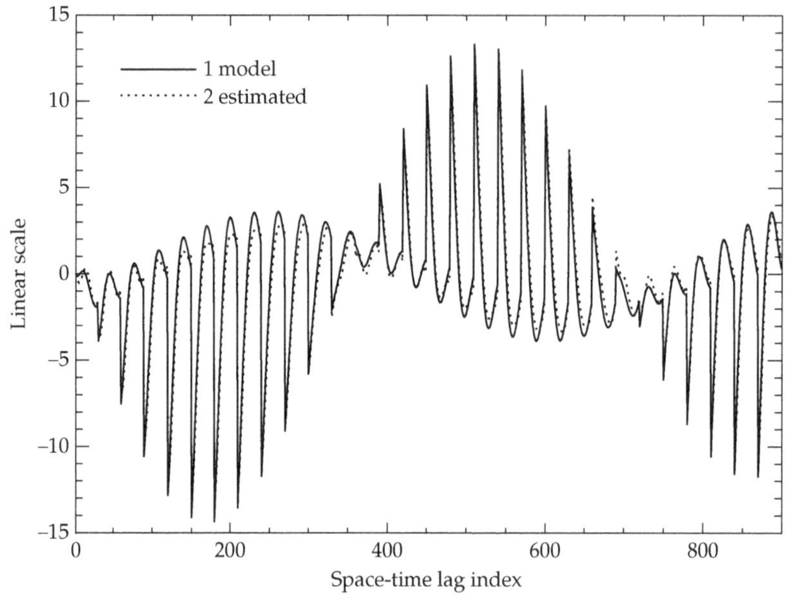

FIGURE 8.3 Imaginary component of the space-time ACS predicted by the model compared with the imaginary component of the space-time ACS estimated from experimental data. © Commonwealth of Australia 2011.

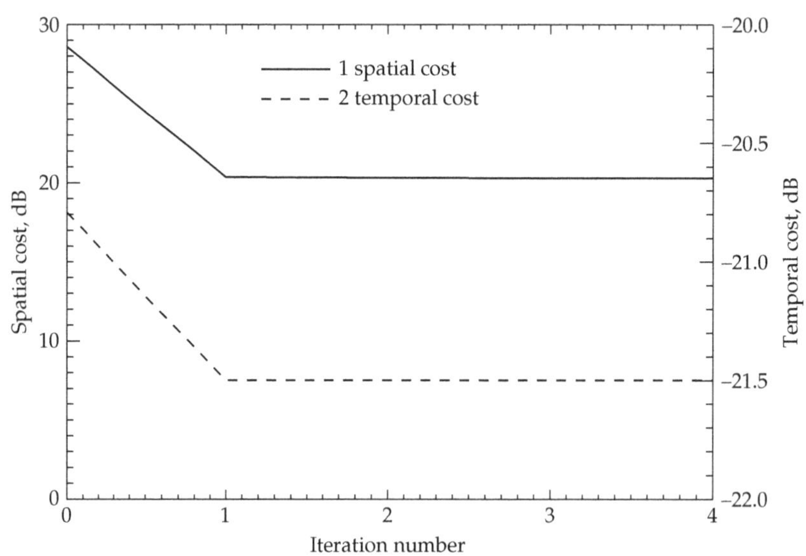

FIGURE 8.4 The spatial cost $\mathrm{Tr}\{\mathbf{P}_w\tilde{\mathbf{T}}^{\dagger}\tilde{\mathbf{T}}\}$ and temporal cost $\mathrm{Tr}\{\mathbf{P}_z\tilde{\mathbf{T}}\tilde{\mathbf{T}}^{\dagger}\}$ as a function of iteration number for the proposed closed-form least squares parameter-estimation technique. The minimum is attained after two iterations, as expected. © Commonwealth of Australia 2011.

evident that only two iterations are required to reach the minimum value in each case.

The reduction between the first and second iteration corresponds to the improvement of the described technique relative to Prony's method, which is traditionally used to estimate the parameters of damped sinusoidal signals in noise. While the need for only two steps has been demonstrated by other investigators based on theoretical analysis and simulated data, the author is not aware of any practical validation of this class of method using real data.

8.4.2 Subspace-Based Approach

The MF-MUSIC algorithm requires specification of the spatial dimension P_s and temporal dimension P_t, which set the size of the data matrix $\mathbf{D}$ in Eqn. (8.32). Like the traditional MUSIC algorithm, larger dimensions are desirable to resolve closely spaced signals, but in practice this often means that fewer data "snapshots" are available for estimation and this increases the variance of the parameter estimates. Values of $P_s = P_t = 6$ were selected for the current application as a compromise between resolution and variance.

Figures 8.5 and 8.6 show the MF-MUSIC spectra $p(\phi)$ plotted in the angle-Doppler domain $(\theta, \Delta f)$ assuming $M = 2$ modes according to Eqn. (8.38) when the parameter vector $\phi = [\alpha(\Delta t), \beta(\Delta d), \theta, \Delta f]^T$ has dampening terms $[\alpha(\Delta t), \beta(\Delta d)]$ set to values estimated for modes 1 and 2 in Table 8.1, respectively. Both spectra exhibit the same peak *coordinates* in mean Doppler frequency and DOA. The MF-MUSIC peak location estimates of $[\Delta f, \theta] = [0.46, 22.2]$ and $[\Delta f, \theta] = [0.39, 21.7]$ are practically identical to values estimated by the least squares technique for modes 1 and 2 in Table 8.1, respectively.

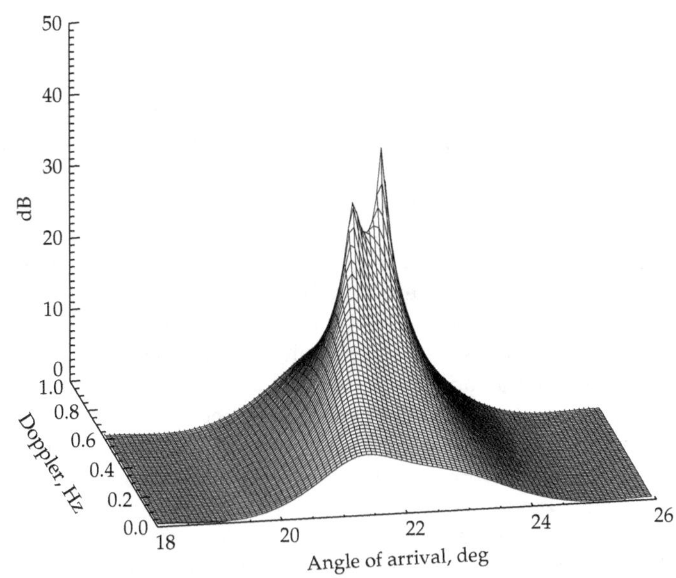

FIGURE 8.5 MF-MUSIC spectrum evaluated over the angle-Doppler domain for $[\alpha, \beta] = [0.998, 0.920]$ (i.e., dampening parameters matched to $m = 1$). The larger peak (40.5 dB) has coordinates $[\Delta f, \theta] = [0.46, 22.2]$, while the smaller peak (33.1 dB) has coordinates $[\Delta f, \theta] = [0.39, 21.7]$. © Commonwealth of Australia 2011.

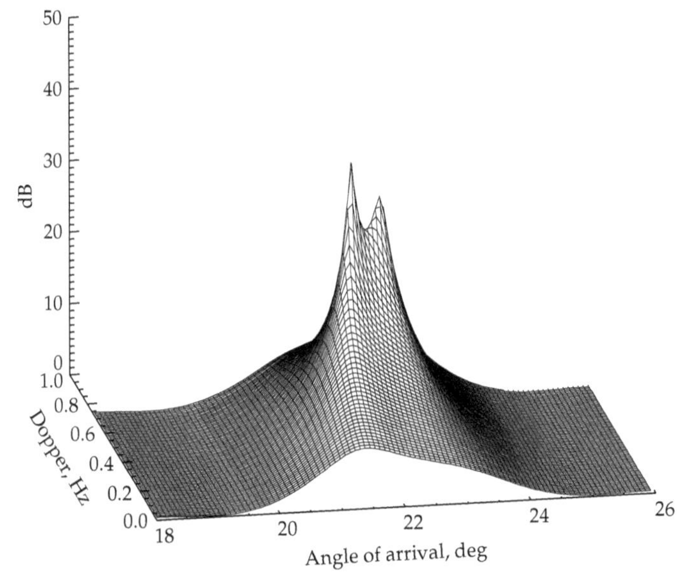

FIGURE 8.6 MF-MUSIC spectrum evaluated over the angle-Doppler domain for $[\alpha, \beta] = [0.998, 0.920]$ (i.e., dampening parameters matched to $m = 2$). The coordinates of both peaks are the same as those quoted for Figure 8.5 but the amplitudes of the peaks have changed from 40.5 dB and 33.1 dB to 31.3 dB and 38.5 dB, respectively.

The *magnitude* of the MF-MUSIC peaks measures the degree of orthogonality between the estimated noise subspace and the modeled signal vector. In other words, a higher peak indicates a tighter match between the hypothesized model ACS on the parameterized manifold and the measured sample ACS. It is emphasized that the peak magnitudes are not related to the strengths of the signals. Rather, the peak magnitudes indicate how close to being orthogonal the estimated noise subspace is to the modeled signal vector. The peak magnitude therefore provides a measure of how well matched the modeled signal vector is to the signal subspace estimated from the real data.

In Figure 8.5, the dampening factors have been set to those estimated for the first mode $[\alpha, \beta] = [0.998, 0.920]$. In this case, the magnitude of the peak at the angle-Doppler coordinates of the first mode is relatively higher (40.5 dB) compared to that of the second mode (33.1 dB). This illustrates the effect of matching the channel spread parameters of the manifold to a particular mode. When the manifold is "focused" onto the channel spread parameters of the second mode $[\alpha, \beta] = [0.997, 0.963]$ in Figure 8.6, the magnitude of the peak at the angle-Doppler coordinates of mode 2 *increases* from 33.1 to 38.5 dB, while that of the mode 1 *reduces* from 40.5 to 31.3 dB. The substantial changes in peak magnitudes with respect to the dampening coefficients provides an additional capability to resolve the modes in the dampening factor domain, which is defined by the channel angle and Doppler *spread* parameters.

An alternative way to plot the MF-MUSIC spectrum is to keep the mean Doppler shift and angle parameters $[\Delta f, \theta]$ constant and to evaluate $p(\phi)$ over a manifold that changes as a function of the Doppler and angular spread parameters $[\alpha, \beta]$. Figures 8.7

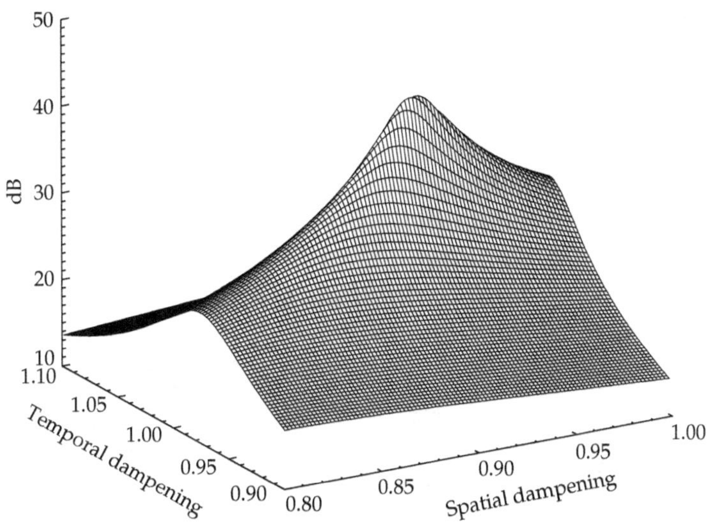

FIGURE 8.7 MF-MUSIC spectrum evaluated for different space-time dampening factors at angle-Doppler coordinates $[\Delta f, \theta] = [0.46, 22.2]$. The peak (43.3 dB) occurs at $[\alpha, \beta] = [0.996, 0.922]$. © Commonwealth of Australia 2011.

and 8.8 illustrate the spectra resulting in the dampening factor domain when $[\Delta f, \theta]$ is set to values of [0.46, 22.2] and [0.39, 21.7] estimated for modes 1 and 2, respectively, in Table 8.1. The physical interpretation is that such spectra attempts to resolve modes that have similar *mean* Doppler shift and cone angle of arrival but are reflected from different ionospheric regions that impose varying amounts of angle and Doppler spread on the modes. In theory, two modes with identical mean Doppler shift and cone angle but different spreads (dampening factors) in space and time due to independent channel fluctuations could be resolved from one another using the MF-MUSIC technique.

In this case, the two sporadic-E modes have sufficiently different Doppler shifts and angles-of-arrival so only a single peak is observed in Figures 8.7 and 8.8. The location of each peak quoted in the captions closely matches the dampening factor values estimated using the least squares method for each mode in Table 8.1. These results indicate that both methods may be used to estimate the distributed signal model parameters from the sample ACS. Moreover, the close agreement between the parameter estimates of mean cone angle and Doppler shift, as well as dampening factors in space and time, reflects the consistency and robustness of the proposed techniques to modeling uncertainty, statistical errors, and additive noise in this experiment.

8.4.3 Summary and Discussion

A space-time parametric signal-processing model, known as the Generalized Watterson model (GWM), was described to represent the statistical characteristics of multipath HF signals received by antenna arrays from a far-field point source via single-hop ionospheric propagation. The GWM adopts a flexible (pole-zero) statistical formulation that can model distributed signals described by a wide variety of power spectral density

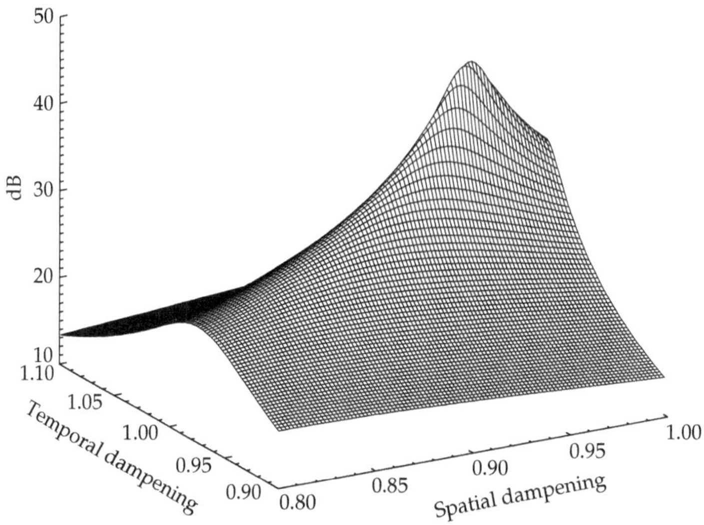

FIGURE 8.8 MF-MUSIC spectrum evaluated for different space-time dampening factors at angle-Doppler coordinates $[\Delta f, \theta] = [0.39, 21.7]$. The peak (45.3 dB) occurs at $[\alpha, \beta] = [0.998, 0.952]$. © Commonwealth of Australia 2011.

functions in angle and Doppler. The simplest (first-order) AR implementation of the GWM is specified by a minimum description set of five physically meaningful parameters per mode. Moreover, the recursive structure of the GWM random process allows statistical realizations of the multi-sensor data to be generated readily for computer simulations. Although the GWM was originally intended for the HF environment, it is envisaged that such a model may be useful for representing distributed signals arising in other practical applications not limited to radar.

The first-order model was adopted in accordance with previous experimental results on isolated signal modes propagated over a relatively quiet mid-latitude ionospheric path. The superposition of M statistically independent AR modes with Lorenzian-shaped angular and Doppler power spectral densities leads to a composite random process described by an ARMA(M, $M-1$) model. This chapter introduced a closed-form parameter-estimation technique that estimates the mode complex parameter tuples (mean and spread parameters) in space and time, providing the best least squares fit to the sample ACS. An alternative (spectral) parameter-estimation method, referred to as matched-field MUSIC, was also introduced. Both techniques have the capability to resolve signal modes in terms of their *mean* (Doppler shift and cone angle) parameters in addition to their (temporal and spatial) *spread* parameters, which depend on the statistical properties of the channel fluctuations.

An experiment was conducted to validate the GWM and proposed space-time parameter-estimation techniques for the case of two distributed signal modes that were unresolved in delay (group range). Specifically, it was found that two ionospheric modes propagated by the sporadic-E layer could not be resolved in range by the antenna array, although their presence was certified by oblique incidence ionogram records. Assuming the presence of two modes, the proposed parameter-estimation techniques were applied and yielded estimates that not only agreed well with each other, but also accurately modeled the real-data space-time sample ACS calculated for the mode mixture. Such results provide a convincing demonstration of the validity of the model as well as the effectiveness of the proposed parameter-estimation techniques applied to real data.

The HF channel simulator described in this chapter may be used in conjunction with a climatological ionospheric model and numerical ray tracing techniques to predict the propagation mode structure over a point-to-point link. This allows the number of dominant modes and their mean parameters (power, Doppler shift, and angle of arrival) to be predicted for a given propagation circuit, time interval, and carrier frequency in order to model a specific practical scenario of interest. Diffuse scattering or micro-multipath effects that produce complex fading of the various propagation modes in space and time (i.e., fine structure) are represented by the GWM in terms of Doppler and angular spread parameters with values that may be deduced from the WBMOD climatological model of naturally occurring small-scale ionization structure (Secan 2004). Physics-based representations of the signal scintillation process using plasma drift velocity models and phase screen diffraction methods, with small-scale ionization structure supplied by WBMOD, have been proposed in Nickisch, St. John, Fridman, and Hausman (2011).

Extension of the one-way ionospheric channel model, which applies directly to HF communication links between two ground stations or modeling radio frequency interference received by an OTH radar system, to the two-way (backscatter) paths for modeling target echoes and clutter is also of significant interest. The transformation from one-way to two-way paths is relatively straightforward for a point scatterer and a monostatic radar configuration, provided that reciprocity may be assumed for the outgoing (transmitter-to-scatterer) and incoming (scatterer-to-receiver) signal paths. In simple terms, the target echo channel model may be derived as the convolution of the one-way channel impulse response function with itself. Modeling clutter backscattered from terrain and/or sea over wide areas requires a surface scattering model and finding all connecting paths for each propagation mode between the transmit and receive antenna elements. With the aid of ray tracing engines, clutter echoes may be simulated as the sum of a large number of two-way point-to-point circuits using an appropriate (land or sea surface) backscattering model, such as that described for deep water at near grazing incidence in Chapter 5.

CHAPTER 9

Interference Cancelation Analysis

The space-time characteristics of HF signals reflected by the ionosphere have been analyzed and modeled from a signal-processing perspective in Chapters 6–8. The physical phenomena that contribute to the observed signal characteristics are clearly of scientific interest, but the more pragmatic question arises as to the practical impact of space-time signal distortions in operational OTH radar systems. The first objective of this chapter is to present the results of a case study that demonstrates the influence of such distortions in a key OTH radar signal-processing application. Specifically, this case study experimentally quantifies the performance of different adaptive beamforming schemes applied to the problem of HF interference cancelation in OTH radar.

The HF channel model described in Chapter 8 was empirically shown to accurately represent the statistical characteristics of signal modes reflected by the ionosphere. While this condition is considered necessary for experimental validation, realizations of the model are also required to accurately portray the performance of signal-processing techniques applied to real data. A simulator with this capability not only helps to guide algorithm design, but also serves to evaluate the relative merits and shortcomings of different signal-processing techniques in a convenient manner. It is therefore of interest to determine the fidelity with which the previously described multi-variate HF channel model and parameter estimation technique can predict the experimental performance of modern OTH radar signal-processing techniques. The second objective of this chapter is to measure and compare the performance of various adaptive beamforming schemes when applied to simulated and real data.

A brief introduction to adaptive beamforming for HF interference rejection is provided in Section 9.1, although readers not familiar with this area are referred to the essential concepts section in Chapter 10, which provides a more detailed introduction to this subject. Section 9.2 describes a number of adaptive beamforming schemes considered suitable for practical implementation in OTH radar, while the experimental performance of these adaptive beamforming schemes is quantified using an actual HF interference source in Sections 9.3 and 9.4.

The accuracy with which the previously described HF channel model can estimate the observed practical performance of the adaptive beamforming schemes is illustrated in Section 9.5. This is carried out by generating realizations of the interference signal and applying the same adaptive beamforming schemes to both simulated and real data. The third and final objective of this chapter is to motivate the development of robust adaptive beamforming algorithms as a lead in to Chapter 10.

9.1 Interference and Noise Mitigation

Although interference-free conditions seldom occur in the HF environment, relatively few studies reported in the open literature have statistically quantified the capabilities and limitations of different adaptive beamforming schemes considered suitable for practical implementation in OTH radar systems. This is particularly so in the context of very wide aperture antenna arrays and the case of interference sources reflected by the ionosphere. This section provides a brief introduction to this topical area and describes a number of experimental investigations where adaptive beamforming has been applied for interference cancelation in HF systems. A level of familiarity with this subject is assumed in the following discussion. The reader may prefer to consult the essential concepts section in Chapter 10 for a more general introduction.

9.1.1 Spatial Processing

It is well known that the conventional beamformer or spatial matched filter is optimum in terms of output SNR when the steering vector of the useful signal is perfectly known and the noise field is spatially white. In OTH radar, the ambient noise field is spatially structured or "colored" due to the anisotropic nature of atmospheric and galactic sources, while directional interference from other users of the HF spectrum is often difficult to avoid completely due to long distance ionospheric propagation. When strong co-channel interference is present, such signals may enter through the antenna pattern sidelobes and significantly contaminate the conventional beam estimate. This so-called "spectral leakage" effect has the potential to mask faint target echoes and degrade the performance of practical OTH radar systems.

Tapering of the array using a window function in conventional beamforming may be used to lower sidelobe levels at the expense of broadening the main lobe width. A wide selection of window functions may be used to shade the array outputs for spectral analysis (Harris 1978). The specific choice of taper weights depends on the tradeoff that yields the best signal detection and estimation performance for the anticipated interference conditions. The tapered or "mismatched" conventional beamformer is generally preferred over the matched filter because of its higher immunity to sidelobe interference and its greater robustness to beam pointing errors. Although this approach is computationally attractive, the sub-optimality of data-independent beamforming may reach intolerable levels when powerful and persistent interference is present, particularly in practical situations where the antenna array is not precisely calibrated.

The alternative is known as data-dependent or adaptive beamforming that can in principle improve signal detection and parameter estimation performance in spatially structured interference environments relative to conventional beamforming. Adaptive beamforming tailors the directional response of the array to the characteristics of the received data so that useful signals may be filtered out more effectively from interference and noise. On the practical implementation side, the higher computational load associated with adaptive beamforming can only be justified when a significant improvement in system performance is attained relative to conventional beamforming. This improvement is often measured in terms of the output signal-to-interference plus noise ratio (SINR).

In essence, the adaptive beamforming problem is to estimate as accurately as possible the optimal set of array-processing weights that provide maximum output SINR for the prevailing signal and interference environment. Simply stated, adaptive beamforming can improve output SINR relative to conventional beamforming by forming deep "nulls" in the directions of strong interferences while maintaining a fixed gain in the desired look direction to receive useful signals. The optimum weight vector is usually synthesized with reference to a known steering vector model for the useful signal and the information contained in the interference-plus-noise spatial covariance matrix, which is estimated from the received data. Typically, adaptive beamforming algorithms do not rely on the plane wave assumption as far as the rejection of *interference* is concerned.

There is a lack of *experimental results* published in the open literature that quantify the interference cancelation performance of adaptive beamforming schemes in very wide aperture HF antenna arrays. In particular, the impact of *time-varying* interference wavefront distortions on adaptive beamformer performance merits more detailed attention. This chapter experimentally quantifies the output SINR improvements achieved by various adaptive beamforming schemes relative to conventional beamforming. The adaptive beamformers are applied to reject a real HF interference source received by very wide aperture array. In addition, the previously described space-time HF channel model is revisited in order to determine the accuracy with which it can predict the experimentally observed output SINR improvements.

9.1.2 Popular Techniques

A vast amount of literature exists on the subject of adaptive beamforming and it is beyond the scope of this chapter to provide a comprehensive list of references. For readers interested in delving further, many of them can be found in the authoritative texts by Monzingo and Miller (1980), Hudson (1981), Compton (1988a), and Li and Stoica (2006) for example, as well as the excellent tutorial papers by Van Veen and Buckley (1988) and Steinhardt and Van Veen (1989). The scope of this section is to identify the broad class of adaptive beamforming techniques considered suitable for OTH radar applications, and to explain the underlying reasons driving this choice.

The minimum variance distortionless response (MVDR) spectrum estimator described in Capon (1969) is frequently adopted as a framework for adaptive beamformer design in a wide variety of practical applications. The MVDR approach derives the optimal array weight vector by minimizing the output power of the processor subject to a linear constraint that provides fixed unity gain to useful signals in the beam steer direction. This procedure is tantamount to maximizing the output SINR. An analytic solution for the optimum weight vector exists in terms of the inverse of the interference-plus-noise spatial covariance matrix and the steering vector of the useful signal. In many respects, this optimum beamforming solution represents the foundation for the development of adaptive algorithms in operational OTH radar systems.

Two fundamental issues must be addressed in order to realize the benefits of adaptive beamforming in practice. The first issue relates to the choice of processor architecture and the allocation of adaptive degrees of freedom[1] (DOF) with due regard to the finite amount of statistically homogeneous interference data available for training. The second critical issue is the choice of adaptive algorithm to estimate the optimum weight vector.

[1]Methods for allocation of adaptive degrees of freedom are suggested in Steinhardt and Van Veen (1989).

The interference-plus-noise spatial covariance matrix is unknown a priori and needs to be estimated from the received data. Furthermore, the useful signal steering vector may deviate from the presumed model due to instrumental and environmental uncertainties. Computational complexity also represents a major consideration that influences the choice of adaptive algorithm for systems required to operate in real-time.

The sample matrix inverse (SMI) technique described in the seminal paper of Reed, Mallet, and Brennan (1974) is a popular approach that directly substitutes the sample covariance matrix for the true covariance matrix in the MVDR criterion function. The SMI approach is often preferred over iterative gradient-descent based techniques such as the least mean square (LMS) algorithm of Widrow et al. (1967), which does not require a matrix inversion but is known to converge slowly when the interference covariance matrix has a large eigenvalue spread. A compelling advantage of the SMI technique is that rapid convergence rate toward the optimal solution is achieved independently of the eigenvalue spread. In addition, the need for a matrix inversion has become less of an issue in current systems than it was in the past due to the availability of faster computers.

It is recalled that the SMI technique coincides with the maximum likelihood (ML) estimate of the optimum weight vector when the interference-plus-noise snapshots are described by a wide-sense stationary multi-variate Gaussian random process. Assuming this interference description, a simple rule of thumb was derived for the SMI technique in (Reed, Mallet, and Brennan 1974) to ensure that average losses in output SINR are within 3 dB of the optimum value. It states that the number of statistically independent interference-plus-noise snapshots used to form the sample covariance matrix needs to be more than twice the adaptive weight-vector dimension. This rule of thumb assumes the useful signal is *absent* from the training data and that it has a perfectly known steering vector. It is therefore often sought to estimate the adaptive beamforming weight vector using data snapshots free of useful signals.

In most adaptive beamforming studies, improvements have been claimed solely on the basis of theoretical analysis and/or computer simulation, where to a large extent, the characteristics of the signal environment and the properties of the hypothesized sensor array are controlled. For example, many works assume that the useful signal and interference sources have a time-invariant plane wave structure. This rather idealized assumption is clearly not satisfied for HF signals reflected by the ionosphere over the time scales commensurate with typical OTH radar CPI. In addition, the advantages of different approaches are often evaluated and compared with respect to at best a subset of the operational, environmental, and instrumental factors that can collectively limit adaptive beamformer performance in a real-world system.

Although numerical analysis of adaptive beamformer performance based on synthetic data provides valuable information for discerning between the potential usefulness of different techniques in a particular application, performance improvements based on simulation results should be interpreted with caution as they may not be representative of those actually encountered in practice. A more direct method of quantifying the output SINR improvement of adaptive beamforming relative to conventional beamforming in a very wide aperture HF antenna array involves processing experimental data acquired by such systems.

9.1.3 HF Applications

The impact of wavefront distortions on useful signal reception using a very wide aperture HF antenna array was analyzed in Sweeney (1970). However, there is an important distinction between this aspect of the problem and the impact of wavefront distortions on the interference *rejection* performance of a very wide aperture HF antenna array. In the former case, signal distortions are observed in the main lobe of the antenna pattern, whereas in the latter case, they are typically observed in the sidelobe region (ignoring the main beam interference scenario). When adaptive beamforming is employed, the interference wavefronts are received close to the deep nulls of the antenna pattern, where the response of the array is much more sensitive to variations in spatial structure.

In addition, useful signals and radio frequency interference (RFI) received by an OTH radar are typically propagated via different ionospheric paths that can have markedly different characteristics. While the operating frequency is usually chosen to optimize the propagation of radar signals to the geographic area of interest, co-channel interference sources are arbitrarily located with respect to the surveillance area and may be propagated by highly perturbed ionospheric regions, such as the equatorial or polar regions. In these circumstances, the temporal variability of the interference spatial structure within the CPI is likely to be more pronounced compared to that of useful signals which propagate over a relatively stable ionospheric path. Fluctuation of the interference wavefronts within the CPI can have a profound effect on the performance of adaptive beamforming.

An online adaptive beamforming capability for HF backscatter radar was developed for a 2.5-km long ULA by Washburn and Sweeney (1976). The ULA consisted of eight 32-element subarrays with each subarray output connected to a digital receiver. The useful signals were aircraft target echoes, while the interference signals were from other users of the HF band as well as signals from a ground-based radar repeater. The performance of a recursive time-domain adaptive beamforming technique that converges to the optimum MVDR solution was compared against the conventional beamformer with −25 dB Dolph taper. The rejection of unwanted signals with the adaptive beamformer was variable, but side-by-side comparisons revealed that adaptive beamforming could reject off-azimuth signals up to 20 dB better than conventional beamforming. However, important quantities such as the mean improvement in output SINR and its variability over different data sets, CPI lengths, and beam steer directions were not reported in Washburn and Sweeney (1976).

Not surprisingly, the interference was canceled more effectively when a temporal filter was used to remove clutter prior to training the adaptive beamforming weights. As Doppler processing is often sufficient to separate clutter from moving target echoes, filtering out the clutter allows all spatial adaptive degrees of freedom to be utilized for the interference rejection task. In Washburn and Sweeney (1976), the weights were adapted in the *time domain* after clutter removal in order to track temporal variations of the interference characteristics over the CPI. The resulting time-varying adaptive weight vectors were stored and then used to process the original data. It was found that the application of time-varying weights to the original data resulted in significant Doppler broadening of the clutter and targets.

A similar phenomenon was encountered in a companion paper (Griffiths 1976). Such broadening can potentially impair target detection after Doppler processing. To prevent this deleterious side effect, Washburn and Sweeney (1976) formed the adaptive beam using a fixed weight vector that resulted at the end of the training interval. Although this eliminated the problem of Doppler broadening, holding the weights fixed over the CPI led to a degradation in interference suppression. The same problem was also cited in Games, Townes, and Williams (1991), where it was concluded that variations in the HF signal environment over time intervals in the order of seconds can significantly impact the interference rejection performance of adaptive beamforming algorithms.

Before considering the use of more sophisticated (and computationally intensive) adaptive beamforming algorithms to address this problem, it is first important to quantify the effectiveness of standard adaptive beamforming techniques, and to assess whether the performance loss warrants the additional effort to implement more advanced routines. The improvement in output SINR achieved by the SMI-MVDR adaptive beamforming technique using fixed adaptive weights was experimentally quantified for different CPI lengths in Fabrizio et al. (1998). This analysis was based on a 1.4-km long ULA with 16 digital receivers. Two different interference sources of opportunity were considered. The first source propagated over a relatively quiet single-hop mid-latitude path, while the second involved multi-hop propagation via the equatorial ionosphere, which is typically more disturbed.

Over very short time intervals (less than 0.1 seconds), performance was limited by finite sample support, i.e., estimation errors due to the small amount of training data. As the CPI was increased from 0.1 to 4 seconds, finite sample effects ceased to limit performance. Over this longer time interval, the source propagated over the mid-latitude path was effectively rejected using fixed adaptive beamforming weights within the CPI. However, a 4–5 dB loss in performance was observed when identical adaptive beamforming was applied to the source propagated via the equatorial ionosphere. Although the loss in output SINR was not dramatic in this example, further analysis is required to better understand the impact of time-varying interference wavefront distortions on adaptive beamformer performance.

9.2 Standard Adaptive Beamforming

This section describes a number of standard adaptive beamforming schemes that may be considered suitable for practical implementation in OTH radar systems. The performance of each scheme is quantified in terms of the output SINR improvement relative to conventional beamforming using real data containing an HF interference source. The improvements are measured by comparing the level of interference-plus-noise rejection achieved with all adaptive beamforming schemes having identical (unit gain) response to ideal useful signals. This method conveniently allows performance to be assessed with the radar system operating in passive mode, i.e., with transmitters switched off such that the antenna array samples HF interference and noise only. This provides an indication of the maximum potential SINR improvement for the various schemes since the presence of clutter and real target echoes is likely to degrade the performance of adaptive beamforming relative to conventional beamforming.

9.2.1 Sample Matrix Inverse Technique

The most common criterion used to define the optimal array weight vector $\mathbf{w}_{opt}$ is that of maximizing the SINR at the beamformer output. Let $\mathbf{x}_k(t)$ be the N-dimensional array snapshot vector received in range bin k of a PRI indexed by t. Assume this vector contains a useful signal $\mathbf{s}_k(t)$ and uncorrelated interference-plus-noise $\mathbf{n}_k(t)$ in Eqn. (9.1).

$$\mathbf{x}_k(t) = \mathbf{s}_k(t) + \mathbf{n}_k(t) = g_k(t)\mathbf{s}(\theta) + \mathbf{n}_k(t) \tag{9.1}$$

The complex scalar $g_k(t)$ is the useful signal waveform and $\mathbf{s}(\theta) = [1, e^{j2\pi d \sin\theta/\lambda}, \ldots, e^{j2\pi(N-1)d \sin\theta/\lambda}]^T$ is the ULA steering vector corresponding to a plane wave incident from cone angle θ, where λ is the wavelength and d is the inter-element separation. The beamformer output $y_k(t)$ is given by the inner product of the weight vector $\mathbf{w}_{opt}$ and data vector $\mathbf{x}_k(t)$ in Eqn. (9.2).

$$y_k(t) = \mathbf{w}_{opt}^{\dagger}\mathbf{x}_k(t) = g_k(t)\mathbf{w}_{opt}^{\dagger}\mathbf{s}(\theta) + \mathbf{w}_{opt}^{\dagger}\mathbf{n}_k(t) \tag{9.2}$$

For an arbitrary weight vector $\mathbf{w}$ satisfying the linear constraint $\mathbf{w}^{\dagger}\mathbf{s}(\theta) = 1$, the output power is given by Eqn. (9.3), where the interference-plus-noise is assumed to be wide-sense stationary process with a statistically expected spatial covariance matrix denoted by $\mathbf{R}_n = \mathrm{E}\{\mathbf{n}_k(t)\mathbf{n}_k^{\dagger}(t)\}$. The optimum weight vector $\mathbf{w}_{opt}$ *minimizes* the output power subject to the unity gain constraint $\mathbf{w}_{opt}^{\dagger}\mathbf{s}(\theta) = 1$, which provides distortionless response for useful signals.

$$\mathrm{E}\{|y_k(t)|^2\} = \sigma_g^2 + \mathbf{w}^{\dagger}\mathrm{E}\{\mathbf{n}_k(t)\mathbf{n}_k^{\dagger}(t)\}\mathbf{w} = \sigma_g^2 + \mathbf{w}^{\dagger}\mathbf{R}_n\mathbf{w} \tag{9.3}$$

In Eqn. (9.3), $\sigma_g^2 = E\{|g_k(t)|^2\}$ is the power of the useful signal in a single receiver (preserved at the beamformer output), while $\mathbf{w}^{\dagger}\mathbf{R}_n\mathbf{w}$ is the residual interference-plus-noise power that must be minimized in order to maximize the output SINR. The minimum variance distortionless response (MVDR) approach finds the optimum weight vector $\mathbf{w}_{opt}$ as the solution of the following linearly constrained optimization problem:

$$\mathbf{w}_{opt} = \arg\min_{\mathbf{w}} \mathbf{w}^{\dagger}\mathbf{R}_n\mathbf{w} \quad \text{subject to}: \ \mathbf{w}^{\dagger}\mathbf{s}(\theta) = 1 \tag{9.4}$$

Using the method of Lagrange multipliers, the solution that maximizes the output SINR under the above-mentioned conditions is given by the closed-form expression in Eqn. (9.5), which is often referred to as the Capon optimum beamformer (Capon 1969).

$$\mathbf{w}_{opt} = \frac{\mathbf{R}_n^{-1}\mathbf{s}(\theta)}{\mathbf{s}^{\dagger}(\theta)\mathbf{R}_n^{-1}\mathbf{s}(\theta)} \tag{9.5}$$

The interference-plus-noise spatial covariance matrix $\mathbf{R}_n$ is unknown a priori and needs to be estimated from the received data in practice. A popular method for estimating the optimum weight vector $\hat{\mathbf{w}}_{opt}$ is the sample matrix inverse (SMI) technique (Reed, Mallet, and Brennan 1974). This involves substitution of $\mathbf{R}_n$ in Eqn. (9.5) with the sample covariance matrix (SCM) $\hat{\mathbf{R}}_n$ in Eqn. (9.6). The SMI technique assumes the availability of training data $\mathbf{n}_k(t)$ containing interference-plus-noise only. Ideally, these vectors are extracted from Δ_k ranges and Δ_p PRI not containing clutter or useful signals. A sufficient number of independent vectors $\Delta_k\Delta_p \geq N$ is required to ensure that $\hat{\mathbf{R}}_n^{-1}$ exists.

$$\hat{\mathbf{R}}_n = \frac{1}{\Delta_k\Delta_p}\sum_{k=1}^{\Delta_k}\sum_{t=1}^{\Delta_p}\mathbf{n}_k(t)\mathbf{n}_k^{\dagger}(t) \tag{9.6}$$

Although this may appear to be an ad hoc approach, the SMI technique has a number of very important properties. For a finite number of available training samples, the SMI method converges more rapidly to the optimal solution than gradient-descent estimation techniques such as the LMS algorithm, particularly when $\mathbf{R}_n$ is poorly conditioned due to the presence of powerful interference with low spatial rank. Moreover, when $\mathbf{n}_k(t)$ is zero-mean complex Gaussian distributed, the SCM is the maximum likelihood (ML) estimate of $\mathbf{R}_n$, and by the invariance principle, $\hat{\mathbf{w}}$ in Eqn. (9.7) is the ML estimate of $\mathbf{w}_{opt}$.

$$\hat{\mathbf{w}} = \frac{\hat{\mathbf{R}}_n^{-1}\mathbf{s}(\theta)}{\mathbf{s}^{\dagger}(\theta)\hat{\mathbf{R}}_n^{-1}\mathbf{s}(\theta)} \tag{9.7}$$

Provided $\hat{\mathbf{w}}$ is computed using $\Delta_k\Delta_p > 2N$ statistically independent training vectors $\mathbf{n}_k(t)$, the average loss in output SINR due to estimation errors in $\hat{\mathbf{w}}$ relative to $\mathbf{w}_{opt}$ is less than 3 dB irrespective of the form of the interference-plus-noise spatial covariance matrix $\mathbf{R}_n$ (Reed, Mallet, and Brennan 1974). The results of Cheremisin (1982) and Abramovich (1981b) show that by appropriate diagonal loading of the sample covariance matrix, the number of independent snapshots required for less than 3 dB average losses in output SINR can under certain circumstances be reduced to $2N_e$, where $N_e < N$ is the interference subspace dimension. The introduction of diagonal loading with appropriate loading factor α in Eqn. (9.8) can therefore significantly improve convergence rate when sample support is limited.

$$\tilde{\mathbf{R}}_n = \hat{\mathbf{R}}_n + \alpha\mathbf{I} \tag{9.8}$$

9.2.2 Practical Implementation Schemes

Operational considerations affect how adaptive beamforming can be applied in practice. A key issue is that during normal operation an OTH radar system receives powerful clutter in addition to useful signals and interference. The presence of radar echoes in the training data needs to be avoided so as to effectively estimate the interference-plus-noise spatial covariance matrix $\mathbf{R}_n$. One approach is to schedule a brief period of passive data collection prior to each active coherent processing interval (CPI). The clutter-free training data collected during this period is used to train the adaptive beamformer which is then applied to the CPI that follows.

Turning the transmitters off temporarily before each CPI represents a possible option. However, this method may not be advisable in certain practical systems due to the risk of damaging hardware. An alternative is to deliberately assign the transmitter to a different carrier frequency for a brief period, such that clutter returns are temporarily out of band and hence not received by the system during the interference-plus-noise sampling interval. This allows the transmitter to be driven continuously at a constant power.

The interference-plus-noise SCM is computed using array snapshots received during the "passive" training interval. This estimate is used to synthesize the adaptive beamforming weight vector according to the SMI or diagonally loaded SMI technique. The resulting spatial filter is then used to process the entire CPI of operational data that immediately follows the interference-plus-noise sample train. The adaptive beamformer does not attempt to cancel clutter in the CPI. Clutter and useful signals can be separated into different frequency bins by Doppler processing after adaptive beamforming. The main role of adaptive beamforming is to reject the interference and pass on the useful signals. This procedure is repeated again to process the next CPI and so on in order to adaptively respond to changes in the spatial properties of the external interference-plus-noise environment.

Such a method is referred to as scheme 1 and has been illustrated in the upper left diagram of Figure 9.1. Assuming the spatial statistics of $\mathbf{n}_k(t)$ are invariant across range

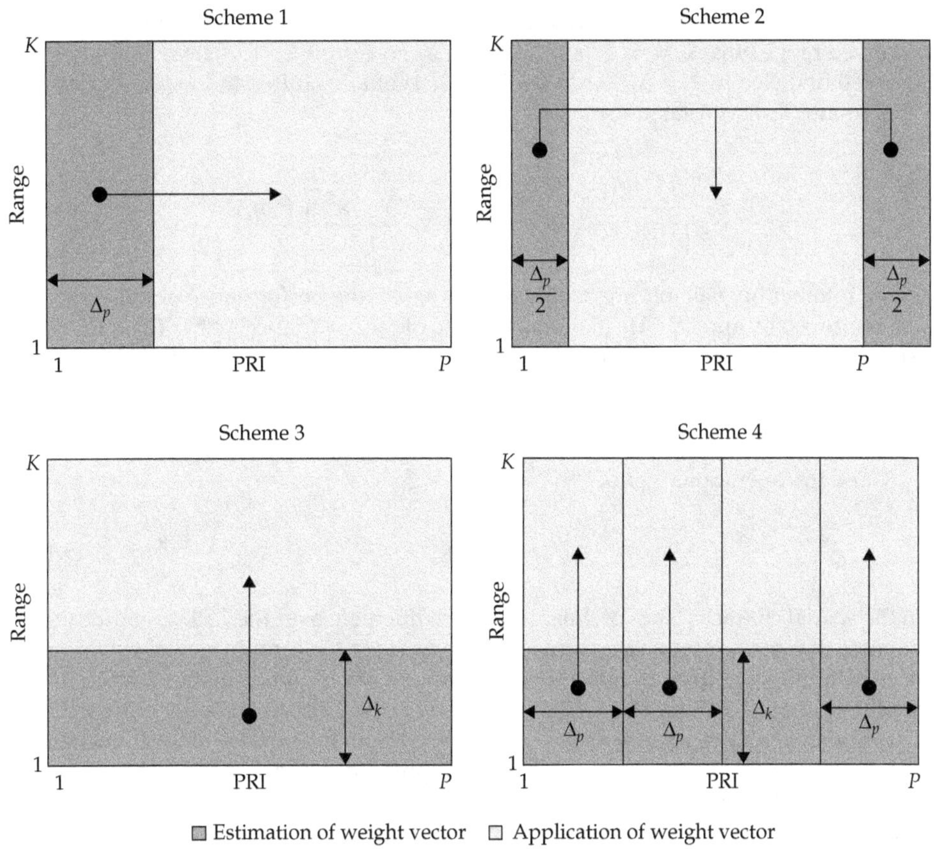

FIGURE 9.1 Conceptual illustration of the four adaptive beamforming schemes. Each diagram shows the range and PRI regions where the array snapshots are used to estimate the adaptive weight vector and the regions where the resulting spatial filter is applied to beamform the radar data. © Commonwealth of Australia 2011.

over a relatively small number of Δ_p consecutive PRI, the received snapshots $\mathbf{n}_k(t)$ for $k = 1, \ldots, \Delta_k$ and $t = 1, \ldots, \Delta_p$ may be regarded as different realizations of a locally stationary random process. These interference-plus-noise realizations may be considered independent, provided the interference bandwidth is larger or equal to the radar bandwidth and the receiver output is sampled at the Nyquist rate. The degree of independence among the different sample vectors in the training set will in general depend on the interference waveform characteristics as well as the window function used for pulse compression, which introduces correlation among neighboring range bins.

In scheme 1, the sample spatial covariance matrix $\hat{\mathbf{R}}_n$ is formed using all available training data acquired during the "listening" period between the end of one CPI and the start of another. This training data is also commonly known as secondary data. The weights for adaptive beamforming scheme 1 are denoted by $\hat{\mathbf{w}}_1$ in Eqn. (9.9). In this case, the interference-plus-noise sample spatial covariance matrix $\hat{\mathbf{R}}_n$ is estimated using all processed range bins $\Delta_k = K$ over a subset of Δ_p consecutive PRI. The remaining PRI in the dwell (indexed by $t = \Delta_p + 1, \Delta_p + 2, \ldots, P$) contain clutter and useful signals in addition to interference-plus-noise.

$$\hat{\mathbf{w}}_1 = \frac{\hat{\mathbf{R}}_n^{-1}\mathbf{s}(\theta)}{\mathbf{s}^\dagger(\theta)\hat{\mathbf{R}}_n^{-1}\mathbf{s}(\theta)}, \quad \hat{\mathbf{R}}_n = \frac{1}{K\Delta_p}\sum_{k=1}^{K}\sum_{t=1}^{\Delta_p}\mathbf{n}_k(t)\mathbf{n}_k^\dagger(t) \tag{9.9}$$

When the number of available snapshots $K\Delta_p$ is small, the performance of this method may be improved by appropriate diagonal loading of the covariance matrix $\tilde{\mathbf{R}}_n = \hat{\mathbf{R}}_n + \alpha\mathbf{I}$, and substituting $\tilde{\mathbf{R}}_n$ for $\hat{\mathbf{R}}_n$ in Eqn. (9.9). The choice of α yielding the best performance is data-dependent. Typically, loading levels above the additive white noise power but well below the interference subspace eigenvalues provide best results. The beamformer output for scheme 1 is given by the scalar signal $y_k^{[1]}(t)$ in Eqn. (9.10) for ranges $k = 1, \ldots, K$ and the operational set of PRI $t = \Delta_p + 1, \Delta_p + 2, \ldots, P$.

$$y_k^{[1]}(t) = \hat{\mathbf{w}}_1^\dagger \mathbf{x}_k(t) \tag{9.10}$$

When the spatial characteristics of the interference fluctuate over the CPI, a modification to scheme 1 that may improve performance is to average the sample spatial covariance matrices resulting at both ends of the CPI. The rationale behind this method, referred to as adaptive scheme 2 in Figure 9.1, is that variations of the interference-plus-noise within the CPI are better captured by a weight vector derived from the average of the two sample covariance matrices. The weight vector for this adaptive beamforming scheme is denoted as $\hat{\mathbf{w}}_2$ in Eqn. (9.11), while the scalar output is given by $y_k^{[2]}(t) = \hat{\mathbf{w}}_2^\dagger \mathbf{x}_k(t)$ for $k = 1, \ldots, K$ and $t = \Delta_p/2 + 1, \ldots, P - \Delta_p/2$ with Δ_p assumed to be even.

$$\hat{\mathbf{w}}_2 = \frac{\hat{\mathbf{R}}_n^{-1}\mathbf{s}(\theta)}{\mathbf{s}^\dagger(\theta)\hat{\mathbf{R}}_n^{-1}\mathbf{s}(\theta)}, \quad \hat{\mathbf{R}}_n = \frac{1}{K\Delta_p}\sum_{k=1}^{K}\left\{\sum_{t=1}^{\Delta_p/2}\mathbf{n}_k(t)\mathbf{n}_k^\dagger(t) + \sum_{t=P-\Delta_p/2+1}^{P}\mathbf{n}_k(t)\mathbf{n}_k^\dagger(t)\right\} \tag{9.11}$$

The matched filter beamformer is defined as $N^{-1}\mathbf{s}(\theta)$, where normalization by N provides unit gain in the beam steer direction, i.e., $N^{-1}\mathbf{s}^\dagger(\theta)\mathbf{s}(\theta) = 1$. A real valued window function is often applied to reduce sidelobe levels. In this case, the mismatched conventional beamformer $\mathbf{v}(\theta)$ may be written in the form of Eqn. (9.12), where the diagonal taper matrix $\mathbf{T} = \text{diag}[w_1, \ldots, w_N]$ contains the window values $\{w_n\}_{n=1}^N$ as its elements. Similar to adaptive schemes 1 and 2, the conventional beamformer in Eqn. (9.12) is also normalized to provide unit gain in the steer direction such that $\mathbf{v}^\dagger(\theta)\mathbf{s}(\theta) = 1$. The conventional output is given by $y_k(t) = \mathbf{v}^\dagger(\theta)\mathbf{x}_k(t)$.

$$\mathbf{v}(\theta) = \frac{\mathbf{T}\mathbf{s}(\theta)}{\mathbf{s}^\dagger(\theta)\mathbf{T}\mathbf{s}(\theta)} \tag{9.12}$$

As all the above-mentioned beamformers share the unit gain response to an ideal useful signal, the SINR improvements of schemes 1 and 2 relative to the conventional beamformer may be estimated as $\hat{q}_1$ and $\hat{q}_2$, respectively, in Eqn. (9.13). In the case of passive mode data, we have that $\mathbf{x}_k(t) = \mathbf{n}_k(t)$, so the quantities in Eqn. (9.13) represent the interference-plus-noise cancelation ratio. The two adaptive beamforming schemes described so far belong to a class of algorithms that we shall refer to as *framing* schemes.

$$\hat{q}_1 = \frac{\sum_{k=1}^{K}\sum_{t=\Delta_p+1}^{P}|y_k(t)|^2}{\sum_{k=1}^{K}\sum_{t=\Delta_p+1}^{P}|y_k^{[1]}(t)|^2}\,, \quad \hat{q}_2 = \frac{\sum_{k=1}^{K}\sum_{t=\Delta_p/2+1}^{P-\Delta_p/2}|y_k(t)|^2}{\sum_{k=1}^{K}\sum_{t=\Delta_p/2+1}^{P-\Delta_p/2}|y_k^{[2]}(t)|^2}\,, \tag{9.13}$$

An alternative method exploits the property that interference and noise are (normally) incoherent with the radar signal and will appear in all range cells processed, while clutter and useful signals often appear over a finite range extent that may only occupy a subset of the processed range cells. Specifically, clutter and useful signals are mainly present in a band of ranges beyond the skip-zone, where sky-wave signal propagation is supported by the ionosphere. Consequently, the signals received in the nearest range cells within the skip-zone, indexed by $k = 1, 2, \ldots, \Delta_k < K$, will be dominated by interference and noise. The skip-zone phenomenon allows the interference-plus-noise spatial covariance matrix to be estimated *within* the CPI, as illustrated in Figure 9.1. This alternative method is referred to as adaptive beamforming scheme 3. The weight vector for scheme 3 is calculated as $\hat{\mathbf{w}}_3$ in Eqn. (9.14).

$$\hat{\mathbf{w}}_3 = \frac{\hat{\mathbf{R}}_n^{-1}\mathbf{s}(\theta)}{\mathbf{s}^\dagger(\theta)\hat{\mathbf{R}}_n^{-1}\mathbf{s}(\theta)}\,, \quad \hat{\mathbf{R}}_n = \frac{1}{P\Delta_k}\sum_{k=1}^{\Delta_k}\sum_{t=1}^{P}\mathbf{n}_k(t)\mathbf{n}_k^\dagger(t) \tag{9.14}$$

The scalar output and SINR improvement factor for scheme 3 are given in Eqn. (9.15). In this case, the operational data to be processed $\mathbf{x}_k(t)$ are the set of range cells $k = \Delta_k + 1, \ldots, K$ immediately beyond the skip-zone, which additionally contain clutter and potentially useful signals. The significant advantage of scheme 3 relative to framing schemes is that the adaptive beamforming weights are derived from an estimate of the interference-plus-noise spatial covariance matrix integrated over the CPI. The presence and number of skip-zone range cells depends on ionospheric conditions and the operating frequency. A potential drawback of scheme 3 is that such cells may not always be available in practice.

$$y_k^{[3]}(t) = \hat{\mathbf{w}}_3^\dagger\mathbf{x}_k(t)\,, \quad \hat{q}_3 = \frac{\sum_{k=\Delta_k+1}^{K}\sum_{t=1}^{P}|y_k(t)|^2}{\sum_{k=\Delta_k+1}^{K}\sum_{t=1}^{P}|y_k^3(t)|^2}\begin{cases} k = \Delta_k+1, \Delta_k+2, \ldots K \\ t = 1, 2, \ldots, P \end{cases} \tag{9.15}$$

Experimental analysis of the relative improvement factors $\hat{q}_{1,2,3}$ enables the cancelation performance of different adaptive beamforming schemes to be compared. It is also of interest to determine whether the performance observed on real data can be accurately predicted by the HF channel simulator described in the previous chapter. For both these purposes, the statistical distribution of the quantities $\hat{q}_{1,2,3}$ will be determined as a function of CPI length. In particular, the mean and deciles of these distributions will be calculated to provide an indication of cancelation performance for the different adaptive beamforming schemes.

9.2.3 Alternative Time-Varying Approach

The spatial structure of HF signals reflected by the ionosphere can vary appreciably over time scales commensurate with the duration of typical OTH radar CPI. However, it has been observed that changes in the signal wavefronts received from individual propagation modes are often highly correlated from one PRI to another. Stated another way, the received wavefronts tend to evolve in relatively smooth manner when viewed at a temporal resolution of less than one-tenth of a second.[2] An HF signal mode reflected from a relatively stable and localized region of the ionosphere will therefore exhibit an almost steady spatial signature over a sufficiently short time interval comparable to the OTH radar PRI. The spatial covariance matrix of such a signal will therefore have a low rank when integrated over the "quasi-instantaneous" PRI. On the other hand, the spatial covariance matrix averaged over the relatively longer CPI converges to its *statistically expected* form, which may grow to full rank as the fluctuations of an angularly spread signal decorrelate in time.

It follows that forming the interference-plus-noise sample spatial covariance matrix over short time segments within the CPI provides a means to reduce the interference subspace dimension and hence improve rejection performance. In other words, readjusting the antenna pattern a number of times within the CPI to adaptively "track" the spatial dynamics of the interference provides greater opportunity to reject such interference effectively. This capability becomes important when: (1) the number of independent interference sources and modes approaches the number of adaptive degrees of freedom, (2) the spatial properties of the interference wavefronts are changing rapidly, and (3) the CPI to be processed is long. In such situations, re-tuning of the adaptive weights a number of times within the CPI is likely to provide more effective interference cancelation.

An apparently reasonable way of adapting the antenna pattern to the changing interference characteristics within the CPI is to form the slow-time varying SCM $\hat{\mathbf{R}}_n(t)$ using the skip-zone range cells integrated over a batch of Δ_p PRI starting at time t, and to calculate the adaptive weight vector $\hat{\mathbf{w}}_4(t)$ in accordance with the well-known rule in Eqn. (9.16). The time-varying weight vector $\hat{\mathbf{w}}_4(t)$ is then used to beamform the operational ranges cells $y^{[4]}(t) = \hat{\mathbf{w}}_4^\dagger(t)\mathbf{x}_k(t)$ for $k = \Delta_k + 1, \Delta_k + 2, \ldots, K$ in the *current* batch of PRI indexed by $t, t+1, \ldots, t+\Delta_p - 1$. This time-varying method, referred to as adaptive scheme 4, is illustrated in Figure 9.1 for the case of non-overlapping batches.

$$\hat{\mathbf{w}}_4(t) = \frac{\hat{\mathbf{R}}_n^{-1}(t)\mathbf{s}(\theta)}{\mathbf{s}^H(\theta)\hat{\mathbf{R}}_n^{-1}(t)\mathbf{s}(\theta)}, \quad \hat{\mathbf{R}}_n(t) = \frac{1}{\Delta_k \Delta_p} \sum_{k=1}^{\Delta_k} \sum_{t'=t}^{t+\Delta_p-1} \mathbf{n}_k(t')\mathbf{n}_k^\dagger(t') \tag{9.16}$$

Once the entire CPI of data is beamformed by the slow-time sequence of adaptive weight vectors, the scalar output needs to be Doppler processed to separate useful signals from the more powerful clutter returns. While the interference is expected to be rejected effectively by this approach, a serious problem arises due to the unwanted interaction between the intra-CPI antenna pattern fluctuations and the unrejected clutter returns that are passed onto the beamformer output. Specifically, the adaptive beampattern readjustments within the CPI temporally modulate the unrejected clutter returns in the beamformer output. This has the deleterious effect of broadening the clutter spectrum after Doppler processing. This operational issue, which can cause target echoes to be masked by clutter energy after Doppler processing, will be described in more detail later.

[2]This applies for single-hop propagation over a relatively quiet mid-latitude ionospheric path.

9.3 Instantaneous Performance Analysis

The interference rejection performance of different adaptive beamforming schemes may be quantified as a function of slow-time to observe how the instantaneous performance changes from one PRI to another within the CPI. Analysis of the instantaneous cancelation ratio sheds light on the intra-CPI performance of different adaptive beamforming schemes applied to interference reflected from the ionosphere. The main purpose of this analysis is to expose a number of important characteristics that are not predicted by simulation results based on traditional models.

9.3.1 Real-Data Collection

The collection of experimental data involved the passive reception of a cooperative HF interference source that radiated a bandlimited white signal. A vertically polarized omnidirectional (whip) antenna was used to emit this radio frequency interference (RFI) signal, which propagated via a one-hop mid-latitude ionospheric path to the Jindalee OTH radar receive system. The cooperative RFI source was located near Darwin in the far field of the 2.8-km long Jindalee receive antenna array. Specifically, the RFI source was at a ground range of 1265 km from the receiver site and offset +22 degrees from the ULA boresight direction. A clear 3 kHz bandwidth channel was exclusively allocated for the experiment at a carrier frequency of 16.050 MHz on 1 April 1998 between 06:22 and 06:32 UT.

It is noted that the channel scattering function (CSF) data analyzed in Chapters 6–8 corresponds to the same mid-latitude ionospheric path and was recorded on the same day immediately prior to this experiment (06:17–06:21 UT) on an adjacent frequency channel (16.110 MHz). The HF channel model parameters estimated from the CSF data are therefore relevant to those experienced by the interference in this experiment.

According to the real-time frequency management system (FMS), the operating frequencies used coincided with near optimum propagation conditions at the time of recording. Hence, experimental results based on this interference data are likely to estimate the impact of time-varying interference wavefront distortions on adaptive beamformer performance conservatively with respect to RFI sources propagated via disturbed ionospheric regions.

The bandlimited white noise signal acquired by the 32 receivers of the Jindalee ULA was observed to spread across the entire range-Doppler map in a significant number of beams after conventional processing. Each CPI consisted of $P = 256$ PRI and $K = 42$ range cells. Although a low PRF (of say 1 Hz) permits interference rejection to be studied over very long CPI (up to 256 seconds), it limits the shortest time interval over which the interference properties are integrated to a minimum of 1 second. On the other hand, increasing the PRF to say 50 Hz allows performance to be observed at a higher temporal resolution of 0.02 seconds, but limits the maximum CPI length that can be analyzed to around 4 seconds.

In this experiment, the selected PRF of 5 Hz represents a compromise between the two competing objectives, yielding a maximum CPI of 50 seconds and a temporal resolution of 0.2 seconds. Based on the previous CSF data analysis, the ionospheric channel under investigation is not expected to vary significantly within a PRI of 0.2 seconds. The RFI source was also switched off momentarily during the experiment in order to record background noise on the same frequency channel. This data serves as a reference to determine how effectively adaptive beamforming is able to reject the interference.

9.3.2 Intra-CPI Performance Analysis

The SINR improvement factor $\hat{q}_m(t)$ for $m = 1, \ldots, 4$ in Eqn. (9.17) measures the relative output SINR improvement (RFI cancelation performance) of adaptive beamforming scheme m as a function of slow-time t. A Hamming taper was used for the conventional beamformer $\mathbf{v}(\theta)$. Note that the instantaneous improvements are computed in $K' = 16$ range cells taken after the first $\Delta_k = 16$ "training" range cells in each PRI (i.e., cross-rejection as opposed to self-rejection).

$$\hat{q}_m(t) = \frac{\mathbf{v}^\dagger(\theta)\hat{\mathbf{R}}_n(t)\mathbf{v}(\theta)}{\hat{\mathbf{w}}_m^\dagger\hat{\mathbf{R}}_n(t)\hat{\mathbf{w}}_m}, \quad \hat{\mathbf{R}}_n(t) = \frac{1}{K'}\sum_{k=\Delta_k+1}^{\Delta_k+K'} \mathbf{n}_k(t)\mathbf{n}_k^\dagger(t) \tag{9.17}$$

The weights $\hat{\mathbf{w}}_1$ are formed using $K = 32$ and $\Delta_p = 6$ in Eqn. (9.9). Curve 1 in Figure 9.2 shows $\hat{q}_1(t)$ as a function of slow-time t over a 50 second CPI when the beam is steered at a cone angle of $\theta = 21.6$ degrees (near the RFI source direction). This curve shows that the filter $\hat{\mathbf{w}}_1$ has a high *initial* effectiveness and yields 30 dB improvement relative to conventional beamforming close to the start of the CPI. However, this improvement drops rapidly and falls to below 5 dB in less than 2 seconds. After about 20 seconds, the filter $\hat{\mathbf{w}}_1$ "ages" to the point where conventional beamforming provides *superior* performance. More than 30 dB variation in performance is not expected under stationary conditions for an adaptive filter trained on $K \times \Delta_p = 6N$ samples. In the traditional model, where the interference modes are assumed to have a time-invariant spatial structure, such a filter would be expected to provide near optimum performance over the whole CPI.

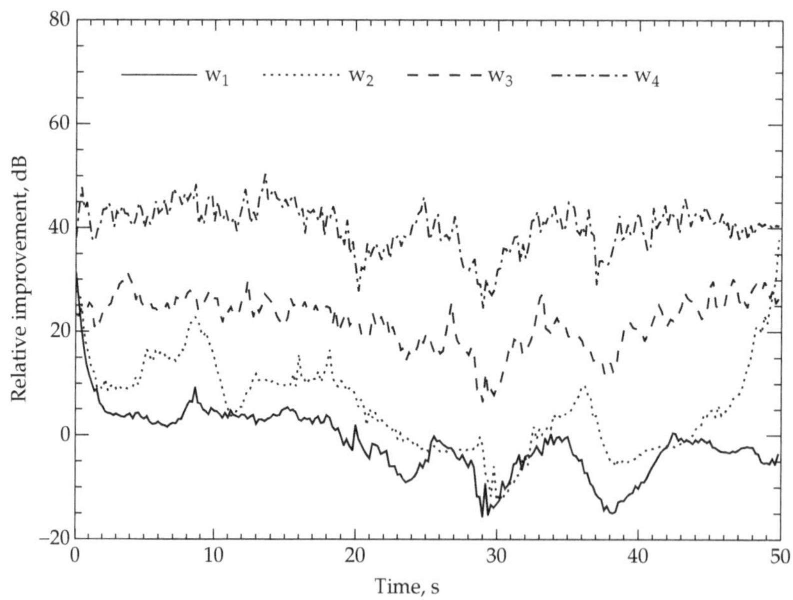

FIGURE 9.2 Relative performance improvements achieved by the four adaptive beamforming schemes as a function of PRI number over the CPI. © Commonwealth of Australia 2011.

The small-scale random variations of $\hat{q}_1(t)$, possibly caused by finite sample effects, are superimposed on a relatively smooth large-scale variation that exhibits a physical (wavelike) characteristic between 20 and 50 seconds. These large-scale changes in rejection performance are not predicted by standard models and can only be attributed to variations in the interference spatial structure over the CPI. Based on this experimental result, it is evident that adaptive scheme 1 is quite ineffective for the purpose of sky-wave HF interference mitigation when the required CPI length exceeds approximately half a second.

Curve 2 in Figure 9.2 shows the results for $\hat{q}_2(t)$, where $\hat{\mathbf{w}}_2$ is formed according to Eqn. (9.11) using identical parameters as adaptive scheme 1 $\{K = 32, \Delta_p = 6\}$. The same number of training vectors are used to estimate $\hat{\mathbf{w}}_1$ and $\hat{\mathbf{w}}_2$. The range cells processed by $\hat{\mathbf{w}}_1$ and $\hat{\mathbf{w}}_2$ are also the same, so $\hat{q}_1(t)$ and $\hat{q}_2(t)$ can be meaningfully compared. Clearly, scheme 2 is a block processing algorithm that requires memory (data storage), whereas scheme 1 is a streaming technique that may be applied as the data is received. A comparison of curves 1 and 2 shows that scheme 2 performs better than scheme 1 over practically all of the CPI.

In particular, scheme 2 performs 20–30 dB better than scheme 1 in the final second of the CPI. This is expected because scheme 2 is based only on an estimate of the covariance matrix that averages the interference spatial statistics immediately before and after the CPI, whereas scheme 1 is based only on an estimate formed prior to the CPI. The performance of scheme 2 in the middle portion of CPI degrades by about 20 dB relative to that observed near the extremities. This is also expected because adaptive scheme 2 has no "knowledge" of the variations in RFI spatial structure over the middle region of the CPI. In general, the spatial properties of time-varying interference in this region of the CPI cannot be deduced or estimated accurately from data received outside the CPI.

Averaging the sample spatial covariance matrices estimated close to both ends of the CPI seems to provide a better estimate of the interference received within the CPI than using only the matrix formed prior to the CPI. This is evidenced by the 10 dB improvement of scheme 2 relative to scheme 1 in the middle region of the CPI. However, the results demonstrate that scheme 2 is also unsuitable for skywave HF interference mitigation in applications where the CPI length exceeds about 1 second. These observations strongly motivate the use of adaptive schemes that operate on estimates of the interference spatial covariance matrix formed *within* the CPI, such as adaptive scheme 3.

Curve 3 in Figure 9.2 shows the relative improvement $\hat{q}_3(t)$ for adaptive scheme 3. The adaptive weight vector $\hat{\mathbf{w}}_3$ in Eqn. (9.14) is estimated using $\Delta_k = 16$ training ranges using the sample covariance matrix averaged over the entire CPI, i.e., integrated over all P PRI. The resulting time-invariant spatial filter is then used to process the "operational" range cells $k = \Delta_k + 1, \ldots, \Delta_k + K'$ in each PRI using $K' = 16$ as before. Unlike the framing schemes, adaptive scheme 3 is based on an estimate of the interference spatial covariance matrix averaged during the CPI to be processed.

The performance of scheme 3 is observed to be 15–20 dB better than scheme 2 over almost all of the CPI. This approach may be quite effective for skywave HF interference mitigation, particularly in aircraft detection OTH radar applications that use a relatively short CPI. However, for a long CPI (e.g. ship detection), or when the interference fluctuations are rapid due to disturbed ionospheric propagation, scheme 3 may not be able to reject the interference effectively.

A remaining option to mitigate against interference spatial structure fluctuations is to update the weight vector within the CPI. Curve 4 in Figure 9.2 shows the improvement

$\hat{q}_4(t)$ corresponding to the slow-time varying weight vector $\hat{\mathbf{w}}_4(t)$. This sequence of spatial filters is computed using Eqn. (9.16), where $\Delta_k = 16$ training cells are used in non-overlapping sub-CPIs each containing $\Delta_p = 4$ PRI. A substantial performance gain in the order of 30–40 dB relative to the conventional beamformer results for scheme 4. A comparison between the time-invariant solution $\hat{\mathbf{w}}_3$ (curve 3) and the time-varying solution $\hat{\mathbf{w}}_4(t)$ (curve 4) demonstrates that re-adapting the weights during the CPI yields an additional 15–20 dB of interference cancelation. It will be shown later that scheme 4 rejects the interference to the background noise floor.

The spatial processors $\hat{\mathbf{w}}_4(t)$ and $\hat{\mathbf{w}}_3$ are clearly of the same dimension $N = 16$, but the number of adaptive degrees of freedom required for effective HF interference rejection tends to grow as the integration time increases. This is primarily caused by the time-varying wavefront distortions imparted on the received interference modes by the ionospheric reflection process. As these distortions evolve in a correlated manner over time, limiting the integration time serves to reduce the effective dimension of the interference subspace. This often allows an adaptive beamformer to cancel the interference more effectively. Exceptions to this may arise in the special case of main beam interference or when the RFI is propagated by the much more stable surface-wave mode.

Figure 9.3, in the same format as Figure 9.2, shows analogous results for a subsequent CPI of data. In this example, the beam steer cone angle of $\theta = 20.8$ degrees is slightly further away from the direction of the RFI source. Although the corresponding curves differ in detail, the qualitative characteristics of the four curves in Figure 9.3 are similar to those of Figure 9.2. To quantify this variability, the statistical performance of the four schemes needs to be evaluated over different beam steer directions and CPI by extended data

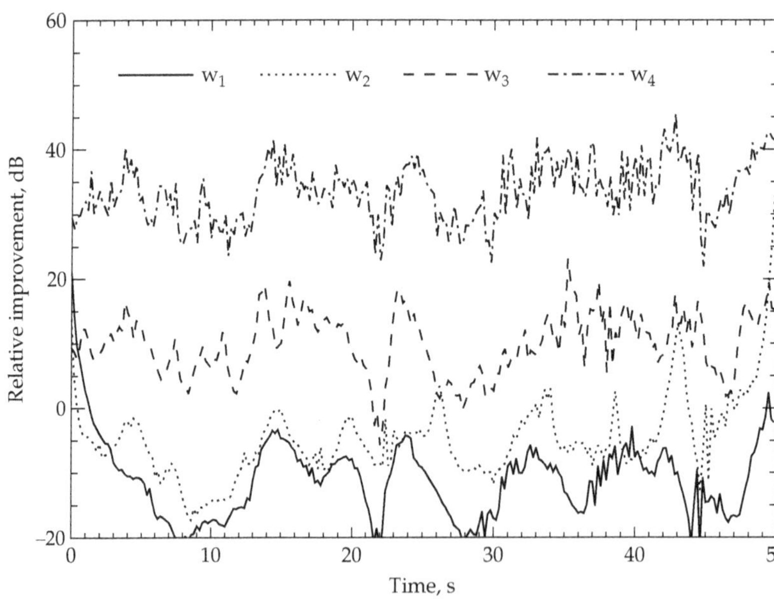

FIGURE 9.3 Relative performance improvements for the same adaptive schemes using a different CPI and beam steer direction. © Commonwealth of Australia 2011.

processing. An estimate of the average or expected performance can then be provided for the different adaptive beamforming schemes as a function of CPI length.

9.3.3 Output SINR Improvement

Before proceeding to the statistical performance analysis, the question arises as to whether adaptive scheme 4 rejects the interference to the background noise level. To answer this question, consider the Doppler spectra in Figure 9.4. Curves 1 and 3 show the Doppler spectra resulting for the conventional beamformer when the interference is present and absent, respectively. An ideal synthetic target has been injected with a normalized Doppler frequency of 0.5 in both cases. The target is clearly visible when only background noise is present (curve 3). Note that this is actual background noise recorded at the same frequency as the interference source when the latter was switched off during the experiment. When the interference is additionally present (curve 1), the target is submerged and cannot be detected in the conventional beamformer output.

Curve 2 in Figure 9.4 shows the Doppler spectrum resulting at the output of adaptive scheme 4 when the interference is *present*. In this case, the adaptive weights $w_4(t)$ were updated from one PRI to another over the CPI with diagonal loading to improve convergence. The slow-time varying weight vectors were then applied to beamform the range cell containing the synthetic target, which was not included in the training data. A comparison of curves 1 and 2 in Figure 9.4 demonstrates that re-adapting the antenna pattern within the CPI attenuates the interference by an additional 40 dB relative to the conventional beamformer.

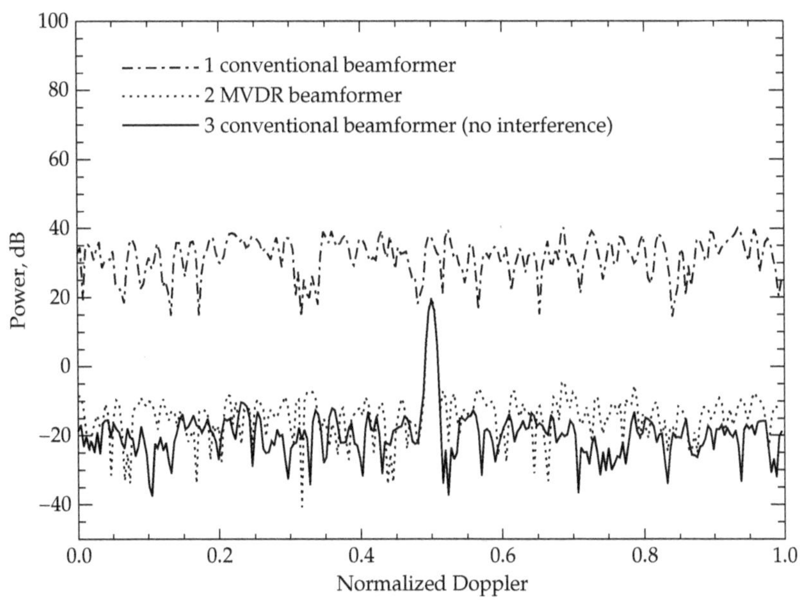

FIGURE 9.4 Conventional Doppler spectra for a simulated target signal in real HF interference (curve 1) and real background noise (curve 3). Curve 2 shows the Doppler spectra resulting when adaptive scheme 4 is applied with interference *present*. © Commonwealth of Australia 2011.

It is evident from curves 2 and 3 in Figure 9.4 that both adaptive and conventional beamformers provide identical response to the target echo. This demonstrates the equivalence between interference cancelation ratio and relative SINR improvement for the considered schemes assuming ideal useful signals. The cancelation ratio translates to an SINR improvement provided the ideal useful signal is not included in the training data. Importantly, curves 2 and 3 also demonstrate that adaptive scheme 4 has practically rejected the interference to the background noise level. Application of scheme 4 therefore has the potential to restore the output SINR to that of the conventional beamformer on a clear frequency channel. From an interference rejection and ideal signal reception viewpoint, the performance of adaptive scheme 4 is as good as can be.

9.4 Statistical Performance Analysis

The previous results illustrated a number of key points in detail, but a more comprehensive performance analysis is required to quantify the relative merits and shortcomings of the different adaptive beamforming schemes. As the relative improvement factors $\{\hat{q}_m\}_{m=1}^{4}$ are random variables, it is of interest to measure the mean and upper/lower deciles of their *distributions* as a function of CPI length by analyzing more data.

9.4.1 Framing Schemes

Figure 9.5 shows the mean and deciles of the distribution of $\hat{q}_1$ for a number of different CPI lengths. For each interrogated CPI length, a total of 10 mutually orthogonal beam

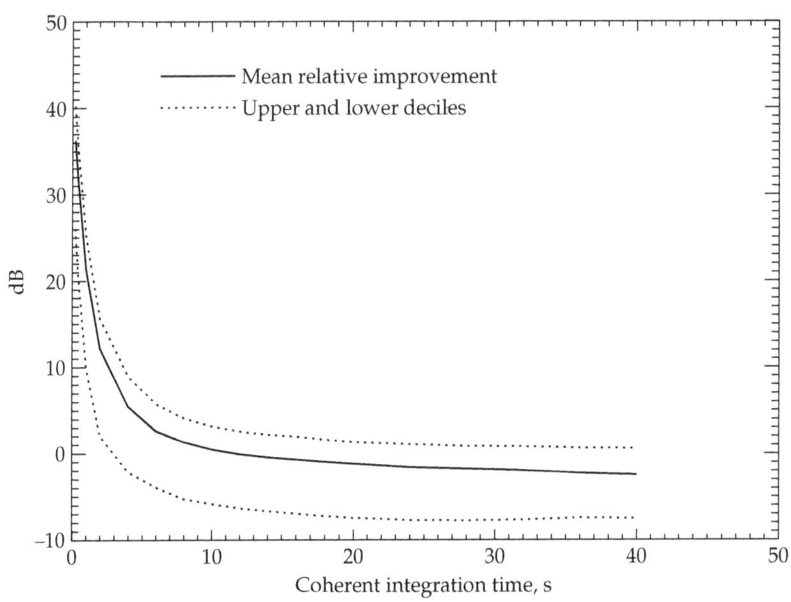

FIGURE 9.5 Mean and deciles of the improvement in interference rejection performance achieved by adaptive scheme 1 ($\hat{q}_1$) over the conventional beamformer as a function of CPI length. © Commonwealth of Australia 2011.

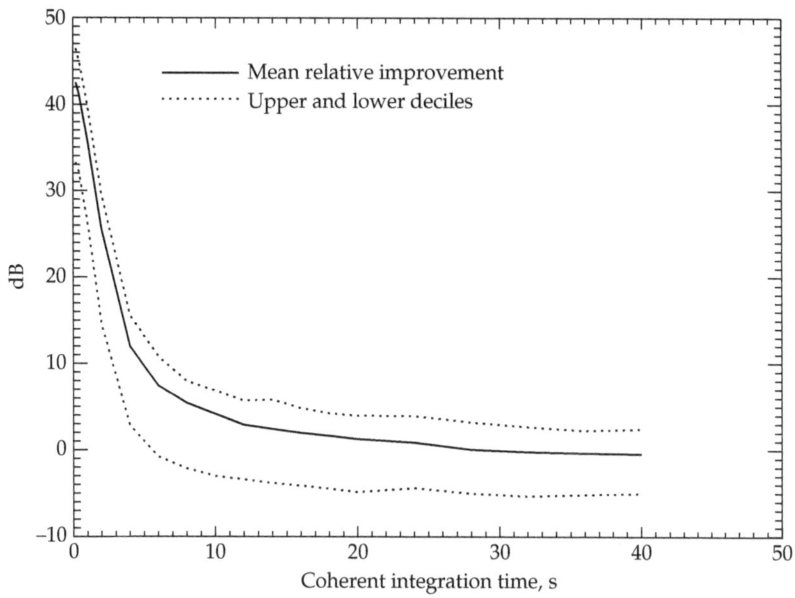

FIGURE 9.6 Mean and deciles of the improvement in interference rejection performance achieved by adaptive scheme 2 ($\hat{q}_2$) over the conventional beamformer as a function of CPI length. © Commonwealth of Australia 2011.

steer directions were processed in 40 different CPIs such that the distribution for each CPI length is based on 400 samples of the improvement factor. Figure 9.5 demonstrates that the mean relative improvement drops from above 30 dB to below 0 dB as the CPI is increased from a fraction of a second to 40 seconds. The fall in performance of adaptive scheme 1 is observed to be very rapid, as over 20 dB of interference rejection is on average lost in less than 2 seconds. The higher computational load of adaptive scheme 1 relative to conventional beamforming is therefore not warranted for OTH radar applications according to these experimental results.

Figure 9.6, in the same format as Figure 9.5, shows the mean and deciles of $\hat{q}_2$ to determine the performance of adaptive scheme 2. The mean relative improvement is above 40 dB for very short CPI, but decays to 25 dB after only two seconds and then down to almost 0 dB when the CPI reaches 30 seconds. Although the performance of scheme 2 decays less rapidly compared to that of scheme 1, the significant performance degradation for CPI lengths greater than about 2 seconds may not be tolerable when powerful interference is present. Based on these experimental results, it is evident that scheme 2 may also not perform sufficiently well in practice to justify operational implementation.

9.4.2 Batch Schemes

Figure 9.7 shows the mean and deciles of the improvement factor $\hat{q}_3$ for scheme 3. The maximum mean value of about 43 dB occurs at a CPI length slightly above 1 second. A degradation of up to 8 dB is observed for shorter CPI due to estimation errors caused by finite sample support. While for longer CPI lengths, the degradation is caused by

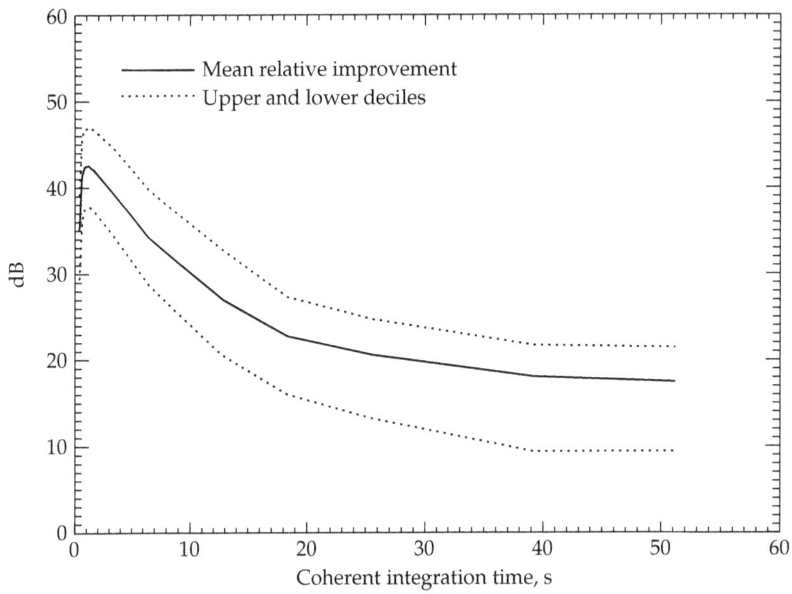

FIGURE 9.7 Mean and deciles of the improvement in interference rejection performance achieved by adaptive scheme 3 ($\hat{q}_3$) over the conventional beamformer as a function of CPI length. © Commonwealth of Australia 2011.

fluctuations in interference spatial structure. Hence, the best performance for scheme 3 occurs at a CPI length that provides an optimal tradeoff between convergence losses due to finite sample support at very short CPI, and dynamic losses caused by interference fluctuations at longer CPI. Adaptive scheme 3 provides quite acceptable performance for CPI lengths shorter than about 4 seconds in the analyzed data set.

Moreover, diagonal loading can be applied to reduce losses due to limited training data. This is illustrated in Figure 9.8, where the statistics have been recalculated using a loading factor of −20 dB ($\alpha = 0.01$). This simple modification increases the mean relative SINR improvement from 35 to 45 dB for CPIs under a second, a remarkable 10 dB increase with respect to the case where no loading is applied. However, the 10 dB performance degradation observed for adaptive scheme 3 as the CPI increases from 1 to 8 seconds cannot be recovered as this is caused by fluctuations of the interference spatial structure. The more significant performance degradation of around 25 dB for CPI approaching 50 seconds also poses a problem in OTH radar applications that require high Doppler resolution.

For this reason it is of interest to determine the potential effectiveness of scheme 4, which adjusts the spatial filter to process the data in batches consisting of relatively few consecutive PRI. The performance of adaptive scheme 4 can also be inferred from Figure 9.7, since every batch of PRI may be considered as a sub-CPI of the total CPI. In other words, if the adaptive weights are updated every second using scheme 4, the mean relative improvement of about 43 dB observed for this update rate in Figure 9.7 can be attained for a CPI that is arbitrarily longer. For example, 43 dB of additional interference

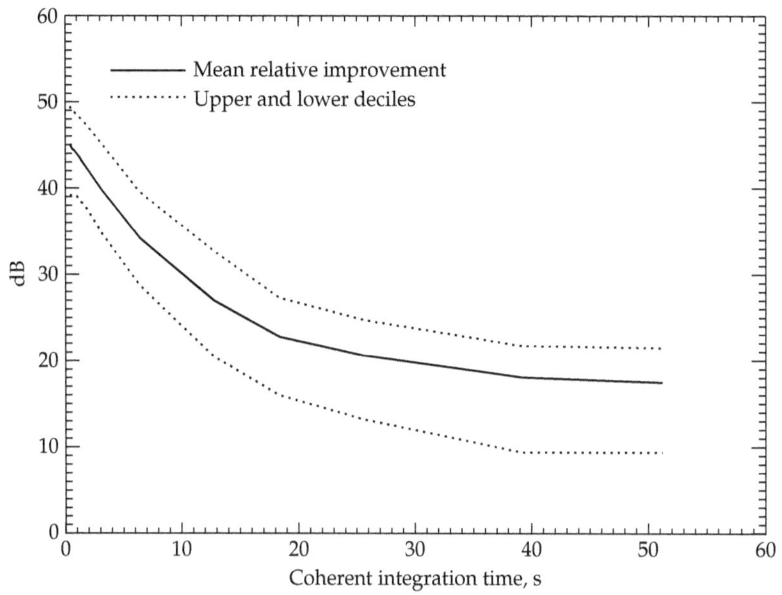

FIGURE 9.8 Mean and deciles of the improvement in interference rejection performance achieved by adaptive scheme 3 *after diagonal loading* relative to the conventional beamformer as a function of CPI length. © Commonwealth of Australia 2011.

cancelation relative to the conventional beamformer could be obtained for a CPI of 50 seconds because the performance of adaptive scheme 4 is determined by the batch length as opposed to the CPI length.

9.4.3 Operational Issues

In relation to adaptive scheme 4, the main operational issue relates to the application of a time-varying array weight vector to a CPI of data that also contains *clutter*. Due to the relatively broad angular coverage of the OTH radar transmit beam, the clutter has a spatially wideband characteristic as the backscatter is incident from a continuum of directions that occupy a significant portion of the receive antenna radiation pattern. Unconstrained fluctuations of the receive antenna pattern during the CPI imposes a temporal modulation on the output clutter signal that causes it to spread across the target velocity search space after Doppler processing.[3] This issue represents the main impediment to the practical application of scheme 4.

Hence, the problem faced is to not only allow adaptive beampattern readjustments within the CPI to enhance interference rejection, but also to simultaneously control these variations so as to preserve the temporal correlation properties of the backscattered clutter at the beamformer output. Alternative approaches that address this problem are

[3] Recall that the response of the adaptive antenna pattern is only constrained in the beam steer direction, where the array provides fixed unity gain response to ideal useful signals. In other directions, the response of the receive antenna pattern is unconstrained for the adaptive schemes considered.

described in the next chapter. Other operational issues that can potentially limit performance include the lack of clutter-free training data and mismatches between the useful signal structure and the presumed steering vector model. These issues will also be discussed in the following chapter. Clearly, the computational complexity of scheme 4 is higher than those of schemes 1 to 3. This may limit scope for real-time implementation. Trading off batch length with performance represents an option to reduce the computational complexity of scheme 4.

9.5 Simulated Performance Prediction

The space-time HF channel model described in the previous chapter may be used to simulate the statistical characteristics of the interference received by the Jindalee antenna array. The temporal parameters of the HF channel model cannot be estimated from the interference because this signal is not coherent with the radar waveform. However, the interference was received over the same propagation path at a similar time and frequency to the channel scattering function (CSF) data analyzed in Chapters 6–8. The channel parameters estimated from the CSF data may therefore be used to simulate the interference. The capability of the HF channel model to predict adaptive beamformer performance results observed on experimental data can then be assessed.

9.5.1 Multi-Channel Model Parameters

The model parameters used for the simulation are listed in Table 9.1. The relative power ratios among the different interference modes are assumed to be the same as those of the CSF data modes, but scaled up in absolute power such that the the total power of the modeled interference equals the total power measured for the actual interference received by the system. The received interference power was estimated by calculating the mean square value of the data in each receiver over the data collection interval and then averaging these results over the different receivers. The average power of the received interference was 30.2 dB.

Note that the mode powers listed in Table 9.1 are not equal to the mode interference-to-noise ratios (INR), since the background noise level is at approximately −27 dB when measured in a single receiver *prior* to Doppler processing. The mode INRs can be therefore

Mode	Power, dB	Temporal Pole, α	Spatial Pole, β
$1E_s(a)$	20.5	$0.998\, e^{jk_1 0.46}$	$0.920\, e^{jk_2 \sin(22.2-\theta_s)}$
$1E_s(b)$	17.9	$0.997\, e^{jk_1 0.39}$	$0.963\, e^{jk_2 \sin(21.7-\theta_s)}$
$1F_2$	28.6	$0.994\, e^{jk_1 0.44}$	$0.988\, e^{jk_2 \sin(20.7-\theta_s)}$
$1F_2(o)$	21.8	$0.997\, e^{jk_1 0.47}$	$0.986\, e^{jk_2 \sin(20.5-\theta_s)}$
$1F_2(x)$	7.9	$0.994\, e^{jk_1 0.53}$	$0.953\, e^{jk_2 \sin(19.9-\theta_s)}$

TABLE 9.1 Interference simulation parameters assuming five propagation modes and the HF channel model estimated from the channel scattering function data. The constants are $k_1 = 2\pi/f_p$ where $f_p = 5$ Hz and $k_2 = 2\pi\,\Delta d/\lambda$, where $\Delta d = 84$ m, $\lambda = c/f_c = 18.7$ m, and the subarray steer direction $\theta_s = 22.0$ degrees.

be calculated by adding 27 dB to the powers listed in Table 9.1. The noise added to the simulated interference is real background noise recorded on the same frequency channel in the absence of interference. Further aspects that need to be addressed in the simulation are the generation of the interference mode waveforms $g_k(t - \tau_m)$ with time delays τ_m, and the correlation among the different modes.

The interference signal has a bandwidth of $f_b = 3$ kHz with a relatively uniform power spectral density. The temporal auto-correlation sequence of the interference waveform is a sinc function with the first null at time delay $\tau = 1/f_b = 0.33$ ms. As the differential mode time-delays are known from the oblique incidence ionogram recorded for the CSF data analysis, each pair of modes can be assigned an inter-mode correlation coefficient according to the interference auto-correlation function. These correlation coefficients $\rho_{i,j}$ for all pairs of modes $i, j = 1, 2, \ldots, M = 5$ are entered into the source covariance matrix $\mathbf{R}_s$ in Eqn. (9.18), where the diagonal elements of $\mathbf{R}_s$ are unity as the mode waveforms $g_k(t - \tau_m)$ are normalized to unit variance by definition in the model.

$$\mathbf{R}_s = E\{\mathbf{g}_k\mathbf{g}_k^\dagger\} = [\rho_{i,j}]_{i,j=1}^M, \quad \mathbf{g}_k = [g_k(t-\tau_1)\cdots g_k(t-\tau_M)]^T \tag{9.18}$$

Once the source covariance matrix has been determined, the interference mode waveforms can be generated simultaneously using Eqn. (9.19), where $\mathbf{R}_s^{1/2}$ is the Hermitian square root of the source covariance matrix. The statistical independence of the zero-mean complex Gaussian vectors $\mathbf{n}_k(t)$ over range k and slow-time t implies that the interference samples generated for a particular mode are white and hence uncorrelated in both of these data dimensions.

$$\mathbf{g}_k(t) = \mathbf{R}_s^{1/2}\mathbf{n}_k(t), \quad E\{\mathbf{n}_{k_1}(t_1)\mathbf{n}_{k_2}^\dagger(t_2)\} = \delta(k_1 - k_2)\delta(t_1 - t_2)\mathbf{I} \tag{9.19}$$

9.5.2 Impact of Wavefront Distortions

In the traditional array-processing model, the interference wavefronts are assumed to remain rigid over the CPI. The absence of time-varying wavefront distortions on the received interference modes implies that the spatial poles of the model lie on the unit circle. It is therefore possible to revert to the traditional model by setting the magnitude of the spatial pole estimated for each interference mode in Table 9.1 to unity. In this case, the interference modes are modeled as plane waves with Doppler shifted and Doppler spread waveforms due to temporal channel fluctuations only. Based on this model, Figure 9.9 shows the mean and deciles of the improvement factor $\hat{\eta}_3$ when the simulated data is processed in identical manner to real data results shown in Figure 9.8.

As the spatial structure of each interference mode is assumed to be time-invariant in the traditional model, the dimension of the interference subspace cannot increase beyond $M = 5$ regardless of the CPI length. This explains why the mean relative improvement in Figure 9.9 remains approximately constant with increasing CPI length. Diagonal loading has been applied here as it was for Figure 9.8. Clearly, the traditional plane wave interference model does not provide an acceptable representation of the experimental results in Figure 9.8. The substantial difference between Figures 9.8 and 9.9 illustrate that results based on this model are of limited value for OTH radar systems, as they do not provide an accurate indication of practical performance in the HF environment.

Figure 9.10 shows the recomputed statistics for $\hat{\eta}_3$ after the time-varying interference wavefront distortions are introduced using the model parameters in Table 9.1. The simulation results in Figure 9.10 show that the mean relative improvement drops from 46 dB

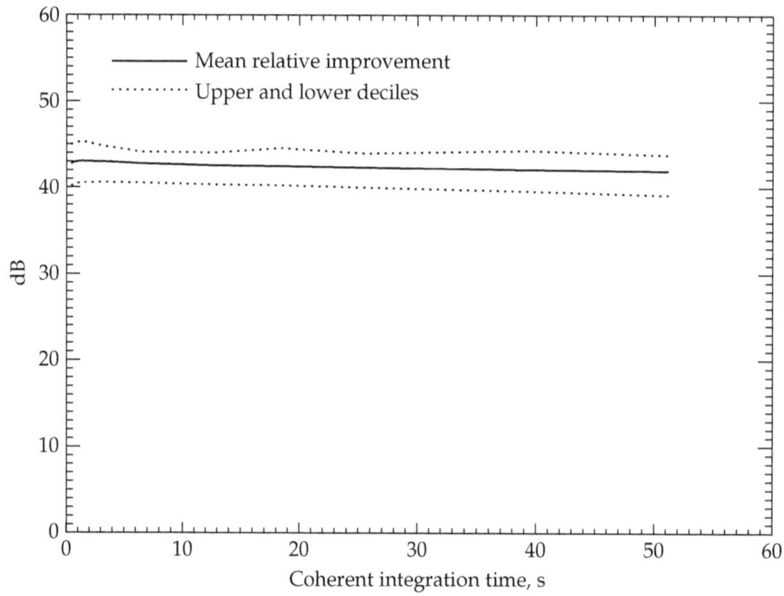

FIGURE 9.9 Relative improvement achieved by adaptive scheme 3 (after diagonal loading) for simulated HF interference with temporal distortions but in the *absence* of spatial distortions. © Commonwealth of Australia 2011.

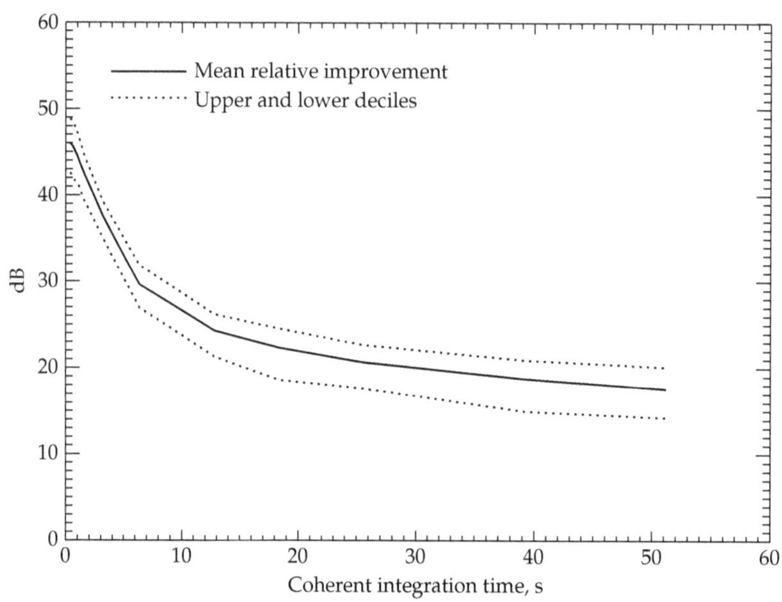

FIGURE 9.10 Relative improvement achieved by adaptive scheme 3 (after diagonal loading) for simulated HF interference with the inclusion of space-time distortions. © Commonwealth of Australia 2011.

for a CPI length of less than 1 second to 18 dB for a CPI of about 50 seconds. This agrees well with the experimental results in Figure 9.8, which report a drop in mean performance from 45 dB to 17 dB over the same range of CPI lengths. The detailed shapes of the curves in Figure 9.8 and Figure 9.10 are clearly not the same, but the described HF channel model and estimated parameters provide a quite accurate description of adaptive beamformer performance in the HF environment. A similar experiment demonstrating the close agreement between simulated and experimental performance results for adaptive beamforming scheme 1 using different CSF and interference data sets can be found in Fabrizio et al. (1998).

9.5.3 Summary and Discussion

The presence and characteristics of wavefront distortions imparted on HF signal modes reflected by the ionosphere were confirmed, analyzed, and modeled in Chapters 6–8. This case study quantified the operational impact of such phenomena on the performance of four standard adaptive beamforming schemes applied to the problem of interference rejection in OTH radar systems. As opposed to useful signals, which are received in the relatively broad main lobe of the beam pattern, interference is typically received near the deep "nulls" of an adaptive antenna pattern, where the array response is much more sensitive to time-varying wavefront distortions. The first important observation is that dynamic wavefront distortions imparted on *interference* signal modes have the potential to dramatically degrade the *rejection* performance of HF systems when time-invariant adaptive beamforming schemes are used to process each CPI.

Second, it has been shown that traditional array signal-processing models poorly represent adaptive beamformer performance in the HF environment. Such models are therefore not suitable for guiding algorithm design or predicting practical performance in OTH radar systems. For applications that require high fidelity, it was also shown that the previously validated HF channel model could be used to simulate HF interference signals received by a very wide aperture antenna array and accurately predict the statistical performance of an adaptive beamformer as a function of CPI length.

The final observation is that the presence of time-varying interference wavefront distortions needs to be accounted for in order to develop effective adaptive beamforming algorithms for operational OTH radar systems. In particular, methods that can stabilize the Doppler spectrum characteristics of the backscattered clutter signals when the adaptive beamforming weights change several times during the CPI to ensure effective interference rejection are required. Without such methods, the potential benefits of updating the spatial filter within the CPI cannot be realized in practice. This motivates the search for alternative adaptive beamforming techniques suitable for real-time implementation in OTH radar.

PART III

Processing Techniques

CHAPTER 10

Adaptive Beamforming

Adaptive processing has the potential to significantly enhance the target detection and tracking performance of OTH radar systems because of the statistically structured or "colored" nature of HF disturbance signals often encountered in practice. Indeed, the application of robust spatial, temporal, or joint space-time adaptive filtering techniques may typically be expected to mitigate strong disturbances such as clutter and interference more effectively than conventional (data-independent) processing. This chapter is concerned with the application of *adaptive beamforming* to improve the output useful-signal-to-disturbance ratio (SDR) in OTH radar surveillance applications.

Specifically, two fundamentally different adaptive beamforming problems will be considered. The first is related to the rejection of ionospherically-propagated HF interference signals that arise from sources independent of the OTH radar waveform, where it is assumed that such signals impinge upon the sidelobes of the receive antenna beam pattern. The special case where interference signals are received in the main lobe of the surveillance beam will be dealt with in the following chapter. A challenge faced by OTH radar systems is that skywave propagation via disturbed ionospheric paths can lead to the reception of interference with time-varying (*non-stationary*) spatial structure over the CPI. The second adaptive beamforming problem considered is the mitigation of clutter echoes, which may be regarded as statistically homogeneous when attention is restricted to a single range cell and CPI, but frequently exhibit spatial statistics that are highly heterogeneous in range due to the variation of scatterer distribution with time delay.

The chapter is divided into four sections. A brief summary of essential concepts in optimum and adaptive filtering is provided in Section 10.1 for readers not familiar with these areas, as well as to introduce terminology and notation. Section 10.2 formulates the spatially non-stationary HF interference rejection problem in the context of OTH radar and highlights the key issues to be resolved. Two alternative time-domain (pre-Doppler) adaptive beamforming algorithms that can effectively address this problem are then described and assessed using experimental data in Section 10.3. The above-mentioned clutter mitigation problem is motivated and illustrated for an HF OTH passive coherent location (PCL) system in Section 10.4. In particular, this section describes the implementation of a frequency-domain (post-Doppler) adaptive beamforming technique and evaluates its practical effectiveness using carefully benchmarked real data that contains clutter returns with range-dependent spatial statistics and useful echoes from a cooperative aircraft target.

10.1 Essential Concepts

The essential concepts of optimum and adaptive filtering may be summarized in rather general terms that are applicable to data processing in one or more dimensions, including adaptive beamforming discussed in this chapter, and space-time adaptive processing (STAP) to be discussed in the following chapter. The main purpose of this section is to provide a broad overview of the key principles in optimum and adaptive filtering that are relevant to OTH radar systems. This not only serves as background, but also defines the terminology and notation used in subsequent sections of this chapter.

As the topic of adaptive filtering represents a vast area, even when attention is restricted solely to radar applications, a detailed description of the general theory (beyond the key principles applicable to OTH radar) is outside the scope of this book. Comprehensive treatments of adaptive filter theory and its application to temporal, spatial, and space-time processing can be readily found in a number of excellent texts, including Haykin (1996), Li and Stoica (2006), and Klemm (2004), in addition to other references that will be provided in due course for readers interested in delving further.

10.1.1 Optimum and Adaptive Filters

The radar is assumed to collect data over a CPI of length T_o seconds that consists of P pulses (sweeps) of the radar waveform transmitted at a pulse repetition frequency (PRF) of f_p pulses per second. The receiving system is assumed to be an array of N sensors, composed of antenna elements or sub-arrays, for example, with each sensor output connected to an individual digital receiver. After down-conversion and baseband filtering, the in-phase and quadrature (I/Q) receiver outputs are sampled at the Nyquist rate of f_r samples per second, with K complex digital samples acquired in each pulse repetition interval (PRI). The K samples acquired in a particular PRI at the rate f_r are known as *fast-time* samples (i.e., range bins), while the sequence of P samples collected in a particular range bin at the rate f_p (i.e., across PRI) are referred to as *slow-time* samples. Based on this description, the radar acquisition during a single CPI may be visualized as a data cube of $N \times P \times K$ complex-valued samples. The CPI of data collected by a repetitive pulse-waveform (PW) or continuous-waveform (CW) radar system may both be described in this manner.

Let the column vector $\mathbf{z} \in \mathcal{C}^N$ in Eqn. (10.1) contain N data samples $\{z(n)\}_{n=1}^N$ to be processed in a particular filtering application. This vector is extracted from the data cube in a manner that depends on the type of filtering to be performed. For example, the set of samples $\{z(n)\}_{n=1}^N$ may correspond to the data received by the N sensors of the array in an element-space beamforming application; whereas in a two-dimensional (space-time) filtering application, $\mathbf{z}$ becomes a stacked vector containing samples extracted from the data cube jointly across reception channels as well as fast-time and/or slow-time. The presence of useful signals is sought in a set of data sample vectors, which are collectively referred to as *primary* data or "test cells" in radar terminology. For the time being, we do not identify the different members of this set and restrict our attention to the single (unnamed) primary-data vector $\mathbf{z}$.

$$\mathbf{z} = [z(1), \ldots, z(N)]^T \tag{10.1}$$

Since N was previously defined as the number of antenna sensors, the discussion may be directly related to element-space beamforming, and reference to this specific application will be made when it is convenient to lend concreteness by practical example. However, the basic concepts to be described are quite general and are equally applicable to filters operating in one or more alternative data-cube dimensions. It is therefore possible to interpret **z** as a sample vector extracted from the data cube in a rather arbitrary manner for the time being. For a particular test cell **z** in the CPI, the statistical quantities of interest here are the mean vector **m** and the covariance matrix **Q**, both defined in Eqn. (10.2).

$$\begin{cases} \mathbf{m} = \mathrm{E}\{\mathbf{z}\} \\ \mathbf{Q} = \mathrm{E}\{(\mathbf{z}-\mathbf{m})(\mathbf{z}-\mathbf{m})^\dagger\} \end{cases} \tag{10.2}$$

The primary vector **z** is assumed to contain a disturbance signal **d** and potentially a useful signal denoted by $\mu e^{j\phi}\mathbf{s}$ in Eqn. (10.3). The two alternative conditions on the useful signal amplitude, $\mu > 0$ and $\mu = 0$, represent the presence and absence of the target echo in the test cell, respectively. The useful signal is assumed to be characterized by a deterministic signature vector **s** (also known as the signal vector), defined to be of unit norm $\mathbf{s}^\dagger\mathbf{s} = 1$, and complex scale $a = \mu e^{j\phi}$.

$$\mathbf{z} = \mu e^{j\phi}\mathbf{s} + \mathbf{d} \tag{10.3}$$

The overall disturbance $\mathbf{d} = \mathbf{e} + \mathbf{n}$ is considered to be a zero-mean multi-variate random process that represents the composite sum of all external unwanted signals **e** (e.g., clutter, interference, ambient noise) plus internal receiver noise of thermal origin **n**. In this case, the primary vector has a mean $\mathbf{m} = a\mathbf{s}$ and covariance matrix $\mathbf{Q} = \mathrm{E}\{\mathbf{d}\mathbf{d}^\dagger\}$. For white thermal noise of power σ_n^2, uncorrelated with all other signals, **Q** takes the form of Eqn. (10.4), where $\mathbf{M} = \mathrm{E}\{\mathbf{e}\mathbf{e}^\dagger\}$ is the external disturbance covariance matrix.

$$\mathbf{Q} = \mathbf{M} + \sigma_n^2\mathbf{I} \tag{10.4}$$

The optimum filter is defined as the complex weight vector $\mathbf{w}_o \in \mathcal{C}^N$ that maximizes the statistically expected SDR at the processor output $y_o = \mathbf{w}_o^\dagger\mathbf{z}$. The optimum filter is given by the well-known formula in Eqn. (10.5), where β is an arbitrary scaling term that does not affect the output SDR. For a Gaussian disturbance process, it can be shown that maximizing the output SDR is equivalent to maximizing the probability of detection P_D for a given false alarm rate P_{FA} at the threshold detector (Brennan and Reed 1973).

$$\mathbf{w}_o = \beta\mathbf{Q}^{-1}\mathbf{s} \tag{10.5}$$

The optimum filter is often scaled to provide a known gain to useful signals. Specifically, the minimum variance distortionless response (MVDR) filter, denoted by $\mathbf{w}_a$, minimizes the output disturbance power while maintaining fixed unit gain for useful signals: $\mathbf{w}_a^\dagger\mathbf{s} = 1$. Mathematically, the MVDR criterion may be cast as the following optimization problem (Capon 1969).

$$\mathbf{w}_a = \arg\min_{\mathbf{w}} \mathbf{w}^\dagger\mathbf{Q}\mathbf{w} \;\; \text{subject to} : \mathbf{w}^\dagger\mathbf{s} = 1 \tag{10.6}$$

The MVDR filter solution is given by Eqn. (10.7) and corresponds to a scaling term of $\beta = \{\mathbf{s}^\dagger\mathbf{Q}^{-1}\mathbf{s}\}^{-1}$ in the optimum filter expression of Eqn. (10.5). Such scaling normalizes the filter response to ensure that $\mathbf{w}_a^\dagger\mathbf{s} = 1$ irrespective of the scale and structure of the disturbance covariance matrix **Q**. This provides distortionless processing of useful signals. Note that if $\mathbf{Q} = \sigma_n^2\mathbf{I}$ (i.e., when only white noise is present), the MVDR optimum filter $\mathbf{w}_a$ in Eqn. (10.7) reduces to the conventional matched filter $\mathbf{s}$.

$$\mathbf{w}_a = \mathbf{Q}^{-1}\mathbf{s}\{\mathbf{s}^\dagger\mathbf{Q}^{-1}\mathbf{s}\}^{-1} \tag{10.7}$$

It can be shown that the residual disturbance power at the MVDR filter output $y_a = \mathbf{w}_a^\dagger\mathbf{z}$ is given by p_a in Eqn. (10.8). Note that p_a is functionally dependent on the disturbance covariance matrix $\mathbf{Q}$. It follows that the false-alarm rate can vary significantly as $\mathbf{Q}$ changes when a given threshold is applied to the envelope of the MVDR filter output for detection. Constant false-alarm rate (CFAR) processing is therefore required prior to threshold detection to ensure that a fixed and predictable P_{FA} is maintained for a given threshold setting independently of the scale and structure of $\mathbf{Q}$.

$$p_a = \mathbf{w}_a^\dagger\mathbf{Q}\mathbf{w}_a = \{\mathbf{s}^\dagger\mathbf{Q}^{-1}\mathbf{s}\}^{-1} \tag{10.8}$$

An extension of the MVDR approach that incorporates multiple linear constraints is known as the linearly constrained minimum variance (LCMV) optimization criterion. LCMV is formulated in Eqn. (10.9). Here, the output disturbance power is minimized subject to M linear constraints expressed as $\mathbf{w}^\dagger\mathbf{C} = \mathbf{f}^\dagger$, where $\mathbf{C} \in \mathcal{C}^{N\times M}$ is the constraint matrix, and $\mathbf{f}^\dagger \in \mathcal{C}^{1\times M}$ is the response vector. The M linear constraints normally include the unity gain constraint $\mathbf{w}^\dagger\mathbf{s} = 1$ that is used to protect the useful signal. In addition, there are $M - 1$ *auxiliary* linear constraints, where $M \ll N$ typically. The specification and purpose of the auxiliary linear constraints depends on the application. An example will be discussed in Section 10.3.

$$\mathbf{w}_b = \arg\min_{\mathbf{w}} \mathbf{w}^\dagger\mathbf{Q}\mathbf{w} \;\; \text{subject to} : \mathbf{w}^\dagger\mathbf{C} = \mathbf{f}^\dagger \tag{10.9}$$

The solution to the LCMV optimization problem may be found using the method of Lagrange multipliers (Frost 1972), and takes the form of $\mathbf{w}_b$ in Eqn. (10.10). In the special case of $M = 1$, the LCMV filter clearly reverts to the MVDR filter when $\mathbf{C} = \mathbf{s}$ and $\mathbf{f} = 1$. However, for $\mathbf{Q} = \sigma_n^2\mathbf{I}$, the LCMV filter $\mathbf{w}_b = \mathbf{C}[\mathbf{C}^\dagger\mathbf{C}]^{-1}\mathbf{f}$ does not revert to the matched filter $\mathbf{s}$ if multiple linear constraints are used ($M > 1$).

$$\mathbf{w}_b = \mathbf{Q}^{-1}\mathbf{C}[\mathbf{C}^\dagger\mathbf{Q}^{-1}\mathbf{C}]^{-1}\mathbf{f} \tag{10.10}$$

The residual disturbance power at the LCMV filter output is given by p_b in Eqn. (10.11), where $p_b \geq p_a$. The inclusion of auxiliary linear constraints can in certain circumstances increase the output disturbance power p_b significantly with respect to p_a. For this reason, the benefit provided by additional linear constraints needs to be weighed up against the potential loss in disturbance suppression due to a reduction in filter degrees of freedom.

$$p_b = \mathbf{f}^\dagger[\mathbf{C}^\dagger\mathbf{Q}^{-1}\mathbf{C}]^{-1}\mathbf{f} \tag{10.11}$$

A performance metric that is commonly used to evaluate the effectiveness of optimum filters is the loss factor $\varrho(\mathbf{w}) \in [0, 1]$ defined in Eqn. (10.12). For $\mathbf{w}^\dagger\mathbf{s} = \mathbf{s}^\dagger\mathbf{s} = 1$, this factor expresses the loss in SDR due to the presence of an external disturbance when the optimum filter is applied (i.e., $\mathbf{w} = \mathbf{w}_a$ or $\mathbf{w} = \mathbf{w}_b$) relative to the SNR arising at the output of the matched filter when only white noise is present. A value of unity indicates that the optimum filter has rejected the external disturbance to the (matched filter) white-noise power level given by σ_n^2.

$$\varrho(\mathbf{w}) = \frac{\sigma_n^2}{\mathbf{w}^\dagger \mathbf{Q} \mathbf{w}} \tag{10.12}$$

Application of the optimum filter requires exact knowledge of the useful signal vector $\mathbf{s}$ and disturbance covariance matrix $\mathbf{Q}$. This is colloquially referred to as the "clairvoyant" case. Obviously, the optimum filter cannot be implemented in practice due to imperfect knowledge of both $\mathbf{s}$ and $\mathbf{Q}$. The unknown quantities $(\mathbf{s}, \mathbf{Q})$ are typically replaced by estimates, denoted by $(\mathbf{v}, \hat{\mathbf{R}})$, which may be substituted into Eqn. (10.5) to approximate the optimum filter as $\hat{\mathbf{w}}_o$ in Eqn. (10.13) for an arbitrary scaling β.

$$\hat{\mathbf{w}}_o = \beta \hat{\mathbf{R}}^{-1} \mathbf{v} \tag{10.13}$$

Estimates of the unknown quantities may be derived from the received data or analytical models based on physical principles. The term *adaptive* refers to techniques that estimate an unknown quantity, such as the disturbance covariance matrix $\mathbf{Q}$, from the received data. As adaptive filters rely on imprecise estimates, such filters experience a loss in the statistically expected output SDR relative to the optimum filter $\mathbf{w}_o$. In Eqn. (10.14), $\mathcal{L}(\mathbf{v}, \hat{\mathbf{R}})$ is defined as the output SDR when the true signal vector $\mathbf{s}$ is replaced by $\mathbf{v}$, and the true disturbance covariance matrix $\mathbf{Q}$ is replaced by $\hat{\mathbf{R}}$, to form the filter estimate $\hat{\mathbf{w}}_o$.

$$\mathcal{L}(\mathbf{v}, \hat{\mathbf{R}}) = \frac{|a \hat{\mathbf{w}}_o^\dagger \mathbf{s}|^2}{\hat{\mathbf{w}}_o^\dagger \mathbf{Q} \hat{\mathbf{w}}_o} = \frac{\mu^2 |\mathbf{v}^\dagger \hat{\mathbf{R}}^{-1} \mathbf{s}|^2}{\mathbf{v}^\dagger \hat{\mathbf{R}}^{-1} \mathbf{Q} \hat{\mathbf{R}}^{-1} \mathbf{v}} \tag{10.14}$$

The *relative loss* in output SDR caused by estimation errors (adaptivity) with respect to the true optimum filter is of considerable interest. The relative SDR loss is defined by the factor $\rho(\mathbf{v}, \hat{\mathbf{R}}) \in [0, 1]$ in Eqn. (10.15). The denominator simplifies to $\mathcal{L}(\mathbf{s}, \mathbf{Q}) = \mu^2 \mathbf{s}^\dagger \mathbf{Q}^{-1} \mathbf{s}$, which is the maximum output SDR corresponding to the clairvoyant case.

$$\rho(\mathbf{v}, \hat{\mathbf{R}}) = \frac{\mathcal{L}(\mathbf{v}, \hat{\mathbf{R}})}{\mathcal{L}(\mathbf{s}, \mathbf{Q})} = \frac{|\mathbf{v}^\dagger \hat{\mathbf{R}}^{-1} \mathbf{s}|^2}{\{\mathbf{v}^\dagger \hat{\mathbf{R}}^{-1} \mathbf{Q} \hat{\mathbf{R}}^{-1} \mathbf{v}\}\{\mathbf{s}^\dagger \mathbf{Q}^{-1} \mathbf{s}\}} \tag{10.15}$$

The estimate of $\mathbf{Q}$ is often formed using an available set of secondary data sample vectors, also known as *reference cells*, which may be considered as the "training data" for the adaptive filter. These vectors are extracted from the data cube in similar fashion to the primary vector, but are selected from a limited number of separate resolution cells that are often located in the vicinity of the test cell for reasons to be described below.

Let $\mathbf{x}_k \in \mathcal{C}^N$ for $k = 1, \ldots, K$ be the set of $K \geq N$ secondary sample vectors used to estimate $\mathbf{Q}$. Ideally, these vectors contain statistically homogeneous realizations of the disturbance process, $\mathbf{x}_k = \mathbf{d}_k$. In this case, it is common to estimate $\mathbf{Q}$ as the sample covariance matrix (SCM) based only on the secondary data, as in Eqn. (10.16). Although K was previously defined as the number of fast-time samples in the data cube, we shall for the moment consider it as the number of secondary sample vectors available for training the adaptive filter, without specific reference to a particular dimension of the data cube.

$$\hat{\mathbf{R}} = \frac{1}{K} \sum_{k=1}^{K} \mathbf{x}_k \mathbf{x}_k^\dagger \tag{10.16}$$

The signal vector $\mathbf{s}$ is traditionally approximated by a deterministic steering vector model that is a function of one or more target parameters, as indicated by Eqn. (10.17). This approach effectively models, often under rather idealized conditions, the expected useful signal signature $\mathbf{v}(\vec{\psi}_0)$ as an analytical function of a parameter vector $\vec{\psi}_0$, which may include the target echo DOA and/or Doppler shift, for example.

$$\mathbf{v} = \mathbf{v}(\vec{\psi}_0) \tag{10.17}$$

The collection of all possible steering vectors $\mathbf{v}(\vec{\psi})$ over the full domain of useful signal parameter values $\vec{\psi} \in \mathcal{D}$ traces out a locus in N-dimensional space known as a *manifold*. A radar system typically searches for target echoes over a finite set or "bank" of L candidate parameter vectors, denoted by $\vec{\psi}_0 = \vec{\psi}_1, \ldots, \vec{\psi}_L$, that sample this manifold at discrete points. This enables the radar to search the target parameter space efficiently, yet with some degree of reliability.

10.1.2 Homogeneous Gaussian Case

We shall now consider the relative SDR loss factor defined by Eqn. (10.15) in more detail to expose several important convergence properties of the previously described adaptive filters. In particular, the relative SDR loss factor is a random variable and probability density functions are described in this section for both supervised and unsupervised training applications, as well as the cases of matched and mismatched useful signals, under the statistically homogeneous Gaussian disturbance assumption.

More specifically, the secondary sample vectors $\{\mathbf{x}_k\}_{k=1}^{K}$ are assumed to be independent and identically distributed (IID) realizations of a zero-mean N-variate Gaussian process. The *supervised* training scenario assumes that the secondary data contains disturbance only $\mathbf{x}_k = \mathbf{d}_k$, and that the covariance matrix $\mathbf{Q}$ of all secondary disturbance vectors is identical to that of the disturbance $\mathbf{d}$ in the primary data. Here, the *unsupervised* training scenario assumes that independent useful signal components, denoted by $\mathbf{s}_k = a_k\mathbf{s}$ for a complex normal distributed amplitude $a_k \sim \mathcal{CN}(0, \sigma_s^2)$, are additionally present in the secondary data, such that $\mathbf{x}_k = \mathbf{s}_k + \mathbf{d}_k$. The supervised and unsupervised secondary data covariance matrices are given by Eqn. (10.18).

$$\begin{cases} \mathrm{E}\{\mathbf{x}_k\mathbf{x}_k^\dagger\} = \mathbf{Q} & \text{Supervised} \\ \mathrm{E}\{\mathbf{x}_k\mathbf{x}_k^\dagger\} = \mathbf{Q} + \sigma_s^2\mathbf{s}\mathbf{s}^\dagger & \text{Unsupervised} \end{cases} \tag{10.18}$$

The presumed signal vector $\mathbf{v} = \mathbf{v}(\psi_0)$ may or may not be equal to the true (actual) signal vector $\mathbf{s}$, which leads to the matched and mismatched useful signal conditions in Eqn. (10.19). To maintain generality, a specific model for the mismatch is not assumed for the moment. In practice, mismatches may arise due to instrumental imperfections, such as array calibration errors, or environmental factors, such as diffuse multipath propagation. The potential impact of matched and mismatched useful signals on output SDR is also of significant interest. Four different combinations of supervised or unsupervised training and matched or mismatched useful signals will now be considered in turn to explain the main features of adaptive algorithms used in practical radar systems.

$$\begin{cases} \mathbf{v} = \mathbf{s} & \text{Matched} \\ \mathbf{v} \neq \mathbf{s} & \text{Mismatched} \end{cases} \tag{10.19}$$

10.1.2.1 Supervised Training with Matched Useful Signal

Restricting attention to the homogeneous Gaussian case, the secondary data SCM $\hat{\mathbf{R}}$ corresponds to the maximum-likelihood estimate (MLE) of $\mathbf{Q}$ in the supervised training scenario. For a disturbance covariance matrix of full rank N, and a number of IID samples $K \geq N$, the inverse $\hat{\mathbf{R}}^{-1}$ is guaranteed to exist such that the adaptive filter $\hat{\mathbf{w}}_1 = \beta\hat{\mathbf{R}}^{-1}\mathbf{s}$ may be formed using the sample matrix inverse (SMI) technique (Reed, Mallet, and Brennan 1974).

The SMI-MVDR adaptive filter sets the scaling as $\beta = \{\mathbf{s}^\dagger\hat{\mathbf{R}}^{-1}\mathbf{s}\}^{-1}$ to provide unity gain to useful signals. In this case, useful signals are assumed to be perfectly matched ($\mathbf{v} = \mathbf{s}$). The ratio between the output SDR of $\hat{\mathbf{w}}_1$ and that of the optimum filter $\mathbf{w}_o$ is given by the SDR loss factor $\rho_1 \in [0, 1]$ in Eqn. (10.20). The loss factor $\rho_1 \to 1$ from below asymptotically, since $\hat{\mathbf{R}} \to \mathbf{Q}$ as $K \to \infty$.

$$\rho_1 = \rho(\mathbf{s}, \hat{\mathbf{R}}) = \frac{\mathcal{L}(\mathbf{s}, \hat{\mathbf{R}})}{\mathcal{L}(\mathbf{s}, \mathbf{Q})} \tag{10.20}$$

In their pioneering paper on the SMI technique, Reed, Mallet, and Brennan (RMB) derived the probability density function of ρ_1 as a beta function. The authors proved the remarkable fact that the distribution of ρ_1 depends only on the dimensional parameters of system, namely, the sample size K and filter length N. Specifically, the density of ρ_1 is given by $f_1(\rho_1)$ in Eqn. (10.21), where the term $B(M, L) = (M-1)!(L-1)!/(M+L-1)!$.

$$f_1(\rho_1) = [B(N-1, K+2-N)]^{-1}\rho_1^{K-N+1}(1-\rho_1)^{N-2}, \quad 0 \leq \rho_1 \leq 1 \tag{10.21}$$

The first and second moments (mean and variance) of the random variable ρ_1 are given in Eqn. (10.22). An important property of the SMI technique is that the convergence rates of the moments of ρ_1 are functionally independent of $\mathbf{Q}$. This is in contrast to the performance of the iterative LMS adaptive algorithm, which converges relatively slowly when $\mathbf{Q}$ is poorly conditioned (i.e., when $\mathbf{Q}$ has a large eigenvalue spread). This important property of the SMI technique provides rapid convergence rate irrespective of the disturbance covariance matrix.

$$\mathrm{E}\{\rho_1\} = \frac{K-N+2}{K+1}, \quad \mathrm{Var}\{\rho_1\} = \frac{(K-N+2)(N-1)}{(K+1)^2(K+2)} \tag{10.22}$$

The probability that ρ_1 has a value less than $1-\delta$ is given by $P[\rho_1 < 1-\delta]$ in Eqn. (10.23), where the term $b(m; K, \delta) = \binom{K}{m}\delta^m(1-\delta)^{K-m}$. For the particular number of samples $K = 2N-3$, the density $f_1(\rho_1)$ is symmetric about $1/2$, such that $P[\rho_1 < 1/2] = 1/2$. It is evident from Eqn. (10.22) that $\mathrm{E}\{\rho_1\} = 1/2$ for $K = 2N-3$, as expected. The frequently quoted RMB rule-of-thumb stems from this relationship; it states that using an IID Gaussian disturbance sample size of $K > 2N$ to train an adaptive filter of length N using the SMI technique yields a mean SDR loss of $\mathrm{E}\{\rho_1\} < 1/2$ relative to the optimum filter (i.e., an average loss of less than 3 dB) under the aforementioned assumptions.

$$P[\rho_1 < 1-\delta] = \sum_{m=0}^{N-2} b(m; K, \delta) \tag{10.23}$$

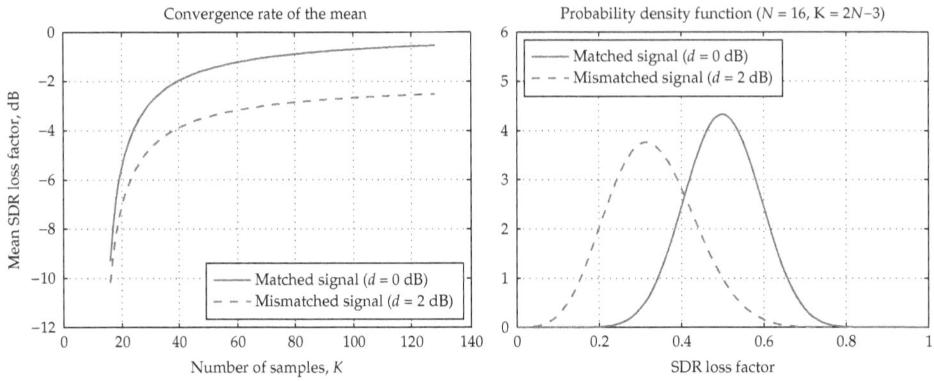

(a) Convergence rate of E{ρ_1} and E{ρ_2}.

(b) Probability density functions of ρ_1 and ρ_2.

FIGURE 10.1 The left panel shows the convergence rate of the mean SDR loss factor for a matched signal E$\{\rho_1\}$ and mismatched signal E$\{\rho_2\}$ with $d = 2$ dB in the *supervised* training scenario. The right panel shows the probability density functions of the SDR loss factors ρ_1 and ρ_2 that result for $K = 2N - 3$ in the *supervised* training scenario. The adaptive filter dimension is $N = 16$ in all cases.

The solid line in Figure 10.1a illustrates the convergence rate of E$\{\rho_1\}$ for $N = 16$, while the solid line in Figure 10.1b shows a plot of $f_1(\rho_1)$ for $K = 2N-3$, which is symmetric about 1/2. It is emphasized that the formulas in this section apply for the *cross-rejection* situation, whereby the adaptive filter is applied to data snapshots independent of those used to calculate the weight vector. Application of the weight vector to the same disturbance vectors used for adaptive filter estimation is known as the *self-rejection* case. For a limited number of samples, an adaptive filter will in the latter case provide *higher* disturbance rejection than the optimum filter (Abramovich, Mikhaylyukov, and Malyavin 1992a). Performance evaluations based on self-rejection analysis should therefore be interpreted with caution.

10.1.2.2 Supervised Training with Useful Signal Mismatch

More often in practice, neither $\mathbf{Q}$ nor $\mathbf{s}$ is known exactly. When useful signal mismatch is present, the SMI-MVDR adaptive filter is estimated as $\hat{\mathbf{w}}_2 = \beta\hat{\mathbf{R}}^{-1}\mathbf{v}$, where $\mathbf{v}$ is the presumed signal vector. If the model $\mathbf{v}$ differs from the actual signal vector $\mathbf{s}$, the estimated adaptive filter $\hat{\mathbf{w}}_2$ will deviate from the optimum filter $\mathbf{w}_o = \beta\mathbf{Q}^{-1}\mathbf{s}$ even in the limit of infinite training samples, as $\hat{\mathbf{w}}_2 \to \beta\mathbf{Q}^{-1}\mathbf{v}$ for large K. The SDR loss factor corresponding to the supervised training scenario under the condition of useful signal mismatch is defined by $\rho_2 \in [0, 1]$ in Eqn. (10.24).

$$\rho_2 = \rho(\mathbf{v}, \hat{\mathbf{R}}) = \frac{\mathcal{L}(\mathbf{v}, \hat{\mathbf{R}})}{\mathcal{L}(\mathbf{s}, \mathbf{Q})} \tag{10.24}$$

The probability density function (PDF) of the scaled SDR loss factor $d\rho_2 \in [0, d]$, where the scale parameter $d = [(\mathbf{s}^\dagger\mathbf{Q}^{-1}\mathbf{s}).(\mathbf{v}^\dagger\mathbf{Q}^{-1}\mathbf{v})]/|\mathbf{v}^\dagger\mathbf{Q}^{-1}\mathbf{s}|^2$, was analytically derived in Boroson (1980) and is given by $f_2(d\rho_2; d)$ in Eqn. (10.25). The inverse of $d \in [1, \infty]$ may be interpreted as a generalized magnitude-squared coherence, which is often written in terms of the direction cosine squared $\cos^2 \Upsilon = 1/d$, where Υ is the angle between the vectors $\mathbf{v}$ and $\mathbf{s}$ in *whitened* data space (i.e., between the vectors $\mathbf{Q}^{-1/2}\mathbf{v}$ and $\mathbf{Q}^{-1/2}\mathbf{s}$).

$$f_2(d\rho_2; d) = \frac{d^{N-2K-1}}{B(N-1, K+2-N)} \sum_{\ell=0}^{K+1-N} \times \left\{ \binom{K+1-N}{\ell}^2 \Big/ \binom{N-2+\ell}{\ell} \right\} (d-1)^{\ell} (d - d\rho_2)^{N-2+\ell} (d\rho_2)^{K+1-N-\ell} \quad (10.25)$$

Boroson (1980) showed that the mean and variance of the random variable ρ_2 are given by Eqn. (10.26). In the case of no useful signal mismatch ($\mathbf{v} = \mathbf{s}$), we have that $d = 1$ and $f_2(\rho_2; 1) = f_1(\rho_1)$, as expected. For $d = 1$, the variance decreases in proportion to $1/K^2$, but as $\mathbf{Q}^{-1/2}\mathbf{v}$ and $\mathbf{Q}^{-1/2}\mathbf{s}$ diverge, $d > 1$ and $\mathrm{Var}\{\rho_2\}$ decreases in proportion to $1/K$, which is a much slower rate for large K.

$$\mathrm{E}\{\rho_2\} = \frac{K+1-N+d}{d(K+1)},$$
$$\mathrm{Var}\{\rho_2\} = \frac{(K-N+2)(N-1) + (d-1)K[d-1+2(K+2-N)]}{d^2(K+1)^2(K+2)} \quad (10.26)$$

From Eqns. (10.22) and (10.26), it can be shown that useful-signal mismatch biases the mean SDR loss of $\hat{\mathbf{w}}_2$ relative to the *adaptive* filter $\hat{\mathbf{w}}_1$ by the multiplicative factor $b \leq 1$ in Eqn. (10.27), where $\xi = d - 1$. As $K \to \infty$, it may be observed that $b \to (1+\xi)^{-1} = 1/d$ from above. Hence, providing $\mathbf{v}$ is a reasonable estimate of $\mathbf{s}$ after transformation by $\mathbf{Q}^{-1/2}$, such that d is not too large, the *additional* mean SDR loss caused by signal mismatch relative to the adaptive filter in the matched condition will not exceed d for all $K \in [N, \infty)$. For example, less than 3-dB mean additional loss for $d < 2$. This alternative rule-of-thumb for the amount of useful signal mismatch that can be tolerated to keep the mean *additional* losses to within 3 dB of the adaptive filter under matched conditions for any sample size K only applies for the *supervised* training scenario.

$$b = \frac{\mathrm{E}\{\rho_2\}}{\mathrm{E}\{\rho_1\}} = \frac{1 + \xi/(K+2-N)}{1+\xi} \quad (10.27)$$

As shown in Boroson (1980), the probability that the random variable $d\rho_2$ has a value less than $1-\delta$ is given by $P[d\rho_2 < 1-\delta; d]$ in Eqn. (10.28). Note that the definition of the SDR loss factor ρ_2 in Eqn. (10.24) is referenced to the SDR of the optimum filter, so it captures the overall relative losses due to both adaptivity and useful signal mismatch. On the other hand, the quantity $d\rho_2$ may be interpreted as the SDR loss factor due to adaptivity only, since its value is referenced to the SDR loss of the asymptotic filter under the mismatched condition. In other words, $d\rho_2 = \mathcal{L}(\mathbf{v}, \hat{\mathbf{R}})/\mathcal{L}(\mathbf{v}, \mathbf{Q})$, as $\rho_2 = \mathcal{L}(\mathbf{v}, \hat{\mathbf{R}})/\mathcal{L}(\mathbf{s}, \mathbf{Q})$ and $d = \mathcal{L}(\mathbf{s}, \mathbf{Q})/\mathcal{L}(\mathbf{v}, \mathbf{Q})$.

$$P[d\rho_2 < 1-\delta; d] = \sum_{\ell=0}^{K+1-N} \sum_{n=0}^{N-2+\ell} b(n; K, \{\delta + d - 1\}/d)\, b(\ell; K+1-N, \{d-1\}/d) \quad (10.28)$$

The dashed line in Figure 10.1a shows the convergence rate for $\mathrm{E}\{\rho_2\}$ with $N = 16$ and $d = 2$ dB. Comparison with the solid line shows that the additional loss with respect to the matched case $\mathrm{E}\{\rho_1\}$ reaches a maximum of $d = 2$ dB for large K. The dashed line in Figure 10.1b shows the density of ρ_2 for $K = 2N-3$. Note that the impact of useful signal mismatch is not only to increase the mean SDR loss, but also to increase the variance of

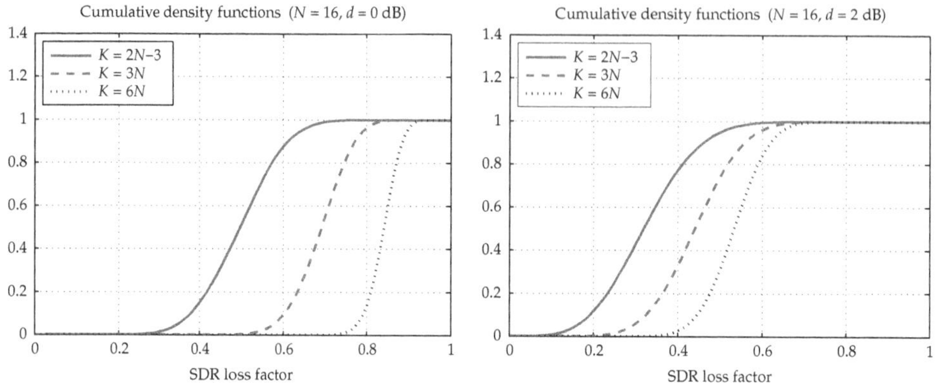

(a) Matched signal condition ρ_1.

(b) Mismatched signal condition ρ_2 ($d = 2$ dB).

FIGURE 10.2 Cumulative density functions for the matched signal SDR loss factor ρ_1 (left panel) and mismatched signal SDR loss factor ρ_2 with $d = 2$ dB (right panel), parameterized by different numbers of sample sizes $K = 2N - 3$, $K = 3N$, and $K = 6N$ in the *supervised* training scenario.

the SDR loss distribution for a given sample size (Boroson 1980). Alternatively, for any mean SDR loss less than d, the mismatched case requires a greater number of samples to achieve the same performance as the matched case (i.e., $\text{E}\{\rho_2\} = \text{E}\{\rho_1\}$) in the *supervised* training scenario. This is illustrated in Figures 10.2a and 10.2b, where the cumulative densities of ρ_1 and ρ_2 are respectively plotted for different sample sizes K.

Although it is true that $\text{E}\{\rho_1\} = 1/2$ for $K = 2N - 3$, it is also true that $\rho_1 < 1/2$ fifty percent of the time (i.e., $P[\rho_1 < 1/2] = 1/2$). A more appropriate performance requirement may be to reduce $P[\rho_1 < 1/2]$ to an acceptably small value. For instance, by setting $K = 3N$, Figure 10.2a shows that $P[\rho_1 < 1/2]$ may be reduced to less than 1 percent. Such a choice of K provides confidence that more than 99 percent of realizations of the SDR loss factor ρ_1 are within 3 dB of the optimum. This example merely serves to point out that analysis based on mean losses only, without regard to the distributional properties of the loss factor, may lead to an optimistic impression of performance.

10.1.2.3 Unsupervised Training with Matched Useful Signal

The adaptive weight vector aims to filter useful signals from the disturbance, so it is not surprising that training data completely free of useful signals is not always available in practical applications. In the unsupervised training scenario, the secondary data that is used for covariance matrix estimation is assumed to contain useful signal components in addition to the disturbance.

To avoid confusion with the previously defined SCM $\hat{\mathbf{R}} = K^{-1}\sum_{k=1}^{K}\mathbf{d}_k\mathbf{d}_k^{\dagger}$, the unsupervised secondary data SCM is defined as $\hat{\mathbf{P}}$ in Eqn. (10.29), where $\mathbf{s}_k = a_k\mathbf{s}$ and the complex amplitude is distributed as $a_k \sim \mathcal{CN}(0, \sigma_s^2)$.

$$\hat{\mathbf{P}} = \frac{1}{K}\sum_{k=1}^{K}[\mathbf{s}_k + \mathbf{d}_k][\mathbf{s}_k + \mathbf{d}_k]^{\dagger} \tag{10.29}$$

Provided that the disturbance $\mathbf{d}_k$ and useful signal $\mathbf{s}_k$ are uncorrelated, the unsupervised secondary-data SCM $\hat{\mathbf{P}}$ tends to its statistically expected form of $\mathbf{P} = \mathbf{Q} + \sigma_s^2\mathbf{s}\mathbf{s}^{\dagger}$ for large K. By applying the matrix inversion lemma, the inverse of $\mathbf{P}$ can be written as in Eqn. (10.30), from which it is readily shown that $\mathbf{P}^{-1}\mathbf{s} = \beta\mathbf{Q}^{-1}\mathbf{s}$ for $\beta = (1+\sigma_s^2\mathbf{s}^\dagger\mathbf{Q}^{-1}\mathbf{s})^{-1}$. Hence, the SMI adaptive filter in the *unsupervised* training scenario also tends to the optimum filter for large K, provided that the useful signal vector $\mathbf{s}$ is perfectly known.

$$\mathbf{P}^{-1} = \mathbf{Q}^{-1} - \frac{\mathbf{Q}^{-1}\mathbf{s}\mathbf{s}^\dagger\mathbf{Q}^{-1}}{1/\sigma_s^2 + \mathbf{s}^\dagger\mathbf{Q}^{-1}\mathbf{s}^{-1}} \tag{10.30}$$

The main question for the unsupervised training scenario in the case of a matched useful signal relates to the convergence rate when the SCM $\hat{\mathbf{P}}$ is used instead of $\mathbf{P}$ to form the adaptive filter $\hat{\mathbf{w}}_3 = \beta\hat{\mathbf{P}}^{-1}\mathbf{s}$. The relative SDR loss factor for this case is defined by ρ_3 in Eqn. (10.31). It was shown by Miller (1976) that this random variable can be expressed as $\rho_3 = \rho'/[1 + p_s(1-\rho')]$, where $\rho' = |\mathbf{s}^\dagger\hat{\mathbf{P}}^{-1}\mathbf{s}|^2/[(\mathbf{s}^\dagger\mathbf{P}^{-1}\mathbf{s}^{-1})(\mathbf{s}^\dagger\hat{\mathbf{P}}^{-1}\mathbf{P}\hat{\mathbf{P}}^{-1}\mathbf{s}^{-1})]$ is the loss factor due to adaptivity for the covariance matrix $\mathbf{P}$, and $p_s = \sigma_s^2\mathbf{s}^\dagger\mathbf{Q}^{-1}\mathbf{s}$ is the power of the signal in whitened data space. The PDF of ρ' is identical to that of ρ_1, since for an IID zero-mean Gaussian sequence, this density depends only on the dimensional parameters of the system (K and N), and is functionally independent of the covariance matrix ($\mathbf{P}$ or $\mathbf{Q}$).

$$\rho_3 = \rho(\mathbf{s}, \hat{\mathbf{P}}) = \frac{\mathcal{L}(\mathbf{s}, \hat{\mathbf{P}})}{\mathcal{L}(\mathbf{s}, \mathbf{Q})} \tag{10.31}$$

The probability that ρ_3 has a value less than $\epsilon = 1-\delta$ is therefore given by $P[\rho_3 < \epsilon]$ in Eqn. (10.32), where $P[\rho_1 < \epsilon(1+p_s)/(1+p_s\epsilon)]$ can be computed from Eqn. (10.23). Note that $p_s = 0$ corresponds to the supervised case.

$$P[\rho_3 < \epsilon] = P[\rho_1 < \epsilon(1+p_s)/(1+p_s\epsilon)] \tag{10.32}$$

Figure 10.3a demonstrates that the relative SDR loss for a fixed sample size increases for higher values of p_s in the unsupervised training scenario. The detrimental effect of

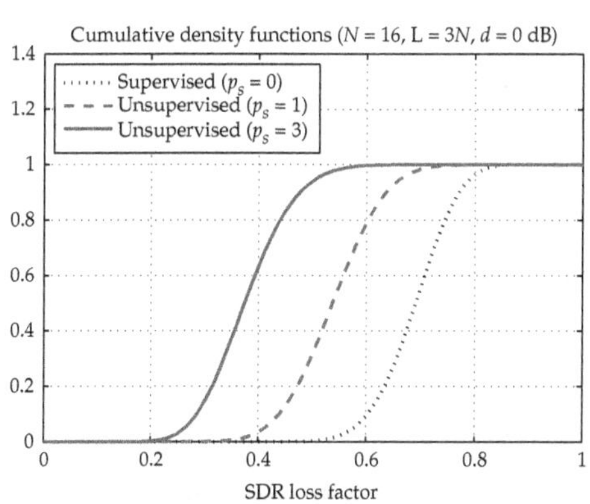

(a) Dependence of ρ_3 on signal power p_s.

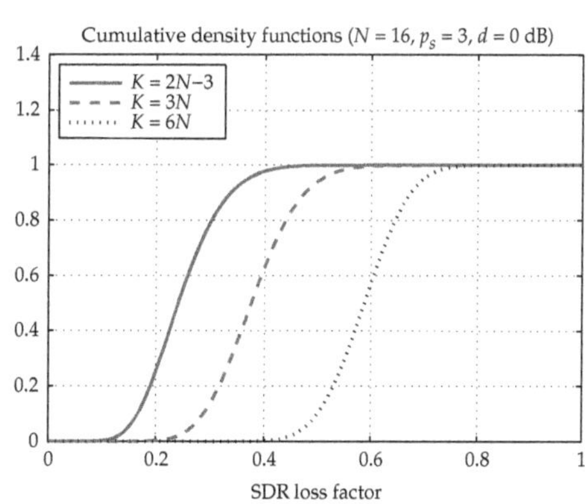

(b) Dependence of ρ_3 on sample size K.

FIGURE 10.3 Cumulative density function of the SDR loss factor ρ_3, which corresponds to the unsupervised training scenario assuming a matched useful signal. The functions are plotted for different signal powers (in whitened data space) p_s and a fixed sample size $K = 3N$ in the left panel, as well as for a fixed signal power $p_s = 3$ and different sample sizes K in the right panel. The adaptive filter dimension is $N = 16$ in all cases.

signal presence in the training data on convergence rate may be appreciated by comparing Figure 10.3b with Figure 10.2a. For example, $P[\rho_3 < 1/2] \approx 0.1$ for $K = 6N$ in Figure 10.3b, whereas for supervised training, $P[\rho_1 < 1/2] \approx 0.01$ for $K = 3N$ in Figure 10.2a.

10.1.2.4 Unsupervised Training with Useful Signal Mismatch

When useful signal mismatch is present, $\mathbf{P}^{-1}\mathbf{v}$ is no longer proportional to $\mathbf{Q}^{-1}\mathbf{s}$ since $\mathbf{Q}$ is assumed to have full rank. Consequently, the adaptive filter $\hat{\mathbf{w}}_4 = \hat{\mathbf{P}}^{-1}\mathbf{v}$ does not tend to the optimum filter for large K. The relative SDR loss factor in the unsupervised training scenario for the case of useful signal mismatch is given by ρ_4 in Eqn. (10.33).

$$\rho_4 = \rho(\mathbf{v}, \hat{\mathbf{P}}) = \frac{\mathcal{L}(\mathbf{v}, \hat{\mathbf{P}})}{\mathcal{L}(\mathbf{s}, \mathbf{Q})} \tag{10.33}$$

Using a similar argument to that of Miller (1976), it was proven in Boroson (1980) that $P[d\rho_4 < \epsilon; d]$ is given by Eqn. (10.34), where $P[d\rho_2 < \epsilon(1 + p_s)\{1 + p_s(d - 1)/d\}/(1 + p_s\epsilon/d); d]$ may be computed from Eqn. (10.28).

$$P[d\rho_4 < \epsilon; d] = P[d\rho_2 < \epsilon(1 + p_s)\{1 + p_s(d - 1)/d\}/(1 + p_s\epsilon/d); d] \tag{10.34}$$

Figure 10.4, in the same format as Figure 10.3, shows the cumulative distributions resulting for ρ_4. For a given sample size $K = 3N$, a comparison of Figure 10.4a with Figure 10.3a shows that useful signal mismatch causes a substantial performance degradation relative to the matched case in the unsupervised training scenario. Importantly, these substantial losses cannot be mitigated by increasing sample support, as shown in Figure 10.4b. This is because a mismatched signal is "actively" suppressed by an amount proportional to p_s in the unsupervised scenario, even in the limit of infinite training samples.

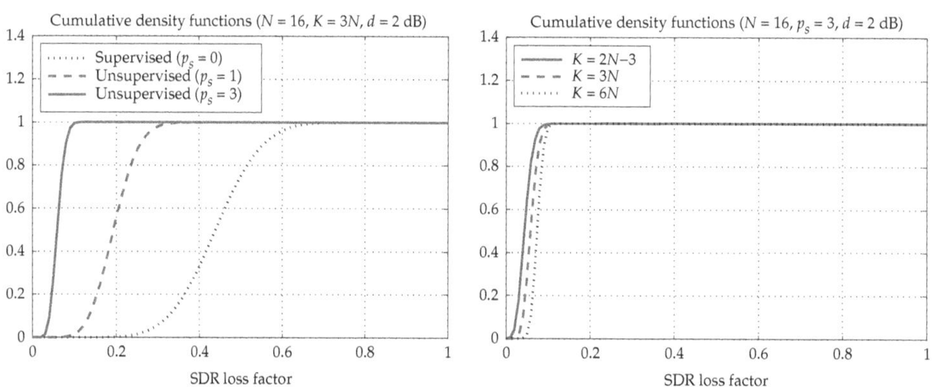

(a) Effect of signal strength p_s on ρ_4.

(b) Effect of sample size K on ρ_4.

FIGURE 10.4 Cumulative density function of the SDR loss factor ρ_4, which corresponds to the unsupervised training scenario in the condition of a mismatched useful signal. The functions are plotted for different signal powers p_s and a fixed sample size $K = 3N$ in the left panel, as well as for a fixed signal power $p_s = 3$ and different sample sizes K in the right panel. The adaptive filter dimension is $N = 16$ in all cases.

It is evident from the collection of results presented in this section that unsupervised training combined with signal mismatch leads to the poorest adaptive filter performance. For this reason, practical measures are often taken to avoid useful signal presence in the training data. For example, this may include the judicious selection and screening of reference cells used to estimate the disturbance covariance matrix. When useful signal corruption cannot be avoided in the training data, minimization of useful signal mismatch becomes critical to avoid significant performance degradation.

Useful signal mismatch is a much less sensitive issue in the supervised training scenario, providing a reasonable estimate of the signal vector is used, as illustrated previously. Clearly, the best performance is most likely to occur in the supervised matched case, assuming a sufficient number of IID secondary vectors are available for disturbance covariance matrix estimation.

The convergence rate and distributional properties of the SDR loss factors $\{\rho_i\}_{i=1}^4$ are directly applicable to the SMI-MVDR adaptive filter implementation, since an arbitrary scaling of the adaptive filters $\{\hat{\mathbf{w}}_i\}_{i=1}^4$ does not affect output SDR. The SMI-LCMV implementation must take the $M-1$ auxiliary linear constraints into account. It has been shown in Abramovich (2000) that the convergence properties of the SMI-LCMV technique may be inferred from those of $\{\rho_i\}_{i=1}^4$ by substituting the true filter dimension N with the effective dimension $N' = N - (M-1)$. This is tantamount to reducing the filter dimension by the number of auxiliary linear constraints $M-1$.

10.1.2.5 Optimum Filter Performance

Thus far, the emphasis has been on assessing adaptive filter performance relative to the corresponding optimum filter in the supervised and unsupervised training scenarios filter as well as for the matched and mismatched useful signal cases. However, the performance of the optimum filter, which has served as the reference for analysis in each case, also needs to be quantified relative to the matched-filter white-noise floor.

The overall performance loss factor may be expressed as the product of the loss factor relative to the optimum filter and the loss factor of the optimum filter relative to the (matched filter) white-noise floor. Using Eqn. (10.12), the overall loss factor may be expressed as $\{\eta_i(\mathbf{w})\}_{i=1}^4$ in Eqn. (10.35), where $\mathbf{w} = \mathbf{w}_a$ for the SMI-MVDR filter, and $\mathbf{w} = \mathbf{w}_b$ for SMI-LCMV filter.

$$\eta_i(\mathbf{w}) = \varrho(\mathbf{w})\rho_i, \quad i = 1, \ldots, 4 \tag{10.35}$$

The density functions of the SMI loss factors $\{\rho_i\}_{i=1}^4$ corresponding to the four considered situations were previously defined for a homogeneous Gaussian disturbance, both with and without the inclusion of auxiliary linear constraints. The question arises as to characteristics of $\varrho(\mathbf{w})$. The loss factor $\varrho(\mathbf{w}_a)$ is highly influenced by the eigenstructure of the external disturbance covariance matrix $\mathbf{M}$ in Eqn. (10.36), particularly the eigenvalue spectrum $\lambda_1, \ldots, \lambda_N$ and the magnitude of the projection of the signal vector $\mathbf{s}$ against the associated eigenvectors $\{\mathbf{q}_n\}_{n=1}^N$.

$$\mathbf{M} = \sum_{n=1}^{N} \lambda_n \mathbf{q}_n \mathbf{q}_n^\dagger \tag{10.36}$$

Broadly speaking, the loss $\varrho(\mathbf{w}_a)$ will be small provided that the *external* disturbance component has an effective rank $N_e < N$ and $\mathbf{s}$ is predominantly spanned by the $N - N_e$ eigenvectors in the "white-noise" subspace (i.e., the strong external disturbance components are received in the "sidelobe" region of the filter and sufficient degrees of freedom are available for effective cancelation).

On the other hand, the loss can be high if $N_e = N$, or if $N_e < N$ but the signal vector $\mathbf{s}$ is predominantly contained in the disturbance subspace (i.e., the powerful external disturbance is effectively full rank or at least one strong disturbance component is received in the "mainlobe" region of the filter). The loss of the LCMV filter $\varrho(\mathbf{w}_b)$ additionally depends, in a rather complex way, on the number and specification of the auxiliary constraints.

Although $\mathbf{M}$ is unknown in practice, its basic structural properties may be qualitatively inferred from the expected physical characteristics of the disturbance type to be mitigated. For example, directional noise-like interference received from a point source typically exhibits a covariance matrix with relatively low spatial rank but full rank in fast or slow-time. Such a disturbance is clearly most amenable to rejection by spatial processing.

Clutter backscattered from extended regions of the Earth's surface will often have a covariance matrix with large or full rank in element or beam space, but relatively low rank in slow-time for a stationary radar platform. Obviously, this type of disturbance is more amenable to rejection by Doppler processing.

Information regarding the expected correlation properties of the external disturbance over different data-cube dimensions, considered either separately or jointly, may therefore be used to guide the selection of filter-processing architecture and allocation of degrees of freedom. While this qualitative description and the assumptions in the previous theoretical analysis may not accurately describe real-world environments, they nevertheless highlight a number of important adaptive filtering principles that serve to guide practical implementations.

10.1.3 Real-World Environments

The assumption of primary and secondary data vectors containing an IID Gaussian disturbance is rather idealistic and seldom encountered in practice. At least some of the secondary data used to train the adaptive filter often exhibits statistical characteristics that are not identical to the disturbance in the cell under test. These heterogeneities or outliers may either be: (1) present among the secondary vectors, (2) confined to the test cell only, or (3) contaminate both the primary and secondary data.

The heterogeneous disturbance case is quite insidious, as it involves systematic as opposed to random errors in the statically expected disturbance covariance matrix, which cannot be alleviated by averaging more samples. In fact, the departure between the secondary data SCM and the statistically expected covariance matrix of the disturbance in the test cell may actually *grow* as the number of reference cells is increased in a heterogeneous environment. The possible causes and effects of different types of heterogeneities are briefly identified and discussed in this section. Some representative methods for training data selection in heterogeneous environments and improved convergence rate in sample-starved conditions are also briefly described.

10.1.3.1 Amplitude and Spectral Mismatch

Heterogeneities that alter the scale or structure of the disturbance covariance matrix among the secondary vectors relative to that of the disturbance in the primary vector are of significant concern in practice. Such phenomena can seriously degrade adaptive filter performance in operational systems. Amplitude heterogeneity refers to differences in the scale of the external disturbance over the secondary vectors relative to that in the test cell. In this case, the disturbance covariance matrix of the secondary vectors takes the form of Eqn. (10.37), where $\mathbf{M}$ is the covariance matrix of the external disturbance in the primary data, and the scalars $\{\gamma_k\}_{k=1}^{K}$ represent the variation of disturbance power over the secondary data $k = 1, \ldots, K$ *relative* to the test cell.

$$\mathrm{E}\{\mathbf{x}_k\mathbf{x}_k^{\dagger}\} = \gamma_k\mathbf{M} + \sigma_n^2\mathbf{I} \tag{10.37}$$

This reflects a situation in which the power level of the external disturbance is changing across different radar resolution cells but its spectral structure remains invariant. A practical example is clutter scattered from a scene with spatially fluctuating reflectivity. Marginal and joint densities for the collection of random variables $\{\gamma_k\}_{k=1}^{K}$ have been proposed in the literature to statistically describe amplitude mismatch. For example, the Weibull, Gamma, and K-distribution have been suggested to represent compound-Gaussian clutter in Gini and Farina (2002a) and references therein.

For a large sample size K, the secondary data SCM will tend to the expected value in Eqn. (10.38), where $\bar{\gamma}$ represents the mean external disturbance scale relative to that in the test cell. In the presence of amplitude heterogeneity, the mean disturbance amplitude will in general not match the disturbance amplitude in any particular test cell ($\bar{\gamma} \neq 1$). This amplitude mismatch causes sub-optimality in the "notch-depth" of the adaptive filter used to mitigate the disturbance.

When the scale is underestimated relative to the test cell (i.e., $\bar{\gamma} \ll 1$), the "notch" is not deep enough, so the external disturbance in the test cell is *under-nulled*. This can raise the false-alarm rate significantly above the anticipated value. On the other hand, when the scale is overestimated relative to the test cell ($\bar{\gamma} \gg 1$), the notch is deeper than required. In this case, the adaptive filter norm, and hence, white-noise gain, may be higher than necessary. In extreme cases, the increase in white-noise gain due to *over-nulling* can lead to target masking and hence missed detections.

$$\mathbf{Q} = \bar{\gamma}\mathbf{M} + \sigma_n^2\mathbf{I} \tag{10.38}$$

Spectral heterogeneity refers to mismatches in the *structure* of the disturbance covariance matrix as opposed to scale only. In the most general case, each realization of the external disturbance in the secondary data is statistically characterized by a different covariance matrix, denoted by $\mathbf{M}_k$ in Eqn. (10.39). Spectral heterogeneity may occur in the secondary data due to varying clutter Doppler-spectrum characteristics, for example, such as when the radar coverage crosses land-sea boundaries in littoral areas. A similar phenomenon arises for spatially non-stationary interference sources when the secondary data is acquired over a relatively long time interval.

$$\mathrm{E}\{\mathbf{x}_k\mathbf{x}_k^{\dagger}\} = \mathbf{M}_k + \sigma_n^2\mathbf{I} \tag{10.39}$$

Although the dominant disturbance can at times be considered locally homogeneous over a limited set of training vectors, the secondary and/or primary data may be corrupted by unwanted coherent signals or clutter "discretes," each being confined to a single radar resolution cell (ignoring spectral leakage effects). For example, such contamination may arise from localized radar cross section (RCS) enhancements in a scattering scene with an underlying homogeneous texture profile.

Clutter discretes from independent scatterers will generally have different signature vectors, denoted by $\mathbf{u}_k$, in the otherwise homogeneous disturbance model of Eqn. (10.40).

Contamination from discretes in the secondary data mainly affects the quality of disturbance suppression in the test cell, which can lead to missed detections. Contamination from discretes in the primary data can cause false alarms, irrespective of whether independent contamination is present or not in the secondary data. This is because the properties of a discrete confined to the test cell cannot be learned from the secondary data used to train the adaptive filter.

$$\mathrm{E}\{\mathbf{x}_k\mathbf{x}_k^\dagger\} = \begin{cases} \mathbf{M} + \sigma_n^2\mathbf{I} & \text{No Discrete} \\ \mathbf{M} + \sigma_n^2\mathbf{I} + \mathbf{u}_k\mathbf{u}_k^\dagger & \text{Discrete} \end{cases} \tag{10.40}$$

10.1.3.2 Training Data Selection

Training data is commonly extracted from a limited number of reference cells that are located in the neighborhood of the test cell in the radar data cube. The test cell, and typically one or more guard cells on either side of it, are usually excluded from the secondary data to reduce the potential for target cancelation by minimizing the amount of useful signal energy leaking into the training data. This attempts to emulate supervised training. The practice of selecting secondary vectors in the vicinity of the test cell in one or more data-cube dimensions is known as *localized training*.

The supposition is that over a relatively small training interval around the test cell, the secondary data is more likely to exhibit a homogeneous disturbance covariance matrix with similar structure to that in the test cell. Localized training is therefore an attempt to minimize amplitude and spectral heterogeneity of the disturbance covariance matrix (i.e., to emulate the homogeneous case). In a heterogeneous clutter environment, this may entail averaging the covariance matrix over a limited number of range cells near the test cell, while for a spatially non-stationary interference signal, this may require averaging sample vectors acquired over a relatively short time interval.

Localized training can often reduce the severity of disturbance heterogeneity, but such a strategy alone will in general not effectively deal with the presence of discretes in the training data. A preprocessing step known as non-homogeneity detection (NHD) may be applied to detect and excise secondary vectors that do not conform well in a statistical sense with the properties of the secondary data viewed collectively.

Once these "outlier" cells are identified using an appropriate detection test, they are screened out from the secondary data, such that they do not contribute to the final SCM computation. A common metric used to rank training cells for NHD is the generalized inner product (GIP) test, denoted by ς_k in Eqn. (10.41). A preassigned number of training cells with the highest values of ς_k may be eliminated from the final SCM computation used to estimate the adaptive filter (Chen, Melvin, and Wicks 1999).

$$\varsigma_k = \mathbf{x}_k^\dagger\hat{\mathbf{R}}^{-1}\mathbf{x}_k \tag{10.41}$$

NHD schemes can be quite effective in identifying secondary cells that contain significant clutter discretes. Sample vectors that pass the NHD test may be pruned further based on power-selected training (PST), which preferentially selects stronger sample vectors to reduce the probability of under-nulling caused by amplitude heterogeneity. A different criterion is to normalize the contribution of each secondary vector in the final SCM computation by its inner product to reduce the likelihood of over-nulling.

The availability of a priori knowledge from databases or other sensors may also be incorporated to improve the quality of secondary data selection; see Melvin, Wicks,

Antonik, Salama, Li, and Schuman (1988) and references therein. An OTH radar application using geographic map-assisted training data selection in a littoral area for enhanced ship detection can be found in Fabrizio, Farina, and De Maio (2006).

10.1.3.3 Sample-Starved Techniques

The quest for homogeneous disturbance-only secondary data often comes at the expense of significantly diminished sample support. In highly heterogeneous environments, the scarcity of suitable training data may effectively "starve" the adaptive processor. Stated simply, the need to use training data from resolution cells close to the test cell may lead to higher estimation errors due to finite sample support. An appropriate tradeoff between these two competing objectives is required to obtain the best performance.

Methods to combat adaptive filter performance loss under conditions of low sample support are of paramount importance in practice. A popular method is known as the diagonally loaded sample-matrix inverse (LSMI) technique, wherein the standard SCM $\hat{\mathbf{R}}$ is replaced by its regularized version $\tilde{\mathbf{R}}$ defined in Eqn. (10.42). The positive scalar α is called the loading factor.

$$\tilde{\mathbf{R}} = \hat{\mathbf{R}} + \alpha\mathbf{I} \tag{10.42}$$

Judicious diagonal loading has the potential to significantly improve convergence rate, particularly when three conditions are satisfied. First, the effective disturbance subspace dimension is much smaller than the adaptive filter dimension, $N_e \ll N$. Second, the dominant disturbance eigenvalues are much higher than the additive white noise eigenvalues, $\{\lambda_n\}_{n=1}^{N_e} \gg \sigma_n^2$. Third, the number of linear constraints M imposed on the adaptive filter leaves spare adaptive degrees of freedom available, $M < N - N_e$.

When all these conditions are met, an appropriate diagonal loading factor in the interval $\sigma_n^2 < \alpha \ll \{\lambda_n\}_{n=1}^{N_e}$ may be used to reduce the number of IID samples needed to achieve better than 3-dB average losses relative to the optimum filter from $K = 2N$ to $K = 2N_e$ for both the MVDR and LCMV LSMI implementations in the supervised training scenario, as described in Abramovich (1981b), Cheremisin (1982), and Gierull (1996).

In addition, diagonal loading is known to stabilize main beam distortion and regularize sidelobe levels, which provides robustness against useful signal mismatch and greater immunity to off-azimuth clutter discretes in the primary data, respectively. The reader is referred to the three citations in the previous paragraph as well as Carlson (1988) for a detailed description of the various benefits of diagonal loading.

Alternative methods to diagonal loading include the use of data-independent filter dimension reduction techniques, such as beam-space or frequency-domain transformations, as well as forward/backward sub-aperture smoothing approaches (Pillai, Kim, and Guerci 2000), which trade off resolution performance against estimation accuracy.

Adaptive rank-reduction techniques based on singular value decomposition (SVD) of the data can also be used to improve convergence rate (Guerci, Goldstein, and Reed 2000). In the clairvoyant case, the performance of a rank-reduced optimum filter will in general be inferior to that of the full-dimension optimum filter. However, the improved convergence rate of the former in the finite sample case may yield superior performance in practice when it comes to an *adaptive* implementation.

Parametric models of the disturbance covariance matrix, based on multi-channel AR processes, for example, have also been proposed to combat finite sample support

(Michels, Himed, and Rangaswamy 2002). Provided that such models are accurate, this type of approach effectively reduces the number of covariance matrix parameters to be estimated to achieve a more rapid convergence rate in sample-starved conditions.

10.2 Problem Formulation

OTH radars operate in the presence of external interference and noise sources that may be received with power levels significantly above the internal (i.e., thermal) noise floor. In the HF band, man-made interference may originate from intentional radiators (e.g., communications and radio broadcasts), or from unintentional radiators (e.g., electrical machinery and industrial processes). Natural HF noise sources include atmospheric noise from lightning discharges, as well as galactic noise from the Sun and other stars.

Real-time spectrum monitoring is performed at the receiver site to aid the selection of unoccupied or "clear" frequency channels for OTH radar operation. This represents the primary method for reducing the interference-plus-noise spectral density in the radar bandwidth, as well as avoiding mutual interference amongst users of the HF spectrum. Besides the performance benefits of operating on a clear frequency channel, OTH radar systems are required to adhere to a policy of noninterference.

The availability of clear frequency channels suitable for OTH radar operation can at times be significantly diminished. High user-congestion in the HF band has the potential to degrade OTH radar performance because it limits the scope which the frequency management system has for optimizing the selection of operating frequency. More specifically, frequency channels that are optimum for target signal reception may contain powerful co-channel interference, or high background noise levels, while clear frequency channels found in electrically "quiet" regions of the spectrum may be inappropriate for radar signal propagation to the geographical area of interest.

Such circumstances arise most often at night, when reduced skywave propagation support at higher frequencies forces users to operate in the lower HF spectrum. In addition, the nighttime ionosphere is relatively prone to propagating interference sources over very long distances due to the absence of D-region absorption. This further compounds the channel availability problem, since it increases the effective number of interference sources that reach the receiver site. At night, it can sometimes be difficult to find a clear channel of suitable frequency and sufficient bandwidth to operate the radar.

Channel occupancy is typically less problematic during the day. However, frequency channels deemed to be clear and suitable for OTH radar operation at a particular time will rarely remain completely free of incidental interference during the entire mission. The rapidly changing spatial and spectral properties of the composite HF signal environment, coupled with the dynamic characteristics of the skywave propagation channel, often means that the reception of incidental radio frequency interference (RFI) is inevitable in practice.

The received RFI may be transient in nature and contaminate small segments of a CPI every now and then, or it may be a time continuous signal that unpredictably appears on the radar channel and then persists over many CPIs. The main point is that the potential for RFI to contaminate radar dwells and mask target echoes can be significantly reduced, but not completely avoided, by frequency agility alone. For this reason, there has always been a strong interest in adaptive signal-processing techniques as a secondary measure for interference and noise reduction in OTH radar.

10.2.1 Interference and Clutter Mitigation

Directional interference received from one or more point sources independent of the radar waveform will typically be highly correlated spatially (across antenna elements), but will often exhibit little or no correlation in "slow time" (over pulses) and "fast time" (over range bins). Provided the interference components to be mitigated are not arriving from the same direction as the target echo (i.e., sidelobe interference), and the number of spatial degrees of freedom available is greater than the effective interference-subspace dimension, adaptive beamforming may in principle be applied in lieu of conventional beamforming to improve the output signal-to-interference-plus-noise ratio (SINR).

On the other hand, OTH radar clutter is typically incident from a broad continuum of directions and therefore exhibits poor correlation across antenna elements. Its spatially broadband structure arises due to the relatively wide main lobe of the transmit antenna beam pattern. However, OTH radar clutter is usually highly correlated from pulse-to-pulse within the CPI, and is therefore best mitigated by slow-time domain filtering or Doppler processing. In many operational OTH radar systems, the detection of useful signals against the composite disturbance, which contains clutter and interference, may be carried out by incorporating an adaptive beamforming step for interference rejection in cascade with a Doppler processing step for clutter mitigation. STAP attempts to combine these two operations in a "joint" processing step. In practice, this is at the expense of substantially higher computational complexity (and sometimes inferior performance) with respect to the cascaded or "factored" space-time processing approach.

In the context of OTH radar, adaptive beamforming may be considered as a means for removing spatially structured interference and noise in frequency channels considered suitable for operation. Such a capability is particularly important for the development of OTH radar systems that treat *adaptive* processing and frequency management as a global optimization problem. It is envisaged that future OTH radar systems equipped with a wideband direct digital receiver per element may routinely analyze the spatial properties of clear frequency channels across the HF band so that the choice of operating frequency and adaptive beamforming algorithm can be optimized jointly to maximize output SINR in the beam-steer directions required for surveillance.

10.2.2 Multi-Channel Data Model

To formulate the problem in mathematical terms, define $\mathbf{z}(t) \in \mathcal{C}^N$ in Eqn. (10.43) as the complex-valued N-dimensional antenna array snapshot vector received by the radar in a particular (unnamed) range cell at pulse t of the CPI. In general, $\mathbf{z}(t)$ contains an additive mixture of backscattered clutter returns $\mathbf{c}(t)$, from sea and/or terrain surfaces, external interference $\mathbf{i}(t)$, representing the sum of the acquired natural and man-made co-channel signals, internal receiver noise $\mathbf{n}(t)$ of thermal origin, and potentially a useful signal $\mathbf{s}(t)$ due to an echo from a point target.

$$\mathbf{z}(t) = \mathbf{s}(t) + \mathbf{c}(t) + \mathbf{i}(t) + \mathbf{n}(t) \tag{10.43}$$

The useful signal received by a well-calibrated narrowband uniform linear array (ULA) of antenna elements from a far-field point target with cone angle-of-arrival θ takes the form of Eqn. (10.44). The scalar waveform $g(t)$ represents the temporal signature of the useful signal in slow-time t, while $\mathbf{v}(\theta) \in \mathcal{C}^N$ denotes its spatial response (i.e., steering vector) on the ULA manifold, which is parameterized by θ. For reasons to become apparent later, θ will be referred to simply as azimuth in this section.

$$\mathbf{s}_k(t) = g_k(t)\mathbf{v}(\theta) \tag{10.44}$$

An ideal target with constant reflectivity and radial velocity during the CPI has a temporal signature given by $g(t) = \mu e^{j(2\pi f_d t+\phi)}$, where f_d is the Doppler frequency shift normalized by the pulse repetition frequency f_p, and $a = \mu e^{j\phi}$ is a complex scalar amplitude. The simplest useful signal model may be defined as Eqn. (10.45), where d is the antenna element spacing of the ULA and λ is the radio wavelength.

$$\mathbf{s}(t) = g(t)\mathbf{v}(\theta) = \mu e^{j(2\pi f_d t+\phi)}[1, e^{j2\pi d \sin\theta/\lambda}, \ldots, e^{2\pi(N-1)d\sin\theta/\lambda}]^T \tag{10.45}$$

The internal receiver noise is assumed to be independent of other signal components and uncorrelated with itself over different receivers and pulses. Stated another way, it is assumed to be spatially and temporally white with correlation properties given by Eqn. (10.46), where σ_n^2 is the thermal noise power and δ_{rs} is the Kronecker delta function.

$$\mathrm{E}\{\mathbf{n}(r)\mathbf{n}^\dagger(s)\} = \delta_{rs}\sigma_n^2\mathbf{I} \tag{10.46}$$

The overall *external* disturbance to be mitigated is given by $\mathbf{e}(t) = \mathbf{c}(t) + \mathbf{i}(t)$. However, the goal of adaptive beamforming is to improve target detection against the interference-plus-noise component, denoted by $\mathbf{j}(t) = \mathbf{i}(t) + \mathbf{n}(t)$. Adaptive beamforming does not attempt to cancel the clutter $\mathbf{c}(t)$, which is filtered out by the subsequent Doppler processing step.

10.2.2.1 External Interference

While the properties of signal propagation from the radar to the surveillance region and back are optimized through the choice of operating frequency, RFI propagation paths are arbitrary and may involve reflections from highly perturbed ionospheric regions, such as the equatorial or auroral plasma. Due to the dynamic nature of the ionosphere, the received interference often exhibits a non-stationary spatial structure that is statistically characterized by a *time-varying* spatial covariance matrix *within* the CPI. A time-varying interference spatial covariance matrix within the CPI may also arise due to the variation in geometry between the radar receiver and interference source(s) (Gershman, Nickel, and Bohme 1997), or the impulsive nature of the sources (Turley and Lees 1987).

To account for this "spatial non-stationarity" phenomenon, define $\mathbf{R}(t) = \mathrm{E}\{\mathbf{j}(t)\mathbf{j}^\dagger(t)\}$ in Eqn. (10.47) as the slow-time varying interference-plus-noise spatial covariance matrix. This assumes the interference wavefronts are essentially frozen over the relatively short pulse-repetition interval (PRI). The variation of $\mathbf{R}(t) = \mathrm{E}\{[\mathbf{i}(t) + \mathbf{n}(t)][\mathbf{i}(t) + \mathbf{n}(t)]^\dagger\}$ from pulse-to-pulse is clearly due to the spatial non-stationarity of the *external* interference component $\mathbf{M}(t) = \mathrm{E}\{\mathbf{i}(t)\mathbf{i}^\dagger(t)\}$. The rate at which the matrices $\mathbf{R}(t)$ evolve on slow-time t, and hence their similarity over different pulses can vary significantly depending on the degree of interference spatial non-stationarity.

$$\mathbf{R}(t) = \mathrm{E}\{\mathbf{i}(t)\mathbf{i}^\dagger(t)\} + \mathrm{E}\{\mathbf{n}(t)\mathbf{n}^\dagger(t)\} = \mathbf{M}(t) + \sigma_n^2\mathbf{I} \tag{10.47}$$

In general, it is observed that interference modes reflected by the ionosphere have rigid underlying plane wavefronts with superimposed gain and phase "corrugations" that fluctuate in a *correlated* manner from one pulse to another over the CPI. This means that the interference spatial covariance matrix $\mathbf{R}(t)$ may be assumed locally homogeneous over a small number of Q consecutive PRIs, as indicated by Eqn. (10.48) where $Q \ll P$, but becomes progressively different (heterogenous) over time as the interference wavefronts change within the CPI. The structure of $\mathbf{R}(t)$ can vary appreciably between the endpoints of a typical OTH radar CPI.

$$\mathbf{R}(t) \approx \mathbf{R}(t+q), \quad q = 1, \ldots, Q \tag{10.48}$$

OTH radar clutter backscattered from the Earth's surface usually occupies a finite range depth within the PRI. On the other hand, interference and noise signals are incoherent with the radar waveform and are therefore present in all range cells of the PRI after pulse compression. Hence, it is often possible to identify a number of secondary range cells in each PRI that are essentially free of clutter and useful signals but contain interference and noise.

These secondary (non-operational) range cells allow for supervised training of the adaptive filter. The unknown quantity $\mathbf{R}(t)$ may therefore be estimated as the sample spatial covariance matrix $\hat{\mathbf{R}}(t)$ in Eqn. (10.49), using interference-plus-noise-only snapshots $\mathbf{j}_k(t)$ extracted from a set of K secondary range cells $k = 1, \ldots, K$. As the interference and noise energy in a particular PRI is spread over all range cells, the RFI spatial properties are assumed to be homogeneous in range, such that $\mathbf{R}(t) = \mathrm{E}\{\mathbf{j}_k(t)\mathbf{j}_k^\dagger(t)\}$ for all k.

$$\hat{\mathbf{R}}(t) = \frac{1}{K}\sum_{k=1}^{K} \mathbf{j}_k(t)\mathbf{j}_k^\dagger(t) \tag{10.49}$$

In conditions of low sample support per pulse, the CPI may be partitioned into M smaller sub-CPI or PRI *batches* each containing Q pulses (i.e., $M = P/Q$), such that the adaptive beamformer weight vector $\mathbf{w}_m \in \mathcal{C}^N$ for $m = 1, \ldots, M$ can be updated from batch-to-batch rather than every pulse. Selection of the batch length Q represents a compromise between fast updates to counter RFI non-stationarity and slow updates to increase secondary data sample support as well as reduce computational load.

For a batch length Q with K secondary samples per pulse, a number of secondary snapshots $KQ > 2N$ is required to provide satisfactory performance using the SMI technique (Reed et al. 1974). The adaptive weight vectors $\mathbf{w}_m$ used to process the batches $m = 1, \ldots, M$ may be synthesized from the associated *integrated* sample covariance matrices $\hat{\mathbf{R}}_m$ in Eqn. (10.50).

$$\hat{\mathbf{R}}_m = \frac{1}{Q}\sum_{t=Q(m-1)}^{Qm-1} \hat{\mathbf{R}}(t) \tag{10.50}$$

In a continuous-wave (CW) OTH radar with well-separated transmit and receive sites, the secondary cells used to train the adaptive filter may be extracted from range bins corresponding to physical distances less than the transmitter-to-receiver distance in a surface-wave radar, or range bins that fall inside the skip-zone when it exists for a skywave radar. Such range cells found near the "start" of the PRI will be effectively clutter free. The former scheme is not feasible in a pulse-waveform (PW) mono-static surface-wave radar, since the closest ranges often contain the most powerful clutter.

However, the high attenuation of the surface wave with distance, which is much more severe than the inverse-square law describing line-of-sight propagation, may allow clutter-free cells to be obtained from nonoperational ranges near the "end" of the PRI. When RFI-only secondary data is not available due to propagation conditions or processing limitations, the clutter must be pre-filtered to obtain suitable training data (Abramovich, Spencer, and Anderson 2000).

10.2.2.2 Clutter Returns

Different clutter models may be proposed for both skywave and surface-wave OTH radar systems. Here, the emphasis is placed on high frequency surface-wave (HFSW) radar because the experimental results to be presented later in this chapter demonstrate the practical effectiveness of adaptive beamforming for non-stationary interference cancelation in such a system. However, the practical methodology adopted is not restricted to HFSW radar, but may also be applied to skywave OTH radar systems. Although experimental results for the skywave case are not reported here, the following chapter will describe a skywave clutter model and illustrate the application of similar concepts on simulated data.

For an HFSW radar, the dominant contribution of sea clutter is produced by first-order scattering from specific spectral components of the ocean surface height wavefield, known as advance and recede Bragg-wave trains. These spectral components have a wavelength $\ell = \lambda/2$ of exactly half the radio wavelength λ and move directly toward and away from the radar, respectively. In water of depth $h > \ell/2$, the Bragg-wave trains move with radial velocities $v_g = \pm\sqrt{g\ell/2\pi}$, where $\ell = \lambda/2$ is the Bragg wavelength and g is acceleration due to gravity. Consequently, radar signal echoes coherently backscattered from Bragg waves at grazing incidence are Doppler shifted by an amount $f_b = \pm\sqrt{g/\lambda\pi}$ relative to the carrier (Lipa and Barrick 1986). When a surface current of mean radial velocity v_s is present, an additional Doppler shift of $f_s = 2v_s/\lambda$ results for both the advance and recede Bragg-wave spectral components.

Let $r_i(t)$ be the ocean clutter contribution received by the first (i.e., reference) *antenna element* in the array due to a scattering "patch" defined by the radar resolution cell size at the location of the unnamed range cell in a narrow *finger-beam* at azimuth θ_i. This contribution to the overall time-domain clutter return is modeled by Eqn. (10.51) in accordance with Hickey, Khan, and Walsh (1995). The advance and recede first-order clutter amplitudes, denoted by A_i and $\tilde{A}_i$, respectively, are proportional to the ocean directional wave-height spectrum at the Bragg wave-vectors, while the surrounding clutter continuum $e_i(t)$ is due to higher-order scatter, arising mainly from the second-order interaction between pairs of waves that constitute the total ocean wavefield.

$$r_i(t) = A_i e^{j\{2\pi(f_s+f_b)t+\phi_i(t)\}} + \tilde{A}_i e^{j\{2\pi(f_s-f_b)t+\phi_i(t)\}} + e_i(t) \tag{10.51}$$

The surface current in the resolution cell may also have a nonuniform or turbulent velocity component relative to the mean value v_s, with different dynamic behavior from one scattering patch to another. Surface-current turbulence imposes the phase modulation $\phi_i(t)$ on the Bragg-wave clutter returns, which broadens the first-order clutter spectral peaks. In practice, the presence of time-varying surface currents during the CPI causes the advance and recede Bragg spectral lines to be smeared in identical shapes by the phase modulation function (Parkinson 1997).

From the viewpoint of time-varying adaptive beamforming, the primary concern is to not destroy the naturally occurring temporal correlation of the first-order clutter components, whose amplitudes are generally two orders of magnitude higher than the surrounding continuum. For this reason, we focus on the properties of the first-order scatter, and momentarily ignore the significantly weaker contribution $e_i(t)$. The overall first-order sea clutter received in the unnamed test cell is the vector addition of the narrow-beam returns from I azimuth cells subtending the intersection of the transmit and receive antenna patterns. The resultant first-order sea clutter snapshots $\mathbf{c}_s(t)$ may be expressed as Eqn. (10.52).

$$\mathbf{c}_s(t) = \sum_{i=1}^{I}[A_i e^{j\{2\pi(f_s+f_b)t+\phi_i(t)\}} + \tilde{A}_i e^{j\{2\pi(f_s-f_b)t+\phi_i(t)\}}]\mathbf{v}(\theta_i) \tag{10.52}$$

A ground-clutter component $\mathbf{c}_g$ with zero Doppler shift may also be present due to backscatter from areas of land or islands in the coverage. The deterministic signature of $\mathbf{c}_g$ assumes that the movement of ground scatterers due to wind, for example, as well as nonideal systems effects can be ignored. Combining terrain and sea contributions, the first-order clutter snapshots $\mathbf{c}(t) = \mathbf{c}_g + \mathbf{c}_s(t)$ received by the array are modeled as Eqn. (10.53), where $\mathbf{c}_a(t) = \sum_{i=1}^{I} A_i e^{j2\pi\phi_i(t)}\mathbf{v}(\theta_i)$ and $\mathbf{c}_r(t) = \sum_{i=1}^{I} \tilde{A}_i e^{j2\pi\phi_i(t)}\mathbf{v}(\theta_i)$ represent the complex-valued modulations imposed by the ocean surface on the advance and recede Bragg waves.

$$\mathbf{c}(t) = \mathbf{c}_g + \mathbf{c}_a(t)e^{j2\pi(f_s+f_b)t} + \mathbf{c}_r(t)e^{j2\pi(f_s-f_b)t} \tag{10.53}$$

Since the instantaneous Bragg frequencies change relatively slowly, they may be regarded quasi-stationary during a sufficiently short sub-CPI. The assumption of constant Bragg frequencies over Q consecutive PRI allows the clutter to be represented as superposition of complex sinusoids in each sub-CPI. Such a model has previously been adopted to parametrically estimate surface currents (Hickey et al. 1995), and to cancel clutter in CPIs shorter than 3 seconds (Root 1998). This assumption implies that $\mathbf{c}_a(t)$ and $\mathbf{c}_r(t)$ can be written in the form of Eqn. (10.54), where $\tau_m = Q(m-1)$ indexes the first pulse in the mth sub-CPI, $t = Q(m-1), \ldots, Qm-1$ indexes the Q consecutive pulses in the mth sub-CPI, and δf_m is the Bragg wave instantaneous frequency that changes from one sub-CPI to another. The slowly changing shift in Bragg wave instantaneous frequency δf_m with m leads to the incoherent integration of the first-order ocean clutter over the CPI.

$$\mathbf{c}_a(t) = \mathbf{c}_a(\tau_m)e^{j2\pi\delta f_m t}, \quad \mathbf{c}_r(t) = \mathbf{c}_r(\tau_m)e^{j2\pi\delta f_m t} \tag{10.54}$$

By defining the $N \times L$ matrix $\mathbf{A}_m = [\mathbf{c}_g, \mathbf{c}_a(\tau_m), \mathbf{c}_r(\tau_m)]$ as the clutter subspace with dimension $L = 3$, and the $L \times 1$ parameter vector $\mathbf{p}(t) = [1, e^{j2\pi(f_m+f_b)t}, e^{j2\pi(f_m-f_b)t}]^T$ with $f_m = f_s + \delta f_m$, the first-order clutter snapshots may be described by the *dynamic spatial subspace* model of Eqn. (10.55). In this example, a value of $L = 3$ captures the ground clutter and the advance/recede Bragg-wave spectral components. The clutter snapshots $\mathbf{c}(t)$ received in sub-CPI of length $Q > L$ are spanned by a subspace $\mathbf{A}_m$ of relatively low-rank $L << N$. This low-rank property reflects the high pulse-to-pulse correlation of the dominant clutter spectral components over a sub-CPI. On the other hand, in the limit of a relatively long CPI, the temporal variation of $\mathbf{A}_m$ with sub-CPI index m reflects the higher rank of the statistically expected clutter spatial covariance matrix, which typically has almost full rank N.

$$\mathbf{c}(t) = \mathbf{A}_m \mathbf{p}(t), \quad t = Q(m-1), \ldots, Qm-1 \tag{10.55}$$

In general, the minimum value of L is set by the number of dominant clutter spectral components resolved in the Doppler spectrum. For example, $L = 3$ corresponds to two resolved Bragg lines and ground clutter, based on the assumption that all dominant components are frequency stable over the sub-CPI. Values of $L > 3$ may be adopted to account for subspace leakage caused by the nonlinear nature of the phase modulation process over a sub-CPI, or to account for some of the second-order clutter. Typically, $L \ll N$. The condition $L \leq \min(Q, N)$ must be satisfied, since Q snapshots of dimension N can always be represented by a subspace of dimension less than or equal to $\min(Q, N)$.

10.2.3 Standard Adaptive Beamforming

Perhaps the simplest approach to account for RFI spatial non-stationarity is to estimate the adaptive beamformer weights using secondary interference-plus-noise snapshots recorded during a short training interval scheduled immediately prior to and after the operational CPI that additionally contains clutter and useful signals (i.e., with the radar transmitter switched off or tuned to a different frequency for a brief period). The resulting adaptive weight vector is then "frozen" and applied to process the operational CPI of data. Experimental investigations confirm that such approaches, referred to as framing schemes 1 and 2 in Chapter 9, are ineffective in practice because their ability to suppress non-stationary interference degrades quickly for data not immediately adjacent to the training periods. Hence, such schemes are of limited utility for the relatively long CPI lengths required in OTH radar.

An alternative method relies on a limited number of (clutter-free) RFI snapshots being available *within* the CPI, such that the adaptive beamformer is trained on the average properties of the interference and noise prevailing in the CPI. While this method, corresponding to the weight vector $\bar{\mathbf{w}}$ in Eqn. (10.56) and called scheme 3 in Chapter 9, is more effective than the "framing" approaches described previously, the practical case study in Chapter 9 revealed that significant losses in RFI rejection are nevertheless experienced for typical OTH radar CPI lengths. In summary, adaptive beamforming procedures that process the CPI based on a *time-invariant* weight vector often fail to effectively remove spatially non-stationary HF interference. The relatively long dwell times required by OTH radars make such approaches susceptible to this frequently observed physical phenomenon.

$$\bar{\mathbf{w}} = [\mathbf{v}^{\dagger}(\theta)\bar{\mathbf{R}}^{-1}\mathbf{v}(\theta)]^{-1}\bar{\mathbf{R}}^{-1}\mathbf{v}(\theta), \quad \bar{\mathbf{R}} = \frac{1}{M}\sum_{m=1}^{M}\hat{\mathbf{R}}_m \tag{10.56}$$

The remaining option is to update the adaptive beamforming weight vector within the CPI. In the traditional time-varying approach, referred to as scheme 4 in Chapter 9, the CPI is processed by a sequence of adaptive weight vectors $\hat{\mathbf{w}}_1, \ldots, \hat{\mathbf{w}}_M$ that maximize the output SINR based on the corresponding sample spatial covariance matrices $\hat{\mathbf{R}}_1, \ldots, \hat{\mathbf{R}}_M$ integrated over M relatively short sub-CPIs. As the interference and noise may be regarded as locally stationary over a sufficiently small number of consecutive PRI, the time-varying adaptive beamformer has greater scope for effective RFI rejection than its time-invariant counterpart. In accordance with the SMI-MVDR criterion, the weight vector used to adaptively beamform the operational range cells processed in PRI batch m is given by $\hat{\mathbf{w}}_m$ in Eqn. (10.57).

$$\hat{\mathbf{w}}_m = [\mathbf{v}^{\dagger}(\theta)\hat{\mathbf{R}}_m^{-1}\mathbf{v}(\theta)]^{-1}\hat{\mathbf{R}}_m^{-1}\mathbf{v}(\theta) \tag{10.57}$$

However, effective RFI cancelation does not automatically imply an improvement in target detection performance. The main problem is that the intra-CPI antenna pattern fluctuations caused by the changing weight vectors $\{\hat{\mathbf{w}}_m\}_{m=1}^{M}$ destroy the temporal coherence of clutter returns in the operational ranges. OTH radar clutter is incident from a broad continuum of directions, while the fixed (unit-gain) constraint only provides distortionless coherent processing for useful signals in the radar look direction. The resulting pulse-to-pulse modulation of off-azimuth signals at the time-varying beamformer output decorrelates the clutter returns and this can severely degrade Doppler processing performance. The consequence of adaptive antenna pattern updates on clutter returns therefore manifests itself as a dramatic reduction in sub-clutter visibility (SCV) after Doppler processing.[1] In simple terms, this may be thought of as the smearing of clutter energy across the Doppler search space.

A standard method to stabilize the sequence of adaptive patterns over the CPI is to diagonally load or *regularize* the sample covariance matrices and to process the data using the LSMI weights defined in Eqn. (10.58). It is evident that as the loading factor α increases, the LSMI weights tend toward the conventional (matched filter) beamformer $\mathbf{v}(\theta)$. The value of α therefore represents a tradeoff between the amount of stabilization required for preserving clutter coherence, and the freedom of adaptation necessary for effective interference rejection. While a small amount of diagonal loading is useful for improving convergence rate in sample-starved conditions, such loading is typically not high enough to prevent significant SCV degradation due to clutter smearing in non-stationary RFI environments. On the other hand, sufficiently heavy loading can steady the adaptive weight vectors to protect against SCV degradation, but the amount of loading required for this will typically spoil the interference rejection capability of the adaptive beamformer. As there is often no value of α that can achieve both objectives simultaneously, the LSMI method does not represent an effective solution for the problem at hand.

$$\hat{\mathbf{w}}_m = [\mathbf{v}^{H}(\theta)\{\hat{\mathbf{R}}_m + \alpha\mathbf{I}\}^{-1}\mathbf{v}(\theta)]^{-1}\{\hat{\mathbf{R}}_m + \alpha\mathbf{I}\}^{-1}\mathbf{v}(\theta) \tag{10.58}$$

If the clutter was produced by a small number of point scatterers, it would in theory be possible to maintain the adaptive pattern fixed in the directions of these scatterers using a set of linear deterministic constraints. The remainder of the pattern would then be free to adapt to the RFI characteristics (Griffiths 1996). However, OTH radar clutter is received by most of the antenna pattern because it arises from a continuum of point scatterers distributed over a relatively wide angular region (i.e., that of the transmitter footprint). It is therefore infeasible to "freeze" the adaptive filter response for all such scatterers and simultaneously maintain sufficient spare degrees of freedom to ensure high quality interference rejection.

[1]The SCV is defined as the power ratio between the maximum clutter peak in the Doppler spectrum and the average disturbance level calculated in the half of the Doppler spectrum that corresponds to the highest radial speeds.

The goal of a robust adaptive beamformer in HF radar applications is to cancel non-stationary RFI and simultaneously preserve the temporal correlation properties of the backscattered clutter in a manner that is computationally efficient and suitable for practical implementation. Standard adaptive beamforming schemes, which may be expected to work well under stationary RFI conditions, are not effective in the HF environment, as they cannot manage to reconcile these two apparently conflicting objectives.

10.3 Time-Varying Approaches

Fortunately, at least two fruitful approaches to address the aforementioned problem have been developed, both of which resolve the apparent "paradox" of allowing the weights to vary within the CPI for effective RFI rejection without affecting the clutter Doppler spectrum characteristics. These two alternative time-domain (i.e., pre-Doppler) adaptive beamforming algorithms, known as the stochastic constraints (SC) method and time-varying spatial adaptive processing (TV-SAP), will be described in this section. The performance of these two candidate operational approaches for OTH radar are evaluated using experimental data in terms of practical performance in an HFSW radar system, computational load for real-time implementation, and flexibility to cater for varying levels of RFI non-stationarity.

10.3.1 Stochastic Constraints

The original scientific breakthrough is represented by the algorithm of Abramovich et al. who pioneered the *stochastic constraints* method. This method has been reported in several landmark papers, including (Abramovich, Mikhaylyukov, and Malyavin 1992b) and (Abramovich, Gorokhov, Mikhaylyukov, and Malyavin 1994) for example, as well as in (Klemm 2004). Hence, only the underlying principles and essential features of the SC method are described here for brevity. The reader is referred to Klemm (2004) for an in-depth treatment of this method and its operational implementation, which is applicable to spatial and space-time adaptive processing. As the extension of the SC approach to space-time adaptive processing will be covered explicitly in the following chapter, the brief description given below avoids the unnecessary duplication of similar details.

The SC-SAP algorithm relies on a relatively low-order scalar-type auto-regressive (AR) model of the random multi-channel clutter process $\mathbf{c}(t)$, such that the AR parameters $\{b_i\}_{i=1}^{L}$ for $L \ll N$ in Eqn. (10.59) determine the Doppler spectrum characteristics of the received clutter. In other words, the clutter snapshot $\mathbf{c}(t)$ in the current pulse t is described as a linear combination of the past L clutter snapshots (in the same range cell) plus innovative noise in accordance with the AR recursive relation of Eqn. (10.59). Here, $\epsilon(t) \in \mathcal{C}^N$ is a zero-mean complex innovative white-noise vector with correlation properties $\mathrm{E}\{\epsilon(i)\epsilon^{\dagger}(j)\} = \delta_{ij}\mathbf{R}_c$, where $\mathbf{R}_c$ is the spatial covariance matrix of the clutter at the unnamed range cell. The AR model order L is assumed to be known, but the AR parameters b_i for $i = 1, .., L$ are unknown. Such a model may be used to represent clutter received by surface-wave and skywave OTH radar systems.

$$\mathbf{c}(t) + \sum_{i=1}^{L} b_i \mathbf{c}(t-i) = \epsilon(t) \tag{10.59}$$

The main idea behind the SC-SAP approach is to preserve the slow-time clutter correlation properties at the beamformer output by imposing L additional "stochastic constraints" on the adaptive weight vector. These auxiliary linear constraints are data-dependent and designed to control the antenna pattern adaptations such that the *output* clutter time-series is statistically described by a temporal AR process with the same scalar parameters $\{b_i\}_{i=1}^{L}$ as the input multi-variate process. The method for achieving this is to estimate a slow-time (and range) dependent adaptive beamforming weight vector $\hat{\mathbf{w}}(t)$ that satisfies the optimization criterion in Eqn. (10.60).

$$\hat{\mathbf{w}}(t) = \arg\min_{\mathbf{w}} \mathbf{w}^{\dagger}\hat{\mathbf{R}}(t)\mathbf{w} \quad \text{subject to} : \mathbf{C}^{\dagger}(t)\mathbf{w} = \mathbf{f}(t) \tag{10.60}$$

The term $\hat{\mathbf{R}}(t)$ is a local estimate of the interference-plus-noise spatial covariance matrix that can effectively reject the RFI at time t, while the linear constraints are defined by the $N \times (L+1)$ matrix $\mathbf{C}(t) = [\mathbf{v}(\theta)\ \mathbf{c}(t-1)\cdots\mathbf{c}(t-L)]$ and the $(L+1)$-dimensional vector $\mathbf{f}(t) = [1\ \mathbf{w}_0^{\dagger}\mathbf{c}(t-1)\cdots\mathbf{w}_0^{\dagger}\mathbf{c}(t-L)]^{\dagger}$. The first weight vector in the sequence $\mathbf{w}_0$ is a reference weight vector derived from the sample covariance matrix averaged over the first $L+1$ pulses (i.e., $t = 1, \ldots, L+1$), *without* the use of L additional stochastic constraints. The SC weights may be computed using Eqn. (10.10), and updated in slow-time t to process the entire CPI. In particular, the beamformed clutter outputs are given by Eqn. (10.61), where $y_c(t) = \mathbf{w}_0^{\dagger}\mathbf{c}(t)$.

$$y_c(t) = \hat{\mathbf{w}}^{\dagger}(t)\mathbf{c}(t) = -\sum_{i=1}^{L} b_i y_c(t-i) + \eta(t) \tag{10.61}$$

The imposed constraints ensure that the current beamformed clutter output is also a linear function of the past L beamformed clutter outputs with the same scalar AR coefficients of the original multi-channel clutter process $\{b_i\}_{i=1}^{L}$ plus innovative noise $\eta(t)$. Hence, it follows that such constraints do in fact preserve the spectral characteristics of the beamformed clutter process, while allowing the adaptive weight vector to change in response to the RFI during the CPI. Naturally, clutter-only snapshots are not available in practice to form the stochastic constraints in Eqn. (10.60). Operational procedures for obtaining close approximations to these constraints are described in Abramovich, Spencer, and Anderson (1998) for supervised training and in Abramovich, Anderson, and Spencer (2000) for unsupervised training. The performance of an operational SC-SAP method using supervised training will be shown later in this section.

10.3.2 Time-Varying Spatial Adaptive Processing

Indeed, the SC-SAP method is based on protecting the AR spectral characteristics of clutter at the output of an adaptive beamformer that fluctuates in response to the changing RFI spatial structure within the CPI. However, in the attempt to stabilize the AR characteristics of the output clutter time series, the SC method forms a new adaptive spatial filter *every* PRI by applying a different set of data-driven constraints. In other words, the SC adaptive beamforming weight vectors are updated in a "sliding-window" fashion, such that for an AR process of order L, the SC weights must be re-adjusted $P-L$ times over the CPI (Abramovich, Gorokhov, Mikhaylyukov, and Malyavin 1994).

Importantly, this rate of adaptation is determined by the need to protect the AR Doppler spectrum properties of the *clutter* at the beamformer output, irrespective of the prevailing level of interference spatial non-stationarity, which ought to be the primary reason for re-adapting the weight vectors. In many situations of practical interest, the adaptive filter may be updated at a much slower rate than once per PRI, without incurring loss in interference rejection performance. In this case, the SC-SAP technique has two major disadvantages.

First, the procedure is computationally expensive, which may prevent its real-time implementation in certain operational HF radar systems. Second, weight-vector estimation errors occur at each update and accumulate with every re-adaptation, which can cause a reduction in SCV after Doppler processing. These reasons strongly motivate further efforts to develop other time-varying adaptive beamforming algorithms that are *computationally efficient* while, at the same time, yield high performance comparable to the SC method.

The TV-SAP method introduced in the next section follows the same basic idea as the SC-SAP technique, but is structured differently to mitigate the aforementioned limitations of the original SC work. The philosophy behind the TV-SAP method is to update the weight vector in non-overlapping batches at a rate that is commensurate with the level of interference spatial non-stationarity. Data-driven constraints are incorporated to protect the dominant clutter spectral components in accordance with the dynamic spatial subspace model described previously. It will be shown that TV-SAP outperforms the SC-SAP technique in the illustrated practical examples with the added advantage of significantly reduced computational complexity.

10.3.2.1 Alternative Algorithm

The TV-SAP approach in Fabrizio, Gershman, and Turley (2004) is based on dividing the CPI into a number of non-overlapping sub-CPIs wherein the dominant *clutter* spectral components are contained in a low-rank spatial subspace. The main idea behind TV-SAP is that *changes* in the adaptive weights from one sub-CPI to another are constrained to be *orthogonal* to the current clutter subspace. In this way, the most powerful clutter returns essentially experience a "stable" spatial filter over the CPI, while the adaptive weight vector varies from one sub-CPI to another to effectively suppress the non-stationary interference.

Specifically, the first adaptive beamformer $\mathbf{w}_1$ in the sequence of weight vectors $\mathbf{w}_1, \ldots, \mathbf{w}_M$ used to process the CPI provides maximum SINR for the sample covariance matrix $\mathbf{R}_1$ integrated over the first sub-CPI in accordance with SMI-MVDR criterion: $\mathbf{w}_1 = [\mathbf{v}^\dagger(\theta)\mathbf{R}_1^{-1}\mathbf{v}(\theta)]^{-1}\mathbf{R}_1^{-1}\mathbf{v}(\theta)$. This weight vector is used to adaptively beamform the array snapshots $y(t) = \mathbf{w}_1^\dagger\mathbf{z}(t)$ received over the first batch of Q pulses, i.e., for $t = 0, \ldots, Q-1$ in Eqn. (10.62). Before proceeding to describe the TV-SAP algorithm, it is instructive to explain why it is *inappropriate* to derive the adaptive weight vectors for the batches $m = 2, \ldots, M$ in analogous fashion: $\mathbf{w}_m = [\mathbf{v}^\dagger(\theta)\mathbf{R}_m^{-1}\mathbf{v}(\theta)]^{-1}\mathbf{R}_m^{-1}\mathbf{v}(\theta)$, and process the CPI as in Eqn. (10.62). In this section, the hat $(\hat{\cdot})$ symbol has been dropped for the estimated weight vector $\mathbf{w}_m$ and SCM $\mathbf{R}_m$ to simplify notation.

$$y(t) = g(t) + \mathbf{w}_m^\dagger\mathbf{c}(t) + \mathbf{w}_m^\dagger\mathbf{j}(t) \qquad \begin{cases} m = 1, \ldots, M \\ t = Q(m-1), \ldots, Qm-1 \end{cases} \tag{10.62}$$

Consider the first two batches of the CPI. The natural slow-time (pulse-to-pulse) coherence of the initial Q *clutter* samples $y_c(t)$ in Eqn. (10.63) is retained over the first batch $t = 0, \ldots, Q-1$ because the data is processed by a *fixed* spatial filter $\mathbf{w}_1$. Using a similar argument, coherence is also maintained among the Q clutter samples $y_c(t)$ in the second batch $t = Q, \ldots, 2Q-1$, which have also been processed by a *fixed* adaptive spatial filter $\mathbf{w}_2$ in Eqn. (10.63). However, the $2Q$ clutter samples beamformed by the first and second adaptive filters constitute a part of the total CPI. Hence, if $\mathbf{w}_2$ is different and formed independently of $\mathbf{w}_1$, the pulse-to-pulse coherence amongst *all* of the $2Q$ clutter samples is no longer preserved. Specifically, the coherence is lost between the last sample of batch $m = 1$ and the first sample of batch $m = 2$ (i.e., across the batch boundary). This is because the instantaneous change in the antenna pattern temporally modulates any unrejected clutter components passed onto the beamformer output. More precisely, $\mathbf{w}_2^\dagger \mathbf{c}(t) \neq \mathbf{w}_1^\dagger \mathbf{c}(t)$ for $t = Q, \ldots, 2Q-1$.

$$y_c(t) = \begin{cases} \mathbf{w}_1^\dagger \mathbf{c}(t) & \text{for } t = 0, \ldots, Q-1 \\ \mathbf{w}_2^\dagger \mathbf{c}(t) & \text{for } t = Q, \ldots, 2Q-1 \end{cases} \tag{10.63}$$

To rectify this coherence "discontinuity" between the first and second batches, without enforcing the condition $\mathbf{w}_2 = \mathbf{w}_1$, the second weight vector $\mathbf{w}_2$ is formed *dependently* of the first $\mathbf{w}_1$, such that the instantaneous *change* in antenna pattern is constrained to be *orthogonal* to the clutter subspace in the second batch. Stated mathematically, the key idea of the TV-SAP algorithm is to maintain the constraint in Eqn. (10.64) as accurately as possible.

$$(\mathbf{w}_2 - \mathbf{w}_1)^\dagger \mathbf{A}_2 = \mathbf{0} \tag{10.64}$$

This permits the vector $\mathbf{w}_2$ to differ from $\mathbf{w}_1$ for non-stationary interference rejection purposes, while ensuring that the change $\Delta\mathbf{w}_{1,2} = \mathbf{w}_2 - \mathbf{w}_1$ does not influence the output for the dominant clutter snapshots $\mathbf{c}(t) = \mathbf{A}_2\mathbf{p}(t)$ received in the second batch. This is achieved because the constraint in Eqn. (10.64) ensures that the scalar clutter outputs $y_c(t)$ for $t = Q, \ldots, 2Q-1$ processed by the adaptive beamformer $\mathbf{w}_2$ are identical to the samples which would have resulted if $\mathbf{w}_1$ was applied to process the clutter snapshots $\mathbf{c}(t) = \mathbf{A}_2\mathbf{p}(t)$ in the second batch. This key point is embodied in Eqn. (10.65).

$$\Delta\mathbf{w}_{1,2}^\dagger \mathbf{c}(t) = (\mathbf{w}_2 - \mathbf{w}_1)^\dagger \mathbf{A}_2\mathbf{p}(t) = 0, \quad t = Q, \ldots, 2Q-1 \tag{10.65}$$

Implementing the constraint in Eqn. (10.64) suggests that the second adaptive filter $\mathbf{w}_2$ should be synthesized according to the criterion in Eqn. (10.66), where $\mathbf{w}_1$ is the standard SMI-MVDR filter used to initialize the sequence. In this way, a distortionless response can be provided for useful signals $\mathbf{w}_1^\dagger \mathbf{v}(\theta) = \mathbf{w}_2^\dagger \mathbf{v}(\theta) = 1$, as well as the dominant clutter returns, since $y_c(t) = \mathbf{w}_1^\dagger \mathbf{c}(t)$ over all $2Q$ samples $t = 0, \ldots, 2Q-1$.

$$\mathbf{w}_2 = \arg\min_{\mathbf{w}} \ \mathbf{w}^\dagger \mathbf{R}_2 \mathbf{w} \quad \text{subject to} \begin{cases} \mathbf{w}^\dagger \mathbf{v}(\theta) = 1, \\ \mathbf{w}^\dagger \mathbf{A}_2 = \mathbf{w}_1^\dagger \mathbf{A}_2 \end{cases} \tag{10.66}$$

In practice, knowledge of the clutter subspace $\mathbf{A}_2$ is not available because the received snapshots $\mathbf{z}(t)$ are contaminated by additive interference. As the clutter subspace is not directly observable due to being immersed in interference, the criterion in Eqn. (10.66) cannot be implemented in the presented form. However, the adaptive filter $\mathbf{w}_1$ will still be effective for interference rejection in the first L pulses of the second batch, where L

is the dimension of the low-rank clutter spatial subspace model. Applying this filter to the first L snapshots of the second batch yields Eqn. (10.67), where the magnitude of the residual interference and noise contribution at the adaptive beamformer output $\mathbf{w}_1^\dagger \mathbf{j}(t)$ is assumed to be negligible compared to that of the unrejected clutter $\mathbf{w}_1^\dagger \mathbf{c}(t)$.

$$\mathbf{w}_1^\dagger \mathbf{z}(t) \approx g(t) + \mathbf{w}_1^\dagger \mathbf{c}(t), \quad t = Q, \ldots, Q + L - 1 \tag{10.67}$$

By the same reasoning, we may write $\mathbf{w}_2^H \mathbf{x}(t) \approx g(t) + \mathbf{w}_2^H \mathbf{c}(t)$ for $t = Q, \ldots, Q + L - 1$, assuming the second weight vector $\mathbf{w}_2$ also provides unity gain response to useful signals. Now, if the second adaptive filter is also made to satisfy the L auxiliary linear *data-driven* constraints defined in Eqn. (10.68), then from Eqn. (10.67), it follows that the clutter contribution at the output of the second filter is as if it were processed by the first filter to a good approximation, since these L data-driven constraints lead to the condition $\mathbf{w}_2^\dagger \mathbf{c}(t) \approx \mathbf{w}_1^\dagger \mathbf{c}(t)$ for $t = Q, \ldots, Q+L-1$. Importantly, the two filters satisfying Eqn. (10.68) may be quite different (i.e., $\mathbf{w}_1 \neq \mathbf{w}_2$) for $L \ll N$, as required for effective non-stationary interference rejection.

$$\mathbf{w}_2^\dagger \mathbf{z}(t) = \mathbf{w}_1^\dagger \mathbf{z}(t), \quad t = Q, \ldots, Q + L - 1 \tag{10.68}$$

Furthermore, since the L clutter snapshots $\mathbf{c}(t) = \mathbf{A}_2 \mathbf{p}(t)$ for $t = Q, \ldots, Q + L - 1$ may be assumed to be *linearly independent*, they collectively span the column space of the matrix $\mathbf{A}_2$. As a consequence, the L data-driven constraints imposed on $\mathbf{w}_2$ also encourage the relative change in antenna pattern to be orthogonal to the clutter subspace, i.e., $(\mathbf{w}_2 - \mathbf{w}_1)^\dagger \mathbf{A}_2 \approx \mathbf{0}$. This condition carries with it the implication that the clutter contribution at the output of the second filter is as if it were beamformed by the first filter over the *whole* of the second batch $t = Q, \ldots, 2Q - 1$. The condition in Eqn. (10.69) therefore holds despite the use of only $L < \min(Q, N)$ constraints.

$$\mathbf{w}_2^\dagger \mathbf{c}(t) \approx \mathbf{w}_1^\dagger \mathbf{c}(t), \quad t = Q, \ldots, 2Q - 1 \tag{10.69}$$

Hence, the problem of synthesizing a second adaptive filter $\mathbf{w}_2$, that may differ from the first $\mathbf{w}_1$ to effectively reject spatially non-stationary interference, while not disturbing the pulse-to-pulse coherence of the clutter output sequence, may be formulated in terms of the optimization criterion in Eqn. (10.70).

$$\mathbf{w}_2 = \arg\min_{\mathbf{w}} \ \mathbf{w}^\dagger \mathbf{R}_2 \mathbf{w} \quad \text{subject to} \begin{cases} \mathbf{w}^\dagger \mathbf{v}(\theta) = 1, \\ \mathbf{w}^\dagger \mathbf{z}(t) = \mathbf{w}_1^\dagger \mathbf{z}(t), \quad t = Q, \ldots, Q + L - 1 \end{cases} \tag{10.70}$$

By defining the $N \times L$ linear constraint matrix as,

$$\mathbf{C}_2 = [\mathbf{v}(\theta), \mathbf{z}(Q), \ldots, \mathbf{z}(Q + L - 1)] \tag{10.71}$$

and the associated $L \times 1$ response vector as,

$$\mathbf{f}_2 = [1, \mathbf{w}_1^\dagger \mathbf{z}(Q), \ldots, \mathbf{w}_1^\dagger \mathbf{z}(Q + L - 1)]^\dagger \tag{10.72}$$

for the second batch $m = 2$, the problem in Eqn. (10.70) can be expressed in the more succinct form of Eqn. (10.73).

$$\mathbf{w}_2 = \arg\min_{\mathbf{w}} \mathbf{w}^\dagger \mathbf{R}_2 \mathbf{w} \qquad \text{subject to}: \quad \mathbf{w}^\dagger \mathbf{C}_2 = \mathbf{f}_2^\dagger \tag{10.73}$$

This is immediately recognized as the linearly constrained minimum variance (LCMV) optimization problem, which has the following closed form solution (Frost 1972).

$$\mathbf{w}_2 = \mathbf{R}_2^{-1}\mathbf{C}_2[\mathbf{C}_2^\dagger \mathbf{R}_2^{-1}\mathbf{C}_2]^{-1}\mathbf{f}_2 \tag{10.74}$$

The remaining batches in the CPI are processed in similar fashion by iterating with respect to batch number $m = 2, \ldots, M$, such that the mth weight vector $\mathbf{w}_m$ is computed as,

$$\mathbf{w}_m = \mathbf{R}_m^{-1}\mathbf{C}_m[\mathbf{C}_m^\dagger \mathbf{R}_m^{-1}\mathbf{C}_m]^{-1}\mathbf{f}_m \tag{10.75}$$

where the constraint matrix and response vector associated with the mth iteration are given by Eqn. (10.76).

$$\begin{aligned} \mathbf{C}_m &= [\mathbf{v}(\theta), \mathbf{z}(Q(m-1)), \ldots, \mathbf{z}(Q(m-1)+L-1)] \\ \mathbf{f}_m &= [1, \mathbf{w}_{m-1}^\dagger \mathbf{z}(Q(m-1)), \ldots, \mathbf{w}_{m-1}^\dagger \mathbf{z}(Q(m-1)+L-1)]^\dagger \end{aligned} \tag{10.76}$$

That is, the iterations are continued for the M batches until the whole CPI is processed. To be clear, the TV-SAP algorithm is *iterative* in the sense that the MVDR weight vector $\mathbf{w}_1$ initializes the sequence at $m = 1$, and the weights for batches $m = 2, \ldots, M$ are computed using the known weights from the *previous* step. The weights at each step are computed analytically using the LCMV solution in Eqn. (10.75).

10.3.2.2 Computational Complexity

Tables 10.1 and 10.2 summarize the computational complexity of the TV- and SC-SAP methods, respectively. The size of the matrix at each processing step is indicated by the square brackets in the first column, while the number of complex multiplications required at each step are shown in the second column. The subtotals S_1, S_2, and S_3 take into account the repetition of steps needed to process a whole CPI of data. The study case parameters mentioned in the caption reflect the experimental data processed in the next section, while the corresponding numerical values are indicated in the rightmost column of each table.

The number of complex multiplications required to process the radar data cube is generally accepted as an appropriate measure of computational load for making comparisons between different algorithms. The computational advantage of TV-SAP relative to the SC method is readily apparent in Tables 10.1 and 10.2. In this case, TV-SAP reduces the number of complex multiplications by a factor of about 15 relative to SC-SAP, without additional memory requirements. From a computational perspective, TV-SAP is much more attractive than SC-SAP, as it provides an *order of magnitude* saving in computational load for real-time practical implementation.

Processing Step	Complex Multiplications	Study Case
Select Training Data $\mathbf{D}_m$ $[N \times KQ]$ Estimate Matrix $\mathbf{R}_m = \mathbf{D}\mathbf{D}^H$ $[N \times N]$ Matrix Inversion $\mathbf{R}_m^{-1}$ $[N \times N]$	- $N^2 \times KQ$ $O(N^3)$	- 64000 32000
Repeated for M batches	$S_1 = M\times$ $\{(N^2 \times KQ) + O(N^3)\}$	1536000
Steering Vector $\mathbf{v}(\theta)$ $[N \times 1]$ Form $\mathbf{z}_1 = \mathbf{R}_1^{-1}\mathbf{v}(\theta)$ $[N \times 1]$ Form $z_1 = \mathbf{v}^H(\theta)\mathbf{z}_1$ $[1 \times 1]$ First Weight Vector $\mathbf{w}_1 = \mathbf{Z}_1/z_1$ $[N \times 1]$ Process $y(t) = \mathbf{w}_1^H\mathbf{x}(t)$ for $t = 0, \ldots, Q-1$	- N^2 N N NQ	- 400 20 20 320
Repeated for N_R ranges and N_B beams	$S_2 = N_R N_B \times$ $\{N^2 + 2N + NQ\}$	729600
Form $\mathbf{C}_m$ $[N \times (L+1)]$ Form $\mathbf{f}_m$ $[(L+1) \times 1]$ Form $\mathbf{M}_1 = \mathbf{R}_m^{-1}\mathbf{C}_m$ $[N \times (L+1)]$ Form $\mathbf{M}_2 = \mathbf{C}_m^H\mathbf{M}_1$ $[(L+1) \times (L+1)]$ Invert Matrix $\mathbf{M}_2^{-1}$ $[(L+1) \times (L+1)]$ Multiply $\mathbf{M}_3 = \mathbf{C}_m\mathbf{M}_2$ $[N \times (L+1)]$ Multiply $\mathbf{M}_4 = \mathbf{R}_m^{-1}\mathbf{M}_3$ $[N \times (L+1)]$ Calculate $\mathbf{w}_m = \mathbf{M}_4\mathbf{f}_m$ $[N \times 1]$ Process m^{th} Batch $y(t) = \mathbf{w}_m^H\mathbf{x}(t)$	- NL $N^2(L+1)$ $N(L+1)^2$ $O((L+1)^3)$ $N(L+1)^2$ $N^2(L+1)$ $N(L+1)$ NQ	- 60 1600 320 96 320 1600 80 320
Repeated $(M-1)$ times for the remaining batches and must also be repeated for the N_R ranges and N_B beams	$S_3 = (M-1) \times N_R N_B \times$ $\{NL + 2N^2(L+1) +$ $2N(L+1)^2 + N(L+1)$ $+NQ + O((L+1)^3)\}$	63302400
Total Complexity	$T = S_1 + S_2 + S_3$	65568000

TABLE 10.1 Computational complexity of the TV-SAP method in terms of the number of complex multiplications required to process a CPI of data. The study case is based on the following parameter values: $N = 20$ receivers, $K = 10$ training ranges per PRI, $Q = 16$ PRIs per batch, $M = 16$ batches ($P = MQ = 256$ pulses in the CPI), and $L = 3$ data-driven constraints. It is assumed that $N_R = 60$ range cells, and $N_B = 16$ beams are to be processed per CPI, with a matrix inversion costing four times matrix multiplication for a dimension of $N = 20$ and six times matrix multiplication for a dimension of $L + 1 = 4$.

10.3.3 Experimental Results

Experimental data collected by the Iluka HFSW radar, previously located near Darwin in north Australia, was used to test the standard and alternative adaptive beamforming techniques described in this section. The Iluka HFSW radar consisted of two transmit sites, namely, a high-power site at Stingray Head (65 km south-west of Darwin), where a 10-kW transmitter fed a log-periodic antenna, and a lower-power site at Lee Point (10 km north-east of Darwin), which was inactive during this experiment. The receiving system at Gunn Point (30 km north-east of Darwin) was based on a 500-m long ULA composed of 32 vertical monopole antenna elements, each connected to an individual high dynamic

Processing Step	Complex Multiplications	Study Case
Select Training Data $\mathbf{D}_m$ $[N \times K(L+1)]$	-	
Estimate Matrix $\mathbf{R}_i = \mathbf{D}\mathbf{D}^H$ $[N \times N]$	$N^2 \times K(L+1)$	16000
Matrix Inversion $\mathbf{R}_i^{-1}$ $[N \times N]$	$O(N^3)$	32000
Repeated for $i = 1, \ldots, P-L$	$S_1 = (P-L)\times$ $\{N^2K(L+1) + O(N^3)\}$	12144
Steering Vector $\mathbf{v}(\theta)$ $[N \times 1]$	-	
Form $\mathbf{z}_1 = \mathbf{R}_1^{-1}\mathbf{v}(\theta)$ $[N \times 1]$	N^2	400
Form $z_1 = \mathbf{v}^H(\theta)\mathbf{z}_1$ $[1 \times 1]$	N	20
First Weight Vector $\mathbf{w}_1 = \mathbf{Z}_1/z_1$ $[N \times 1]$	N	20
Process $y(t) = \mathbf{w}_1^H\mathbf{x}(t)$ for $t = 0, \ldots, L$	$N(L+1)$	80
Repeated for N_R ranges and N_B beams	$S_2 = N_R N_B \times$ $\{N^2 + 2N + N(L+1)\}$	499200
Form $\mathbf{C}_i$ $[N \times (L+1)]$	-	-
Form $\mathbf{f}_i$ $[(L+1) \times 1]$	NL	60
Form $\mathbf{M}_1 = \mathbf{R}_i^{-1}\mathbf{C}_i$ $[N \times (L+1)]$	$N^2(L+1)$	1600
Form $\mathbf{M}_2 = \mathbf{C}_i^H\mathbf{M}_1$ $[(L+1) \times (L+1)]$	$N(L+1)^2$	320
Invert Matrix $\mathbf{M}_2^{-1}$ $[(L+1) \times (L+1)]$	$O((L+1)^3)$	96
Multiply $\mathbf{M}_3 = \mathbf{C}_i\mathbf{M}_2$ $[N \times (L+1)]$	$N(L+1)^2$	320
Multiply $\mathbf{M}_4 = \mathbf{R}_i^{-1}\mathbf{M}_3$ $[N \times (L+1)]$	$N^2(L+1)$	1600
Calculate $\mathbf{w}_i = \mathbf{M}_4\mathbf{f}_i$ $[N \times 1]$	$N(L+1)$	80
Process i^{th} window $y(t) = \mathbf{w}_i^H\mathbf{x}(t)$	$N(L+1)$	80
Repeated $(P-L)$ times for windows remaining and also repeated for the N_R ranges and N_B beams	$S_3 = (P-L) \times N_R N_B \times$ $\{NL + 2N^2(L+1) +$ $2N(L+1)^2 + 2N(L+1) +$ $O((L+1)^3)\}$	1009409280
Total Complexity	$T = S_1 + S_2 + S_3$	1009920624

TABLE 10.2 Computational complexity of the SC-SAP method in terms of the number of complex multiplications required to process a CPI of data. The study case is based on the following parameter values: $N = 20$ receivers, $K = 10$ training ranges per PRI, $P = 256$ pulses in the CPI, and a third order ($L = 3$) AR clutter model. It is assumed that $N_R = 60$ range cells and $N_B = 16$ beams are to be processed per CPI, with a matrix inversion costing four times matrix multiplication for a dimension of $N = 20$ and six times matrix multiplication for a dimension of $L + 1 = 4$.

range receiver. The separation between transmit and receive sites permitted operation using a waveform with unit duty cycle. In this case, the radar transmitted a repetitive linear frequency modulated continuous waveform (FMCW) at a center frequency of $f_c = 7.719$ MHz.

Data were acquired in a daytime campaign, during which the interference background was unknown possibly due to a multiplicity of man-made and natural sources. Each CPI was 32 seconds long and contained $P = 256$ linear FMCW pulses or sweeps with bandwidth $B = 50$ kHz and pulse repetition frequency (PRF) $f_p = 8$ Hz. The analysis was performed on a smaller ULA of $N = 20$ well-calibrated receivers, as initial data screening revealed that certain receivers produced unreliable outputs so these receivers were excluded from further processing. A total of 60 range cells were retained after pulse compression. The first 12 range cells were free from clutter as they sampled distances less than the transmitter-to-receiver separation. A subset of these range cells provided secondary interference-plus-noise-only snapshots for training the adaptive beamformers.

10.3.3.1 Conventional and Time-Varying SMI-MVDR Beamforming

The standard time-varying SMI-MVDR beamformer in Eqn. (10.57) was implemented using an intra-CPI update interval of $Q = 16$ PRI (i.e., 2 seconds) with $M = 16$ batches in the CPI. Secondary data were extracted from the first $K = 10$ range cells in every PRI, providing $KQ = 160 = 8N$ training snapshots per batch to form the integrated interference-plus-noise sample spatial covariance matrices $\mathbf{R}_1, \ldots \mathbf{R}_M$.

Figure 10.5 compares the Doppler spectra of *clutter-free* range cell 12 at the output of the conventional beamformer tapered by a Hamming window, and the time-varying SMI-MVDR adaptive beamformer of Eqn. (10.57) with *no data-driven constraints*. This range cell contains interference and noise only, but is *not* part of the secondary data used to train the adaptive beamformer (i.e., the cross-rejection scenario). Both the conventional and adaptive beamformers are normalized to unit gain in the steer direction. Note

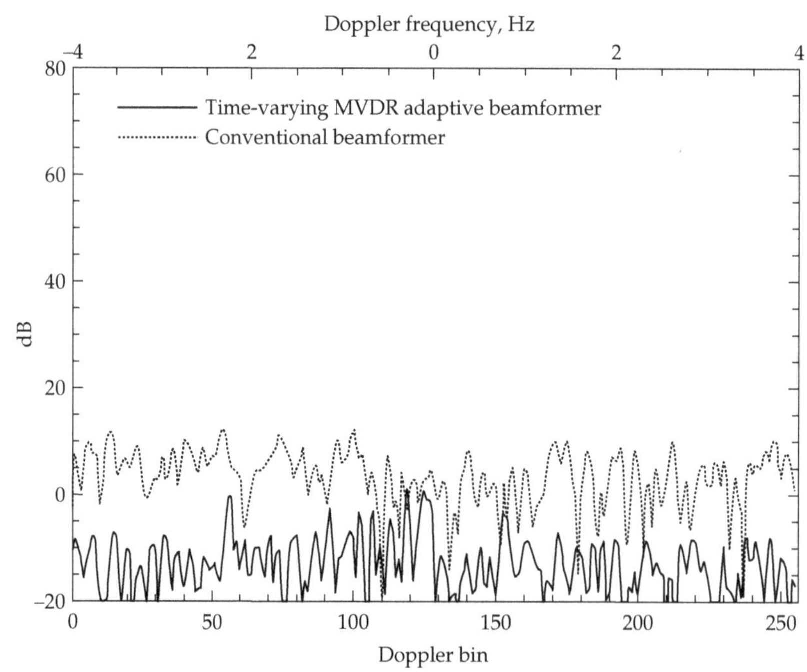

FIGURE 10.5 Interference-plus-noise Doppler spectra for conventional beamformer (with Hamming taper) and time-varying SMI-MVDR adaptive beamformer using *no data-driven constraints*. The mean relative improvement in cross-range RFI rejection is approximately 15–20 dB. © Commonwealth of Australia 2011.

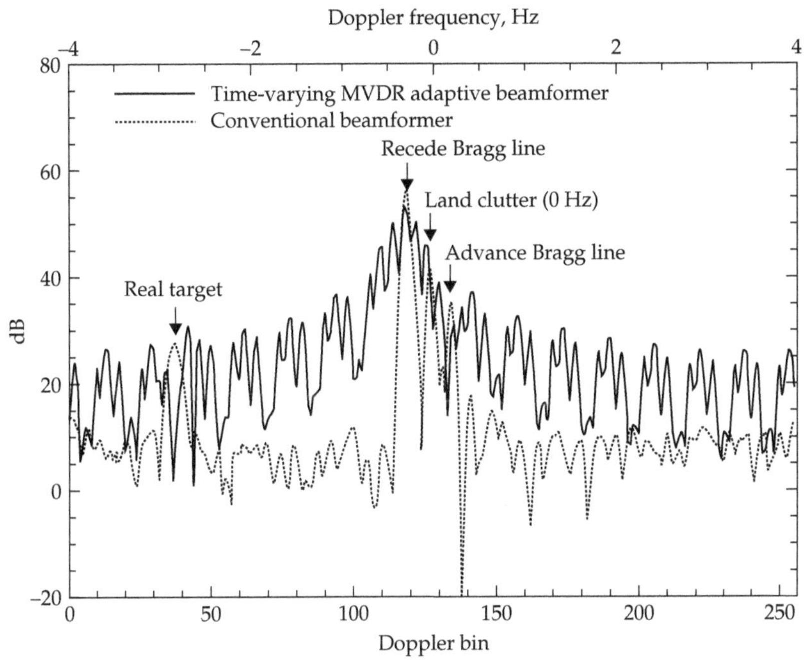

FIGURE 10.6 Doppler spectra for the conventional beamformer (with Hamming taper) and time-varying SMI-MVDR adaptive beamformer using *no data-driven constraints* in an operational range cell. The changing adaptive antenna patterns within the CPI cause the clutter returns to smear across Doppler space and mask a real aircraft target echo near Doppler bin 38. © Commonwealth of Australia 2011.

that the time-varying adaptive beamformer improves interference and noise rejection by 15–20 dB across all of Doppler space.

Figure 10.6 illustrates the application of the same conventional and adaptive weight vectors to range cell 16, which additionally contains clutter and a useful echo from a real aircraft target in the vicinity of Doppler bin 38. The ground clutter component at Doppler bin 128 (0 Hz), and the two Bragg lines resolved and labeled either side of it in Figure 10.6 are clearly evident in the conventional spectrum (dotted line). Although the time-varying SMI-MVDR beamformer effectively rejects interference in Figure 10.5, the solid line in Figure 10.6 shows that it severely degrades target detection due to the smearing of clutter across Doppler space. This occurs because the changing antenna patterns destroy the temporal coherence of the clutter as well as any useful signals received with a spatial signature that does not coincide *exactly* with the presumed array steering vector model.

The effect of changing adaptive antenna patterns on matched and mismatched useful signal reception is illustrated in Figure 10.7. Here, two equal-power synthetic target echoes with different Doppler shifts are injected into clutter-free range cell 12 prior to applying the same conventional and adaptive weight vectors as in Figures 10.5 and 10.6. The spatial signature of the ideal target echo on the left exactly matches the steering vector in the look direction, while the mismatched target on the right is characterized by

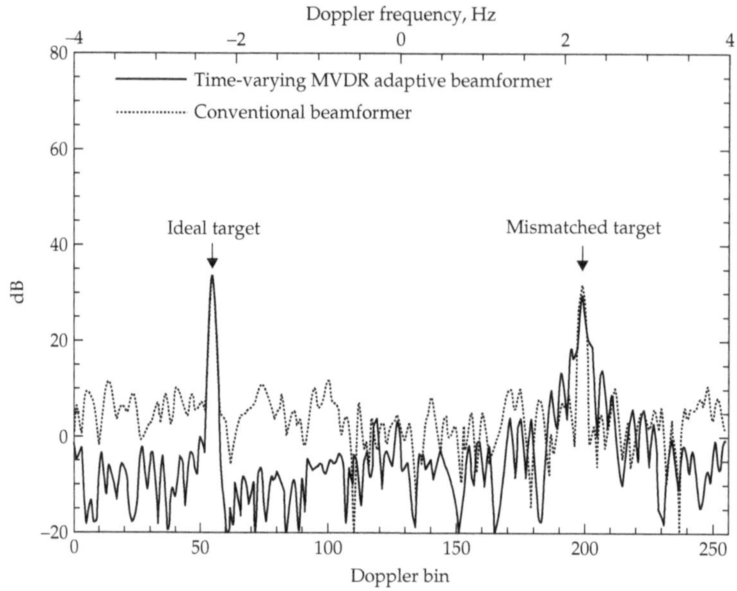

FIGURE 10.7 Interference-plus-noise Doppler spectra for conventional beamformer (with Hamming taper), and time-varying MVDR adaptive beamformer with ideal and mismatched synthetic targets injected. The changing antenna patterns modulate the mismatched useful signal, resulting in a lower peak value and higher Doppler sidelobes compared to the matched useful signal. © Commonwealth of Australia 2011.

a plane wave that has a DOA shifted by half the Rayleigh beam-width away from the steer direction (i.e., precisely between two adjacent orthogonal conventional beams).

As expected, the ideal target echo experiences no distortion in Doppler due to the fixed unity gain of the beamformers in the steer direction. However, the mismatched target has been modulated by the time-variation of the adaptive weights, causing its energy to spread in Doppler. The peak of this target echo is lower compared to the conventional beamformer, and the higher Doppler sidelobes causes the apparent "disturbance" level to rise in the vicinity of the mismatched target.

10.3.3.2 Time-Varying LSMI-MVDR Beamforming

The eigenvalues of the interference-plus-noise sample covariance matrix may be used to guide selection of the diagonal loading factor α in the LSMI-MVDR technique of Eqn. (10.58). Figure 10.8 shows the eigenvalue spectra of the sample covariance matrix $\mathbf{R}_0 = M^{-1}\sum_{m=1}^{M}\mathbf{R}_m$ averaged over the whole CPI (solid line), and the sample covariance matrix $\mathbf{R}_1$ integrated over a single batch (dotted line). The sample sizes used to form these estimates is large (i.e., much greater than $2N$) in both cases. In the hypothetical case of a stationary interference-plus-noise process, the large sample size implies that $\mathbf{R}_0$ and $\mathbf{R}_1$ have practically converged to the same asymptotic form. The significant difference between the two eigenvalue spectra in Figure 10.8 is a clear manifestation of the interference spatial non-stationarity. As the integration time increases, the interference is not

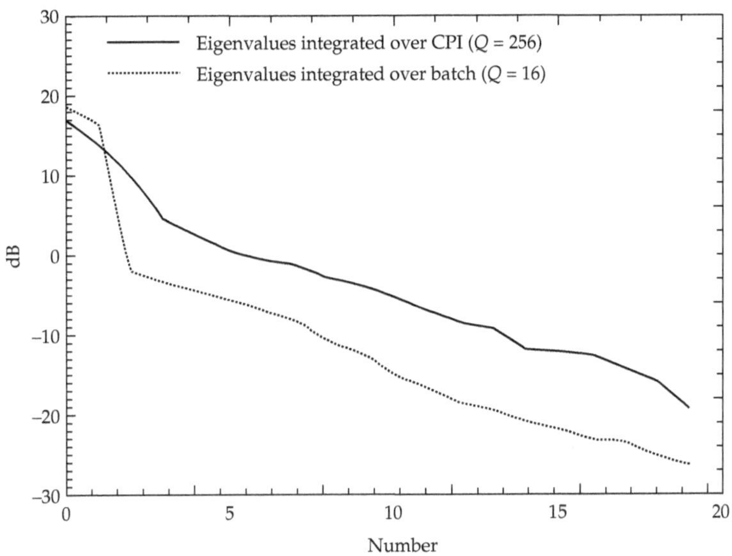

FIGURE 10.8 Eigenvalue spectra for the interference-plus-noise SCM integrated over 2 seconds (dotted line) and the entire 32 second CPI (solid line). The effective interference subspace dimension grows due to spatial non-stationarity as the integration time increases from 2 to 32 seconds. © Commonwealth of Australia 2011.

confined to the same subspace, so $\mathbf{R}_0$ has a higher effective rank than $\mathbf{R}_1$. This "subspace leakage" effect is expected to make the interference rejection task more difficult for a time-invariant (compared with a time-varying) adaptive filter that is tuned on the matrix $\mathbf{R}_0$ averaged over the whole CPI.

With reference to the eigenvalue spectrum shown as the dotted line in Figure 10.8, the time-varying LSMI weights were calculated using Eqn. (10.58) for the cases of moderate loading $\alpha = 1$ (0 dB), and heavy loading $\alpha = 10$ (10 dB). Figure 10.9 shows the application of the two resulting time-varying LSMI-MVDR weight vector sequences to the same data processed in Figure 10.6. In this case, the moderate loading factor is clearly unable to prevent the clutter from smearing over Doppler space and obscuring the target echo (dotted line). Heavy loading helps to restore the gross features of the clutter Doppler spectrum, and makes the target echo somewhat distinguishable, but appreciable distortion remains even in this case.

Figure 10.10 shows the effect of applying the same time-varying LSMI-MVDR weight vector sequences to the interference-plus-noise-only data containing the two synthetically injected targets. A comparison of these results with the solid line in Figure 10.7 demonstrates that the amount of distortion on the mismatched target is significantly reduced as the loading level increases. However, in the negative Doppler bins of Figure 10.7, it can be seen that the penalty paid for loading is a 5–10 dB loss in interference rejection performance. Hence, although diagonal loading can improve convergence rate in cases of limited sample support (i.e., when estimation errors are significant and dominate performance loss), it can *degrade* interference rejection performance when the

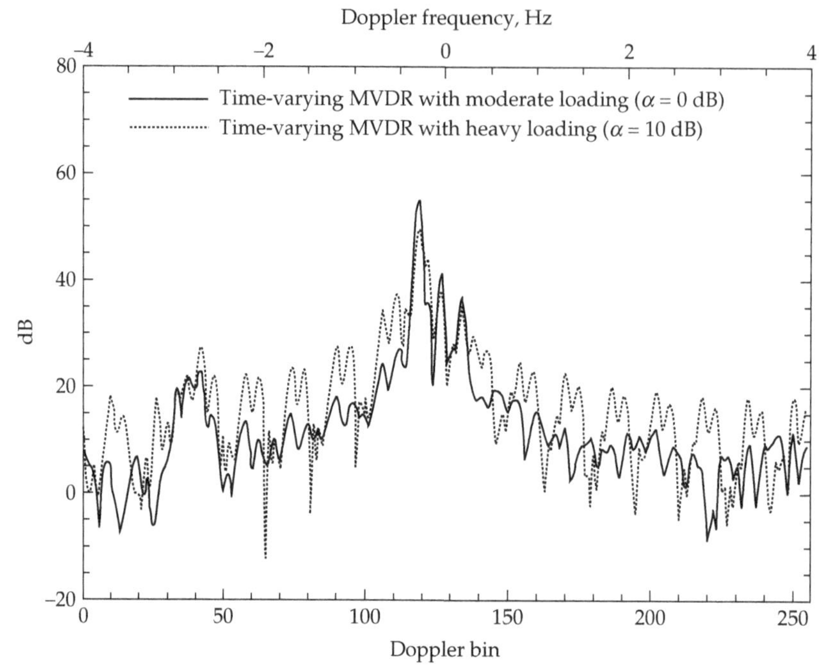

FIGURE 10.9 Operational Doppler spectra for the time-varying LSMI-MVDR adaptive beamformer with moderate and heavy diagonal loading levels. *Moderate* diagonal loading (dotted line) has little effect on protecting SCV. Although heavy loading improves the clutter spectrum, the Doppler sidelobes are still significantly disturbed. © Commonwealth of Australia 2011.

sample covariance matrices are well estimated because it causes the LSMI weight vector to deviate further than its SMI counterpart from the optimal value.

10.3.3.3 TV-SAP Method with Data-Driven Constraints

Figure 10.11 shows the results for operational range cell 16 processed by TV-SAP using $L = 1$ and $L = 2$ data-driven constraints. The choice $L = 1$ performs poorly, while $L = 2$ makes the target echo more visible, although the Doppler spectrum characteristics of the first-order clutter echoes remain severely perturbed compared to the conventional beamformer output.

TV-SAP was then implemented with $L = 3$ data-driven constraints, as this number of constraints matches the three well-resolved Doppler spectrum components that are associated with the dominant ocean and ground clutter returns. It can be appreciated from Figure 10.12 that the SINR of the real target echo has been improved by approximately 20 dB with respect to the conventional beamformer. For $L = 3$, there is no noticeable degradation in sub-clutter visibility at the output of TV-SAP. This represents a substantial improvement with respect to the LSMI results in Figure 10.9. The time-sequence of TV-SAP weight vectors effectively filters out the spatially non-stationary interference, but the changes in antenna pattern remain orthogonal to the processed clutter subspace

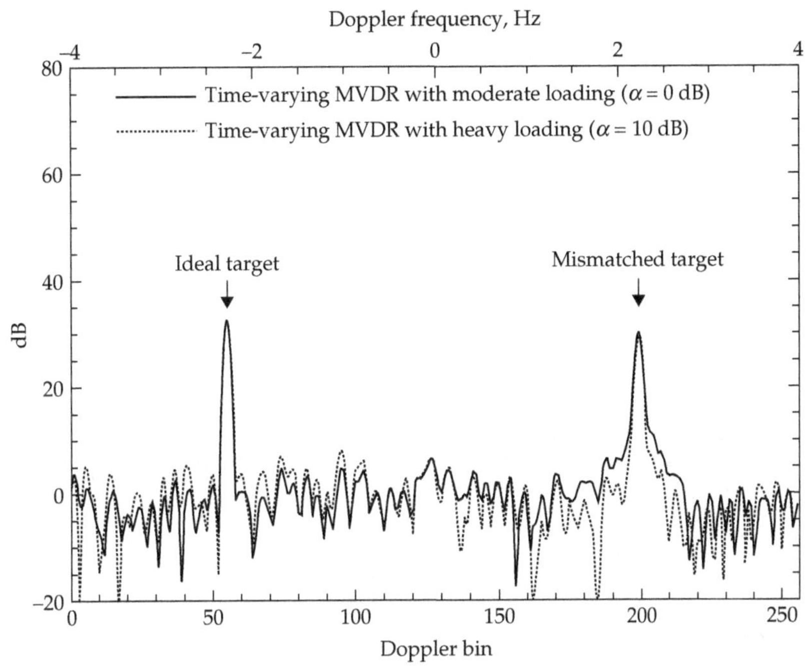

FIGURE 10.10 Interference-plus-noise Doppler spectra for the time-varying LSMI-MVDR adaptive beamformer using moderate and heavy diagonal loading with synthetic targets. The mismatched target response is stabilized by the adaptive filter with heavy loading, but this comes at the expense of a 5 dB loss in interference and noise rejection performance compared to the unloaded adaptive filter in Figure 10.7. © Commonwealth of Australia 2011.

(or approximately so), such that the latter experiences a virtually "stable" spatial filter over the entire CPI.

For targets with perfectly matched steering vectors, TV-SAP is insensitive to the form of the temporal signature $g(t)$, which may differ from a complex sinusoid in the case of a target with Doppler spread (i.e., due to accelerations or changes in reflectivity within the CPI). However, the question may be asked as to the performance of TV-SAP for targets with mismatched steering vectors. Figure 10.13 shows that for a mismatched target, the TV-SAP approach not only controls the Doppler spreading better than the LSMI approach with *heavy* diagonal loading, but also achieves a 5–10 dB relative improvement in interference rejection. This demonstrates that TV-SAP is a superior approach to LSMI, for the considered problem.

To demonstrate performance on weaker useful-signals as well as those close to the dominant clutter spectral components, three ideal synthetic targets denoted by "T1," "T2," and "T3" were injected into operational range cell 33. Figure 10.14 compares the performance of TV-SAP ($Q = 16$ and $L = 3$) with the time-invariant or "fixed" adaptive beamformer trained on the same $K = 10$ range cells but integrated over the *whole* CPI (i.e., steady SAP with $Q = 256$ and $L = 0$). Only TV-SAP is capable of rejecting the non-stationary interference sufficiently well to detect the weaker high velocity target

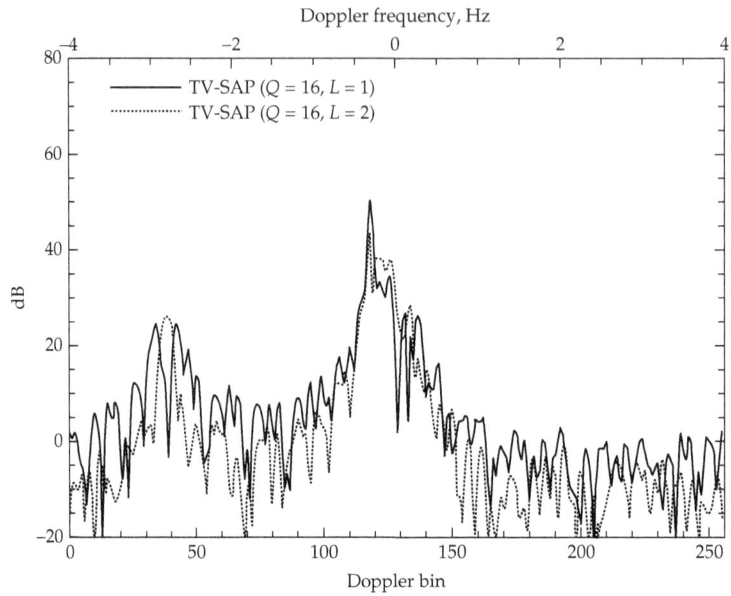

FIGURE 10.11 Operational Doppler spectra for TV-SAP illustrating the use of $L = 1$ and $L = 2$ data-driven constraints. These numbers of data-driven constraints are insufficient to protect the three dominant clutter spectral components. © Commonwealth of Australia 2011.

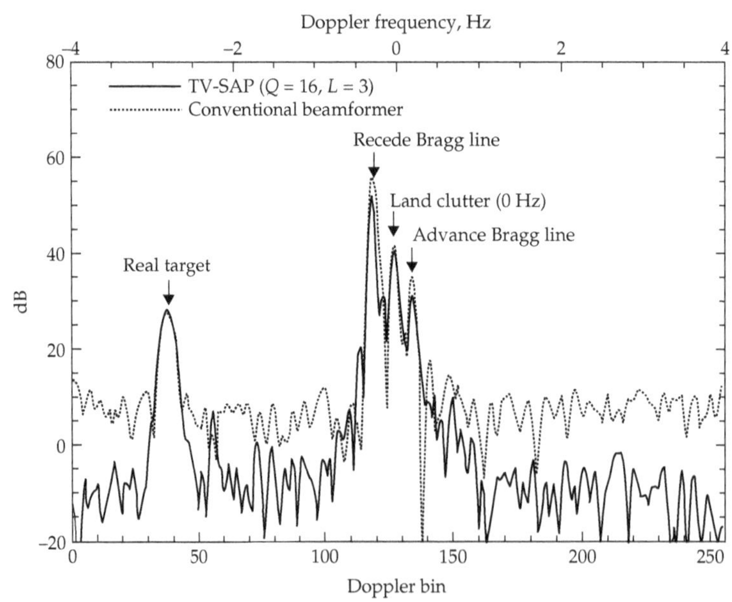

FIGURE 10.12 Operational Doppler spectra for the conventional beamformer (with Hamming taper) and TV-SAP using $L = 3$ data-driven constraints. TV-SAP yields a dramatic 20 dB SINR improvement relative to the conventional beamformer on a real aircraft target echo. © Commonwealth of Australia 2011.

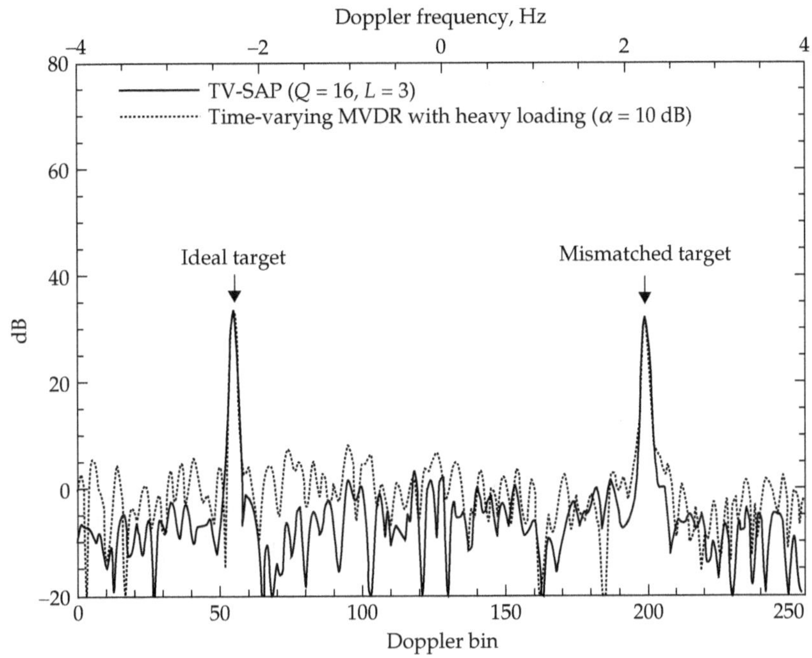

FIGURE 10.13 Interference-plus-noise Doppler spectra with synthetic targets for TV-SAP and the time-varying LSMI-MVDR adaptive beamformer using *heavy* diagonal loading. Not only does TV-SAP protect the Doppler sidelobes of the mismatched target effectively, but it also improves the SINR of both targets by approximately 5 dB relative to the LSMI-MVDR technique. © Commonwealth of Australia 2011.

(T3). This is achieved without compromising the detection of stronger targets, either in the interference and noise region (T1), or near the Bragg line return (T2). The 15–20 dB SINR improvement of TV-SAP relative to the time-invariant adaptive beamformer may be expected from the eigenvalue spectra in Figure 10.8. This is because the sample covariance matrix averaged over the whole CPI clearly has a higher effective rank compared to that integrated over a single sub-CPI.

10.3.3.4 Comparison with Stochastic Constraints

The fundamental difference between TV-SAP and SC-SAP is that the former is free to update the adaptive weight vectors at a rate that is determined by the prevailing characteristics of the interference, while the latter is updated every PRI regardless of the level of interference spatial non-stationarity in an attempt to protect the AR properties of the clutter Doppler spectrum. Figure 10.15 compares the interference-plus-noise Doppler spectra for TV-SAP with parameters ($Q = 16$, $L = 3$) and the SC-SAP method using a third-order AR clutter model (Abramovich, Gorokhov, Mikhaylyukov, and Malyavin 1994). The interference rejection performance is practically identical, so evidently, the faster updates of the SC method do not provide any RFI cancelation benefit relative to TV-SAP in this practical example. This suggests that the significantly greater computational effort of the SC technique is in this case not warranted from an interference rejection viewpoint. The

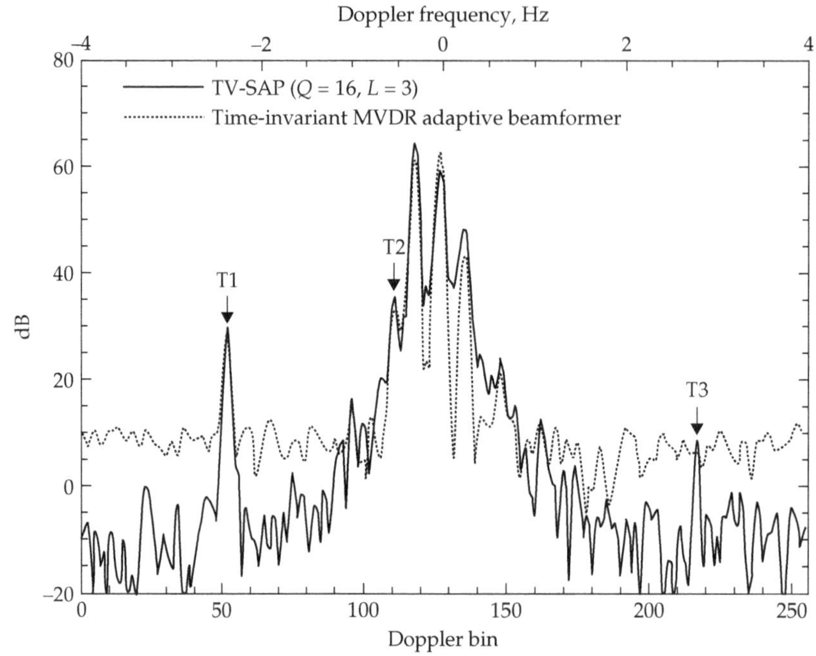

FIGURE 10.14 Doppler spectra with synthetically injected targets for TV-SAP and the *time-invariant* MVDR adaptive beamformer (i.e., using fixed weights over the *whole* CPI). TV-SAP additionally enables the detection of a weak target "T3" that is well resolved from the clutter and competes for detection against interference-plus-noise. © Commonwealth of Australia 2011.

question then arises as to how well TV-SAP, based on an alternative clutter model and a differently structured set of constraints, protects the HFSW radar clutter compared with the SC technique.

Figure 10.16 compares the performance of TV-SAP using the same parameters ($Q = 16$, $L = 3$) against the SC-SAP method based on a third-order AR clutter model. TV-SAP is observed to be more robust and produces sharper Bragg lines that enable the low-velocity target near the recede Bragg line (T2) to be detected. The unnecessarily fast update rate of the SC-SAP method (i.e., every PRI or 0.125 seconds) not only widens the Bragg lines relative to TV-SAP, but also smears clutter energy into other regions of Doppler space. This makes a weak higher velocity target (T3) more difficult to distinguish against the competing background. In order to quantify the robustness of TV-SAP relative to SC-SAP from a clutter Doppler spectrum contamination viewpoint, it is required to process a number of range cells with quasi-homogenous clutter Doppler spectra that can be meaningfully compared in a statistical sense.

A total of 10 contiguous range cells not containing useful signals (target echoes) were processed by TV-SAP and SC-SAP in a manner identical to Figure 10.16, but without the injected targets. Figure 10.17 shows the median of the Doppler spectrum values calculated over this set of range cells. It is clear that TV-SAP increases the median SCV by 3–5 dB (i.e., enhancing the detection of targets with high radial speed) and produces

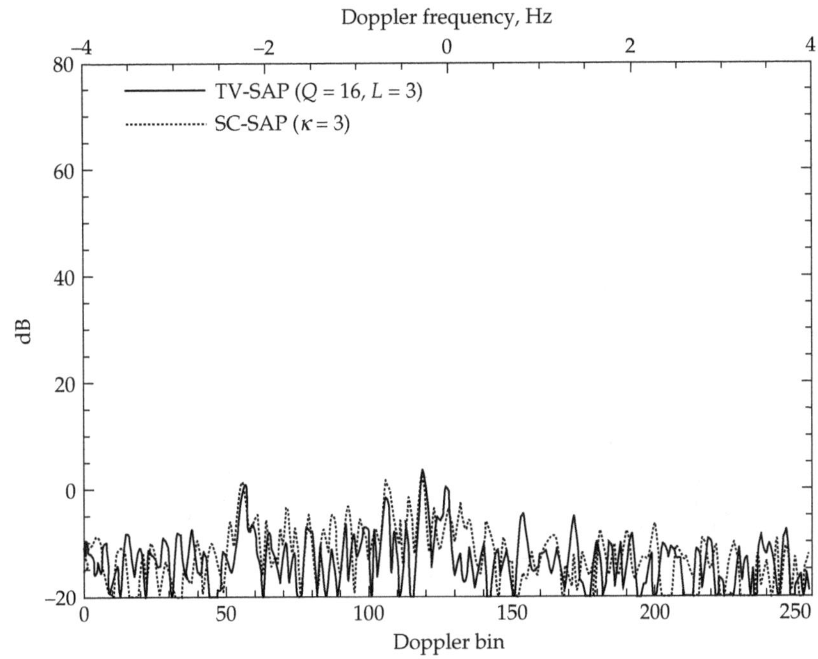

FIGURE 10.15 Interference-plus-noise Doppler spectra for TV-SAP and the SC-SAP method. The level of rejection is practically equivalent, but SC-SAP updates the weights every PRI (i.e., 0.125 seconds), while TV-SAP achieves similar performance with updates every 2.0 seconds. © Commonwealth of Australia 2011.

slightly sharper Bragg lines (i.e., enhancing the detection of targets with low radial speed). This robustness to the accumulation of weight estimation errors is also apparent in Figures 10.18 and 10.19, which show the upper and lower deciles of the clutter Doppler spectrum values, respectively.

These experimental results demonstrate that TV-SAP can outperform the SC-SAP method in practice with the added advantage of much lower computational load. Hence, from both performance and computational points of view, the TV-SAP algorithm represents an attractive candidate for real-time implementation in practical systems. These advantages stem from the alternative philosophy of updating the weight vector at a rate commensurate with the interference spatial non-stationarity instead of every PRI to protect the (AR) clutter characteristics. Such a rate was shown to be faster than necessary for effective interference suppression, while the larger number of updates within the CPI performed by SC-SAP also reduced the SCV by 3–5 dB relative to TV-SAP.

The benefits of TV-SAP with respect to existing adaptive beamforming approaches have been demonstrated using HFSW OTH radar data. However, TV-SAP is also applicable in HF skywave OTH radars and potentially other types of radar. The approach may also be extended to space-time adaptive processing (i.e., TV-STAP), which will be described in the next chapter.

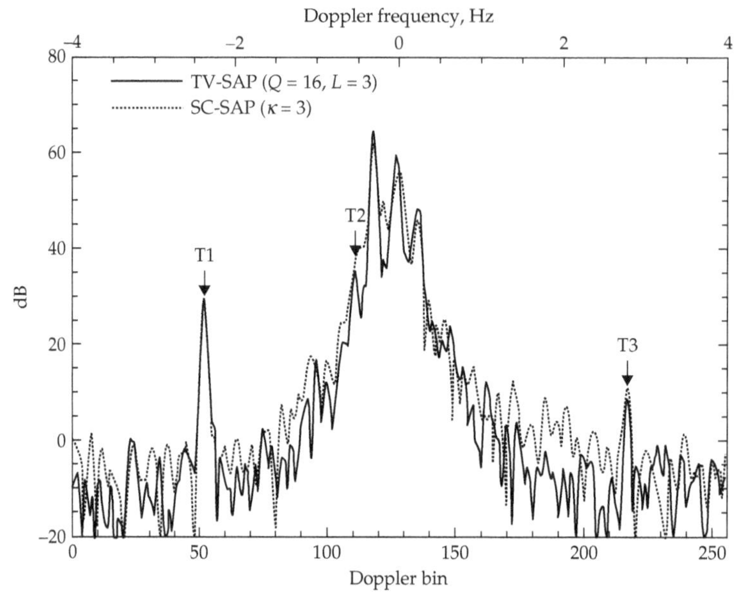

FIGURE 10.16 Doppler spectra with synthetically injected targets for TV-SAP and the SC-SAP method. TV-SAP provides an order of magnitude saving in computational load relative to SC-SAP, and is more robust in protecting SCV. This robustness allows a target close to the recede Bragg line ("T2") to be detected, and also makes the weaker target ("T3") easier to distinguish against the competing background. © Commonwealth of Australia 2011.

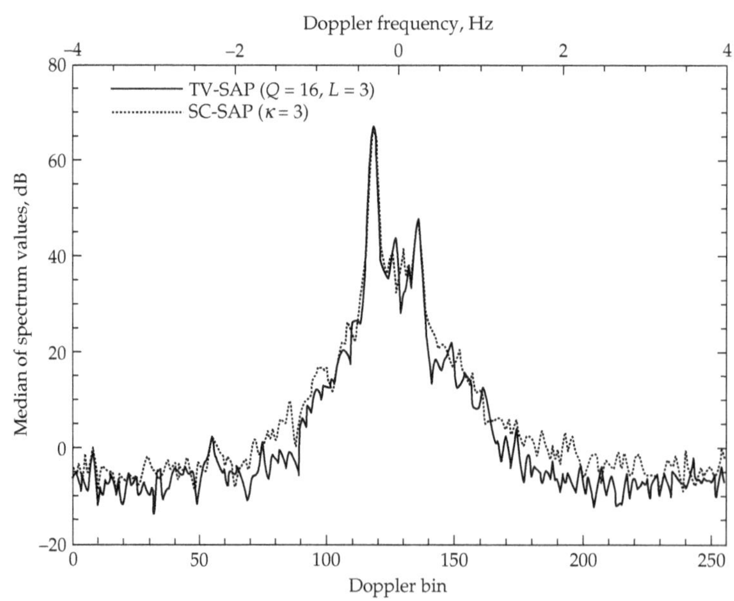

FIGURE 10.17 Median Doppler spectrum values for TV-SAP and the SC-SAP method over 10 consecutive range cells. TV-SAP is more robust, leading to a 3–5 dB improvement in median SCV as well as sharper Bragg lines. © Commonwealth of Australia 2011.

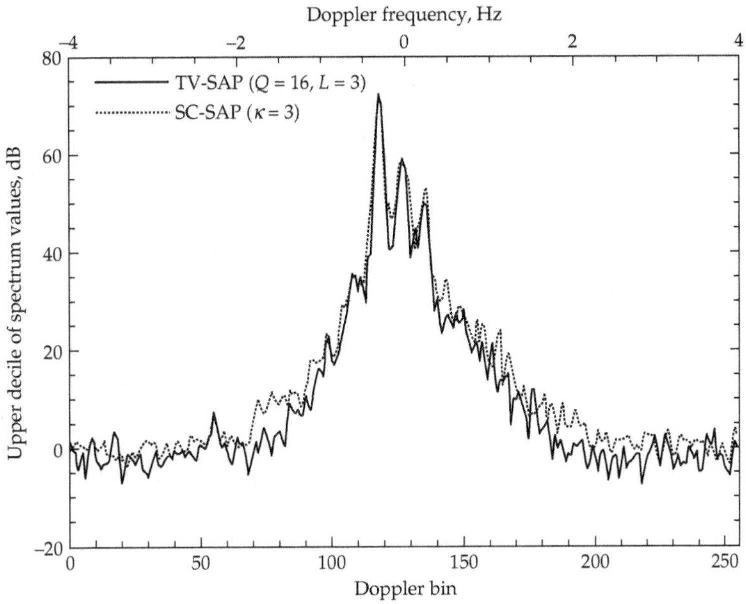

FIGURE 10.18 *Upper* decile of the Doppler spectrum values for the TV-SAP and the SC-SAP method over 10 consecutive range cells. The TV-SAP values are 3–5 dB lower over essentially all parts of the Doppler spectrum where targets are sought. © Commonwealth of Australia 2011.

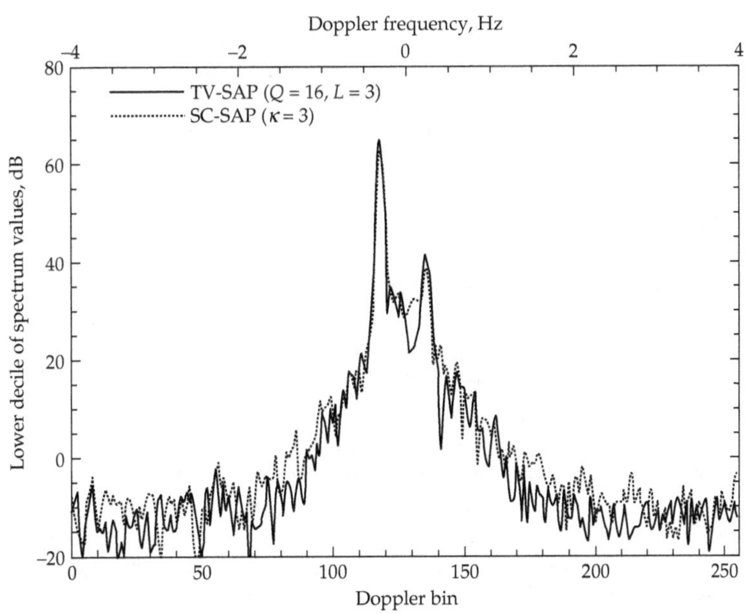

FIGURE 10.19 *Lower* decile of the Doppler spectrum values for the TV-SAP and the SC-SAP method over 10 consecutive range cells. The TV-SAP values are 3–5 dB lower over essentially all parts of the Doppler spectrum where targets are sought. © Commonwealth of Australia 2011.

10.4 Post-Doppler Techniques

Thus far, adaptive beamforming has been applied to cancel interference and noise signals that are not coherent with the radar waveform. In this case, the disturbance energy is generally spread over the entire range-Doppler search space after matched filtering, regardless of the actual group range and Doppler shift of the signal path(s) linking the source(s) to the receiver.

As the matched filter output does not measure the energy distribution of an *incoherent* signal in physical group-range and Doppler shift, each individual resolution cell will typically contain a superposition of all disturbance contributions. This situation justifies forming the interference-plus-noise sample spatial covariance matrix by averaging array snapshots acquired over different range cells, as described in the previous section.

However, when the disturbance signal to be mitigated is *coherent* with the radar waveform, the energy distribution arising from spatially separate (scattering) sources can in principle be resolved into different time delay and Doppler-frequency bins after matched filtering. As the time-delay and Doppler shift are physically related to the signal propagation path in this case, each (range-Doppler) resolution cell will contain a subset of disturbance contributions that collectively make up the overall disturbance signal. In particular, the disturbance spatial covariance matrix may vary significantly as a function of range and Doppler as the contributions present in each resolution cell may originate from scattering sources with different DOAs. The spatial properties of such a disturbance are said to be statistically heterogeneous across different resolutions cells.

This phenomenon may occur in active and passive radar systems. The latter utilize emitters of opportunity instead of a purpose-built radar transmitter for target detection and tracking. Similar to active radars, a passive radar will in general receive clutter from a spatial distribution of scatterers in the environment that are illuminated by the transmitter. Importantly, scatterers located at different ranges can give rise to echoes with different DOAs, which causes the spatial statistics of the clutter disturbance to become heterogeneous in range. From an adaptive beamforming perspective, the mitigation of such a disturbance requires a fundamentally different approach to that described in the previous section.

10.4.1 Motivating Practical Application

The spatial covariance matrix of the overall clutter signal scattered from transmitter to receiver may be represented as the sum of different spatial covariance matrices related to scattering sources that have been "resolved" into different range and Doppler cells after matched filtering. Importantly, the spatial covariance matrix of the clutter components received in a particular resolution cell often has a lower effective rank than the spatial covariance matrix of the composite clutter signal integrated over all resolution cells.

This is *conceptually* similar to the spatial non-stationarity problem discussed earlier, where the interference spatial covariance matrix integrated over a relatively long CPI may be viewed as the sum of many "quasi-instantaneous" spatial covariance matrices of relatively lower effective rank that change over time. In that case, interference spatial non-stationarity was mitigated by filter readjustments in time. Analogously, the heterogeneous clutter problem motivates the application of adaptive beamforming after matched filtering, such that the adaptive weight vectors can be updated across range and Doppler cells.

A post-Doppler adaptive beamforming scheme enables the processor to exploit the lower spatial rank of a coherent disturbance by partitioning the CPI of data into range and Doppler batches. In other words, range and Doppler localized training strategies may be employed for adaptive beamforming after matched filtering is applied in all receivers of the array. A range- and Doppler-dependent adaptive beamforming approach for mitigating statistically heterogeneous clutter disturbances in passive radar systems will be developed and experimentally evaluated in this section.

Although the practical performance of the post-Doppler adaptive beamforming approach will be illustrated on an HF passive radar system, it should be mentioned that the notion of range and Doppler-dependent adaptive processing with localized training after matched filtering has also been applied to mitigate spatially heterogeneous disturbances in active radar systems. In particular, such techniques have been used to mitigate ionospheric clutter in HFSW radar, as described in Abramovich, Anderson, Gorokhov, and Spencer (2004), Fabrizio and Holdsworth (2008), and references therein.

10.4.1.1 Passive Radar Background

International initiatives to develop aircraft target detection and tracking systems that exploit waveforms transmitted by uncooperative sources of opportunity have resulted in the fielding of many diverse passive radar systems, also commonly referred to as passive bistatic radar (PBR), passive covert radar (PCR), or passive coherent location (PCL). The inherent advantages of such systems, with respect to conventional monostatic (active) radars, include significantly lower capital cost, smaller physical size, higher immunity to ECM threats, no additional demands on spectrum resources, no requirement for licenses to radiate, and potentially higher target RCS due to the combined use of bistatic configurations and often lower frequency bands relative to microwave radars.

Although the passive radar concept was conceived many decades ago (circa 1935), there continues to be a significant interest in PCL systems within industry, defence, and academic communities; see Griffiths and Baker (2005) and Baker, Griffiths, and Papoutsis (2005). This is due to a number of contributing factors, including: (1) the development of high dynamic-range digital receivers and advances in computing technology for real-time operation of practical PCL systems, (2) the availability of a wider range of illuminators that combine high bandwidth and power to provide more suitable waveforms for PCL systems (mainly arising from the trend towards digital broadcasting), and (3) the progress and maturity of *robust* adaptive processing techniques, particularly array processing algorithms for target detection and parameter estimation in heterogeneous disturbance environments.

However, the geographic locations and antenna pattern properties of the radiators used by PCL systems may be far from optimal, and importantly, the characteristics of the transmitted waveform cannot be controlled. This leads to a number of significant technical challenges, which need to be overcome before the PCL concept can be turned into an effective and reliable surveillance tool. First, it is necessary to obtain a "clean" copy of the unknown source waveform in the so-called *reference* channel.[2] This copy is used as a reference signal to perform coherent processing in the *surveillance* channel[3] by

[2]This often refers to a spatial channel or beam used to receive a clean copy of the source waveform.

[3]The surveillance channel beam is formed to receive the target echo in a different direction.

matched filtering in the dimensions of delay and Doppler. Ideally, the reference signal should be free of multipath contamination, useful signals (target echoes), and co-channel interference from other emitters that may overlap the same frequency band as the source of opportunity.

A second key issue is to prevent strong direct-wave interference (DWI), due to the signal received directly from the transmitter, and other strong clutter echoes, from leaking excessively into the surveillance channel where target echoes are sought. This is often required because the clutter sidelobes at the matched-filter output may be substantially higher than those corresponding to waveforms designed for radar use. The sub-optimal ambiguity function characteristics of the source waveform for coherent processing has the potential to mask much weaker target echoes, even when such echoes are largely displaced in delay (range) and Doppler from the main clutter peak.

PCL has emerged as a candidate sensor for air surveillance on both fixed and mobile platforms, particularly to augment the conventional radar coverage for low-altitude targets in an economical manner, as described by Kuschel, Heckenbach, Muller, and Appel (2008), or to perform surveillance in conditions where radio silence is necessary. While the operation of practical PCL systems has been successfully demonstrated in various frequency bands, the great majority of current systems utilize emitters in the VHF and UHF bands. This includes systems based on analogue FM radio [(Howland, Maksimiuk, and Reitsma 2005), (Di Lallo, Farina, Fulcoli, Genovesi, Lalli, and Mancinelli 2008), (Bongioanni, Colone, and Lombardo 2008)] and analogue TV transmissions [(Griffiths and Long 1986), (Howland 1999)], for example.

In more recent times, Digital Audio Broadcasts (DAB) with 1.5 MHz bandwidth at frequencies slightly above 200 MHz, and Digital Video Broadcasts Terrestrial (DVB-T) at UHF with 7 MHz bandwidth represent attractive alternatives that provide fine-range resolution; see Poullin (2005), and Saini and Cherniakov (2005). Coded orthogonal frequency division multiplex (COFDM) signals used for commercial digital broadcasting in multistatic single-frequency networks (SFN), at power levels of 1–10 kW, are also attractive for PCL purposes because such signals may be readily decoded to obtain accurate copies of the source waveform. A host of other PCL systems utilizing mobile communication signals from GSM base stations (Tan, Sun, Lui, Lesturgie, and Chan 2005), signals from the Global Navigation Satellite System (GNSS) (Cherniakov, Zeng, and Plakidis 2003), and signals from satellite communication systems (Cherniakov, Nezlin, and Kubin 2002) have also been developed.

A great deal less work has been done on PCL in the HF band (3–30 MHz), since the famous “Daventry Experiment” in February 1935, where short-wave BBC Empire broadcasts were used to detect a bomber aircraft in what is recognized as the first passive radar experiment (Willis 1991). An operational HF-OTH PCL system called “Sugar Tree” was developed for the US army between 1962 and 1970 to detect missile launch (Doppler-time) signatures using uncooperative AM radio broadcast transmitters located near the suspected launch sites. In this particular implementation, the illumination path was via the surface-wave or line-of-sight propagation, while the “pilot signal” path (transmitter-to-receiver), and “target echo” path from the missile launch area to the receiver, was via skywave propagation to a remote phased-array of monopole antennas. Performance details of Sugar Tree are not readily available, despite the system being dismantled over three decades ago (Willis and Griffiths 2007).

In the last few years, the emergence of Digital Radio Mondiale (DRM) has promised new potential for PCL systems in the HF band; see Thomas, Griffiths, and Baker (2006), Thomas, Baker, and Griffiths (2007), and Coleman, Watson, and Yardley (2008). However, the ambiguity function analysis of DRM signals conducted thus far do not provide convincing experimental evidence that HF PCL systems can indeed be used to reliably detect airborne targets at long ranges with reasonable false-alarm rates.

A preliminary feasibility study based on AM broadcasts of opportunity was undertaken in Durbridge (2004) using *conventional* processing, but the poor ambiguity-function properties of such signals limited system performance (Ringer and Frazer 1999). The more recent paper of Fabrizio, Colone, Lombardo, and Farina (2009) aimed to redress the lack of PCL research in the HF band, particularly in regards to the application of *adaptive beamforming* for mitigating direct-wave clutter.

10.4.1.2 HF-OTH PCL System

The principle of operation of the HF-OTH PCL system illustrated in Figure 10.20 differs from the Sugar Tree system in that the transmitter of opportunity illuminates the surveillance region via the *skywave* propagation mode, while the target echoes propagate to a receiving antenna array located in the line-of-sight (LOS). Figure 10.20 also illustrates multipath propagation for both the direct-wave interference and target echoes due to the presence of multiple reflecting layers in the ionosphere. With respect to the Sugar Tree

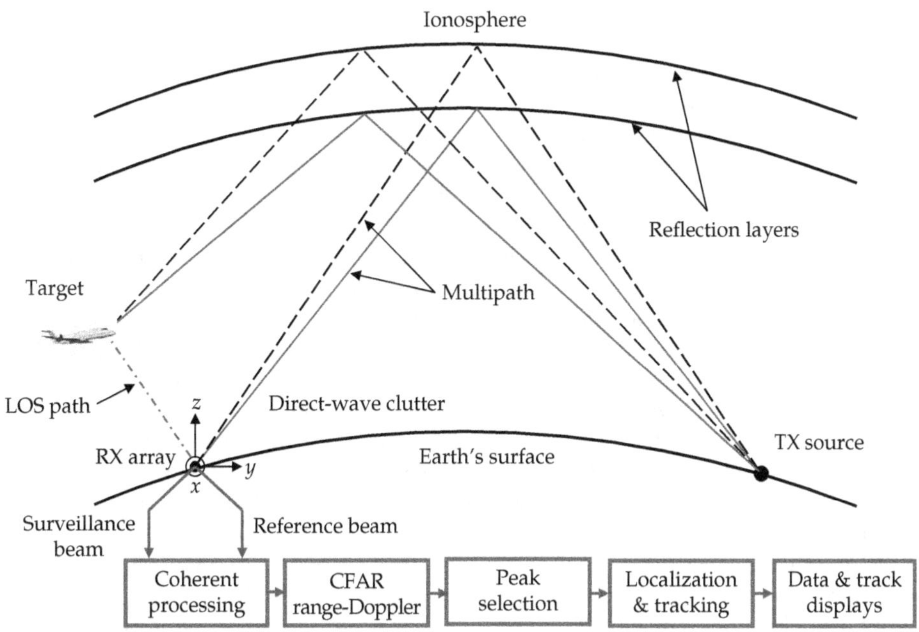

FIGURE 10.20 Illustration of operational concept and signal processing chain for a PCL system in the HF band utilizing emitters of opportunity that are located over the horizon. Although shown as a vertical line, the z-axis of the coordinate system is oriented radially toward the center of the Earth, such that the antenna elements of the ground-based receiving array are deployed in the xy-plane. © Commonwealth of Australia 2011.

(a) One ULA arm of line-of-sight HF receive antenna array. (b) Monopole antenna geometry and orientation.

FIGURE 10.21 Two-dimensional (L-shaped) LOS receiver array showing layout of monopole elements and orientation in degrees clockwise from true north. © Commonwealth of Australia 2011.

architecture, placing the receiving station in the surveillance region has the advantage that the transmitted signal and target echoes may be incident from significantly different DOAs. This can allow useful signals to be discriminated against clutter more effectively by spatial processing.

At the receive site, a multi-channel antenna array digitally forms a *reference beam* in the transmitter direction to estimate a copy of the radiated waveform over the CPI. Ideally, the reference beam minimizes corruption in the source waveform estimate due to multipath propagation, target echoes, as well as interference and noise signals. One or more *surveillance beams* are also simultaneously formed and steered in different target search direction(s). The surveillance beams aim to maximize the useful signal-to-disturbance ratio by canceling clutter, interference, and noise as much as possible.

The reference beam output is cross-correlated with each surveillance beam output in turn using a matched filter bank to produce a bistatic range-Doppler map for each surveillance beam. The magnitude envelopes of the matched filter output samples are then passed on for constant false-alarm rate (CFAR) processing and threshold detection. The peak information extracted from each CPI is input to a tracking filter and the resulting tracks displayed for the operator. The tracking component is not considered here.

The HF PCL receiving system is based on a two-dimensional L-shaped antenna array located near Darwin, north Australia. Figure 10.21a shows one arm of the receiving antenna array, which is configured as eight vertical monopole elements spaced 8 m apart on each arm. This results in 16 spatial channels overall with a digital receiver-per-element. A dummy element is added at the end of each arm to reduce the effects of mutual coupling. Figure 10.21b defines the two-dimensional element geometry and orientation of the L-shaped array, which permits independent digital beam steering in azimuth and elevation. During the experiment, each reception channel acquired both in-phase and quadrature (I/Q) signal components at a sampling rate of 62.5 kHz (equal to the receiver passband). The sampled data was processed as a set of consecutive CPI of 2 seconds duration, which is typical for OTH radar aircraft detection.

A cooperative aircraft was used as a dedicated test target for a trial carried out in April 2004. Figure 10.22a shows a picture of the Westwind IL24 Learjet aircraft target used for this experiment. The cooperative target was equipped with an on-board GPS data

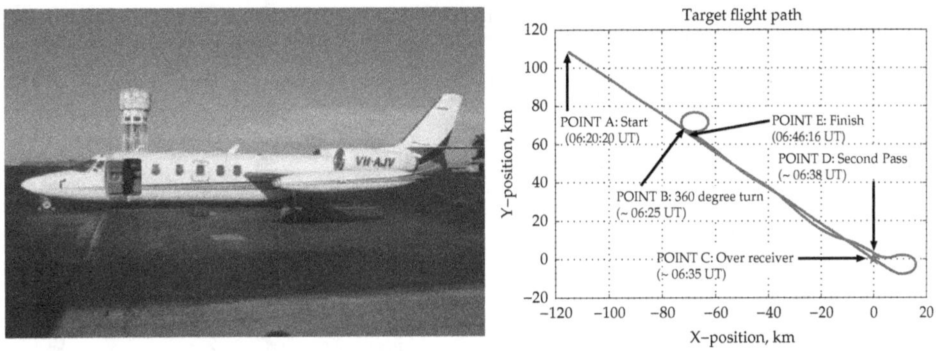

(a) Cooperative Westwind IL24 Learjet aircraft.

(b) Plan view of target flight path from GPS data.

FIGURE 10.22 The private charter aircraft used as a cooperative target for the trial, and its flight path relative to the LOS receiver location measured by the on-board GPS data logger.

logger to record its position, altitude, and attitude every 2 seconds. At its known cruising altitude of 31000 feet, the target falls below the horizon at a range of about 350 km. The aircraft was requested to complete a 360 degree turn every 10–15 minutes to mark the radar return with a known Doppler-time signature for identification purposes. The GPS data plotted in Figure 10.22b shows a plan view of the analyzed part of the target flight path with respect to the LOS receiver location.

The GPS data from the aircraft were transformed to radar coordinates in terms of full DOA (azimuth/elevation) and differential bistatic range/Doppler versus time profiles. Here, the term "differential" refers to the *additional* path length traveled by the target skin-echo relative to the signal received directly from the transmitter, which is used as the zero time-delay reference. The points labeled A to D in Figure 10.22b are associated with the corresponding times in the azimuth and elevation profiles of Figure 10.23. The

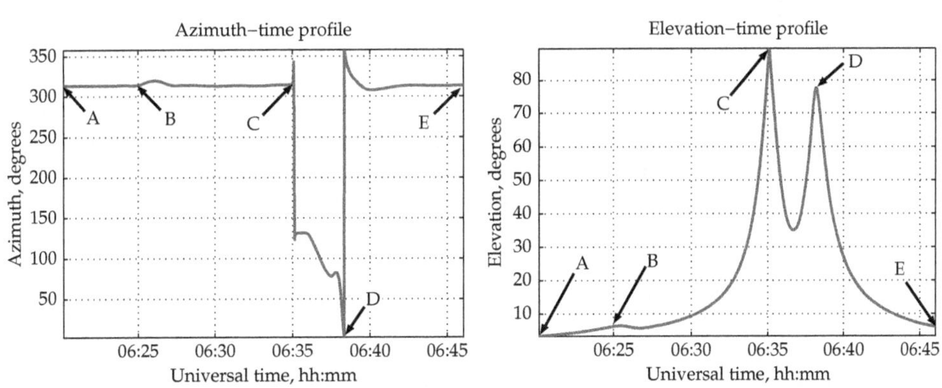

(a) Bearing versus time profile.

(b) Elevation versus time profile.

FIGURE 10.23 Direction of cooperative aircraft in azimuth and elevation with respect to the line-of-sight receiver calculated using GPS data recorded on-board the aircraft.

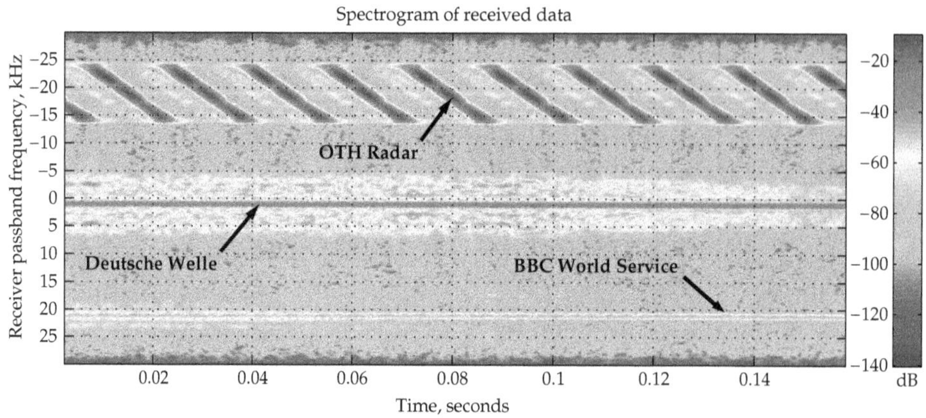

FIGURE 10.24 Raw data spectrogram at a single receiver output showing three waveforms emitted from different sources of opportunity in the receiver passband.

bistatic range and Doppler profiles will be shown later. The 360-degree turn commences at approximately 06:25 UT, while a pass over the receiver occurs at approximately 06:36 UT.

Three different sources were recorded in the receiver bandwidth at a center frequency of 21.639 MHz. The spectrogram in Figure 10.24 shows the signals acquired by a single receiver. The source locations and waveform types were identified and listed in Table 10.3. Figure 10.25 shows the extraction of the 10-kHz bandwidth linear FMCW signal by digital-down conversion to baseband and low-pass filtering. This signal has the most desirable ambiguity function characteristics for PCL purposes. For this reason, further analysis focuses on using this signal, which may be frequently found in the HF spectrum due to the presence of radar systems used for surveillance and remote sea-state sensing.

When the parametric form of the source waveform is known, as it is for an LFMCW signal, this information may be exploited for reference estimation in a practical HF-OTH radar system. However, as a preliminary step toward the development and assessment of post-Doppler adaptive beamforming techniques for a general HF PCL system, no prior information about the structure of the transmitted signal will be assumed. While this does not make best use of the available knowledge in this case, it allows insights to be gained regarding the relative merits and shortcomings of different adaptive beamforming techniques. Such insights may be used to guide the choice of adaptive beamforming approach

Illumination Source	Frequency, MHz	Ground Range, km	Bearing, deg.	Power, kW
HF Radar	21.620	1,851	134	200
Deutsche Welle	21.640	5,980	290	250
BBC World Service	21.660	3,400	295	100

TABLE 10.3 Main characteristics of the identified illumination sources. A HF radar source from Longreach, Australia, is present along with two AM radio broadcasts; Deutsche Welle, from Trincomalee, Sri Lanka, and BBC World Service, from Kranji, Singapore.

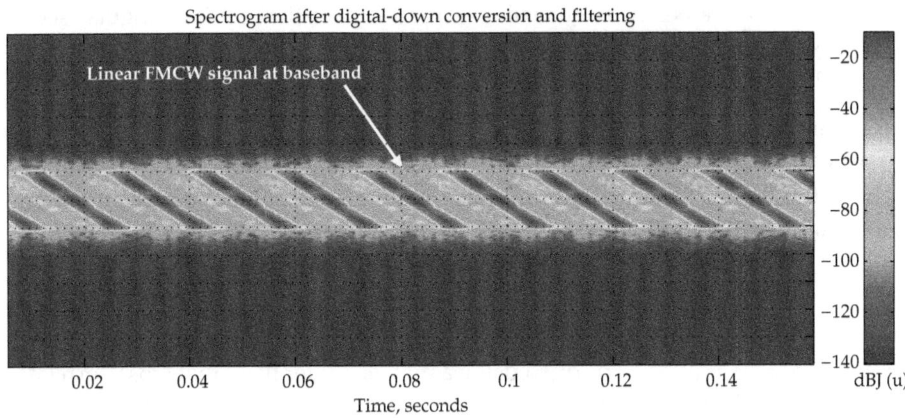

FIGURE 10.25 Spectrogram of the same data in Figure 10.24 after digital-down conversion to baseband and low-pass filtering of the LFMCW signal. © Commonwealth of Australia 2011.

in passive radar systems based on more general HF-OTH signals of opportunity, where no knowledge about the source signal structure is available.

10.4.1.3 Conventional Processing Scheme

For a narrowband antenna array of arbitrary element geometry, the plane-wave steering vector for an azimuth θ and elevation ϕ is given by Eqn. (10.77), where $\mathbf{r}_n = [x_n, y_n, z_n]^T$ is the position vector of the nth antenna element relative to the phase reference (taken as the origin of the coordinate system), and $\mathbf{k}(\theta, \phi) = \frac{2\pi}{\lambda}[\cos(\phi)\sin(\theta), \cos(\phi)\cos(\theta), \sin(\phi)]^T$ is the wave vector. The position of the first receiver is assumed to be at the origin ($\mathbf{r}_1 = \mathbf{0}$), while the azimuth angle θ (great-circle bearing) is defined in conventional manner with respect to true north.

$$\mathbf{s}(\theta, \phi) = [e^{j\mathbf{k}(\theta,\phi)\cdot\mathbf{r}_1}, e^{j\mathbf{k}(\theta,\phi)\cdot\mathbf{r}_2}, \ldots, e^{j\mathbf{k}(\theta,\phi)\cdot\mathbf{r}_N}]^T \tag{10.77}$$

The reference beam is pointed in the direction $\{\theta_r, \phi_r\}$ that maximizes the power received from the source (i.e., the DOA coordinates of the peak value in the spatial spectrum). As such, $\{\theta_r, \phi_r\}$ represents the conventional DOA estimate of the dominant direct-wave signal mode. Let $\mathbf{x}(t)$ be the complex N-dimensional snapshot vector received by the antenna array at A/D sample index $t = 0, 1, \ldots, M-1$. The reference waveform is estimated as the scalar output time-series $y_c(t)$ in Eqn. (10.78), where $\mathbf{v}^\dagger(\theta_r, \phi_r) = N^{-1}\mathbf{s}^\dagger(\theta_r, \phi_r)$ provides unit gain in the steer direction $\mathbf{v}^\dagger(\theta_r, \phi_r)\mathbf{s}(\theta_r, \phi_r) = 1$. The conventional reference estimate $y_c(t)$ may not resolve multipath direct-wave components that are closely spaced in DOA. The main consequence of not resolving multipath with similar DOA is that the reference estimate will be contaminated by a superposition of source waveform copies with different delays and possibly Doppler shifts. This can lead to the appearance of more resolved echoes than multipath components from a single physical target after matched filtering.

$$y_c(t) = \mathbf{v}^\dagger(\theta_r, \phi_r)\mathbf{x}(t), \quad t = 0, 1, \ldots, M-1. \tag{10.78}$$

A set of surveillance beams are usually steered to cover the required angular sector quickly, yet with some degree of reliability. Denoting the radar search direction by $\{\theta_s, \phi_s\}$, the surveillance channel output $z_c(t)$ may also be generated using a conventional beamformer, as in Eqn. (10.79). Clearly, the reference and surveillance outputs are formed using the same CPI of data, but the surveillance beam-steer direction will generally differ to that of the reference beam. A potential drawback of forming a conventional surveillance beam is that powerful DWI may not be sufficiently attenuated. Strong DWI entering the surveillance beam through the conventional antenna pattern sidelobes can degrade target detection performance after matched filtering.

$$z_c(t) = \mathbf{v}^\dagger(\theta_s, \phi_s)\mathbf{x}(t)\,, \quad t = 0, 1, \ldots, M-1. \tag{10.79}$$

Coherent processing is often performed using a matched filter bank that cross-correlates the surveillance and reference beam outputs, $z_c(t)$ and $r_c(t)$, with the latter being delayed and frequency shifted over a set of bistatic range and Doppler bins. The time delays are usually integer multiples of the sampling interval, while the interrogated Doppler shifts are usually defined by the FFT frequency bins. For arbitrary reference and surveillance channel outputs, referred to simply as $y(t)$ and $z(t)$, respectively, the matched filter output is given by Eqn. (10.80), where $f(t)$ is a taper function, and the complex scalar $c(\ell, k)$ represents a resolution cell in the bistatic range (ℓ) and Doppler (k) map. The symbol $*$ in Eqn. (10.80) denotes complex conjugate.

$$c(\ell, k) = \sum_{t=0}^{M-\ell-1} f(t).z(t).y^*(t-\ell).e^{j2\pi kt/M} \tag{10.80}$$

The matched filter makes use of a uniform taper function $f(t) = 1$. The intensity-modulated range-Doppler display in Figure 10.26a shows the envelope of the matched filter outputs $|c(\ell, k)|$ for a CPI recorded at 06:40:44 UT. In this case, the surveillance beam is steered in the cooperative target direction of $\{\theta_s = 309°, \phi_s = 20°\}$, which was known at

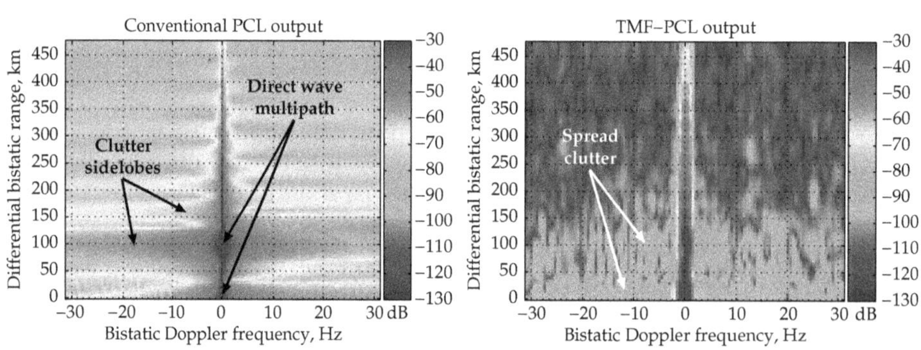

(a) Conventional matched filter output.

(b) Tapered matched filter output.

FIGURE 10.26 Bistatic range-Doppler map formed using the conventional and tapered matched filter for a surveillance beam steered in the target direction. The tapered matched filter significantly reduces clutter sidelobes, but is not able to detect the target echo.

this time from the GPS data in Figures 10.23a and 10.23b. The Learjet is expected to have a differential bistatic range of $\Delta r = 61$ km, and a bistatic Doppler shift of $\Delta f = -29$ Hz at 06:40:44 UT.

The cooperative target echo is not visible in Figure 10.26a because it is masked by the high range-Doppler sidelobes of the more powerful clutter. Using the same data, Figure 10.26b shows the effect of applying a Blackman-Harris −92-dB sidelobe window function defined by Eqn. (10.81). Although the sub-clutter visibility has greatly improved, the target echo cannot be distinguished from the residual "spread-Doppler" clutter that is observed to contaminate ranges in the interval 0–125 km.

$$f(t) = 0.35875 - 0.48829\cos\left(\frac{2\pi t}{M-1}\right) + 0.14128\cos\left(\frac{4\pi t}{M-1}\right) - 0.01168\cos\left(\frac{6\pi t}{M-1}\right) \tag{10.81}$$

To evaluate PCL system performance, the matched filter envelope $|c(\ell, k)|$ was passed on for CFAR processing using a cell-averaging method (Turley 1997). Peaks were selected from the CFAR output as samples greater than their immediate neighbors in beam, range, and Doppler. The coordinates of each peak were estimated using a three-dimensional quadratic interpolation technique that best fits the peak and its immediate neighbors in each dimension. Peak estimates derived from more than 25 minutes of experimental data were stored for the described conventional processing scheme. This was also carried out for various adaptive schemes described below, in order to compare detection performance in the final part of this section.

10.4.2 Range-Dependent Adaptive Beamforming

Temporal adaptive processing techniques have been proposed for the removal of direct signal interference in the surveillance channel. This includes the Extensive Cancelation Algorithm (ECA) based on Least Squares (LS) applied across different batches and stages (Lombardo, Colone, Bongioanni, Lauri, and Bucciarelli 2008), and other methods described in Kulpa and Czekala (2005), Gunner, Temple, and Claypoole (2003), and Saini, Cherniakov, and Lenive (2003), for example.

However, if the reference signal is contaminated by multipath, as in the present case, it is known that the cancelation capability of these algorithms can be severely limited (Lombardo et al. 2008). Multipath removal and waveform estimation based on a priori knowledge about the source signal may also be performed (Treichler and Agee 1983), but this often restricts operation to certain classes of signals in practice. These different issues motivate the use of adaptive beamforming to attenuate direct-wave interference in the surveillance channel of an HF-OTH PCL system.

10.4.2.1 Traditional Adaptive Beamforming

Figure 10.27 schematically depicts the time-domain implementation of the diagonally loaded LSMI-MVDR technique in the PCL processing chain. Specifically, the reference channel $y_c(t)$ is generated by conventional beamforming, as before, whereas the surveillance beam is formed using the adaptive weight vector $\mathbf{w}(\theta_s, \phi_s)$ given by Eqn. (10.82). As described previously, $\hat{\mathbf{R}} = \frac{1}{M}\sum_{t=1}^{M} \mathbf{x}(t)\mathbf{x}^{\dagger}(t)$ is the received data sample spatial covariance matrix, and α is the diagonal loading factor. The integration of all time-domain

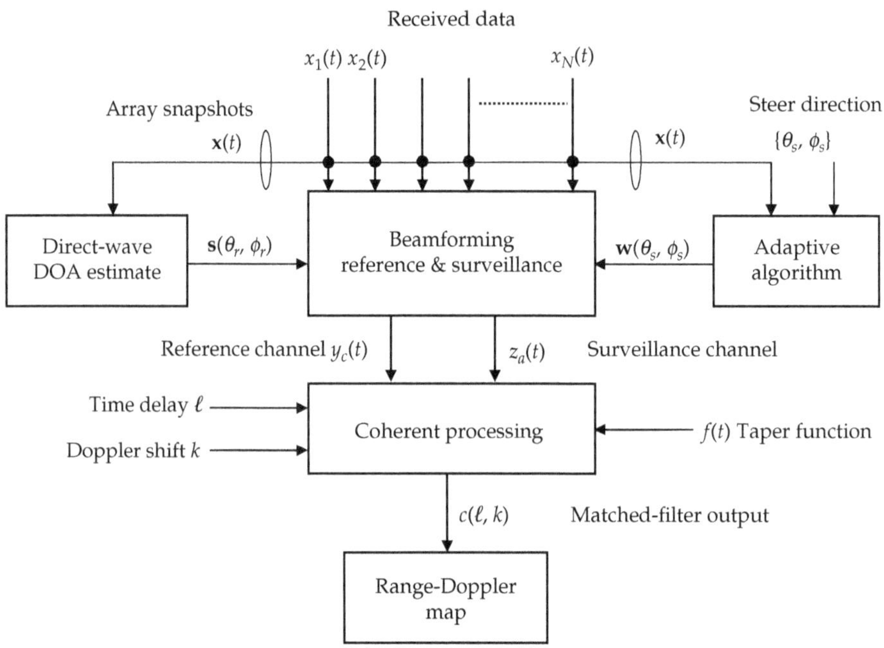

FIGURE 10.27 Traditional PCL signal-processing architecture incorporating conventional beamforming for the reference channel and standard (LSMI) adaptive beamforming on the surveillance channel in the time domain. © Commonwealth of Australia 2011.

snapshots in the CPI is equivalent to forming the estimate of $\hat{\mathbf{R}}$ by averaging the snapshots in all range-Doppler cells after matched filtering (neglecting the influence of the taper function).

$$\mathbf{w}(\theta_s, \phi_s) = [\hat{\mathbf{R}} + \alpha \mathbf{I}]^{-1}\mathbf{s}(\theta, \phi)\{\mathbf{s}^\dagger(\theta_s, \phi_s)[\hat{\mathbf{R}} + \alpha \mathbf{I}]^{-1}\mathbf{s}(\theta_s, \phi_s)\}^{-1} \tag{10.82}$$

This weight vector is applied to produce the surveillance channel output $z_a(t)$ in Eqn. (10.83). Such a procedure is straightforward, but has two potential shortcomings. First, it attempts to cancel the *global* disturbance energy, even though almost all of it is concentrated around zero Doppler frequency after matched filtering, where it does not impede the detection of Doppler-shifted targets. This "over-nulling" comes at the expense of higher white-noise gain, which actually *does* impede the detection of Doppler-shifted targets. Second, the target is included in the training data, and is therefore prone to self-cancelation. A well-chosen loading factor α may reduce susceptibility to these side effects. A loading factor empirically found to perform most favorably for this scheme was adopted in the analysis.

$$z_a(t) = \mathbf{w}(\theta_s, \phi_s)^\dagger \mathbf{x}(t)\,, \quad t = 0, 1, \ldots, M-1 \tag{10.83}$$

Before we move on to discuss the post-Doppler processing scheme (Figure 10.28), it is noted that Figure 10.29a shows the range-Doppler map resulting for the same CPI of data as in Figure 10.26b when the LSMI-MVDR adaptive beamformer is applied in lieu of the conventional beamformer to generate the surveillance channel. Since the direct-wave clutter near zero Doppler frequency has a different DOA to the surveillance beam steer direction, it has been strongly attenuated by the adaptive beamformer in comparison to Figure 10.26b. The disturbance level at Doppler frequencies well away from 0 Hz is also reduced by the LSMI-MVDR scheme, particularly in the 0–125-km range band. Despite this improvement, the target is still not clearly visible at its expected location. The reasons for this will become evident later in this section.

10.4.2.2 Post-Doppler RD-SAP Method

Alternatively, the multi-channel data can be transformed to the range-Doppler domain prior to adaptive beamforming by applying a matched filter to each *receiver output* in an element-wise manner. This yields the array snapshot vectors $\mathbf{c}(\ell, k)$ indexed by range cells ℓ and Doppler bins k in Eqn. (10.84). In this scheme, the N receiver outputs are coherently processed using the conventional reference waveform estimate $y(t) = y_c(t)$ to produce N range-Doppler maps, denoted by $c_n(\ell, k)$ for $n = 1, \ldots, N$. At a particular range-Doppler cell, the N outputs across different receivers may be assembled into the array *snapshot vector* $\mathbf{c}(\ell, k) = [c_1(\ell, k), \ldots, c_N(\ell, k)]^T$. Figure 10.28 illustrates this alternative processing scheme.

$$\mathbf{c}(\ell, k) = [c_1(\ell, k), \ldots, c_N(\ell, k)]^T = \sum_{t=0}^{M-\ell-1} f(t).\mathbf{x}(t).y^*(t-\ell).e^{j2\pi\frac{tk}{M}} \tag{10.84}$$

The matched filter not only concentrates the strongest clutter components near zero Doppler frequency, but it also effectively decomposes the total angular occupancy of the clutter over different range and Doppler cells. This enables *localized* sample spatial covariance matrices to be estimated from carefully selected secondary range-Doppler bins that are in the vicinity of test cells where target echoes are sought. This training philosophy is more likely to provide locally homogeneous estimates of the disturbance covariance matrix.

Besides the lower effective rank of the statistically expected local disturbance covariance matrix relative to the global one (integrated over all processed range and Doppler cells), such estimates may also be expected to be more representative of the disturbance present in nearby test cells. In principle, this localized processing strategy may be expected to lead to more effective disturbance cancelation, and hence improved detection performance. This approach forms the basis of the adaptive beamforming technique that will now be described.

Specifically, the adaptive beamforming technique strives to fulfill three main objectives, namely: (1) that the sufficient training data selected is as statistically homogeneous as possible with the disturbance present in the cells under test (CUT), (2) that useful signals hypothetically present in the CUT are excluded from the training data to avoid the potential problem of target self-cancelation, and (3) that the adaptive algorithm is not overly complicated to facilitate performance diagnosis and real-time operation. Based on these objectives, a range-dependent spatial adaptive processing (RD-SAP) algorithm may be proposed in terms of the following three steps.

STEP 1: Partition the total number of ranges cells L into N_b non-overlapping batches, where each batch contains an equal number Q of consecutive range cells, such that all ranges are accounted for by an integer number of batches, i.e., $N_b = L/Q$. The adaptive algorithm can be readily generalized and is not restricted to such assumptions, which

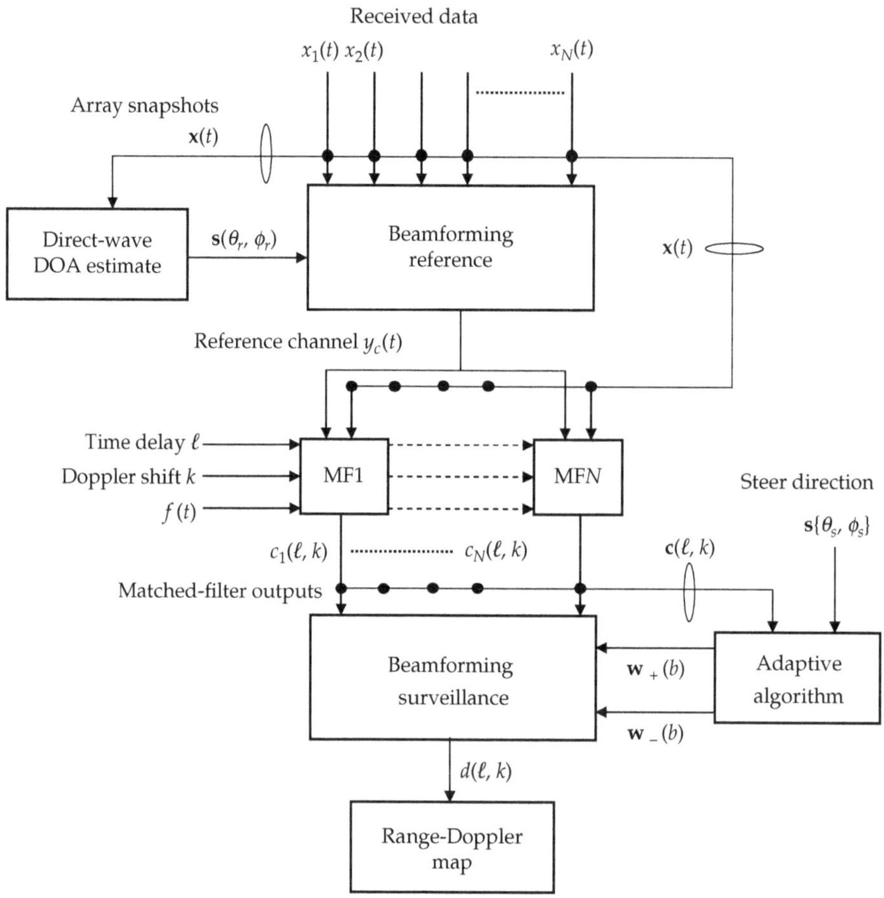

FIGURE 10.28 Alternative signal-processing architecture implementing post-Doppler range-dependent adaptive beamforming for the surveillance channel after matched filtering in each receiver. © Commonwealth of Australia 2011.

only serve to simplify its description. In each batch $b = 1, 2, \ldots, N_b$, define a narrow clutter guard-band that contains most of the direct-wave energy centered around zero Doppler frequency (i.e., specify the left and right "clutter edges"). These edges, denoted by the positive integers g_l and g_r, respectively, are assumed invariant with respect to batch number for convenience of description. This gives rise to two training data regions, one indexed by a set of positive Doppler frequency bins: $k \in G^+ \equiv \{g_r, g_r + 1, \ldots, K\}$, and the other indexed by a set of negative Doppler frequency bins: $k \in G^- \equiv \{-g_l, -g_l - 1, \ldots, -K\}$. Typically, $\{g_l, g_r\} \ll K$ to exclude the most powerful clutter returns. For batch b, training data from the positive and negative Doppler intervals is used to form two sample spatial covariance matrices in Eqn. (10.85). Note that these matrices contain *range-local* disturbance statistics. Proper estimation requires a statistically sufficient number of training snapshots $Q[K - \max(g_r, g_l)] > 2N$ to avoid significant convergence losses (Reed et al. 1974). For $Q = 1$, $K = 62$, and $g_r = g_l = 4$, we have $K - \max(g_r, g_l) = 58$, which is greater than $3N$ for $N = 16$. In the current application, this makes adaptation on a single range-by-range basis feasible from a finite sample support perspective.

$$\mathbf{R}_{+}(b) = \sum_{\ell=(b-1)Q}^{bQ-1} \sum_{k \in G^{+}} \mathbf{c}(\ell, k)\mathbf{c}(\ell, k)^{\dagger}, \quad \mathbf{R}_{-}(b) = \sum_{\ell=(b-1)Q}^{bQ-1} \sum_{k \in G^{-}} \mathbf{c}(\ell, k)\mathbf{c}(\ell, k)^{\dagger} \tag{10.85}$$

STEP 2: Form the batch-dependent weight vectors defined by $\mathbf{w}_{+}(b)$ and $\mathbf{w}_{-}(b)$ in Eqn. (10.86) for adaptive beamforming in the surveillance direction $\{\theta_s, \phi_s\}$, note that steer angle dependence on the weights has been dropped to simplify notation. The weight vectors are updated with respect to batches $b = 0, 1, \ldots, N_b - 1$ to minimize performance loss due to statistical heterogeneity of the disturbance spatial covariance matrix. Relative to the standard LSMI method described previously, the additional computational load of this routine is dominated by two $N \times N$ matrix inversions per batch in Eqn. (10.86). As these matrices are of moderate dimension $N = 16$, and all the surveillance beams steered in different directions can be formed without further inversions,[4] the computational load of approach can be scaled for real-time operation by an appropriate choice of batch size Q.

$$\begin{cases} \mathbf{w}_{+}(b) = \mathbf{R}_{+}^{-1}(b)\mathbf{s}(\theta_s, \phi_s)\{\mathbf{s}^{\dagger}(\theta_s, \phi_s)\mathbf{R}_{+}^{-1}(b)\mathbf{s}(\theta_s, \phi_s)\}^{-1} \\ \mathbf{w}_{-}(b) = \mathbf{R}_{-}^{-1}(b)\mathbf{s}(\theta_s, \phi_s)\{\mathbf{s}^{\dagger}(\theta_s, \phi_s)\mathbf{R}_{-}^{-1}(b)\mathbf{s}(\theta_s, \phi_s)\}^{-1} \end{cases} \tag{10.86}$$

STEP 3: The critical point is that the weight vector $\mathbf{w}_{+}(b)$ is applied to beamform the *negative* Doppler bins, and $\mathbf{w}_{-}(b)$ is applied to beamform the *positive* Doppler bins, such that the scalar outputs $d(\ell, k)$ given by Eqn. (10.87) yield the range-Doppler map for the surveillance beam. This positive-to-negative "swap" ensures that the test cells processed by the adaptive filter are *not* included in the training data used to synthesize it. This procedure significantly reduces the likelihood of target cancelation. Moreover, since the spatial structure of the direct-wave clutter in a particular range cell spreads over both positive and negative Doppler frequencies after matched filtering, this training methodology provides a reasonable estimate of the local disturbance characteristics in the alternate part of the Doppler spectrum. The RD-SAP algorithm is relatively simple to interpret and implement, with these three steps being repeated until all range cells $\ell = (b - 1)Q, \ldots, bQ - 1$ and batches $b = 1, \ldots, N_b$ are processed.

$$d(\ell, k) = \begin{cases} \mathbf{w}_{-}(b)^{\dagger}\mathbf{c}(\ell, k), & k = 1, \ldots, K \\ \mathbf{w}_{+}(b)^{\dagger}\mathbf{c}(\ell, k), & k = 0, -1, \ldots, -K \end{cases} \tag{10.87}$$

Figure 10.29b shows the range-Doppler map for the same example CPI using RD-SAP with $Q = 1$ and $g_r = g_l = 4$. The target echo can now be clearly distinguished above the local background at the aircraft's expected range-Doppler coordinates. Two weaker echo components at similar Doppler shift but closer and further ranges are also visible in the RD-SAP output. These returns arise due to the presence of multipath in the reference and surveillance channels. Based on the presence of two dominant ionospheric modes, the "early echo" (labeled A in Figure 10.29b) is due to the temporal alignment of the target echo with *smaller* delay in the surveillance channel and the direct-wave with *larger* delay in the reference channel. Conversely, the "late echo" (labeled B in Figure 10.29b) is

[4] Once $\mathbf{R}_{+}(b)$ and $\mathbf{R}_{-}(b)$ have been inverted, they may be substituted in Eqn. (10.86) to steer the adaptive beam in all required surveillance directions $\{\theta_s, \phi_s\}$.

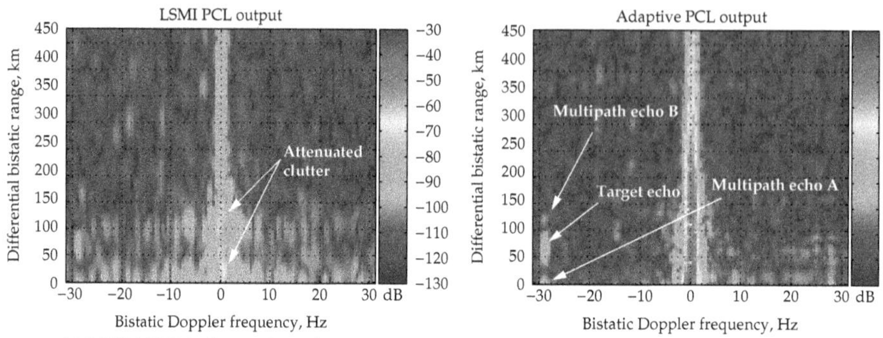

(a) LSMI-MVDR adaptive beamforming.

(b) Post-Doppler RD-SAP method.

FIGURE 10.29 Bistatic range-Doppler maps resulting when the LSMI-MVDR adaptive beamformer and post-Doppler range-dependent spatial adaptive processor (RD-SAP) are applied to generate the surveillance channel, respectively. © Commonwealth of Australia 2011.

caused by the alignment of the target echo with *larger* delay in the surveillance channel and the direct-wave with *smaller* delay in the reference channel. This example illustrates the multipath effects at the matched filter output due to the presence of two dominant propagation modes in both the reference waveform estimate and for the target echoes in the surveillance beam.

Figure 10.30 compares the Doppler spectra at the true target range (labeled "target echo" in Figure 10.29b) for the various processing schemes. It is evident that conventional

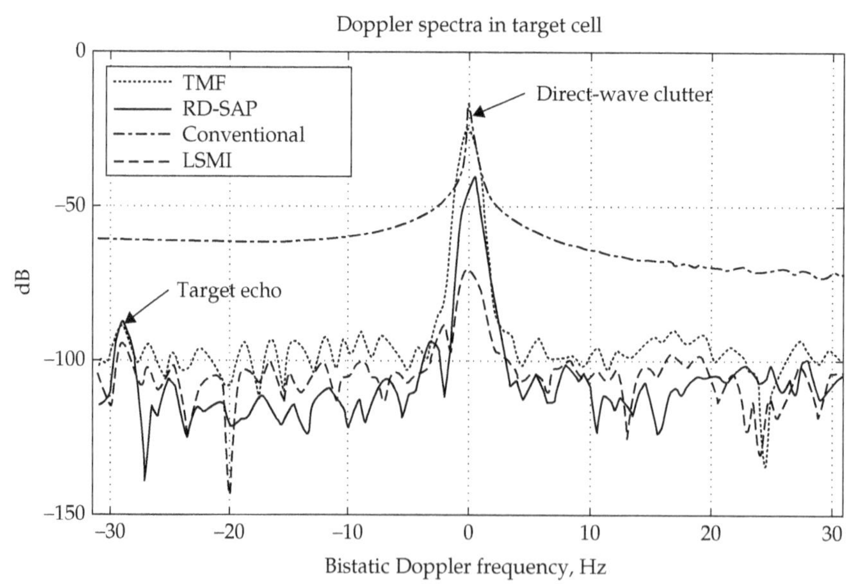

FIGURE 10.30 Doppler spectra for different PCL processing schemes at the true target range showing that the target echo can be clearly detected against the local background only when the proposed adaptive beamformer (RD-SAP) is applied in this example.
© Commonwealth of Australia 2011.

processing methods do not provide sufficient suppression of the direct-wave clutter to enable target detection above the residual background disturbance level (i.e., the dash-dot and dotted curves in Figure 10.30). The standard LSMI-MVDR adaptive beamformer, which integrates the disturbance spatial structure over the whole CPI (i.e., over the entire range-Doppler map), significantly cancels the strongest direct-wave clutter energy near zero Doppler frequency, as shown by the dashed curve in Figure 10.30.

However, these powerful components seduce the adaptive processor into consuming extra degrees of freedom for canceling clutter in a very confined band of Doppler frequencies. The high level of clutter rejection achieved in this small Doppler interval is largely ineffective for target detection because it comes at the expense of poorer rejection at the higher Doppler frequencies of interest. In addition, a 5-dB loss in target echo peak due to self-cancelation is evident for the LSMI method relative to the RD-SAP technique (solid line in Figure 10.30).

RD-SAP rejects much less direct-wave clutter near 0 Hz compared to the LSMI scheme. Importantly, it lowers the background disturbance level in a larger number of Doppler bins compared to the LSMI, such that target echoes can be more easily distinguished. The key point is that RD-SAP leads to an increase in the probability of target detection across a much greater section of Doppler space in a particular range cell, despite the total output disturbance power being substantially higher than for the LSMI scheme due to inferior rejection of the direct wave in the vicinity of 0-Hz Doppler frequency.

In summary, RD-SAP leads to more fruitful use of adaptive degrees of freedom, and yields higher disturbance cancelation in regions of the search space *where it is necessary* for target detection. In addition, RD-SAP strives to protect the target echo from being canceled by interchanging the adaptive filters used to process the positive and negative Doppler bins.

Although more sophisticated adaptive schemes could be proposed, the presented RD-SAP version serves to illustrate the main concepts and advantages of post-Doppler adaptive beamforming using localized training in a heterogenous disturbance environment. The described method attempts to strike a balance between performance (evaluated more formally in the next section), simplicity for implementation and interpretation, as well as scalable computational load for real-time operation.

The computational load may be reduced in more homogeneous disturbance environments by increasing the range batch size ($Q > 1$), as the number of matrix inversions is directly proportional to Q. The clutter edges g_l and g_r may be preset based on empirical observations, or estimated online, and only need to exclude the most powerful clutter components often contained in a small band of Doppler frequencies centered near 0 Hz.

10.4.3 Extended Data Analysis

While the analysis of a single CPI of data enables very detailed performance comparisons to assist with algorithm development and testing, statistics that indicate PCL system performance over an extended data set are ultimately of more interest to an operator. This section quantifies the performance of the previously described conventional and adaptive processing schemes over an extended data set containing 322 CPIs.

Specifically, performance statistics are computed in two range bands, namely, a "short" range band with differential bistatic ranges in the interval $\Delta r \in \mathcal{R}_1 \equiv [0, 125)$ km, and a "long" range band defined by the interval $\Delta r \in \mathcal{R}_2 \equiv [125, 250)$ km. Using the cooperative Learjet aircraft as a test target, as well as the known GPS truth information on the flight path, performance statistics were evaluated for the short and long range bands using 153 and 169 CPIs, respectively.

For both range bands, experimental results showing PCL system performance are presented as: (1) empirical receiver operating characteristic (ROC) curves, which show the relative frequency of target detections against the density of false alarms, (2) scatter plots of target detections (and missed detections) over time, which also compare the target coordinates estimated by the PCL system with the GPS truth data, and (3) a first-order indication of the probability of target track initiation in terms of the observed likelihood of at least m target detections out of n consecutive CPI.

10.4.3.1 Empirical Receiver-Operating Characteristics

Figures 10.31a and 10.32a show empirical ROC curves resulting for different processing schemes in the short- and long-range bands, respectively. The method used to produce these curves is as follows. For each CPI, a target detection was declared whenever a peak in the CFAR output that exceeded the threshold value was found at the expected radar coordinates of the target (calculated from the GPS data) plus a small tolerance in range and Doppler. Any other peaks found above this threshold in the range-Doppler map (i.e., over a total of $K \times L = 12500$ cells per CPI) were considered false alarms. This represents a conservative estimate of the false alarm density since other genuine but unconfirmed targets may be present.

For each range band, the percentage number of CPI in which the target could be detected was recorded for a number of different threshold settings that varied in steps of 0.5 dB down from a maximum of 25 dB above the mean background level. For each threshold setting, the false-alarm density in each CPI was also recorded as a percentage value, and the resulting values averaged over all the CPIs processed. The large amount of data processed enables robust false-alarm density estimates to be made down to 10^{-4} (0.01%). Hence, a single point on the ROC curves in Figures 10.31a and 10.32a indicates the relative frequency of target detections and false-alarm density that can be simultaneously achieved with a particular threshold value. The ROC curve is a collection of such points, which are plotted for different threshold values.

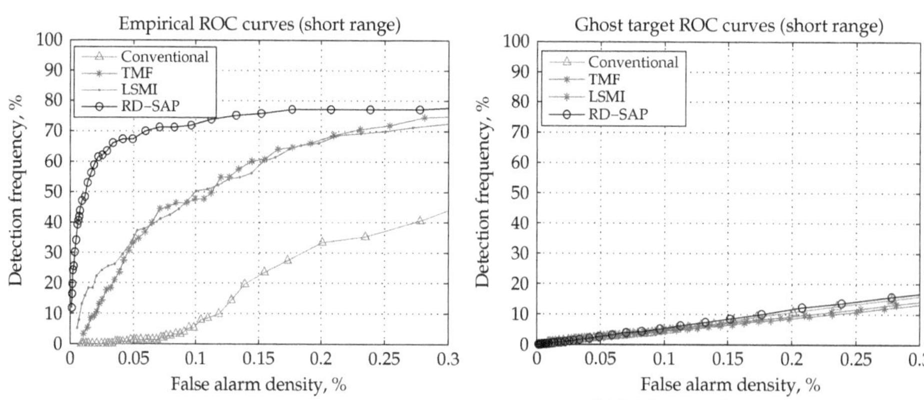

(a) ROC curves for real target.
(b) ROC curves for "ghost" target.

FIGURE 10.31 Empirical receiver operating characteristic curves for the real target in the short-range band $\mathcal{R}_1 \in [0\text{–}125)$ km, as well as "ghost targets" with trajectory similar to the real target but slightly displaced in Doppler frequency. © Commonwealth of Australia 2011.

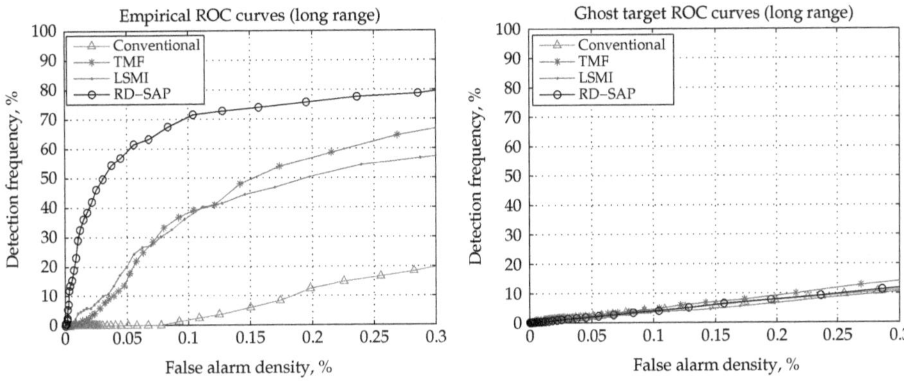

(a) ROC curves for real target.

(b) ROC curves for "ghost" target.

FIGURE 10.32 Empirical receiver operating characteristic curves for the real target in the long-range band $\mathcal{R}_2 \in [125\text{–}250)$ km, as well as "ghost targets" with trajectory similar to the real target but slightly displaced in Doppler frequency. © Commonwealth of Australia 2011.

The ROC curves empirically indicate the "probability of detection" for the cooperative target within each range band at a number of operationally feasible "false-alarm rates." For a false alarm rate of 10^{-3} (0.1%), it is possible to detect the target 39% of the time using conventional processing in the long-range band, compared to 72% of the time using the RD-SAP method. This is a substantial improvement relative to the tapered matched filter (TMF) at times when the target echo is weaker due to the longer range. Although not as substantial as for longer ranges, a noteworthy improvement also results for RD-SAP over conventional processing in the short-range band $\mathcal{R}_1$, as reported in Table 10.4.

To benchmark the ROC curves, the expected target coordinates were deliberately mistuned from the true ones by displacing the search for the target peak in Doppler relative to the GPS predictions. The system then searches for a "ghost target" with similar trajectory to the true target over the same data set. The statistics were recalculated for the ghost target in identical manner, and averaged over a series of different ghost targets. The resulting ROC curves for the two range bands are shown in Figures 10.31b and Figure 10.32b. These benchmark curves demonstrate that the percentage of detections recorded in the *absence* of the true target is small and nearly identical for all applied processing schemes.

10.4.3.2 Distribution of Detections

Figure 10.33 shows the distribution of target detections over time and their location in differential bistatic range relative to GPS data for conventional processing (TMF). Open

Processing	Threshold	$P_D \in \mathcal{R}_1$	$P_{FA} \in \mathcal{R}_1$	$P_D \in \mathcal{R}_2$	$P_{FA} \in \mathcal{R}_2$
RD-SAP	11 dB	76%	0.15%	72%	0.10%
Conventional TMF	9.5 dB	66%	0.19%	39%	0.10%

TABLE 10.4 Detection performance for the conventional TMF and RD-SAP processing schemes in the long- and short-range bands using a fixed threshold value in each band.

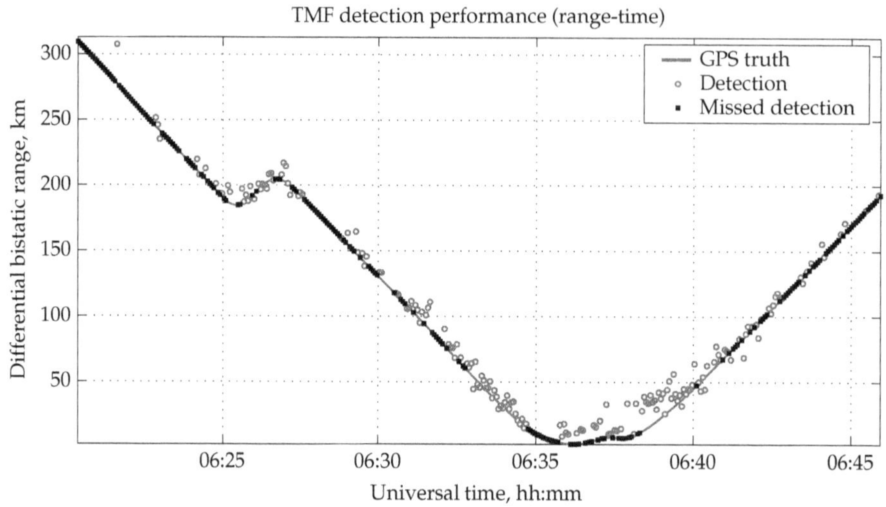

FIGURE 10.33 Distribution of cooperative aircraft detections and missed detections as function of target *differential bistatic range* for conventional TMF processing using a threshold of 9.5 dB (i.e., false alarm rate of 0.19% in $\mathcal{R}_1$ and 0.1% in $\mathcal{R}_2$). © Commonwealth of Australia 2011.

circles mark the times and coordinates of target detections relative to the GPS data, which is shown as a solid line, while the solid squares plotted on top of the GPS truth indicate the missed detections. Figure 10.34, in the same format as Figure 10.33, shows the corresponding results for RD-SAP. The detection thresholds listed in Table 10.4 have been used to produce these figures. Such threshold values provide an identical false

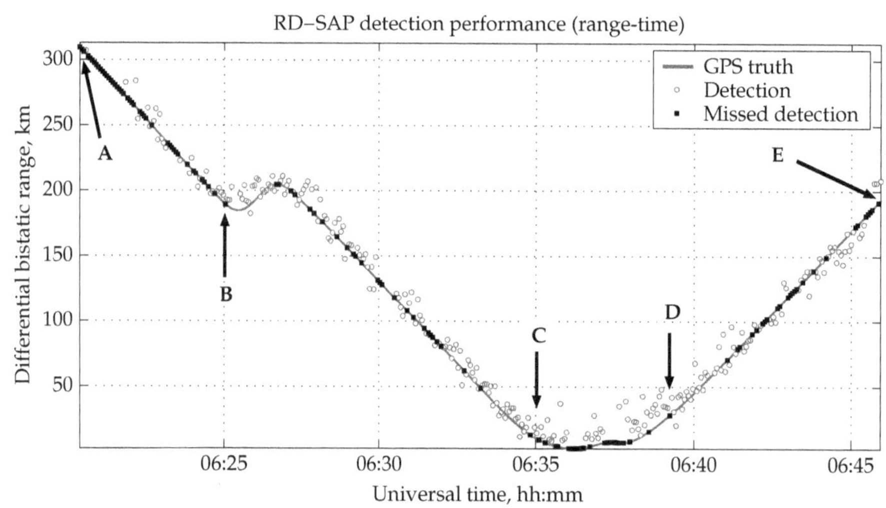

FIGURE 10.34 Distribution of cooperative aircraft detections and missed detections as function of target *differential bistatic range* using RD-SAP and a threshold of 11 dB (i.e., false alarm rate of 0.15% in $\mathcal{R}_1$ and 0.1% in $\mathcal{R}_2$). © Commonwealth of Australia 2011.

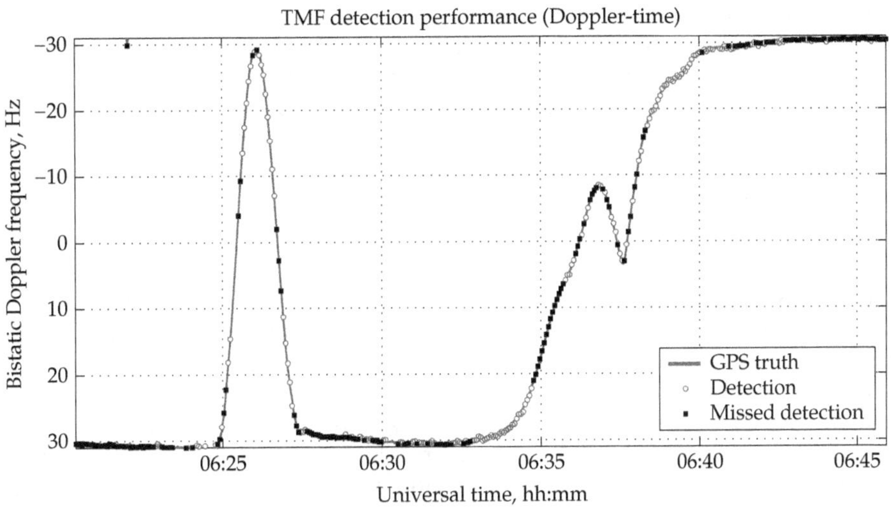

FIGURE 10.35 Distribution of cooperative aircraft detections and missed detections as function of target *bistatic Doppler frequency* for conventional TMF processing using a threshold of 9.5 dB (i.e., false alarm rate of 0.19% in $\mathcal{R}_1$ and 0.1% in $\mathcal{R}_2$). © Commonwealth of Australia 2011.

alarm rate of 0.1% for both conventional and adaptive processing schemes in $\mathcal{R}_2$, but *favor* conventional processing in $\mathcal{R}_1$, due to the relatively higher false-alarm rate of 0.19%.

Similarly, Figures 10.35 and 10.36 demonstrate the bistatic Doppler versus time results, where the high accuracy of detected peak locations relative to the expected GPS coordinates may be appreciated. Note that a number of detections are missed for both

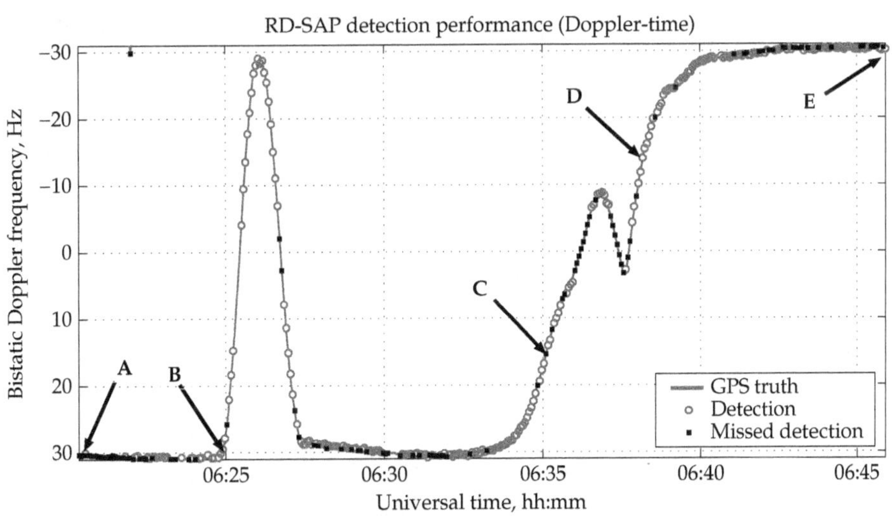

FIGURE 10.36 Distribution of cooperative aircraft detections and missed detections as function of target *bistatic Doppler frequency* using RD-SAP and a threshold of 11 dB (i.e., false alarm rate of 0.15% in $\mathcal{R}_1$ and 0.1% in $\mathcal{R}_2$). © Commonwealth of Australia 2011.

schemes when the target bistatic Doppler frequency is close to 0 Hz (i.e., near the strong clutter ridge), and at differential bistatic ranges greater than 250 km, due to the reduced gain of the antennas at low elevation angles, which effectively limits system performance at very long ranges. In almost all segments of the cooperative target flight path, use of RD-SAP is generally observed to significantly improve performance relative to conventional TMF processing in terms of the frequency of target "hits" for an equivalent or lower false-alarm rate.

10.4.3.3 Indication of Track Initiation

A method commonly used to provide a *first-order* indication of target track-initiation probability is to determine the frequency that at least m detections are made in n consecutive CPI. This "m-out-of-n" detection logic provides valuable information about the continuity of detections made by a particular processing scheme over time. Although this measure does not assess tracking performance, as it is independent of a track filter, detection continuity is of fundamental importance for establishing and maintaining tracks. Figures 10.37 and 10.38 show the results for *at least* $m = 3$ target

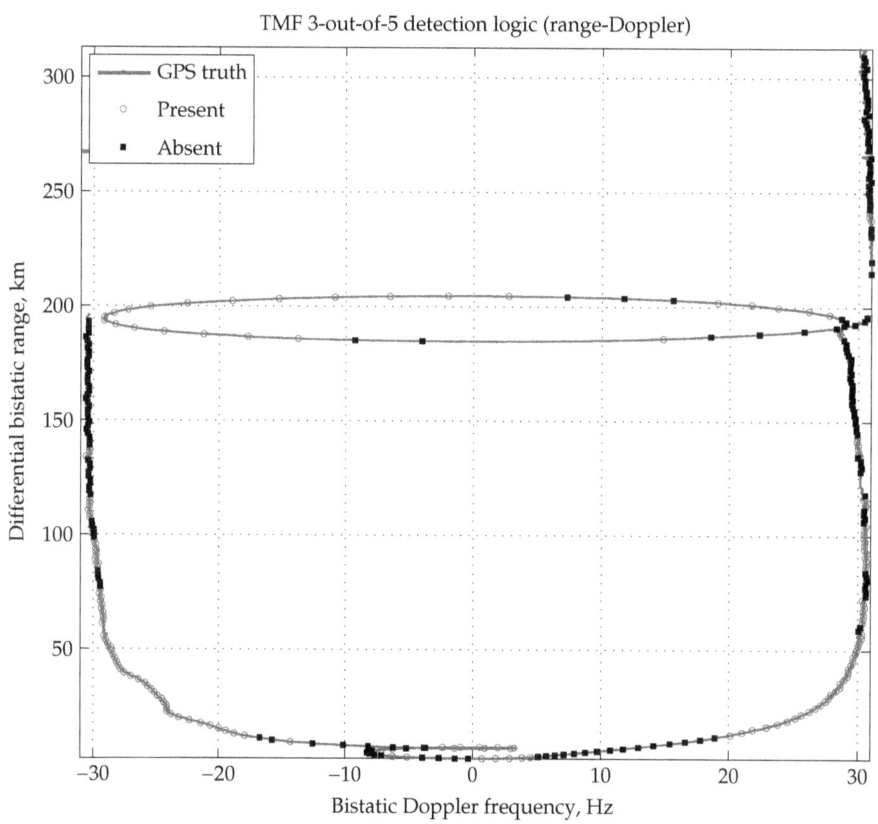

FIGURE 10.37 Distribution of occurrences satisfying the 3-out-of-5 detection logic overlayed on the target GPS profile in range-Doppler space for conventional TMF processing using a threshold of 9.5 dB. © Commonwealth of Australia 2011.

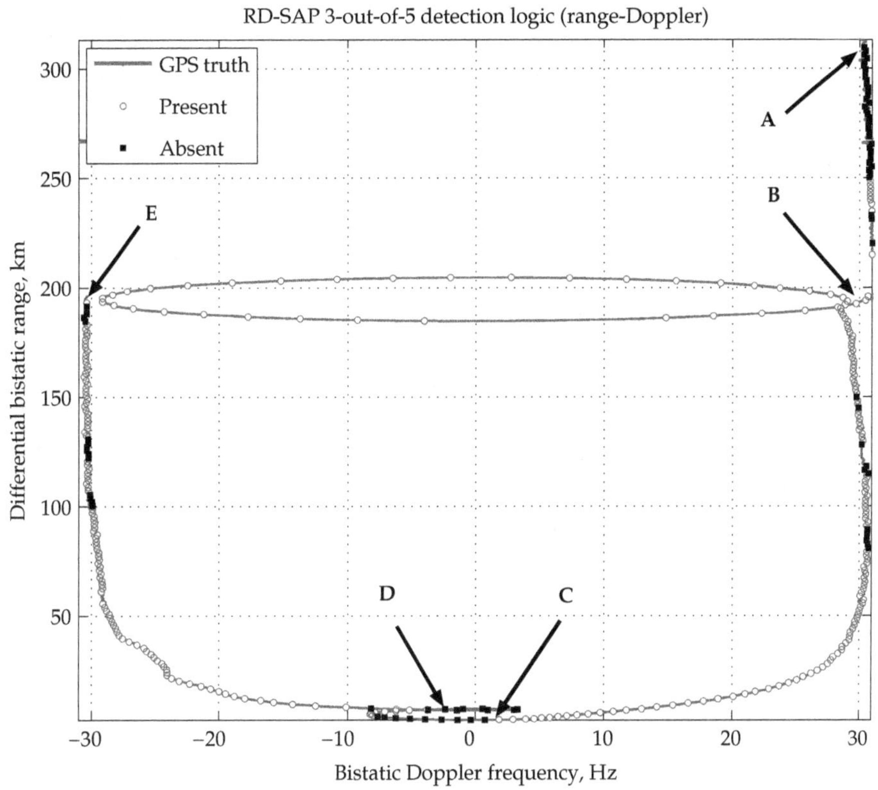

FIGURE 10.38 Distribution of occurrences satisfying the 3-out-of-5 detection logic overlayed on the target GPS profile in range-Doppler profile for RD-SAP with a threshold of 11 dB. © Commonwealth of Australia 2011.

detections in $n = 5$ consecutive CPI using the TMF and RD-SAP processing schemes, respectively.

The solid line represents the GPS locus of target bistatic range-Doppler coordinates during the trial. An open circle indicates the time centroid of an $n = 5$ CPI window in which at least $m = 3$ target detections were made, while the solid squares indicate cases where the number of detections made in the CPI window was less than $m = 3$. The detection thresholds used to produce these results are the same as those listed in Table 10.4. Table 10.5 expresses the number of occurrences satisfying the 3-out-of-5

Processing Scheme	Threshold (dB)	$P_{TI} \in \mathcal{R}_1$ (%)	$P_{TI} \in \mathcal{R}_2$ (%)
RD-SAP	11	84	83
Conventional TMF	9.5	69	32

TABLE 10.5 Indication of cooperative aircraft target track initiation probability (P_{TI}) based on a 3-out-of-5 detection logic using conventional TMF processing and RD-SAP.

detection logic as a percentage for both processing schemes. It is evident from these results, as well as the corresponding visual interpretation in Figures 10.37 and 10.38, that RD-SAP significantly increases the probability of track initiation relative to conventional TMF processing in the long- and short-range bands.

The improvement is especially notable in the long-range band $\mathcal{R}_2$, where the percentage of time the detection logic satisfied is 83% for RD-SAP, but only 32% for conventional TMF processing. It is also noted that only RD-SAP is able to satisfy this detection logic over the full 360-degree maneuver made by the target at a differential bistatic range of approximately 200 km. The false track performance is more difficult to gain insight on without implementing a tracking filter, but the considered false-alarms rates (0.1%) are expected to be small enough to produce acceptable false track results for the studied HF-OTH PCL application.

Taking into account the high density (i.e., continuity) of detections in time, and the accuracy of peak locations estimates relative to the GPS-measured target coordinates, particularly in the Doppler dimension, it is likely that the RD-SAP adaptive beamformer would allow effective tracking of the Learjet aircraft target to considerable ranges (greater than 100 km from the line-of-sight receiver).

However, this result should not be extrapolated to infer the performance of the system for signals other than the one analyzed; further analysis on different signals of opportunity will be required to make a more general statement about the capability of this HF-OTH PCL system. While this analysis indicates the benefits of post-Doppler adaptive beamforming for real-target detection in a practical system, and may serve to guide the choice of training strategy for HF-OTH PCL systems, there remain several issues to consider. This includes evaluating performance for other signal classes in the HF band.

CHAPTER 11

Space-Time Adaptive Processing

Space-time adaptive processing (STAP) refers to a class of multi-dimensional adaptive filtering techniques which are used in radar to simultaneously combine data received across the elements of an antenna array with samples acquired in the dimensions of slow- and/or fast-time over a coherent processing interval to produce a filtered scalar output. The primary goal of the STAP filter is to maximize the output signal-to-disturbance ratio (SDR) by suppressing clutter and interference in each processed azimuth-range-Doppler cell. In the presence of powerful disturbance signals, STAP can significantly improve the target detection performance of a radar system with respect to conventional processing.

STAP is a topic which has received enormous attention in the literature. The intense interest over the last two decades in particular has led to the development of a wide variety of STAP techniques for different radar systems, practical applications, and operational scenarios. The scope of this chapter is not to review the extensive collection of theoretical and experimental works on STAP, but rather to focus on the specific STAP implementations that hold most promise for OTH radar systems.

For a general introduction to the subject of STAP, the reader is referred to a number of authoritative treatments, such as the definitive texts by Klemm (2002) and Guerci (2003), and the comprehensive review articles of Melvin (2004), Wicks, Rangaswamy, Adve, and Hale (2006), and references therein. The technical report by Ward (1994) and seminal paper of Brennan and Reed (1973) are also highly recommended.

In certain situations of practical interest to radar operators, STAP offers the potential for more effective disturbance cancelation than separate spatial and temporal adaptive processing. Depending on the configuration employed, STAP can involve the use of antenna elements or beams as spatial channels, and either time or frequency domain samples in each spatial channel. Restricting attention to antenna-element and time-domain architectures, STAP implementations may incorporate slow-time and/or fast-time samples to combat disturbances such as clutter and/or interference, which inherently possess different correlation properties.

Areas in which STAP offers significant benefits relative to sequential or "factored" space-time processing include the rejection of backscattered surface clutter for radars mounted on moving platforms (slow-time STAP), and the rejection of diffusely scattered multipath interference, which may be received through the main lobe of the antenna pattern (fast-time STAP). The most general "fully adaptive" STAP approach simultaneously combines signals from antenna elements, slow-time samples, and fast-time samples

(3D-STAP). This architecture has been proposed for the problem of jointly mitigating surface clutter and terrain-scattered interference in airborne radar (Fante and Torres 1995).

Important practical issues for STAP include the need for sufficient and statistically homogeneous training data, as well as low computational load for real-time processing. Such considerations have led to a taxonomy of partially adaptive algorithms with reduced dimension or rank (Goldstein and Reed 1997). Self-configuring STAP techniques that are robust to instrumental imperfections and changing environmental conditions (without requiring operator intervention) are also highly desirable in practice.

The first section of this chapter describes three different STAP architectures implemented in the antenna-element/time domain and discusses the motivation for each architecture in connection with the characteristics of the disturbance to be mitigated. The main purpose of this section is to identify the STAP typologies that are most suitable for OTH radar, and to delineate the peculiarities of the STAP problem for OTH radar with respect to that encountered in airborne microwave radar systems.

Data models are formulated in the second section to describe the characteristics of surface-scattered clutter and diffuse multipath interference signals received by OTH radar after reflection from the ionosphere. The third section presents standard and alternative STAP techniques to address the problem of rejecting non-stationary diffuse multipath interference or "hot clutter" in OTH radar. The performance of these algorithms is illustrated using simulated data based on the formulated data models.

The final section describes a post-Doppler STAP technique for canceling a mixture of narrowband interference and spread-Doppler clutter that is statistically heterogeneous in range. With practical applications in mind, this method incorporates a reduced dimension beam-space architecture to ease demands on sample support and computational load. An important theme relevant to the entire discussion is that the added sophistication and computational complexity of STAP algorithms needs to be justified in terms of the practical performance benefits relative to processing schemes that operate separately on a single radar data-cube dimension at a time.

11.1 STAP Architectures

To explain and motivate the different antenna-element/time-domain STAP architectures, it is useful to recall the "anatomy" of the radar data cube. The radar system operates by collecting data over a coherent processing interval (CPI), which consists of P transmitted pulses or "sweeps" emitted at a pulse repetition frequency of f_p pulses per second. The receiving system is composed of N spatial channels, consisting of antenna elements or sub-arrays, for example, with each reception channel being connected to an individual digital receiver.

After down-conversion and filtering, the received in-phase and quadrature (I/Q) components of the baseband signal are sampled at the Nyquist rate of f_t samples per second, such that K complex samples are acquired in each pulse repetition interval (PRI). The raw data cube collected in this manner over a single CPI therefore consists of $N \times P \times K$ complex samples. Increments acquired at the Nyquist rate within a particular PRI are referred to as "fast-time" samples or range bins, while those resulting across different PRI intervals over the CPI are termed "slow-time" samples (Griffiths 1996).

This section is divided into three parts, which describe the main characteristics of slow-time, fast-time, and fully adaptive 3D-STAP. These different STAP techniques are described with a view to explaining the potential application of each to OTH radar.

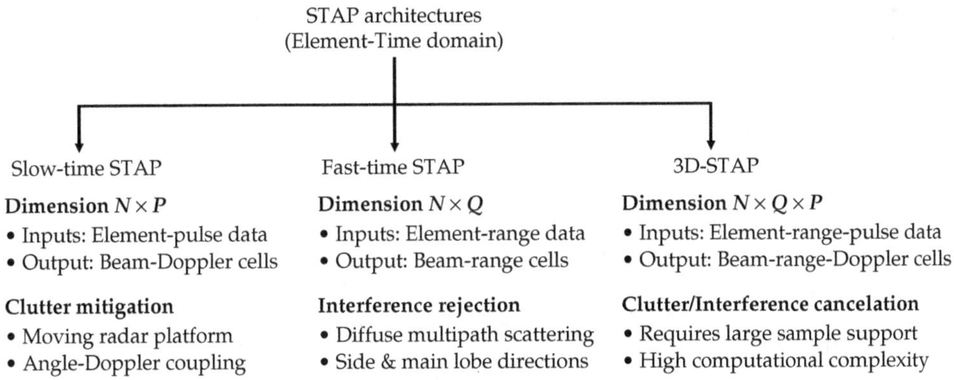

FIGURE 11.1 Different STAP architectures for processing in the antenna-element/time domain with a representative application in each case. The number of fast-time samples used is generally less than the maximum number available, i.e., $Q < K$.

Figure 11.1 summarizes the filter dimensions and input/output data formats of the three STAP typologies considered, along with representative applications, which will be described in more detail below.

11.1.1 Slow-Time STAP

Slow-time STAP operates on a single range cell in turn and produces an output that is a weighted linear combination of spatial samples collected by different antenna elements in the array, and temporal samples acquired over multiple coherent pulses of the radar waveform during the CPI. The filter dimension corresponding to the full slow-time STAP architecture is therefore $N \times P$. At a given range cell, the STAP weights are adjusted to provide an output for each combination of beam steer direction and echo Doppler frequency at which targets are sought. Ideally, the STAP weights are synthesized in a manner that maximizes the signal-to-disturbance ratio in the output beam-range-Doppler cells. Here, *useful signals* refer to target echoes matched to the interrogated steer direction and Doppler frequency, while *disturbance* refers to both clutter and interference in general.

When it is necessary to reduce the number of degrees of freedom (DOFs) in the STAP processor, partially adaptive slow-time STAP configurations may be implemented post-Doppler or in beam space, for example. Alternatively, more sophisticated rank-reduction transforms based on singular-value decomposition may be used. In any case, slow-time STAP techniques operate on all or part of the information contained in the receivers and pulses of the radar data cube. For the time being, we shall not distinguish between full and partially adaptive slow-time STAP, or concern ourselves with issues regarding training data and computational load. These aspects are of course very important for practical implementation, but peripheral in the sense of identifying the main objective of the processor itself.

The slow-time STAP approach has received considerable attention in the context of its application to airborne microwave radar systems. The question arises as to the driving factors which motivate this two-dimensional processing architecture for such systems, and the conditions in which slow-time STAP may be expected to provide performance benefits relative to the application of beamforming and Doppler processing separately.

For the case of a moving radar platform, a large part of the answer resides in the strong coupling that exists between the direction of arrival and Doppler frequency of radar signals backscattered from extended regions of the Earth's surface. More specifically, relatively faint target echoes received in the main beam can be masked by strong surface clutter returns that share the same Doppler frequency and group range (either coincident or ambiguous) as the target echo, but are incident from directions other than the radar look direction.

An important objective of slow-time STAP in airborne radar systems is to mitigate sidelobe clutter, which is distributed in Doppler frequency due to platform motion. It is often difficult for a conventional beamformer to achieve extremely low sidelobes in practice due to array calibration uncertainties and local scattering effects arising from the presence of the aircraft. In addition, conventional processing typically achieves relatively low sidelobes at the expense of an increase in main-lobe width. As adaptive beamforming may alleviate some of these issues, it is reasonable to ask why slow-time STAP is applied in preference to the combination of adaptive beamforming and Doppler processing.

The main advantage of slow-time STAP with respect to adaptive beamforming stems from the fact that the surface clutter is incident from a continuum of directions and therefore tends to have full spatial rank. This makes clutter rejection via pure spatial processing ineffective in general. However, due to the strong DOA-Doppler coupling of the backscattered clutter, the energy contained in such signals has the potential to be concentrated in a relatively low dimensional subspace of the space-time covariance matrix formed in the joint antenna-element/slow-time domain. This property opens up the possibility for more effective sidelobe-clutter rejection using slow-time STAP.

Slow-time STAP provides the possibility to simultaneously cancel a limited number of jamming sources that may occupy all range-Doppler bins (i.e., broadband interference), but are not incident from the same direction as the target echo (i.e., not entering through the main beam). For an airborne radar system, this typically includes the direct and specularly reflected interference paths but not diffusely scattered components incident from the main beam direction. The advantage of slow-time STAP over adaptive beamforming when sidelobe interference is present is that such interference can in principle be rejected effectively even when clutter-free training data is not available.

OTH radars operate with fixed land-based receive and transmit systems; an exception to this is HF surface-wave radars installed on moving ship-borne platforms, but this case will not be considered further. In general, backscattered surface clutter received by OTH radars exhibits relatively weak (if any) ionospherically-induced DOA-Doppler coupling.[1] In a given range cell, both main-lobe and sidelobe clutter backscattered from the OTH radar transmitter footprint often occupy a similar Doppler frequency band typically near 0-Hz. In other words, target echoes are often Doppler-shifted by a similar amount relative to both the main-lobe and sidelobe clutter in OTH radar. In this situation, slow-time STAP provides minimal or no advantages with respect to factored space-time processing.

As standard Doppler processing is often quite effective for separating moving target echoes and quasi-stationary surface-clutter returns (incident from all directions) into

[1]This is often the case for a relatively quiet mid-latitude ionosphere. However, in auroral regions, e.g., the occurrence of high plasma drift velocities due to convection can lead to appreciable DOA-Doppler coupling on the scattered skywave signals. In this case, the radar platform is stationary but the propagation medium is moving.

different Doppler bins, there is often no strong motivation to apply slow-time STAP in OTH radar systems. A possible reason to justify the additional computational complexity of slow-time STAP relative to the application of adaptive beamforming and Doppler processing is when supervised training data containing only interference and noise contributions is difficult to obtain. Clutter contamination in the training data used for adaptive beamforming can bias the weight estimates and degrade interference-plus-noise rejection performance. Since there are different methods for obtaining such training data in practice, slow-time STAP has not found widespread use in OTH radar systems.

11.1.2 Fast-Time STAP

The schematic diagram in Figure 11.2 illustrates the fast-time STAP architecture which operates on data from a single radar pulse in turn. Here, the output is a weighted linear combination of spatial samples acquired by the receiving elements of the antenna array and multiple fast-time samples corresponding to different range-gates in a PRI. Note that the fast-time samples are delayed by $T_s = 1/f_s$ seconds, where for a receiver bandwidth B, the time-bandwidth product is typically less than or equal to unity $BT_s \leq 1$.

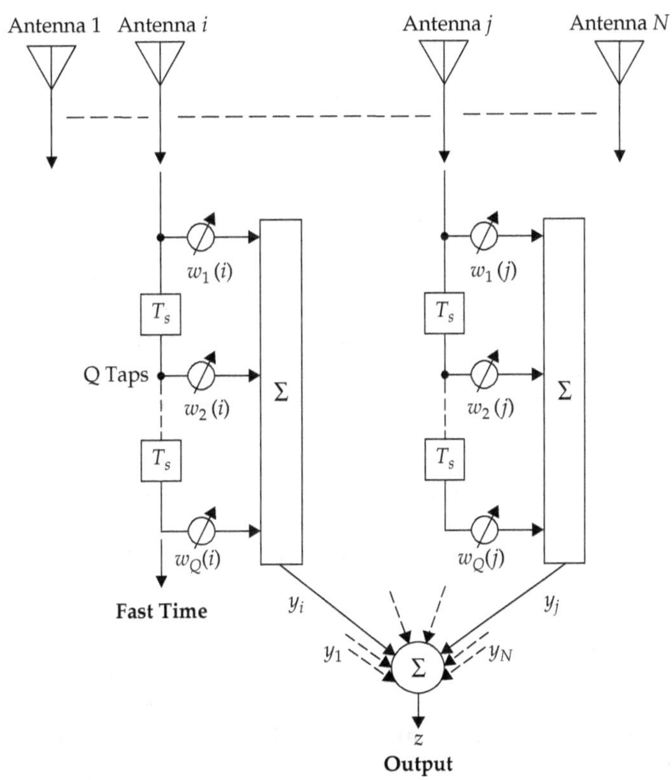

FIGURE 11.2 The fast-time STAP architecture implemented in the element-time domain combines data from N antenna sensors and Q fast-time samples.

The dimensionality of the full fast-time STAP architecture would be $N \times K$, which can be large when many range cells are processed. To train such a filter in practice, only a small subset of range taps $Q \ll K$ are used such that the filter dimension is reduced to $N \times Q$. The fast-time STAP weights are ideally adjusted to maximize the output SDR for each beam steer direction and range bin processed. In this sense, fast-time STAP may be viewed as an extension of adaptive beamforming, since the output samples correspond to beam-range-pulse data.

Unlike slow-time STAP which generates beam-Doppler outputs for each range cell processed, the fast-time STAP outputs are in the PRI domain and must be subsequently Doppler processed to coherently integrate the pulses in the CPI. For additional information pertaining to the fast-time STAP architecture, the reader is reffered to Fante and Torres (1995), Kogon, Williams, and Holder (1998), Jouny and Culpepper (1995), and Griffiths (1997) and references therein.

The disturbance type which motivates the fast-time STAP architecture in airborne radar systems is known as terrain-scattered jamming (TSJ) or "hot clutter." A jamming signal is generally not received as a simple rank-one spatial interferer incident on the sidelobes of the antenna radiation pattern. This would require the transmit antenna of the jammer to have a fictitious (unrealistic) "pencil beam" with no sidelobes such that only the "direct-path" interference is received by the radar. Real antennas have sidelobes, and consequently, significant amounts of jammer energy may be scattered from the Earth's surface into the radar, resulting in both in-plane and out-of-plane multipath interference.

Since terrain and sea surfaces are never perfectly smooth, the jamming signal is not specularly reflected but rather diffusely scattered, possibly over an extended area that spans a wide angular region. As a result, high levels of jammer energy can enter through both the mainbeam and sidelobes of the receive antenna pattern. In general, spatial-only adaptive processing (SAP) does not provide a solution for the hot-clutter problem, since this type of interference cannot be canceled effectively by simply placing "nulls" in the receive antenna pattern alone.

A consequence of diffuse multipath scattering is that the total number of interference paths summed over the number of independent sources can significantly exceed the number of spatial DOF available (i.e., the number of antenna elements N). In other words, the effective number of linearly independent interference components to be canceled may overwhelm the processor in the sense that the resulting interference spatial covariance matrix is of full rank N. Moreover, while the direct and specularly reflected jammer paths may be received from sidelobe directions, diffusely scattered multipath components can potentially enter through the main beam, particularly when scattering is from rough surfaces that act to spatially distribute the signal over a broad continuum of angles. The presence of main beam interference poses a problem for SAP even when the interference spatial covariance matrix has low rank.

Adaptive beamforming cannot be expected to effectively mitigate the composite hot-clutter signal when the condition of full rank or main-beam interference arises. Jamming signals are often assumed to emit waveforms that are uncorrelated with that of the radar and to have a bandwidth that is comparable with the receiver bandwidth. It follows that temporal DOF available in the slow-time STAP architecture cannot be used to mitigate such signals due to the long interval between pulses relative to the jammer waveform correlation time. However, the hot-clutter multipath components may be highly correlated with each other over time intervals in the order of the inverse of the system bandwidth. Thus, a STAP architecture that exploits fast-time taps can be effective for removing hot clutter.

The primary goal of fast-time STAP in airborne radar systems is therefore not to cancel the backscattered clutter signal, but rather to mitigate diffuse multipath interference received in the main beam and sidelobes of the antenna pattern. During the relatively short PRI, hot clutter from a particular source may be described as a linear combination of complex weighted and delayed replicas of the source waveform. In the fast-time STAP architecture, the finite impulse response (FIR) tap-delay-line filter behind each antenna element can in principle reverse the TSJ formation process. System identification requires the length of the FIR filter to be commensurate with the maximum impulse response duration of the propagation channel for the case of a single source. As we shall see later, the rejection of hot clutter from one or more sources may be achieved with less restrictive conditions on the number of required fast-time taps.

A key point is that the time dispersion of the hot-clutter channel needs to be acquired by each fast-time delay line at a temporal resolution that ensures the interference is not undersampled, i.e., $BT_s < 1$ (Fante and Torres 1995). Essentially, the idea behind this architecture is that interference components received from scatterers located in the direction of the main beam may be canceled using multipath versions of the same signal received from highly correlated scatterers of the same source that are located outside the main beam but which are simultaneously captured within the fast-time tap-delay lines. This fast-time STAP concept will be described in more detail later.

The use of fast-time STAP for OTH radar applications was investigated in Anderson, Abramovich, and Fabrizio (1997), Abramovich, Anderson, Gorokhov, and Spencer (1998), and Abramovich, Anderson, and Spencer (2000). In this application, the different ionospheric layers which reflect the HF interference signal are responsible for creating the "diffuse multipath" phenomenon. The various ionospheric layers may be regarded as irregular reflection surfaces that diffusely scatter the interference signal from source to receiver along multiple propagation paths. In addition, the constant electron-density contours defining these scattering surfaces do not maintain a rigid structure over the relatively long (OTH radar) CPI. For example, the received interference modes are typically Doppler-shifted due to the mean or regular component of ionospheric layer motion. Importantly, any differential Doppler shift between these modes causes the hot-clutter channel, and hence the optimum fast-time STAP filter, to become time dependent over the CPI.

If the propagation paths involve reflections from highly perturbed ionospheric regions, such as those often encountered at low and high magnetic latitudes, significant random fluctuations of the channel can also contribute to the "non-stationarity" of the interference space/fast-time covariance matrix over the CPI. This motivates the use of fast-time STAP in OTH radar with the filter weights being updated a number of times within the CPI to counter non-stationary multipath interference.

A large portion of this chapter is devoted to the description of time-varying fast-time STAP algorithms that can effectively deal with this practical problem. An analogous effect may be observed in airborne radar systems due to the relative motion between the radar platform and jamming source(s). However, the airborne radar case differs in some important respects, and only algorithms appropriate for OTH radar applications will be discussed in this chapter.

11.1.3 3D-STAP

The most general "fully adaptive" STAP operates simultaneously on all three data-cube dimensions of elements, ranges, and pulses. Figure 11.3 illustrates this processor which was motivated and analyzed for the case of airborne radar systems in Fante and Vacarro

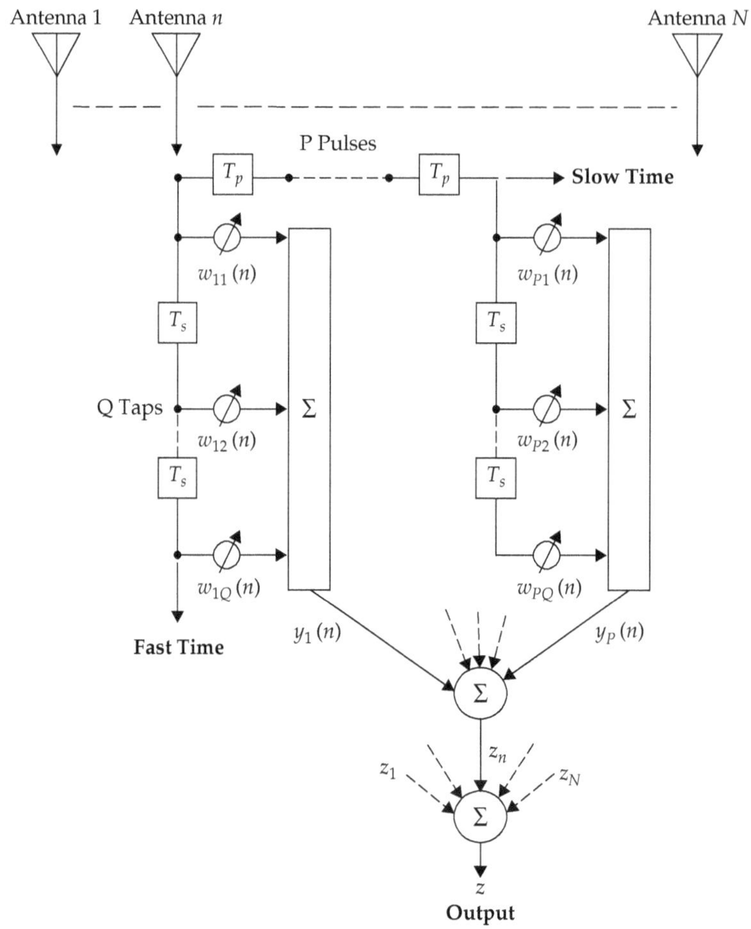

FIGURE 11.3 The general fully adaptive STAP architecture implemented in the element-time domain combines data from N antenna sensors, Q fast-time samples (range bins), and P slow-time samples (coherent pulses). The tap delays are such that $BT_s < 1$ where $T_s \ll T_p = 1/f_p$.

(1998), Rabideau (2000), and Seliktar, Williams, and Holder (2000), for example. This type of approach can in theory solve the problem of joint hot- and cold-clutter suppression, where the latter refers to ordinary backscattered radar signal clutter. To jointly mitigate hot and cold clutter, both fast-time and slow-time temporal DOF are needed in addition to spatial DOFs. Simultaneous processing of all three data-cube dimensions is also called 3D STAP in the radar nomenclature. However, it is rarely possible to effectively implement the full 3D-STAP architecture in practice due to problems associated with the large processor dimension. In the 3D-STAP processor, the number of adaptive DOFs grows to $N \times Q \times P$, where Q is the number of fast-time samples or range bins used in the tap delay-line. For OTH radar systems with typical parameters of $N = 32$ and $P = 128$, this results in a prohibitively large weight-vector dimension of $N \times Q \times P = 16384$ for $Q = 4$

fast-time taps. The major concern for STAP architectures with many adaptive degrees of freedom is the lack of statistically homogeneous training data to effectively estimate the adaptive filter coefficients. Another important issue is the high computational load associated with computing the fully adaptive STAP solution. In airborne moving target indicator (MTI) applications, the relatively low-system dimensions (e.g., $P \leq 3$ and $N \leq 16$) may allow 3D-STAP to be implemented effectively (Abramovich et al. 1998).

Rank-reduction transforms such as those proposed by Guerci, Goldstein, and Reed (2000) may be used to reduce the number of adaptive DOF. A rank-reduced 3D STAP could be proposed for OTH radar, but the underlying basis for such an approach would be poorly motivated for two reasons. First, ordinary clutter does not often exhibit significant angle-Doppler coupling in OTH radar, so the combination of space/slow-time processing is unlikely to provide significant cold-clutter mitigation benefits relative to standard Doppler processing. Second, the cold-clutter is usually received via relatively stable ionospheric propagation paths, which are often optimized by the choice of operating frequency, but the hot clutter typically originates from sources that are arbitrarily located with respect to the surveillance region, and may therefore propagate to the radar via reflections from highly perturbed ionospheric regions. This gives rise to a situation where the hot-clutter statistical properties are non-stationary over the CPI and time-dependent adaptation of the STAP filter is required for effective mitigation, while those of the cold clutter may be relatively stationary and not require the adaptive filter to be updated during the CPI.

The joint removal of hot and cold clutter is often not considered for OTH radar. One possible exception is for cases where hot-clutter-only training data cannot be obtained due to the presence of cold-clutter in all available ranges. This particular situation, which motivates 3D STAP in OTH radar, will not be considered here but has been treated in Abramovich, Anderson, and Spencer (2000). In summary, the fast-time STAP category represents the most well-motivated candidate out of all STAP architectures for practical application in OTH radar. For this reason, this chapter focuses on STAP techniques within this class.

11.2 Data Model

OTH radar systems are required to operate in signal environments where the composite disturbance is generally the sum of surface-scattered clutter, diffuse multipath interference (from one or more sources), and additive noise, all of which compete for detection against relatively faint target echoes. This section describes models for the various signal components received by an OTH radar in all three data-cube dimensions. Relatively simple models are described for the target and additive noise, while more detailed attention is paid to modeling the surface clutter and diffuse multipath interference, which are the dominant components to be mitigated in the received signal.

An important aspect of the formulated data models is that they allow realizations of OTH radar data to be readily generated for evaluating signal processing performance in computer simulations. The simulation results showing the performance of various fast-time STAP algorithms in Section 11.3.3 are based on synthetic data generated using the data model presented in this section.

11.2.1 Composite Signal

Let $\mathbf{x}_k(t) \in \mathcal{C}^N$ be the N-variate spatial snapshot vector received by an array of N antenna sensors at fast-time sample $k = 1, \ldots, K$ and slow-time sample $t = 1, \ldots, P$ within the CPI. In general, $\mathbf{x}_k(t)$ may be written as in Eqn. (11.1), where $\mathbf{c}_k(t)$ is ordinary radar clutter backscattered from the Earth's surface (cold clutter), $\mathbf{j}_k(t)$ is the superposition of diffusely scattered multipath interference components from all jamming sources (hot clutter), and $\mathbf{n}_k(t)$ is the sum of internal and external additive noise from other sources. The potential presence of a point-target echo is represented by the term $\mathbf{s}_k(t)$, which is the *useful* signal.

$$\mathbf{x}_k(t) = \mathbf{s}_k(t) + \mathbf{c}_k(t) + \mathbf{j}_k(t) + \mathbf{n}_k(t) \tag{11.1}$$

For a uniform linear array (ULA) steered to a cone angle-of-arrival φ_0, a useful signal incident from the radar look direction may be expressed in the form of Eqn. (11.2). Here, a is a complex scalar amplitude, ψ_k is the signal waveform, f_d is the target Doppler-shift normalized by the PRF, $\mathbf{s}(\varphi_0)$ is the steering vector on the ULA manifold, and γ_k is a range-dependent phase. A more complex model incorporating spatial spreading and temporal fading could be proposed, while other practical issues such as range straddling and range sidelobes could also be accounted for. Similarly, it would be possible to model mismatches in useful signal DOA, or extend the model to 2D arrays steered independently in azimuth θ_0 and elevation ϕ_0. However, such generalizations detract from the main intent of describing the key points, so the simplest target model in Eqn. (11.2) may be adopted for this purpose. For example, a signal matched to range bin k_0 has a fast-time signature $\psi_k = \delta(k - k_0)$ for a pulsed-waveform (PW) system, whereas for a continuous-wave (CW) system, $\psi_k = u_p(k - k_0)$, where $u_p(k)$ is the transmitted signal pulse. It is convenient to consider the former case as it enables us to directly interpret fast-time samples as range bins. However, the main concepts illustrated in the following discussion are equally applicable to CW systems, as explained later.

$$\mathbf{s}_k(t) = a\,\psi_k \mathbf{s}(\varphi_0) \exp\{j2\pi f_d t + \gamma_k\} \tag{11.2}$$

The additive noise $\mathbf{n}_k(t)$ is in general a mixture of internal receiver noise (i.e., thermal noise) and naturally occuring external noise (i.e., ambient noise). For the purpose of the following analysis, we avoid delving into the fine detail of physical noise models and simply assume that this process is complex circular Gaussian distributed and white across all radar data-cube dimensions. In other words, the additive noise has spatial and temporal correlation properties given by Eqn. (11.3), where σ_n^2 is the noise power per antenna element, and $\mathbf{I}_N$ is the N-dimensional identity matrix. In Eqn. (11.3), E{·} denotes statistical expectation, $\dagger$ is the Hermitian (conjugate transpose) operator, and $\delta_{kk'}$ is shorthand notation for the delta function $\delta(k - k')$. As the cold- and hot-clutter disturbances are much more powerful than the additive noise, the structure or "color" of the additive noise is largely inconsequential because $\mathbf{n}_k(t)$ is not the signal contribution that limits performance when hot clutter is present.

$$\mathrm{E}\{\mathbf{n}_k(t)\mathbf{n}_{k'}^{\dagger}(t')\} = \delta_{kk'}\delta_{tt'}\mathbf{I}_N \tag{11.3}$$

The fast-time STAP architecture considered here jointly operates on data acquired by the N antenna sensors of the array (i.e., element-space) and Q successive fast-time samples or range bins. Recall that Q is the number of fast-time taps in the delay line for each antenna sensor. The collection of these complex samples are conveniently assembled into the NQ-variate "stacked" data vector $\tilde{\mathbf{x}}_k(t)$ defined in Eqn. (11.4). The stacked vectors for the useful signal $\tilde{\mathbf{s}}_k(t)$, cold clutter $\tilde{\mathbf{c}}_k(t)$, hot clutter $\tilde{\mathbf{j}}_k(t)$, and additive noise $\tilde{\mathbf{n}}_k(t)$ are constructed in analogous manner.

$$\tilde{\mathbf{x}}_k(t) = \begin{bmatrix} \mathbf{x}_k(t) \\ \mathbf{x}_{k-1}(t) \\ \vdots \\ \mathbf{x}_{k-Q+1}(t) \end{bmatrix} = \tilde{\mathbf{s}}_k(t) + \tilde{\mathbf{c}}_k(t) + \tilde{\mathbf{j}}_k(t) + \tilde{\mathbf{n}}_k(t) \tag{11.4}$$

The scalar output $z_k(t)$ processed by the NQ-variate STAP filter $\tilde{\mathbf{w}}_k(t)$ is then given by Eqn. (11.5), where $s_k(t) = \tilde{\mathbf{w}}_k^\dagger(t)\tilde{\mathbf{s}}_k(t)$ is the useful signal component. The other components are defined in similar manner. The fast-time STAP weight vector aims to protect the useful signal while attenuating the hot-clutter-plus-noise as much as possible. This filter does *not* attempt to cancel the cold clutter. In OTH radar applications, the scalar sequence $z_k(t)$ corresponds to a "finger beam" output, which subsequently requires coherent processing over a sequence of PRI to isolate moving target echoes from the residual cold clutter signal at the fast-time STAP filter output (i.e., Doppler spectrum analysis).

$$z_k(t) = \tilde{\mathbf{w}}_k^\dagger(t)\tilde{\mathbf{x}}_k(t) = s_k(t) + c_k(t) + j_k(t) + n_k(t) \tag{11.5}$$

11.2.2 Cold Clutter

In the work of Abramovich et al. (1998), the backscattered cold-clutter snapshots $\mathbf{c}_k(t)$ received in a particular CPI are modeled as realizations of a multi-variate stationary Gaussian random process with second-order statistics given by Eqn. (11.6). Here, $\mathbf{R}_c(\tau)$ is the $N \times N$ cold clutter spatial covariance matrix at slow-time lag $\tau = t' - t$. In this representation, the clutter snapshots $\mathbf{c}_k(t)$ received in different range bins k are assumed to be statistically independent (ignoring range sidelobes).

$$\mathrm{E}\{\mathbf{c}_k(t)\mathbf{c}_{k'}^\dagger(t')\} = \delta_{kk'}\mathbf{R}_c(\tau) \tag{11.6}$$

In contrast to airborne radar, where the rapidly moving antenna platform creates a clutter power spectrum with angle-Doppler coupling, the Doppler spectrum characteristics of clutter received by an OTH radar tends to exhibit a weaker dependence on beam-steer direction within the transmitter footprint. Provided that this angle-Doppler dependence may be considered negligible, the slow-time lagged clutter spatial covariance matrix $\mathbf{R}_c(\tau)$ may be represented in the special form of Eqn. (11.7). In words, $\mathbf{R}_c(\tau)$ becomes separable and may be factored into the clutter spatial covariance matrix $\mathbf{R}_c$ and the scalar function $r(\tau)$, which represents the clutter slow-time auto-correlation coefficients. Note that $r(0) = 1$ by definition.

$$\mathbf{R}_c(\tau) = \mathbf{R}_c r(\tau) \tag{11.7}$$

From Eqn. (11.7), it follows that the clutter cross-spectral matrix $\mathbf{S}_c(f)$ can be expressed as in Eqn. (11.8), where $\mathbf{R}_c$ defines the spatial distribution of the cold clutter, and $r(\tau)$ determines its Doppler power spectrum structure according to the scalar function $s_c(f)$. The clutter Doppler power spectrum at the conventional beamformer output, is given by $p(\varphi_0, f) = \mathbf{s}^\dagger(\varphi_0)\mathbf{S}_c(f)\mathbf{s}(\varphi_0) = [\mathbf{s}^\dagger(\varphi_0)\mathbf{R}_c\mathbf{s}(\varphi_0)]s_c(f)$. The scale of this spectrum may change with radar look direction φ_0 since $\mathbf{R}_c$ is typically not equal to the identity matrix. However, the *structure* of this spectrum has the same form $s_c(f)$ independent of the steer angle φ_0. From a physical viewpoint, this implies that there is no angle-Doppler coupling in the cold-clutter spectrum to within a complex scalar. While such a model is often quite appropriate for OTH radar in an approximate sense,[2] it is clearly not valid for the airborne microwave radar case. Despite the analogies drawn between these two radar systems with respect to the hot-clutter problem in Abramovich et al. (1998), it may be expected that any fast-time STAP approach based strongly on this type of clutter model will probably not be directly appropriate for airborne radar.

$$\mathbf{S}_c(f) = \mathbf{R}_c \sum_{\tau=-\infty}^{\infty} r(\tau)e^{-j2\pi f\tau} = \mathbf{R}_c s_c(f) \tag{11.8}$$

Once it has been accepted that the clutter correlation properties can be written as Eqn. (11.7), or equivalently in Eqn. (11.8), statistical realizations of the Gaussian distributed clutter process $\mathbf{c}_k(t)$ may be generated using the *scalar* multi-variate auto-regressive (AR) model defined in Eqn. (11.9). The order of this model κ depends on the characteristics of the clutter slow-time auto-correlation coefficient function $r(\tau)$. Based on experimental observations of skywave OTH radar clutter, empirical analysis suggests that the snapshots $\mathbf{c}_k(t)$ may be statistically modeled quite accurately using a relatively low-order AR model, where typically $\kappa \ll N$, as described in (Abramovich et al. 1998).

$$\mathbf{c}_k(t) + \sum_{i=1}^{\kappa} b_i \mathbf{c}_k(t-i) = \boldsymbol{\varepsilon}_k(t) \tag{11.9}$$

In this clutter model, the complex scalar coefficients $\{b_i\}_{i=1}^{\kappa}$ are the temporal AR model parameters (for range cell k), while $\boldsymbol{\varepsilon}_k(t) \in \mathcal{C}^N$ is a temporally white innovative noise vector with correlation properties given by Eqn. (11.10). The terms σ_ε^2 and $\mathbf{R}_c$ will now be discussed. Due to the relatively broad transmit beam used in OTH radar to floodlight the surveillance region, clutter received at a certain group range is returned by a spatially extended area of the Earth's surface. This area is defined by the set of all scatterers on the locus of constant path delay within the range resolution cell limits after propagation through the ionosphere. As a result, the backscattered clutter received in any given range cell will be spatially distributed over a relatively wide angular region, such that $\mathbf{R}_c$ effectively has full rank. For a spatially stationary clutter process, $\mathbf{R}_c$ is a Toeplitz matrix with diagonal elements equal to the clutter power σ_c^2 received by each antenna element.

$$\mathbf{E}\{\boldsymbol{\varepsilon}_k(t)\boldsymbol{\varepsilon}_{k'}^{\dagger}(t')\} = \delta_{kk'}\delta_{tt'}\sigma_\varepsilon^2\mathbf{R}_c \tag{11.10}$$

Recall that the scalar AR parameters $\{b_i\}_{i=1}^{\kappa}$ satisfy the $(\kappa+1)$-variate Yule-Walker equations in Eqn. (11.11), where the conjugate-symmetry property $r(-\tau) = r^*(\tau)$ of a wide-sense stationary process is used to define the matrix on the left-hand side. Since

[2]In slow-moving target detection applications, angle-Doppler coupling effects can become more significant in OTH radar as useful signals are often located close to the main clutter "ridge."

the zero-lag correlation coefficient $r(0) = 1$, by definition, σ_ε^2 in Eqn. (11.11) is defined as the innovative noise power corresponding to an AR process output of *unit* variance.

$$\begin{bmatrix} r(0) & r(1) & \cdots & r(\kappa) \\ r^*(1) & r(0) & & \\ \vdots & & \ddots & \vdots \\ r^*(\kappa) & & \cdots & r(0) \end{bmatrix} \begin{bmatrix} 1 \\ b_1 \\ \vdots \\ b_\kappa \end{bmatrix} = \begin{bmatrix} \sigma_\varepsilon^2 \\ 0 \\ \vdots \\ 0 \end{bmatrix} \tag{11.11}$$

The $(\kappa+1)$-variate Toeplitz matrix constructed from the clutter temporal auto-correlation coefficients in Eqn. (11.11) is denoted by $\mathbf{R}_\tau = \text{Toep}[r(0), r(1), \ldots, r(\kappa)]$. The associated AR parameter vector $\mathbf{b} = [1, b_1, \ldots, b_\kappa]^T$ and the innovative noise power scaling term σ_ε^2 are given by the solution of the Yule-Walker equations in Eqn. (11.12), where $\mathbf{u}_1 = [1, 0, \ldots, 0]^T$ is the $(\kappa+1)$-dimensional unit vector (with the first element equal to unity), and T denotes transpose.

$$\mathbf{b} = \sigma_\varepsilon^2 \mathbf{R}_\tau^{-1} \mathbf{u}_1\,, \quad \sigma_\varepsilon^2 = \left\{ \mathbf{u}_1^T \mathbf{R}_\tau^{-1} \mathbf{u}_1 \right\}^{-1} \tag{11.12}$$

The structure of the AR Doppler power spectrum is given by the parametric model $s_c(f) = \{|B(e^{j2\pi f})|^2\}^{-1}$, where $B(z) = 1 + \sum_{i=1}^{\kappa} b_i z^{-i}$ is the characteristic polynomial in $z = e^{j2\pi f}$ of order κ with no roots outside the unit circle. This polynomial may also be expressed as $B(z) = \prod_{i=1}^{\kappa}(1 - p_i z^{-1})$, where the poles p_i for $i = 1, \ldots, \kappa$ have magnitudes less than or equal to unity. Inserting this parametric description into Eqn. (11.8) yields the clutter cross-spectral matrix model of Eqn. (11.13), where $f \in [-1/2, 1/2)$ is the Doppler frequency normalized by the PRF.

$$\mathbf{S}_c(f) = \frac{\mathbf{R}_c}{\left|1 + \sum_{i=1}^{\kappa} b_i e^{-j2\pi i f}\right|^2} = \frac{\mathbf{R}_c}{\prod_{i=1}^{\kappa} |1 - p_i e^{-j2\pi i f}|^2} \tag{11.13}$$

In Abramovich et al. (1998), it is assumed that the stacked clutter snapshots $\tilde{\mathbf{c}}_k(t)$ may also be described by a scalar-type AR process. Provided that the AR clutter parameters in Eqn. (11.9) are locally homogeneous over a limited number of Q fast-time taps, the stacked clutter snapshots $\tilde{\mathbf{c}}_k(t)$ will also obey the recursive relation of Eqn. (11.14). In summary, such a model is appropriate when the clutter Doppler spectrum structure may be considered invariant over Q adjacent ranges and spatially homogeneous in cone angle to within a complex scale factor over the radar footprint.

$$\tilde{\mathbf{c}}_k(t) + \sum_{i=1}^{\kappa} b_i \tilde{\mathbf{c}}_k(t-i) = \tilde{\varepsilon}_k(t) \tag{11.14}$$

It is convenient to define the N-dimensional innovative noise vector $\eta_k(t)$ such that $\varepsilon_k(t) = \sigma_\varepsilon \eta_k(t)$. The NQ-variate stacked innovative noise vector may then be written as $\tilde{\varepsilon}_k(t) = \sigma_\varepsilon \tilde{\eta}_k(t)$, where $\tilde{\eta}_k(t)$ is a stacked vector of Q independent innovative noise vectors $\{\eta_k(t), \eta_{k-1}(t), \ldots, \eta_{k-Q+1}(t)\}$ with identical covariance matrix $\mathbf{R}_c = \mathrm{E}\{\eta_k(t)\eta_k^\dagger(t)\}$. The second-order statistics of $\tilde{\varepsilon}_k(t)$ can be expressed in the form of Eqn. (11.15), where the $NQ \times NQ$ block diagonal matrix $\tilde{\mathbf{R}}_c = \mathrm{E}\{\tilde{\eta}_k(t)\tilde{\eta}_k^\dagger(t)\}$ is given by $\tilde{\mathbf{R}}_c = \text{diag}[\mathbf{R}_c, \ldots, \mathbf{R}_c]$.

$$\mathrm{E}\{\tilde{\varepsilon}_k(t)\tilde{\varepsilon}_{k'}^\dagger(t')\} = \delta_{tt'}\delta_{kk'}\sigma_\varepsilon^2 \tilde{\mathbf{R}}_c \tag{11.15}$$

The simplest first-order ($\kappa = 1$) AR model in Eqn. (11.16) may be used to represent terrain-scattered clutter. Using Eqns. (11.11) through to (11.14), it is readily verified that the inter-pulse clutter correlation coefficient $\rho_t = r(1) = -b_1 = p_1$. For a stable AR process, the parameter ρ_t lies inside the unit circle. The modulus $|\rho_t| < 1$ determines the width of the clutter spectrum (Doppler frequency spread) due to ionospheric propagation. For $\kappa = 1$, this spectrum is parameterized by a Lorentzian profile. The argument $\measuredangle \rho_t$ determines the centroid of this spectrum in frequency to reflect the mean ionospheric Doppler shift. For high PRF applications (aircraft detection missions), $|\rho_t| \to 1$ in stable ionospheric conditions, with values of about 0.999 being typical for a PRF of 50-Hz. From Eqn. (11.12), we have that $\sigma_\varepsilon = \sqrt{1-|\rho_t|^2}$ for a first-order AR process.

$$\tilde{\mathbf{c}}_k(t) = \rho_t \tilde{\mathbf{c}}_k(t-i) + \sqrt{1-|\rho_t|^2}\,\tilde{\boldsymbol{\eta}}_k(t) \tag{11.16}$$

A simple model for the spatial distribution of the backscattered cold clutter assumes the covariance matrix $\mathbf{R}_c = \sigma_c^2 \text{Toep}[1, \rho_s, \ldots, \rho_s^{N-1}]$, where the complex scalar ρ_s is the inter-sensor spatial correlation coefficient of the clutter. This parameter determines the angular width of the (Lorentzian shaped) spatial spectrum and its mean DOA relative to broadside. For example, a value of $\rho_s = 0.5$ was assumed for an OTH radar footprint steered at broadside in Abramovich, Gorokhov, Mikhaylyukov, and Malyavin (1994), and Abramovich (1992). The Q independent innovative noise vectors $\{\boldsymbol{\eta}_k(t), \boldsymbol{\eta}_{k-1}(t), \ldots, \boldsymbol{\eta}_{k-Q+1}(t)\}$ used to construct $\tilde{\boldsymbol{\eta}}_k(t)$ in Eqn. (11.16) may be generated by the element-space AR(1) process in Eqn. (11.17), where the superscript $[n]$ for $n = 1, \ldots, N$ denotes the elements of $\boldsymbol{\eta}_k(t)$. Here, $\gamma_n(t,k)$ is complex driving white Gaussian noise with correlation properties given by $\mathrm{E}\left\{\gamma_n(t,k)\gamma_{n'}(t',k')^*\right\} = \sigma_c^2 \delta_{nn'}\delta_{tt'}\delta_{kk'}$.

$$\eta_k^{[n]}(t) = \rho_s \eta_k^{[n-1]}(t) + \sqrt{1-|\rho_s|^2}\,\gamma_n(t,k) \tag{11.17}$$

For the case of sea-surface scattering, a second-order ($\kappa = 2$) AR model may be proposed to represent the two dominant Bragg lines in the clutter Doppler spectrum. The values of $\kappa = 1$ and $\kappa = 2$ (corresponding to the simplest terrain and sea-clutter AR models, respectively) are therefore minimum model order requirements. In practice, higher-order models are often required to capture the received clutter Doppler spectra more accurately. A basic parameter set for modeling sea-surface clutter at low (ship-detection) PRFs of about 5-Hz has been specified as $b_1 = -1.9359$, $b_2 = 0.998$, $\sigma_\varepsilon^2 = 0.009675$ in Abramovich (1992) and Abramovich (1994).

While the value of κ may be selected a priori based on the expected characteristics of the clutter, the model parameters $\{b_i\}_{i=1}^{\kappa}$ will be unknown in general. Moreover, cold-clutter signals may be partially or fully submerged by the hot clutter in an operational system. In this case, access to cold clutter-only snapshots is not directly available in practice for identifying (estimating) the AR model parameters.

11.2.3 Hot Clutter

Hot clutter is assumed to arise from a convolutive mixture of M external interference sources emitting independent complex scalar waveforms denoted by $g_{mk}(t)$ for $m = 1, \ldots, M$. The received hot-clutter spatial snapshot $\mathbf{j}_k(t)$ may be written as the multi-channel discrete convolution in Eqn. (11.18), where L is the maximum duration of the hot-clutter channel impulse response in fast-time samples for the source with the largest multipath time dispersion. In other words, the multipath components received from the M hot-clutter sources are contained within a differential group-range interval of

$\Delta R = cL/f_s$. Although L may loosely be referred to as the maximum number of paths or ionospheric modes over all M sources, it is more accurate to interpret L as the maximum fast-time sample interval between loci of constant path-delay in the case of continuously distributed scatterers. The complex multi-channel FIR function that links source m to the N antenna elements is denoted by the N-variate vector $\mathbf{h}_{m\ell}(t)$ for $\ell = 1, \ldots, L$. The channel impulse response coefficients in $\mathbf{h}_{m\ell}(t)$ may be considered essentially frozen in fast-time k (i.e., over the relatively short PRI), but they may fluctuate with respect to slow-time t over the relatively long CPI. The rate of channel "non-stationarity" is related the highest differential Doppler shift between the scatterers on each loci of constant path delay (Fante and Torres 1995).

$$\mathbf{j}_k(t) = \sum_{m=1}^{M}\sum_{\ell=1}^{L} \mathbf{h}_{m\ell}(t) g_{mk-\ell+1}(t) \tag{11.18}$$

The hot-clutter array snapshot vector $\mathbf{j}_k(t)$ is more conveniently expressed in the form of Eqn. (11.19). Here, the M-dimensional signal vector $\mathbf{g}_k(t) = [g_{1k}(t), \ldots, g_{Mk}(t)]^T$ contains the complex source waveforms received at fast-time k and slow-time t, while the $N \times M$ matrix $\mathbf{H}_\ell(t) = [\mathbf{h}_{1\ell}(t), \ldots, \mathbf{h}_{M\ell}(t)]$ represents the instantaneous total impulse response of the hot-clutter channel at fast-time delay ℓ. This matrix remains constant during the "quasi-instantaneous" PRI but changes as the channel evolves in slow-time t over the CPI. More specifically, the $(n, m)^{\text{th}}$ element of $\mathbf{H}_\ell(t)$ contains the complex scalar channel coefficient that transfers source m to receiver n at relative delay ℓ in repetition period t. In the hypothetical case of no multipath $L = 1$, Eqn. (11.19) reverts back to the familiar instantaneous mixture model $\mathbf{j}_k(t) = \mathbf{H}(t)\mathbf{g}_k(t)$, where the columns of the mixing matrix $\mathbf{H}(t) = [\mathbf{h}_1(t), \ldots, \mathbf{h}_M(t)]$ may be regarded as the M interference wavefronts received at slow-time t.

$$\mathbf{j}_k(t) = \sum_{\ell=1}^{L} \mathbf{H}_\ell(t)\mathbf{g}_{k-\ell+1}(t) \tag{11.19}$$

The source waveforms $g_{mk}(t)$ are assumed to be mutually independent with correlation properties in Eqn. (11.20), where $r_m(k)$ is the fast-time auto-correlation function of the mth hot-clutter source and $*$ denotes complex conjugate. As the power of each hot-clutter signal (mode) will be accounted for in the channel impulse response definition later on, the source waveforms may be scaled to unit variance ($r_m(0) = 1$) without loss of generality. Unless otherwise stated, we shall assume broadband sources with an essentially flat power spectral density over the jammer bandwidth $B_m > f_s = 1/T_s$, such that $r_m(k) = \delta(k)$. In the final section of this chapter, interference sources with narrow bandwidths in the interval $1/T_p = f_p \gg B_m < f_s$ will be considered, such that $|r_m(k)| \to 1$ for $kT_s \ll T_p$.

$$\mathrm{E}\{g_{mk}(t)g^*_{m'k'}(t')\} = \delta_{mm'}\delta_{tt'}r_m(k-k') \tag{11.20}$$

Now consider the NQ-variate stacked hot-clutter vector $\tilde{\mathbf{j}}_k(t)$, which may be expressed in the compact form of Eqn. (11.21). Here, $\tilde{\mathbf{g}}_k(t)$ is the $M(L+Q-1)$-variate stacked vector of $\{\mathbf{g}_k(t), \mathbf{g}_{k-1}(t), \ldots \mathbf{g}_{k-L+1-Q+1}(t)\}$, while $\tilde{\mathbf{H}}(t)$ is an $NQ \times M(L+Q-1)$ block-Sylvester matrix constructed from the matricies $\{\mathbf{H}_1(t), \ldots, \mathbf{H}_L(t)\}$. It may be readily verified that the expression for $\tilde{\mathbf{j}}_k(t)$ in Eqn. (11.21) is consistent with the definition of the individual spatial snapshots $\{\mathbf{j}_k(t), \mathbf{j}_{k-1}(t), \ldots, \mathbf{j}_{k-Q+1}(t)\}$ in Eqn. (11.19), which form the stacked hot-clutter vector.

$$\tilde{\mathbf{j}}_k(t) = \tilde{\mathbf{H}}(t)\tilde{\mathbf{g}}_k(t) = \begin{bmatrix} \mathbf{H}_1(t) & \cdots & \mathbf{H}_L(t) & \mathbf{0} & \cdots & \mathbf{0} \\ \mathbf{0} & \mathbf{H}_1(t) & \cdots & \mathbf{H}_L(t) & \ddots & \vdots \\ \vdots & \ddots & \ddots & & \ddots & \mathbf{0} \\ \mathbf{0} & \cdots & \mathbf{0} & \mathbf{H}_1(t) & \cdots & \mathbf{H}_L(t) \end{bmatrix} \times \begin{bmatrix} \mathbf{g}_k(t) \\ \mathbf{g}_{k-1}(t) \\ \vdots \\ \mathbf{g}_{k-L+1-Q+1}(t) \end{bmatrix} \tag{11.21}$$

Using Eqn. (11.21), the $NQ \times NQ$ hot-clutter covariance matrix $\tilde{\mathbf{R}}_h(t) = \mathrm{E}\{\tilde{\mathbf{j}}_k(t)\tilde{\mathbf{j}}_k^\dagger(t)\}$ is given by Eqn. (11.22), where $\tilde{\mathbf{G}} = \mathrm{E}\{\tilde{\mathbf{g}}_k(t)\tilde{\mathbf{g}}_k^\dagger(t)\}$ is the associated source covariance matrix. The matrix $\tilde{\mathbf{R}}_h(t)$ is slow-time varying due to the dynamic channel impulse responses over the CPI. However, it may be regarded essentially constant in fast-time over a relatively short PRI, which effectively observes a "quasi-instantaneous" snapshot of the channel fluctuations.

For independent broadband jamming signals, the source covariance matrix has full rank given by $\mathcal{R}\{\tilde{\mathbf{G}}\} = M(L+Q-1)$, where the operator $\mathcal{R}\{\cdot\}$ returns the rank of a matrix. As the dimensions of the system matrix $\tilde{\mathbf{H}}(t)$ are $NQ \times M(L+Q-1)$, it follows that the (noise-free) $NQ \times NQ$ hot-clutter covariance matrix $\tilde{\mathbf{R}}_h(t)$ is guaranteed to be rank deficient when $NQ > M(L+Q-1)$, where we recall that L is defined as the maximum impulse response duration over all M sources.

$$\tilde{\mathbf{R}}_h(t) = \tilde{\mathbf{H}}(t)\mathrm{E}\{\tilde{\mathbf{g}}_k(t)\tilde{\mathbf{g}}_k^\dagger(t)\}\tilde{\mathbf{H}}^\dagger(t) = \tilde{\mathbf{H}}(t)\tilde{\mathbf{G}}\tilde{\mathbf{H}}^\dagger(t) \tag{11.22}$$

The rank deficiency of $\tilde{\mathbf{R}}_h(t)$ for dimensional parameters satisfying the condition in Eqn. (11.23) has important implications for hot-clutter rejection. Specifically, it means that a fast-time STAP architecture of dimension NQ has sufficient DOFs to cancel the hot clutter effectively when the filter is updated on a pulse-by-pulse basis. In conditions of no multipath $L = 1$, Eqn. (11.23) holds only if the number of antenna elements is greater than the number of independent sources $N > M$, which confirms that fast-time STAP has no scope to outperform SAP in the absence of multipath. Indeed, the condition in Eqn. (11.23) holds only if $N > M$, irrespective of the values of L and Q, i.e., the maximum number of independent sources that can be effectively canceled by both SAP and fast-STAP must be less than the number of antenna elements. In the case of pure SAP ($Q = 1$), Eqn. (11.23) suggests that hot clutter can be effectively canceled when $N > ML$. As not all sources will have the maximum impulse response length L in practice, the milder condition $N > \sum_{m=1}^{M} L_m$ applies, where L_m is the duration of source m. There is another subtle point; while Eqn. (11.23) indicates the potential for effective hot clutter cancelation using fast-time STAP filters updated in slow-time, a generalized "main beam" scenario may unfortunately result if the stacked useful signal vector $\tilde{\mathbf{s}}_k(t)$ is accurately spanned by the hot-clutter subspace, i.e., as a linear combination of the columns of $\tilde{\mathbf{H}}(t)$. In this case, hot-clutter rejection is still possible, but the signal-to-hot-clutter ratio will be degraded.

$$NQ > M(L+Q-1) \tag{11.23}$$

The condition in Eqn. (11.23) may be interpreted as a fast-time STAP generalization of the spatial-only condition $N > M$ necessary for the effective rejection of independent interference sources by SAP. This expression may be recast in the form of Eqn. (11.24), which allows the designer to determine the minimum number of fast-time taps necessary to ensure that the (quasi-instantaneous) stacked hot clutter-only covariance matrix $\tilde{\mathbf{R}}_h(t)$ is rank deficient for a given number of sources M and maximum impulse response duration L. The number of taps $Q = L$, typically adopted as a rule-of-thumb, can only guarantee rank deficiency of $\tilde{\mathbf{R}}_h(t)$ for the case $M < N/2$, i.e., when the number of independent sources is less than half the number of antenna elements. For $M \ll N/2$, values of $Q < L$ are sufficient, which is a less restrictive condition than the rule-of-thumb. Whereas for the maximum number of independent sources that can possibly be canceled $M_{max} = N - 1$, the number of fast-time taps required to ensure rank deficiency of $\tilde{\mathbf{R}}_h(t)$ is $Q_{max} = M_{max}(L - 1)$ which is typically greater than L in practical scenarios. Clearly, SAP will be ineffective for cases where $Q > 1$ taps are required for rank deficiency in Eqn. (11.24).

$$Q > \frac{M(L-1)}{N-M} \tag{11.24}$$

Having described the underlying conditions for which fast-time STAP has the potential to effectively cancel hot clutter and outperform SAP, we may now consider specific models that may be used to simulate the hot-clutter signal. Based on the work of Abramovich, Spencer, and Anderson (1998), and real-data processing results presented for the ionospheric HF channel in the second part of this text, a modified version of the generalized Watterson model (GWM) may be used to simulate the channel vectors $\mathbf{h}_{m\ell}(t)$ that give rise to the "non-stationary" hot-clutter phenomenon. In array processing terminology, $\mathbf{h}_{m\ell}(t)$ in the model of Eqn. (11.25) may be regarded as the hot-clutter "wavefront" received in PRI t from source m for the mode with relative delay ℓ. Note that the sum is over the maximum number of paths $\ell = 1, \ldots, L$ for all sources, but it is clear that $\mathbf{h}_{m\ell}(t) = 0$ for $\ell > L_m$, i.e., when the fast-time delay exceeds the impulse response duration of source m.

$$\mathbf{j}_k(t) = \sum_{m=1}^{M} \sum_{\ell=1}^{L} \mathbf{h}_{m\ell}(t) g_{mk-\ell+1}(t) \tag{11.25}$$

The slow-time varying channel vectors $\mathbf{h}_{m\ell}(t)$ are assumed to be random and statistically independent for different sources and modes. In the GWM model of Eqn. (11.26), the terms $A_{m\ell}$ and $f_{m\ell}$ denote the RMS amplitude and Doppler shift of mode ℓ from source m, respectively, while the $N \times N$ matrix $\mathbf{S}_{m\ell}$ represents the mean synthetic wavefront of this mode over the CPI (as described below). The multi-variate complex Gaussian distributed N-dimensional vector $\mathbf{c}_{m\ell}(t)$ encapsulates the random space-time fluctuations of the received hot-clutter wavefronts. This accounts for the DOA and Doppler spread imposed on the various sources and modes. In Chapter 8, the simplest model for $\mathbf{c}_{m\ell}(t)$ was described as a Markov chain defined by two parameters, namely, a temporal correlation coefficient $\alpha_{m\ell}$, and a spatial correlation coefficient $\beta_{m\ell}$, both real-valued quantities in the interval between zero and one. Lower values of $\alpha_{m\ell}$ and $\beta_{m\ell}$ correspond to diffusely scattered modes with faster temporal fluctuations and greater wavefront variability. In other words, the parameters $\alpha_{m\ell}$ and $\beta_{m\ell}$ represent the prevailing characteristics of the different ionospheric paths responsible for producing the hot-clutter signal. Specific values of $\alpha_{m\ell}$ and $\beta_{m\ell}$ will be quoted in Section 11.3.3 for individual hot-clutter modes.

$$\mathbf{h}_{m\ell}(t) = A_{m\ell} \mathbf{S}_{m\ell} \mathbf{c}_{m\ell}(t) \exp\{j2\pi f_{m\ell} t\} \tag{11.26}$$

The only modification to the GWM described in Chapter 8 relates to the definition of $\mathbf{S}_{m\ell}$. When diffuse scattering occurs from a rough ionospheric surface, the resulting hot-clutter "mode" is not reflected from a single point hence the mean wavefront will not be planar. In this case, signal components arriving with similar path delay ℓ may be due to a continuum of point scatterers that are spatially distributed over an extended region. These "micro-multipath" components received from scatterers along a locus of constant path delay superimpose to produce a synthetic wavefront, which may deviate significantly from a plane wave. The assumption of a mean plane-wavefront model is therefore not suitable for these hot-clutter modes.

In Abramovich et al. (1998), it is proposed to define $\mathbf{S}_{m\ell}$ based on the Karhunen-Loève expansion of the hot-clutter mode spatial covariance matrix $\mathbf{F}_{m\ell}$ averaged over an infinite time interval in Eqn. (11.27). While the spatial rank of a single hot-clutter mode is assumed to be unity in any given repetition period, it is noted that the time-averaged spatial covariance matrix of a single hot-clutter mode tends to full rank when it is integrated over a relatively long CPI because of the mode wavefront structure variations embodied in the slow-time varying channel vector $\mathbf{h}_{m\ell}(t)$.

$$\mathbf{F}_{m\ell} = \mathbf{S}_{m\ell}\mathbf{S}_{m\ell}^{\dagger} = \lim_{P\to\infty} \sum_{t=1}^{P} \mathbf{h}_{m\ell}(t)\mathbf{h}_{m\ell}^{\dagger}(t) \tag{11.27}$$

The $NQ \times NQ$ space/fast-time covariance matrix $\tilde{\mathbf{R}}(t) = \mathrm{E}\{\tilde{\mathbf{i}}_k(t)\tilde{\mathbf{i}}_k^{\dagger}(t)\}$, defined in terms of the stacked hot-clutter-plus-noise vector $\tilde{\mathbf{i}}_k(t) = \tilde{\mathbf{j}}_k(t) + \tilde{\mathbf{n}}_k(t)$, has the block-Toeplitz structure in Eqn. (11.28), where the $N \times N$ blocks are given by the fast-time lagged hot-clutter-plus-noise spatial covariance matrices $\mathbf{R}_q(t) = \mathrm{E}\{\mathbf{i}_k(t)\mathbf{i}_{k-q}^{\dagger}(t)\}$ for sample lags $q = 0, \ldots, Q-1$ in the fast-time tap delay line.

$$\tilde{\mathbf{R}}(t) = \mathrm{Toep}[\mathbf{R}_0(t), \mathbf{R}_1(t), \ldots, \mathbf{R}_{Q-1}(t)] \tag{11.28}$$

Specifically, the matrix blocks are given by $\mathbf{R}_q(t) = \mathrm{E}\{[\mathbf{j}_k(t) + \mathbf{n}_k(t)][\mathbf{j}_{k-q}(t) + \mathbf{n}_{k-q}(t)]^{\dagger}\}$, and expanded further in Eqn. (11.29), where the assumption of independence between the composite hot-clutter signal and additive noise is invoked to separate the two expectations.

$$\mathbf{R}_q(t) = \mathrm{E}\left\{\left(\sum_{m=1}^{M}\sum_{\ell=1}^{L} \mathbf{h}_{m\ell}(t) g_{mk-\ell+1}(t)\right)\left(\sum_{m=1}^{M}\sum_{\ell=1}^{L} \mathbf{h}_{m\ell}(t) g_{mk-\ell-q+1}(t)\right)^{\dagger}\right\} + \mathrm{E}\{\mathbf{n}_k(t)\mathbf{n}_{k-q}^{\dagger}(t)\} \tag{11.29}$$

The mutual independence of the individual hot-clutter sources and modes means that all the cross-terms in Eqn. (11.29) cancel. Eliminating these terms, and substituting $\mathbf{h}_{m\ell}(t)$ for its definition in Eqn. (11.26), the standard spatial covariance matrix corresponding to zero fast-time lag ($q = 0$) is given by Eqn. (11.30). Recall that the additive noise $\mathbf{n}_k(t)$ is assumed to be spatially white of power σ_n^2, such that $\mathrm{E}\{\mathbf{n}_k(t)\mathbf{n}_k^{\dagger}(t)\} = \sigma_n^2 \mathrm{I}_N$.

$$\mathbf{R}_0(t) = \sum_{m=1}^{M}\sum_{\ell=1}^{L} A_{m\ell}^2 \mathbf{S}_{m\ell}\mathbf{c}_{m\ell}(t)\mathbf{c}_{m\ell}^{\dagger}(t)\mathbf{S}_{m\ell}^{\dagger} + \sigma_n^2 \mathbf{I}_N \tag{11.30}$$

For $q = 1, \ldots, Q-1$, the fast-time lagged spatial covariance matrices $\mathbf{R}_q(t)$ are given by Eqn. (11.31). The additive noise component doesn't appear because $\mathrm{E}\{\mathbf{n}_k(t)\mathbf{n}_{k-q}(t)^{\dagger}\} = \mathbf{0}$

for $q > 0$. Since the hot-clutter source waveforms are assumed to be temporally white in the broadband case, the only hot-clutter contributions in $\mathbf{R}_q(t)$ are from pairs of modes that have a differential path delay of q fast-time samples. Clearly, $\mathbf{R}_q(t) = \mathbf{0}$ for $q \geq L$ since there is no pair of modes with a differential delay exceeding the maximum impulse response duration of the hot-clutter channel.

$$\mathbf{R}_q(t) = \sum_{m=1}^{M} \sum_{\ell=1}^{L-q} A_{m\ell} A_{m\ell+q} \mathbf{S}_{m\ell} \mathbf{c}_{m\ell}(t) \mathbf{c}^{\dagger}_{m\ell+q}(t) \mathbf{S}^{\dagger}_{m\ell+q} \tag{11.31}$$

11.3 Mitigation Techniques

Most schemes for hot- and cold-clutter mitigation are based on a cascaded processing approach in which a hot-clutter canceler, typically a fast-time STAP technique, precedes a cold-clutter suppression stage, either implemented as a slow-time STAP technique or standard Doppler processing. In airborne radar applications, this processing sequence is largely motivated by practical considerations, including training strategies and computational complexity. In OTH radar, such processing is additionally motivated by the cold-clutter properties, which in general allow it to be mitigated effectively by standard Doppler processing. For this reason, the cascaded approach in which fast-time STAP for hot-clutter cancelation is followed by standard Doppler processing for cold-clutter suppression is considered.

Standard fast-time STAP techniques may be broadly distinguished in terms of whether the filter weights are held fixed or updated within the CPI. Intra-CPI filter adaptations are primarily motivated by the need to track the time-dependence or "non-stationarity" of the hot-clutter space/fast-time covariance matrix over the relatively long radar dwell. However, the deleterious impact of a STAP filter that changes during the CPI on the subsequent cold-clutter suppression stage is not always taken into account. The scope of this section is to describe fast-time STAP techniques applicable for hot-clutter cancelation in pulse-waveform (PW) and continuous waveform (CW) OTH radar systems. Methods required to tailor such techniques to the peculiarities of airborne radar systems will not be discussed here.

Only a handful of studies have specifically addressed the design of the cascaded approach in which the slow-time varying STAP filter for hot-clutter rejection is synthesized with due regard to the effect on the performance of the subsequent cold-clutter canceler. The first part of this section recalls standard fast-time STAP techniques based on static and dynamic filters over the CPI. Alternative fast-time STAP techniques that overcome the limitations of these standard approaches are then described in the second part of this section. The third part of this section compares the performance of standard and alternative processing schemes using simulated data.

11.3.1 Standard Schemes

Two basic types of standard fast-time STAP filters may be identified. The first adopts a fixed weight vector $\tilde{\mathbf{w}}$ to process all ranges and pulses in the CPI. This filter varies only with radar look direction φ_0, but this dependence is considered implicit and omitted here for notational convenience. The second is based on a slow-time dependent filter $\tilde{\mathbf{w}}(t)$ that processes all range bins in pulse t for $t = 1, \ldots, P$. The most general STAP filter, denoted by $\tilde{\mathbf{w}}_k(t)$, is both slow-time varying and range dependent. This type of filter will be discussed in the second part of this section, which describes alternative STAP techniques.

The majority of STAP techniques described in the open literature are based on the solution of a linearly constrained minimum variance (LCMV) optimization problem, which incorporates multiple linear constraints. The LCMV formulation is a generalization of the more familiar minimum variance distortionless response (MVDR) approach, which incorporates a single linear constraint. The LCMV optimization problem may be written as Eqn. (11.32). Here, the argument $\mathbf{w} \in \mathcal{C}^{NQ}$ is the weight vector of the (generic) fast-time STAP filter, $\mathbf{R}$ normally represents the $NQ \times NQ$ hot-clutter-plus-noise covariance matrix, while the $NQ \times q$ constraint matrix $\mathbf{C}$, and the q-dimensional response vector $\mathbf{f}$, define the q linear constraints imposed on the filter $\mathbf{w}$.

$$\mathbf{w}_o = \arg\min_{\mathbf{w}} \mathbf{w}^\dagger \mathbf{R}\mathbf{w} \quad \text{subject to} : \mathbf{w}^\dagger \mathbf{C} = \mathbf{f}^\dagger \tag{11.32}$$

The optimum solution $\mathbf{w}_o$ is the weight vector that minimizes the interference power $\mathbf{w}^\dagger\mathbf{R}\mathbf{w}$ at the STAP output subject to the q linear constraints $\mathbf{w}^\dagger\mathbf{C} = \mathbf{f}^\dagger$. This filter is given by Eqn. (11.33), where the matrix $\mathbf{R}$ is assumed to be positive definite, so that $\mathbf{R}^{-1}$ exists. The derivation of this solution using the method of Lagrange multipliers can be found in Frost (1972). We shall find it useful to refer back to this general expression and assign specific definitions to the various terms in Eqn. (11.33). The terms $\mathbf{R}$, $\mathbf{C}$, and $\mathbf{f}$ conveniently serve as "place-holders" for now. It is mainly with regard to these definitions, and the manner in which filter is implemented, that the standard and alternative fast-time STAP approaches differ.

$$\mathbf{w}_o = \mathbf{R}^{-1}\mathbf{C}[\mathbf{C}^\dagger\mathbf{R}^{-1}\mathbf{C}]^{-1}\mathbf{f} \tag{11.33}$$

11.3.1.1 Linear Deterministic Constraints

Traditionally, the main purpose of the linear constraints is to ensure that the filter provides fixed gain and distortionless processing of useful signals incident from the look direction φ_0. Standard fast-time STAP techniques typically employ *deterministic* constraints for this purpose. Different approaches have been proposed to protect the useful signal from attenuation and distortion at the fast-time STAP output. To motivate these approaches, it is instructive to express the stacked useful signal vector $\tilde{\mathbf{s}}_k(t)$ in the form of Eqn. (11.34). With reference to the spatial snapshot model in Eqn. (11.2), the Q-variate fast-time vector $\boldsymbol{\psi}_k = [\psi_k, \psi_{k-1}, \ldots, \psi_{k-Q+1}]^T$ contains the useful signal samples in the Q-tap delay line of the STAP filter, while the $NQ \times Q$ matrix $\mathbf{A}_Q(\varphi_0)$ is given by $\mathbf{A}_Q(\varphi_0) = \mathbf{s}(\varphi_0) \otimes \mathbf{I}_Q$, where $\otimes$ denotes Kronecker product.

$$\tilde{\mathbf{s}}_k(t) = a \exp\{j2\pi f_d t + \gamma_k\}\mathbf{A}_Q(\varphi_0)\boldsymbol{\psi}_k \tag{11.34}$$

Now consider the case of $q = Q$ linear deterministic constraints with the constraint matrix defined as $\mathbf{C} = \mathbf{A}_Q(\varphi_0)$. To determine the effect on the useful signal, assume that a generic fast-time STAP filter $\mathbf{w}$ satisfying the condition $\mathbf{w}^\dagger\mathbf{C} = \mathbf{f}^\dagger$ processes the useful signal vector $\tilde{\mathbf{s}}_k(t)$. Using Eqn. (11.34), it is readily determined that the scalar output $s_k(t) = \mathbf{w}^\dagger\tilde{\mathbf{s}}_k(t)$ is given by Eqn. (11.35), since the condition $\mathbf{w}^\dagger\mathbf{A}(\varphi_0) = \mathbf{f}^\dagger$ is enforced by the constraints matrix $\mathbf{C} = \mathbf{A}(\varphi_0)$. In this case, it is observed that the STAP output $s_k(t)$ depends on the Q-variate response vector $\mathbf{f}$ through the inner product $\mathbf{f}^\dagger\boldsymbol{\psi}_k$. As pointed out in Griffiths (1996), the complex scalar $\mathbf{f}^\dagger\boldsymbol{\psi}_k$ may be interpreted as the output of a correlation receiver applied in the fast-time sample domain, where the filter coefficients of this receiver are given by the elements of the response vector $\mathbf{f}$.

$$s_k(t) = a \exp\{j2\pi f_d t + \gamma_k\}\mathbf{f}^\dagger \psi_k \tag{11.35}$$

Two cases are of particular interest. The first is the matched-filter receiver, given by $\mathbf{f} = \alpha\psi_k$ for an arbitrary constant α, which maximizes the signal-to-white-noise ratio in the output $s_k(t)$. The second is the receiver that provides fixed unity gain and distortionless coherent processing to yield the output $s_k(t)$ in Eqn. (11.36). The latter is clearly obtained by setting the first element of $\mathbf{f}$ to unity and the other $Q-1$ elements to zero, i.e., $\mathbf{f} = \mathbf{e}_Q = [1, 0\ldots, 0]^T$, such that $\mathbf{f}^\dagger\psi_k = \psi_k$. Ideally, a pulse-compressed useful signal is impulsive in fast-time when range sidelobes are ignored. When such a signal is matched to the current range k, we have that $\psi_k = [1, 0, \ldots, 0]^T$. In this case, the matched filter receiver coincides with the distortionless response receiver $\mathbf{f} = \mathbf{e}_Q$.

$$s_k(t) = a \exp\{j2\pi f_d t + \gamma_k\}\psi_k \tag{11.36}$$

Hence, the Q linear deterministic constraints defined by $\mathbf{C} = \mathbf{A}_Q(\varphi_0)$ and $\mathbf{f} = \mathbf{e}_Q$ represent a minimum requirement to ensure fixed unit gain and distortionless processing of an ideal useful signal at the output of a fast-time STAP filter. If it is desired to make the output useful signal $s_k(t)$ more robust to pointing errors, a corresponding constraint on derivatives might also be imposed, e.g., by setting $\mathbf{C} = \mathbf{A}_{2Q}(\varphi_0)$ defined in Eqn. (11.37), and $\mathbf{f} = \mathbf{e}_{2Q}$, where $\dot{\mathbf{s}}(\varphi) = \partial\mathbf{s}(\varphi)/\partial\varphi$. We may write $\mathbf{C} = \mathbf{A}_q(\varphi_0)$ and $\mathbf{f} = \mathbf{e}_q$ for the general case of $q > Q$ linear deterministic constraints.

$$\mathbf{A}_{2Q}(\varphi_0) = \begin{bmatrix} \mathbf{s}(\varphi_0) & \dot{\mathbf{s}}(\varphi_0) & 0 & 0 & \cdots & 0 & 0 \\ 0 & 0 & \mathbf{s}(\varphi_0) & \dot{\mathbf{s}}(\varphi_0) & & & \\ \vdots & & & & \ddots & & \\ 0 & 0 & & & & \mathbf{s}(\varphi_0) & \dot{\mathbf{s}}(\varphi_0) \end{bmatrix} \tag{11.37}$$

The use of a single linear deterministic constraint has been advocated in certain fast-time STAP studies. To motivate this concept, it is more convenient to express the stacked useful signal vector $\tilde{\mathbf{s}}_k(t)$ in the alternative form of Eqn. (11.38), where the NQ-variate vector $\psi_k \otimes \mathbf{s}(\varphi_0)$ is substituted for the equivalent matrix multiplication $\mathbf{A}_Q(\varphi_0)\psi_k$ in Eqn. (11.34). For an ideal target echo matched to the current range cell k, we impose the condition $\psi_k = \mathbf{e}_Q$ in Eqn. (11.38) such that $\tilde{\mathbf{s}}_k(t) = a \exp\{j2\pi f_d t + \gamma_k\}\mathbf{v}(\varphi_0)$, where the vector $\mathbf{v}(\varphi_0) = \mathbf{e}_Q \otimes \mathbf{s}(\varphi_0)$ is regarded as the space/fast-time steering vector.

$$\tilde{\mathbf{s}}_k(t) = a \exp\{j2\pi f_d t + \gamma_k\}\psi_k \otimes \mathbf{s}(\varphi_0) \tag{11.38}$$

It would appear that the single linear constraint defined in Eqn. (11.39) suffices in this case. Indeed, this constraint provides unit gain to a useful signal when the location of the impulse is matched to the current range k. However, as the fast-time STAP filter "slides over" different fast-time samples to process different range cells k, the location of this impulse moves into a subsquent tap of the delay line that trails the current range cell processed. In these trailing taps, the spatial response of the antenna weights in the direction φ_0 is unconstrained when only the constraint in Eqn. (11.39) is imposed. The spatial response of the processor to the useful signal in these taps will in general be non-zero and may fluctuate over pulses if the weight vector is updated within the CPI. This causes a degradation in the range-sidelobe structure of useful signals at the fast-time

STAP output. Visually, the processed target echo may appear spread or "smeared" in range over the full length Q of the fast-time tap delay line. In the case of a dynamic filter that is updated from pulse to pulse, temporal variations in the range sidelobe structure over the PRI will additionally cause these sidelobes to appear spread in Doppler.

$$\mathbf{w}^{\dagger}\mathbf{v}(\varphi_0) = \mathbf{w}^{\dagger}\left[\mathbf{e}_Q \otimes \mathbf{s}(\varphi_0)\right] = 1 \tag{11.39}$$

In summary, a single linear constraint can provide unit gain to matched useful signals but cannot ensure distortionless processing of such signals. For this reason, the set of Q linear deterministic constraints defined previously is recommended for fast-time STAP. The first of these linear constraints provides fixed unity gain to useful signals incident from the radar look direction. This not only protects a useful signal matched to the current range from being inadvertently attenuated, but also ensures Doppler coherence of the target echo across the different repetition periods in the CPI. On the other hand, the remaining $Q-1$ constraints ensure the fast-time STAP filter has zero response in the look direction over the trailing taps of the delay line to avoid smearing the output useful signal energy in range (and in Doppler for a time-varying filter).

11.3.1.2 Time-Invariant STAP

The first standard fast-time STAP approach to be described is based on a time-invariant weight vector held fixed over the CPI. The optimum (time-invariant) STAP filter is given by $\tilde{\mathbf{w}}$ in Eqn. (11.40), where $\tilde{\mathbf{R}} = \sum_{t=1}^{P} \tilde{\mathbf{R}}(t)$ is the hot-clutter-plus-noise covariance matrix averaged over the CPI. To arrive at the solution $\tilde{\mathbf{w}}$, the term $\mathbf{R}$ in Eqn. (11.33) is substituted for $\tilde{\mathbf{R}}$, while the linear deterministic constraints are defined by $\mathbf{C} = \mathbf{A}_q(\varphi_0)$ and $\mathbf{f} = \mathbf{e}_q$. The filter $\tilde{\mathbf{w}}$ is optimum in terms of output signal-to-hot-clutter-plus-noise ratio when conditioned on the set of all *time-invariant* filters.

$$\tilde{\mathbf{w}} = \tilde{\mathbf{R}}^{-1}\mathbf{A}_q(\varphi_0)[\mathbf{A}_q(\varphi_0)^{\dagger}\tilde{\mathbf{R}}^{-1}\mathbf{A}_q(\varphi_0)]^{-1}\mathbf{e}_q \tag{11.40}$$

Stacked hot-clutter-plus-noise training snapshots are required to estimate the unknown matrix $\tilde{\mathbf{R}}$. In an OTH skywave radar system, a limited number of practically clutter-free snapshots may be found near the start of the PRI (i.e., at short ranges) due to the skip-zone phenomenon. Whereas in an HFSW radar system, the high attenuation of the surface-wave at long ranges often permits clutter-free snapshots to be obtained near the end of the PRI. In any case, this allows for *supervised* training using N_k hot-clutter-plus-noise-only snapshots $\tilde{\mathbf{x}}_k(t) = \tilde{\mathbf{j}}_k(t) + \tilde{\mathbf{n}}_k(t)$ available in each PRI, where $N_k < K$. Using the first N_k range cells, e.g., the unknown matrix may be estimated as $\hat{\mathbf{R}}$ in Eqn. (11.41).

$$\hat{\mathbf{R}} = \sum_{t=1}^{P}\sum_{k=1}^{N_k} \tilde{\mathbf{x}}_k(t)\tilde{\mathbf{x}}_k(t) \tag{11.41}$$

Diagonal loading is often not necessary when the sample covariance matrix is averaged over the whole CPI, as typically $PN_k \gg 2NQ$. The main issue relates to the rank-expansion of $\hat{\mathbf{R}}$ over the relatively long CPI and the inability of the associated STAP filter to effectively cancel non-stationary hot clutter. Specifically, the adaptive implementation of this first standard approach is denoted by the time-invariant STAP filter $\hat{\mathbf{w}}$ in Eqn. (11.42), which is used to process all ranges and pulses in the CPI. This approach shall be referred to as *time-invariant* STAP hereafter. As discussed previously, the number of deterministic constraints may be $q = Q$ or $q = 2Q$, depending on whether robustness to beam-pointing errors is deemed neccessary.

$$\hat{\mathbf{w}} = \hat{\mathbf{R}}^{-1}\mathbf{A}_q(\varphi_0)[\mathbf{A}(\varphi_0)^{\dagger}\hat{\mathbf{R}}^{-1}\mathbf{A}_q(\varphi_0)]^{-1}\mathbf{e}_q \tag{11.42}$$

Similar concepts to those described above for PW systems also apply to CW systems after range processing is performed. It is shown in Abramovich et al. (2000) that range processing by FMCW deramping and FFT-based spectral analysis does not effect the hot-clutter-plus-noise covariance matrix model in Eqn. (11.22) under relatively mild assumptions that normally hold in practice. Hence, fast-time STAP techniques described in this and the following sections are applicable to both PW and CW OTH radars when supervised training is possible.

Depending on system characteristics and propagation conditions, it may occur that all range bins contain significant cold-clutter contributions. In this *unsupervised* training scenario, it is desirable to perform pre-processing to attenuate the cold-clutter signal prior to hot-clutter covariance matrix estimation. A scheme that makes use of an MTI clutter removal filter to obtain suitable training data is described in Abramovich et al. (2000).

11.3.1.3 Unconstrained STAP

Now let's turn our attention to the specification of the optimum slow-time varying STAP filter denoted by $\tilde{\mathbf{w}}(t)$. In this case, the term $\mathbf{R}$ in Eqn. (11.33) is substituted for the quasi-instantaneous hot-clutter-plus-noise covariance matrix $\tilde{\mathbf{R}}(t)$ defined in Eqn. (11.28), while the linear deterministic constraints are defined by $\mathbf{C} = \mathbf{A}_q(\varphi_0)$ and $\mathbf{f} = \mathbf{e}_q$, as before. The slow-time dependent optimum filter $\tilde{\mathbf{w}}(t)$ is given by Eqn. (11.43).

$$\tilde{\mathbf{w}}(t) = \tilde{\mathbf{R}}^{-1}(t)\mathbf{A}_q(\varphi_0)[\mathbf{A}_q(\varphi_0)^{\dagger}\tilde{\mathbf{R}}^{-1}(t)\mathbf{A}_q(\varphi_0)]^{-1}\mathbf{e}_q \tag{11.43}$$

Since the hot clutter is assumed to be stationary over the quasi-instantaneous PRI, the STAP filter $\tilde{\mathbf{w}}(t)$ is optimum in terms of output signal-to-hot-clutter-plus-noise ratio. In practical applications, the hot-clutter-plus-noise covariance matrix $\tilde{\mathbf{R}}(t)$ is unknown, but may be estimated as $\hat{\mathbf{R}}(t)$ in Eqn. (11.44). Here, only the training range cells in the *current* PRI are used. Diagonal loading at an appropriate level σ^2 is often applied to improve convergence rate in conditions of low sample support when $N_k < 2NQ$.

$$\hat{\mathbf{R}}(t) = \sum_{k=1}^{N_k} \tilde{\mathbf{x}}_k(t)\tilde{\mathbf{x}}_k(t) + \sigma^2\mathbf{I}_{NQ} \tag{11.44}$$

The true covariance matrix $\tilde{\mathbf{R}}(t)$ may be substituted for the regularized sample estimate $\hat{\mathbf{R}}(t)$ in the optimum filter expression of Eqn. (11.43) to yield the adaptive STAP filter $\hat{\mathbf{w}}(t)$ in Eqn. (11.45). This practical filter is used to process the operational range cells $k = N_k + 1, \ldots, K$ in the current PRI t. This approach is characterized by relatively higher computational complexity compared to time-invariant STAP, but is potentially more effective for non-stationary hot-clutter cancelation provided that the condition $NQ > M(L+Q-1)$ is met. This second standard approach is referred to as *unconstrained* STAP in the sense that the weights may vary arbitrarily in slow-time t aside from the q linear deterministic constraints.

$$\hat{\mathbf{w}}(t) = \hat{\mathbf{R}}^{-1}(t)\mathbf{A}_q(\varphi_0)[\mathbf{A}_q(\varphi_0)^{\dagger}\hat{\mathbf{R}}^{-1}(t)\mathbf{A}(\varphi_0)]^{-1}\mathbf{e}_q \tag{11.45}$$

Although the deterministic linear constraints protect the gain and Doppler spectrum of signals incident from the radar look direction, the (otherwise unconstrained) changes in $\tilde{\mathbf{w}}(t)$ over the CPI will temporally modulate cold-clutter returns incident from other directions. This is because the response of $\tilde{\mathbf{w}}(t)$ is unconstrained in all directions but the look direction, and is therefore free to fluctuate in an uncontrolled manner from pulse to pulse.

11.3.1.4 Relative Merits and Shortcomings

In summary, two standard operational fast-time STAP techniques have been motivated and described. The first is referred to as the standard *time-invariant* STAP scheme, which corresponds to the static adaptive filter $\hat{\mathbf{w}}$ in Eqn. (11.42). This approach has the advantage of relatively low computational complexity and reduced demands on sample support for practical implementation. Importantly, use of a fixed filter to process the CPI also preserves the temporal inter-pulse correlation properties of the cold clutter at the fast-time STAP output. This is vital for effective cold-clutter mitigation in the subsequent Doppler processing step.

The main problem with this approach is that it often cannot effectively reject non-stationary hot clutter because the sample covariance matrix $\hat{\mathbf{R}}$ has large or full rank when averaged over a typical OTH radar CPI. In practice, this scheme may be regarded appropriate when the non-stationarity of the hot clutter is not significant on the scale of the CPI. Real-data processing results suggest that this typically corresponds to CPI lengths which are too short for OTH radar purposes, particularly for (but not limited to) ship-detection applications.

The other standard method, referred to as *unconstrained* STAP, is based on the slow-time varying adaptive filter $\hat{\mathbf{w}}(t)$ in Eqn. (11.45). While this scheme is typically able to reject the hot clutter effectively, the slow-time modulation induced on the cold-clutter output by the STAP filter updates will in general preclude effective cold-clutter suppression via Doppler processing. Indeed, application of the standard unconstrained STAP filter often has a devastating effect on SCV after Doppler processing. Diagonal loading heavier than required to improve the convergence rate may be applied to stabilize the fluctuations of $\hat{\mathbf{w}}(t)$. This method can improve the inter-PRI correlation properties of the cold clutter at the STAP output. However, for sufficiently non-stationary hot clutter, no level of diagonal loading can simultaneously provide effective hot-clutter mitigation and distortionless processing of the cold-clutter signal.

If the cold clutter were generated by a small number of point scatterers, its Doppler spectrum properties could be preserved by imposing a limited number of additional linear deterministic constraints that freeze the STAP filter response in the direction of each point scatterer (Griffiths 1996). As explained in Abramovich et al. (2000), the spatial distribution of the cold clutter is generally quite broad, so the backscattered radar signal enters through a large part of the receiving antenna pattern rather than from a few discrete directions only. Hence, the addition of deterministic linear constraints is not a feasible solution since it is not possible to hold constant all or most of the antenna pattern without incurring a dramatic degradation in non-stationary hot-clutter rejection.

11.3.2 Alternative Procedures

Only a few studies have specifically addressed the problem of preserving the natural slow-time correlation properties of the spatially broadband cold clutter at the STAP filter output while simultaneously updating the weight vector in slow-time to counter the non-stationarity of hot clutter. Perhaps the first attempt to provide a solution to this problem was reported in Anderson et al. (1997) and Abramovich, Spencer, and Anderson (1998) for supervised training scenarios, and subsequently in Abramovich et al. (2000) for unsupervised training. The pioneering work of Abramovich in this area represents a generalization of specific methodologies described in Abramovich et al. (1994) for spatial-only adaptive processing. The main objective of this approach, known as the stochastically constrained fast-time STAP or SC-STAP method, is to effectively cancel non-stationary hot clutter while preserving the slow-time correlation properties of the clutter output for coherent Doppler processing. The SC-STAP method is briefly recalled in this section. Further details can be found in Klemm (2004).

Motivated by practical considerations and the high computational load associated with the SC-STAP method, this section introduces a new approach called time-varying STAP or TV-STAP, which provides an attractive alternative for real-time implementation. With respect to the SC-STAP method, the TV-STAP approach allows the computational load to be significantly reduced in situations where the hot clutter can be effectively canceled by STAP filters updated at a rate slower than every PRI. Importantly, the reduced computational load is not traded-off against hot-clutter cancelation performance in this case.

Both SC-STAP and TV-STAP have been designed with OTH radar applications in mind, as they both rely in an approximate sense on the aforementioned cold-clutter model. Consequently, the SC-STAP and TV-STAP techniques described below are not directly applicable to airborne radar systems, where cold clutter exhibits strong angle-Doppler coupling due to platform motion and is more accurately represented by a multi-variate (as opposed to a scalar-type) AR model. A slow-time dependent STAP filter that can address the airborne radar problem has been proven to exist in theory; see Abramovich et al. (1998). However, operational versions of this approach will not be discussed here.

11.3.2.1 Stochastically-Constrained STAP (SC-STAP)

The full details of the stochastic constraints (SC) technique for fast-time STAP are not repeated here for brevity, but an overview of the approach is given for two main reasons: (1) to provide motivation for the development of the subsequent TV STAP algorithm, and (2) to explain the similarities and differences between TV-STAP and the SC-STAP method.

The key idea behind SC-STAP is to apply a PRI-varying and *range-dependent* filter $\tilde{\mathbf{w}}_k(t)$ that is able to preserve SCV by statistically approximating the condition $\tilde{\mathbf{w}}_k^\dagger(t)\tilde{\mathbf{c}}_k(t) \approx \tilde{\mathbf{w}}_0^\dagger\tilde{\mathbf{c}}_k(t)$ at the cold-clutter output, where $\tilde{\mathbf{w}}_0$ is a fixed reference STAP weight vector that provides distortionless cold-clutter processing. Clearly, the SC-STAP filter $\tilde{\mathbf{w}}_k(t)$ differs from $\tilde{\mathbf{w}}_0$ and changes over the CPI to effectively reject non-stationary hot clutter. Utilizing the κth order scalar-type AR model for the stacked cold-clutter snapshots $\tilde{\mathbf{c}}_k(t)$ in Eqn. (11.14), the scalar cold-clutter output $c_k(t)$ of the SC-STAP filter is given by Eqn. (11.46).

$$c_k(t) = \tilde{\mathbf{w}}_k^\dagger(t)\tilde{\mathbf{c}}_k(t) = \tilde{\mathbf{w}}_k^\dagger(t)\tilde{\boldsymbol{\varepsilon}}_k(t) - \sum_{i=1}^{\kappa} b_i \tilde{\mathbf{w}}_k^\dagger(t)\tilde{\mathbf{c}}_k(t-i) \tag{11.46}$$

Inspection of Eqn. (11.46) reveals that the terms in the summation may be made identical to those processed by a stable filter $\tilde{\mathbf{w}}_0$ provided that the weight vector $\tilde{\mathbf{w}}_k(t)$ satisfies κ

linear stochastic constraints in Eqn. (11.47). If this weight vector also satisfies the quadratic constraint $\tilde{\mathbf{w}}_k^\dagger(t)\tilde{\mathbf{R}}_c\tilde{\mathbf{w}}_k(t) = \tilde{\mathbf{w}}_0^\dagger\tilde{\mathbf{R}}_c\tilde{\mathbf{w}}_0$, where $\tilde{\mathbf{R}}_c = \mathrm{E}\{\tilde{\varepsilon}_k(t)\tilde{\varepsilon}_k^\dagger(t)\}$, the power of the output innovative noise in Eqn. (11.46) is also equalized. The scalar innovations $\tilde{\mathbf{w}}_k^\dagger(t)\tilde{\varepsilon}_k(t)$ and $\tilde{\mathbf{w}}_0^\dagger\tilde{\varepsilon}_k(t)$ will correspond to different white-noise realizations, i.e., $\tilde{\mathbf{w}}_k^\dagger(t)\tilde{\varepsilon}_k(t) \neq \tilde{\mathbf{w}}_0^\dagger\tilde{\varepsilon}_k(t)$, but the stochastic constraints ensure that the cold-clutter processes $\tilde{\mathbf{w}}_k^\dagger(t)\tilde{\mathbf{c}}_k(t)$ and $\tilde{\mathbf{w}}_0^\dagger\tilde{\mathbf{c}}_k(t)$ are statistically equivalent under these conditions. In this case, the scalar cold-clutter output $c_k(t)$ processed by the filter $\tilde{\mathbf{w}}_k(t)$ is statistically identical to the cold-clutter output of the stable filter $\tilde{\mathbf{w}}_0$. Consequently, $c_k(t)$ is described by a stationary κ^{th} order AR model with the same parameters $\{b_i\}_{i=1}^{\kappa}$ as those of the input slow-time sequence of cold-clutter stacked vectors defined in Eqn. (11.14). The important point is that the requirement to preserve the auto-regressive correlation properties of the cold clutter at the STAP filter *output* does not necessarily imply that the STAP filter needs to be time invariant.

$$\tilde{\mathbf{w}}_k^\dagger(t)\tilde{\mathbf{c}}_k(t-i) = \tilde{\mathbf{w}}_0^\dagger\tilde{\mathbf{c}}_k(t-i) \quad i = 1, .., \kappa \tag{11.47}$$

These fundamental observations made in Abramovich et al. (1998), as well as in earlier references therein for the case of pure spatial filtering, provide scope for the weight vector $\tilde{\mathbf{w}}_k(t)$ to change over the CPI in response to hot-clutter non-stationarity, while simultaneously stabilizing the auto-regressive cold-clutter characteristics at the output of the fast-time STAP filter. It follows that the optimum SC-STAP filter may be synthesized according to Eqn. (11.48) using the matrix $\tilde{\mathbf{R}}_\kappa(t)$ defined in Eqn. (11.49).

$$\tilde{\mathbf{w}}_k(t) = \arg\min_{\mathbf{w}} \ \mathbf{w}^\dagger\tilde{\mathbf{R}}_\kappa(t)\mathbf{w} \qquad \text{subject to} \begin{cases} \mathbf{w}^\dagger\mathbf{A}_q(\theta_0) = \mathbf{e}_q^T, \\ \mathbf{w}^\dagger\tilde{\mathbf{c}}_k(t-i) = \tilde{\mathbf{w}}_0^\dagger\tilde{\mathbf{c}}_k(t-i), \quad i = 1, \ldots, \kappa \end{cases} \tag{11.48}$$

The linear deterministic constraints $\mathbf{w}^\dagger\mathbf{A}_q(\theta_0) = \mathbf{e}_q^T$ are applied for the same reasons as in the two standard STAP schemes. In Eqn. (11.49), $\tilde{\mathbf{R}}_\kappa(t)$ is defined as the hot-clutter-plus-noise covariance matrix averaged over $\kappa + 1$ adjacent repetition periods. This sliding-window average terminates at pulse t and moves forward one pulse at a time until the end of the CPI is reached, i.e., $t = \kappa + 1, \ldots, P$. The first matrix in the sequence is denoted by $\tilde{\mathbf{R}}_\kappa = \tilde{\mathbf{R}}_\kappa(\kappa + 1)$. Since $\kappa \ll N$, the average is performed over a relatively small number of adjacent pulses in which the true hot-clutter-plus-noise covariance matrix may be regarded as locally stationary, i.e., $\tilde{\mathbf{R}}_\kappa(t-\kappa) \approx \tilde{\mathbf{R}}_\kappa(t-\kappa+1)\ldots \approx \tilde{\mathbf{R}}_\kappa(t)$.

$$\tilde{\mathbf{R}}_\kappa(t) = \frac{1}{\kappa+1}\sum_{i=t-\kappa}^{t}\tilde{\mathbf{R}}(i) \tag{11.49}$$

The *initial* filter of the SC-STAP sequence $\tilde{\mathbf{w}}_0$ is formed *without* stochastic constraints, as in Eqn. (11.50), using the matrix $\tilde{\mathbf{R}}_\kappa$ averaged over the first $\kappa + 1$ PRI. This filter is independent of k and is used to process all operational range cells in the first $\kappa + 1$ PRIs, i.e., $\tilde{\mathbf{w}}_k(t) = \tilde{\mathbf{w}}_0$ for $t = 1, \ldots, \kappa+1$ and $k = N_k+1, \ldots, K$. Local averaging over κ adjacent PRIs does not preclude the initial filter $\tilde{\mathbf{w}}_0$ from effectively canceling the hot clutter in the first $\kappa + 1$ pulses provided that $NQ > (\kappa + 1)M(L + Q - 1)$. This is because the hot-clutter subspace dimension strictly grows by $M(L + Q - 1)$ with every pulse averaged. Less restrictive conditions apply on the *effective* rank of the hot-clutter covariance matrix when the mode wavefronts vary in a highly correlated manner over adjacent PRI. The condition $N > (\kappa + 1)P$ is also necessary because the apparent number of independent sources is multiplied by averaging over $\kappa + 1$ PRIs (Abramovich et al. 1998).

$$\tilde{\mathbf{w}}_0 = \tilde{\mathbf{R}}_\kappa^{-1}\mathbf{A}_q(\varphi_0)[\mathbf{A}_q(\varphi_0)^\dagger\tilde{\mathbf{R}}_\kappa^{-1}\mathbf{A}_q(\varphi_0)]^{-1}\mathbf{e}_q \tag{11.50}$$

The remaining SC-STAP filters $\tilde{\mathbf{w}}_k(t)$ in the slow-time sequence $t = \kappa+2, \ldots, P$ are *range dependent* and *stochastically constrained*. These SC-STAP filters are given by Eqn. (11.51), where the constraint matrix is augmented to incorporate the κ stochastic constraints $\mathbf{A}_{q+\kappa}(\varphi_0) = [\mathbf{A}_q(\varphi_0), \tilde{\mathbf{c}}_k(t-1), \ldots, \tilde{\mathbf{c}}_k(t-\kappa)]$, while the response vector is extended to the form $\mathbf{e}_{q+\kappa} = [e_q^T, \tilde{\mathbf{w}}_0^\dagger\tilde{\mathbf{c}}_k(t-1), \ldots, \tilde{\mathbf{w}}_0^\dagger\tilde{\mathbf{c}}_k(t-\kappa)]^T$. The quadratic constraint mentioned earlier is not implemented in the SC-STAP approach. It has been shown that its impact is negligible in most cases, since the cold-clutter covariance matrix is generally well conditioned and the effect of weight vector fluctuations on AR innovative noise power is small (Abramovich 1992).

$$\tilde{\mathbf{w}}_k(t) = \tilde{\mathbf{R}}_\kappa^{-1}\mathbf{A}_{q+\kappa}(\varphi_0)[\mathbf{A}_{q+\kappa}(\varphi_0)^\dagger\tilde{\mathbf{R}}_\kappa^{-1}(t)\mathbf{A}_{q+\kappa}(\varphi_0)]^{-1}\mathbf{e}_{q+\kappa} \tag{11.51}$$

The question then becomes how to synthesize such a filter in an operational system, particularly as the cold-clutter only snapshots $\tilde{\mathbf{c}}_k(t)$ are not directly available to form the stochastic constraints when hot clutter is present. An operational algorithm that can closely approximate the ideal solution is derived according to the SC-STAP criterion in Eqn. (11.52). The slow-time sequence of SC-STAP filters for range cell k are formed in turn for pulses in the CPI with the current filter $\hat{\mathbf{w}}_k(t)$ being dependent on the filter from the previous pulse $\hat{\mathbf{w}}_k(t-1)$ through the κ stochastic constraints in Eqn. (11.52).

$$\hat{\mathbf{w}}_k(t) = \arg\min_{\mathbf{w}} \; \mathbf{w}^\dagger\hat{\mathbf{R}}_\kappa(t)\mathbf{w}$$
$$\text{subject to} \begin{cases} \mathbf{w}^\dagger\mathbf{A}_q(\theta_0) = \mathbf{e}_q^T, \\ \mathbf{w}^\dagger\tilde{\mathbf{x}}_k(t-i) = \hat{\mathbf{w}}_k(t-1)^\dagger\tilde{\mathbf{x}}_k(t-i), \quad i = 1, \ldots, \kappa \end{cases} \tag{11.52}$$

In Eqn. (11.53), $\hat{\mathbf{R}}_\kappa(t)$ is defined as a regularized *sliding-window* estimate of the hot-clutter-plus-noise covariance matrix averaged over a length of $\kappa + 1$ repetition periods using the first N_k ranges cells as training data. This sliding-window average terminates at pulse t and moves forward one pulse at a time until the end of the CPI is reached, as before. The first matrix of the sequence is denoted by $\hat{\mathbf{R}}_\kappa = \hat{\mathbf{R}}_\kappa(\kappa + 1)$. Similarly, the *initial* filter of the SC-STAP sequence is formed *without* stochastic constraints as $\hat{\mathbf{w}}_0 = \hat{\mathbf{R}}_\kappa^{-1}\mathbf{A}_q(\varphi_0)[\mathbf{A}_q(\varphi_0)^\dagger\hat{\mathbf{R}}_\kappa^{-1}\mathbf{A}_q(\varphi_0)]^{-1}\mathbf{e}_q$. This adaptive filter is independent of range bin k and is used to process the first $\kappa + 1$ PRIs of the CPI, i.e., $\hat{\mathbf{w}}_k(t) = \hat{\mathbf{w}}_0$ for $t = 1, \ldots, \kappa + 1$ and $k = N_k + 1, \ldots, K$.

$$\hat{\mathbf{R}}_\kappa(t) = \frac{1}{(\kappa+1)N_k}\sum_{i=t-\kappa}^{t}\sum_{k=1}^{N_k}\tilde{\mathbf{x}}_k(i)\tilde{\mathbf{x}}_k(i) + \sigma^2\mathbf{I}_{NQ} \tag{11.53}$$

The remaining SC-STAP filters $\hat{\mathbf{w}}_k(t)$ in the slow-time sequence $t = \kappa + 2, \ldots, P$ are *range-dependent* and *stochastically constrained*. These SC-STAP filters are given by Eqn. (11.54), where the constraint matrix is augmented to incorporate the κ stochastic constraints $\hat{\mathbf{A}}_{q+\kappa}(\varphi_0) = [\mathbf{A}_q(\varphi_0), \tilde{\mathbf{x}}_k(t-1), \ldots, \tilde{\mathbf{x}}_k(t-\kappa)]$, while the response vector is extended accordingly $\hat{\mathbf{e}}_{q+\kappa} = [e_q^T, \tilde{\mathbf{w}}_k(t-1)^\dagger\tilde{\mathbf{x}}_k(t-1), \ldots, \tilde{\mathbf{w}}_k(t-1)^\dagger\tilde{\mathbf{x}}_k(t-\kappa)]^T$. With respect to the optimum SC-STAP filter in Eqn. (11.51), the adaptive implementation in Eqn. (11.54) involves a regularized sample estimate of the locally integrated hot-clutter-plus-noise covariance matrix $\hat{\mathbf{R}}_\kappa(t)$, as well as the estimates of the ideal stochastic constraints given by $\hat{\mathbf{A}}_{q+\kappa}(\varphi_0)$ and $\hat{\mathbf{e}}_{q+\kappa}$.

$$\hat{\mathbf{w}}_k(t) = \hat{\mathbf{R}}_\kappa^{-1}(t)\hat{\mathbf{A}}_{q+\kappa}(\varphi_0)[\hat{\mathbf{A}}_{q+\kappa}(\varphi_0)^\dagger\hat{\mathbf{R}}_\kappa^{-1}(t)\hat{\mathbf{A}}_{q+\kappa}(\varphi_0)]^{-1}\hat{\mathbf{e}}_{q+\kappa}\ , \quad t = \kappa+2, \ldots, P \tag{11.54}$$

The operational stochastic constraints in Eqn. (11.52) are not the same as the ideal ones in Eqn. (11.47), but they can approximate them well when the hot-clutter component is effectively rejected by the weight vectors $\hat{\mathbf{w}}_k(t)$ over a window length of κ PRI. Provided that the residual hot clutter plus noise is small compared to the processed cold clutter contribution, the adaptive SC-STAP filter output may be approximated as Eqn. (11.55). By setting $\hat{\mathbf{w}}_k^\dagger(t)\tilde{\mathbf{x}}_k(t-i) = \hat{\mathbf{w}}_0^\dagger\tilde{\mathbf{x}}_k(t-i)$ for $i = 1, \ldots, \kappa$ and $t = \kappa + 2$, the second weight vector in the SC-STAP sequence provides a good approximation to the ideal condition in Eqn. (11.47). The same follows for the remaining SC-STAP filters in the slow-time sequence $t = \kappa + 3, \ldots, P$. This is because all weight vectors in the chain are ultimately referenced to $\hat{\mathbf{w}}_0$ by virtue of the iterations. As far as useful signals are concerned, the deterministic linear constraints ensure that $\tilde{\mathbf{w}}_k^\dagger(t)\tilde{\mathbf{s}}_k(t-i) = \tilde{\mathbf{w}}_0\tilde{\mathbf{s}}_k(t-i)$ for all SC-STAP weight vectors.

$$\hat{\mathbf{w}}_k^\dagger(t)\tilde{\mathbf{x}}_k(t-i) \approx \hat{\mathbf{w}}_k^\dagger(t)\tilde{\mathbf{c}}_k(t-i)\ , \quad i = 1, \ldots, \kappa \tag{11.55}$$

The main point is that clutter-only snapshots are not necessary to approximate the stochastic constraints in this operational routine. Moreover, the unknown AR clutter parameters are only required to be locally stationary over $\kappa + 1$ successive pulses for the stochastic constraints to preserve the output cold clutter correlation properties.

11.3.2.2 Time-Varying STAP (TV-STAP)

In an attempt to stabilize the AR spectral characteristics of the cold clutter output, the SC method forms a new adaptive filter every PRI in a "sliding window" fashion using a different set of linear stochastic constraints, such that for an AR process of order κ, the SC-STAP weights are updated a total of $P - \kappa$ times over the CPI. This effectively means that the filter coefficients are updated at a rate equal to the PRF, irrespective of the physical duration of the waveform repetition period or the prevailing hot-clutter characteristics. This rate of adaptation is mainly determined by the need to protect the AR properties of the cold clutter regardless of whether or not such fast adaptation is actually necessary for effective hot-clutter cancelation, which ought to be the primary reason for updating the weight vectors.

In high PRF (aircraft detection) applications, or when the ionospheric propagation channels are not changing rapidly, the hot clutter may have a local stationarity interval which significantly exceeds a single PRI. This can be exploited to update the weights less frequently over the CPI with practically no loss in hot-clutter rejection performance. In this situation, the SC-STAP technique has at least one major drawback. That is, the procedure is computationally expensive as it requires the calculation of a separate weight solution for every PRI and range cell processed in the CPI. It is not surprising that such a scheme typically prevents real-time implementation in operational systems, particularly when the full chain of signal processing steps includes other intensive algorithms that must also be performed over a time interval less than the CPI. This issue strongly motivates the search for computationally efficient fast-time STAP techniques that can effectively address the non-stationary hot-clutter cancelation problem.

A second possible limitation, illustrated by practical examples for the case of pure spatial adaptive processing in the previous chapter, is that weight-estimation errors relative to the ideal solution occur at each filter update and accumulate over the CPI. Since these estimation errors ultimately cause a reduction in SCV after Doppler processing, minimizing the number of filter adaptations is generally preferable from the viewpoint of computational load as well as robustness to clutter Doppler spectrum broadening.

The TV-STAP method introduced in this section follows the same basic idea as the SC technique, but is structured differently to mitigate the aforementioned limitations of the original SC work. The philosophy behind TV-STAP is to update the weight vector in non-overlapping batches at a rate commensurate with the prevailing level of hot-clutter non-stationarity. Data-driven constraints are incorporated to protect the spectral integrity of the cold clutter output in accordance with an alternative clutter model to be formulated below. The main motivation of TV-STAP is to achieve comparable performance to SC-STAP, while at the same time breaking the bottleneck of real-time processing when conditions permit.

TV-STAP partitions the CPI into N_b smaller sub-CPI or *batches* each containing N_p pulses, and then updates the fast-time STAP weights from batch to batch rather than from PRI to PRI. For simplicity, assume that the number of batches $N_b = P/N_p$ is an integer. Selection of the batch length N_p represents a compromise between smaller values (rapid updates) to counter hot-clutter non-stationarity and larger values (slower updates) to reduce computational load.

More specifically, the TV-STAP weights are adjusted using batch-integrated hot clutter plus noise covariance matrices $\tilde{\mathbf{R}}_b$, which are defined in Eqn. (11.56) for $b = 1, \ldots, N_b$. A salient feature of Eqn. (11.56) is that the first κ pulses in the batch that follows the current one are also included in the summation. For reasons to become apparent later, this ensures the TV-STAP filter is effective for hot-clutter rejection in the first κ pulses of the next batch. Such a modification is clearly not applied to the final batch $b = N_b$, as no such pulses are available (i.e., the end of the CPI is reached).

$$\tilde{\mathbf{R}}_b = \frac{1}{N_p} \sum_{t=(b-1)N_p+1}^{bN_p+\kappa} \tilde{\mathbf{R}}(t) \tag{11.56}$$

The TV-STAP algorithm is based on a dynamic subspace cold clutter model in the form of Eqn. (11.57). In this representation, the N_p stacked cold clutter snapshots $\tilde{\mathbf{c}}_k(t)$ received in batch b at range k are assumed to be spanned by a stacked vector subspace of rank κ, denoted by the $NQ \times \kappa$ matrix $\tilde{\mathbf{Q}}_k(b)$ in Eqn. (11.57). Here, κ is the number of dominant spectral components resolved in the cold clutter Doppler spectrum. The N_p received cold clutter snapshots are assumed to lie in the range-space of $\tilde{\mathbf{Q}}_k(b)$ to a good approximation, where typically $P > N_p > \kappa$. The κ-variate parameter vector $\tilde{\mathbf{p}}_k(t)$ contains the linear combination coefficients that define the instantaneous structure of the synthetic clutter wavefront at range k and pulse t.

$$\tilde{\mathbf{c}}_k(t) = \tilde{\mathbf{Q}}_k(b)\tilde{\mathbf{p}}_k(t)\,, \quad t = (b-1)N_p + 1, \ldots, bN_p \tag{11.57}$$

The physical motivation behind the dynamic subspace model is that the instantaneous frequencies of the dominant cold clutter Doppler spectrum components often vary in a highly correlated manner across adjacent pulses such that the changes in phase-path for each component in slow-time may be regarded close to linear during a sufficiently short sub-CPI (batch). When the ionospheric paths propagating the cold clutter are frequency-stable, such a model may be expected to provide a fairly accurate approximation of the received clutter snapshots over a considerable number of pulses N_p, i.e., $N_p \gg \kappa$. Alternatively, when the phase-path fluctuations are caused by rapid and irregular ionospheric variations that induce significant Doppler spread on the κ dominant cold clutter spectral components, the approximation will be accurate only for a relatively small number of pulses (that may not greatly exceed κ).

It is of interest to discuss the connection between the dynamic subspace representation in Eqn. (11.57) and the stationary scalar-type AR cold clutter model adopted for skywave OTH radar in Eqn. (11.58). The two may be regarded as equivalent in the absence of innovative noise $\tilde{\varepsilon}_k(t)$, since any snapshot resulting from the recursive sequence without such innovations can be expressed exactly as a linear combination of any other κ (linearly independent) snapshots in the same sequence.

The presence of full-rank innovative noise in the AR model implies that Eqn. (11.57) can at best only approximate the cold clutter snapshots generated by Eqn. (11.58) when $N_p > \kappa$. As the innovative noise distinguishes the AR model and the dynamic subspace description, the accuracy of the approximation for $N_p > \kappa$ will tend to improve as the innovative noise power decreases and/or the batch length N_p becomes smaller. The question arises as to whether the use of Eqn. (11.57), when the clutter actually obeys the AR model in Eqn. (11.58), results in weight-estimation errors that are comparable to those arising from the operational stochastic constraints. This issue is investigated in the simulation analysis of Section 11.3.3.

$$\tilde{\mathbf{c}}_k(t) = -\sum_{i=1}^{\kappa} b_i \tilde{\mathbf{c}}_k(t-i) + \tilde{\boldsymbol{\varepsilon}}_k(t)\,, \quad t = 1, \ldots, P \tag{11.58}$$

For the moment, let's assume the cold clutter model in Eqn. (11.57) is appropriate. If $\tilde{\mathbf{Q}}_k(b)$ were known, it would be possible to fix the response of the TV-STAP filter $\tilde{\mathbf{w}}_k(b)$ to the clutter *synthetic wavefronts* received in batch b by introducing a set of auxiliary linear constraints $\tilde{\mathbf{w}}_k^\dagger(b)\tilde{\mathbf{Q}}_k(b) = \tilde{\mathbf{w}}(1)^\dagger \tilde{\mathbf{Q}}_k(b)$ for $b = 2, \ldots, N_b$. Here, $\tilde{\mathbf{w}}(1)$ is the first weight vector in the TV-STAP sequence given by Eqn. (11.59). This weight vector is used to process the operational range cells in the first batch ($b = 1$) of N_p pulses.

$$\tilde{\mathbf{w}}(1) = \tilde{\mathbf{R}}_1^{-1}\mathbf{A}_q(\varphi_0)[\mathbf{A}_q(\varphi_0)^\dagger \tilde{\mathbf{R}}_1^{-1}\mathbf{A}_q(\varphi_0)]^{-1}\mathbf{e}_q \tag{11.59}$$

Such a system of constraints ensures that the dominant clutter components experience a stable filter $\tilde{\mathbf{w}}(1)$ over the entire CPI as far as the output cold clutter contribution is concerned, while leaving $NQ-(q+\kappa)$ spare adaptive DOFs for effective non-stationary hot-clutter rejection at each weight update. The optimum TV-STAP weight vector $\tilde{\mathbf{w}}_k(b)$ for batch b is derived according to Eqn. (11.60)

$$\tilde{\mathbf{w}}_k(b) = \arg\min_{\mathbf{w}} \ \mathbf{w}^\dagger \tilde{\mathbf{R}}_b \mathbf{w} \qquad \text{subject to} \begin{cases} \mathbf{w}^\dagger \mathbf{A}_q(\theta) = \mathbf{e}_q^T, \\ \mathbf{w}^\dagger \tilde{\mathbf{Q}}_k(b) = \tilde{\mathbf{w}}_k^\dagger(b-1)\tilde{\mathbf{Q}}_k(b) \end{cases} \tag{11.60}$$

This leads to the weight solution in Eqn. (11.61) for batches $b = 2, \ldots, N_b$, where the constraint matrix and response vector are defined as $\mathbf{C}_k(b) = [\mathbf{A}_q(\theta_0), \tilde{\mathbf{Q}}_k(b)]$ and $\mathbf{f}_k(b) =$

$[\mathbf{e}_q^T, \tilde{\mathbf{w}}_k^\dagger(b-1)\tilde{\mathbf{Q}}_k(b)]^T$, respectively. Similar to the SC-STAP method, this set of data-driven constraints implies that all TV-STAP filters except the first one are range dependent.

$$\tilde{\mathbf{w}}_k(b) = \tilde{\mathbf{R}}_b^{-1}\mathbf{C}_k(b)[\mathbf{C}_k^\dagger(b)\tilde{\mathbf{R}}_b^{-1}\mathbf{C}_k(b)]^{-1}\mathbf{f}_k(b)\ , \quad b = 2, \ldots, N_b \tag{11.61}$$

As stated previously, the quasi-instantaneous cold clutter subspace $\tilde{\mathbf{Q}}_k(b)$ is not directly accessible in practice due to the presence of hot clutter, so the ideal auxiliary constraints cannot be implemented. Secondly, the batch-integrated covariance matrix $\tilde{\mathbf{R}}_b$ is unknown and must be estimated using the training range cells in each PRI. The operational TV-STAP procedure replaces the unknown matrix $\tilde{\mathbf{R}}_b$ by its regularized sample estimate $\hat{\mathbf{R}}_b$ in Eqn. (11.62).

$$\hat{\mathbf{R}}_b = \frac{1}{N_p N_k} \sum_{t=(b-1)N_p+1}^{bN_p+\kappa} \sum_{k=1}^{N_k} \tilde{\mathbf{x}}_k(t)\tilde{\mathbf{x}}_k^\dagger(t) + \sigma^2 \mathbf{I}_{NQ} \tag{11.62}$$

As far as the constraints are concerned, development of the operational routine is based on two key observations. First, the linear independence of the cold clutter snapshots received in the first κ pulses of batch b implies that the range-space of the $NQ \times \kappa$ matrix $\tilde{\mathbf{Q}}_k(b)$ is spanned by the set of vectors $[\tilde{\mathbf{c}}_k(N_p(b-1)), \ldots, \tilde{\mathbf{c}}_k(N_p(b-1)+\kappa-1)]$. The second is that for the stacked vectors in the data matrix $\tilde{\mathbf{D}}_k(b) = [\tilde{\mathbf{x}}_k(N_p(b-1)), \ldots, \tilde{\mathbf{x}}_k(N_p(b-1)+\kappa-1)]$, the approximation $\hat{\mathbf{w}}_k^\dagger(b)\tilde{\mathbf{D}}_k(b) \approx \tilde{\mathbf{w}}_k^\dagger(b)\tilde{\mathbf{Q}}_k(b)$ is accurate provided that the TV-STAP filter effectively rejects the hot clutter in batch b, i.e., $\hat{\mathbf{w}}_k^\dagger(b)\tilde{\mathbf{x}}_k(t) \approx \hat{\mathbf{w}}_k^\dagger(b)\tilde{\mathbf{c}}_k(t)$. Hence, the operational TV-STAP algorithm may be formulated in terms of q deterministic linear constraints, and κ data-driven auxiliary linear constraints, as in Eqn. (11.63).

$$\hat{\mathbf{w}}_k(b) = \arg\min_{\mathbf{w}} \ \mathbf{w}^\dagger\hat{\mathbf{R}}_b\mathbf{w}$$
$$\text{subject to} \begin{cases} \mathbf{w}^\dagger\mathbf{A}_q(\theta) = \mathbf{e}_q^T, \\ \mathbf{w}^\dagger\tilde{\mathbf{x}}_k(t) = \hat{\mathbf{w}}_k^\dagger(b-1)\tilde{\mathbf{x}}_k(t), \quad t = N_p(b-1), \ldots, N_p(b-1)+\kappa-1 \end{cases} \tag{11.63}$$

This leads to the operational TV-STAP solution in Eqn. (11.64) for batches $b = 2, \ldots, N_b$, where the constraint matrix $\hat{\mathbf{C}}_k(b) = [\mathbf{A}_q(\theta_0), \tilde{\mathbf{x}}_k(N_p(b-1)), \ldots, \tilde{\mathbf{x}}_k(N_p(b-1)+\kappa-1)]$ and response vector $\hat{\mathbf{f}}_k(b) = [\mathbf{e}_q^T, \hat{\mathbf{w}}_k^\dagger(b-1)\tilde{\mathbf{x}}_k(N_p(b-1)), \ldots, \hat{\mathbf{w}}_k^\dagger(b-1)\tilde{\mathbf{x}}_k(N_p(b-1)+\kappa-1)]^T$.

Analogously to SAP in the previous chapter, TV-STAP reduces the number of complex multiplications relative to SC-STAP by a factor that is closely approximated by the batch length N_p. For example, if the hot clutter can be considered locally stationary over a typical surface-mode PRI with repetition frequency $f_p = 4$ Hz, the same signal may be considered stationary over 15 consecutive air-mode PRIs when $f_p = 60$ Hz, as the physical time interval is unchanged. In this hypothetical example, TV-STAP using $N_p = 16$ may be expected to cancel the hot clutter as effectively as SC-STAP with an order of magnitude reduction in computational load.

$$\hat{\mathbf{w}}_k(b) = \hat{\mathbf{R}}_b^{-1}\hat{\mathbf{C}}_k(b)[\hat{\mathbf{C}}_k^\dagger(b)\hat{\mathbf{R}}_b^{-1}\hat{\mathbf{C}}_k(b)]^{-1}\hat{\mathbf{f}}_k(b)\ , \quad b = 2, \ldots, N_b \tag{11.64}$$

It may be verified that for the parameter choice $N_p = 1$, the TV-STAP algorithm reverts back to the SC-STAP method, albeit applied to the data in reverse slow-time order. In

Fast-Time STAP Technique	Optimum Weights	Adaptive Weights
Time-invariant	$\tilde{\mathbf{w}}$ Eqn. (11.40)	$\hat{\mathbf{w}}$ Eqn. (11.42)
Unconstrained	$\tilde{\mathbf{w}}(t)$ Eqn. (11.43)	$\hat{\mathbf{w}}(t)$ Eqn. (11.45)
Stochastic constraints	$\tilde{\mathbf{w}}_k(t)$ Eqn. (11.51)	$\hat{\mathbf{w}}_k(t)$ Eqn. (11.54)
Time-varying	$\tilde{\mathbf{w}}_k(b)$ Eqn. (11.61)	$\hat{\mathbf{w}}_k(b)$ Eqn. (11.64)

TABLE 11.1 Summary of the considered fast-time STAP techniques. Time-invariant and unconstrained STAP are standard schemes, while SC-STAP and TV-STAP are regarded as the alternative techniques. TV-STAP is most general in the sense that other techniques can be represented as special cases of TV-STAP for appropriate parameter choices.

this case, TV-STAP is able to protect the cold clutter AR spectral characteristics in similar fashion to the SC-STAP method. Hence, TV-STAP may be regarded as a generalization of SC-STAP, which provides additional flexibility to reduce the computational burden when the prevailing interference environment permits (i.e., when the effective stationarity interval of the received hot clutter exceeds a single PRI). Clearly, the standard time-invariant STAP corresponds to $N_p = P$, while the standard time-dependent STAP scheme results for $N_p = 1$ and $\kappa = 0$. Through appropriate choices of parameters N_p and κ, the TV-STAP algorithm encapsulates all STAP schemes described above as special cases, namely; the two standard STAP schemes and SC-STAP. A summary of the fast-time STAP techniques described in this section is provided in Table 11.1.

11.3.3 Simulation Results

The simulations reported in Abramovich et al. (1998) are performed here to reproduce the numerical results described therein. In addition to ensuring consistency with the benchmark results of Abramovich et al. (1998), use of the same hot- and cold-clutter model parameters provides a fair basis on which to compare the newly proposed TV-STAP algorithm with the SC-STAP method in cases where the two techniques are not equivalent.

Once the hot and cold clutter models are defined, and the dimensions of the OTH radar data-cube are specified, the numerical analysis proceeds in three stages. The first stage examines the maximum potential effectiveness of each processing scheme in terms of hot-clutter rejection. These schemes include conventional processing, spatial-only adaptive processing (SAP), and the four previously described fast-time STAP techniques. This analysis of rejection performance is based on clairvoyant knowledge of the true hot-clutter plus noise covariance matrix. The main objective is to quantify the *upper limit* on hot-clutter rejection achieved by the various schemes.

The second stage investigates the impact on cold clutter processing for the schemes capable of effective hot-clutter rejection, particularly with respect to preserving SCV after Doppler processing. The synthesis of first and second stage results enables schemes that have the potential to be appropriate for hot- and cold-clutter mitigation in OTH radar to be identified. The final stage compares the performance of fully operational routines for these identified schemes. In this case, training data is used to estimate the hot-clutter covariance matrices, while cold-clutter samples are not directly available to form data-dependent constraints.

Hot Clutter	$\theta_{m\ell}$, deg.	$\alpha_{m\ell}$, f_p = 5 Hz	$\alpha_{m\ell}$, f_p = 50 Hz	$\beta_{m\ell}$	HCNR, dB
Mode 1	0.5	1.00	1.00	1.00	30
Mode 2	20.5	0.90	0.98	0.91	25
Mode 3	39.3	0.88	0.97	0.90	20
Mode 4	44.9	0.91	0.99	0.90	35

TABLE 11.2 Hot-clutter parameters for a single ($M = 1$) source and four ($L = 4$) propagation modes.

A uniform linear array (ULA) composed of $N = 16$ identical antenna sensors with half-wavelength inter-element spacing is considered for the simulations. The beam is steered in the broadside direction, i.e., $\varphi_0 = 0$. As in Abramovich et al. (1998), a single far-field hot clutter source is assumed with four propagation modes, i.e., $M = 1$ and $L = 4$. The spatial and temporal parameters describing the four hot clutter modes are listed in Table 11.2. The DOA of the first hot clutter mode is intentionally chosen to lie in the main beam so as to demonstrate the benefit of fast-time STAP relative to SAP in such a scenario. Temporal parameters have been specified for low and high PRF modes in Table 11.2, where $f_p = 5$ Hz and $f_p = 50$ Hz are typical for OTH radar ship and aircraft detection missions, respectively.

The radar is assumed to transmit $P = 256$ pulses over the CPI. The model parameters corresponding to the cold clutter are listed in Table 11.3 for terrain ($\kappa = 1$) and sea surface ($\kappa = 2$) scattering. Recall that the rank of the hot clutter space/fast-time covariance matrix averaged over $\kappa + 1$ repetition periods is given by $(\kappa + 1)M(L + Q - 1)$. A number of fast-time taps $Q \geq 2$ is needed to cope with main beam hot clutter condition. Selecting $Q = 3$ results in $NQ = 48 \gg 18 = (\kappa + 1)M(L + Q - 1)$ for the case of sea-surface scattering. This number of fast-time taps is sufficient to ensure that fast-time STAP has the potential to effectively reject the hot clutter.

11.3.3.1 Hot-Clutter Rejection

A standard metric for comparing the performance of different processing techniques is the output signal-to-hot clutter plus noise ratio (SHCR). For an ideal useful signal with unit power, the conventional matched filter beamformer $\mathbf{s}(\varphi_0)$ yields an output SHCR given by Eqn. (11.65), where $\mathbf{R}_0(t)$ is the statistically expected hot clutter plus noise *spatial* covariance matrix for pulse t. Note that the steering vector is normalized such that $\mathbf{s}(\varphi_0)^\dagger\mathbf{s}(\varphi_0) = N = 16$.

$$q_{CBF}(t) = \frac{|\mathbf{s}(\varphi_0)^\dagger\mathbf{s}(\varphi_0)|^2}{\mathbf{s}(\varphi_0)^\dagger\mathbf{R}_0(t)\mathbf{s}(\varphi_0)} \tag{11.65}$$

Cold Clutter	b_1	b_2	σ_ε^2	ρ_s	CNR, dB
Sea-surface, $f_p = 5$ Hz	−1.9359	0.998	0.009675	0.5	50
Terrain, $f_p = 50$ Hz	0.999	0	0.002	0.5	50

TABLE 11.3 Cold-clutter AR model parameters for terrain ($\kappa = 1$) and sea-surface ($\kappa = 2$) scattering.

In the clairvoyant case, $\mathbf{R}_0(t)$ is known and the optimum time-varying SAP filter takes the form of $\mathbf{w}(t) = \{\mathbf{s}(\varphi_0)^\dagger \mathbf{R}_0^{-1}(t)\mathbf{s}(\varphi_0)\}^{-1}\mathbf{R}_0^{-1}(t)\mathbf{s}(\varphi_0)$. This shall be referred to as the standard unconstrained SAP filter. For an ideal useful signal with unit power per element, the optimum SAP filter yields a maximum instantaneous output SHCR given by Eqn. (11.66). The highest potential effectiveness is bounded above by the signal-to-white noise ratio (SWNR) at the matched-filter output when only white additive noise is present. For additive white noise of unit power per element, we have $\mathbf{R}_0(t) = \mathbf{I}_N$ in the *absence* of hot clutter, so the upper bound for $q_{SAP}(t)$ and $q_{CBF}(t)$ is $10\log_{10} N = 12$ dB.

$$q_{SAP}(t) = \mathbf{s}(\varphi_0)^\dagger \mathbf{R}_0^{-1}(t)\mathbf{s}(\varphi_0) \tag{11.66}$$

The output SHCR for the optimum fast-time STAP filter associated with the time-varying weight vector $\tilde{\mathbf{w}}(t) = \tilde{\mathbf{R}}^{-1}(t)\mathbf{A}_q[\mathbf{A}_q^\dagger \tilde{\mathbf{R}}(t)^{-1}\mathbf{A}_q]^{-1}\mathbf{e}_q$ that employs only linear deterministic constraints is given by Eqn. (11.67). Recall that this filter provides fixed unit gain and distortionless response for ideal useful signals in the look direction φ_0, and that $\tilde{\mathbf{R}}(t)$ is the true hot clutter plus noise covariance matrix at pulse t. This standard technique was previously referred to as *unconstrained* STAP.

$$q_{STAP}(t) = \{\tilde{\mathbf{w}}^\dagger(t)\tilde{\mathbf{R}}(t)\tilde{\mathbf{w}}(t)\}^{-1} = \{\mathbf{e}_Q^T[\mathbf{A}_Q^\dagger \tilde{\mathbf{R}}(t)^{-1}\mathbf{A}_Q]^{-1}\mathbf{e}_Q\}^{-1} \tag{11.67}$$

On the other hand, the output SHCR for the optimum fast-time STAP technique based on the time-invariant weight vector $\tilde{\mathbf{w}} = \tilde{\mathbf{R}}^{-1}\mathbf{A}_q[\mathbf{A}_q^\dagger \tilde{\mathbf{R}}^{-1}\mathbf{A}_q]^{-1}\mathbf{e}_q$, derived from the hot clutter plus noise covariance matrix $\tilde{\mathbf{R}} = \sum_{t=1}^{P}\tilde{\mathbf{R}}(t)$ *averaged* over the whole CPI, is given by Eqn. (11.68). This standard technique was previously referred to as *time-invariant* STAP.

$$q_{AVE}(t) = \left\{\tilde{\mathbf{w}}^\dagger \tilde{\mathbf{R}}(t)\tilde{\mathbf{w}}\right\}^{-1} \tag{11.68}$$

The simulation results in Figure 11.4 illustrate the quasi-instantaneous output SHCR for the conventional beamformer, the optimum unconstrained SAP filter $q_{SAP}(t)$, the optimum unconstrained STAP filter $q_{STAP}(t)$, and the optimum time-invariant STAP filter $q_{AVE}(t)$, as a function of pulse number t over the CPI. Recall that the term unconstrained refers to the use of standard deterministic constraints only, i.e., without the auxiliary data-driven linear constraints used by SC-STAP and TV-STAP.

The curves in Figure 11.4 indicate the maximum potential effectiveness of the various *standard* techniques when estimation errors due to finite sample support are neglected. As expected, unconstrained STAP performs best. It provides an output SHCR improvement of about 60 dB with respect to the conventional beamformer $q_{CBF}(t)$. Moreover, it yields an output SHCR that is within about 6 dB of the upper bound on performance in this example (i.e., 12 dB).

On the other hand, unconstrained SAP is rather ineffective due to the presence of a main-beam hot-clutter component. This illustrates the susceptibility of a slow-time varying SAP filter to hot clutter, even when the number of antennas exceeds the number of independent interference components (i.e., $N = 16 > 4 = ML$). In this example, unconstrained SAP leads to a loss in output SHCR of around 15 dB with respect to unconstrained STAP.

Greater relative losses of approximately 30 dB are observed for time-invariant STAP, which cannot cancel non-stationary hot clutter effectively over the CPI. Among the considered standard approaches, only the fast-time STAP approach with slow-time-varying weight vectors can provide the possibility to effectively cancel non-stationary hot clutter.

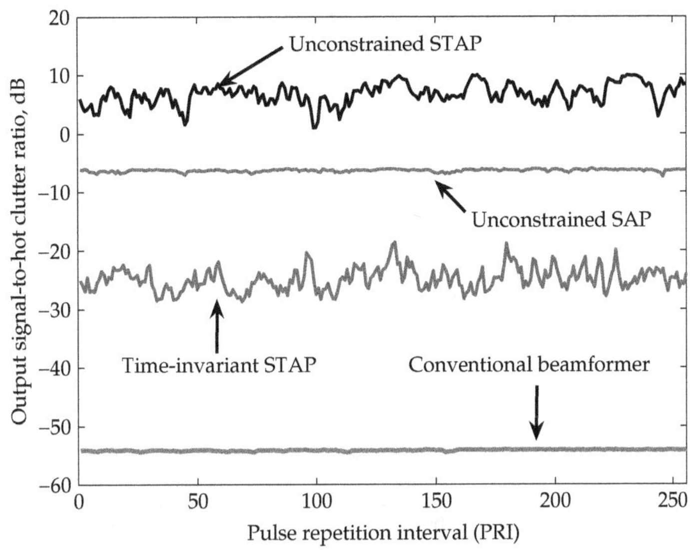

FIGURE 11.4 Optimum output SHCR of standard fast-time STAP and pure SAP approaches as a function of slow-time over the CPI. © Commonwealth of Australia 2011.

The other standard schemes are clearly inappropriate, even when the hot clutter statistics are perfectly known. These numerical results agree closely with those reported for an independent realization of this simulation in Abramovich et al. (1998).

11.3.3.2 Cold-Clutter Processing

From a hot-clutter rejection perspective, the standard unconstrained STAP filter performs very well. The main problem arises when this slow-time sequence of weight vectors $\tilde{\mathbf{w}}(t)$ is used to process range cells that additionally contain cold clutter. To observe the impact on processing the cold clutter only, the hot clutter is momentarily removed from the operational range cells. Figure 11.5 illustrates the cold-clutter Doppler spectrum at the output of the standard unconstrained STAP filter compared to that processed by the standard time-invariant STAP filter using the second-order AR (sea-scattering) model. The dramatic degradation in SCV is obvious when unconstrained STAP is applied. Although $\tilde{\mathbf{w}}(t)$ can effectively remove the non-stationary hot clutter, it is apparent that this filter is completely inappropriate for processing the cold clutter. The unconstrained fluctuations of $\tilde{\mathbf{w}}(t)$ destroy the pulse-to-pulse correlation properties of the scalar cold-clutter output.

Typical target echoes have peaks that may be 40 dB below the main clutter peak in the OTH radar Doppler spectrum. Such targets are undetectable when the SCV falls to about 30 dB, as it does for the unconstrained STAP approach in Figure 11.5. In contrast, time-invariant STAP is suitable as far as cold clutter processing is concerned. This is because a static filter over the CPI avoids any degradation in SCV. However, this method relies on averaging the hot-clutter covariance matrix over the whole CPI, which leads to a degradation in rejection performance of approximately 30 dB in Figure 11.4. In this

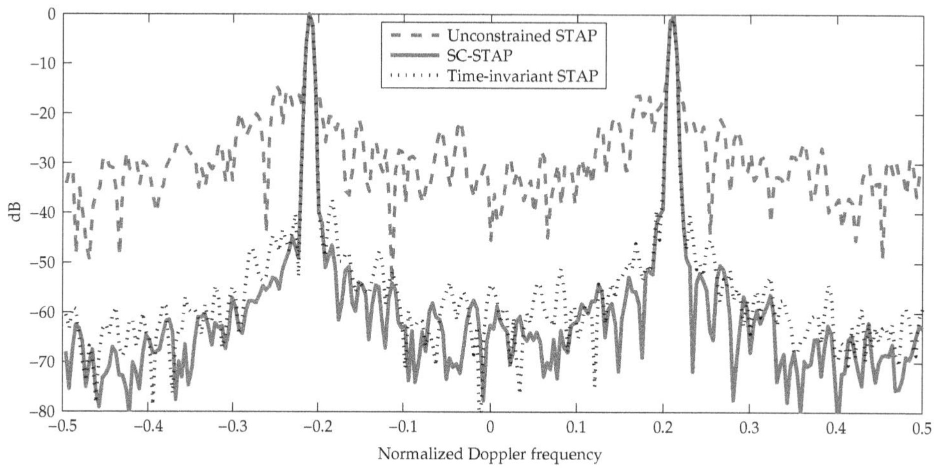

FIGURE 11.5 Doppler spectra showing the sub-clutter visibility at the output of two standard STAP techniques and SC-STAP when these filters are applied to process the cold clutter. © Commonwealth of Australia 2011.

case, target echoes may not be obscured by clutter after Doppler processing, but are likely to be buried under the high residual hot-clutter level.

The simulation results presented thus far demonstrate that both standard fast-time STAP approaches (i.e., unconstrained and time-invariant) are ineffective for the considered problem, as neither can simultaneously yield effective hot-clutter rejection and distortionless cold-clutter processing.

The cold-clutter Doppler spectrum processed by the optimum SC-STAP technique using two stochastic constraints for the AR(2) sea scattering model is also illustrated in Figure 11.5. Ideal stochastic constraints were constructed using pure cold clutter snapshots and are therefore non-operational in this sense. The sequence of optimum SC-STAP filters $\tilde{\mathbf{w}}_k(t)$ can clearly preserve SCV after Doppler processing (similar to time-invariant STAP). The question arises as to the impact of the stochastic constraints on hot clutter rejection.

The top curve in Figure 11.6 shows the maximum potential hot-clutter rejection achieved by SC-STAP based on the clairvoyant hot-clutter-plus-noise covariance matrix and the incorporation of two ideal stochastic constraints. It is observed that the effect of these additional linear constraints on hot-clutter rejection performance is negligible. In other words, optimum SC-STAP achieves practically the same hot clutter rejection performance as the unconstrained STAP approach, which only uses deterministic constraints.

Figure 11.6 also shows the maximum hot-clutter rejection effectiveness of TV-STAP with two ideal data-driven constraints (using pure cold-clutter snapshots), and different batch lengths N_p. Using the simulation parameters in the ship-detection example ($f_p =$ 5 Hz), it is noticed that updating the STAP filter every $N_p = 2$ pulses yields a negligible loss in hot-clutter rejection relative to SC-STAP, which updates the filter every PRI. In this case, TV-STAP demands approximately half the computational load of SC-STAP. Increasing the batch size to $N_p = 16$ pulses leads to rejection losses of about 10–15 dB relative to SC-STAP due to hot-clutter non-stationarity over the batch length. In this

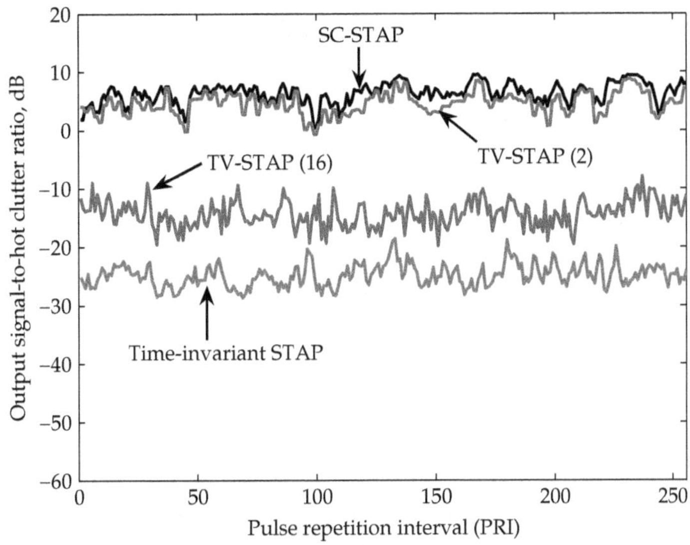

FIGURE 11.6 Optimum output SHCR of SC-STAP and TV-SAP using two ideal auxiliary constraints, and different batch lengths N_p for TV-STAP. The performance of standard time-invariant STAP is shown again here for reference. © Commonwealth of Australia 2011.

case, TV-STAP trades this loss in output SHCR for an order of magnitude reduction in computational complexity, which may be necessary to allow real-time implementation.

It is observed that TV-STAP with $N_p = 16$ still provides a 10–15-dB improvement in hot-clutter rejection performance with respect to standard time-invariant STAP. The other issue for TV-STAP is the impact on cold-clutter processing. Figure 11.7 confirms that the cold-clutter Doppler spectra at the output of TV-STAP and SC-STAP have practically identical SCV. This illustrates that TV-STAP with two data-driven constraints can protect SCV when the weights are updated over batches, and the cold-clutter signal is strictly described by a scalar-type AR model.

Based on the model parameters used to simulate the ship detection example, it may be anticipated that TV-STAP can yield comparable performance to SC-STAP for about half the computational cost (i.e., with a batch length of $N_p = 2$ pulses). Among the fast-time STAP techniques discussed in this chapter, SC-STAP and TV-STAP represent the only promising candidates in terms of offering an effective solution to the problem at hand.

11.3.3.3 Fully Operational Schemes

Figures 11.8 and 11.9 show the performance of operational STAP schemes, where the hot-clutter covariance matrix is estimated from training data, and data snapshots containing a mixture of hot and cold clutter are used to generate the auxiliary data-driven linear constraints. In all examples, the first $N_k = 50$ range cells in each PRI were deemed to be free of cold clutter. These cells were used as training data to estimate the unknown hot-clutter-plus-noise covariance matrices.

Figure 11.8 illustrates the output of the operational SC-STAP filter $\hat{\mathbf{w}}_k(t)$, which can detect a synthetic target 30 dB above the local background disturbance level. In contrast,

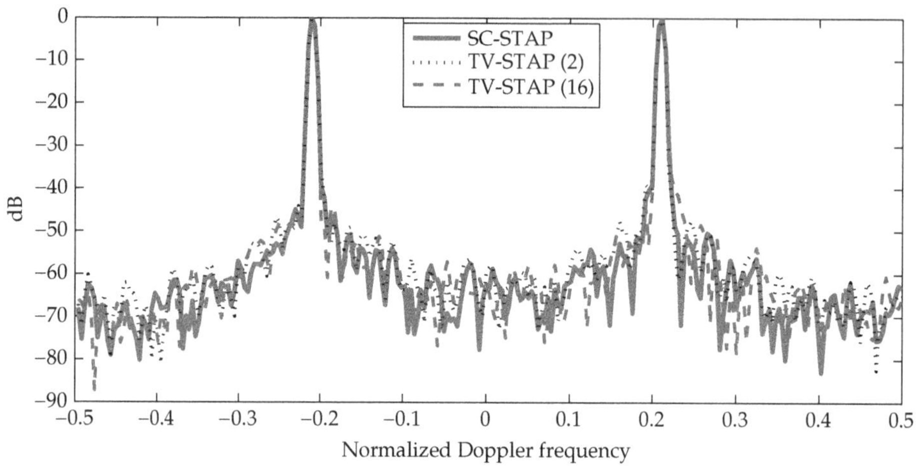

FIGURE 11.7 Doppler spectra showing the sub-clutter visibility at the output of SC-STAP and TV-STAP using different batch lengths when these filters are applied to process the cold clutter. © Commonwealth of Australia 2011.

the time-invariant STAP filter $\hat{\mathbf{w}}$ was not able to distinguish this useful signal due to masking from inadequately rejected hot clutter. Figure 11.9 compares the performance of SC-STAP and TV-STAP using a batch length of $N_p = 2$ PRIs. The curves indicate quite similar performance, which confirms that TV-STAP can offer a useful computational advantage for negligible performance loss in the simulated ship-detection example.

Now consider the high PRF (air-detection) scenario, with terrain-scattered cold clutter. In this case, the PRI has one-tenth of the duration assumed for the previous (ship-detection) simulation. The inter-PRI temporal correlation coefficients of the hot-clutter

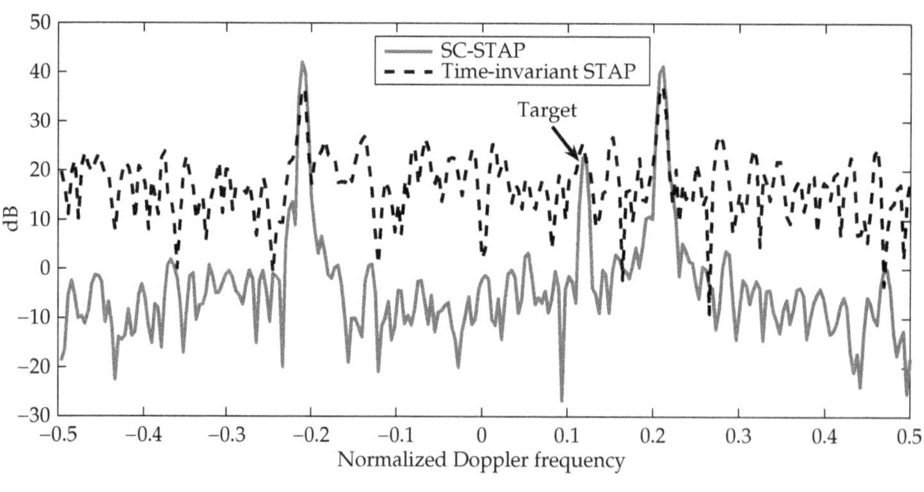

FIGURE 11.8 Doppler spectra for the operational time-invariant STAP and SC-STAP routines in the ship detection example with an injected useful signal. © Commonwealth of Australia 2011.

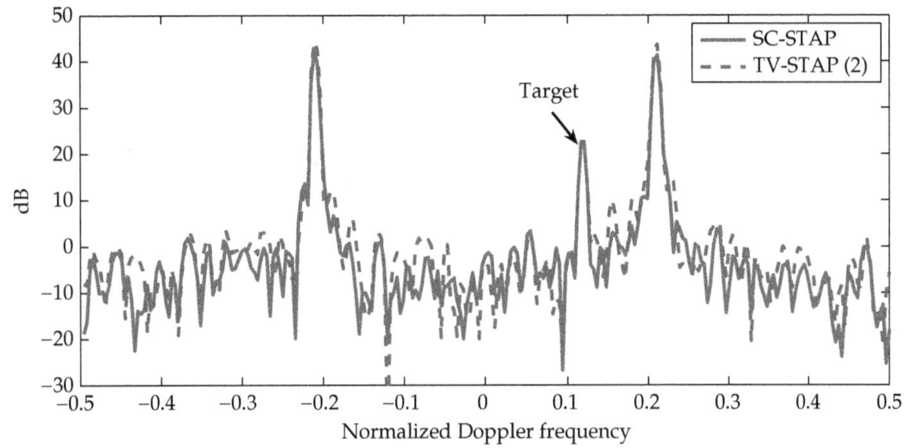

FIGURE 11.9 Doppler spectra for the operational SC-STAP and TV-STAP routines in the ship detection example with an injected useful signal. © Commonwealth of Australia 2011.

modes may be adjusted to account for the higher PRF of $f_p = 50$ Hz in accordance with the decaying exponential model $\alpha(f_p) = e^{-B_t/f_p}$, where the bandwidth B_t is computed from the hot-clutter parameters listed for the low PRF mode ($f_p = 5$ Hz) in Table 11.2. This procedure is used to adjust the hot-clutter coefficients α_{ml} for the $f_p = 50$ Hz case in Table 11.2. The first-order AR model is adopted to describe terrain-scattered cold clutter using the parameters listed in Table 11.3.

Figures 11.10 and 11.11 show the Doppler spectra resulting for operational STAP schemes in the high PRF example. In this case, both SC-STAP and TV-STAP employ

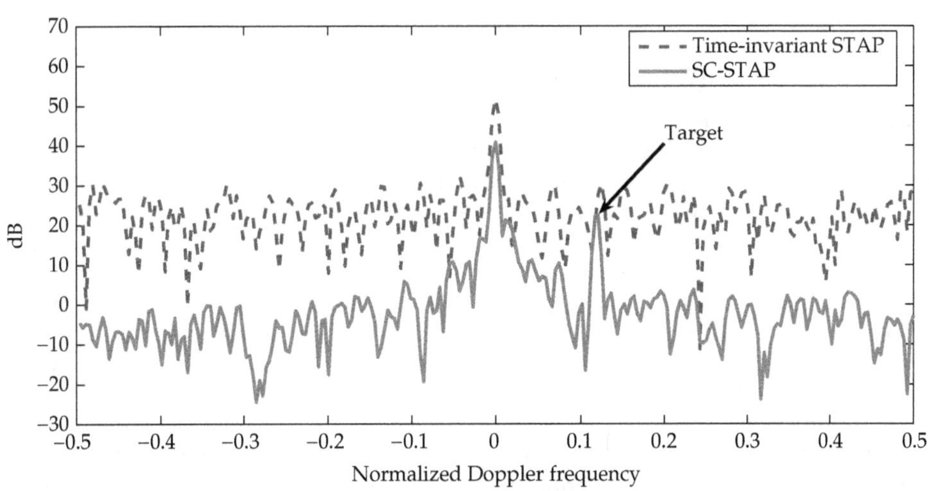

FIGURE 11.10 Doppler spectra for operational time-invariant STAP and SC-STAP routines in the aircraft detection example with an injected target. © Commonwealth of Australia 2011.

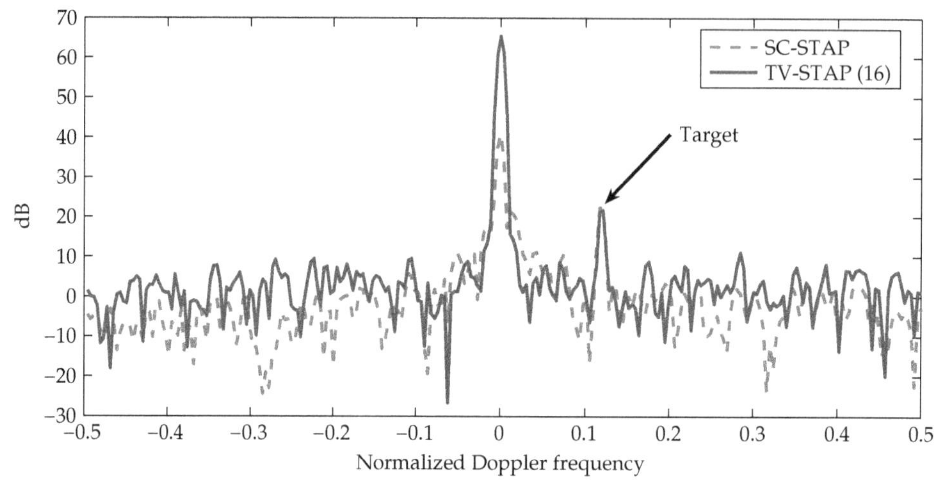

FIGURE 11.11 Doppler spectra for operational SC-STAP and TV-STAP routines in the aircraft detection example with an injected target. © Commonwealth of Australia 2011.

a single auxiliary linear constraint to protect SCV, which is appropriate for the assumed first-order AR cold-clutter model. It is clear from Figure 11.10 that time-invariant STAP leads to a degradation in hot clutter rejection of about 20 dB relative to SC-STAP. This degradation is severe enough to mask the useful signal, which is readily detected by SC-STAP in Figure 11.10. Figure 11.11 compares SC-STAP with TV-STAP using a batch length of $N_p = 16$ pulses. This batch length provides an order of magnitude reduction in computational load with respect to SC-STAP for negligible loss in hot-clutter rejection. Clearly, both SC-STAP and TV-STAP detect the useful signal in Figure 11.11.

11.4 Post-Doppler STAP Implementation

In the previous section, fast-time STAP was used to mitigate hot clutter signals that were time-continuous over the CPI, broadband with respect to the sampling rate, incoherent with the radar waveform, and received as multiple propagation modes, one or more of which may enter through the main lobe of the radar beam. The correlation properties of these *active* interferences were assumed to be statistically homogeneous in fast-time, which allowed STAP filters to be trained effectively using secondary data obtained from a limited number range bins free of cold clutter in each repetition period. Unwanted signals from *passive* sources that reflect or scatter the radar signal, as opposed to radiating an independent waveform, give rise to coherent disturbances, which pose a quite different problem as far as mitigation by means of STAP is concerned.

For example, radar echoes from meteor trails, or highly dynamic electron-density irregularities in the ionosphere, may produce echoes that are significantly spread in Doppler. Such returns have the potential to mask useful signals over much of the target velocity search space. Perhaps more importantly, these coherent disturbance signals tend to have spatial characteristics that are highly heterogeneous in range. This is because the energy of a coherent signal received from a particular scatterer is localized in range after pulse compression. In general, this means that each range cell samples a different spatial distribution of scatterers. The main implication of such heterogeneity is that adaptive filters need to be trained on one range cell at a time. In this case, secondary data is taken from slow-time samples or Doppler bins in the range cell under test.

In the previous chapter, a post-Doppler SAP method with range-dependent weights was described. When main-beam interference from active sources is additionally present, this scheme will attempt to jointly cancel the coherent disturbance and main-beam interference, which typically leads to performance degradation. Moreover, the main-beam interference may be received via a single dominant mode instead of multiple propagation paths. In this case, fast-time STAP is not effective for broadband sources as it relies on multipath to cancel the main-beam interference. However, if the main-beam interference has an effective bandwidth that is small compared to the radar bandwidth, the significant correlation existing between fast-time samples (acquired at the Nyquist rate f_s) can be exploited for cancelation. Typically, narrowband interference in the HF band has an effective bandwidth that is higher than the pulse repetition frequency f_p, so its energy spreads across Doppler. Narrowband interference therefore has the potential to mask targets in all range-Doppler cells.

In this section, a post-Doppler STAP approach is described for jointly mitigating range-heterogeneous spread-Doppler clutter and main-beam narrowband (i.e., range-correlated) interference. The proposed architecture simultaneously processes radar data across a number of auxiliary beams and ranges. The former provides spatial DOFs for canceling sidelobe signals, while the latter provides fast-time DOFs for canceling main-beam interference that is correlated in range. An important point is that training of the STAP filter is performed on a range-by-range basis in the Doppler domain for the reasons stated above. Another key point is that the use of a fast-time tap delay line behind each spatial DOF is dispensed with in this application. An alternative reduced-dimension post-Doppler STAP architecture in which the number of spatial and temporal DOFs *add* instead of *multiply* will be proposed in Section 11.4.1. Experimental results are illustrated for a real-time practical implementation of this alternative STAP architecture. These results demonstrate the performance of an operational routine in a trial involving a cooperative aircraft target where on-board GPS data is used to provide ground-truth on the flight path.

11.4.1 Algorithm Description

Some of the symbols used in the first part of this chapter will be redefined in this section to avoid cumbersome notation. In traditional STAP architectures, the use of K spatial channels (receivers or beams) and L temporal taps (fast-time samples in this case) results in a space-time filter dimension equal to the product $K \times L$. Even for modest parameter values, e.g., $K = 16$ and $L = 8$, this architecture results in a processor with $K \times L = 128$ DOFs. The problem with such a large filter dimension is that adaptation on a range-by-range basis needs to be achieved by training over Doppler bins, which are typically limited to $P = 256$ or less in practical OTH radar applications. Moreover, the effective number of independent Doppler bins is usually about half this amount due to the low sidelobe windows often used for Doppler processing.

To counter the finite sample-support problem, rank reduction based on singular value decomposition may be performed on the data, but such procedures are computationally expensive to apply for each range cell. Diagonal loading provides an alternative to improve convergence rate, but the processor dimension and hence computational load is not reduced. A different option is to identify a reduced dimension post-Doppler STAP architecture to address the practical issues of finite sample support and computational load.

This section describes an alternative post-Doppler STAP formulation that uses K beams for spatial adaptation and L fast-time taps for temporal adaptation with the latter taken only from the *reference beam*. In this way, the STAP filter dimension is given by the sum $Q = K + L$, which is typically much less than the product $Q \ll KL$. The main benefits of this architecture include reduced demands on statistically homogeneous training data and lower computational load for real-time implementation.

Figure 11.12 illustrates the reduced-dimension STAP processing architecture using $K + L$ degrees of freedom. In our previous example, where $K = 16$ and $L = 8$, we have $Q = 24$. This is a significant reduction on $K \times L = 128$. In addition, $Q > 5P$ for $P = 128$ Doppler cells, which means that sufficient training is available for range-by-range adaptation. In this architecture, the main and auxiliary beams are formed conventionally. The weighted combination of auxiliary beam outputs provides a means to cancel *sidelobe* disturbances. This includes unwanted (sidelobe) signals from coherent and incoherent sources.

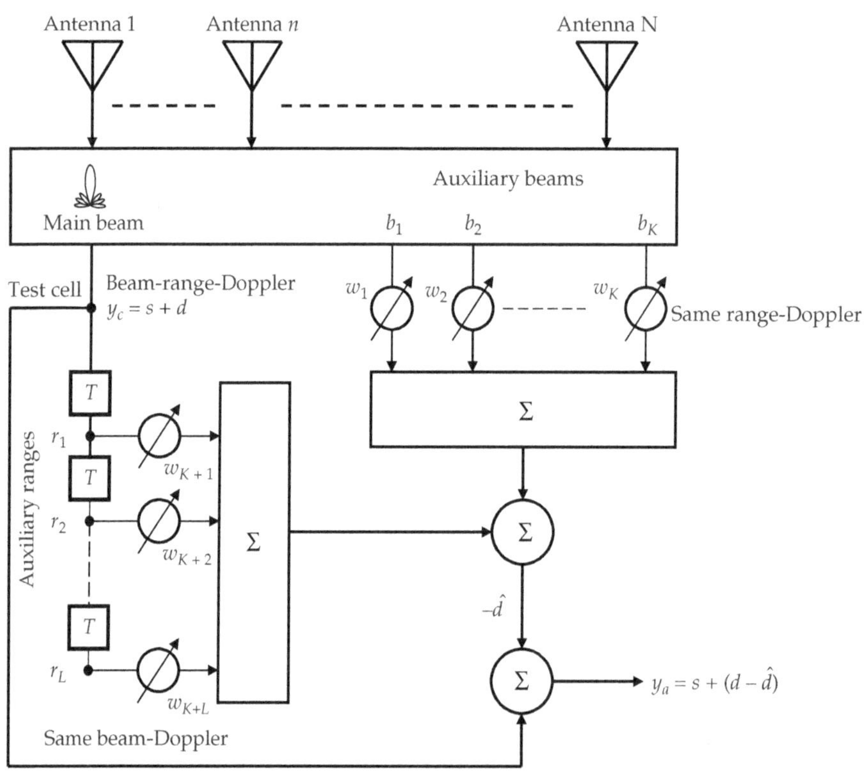

FIGURE 11.12 Reduced-dimension STAP processing scheme with K spatial taps (auxiliary beams) and L temporal taps (auxiliary ranges) that results in a processor dimension of $K + L$ instead of $K \times L$ (degrees of freedom). © Commonwealth of Australia 2011.

On the other hand, the fast-time taps behind the main beam output provide a means to reject range-correlated interference that may be received from the useful signal direction. This may include interference from transient (impulsive) signals, or long-lived (narrowband) signals on the scale of the CPI. Naturally, impulsive interference that corrupts only a few PRIs can also be mitigated by excision and linear prediction techniques applied in the slow-time domain. However, exploiting the range-correlation structure of such interferences can have advantages when a significant number of PRIs in the CPI are affected.

As the spatial and temporal taps in this STAP architecture are used to mitigate different interference types, it can be argued that a factored approach involving SAP followed by range-only adaptive processing (RAP), or vice versa, could be applied to further reduce sample-support requirements and computational load. The main issue is that supervised training data containing only one interference type in isolation from the other is generally not available for the described application. For example, the presence of main-beam interference in the SAP training data can bias the filter weights needed to cancel sidelobe disturbances, as well as distort the main lobe of the antenna pattern. In such cases, STAP provides an avenue for joint cancelation based on training data that contains a mixture of the different disturbance types.

However, turning the STAP architecture of Figure 11.12 into a robust and practical technique requires a number of significant technical challenges to be overcome. Performance-related aspects include: (1) the quality of disturbance suppression, which depends on the allocation and relative distribution of adaptive DOFs in space and time, as well as the training-data selection strategy, and (2) the impact on useful signals, where it is important to avoid target echo self-cancelation and copying effects that can produce false alarms or degrade parameter estimation accuracy for target tracking.

Robustness to the wide diversity of operational conditions encountered by practical OTH radar systems in the unpredictable HF environment is also essential. All of these prerequisites need to be achieved with minimal or no requirement for human intervention, while the algorithm needs to be computationally efficient to permit real-time operation on existing platforms without severely consuming radar resources. Satisfying this combination of inter-dependent objectives is typically not straightforward, but necessary in order to claim an effective real-time STAP capability. At this point, the proposed post-Doppler STAP method, referred to as range-dependent (RD)-STAP, is described.

Let the N-dimensional complex vector $\mathbf{x} = [x_1, x_2, \ldots, x_N]^T$ be the *element-space* spatial snapshot resulting at a particular *range-Doppler* cell after matched filtering is performed in each receiver. For a two-dimensional array, the radar look direction is defined by steering angles in azimuth θ_o and elevation ϕ_o. The conventional beamformer output for the main beam steered in this direction is given by the complex scalar $y_c = \mathbf{s}^\dagger(\theta_o, \phi_o)\mathbf{x}$, where $\mathbf{s}^\dagger(\theta_o, \phi_o)$ is the array steering vector.

The array steering vector is parametrically defined by Eqn. (11.69), where the term $\mathbf{k}(\theta, \phi) = \frac{2\pi}{\lambda}[\cos\phi\cos\theta,\ \cos\phi\sin\theta,\ \sin\phi]^T$ is the signal wave vector, and $\mathbf{r}_n = [x_n, y_n, z_n]^T$ is the antenna position vector in Cartesian coordinates relative to the phase reference for the elements $n = 1, 2, \ldots, N$. Conventional beamforming of each range-Doppler cell processed by the system over a set of beams steered in different directions that (at least) cover the surveillance region, results in a beam-range-Doppler (BRD) data cube. In other words, the raw data cube with dimensions of antenna element, fast-time and slow-time is transformed by conventional processing to a data cube with dimensions of steer direction, range cell, and Doppler bin. This conventionally processed data cube is normally passed on for CFAR processing followed by peak detection-estimation and tracking.

$$\mathbf{s}(\theta, \phi) = [e^{j\mathbf{k}(\theta,\phi)\cdot\mathbf{r}_1}, e^{j\mathbf{k}(\theta,\phi)\cdot\mathbf{r}_2}, \ldots, e^{j\mathbf{k}(\theta,\phi)\cdot\mathbf{r}_N}]^T \tag{11.69}$$

The proposed RD-STAP method is applied directly to the conventional BRD data cube, and may therefore be "switched in" as the final stage of the signal-processing chain, i.e., immediately prior to the CFAR processing step. A well-designed STAP procedure is able to substantially improve detection performance when interference is present, but should automatically revert back to what is effectively conventional processing under quiescent conditions. In this case, a radar operator can keep STAP operating all the time, providing the computational burden is low enough so as not to compromise other radar functions or performance. In essence, the purpose of RD-STAP is to remove residual interference that contaminates the conventionally processed BRD map before this data is passed on for subsequent CFAR processing.

Define $\mathbf{z} \in \mathcal{C}^{K+L}$ in Eqn. (11.70) as the primary data vector to be processed by the STAP filter. The first scalar element of this data vector y_c is the conventional output at a particular (unnamed) location in the BRD data cube. The RD-STAP procedure assigns each cell in the BRD map as the current test cell in turn. The second component of the primary data vector contains a set of K auxiliary beam outputs, contained in the vector $\mathbf{b} = [b_1, b_2, \ldots, b_K]^T$, where each auxiliary beam sample b_k is extracted from the *same* range-Doppler coordinate as the test cell y_c, but from an auxiliary beam steered in a different direction to the main beam. The third component of $\mathbf{z}$ contains a set of L auxiliary range cells denoted by $\mathbf{r} = [r_1, r_2, \ldots, r_L]^T$, where each fast-time sample r_ℓ is taken from the *same* beam-Doppler coordinate as the test cell y_c, but from a fast-time sample different to the current range.

$$\mathbf{z} = [y_c, \mathbf{b}^T, \mathbf{r}^T]^T \tag{11.70}$$

Figure 11.12 illustrates the conventional beam-range-Doppler sample y_c, the L range samples $r_1, \ldots, r_L$ from the same beam and Doppler cell as y_c, and the K auxiliary beams $b_1, \ldots, b_L$ from the same range and Doppler cell as y_c. Naturally, the useful signal vector needs to be modified in accordance with the beam and range transformations to reflect the chosen set of auxiliary spatial channels and range taps in the primary data vector $\mathbf{z}$.

Let $\mathbf{T}_b$ be the $K \times N$ matrix that transforms the N receiver outputs to the K selected auxiliary beams, such that $\mathbf{b} = \mathbf{T}_b\mathbf{x}$. For example, if the K selected auxiliary beams are conventionally formed in steer directions $\{\theta_k, \phi_k\}_{k=1}^K$, the transformation matrix would be defined as $\mathbf{T}_b = [\mathbf{s}(\theta_1, \phi_1), \ldots, \mathbf{s}(\theta_K, \phi_K)]^\dagger$. If the auxiliary beams are formed using a taper function $\{w_n\}_{n=1}^N$ to lower the antenna pattern sidelobes, the transformation matrix becomes $\mathbf{T}_b = [\mathbf{t}(\theta_1, \phi_1), \ldots, \mathbf{t}(\theta_K, \phi_K)]^\dagger$, where $\mathbf{t}(\theta_k, \phi_k) = \mathbf{D}\mathbf{s}(\theta_k, \phi_k)$ and $\mathbf{D} = \text{diag}[w_1, \ldots, w_n]$. In any case, the *spatial* transformation of the steering vector is given by

$$\mathbf{v}_b = \mathbf{T}_b\mathbf{s}(\theta_o, \phi_o) \tag{11.71}$$

Similarly, define $\mathbf{T}_r$ as the $L \times M$ matrix that transforms the M fast-time samples in the PRI to the L selected auxiliary range cells. For a CW-OTH radar system, each row of $\mathbf{T}_r = [\mathbf{g}(\tau_1), \ldots, \mathbf{g}(\tau_L)]^\dagger$ is an M-dimensional vector $\mathbf{g}^\dagger(\tau_\ell)$ that contains the matched filter coefficients used for pulse compression, i.e., to form the range bin at fast-time sample delay τ_ℓ. Typically, the elements of $\mathbf{g}(\tau_\ell)$ correspond to the transmitted radar waveform delayed by τ_ℓ fast-time samples and shaded in amplitude to reduce range sidelobes. For a PW-OTH radar system, where fast-time samples correspond directly to range bins, $\mathbf{g}(\tau_\ell)$ may be considered a vector with a unit element in position ℓ and zeros elsewhere. The modification to the *temporal* component of the steering vector is given by Eqn. (11.72), where $\mathbf{g}(\tau_o)$ is the radar waveform with delay τ_o corresponding to the range bin of the current output y_c.

$$\mathbf{v}_r = \mathbf{T}_r \mathbf{g}(\tau_o) \tag{11.72}$$

The space-time steering vector for the proposed RD-STAP architecture may then be constructed as $\mathbf{v}$ in Eqn. (11.73). Note that $\mathbf{v}$ involves a concatenation of the transformed spatial and temporal steering vectors rather than a Kronecker product. Stated simply, $\mathbf{v}$ is the useful signal vector for the proposed STAP architecture. This vector represents the space-time signature of the target echo that is sought in the primary data $\mathbf{z}$. When a detection results, target presence is declared at the beam-range-Doppler coordinates of the test sample y_c. Note that the signal vector $\mathbf{v}$ is valid for all Doppler frequency bins, and only changes when different auxiliary beams or ranges are selected. The vectors $\mathbf{v}_b$ and $\mathbf{v}_r$ effectively reduce to zero when the beam and range (sidelobe) responses to the target echo fall below the thermal noise floor in the selected auxiliary beams and ranges, respectively. In this case, any useful signal energy leaking into the auxiliary beams and ranges is small enough to be neglected.

$$\mathbf{v} = [1, \mathbf{v}_b^T, \mathbf{v}_r^T]^T \tag{11.73}$$

When such conditions do not hold, either because of the coordinates of the selected auxiliary beams or ranges relative to the test cell, or the manner in which the auxiliary beams and ranges are formed (with due regard to the strength of the useful signal), the vectors $\mathbf{v}_b$ and $\mathbf{v}_r$ will contain significant nonzero elements. In this case, it is advisable to normalize the useful signal signature vector, such that it has fixed unit norm $\mathbf{v}^\dagger\mathbf{v} = 1$. If the statistically expected covariance matrix $\mathbf{R} = \mathrm{E}\{\mathbf{z}\mathbf{z}^H|H_0\}$ for the primary data vector were known under the null hypothesis (i.e., the disturbance-only covariance matrix), the optimal STAP weight vector $\mathbf{w}_{opt}$ for processing the data vector would be given by the well-known rule in Eqn. (11.74).

$$\mathbf{w}_{opt} = \frac{\mathbf{R}^{-1}\mathbf{v}}{\mathbf{v}^H\mathbf{R}^{-1}\mathbf{v}} \tag{11.74}$$

In practice, this matrix is not known a priori and must be estimated from a set of D secondary data vectors $\mathbf{z}_d$ indexed by $d = 1, 2, \ldots, D$. These secondary vectors are assumed to be free of the useful signal and to share the same statistical characteristics as the *disturbance* in the primary data $\mathbf{z}$. In the current application, the disturbance statistics are presumed to be heterogeneous in range. For this reason, the secondary data is extracted from P Doppler bins available in the same range cell as the sample under test. The set of $D < P$ Doppler bins utilized for training typically exclude the test cell plus a guard cell either side of it. Doppler bins deemed to contain "outliers" are also screened out from the training set. A disturbance sample covariance matrix with appropriate regularization $\hat{\mathbf{R}}$ is formed using the selected secondary data, as in Eqn. (11.75). The operational STAP weights $\hat{\mathbf{w}}$ are then computed by replacing the unknown true covariance matrix $\mathbf{R}$ in Eqn. (11.74) with its estimate $\hat{\mathbf{R}}$ in Eqn. (11.75).

$$\hat{\mathbf{w}} = \frac{\hat{\mathbf{R}}^{-1}\mathbf{v}}{\mathbf{v}^H\hat{\mathbf{R}}^{-1}\mathbf{v}}, \quad \hat{\mathbf{R}} = \frac{1}{D}\sum_{d=1}^{D} \mathbf{z}_d\mathbf{z}_d^{\dagger} + \sigma^2\mathbf{I} \tag{11.75}$$

The test and guard cells are excluded from the training data to avoid attenuation of target echoes. These echoes are assumed to have a steady Doppler shift over the CPI, such that significant useful signal energy is not present outside of the guard cells. On the other hand, disturbances such as ionospheric clutter and incoherent interference are often more spread in Doppler than target echoes and will therefore be captured in the training data. The resulting adaptive filter is applied to the primary data to obtain the RD-STAP output y_a given by Eqn. (11.76). This output may be directly compared with the conventional output y_c.

$$y_a = \hat{\mathbf{w}}^{\dagger}\mathbf{z} \tag{11.76}$$

At this point, a number of questions arise. First, how are the number and coordinates of auxiliary beams and ranges chosen? Second, how are number and coordinates of the Doppler training bins selected for a particular test cell? These decisions need to strike a balance between performance and computational complexity. As far as convergence rate for the the sample matrix inverse (SMI) technique is concerned, it is recommended that the number of independent training vectors is *at least* twice the STAP filter dimension (Reed, Mallet, and Brennan 1974). This rule of thumb is expressed in Eqn. (11.77). Given that adjacent Doppler cells are not strictly independent due to the window used for controlling spectral leakage, the requirement $D > 4Q$ may be more appropriate if neighboring Doppler bins are used and regularization is not applied. Appropriate diagonal loading can under certain circumstances significantly reduce the number of independent samples required (Cheremisin 1982). In general, the choice of processor DOFs Q should ensure that sufficient training data is available for adaptation given the number of Doppler bins is limited to P.

$$D > 2Q \tag{11.77}$$

From a computational perspective, updating the STAP weights every Doppler cell may be prohibitive. The matrix inversion lemma (Woodbury's identity) can be used to reduce computational load for low-rank updates of the covariance matrix, but a much faster method is to process a set of Doppler cells using a single weight vector. Similar to the purely spatial approach described in the previous chapter, the positive and negative Doppler frequency bins are processed by independent STAP filters. This requires two matrix inversions of dimension $Q + 1$ for processing all Doppler bins in a beam-range resolution cell. While this approach clearly avoids target contamination in the training data, the number of training cells available for each estimate is half the total number of Doppler bins minus the number of outliers.

The outliers are excluded from the training data because their statistical characteristics are deemed to be not sufficiently representative of the disturbance to be mitigated. In this application, outliers are mainly due to powerful surface clutter components concentrated near zero Doppler frequency, as well as Doppler-shifted target-like returns. For a training data volume of at least four times the STAP filter dimension ($F = 4$), the described rationale imposes an upper limit on the value of Q given by Eqn. (11.78), where J is the maximum number of discarded outlier samples in the positive or negative half of the Doppler spectrum, and $\lfloor\cdot\rfloor$ denotes integer truncation (the floor function).

$$Q \leq \lfloor (P/2 - J)/F \rfloor \tag{11.78}$$

This may be interpreted as an upper limit on Q based on adaptive filter performance considerations. As computational load grows with Q, the upper limit on the value of Q for real-time processing may be lower than that of Eqn. (11.78). Hence, Q is selected as the highest value that satisfies the finite sample-support constraints in Eqn. (11.78), and the restrictions on real-time processing capacity. The processing load may be reduced if the set of beams and ranges contained in the data vector $\mathbf{z}$ remains the same as the sample under test changes, i.e., when one of the auxiliary cells is interchanged with the sample under test and vice versa without introducing new beams or ranges in $\mathbf{z}$. It is then possible to process a block of $K + 1$ beams and $L + 1$ ranges with no further matrix inversions simply by modifying the useful signal signature vector to reflect the new position of the current sample under test.

For example, if the previous test cell becomes the first auxiliary beam and vice versa, the signal vector may be changed to Eqn. (11.79), where unity defines the current position of the main beam and the elements b_k are calculated accordingly. The resulting signal vector is normalized to $\mathbf{v}^\dagger\mathbf{v} = 1$, as before. The weights are then computed by substituting the new signal vector into Eqn. (11.75) with the same matrix inverse for either the positive or negative Doppler bins. An analogous modification is made when swapping one of the auxiliary range cells for the sample under test. Significant computational savings can be achieved when a fixed set of beams and ranges are processed as a block in this manner. The compromise here is that the choice of auxiliary beams and ranges cannot be tailored to specifically suit the test cells on an individual basis.

$$\mathbf{v} = [b_1, 1, b_2, \ldots, b_K, \mathbf{v}_r^T]^T \tag{11.79}$$

Once Q is selected based on the aforementioned considerations, the next issue relates to the allocation of DOFs among spatial channels and fast-time taps, as well as the selection of specific auxiliary beams and ranges from the conventional BRD data cube. The latter requires a method for preferentially ranking the auxiliary beams and ranges. A reasonable approach is to rank auxiliary beams according to an estimate of the received disturbance power. For example, the median value of the Doppler spectrum may be used to provide a robust measure of disturbance level. Auxiliary beams with higher disturbance levels are chosen in preference to those containing lower power. On the other hand, the auxiliary ranges are selected in the immediate neighborhood of the test range, since the disturbance in these "local" cells is likely to be more highly correlated with that in the test cell. Such criteria may be used to pick the auxiliary beams and ranges once K and L are specified from the value of Q.

The partitioning of Q into K spatial channels and L fast-time taps is a more complex issue, as it depends on the prevailing correlation properties of the disturbance relative to the structure of the useful signal vector. This issue was investigated in Holdsworth and Fabrizio (2008). A reasonable starting point is to calculate the relative disturbance cancelation ratio using spatial and temporal DOF for adaptation separately. The relative gains may be compared and the values of K and L allocated based on an estimate of the relative benefits from adaptation in each dimension. An optimum strategy has not been developed excluding the brute force analysis of all possible combinations. This aspect merits further study. In practice, some degree of experimentation may be required to empirically derive suitable rules for determining the values of K and L for a given Q.

Once the RD-STAP system dimensions are selected, a reliable method for identifying outlier cells in each half of Doppler space is required. A preliminary step involves identifying the Doppler bins occupied by relatively strong surface clutter typically near zero Doppler frequency. Once the edges defining the clutter Doppler band to be excluded have been determined, an initial SCM may be formed using all remaining Doppler bins. A nonhomogeneity detector based on the Generalized Inner Product (GIP) can then be applied to identify the presence of outliers in the training data. A small but fixed percentage of samples with the highest level of heterogeneity may then be discarded from the final covariance matrix estimates used to derive the RD-STAP filters for each half of Doppler space. This ensures a that known number of snapshots remain for training, while reducing the influence of the strongest outliers that could potentially bias the filter estimate.

In summary, the algorithm may be broken down into three main steps: (1) select the total number of DOFs Q to be allocated for adaptive processing, (2) partition the available DOFs into K spatial channels and L temporal taps and select these from the conventional BRD data cube based on a ranking system, and (3) apply the training-data selection strategy to estimate the RD-STAP filters for each resolution cell in the radar coverage. When the rules for these three main procedures are established and automatically implemented, the STAP algorithm is essentially self-configuring with no need for human intervention.

11.4.2 Experimental Results

The experimental data analyzed in this study were collected between 04:45 and 05:15 UT on 17 April 2004 using a two-dimensional (L-shaped) antenna array located near Darwin in Northern Australia. The array consisted of 16 vertically polarized "whip" antenna elements with 8 elements on each arm uniformly spaced 8 m apart (a dummy element was included at the end of each arm to avoid mutual coupling). The output of each antenna element was connected to an individual HF receiver. The main features of the receiving system have been discussed in the previous chapter. An OTH radar transmitter located approximately 1850 km to the south-east of Darwin illuminated the region around the receiver via the ionosphere. This allowed the forward-based receiver array to acquire echoes from targets in the line of sight via a direct path.

The radar signal was a linear frequency modulated continuous waveform (FMCW) with carrier frequency $f_c = 19.380$ MHz, bandwidth $f_b = 20$ kHz, and pulse repetition frequency $f_p = 62.5$ Hz. The coherent integration time (CIT) consisted of $P = 248$ PRIs and was approximately 4 seconds long. An HF spectrum analyzer was used to monitor channel occupancy and indicated that the carrier frequency was clear of other users at the time of the experiment. The trial involved a cooperative Learjet aircraft target (see Fig 10.22) that was chartered to fly out from Darwin in a north-west direction to a range of approximately 400 km at a cruising altitude of about 31000 ft.

GPS logging equipment on board the aircraft enabled the target's range, bearing, and bi-static Doppler shift to be determined during the flight. At predetermined locations, the aircraft performed a 360 degree turn to clearly distinguish its echo from other echoes on the radar display by virtue of its unique Doppler-time signature. Range and Doppler profiles of the flight path derived from on-board GPS data logs appear in Figures 11.13 and 11.14 over the 15-minute period of interest (04:25–04:40 UT).

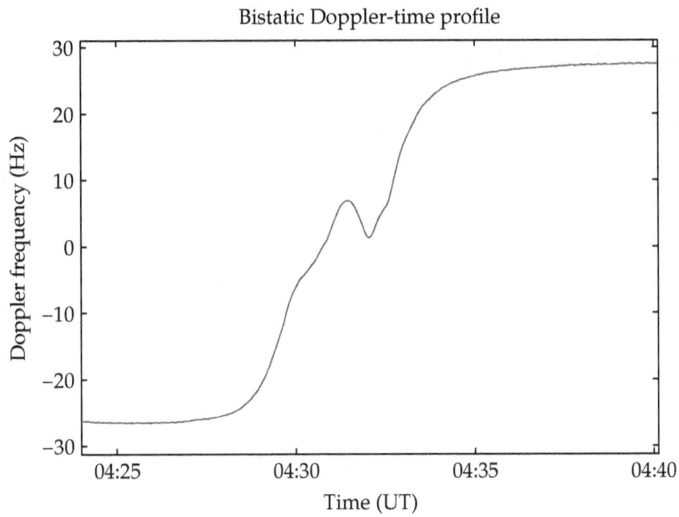

FIGURE 11.13 Bistatic Doppler-time profile of the aircraft target flight path computed from the GPS data. © Commonwealth of Australia 2011.

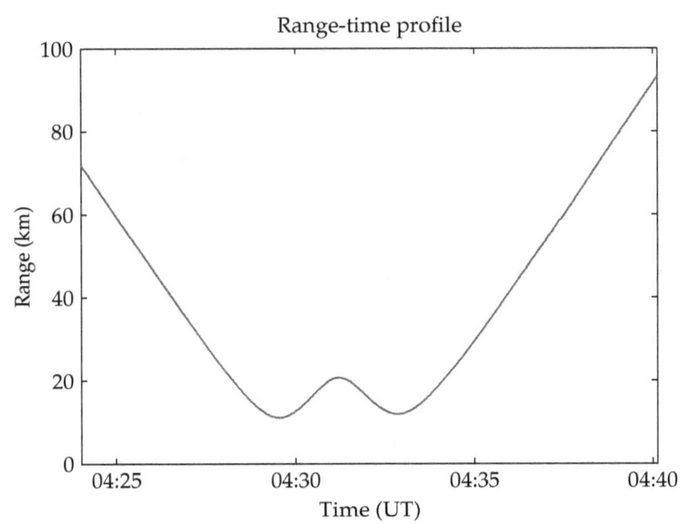

FIGURE 11.14 Range-time profile of the aircraft target flight-path (relative to the LOS receiver) computed from the GPS data. © Commonwealth of Australia 2011.

During this time, a 360-degree maneuver commenced at about 04:29 UT. The maneuver is quite noticeable in the range and Doppler versus time profiles.

11.4.2.1 Range-Doppler Displays

The target maneuver occurs when the echo is in range cells contaminated by relatively strong spread-Doppler clutter and occasionally strong RFI that affects the main beam but is correlated in range. Figure 11.15 shows a range-Doppler map for a conventional beam steered in the target direction. The azimuth and elevation of the target were determined from the known position of the aircraft (using GPS data) at the time when this CPI was recorded. In this display, range bins are ordered vertically with the nearest range at the bottom. This CPI is representative of the data collected during intervals deemed to be free of strong RFI.

In addition to the main clutter "ridge," which is apparent in all ranges near zero Doppler frequency, spread-Doppler clutter is also present and contributes to raising the disturbance level over the entire velocity search space in a band of range bins with indices roughly between 10 and 25. The expected target position in radar coordinates calculated from the GPS data is circled on this display at about 14 Hz Doppler frequency and near range bin 21. However, conventional processing cannot clearly distinguish the faint target echo, which is masked by the more powerful spread-Doppler clutter in this CPI.

The proposed RD-STAP technique was implemented with $K = 8$ auxiliary beams and $L = 4$ auxiliary ranges. The auxiliary beams were steered at the same elevation as the main beam and equally spaced 10 degrees apart in azimuth, such that four beams were formed either side of the main beam. Once the clutter-contaminated Doppler bins near 0 Hz and the outliers were removed from the training data, the number of samples remaining in the positive and negative halves of Doppler space was $D = 96$. For $Q = K + L = 12$, this corresponds to $D = 8Q$. Figure 11.16 shows the RD-STAP output for the same data

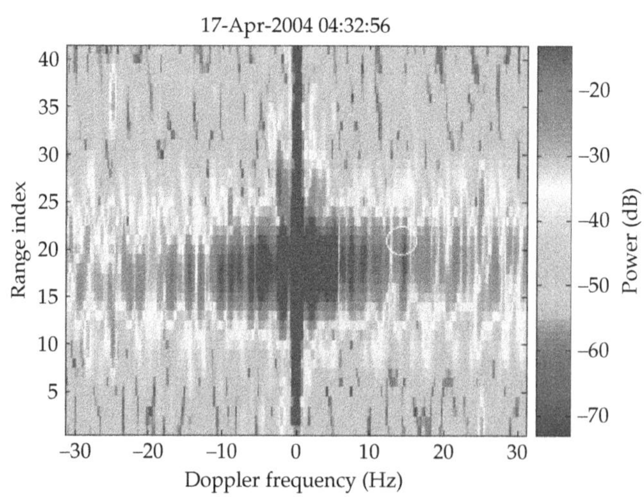

FIGURE 11.15 Conventional range-Doppler map for a beam steered in the aircraft target direction. The target echo expected at the circled location is submerged under powerful spread-Doppler clutter and cannot be detected. © 2007 IEEE. Reprinted with permission.

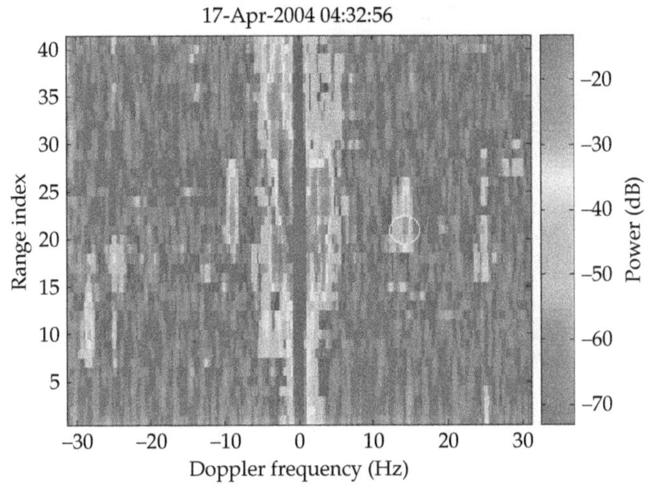

FIGURE 11.16 Range-Doppler map after RD-STAP showing that the spread-Doppler clutter has been significantly attenuated, and that the target echo can be clearly detected at the expected location. © 2007 IEEE. Reprinted with permission.

as Figure 11.15. The offending spread-Doppler clutter has been effectively removed and the target can be easily detected at the expected location.

In identical format, Figures 11.17 and 11.18 show the conventional and RD-STAP results for a CPI recorded 7 minutes later, when the target had moved out of the spread-Doppler clutter affected ranges, but was masked by strong RFI from an unknown source.

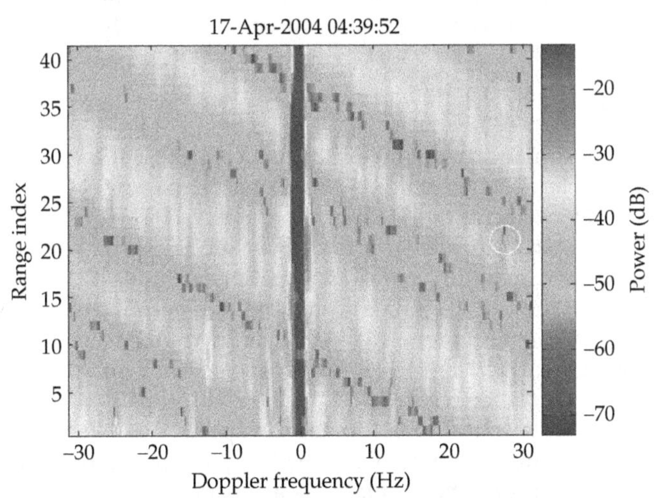

FIGURE 11.17 Conventional range-Doppler map when the target is outside the range interval occupied by spread-Doppler clutter, but at a time when strong RFI is present. © 2007 IEEE. Reprinted with permission.

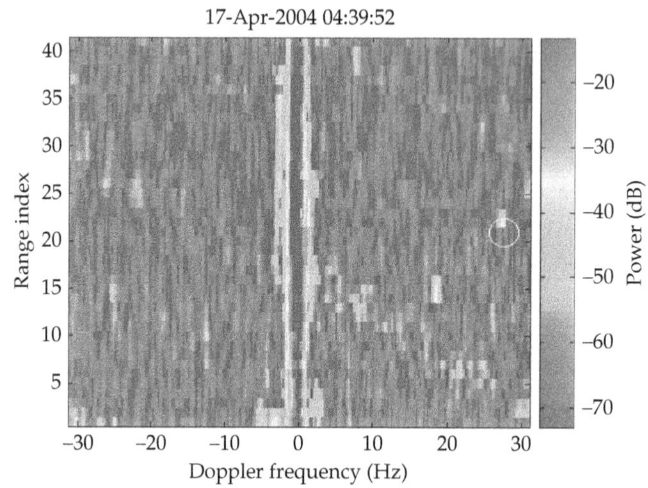

FIGURE 11.18 Range-Doppler map illustrating the effective removal of RFI by RD-STAP, which allows the target echo to be detected at the expected location. © 2007 IEEE. Reprinted with permission.

This source was received within the main beam and effectively precluded target detection over practically all range-Doppler bins when conventional processing was applied. Figure 11.17 indicates that the RFI exhibits significant structure in range, which may be exploited for cancelation. Indeed, Figure 11.18 shows that STAP cancels the RFI and clearly detects the target at the expected location near range bin 22 with a Doppler shift of about 28 Hz. A potential target of opportunity is also distinguished near range bin 15 at a Doppler shift of about 19 Hz, but this candidate is unconfirmed and will not be considered further.

11.4.2.2 Doppler-Time Signature

A synoptic view of performance over a collection of CPIs is most effectively shown as the time-evolution of Doppler spectra resulting in the target beam-range cell. Displays of this kind are sometimes referred to as *waterfall* or *scroll* displays. Since the target migrates to different range and beam coordinates over time, the waterfall display is populated only with Doppler spectra corresponding to the radar spatial resolution cell known to contain the target at the time of the CPI. GPS data were used to determine the range-beam cell containing the target for each CPI processed, such that the appropriate Doppler spectrum could be extracted. Figures 11.19 and 11.20 show the waterfall displays for conventional processing and RD-STAP, respectively.

In these intensity-modulated displays, the Doppler spectrum is plotted vertically, with a single-line spectrum being shown per CPI. The Doppler spectra from successive CPI are stacked next to one another horizontally as a function of the CPI start time. The 15 minutes of data analyzed to produce these displays contains more than 200 CPI. The powerful horizontal clutter trace near zero Doppler frequency is clearly evident in both displays. As the target initially moves toward the radar in range, it enters the spread-Doppler clutter effected region around 04:32 UT. At this time, the disturbance commences

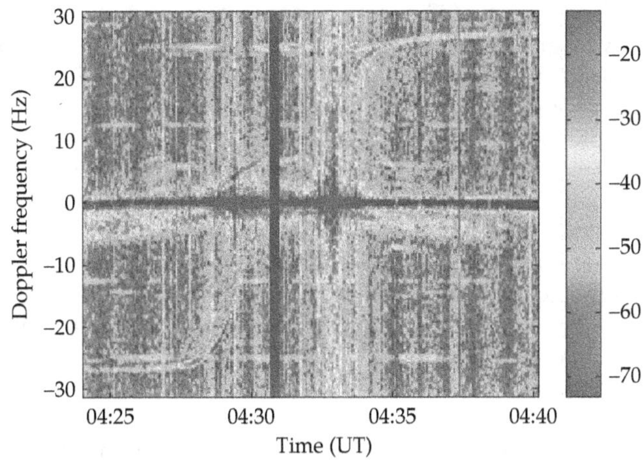

FIGURE 11.19 Waterfall display after conventional processing showing that the target can be seen before and after the 360-degree turn, but not during its passage from negative to positive Doppler frequencies as the target is obscured by the spread-Doppler clutter during the maneuver. © 2007 IEEE. Reprinted with permission.

to obscure the Doppler-time profile of the target maneuver in Figure 11.19. An example of such obscuration is shown in the conventional range-Doppler display of Figure 11.15.

The 360-degree turn results in a passage of the target trace from one side of Doppler space to the another. Since this occurs at a time when the target range coincides with that of the spread-Doppler clutter, the target maneuver is not clearly observed in the

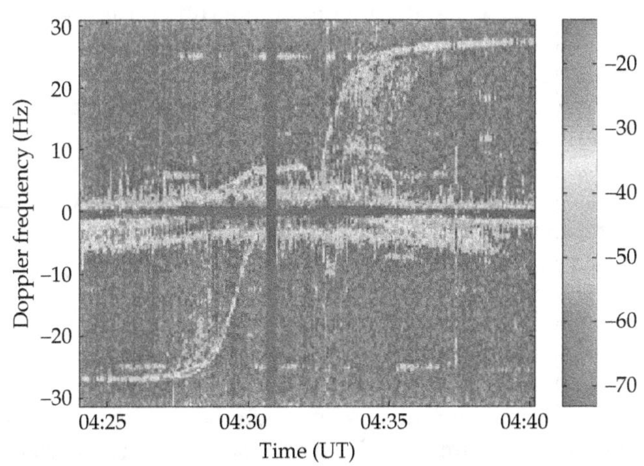

FIGURE 11.20 Waterfall display after RD-STAP showing a clearly visible target Doppler-time signature that can be easily reconciled with the expected profile shown in Figure 11.13. © 2007 IEEE. Reprinted with permission.

conventionally processed output of Figure 11.19. On the other hand, Figure 11.20 shows that RD-STAP effectively and consistently removes the spread-Doppler clutter to make the target maneuver clearly visible. The RD-STAP waterfall display in Figure 11.20 reveals a sharp and continuous Doppler-time target signature that agrees remarkably well with the Doppler-time profile predicted from the GPS data in Figure 11.13. Before and after the target maneuver (i.e., toward the left and right extremities of the waterfall display), the echo has an essentially "steady-state" bistatic Doppler shift of around ± 28 Hz.

In these sections of the flight path the target range is not within the spread-Doppler clutter effected region and conventional processing performs rather well except for the occasional CPI contaminated by strong RFI, as shown in Figure 11.17 for example. RD-SAP is seen to remove the interference due to these sporadic events, as illustrated by Figure 11.18. In this data set, the RD-STAP technique is shown to provide the radar system with greater immunity against spread-Doppler clutter and RFI that is correlated in range. The former is particularly relevant for detection at ranges close to the direct-wave clutter, while latter may be important in operational situations where a frequency change is not practical at a critical time in the mission and the presence of *persistent* range-correlated RFI threatens to preclude target detection across all range-Doppler cells over many CPIs.

11.4.3 Discussion

The first section of this chapter discussed three different STAP architectures referred to as slow-time STAP, fast-time STAP, and 3D-STAP. The disturbance characteristics motivating each STAP architecture were briefly identified in the context of both airborne microwave radar and HF OTH radar. It was argued that fast-time STAP is applicable in OTH radar for the problem of hot clutter mitigation, while slow-time STAP for cold clutter mitigation is not as well motivated for OTH radar as it is for airborne radar. In unsupervised training scenarios, 3D-STAP for joint hot- and cold-clutter mitigation may have applications for OTH radar, but practical issues such as finite sample support and computational complexity generally become more difficult to overcome for such schemes. For these reasons, only fast-time STAP was considered further and appropriate data models were developed in the second section of this chapter.

Fast-time STAP offers two important advantages with respect to spatial adaptive processing (SAP). The first situation of practical interest arises when the number of diffusely scattered hot clutter paths summed over the number of independent sources exceeds the number of antenna elements or spatial channels. In this case, SAP is exhausted of degrees of freedom and cannot effectively cancel the composite disturbance. Fast-time STAP provides a means to extend the number of adaptive DOFs beyond the rank of the hot clutter subspace and is therefore potentially able to cancel the composite disturbance more effectively than SAP. Second, SAP is not suitable for canceling disturbances received in the main-beam, as effective rejection often comes at the expense of significant main-beam distortion in this scae. Fast-time STAP is able to exploit multipath scattering from different angles to provide immunity against main-beam disturbance components. In the absence of multipath, fast-time STAP can also exploit range correlation to cancel main-beam interference in the special case of narrowband interference or impulsive noise sources.

In the general case of broadband interference sources, the non-stationarity of the hot-clutter covariance matrix over the relatively long (OTH radar) CPI creates a problem for standard fast-time STAP procedures. Specifically, the standard approach based on a fast-time STAP filter that is held fixed over the CPI fails to effectively reject the hot clutter because the integrated covariance matrix (averaged over the whole CPI) typically has full rank as a consequence of the non-stationarity hot-clutter propagation paths. On the other hand, the standard approach based on a time-varying weight vector that changes from one PRI to another can effectively reject non-stationary hot clutter, but typically has a devastating effect on sub-clutter visibility, as such filters destroy the slow-time correlation properties of the processed cold-clutter output.

In the third section of this chapter, two alternative fast-time STAP techniques were described to mitigate hot-clutter non-stationarity over the CPI, while simultaneously preserving the Doppler spectrum characteristics of the output cold-clutter signal. The SC-STAP method updates the weight vector every PRI but employs auxiliary linear stochastic constraints to stabilize the auto-regressive characteristics of the output cold-clutter signal. Motivated by the need to reduce computational load for practical implementation in real-time operational systems, the TV-STAP algorithm was subsequently introduced.

By updating the weight vector at a rate commensurate with the prevailing rate of hot-clutter non-stationarity, simulations demonstrated that TV-STAP can achieve computational advantages relative to SC-STAP while yielding similar performance to the SC-STAP method. Importantly, it was also shown that TV-STAP can protect the cold-clutter Doppler spectrum when it is described by a scalar-type AR process. Indeed, the TV-STAP algorithm reverts back to the SC-STAP technique as a special case for a particular choice of algorithm parameters. As such, TV-SAP may be viewed as a generalization of the SC-STAP technique that provides scope to jointly optimize performance and computational load.

The mitigation of disturbance signals coherent with the radar waveform via STAP was considered in the final section of this chapter. Such signals can have statistical characteristics that are significantly heterogeneous in range. In this case, effective mitigation requires the adaptive filter to be updated on a range-by-range basis.

A post-Doppler and beam-space fast-time STAP architecture for the joint cancelation of sidelobe spread-Doppler clutter and range-correlated interference (potentially received in the main beam) was described and tested in the final section. A peculiarity of this reduced-dimension RD-STAP technique is that the adaptive filter dimension is the sum rather than the product of the number of spatial and temporal DOFs. This enables the processor to cope well with low sample support and keep the computational load to modest levels for real-time implementation. Both aspects are of prime importance and must be adequately addressed for operational systems.

The performance of the proposed RD-STAP method, which is capable of running in real-time on contemporary computing platforms, was demonstrated and compared against conventional processing using over 200 CPIs of experimental data. The advantages provided by this STAP architecture relative to conventional processing were observed in cases where spread-Doppler clutter and intermittent RFI masked echoes from a cooperative (Learjet) aircraft target. STAP clearly revealed the Doppler-time signature of the cooperative target during a 360-degree turn maneuver, which was not easily visible at the conventional processing output. The target trace seen at the STAP output agreed remarkably well with the flight path predictions derived from GPS data recorded on-board the cooperative aircraft.

CHAPTER 12

GLRT Detection Schemes

In many radar systems, target detection is carried out by the sequential application of adaptive filtering to the input data, followed by constant false-alarm rate (CFAR) processing on the amplitude of the filter output, as two separate steps prior to sample thresholding. The former attempts to maximize the signal-to-disturbance ratio (SDR) of the scalar samples at the filter output, while the latter aims to normalize the envelope of the output sequence based on estimates of the mean disturbance level, such that a fixed threshold setting may be used to maintain a constant and predictable false-alarm rate.

An alternative approach is to formally cast the detection problem as a binary hypothesis test directly in terms of the input data. Ideally, a likelihood-ratio test (LRT) is applied to discriminate between the null hypothesis (disturbance-only) and the alternative hypothesis (signal-plus-disturbance). The LRT is optimum in the Neyman-Pearson sense of maximizing the probability of detection for a given probability of false alarm (PFA). However, the LRT cannot be implemented in practice due to the unknown distributional parameters of the input data. In such situations, it is possible to resort to the generalized likelihood-ratio test (GLRT), which substitutes these unknown "nuisance" parameters by their maximum-likelihood estimates calculated from the received data.

Simply stated, the GLRT may be regarded as an adaptive implementation of the LRT. Specifically, the GLRT transforms one or more input data vectors directly into a test statistic for deciding between the two hypotheses. In this sense, the GLRT represents an explicit decision rule for declaring target presence or absence directly as a function of the received data vectors. This differs from adaptive filtering, where the detection task remains to be accomplished.

In the seminal paper of Kelly (1986) and technical report of Kelly and Forsythe (1989), no optimality properties are claimed for the GLRT. However, the form of the test allows the detection and false-alarm probabilities to be determined for performance evaluation. In addition, the GLRT formulism often yields test statistics with the desirable CFAR property; i.e., the distribution of the test statistic under the null hypothesis is independent of the disturbance-model parameters. The GLRT may be qualitatively interpreted as combining the functions of adaptive filtering and CFAR processing into one procedure.

The performance of the GLRT for a particular hypothesis test may be quantified in terms of receiver operating characteristics (i.e., probability of detection P_D versus the probability of false alarm P_{FA}), or other metrics (e.g., P_D versus input SDR curves for a given P_{FA}), based on the statistical models assumed for the input data vectors under both hypotheses. In radar, GLRT-based detectors may be applied in one or more data-cube dimensions. For example, such processing may be applied to snapshot vectors acquired by multiple spatial channels of an antenna array to improve target detection in directional noise fields, which may include jamming. Alternatively, such processing may be applied to the sequence of pulses in the radar CPI for mitigating disturbances with structure in slow time, including clutter or impulsive interference. GLRT-based detectors may also be implemented in two dimensions (i.e., space/slow-time) for disturbances with angle-Doppler coupling.

An objective of this chapter is to provide a broad summary of the essential concepts related to the formulation of binary hypothesis tests and the derivation of adaptive detectors based on the GLRT. The other objective is to demonstrate the experimental performance of GLRT-based processing schemes in real-world OTH radar systems to complement the many existing numerical studies, which analyze and compare the performance of different techniques under controlled conditions by computer simulation.

The first section provides background and explains the connection between several hypothesis tests and realistic detection problems in radar. Measurement models for representing the useful signal, disturbances, as well as coherent interference are discussed in the second section, with particular attention paid to robust models that can account for partial uncertainty in the received data. The third section describes a number of popular GLRT-based detection schemes and derives some less well-known tests that can potentially provide performance advantages in situations of practical interest. The final section focuses on the experimental application of the GLRT design principle for spatial and temporal processing of real data from skywave and surface-wave OTH radar systems.

12.1 Problem Description

Adaptive signal detection theory, and its application to practical systems including radar, represents a very rich and mature field of research. The first part of this section provides a brief (non-exhaustive) synopsis of several research threads that have been followed in this area since the pioneering contribution of Kelly (1986). A relatively small number of representative works on selected topics are discussed to highlight the main themes, as well as to motivate the directions followed in the remainder of this chapter. A detailed literature review on the subject of adaptive detection and the GLRT is beyond the scope of this chapter. For a more comprehensive account of such developments, the reader is referred to the authoritative texts of Kay (1998), Scharf (1991), and Van Trees (2002), for example, all of which provide an excellent coverage of detection theory.

The second part of this section describes the traditional binary hypothesis testing problem that is often considered for radar, and introduces the fundamental precepts upon which the LRT and GLRT are based. This traditional formulation is then compared with the operating conditions frequently encountered by practical OTH radar systems. This serves to identify a number of factors that drive the development of alternative hypothesis testing problems relative to the traditional formulation. The third part of this section describes these alternative hypothesis tests, which not only differ in the measurement models adopted for the various signal components, but also in terms of the construction of the different hypotheses. A summary of the modifications to the traditional hypothesis test, which may be incorporated to represent more realistic signal environments in systems not limited to OTH radar, is provided at the end of this section.

12.1.1 Background and Motivation

Many radar systems are required to detect (and estimate the parameters of) useful signals (i.e., target echoes) embedded in powerful disturbances, which generally consist of an additive mixture of clutter, interference, and noise. In a "white-noise" environment, the conventional matched filter (MF) provides an effective and computationally efficient method for signal detection and parameter estimation. However, such a noise environment seldom arises in practice, even under benign or "quiescent" conditions, particularly for OTH radar systems, which are limited by disturbance signals from external sources as opposed to internal noise of thermal origin. In the presence of a powerful but statistically structured (colored) disturbance, the sub-optimality of conventional (data-independent) processing can at times lead to serious degradations in detection performance. This scenario motivates the use of adaptive processing for useful signal detection and parameter estimation.

Fundamentally, the design of an appropriate adaptive detection strategy involves at least two key steps. The first is the formulation of an appropriate hypothesis test based on signal models that accurately portray the properties of the input data vectors under the hypotheses of signal presence and absence, respectively. The construction of the hypothesis test, and the definition of its various components, needs to strike a balance between representing the realistic scenario as faithfully as possible, while retaining some degree of mathematical tractability to facilitate the derivation of an adaptive detection scheme suitable for practical implementation. The GLRT design methodology provides a theoretical framework to derive the detection statistic for a particular hypothesis test as the second step. Alternative techniques for deriving detection statistics, such as the Rao and Wald tests, may also be proposed, but these will not be considered here.

The seminal contributions of Reed, Mallet, and Brennan (1974) and of Kelly (1986) represent breakthroughs in the fields of adaptive filtering and adaptive detection, respectively. In these works to be described in more detail later, the possible presence of a useful signal with a known N-dimensional signature vector, but unknown complex scale, is accepted for a primary data vector, which is also assumed to contain an additive disturbance signal described by a zero-mean multi-variate complex Gaussian random process with unknown covariance matrix. The availability of a sufficient number $K \geq N$ of secondary data vectors with independent and identically distributed (IID) disturbance components to that in the primary vector is also assumed. The statistically homogeneous secondary data, also known as training data, is used to form the maximum-likelihood (ML) estimate of the unknown disturbance covariance matrix. The inverse of this sample covariance matrix (SCM), and the known useful signal signature vector, are key elements of the Reed, Mallet, and Brennan (RMB) sample matrix inverse (SMI) adaptive filter and Kelly's GLRT.

The convergence rate of the SMI technique was formally analyzed in terms of the output signal-to-disturbance ratio (Reed et al. 1974), but a decision rule for declaring target presence (or absence) was not explicitly specified in terms of the sample sequence at the adaptive filter output. Kelly formally considered the problem of adaptive detection as an exercise in hypothesis testing, and derived the GLRT by maximizing the likelihood function under each hypothesis separately over the set of unknown distributional parameters. Based on the same problem formulation as Kelly, the adaptive matched filter (AMF) was subsequently introduced in Robey, Fuhrmann, Kelly, and Nitzberg (1992), Chen and Reed (1991), and Cai and Wang (1990). The AMF is known as a "two-step" GLRT because it derives a GLRT expression assuming the disturbance covariance matrix is known in the first step, and then substitutes the SCM formed on secondary data into this expression to arrive at the AMF statistic in the second step.

Kelly's GLRT and the AMF represent two well-known detection statistics that can effectively test for target presence in structured disturbance environments. Both adaptive detection algorithms have the desirable CFAR property, which implies that the false-alarm rate of the test statistic is functionally *independent* of the disturbance covariance matrix in the problem considered. Indeed, the SMI adaptive filter represents an important component of Kelly's GLRT and the AMF detection statistic. A significant difference with respect to the adaptive filtering application considered by RMB is that the performance of Kelly's GLRT and the AMF can be quantified in terms of detection and false-alarm probabilities. Since the tracker output can be sensitive to the values of P_D and P_{FA}, it follows that a knowledge of these performance metrics is vital to the successful operation of a radar system.

Despite the widespread appeal of Kelly's GLRT and the AMF, which are often regarded as benchmark adaptive detectors for performance comparisons, their implementation in a practical radar system can be met with unexpected results. This may occur for a number of environmental and instrumental reasons that cause departures between the rather ideal conditions under which the hypothesis testing problem is defined and those which prevail in the real world. For example, the spatial, temporal, or space-time signature of a target echo received by a practical system may deviate significantly from the presumed model due to multipath propagation effects (Fabrizio, Gray, and Turley 2000), array-manifold uncertainties and receiver channel mismatches (Farina 1992), as well as pointing (steering) errors in radar systems that search over a discrete bank of relatively coarsely spaced target angle and/or Doppler frequency bins.

The traditional steering vector model has limited ability to capture such uncertainties as it assumes that the useful signal vector is perfectly known up to a complex scale. This model is typically described in terms of a parametric signature vector, which is assumed to lie on a manifold to be defined in due course. Differences between the interrogated point on this manifold and the true signature of the target echo can have a detrimental effect on the performance of the aforementioned detectors because such differences are not accounted for in the construction of the hypothesis test. This has stimulated the use of multi-rank subspace models, where the useful signal vector is described by an unknown linear combination of known basis vectors. The amazingly general work of Kelly and Forsythe (1989) incorporates the possibility of multi-rank or *subspace* signal models, which provide greater flexibility to model partial uncertainty in the signature vector of a target echo.

In addition to signal modeling uncertainties, the disturbance in the test cell or *primary data*, where signal presence is sought, and that in the reference cells or *secondary data*, used to train the adaptive detector, may not be statistically homogeneous or Gaussian distributed. For instance, the received disturbance may be more accurately described by an alternative statistical model, such as the compound-Gaussian process. Indeed, analysis of radar data collected in the field confirms that clutter is often non-Gaussian and better represented by a Weibull or K-distribution, both of which are compatible with the compound-Gaussian description. The problem of adaptive signal detection in compound-Gaussian clutter has been considered extensively in the radar literature. For example, this problem has been addressed in the context of temporal (pulse-to-pulse) processing in Sangston and Gerlach (1994), Conte, Lops, and Ricci (1995), Gini (1997), Gini and Farina (2002a), Gini and Farina (2002b), and references therein.

In many studies, the non-Gaussian clutter vector is described by a special case of the compound Gaussian model known as a spherically invariant random process (SIRP).

In the context of spatial processing, the SIRP was also shown to be more appropriate than the traditional homogeneous Gaussian model for representing *interference* vectors received by an HF antenna array (Fabrizio, Farina, and Turley 2003a). Importantly, Kelly's GLRT and the AMF are known to lose their CFAR property when the homogeneous IID Gaussian disturbance assumption is not valid. In this case, the false-alarm rate is no longer solely a function of the dimensional parameters of the system and the threshold setting, and may vary unpredictably when Kelly's GLRT or the AMF are applied in such disturbance environments. This has encouraged the development of GLRTs that explicitly account for compound-Gaussian or other non-Gaussian disturbances by incorporating such disturbance models in the definition of the hypothesis test (i.e., at the design stage).

The normalized matched filter (NMF) was derived as a GLRT in Conte et al. (1995) for the case of a SIRP disturbance with known covariance matrix, but unknown texture marginal density function.[1] It was shown that the NMF is an asymptotically optimal CFAR detector for a SIRP disturbance, where the term "asymptotically" refers here to an infinite number of integrated samples (i.e., large data vector length), "optimal" refers to maximizing the probability of detection for a given PFA, and CFAR refers to the invariance of the NMF test statistic distribution under the null hypothesis with respect to the SIRP disturbance power and texture marginal density function. The adaptive version of this detector, which replaces the true disturbance covariance matrix by its sample estimate calculated from the secondary data, coincides with the adaptive coherence estimator (ACE) independently derived by Scharf and Mc Whorter (1996).

In Kraut and Scharf (2001), ACE was shown to be a GLRT for a Gaussian disturbance model with unknown scale change between the training and test data. For this disturbance model, ACE was shown to be a GLRT for measurement vectors of finite length with the desirable (scale-invariant) CFAR property. In particular, the CFAR property of ACE is retained for the class of SIRP disturbance process that satisfies the "partially homogeneous" texture model described in Conte, De Maio, and Ricci (2001), wherein the texture is assumed to be fully correlated over the secondary data, but may be independent of that in the primary data. ACE also retains the CFAR property for other classes of non-Gaussian disturbances described in Richmond (1996), and Pulsone and Raghavan (1999).

A multi-rank extension of ACE, known as the adaptive subspace detector (ASD), was derived as a GLRT for the partially homogeneous disturbance environment in Kraut and Scharf (1999). Importantly, the ASD may be configured to handle partial uncertainty in the target signature vector by virtue of a subspace signal representation. The ASD therefore provides more flexibility to model useful signal uncertainty than ACE by allowing for subspace dimensions greater than unity, and unlike Kelly's GLRT or the AMF, its CFAR property remains invariant to *independent* scaling of the disturbance in the primary and secondary data. These additional features of the ASD with respect to ACE, the AMF, and Kelly's GLRT have significant robustness implications for operational OTH radar systems, making the ASD an attractive candidate for practical implementation.

Another issue of concern to radar systems relates to the problem of unwanted signals arising from passive (scattering) or active (repeater-like) sources of coherent interference. These unwanted signals may be present along with useful signals in the primary data, but due to their coherence with the radar waveform, these unwanted signals are essentially confined to a single resolution cell. Hence, the characteristics of unwanted signals in the

[1] The meaning of the term texture will be explained in Section 12.2.

primary data cannot be gleaned from the secondary data. The GLRT-based adaptive detection schemes discussed so far were derived for hypothesis tests that do not explicitly account for the possible presence of unwanted signals. Not surprisingly, the presence of such signals has the potential to degrade the performance of all the aforementioned detectors. For Kelly's GLRT and the AMF, mismatched signals due to coherent interference have the potential to trigger unwanted detections and hence raise the false-alarm rate. On the other hand, the application of ACE or the ASD, which are very intolerant to signal mismatch, can cause useful signals to be masked by coherent interference present in the same resolution cell, but at different DOA and/or Doppler coordinates. This "blanking" phenomena effectively lowers the probability of detection.

Early attempts to incorporate unwanted signals directly into the hypothesis test appeared in Pulsone and Rader (2001) for unit-rank (steering vector) useful signals, and subsequently in Fabrizio et al. (2003a) for the case of both useful signals and coherent interference described by subspace models. In the former case, the GLRT for a homogeneous Gaussian disturbance led to the ABORT detector, while the GLRT derived in the latter case possesses the scale-invariant CFAR property for the partially homogeneous disturbance environment. Tunable receivers based on Kelly's GLRT, e.g., Kalson (1992), and two-stage receivers that regulate the trade off between robustness to *slightly* mismatched useful signals, and selectivity for rejecting significantly mismatched unwanted signals, have also been proposed in Bandiera, De Maio, De Nicola, Farina, Orlando, and Ricci (2010) and references therein.

In the majority of works cited above, the relative performance advantages of each proposed detector have been claimed solely on the basis of theoretical analysis and/or computer simulation results, where to a large extent, the characteristics of the signal environment and hypothesized sensor are controlled by making certain assumptions. While such studies provide valuable insights for discerning between the potential usefulness of different adaptive detection schemes, the actual performance improvements should be interpreted with caution, as they may not be typical of those encountered in practice. There is currently a lack of *experimental* results published in the open literature to indicate the relative performance of different adaptive detection schemes in order to justify their practical implementation. Another aspect which has received insufficient attention is the detection of target echoes that are spread over *multiple* primary data vectors, as in Conte et al. (2001) for example.

The main purpose of this chapter is to explore a number of selected topics in relation to the four key issues of: (1) useful signal uncertainty, (2) non-Gaussian disturbance processes, (3) unwanted coherent signals, and (4) target echo spread over multiple primary data vectors, which are relevant for practical applications of adaptive detection in radar systems. Alternative options for constructing the hypothesis test, and modeling the various signal components under each hypothesis, will also be described to derive robust adaptive detection schemes based on the GLRT methodology. The performance of several adaptive detection schemes will then be compared using real data collected by surface-wave and skywave OTH radar systems.

12.1.2 Traditional Hypothesis Test

Suppose the radar transmits a coherent burst of N_p identical radio frequency (RF) pulses in a coherent processing interval (CPI), and that N_g complex range-gated samples are acquired per pulse in each of N_a spatial reception channels. Each reception channel consists of an antenna sensor (e.g., a subarray), connected to a receiver "front end," which carries out amplification, filtering, conversion to baseband and pulse compression of the received in-phase (I) and quadrature (Q) signal components. These components are digitized at the Nyquist rate to produce a CPI data cube of $N_a \times N_g \times N_p$ complex samples. The *primary* inputs to the detection processor consist of N complex samples, denoted by $z(n) = z_I(n) + jz_Q(n)$ for $n = 1, \ldots, N$, where $z_I(n)$ and $z_Q(n)$ correspond to the digitized I and Q outputs, respectively. The input samples are conveniently arranged into an N-dimensional column vector $\mathbf{z}$ in Eqn. (12.1), referred to as the primary data vector, where T denotes transpose.

$$\mathbf{z} = [z(1), \ldots, z(N)]^T \tag{12.1}$$

Depending on the application, the samples in $\mathbf{z}$ may be extracted from the data cube in the dimensions of space, time, or space-time. For example, the vector $\mathbf{z}$ has length $N = N_a$ for antenna element spatial processing, $N = N_p$ for pulse-to-pulse temporal processing, or $N = N_a \times N_p$ for space-time processing. Detection processing may also be performed in beam-space or in the Doppler domain, but we shall not confine ourselves to specific applications for the moment. The detection procedure is typically performed on one primary data vector at a time until all such cells in the data cube are processed. We may restrict our attention to the (unnamed) vector $\mathbf{z}$ for the time being. In the binary hypothesis testing problem, the primary data vector $\mathbf{z}$ is traditionally assumed to take the form of Eqn. (12.2) under the null (H_0), and alternative (H_1), hypotheses.

$$\begin{cases} \mathbf{z} = \mathbf{d}, & H_0 : \text{disturbance-only} \\ \mathbf{z} = \mathbf{s} + \mathbf{d}, & H_1 : \text{signal-in-disturbance} \end{cases} \tag{12.2}$$

Under H_0, $\mathbf{z}$ is assumed to contain a disturbance component only, while in the alternative hypothesis H_1, $\mathbf{z}$ is assumed to contain a useful signal vector $\mathbf{s}$ in addition to the disturbance vector $\mathbf{d}$. In general, the overall disturbance may be expressed as the sum of internal receiver noise $\mathbf{n}$, as well as external disturbance contributions arising from clutter $\mathbf{c}$ and interference $\mathbf{i}$ received by the radar, as in Eqn. (12.3). One of the two external disturbance contributions may dominate, or may be identified as the disturbance type to be mitigated when processing in a particular dimension. For example, spatial processing is often applied to mitigate directional interferences $\mathbf{i}$ from point sources, whereas temporal (pulse-to-pulse) processing may be used to mitigate backscattered clutter $\mathbf{c}$. Space-time processing can in principle mitigate clutter received by a moving radar platform, or both clutter and interference simultaneously, particularly when neither of them dominate.

$$\mathbf{d} = \mathbf{c} + \mathbf{i} + \mathbf{n} \tag{12.3}$$

The disturbance vector $\mathbf{d}$ is typically assumed to be a zero-mean (i.e., purely statistical) vector, which is modeled as a realization of an N-variate circular-symmetric complex Gaussian process with an unknown positive-definite (full-rank) Hermitian covariance matrix $\mathbf{R}$, as defined in Eqn. (12.4). In shorthand notation, $\mathbf{d} \sim \mathcal{CN}(\mathbf{0}, \mathbf{R})$. The operator $\mathbf{E}\{\cdot\}$ denotes statistical expectation, while $\dagger$ is the Hermitian or conjugate-transpose operator. Assuming the various disturbance components are mutually uncorrelated, it follows that $\mathbf{R} = \mathbf{R}_c + \mathbf{R}_i + \mathbf{R}_n$, where $\mathbf{R}_c = \mathbf{E}\{\mathbf{c}\mathbf{c}^\dagger\}$ and $\mathbf{R}_i = \mathbf{E}\{\mathbf{i}\mathbf{i}^\dagger\}$ are the clutter and interference covariance matrices, respectively, whereas $\mathbf{R}_n = \sigma_n^2 \mathbf{I}_N$ for white noise of power σ_n^2 per element.

$$\mathbf{R} = \mathrm{E}\{\mathbf{d}\mathbf{d}^\dagger\} = \mathbf{R}_c + \mathbf{R}_i + \mathbf{R}_n \tag{12.4}$$

On the other hand, the useful signal $\mathbf{s}$ is considered to be a purely *deterministic* vector, such that the mean of $\mathbf{z}$ conditioned on H_1 has the form of Eqn. (12.5). The complex scale $a = \mu e^{j\vartheta}$ is assumed to be unknown because of the lack of knowledge regarding the target amplitude μ and phase ϑ due to scattering and propagation phenomena. The term $\mathbf{v} \in \mathcal{C}^N$ is referred to as the target signature vector. In the traditional hypothesis test, its *structure* is assumed to be perfectly known. The norm of $\mathbf{v}$ is clearly arbitrary as a is a free parameter, but a convenient choice is $\mathbf{v}^\dagger \mathbf{v} = N$.

$$\mathbf{m} = \mathrm{E}\{\mathbf{z}|H_1\} = \mathbf{s} = a\mathbf{v} \tag{12.5}$$

In practice, a radar system searches for targets by assuming that the signature vector $\mathbf{v}$ lies on a known manifold $\mathcal{M}$ in N-dimensional space. This manifold is often defined in terms of a steering vector model $\mathbf{v}(\omega)$, which depends on one or more target echo parameters, such as direction of arrival and/or Doppler shift, that are collectively represented by its argument ω. The manifold $\mathcal{M}$ is defined as the set of steering vectors $\mathbf{v}(\omega)$ corresponding to all possible target parameter values ω in the feasible domain $\mathcal{D}$, as in Eqn. (12.6). The condition $\mathbf{v} \in \mathcal{M}$ is typically based on a point-target model that relies on rather idealized physical assumptions about the scattering process, propagation channel, and receiving instrument.

$$\mathbf{v} \in \mathcal{M} \equiv \{\mathbf{v}(\omega)\ ;\ \omega \in \mathcal{D}\} \tag{12.6}$$

The steering vector $\mathbf{v}(\omega)$ often takes the form of Eqn. (12.7). The parameter ω may correspond to a temporal angular frequency $\omega = 2\pi f$ in pulse-to-pulse processing (where f is the target Doppler shift normalized by the PRF), or a spatial angular frequency $\omega = 2\pi d \sin\varphi/\lambda$ in array processing for a uniform linear array (ULA).[2] For space-time processing, the steering vector is given by the Kronecker product of the temporal and spatial steering vectors. In general, the functional form of the steering vector is not restricted to the class in Eqn. (12.7). For convenience, we shall simply refer to the steering vector as $\mathbf{v}(\omega)$, with the understanding that its definition may vary depending on system characteristics, target model, and processing dimension(s).

$$\mathbf{v}(\omega) = [1, e^{j\omega}, e^{j2\omega}, \ldots, e^{j(N-1)\omega}]^T \tag{12.7}$$

The optimum detector in the Neyman-Pearson sense of maximizing the probability of detection for a given probability of false alarm is the likelihood-ratio test (LRT). According to the described data model, the distribution of the primary vector $\mathbf{z}$ under the hypothesis H_δ is given by $p_\delta[\mathbf{z}]$ for $\delta = 0, 1$ in Eqn. (12.8), where $\|\cdot\|$ denotes the determinant of a matrix.

$$p_\delta[\mathbf{z}] = \frac{1}{\pi^N \|\mathbf{R}\|} e^{-(\mathbf{z}-\delta\mathbf{m})^\dagger \mathbf{R}^{-1} (\mathbf{z}-\delta\mathbf{m})} \tag{12.8}$$

[2]Here, φ is the target cone-angle-of-arrival, d is the inter-sensor spacing, and λ is the carrier wavelength.

The LRT compares the statistic $\ell(\mathbf{z})$ in Eqn. (12.9) with a threshold for an input vector $\mathbf{z}$. A signal is declared to be present if $\ell(\mathbf{z})$ exceeds the threshold, otherwise the signal is declared absent. The probabilities of detection P_D and false alarm P_{FA} are those of declaring a signal present when H_1 and H_0 are actually true, respectively. Based on the Neyman-Pearson criterion, the LRT is known to be the uniformly most powerful (UMP) test for deciding between two simple hypotheses H_0 and H_1.

$$\ell(\mathbf{z}) = \frac{p_1[\mathbf{z}]}{p_0[\mathbf{z}]} \tag{12.9}$$

The detection statistic $\ell(\mathbf{z})$ is often expressed as a log-likelihood ratio $L(\mathbf{z}) = \ln\{\ell(\mathbf{z})\}$. Using Eqn. (12.8), the decision rule based on the log-likelihood ratio may be written as Eqn. (12.10), where γ is the detection threshold. Unfortunately the test $L(\mathbf{z})$ cannot be implemented in practice because of the unknown distributional parameters of $\mathbf{z}$ under each hypothesis; namely, the disturbance covariance matrix $\mathbf{R}$ under H_0, in addition to the complex scale a in the mean vector $\mathbf{m} = a\mathbf{v}$ under H_1.

$$L(\mathbf{z}) = \mathbf{z}^\dagger \mathbf{R}^{-1}\mathbf{z} - (\mathbf{z}-\mathbf{m})^\dagger \mathbf{R}^{-1}(\mathbf{z}-\mathbf{m}) \underset{H_0}{\overset{H_1}{\gtrless}} \gamma \tag{12.10}$$

Now suppose that the processor also has a set of $K_s \geq N$ secondary data vectors, denoted by $\mathbf{z}_k \in \mathcal{C}^N$ for $k \in \Omega_s \equiv [1, \ldots, K_s]$. These vectors are extracted from the data cube in similar fashion to the primary vector $\mathbf{z}$, but from different resolution cells, which are often chosen in the proximity of the cell under test (CUT). For example, in the case of joint space-time processing, where $N = N_a \times N_p$, the secondary vectors $\mathbf{z}_k$ can be extracted from a subset of the N_g available range cells, excluding the range of the CUT, and typically a neighboring "guard" cell either side of it. The secondary data matrix $\mathbf{Z}_s \in \mathcal{C}^{N\times K_s}$ is defined by Eqn. (12.11).

$$\mathbf{Z}_s = [\mathbf{z}_1, \ldots, \mathbf{z}_{K_s}] \tag{12.11}$$

The secondary vectors are assumed to contain disturbance-only realizations $\mathbf{z}_k = \mathbf{d}_k$ that are statistically independent and identically distributed with each other as well as the disturbance in the primary data $\mathbf{d}$. These secondary vectors, also referred to as training data or reference cells, therefore have the same covariance matrix $\mathbf{R}$ in Eqn. (12.12), which is identical to that of the primary data. This disturbance model is known as the homogeneous Gaussian case.

$$\mathbf{E}\{\mathbf{z}_k\mathbf{z}_k^\dagger\} = \mathbf{R}, \quad k \in \Omega_s \tag{12.12}$$

The traditional binary hypothesis test including the secondary data may therefore be formulated as Test $\mathcal{A}$ in Eqn. (12.13). For convenience, we define the matrix $\mathbf{Z} = [\mathbf{z}, \mathbf{Z}_s]$, which contains the total input data set (i.e., primary and secondary) made available to the processor. The assumption of signal-free secondary data can be quite unrealistic in practice, although careful training data selection and screening can reduce the incidence of signal contamination in the secondary data. In this chapter, the issue of unwanted signal contamination is ignored as far as the *secondary* data is concerned.

$$\text{Test } \mathcal{A}: \begin{cases} H_0: \begin{cases} \mathbf{z} &= \mathbf{d}, \\ \mathbf{z}_k &= \mathbf{d}_k, \quad k \in \Omega_s \end{cases} \\ \\ H_1: \begin{cases} \mathbf{z} &= \mathbf{s} + \mathbf{d}, \\ \mathbf{z}_k &= \mathbf{d}_k, \quad k \in \Omega_s \end{cases} \end{cases} \tag{12.13}$$

As proposed by Kelly (1986), a way forward is to resort to the GLRT for Test $\mathcal{A}$, which is based on the joint distribution of all primary- and secondary-data vectors under each hypothesis. By exploiting the IID property of the disturbance vectors, it is readily shown that the joint density of the data vectors in $\mathbf{Z}$ conditioned on the unknown distributional parameters $(a, \mathbf{R})$ is given by $f_\delta[\mathbf{Z}; a, \mathbf{R}]$ in Eqn. (12.14) under H_δ, where $\text{Tr}\{\cdot\}$ is the trace operator. The unknown parameters in this density are $(a, \mathbf{R})$ under H_1, and $\mathbf{R}$ only under H_0.

$$f_\delta[\mathbf{Z}; a, \mathbf{R}] = \frac{1}{\pi^{N(K_s+1)} \|\mathbf{R}\|^{K_s+1}} \exp -\text{Tr}\left\{ \mathbf{R}^{-1} \left[(\mathbf{z} - \delta a \mathbf{v})(\mathbf{z} - \delta a \mathbf{v})^\dagger + \sum_{k=1}^{K_s} \mathbf{z}_k \mathbf{z}_k^\dagger \right] \right\} \tag{12.14}$$

Derivation of the GLRT requires the joint density $f_\delta[\mathbf{Z}; a, \mathbf{R}]$ to be separately maximized over all the unknown distributional parameters under each hypothesis $\delta = 0, 1$. The maximizing parameter values are, by definition, the maximum-likelihood (ML) estimates of these parameters, so the maximized densities are obtained by replacing the unknown parameters a and $\mathbf{R}$ by their ML estimates. In particular, it is useful to define the disturbance sample covariance matrix (SCM) $\mathbf{S}$ computed from the secondary data in Eqn. (12.15), which is the ML estimate of the disturbance covariance matrix based on the secondary data alone.

$$\mathbf{S} = K_s^{-1} \mathbf{Z}_s \mathbf{Z}_s^\dagger = K_s^{-1} \sum_{k=1}^{K_s} \mathbf{z}_k \mathbf{z}_k^\dagger \tag{12.15}$$

The ratio of the maximized densities under the two hypotheses, defined by $\mathcal{G}(\mathbf{Z})$ in Eqn. (12.16), is the GLRT for the problem at hand. The alternative hypothesis H_1 whose density appears in the numerator is accepted as true when $\mathcal{G}(\mathbf{Z})$ exceeds a predetermined threshold.

$$\mathcal{G}(\mathbf{Z}) = \frac{\max_{\mathbf{R},a} f_1[\mathbf{Z}; a, \mathbf{R}]}{\max_{\mathbf{R}} f_0[\mathbf{Z}; \mathbf{R}]} \tag{12.16}$$

This procedure results in Kelly's GLRT, which may be written in the elegant form of $\mathcal{L}(\mathbf{Z}) = (\mathcal{G}'(\mathbf{Z}) - 1)/\mathcal{G}'(\mathbf{Z})$ in Eqn. (12.17), where $\mathcal{G}'(\mathbf{Z})$ is the $(K_s + 1)^{\text{th}}$ root of the ratio $\mathcal{G}(\mathbf{Z})$ in Eqn. (12.16), and $(\mathcal{G}'(\mathbf{Z}) - 1)/\mathcal{G}'(\mathbf{Z})$ is a monotonic function that does not effect the performance of the test. Details of the derivation of $\mathcal{L}(\mathbf{Z})$, and proof of its CFAR property, can be found in Kelly (1986).

$$\mathcal{L}(\mathbf{Z}) = \frac{|\mathbf{v}^\dagger \mathbf{S}^{-1} \mathbf{z}|^2}{\mathbf{v}^\dagger \mathbf{S}^{-1} \mathbf{v}(1 + K_s^{-1} \mathbf{z}^\dagger \mathbf{S}^{-1} \mathbf{z})} \underset{H_0}{\overset{H_1}{\gtrless}} K_s \gamma \tag{12.17}$$

12.1.3 Alternative Binary Tests

In practice, the target parameter ω is unknown a priori, and therefore needs to be guessed by the system to compute the GLRT using $\mathbf{v} = \mathbf{v}(\omega)$. Typically, the radar searches over a discrete set of Q values for ω defined by ω_s in the target parameter bank $\mathcal{B}$ of Eqn. (12.18). The actual target parameter ω takes values in the continuous domain $\omega \in \mathcal{D}$, and may be matched or mismatched to the current search parameter $\omega_s \in \mathcal{B}$. For the moment, assume that one value of ω_s in the bank perfectly matches the true target parameter ω.

$$\omega_s \in \mathcal{B} \equiv \{\omega_1, \ldots, \omega_Q\} \tag{12.18}$$

As the radar is steered to different target parameters $\omega_s \in \mathcal{B}$, the aim of the search is not only to detect a target when it is present, but also to report its approximate angle of arrival and/or Doppler frequency, as the case may be. Hence, when a detection is declared using a particular value of ω_s in Eqn. (12.19), the target parameter is implicitly estimated as the current search parameter ω_s. The search must therefore be considered unsuccessful if the presence of a target with actual parameter ω is detected while the radar is "looking" elsewhere, i.e., $\omega \neq \omega_s$. For example, a strong target echo located in the sidelobe region of a spatial processor may trigger a detection and be (erroneously) interpreted as a useful signal incident from the mainlobe direction.

$$\mathcal{L}(\mathbf{Z}, \omega_s) = \frac{|\mathbf{v}(\omega_s)^\dagger \mathbf{S}^{-1}\mathbf{z}|^2}{\mathbf{v}(\omega_s)^\dagger \mathbf{S}^{-1}\mathbf{v}(\omega_s)(1 + K_s^{-1}\mathbf{z}^\dagger \mathbf{S}^{-1}\mathbf{z})} \underset{H_0}{\overset{H_1}{\gtrless}} K_s\gamma, \quad \omega_s \in \mathcal{B} \tag{12.19}$$

With this in mind, we follow the usual convention and define the probability of false alarm P_{FA} as that of accepting H_1 when H_0 is true (in force), and the probability of detection P_D as that of accepting H_1 when H_1 true. However, P_D will have different meanings depending on whether a matched or mismatched signal is present under H_1.

Borrowing terminology from Kalson (1995), we call P_{DS} the probability of *sidelobe* detection when the steer parameter ω_s is mismatched to the signal parameter ω. On the other hand, if $\omega_s = \omega$, then we call P_{DM} the probability of *mainlobe* detection for a matched signal. These two probabilities are defined in Eqn. (12.20). A detector that achieves a low value of P_{DS} is said to have high *selectivity*, while a detector that achieves a high value of P_{DS} is said to have high *sensitivity*. In Section 12.2, we shall introduce the notion of *robustness* for the case of slightly mismatched signals.

$$P_{DM} = P[\mathcal{L}(\mathbf{Z}, \omega_s) > K_s\gamma \,|H_1;\ \omega_s = \omega]\,, \quad P_{DS} = P[\mathcal{L}(\mathbf{Z}, \omega_s) > K_s\gamma \,|H_1;\ \omega_s \neq \omega] \tag{12.20}$$

Now consider the situation where H_1 is in force due to the presence of a signal $a\,\mathbf{v}(\omega)$ in the primary data. The system checks for targets by steering over the Q candidate target parameters $\omega_s \in \mathcal{B}$ in turn, only one of which can perfectly match the signal parameter ω. If $\omega_s = \omega$, then $a\,\mathbf{v}(\omega)$ is clearly a useful signal $\mathbf{s}$ to be detected by the system. However, for any other $\omega_s \neq \omega$, the signal $a\,\mathbf{v}(\omega)$ residing in the primary data vector will effectively change status from a useful signal $\mathbf{s}$ to an unwanted signal $\mathbf{r}$ in Eqn. (12.21). An unwanted signal needs to be rejected or "blanked" by the system to prevent it from triggering a false detection, i.e., one that falsely leads the system to believe that a target with parameter ω_s is present.

$$a\mathbf{v}(\omega) = \begin{cases} \mathbf{s} \text{ for } \omega = \omega_s \\ \mathbf{r} \text{ for } \omega \neq \omega_s \end{cases} \tag{12.21}$$

Test $\mathcal{A}$ in Eqn. (12.13) therefore includes a valid hypothesis for the primary data only if a perfectly matched signal is present, or if the signal is absent. However, when a signal is present, the hypothesis H_1 in test $\mathcal{A}$ can only be valid for one of the Q guessed parameter values. For the remaining $Q-1$ guesses, none of the hypotheses in Test $\mathcal{A}$ properly reflects the actual state of the primary data. Detection statistics based on Test $\mathcal{A}$ are not designed to discriminate between matched and mismatched signals, and can give rise to undesirably high values of P_{DS}.

To derive selective detectors, the possibility of mismatched signal presence may be accounted for directly in the hypothesis test. As in Pulsone and Rader (2001), hypothesis test $\mathcal{B}$ in Eqn. (12.22) may be formulated to accept the presence of mismatched signal $\mathbf{r}$ under H_0, and a matched signal $\mathbf{s}$ under H_1. When a signal is present in the primary data, test $\mathcal{B}$ always has one valid hypothesis for all values of $\omega_s \in \mathcal{B}$. Methods for modeling the unwanted signal $\mathbf{r}$ will be described in the next section.

$$\text{Test } \mathcal{B}: \begin{cases} H_0: \begin{cases} \mathbf{z} \;= \mathbf{r}+\mathbf{d}, & \text{unwanted signal in disturbance} \\ \mathbf{z}_k = \mathbf{d}_k, & k \in \Omega_s. \end{cases} \\ \\ H_1: \begin{cases} \mathbf{z} \;= \mathbf{s}+\mathbf{d}, & \text{useful signal in disturbance} \\ \mathbf{z}_k = \mathbf{d}_k, & k \in \Omega_s. \end{cases} \end{cases} \tag{12.22}$$

Another case of interest arises when multiple coherent signals with different parameters are received in the primary data. As the system switches the search parameter through the bank of candidates $\omega_s \in \mathcal{B}$, the value of ω_s can only ever match the parameter of one of these signals at a time, or none of them. In this case, the two possible options for any choice of $\omega_s \in \mathcal{B}$ are represented by the binary hypothesis test formulated as Test $\mathcal{C}$ in Eqn. (12.23). The main point is that the fidelity with which a hypothesis test can represent the actual characteristics of the primary data depends not only on whether the signal is present or absent, but also on the number of such signals and their parameters with respect to the search parameter the system is currently steered to.

$$\text{Test } \mathcal{C}: \begin{cases} H_0: \begin{cases} \mathbf{z} \;= \mathbf{r}+\mathbf{d}, & \text{unwanted signal in disturbance} \\ \mathbf{z}_k = \mathbf{d}_k, & k \in \Omega_s. \end{cases} \\ \\ H_1: \begin{cases} \mathbf{z} \;= \mathbf{s}+\mathbf{r}+\mathbf{d}, & \text{useful and unwanted signals in disturbance} \\ \mathbf{z}_k = \mathbf{d}_k, & k \in \Omega_s. \end{cases} \end{cases} \tag{12.23}$$

It is evident that each binary hypothesis test can only cover two of the four possible hypotheses appearing in the three tests $\mathcal{A}$, $\mathcal{B}$, and $\mathcal{C}$. Specifically, when one signal $a\mathbf{v}(\omega)$ is known to be present in the primary data, only Test $\mathcal{B}$ can provide a valid hypothesis for all values of $\omega_s \in \mathcal{B}$. When multiple signals with different parameters are known to be present in the primary data, then only Test $\mathcal{C}$ can provide a valid hypothesis for all values of $\omega_s \in \mathcal{B}$. Clearly, only Test $\mathcal{A}$ provides a valid null hypothesis when the primary data contains disturbance only. A multi-hypothesis test formulation is therefore natural in the sense that at least one of the four structurally different hypotheses will correspond to the actual state of the primary vector over the full range of considered conditions, regarding the number of signals present in the primary data, and their parameters with respect to the current system search parameter.

12.1.3.1 Generalized Multi-Hypothesis Tests

Thus far, the presence of a useful signal has been sought in a *single* primary data vector at a time. This is because an echo from a point target is essentially localized to a single resolution cell or measurement vector (ignoring spectral leakage effects). This assumption is valid when the radar has a range-resolution cell size that exceeds the physical dimensions of the target. In the case of high-resolution radars, a relatively large target may be spread over several range cells, such that useful signal energy is present across *multiple* primary data vectors simultaneously.

Hence, in addition to the multi-hypothesis test eluded to in the previous section, the possibility for the detection of targets that are spread or *distributed* across multiple primary vector should also be considered as an option. To this end, a more general multi-hypothesis test that accounts for these extensions can be formulated as in Eqn. (12.24), where the K_p primary data vectors are denoted by $\mathbf{z}_k$ for $k \in \Omega_p \equiv \{K_s + 1, \ldots, K\}$ and $K = K_s + K_p$. This generalized multi-hypothesis tests (GMHT) naturally divides into two classes of hypotheses, namely, H_0 or H_1 (target absence), and H_2 or H_3 (target presence).

$$\text{GMHT:} \begin{cases} \text{Target Absent:} \begin{cases} H_0 : \begin{cases} \mathbf{z}_k = \mathbf{d}_k, & k \in \Omega_p, \\ \mathbf{z}_k = \mathbf{d}_k, & k \in \Omega_s. \end{cases} \\ H_1 : \begin{cases} \mathbf{z}_k = \mathbf{r}_k + \mathbf{d}_k, & k \in \Omega_p, \\ \mathbf{z}_k = \mathbf{d}_k, & k \in \Omega_s. \end{cases} \end{cases} \\ \text{Target Present:} \begin{cases} H_2 : \begin{cases} \mathbf{z}_k = \mathbf{s}_k + \mathbf{d}_k, & k \in \Omega_p, \\ \mathbf{z}_k = \mathbf{d}_k, & k \in \Omega_s. \end{cases} \\ H_3 : \begin{cases} \mathbf{z}_k = \mathbf{s}_k + \mathbf{r}_k + \mathbf{d}_k, & k \in \Omega_p, \\ \mathbf{z}_k = \mathbf{d}_k, & k \in \Omega_s. \end{cases} \end{cases} \end{cases} \tag{12.24}$$

The majority of GLRT schemes described in the literature are based on $K_p = 1$, $K_s \geq N$, and a binary hypothesis test, where with reference to the GMHT in Eqn. (12.24), one hypothesis is structured as H_0 or H_1 (target absent), and the other is structured as H_2 or H_3 (target present). For example, Test $\mathcal{A}$ corresponds to H_0 versus H_2, Test $\mathcal{B}$ corresponds to H_1 versus H_2, and Test $\mathcal{C}$ corresponds to H_1 versus H_3, with $K_p = 1$ and $K_s \geq N$ in all cases. GLRT expressions for H_0 versus H_2 with $K_p > 1$ and $K_s \geq N$ were derived in Kelly and Forsythe (1989), and in Conte et al. (2001), for example.

Specifically, the GLRTs derived in Conte et al. (2001) assume that the target components $\mathbf{s}_k$, which are present only in the primary data, share a common and known structure $\mathbf{v}$, but have an unknown and arbitrary complex scale a_k, as in Eqn. (12.25). This is referred to as an *incoherently distributed* target. The detection of *coherently distributed* targets, wherein the complex scalers a_k are related by a known function was in fact addressed in the remarkable work of Kelly and Forsythe (1989) for the homogeneous Gaussian case assuming no unwanted signals. More will be said on this in section 12.3.

$$\mathbf{s}_k = a_k \mathbf{v}, \quad k \in \Omega_p \tag{12.25}$$

While the GMHT framework of Eqn. (12.24) encompasses the structural elements of many hypothesis tests considered for radar applications, generalizations can also be made in terms of the models specified for the various signal components. A number of important modifications have been advocated in this respect since the original paper of Kelly (1986). These key advances fall under three main categories: (1) generalization of the homogeneous Gaussian case to account for more realistic disturbance models with heterogeneity in scale and possibly structure, (2) extension of the "steering-vector" target model to capture partial uncertainty in the useful signal by means of multi-rank (i.e., subspace) representations, (3) incorporation of unwanted signals due to coherent returns in the primary and possibly secondary data (the latter case is not considered here). Measurement models for the disturbance, useful signal, and coherent interference will be discussed in Section 12.2.

To arrive at a detection statistic for the GMHT, the multi-hypothesis testing problem is typically broken down into a sequence of binary tests. For example, the first binary test may decide if $\mathbf{z}$ contains disturbance-only (H_0), or signal-plus-disturbance, irrespective of whether it is a useful or unwanted signal. If a signal is deemed to be present (H_1, H_2, H_3), then the next test in the chain aims to discriminate between mismatched (H_1) and matched (H_2) signal hypotheses. If the former is accepted, a further test may be performed to decide if $\mathbf{z}$ contains only a mismatched signal (H_1), or whether it additionally contains a matched signal (H_3) that has been masked by a stronger mismatched signal. The focus in this chapter is on GLRT-based detectors for different binary hypothesis tests, since these statistics form the building blocks to address the multi-hypothesis testing problem.

12.2 Measurement Models

In order to completely define a detection problem, measurement models for the various signal components appearing in each hypothesis of the test must be properly specified. This section describes a number of traditional and alternative models for the disturbance, useful signal and unwanted signal components. These models may be assumed for the binary hypothesis tests $\mathcal{A}$, $\mathcal{B}$, and $\mathcal{C}$, or in the GMHT of Eqn. (12.24). The combination of different hypothesis tests (described in the previous section), and measurement models for the individual components of these tests (to be described in this section), give rise to a wide diversity of detection problems, many of which have been considered in the literature.

In realistic scenarios, precise knowledge about the signal environment is typically not available, and standard models based on idealistic assumptions may at times be deemed over-simplistic or inappropriate in certain applications. This motivates the use of alternative disturbance and signal models that attempt to account for a higher degree of uncertainty in the received data characteristics. Such models can potentially lead to detection schemes that are more robust in practice. This section presents measurement models with varying levels of complexity, some of which will be adopted to derive GLRT-based detections schemes in Section 12.3.

12.2.1 Disturbance Process

The multi-variate (circular-symmetric) complex Gaussian distribution has traditionally been used to model the disturbance received in the primary and secondary data. However, analysis of collected data suggests that disturbance vectors received by real radar systems are not always adequately described by an identical Gaussian distribution across all primary and secondary resolution cells. A prime example is clutter samples received over different pulses of the CPI by high-resolution microwave radars, particularly at low grazing angles, as well as land and sea clutter received by HF radar systems. Temporal clutter vectors acquired in different range and beam resolution cells can exhibit significant departures from the IID Gaussian assumption.

The compound-Gaussian distribution represents an alternative model with inherently greater flexibility to describe the experimentally observed clutter statistics (Gini and Farina 2002a). Probability densities compatible with the compound-Gaussian process include the Weibull and K-distributions. An important class of compound-Gaussian model is known as the spherically invariant random process (SIRP). In radar applications, the SIRP is well motivated from a physical viewpoint and is widely adopted as an alternative model for temporal (pulse-train) clutter vectors. A SIRP characterization may also be used in lieu of the IID Gaussian model for incoherent disturbance signals in spatial processing applications, such as interference snapshots received by the elements of an HF antenna array (Fabrizio et al. 2003a).

12.2.1.1 Homogeneous Gaussian Case

In the statistically homogeneous model, the $K = K_s + K_p$ disturbance vectors in the primary and secondary data $\mathbf{d}_k$ for $k \in \Omega \equiv [\Omega_s, \Omega_p]$ are assumed to have zero mean and share a common positive-definite Hermitian covariance matrix $\mathbf{R}$.

$$\mathbf{R} = \mathrm{E}\{\mathbf{d}_k \mathbf{d}_k^\dagger\}, \quad \forall k \in \Omega \tag{12.26}$$

The disturbance vectors $\mathbf{d}_k$ are assumed to be identically distributed with a marginal density defined by the N-variate (circular-symmetric) complex normal distribution $p_d[\mathbf{d}_k]$ in Eqn. (12.27).

$$p_d[\mathbf{d}_k] = \frac{1}{\pi^N \|\mathbf{R}\|} e^{-\mathbf{d}_k^\dagger \mathbf{R}^{-1} \mathbf{d}_k} \tag{12.27}$$

The K disturbance vectors are also assumed to be (mutually) statistically independent with a joint density function $f_d[\mathbf{d}_1, \ldots, \mathbf{d}_K]$ given by the product of the marginal densities in Eqn. (12.28).

$$f_d[\mathbf{d}_1, \ldots, \mathbf{d}_K] = \prod_{k=1}^{K} p_d[\mathbf{d}_k] = \frac{1}{\pi^{NK} \|\mathbf{R}\|^K} \exp\left(-\mathrm{Tr}\left\{\mathbf{R}^{-1} \sum_{k=1}^{K} \mathbf{d}_k \mathbf{d}_k^\dagger\right\}\right) \tag{12.28}$$

This disturbance description, originally proposed in Reed et al. (1974), and later adopted in Kelly (1986), is commonly known as the homogeneous Gaussian disturbance model. The disturbance covariance matrix $\mathbf{R}$ represents the only unknown parameter in this statistical model.

12.2.1.2 Compound-Gaussian Model

The compound-Gaussian disturbance model is often assumed in cases where external disturbance signals dominate the internal noise. To lend concreteness by way of example, suppose that interference is not present and that clutter dominates internal noise, such that the latter may be neglected. In this case, the $N = N_p$ dimensional temporal (pulse-to-pulse) disturbance vector $\mathbf{d}_k$ is well approximated by the clutter-only contribution $\mathbf{c}_k$

in Eqn. (12.29). The influence of additive thermal noise will be discussed later in this section.

$$\mathbf{d}_k = \mathbf{c}_k \tag{12.29}$$

In the compound-Gaussian model, the disturbance vector $\mathbf{d}_k = [d_k(1), \ldots, d_k(N)]^T$ has elements $d_k(n)$ that are deemed to be a product of two mutually independent random processes; namely, a possibly correlated wide-sense stationary (WSS) zero-mean complex circular Gaussian process, referred to as *speckle* $x_k(n)$, multiplied by a real-valued modulating component drawn from samples of a non-negative possibly correlated WSS random process with finite mean-square value, referred to as *texture* $\tau_k(n)$. This two-scale product model may be written as Eqn. (12.30). Without loss of generality, the texture is assumed to have marginal density function $p_\tau(\tau)$ with unit second moment.

$$d_k(n) = \tau_k(n) x_k(n), \quad n = 1, \ldots, N \tag{12.30}$$

The compound-Gaussian model is evidently a doubly-stochastic process. Note that the K-distribution is fully compatible with such a model, whereas the Weibull probability density function (PDF) is amenable to a compound-Gaussian description for shape parameter values in the range (0, 2]. The disturbance vector $\mathbf{d}_k$ can be expressed as the element-wise product in Eqn. (12.31), where $\mathbf{x}_k = [x_k(1), \ldots, x_k(N)]^T$ is the N-variate speckle vector, and $\boldsymbol{\tau}_k = [\tau_k(1), \ldots, \tau_k(N)]^T$ is N-variate texture vector, which modulates the disturbance envelope.

$$\mathbf{d}_k = \boldsymbol{\tau}_k \odot \mathbf{x}_k \tag{12.31}$$

The mutual independence between the speckle and texture means that the disturbance covariance matrix $\mathbf{R} = \mathrm{E}\{\mathbf{d}_k \mathbf{d}_k^\dagger\}$ can be expressed in the form of Eqn. (12.32), where $\mathbf{M}_\tau = \mathrm{E}\{\boldsymbol{\tau}_k \boldsymbol{\tau}_k^\dagger\}$ is the texture covariance matrix with unit diagonal elements by definition, and $\boldsymbol{\Sigma} = \mathrm{E}\{\mathbf{x}_k \mathbf{x}_k^\dagger\}$ is the positive-definite Hermitian speckle covariance matrix.

$$\mathbf{R} = \mathbf{M}_\tau \odot \boldsymbol{\Sigma} \tag{12.32}$$

The assumption of a compound-Gaussian model with a common *statistically expected* disturbance covariance matrix $\mathbf{R}$ over all primary and secondary data vectors is often justified in practice providing the two data sets are extracted from a limited number of resolution cells in a localized region of the data cube, and that the training data is properly vetted to screen out any outliers.

12.2.1.3 Spherically Invariant Random Process (SIRP)

Analysis of real clutter data indicates that the modulating texture component is slowly varying compared to the Gaussian speckle component. When the coherence time of the texture is greater than the radar CPI, the texture samples $\tau_k(n)$ may be assumed to be fully correlated over the observation interval $n = 1, \ldots, N$. This condition allows the texture vector $\boldsymbol{\tau}_k$ to be replaced by a scalar random variable τ_k that multiplies all samples of the speckle vector $\mathbf{x}_k$ in Eqn. (12.33). In such situations, the compound-Gaussian model in Eqn. (12.31) degenerates into the zero-mean complex-valued spherically invariant random process (SIRP).

$$\mathbf{d}_k = \tau_k \mathbf{x}_k \tag{12.33}$$

Since the texture has unit variance $\mathbf{E}\{\tau_k^2\} = 1$, by definition, the disturbance covariance matrix for the SIRP model in Eqn. (12.33) is given by $\mathbf{R} = \boldsymbol{\Sigma}$. The representation theorem in Yao (1973) may be used to write the marginal density function of the SIRP disturbance vectors $\mathbf{d}_k$ as $p_d[\mathbf{d}_k]$ in Eqn. (12.34).

$$p_d[\mathbf{d}_k] = \frac{1}{\pi^N \|\mathbf{\Sigma}\|} h_N(\mathbf{d}_k^\dagger \mathbf{\Sigma}^{-1} \mathbf{d}_k) \tag{12.34}$$

The function $h_N(y)$ is defined in Eqn. (12.35). Both $\mathbf{\Sigma}$ and the texture marginal density $p_\tau(\tau)$ are assumed to be unknown. The SIRP description reverts back to the homogeneous Gaussian model with $\mathbf{R} = \mathbf{\Sigma}$ in the special case of $p_\tau(\tau) = \delta(\tau - 1)$.

$$h_N(y) = \int_0^\infty \tau^{-2N} \exp\left(-\frac{y}{\tau^2}\right) p_\tau(\tau)\, \mathrm{d}\tau \tag{12.35}$$

The texture auto-correlation function depends on the joint texture PDF $f_\tau(\tau_1, \ldots, \tau_K)$ over the primary and secondary data $k \in \Omega$. At one extreme, the texture samples τ_k may be considered mutually independent over all primary and secondary cells, such that the joint PDF is given by Eqn. (12.36). Such a model is perhaps most suitable for high-resolution systems that involve less "averaging" of the texture contributions in a single cell, or if the reflective properties of the scattering surfaces vary rapidly on the spatial scale of the radar resolution cell.

$$f_\tau(\tau_1, \ldots, \tau_K) = \prod_{k=1}^{K} p_\tau(\tau_k) \tag{12.36}$$

At the other extreme, the texture samples may be considered fully dependent, such that $\mathbf{d}_k = \tau \mathbf{x}_k$ for all $k \in \Omega$. This effectively results in a homogeneous Gaussian model with a randomly distributed disturbance power. The disturbance covariance matrix conditioned on the texture value τ is given by Eqn. (12.37). It is evident from this expression that the disturbance *structure* or spectral form is embedded in the speckle covariance matrix $\mathbf{\Sigma}$, while the disturbance power or *scale* is determined by the texture component τ.

$$\mathbf{E}\{\mathbf{d}_k \mathbf{d}_k^\dagger | \tau\} = \tau^2 \mathbf{\Sigma} \tag{12.37}$$

An intermediate case arises when the texture is fully dependent over the secondary data, and separately over the primary data if $K_p > 1$, but not over both, as in Eqn. (12.38). This situation corresponds to a case where the maximum spacing between any pair of resolution cells in the primary or secondary data is small compared to the scale over which the texture decorrelates, but where the secondary vectors are not in the immediate vicinity of the primary data, such that the texture component τ_s will in general differ in a random manner from τ_p.

$$\begin{cases} \mathbf{d}_k = \tau_s \mathbf{x}_k & k \in \Omega_s \\ \mathbf{d}_k = \tau_p \mathbf{x}_k & k \in \Omega_p \end{cases} \tag{12.38}$$

For the model in Eqn. (12.38), define $\mathbf{R}_s = \mathbf{E}\{\mathbf{d}_k \mathbf{d}_k^\dagger | \tau_s\}$ as the secondary data disturbance covariance matrix conditioned on τ_s in Eqn. (12.39). Similarly, the primary data disturbance covariance matrix conditioned on τ_p is given by $\mathbf{R}_p$ in Eqn. (12.39). The *relative* scaling factor $\nu = (\tau_p/\tau_s)^2$ between the primary and secondary data disturbance covariance matrices is a random variable. This simplified SIRP model is referred to as the partially homogeneous case (Conte et al. 2001), also known as the Gaussian disturbance model with unknown scale change between the training and test data. In certain studies, ν is regarded as an unknown deterministic nuisance parameter (Kraut and Scharf 2001).

$$\begin{cases} \mathbf{R}_s = \tau_s^2 \boldsymbol{\Sigma} & k \in \Omega_s \\ \mathbf{R}_p = \tau_p^2 \boldsymbol{\Sigma} = \nu \mathbf{R}_s & k \in \Omega_p \end{cases} \tag{12.39}$$

An implicit assumption is that the external disturbance dominates the internal additive noise. In practice, this approximation is accurate when the disturbance-to-noise ratio is high and the texture profile is not too "spikey," as described in Michels, Himed, and Rangaswamy (2000), Michels, Himed, and Rangaswamy (2002), and Liu, Chen, and Michels (2002). When additive white noise cannot be neglected, it is clear from associated covariance matrices in Eqn. (12.40) that $\mathbf{R}_p \neq \nu \mathbf{R}_s$ for $\tau_s \neq \tau_p$.

$$\begin{cases} \mathbf{R}_s = \tau_s^2 \boldsymbol{\Sigma} + \sigma_n^2 \mathbf{I}_N & k \in \Omega_s \\ \mathbf{R}_p = \tau_p^2 \boldsymbol{\Sigma} + \sigma_n^2 \mathbf{I}_N & k \in \Omega_p \end{cases} \tag{12.40}$$

In the remainder of this chapter, the emphasis will be on the use of the homogeneous Gaussian disturbance model described previously, and the partially homogeneous SIRP disturbance model. These two popular models have been extensively used to derive adaptive detection tests based on the GLRT philosophy.

12.2.2 Useful Signal

The scope of this section is not to provide an exhaustive taxonomy of useful signal models for radar applications, but rather to describe and contrast the main approaches used to model $\mathbf{s}$ in the binary hypothesis tests $\mathcal{A}$, $\mathcal{B}$, and $\mathcal{C}$, or $\mathbf{s}_k$ in the GMHT. Traditionally, the useful signal model $\mathbf{s} = a\mathbf{v}$ consists of a *known* deterministic signature vector $\mathbf{v}$, which describes the structure of the target echo, multiplied by an *unknown* deterministic complex scale a, which determines its amplitude and phase due to uncertain scattering and propagation effects.

Although the primary focus is on deterministic useful signal models, the characteristics of $\mathbf{s}$ may be described in rather general terms, which includes the possibility of statistical signals. In Farina and Russo (1986), for example, the target signal is modeled as a multi-variate complex Gaussian vector with mean vector $\mathbf{m}_s$ and covariance matrix $\mathbf{M}_s$ defined in Eqn. (12.41).

$$\begin{cases} \mathbf{m}_s = \mathrm{E}\{\mathbf{s}\} \\ \mathbf{M}_s = \mathrm{E}\{(\mathbf{s} - \mathbf{m}_s)(\mathbf{s} - \mathbf{m}_s)^\dagger\} \end{cases} \tag{12.41}$$

Specifically, a signal vector $\mathbf{s}$ with both mean and fluctuating components may be statistically represented by the multi-variate circular-symmetric complex-normal density function $f_s[\cdot]$ in Eqn. (12.42), where $\mathbf{M}_s$ is a positive definite Hermitian matrix. In short-hand notation, $\mathbf{s} \sim \mathcal{CN}(\mathbf{m}_s, \mathbf{M}_s)$. Although non-Gaussian multi-variate density functions, such as compound-Gaussian or elliptically contoured distributions, may also be proposed, such models will not be considered here as far as the useful signal is concerned. The great majority of adaptive signal detection problems assume a purely deterministic model for the useful signal, and this convention will be followed for the most part in this chapter.

$$f_s[\mathbf{s}] = \frac{1}{\pi^N \|\mathbf{M}_s\|} e^{-(\mathbf{s}-\mathbf{m}_s)^\dagger \mathbf{M}_s^{-1} (\mathbf{s}-\mathbf{m}_s)} \tag{12.42}$$

12.2.2.1 Discrete-Target Echo

The discrete-target model is based on the notion of a steering vector manifold for the useful signal structure. This manifold is typically defined as an analytic function of one or more target parameters, denoted here by ω. As ω takes all possible values in the target parameter domain $\omega \in \mathcal{D}$, the steering vector $\mathbf{v}(\omega)$ traces out a locus in N-dimensional space. The set of all points on this locus, designated $\mathcal{M}$ in Eqn. (12.43), is known as the steering vector manifold.

$$\mathcal{M} \equiv \{\mathbf{v}(\omega);\ \omega \in \mathcal{D}\} \tag{12.43}$$

A discrete-target echo is defined as a useful signal $\mathbf{s} = a\mathbf{v}$ with a signature vector $\mathbf{v}$ on the manifold $\mathcal{M}$. In other words, $\mathbf{v} = \mathbf{v}(\omega_s)$ for some $\omega_s \in \mathcal{D}$. The discrete-target model in Eqn. (12.44) does not necessarily assume that the parameter ω_s is known *a priori*. In practical surveillance radar applications, the target parameter ω_s is unknown and is searched for over a bank of preassigned candidates; e.g., a finite set of beam-steer directions and/or Doppler frequency bins.

$$\mathbf{s} = a\mathbf{v}(\omega_s) \tag{12.44}$$

The signal model for a discrete target can be deterministic or statistical, depending on the properties of the complex scalar a. The deterministic description represents the target signal $\mathbf{s}$ by its mean vector $\mathbf{m}_s$ in Eqn. (12.45). In this case, the complex scale $a = \mu e^{j\vartheta}$ is unknown but fixed, and represents incomplete knowledge about the target echo amplitude μ and phase ϑ due to (invariant) scattering and propagation effects from scan to scan. The signature vectors in $\mathcal{M}$ are typically assumed to have constant norm, e.g., $\mathbf{v}(\omega_s)^\dagger\mathbf{v}(\omega_s) = N$.

$$\text{deterministic} \begin{cases} \mathbf{m}_s = a\mathbf{v}(\omega_s) \\ \mathbf{M}_s = \mathbf{0} \end{cases} \tag{12.45}$$

The *purely* statistical description of $\mathbf{s}$ assumes that the mean vector $\mathbf{m}_s$ is zero and the covariance matrix is given by $\mathbf{M}_s = \mathbf{E}\{\mathbf{s}\mathbf{s}^\dagger\}$. For the discrete-target model, this corresponds to a zero-mean fluctuating complex amplitude a and a deterministic signature vector $\mathbf{v}(\omega_s)$. When a has IID Gaussian real and imaginary parts that vary independently from scan to scan to produce Rayleigh fading, $\mathbf{s}$ is described by the Swerling I model of Eqn. (12.46), where $\sigma_s^2 = \mathbf{E}\{|a|^2\}$ is the variance of the scale fluctuations.

$$\text{Swerling I} \begin{cases} \mathbf{m}_s = \mathbf{0} \\ \mathbf{M}_s = \sigma_s^2\mathbf{v}(\omega_s)\mathbf{v}(\omega_s) \end{cases} \tag{12.46}$$

Although multi-rank models are covered in the next section, a special case is discussed here, wherein the scale and structure of $\mathbf{s}$ fluctuates from scan to scan. The Swerling II model in Eqn. (12.47) results when all elements in the signal vector $\mathbf{s} = [s_1, s_2, \ldots, s_N]^T$ are zero-mean IID Gaussian random variables that produce Rayleigh fading. This leads to the statistical description in Eqn. (12.47), where $\sigma_s^2 = \mathbf{E}\{|s_n|^2\}$ for $n = 1, 2\ldots, N$.

$$\text{Swerling II} \begin{cases} \mathbf{m}_s = \mathbf{0} \\ \mathbf{M}_s = \sigma_s^2\mathbf{I}_N \end{cases} \tag{12.47}$$

A non-zero mean vector ($\mathbf{m}_s \neq \mathbf{0}$) in this IID Gaussian model produces useful signal samples $\{s_n\}_{n=1}^N$ with a Ricean distributed amplitude envelope. The Swerling III and IV models result for the case of $\mathbf{m}_s \neq \mathbf{0}$ in Eqns. (12.46) and (12.47), respectively.

12.2.2.2 Extended Useful Signals

An extended useful signal is defined as a target echo $\mathbf{s} = a\mathbf{v}$ with a signature vector $\mathbf{v}$ that does not lie on the manifold $\mathcal{M}$. In other words, $\mathbf{v} \neq \mathbf{v}(\omega_s)$ for any $\omega_s \in \mathcal{D}$. The causes for this may be environmental or instrumental, but we shall not distinguish between them. An extended signal $\mathbf{s}$ is not aligned with any single point on the manifold, but can be expressed as the sum of a potentially large number N_r of closely clustered points on the manifold, as in Eqn. (12.48), where ρ_r are the ray complex amplitudes and δ_r are small displacements relative to the nominal target parameter ω_s (i.e., centroid of the cluster).

$$\mathbf{s} = \sum_{r=1}^{N_r} \rho_r \mathbf{v}(\omega_s + \delta_r) \tag{12.48}$$

Such a model tends to be most appropriate when the deviation between the actual signal structure $\mathbf{v}$ and the nominal steering vector $\mathbf{v}(\omega_s)$ is not large. The complex amplitudes ρ_r and parameter displacements δ_r are assumed unknown. For small displacements δ_r, the steering vectors $\mathbf{v}(\omega_s + \delta_r)$ may be accurately approximated by a truncated Taylor series expansion with M terms, as in Eqn. (12.49). The manifold $\mathcal{M}$ is assumed to be continuously differentiable with respect to ω, and the vectors $\mathbf{d}_m(\omega_s) = \partial^m \mathbf{v}(\omega)/\partial\omega^m|_{\omega=\omega_s}$ for $m = 1, \ldots, M-1 \ll N$ are assumed to be linearly independent, n.b., $\mathbf{d}_0(\omega_s) = \mathbf{v}(\omega_s)$.

$$\mathbf{s} = \sum_{r=1}^{N_r} \rho_r \left\{ \mathbf{v}(\omega_s) + \sum_{m=1}^{M-1} \frac{\rho_r(\delta_r)^m}{m!} \left[\frac{\partial^m \mathbf{v}(\omega)}{\partial\omega^m}\right]_{\omega=\omega_s} \right\} \tag{12.49}$$

Once the model in Eqn. (12.49) is accepted for $\mathbf{s}$, the useful signal may be written as Eqn. (12.50), where $\mathbf{d}_m(\omega_s)$ for $m = 0, \ldots, M-1$ are linearly independent basis vectors or "modes." The form of each basis vector is a known function of the nominal search parameter ω_s. The unknown linear combination coefficients are given by the complex weights $\theta_m = \sum_{r=1}^{N_r} \rho_r(\delta_r)^m/m!$ for $m = 0, \ldots, M-1$. For example, a first-order ($M = 1$) Taylor series expansion was used in Astely, Ottersten, and Swindelhurst (1998) to model the spatial signatures of locally scattered sources.

$$\mathbf{s} = \theta_0 \mathbf{v}(\omega_s) + \sum_{m=1}^{M-1} \theta_m \mathbf{d}_m(\omega_s) \tag{12.50}$$

This model is less appropriate for an extended signal $\mathbf{s}$ where the component rays deviate significantly from the nominal steering vector $\mathbf{v}(\omega_s)$. An alternative is to represent $\mathbf{s}$ as the superposition of a relatively small number $M \ll N$ of more widely spaced steering vectors $\mathbf{v}(\omega_s + \Delta_m)$ in Eqn. (12.51). The different positive or negative displacements Δ_m for $m = 1, \ldots, M-1$ may take values between ω_s and its immediate neighbors in the target search parameter bank, for example. The Vandermonde structure of the manifold $\mathcal{M}$ ensures that M basis vectors or modes are linearly independent. This was referred to as the wave-interference model in Fabrizio et al. (2003a).

$$\mathbf{s} = \theta_0 \mathbf{v}(\omega_s) + \sum_{m=1}^{M-1} \theta_m \mathbf{v}(\omega_s + \Delta_m) \tag{12.51}$$

In both cases, $\mathbf{s}$ can be represented by the multi-rank (subspace) model of Eqn. (12.52), where the mode matrix $\mathbf{V}(\omega_s)$ may be defined as $\mathbf{V}(\omega_s) = [\mathbf{v}(\omega_s), \mathbf{d}_1(\omega_s), \ldots, \mathbf{d}_{M-1}(\omega_s)]$, or in the alternative form of $\mathbf{V}(\omega_s) = [\mathbf{v}(\omega_s), \mathbf{v}(\omega_s + \Delta_1), \ldots, \mathbf{v}(\omega_s + \Delta_{M-1})]$, for example. Both representations offer greater flexibility than searching over a series of finely spaced steering vectors individually, as subspace models can represent extended signals that are not on the manifold. Clearly, other models may be postulated for $\mathbf{V}(\omega_s)$. Just as $\mathbf{V}(\omega_s)$ is the subspace generalization of the signature vector $\mathbf{v}(\omega_s)$, the M-variate coordinate vector $\underline{\theta} = [\theta_0, \theta_1, \ldots, \theta_{M-1}]^T$ is the multi-rank extension of the complex scale a in going from the discrete target echo model to an extended useful signal description. For a purely deterministic signal, $\mathbf{V}(\omega_s)$ is assumed to be known, whereas $\underline{\theta}$ is assumed unknown but *invariant* from scan to scan.

$$\mathbf{s} = \mathbf{V}(\omega_s)\underline{\theta}, \quad \mathbf{V}(\omega_s) \in \mathcal{C}^{N\times M}, \quad \underline{\theta} \in \mathcal{C}^M \tag{12.52}$$

A statistical signal $\mathbf{s} = a\mathbf{v}$ may be written in the form of Eqn. (12.53), where $\mathbf{v}(\omega_s)$ is the steering vector most closely matched to $\mathbf{v}$ in a minimum mean square error (MMSE) sense, and $\mathbf{b} \in \mathcal{C}^N$ is a zero-mean (complex Gaussian distributed) multiplicative distortion vector. The symbol $\odot$ denotes element-wise product. In simple terms, $\mathbf{s}$ may be thought of as possessing a "gross" structure determined by the nominal steering vector $\mathbf{v}(\omega_s)$, and an uncertain "fine" structure determined by the modulating *random* vector $\mathbf{b}$.

$$\mathbf{s} = \mathbf{b} \odot \mathbf{v}(\omega_s) \tag{12.53}$$

In the purely statistical case, the random vector $\mathbf{b}$ has zero mean, such that $\mathbf{m}_s = \mathbf{0}$. The signal covariance $\mathbf{M}_s$ is given by Eqn. (12.54), where $\mathrm{E}\{\mathbf{b}\mathbf{b}^\dagger\} = \sigma_s^2\mathbf{B}$. The spreading matrix $\mathbf{B}$ has unit diagonal elements $[\mathbf{B}]_{ii} = 1$ by definition, and determines the spectral width of the statistical signal, while σ_s^2 represents the signal power. The dependence of $\mathbf{M}_s$ on ω_s is implicit and has been dropped for notational convenience.

$$\mathbf{M}_s = \sigma_s^2\mathbf{B} \odot \mathbf{v}(\omega_s)\mathbf{v}^\dagger(\omega_s) \tag{12.54}$$

Commonly adopted models for element (i, j) of $\mathbf{B}$, denoted by $\mathbf{B}_{ij}$, are the Gaussian or exponentially decaying functions of the one-lag correlation coefficient $\rho \in [0, 1]$, as defined in Eqn. (12.55). Such models parametrically characterize the "coherence loss" in the signal vector $\mathbf{s}$ by power spectral density functions with Gaussian and Lorentzian profiles, respectively. See Paulraj and Kailath (1988) for further details regarding this statistical signal model.

$$\mathbf{B}_{ij} = \rho^{|i-j|^2} \text{ (Gaussian)}, \quad \mathbf{B}_{ij} = \rho^{|i-j|} \text{ (decaying exponential)} \tag{12.55}$$

In the limiting case of $\rho \to 1$, $\mathbf{M}_s$ approaches the unit ($M = 1$) rank model in Eqn. (12.56). This is because $\mathbf{B}$ tends to a matrix of unit entries, and the target spectrum approaches a delta function centered on ω_s. It is evident that the unit-rank model of $\mathbf{M}_s$ is equivalent to the Swerling I model, which corresponds to a fully coherent target of completely known form up to a random complex constant.

$$\lim_{\rho\to 1} \mathbf{M}_s = \sigma_s^2\mathbf{v}(\omega_s)\mathbf{v}^\dagger(\omega_s) \tag{12.56}$$

As $\rho \to 0$, $\mathbf{M}_s$ approaches the full ($M = N$) rank model in Eqn. (12.56) since $\mathbf{B} \to \mathbf{I}_N$, and the target spectrum approaches a uniform profile. The full-rank model of $\mathbf{M}_s$ in Eqn. (12.56) is equivalent to the Swerling II model of a useful signal vector with completely uncorrelated elements. Values in the range of $0 < \rho < 1$ correspond to useful signal vectors with partially correlated elements (i.e., a useful signal of finite spectral width), and covariance matrices $\mathbf{M}_s$ of intermediate effective rank $1 < M < N$. Such values of ρ therefore represent a continuous transition between the Swerling I and II models, which represent the two extreme cases.

$$\lim_{\rho \to 0} \mathbf{M}_s = \sigma_s^2 \mathbf{I}_N \tag{12.57}$$

The positive semi-definite Hermitian matrix $\mathbf{M}_s$ may be approximated by a rank-M truncation of its eigen-decomposition in Eqn. (12.54), where the $N \times M$ unitary matrix $\mathbf{Q}_s = [\mathbf{q}_1, \ldots, \mathbf{q}_M]$ contains the eigenvectors $\mathbf{q}_m$ associated with the M-dominant positive eigenvalues λ_m in the $M \times M$ diagonal matrix $\mathbf{\Lambda}_s = \text{diag}[\lambda_1, \ldots, \lambda_M]$. The accuracy of the approximation depends on M and the eigenvalue spectrum of $\mathbf{M}_s$, which is determined by ρ in Eqn. (12.54).

$$\mathbf{M}_s = \mathbf{Q}_s \mathbf{\Lambda}_s \mathbf{Q}_s^\dagger + \mathbf{Q}_n \mathbf{\Lambda}_n \mathbf{Q}_n^\dagger \tag{12.58}$$

As ρ approaches unity, $\mathbf{M}_s$ is close to being rank deficient as all but the first eigenvalue drop rapidly toward zero. This defines a relatively low *effective* rank $M \ll N$, which contains almost all of the useful signal energy. In this case, the statistical signal $\mathbf{s}$ lies in a relatively low-rank subspace $\mathbf{Q}_s$ to a good approximation. This allows it to be accurately represented by a subspace model in the form of Eqn. (12.52), but in this case, the mode matrix contains the M principle eigenvectors of $\mathbf{M}_s$, as in Eqn. (12.59). In the purely statistical case, the mode matrix in Eqn. (12.59) is assumed to be known, whereas $\underline{\theta}$ is a *fluctuating* unknown zero-mean vector.

$$\mathbf{s} = \mathbf{Q}_s \underline{\theta} \tag{12.59}$$

In contrast to the extended signal resulting for the deterministic subspace model, where $\underline{\theta}$ is assumed to be fixed, the random variation of $\underline{\theta}$ in the purely statistical subspace model gives rise to an extended signal with a changing scale and structure. In accordance with Eqns. (12.58) and (12.59), statistical fluctuations of the parameter vector $\underline{\theta}$ are characterized by the covariance matrix $\mathbf{M}_\theta$ in Eqn. (12.60).

$$\mathbf{M}_\theta = \mathbf{E}\{\underline{\theta}\underline{\theta}^\dagger\} = \mathbf{\Lambda}_s \tag{12.60}$$

In a Ricean channel model, with non-zero mean $\mathbf{m}_s = \theta_0 \mathbf{v}(\omega_s)$ and covariance matrix $\mathbf{M}_s = \sigma_s^2 \mathbf{B} \odot \mathbf{v}(\omega_s)\mathbf{v}^\dagger(\omega_s)$, the Gaussian distributed signal $\mathbf{s}$ has the density $f_s[\mathbf{s}]$ in Eqn. (12.42) for a positive-definite Hermitian matrix $\mathbf{M}_s$. Such a signal may be approximated by Eqn. (12.61) using the subspace model $[\mathbf{v}(\omega_s), \mathbf{Q}_s]$, where the first element of $\underline{\theta}$ is fixed, but the other elements are fluctuating. The main point is that partial uncertainty in the scale and structure of a deterministic or statistical extended signal may be modeled by assuming that $\mathbf{s}$ lies in a given linear subspace of relatively low rank.

$$\mathbf{s} = [\mathbf{v}(\omega_s), \mathbf{Q}_s]\underline{\theta} \tag{12.61}$$

The use of subpace models for the useful signal may be summarized as follows. The degree of available information about the useful signal is embodied in a mode matrix $\mathbf{V} = [\mathbf{v}_1, \ldots, \mathbf{v}_M]$ defined by the M known linearly independent basis vectors $\mathbf{v}_m \in \mathcal{C}^N$ for $M \ll N$; whereas $\underline{\theta}$ is a deterministic (or statistical) unknown M-dimensional parameter vector that represents the partial uncertainty about the useful signal.

The case of *partial* uncertainty corresponds to subspace dimension values in the range $1 < M < N$. The value $M = 1$ corresponds to a signal vector that is completely known to within a complex multiplicative constant, while $M = N$ reflects the case where the signal to be detected is a totally unknown N-dimensional vector. For simplicity, we shall henceforth refer to the subspace model more simply as $\mathbf{s} = \mathbf{V}\underline{\theta}$, with specific definitions for $\mathbf{V}$ and $\underline{\theta}$ to be stated, as required.

12.2.2.3 Distributed Useful Signals

Thus far, we have considered the possibility of useful signal presence in a single primary data vector. In certain detection problems, echoes from a particular target are not confined to a single resolution cell, but are distributed over a group of cells. The so-called "range-spread" target echoes received by high range-resolution radars represents a well-known example (Cone et al. 2001). We shall use the terms *point* and *distributed* to respectively distinguish between cases where the target echo energy is confined to a single primary vector (ignoring spectral leakage), or is present in multiple primary data vectors that are to be processed jointly for detection.

The distinction between a discrete and point target is worth clarifying, as these terms are often used interchangeably in the literature, and this can lead to confusion. Recall that the detection process requires the designation of a processing dimension in the data cube, over which the acquired data samples are assembled to form the primary and secondary data vectors (e.g., snapshots in space or space-time), and one or more complementary data cube dimensions, over which different resolution cells give rise to different primary and secondary data vectors (e.g., range bins).

As defined here, a point-target echo has its energy confined to a single resolution cell (ignoring spectral leakage effects), otherwise it is deemed to be distributed. On the other hand, a discrete-target echo is one that possesses a structure coinciding with a particular member of the steering vector manifold assumed for the useful signal in the processing dimension, otherwise it is said to be extended. Although these are not standard definitions, they serve to differentiate between various types of models described for the useful signal.

Models for point target echoes were discussed in the previous subsections. Here, we discuss distributed target echoes, not limited to the range-spread example. The possible presence of distributed useful signal components $\mathbf{s}_k$ is accepted only for the primary data vectors indexed by $k = K_s + 1, \ldots, K = K_s + K_p$. The vectors $\mathbf{s}_k$ in Eqn. (12.62) may be assumed to have a static signature vector, wherein only the target complex scale a_k may change, or a dynamic signature vector, which potentially involves changes in target scale and structure across the primary data $k \in \Omega_p$. In radar applications, attention is mainly restricted to the static signature model described by Eqn. (12.62).

$$\mathbf{s}_k = \begin{cases} a_k\mathbf{v} & k \in \Omega_p \\ \mathbf{0} & k \in \Omega_s \end{cases} \tag{12.62}$$

For this type of distributed target model, it is of interest to describe the properties of the K_p-dimensional spreading vector $\mathbf{a} = [a_{K_s+1}, a_{K_s+2}, \ldots, a_K]^T$. In several previous works dealing with range-spread targets, the elements of $\mathbf{a}$ are assumed to be deterministic and completely unknown. This special case may be referred to as "unstructured" spread, since it corresponds to an arbitrary vector $\mathbf{a} \in \mathcal{C}^{K_p}$. In certain applications, the spreading vector exhibits a structure $\mathbf{a} = a\,\mathbf{e}(\psi_s)$ aligned with a template model $\mathbf{e}(\psi_s)$ of unit norm, which may be analytically defined in terms of one or more target parameters ψ_s in the complementary data-cube dimension(s). In this case, the spreading may be considered parameterically known up to a complex multiplicative constant a. This is referred to as "structured" spread. Both scenarios are summarized in Eqn. (12.63).

$$\begin{cases} \mathbf{a} = a\,\mathbf{e}(\psi_s) & \text{structured} \\ \mathbf{a} \in \mathcal{C}^{K_p} & \text{unstructured} \end{cases} \tag{12.63}$$

A deterministic unstructured model for **a** was adopted for range-spread targets in Conte et al. (2001), and in other related works concerned with distributed targets, such as Bandiera, Orlando, and Ricci (2006). The structured spread model has received less attention in the literature. Furthermore, the notion of distributed targets has often been cast in the range dimension, where the amplitudes of echoes from various resolved "scattering centers" of a target may be considered independent, or at least not to have any a priori known structure that can be exploited for detection. A practical adaptive signal detection problem which motivates the structured spread model in radar applications was proposed in Fabrizio, De Maio, Farina, and Gini (2007) and will be discussed in the next section.

For example, consider the spatial processing application and assume that a target echo is confined to a particular range cell. Now consider the slow-time sequence of spatial snapshots recorded at this range cell over the different pulses of the radar CPI. The target echo is present in all pulses, and therefore all of these spatial snapshots, which may be regarded as multiple primary data vectors in the range cell under test.

Using the previous nomenclature, the processing dimension is space (i.e., $N = N_a$), while the complementary dimension over which the target spreads is slow time (i.e., $K_p = N_p$). The secondary data may be extracted from the $N_g - 1$ remaining ranges in the data cube, or a subset of them. In this example, **v** in Eqn. (12.62) represents the spatial signature vector of the target, assumed static over the CPI, while a_k represents the temporal progression of the target complex amplitudes from one pulse to another. For a steadily moving target with normalized angular Doppler frequency $\psi_s = 2\pi f_s$, it follows that $a_k = a\,\exp[j(k-1-K_s)\psi_s]$, where a is the target complex amplitude in the first pulse, as per Eqn. (12.64).

$$\mathbf{e}(\psi_s) = \frac{1}{\sqrt{K_p}}[1, e^{j\psi_s}, \ldots, e^{j\psi_s(K_p-1)}]^T \tag{12.64}$$

Interference-plus-noise array snapshots received in different radar pulses may be considered independent over the K_p primary data vectors. As we shall see in Section 12.3, this property facilitates the GLRT derivation. This is a *spatial-only* processing example involving multiple primary vectors. An analogous example may be proposed for clutter mitigation in a *temporal-only* processing application. The mutual independence of disturbance vectors does not apply in general for the sub-aperture STAP scheme described in Aboutanios and Mulgrew (2005), where multiple primary vectors are extracted in sliding window fashion from the range cell under test. In this scheme, the independence of the disturbance vectors holds strictly only for the special case of spatially and temporally white noise.

12.2.3 Coherent Interference

Besides the disturbance process, which is present in both the primary and secondary data, radar systems may also receive *unwanted signals* due to coherent scatter. Unwanted signals may be present in the CUT independently of whether the useful signal is present or not. However, unlike the disturbance process $\mathbf{d}_k$, the characteristics of a coherent interference signal $\mathbf{r}_k$ in the primary data cannot be estimated or "learnt" from the secondary data. This property is reflected by Eqn. (12.65), where b_k and $\mathbf{u}$ denote the complex scale and structure of the coherent interference $\mathbf{r}_k$, respectively.

$$\mathbf{r}_k = \begin{cases} b_k\mathbf{u} & k \in \Omega_p \\ \mathbf{0} & k \in \Omega_s \end{cases} \tag{12.65}$$

Unwanted signals usually originate from passive (scattering) sources, such as echoes reflected from "sidelobe targets," which have angle and/or Doppler coordinates not matched to the assumed useful signal. In spatial processing applications, this phenomenon can arise in OTH and airborne surveillance radars that illuminate the search area using a relatively wide transmit beam, and then electronically steer narrower "finger" beams within the transmit pattern using a receiver array with wider aperture. In airborne radars, discrete "non-homogeneities" can also arise due to corner reflectors located in the primary range cell, but not in the beam-steer direction, see Adve, Hale, and Wicks (2000b) and Adve, Hale, and Wicks (2000c). Active sources of unwanted signals may include beacons (i.e., repeaters or transponders) or electronic counter measures (i.e., deception jamming).

The unexpected presence of unwanted signals in the primary data may adversely impact performance in two ways: (1) by masking weaker target echoes in the same resolution cell, or (2) by triggering false alarms that possibly lead to false tracks. For these reasons, the incorporation of unwanted signals explicitly into the hypothesis testing problem has recently received more attention. Accepting the possibility of unwanted signal presence at the problem formulation stage enables the derivation of detection schemes with greater immunity against target masking and false alarms.

Similar to the useful signal, the coherent interference $\mathbf{r}_k$ is assumed to belong to a known linear subspace $\mathbf{U} \in \mathcal{C}^{N\times P}$ of rank $P \leq N - M$, where M is the rank of the useful signal subspace. The P basis vectors $\mathbf{u}_p$ in the columns of $\mathbf{U} = [\mathbf{u}_1, \ldots, \mathbf{u}_P]$ are assumed linearly independent with each other and the modes $\mathbf{v}_m$ in the target subspace $\mathbf{V} = [\mathbf{v}_1, \ldots, \mathbf{v}_M]$. In Eqn. (12.66), $\underline{\phi}_k$ is an unknown P-dimensional parameter vector, which may be modeled as a deterministic or statistical quantity, as described previously for the useful signal.

$$\mathbf{r}_k = \mathbf{U}\underline{\phi}_k, \qquad \mathbf{U} \in \mathcal{C}^{N\times P}, \;\; \underline{\phi}_k \in \mathcal{C}^P \tag{12.66}$$

While the subspace model $\mathbf{V}$ for the useful signal is defined on the basis of a nominal steer direction ω_s, a subspace model for $\mathbf{r}_k$ is more difficult to define in practice because an unwanted signal may arrive from any "direction" (that is typically unknown a priori). Moreover, the structure of $\mathbf{r}_k$ depends on the unwanted signal source coordinates, which may differ from one resolution cell to another. In the absence of a specific subspace model for $\mathbf{r}_k$, one approach is to model $\mathbf{r}_k$ as orthogonal to $\mathbf{s}_k$ by assuming $P = N - M$. This includes all possible unwanted signals and has the advantage of being universally applicable to every resolution cell. Hence, if $\mathbf{r}_k$ is present, it can be assumed to lie in the range space of $\mathbf{U}$ that satisfies the orthogonality condition $\mathbf{U}^{\dagger}\mathbf{V} = \mathbf{0}$, where the augmented matrix $[\mathbf{U}, \mathbf{V}]$ spans all of N-dimensional space.

If the test vector is to be transformed to "whitened" disturbance space, then the transformed desired and undesired signal subspaces may be assumed orthogonal in whitened data space, as pointed out in Fabrizio et al. (2003a). For example, given the model $\mathbf{V}$, the subspace $\mathbf{U}$ may be defined such that it satisfies Eqn. (12.67), where the disturbance SCM $\mathbf{S}$ is used to "quasi-whiten" the data. Alternatively, the true disturbance covariance matrix $\mathbf{R}$ may be substituted for $\mathbf{S}$ in Eqn. (12.67), as suggested in Bandiera, Besson, and Ricci (2008). The operator $< \cdot >$ denotes the range-space of a matrix:

$$< \mathbf{S}^{-1/2}\mathbf{V} >^{\perp} = < \mathbf{S}^{-1/2}\mathbf{U} > \tag{12.67}$$

The modes of $\mathbf{U} = [\mathbf{u}_1, \ldots, \mathbf{u}_P]$ may in some cases be known or accurately modeled using a priori information about the source of the unwanted signal. For example, a transponder signal may arrive from a known direction or have known Doppler offsets, while resonant (first-order) sea clutter in a high frequency surface-wave radar is composed of Bragg lines with Doppler shifts that are a function of the operating frequency (in the absence of surface currents). In such situations, the modes can be modeled as $\mathbf{u}_p = \mathbf{v}(\omega_p)$, where the parameters ω_p for $p = 1, \ldots, P$ are the coordinates of point scatterers causing the coherent interference ($\omega_p = \omega_s$). A practical method for obtaining such coordinates is illustrated in Section 12.4.

12.3 Processing Schemes

Once the structure of the hypothesis test is specified, and models for the signal components are defined, the detection statistic may be derived using the GLRT method. The first part of this section briefly recalls some well-known adaptive detection schemes based on (one- and two-step) GLRTs for the traditional hypothesis test assuming a homogeneous Gaussian disturbance. This is followed by a description of adaptive subspace detectors, which account for the possibility of multi-rank useful signals, and a partially homogeneous disturbance. Some less well-known GLRTs that additionally include the possibility of unwanted subspace signals in the primary data are also described in the second part of this section.

GLRTs are first presented for the case where echoes from individual targets are localized to a single resolution cell (i.e., point targets). GLRT procedures for the detection of distributed targets using multiple primary and secondary data vectors are derived for the homogeneous and partially homogeneous disturbance cases in the third part of this section. Here, particular attention is paid to the derivation of one- and two-step GLRTs for distributed targets that exhibit *structured* spread across the primary data vectors. The performance of various GLRTs described in this section will be demonstrated by real-data processing in Section 12.4.

12.3.1 One- and Two-Step GLRT

To motivate the discussion, we recall the original problem formulation of Reed et al. (1974), where the *primary data* vector $\mathbf{z} = [z(1), \ldots, z(N)]^T$ is processed by an adaptive filter $\hat{\mathbf{w}}$ estimated using the sample matrix inverse (SMI) technique in Eqn. (12.68). Here, $\mathbf{v}$ is the known useful signal signature vector, $\mathbf{S}$ is the disturbance sample covariance matrix, and ρ is an arbitrary scalar.

$$\hat{\mathbf{w}} = \rho \mathbf{S}^{-1}\mathbf{v} \tag{12.68}$$

The RMB procedure replaces the unknown *true* positive-definite Hermitian disturbance covariance matrix $\mathbf{R}$ in the optimum filter expression $\mathbf{w} = \rho\mathbf{R}^{-1}\mathbf{v}$ by its sample estimate $\mathbf{S}$ formed on K_s *secondary* data vectors $\mathbf{z}_1, \ldots, \mathbf{z}_k$ that contain IID Gaussian disturbance realizations with the same statistical properties as the disturbance in the primary vector. The assumption $K_s \geq N$ is made such that $\mathbf{S}$ is non-singular.

$$\mathbf{S} = K_s{}^{-1} \sum_{k=1}^{K_s} \mathbf{z}_k \mathbf{z}_k^{\dagger} \tag{12.69}$$

The minimum variance distortionless response (MVDR) filter adopts the normalization $\rho = (\mathbf{v}^{\dagger}\mathbf{S}^{-1}\mathbf{v})^{-1}$ to ensure fixed unit-gain response to useful signals (i.e., $\hat{\mathbf{w}}^{\dagger}\mathbf{v} = 1$). In this case, the filtered scalar output sample $z = \hat{\mathbf{w}}^{\dagger}\mathbf{z}$ is given by Eqn. (12.70). It is well known that the SMI filter asymptotically maximizes the signal-to-disturbance ratio in the output sample z as $K \to \infty$. The convergence properties of the SMI technique are analyzed in Reed et al. (1974).

$$z = \frac{\mathbf{v}^{\dagger}\mathbf{S}^{-1}\mathbf{z}}{\mathbf{v}^{\dagger}\mathbf{S}^{-1}\mathbf{v}} \tag{12.70}$$

The implication is that the amplitude envelope or modulus squared of the filtered output z is compared against a threshold η for signal detection, as in Eqn. (12.71). As stated in Kelly (1986), no predetermined threshold η can be assigned to achieve a given probability of false alarm (PFA), since the detector in Eqn. (12.71) is supposed to operate in a disturbance environment of unknown form and intensity.

$$|z|^2 \underset{H_0}{\overset{H_1}{\gtrless}} \eta \tag{12.71}$$

For example, in the simple case of white noise $\mathbf{R} = \sigma_n^2\mathbf{I}$ with unknown power σ_n^2, the optimum filter is the conventional matched filter $\mathbf{v}$, and $\hat{\mathbf{w}} \to (\mathbf{v}^{\dagger}\mathbf{v})^{-1}\mathbf{v}$ as $K \to \infty$. For $\mathbf{v}^{\dagger}\mathbf{v} = N$, detection applied directly to the matched filtered output results in Eqn. (12.72). The distribution of $|\mathbf{v}^{\dagger}\mathbf{z}|^2$ in the noise-only hypothesis clearly depends on the value of σ_n^2. For this reason, no predetermined threshold η can be set in Eqn. (12.72) to achieve a constant and predictable PFA, which leaves the performance of the test undetermined.

$$|\mathbf{v}^{\dagger}\mathbf{z}|^2 \underset{H_0}{\overset{H_1}{\gtrless}} N\eta \tag{12.72}$$

12.3.1.1 Kelly's GLRT

Many practical systems ultimately require a *decision rule* for declaring target presence, where the PFA of the procedure can be determined for a given threshold setting in a disturbance environment of unknown form and intensity. Kelly (1986) reconsidered the problem formulated by RMB and cast it as a hypothesis test, i.e., Test A in Eqn. (12.13). As described earlier, this led to Kelly's generalized likelihood-ratio test (GLRT), reproduced for convenience in Eqn. (12.73). Kelly's GLRT is generally regarded as a benchmark adaptive detection algorithm for performance comparisons.

$$\frac{|\mathbf{v}^\dagger \mathbf{S}^{-1}\mathbf{z}|^2}{\mathbf{v}^\dagger \mathbf{S}^{-1}\mathbf{v}(1 + K_s^{-1}\mathbf{z}^\dagger \mathbf{S}^{-1}\mathbf{z})} \underset{H_0}{\overset{H_1}{\gtrless}} K_s \gamma \tag{12.73}$$

Kelly's GLRT possesses the desirable CFAR property for the problem at hand. This means that the false-alarm rate is functionally *independent* of the form and intensity of the disturbance covariance matrix **R**. The false alarm and detection probabilities of Kelly's GLRT are analyzed in Kelly (1986).

A higher dimensional *subspace* model may be used to improve robustness against slightly mismatched signals. As opposed to the unit-rank useful signal model $\mathbf{s} = a\mathbf{v}$, the multi-rank model is defined as $\mathbf{s} = \mathbf{V}\underline{\theta}$, where $\mathbf{V} \in \mathcal{C}^{N \times M}$ is a known mode matrix of relatively low rank $M \ll N$, and $\underline{\theta} \in \mathcal{C}^M$ is an unknown parameter vector. The GLRT for a multi-rank signal model was derived in Kelly and Forsythe (1989) and appears in Eqn. (12.74). Clearly, Eqn. (12.74) reverts back to Eqn. (12.73) for $M = 1$ where $\mathbf{V} = \mathbf{v}$.

$$\frac{\mathbf{z}^\dagger \mathbf{S}^{-1}\mathbf{V}(\mathbf{V}^\dagger \mathbf{S}^{-1}\mathbf{V})^{-1}\mathbf{V}^\dagger \mathbf{S}^{-1}\mathbf{z}}{1 + K_s^{-1}\mathbf{z}^\dagger \mathbf{S}^{-1}\mathbf{z}} \underset{H_0}{\overset{H_1}{\gtrless}} K_s \gamma \tag{12.74}$$

Kelly (1986) analyzed the SNR penalty paid to achieve a specified detection probability at a given false-alarm rate when the disturbance covariance matrix is unknown compared to when it is known a priori, i.e., the SNR penalty incurred for having to estimate the disturbance covariance matrix. It was shown that this penalty has two components, one being due to the CFAR character of the decision rule, and the other being an effective SNR loss factor similar to the quantity analyzed for the SMI technique in Reed et al. (1974). The CFAR loss contribution depends mainly on the number of available secondary vectors, and decreases as K_s increases. On the other hand, the SNR loss factor depends approximately on the ratio of K_s to N, and this contribution decreases as K_s/N increases.

12.3.1.2 Adaptive Matched Filter (AMF)

An alternative adaptive detection algorithm for the same problem was developed in Robey et al. (1992). The first step derived the GLRT assuming the disturbance covariance matrix **R** is known. After this test statistic was derived, the maximum-likelihood estimate of the covariance matrix **S** computed from the secondary data was substituted for **R** in a second step. The resulting *two-step* GLRT in Eqn. (12.75) has the form of a normalized adaptive matched filter (AMF).

$$\frac{|\mathbf{v}^\dagger \mathbf{S}^{-1}\mathbf{z}|^2}{\mathbf{v}^\dagger \mathbf{S}^{-1}\mathbf{v}} \underset{H_0}{\overset{H_1}{\gtrless}} \eta \tag{12.75}$$

Similar to Kelly's GLRT, this test also possesses the CFAR property. However, the AMF statistic does not contain the factor in brackets appearing in the denominator of Eqn. (12.73). As this term tends to unity when K_s is large, Kelly's GLRT converges to the AMF statistic as $K_s \rightarrow \infty$. Inspection of Eqn. (12.75) reveals that the AMF detector can equivalently be written in the form of $\hat{\mathbf{w}}^\dagger \mathbf{z}$, where $\hat{\mathbf{w}} = \rho \mathbf{S}^{-1}\mathbf{v}$ is the SMI adaptive weight vector normalized by $\rho = (\mathbf{v}^\dagger \mathbf{S}^{-1}\mathbf{v})^{-1/2}$ instead of the MVDR normalization $\rho = (\mathbf{v}^\dagger \mathbf{S}^{-1}\mathbf{v})^{-1}$. This modification ensures the CFAR behavior of the AMF detector in the homogeneous Gaussian case. The multi-rank version of the AMF is given by Eqn. (12.76).

$$\mathbf{z}^\dagger \mathbf{S}^{-1}\mathbf{V}(\mathbf{V}^\dagger \mathbf{S}^{-1}\mathbf{V})^{-1}\mathbf{V}^\dagger \mathbf{S}^{-1}\mathbf{z} \underset{H_0}{\overset{H_1}{\gtrless}} \eta \tag{12.76}$$

For a finite set of secondary data, the performance of the AMF and Kelly's GLRT exhibit different behavior in both the matched and mismatched useful signal cases. In the case of matched signals, it is found in Robey et al. (1992) that the performance of the AMF exhibits a small loss in P_D compared to Kelly's GLRT at low SNRs, but a greater P_D than Kelly's GLRT at high SNRs. The GLRT has no known optimality property, and the results in Robey et al. (1992) confirm that Kelly's GLRT is not optimal in the Neyman-Pearson sense of maximizing the probability of detection for a given probability of false alarm.

The AMF is generally regarded as a more *sensitive* detector than Kelly's GLRT since the low SNR performance is comparable. Another practical advantage stated in Robey et al. (1992) is that the generalized inner product $\mathbf{z}^\dagger \mathbf{S}^{-1}\mathbf{z}$ appearing in Kelly's GLRT may be computationally intensive for real-time systems as it must be calculated for each new input vector $\mathbf{z}$. However, the absence of this denominator term causes the AMF to be much more sensitive to mismatched signals. In other words, the AMF has a higher probability of sidelobe detection and is therefore not as *selective* as Kelly's GLRT, which tends to reject mismatched signals more strongly (Kelly 1989).

12.3.1.3 Alternative GLRT Expressions

Before proceeding to the GLRT expressions for tests $\mathcal{B}$ and $\mathcal{C}$ in the homogeneous case, it is noted that Kelly's expression in Eqn. (12.74) can be written in terms of the test statistic $\ell_\mathcal{A}$ in Eqn. (12.77). Here, we have defined the "quasi-whitened" quantities $\tilde{\mathbf{z}} = \mathcal{S}^{-1/2}\mathbf{z}$ and $\tilde{\mathbf{V}} = \mathcal{S}^{-1/2}\mathbf{V}$, where $\mathcal{S} = K_s\mathbf{S}$, as well as the projection matrix $\mathbf{P}_{\tilde{\mathbf{V}}} = \tilde{\mathbf{V}}(\tilde{\mathbf{V}}^\dagger\tilde{\mathbf{V}})^{-1}\tilde{\mathbf{V}}^\dagger$. The orthogonal projection matrix is $\mathbf{P}_{\tilde{\mathbf{V}}}^\perp = \mathbf{I} - \mathbf{P}_{\tilde{\mathbf{V}}}$.

$$\ell_\mathcal{A} = \frac{1 + \tilde{\mathbf{z}}^\dagger\tilde{\mathbf{z}}}{1 + \tilde{\mathbf{z}}^\dagger \mathbf{P}_{\tilde{\mathbf{V}}}^\perp \tilde{\mathbf{z}}} \tag{12.77}$$

Specifically, Kelly's expression in Eqn. (12.74), which is the GLRT for Test $\mathcal{A}$ assuming multi-rank useful signals and the homogeneous Gaussian case, is equivalent to the test statistic $\eta_\mathcal{A} = (\ell_\mathcal{A} - 1)/\ell_\mathcal{A}$, which simplifies to Eqn. (12.78).

$$\eta_\mathcal{A} = \frac{\tilde{\mathbf{z}}^\dagger \mathbf{P}_{\tilde{\mathbf{V}}}\tilde{\mathbf{z}}}{1 + \tilde{\mathbf{z}}^\dagger\tilde{\mathbf{z}}} \underset{H_0}{\overset{H_1}{\gtrless}} \gamma \tag{12.78}$$

The GLRT for Test $\mathcal{B}$, which assumes the presence of a mismatched signal $\mathbf{r} = \mathbf{U}\underline{\phi}$ in addition to the disturbance under the null hypothesis, is given by the more general expression in Eqn. (12.79) for the homogeneous case. Here, we have defined $\tilde{\mathbf{U}} = \mathcal{S}^{-1/2}\mathbf{U}$,

$\mathbf{P}_{\tilde{\mathbf{u}}} = \tilde{\mathbf{U}}(\tilde{\mathbf{U}}^\dagger\tilde{\mathbf{U}})^{-1}\tilde{\mathbf{U}}^\dagger$ and $\mathbf{P}_{\tilde{\mathbf{u}}}^\perp = \mathbf{I} - \mathbf{P}_{\tilde{\mathbf{u}}}$. Note that the GLRTs $\ell_{\mathcal{A}}$ and $\ell_{\mathcal{B}}$ differ in the numerator only.

$$\ell_{\mathcal{B}} = \frac{1 + \tilde{\mathbf{z}}^\dagger \mathbf{P}_{\tilde{\mathbf{u}}}^\perp \tilde{\mathbf{z}}}{1 + \tilde{\mathbf{z}}^\dagger \mathbf{P}_{\tilde{\mathbf{v}}}^\perp \tilde{\mathbf{z}}} \tag{12.79}$$

Following the lead of Kelly, we define the statistic for Test $\mathcal{B}$ as $\eta_{\mathcal{B}} = (\ell_{\mathcal{B}} - 1)/\ell_{\mathcal{B}}$. It can be shown that the decision rule based on this test statistic is given by Eqn. (12.80). Note that for $\mathbf{U} = \mathbf{0}$ (i.e., no mismatched signals), the test in Eqn. (12.80) is equivalent to Kelly's GLRT in Eqn. (12.74).

$$\eta_{\mathcal{B}} = \frac{\tilde{\mathbf{z}}^\dagger \mathbf{P}_{\tilde{\mathbf{v}}} \tilde{\mathbf{z}} - \tilde{\mathbf{z}}^\dagger \mathbf{P}_{\tilde{\mathbf{u}}} \tilde{\mathbf{z}}}{1 + \tilde{\mathbf{z}}^\dagger \tilde{\mathbf{z}} - \tilde{\mathbf{z}}^\dagger \mathbf{P}_{\tilde{\mathbf{u}}} \tilde{\mathbf{z}}} \underset{H_0}{\overset{H_1}{\gtrless}} \gamma \tag{12.80}$$

An alternative transformation is $\eta'_{\mathcal{B}} = \ell_{\mathcal{B}}/(\ell_{\mathcal{B}} + 1)$. It can be shown that the decision rule based on $\eta'_{\mathcal{B}}$ is given by Eqn. (12.81). If the useful and unwanted signals are assumed to be orthogonal in quasi-whitened data space (i.e., $\mathbf{P}_{\tilde{\mathbf{u}}}^\perp = \mathbf{P}_{\tilde{\mathbf{v}}}$), the test statistic in Eqn. (12.81) simplifies to $\eta'_{\mathcal{B}} = (1 + \tilde{\mathbf{z}}^\dagger \mathbf{P}_{\tilde{\mathbf{v}}} \tilde{\mathbf{z}})/(2 + \tilde{\mathbf{z}}^\dagger \tilde{\mathbf{z}})$. This is the multi-rank version of the adaptive beamformer orthogonal rejection test (ABORT), which was proposed for the unit-rank signal model in Pulsone and Rader (2001).

$$\frac{1 + \tilde{\mathbf{z}}^\dagger \mathbf{P}_{\tilde{\mathbf{u}}}^\perp \tilde{\mathbf{z}}}{2 + \tilde{\mathbf{z}}^\dagger \mathbf{P}_{\tilde{\mathbf{u}}}^\perp \tilde{\mathbf{z}} + \tilde{\mathbf{z}}^\dagger \mathbf{P}_{\tilde{\mathbf{v}}}^\perp \tilde{\mathbf{z}}} \underset{H_0}{\overset{H_1}{\gtrless}} \gamma \tag{12.81}$$

The GLRT for Test $\mathcal{C}$, which assumes the presence of a mismatched signal $\mathbf{r} = \mathbf{U}\underline{\phi}$ in addition to the homogeneous disturbance under both the null and alternative hypotheses, is given by the compact expression in Eqn. (12.82), where $\tilde{\mathbf{W}} = [\tilde{\mathbf{V}}, \tilde{\mathbf{U}}]$ is the concatenation of the (quasi-whitened) useful and unwanted signal subspaces.

$$\ell_{\mathcal{C}} = \frac{1 + \tilde{\mathbf{z}}^\dagger \mathbf{P}_{\tilde{\mathbf{u}}}^\perp \tilde{\mathbf{z}}}{1 + \tilde{\mathbf{z}}^\dagger \mathbf{P}_{\tilde{\mathbf{w}}}^\perp \tilde{\mathbf{z}}} \tag{12.82}$$

The related statistic $\eta_{\mathcal{C}} = (\ell_{\mathcal{C}} - 1)/\ell_{\mathcal{C}}$ is given by Eqn. (12.83). The performance of the ℓ and η versions of the GLRTs for hypothesis tests $\mathcal{A}$, $\mathcal{B}$, and $\mathcal{C}$ are identical in terms of P_D versus P_{FA}. In summary, the expressions $\ell_{\mathcal{A}}$, $\ell_{\mathcal{B}}$, and $\ell_{\mathcal{C}}$ (or $\eta_{\mathcal{A}}$, $\eta_{\mathcal{B}}$, and $\eta_{\mathcal{C}}$) correspond to the GLRTs for the three associated binary hypothesis tests assuming multi-rank signal models and a homogeneous Gaussian disturbance.

$$\eta_{\mathcal{C}} = \frac{\tilde{\mathbf{z}}^\dagger \mathbf{P}_{\tilde{\mathbf{w}}} \tilde{\mathbf{z}} - \tilde{\mathbf{z}}^\dagger \mathbf{P}_{\tilde{\mathbf{u}}} \tilde{\mathbf{z}}}{1 + \tilde{\mathbf{z}}^\dagger \tilde{\mathbf{z}} - \tilde{\mathbf{z}}^\dagger \mathbf{P}_{\tilde{\mathbf{u}}} \tilde{\mathbf{z}}} \underset{H_0}{\overset{H_1}{\gtrless}} \gamma \tag{12.83}$$

12.3.2 Partially Homogeneous Case

The AMF and Kelly's GLRT are known to lose their CFAR property when the disturbance does not obey the homogeneous Gaussian model. The same is true for the other GLRTs described in the previous section. This section describes alternative GLRT-based detection schemes which preserve their CFAR property in the partially homogeneous SIRP disturbance environment, which is characterized by an unknown scale change between the primary and secondary data.

These scale-invariant CFAR adaptive detectors may be configured with unit-rank (steering vector) or multi-rank (subspace) useful signal models. A further generalization is the incorporation of unwanted signals in the hypothesis test, either under H_0 only, or under both H_0 and H_1. The partially homogeneous case leads to three fundamentally different GLRTs for tests $\mathcal{A}$, $\mathcal{B}$, and $\mathcal{C}$, which will now be discussed.

12.3.2.1 Adaptive Coherence Estimator (ACE)

In the special case of the SIRP disturbance model with a partially homogeneous texture, also known as the Gaussian disturbance model with scale change between the primary and secondary data, the GLRT for test $\mathcal{A}$ is given by the adaptive coherence estimator (ACE) derived in Kraut and Scharf (1999).

In the partially homogeneous case, the ACE detector coincides with the NAMF receiver described in Conte et al. (1995), wherein the true disturbance covariance is replaced by the SCM calculated from the secondary data. The ACE or NAMF test appears in Eqn. (12.84).

$$\frac{|\mathbf{v}^\dagger \mathbf{S}^{-1}\mathbf{z}|^2}{(\mathbf{v}^\dagger \mathbf{S}^{-1}\mathbf{v})(\mathbf{z}^\dagger \mathbf{S}^{-1}\mathbf{z})} \underset{H_0}{\overset{H_1}{\gtrless}} \eta \tag{12.84}$$

With respect to Kelly's GLRT, ACE maintains its CFAR property for the partially homogeneous case. It has also been shown that ACE is a uniformly most powerful invariant (UMPI) adaptive detection statistic. This means that it has a claim to optimality, having the largest detection probability for a given false-alarm rate out of the class of detectors that are invariant to a group of transformations on the data set which leave its statistical distribution under the two hypotheses unchanged. This claim to optimality for ACE is based on maximal invariant arguments in Kraut, Scharf, and Butler (2005).

12.3.2.2 Adaptive Subspace Detector (ASD)

The scale-invariant CFAR property of ACE is associated with this test being highly intolerant to signal mismatch.[3] Recall that a test whose response falls off rapidly as the degree of mismatch between the received useful signal and the postulated target model increases is said to be highly selective. Such a test is desirable in general since this property lowers the probability of sidelobe detection P_{DS}. However, the radar system also needs to maintain an acceptable detection loss for *slightly* mismatched or "mainlobe" target signals. Such signals may be thought of as having a signature that is best matched to the current steering vector out of all the candidate steering vectors that the radar searches for on the manifold, but is not perfectly matched to this steering vector.

A mainlobe target signature model with a greater capability to capture distortions imposed on the received signal can provide a degree of protection against losses in the probability of mainlobe detection P_{DM} when using a highly selective detector. A higher dimensional *subspace* model effectively opens up the "acceptance bandwidth" of the system to useful signals in the spatial and/or temporal domain, as the case may be. As described earlier, the multi-rank useful signal model may be defined as $\mathbf{s} = \mathbf{V}\underline{\theta}$. The

[3] ACE can strongly suppress slightly mismatched signals, which may prevent registering a detection.

scale-invariant CFAR adaptive subspace detector (ASD) is given as a generalized cosine-squared statistic in Eqn. (12.85). This is the multi-rank GLRT for test $\mathcal{A}$ in the partially homogeneous case.

$$\hat{\cos}^2 = \frac{\mathbf{y}^\dagger \mathbf{P}_\Psi \mathbf{y}}{\mathbf{y}^\dagger \mathbf{y}} \underset{H_0}{\overset{H_1}{\gtrless}} \eta \tag{12.85}$$

In Eqn. (12.85), $\mathbf{y} = \mathbf{S}^{-1/2}\mathbf{z}$ and $\Psi = \mathbf{S}^{-1/2}\mathbf{V}$ are the "quasi-whitened" data vector and mode matrix, respectively, while $\mathbf{P}_\Psi = \Psi(\Psi^\dagger\Psi)^{-1}\Psi^\dagger$ is the projection matrix onto the range space of Ψ. As expected, Eqn. (12.85) reverts to Eqn. (12.84) for the unit-rank signal model ($M = 1$), where $\mathbf{V} = \mathbf{v}$.

The CFAR property of the ASD is invariant to independent scaling between the disturbance in the training data and test data. This is in contrast to Kelly's GLRT and the AMF detector, whose CFAR property is invariant to arbitrary but uniform scaling of the training and test data (as a whole). With respect to ACE, the ASD trades greater robustness to slightly mismatched signals for a slight loss in sensitivity to perfectly matched signals. The ASD can also be expressed in the equivalent form of $\mathcal{L}_\mathcal{A} = 1/(1 - \hat{\cos}^2)$ in Eqn. (12.86), where $\mathbf{P}_\Psi^\perp = \mathbf{I} - \mathbf{P}_\Psi$ and $\gamma = 1/(1 - \eta)$.

$$\mathcal{L}_\mathcal{A} = \frac{\mathbf{y}^\dagger \mathbf{y}}{\mathbf{y}^\dagger \mathbf{P}_\Psi^\perp \mathbf{y}} \underset{H_0}{\overset{H_1}{\gtrless}} \gamma \tag{12.86}$$

To summarize, the ASD in Eqn. (12.85) may be interpreted as an extension of Kelly's GLRT in two respects: (1) the target *signature* vector is not fully specified, but is rather parameterized as a linear combination of M known mode vectors (the columns of $\mathbf{V}$), with the M weights given by the unknown coefficients in $\underline{\theta}$ (i.e., the multidimensional generalization of the complex amplitude $a = \mu e^{j\vartheta}$), and (2) the CFAR property is not only maintained with respect to a disturbance covariance matrix $\mathbf{R}$ of unknown form and intensity statistically characterizing the secondary data, but also with respect to a relative scale (power level) difference ν between the disturbance in the primary and secondary data.

12.3.2.3 Generalized Subspace Detectors (GSD)

It is now of interest to derive GLRT expressions for hypothesis tests $\mathcal{B}$ and $\mathcal{C}$ in the partially homogenous case. The GLRT expressions derived for tests $\mathcal{B}$ and $\mathcal{C}$ assuming multi-rank useful and unwanted signal models in a partially homogeneous disturbance environment will be referred to as generalized subspace detectors (GSD). The GLRT for test $\mathcal{B}$ was described in Fabrizio et al. (2003a), while that for test $\mathcal{C}$ appears in Fabrizio and Farina (2007a). Sketch derivations, reproduced below for convenience, adhere closely to the procedure of Kraut and Scharf (2001), which in turn leans heavily on the original paper of Kelly (1986).

Commencing with test $\mathcal{B}$, recall that the primary data vector $\mathbf{z}$ is distributed as in Eqn. (12.87) under the null H_0 and alternative H_1 hypotheses, respectively, where $\mathbf{s} = \mathbf{V}\underline{\theta}$ is a deterministic multi-rank useful signal, $\mathbf{r} = \mathbf{U}\underline{\phi}$ is a deterministic multi-rank unwanted signal, and $\mathbf{R}$ is the disturbance covariance matrix of the secondary data.

$$H_0 : \mathbf{z} \sim \mathcal{CN}(\mathbf{U}\underline{\phi}, \nu\mathbf{R}), \quad H_1 : \mathbf{z} \sim \mathcal{CN}(\mathbf{V}\underline{\theta}, \nu\mathbf{R}) \tag{12.87}$$

The unknowns are the parameter vectors $\underline{\theta}$ and $\underline{\phi}$, the covariance matrix of the secondary data vectors $\mathbf{R}$, and the relative scale $\nu > 0$ of the disturbance in the primary data. The density of the primary vector $p_{\mathbf{z}|H_\delta}(\mathbf{z})$ conditioned on the two hypotheses H_δ for $\delta = 0, 1$ is given by Eqn. (12.88), where $\mathbf{m}_0 = \mathbf{U}\underline{\phi}$ and $\mathbf{m}_1 = \mathbf{V}\underline{\theta}$. The dependence of the density $p_{\mathbf{z}|H_\delta}(\mathbf{z})$ on the unknown nuisance parameters ν, $\mathbf{R}$, and $\underline{\theta}$ or $\underline{\phi}$, is omitted to simplify notation.

$$p_{\mathbf{z}|H_\delta}(\mathbf{z}) = \frac{1}{(\pi\nu)^N\|\mathbf{R}\|}\exp\left\{-\frac{1}{\nu}(\mathbf{z}-\mathbf{m}_\delta)^\dagger\mathbf{R}^{-1}(\mathbf{z}-\mathbf{m}_\delta)\right\} \tag{12.88}$$

The GLRT substitutes the unknown parameters of the likelihood ratio by their ML estimates, under each hypothesis separately, using the joint density $f_\delta(\mathbf{Z})$ of all the primary and secondary data vectors in $\mathbf{Z} = [\mathbf{z}, \mathbf{z}_1, \ldots, \mathbf{z}_{K_s}]$. Since the disturbance vectors are assumed to be mutually independent, the joint PDF $f_\delta(\mathbf{Z})$ under each hypothesis is given by the product of the secondary-data marginal densities and that of the primary data conditioned on each hypothesis, as in Eqn. (12.89).

$$f_\delta(\mathbf{Z}) = p_{\mathbf{z}|H_\delta}(\mathbf{z})\prod_{k=1}^{K_s}\frac{1}{\pi^N\|\mathbf{R}\|}\exp\{-\mathbf{z}_k^\dagger\mathbf{R}^{-1}\mathbf{z}_k\} \tag{12.89}$$

The GLRT for hypothesis test $\mathcal{B}$ is given by Eqn. (12.90). The denominator of this expression differs from test $\mathcal{A}$ due to the additional maximization that needs to be performed over the unknown unwanted signal parameter $\underline{\phi}$. The purpose of the additional maximization over $\underline{\phi}$ is to yield a test with greater ability to discriminate between $\mathbf{s} = \mathbf{V}\underline{\theta}$, and $\mathbf{r} = \mathbf{U}\underline{\phi}$.

$$\frac{\max_{\nu,\mathbf{R},\underline{\theta}}\ f_1(\mathbf{Z})}{\max_{\nu,\mathbf{R},\underline{\phi}}\ f_0(\mathbf{Z})} \underset{H_0}{\overset{H_1}{\gtrless}} \gamma \tag{12.90}$$

Maximization of the joint densities $f_\delta(\mathbf{Z})$ under the null and alternative hypotheses with respect to the (common) unknown parameters $\mathbf{R}$ and ν leads to the following expression, which was derived in Kraut and Scharf (2001).

$$\max_{\mathbf{R},\,\nu} f_\delta(\mathbf{Z}) = \frac{(K_s+1)^{(N-1)(K_s+1)}(K_s-N+1)^{K_s-N+1}(K_sN)^N}{(e\pi K_s)^{N(K_s+1)}\|\mathbf{S}\|^{K_s+1}[(\mathbf{z}-\mathbf{m}_\delta)^\dagger\mathbf{S}^{-1}(\mathbf{z}-\mathbf{m}_\delta)]^N} \tag{12.91}$$

Hence, for the problem at hand, we may write the *intermediate* GLRT as Eqn. (12.92), where $\mathbf{m}_0 = \mathbf{U}\underline{\phi}$ and $\mathbf{m}_1 = \mathbf{V}\underline{\theta}$. The numerator and denominator of the intermediate GLRT must be *minimized* separately with respect to $\underline{\phi}$ and $\underline{\theta}$ in order to arrive at the final GLRT. Note that minimizing the quadratic forms $(\mathbf{z}-\mathbf{m}_\delta)^\dagger\mathbf{S}^{-1}(\mathbf{z}-\mathbf{m}_\delta)$ for $\delta = 0, 1$ in Eqn. (12.92) with respect to the unknown parameters *maximizes* the likelihood function associated with H_γ in Eqn. (12.91).

$$\frac{\max_{\nu,\mathbf{R}}\ f_1(\mathbf{Z})}{\max_{\nu,\mathbf{R}}\ f_0(\mathbf{Z})} = \left[\frac{(\mathbf{z}-\mathbf{m}_0)^\dagger\mathbf{S}^{-1}(\mathbf{z}-\mathbf{m}_0)}{(\mathbf{z}-\mathbf{m}_1)^\dagger\mathbf{S}^{-1}(\mathbf{z}-\mathbf{m}_1)}\right]^N \tag{12.92}$$

By defining $\mathbf{y} = \mathbf{S}^{-1/2}\mathbf{z}$, $\mathbf{\Psi} = \mathbf{S}^{-1/2}\mathbf{V}$, and $\mathbf{\Omega} = \mathbf{S}^{-1/2}\mathbf{U}$, it is readily shown that the minimizing arguments of the intermediate GLRT are given by the two ML estimates $\hat{\underline{\theta}}$ and $\hat{\underline{\phi}}$ of the unknown parameter vectors in Eqn. (12.93).

$$\begin{aligned}\hat{\underline{\theta}} &= (\mathbf{V}^\dagger\mathbf{S}^{-1}\mathbf{V})^{-1}\mathbf{V}^\dagger\mathbf{S}^{-1}\mathbf{z} = (\mathbf{\Psi}^\dagger\mathbf{\Psi})^{-1}\mathbf{\Psi}^\dagger\mathbf{y}\\ \hat{\underline{\phi}} &= (\mathbf{U}^\dagger\mathbf{S}^{-1}\mathbf{U})^{-1}\mathbf{U}^\dagger\mathbf{S}^{-1}\mathbf{z} = (\mathbf{\Omega}^\dagger\mathbf{\Omega})^{-1}\mathbf{\Omega}^\dagger\mathbf{y}\end{aligned} \tag{12.93}$$

Inserting these two estimates into the intermediate GLRT of Eqn. (12.92) and taking the Nth root of the resulting expression yields the final GLRT in Eqn. (12.94). This GLRT can be written in compact form by observing that $\mathbf{q} = \mathbf{S}^{-1/2}(\mathbf{z} - \mathbf{V}\hat{\underline{\theta}}) = \mathbf{y} - \mathbf{\Psi}\hat{\underline{\theta}} = (\mathbf{I} - \mathbf{P}_\Psi)\mathbf{y}$ where $\mathbf{P}_\Psi = \mathbf{\Psi}(\mathbf{\Psi}^\dagger\mathbf{\Psi})^{-1}\mathbf{\Psi}^\dagger$, such that $\mathbf{q}^\dagger\mathbf{q} = (\mathbf{z} - \mathbf{V}\hat{\underline{\theta}})^\dagger\mathbf{S}^{-1}(\mathbf{z} - \mathbf{V}\hat{\underline{\theta}}) = \mathbf{y}^\dagger(\mathbf{I} - \mathbf{P}_\Psi)\mathbf{y}$ for the denominator, and similarly for the numerator. The idempotent property of projection operators is also used to simplify the final GLRT expression.

$$\mathcal{L}_B = \frac{\mathbf{y}^\dagger(\mathbf{I} - \mathbf{P}_\Omega)\mathbf{y}}{\mathbf{y}^\dagger(\mathbf{I} - \mathbf{P}_\Psi)\mathbf{y}} \underset{H_0}{\overset{H_1}{\gtrless}} \gamma \tag{12.94}$$

In summary, the GLRT in Eqn. (12.94) has been derived for Test $\mathcal{B}$ under the following conditions:

1. The disturbance in the primary data $\mathbf{d}$ and secondary data $\mathbf{d}_1, \ldots, \mathbf{d}_{K_s}$ is described by an SIRP with partially homogeneous texture, also referred to as the Gaussian disturbance model with *unknown* covariance matrix $\mathbf{R} = E\{\mathbf{z}_k\mathbf{z}_k^\dagger\}$ and scale change ν between the training and test data.
2. The useful and unwanted signals are assumed to be deterministic and described by the subspace models $\mathbf{s} = \mathbf{V}\underline{\theta}$ and $\mathbf{r} = \mathbf{U}\underline{\phi}$, respectively. The mode matrices $\mathbf{V}$ and $\mathbf{U}$ are assumed known but the parameter vectors $\underline{\theta}$ and $\underline{\phi}$ are arbitrary and unknown.
3. After transforming the mode matrices to "quasi-whitened" data space using the positive-definite Hermitian SCM computed from the secondary data only $\mathbf{S}$, the resulting subspaces $\mathbf{\Psi} = \mathbf{S}^{-1/2}\mathbf{V}$ and $\mathbf{\Omega} = \mathbf{S}^{-1/2}\mathbf{U}$ may be linearly independent, but are *not* necessarily orthogonal.

The rapprochement of this GLRT with the ASD in Eqn. (12.85) can be seen by setting the unwanted signal subspace to zero $\mathbf{U} = \mathbf{0}$, whereupon $\mathbf{P}_\Omega = \mathbf{0}$ and the test in Eqn. (12.94) reduces to the simpler form of Eqn. (12.95). If the numerator and denominator of this expression are divided by $\mathbf{y}^\dagger\mathbf{y}$, the result is a monotonic function of a generalized cosine-squared statistic $\hat{\cos}^2 = \mathbf{y}^\dagger\mathbf{P}_\Psi\mathbf{y}/\mathbf{y}^\dagger\mathbf{y}$, identical to the scale-invariant CFAR ASD in Eqn. (12.85).

$$\frac{\mathbf{y}^\dagger\mathbf{y}}{(\mathbf{y}^\dagger\mathbf{y} - \mathbf{y}^\dagger\mathbf{P}_\Psi\mathbf{y})} = \frac{1}{1 - \hat{\cos}^2} \tag{12.95}$$

Alternatively, when the quasi-whitened subspaces $\mathbf{\Omega}$ and $\mathbf{\Psi}$ are orthogonal to each other (i.e., $\mathbf{\Omega}^\dagger\mathbf{\Psi} = \mathbf{0}$), and collectively span all of N-dimensional space (i.e., $M + P = N$), we may write $\mathbf{P}_\Omega = \mathbf{I} - \mathbf{P}_\Psi = \mathbf{P}_\Psi^\perp$. In this case, the GLRT in Eqn. (12.94) takes the special form of Eqn. (12.96), which is also a monotonic function of the generalized cosine-squared statistic $\hat{F} = \hat{\cos}^2/(1 - \hat{\cos}^2)$. In this special case, the GLRT in Eqn. (12.94) is equivalent to the ASD in Eqn. (12.85). This explains why the ASD, or ACE for unit-rank signals, is a more selective detector than Kelly's GLRT. However, in the general case where $\mathbf{U} \neq \mathbf{0}$ and $M+P < N$, the GLRT in Eqn. (12.94) differs from the ASD. In this respect, the GLRT in Eqn. (12.94) is referred to as a generalized subspace detector (GSD) as it encompasses the ACE and ASD tests for particular parameter choices.

$$\hat{F} = \frac{\mathbf{y}^\dagger \mathbf{P}_\Psi \mathbf{y}}{\mathbf{y}^\dagger \mathbf{P}_\Psi^\perp \mathbf{y}} \underset{H_0}{\overset{H_1}{\gtrless}} \gamma \tag{12.96}$$

If no prior information is available regarding the structure of $\mathbf{U}$, it is quite reasonable to assume that the unwanted signal lies in a subspace orthogonal to the useful signal in quasi-whitened data space, such that $\mathbf{P}_\Psi + \mathbf{P}_\Omega = \mathbf{I}$. In this case, the ASD becomes the GLRT. However, when prior information is available about $\mathbf{U}$, the GLRT is given by the GSD in Eqn. (12.94), which provides an alternative option to the ASD.

It is also worth pointing out that the GLRT in Eqn. (12.94) generalizes the result of Pulsone and Rader (2001) in several respects. In contrast to ABORT, the GSD possesses the scale-invariant CFAR property and adopts a subspace signal model for the useful signal to improve robustness for the detection of slightly mismatched mainlobe targets. Moreover, the unwanted signal subspace is not pre-configured as the orthogonal subspace to that of the useful signal, which has rank $N-1$ for a unit-rank signal model. As shown in the subsequent experimental results, an effective method exists for configuring the GSD with an unwanted signal subspace of relatively low rank (i.e., comparable to that of the useful signal).

When a significantly mismatched signal is additionally present, the action of the ASD detector is reminiscent of a classical sidelobe blanker that suppresses an echo when the ratio between signals in a main and auxiliary antenna is greater than a chosen threshold. In other words, a faint mainlobe target can be rejected when a stronger sidelobe signal is also present. In this scenario, the detection probability can be reduced dramatically, even though the useful signal embedded in the disturbance could have been detected by the ASD in the absence of the unwanted signal (Kalson 1992). In test $\mathcal{C}$, the useful and unwanted signal are simultaneously present under H_1, so the GLRT for test $\mathcal{C}$ involves solving the problem in Eqn. (12.97).

$$\frac{\max_{\nu,\mathbf{R},\underline{\theta},\underline{\phi}} f_1(\mathbf{Z})}{\max_{\nu,\mathbf{R},\underline{\phi}} f_0(\mathbf{Z})} \underset{H_0}{\overset{H_1}{\gtrless}} \gamma \tag{12.97}$$

Before evaluating the left hand side of Eqn. (12.97), the signal "masking" phenomenon, is perhaps most readily appreciated by considering the $\hat{F}$-version of the ASD in Eqn. (12.96). The test statistic F will have low value and therefore not exceed the detection threshold if the energy contained in the subspace orthogonal to the useful signal in quasi-whitened space (i.e., the denominator) is much higher than that contained in the signal subspace (i.e., the numerator). The ASD is therefore highly susceptible to missed detections when a strong sidelobe signal is present along with a weaker mainlobe target. The intermediate GLRT expression for Test $\mathcal{C}$ takes the same form as Eqn. (12.92), except that $\mathbf{m}_0 = \mathbf{U}\underline{\phi}$ and $\mathbf{m}_1 = [\mathbf{V}, \mathbf{U}][\underline{\theta}^T, \underline{\phi}^T]^T$. By defining $\mathbf{\Gamma} = \mathbf{S}^{-1/2}[\mathbf{V}, \mathbf{U}]$ and the projection $P_\Gamma = \mathbf{\Gamma}(\mathbf{\Gamma}^\dagger\mathbf{\Gamma})^{-1}\mathbf{\Gamma}^\dagger$ the GLRT for Test $\mathcal{C}$ takes the form of Eqn. (12.98), as shown in Fabrizio, Scharf, Farina, and Turley (2004b).

$$\mathcal{L}_\mathcal{C} = \frac{\mathbf{y}^\dagger(\mathbf{I} - \mathbf{P}_\Omega)\mathbf{y}}{\mathbf{y}^\dagger(\mathbf{I} - \mathbf{P}_\Gamma)\mathbf{y}} \underset{H_0}{\overset{H_1}{\gtrless}} \gamma \tag{12.98}$$

This detector represents a generalization of the ASD in Kraut and Scharf (1999) for an overall disturbance modeled by a deterministic low-rank subspace component $\mathbf{r} = \mathbf{U}\underline{\phi}$ plus a statistical (possibly "colored") full-rank component $\mathbf{d}$.

The GSD in Eqn. (12.98) assumes the subspace $\mathbf{U}$ is known, but in practice, it must be estimated. It is proposed to make a first pass assuming $\mathbf{U} = \mathbf{0}$, for which the GLRT corresponds to the ACE ($M = 1$) or ASD ($M > 1$) test. Any detections made during the first pass are recorded and their coordinates (e.g., DOA and/or Doppler shift) are stored for reference in the second pass. If the detections observed in the first pass occur at coordinates different to the useful signal coordinate currently being interrogated in the second pass, the GSD in Eqn. (12.98) is used to process the data using an unwanted signal model that accounts for the signals detected during the first pass. The aim of the second pass is therefore to uncover any new targets that may have been masked by stronger signals at different coordinates that were detected during the first pass. The operation of this procedure to alleviate the signal-masking problem is illustrated using experimental data in Section 12.4.2.

12.3.3 Joint Data-Set Detection

The aforementioned detection schemes operate on a single primary data vector at a time. In this section, we consider the possibility of detecting distributed targets, where the echo from a single target has components spread over multiple primary vectors. Traditional techniques that utilize a set of training data to implement the adaptive detector applied to one primary vector at a time are referred to as two-data set (TDS) algorithms. On the other hand, single-data set (SDS) algorithms have been proposed to address the problem of disturbance statistical heterogeneity by operating on the test data only (i.e., no secondary data is used). In order to distinguish against TDS techniques that operate on a single primary data vector, we shall refer to detectors that operate on multiple secondary and primary data vectors simultaneously as hybrid or joint-data set (JDS) techniques.

Examples of JDS detectors for range-spread targets appear in Conte et al. (2001), where the GLRT was derived for both the homogeneous and partially homogeneous cases. This work was extended in Bandiera, De Maio, Greco, and Ricci (2007) to include deterministic subspace interference. JDS adaptive detection was also investigated in Aboutanios and Mulgrew (2010) for space-time adaptive processing (STAP) assuming a statistically homogeneous Gaussian disturbance and no unwanted signals. The GLRT was derived on the assumption of independent disturbance vectors, which does not strictly hold for the procedure described in Aboutanios and Mulgrew (2010), except for the special case of temporally and spatially white noise.

The first part of this section examines the JDS detection problem in the context of purely spatial or temporal processing for the case of distributed targets with *structured* spread. The practical formulation of this problem differs in detail to the one described in Aboutanios and Mulgrew (2010). Specifically, the alternative formulation proposed herein allows the GLRT to be derived based on a more realistic assumption of statistical independence, which can hold for disturbance vectors with arbitrary covariance matrix. The considered problem also differs from those studied in Conte et al. (2001) and Bandiera et al. (2007), wherein the GLRT does not assume or incorporate any knowledge regarding the structure of the target complex amplitudes over the primary data.

The JDS problem for distributed targets in a homogenous Gaussian disturbance with no unwanted signals was treated in Kelly and Forsythe (1989). As the GLRT resulting for this problem is not widely known, it will be briefly described and discussed in the first part of this section. The second part of this section remains on the theme of JDS detection for distributed targets with structured spread, but includes subspace useful signals, unwanted signals, and a partially homogeneous disturbance model in the problem formulation. The latter two extensions permit the derivation of one- and two-step GLRT detectors that do not appear in the pioneering work of Kelly and Forsythe (1989).

12.3.3.1 Homogeneous Case

The traditional hypothesis testing problem for a deterministic useful signal distributed over multiple primary data vectors can be represented by Eqn. (12.99). The useful signals are assumed to have unit-rank, as in Eqn. (12.62).

$$\begin{cases} H_0 : \begin{cases} \mathbf{z}_k = \mathbf{d}_k, & k \in \Omega_p \\ \mathbf{z}_k = \mathbf{d}_k, & k \in \Omega_s \end{cases} \\ H_1 : \begin{cases} \mathbf{z}_k = \mathbf{s}_k + \mathbf{d}_k, & k \in \Omega_p \\ \mathbf{z}_k = \mathbf{d}_k, & k \in \Omega_s \end{cases} \end{cases} \tag{12.99}$$

In the homogeneous case, the set of K_p primary data vectors denoted by $\mathbf{z}_k \in \mathcal{C}^N$ for $k \in \Omega_p \equiv [K_s + 1, \ldots, K_s + K_p]$ are assumed to share the same disturbance covariance matrix $\mathbf{R}$ and to have a marginal density given by $p_\delta[\mathbf{z}_k]$ in Eqn. (12.100), where $\delta = 0, 1$ signifies the distribution under hypothesis H_δ.

$$p_\delta[\mathbf{z}_k] = \frac{1}{\pi^N \|\mathbf{R}\|} \exp\left\{-(\mathbf{z}_k - \delta a_k \mathbf{v})^\dagger \mathbf{R}^{-1} (\mathbf{z}_k - \delta a_k \mathbf{v})\right\}, \quad k \in \Omega_p \tag{12.100}$$

For convenience, we shall also define $\mathbf{a} \in \mathcal{C}^{K_p}$ in Eqn. (12.101) as the vector of unknown target complex amplitudes a_k over the primary data $k \in \Omega_p$.

$$\mathbf{a} = [a_{K_s+1}, \ldots, a_{K_s+K_p}]^T \tag{12.101}$$

The K_s secondary data vectors $\mathbf{z}_k \in \mathcal{C}^N$ for $k \in \Omega_s \equiv [1, \ldots, K_s]$ are assumed to contain an IID Gaussian disturbance $\mathbf{z}_k = \mathbf{d}_k$ with covariance matrix $\mathbf{R}$ identical to that of the primary data. Using Eqn. (12.100), the marginal density of the secondary vectors $\mathbf{z}_k$ is given by $p_0[\mathbf{z}_k]$ for $k \in \Omega_s$. It follows that the joint PDF $f_\delta[\mathbf{Z}; \mathbf{R}, \mathbf{a}]$ of all the $K = K_s + K_p$ input vectors $\mathbf{Z} = [\mathbf{z}_1, \ldots, \mathbf{z}_K]$ is given by the product of the marginal densities, as in Eqn. (12.102). Clearly, the density under H_0 only depends on $\mathbf{R}$.

$$f_\delta[\mathbf{Z}; \mathbf{R}, \mathbf{a}] = \prod_{k=1}^{K_s} p_0[\mathbf{z}_k] \; \times \prod_{k=K_s+1}^{K} p_\delta[\mathbf{z}_k], \quad \begin{cases} H_0 : & \delta = 0 \\ H_1 : & \delta = 1 \end{cases} \tag{12.102}$$

The expression for the joint density of all the data can be simplified by making use of the equivalence: $\mathbf{t}^\dagger \mathbf{M} \mathbf{t} = \text{Tr}\{\mathbf{M}\mathbf{t}\mathbf{t}^\dagger\} = \text{Tr}\{\mathbf{M}\mathbf{T}\}$ for any matrix $\mathbf{M}$ and vector $\mathbf{t}$, where $\text{Tr}\{\cdot\}$ is the trace operator. Applying this equivalence to all the factors of the joint PDF in Eqn. (12.102), it is readily shown that $f_\delta[\mathbf{Z}; \mathbf{R}, \mathbf{a}]$ may be written as in Eqn. (12.103),

$$f_\delta[\mathbf{Z}; \mathbf{R}, \mathbf{a}] = \left\{\frac{1}{\pi^N \|\mathbf{R}\|} \exp\left[-\text{Tr}(\mathbf{R}^{-1}\mathbf{T}_\delta)\right]\right\}^K \tag{12.103}$$

where,

$$\mathbf{T}_\delta = \frac{1}{K}\left(\sum_{k=1}^{K_s} \mathbf{z}_k\mathbf{z}_k^\dagger + \sum_{k=K_s+1}^{K} (\mathbf{z}_k - \delta a_k\mathbf{v})(\mathbf{z}_k - \delta a_k\mathbf{v})^\dagger\right) \tag{12.104}$$

Under H_0, the density $f_0[\mathbf{Z}; \mathbf{R}]$ is maximized by replacing the unknown covariance matrix $\mathbf{R}$ by its ML estimate $\mathbf{T}_0$ (i.e., the sample covariance matrix formed on the primary and secondary data). Similarly, maximizing $f_1[\mathbf{Z}; \mathbf{R}, \mathbf{a}]$ with respect to $\mathbf{R}$ involves replacing this matrix by $\mathbf{T}_1$. This results in Eqn. (12.105), where the density under H_0 is clearly independent of $\mathbf{a}$.

$$\max_{\mathbf{R}} f_\delta[\mathbf{Z}; \mathbf{R}, \mathbf{a}] = f_\delta[\mathbf{Z}; \mathbf{T}_\delta, \mathbf{a}] = \left\{\frac{1}{(e\pi)^N \|\mathbf{T}_\delta\|}\right\}^K \tag{12.105}$$

Taking the Kth root of the ratio between the results for H_1 and H_0 yields the intermediate GLR expression in Eqn. (12.106), where the matrix $\mathbf{T}_1$ is implicitly dependent on $\mathbf{a}$.

$$\lambda(\mathbf{Z}; \mathbf{a}) = \sqrt[K]{\frac{f_1[\mathbf{Z}; \mathbf{T}_1, \mathbf{a}]}{f_0[\mathbf{Z}; \mathbf{T}_0]}} = \frac{\|\mathbf{T}_0\|}{\|\mathbf{T}_1\|} \tag{12.106}$$

To arrive at the final GLRT, denoted by $\lambda(\mathbf{Z})$, it remains to maximize $\lambda(\mathbf{Z}; \mathbf{a})$ with respect to the unknown parameters in the vector $\mathbf{a}$, as in Eqn. (12.107). This final step requires the minimization of the determinant $\|\mathbf{T}_1\|$ with respect to the elements of $\mathbf{a}$.

$$\lambda(\mathbf{Z}) = \max_{\mathbf{a}} \lambda(\mathbf{Z}; \mathbf{a}) = \frac{\|\mathbf{T}_0\|}{\min_{\mathbf{a}} \|\mathbf{T}_1\|} \tag{12.107}$$

To perform this minimization, it is convenient to expand the terms in Eqn. (12.104) and write $K\mathbf{T}_1$ in the form of Eqn. (12.108), where it is recalled that $K\mathbf{T}_0 = \sum_{k=1}^{K} \mathbf{z}_k\mathbf{z}_k^\dagger$.

$$K\mathbf{T}_1 = K\mathbf{T}_0 + \sum_{k=K_s+1}^{K} \left(|a_k|^2\mathbf{v}\mathbf{v}^\dagger - a_k\mathbf{v}\mathbf{z}_k^\dagger - a_k^*\mathbf{z}_k\mathbf{v}^\dagger\right) \tag{12.108}$$

For a distributed target with spreading vector $\mathbf{a}$ of known structure, we may write $\mathbf{a} = a\mathbf{e}$, where $a = \mu e^{j\vartheta}$ is an unknown complex amplitude, and $\mathbf{e} = [e_{K_s+1}, \ldots, e_K]^T$ is a known template vector of unit length $\mathbf{e}^\dagger\mathbf{e} = 1$. At this stage, it is convenient to define the N-dimensional vector $\mathbf{g}$ in Eqn. (12.109).

$$\mathbf{g} = \sum_{k=K_s+1}^{K} e_k^*\mathbf{z}_k \tag{12.109}$$

Using this definition, the right hand side of Eqn. (12.108) is given by Eqn. (12.110).

$$K\mathbf{T}_1 = K\mathbf{T}_0 + |a|^2\mathbf{v}\mathbf{v}^\dagger - a\mathbf{v}\mathbf{g}^\dagger - a^*\mathbf{g}\mathbf{v}^\dagger \tag{12.110}$$

By completing the square in Eqn. (12.110), the above expression can be written as Eqn. (12.111).

$$K\mathbf{T}_1 = K\mathbf{T}_0 + (a\mathbf{v} - \mathbf{g})(a\mathbf{v} - \mathbf{g})^\dagger - \mathbf{g}\mathbf{g}^\dagger \tag{12.111}$$

Next, we make the substitution $\mathbf{M} = K\mathbf{T}_0 - \mathbf{g}\mathbf{g}^\dagger$, to write Eqn. (12.111) as Eqn. (12.112).

$$K\mathbf{T}_1 = \mathbf{M} + (a\mathbf{v} - \mathbf{g})(a\mathbf{v} - \mathbf{g})^\dagger \tag{12.112}$$

In addition, we note that $K\mathbf{T}_0$ may also be written in terms of $\mathbf{M}$ in Eqn. (12.113).

$$K\mathbf{T}_0 = \mathbf{M} + \mathbf{g}\mathbf{g}^\dagger \tag{12.113}$$

A well-known matrix determinant lemma may be invoked to evaluate $\|\mathbf{T}_0\|$ in Eqn. (12.114)

$$K^N\|\mathbf{T}_0\| = \|\mathbf{M}\|(1 + \mathbf{g}^\dagger\mathbf{M}^{-1}\mathbf{g}) \tag{12.114}$$

Application of the same lemma to Eqn. (12.112) yields a similar expression for $\|\mathbf{T}_1\|$ in Eqn. (12.115).

$$K^N\|\mathbf{T}_1\| = \|\mathbf{M}\|[1 + (a\mathbf{v} - \mathbf{g})^\dagger\mathbf{M}^{-1}(a\mathbf{v} - \mathbf{g}]\} \tag{12.115}$$

We may now write the likelihood ratio as

$$\lambda(\mathbf{Z}) = \frac{1 + \mathbf{g}^\dagger\mathbf{M}^{-1}\mathbf{g}}{\min_a[1 + (a\mathbf{v} - \mathbf{g})^\dagger\mathbf{M}^{-1}(a\mathbf{v} - \mathbf{g})]} \tag{12.116}$$

In the following, we shall refer to $\lambda(\mathbf{Z})$ simply by λ. The minimizing argument $\hat{a}$ is given by Eqn. (12.117).

$$\hat{a} = \frac{\mathbf{v}^\dagger\mathbf{M}^{-1}\mathbf{g}}{\mathbf{v}^\dagger\mathbf{M}^{-1}\mathbf{v}} \tag{12.117}$$

Substituting this ML estimate into the denominator of Eqn. (12.116) yields the GLRT in Eqn. (12.118).

$$\lambda = \frac{1 + \mathbf{g}^\dagger\mathbf{M}^{-1}\mathbf{g}}{1 + \mathbf{g}^\dagger\mathbf{M}^{-1}\mathbf{g} - |\mathbf{v}^\dagger\mathbf{M}^{-1}\mathbf{g}|^2(\mathbf{v}^\dagger\mathbf{M}^{-1}\mathbf{v})^{-1}} \tag{12.118}$$

Following the lead of Kelly (1986), we can introduce the quantity η in Eqn. (12.119).

$$\eta = \frac{\lambda - 1}{\lambda} = \frac{|\mathbf{v}^\dagger\mathbf{M}^{-1}\mathbf{g}|^2}{\mathbf{v}^\dagger\mathbf{M}^{-1}\mathbf{v}(1 + \mathbf{g}^\dagger\mathbf{M}^{-1}\mathbf{g})} \tag{12.119}$$

A test on λ using a threshold λ_0 is equivalent to a test on η using a threshold $\eta_0 = (\lambda_0 - 1)/\lambda_0$.

$$\frac{|\mathbf{v}^\dagger\mathbf{M}^{-1}\mathbf{g}|^2}{\mathbf{v}^\dagger\mathbf{M}^{-1}\mathbf{v}(1 + \mathbf{g}^\dagger\mathbf{M}^{-1}\mathbf{g})} \underset{H_0}{\overset{H_1}{\gtrless}} \eta_0 \tag{12.120}$$

To facilitate the interpretation of this GLRT, note that $\mathbf{M}$ can be written as Eqn. (12.121), where $\mathbf{S}$ and $\mathbf{P}$ are the sample covariance matrices computed from the secondary and primary data sets, respectively.

$$\mathbf{M} = K_s\mathbf{S} + K_p\mathbf{P} - \mathbf{g}\mathbf{g}^\dagger \tag{12.121}$$

Assume a template model of the form $\mathbf{e} = [1, e^{j\psi}, \ldots, e^{j\psi(K_p-1)}]/\sqrt{K_p}$. Here, ψ may be regarded as the target spreading parameter. The vector $\mathbf{g}$ is then given by Eqn. (12.122).

$$\mathbf{g} = \frac{1}{\sqrt{K_p}} \sum_{k=K_s+1}^{K_s+K_p} e^{-j\psi(k-K_s-1)} \mathbf{z}_k \tag{12.122}$$

In the TDS problem, $K_s \geq N$ and $K_p = 1$, so $\mathbf{g} = \mathbf{z}_{K_s+1}$ is equal to the only primary vector. In addition, $\mathbf{M} = K_s\mathbf{S}$ for $K_p = 1$. Substitution of these values into Eqn. (12.120) and Eqn. (12.117) leads to Kelly's GLRT in Eqn. (12.123), as expected. It also noted that the ML estimate of the target amplitude $\hat{a} = \hat{\mathbf{w}}^\dagger \mathbf{z}$ is derived from the RMB-SMI filter $\hat{\mathbf{w}} = \mathbf{S}^{-1}\mathbf{v}/(\mathbf{v}^\dagger\mathbf{S}^{-1}\mathbf{v})$ in this case.

$$\frac{|\mathbf{v}^\dagger\mathbf{S}^{-1}\mathbf{z}|^2}{\mathbf{v}^\dagger\mathbf{S}^{-1}\mathbf{v}(1 + K_s^{-1}\mathbf{z}^\dagger\mathbf{S}^{-1}\mathbf{z})} \underset{H_0}{\overset{H_1}{\gtrless}} K_s\eta_0 \tag{12.123}$$

Now consider the SDS problem, where $K_s = 0$ and $K_p \geq N$. In this case, $\mathbf{M}$ is given in Eqn. (12.124), where we have defined the matrix $\mathbf{Q} = \mathbf{P} - \mathbf{g}\mathbf{g}^\dagger/K_p$. It may be observed that the ML estimate of the target amplitude $\hat{a} = \hat{\mathbf{w}}^\dagger\mathbf{z}$ is derived from the amplitude and phase estimation (APES) filter given by $\hat{\mathbf{w}} = \mathbf{Q}^{-1}\mathbf{v}/(\mathbf{v}^\dagger\mathbf{Q}^{-1}\mathbf{v})$, as opposed to the SMI filter estimate.

$$\mathbf{M} = K_p\mathbf{P} - \mathbf{g}\mathbf{g}^\dagger = K_p\mathbf{Q} \tag{12.124}$$

When $\hat{a} = \mathbf{v}^\dagger\mathbf{Q}^{-1}\mathbf{g}/(\mathbf{v}^\dagger\mathbf{Q}^{-1}\mathbf{v})$ is substituted for a, the result is the GLRT for the SDS problem in Eqn. (12.125). This may be interpreted as the GLRT in which the APES filter described by Li and Stoica (1996) plays a prominent role. This is analogous to Kelly's GLRT, in which the RMB-SMI adaptive filter represents a significant component.

$$\frac{|\mathbf{v}^\dagger\mathbf{Q}^{-1}\mathbf{g}|^2}{\mathbf{v}^\dagger\mathbf{Q}^{-1}\mathbf{v}(1 + K_p^{-1}\mathbf{g}^\dagger\mathbf{Q}^{-1}\mathbf{g})} \underset{H_0}{\overset{H_1}{\gtrless}} K_p\eta_0 \tag{12.125}$$

For large K_p, the GLRT in Eqn. (12.125) tends to the test statistic in Eqn. (12.126). This may be recognized as the AMF equivalent of the TDS problem for the SDS problem. The test in Eqn. (12.125) is referred to here as the normalized APES or NAPES filter, which may also be derived as a two-step GLRT for the SDS problem, in similar fashion to the AMF for the SDS problem.

$$\frac{|\mathbf{v}^\dagger\mathbf{Q}^{-1}\mathbf{g}|^2}{\mathbf{v}^\dagger\mathbf{Q}^{-1}\mathbf{v}} \underset{H_0}{\overset{H_1}{\gtrless}} K_p\eta_0 \tag{12.126}$$

The GLRT for the more general JDS problem in Eqn. (12.120), where $K_s > 1$, $K_p > 1$, and $K \geq N$, is less known than Kelly's GLRT for the TDS problem in Eqn. (12.123). The same can be said for the SDS GLRT described above, where the APES filter plays a prominent role. The JDS GLRT is well motivated for problems where the distributed target is spread in a structured manner over the primary data.

12.3.3.2 Multi-Rank Signals and Partially Homogeneous Disturbance

To motivate the JDS problem, consider a multi-channel radar system that transmits a series of N_p pulses during the CPI. The echoes from each pulse are received in N_g range cells after pulse compression and in N_a spatial reception channels. At a particular range cell r, we define a set of $K_p = N_a$ primary vectors $\mathbf{z}_k = [z_k(1), \ldots, z_k(N_p)]^T$ for $k \in \Omega_p$. The primary vectors $\mathbf{z}_k$ have dimension $N = N_p$, and contain the pulse-train outputs acquired in range cell r by the $K_p = N_a$ receivers. In addition to these K_p primary vectors, we define a set of K_s secondary vectors $\mathbf{z}_k$ for $k = 1, \ldots, K_s$ that contain pulse-train outputs extracted from "training" range cells near to the cell under test (CUT).

Spatially broadband OTH radar clutter will give rise to practically independent disturbance pulse-train vectors from one receiver to another, while the return from a target with fixed DOA will have a structure across the array determined by a phase progression over receivers. When unwanted coherent signals are included under H_0 and H_1, the JDS hypothesis test may be formulated as in Eqn. (12.127). We shall introduce subspace models for the useful signal $\mathbf{s}_k$ and the unwanted signal $\mathbf{r}_k$ in a moment. It is assumed that the spreading of these signals over the primary data is *structured* in both cases. A GLRT is derived for the homogeneous case, and a two-step GLRT is proposed for the partially homogeneous case.

$$\begin{cases} H_0 : \begin{cases} \mathbf{z}_k = \mathbf{r}_k + \mathbf{d}_k, & k \in \Omega_p \\ \mathbf{z}_k = \mathbf{d}_k, & k \in \Omega_s \end{cases} \\ H_1 : \begin{cases} \mathbf{z}_k = \mathbf{s}_k + \mathbf{r}_k + \mathbf{d}_k, & k \in \Omega_p \\ \mathbf{z}_k = \mathbf{d}_k, & k \in \Omega_s \end{cases} \end{cases} \tag{12.127}$$

Here, $\mathbf{E}\{\mathbf{d}_k\mathbf{d}_k^\dagger\} = \mathbf{R}$ for $k \in \Omega_p$ and $\mathbf{E}\{\mathbf{d}_k\mathbf{d}_k^\dagger\} = \nu\mathbf{R}$ for $k \in \Omega_s$ with both $\mathbf{R}$ and ν unknown.[4] A subspace target model for $\mathbf{s}_k$ is adopted in Eqn. (12.128) for a known rank-M mode matrix $\mathbf{V} = [\mathbf{v}_1, \ldots \mathbf{v}_M]$ but unknown coordinates $\underline{\theta}_k$. Similarly, a subspace model is also proposed for the unwanted signal $\mathbf{r}_k$ in Eqn. (12.128), with a known rank-P mode matrix $\mathbf{U} = [\mathbf{u}_1, \ldots \mathbf{u}_P]$ and an unknown parameter vector $\underline{\phi}_k$. An imposed condition is that the collective subspace $[\mathbf{V}, \mathbf{U}]$ has full rank $M + P$ where $M + P \ll N$.

$$\mathbf{s}_k = \mathbf{V}\underline{\theta}_k, \quad \mathbf{r}_k = \mathbf{U}\underline{\phi}_k \tag{12.128}$$

For a target with structured spread over different primary vectors, the matrix of target components $\mathbf{F} = [\mathbf{s}_{K_s+1}, \ldots, \mathbf{s}_K]$ can be written as $\mathbf{F} = \mathbf{V}\Theta$, where $\Theta = [\underline{\theta}_{K_s+1}, \ldots, \underline{\theta}_K]$. For a target echo incident from azimuth α and elevation β, the steering vector can be expressed as: $\mathbf{e} = [1, e^{j\mathbf{k}(\alpha,\beta)\cdot\mathbf{r}_2}, \ldots, e^{j\mathbf{k}(\alpha,\beta)\cdot\mathbf{r}_{N_a}}]^\dagger$, where $\mathbf{k}(\alpha, \beta)$ is the wavevector and $\mathbf{r}_n$ for $n = 1, \ldots, N_a$ is the position vector of the nth antenna sensor relative to the first ($n = 1$), assigned as the phase reference. Exploiting this relationship, and defining $\underline{\theta} = \underline{\theta}_{K_s+1}$, we have that $\Theta = \underline{\theta}\mathbf{e}^\dagger$ and $\mathbf{F} = \mathbf{V}\underline{\theta}\mathbf{e}^\dagger$. In similar fashion, an unwanted signal in the same

[4] It is convenient to define $\mathbf{R}$ and $\nu\mathbf{R}$ as the primary and secondary data covariance matrices, respectively, in this section.

resolution cell as the target but having different Doppler characteristics (e.g., a "crossing" target) may be expressed as $\mathbf{G} = [\mathbf{r}_{K_s+1}, \ldots, \mathbf{r}_K] = \mathbf{U}\underline{\phi}\mathbf{e}^\dagger$.

$$\begin{cases} \mathbf{F} = [\mathbf{s}_{K_s+1}, \ldots, \mathbf{s}_K] = \mathbf{V}\underline{\theta}\mathbf{e}^\dagger \\ \mathbf{G} = [\mathbf{r}_{K_s+1}, \ldots, \mathbf{r}_K] = \mathbf{U}\underline{\phi}\mathbf{e}^\dagger \end{cases} \tag{12.129}$$

This JDS problem is not captured in the formulation of Kelly and Forsythe (Kelly 1989), which does not consider the partially homogeneous case, or the inclusion of unwanted signals. The same differences apply with respect to the JDS problem considered in Aboutanios and Mulgrew (2010), which additionally restricts attention to a unit-rank useful signal model. The problem also differs in detail to the ones considered in Conte et al. (2001) and Bandiera et al. (2007), as it exploits structure in the target complex amplitudes over the primary data.

Due to the various unknowns in the hypothesis test, we resort to the GLRT to decide between H_0 and H_1, given the joint PDFs $f_\delta[\cdot]$ under the two hypotheses H_δ for $\delta = 0, 1$. The GLRT $\Lambda(\mathbf{Z})$ for this problem is based on the decision rule in Eqn. (12.130), where $\mathbf{Z} = [\mathbf{Z}_p, \mathbf{Z}_s]$ for $\mathbf{Z}_p = [\mathbf{z}_{K_s+1}, \ldots, \mathbf{z}_K]$, $\mathbf{Z}_s = [\mathbf{z}_1, \ldots, \mathbf{z}_{K_s}]$, and γ is a threshold that maintains an acceptable probability of false alarm. The test statistic $\Lambda(\mathbf{Z})$ was derived in Fabrizio et al. (2007), with the maximization over ν performed as a two-step GLRT. The main elements of the derivation are reviewed below.

$$\Lambda(\mathbf{Z}) = \frac{\max_{\mathbf{R},\underline{\theta},\underline{\phi},\nu} f_1[\mathbf{Z}; \mathbf{R}, \underline{\theta}, \underline{\phi}, \nu]}{\max_{\mathbf{R},\underline{\phi},\nu} f_0[\mathbf{Z}; \mathbf{R}, \underline{\phi}, \nu]} \underset{H_0}{\overset{H_1}{\gtrless}} \gamma \tag{12.130}$$

The joint PDFs in Eqn. (12.131) can be defined in terms of $\mathbf{T}_0 = [\frac{1}{\nu}\mathbf{Z}_s\mathbf{Z}_s^\dagger + (\mathbf{Z}_p - \mathbf{G})(\mathbf{Z}_p - \mathbf{G})^\dagger]$ and $\mathbf{T}_1 = [\frac{1}{\nu}\mathbf{Z}_s\mathbf{Z}_s^\dagger + (\mathbf{Z}_p - \mathbf{H})(\mathbf{Z}_p - \mathbf{H})^\dagger]$, where $\mathbf{H} = \mathbf{F} + \mathbf{G}$, recalling that $\mathbf{F} = \mathbf{V}\underline{\theta}\mathbf{e}^\dagger$ and $\mathbf{G} = \mathbf{U}\underline{\phi}\mathbf{e}^\dagger$.

$$f_\delta[\cdot] = \left[\frac{1}{\pi^N ||\mathbf{R}||}\right]^K \left(\frac{1}{\nu}\right)^{NK_s} \exp\{-\mathrm{tr}(\mathbf{R}^{-1}\mathbf{T}_\delta)\} \tag{12.131}$$

Maximization with respect to $\mathbf{R}$ under H_0 involves substituting true matrix $\mathbf{R}$ with the sample covariance matrix $\hat{\mathbf{R}} = K^{-1}\mathbf{T}_0$. This results in Eqn. (12.132), when also making substitutions for $\mathbf{S} = K_s^{-1}\mathbf{Z}_s\mathbf{Z}_s^\dagger$ and $\mathbf{G} = \mathbf{U}\underline{\phi}\mathbf{e}^\dagger$.

$$f_0[\mathbf{Z}; \hat{\mathbf{R}}, \underline{\phi}, \nu] = \left[\frac{K}{\pi e}\right]^{NK} \left(\frac{1}{\nu}\right)^{NK_s} \left\{\frac{1}{||\frac{1}{\nu}K_s\mathbf{S} + (\mathbf{Z}_p - \mathbf{U}\underline{\phi}\mathbf{e}^\dagger)(\mathbf{Z}_p - \mathbf{U}\underline{\phi}\mathbf{e}^\dagger)^\dagger||}\right\}^K \tag{12.132}$$

By expanding the terms in the determinant of Eqn. (12.132) and defining $\mathbf{P} = K_p^{-1}\mathbf{Z}_p\mathbf{Z}_p^\dagger$, $\mathbf{g} = \mathbf{Z}_p\mathbf{e}/|\mathbf{e}|$, and $\mathbf{M} = \frac{1}{\nu}K_s\mathbf{S} + K_p\mathbf{P} - \mathbf{g}\mathbf{g}^\dagger$, the determinant can be expressed as

$||\mathbf{M} + (|\mathbf{e}|\mathbf{U}\underline{\phi} - \mathbf{g})(|\mathbf{e}|\mathbf{U}\underline{\phi} - \mathbf{g})^\dagger||$. Using the same matrix determinant lemma as above, this expression may be further simplified to yield:[5]

$$f_0[\mathbf{Z}; \hat{\mathbf{R}}, \underline{\phi}, \nu)] = \left[\frac{K}{\pi e}\right]^{NK} \left(\frac{1}{\nu}\right)^{NK_s} \left\{\frac{1}{||\mathbf{M}||[1 + (|\mathbf{e}|\mathbf{U}\underline{\phi} - \mathbf{g})^\dagger \mathbf{M}^{-1}(|\mathbf{e}|\mathbf{U}\underline{\phi} - \mathbf{g})]}\right\}^K \tag{12.133}$$

Maximization of $f_0[\cdot]$ with respect to $\underline{\phi}$ is achieved by substituting the minimizer of the determinant $\hat{\underline{\phi}} = \frac{1}{|\mathbf{e}|}(\mathbf{U}^\dagger\mathbf{M}^{-1}\mathbf{U})^{-1}\mathbf{U}^\dagger\mathbf{M}^{-1}\mathbf{g}$ into (12.133). This yields (12.134), written in terms of the "whitened" vector $\tilde{\mathbf{g}} = \mathbf{M}^{-1/2}\mathbf{g}$, the "whitened" unwanted signal subspace $\tilde{\mathbf{U}} = \mathbf{M}^{-1/2}\mathbf{U}$, and the orthogonal projection matrix $\mathbf{P}_{\tilde{\mathbf{U}}}^{\perp} = \mathbf{I} - \tilde{\mathbf{U}}\tilde{\mathbf{U}}^{+}$, where $\tilde{\mathbf{U}}^{+} = (\tilde{\mathbf{U}}^\dagger\tilde{\mathbf{U}})^{-1}\tilde{\mathbf{U}}^\dagger$ is the Moore-Penrose pseudo-inverse of $\tilde{\mathbf{U}}$.

$$f_0[\mathbf{Z}; \hat{\mathbf{R}}, \hat{\underline{\phi}}, \nu] = \left[\frac{K}{\pi e}\right]^{NK} \left(\frac{1}{\nu}\right)^{NK_s} \left\{\frac{1}{||\mathbf{M}||(1 + \tilde{\mathbf{g}}^\dagger \mathbf{P}_{\tilde{\mathbf{U}}}^{\perp}\tilde{\mathbf{g}})}\right\}^K \tag{12.134}$$

An analogous procedure can be used to maximize the H_1 density $f_1[\cdot]$, except in this case we combine the maximization with respect to $\underline{\psi} = [\underline{\theta}, \underline{\phi}]$ for an augmented subspace $\mathbf{W} = [\mathbf{V}, \mathbf{U}]$. By defining $\tilde{\mathbf{W}} = \mathbf{M}^{-1/2}\mathbf{W}$ and its orthogonal projection $\mathbf{P}_{\tilde{\mathbf{W}}}^{\perp} = \mathbf{I} - \tilde{\mathbf{W}}\tilde{\mathbf{W}}^{+}$, we arrive at Eqn. (12.135).

$$f_1[\mathbf{Z}; \hat{\mathbf{R}}, \hat{\underline{\theta}}, \hat{\underline{\phi}}, \nu] = \left[\frac{K}{\pi e}\right]^{NK} \left(\frac{1}{\nu}\right)^{NK_s} \left\{\frac{1}{||\mathbf{M}||(1 + \tilde{\mathbf{g}}^\dagger \mathbf{P}_{\tilde{\mathbf{W}}}^{\perp}\tilde{\mathbf{g}})}\right\}^K \tag{12.135}$$

At this point, we assume that ν is known and evaluate the Kth root of the GLR as Eqn. (12.136), noting that the remaining dependence on the scaling factor ν resides in the "whitening" matrix $\mathbf{M} = \frac{1}{\nu}\mathcal{S} + \mathcal{P} - \mathbf{g}\mathbf{g}^\dagger$, which is written more compactly in terms of $\mathcal{S} = K_s\mathbf{S}$ and $\mathcal{P} = K_p\mathbf{P}$.

$$\Lambda(\mathbf{Z}, \nu) = \frac{1 + \tilde{\mathbf{g}}^\dagger \mathbf{P}_{\tilde{\mathbf{U}}}^{\perp}\tilde{\mathbf{g}}}{1 + \tilde{\mathbf{g}}^\dagger \mathbf{P}_{\tilde{\mathbf{W}}}^{\perp}\tilde{\mathbf{g}}} \tag{12.136}$$

This is the GLRT expression for known ν. For the homogeneous case, we may simply insert $\nu = 1$ to compute $\mathbf{M} = \mathcal{S} + \mathcal{P} - \mathbf{g}\mathbf{g}^\dagger$. For a unit-rank useful signal $\mathbf{V} = \mathbf{v}$, no unwanted signals $\mathbf{U} = \mathbf{0}$, and the homogeneous case $\nu = 1$, this expression reduces to $\Lambda(\mathbf{Z}, 1)$ in Eqn. (12.137), which is precisely the GLRT λ derived in Eqn. (12.118).

$$\Lambda(\mathbf{Z}, 1) = \frac{1 + \tilde{\mathbf{g}}^\dagger\tilde{\mathbf{g}}}{1 + \tilde{\mathbf{g}}^\dagger \mathbf{P}_{\tilde{\mathbf{v}}}^{\perp}\tilde{\mathbf{g}}} = \frac{1 + \mathbf{g}^\dagger\mathbf{M}^{-1}\mathbf{g}}{1 + \mathbf{g}^\dagger\mathbf{M}^{-1}\mathbf{g} - |\mathbf{v}^\dagger\mathbf{M}^{-1}\mathbf{g}|^2(\mathbf{v}^\dagger\mathbf{M}^{-1}\mathbf{v})^{-1}} \tag{12.137}$$

With respect to Eqn. (12.137), the more general GLRT in Eqn. (12.136) is expected to provide greater robustness against useful signal uncertainty and unwanted signals. If invariance to disturbance scale is also deemed necessary, a possible approach is to insert an estimate for $\hat{\nu}$ into Eqn. (12.136). The secondary data disturbance sample covariance matrix $\mathbf{S}$ can be used to "whiten" the primary vectors $\bar{\mathbf{z}}_k = \mathbf{S}^{-1/2}\mathbf{z}_k$. Any deterministic components in the transformed signal subspace $\bar{\mathbf{W}} = \mathbf{S}^{-1/2}\mathbf{W}$ may be removed by

[5]Note that $\mathbf{M}$ has been redefined to account for the relative scale ν in the partially homogeneous case.

applying an orthogonal projection $\mathbf{P}_{\bar{\mathbf{W}}}^{\perp}$ to the vectors $\bar{\mathbf{z}}_k$. The resulting vectors are then used to form the estimate $\hat{\nu}$ in Eqn. (12.138).

$$\hat{\nu} = \left\{ \frac{1}{(N-M-P)K_p} \sum_{k=K_s+1}^{K} \bar{\mathbf{z}}_k^{\dagger} \mathbf{P}_{\bar{\mathbf{W}}}^{\perp} \bar{\mathbf{z}}_k \right\}^{-1} \tag{12.138}$$

The estimate of $\hat{\nu}$ is consistent in the sense that $\hat{\nu} \to \nu$ as $K_s \to \infty$ and $K_p \to \infty$. The two-step GLRT for the partially homogeneous case, which first assumes ν is known, and then substitutes the estimate $\hat{\nu}$ for this unknown quantity in the resulting GLRT expression, is given by the test statistic $\Lambda(\mathbf{Z}, \hat{\nu})$. Denoting this statistic simply as Λ, and defining $\chi = (\Lambda - 1)/\Lambda$, then a test on Λ with threshold Λ_0 is equivalent to the test in Eqn. (12.139), where $\chi_0 = (\Lambda_0 - 1)/\Lambda_0$. The quasi-whitening operation to form the various terms in Eqn. (12.139) is performed using the matrix $\mathbf{M} = \frac{1}{\hat{\nu}}\mathcal{S} + \mathcal{P} - \mathbf{g}\mathbf{g}^{\dagger}$ in this case.

$$\chi = \frac{\tilde{\mathbf{g}}^{\dagger}\mathbf{P}_{\tilde{\mathbf{W}}}\tilde{\mathbf{g}} - \tilde{\mathbf{g}}^{\dagger}\mathbf{P}_{\tilde{\mathbf{U}}}\tilde{\mathbf{g}}}{1 + \tilde{\mathbf{g}}^{\dagger}\tilde{\mathbf{g}} - \tilde{\mathbf{g}}^{\dagger}\mathbf{P}_{\tilde{\mathbf{U}}}\tilde{\mathbf{g}}} \underset{H_0}{\overset{H_1}{\gtrless}} \chi_0 \tag{12.139}$$

12.4 Practical Applications

This section illustrates the practical application of several GLRT processing schemes to real data from experimental skywave and surface-wave OTH radar systems. Specifically, *spatial* processing is applied in the first part of this section for the problem of detecting targets submerged in directional interference due to far-field sources that are incoherent with the radar waveform. In the second part of this section, *temporal* processing is applied to detect targets masked by clutter, which is backscattered from terrain and sea surfaces. The final part of this section illustrates the application of a JDS adaptive detection technique that simultaneously processes multiple primary data vectors.

12.4.1 Spatial Processing

The spatial processing problem is of significant interest for the detection of target echoes that are not masked by clutter after Doppler processing, due to their significant Doppler frequency shift, but are effectively submerged in interference and noise after such filtering as a result of incoherent sources that fully or partially overlap the radar bandwidth. From the viewpoint of Doppler processing, the detection of target echoes obscured by this type of interference, which is directional and often occupies the entire range-Doppler search space, is ultimately a spatial processing task.

The Jindalee skywave OTH radar and Iluka HF surface-wave radar, located near Alice Springs (central Australia) and Darwin (northern Australia), respectively, were used to collect real data in two separate experiments for this spatial processing study. The procedure used to collect the experimental data is described next, after which the practical performance of Kelly's GLRT, the AMF, as well as ACE and the ASD detectors will be illustrated and compared.

12.4.1.1 Jindalee Experiment

The receiving system of the Jindalee skywave OTH radar is based on a 2.8-km long ULA composed of 462 dual-fan antenna elements spaced 6 m apart. The aperture is partitioned into $N = 32$ subarrays of 28 dual-fan elements that overlap with neighboring subarrays by 50 percent. The individual subarrays are steered toward the surveillance region by analog (delay-line) beamforming networks. Each subarray output is connected to an individual HF receiver, where the RF signal is filtered, down-converted, mixed with a linearly swept repetitive frequency-modulated continuous waveform (FMCW) and digitized at the Nyquist rate. Prior to data acquisition, all 32 receiver passbands were equalized using internal calibration signals. More information about the Jindalee system can be found in Chapter 6.

To measure the signal detection and azimuthal localization performance, as well as false-alarm rate, of different processing schemes in conditions of severe interference, the Jindalee radar was operated in *passive* mode (i.e., with the transmitters turned off) to receive interference and noise in addition to a useful (target-like) signal emitted by a far-field cooperative source at a known location. The *absence* of clutter allows the false-alarm rate arising from the interference and noise to be observed in isolation from coherent backscatter, while the presence of a target-like signal enables the signal detection and azimuthal localization performance to be evaluated and compared with the known geometry of the experiment.

Apart from permitting the spatial processing performance of different detection schemes to be assessed with a high degree of confidence, such data also enables the effect of clutter to be appreciated in the second experiment, where the Iluka HF surface-wave radar was operated in active mode to detect real target echoes against clutter, interference, and noise.

The useful signal here is an FMCW with center frequency $f_c = 15.960$ MHz, bandwidth $B = 20$ kHz, and waveform repetition frequency (WRF) $f_p = 60$ Hz, which was received by the Jindalee radar from a cooperative emitter located near Darwin, approximately 1265 km north of the receiver site, and 22 degrees off array boresight. At the receiver site, located about 40 km north-west of Alice Springs in central Australia, the FMCW used for demodulation was a time-synchronized replica of the transmitted signal, except for a positive Doppler shift of 15 Hz imposed on the latter. This Doppler shift is representative of an inbound target moving at approximately 540 km/h.

The useful signal was received after reflection from the ionosphere in all $N_a = 32$ subarrays on 31 March 1998 between 05:52 and 06:01 UT. The interference was recorded separately during this period at $f_c = 16.052$ MHz and superimposed on the useful signal. The interference signal was noise-like (incoherent with the local FMCW) and relatively strong, with a power level more than 20 dB above the background noise in a single receiver. This interference was emitted from a spatially separate source in the Darwin region with a DOA approximately 1 degree closer to boresight than that of the useful signal.

Each receiver processed and retained $N_g = 20$ complex range cells between distances of 1162 and 1447 km in $N_p = 128$ waveform repetition intervals (PRIs). The range cells collected in the CPI of 2.1 seconds were then Doppler processed. Since the first $L' = 6$ range cells correspond to group ranges (1162–1237 km) less than the ground distance between the useful signal source and receiver array (1265 km), the associated $PL' = 768$ range-Doppler bins were deemed signal-free and used as secondary data for training the adaptive processors.

To assist with data interpretation, an oblique sounder was used to monitor the multipath characteristics of the ionospheric channel that propagated the useful signal. This sounder is part of the Jindalee Frequency Management System (FMS) (Earl and Ward 1986), and routinely measures ionograms to identify the number of propagation modes, and the ionospheric layers that reflected them, as a function of operating frequency.

Figure 12.1 shows the Jindalee azimuth-range-Doppler (ARD) map resulting for the conventional tapered beamformer (CTB) with Hann window (left panel), the AMF with uniform taper (central panel), and a *normalized* ASD statistic to be described in the following paragraphs (right panel). In these intensity-modulated displays, Doppler bins are arranged along the horizontal axis, while beams are stacked vertically, with nested range cells. A total of 16 beams were steered at the angles indicated in Figure 12.1 to cover the angular arc of interest (i.e., near where the useful signals and interference were located). White depicts high power and black depicts low power.

The directional interference is most evident in beams 6–9 of the CTB output, where no useful signals are apparent due to the interference masking all range-Doppler cells. Variation of the interference power is also evident over different ranges, with lower ranges having less power than higher ranges. This power variation is relatively smooth (non-spiky), and most discernable in beams less contaminated by the interference in the CTB output (e.g., beams 5 or 10).

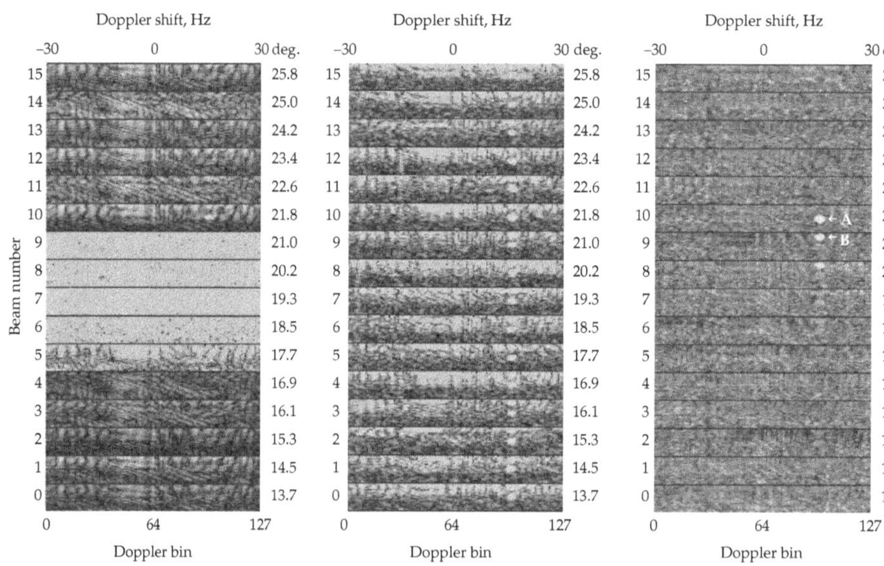

FIGURE 12.1 ARD output of conventional tapered beamformer using a Hann window (left panel), adaptive matched filter using range cells 0–5 inclusive for adaptive training (middle panel), and adaptive subspace detector using identical training data (right panel). The AMF output has a false-alarm rate that changes as a function of range, and the two useful signal modes are detected in all antenna beam-steer directions. The two main features of the ASD output are (1) the noise background is remarkably uniform, which allows CFAR detection to be carried out using a fixed threshold, and (2) both useful signal modes are detected and correctly localized close to their expected azimuths without false detections in other beams.

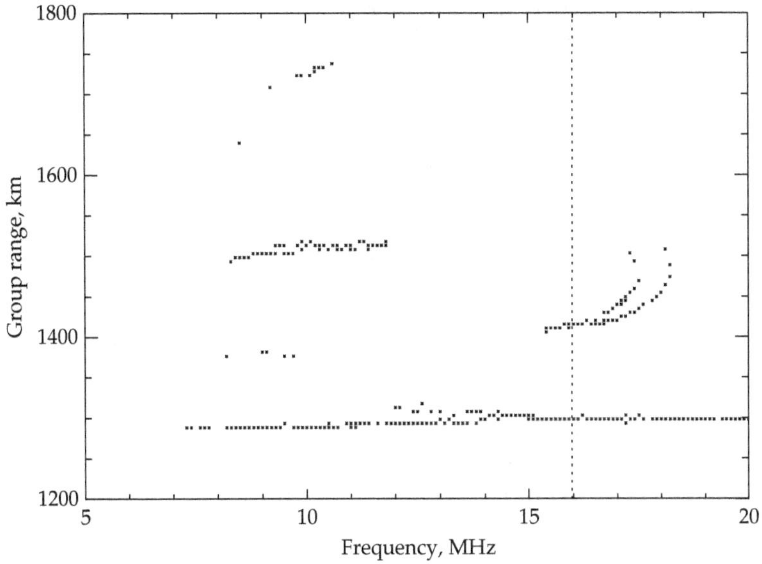

FIGURE 12.2 Oblique incidence ionogram for the Darwin to Alice Springs ionospheric propagation path (i.e., similar to that of the useful signal) on 31 March 1998 at 06:34 UT. At a frequency of 15.960 MHz, two propagation modes are resolved by the ionogram, namely; a one-hop sporadic-E reflection at 1300-km group range (mode A), and a one-hop F-layer reflection at 1420-km group range (mode B). © Commonwealth of Australia 2011.

Before discussing the other two ARDs in detail, let us consider Figure 12.2 which shows the oblique incidence ionogram for the path propagating the useful signal. The mode content is estimated as the point(s) of intersection between the ionogram trace and a line drawn vertically at the operating frequency. At $f_c = 15.960$ MHz, the ionogram resolves two distinct propagation modes at group ranges of 1300 and 1420 km. These group ranges correspond to virtual ionospheric heights of approximately 122 and 292 km respectively. The mode with the shorter path length (referred to as mode A) is a one-hop reflection from the sporadic E layer, while the mode with longer path length (mode B) is a one-hop reflection from the F2 layer (Davies 1990).

Assuming a spherical earth model, and ignoring any large-scale gradients or tilts at the ionospheric reflection points, modes A and B are expected to be impinge upon the array at cone steering angles close to $\varphi_A = 21.7$ degrees (for azimuth and elevation angles of 22 and 10.3 degrees, respectively), and $\varphi_B = 20.3$ degrees (for azimuth and elevation angles of 22 and 22.4 degrees, respectively), due to the coning ambiguity associated with a ULA (Gething 1991). The performance of the ASD and AMF will now be discussed.

A deterministic (wave-interference) subspace signal model with $M = 3$ rays was adopted for the ASD. A model of this kind was validated for individual HF signal modes in Fabrizio et al. (2000). Stated mathematically, $\mathbf{s} = \mathbf{V}\underline{\theta}$, where $\mathbf{V} = [\mathbf{v}(\varphi), \mathbf{v}(\varphi + \Delta), \mathbf{v}(\varphi - \Delta)]$, $\mathbf{v}(\varphi)$ is the ULA steering vector on the array manifold, $\varphi = \varphi_1, .., \varphi_N$ are the nominal steer directions, and Δ is an angular offset equivalent to 0.2 of the main-lobe width.[6]

[6]The main-lobe width is defined here as the angle between the nominal steer direction φ and the first null of the antenna pattern resulting for a conventional beam with uniform taper.

This multi-rank signal model is capable of representing a level of mismatch between the useful signal and the ideal steering vector. Such mismatch may be caused by slight beam-pointing errors, antenna array imperfections, or the non-specular ionospheric reflection process.

As the value of Δ is constrained to lie inside the main-lobe of the beam pattern, we may assign the nominal cone angle-of-arrival φ to any signal detected when the array is steered in this direction. The ray spacing Δ depends on the anticipated amount of spatial spreading encountered. The choice made here is based on quantitative results in Fabrizio et al. (2000), where the angular separation of rays resolved due to a single ionospheric reflection using super-resolution techniques was typically less than 20% of the main-lobe width.

The number of rays M also requires consideration; although using more rays is likely to improve signal modeling accuracy, it also increases the rank of the signal subspace, which opens up the spatial bandwidth and therefore reduces gain against interference and noise. On the other hand, selecting too few rays may lead to significant mismatch between the assumed signal model and the actual signal received by the system. This can result in partial rejection of the desired signal due to the high selectivity of the ASD. The choice $M = 3$ represents a tradeoff between these conflicting objectives in this application.

The ARD in the rightmost panel of Figure 12.1 corresponds to the F-version of the ASD statistic in Eqn. (12.96) normalized by $(N - M)/M$ (i.e., a per subspace-dimension normalization applied to the numerator and denominator of the ASD). This display shows that modes A and B are clearly detected by the ASD. The former appears in beam 10 at 21.8 degrees (compared with the expected value of $\varphi_A = 21.7$ degrees), while the latter appears to straddle beams 8 and 9 at 20.2–21.0 degrees (compared to $\varphi_B = 20.3$ degrees). These results agree well with the geometry of the experiment and demonstrate that the wave-interference model adopted for the ASD can be used to accurately localize both useful signal modes in azimuth. The range spectra shown in Figure 12.3 illustrate the strength of these detections as 24 dB for mode A, and 21 dB for mode B. Such spectra also demonstrate the high spatial selectivity of the ASD, as there is effectively no leakage of mode A into the range spectrum of mode B in the adjacent beam and vice versa. Note that the group ranges of the two modes are also in good agreement with those predicted by the ionogram (i.e., 1300 and 1420 km for modes A and B, respectively).

The noise background in the ASD output is remarkably uniform in ARD cells where no signals are expected. This includes all of the secondary data (ranges 0–5 inclusive), and all the primary data (ranges 6–19 inclusive) where there are no useful signals. In range cells 13 and 14 of Figure 12.3, the range sidelobes of modes A and B fall well below the noise level. These two range cells may therefore be used to analyze the false-alarm rate in the beam containing the useful signal. For mode A in beam 10, a total of $2N_p = 256$ samples are available in these two ranges to compute the cumulative distribution of the primary ASD outputs under the disturbance-only hypothesis. Figure 12.4 compares the cumulative distributions of the $N_p \times L' = 768$ *secondary* ASD outputs (using range cells 0–5), and the $2N_p = 256$ primary ASD outputs (using range cells 13–14), with the central F-distribution using $2M = 6$ and $2(N - M) = 58$ degrees of freedom. This distribution is theoretically expected for the complex data ASD in the clairvoyant (large-sample) case under the disturbance-only hypothesis (Scharf 1991).

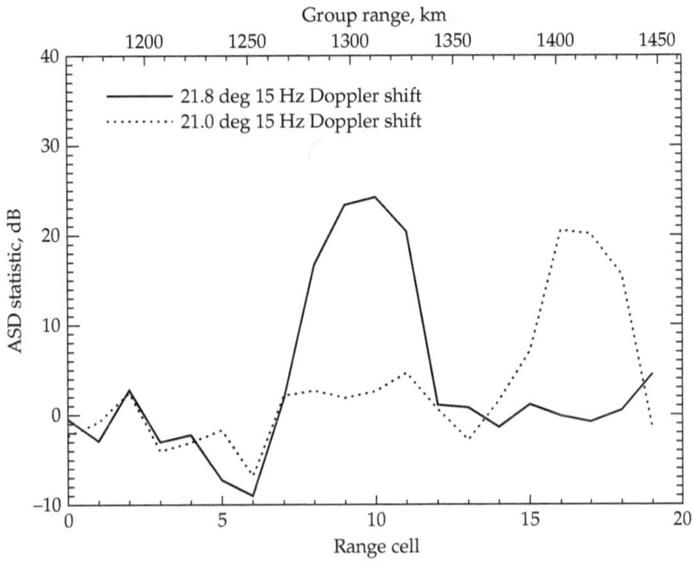

FIGURE 12.3 Range spectra from the ASD output at 15-Hz Doppler shift in beam 10 (solid line) to show that mode A is detected at about 1300 km, and beam 9 (dotted line) to show that mode B is detected at about 1400 km. The range spectra also illustrate the high selectivity of the ASD which suppresses the azimuthal response of each mode in beams that do not point in the useful signal direction. © Commonwealth of Australia 2011.

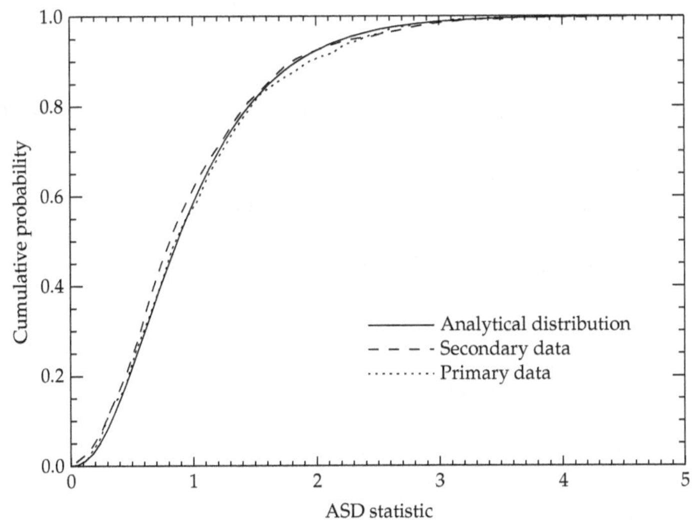

FIGURE 12.4 Cumulative density functions of ASD statistic for the secondary data (dashed line) and primary data (dotted line) in regions of the ARD containing no useful signals, as well as the analytical central F-distribution (solid line) with 6 and 58 degrees of freedom, which is theoretically expected for this statistic under H_0 in the clairvoyant (large-sample) case. © Commonwealth of Australia 2011.

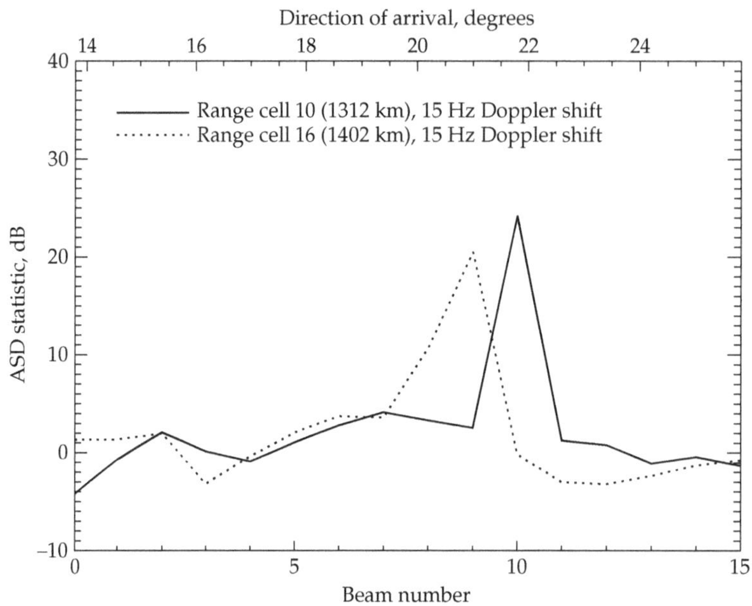

FIGURE 12.5 Azimuthal response of the ASD for mode A (solid line) and mode B (dotted line). The peak for mode A occurs in beam 10 (21.8 degrees) while the peak for mode B is between beams 8 and 9 (20.2–21.0 degrees). These results are consistent with the coning effect for a source at great-circle bearing of 22 degrees with elevation angles of 10.3 and 22.4 degrees calculated for modes A and B, respectively, ignoring the possibility of ionospheric tilts.

Figure 12.4 shows that the real-data false-alarm rate at any given threshold is practically identical to the theoretical prediction over both the secondary and primary ASD outputs. This implies that a fixed and known threshold may be applied directly to the ASD outputs for detection with a predictable and constant false-alarm rate. In addition, the close agreement between the false-alarm rate characteristics of both the primary and secondary ASD outputs with the analytical (large-sample) distribution suggests that the interference sample covariance matrix has practically converged to its expected value using the available secondary-data, and that the primary data is statistically described by a possibly scaled version of this secondary-data covariance matrix. The azimuthal response of the ASD is shown for each mode in Figure 12.5. In each case, only one peak for each signal is detected above a threshold of $\gamma = 5$ dB ($\gamma = 3.16$ in the linear scale of Figure 12.4), which corresponds to a P_{FA} of approximately 1 percent.

The AMF output is shown in the central panel of Figure 12.1, where the varying intensity of the noise background as a function of range is clearly evident. The significant variation in false-alarm rate between the primary and secondary data is confirmed by the cumulative distributions in Figure 12.6, derived similarly to those of Figure 12.4. The analytical (large-sample) distribution of the AMF output using complex-valued input data is a central chi-squared distribution with two degrees of freedom (Scharf 1991). Note that for complex data it is required to scale the argument of this distribution by a factor of two, as the *real* and *imaginary* parts of the (whitened) array snapshot vector elements have an asymptotic variance of 0.5 rather than unity.

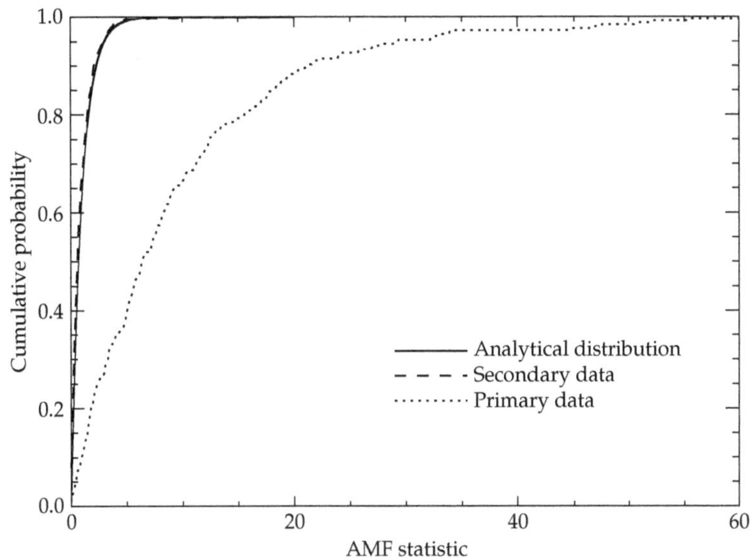

FIGURE 12.6 Cumulative densities of the AMF output for the primary data (dotted line), secondary data (dashed line), and the analytical central chi-squared distribution (solid line) with two degrees of freedom. The horizontal axis was compressed by a factor of two for the analytical distribution as the complex whitening operation produces real and imaginary parts with an asymptotic variance of 0.5 rather than unity. © Commonwealth of Australia 2011.

The close match between the analytical distribution and that of the *secondary* AMF outputs is consistent with a fully correlated interference texture model over the secondary cells. However, the large discrepancy between this distribution and that of the primary AMF outputs indicates that the interference texture in the primary data is not fully correlated with that in the secondary data. The significant variation of the interference power in range causes a large difference in false-alarm rate between the primary and secondary AMF outputs.

The combination of this observation for the AMF and the practically CFAR output observed previously for the ASD over both primary and secondary data suggests that: (1) an SIRP disturbance model with fully correlated texture over the secondary data, but different texture in the primary data, is more appropriate than the homogeneous Gaussian model traditionally adopted in Reed et al. (1974), Kelly (1986), and Robey et al. (1992), and (2) the assumption that external interference dominates additive noise is reasonable from a false-alarm rate perspective in this case, since the ASD output is not expected to be CFAR (or nearly so) if the additive noise contribution were significant.

To illustrate the false-alarm rate properties of the AMF, which is known to be CFAR in a homogeneous Gaussian environment, a detection threshold of $\gamma = 7$ dB ($\gamma = 5$ on the linear scale of Figure 12.6) yields a false-alarm rate of 1 percent in the secondary data, but an intolerable 60 percent in the primary data (at range cells 13 and 14). For a 1-percent false-alarm rate in this primary data, the detection threshold needs to be raised to approximately $\gamma = 17$ dB (i.e., to $\gamma = 50$ on the linear scale of Figure 12.6).

Besides being far from CFAR in this practical interference environment, the AMF also detects both useful signal modes when the antenna beam is not steered in the actual

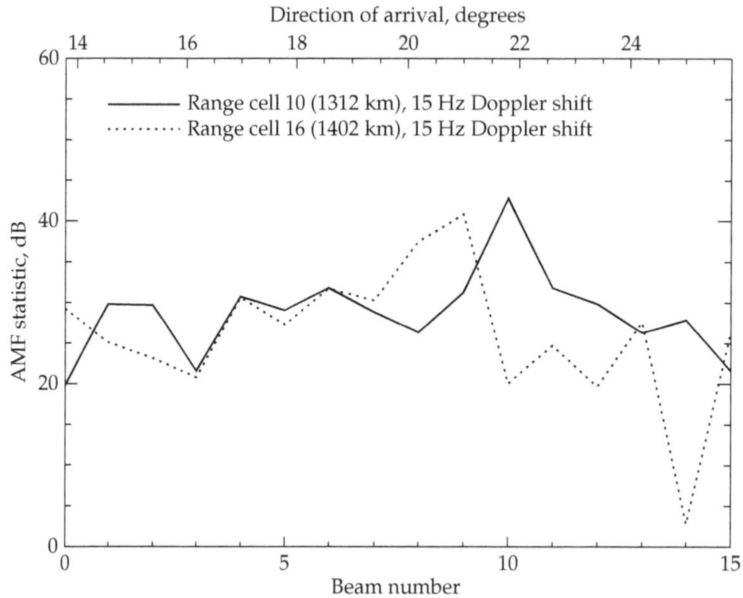

FIGURE 12.7 Azimuthal response of the AMF for mode A (solid line) and mode B (dotted line). The AMF detects both modes but its azimuthal response is poor and false peaks appear significantly above the noise level at angles not corresponding to the useful signal direction (i.e., compare the output for mode A in beam 4 of about 30 dB with the cumulative density of the noise in the primary AMF outputs above a threshold of $\gamma = 50$ or 17 dB).
© Commonwealth of Australia 2011.

directions of these signals. This is evident in the middle panel of Figure 12.1 and is quantitatively shown by the azimuthal response of AMF in the range-Doppler cells containing modes A and B in Figure 12.7. The presence of several false peaks well above the $\gamma = 17$ dB threshold (e.g., solid line in beam 6) demonstrates the poor selectivity of the AMF with respect to the ASD.

However, when the beam is steered in the actual directions of modes A and B, the AMF output is 20–25 dB higher than the 1 percent false-alarm threshold of $\gamma = 17$ dB. The ASD detections are only 15–20 dB higher than a threshold of $\gamma = 5$ dB, which yields the same false-alarm rate of 1 percent in the primary data. This practical example confirms the higher sensitivity of the AMF, which is predicted in simulations, and the tradeoff made by the ASD for the scale-invariant CFAR property.

It is important to emphasize that these observations are *not* related to finite-sample support effects, but rather to the inherent properties of the AMF and ASD. In particular, the unwanted detection of sidelobe signals is known to be a potential problem for the AMF when all adaptive degrees of freedom are absorbed in the task of interference rejection. Importantly, the AMF results presented here are also representative of Kelly's GLRT, since the ratio of dimensional parameters K-to-N of $768/32 = 24$ is large and the two detectors have practically converged.[7]

[7]The output of Kelly's GLRT converges to the AMF as the number of secondary vectors $K \to \infty$.

12.4.1.2 Iluka Experiment

The Iluka OTH radar operates by normal line-of-sight propagation (the standard mode for a microwave radar), as well as *surface-wave* propagation, which is enabled by the high conductivity of sea water in the HF band. The surface-wave propagation mode experiences an attenuation with distance that is significantly more severe than the inverse-square law describing line-of-sight propagation. Iluka's range coverage over sea water extends out from the coast to perhaps 200–300 km for large steel-hulled surface vessels of ocean-going size (e.g., cargo ships). More information about HF surface-wave radar and its capabilities can be found in Chapter 5.

The Iluka radar previously consisted of two transmit sites, a high-power site at Stingray Head (65 km south-west of Darwin), and a low-power site at Lee Point (10 km north-east of Darwin). The receiving system at Gunn Point (30 km north-east of Darwin) was sufficiently isolated from each transmitter to allow radar operation with a linear FMCW. The receiving system was based on a 500-m-long ULA of 32 vertical monopoles. Each antenna element was connected to a well-calibrated HF receiver, and two dummy elements were added at either end of the array to reduce the effects of mutual coupling. The array boresight is oriented in a westerly direction, approximately perpendicular to the coastline.

At the low-power site, a 250-W amplifier drove a pair of monopoles spaced 20 m apart and fed in anti-phase to radiate at endfire. In this experiment, the system radiated a linear FMCW with center frequency $f_c = 11.880$ MHz, bandwidth $B = 50$ kHz, and WRF $f_p = 4$ Hz. A total of $N_p = 128$ waveform repetition intervals per dwell were recorded over a CPI of $N_p/f_p = 32$ seconds. After range-Doppler processing, $N_g = 40$ range cells were retained to cover a range depth of 3–123 km. The relative velocity ambiguities were at ± 91 km/h (equivalent to ± 2 Hz bistatic Doppler shift at $f_c = 11.880$ MHz).

During this experiment, a fast boat target was undertaking 2-km long sorties on a straight inbound/outbound leg approximately 20 km offshore at a nominal cruising speed of 20 knots (37 km/h). For the radar range resolution of 3–4 km, and azimuthal resolution of 4–5 degrees, this fast boat target was essentially confined to a single range-azimuth cell. A radar calibration device (RCD) and stationary support boat were also known to be present in the same range-azimuth cell as the fast boat target.

At the nominal cruising speed, the echo from the target is expected to experience a bistatic Doppler shift of around 0.8 Hz, which is well separated from the dominant sea-clutter returns (first-order Bragg lines) that theoretically appear at Doppler shifts *less than* the monostatic value given by $f_d = \pm\sqrt{g/(\pi\lambda_c)\cos\theta_g} = \pm\ 0.35$ Hz. In this expression, g is acceleration due to gravity, λ_c is the operating wavelength, and θ_g is the grazing angle of incidence to the sea surface. The maximum separation of $\pm$ 0.35 Hz is calculated for near zero grazing angle. Doppler processing can therefore isolate the fast-moving boat target echo from the first-order sea-clutter returns. This allows the target echo to be detected in a region of Doppler space where external interference and noise signals dominate the disturbance background.

The Iluka experiment enables the performance of the CTB, AMF, and scale-invariant ASD spatial processing schemes to be evaluated on *real* targets with available truth information about their position and velocity, in the presence of unknown interference, noise, and clutter. In this experiment, interference-plus-noise-only range cells were not available due to the presence of clutter in all range cells, so an alternative means for selecting the secondary data was required. A method to obtain secondary data that is essentially free of clutter and echoes from targets of interest is to select Doppler bins corresponding to Doppler shifts that exceed those expected for clutter and target echoes that need to be excluded from the training set.

For example, targets of interest with a cruising speed of less than ±25 knots (±46 km/h) produce bistatic Doppler shifts with a magnitude of less than 1 Hz at 11.880 MHz. As the *dominant* clutter returns are expected to be contained in the ±1 Hz Doppler frequency band, all range-Doppler bins that correspond to Doppler shifts of magnitude greater than 1 Hz may be used as secondary data. In this experiment, the secondary data is taken from 64 Doppler bins (half the Doppler spectrum) and 20 range cells to yield $64 \times 20 = 1280$ training snapshots outside of the ±1 Hz Doppler frequency band. These range-Doppler bins contain interference, but not echoes from targets of interest or dominant clutter returns. Apart from this alternative reference cell selection method, all other data processing remains identical to that described previously for the Jindalee experiment.

The CTB ARD output, shown in the leftmost panel of Figure 12.8, exhibits significant contamination due to interference in several beams, which prevents the fast boat target from being detected at its expected position (range cell 6 and beam 3). The ASD output (rightmost display) clearly shows the outbound fast boat target detected at the base of "T,"

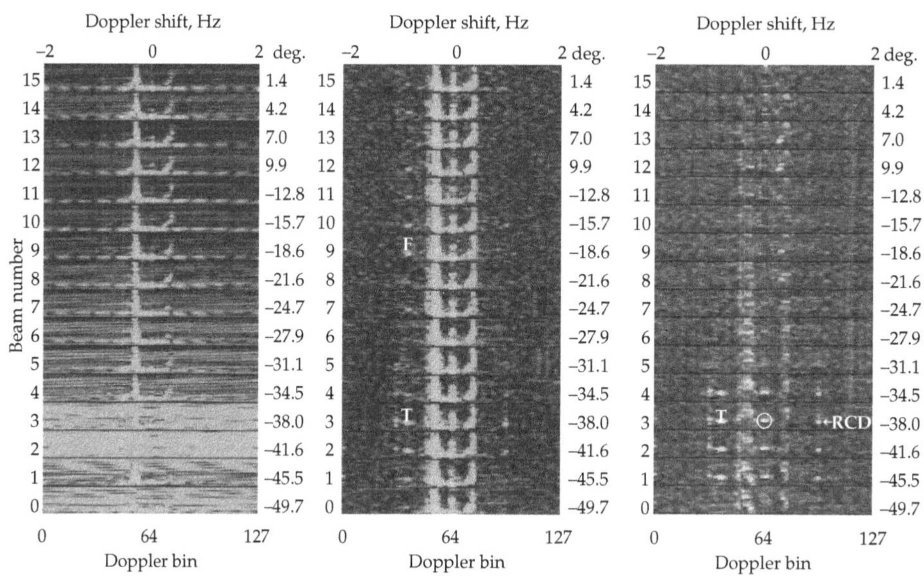

FIGURE 12.8 Iluka radar ARD data after spatial processing with the CTB using a Hann window (left panel), the AMF (middle panel), and the ASD (right panel). In the CTB output, the interference masks the fast boat target echo, the RCD signal sidebands at ±1 Hz Doppler shift, and a target-like return near zero Doppler frequency in beams 2 and 3. The fast boat target is clearly detected by the ASD at the base of "T" in beam 3 and range cell 6. The RCD sidebands, and a target-like return near zero Doppler shift, are also detected in the same range-azimuth cell. The AMF detects the target labeled "T" in beam 3, but produces a strong false alarm marked as "F" in beam 6, due to the same target echo entering through the sidelobes of the AMF. The dominant clutter returns from the resonant (first-order) Bragg lines at approximately ±0.3 Hz Doppler shift are clearly evident in the AMF output. © Commonwealth of Australia 2011.

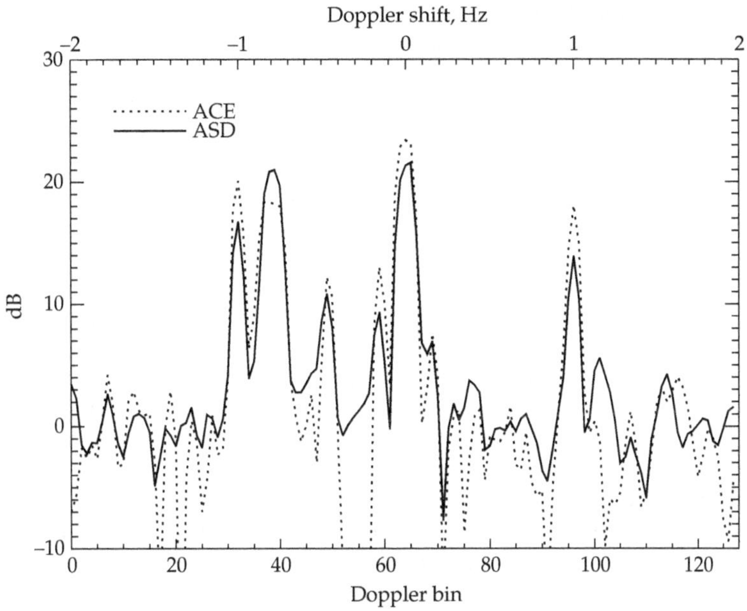

FIGURE 12.9 Doppler spectra at the target range-azimuth cell for the ASD and ACE. Although the ASD provides a 3-dB improvement for the target of interest, both the RCD signal and stationary target-like return exhibit a loss of 2–4 dB relative to ACE. This demonstrates the tradeoff between matched filter gain (better at low model order) and coherent gain for slightly mismatched signals (better at high model order). © Commonwealth of Australia 2011.

along with the RCD sidebands near ±1 Hz, with the right (positive Doppler) sideband labeled as "RCD," and a stationary target (circled) at range cell 6 and beam 3. The circled detection near zero Doppler frequency may correspond to the stationary support boat known to be located in the same range-azimuth cell as the target and RCD. The solid line in Figure 12.9 represents the Doppler response of the ASD at the target range-azimuth cell and shows that the target is detected at approximately −0.78 Hz (Doppler bin 38). This Doppler shift corresponds to an outbound target moving at about 19.2 knots (35.5 km/h), which closely matches the expected nominal cruising speed (20 knots) of the fast boat target.

The dotted line in Figure 12.9 represents the Doppler response of the ACE receiver in the same range-azimuth cell. A comparison of these two curves reveals that the fast boat target detection is about 3 dB stronger at the ASD output compared to the ACE receiver. On the other hand, ACE detects the stationary target and RCD sidebands with a strength of about 3 dB greater than the ASD. This demonstrates that while the higher-dimensional signal subspace model in the ASD may provide higher gain on a slightly mismatched useful signal relative to ACE, it simultaneously trades this benefit against the higher signal-to-disturbance ratio (SDR) achieved for well-matched targets by ACE, which adopts a unit-rank (steering vector) useful signal model. Care must therefore be taken not to over-model the useful signal by selecting an excessively high model order in the ASD. As different targets may have spatial signatures with varying degrees of mismatch to the (best matched) steering vector in the filter bank, the relative benefits of ACE and ASD may be expected to depend on the particular useful signal being processed, as exemplified by Figure 12.9.

The AMF output in Figure 12.8 removes interference more effectively than the CTB. This exposes the clutter structure in the ARD, which is dominated by the first-order Bragg lines near ±0.3 Hz. Figure 12.10 shows the Doppler spectra of the AMF (solid line) and CTB (dotted line) at the target range-azimuth cell. The AMF increases the sub-clutter visibility (SCV) by about 20 dB relative to the CTB, where SCV is defined as the difference between the highest clutter peak in the spectrum to the (average) background level. This improvement enables the target(s) and RCD returns to be detected by the AMF. These signals could not be detected by the CTB at an acceptable false-alarm rate. The AMF detects the target marked "T" in the correct beam, but produces a strong false alarm marked "F" at the same range-Doppler cell in different beam, which is well displaced in steer direction from the actual target azimuth. This is due to the same target echo entering through the sidelobes of the AMF response as an unwanted signal in this case.

The AMF cannot actively suppress this sidelobe "target" echo because its characteristics cannot be learnt from the secondary data, which excludes cells in ±1 Hz Doppler frequency band. Further processing is therefore required to edit out this unwanted detection. The presence of *clutter* may also produce false alarms in the AMF (and ASD)

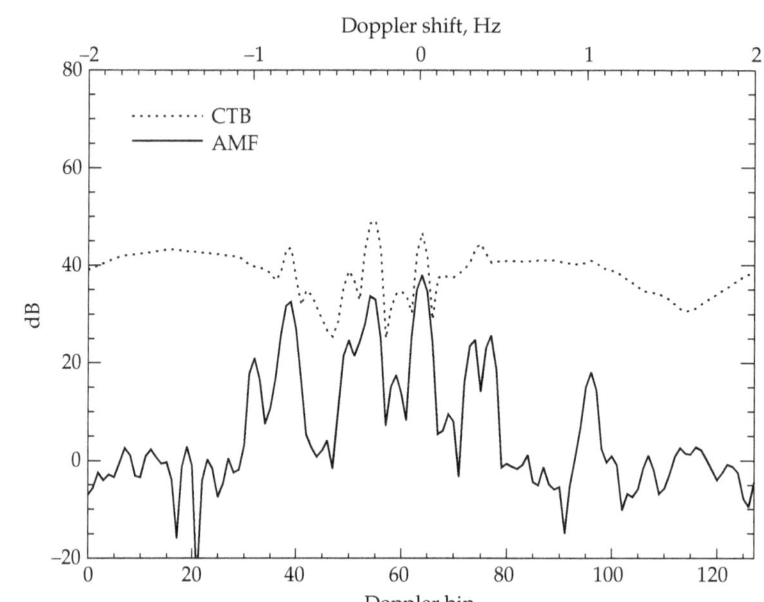

FIGURE 12.10 Doppler spectra at the target range-azimuth cell for the CTB and AMF. The SCV for the CTB is only 10 dB, which precludes the detection of the target at a reasonable false-alarm rate. The corresponding SCV for the AMF is approximately 30 dB, which enables the boat target, the RCD, and perhaps the support boat to be detected. © Commonwealth of Australia 2011.

outputs, which needs to be dealt with via subsequent CFAR processing step based on cell averaging or ordered statistics of the amplitude samples in the range-Doppler map.

It may be noticed from Figure 12.8 that the clutter Bragg lines are more highly attenuated by the ASD compared to the AMF. This apparent benefit is only illusory, as masking of useful signals due to strong clutter echoes not included in the training data arises for ACE and the ASD (i.e., both the strong clutter echoes and useful signals at the same Doppler frequency are attenuated in this case). This "blanking" phenomenon is most evident for the ACE receiver near Doppler cell 55 in Figure 12.9. The susceptibility of ACE and the ASD to target masking is illustrated and addressed next.

12.4.2 Temporal Processing

In this section, we turn our attention to temporal pulse-to-pulse processing for detecting targets that would otherwise be embedded in clutter, as opposed to interference and noise only, after conventional Doppler processing. Moreover, we utilize this application as a vehicle to expose some latent shortcomings of the ACE and ASD tests, which were used to good effect for spatial processing in the previous section. In particular, we investigate the problem of target-echo masking due to strong unwanted signals in the primary data, and highlight the potential advantages of the generalized subspace detection (GSD) tests derived in Section 12.3. The results described below also illustrate the possibility of applying adaptive detection schemes in the slow-time domain to counter a superposition of clutter and transient (impulsive) interferences on the time scale of the radar CPI.

To motivate the problem in specific terms, a significant signal processing challenge for modern HF radar systems is the simultaneous detection and tracking of ship and aircraft targets using different carrier frequencies. Ship detection often requires long CPIs to resolve slow-moving target echoes against clutter using conventional FFT-based Doppler processing (typically 30–60 seconds). Such long CPIs heavily consume radar resources and prevent the radar from performing other functions, such as revisiting aircraft targets frequently enough for effective tracking. For this reason, there is currently great interest in performing ship detection using shorter CPIs of perhaps 10–20 seconds.

Besides the limited frequency resolution of tapered FFT Doppler processing, which may preclude ship detection with such short CPI, conventional processing is also suboptimal for temporally structured disturbances. When the composite disturbance is highly structured in slow-time, there is scope for an adaptive detector to enhance performance with respect to the tapered FFT in Doppler processing. This section is concerned with the application of adaptive detection in the pulse-to-pulse domain to enhance the detection of slow-moving targets against clutter and impulsive interference using short CPIs.

Unlike the adaptive matched filter (AMF) and Kelly's GLRT, the scale-invariant ACE and ASD detectors in Kraut and Scharf (1999) maintain the CFAR property when the disturbance in the test cell and training data share the same covariance matrix structure, but possibly have different scale. This additional CFAR invariance is appealing in practice as it provides a further degree of protection in scenarios where the CFAR property may otherwise be lost due to this phenomenon (as experimentally demonstrated in the previous section).

However, ACE is known to discriminate strongly against mismatched signals (highly selective), and this can be a problem for useful signals that are not exactly aligned with the presumed target signature model. The extension of ACE to a multi-rank (subspace) signal representation in the ASD provides a degree of robustness for a partially unknown target response vector, as shown for a real (fast-boat) target echo in the previous section.

Despite the widespread appeal of ACE and the ASD, particularly in terms of their CFAR invariance to disturbance scale, and the capability of the ASD to cater for slightly mismatched useful signals, a significant shortcoming of both these tests is that they are susceptible to unwanted signals (coherent interference) present in the primary data but not in the secondary data. Specifically, the presence of such signals can cause undesirable blanking of (weaker) useful signals in the same resolution cell and preclude their detection in realistic environments.

This rather latent but significant side effect has received considerably less attention in the literature. As a possible remedy for the useful signal masking problem, the performance of an alternative GSD receiver, which was derived as a GLRT in section 12.3 to address this particular problem, will also be illustrated. The less known GSD not only possesses the valuable CFAR property with invariance to disturbance scale between the primary and secondary data, as well as the ability to model useful signal uncertainty for greater robustness to slightly mismatched signals, but also exhibits higher immunity to target blanking that can arise due to the presence of one or more strong unwanted signals in the primary data at different DOA or Doppler coordinates to the useful signal. The distinct advantages of the GSD over ACE and the ASD, in this respect, will be demonstrated below using real data.

The experimental data were collected by 16 reception channels of the Iluka HFSW radar described in the preceding part of this section. The transmitted linear FMCW signal had a center frequency $f_c = 7.719$ MHz, bandwidth $B = 50$ kHz, and PRF $f_p = 2$ Hz. The short CPIs were approximately 16 seconds long and consisted of $N_p = 32$ sweeps. High Doppler resolution CPIs with $N_p = 128$ sweeps (64 seconds long) were also recorded for identifying targets of opportunity to provide "ground-truth" information. A total of $N_g = 40$ range cells and 16 conventional beams were processed in both the short and long CPI cases.

Figure 12.11, reproduced from Fabrizio and Farina (2007a), shows an intensity modulated range-Doppler map for a beam containing three targets of opportunity when $N_p = 128$ pulses in the long CPI are integrated. Conventional Doppler processing has

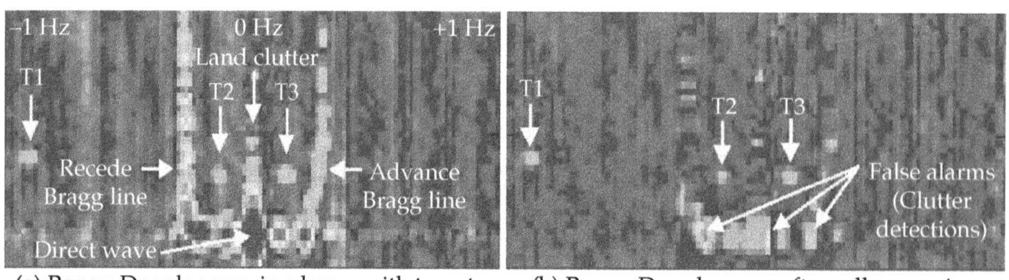

(a) Range-Doppler map in a beam with targets.

(b) Range-Doppler map after cell-averaging CFAR processing.

FIGURE 12.11 Conventionally processed range-Doppler map and CFAR output for a high-Doppler resolution CPI. Different types of clutter (Bragg lines, land clutter, and direct wave) are indicated along with three real target echoes. © 2007 IEEE. Reprinted with permission.

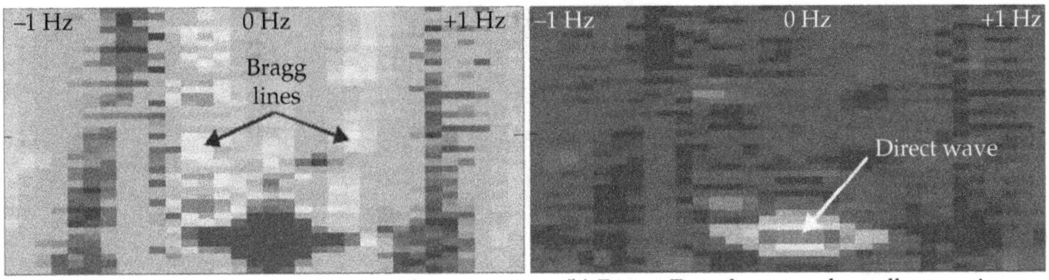

(a) Range-Doppler map in a beam with targets. (b) Range-Doppler map after cell-averaging CFAR processing.

FIGURE 12.12 Conventionally processed range-Doppler map and CFAR output for a short CPI in which impulsive interference is additionally present. © 2007 IEEE. Reprinted with permission.

been applied to this 64-second long CPI using a Blackman-Harris window. A number of features are indicated in this display. In particular, the existence of three targets, labeled T1, T2, and T3, respectively. Note that T2 and T3 are in the *same* range-azimuth cell but have different Doppler shifts, while T1 is in a different range cell to T2 and T3. In addition, T1 is the only visible target in this range cell. A cell-averaging (CA) constant false-alarm rate processing step is applied to this range-Doppler map prior to threshold detection. The output is shown in Figure 12.11, where all three targets can be detected using the long CPI. Some false alarms due to clutter near the direct wave from the transmitter are also indicated.

Figure 12.12, in the same format, shows the conventionally processed output for a short CPI ($N_p = 32$, CPI = 16 seconds) that was additionally contaminated by impulsive interference.[8] In this case, FFT-based processing is not able to detect any targets after CA-CFAR processing. Only the direct-wave return is visible at the CFAR output in Figure 12.12. Using the same data, Figure 12.13 shows the output of the ACE and ASD receivers. The ASD was based on $M = 3$ modes, with one matching the nominal search frequency, and the other two equally spaced either side of this frequency with a half Doppler bin displacement ($\Delta = \pi/N_p$). The disturbance sample covariance matrix $\mathbf{S}$ is formed using $4N_p = 128$ secondary vectors extracted from range and beam cells neighboring the CUT (with one guard cell inserted in both dimensions).

ACE can detect T1 and T2 quite easily, but misses T3 in Figure 12.13. Recall that T3 is in the same range cell as T2. When ACE searches for T3, the signal from T2 is at a different Doppler frequency to the current search frequency. In this case, T2 effectively represents an unwanted signal that blanks T3, preventing T3 from being detected. The ASD output in Figure 12.13 also misses T3 for the same reason. However, the ASD performs better than ACE on T1, which demonstrates once again the advantage of a subspace model in practice. The detection of T3 remains a problem for both ACE and the ASD. This is because both of these detectors are derived as GLRTs for a hypothesis test that does not explicitly account for the presence of unwanted signals, particularly one that is present in *addition* to the useful signal under H_1. The ASD also exhibits slightly inferior clutter suppression performance compared to ACE near the direct-wave signal. This example

[8]The short CPI in Figure 12.12 was recorded a little earlier than the long CPI, which was not contaminated by impulsive interference in Figure 12.11.

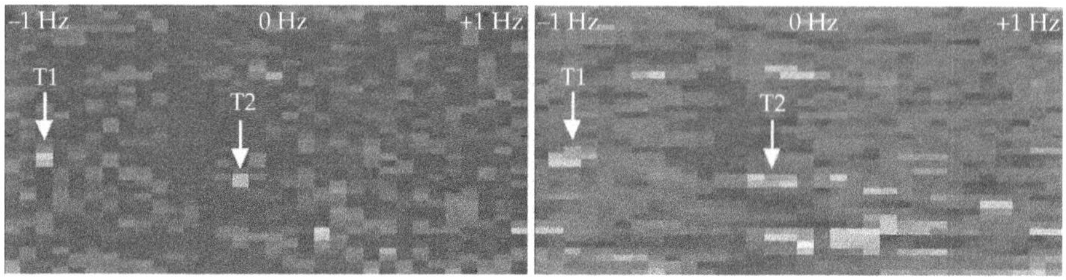

(a) Adaptive coherence estimator (ACE). (b) Adaptive subspace detector (ASD).

FIGURE 12.13 ACE and ASD outputs for the short CPI containing interference.

shows that robustness to slightly mismatched useful signals comes at the expense of reduced detector selectivity.

The ACE or ASD output represent the first pass of the proposed GSD scheme, denoted here by G-ACE and G-ASD, respectively, in order to distinguish between the unit-rank and multi-rank implementations of the GSD. Specifically, the detections made during the first pass from the ACE and ASD outputs in Figures 12.12 and 12.13, respectively, were used to form the unwanted signal model for G-ACE and G-ASD applied in the second pass. The aim of the second pass is to uncover useful signals that were not detected by ACE or the ASD in the first pass (T3) due to masking or "blanking" by another useful signal with a different Doppler shift in the same resolution cell (T2). Figure 12.14 shows that G-ACE was unsuccessful in revealing T3. This is probably due to the high selectivity of this detector combined with the unit-rank (single complex sinusoid) model used for both T3 and the unwanted signal T2, which may not be sufficiently accurate to represent these signals.

On the other hand, it is evident from Figure 12.14 that the G-ASD detector detects all the targets (T1, T2, and T3). Its ability to outperform G-ACE may be due to the more robust subspace model used for both the useful and unwanted signals in the cell under test. The Doppler profiles in Figure 12.15 clearly show the advantage of G-ASD over the ASD in the two range cells that contain targets. While the ASD (left panel) performs very well when the CUT contains a single target (range 18), it fails to detect both targets in

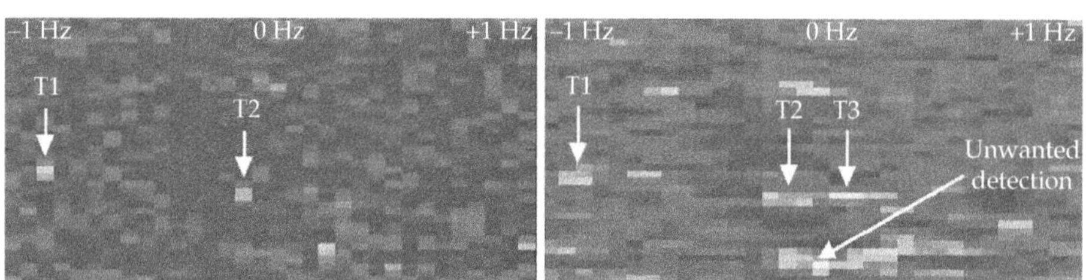

(a) Generalized ACE (G-ACE). (b) Generalized ASD (G-ASD).

FIGURE 12.14 Generalized subspace detector outputs for the short CPI containing interference.

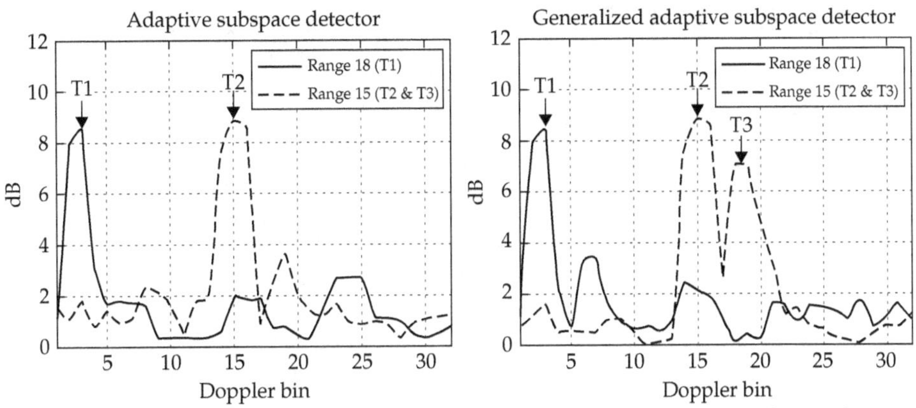

(a) Adaptive subspace detector.

(b) Generalized adaptive subspace detector.

FIGURE 12.15 Doppler profiles for the two range cells containing targets T1, T2, and T3. © 2007 IEEE. Reprinted with permission.

range 15. The G-ASD explicitly takes unwanted signals into account and is able to detect T1, T2, and T3. In particular, the generalized G-ASD detector clearly showed its ability to detect two targets in the same CUT, unlike the ACE or ASD receivers. The G-ACE version was not as effective as G-ASD in this case, perhaps due to its inability to model the unwanted and desired signals well enough, combined with the high selectivity of the test. The practical benefits of the GSD technique are not limited to HFSW radar (or Doppler processing). It is envisaged that this technique can be applied to skywave OTH radar, and possibly other type of radar systems, also for spatial processing and space-time adaptive processing (STAP).

12.4.3 Hybrid Technique

The JDS adaptive detection approach is applied here to a different dwell of the Iluka HFSW radar data previously used for the spatial processing analysis described in Section 12.4.1.2. In this case, the slow-time domain was designated as the processing dimension (i.e., $N = N_p = 32$ pulses), while the spatial channels of the receiver array were designated as the complementary dimension for extracting the multiple primary data vectors (i.e., $K_p = N_a = 32$ elements). The processor is applied to each of the $N_g = 40$ range resolution cells separately (in turn). The high Doppler resolution mode containing $N_p = 128$ pulses per CPI was used to provide ground truth, as before.

At the cruising speed of 20 knots (37 km/h), the echo from the cooperative inbound target is expected to experience a bistatic Doppler shift magnitude of around 0.8 Hz, while the RCD signal sidebands appear at ±1 Hz. The various signal components are identified in Figure 12.16, which shows a conventionally processed range-Doppler map for the beam containing the target in a long (32 second) CPI with $N_p = 128$ integrated pulses. The data in Figure 12.16 is displayed for reference to clearly illustrate the different signals present. A CPI of 32 seconds significantly consumes radar resources, so there is significant interest to shorten the CPI for surface target detection.

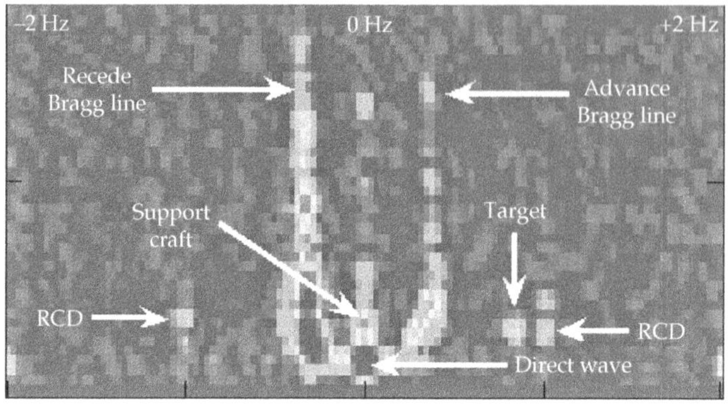

FIGURE 12.16 Range-Doppler map after conventional processing for the high-resolution CPI. The target echo, sea clutter (Bragg lines), and unwanted signals (e.g., from the RCD) are indicated, along with the direct-wave signal from transmitter. © Commonwealth of Australia 2011.

Figure 12.17 shows the output of conventional processing after cell-averaging CFAR is applied to a shorter CPI of $N_p = 32$ pulses (i.e., 8 seconds). In this case, the target known to be present is not detected by conventional processing. Using the same data as Figure 12.16, Figure 12.17 shows the output of the proposed two-step GLRT procedure in Eqn. (12.139). This adaptive processing scheme has enabled the detection of the target that would be missed by conventional processing.

To implement this detector, a rank $M = 3$ subspace model $\mathbf{V}$ was used. This model includes the ideal target response for the search Doppler frequency and the response vectors expected for half a Doppler bin displacement either side of it. The unwanted signal subspace $\mathbf{U}$ was also of rank $P = 3$. In this case, $\mathbf{U}$ included the two RCD components known to be at ± 1 Hz Doppler shift, and a zero Doppler frequency component that accounts for land clutter.

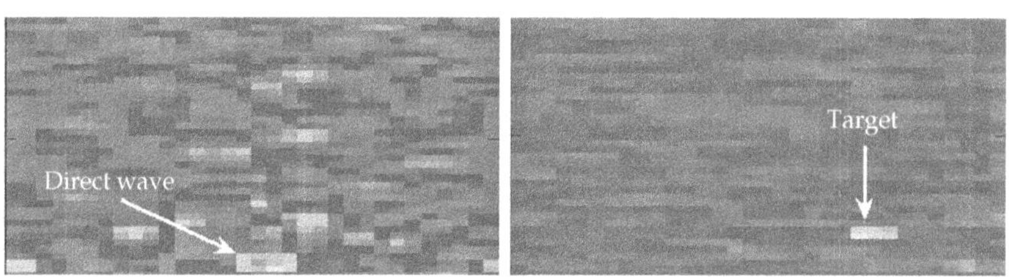

(a) Conventional processing (b) Two-step GLRT

FIGURE 12.17 Range-Doppler map resulting for the short CPI, in which a total of $N = 32$ pulses were integrated. In the left panel, the target is not distinguishable after conventional (tapered FFT) Doppler processing. In the right panel, the target is clearly detected by the proposed two-step GLRT. In addition, the clutter and unwanted signals are well suppressed at the output of this adaptive detector. © Commonwealth of Australia 2011.

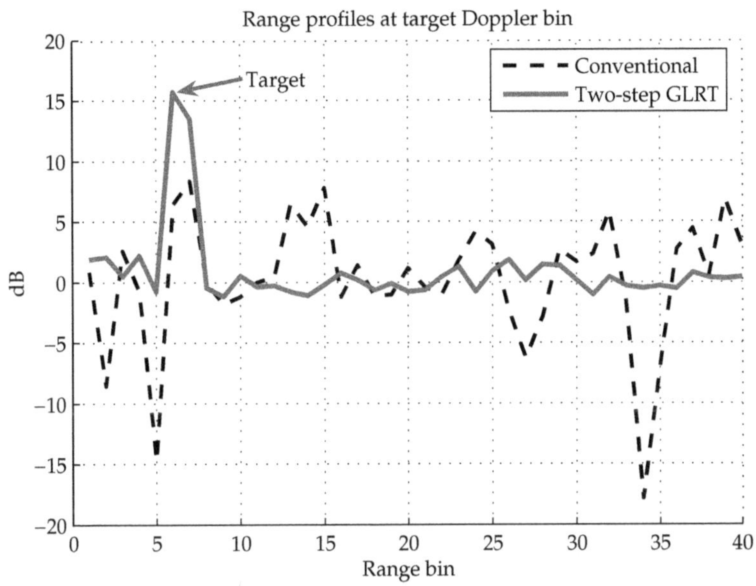

FIGURE 12.18 Range-profiles extracted at the target Doppler frequency for conventional FFT-based processing (dashed line), and the two-step GLRT (solid line). By using a threshold of 10 dB, the target can be detected with no false alarms at the output of the proposed adaptive detector. The same target would not be detected by conventional processing using this threshold. A lower threshold of 5 dB would allow the target to be detected at the conventional output, but only at the expense of multiple false alarms. © Commonwealth of Australia 2011.

For the short CPI, we have $N = 32$ pulses in a primary vector and a total of $K_p = 32$ primary vectors (receivers) in the range-bin CUT. The training data were taken from two range cells, one either side of the CUT, excluding a guard cell immediately adjacent to the CUT. This yielded a total of $K_s = 2N = 64$ secondary vectors. Figure 12.18 shows range-profiles taken at the target Doppler frequency for conventional processing (dashed line), and the two-step GLRT in Eqn. (12.139) (solid line).

By using a threshold of 10 dB, the target can be detected with no false alarms in the two-step GLRT range spectrum. A threshold of 5 dB may be used to detect the target in the conventional range spectrum, but this would produce four false alarms over the $N_g = 40$ processed range cells. At the time of writing, this is perhaps the first illustration of the successful application of a multiple primary and secondary data GLRT-based adaptive detector to experimental HF radar data containing a real target echo.

CHAPTER 13

Blind Waveform Estimation

Sensor arrays are often used in practice to separate and estimate the waveforms of superimposed signals that share similar frequency spectra but have different spatial structure. Source waveform recovery is traditionally a data communication problem, but the signal-processing techniques developed in this field also have a number of practical radar-related applications. For example, passive radar systems require a "clean copy" of the waveform transmitted by an uncooperative source to perform effective matched filtering. Such systems often receive signals from an emitter of opportunity that are contaminated by multipath arrivals and interference from independent sources. The ability to accurately estimate the signal emitted by an unknown source in the presence of multipath and other sources is not only relevant for passive radar, but is also of interest to active radar systems for interference cancelation.

In many problems of practical interest, the interfering signals do not originate from independent sources radiating on the same frequency channel, but rather arise from a single source due to multipath propagation. This gives rise to a resultant signal that is an additive mixture of amplitude-scaled, time-delayed, and possibly Doppler-shifted versions of the source waveform, which are typically incident from different directions of arrival (DOAs). In real-world environments, multipath propagation from source to receiver often occurs due to diffuse scattering from spatially extended regions of an irregular medium as opposed to ideal specular reflection. This scenario is often encountered in mobile communications, underwater acoustics, and radar systems, for example.

In particular, HF skywave signals usually consist of a relatively small number of dominant multipath components or "modes" that propagate from source to receiver along distinct paths. Each signal mode is in turn composed of many locally scattered "rays" that are clustered in some manner about the nominal mode propagation path. This phenomenon, often referred to as "micro-multipath," gives each propagating mode its fine structure, and is a feature of so-called doubly-spread channels. The large-scale delay, Doppler, and DOA spread of the channel is due to the well-separated nominal paths of the dominant modes, while on a smaller scale, the spread is also due to the continuum of diffusely scattered rays distributed about each of these nominal paths.

Mutual interference among the different propagation modes may cause significant frequency-selective fading or signal envelope distortions at a single receiver output. This can significantly degrade or even impair the performance of systems that rely on accurate source-waveform estimation. The ability to separate the individual modes by spatial filtering in a narrowband single-input multiple-output (SIMO) system not only isolates the output of one or more useable paths for high fidelity waveform estimation, but may also enable the various signal modes to be combined constructively to benefit from the additional energy that each path provides. The waveform estimation problem may be generalized to multiple sources by considering a multiple-input multiple-output (MIMO) system. In this case, it is necessary to separate the different sources as well as the multipath components of each source for effective waveform estimation.

However, neither the propagation channel characteristics nor the signal properties may be known in practice. Moreover, parametric models that can accurately describe the received signal wavefronts may not be available due to environmental and instrumental uncertainties, including diffuse multipath scattering and array calibration errors, for example. The lack of a priori knowledge regarding both the source signal(s) and propagation channel(s) poses a major challenge for the task at hand. This situation calls for processing techniques that can recover the source waveforms from the received signal mixture in a strictly blind manner.

This chapter discusses blind waveform estimation techniques based on relatively mild assumptions regarding the source signal(s), propagation channel(s), and sensor array. Specifically, a new technique referred to as the Generalized Estimation of Multipath Signals (GEMS) algorithm is introduced. The ability of GEMS to estimate arbitrarily modulated source waveforms in narrowband finite-impulse-response (FIR) SIMO and MIMO systems is experimentally demonstrated and compared against benchmark approaches.

The first section formulates the problem by describing the data model, processing objectives, and main assumptions. The second section explains the relationship between existing blind signal-processing approaches and the specific problem considered to provide motivation for the GEMS technique. The third section introduces the GEMS algorithm and compares its computational complexity with a benchmark approach. The remaining sections present experimental results to illustrate the potential applications of GEMS for blind source waveform and propagation channel estimation in practical HF systems.

13.1 Problem Formulation

Figure 13.1 conceptually illustrates the problem considered. On the left, a number of independent sources are assumed to emit narrowband waveforms with overlapping power spectral densities. These waveforms propagate via different multipath channels before being received by a sensor array in the far-field. Multipath is due to a relatively small number of dominant signal modes, each being composed of possibly a large number of diffusely scattered rays that superimpose to produce distorted (non-planar) wavefronts with path-dependent "crinkles."

It is assumed that the sensor array is connected to a multi-channel digital receiver that samples the incident signal mixture and additive noise in space and time. The objective of the processor is to recover a clean copy of each transmitted source waveform. Ideally, each waveform estimate is as free as possible of contamination from multipath, other signals, and noise.

The first part of this section describes the physical significance of this problem to HF systems that receive signals via skywave propagation, and develops a mathematical model for the space-time samples acquired by the receiver array as inputs to the processor. The data model is described in relatively general terms, and may also be appropriate in other applications not restricted to HF systems or electromagnetic signals. The second

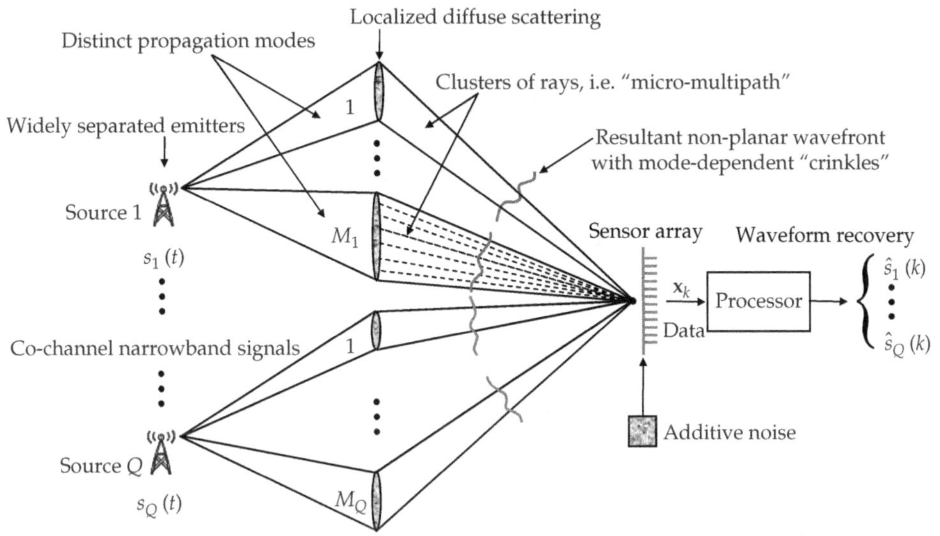

FIGURE 13.1 Conceptual illustration of the blind waveform estimation problem. The illustration depicts a finite impulse response (FIR) multiple-input multiple-output (MIMO) system, which is the most general case considered. The FIR single-input multiple-output (SIMO) system, where there is only one source and multiple echoes is also of interest in a number of practical applications. © Commonwealth of Australia 2011.

part of this section defines the two main tasks of the processor, namely, source waveform recovery and channel parameter estimation. The main assumptions are also summarized for convenience. The final part of this section provides a simple motivating example to show that wavefront distortions caused by diffuse multipath scattering can be exploited to spatially resolve signal modes with closely spaced nominal DOAs.

13.1.1 Multipath Model

Figure 13.2 illustrates the reception of HF signals from distant sources on a ground-based antenna array via reflection from the ionosphere. Multipath arises due to a number of different "layers" or regions in the ionosphere that propagate the HF signal from source to receiver along well-separated (distinct) paths. However, each ionospheric layer does not act as a perfectly smooth specular reflection surface to the incident signal, but rather presents a spatially extended scattering region that transforms the far-field point source into a distributed signal mode at the receiver. The experimental analysis in Part II of this text confirms this characteristic of the ionospheric reflection process for individual HF signal modes.

Specifically, Figure 13.2 depicts two localized scattering regions at E and F layer heights in the ionosphere. For a particular HF source, each localized scattering region gives rise to an individual signal mode. The physical dimensions of these regions effectively depends on the "roughness" of the isoionic contours that scatter the source signal into the receiver. In practice, there is usually a relatively small number of dominant signal modes that are received from physically separated scattering regions. However, the sum of two or more

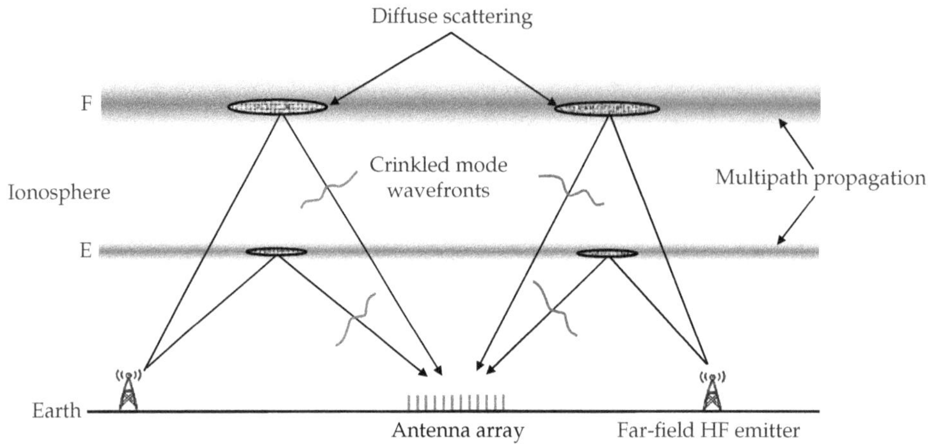

FIGURE 13.2 Notional diagram showing the reception of HF signals via multiple skywave paths. The dominant modes arise from a relatively small number of localized scattering volumes in the E and F regions of the ionosphere. In practice, a layer may have one or more different scattering regions due to the presence of irregularities/disturbances, low- and high-angle modes, as well as magneto-ionic splitting, which produces ordinary and extraordinary waves. This simple diagram illustrates two dominant modes diffusely scattered by the E and F layers, for each source. © Commonwealth of Australia 2011.

propagation modes with comparable strength, but different time delays, Doppler shifts, and angles of arrival, can lead to significant frequency-selective fading of the signal received by the system.

The cone of diffusely scattered rays that emanate toward the receiver from each localized scattering region may combine coherently or incoherently with each other. The former case gives rise to a signal mode with a time-invariant crinkled wavefront that carries an amplitude-scaled, time-delayed, and possibly Doppler-shifted copy of the source waveform. In this case, the aim of signal processing at the receiver is to isolate a single ionospheric mode per source, as this recovers a suitable waveform estimate. For example, a spatial filter may be applied to preserve a selected mode from a certain source, while using spare degrees of freedom to null or attenuate other signals received by the array.

In a dynamic propagation medium, the magnitude and phase relationship between the rays diffusely scattered from a localized region will change over time. When the time scale of such changes is long compared to the observation interval, the signal mode impinges on the array as a crinkled wavefront that exhibits an effectively "frozen" spatial structure (approaching the coherently distributed case). However, if the time scale of such changes is short compared to the observation interval, the result is an incoherently distributed signal mode that is characterized by a time-varying crinkled wavefront.

The data model assumed for a multipath signal received by a sensor array will be described in four steps. The first derives a relatively general expression for the analytic continuous-time signal received from a single source. The second incorporates coherently distributed (CD) and incoherently distributed (ID) ray descriptions into this expression to derive a CD and ID signal model. The third converts the continuous-time model to a discrete-time model, which represents the digital samples input to the signal processor.

The fourth generalizes this data model to the case of multiple sources. Although the end result is a signal-processing model that can be expressed in a relatively familiar mathematical form, several important steps in the derivation are included to highlight the underlying assumptions required for such a model to be valid in practice.

13.1.1.1 Received Signal

Let the scalar signal $g(t)$ in Eqn. (13.1) be the analytic representation of the narrowband waveform emitted by a source of interest. Here, f_c is the carrier frequency, and $s(t)$ is a baseband complex envelope with effective bandwidth B. The narrowband assumption implies a small fractional bandwidth $B/f_c \ll 1$. An alternative definition of narrowband relevant to the development of the model is specified later. Attention is restricted to the single-source case first; the extension to multiple sources is considered later.

$$g(t) = s(t)e^{j2\pi f_c t} \tag{13.1}$$

Define $h_n(t, \tau)$ as the time-varying FIR function of the channel that links the source to receiving element $n = 1, \ldots, N$ of the sensor array at time $t \in [0, T_o)$, where T_o is the observation interval. Note that the variable t denotes continuous time, while τ is the delay variable of the impulse response. For example, the impulse response at time t_0 is $h_n(t_0, \tau)$ for $\tau \in [0, T_n)$, where T_n is the impulse response duration of channel n. The vector $\mathbf{h}(t, \tau) \in \mathcal{C}^N$ in Eqn. (13.2) represents the multi-channel FIR system function with support over the delay interval $\tau \in [0, T_c)$, where $T_c = \max\{T_n\}_{n=1}^N$ is the maximum impulse response duration over all N channels.

$$\mathbf{h}(t, \tau) = [h_1(t, \tau), \ldots, h_N(t, \tau)]^T \tag{13.2}$$

The set of complex scalar signals $\{x_n(t)\}_{n=1}^N$ received by the N elements of the sensor array may be assembled into a spatial snapshot vector $\mathbf{x}(t) = [x_1(t), \ldots, x_N(t)]^T$. In Eqn. (13.3), this vector is defined as the convolution of the source signal with the multi-channel impulse response function, plus measurement noise $\mathbf{n}(t) \in \mathcal{C}^N$. The blind waveform estimation problem assumes that the observations $\mathbf{x}(t)$ are accessible, but the system function $\mathbf{h}(t, \tau)$, source waveform $g(t)$, and additive noise $\mathbf{n}(t)$ are not.

$$\mathbf{x}(t) = \int_0^{T_c} \mathbf{h}(t, \tau)\, g(t - \tau)\; d\tau + \mathbf{n}(t) \tag{13.3}$$

Figure 13.3 illustrates the nominal propagation path and cone of diffusely scattered rays for a single dominant mode referred to by the index m. The meaning of narrowband needs to be more carefully defined in two respects. First, it is assumed that the maximum separation between sensors in the array D_a is such that the time-bandwidth product condition in Eqn. (13.5) is satisfied, where c is the speed of light in free space. In other words, the elements of the sensor array present a coherent aperture to the rays in a signal mode. Each of these rays is incident as a plane-wave component.

$$BD_a/c \ll 1 \tag{13.4}$$

Second, it is assumed that diffuse scattering for all dominant modes $m = 1, \ldots, M$ occurs within a localized region such that the maximum time-dispersion of the rays Δ_m relative to the nominal time delay τ_m of the mode propagation path is much less than the reciprocal

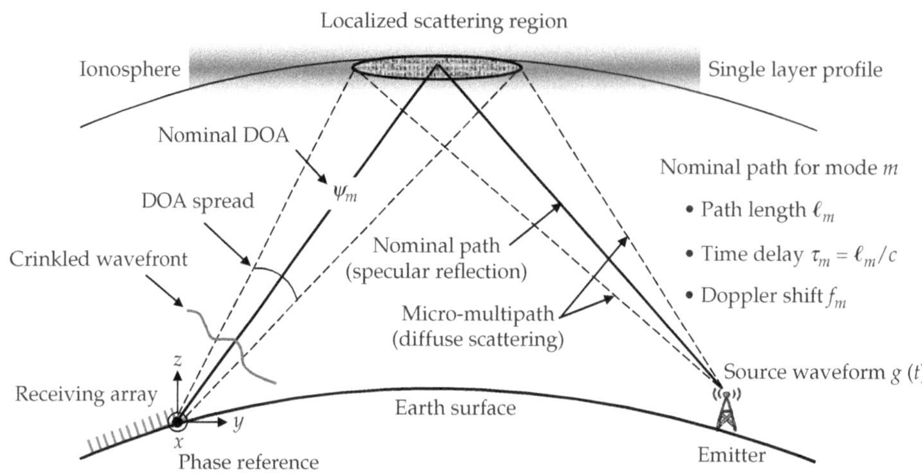

FIGURE 13.3 Schematic diagram showing the nominal path and cone of diffusely scattered rays for a single dominant propagation mode. The time delay, Doppler shift, and DOA of the nominal mode propagation path are indicated. A Doppler shift may arise due to the regular component of large-scale motion of the scattering region, and/or the movement of the source.

of the signal bandwidth B. In other words, it is assumed that the condition in Eqn. (13.5) is also satisfied.

$$B\Delta_m \ll 1 \tag{13.5}$$

Provided these conditions are met, $\mathbf{h}(t, \tau)$ may be written in the form of Eqn. (13.6), where $\delta(\cdot)$ is the Dirac delta function. Here, the impulse response is described as the sum of M dominant signal modes with distinct nominal time delays $\{\tau_m\}_{m=1}^{M}$, where M is assumed to be less than N. Each mode is transferred from source to receiver by its own time-varying channel vector $\mathbf{c}_m(t)$. This vector represents the instantaneous summation of a large number of rays diffusely scattered from a localized region. In other words, $\mathbf{c}_m(t)$ represents the time-varying crinkled wavefront of mode m.

$$\mathbf{h}(t, \tau) = \sum_{m=1}^{M} \delta(\tau - \tau_m)\mathbf{c}_m(t) \tag{13.6}$$

Substituting Eqn. (13.6) into Eqn. (13.3) yields the received signal model of Eqn. (13.7). In this model, within-mode ray interference gives rise to relatively "slow" flat-fading. The flat-fading process is embodied in the time-variation of the channel vector $\mathbf{c}_m(t)$. On the other hand, the summation of M modes with time-delay differences that may significantly exceed $1/B$ produces relatively "fast" frequency-selective fading. The latter has the potential to significantly distort the temporal signature of the source waveform. Hence, the main objective is to remove frequency-selective fading by isolating one of the dominant signal modes for waveform estimation at the processor output.

$$\mathbf{x}(t) = \sum_{m=1}^{M} g(t - \tau_m)\mathbf{c}_m(t) + \mathbf{n}(t) \tag{13.7}$$

For an infinite number (i.e., a continuum) of diffusely scattered rays, the channel vector may be expressed in the form of Eqn. (13.8). The complex scalar function $f_m(\psi, t)$ is the time-varying angular spectrum of mode m, where the DOA parameter vector $\psi = [\theta, \phi]$ includes azimuth θ and elevation ϕ. The vector $\mathbf{v}(\psi) \in \mathcal{C}^N$ denotes the plane-wave array steering vector. This vector represents the spatial response of the sensor array to a single ray that emanates from a far-field point and is incident as a plane wave with DOA ψ.

$$\mathbf{c}_m(t) = \int f_m(\psi, t)\ \mathbf{v}(\psi)\ \mathrm{d}\psi \tag{13.8}$$

In the presence of near-field scattering effects and/or array calibration errors, $f_m(\psi, t)$ may be interpreted as an equivalent angular spectrum which gives rise to the received channel vector $\mathbf{c}_m(t)$. However, in this case, $f_m(\psi, t)$ loses its physical meaning as the angular spectrum of the scattered rays. Stated another way, this function would then no longer represent the received complex amplitudes of the downcoming rays that impinge on the sensor array from different DOAs.

13.1.1.2 Localized Scattering

A coherently distributed (CD) signal arises due to a scattering process that is effectively "frozen" or deterministic over the observation interval, such that $f_m(\psi, t) \rightarrow f_m(\psi)$ for $t \in [0, T_o)$. A limitation of this description is that it does not capture Doppler frequency shifts, which are often significant in practice. Different mode Doppler shifts $\{f_m\}_{m=1}^{M}$ may be incorporated into the CD model by assuming the temporal variation of the ray angular spectrum is separable and can be modeled as in Eqn. (13.9).

$$f_m(\psi, t) = e^{j2\pi f_m t} f_m(\psi) \tag{13.9}$$

The presence of a Doppler shift captures the regular component of large-scale movement of the scattering region and source over the observation interval by introducing a linear phase-path variation common to all rays. From Eqn. (13.8), the channel vector $\mathbf{c}_m(t)$ may then be expressed in terms of a time-invariant mode wavefront $\mathbf{a}_m$, and nominal mode Doppler shift f_m in Eqn. (13.10). The vector $\mathbf{a}_m$ may be interpreted as a crinkled wavefront that does not lie on the plane-wave array manifold in general.

$$\mathbf{c}_m(t) = e^{j2\pi f_m t}\mathbf{a}_m, \quad \mathbf{a}_m = \int f_m(\psi)\mathbf{v}(\psi)\ \mathrm{d}\psi \tag{13.10}$$

The mode wavefront may alternatively be written as $\mathbf{a}_m = \alpha_m \mathbf{v}_m$, where α_m is a complex amplitude, and $\mathbf{v}_m \in \mathcal{C}^N$ is a spatial signature vector with fixed L_2-norm $\|\mathbf{v}_m\|^2 = N$. The spatial signature may in turn be expressed as the Hadamard (element-wise) product in Eqn. (13.11), where $\mathbf{d}_m \in \mathcal{C}^N$ is a multiplicative distortion vector that modulates an underlying plane wavefront $\mathbf{v}(\psi_m)$ parameterized by the nominal DOA $\psi_m = [\theta_m, \phi_m]$ of mode m.

$$\mathbf{a}_m = \alpha_m \mathbf{v}(\psi_m) \odot \mathbf{d}_m \tag{13.11}$$

Substituting Eqn. (13.10) and Eqn. (13.11) into Eqn. (13.7) yields the CD diffuse multipath model in Eqn. (13.12), which has been extended to include mode Doppler shifts. The assumption of linearly independent mode wavefronts $\{\mathbf{a}_m\}_{m=1}^{M}$ is often justified for $M < N$

dominant modes due to the different diffuse scattering processes involved, as well as the differences in the nominal mode DOAs.

$$\mathbf{x}(t) = \sum_{m=1}^{M} \alpha_m g(t - \tau_m) e^{j2\pi f_m t} \mathbf{v}(\psi_m) \odot \mathbf{d}_m + \mathbf{n}(t) \tag{13.12}$$

The incoherently distributed (ID) signal model gives rise to channel vector variations that are not separable in space and time. After the nominal mode Doppler shift is factored out, this leaves a time-varying mode wavefront, denoted by $\mathbf{a}_m(t)$. If the channel is assumed Gaussian and wide-sense stationary over the observation interval, for example, the mode wavefronts may be statistically described by a mean vector $\mathbf{a}_m$ and covariance matrix $\mathbf{R}_m$, as in Eqn. (13.13). Recall that $\sim \mathcal{CN}$ denotes the complex normal distribution.

$$\mathbf{a}_m(t) \sim \mathcal{CN}(\mathbf{a}_m, \mathbf{R}_m) \tag{13.13}$$

Although $\mathbf{R}_m$ may have full rank N, most of the energy in the wavefront fluctuations is typically contained in a small number of $\mathcal{I}_m < N$ eigenvalues. Defining the *effective* subspace as $\mathbf{Q}_m \in \mathcal{C}^{N \times \mathcal{I}_m}$, where the columns of $\mathbf{Q}_m$ are the $\mathcal{I}_m$ dominant eigenvectors of $\mathbf{R}_m$, the dynamic component of the mode wavefront is well-approximated by $\mathbf{Q}_m \underline{\varsigma}_m(t)$, where $\underline{\varsigma}_m(t) \in \mathcal{C}^{\mathcal{I}_m}$ is a time-varying coordinate vector. In this case, the channel vector takes the form in Eqn. (13.14).

$$\mathbf{c}_m(t) = e^{j2\pi f_m t} \{\mathbf{a}_m + \mathbf{Q}_m \underline{\varsigma}_m(t)\} \tag{13.14}$$

This description incorporates purely statistical ID signals with zero-mean (i.e., $\mathbf{a}_m = \mathbf{0}$), and partially correlated distributed (PCD) signals, where $\underline{\varsigma}_m(t)$ changes smoothly over time as correlated (dependent) realizations. In the following, there is no requirement to invoke a specific model for $\mathbf{a}_m$ and $\mathbf{Q}_m$, such as the Gaussian model alluded to previously. The main assumption is that $\mathbf{a}_m$ is linearly independent of the columns of $\mathbf{Q}_m$. As stated earlier, linear independence is also required amongst the different modes. The ID signal model of $\mathbf{x}(t)$ is given by Eqn. (13.15).

$$\mathbf{x}(t) = \sum_{m=1}^{M} g(t - \tau_m)\, e^{j2\pi f_m t} \{\mathbf{a}_m + \mathbf{Q}_m \underline{\varsigma}_m(t)\} + \mathbf{n}(t) \tag{13.15}$$

In this chapter, particular emphasis is on waveform estimation using the CD model that was extended to include Doppler shifts in Eqn. (13.12). The additive noise $\mathbf{n}(t)$ may be of ambient or thermal origin, and is notionally considered to be spatially and temporally white. However, arguments are made later to justify the robustness of the developed approach for signals described by the ID model of Eqn. (13.15), and additive noise that is structured or "colored."

13.1.1.3 Acquired Data

After down-conversion and baseband filtering of the received signal, the in-phase and quadrature (I/Q) outputs are uniformly sampled at time instants $\{t = kT_s\}_{k=1}^{K}$. From Eqn. (13.1), it is straightforward to show that the samples of the source waveform are given by Eqn. (13.16). By ignoring the immaterial phase term $e^{-j2\pi f_c \tau_m}$, we may replace the continuous time signal $g(t)$ by the sampled baseband sequence $s(kT_s)$. For the single-source case, the contracted notation $s_k = s(kT_s)$ will be used.

$$\left\{g(t - \tau_m)e^{-j2\pi f_c t}\right\}_{t=kT_s} = s(kT_s - \tau_m)e^{-j2\pi f_c \tau_m} \tag{13.16}$$

Based on the CD model in Eqn. (13.12), the array snapshots $\mathbf{x}_k$ are given by Eqn. (13.17), where $\nu_m = f_m/f_s$ is the Doppler shift normalized by the sampling frequency ($f_s = 1/T_s$), $\ell_m = \tau_m/T_s$ is the time delay normalized by the sampling period, $\mathbf{a}_m = \alpha_m \mathbf{v}(\psi_m) \odot \mathbf{d}_m$ is the received mode wavefront, and $\mathbf{n}_k$ is additive noise. Provided the time-bandwidth product BT_s is smaller than unity (i.e., the signal is not under-sampled), τ_m is not required to coincide exactly with a time delay bin such that ℓ_m is an integer). This case is adopted only to simplify the description of the model.

$$\mathbf{x}_k = \sum_{m=1}^{M} s_{k-\ell_m}\, e^{j2\pi\nu_m k}\, \mathbf{a}_m + \mathbf{n}_k \tag{13.17}$$

The spatial snapshots $\mathbf{x}_k$ may be described by the familiar array signal processing model in Eqn. (13.18), where the columns of the *multipath* mixing matrix contain the M mode wavefronts $\mathbf{A} = [\mathbf{a}_1, \ldots, \mathbf{a}_M]$, and the M-variate *multipath* signal vector $\mathbf{s}_k = [s_{k-\ell_1} e^{j2\pi\nu_1 k}, \ldots, s_{k-\ell_M} e^{j2\pi\nu_M k}]^T$ contains the source waveform propagated by the different modes. Note that the definition of this vector incorporates the nominal mode time delays and Doppler shifts.

$$\mathbf{x}_k = \mathbf{A}\, \mathbf{s}_k + \mathbf{n}_k \tag{13.18}$$

The overall data set consisting of K array snapshots acquired over a processing interval of T_o seconds may be represented using the matrix notation in Eqn. (13.19), where the $N \times K$ data matrix is $\mathbf{X} = [\mathbf{x}_1, \ldots, \mathbf{x}_K]$, the $M \times K$ signal matrix is $\mathbf{S} = [\mathbf{s}_1, \ldots, \mathbf{s}_K]$, and $\mathbf{N} = [\mathbf{n}_1, \ldots, \mathbf{n}_K]$ is the noise matrix.

$$\mathbf{X} = \mathbf{A}\,\mathbf{S} + \mathbf{N} \tag{13.19}$$

By defining $\mathbf{h}(k, \ell) \in \mathcal{C}^N$ as the discrete-time multi-channel impulse response function, and $L = \lceil T_c/T_s \rceil$ as the FIR model order determined by the maximum duration of the channel $T_c = \max\{\tau_m\}_{m=1}^{M}$, the array snapshots may be represented in the alternative form of Eqn. (13.20).

$$\mathbf{x}_k = \sum_{\ell=0}^{L} \mathbf{h}(k, \ell)\, s_{k-\ell} + \mathbf{n}_k \tag{13.20}$$

It is readily shown that $\mathbf{h}(k, \ell) = [h_1(k, \ell), \ldots, h_N(k, \ell)]^T$ is given by Eqn. (13.21), where by analogy with the time-continuous FIR channel function $h_n(t, \tau)$, the complex scalar $h_n(k, \ell)$ denotes the impulse response that links the source to receiving element n at time k with relative delay ℓ. This model is representative of an FIR-SIMO system with *time-varying* channel coefficients.

$$\mathbf{h}(k, \ell) = \sum_{m=1}^{M} \delta(\ell - \ell_m)\mathbf{a}_m e^{j2\pi\nu_m k} \tag{13.21}$$

For a total number of samples K, a time-invariant FIR-SIMO model arises only for an observation interval $T_o = KT_s$ that is sufficiently short to negate the effect of mode Doppler shifts. In other words, the condition $\nu_m K \ll 1$ is required for all modes m, such that $\mathbf{h}(k, \ell) \rightarrow \mathbf{h}(\ell)$. In this case, the time-invariant multi-channel FIR function is given by Eqn. (13.22).

$$\mathbf{h}(\ell) = \sum_{m=1}^{M} \delta(\ell - \ell_m)\mathbf{a}_m \tag{13.22}$$

Use of the CD model extended by Doppler shifts in Eqn. (13.18) suffices for describing the key elements of the waveform estimation problem in the following parts of this section. This model has been presented in the alternative mathematical forms of Eqns. (13.18) and (13.20) to explain its relationship to blind system identification (BSI) and blind signal separation (BSS) techniques described in Section 13.2. The ID version of this model will be considered in Section 13.3, where the GEMS algorithm is described.

13.1.1.4 Multiple Sources

In the case of Q independent sources, the FIR-MIMO system data model of Eqn. (13.23) is a straightforward generalization of the FIR-SIMO model presented in Eqn. (13.3). The Q co-channel sources are assumed to emit different narrowband waveforms $g_q(t)$ for $q = 1, \ldots, Q$. The multi-sensor channel impulse response function for source q is $\mathbf{h}_q(t, \tau) = [h_{q1}(t, \tau), \ldots, h_{qN}(t, \tau)]^T$ with maximum time duration T_q. The sources are assumed to be widely separated, such that the FIR channels $h_{qn}(t, \tau)$ for all sources q and receivers n are sufficiently diverse to be identifiable. More will be said on the topic of identifiability in Section 13.2.

$$\mathbf{x}(t) = \sum_{q=1}^{Q} \int_0^{T_q} \mathbf{h}_q(t, \tau)\, g_q(t - \tau)\ \mathrm{d}\tau + \mathbf{n}(t) \tag{13.23}$$

The same steps taken previously for the single-source case lead us to the discrete-time model of the received array data $\mathbf{x}_k \in \mathcal{C}^N$ in Eqn. (13.24). The scalar waveform $s_q(k)$ to be recovered is the baseband-sampled version of $g_q(t)$, where the sample index k is left inside the brackets for the multiple-source case to avoid confusing notation. The number of dominant modes propagated along distinct paths for source q is denoted by M_q. The terms ℓ_{mq}, ν_{mq}, and $\mathbf{a}_{mq}$ are the mode time delays, Doppler shifts, and crinkled wavefronts, respectively. The additive noise $\mathbf{n}_k$ is assumed to be uncorrelated with all sources.

$$\mathbf{x}_k = \sum_{q=1}^{Q} \sum_{m=1}^{M_Q} s_q(k - \ell_{mq})\, e^{j2\pi \nu_{mq} k}\ \mathbf{a}_{mq} + \mathbf{n}_k \tag{13.24}$$

In the adopted FIR-MIMO model, the total number of signal components is $R = \sum_{q=1}^{Q} M_q$. It is further assumed that multipath is present for all sources, i.e., $M_q > 1$. With reference to Eqn. (13.18), the array snapshots $\mathbf{x}_k$ resulting for Q sources may be expressed as in Eqn. (13.25). Here, $\mathbf{A}_q \in \mathcal{C}^{N \times M_q}$ and $\mathbf{s}_q(k) \in \mathcal{C}^{M_q}$ are the multipath mixing matrix and multipath signal vector for source q, respectively, defined analogously to the single-source case.

$$\mathbf{x}_k = \sum_{q=1}^{Q} \mathbf{A}_q \mathbf{s}_q(k) + \mathbf{n}_k \tag{13.25}$$

The data may be expressed as in Eqn. (13.26), where the $N \times R$ matrix $\mathbf{H} = [\mathbf{A}_1, \ldots, \mathbf{A}_Q]$ contains the wavefronts of all R signal components, while the augmented (source and multipath) signal vector $\mathbf{p}_k \in \mathcal{C}^R$ is a stacked vector of $\{\mathbf{s}_1(k), \ldots, \mathbf{s}_Q(k)\}$. Recall that the multipath signal vector $\mathbf{s}_q(k)$ incorporates the time delays $\{\ell_{mq}\}$ and Doppler shifts $\{\nu_{mq}\}$ for all modes $m = 1, \ldots, M_q$ of source q.

$$\mathbf{x}_k = \mathbf{H}\mathbf{p}_k + \mathbf{n}_k \tag{13.26}$$

In this case, the totality of the data acquired during the processing interval may be represented in the matrix form of Eqn. (13.27), where the $R \times K$ augmented signal matrix $\mathbf{P} = [\mathbf{p}_1, \ldots, \mathbf{p}_K]$. This multiple-source model will not be considered in the following parts of this section, but will be used in Sections 13.2 and 13.3.

$$\mathbf{X} = \mathbf{H}\mathbf{P} + \mathbf{N} \tag{13.27}$$

In anticipation of the discussion in Section 13.2, the FIR-MIMO model is also presented here in an alternative mathematical form. By defining $\mathbf{h}_q(k, \ell) \in \mathcal{C}^N$ as the discrete-time multi-channel impulse response function for source q, and $L_q = \lceil T_q/T_s \rceil$ as the associated FIR model determined by $T_q = \max\{\tau_{mq}\}_{m=1}^{M_q}$, the array snapshots may be represented by the convolutive mixture model in Eqn. (13.28).

$$\mathbf{x}_k = \sum_{q=1}^{Q} \sum_{\ell=0}^{L_q} \mathbf{h}_q(k, \ell)\, s_q(k - \ell) + \mathbf{n}_k \tag{13.28}$$

It is readily shown that $\mathbf{h}_q(k, \ell) = [h_{q1}(k, \ell), \ldots, h_{qN}(k, \ell)]^T$ is given by Eqn. (13.29), where by analogy with the time-continuous FIR channel function $h_{qn}(t, \tau)$, the complex scalar $h_{qn}(k, \ell)$ denotes the impulse response component that links source q to receiver n at time k with relative delay ℓ. This constitutes an FIR-MIMO system with *time-varying* channel coefficients.

$$\mathbf{h}_q(k, \ell) = \sum_{m=1}^{M_q} \delta(\ell - \ell_{mq})\mathbf{a}_{mq} e^{j2\pi \nu_{mq} k} \tag{13.29}$$

The time-invariant FIR-MIMO model arises for observation intervals that are sufficiently short to negate the effect of the largest signal Doppler shift over all sources and modes. In other words, the condition $K\nu_{mq} \ll 1$ needs to be met for all (m, q), such that $\mathbf{h}_q(k, \ell) \rightarrow \mathbf{h}_q(\ell)$ given by Eqn. (13.30).

$$\mathbf{h}_q(\ell) = \sum_{m=1}^{M_q} \delta(\ell - \ell_{mq})\mathbf{a}_{mq} \tag{13.30}$$

13.1.2 Processing Objectives

In the blind signal-processing problem considered, the objective of the processor is to estimate the source input sequences and propagation channel parameters from noisy measurements of the received signals. In general, the sensor array receives signals from a number of different sources where there are multiple propagation paths per source. To maintain simplicity in the first instance, the processing objectives are described for the case of a single source and multiple echoes (modes) in this section. The more general case involving multiple sources will be dealt with in subsequent sections of this chapter.

Depending on the type of system in question, the primary objective of the processor is often either to estimate the source waveform or channel parameters, depending which of these unknowns is of more interest. However, once either has been estimated through the use of blind signal processing, it is usually straightforward to estimate the other in a subsequent (non-blind) processing step. The signal and channel estimation problems addressed in this chapter are described below. Several practical applications of blind signal and channel estimation not limited to radar are also described. Finally, the main assumptions related to the data model and processing task are summarized to complete the problem formulation.

13.1.2.1 Waveform Estimation

The waveform estimation task is to obtain a clean copy of the signal emitted from the source of interest, where this may be an amplitude-scaled, time-delayed, and possibly Doppler-shifted version of the transmitted baseband modulation envelope. Recovering the source waveform may be regarded as a complementary problem to those of signal detection and source localization. From an array processing perspective, the main aim of spatial filtering is to pass the dominant propagation mode for waveform estimation, and to reject all other interfering multipath signals based on differences in received wavefront structure. According to the signal model of Eqn. (13.18), waveform estimation is tantamount to a multipath separation problem for the single-source case.

Knowledge of the source waveform may be useful for several different reasons depending on the relationship between the signal of interest and system function. Examples of practical systems that can benefit from high-fidelity waveform estimation are mentioned below. Although the significance of the signal to the system is quite different in each case, a common thread is the underlying requirement for an accurate estimate of the source waveform. Fundamental techniques to address this problem therefore have a variety of practical applications, not limited to those described here.

- In communication systems, the primary interest is to extract the information that is encoded in the modulation of the transmitted signal. Minimizing frequency-selective fading caused by multipath at the receiver can reduce signal envelope distortions and significantly improve link performance.
- In passive radar systems, the source waveform is used as a reference signal for matched filtering. The data carried by a signal of opportunity is not of direct interest here, but a clean copy of the source waveform is required to effectively detect and localize echoes from man-made targets.
- In active radar systems, a co-channel source may represent an unwanted signal that can potentially mask useful signals. Knowledge of the source waveform can facilitate the mitigation of such interference, particularly when it is received through the main lobe of the antenna beam pattern.

The clairvoyant[1] signal-copy weight vector $\mathbf{w}_m \in \mathcal{C}^N$ that perfectly isolates mode m from all other modes at the spatial processor output is given by the well-known *minimum-norm* solution in Eqn. (13.31). Recall that $\mathbf{A}$ is the multipath mixing matrix defined previously. The vector $\mathbf{u}_m \in \mathcal{C}^M$ has unity in position m and zeros elsewhere,

[1]The term "clairvoyant" implies that the signal mixing matrix $\mathbf{A}$ is known a priori.

while β is an arbitrary complex scalar that does not affect output signal-to-noise ratio (SNR). The symbol $\dagger$ denotes the Hermitian (conjugate-transpose) operator.

$$\mathbf{w}_m = \beta\mathbf{A}(\mathbf{A}^\dagger\mathbf{A})^{-1}\mathbf{u}_m \tag{13.31}$$

From the model in Eqn. (13.18), the clairvoyant signal-copy weights $\mathbf{w}_m$ yield the output z_k in Eqn. (13.32), where $\beta^* s_{k-\ell_m} e^{j2\pi\nu_m k}$ is the desired estimate of the source waveform, and $n_{km} = \mathbf{w}_m^\dagger \mathbf{n}_k$ is a residual noise contribution. This deterministic "null-steering" spatial filter estimates a copy of the source signal that is free of multipath contamination. Among all the linear combiners that produce a multipath-free estimate, $\mathbf{w}_m$ maximizes the output SNR in the case of spatially white additive noise.

$$z_k = \mathbf{w}_m^\dagger \mathbf{z}_k = \beta^* s_{k-\ell_m} e^{j2\pi\nu_m k} + n_{km} \tag{13.32}$$

The optimum filter maximizing the output signal-to-interference-plus-noise ratio (SINR) is given by Eqn. (13.33), where $\mathbf{Q}_m$ is the statistically expected spatial covariance matrix of all the unwanted signal modes plus noise. The optimum filter is in general different to $\mathbf{w}_m$, but tends to the expression in Eqn. (13.31) when the interfering modes are not coherent and much more powerful than the additive noise.

$$\mathbf{w}_m^{\text{opt}} = \beta\mathbf{Q}_m^{-1}\mathbf{a}_m \tag{13.33}$$

In the spatial processing context, the objective of blind source waveform recovery is to estimate the weights $\mathbf{w}_m$ for a multipath-free output, or $\mathbf{w}_m^{\text{opt}}$ for an output with minimum mean-square error, depending on which criterion is deemed most desirable. Application of the weight vector to the received array snapshots yields an amplitude-scaled, time-delayed, and possibly Doppler-shifted estimate of the source waveform at the spatial filter output, which fulfils the processing objective.

The key point is that the mixing matrix $\mathbf{A} = [\mathbf{a}_1, \ldots, \mathbf{a}_M]$ is unknown a priori, and supervised training to estimate $\mathbf{Q}_m$ is not possible in the considered problem. The model order M and wavefronts $\mathbf{a}_m$ can be estimated under certain conditions, as described in Section 13.2. However, errors in the reconstruction of $\mathbf{A}$ due to estimation uncertainty or model-mismatch will lead to the output being corrupted by residuals of the interfering modes, which can significantly reduce the SINR, and hence quality of the waveform estimate.

13.1.2.2 Channel Estimation

The channel impulse response or system function is of more direct interest in certain applications than the source input sequence. In the model of Eqn. (13.20), the channel parameters are the number of modes M, the nominal delay, and Doppler shift of each mode $\{\tau_m, f_m\}$, and the mode wavefronts $\mathbf{a}_m = \alpha_m \mathbf{v}(\psi_m) \odot \mathbf{d}_m$ from which the nominal mode DOAs ψ_m may be inferred.

In practice, it is not always possible to determine the absolute (as opposed to relative) values of certain channel parameters. This is due to the ambiguity in attributing absolute values uniquely to the channel or source. In addition to the obvious ambiguity in complex scale α_m, the absolute time delays τ_m and Doppler shifts f_m of the channel are unobservable from $\mathbf{x}_k$ without further information about the source. Hence, the objective of channel parameter estimation is to estimate M and the relative mode complex-scales, time-delays, and Doppler-shifts, in addition to the wavefront structure and nominal DOA of each mode.

Knowledge of the propagation channel parameters can have a number of practical uses. In communication systems, traditional methods for multipath channel equalization require the transmission of training symbols or pilot sequences prior to the data frame. This enables the channel parameters to be estimated so that compensation for multipath can be applied to the information-carrying signals. However, training signals may be not be available in some cases, for instance when the source is uncooperative or occurs due to natural phenomena. This has motivated the development of blind system identification techniques that aim to estimate the channel parameters without a requirement for training sequences. Such techniques will be discussed in Section 13.2.

In the HF band, information about the nominal mode DOAs ψ_m and relative time delays τ_m may be used in conjunction with an ionospheric model to estimate the position of an uncooperative source. Geolocation of unknown HF sources from a single site is a problem of interest to the HF direction-finding community. The inverse problem of using uncooperative HF sources at known locations as reference points to estimate ionospheric reflection heights, and possibly tilts, is of significant interest for coordinate registration in OTH radar. Although the emphasis in this chapter is to address the blind source waveform recovery problem, blind estimation of the propagation channel parameters will be illustrated for an HF single-site location (SSL) application in Section 13.6.

13.1.2.3 Main Assumptions

The main assumptions related to the model of Eqn. (13.18) are summarized here to complete the problem formulation. With reference to Eqn. (13.18), these assumptions pertain to the source waveform s_k, the mixing matrix $\mathbf{A}$, the nominal mode delays τ_m, and Doppler shifts f_m. The generalization of these assumptions for the multiple-source case will be described in Section 13.3.

1. *Source Complexity:* The narrowband assumption described earlier is a necessary but not sufficient condition as far as the source waveform is concerned. In order to ensure the blind signal estimation problem is identifiable, the source waveform is required to have a finite bandwidth, such that s_k is not a constant or sinusoid, for example. Specifically, the input sequence is required to have a linear complexity $P > 2L$, where L is the maximum FIR model order over all N channels. Linear complexity of a finite-length deterministic sequence $\{s_k\}_{k=1}^{K}$ is defined as the smallest integer P for which there exist coefficients $\{\lambda_p\}_{p=1}^{P}$ that satisfy Eqn. (13.34) for all $k = 1, \ldots, K$.

$$s_k = -\sum_{p=1}^{P} \lambda_p s_{k-p} \tag{13.34}$$

As virtually all finite bandwidth signals of interest satisfy this linear complexity condition for realistic HF channels, the source waveform may be considered to have a practically arbitrary temporal signature. Importantly, no further information is assumed regarding the deterministic structure or statistical properties of the source waveform, which is therefore not restricted to having any particular modulation format. Clearly, the source is assumed to be uncooperative in the sense that training symbols or pilot signals are not available for estimation purposes.

2. *Channel Diversity:* The source waveform is assumed to be received as an *unknown* number M of diffusely scattered dominant modes, where $1 < M < N$. In other words, multipath propagation is assumed to exist, but the number of dominant modes is less than the number of receiving elements. In addition, the mode wavefront vectors $\mathbf{a}_m \in \mathcal{C}^N$

for $m = 1, \ldots, M$ are assumed to form a linearly independent set, such that the multipath mixing matrix has full rank. These requirements are expressed in Eqn. (13.35), where the operator $\mathcal{R}\{\cdot\}$ returns the rank of a matrix.

$$\mathcal{R}\{\mathbf{A}\} = M\,, \quad 1 < M < N \tag{13.35}$$

Apart from satisfying these conditions, which ensure the system is identifiable and that the problem is not ill-posed, no other information is assumed about the mixing matrix. This implies that $\{\mathbf{a}_1, \ldots, \mathbf{a}_M\}$ are otherwise arbitrary vectors, which are not confined to lie on or close to a parametrically defined spatial signature manifold, such as the plane-wave steering vector model. It follows that the sensor array is not restricted to a particular geometry, and that array *manifold* uncertainties due to nonidentical element gain and phase responses, sensor position errors, and mutual coupling can be tolerated.

3. *Sample Support:* This assumption relates to the acquisition of sufficient data in the processing interval, such that enough samples exist to determine the number of system unknowns. From Abed-Meraim, Qui, and Hua (1997), the number of samples K required to ensure identifiability needs to satisfy the condition in Eqn. (13.36), where L is the maximum channel impulse response duration defined previously. More will be said on the subject of identifiability in Section 13.2.

$$K > 3L\,, \quad L = \max\{\ell_m\}_{m=1}^{M} \tag{13.36}$$

4. *Distinct Modes:* Besides the quite mild conditions assumed for $\mathbf{A}$, s_k, and K above, the number of modes M and the associated delay-Doppler parameters $\{\ell_m, \nu_m\}_{m=1}^{M}$ are also assumed to be unknown. The differences between the mode parameter tuples $\underline{\rho}_m = [\ell_m, \nu_m]$ are assumed to be distinct, as in Eqn. (13.37). Note that this condition applies to the relatively small number of dominant signal modes and not the diffusely scattered rays within each mode.

$$\{\rho_i - \rho_i\}_{i,j=1}^{M} \neq \{\rho_n - \rho_m\}_{n,m=1}^{M} \quad \text{for } \{i, j\} \neq \{n, m\} \tag{13.37}$$

Except for certain contrived scenarios, the differential delay *and* Doppler between two distinct modes $(i \neq j)$ will in general not be *identical* to that of another pair of modes $(n \neq m)$ when attention is restricted to the few dominant modes.

13.1.3 Motivating Example

Diffusely scattered signals are present in a variety of fields. In wireless communications, they are produced by "local scattering" in the vicinity of mobile transmitters, particularly where there is no line-of-sight between the transmitter and receiving base-station, see Zetterberg and Ottersten (1995); Pedersen, Mogensen, Fleury, Frederiksen, Olesen, and Larsen (1997); Adachi, Feeney, Williamson, and Parsons (1986); and Ertel, Cardieri, Sowerby, Rappaport, and Reed (1998) for example.

In sonar, large hydrophone arrays are used to localize spatially distributed acoustic sources that have distorted wavefronts due to propagation in the heterogeneous underwater channel (Owsley 1985). Distributed signals with non-stationary wavefront amplitude and phase "abberations" are also observed in ultrasonics due to irregular propagation through tissue, where wavefronts received over different paths are noticed to experience different distortions; see Liu and Waag (1995, 1998) and Flax and O'Donnell (1988). A similar phenomenon is encountered in radio astronomy due to "scintillation" of signals as they pass through nonuniform plasma profiles (Yen 1985). Blind spatial processing techniques based on distributed multipath signal models may, therefore, find uses in diverse applications.

Before proceeding to a brief review of existing techniques for BSI and BSS, a simple example is used to illustrate how wavefront distortions may assist to separate signal modes with closely spaced nominal DOAs. This is a common problem due to the limited spatial resolution of practical sensor arrays, the fact that multipath components are often received from very similar directions, and that super-resolution techniques for DOA estimation are highly sensitive to model mismatch. A notorious example in microwave radar is the low-elevation multipath encountered over seawater between an airborne source and the receiving antenna array on a surface vessel.

The motivating example described below is relevant to HF radar and directly related to a real-world scenario where data has been collected and processed in Section 13.4. Suppose there are $M = 2$ signal modes, and the goal is to pass mode $m = 1$ and perfectly cancel mode $m = 2$. The clairvoyant minimum-norm signal-copy weight vector that accomplishes this task is given by $\mathbf{w}_1 = \beta \mathbf{A}(\mathbf{A}^\dagger \mathbf{A})^{-1}\mathbf{u}_1$, where $\mathbf{A} = [\mathbf{v}_1, \mathbf{v}_2]$ may be defined in terms of the spatial signatures $\mathbf{v}_m$ instead of the mode wavefronts $\mathbf{a}_m$ without loss of generality.

The SNR gain in white noise, defined as the SNR of mode 1 at the linear combiner output SNR_o relative to that in the reference (first) receiver of the array SNR_i, is given by $\text{SNR}_g = \text{SNR}_o/\text{SNR}_i = |\beta|^2/\|\mathbf{w}_1\|^2$. The L_2-norm $\|\mathbf{w}_1\|^2$ may be evaluated by substituting $\mathbf{A} = [\mathbf{v}_1, \mathbf{v}_2]$ and $\mathbf{u}_1 = [1, 0]^T$ into Eqn. (13.38), and noting that the determinant of $(\mathbf{A}^\dagger \mathbf{A})$ is given by $\nabla = ad - bc$, where $a = \|\mathbf{v}_1\|^2$, $b = \mathbf{v}_1^\dagger \mathbf{v}_2$, $c = \mathbf{v}_2^\dagger \mathbf{v}_1$, and $d = \|\mathbf{v}_2\|^2$.

$$\|\mathbf{w}_1\|^2 = |\beta|^2 \mathbf{u}_1^\dagger (\mathbf{A}^\dagger \mathbf{A})^{-1} \mathbf{u}_1 = |\beta|^2 \|\mathbf{v}_2\|^2 \nabla^{-1} \tag{13.38}$$

From Eqn. (13.38), $\text{SNR}_g = \nabla/\|\mathbf{v}_2\|^2$, and after evaluating the determinant $\nabla = \|\mathbf{v}_1\|^2 . \|\mathbf{v}_2\|^2 - \|\mathbf{v}_1^\dagger \mathbf{v}_2\|^2$, it is straightforward to show that SNR_g is given by Eqn. (13.39). The maximum SNR gain with which mode 1 can be estimated with no contamination from mode 2 in white noise depends on the magnitude-squared coherence (MSC) between the spatial signatures: $\cos^2 \Upsilon = \|\mathbf{v}_1^\dagger \mathbf{v}_2\|^2 (\|\mathbf{v}_2\|^2 \|\mathbf{v}_1\|^2)^{-1} = \mathcal{F}\{\mathbf{v}_1, \mathbf{v}_2\}$, where Υ is the angle between $\mathbf{v}_1$ and $\mathbf{v}_2$ in N-dimensional space. The highest SNR gain subject to the constraint of no multipath interference at the output is $\|\mathbf{v}_1\|^2$. This is precisely the matched filter gain, attainable *only* when the spatial signatures are orthogonal $\mathbf{v}_1^\dagger \mathbf{v}_2 = 0$, such that $\cos^2 \Upsilon = 0$. For all other cases, the SNR gain is smaller and tends to zero as the spatial signatures align, i.e., as $\cos^2 \Upsilon \to 1$.

$$\text{SNR}_g = \|\mathbf{v}_1\|^2 \left(1 - \frac{\|\mathbf{v}_1^\dagger \mathbf{v}_2\|^2}{\|\mathbf{v}_1\|^2 \|\mathbf{v}_2\|^2}\right) = ||\mathbf{v}_1||^2 (1 - \cos^2 \Upsilon) \tag{13.39}$$

The improvement in SNR gain due to the presence of "crinkled wavefronts," relative to the hypothetical case of specular reflection which gives rise to plane waves with the same DOAs (nominal DOAs for diffuse scattering), is denoted by SNR_{IF} in Eqn. (13.40). This is simply the ratio of SNR gains in Eqn. (13.39) occurring for the two cases. Here, $\cos^2 \Phi = \mathcal{F}\{\mathbf{v}(\psi_1), \mathbf{v}(\psi_2)\}$ is the MSC for plane waves at the nominal mode DOAs, and $\cos^2 \Upsilon$ is the MSC for the mode spatial signatures $\mathbf{v}_m = \mathbf{v}(\psi_m) \odot \mathbf{d}_m$ with multiplicative distortions $\{\mathbf{d}_m\}_{m=1,2}$. When the nominal mode DOAs are closely spaced ($\psi_1 \to \psi_2$), we have that $\cos^2 \Phi \to 1$ and hence the denominator $\Delta = (1 - \cos^2 \Phi) \to 0$. As $\psi_1 \to \psi_2$,

$\cos^2 \Upsilon \rightarrow \cos^2 \Theta$, where $\cos^2 \Theta = \mathcal{F}\{\mathbf{d}_1, \mathbf{d}_2\}$ is the MSC of the multiplicative distortion vectors. Hence, $\text{SNR}_{\text{IF}} \rightarrow \sin^2 \Theta / \Delta$, where $\Delta \rightarrow 0$, as $\psi_1 \rightarrow \psi_2$. Large improvements in output SNR can result provided that the distortions $\{\mathbf{d}_m\}_{m=1,2}$ are sufficiently different, so that $\sin^2 \Theta \gg \Delta$.

$$\text{SNR}_{\text{IF}} = \frac{1 - \cos^2 \Upsilon}{1 - \cos^2 \Phi} \rightarrow \frac{\sin^2 \Theta}{\Delta} \quad \text{as } \psi_1 \rightarrow \psi_2 \tag{13.40}$$

Thus, when the mode nominal DOAs are closely spaced, different distortions caused by independent diffuse scattering processes can be exploited to estimate the waveform with no multipath contamination at higher output SNR.[2] This situation is particularly relevant for an HF source located close to boresight of a linear array, where ionospheric modes reflected from different layers share similar nominal cone angles, but are likely to exhibit different wavefront distortions due to the independent diffuse scattering processes.

This illustrative example motivates the use of manifold-free procedures that can take advantage of wavefront distortions for separating signal modes with closely spaced nominal DOAs. Moreover, reliance on the plane-wave model when distortions are actually present would lead to higher SINR improvements than those predicted by Eqn. (13.40).

13.2 Standard Techniques

Standard techniques relevant to the formulated problem fall under two main classes, namely, multi-channel blind system identification (BSI), and blind signal separation (BSS). The former typically assumes the presence of a single source and models the propagation channel linking the source to each receiving element of the array by a different finite impulse response (FIR) function. In the standard multi-channel BSI problem, the source input sequence and FIR system function are both assumed to be unknown. At times, the main intent is to estimate the system function, as opposed to the input sequence, which may be viewed merely as a probing signal. However, the problem can be easily recast to estimate the input sequence directly. In any case, multi-channel BSI techniques often involve the joint processing of a block of space-time data to estimate the unknown parameters.

As its acronym suggests, BSS techniques assume the presence of multiple sources, where the emitted signals cannot be separated by simple operations, such as bandpass filtering. In the array processing context, many BSS techniques assume an instantaneous mixture model, wherein the signals propagate directly from source to receiver without multipath reflections. This is called an instantaneous multiple-input multiple-output (I-MIMO) system. Such a system is appropriate for applications that involve line-of-sight propagation, for example. Alternative BSS techniques address the problem of convolutive signal mixtures, where the presence of multipath gives rise to an FIR-MIMO system. In any case, the aim of multi-channel BSS techniques is to separate and estimate the different source signals from the received array data. This is commonly achieved by spatial-only processing, although space-time processing may be used.

The first purpose of this section is to provide background and reference material on the subjects of multi-channel BSI and BSS. This includes a description of the data models

[2] In the case of signal modes with widely separated nominal DOAs (i.e., beyond the Rayleigh resolution limit), the presence of wavefront distortions also has the potential to degrade performance.

often adopted in standard BSI and BSS approaches, with both the I-MIMO and FIR-MIMO systems considered in the latter case. The second purpose is to provide an overview of the underlying assumptions upon which many existing BSI and BSS techniques are based. This is undertaken with a view to determining whether such techniques are applicable to the previously formulated problem. A discussion at the end of this section motivates the development of a new blind waveform estimation technique, referred to as Generalized Estimation of Multipath Signals (GEMS).

13.2.1 Blind System Identification

Traditional approaches for multipath equalization require the scheduling of training data sequences on transmit, which can significantly consume channel capacity and system resources in a time-varying environment (Paulraj and Papadias 1997). Moreover, training data is obviously unavailable when the signal of interest is transmitted by an uncooperative source. These factors have led to the development of BSI techniques, also known as blind channel equalization or blind deconvolution. A detailed description of BSI techniques is beyond the scope of this text, but comprehensive treatments can be found in the excellent review articles of Abed-Meraim et al. (1997), Tong and Perreau (1998), as well as the text of Haykin (1994), for example.

Figure 13.4 shows a discrete-time representation of the standard multi-channel BSI problem. The source produces a scalar input sequence s_k, which is received at sensors $n = 1, \ldots, N$ after propagation through a linear channel with an impulse response function

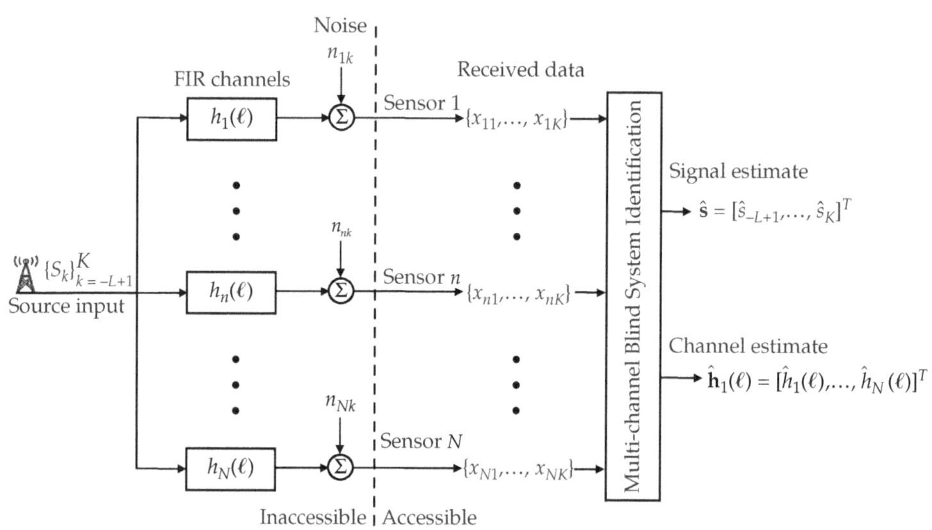

FIGURE 13.4 Representation of the standard multi-channel BSI architecture as a time-invariant FIR-SIMO system. The FIR model order is L, the number of sensors is N, and the number of data samples received by each sensor is K. The source input sequence, channel impulse responses, and additive noise processes are assumed to be inaccessible. The objective of the BSI processor is to jointly identify the source signal and channel coefficients from the observed space-time data to within a complex scale ambiguity. © Commonwealth of Australia 2011.

$h_n(\ell)$. Multi-channel BSI is traditionally based on a *time-invariant* FIR model of each channel $h_n(\ell)$ with support over a delay interval $\ell = 0, \ldots, L$. Recall that L is defined as the maximum FIR model order for the N-channel system. According to this model, the complex data sample x_{nk} received by sensor n at time k is given by Eqn. (13.41), where n_{nk} is additive noise independent of the signal.

$$x_{nk} = \sum_{\ell=0}^{L} h_n(\ell) s_{k-\ell} + n_{nk} \tag{13.41}$$

The source input sequence s_k and multi-channel system function $\mathbf{h}(\ell) = [h_1(\ell), \ldots, h_N(\ell)]^T$ are assumed to be inaccessible. Only the time-series data $\mathbf{y}_n = [x_{n1}, \ldots, x_{nK}]^T$ received by the N-sensor array is deemed to be observable. The vector $\mathbf{y}_n \in \mathcal{C}^K$ may be expressed in the form of Eqn. (13.42).

$$\mathbf{y}_n = \mathcal{H}_n \mathbf{s} + \underline{\epsilon}_n \tag{13.42}$$

In Eqn. (13.42), $\mathcal{H}_n \in \mathcal{C}^{K\times(K+L)}$ is the Sylvester matrix containing the impulse response of channel n, as defined in Eqn. (13.43). For K output samples in $\mathbf{y}_n$, it follows that $\mathbf{s} = [s_{-L+1}, \ldots, s_K]^T$ is the $(K+L)$-dimensional vector of input samples extended by the maximum FIR model order L, and $\underline{\epsilon}_n = [n_{n1}, \ldots, n_{nK}]^T$ is the K-dimensional vector of additive noise.

$$\mathcal{H}_n = \begin{bmatrix} h_n(L) & \cdots & h_n(0) & 0 & \cdots & 0 \\ 0 & h_n(L) & \cdots & h_n(0) & \ddots & \vdots \\ \vdots & \ddots & \ddots & & \ddots & 0 \\ 0 & \cdots & 0 & h_n(L) & \cdots & h_n(0) \end{bmatrix} \tag{13.43}$$

Alternatively, the data vector $\mathbf{y}_n$ may be written in the form of Eqn. (13.44), where $\mathcal{S} \in \mathcal{C}^{K\times(L+1)}$ is the Toeplitz matrix of the input sequence extended by the maximum FIR model order L, and $\mathbf{h}_n = [h_n(0), \ldots, h_n(L)]^T$ is the $(L+1)$-dimensional vector of the impulse response coefficients of channel n.

$$\mathbf{y}_n = \mathcal{S}\mathbf{h}_n + \underline{\epsilon}_n \tag{13.44}$$

To be clear, $\mathcal{S} \in \mathcal{C}^{K\times(L+1)}$ is defined in Eqn. (13.45). This representation is relevant to supervised channel estimation using a known input sequence. In this case, $\mathcal{S}$ is known during the training interval, and $\mathcal{R}\{\mathcal{S}\} = L+1$ by design. The channel coefficients may be estimated from the received data as $\hat{\mathbf{h}}_n = \mathcal{S}^+\mathbf{y}_n = \mathbf{h}_n + \mathcal{S}^+\underline{\epsilon}_n$, where $\mathcal{S}^+ = (\mathcal{S}^\dagger\mathcal{S})^{-1}\mathcal{S}^\dagger$.

$$\mathcal{S} = \begin{bmatrix} s_1 & \cdots & s_{1-L} \\ s_2 & \cdots & s_{2-L} \\ \vdots & & \vdots \\ s_K & \cdots & s_{K-L} \end{bmatrix} \tag{13.45}$$

The total data acquired by the time-invariant FIR-SIMO system over the processing interval may be written as Eqn. (13.46), where $\mathbf{y} = [\mathbf{y}_1^T, \ldots, \mathbf{y}_N^T]^T$ is the NK-dimensional stacked vector of space-time samples, $\mathcal{H} \in \mathcal{C}^{NK\times(K+L)}$ is the generalized Sylvester matrix formed by stacking $\{\mathcal{H}_1, \ldots, \mathcal{H}_N\}$, and $\underline{\varepsilon} \in \mathcal{C}^{NK}$ is the stacked vector of additive noise, constructed similar to $\mathbf{y}$.

$$\mathbf{y} = \mathcal{H}\,\mathbf{s} + \underline{\varepsilon} \tag{13.46}$$

Using Eqn. (13.44), it is also possible to write $\mathbf{y}$ in the form of Eqn. (13.47), where $\mathbf{h} = [\mathbf{h}_1^T, \ldots, \mathbf{h}_N^T]^T$ is the stacked vector of the N channel impulse response functions, and the $NK \times N(L+1)$ matrix $\mathcal{S}_N = \text{diag}\{\mathcal{S}, \ldots, \mathcal{S}\}$ is formed as N diagonal blocks each containing the Toeplitz matrix of the input sequence $\mathcal{S}$ defined in Eqn. (13.45).

$$\mathbf{y} = \mathcal{S}_N\,\mathbf{h} + \underline{\varepsilon} \tag{13.47}$$

When training data is available to estimate the channel coefficients as $\hat{\mathcal{H}}$, linear space-time equalization may be applied to estimate the input sequence from the received data as in Eqn. (13.48). This assumes $NK > (K+L)$ and that $\mathcal{H}$ has full column-rank $(K+L)$. However, when the channel and source are both unknown, the variables $\mathcal{H}$ and $\mathbf{s}$ need to be estimated jointly from the observations $\mathbf{y}$ (i.e., blindly).

$$\hat{\mathbf{s}} = \hat{\mathcal{H}}^{+}\mathbf{y} = (\hat{\mathcal{H}}^{\dagger}\hat{\mathcal{H}})^{-1}\hat{\mathcal{H}}^{\dagger}\mathbf{y} \tag{13.48}$$

For additive white Gaussian noise, application of the maximum-likelihood (ML) criterion for the joint estimation of the system matrix $\mathcal{H}$ and input sequence $\mathbf{s}$, when neither is known, requires finding the solution to Eqn. (13.49), which is a nonlinear optimization problem. An elegant two-step ML method for calculating $(\hat{\mathcal{H}}, \hat{\mathbf{s}})_{\text{ML}}$ is given in Hua (1996). If the FIR-SIMO system is time-varying, because of Doppler shifts for example, frequent updates of this procedure are needed to counter channel variations.

$$(\hat{\mathcal{H}}, \hat{\mathbf{s}})_{\text{ML}} = \arg\min_{\mathcal{H},\mathbf{s}} ||\mathbf{y} - \mathcal{H}\,\mathbf{s}||^2 \tag{13.49}$$

An attractive feature of the BSI approach is that the source input sequence and polynomial channel coefficients can be estimated under relatively general identifiability conditions. The FIR-SIMO system described by Eqn. (13.46) is considered identifiable when a given output $\mathbf{y}$ implies a unique solution for the system matrix $\mathcal{H}$ and the input sequence $\mathbf{s}$ up to an unknown complex scalar in the absence of noise. In accordance with Hua and Wax (1996), the sufficient identifiability conditions for a time-invariant FIR-SIMO system of known model order L are

- The N polynomial sub-channels do not share a common zero. This condition reflects the need for coprime FIR sub-channels (i.e., sufficient channel diversity).
- The input sequence has linear complexity $P > 2L$. This implies that the input cannot be a constant or sinusoid for example (i.e., sufficient signal complexity).
- The total number of samples $K > 3L$. This reflects the need for enough data to determine the number of system unknowns (i.e., sufficient sample support).

However, standard BSI approaches also have some known limitations. For example, the FIR model order is often unknown in practice, and its estimation is a challenging problem

(Abed-Meraim et al. 1997). BSI performance can be sensitive to poor estimation of L. Moreover, sparse channels that are highly time-dispersive may require large values of L for equalization, despite the possible presence of few dominant propagation modes, i.e., $L \gg M$. This leads to multi-channel equalizers of large dimension $N(L+1)$, which increases demands on finite sample support, not to mention computational load.

In addition, large Doppler shifts may stipulate the use of very short data frames to satisfy the time-invariant channel assumption. This restricts the observation interval that can be coherently processed (Hua 1996). The requirement for fast updates may also reduce sample support and increase computational load. These factors can limit the practical performance of standard BSI approaches. This motivates the search for alternative methods that are less prone to these drawbacks, yet strive to retain the general identifiability conditions of the BSI problem formulation.

It is evident from Figure 13.4 that the spatial snapshot data vector $\mathbf{x}_k = [x_{1k}, \ldots, x_{Nk}]^T$ is given by Eqn. (13.50), where $\mathbf{h}(\ell) = [h_1(\ell), \ldots, h_N(\ell)]^T$. The standard BSI model may be reconciled with the CD data model in Eqn. (13.20) by substituting $\mathbf{h}(\ell) = \sum_{m=1}^{M} \delta(\ell - \ell_m)\mathbf{a}_m$ into Eqn. (13.50). This impulse response function model was derived in Eqn. (13.22) for the case of a narrowband source, local scattering, and M dominant modes received by a coherent sensor array, where the length of the data record is sufficiently short to neglect the mode Doppler shifts.

$$\mathbf{x}_k = \sum_{\ell=0}^{L} \mathbf{h}(\ell) s_{k-\ell} + \mathbf{n}_k \tag{13.50}$$

Two comparative remarks are made. First, the multi-channel impulse response function $\mathbf{h}(\ell)$ is assumed to be time-invariant in the standard BSI model, whereas Doppler shifts give rise to a time-varying impulse response function $\mathbf{h}(k, \ell)$. Importantly, standard BSI approaches are not designed to handle a time-varying impulse response function during the observation interval. Recall that the CD model was extended to include the effect of mode Doppler shifts, which may be significant in the problem considered.

Second, standard BSI approaches are not designed for the case of multiple sources, where it is required to estimate the waveform of each source. On the other hand, the FIR model of $\mathbf{h}(\ell)$ in the BSI problem is general in the sense that it is also applicable to broadband signals, extended scattering, and receiving arrays with widely spaced sensors. In other words, the FIR structure of $\mathbf{h}(\ell)$ in Figure 13.4 is not restricted to the form of Eqn. (13.22).

13.2.2 Blind Signal Separation

Multi-channel BSS methods are applied to separate and recover the waveforms of multiple sources received by a sensor array, where the different signals cannot be discriminated readily in time or frequency. In the spatial processing context, the vast majority of BSS techniques, also referred to as "unsupervised" or "self-recovering" methods, fall into two main categories: (1) those which make certain assumptions regarding the propagation channel and sensor array in order to characterize the spatial signatures of the signals to be estimated, i.e., manifold-based methods, and (2) those which are not based on a manifold model but instead utilize a priori information regarding some known deterministic or statistical properties of the waveforms to be separated, i.e., source-based methods.

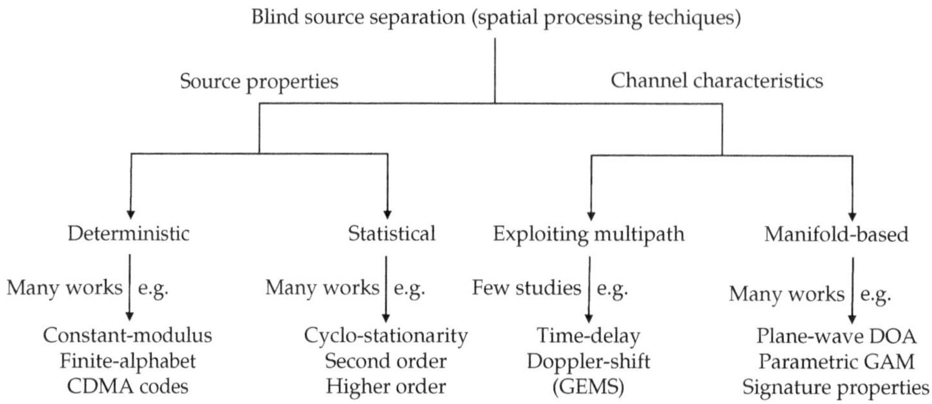

FIGURE 13.5 Taxonomy of standard BSS approaches in terms of underlying assumptions used as the basis for separation. The lack of information about the source signals has in many works been remedied by assuming partial knowledge about the propagation channel and sensor array, such that parametric spatial signature models may be used for separation. On the other hand, alternative approaches assume that certain deterministic or statistical properties are known about the source waveforms in order to compensate for the lack of knowledge regarding the propagation channel and sensor array characteristics. In the spatial processing context, relatively few works have directly exploited multipath as the physical mechanism to enable BSS. The GEMS algorithm is based on the concept that this alternative approach may be used to relax the assumptions required for the source waveforms and array manifold, jointly, rather than separately. © Commonwealth of Australia 2011.

In-depth treatments of manifold- and source-based BSS methods can be found in Van Der Veen (1998) and Cardoso (1998), respectively.

BSS using spatial processing is a topic that has received enormous attention in the literature. Figure 13.5 shows a top-level breakdown of BSS techniques, including source-based methods, manifold-based methods, and the alternative of exploiting multipath as the enabling physical mechanism. A brief overview of the many works existing on manifold- and source-based BSS methods is provided in this section. By comparison, relatively few works in the open literature have directly exploited multipath as the mechanism for enabling blind source separation.

Importantly, it will be shown that this approach allows certain assumptions regarding both the source and manifold to be relaxed. The GEMS algorithm exploits multipath for signal separation under relatively mild assumptions regarding both the mode wavefront and source waveform properties. It is precisely the mildness of these assumptions that makes the GEMS approach noteworthy and robust.

The I-MIMO system is schematically depicted in Figure 13.6, where source q is linked to channel n by a complex scalar transfer coefficient a_{qn}. The received array snapshot $\mathbf{x}_k = [x_{1k}, \ldots, x_{Nk}]^T$ can be represented in the standard form of Eqn. (13.51) where the $N \times Q$ instantaneous source mixing matrix $\mathcal{A} = [\mathbf{a}_1, \ldots, \mathbf{a}_Q]$ contains the Q channel vectors, denoted by $\mathbf{a}_q = [a_{q1}, \ldots, a_{qN}]^T$ for $q = 1, \ldots, Q$, while the Q-dimensional source signal vector $\mathbf{s}(k) = [s_1(k), \ldots, s_Q(k)]^T$ contains the different input sequences.

$$\mathbf{x}_k = \mathcal{A}\mathbf{s}(k) + \mathbf{n}_k \tag{13.51}$$

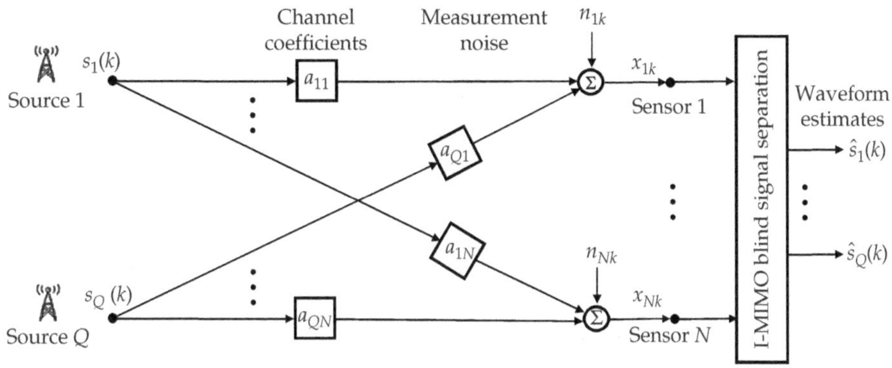

FIGURE 13.6 Illustration of the instantaneous multiple-input multiple-output (I-MIMO) system model. The source input sequences, complex-scalar channel transfer coefficients, and additive noise processes are inaccessible. The objective of the BSS processor is to jointly estimate all of the source waveforms to within an unknown complex scale by spatially weighing and combining the received array data. © Commonwealth of Australia 2011.

The source signal vector $\mathbf{s}(k)$ should not be confused with the multipath signal vector $\mathbf{s}_k$ in Eqn. (13.18). The instantaneous source mixing matrix $\mathcal{A}$ also has a different physical interpretation to the multipath mixing matrix $\mathbf{A}$ in Eqn. (13.18). However, it is apparent that Eqn. (13.51) has an equivalent mathematical form to Eqn. (13.18) in the *special case* where the Q sources are assumed to emit signals that are time-delayed and Doppler-shifted versions of a common input sequence, as in Eqn. (13.52). From the BSS viewpoint, there is clearly no distinction between Eqns. (13.51) and (13.18) in this special case.

$$s_q(k) = s_{k-\ell_q} e^{j2\pi \nu_q k} \tag{13.52}$$

The FIR-MIMO system model illustrated in Figure 13.7 is of more direct interest for multiple sources propagated over multipath channels. Indeed, the FIR-MIMO framework generalizes the BSI problem in Figure 13.4 to the multiple-source case. In Figure 13.7, $h_{qn}(\ell)$ for $\ell = 0, \ldots, L$ denotes the FIR function of the channel that links source q to sensor n. It follows from the single-source expression in Eqn. (13.50) that the FIR-MIMO spatial snapshots $\mathbf{x}_k$ are given by Eqn. (13.53), where $\mathbf{h}_q(\ell) = [h_{q1}(\ell), \ldots, h_{qN}(\ell)]^T$ and $L = \max\{L_q\}_{q=1}^{Q}$ is the maximum FIR model order.

$$\mathbf{x}_k = \sum_{q=1}^{Q} \sum_{\ell=0}^{L} \mathbf{h}_q(\ell) s_q(k-\ell) + \mathbf{n}_k \tag{13.53}$$

The connection between Eqn. (13.53) and the multiple-source model of Eqn. (13.51) developed in the previous subsection becomes evident if we substitute the time-varying impulse response $\mathbf{h}_q(k,\ell) = \sum_{m=1}^{M_q} \delta(\ell - \ell_{mq}) \mathbf{a}_{mq} e^{j2\pi \nu_{mq} k}$ for $\mathbf{h}_q(\ell)$ in Eqn. (13.53). Making this substitution yields the received data snapshots in Eqn. (13.54), where $s'_q(k - \ell_{mq}) = s_q(k - \ell_{mq}) e^{j2\pi \nu_{mq} k}$. This expression is consistent with the model of Eqn. (13.24). Note that the definition of $s'_q(k - \ell_{mq})$ incorporates the Doppler shift of each mode in the waveform

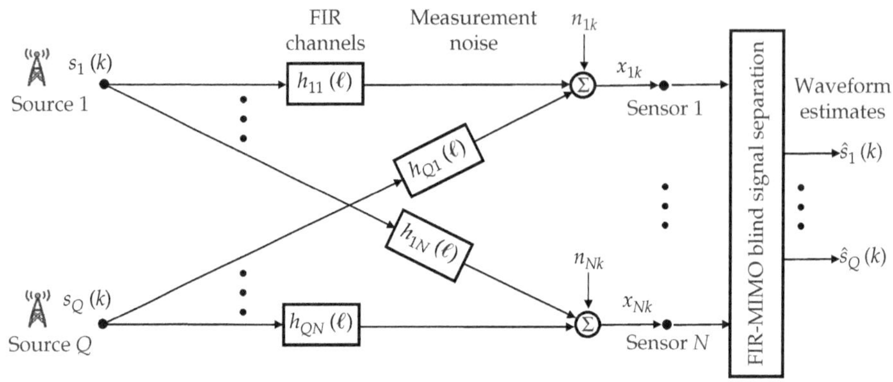

FIGURE 13.7 Illustration of the FIR multiple-input multiple-output (FIR-MIMO) system model. The source input sequences, channel impulse responses, and additive noise processes are inaccessible. The objective of the BSS processor is to jointly estimate all of the source waveforms to within an unknown complex scale by spatially weighting and combining the received array data. © Commonwealth of Australia 2011.

to be estimated. This effectively accounts for channel variations by modifying the waveform of each mode by a different Doppler shift.

$$\mathbf{x}_k = \sum_{q=1}^{Q} \sum_{m=1}^{M_Q} s'_q(k - \ell_{mq})\, \mathbf{a}_{mq} + \mathbf{n}_k \tag{13.54}$$

The form of Eqn. (13.54) indicates that BSS techniques based on a time-invariant channel model are in principle applicable to the formulated problem since each mode waveform may be considered as a different signal to be estimated. The critical issue here is whether the assumptions required by standard BSS techniques to recover suitable waveform estimates in such a situtation are compatable with those previously set out in the problem formulation. This point will be considered in Section 13.2.3 with reference to the multiple-source model reproduced in Eqn. (13.55), and the single-source model $\mathbf{X} = \mathbf{AS} + \mathbf{N}$ described in the previous section.

$$\mathbf{X} = \mathbf{HP} + \mathbf{N} \tag{13.55}$$

In the single-source case, the identifiability condition relating to channel diversity implies that the mode wavefronts in the mixing matrix $\mathbf{A}$ are linearly independent so that $\mathbf{A}$ has full *column-rank* M, while the identifiability condition relating to input sequence linear complexity implies that the signal matrix $\mathbf{S}$ has full *row-rank* M. This leads to the fundamental property upon which nearly all BSS techniques are based, namely, that the column span of $\mathbf{X}$ provides a basis for the column span of $\mathbf{A}$, and that the row span of $\mathbf{X}$ provides a basis for the row span of $\mathbf{S}$. Similar concepts apply for the multiple-source case.

13.2.2.1 Manifold-Based Methods

A popular manifold-based method is to discriminate the signals on the basis of differences in DOA by assuming a plane-wave model. Such a model is valid for narrowband signals and point sources in the far-field of a well-calibrated sensor array. This approach may be used to resolve a number of independent sources, or multiple propagation modes from a single source, that impinge on the array as plane waves. In the latter case, specular reflection of the multipath components is often assumed.

Super-resolution techniques such as MUSIC (Schmidt 1981), ESPRIT (Roy and Kailath 1989), MODE (Stoica and Sharman 1990b), WSF (Viberg, Ottersten, and Kailath 1991), ML (Stoica and Sharman 1990a), and their variants described in Krim and Viberg (1996), may be used to resolve signals with closely spaced DOAs. These techniques may be considered "blind" in the sense that the source properties and mixing matrix are not known a priori. In this case, lack of knowledge regarding the source properties is compensated for by assuming the spatial signatures in the mixing matrix lie on a manifold with a known parametric form. Specifically, the plane-wave manifold is defined by the DOA parameter alone.

DOA estimates of the incident signals are used to reconstruct the mixing matrix, which allows a deterministic null-steering weight vector to estimate the individual source waveforms with reduced contamination from multipath components and other signal sources. This is the classic "signal-copy" procedure. Ideally, the first step estimates the exact DOAs of all signal components, while the second step adjusts the weights of the linear combiner to perfectly null all interfering signals, leaving only the desired source waveform and measurement noise at the output. This signal-copy procedure is effective provided that the number of plane-wave signals is less than the number of receivers, the signal DOAs are not too closely spaced, the model order is selected appropriately, and the SNR is adequate for the amount of training data available. In this event, performance is limited mainly by statistical errors.

However, the plane-wave assumption is rather strong and seldom holds in practical scenarios. In particular, diffuse scattering caused by an irregular propagation medium, combined with the presence of array calibration errors, may lead to significant deviations between the spatial signatures of the signals received by the system and the presumed plane-wave manifold model. Diffuse scattering has been observed and analyzed in a number of different fields not limited to wireless communications (Zetterberg and Ottersten 1995), radio astronomy (Yen 1985), underwater acoustics (Gershman, Turchin, and Zverev 1995), speech recognition (Juang, Perdue, and Thompson 1995), medical imaging (Flax and O'Donnell 1988), seismology (Wood and Treitel 1975), and radar (Barton 1974).

In the HF band, point-to-point communication systems and OTH radars that rely on skywave propagation experience diffuse scattering from different horizontally stratified regions or layers in the ionosphere (Fabrizio, Gray, Turley 2000a). In such applications, a spatially distributed signal representation is often more appropriate than a point-source model. Super-resolution methods for DOA estimation may be applied, but even small departures from the plane-wave model can seriously degrade unwanted signal rejection, and hence waveform estimation quality. Performance degradations tend to be most pronounced for small array apertures, closely spaced sources, and powerful signal components; see Swindlehurst and Kailath (1992) and Friedlander and Weiss (1994), for example.

For DOA-based signal-copy procedures, the problem of spatially distributed signals amounts to decomposing the received (non-planar) wavefronts as a sum of vectors on the plane-wave manifold (Van Der Veen 1998). Estimating the DOAs of possibly a very large number of diffusely scattered rays for each distributed signal component represents a formidable task. In many cases, this task is infeasible due to the limited number of receivers and resolution available. Consequently, generalized array manifolds (GAM) not confined to the plane-wave model have attracted significant attention for spatial signature estimation as well as the parametric localization of coherently and incoherently distributed sources. For more information on this topic, the reader is referred to the works of Swindlehurst (1998), Jeng, Lin, Xu, and Vogel (1995), Lee, Choi, Song, and Lee (1997), Raich, Goldberg, and Messer (2000), Meng, Stoica, and Wong (1996), Astely, Ottersten, and Swindelhurst (1998), Trump and Ottersten (1996), Valaee and Champagne (1995), Fabrizio, Gray, and Turley (2000), Besson, Vincent, Stoica, and Gershman (2000), Besson and Stoica (2000), Astely, Swindlehurst, and Ottersten (1999), Weiss and Friedlander (1996), and Stoica, Besson, and Gershman (2001).

Although more flexible than the plane-wave manifold, many GAM models are nevertheless based on certain assumptions. For example, the GAM proposed in Astely et al. (1998) is based on a first-order Taylor series expansion that requires the angular spread of the signal to be small for accurate modeling. On the other hand, DOA-independent and amplitude-only wavefront distortions were assumed in Weiss and Friedlander (1996) and Stoica et al. (2001), respectively. In Valaee et al. (1995), angular spectrum profiles that are known analytic functions of a nominal DOA and spatial spread parameter are presumed to be available. Despite broadening the domain of applicability with respect to the plane-wave manifold, such models may not be general enough to accurately capture an *arbitrary* set of linearly independent spatial signatures.

In many real-world environments, the spatial signatures may have large angular spreads. Moreover, the received wavefronts typically exhibit a combination of gain and phase distortions relative to the plane-wave model that may differ significantly from one signal component to another. Such distortions can be very difficult to accurately characterize due to the complex diffuse scattering processes involved. Although DOA- and GAM-based techniques are applicable to particular classes of problems, where the assumptions made can be physically justified, they may not allow an arbitrary set of linearly independent spatial signatures to be effectively modeled and resolved for waveform estimation purposes. This restriction can limit the performance of manifold-based methods in practical applications.

13.2.2.2 Source-Based Methods

Performance degradations caused by the sensitivity of manifold-based methods to model mismatch between the actual and presumed signal spatial signatures may be avoided by resorting to a different class of BSS techniques, which are based on source properties as opposed to channel and array characteristics. The purported advantage of source-based BSS techniques stems from the fact that many human-made signals have certain properties that are accurately known a priori in a number of applications. This leads to more robust algorithms, which do not depend on reliable array calibration, or well-understood channel characteristics (Van Der Veen 1998).

For example, BSS may be based on the constant-modulus property of frequency modulated or phase-coded signals, as in Treichler and Agee (1983), van der Veen and Paulraj (1996), and Papadias and Paulraj (1997), or on finite-alphabet signals with known constellations, as in Yellin and Porat (1993), Anand, Mathew, and Reddy (1995), and Talwar, Viberg, and Paulraj (1994), for example. BSS may also exploit known properties regarding the signal second-order statistics (SOS), such as cyclostationarity, which is often encountered in digital communications due to the bauded nature of the transmissions; see Agee, Schell, and Gardner (1990), Xu and Kailath (1992), and Wu and Wong (1996), for example. BSS methods based on the joint diagonalization of spatial covariance matrices have been developed to separate uncorrelated signals in Belouchrani, Abed-Meraim, Cardoso, and Moulines (1997), but such techniques may not be appropriate in multipath environments, where the signals to be separated are often correlated.

Deterministic properties, such as parametrically known templates of periodic radar signals, or the known pulse-shape and code-vectors of CDMA signals, may also be utilized for BSS. Examples of the latter include Liu and Xu (1996), and Liu and Zoltowski (1997). For non-Gaussian signals, the joint distributional properties of the source signals, such as mutual independence, may be used to separate signals based on higher order statistics (HOS), as in Cardoso and Souloumiac (1993), Porat and Friedlander (1991), Dogan and Mendel (1994), Gonen and Mendel (1997), and Yuen and Friedlander (1996). However, the slow convergence rate of the higher order moment sample estimates often poses significant limitations for small data volumes. Perhaps more importantly, cumulant-based methods are not applicable to Gaussian signals, which often emerge from a large number of superimposed rays with random amplitudes and uniformly distributed phases.

Many waveforms emitted by natural and human-made sources of potential interest do not belong to a class of signal with known deterministic or statistical properties that can be utilized for BSS. Specific assumptions regarding the modulation format of a signal can be restrictive in some BSS applications. For this reason, there is value in broadening the scope of BSS techniques by allowing the waveforms to have a practically arbitrary temporal signatures, subject to satisfying the condition required for identifiability (Hua and Wax 1996). Apart from satisfying identifiability, no further knowledge may be available regarding the deterministic or statistical properties of the waveforms to be estimated. Within this more general framework, it would appear that none of the above-mentioned source-based methods are designed for such applications due to the additional information they assume about the source waveforms.

13.2.3 Discussion

There is a perceived lack of blind spatial processing techniques in situations where neither of the premises relied upon by manifold- or source-based BSS methods apply. Specifically, manifold-based methods that assume a plane-wave model or GAM characterization of the received wavefronts may not be capable of describing a completely arbitrary set of M linearly independent spatial signatures using a relatively small number of parameters. On the other hand, source-based methods that depend on certain deterministic or statistical properties of the M signals to be separated may not be flexible enough to deal with practically arbitrary waveforms that are only required to satisfy the linear complexity condition needed for identifiability.

The motivation behind GEMS is to exploit multipath propagation to significantly relax assumptions with regard to both the source signals and spatial signatures. In addition, rather than considering diffuse scattering as a nuisance or complicating factor in the BSS problem, the contrary idea is espoused, in that wavefront distortions imposed on the signals by the propagation medium may actually *help* to separate them. Although this is well-known for source-based BSS methods, the notion that wavefront distortions can be exploited is less appreciated in the context of manifold-based BSS techniques which are agnostic to the properties of the sources. This alternative viewpoint calls for quite different BSS approaches that will be explained and experimentally validated in the remainder of this chapter.

13.3 GEMS Algorithm

GEMS is based on a deterministic optimization problem with a unique algebraic solution that identifies a complex-scaled, time-delayed, and possibly Doppler-shifted copy of the source waveform *exactly* using a finite amount of data when noise is absent. The noiseless case is considered in the first part of this section to lay the foundations for the practical GEMS routine. The noiseless optimization criterion is introduced and motivated for the single- and multiple-source scenarios, as well as for the case of time-varying mode wavefronts.

The practical GEMS filter operating in the presence of full-rank noise is derived in the second part of this section as a least-squares version of the noiseless optimization criterion, as in Fabrizio and Farina (2011a), and Fabrizio and Farina (2011b). This approach is often followed in deterministic blind beamforming (Van Der Veen 1998). The final part of this section discusses the aspect of computational complexity and proposes alternative approaches for implementing the GEMS procedure.

13.3.1 Noiseless Case

The GEMS optimization criterion for the noiseless case is nominally based on the CD signal model described in Section 13.1. The applicability of this optimization criterion for waveform estimation is considered for three scenarios. The first is the case of a single source with multiple echoes. It is shown that multipath itself can, in principle, be exploited to estimate a multipath-free copy of the source waveform. This is the essence of the GEMS algorithm, i.e., the problem of signal distortion caused by multipath and the physical mechanism enabling this distortion to be removed are one and the same.

The second considered scenario is concerned with showing that the developed approach can be extended to cater for multiple sources, in which case it is required to estimate a number of different waveforms. Finally, arguments are provided to justify the robustness of the developed approach to the ID signal model that is characterized by time-varying wavefronts.

13.3.1.1 Single Source

Denote the noiseless array data by the snapshot vector $\bar{\mathbf{x}}_k \in \mathcal{C}^N$ in Eqn. (13.56). From Eqn. (13.18), we recall that $\mathbf{A} \in \mathcal{C}^{N\times M}$ is the multipath mixing matrix of full rank M, and $\mathbf{s}_k \in \mathcal{C}^M$ is the multipath signal vector, which contains M copies of the input sequence $\{s_k\}_{k=1}^K$ with Doppler shifts $\{\nu_m\}_{m=1}^M$ and time delays $\{\ell_m\}_{m=1}^M$. Here, M is the number of dominant modes, with $M < N$, and $\ell_m \in (0, L]$, where L is the maximum FIR channel length.

$$\bar{\mathbf{x}}_k = \mathbf{A}\mathbf{s}_k = \mathbf{A}[s_{k-\ell_1}e^{j2\pi\nu_1 k}, \ldots, s_{k-\ell_M}e^{j2\pi\nu_M k}]^T \tag{13.56}$$

Now define the noiseless auxiliary data vector $\bar{\mathbf{u}}_k$ in Eqn. (13.57), where $\bar{\mathbf{x}}_k$ is delayed by ℓ samples and frequency offset by ν. Henceforth, we shall denote $\{\bar{\mathbf{x}}_k\}_{k=1}^K$ and $\{\bar{\mathbf{u}}_k\}_{k=1}^K$ as the reference and auxiliary noiseless snapshots, respectively. The notation $\tilde{\mathbf{s}}_k = \mathbf{s}_{k-\ell}e^{j2\pi\nu k}$ is used for the time-delayed and Doppler-shifted version of the signal vector $\mathbf{s}_k$.

$$\bar{\mathbf{u}}_k = \bar{\mathbf{x}}_{k-\ell}e^{j2\pi\nu k} = \mathbf{A}\mathbf{s}_{k-\ell}e^{j2\pi\nu k} = \mathbf{A}\tilde{\mathbf{s}}_k \tag{13.57}$$

Now define $\mathbf{w} \in \mathcal{C}^N$ as the reference weight vector, and let $\bar{z}_k$ be the scalar reference data output given by Eqn. (13.58). Also, by defining the M-dimensional vector $\mathbf{f} = \mathbf{A}^\dagger \mathbf{w}$, the complex scalar $\bar{z}_k$ may be written in terms of $\mathbf{f}$ in Eqn. (13.58).

$$\bar{z}_k = \mathbf{w}^\dagger \bar{\mathbf{x}}_k = \mathbf{f}^\dagger \mathbf{s}_k \tag{13.58}$$

Similarly, define an auxiliary weight vector $\mathbf{r} \in \mathcal{C}^N$ and the auxiliary data output $\bar{y}_k$ in Eqn. (13.59). This output is similarly written in terms of the M-dimensional vector $\mathbf{g} = \mathbf{A}^\dagger \mathbf{r}$ in Eqn. (13.59).

$$\bar{y}_k = \mathbf{r}^\dagger \bar{\mathbf{u}}_k = \mathbf{g}^\dagger \tilde{\mathbf{s}}_k \tag{13.59}$$

For an equal number K of output samples $\bar{z}_k$ and $\bar{y}_k$, and $\ell \in (0, L]$, it follows that a total of $K + L$ data snapshots are assumed to be available. For system identifiability, the previously stated condition on the total number of samples is $K + L > 3L$, which implies Eqn. (13.60).

$$K > 2L \tag{13.60}$$

Let the K-dimensional error vector be $\bar{\mathbf{e}} = [\bar{e}_1, \ldots, \bar{e}_K] = \bar{\mathbf{z}} - \bar{\mathbf{y}}$, where $\bar{\mathbf{z}} = [\bar{z}_1, \ldots, \bar{z}_K]$ is the reference output and $\bar{\mathbf{y}} = [\bar{y}_1, \ldots, \bar{y}_K]$ is the auxiliary output. Using Eqns. (13.58) and (13.59), $\bar{\mathbf{e}}$ may be expressed in terms of the matrices $\mathbf{S} = [\mathbf{s}_1, \ldots, \mathbf{s}_K]$ and $\tilde{\mathbf{S}} = [\tilde{\mathbf{s}}_1, \ldots, \tilde{\mathbf{s}}_K]$ in Eqn. (13.61).

$$\bar{\mathbf{e}} = \mathbf{f}^\dagger \mathbf{S} - \mathbf{g}^\dagger \tilde{\mathbf{S}} \tag{13.61}$$

The GEMS algorithm is based on a non-trivial solution for $\{\mathbf{f}, \mathbf{g}\}$ and $\{\ell, \nu\}$ that satisfies $\bar{\mathbf{e}} = \mathbf{0}$ in Eqn. (13.62). The constraint $\|\mathbf{f}^\dagger \mathbf{S}\|^2 = 1$ prevents the trivial solution $\mathbf{f} = \mathbf{g} = \mathbf{0}$, which yields $\bar{\mathbf{e}} = \mathbf{0}$ for any $\{\ell, \nu\}$. The condition $\ell \in (0, L]$ avoids the trivial solution at $\{\ell = 0, \nu = 0\}$, where $\mathbf{S} = \tilde{\mathbf{S}}$, and any nonzero vector $\mathbf{f} = \mathbf{g} \in \mathcal{C}^M$ yields $\bar{\mathbf{e}} = \mathbf{0}$.

$$\bar{\mathbf{e}} = \mathbf{0} \quad \text{s.t.} \quad \begin{cases} \|\mathbf{f}^\dagger \mathbf{S}\|^2 = 1 \\ \ell \in (0, L] \end{cases} \tag{13.62}$$

The implication of a non-trivial solution satisfying Eqn. (13.62) will now be described. Defining $\underline{\upsilon} \in \mathcal{C}^{2M}$ as the stacked vector of $\{\mathbf{f}, -\mathbf{g}\}$, and $\bar{\mathbf{s}}_k \in \mathcal{C}^{2M}$ as the stacked vector of $\{\mathbf{s}_k, \tilde{\mathbf{s}}_k\}$, the errors $\bar{e}_k = \underline{\upsilon}^\dagger \bar{\mathbf{s}}_k$ can be written in the compact form of Eqn. (13.63), where $\bar{\mathbf{S}} = [\bar{\mathbf{s}}_1, \ldots, \bar{\mathbf{s}}_K]$ is the $2M \times K$ stacked matrix of $\{\mathbf{S}, \tilde{\mathbf{S}}\}$.

$$\bar{\mathbf{e}} = \underline{\upsilon}^\dagger [\bar{\mathbf{s}}_1, \ldots, \bar{\mathbf{s}}_K] = \underline{\upsilon}^\dagger \bar{\mathbf{S}} \tag{13.63}$$

From Eqn. (13.64), each error term $\bar{e}_k = \underline{\upsilon}^\dagger \bar{\mathbf{s}}_k$ is a linear combination of $2M$ time-delayed and Doppler-shifted samples of the input sequence $\{s_k\}_{k=1}^K$. The time delays $\{\ell_m\}_{m=1}^M \in (0, L]$ are distinct by definition, so for a fixed displacement $\ell \in (0, L]$, the time delays $\{\ell_m + \ell\}_{m=1}^M \in (0, 2L]$ are also distinct.

$$\bar{\mathbf{s}}_k = \begin{bmatrix} \mathbf{s}_k \\ \tilde{\mathbf{s}}_k \end{bmatrix} = [s_{k-\ell_1} e^{j2\pi\nu_1 k}, \ldots, s_{k-\ell_M} e^{j2\pi\nu_M k}, s_{k-\ell-\ell_1} e^{j2\pi(\nu_1+\nu)k}, \ldots, s_{k-\ell-\ell_M} e^{j2\pi(\nu_M+\nu)k}]^T \tag{13.64}$$

Consider a null hypothesis H_0 for which the displacement ℓ is such that all $2M$ time delays are distinct, as in Eqn. (13.65). In other words, the delay applied to generate the auxiliary data does not match the differential time delay between any pair of modes. Momentarily ignoring the Doppler shifts, this implies that each error term $\bar{e}_k$ is a linear combination of $2M$ samples of the input sequence s_k with distinct time delays in the interval $(0, 2L]$.

$$H_0: \ \ell_i \neq \ell + \ell_j, \quad \forall\, i, j \in [1, M] \tag{13.65}$$

Recall that for an input sequence $\{s_k\}_{k=1}^{K}$ of linear complexity $P > 2L$, there exists no solution for the $(2L+1)$-dimensional vector $\underline{\lambda} = [\lambda_{2L}, \ldots, \lambda_1, 1]^T$ that satisfies Eqn. (13.66) for $K > 2L$ according to the definition in Eqn. (13.34), where the $(2L+1)$-dimensional vector $\dot{\mathbf{s}}_k = [s_{k-2L}, \ldots, s_{k-1}, s_k]^T$ contains samples of the input sequence at all possible distinct time delays in the interval $\ell = (0, 2L]$.

$$\dot{\mathbf{e}} = \underline{\lambda}^\dagger[\dot{\mathbf{s}}_1, \ldots, \dot{\mathbf{s}}_K] = \mathbf{0} \tag{13.66}$$

Stated another way, the $(2L+1) \times K$ Hankel matrix of the input sequence $\dot{\mathbf{S}} = [\dot{\mathbf{s}}_1, \ldots, \dot{\mathbf{s}}_K]$ has full rank $(2L+1)$ under the conditions $P > 2L$ and Eqn. (13.60). It follows that the $(2L+1) \times (2L+1)$ sample covariance matrix $\dot{\mathbf{M}} = \dot{\mathbf{S}}\dot{\mathbf{S}}^\dagger$ is a positive definite Hermitian matrix and the L_2-norm of $\dot{\mathbf{e}} = \underline{\lambda}^\dagger\dot{\mathbf{S}}$ must be greater than zero for all nonzero vectors $\underline{\lambda}$, irrespective of the Doppler shifts, as in Eqn. (13.67).

$$\|\dot{\mathbf{e}}\|^2 = \underline{\lambda}^\dagger\dot{\mathbf{M}}\underline{\lambda} > 0 \tag{13.67}$$

Since the $2M$ distinct time delays in $\bar{\mathbf{s}}_k$ constitute a subset of all $2L$ possible distinct time delays in $\dot{\mathbf{s}}_k$, it follows that a non-trivial solution for $\underline{\upsilon}^\dagger = [\mathbf{f}^\dagger, -\mathbf{g}^\dagger]$ that satisfies $\bar{\mathbf{e}} = \mathbf{0}$ in Eqn. (13.63) does not exist under H_0. In other words, when all $2M$ time delays are distinct, we have

$$\bar{\mathbf{e}} = \underline{\upsilon}^\dagger\bar{\mathbf{S}} = \mathbf{0} \iff \underline{\upsilon} = \mathbf{0} \ \text{(under } H_0) \tag{13.68}$$

Consider the alternative hypothesis H_1 where displacements ℓ and ν match the differential time delay and Doppler shift existing between an arbitrary pair of modes, denoted by modes i and j in Eqn. (13.69).

$$H_1: \ \{\ell = \ell_i - \ell_j, \ \nu = \nu_i - \nu_j\}, \ \ (i, j) \in [1, M], \ \ i \neq j \tag{13.69}$$

Due to the assumption of distinct differential coordinates in Eqn. (13.37), not more than one pair of modes can be simultaneously matched by a particular value of ℓ and ν. The number of different matched pair combinations for M modes is given by Eqn. (13.70), hence $\mathcal{L}$ unique values of $\{\ell, \nu\}$ fall under H_1.

$$\mathcal{L} = \frac{M!}{2(M-2)!} \tag{13.70}$$

Without loss of generality, let $i = M$ and $j = 1$, such that $\ell = \ell_M - \ell_1$ and $\nu = \nu_M - \nu_1$ provides a match between the first and last modes (the indexing is arbitrary). From Eqn. (13.64), this yields a stacked vector $\bar{\mathbf{s}}_k$ with two identical elements in Eqn. (13.71). Specifically, the last element of the vector $\mathbf{s}_k$ in position M of $\bar{\mathbf{s}}_k$ is identical to the first element of the vector $\tilde{\mathbf{s}}_k$ in position $M+1$ of $\bar{\mathbf{s}}_k$. In general, the condition in Eqn. (13.69) causes element i of $\mathbf{s}_k$ to match element j of $\tilde{\mathbf{s}}_k$ for all $k = 1, \ldots, K$.

$$\bar{\mathbf{s}}_k = \left[s_{k-\ell_1}e^{j2\pi\nu_1 k}, \ldots, s_{k-\ell_M}e^{j2\pi\nu_M k}, s_{k-\ell_M}e^{j2\pi\nu_M k}, \ldots, s_{k-2\ell_M+\ell_1}e^{j2\pi(2\nu_M-\nu_1)k}\right]^T \tag{13.71}$$

Under H_1, row i of $\mathbf{S} = [\mathbf{s}_1, \ldots, \mathbf{s}_K]$ becomes identical to row j of $\tilde{\mathbf{S}} = [\tilde{\mathbf{s}}_1, \ldots, \tilde{\mathbf{s}}_K]$, such that the stacked matrix $\bar{\mathbf{S}}$ of $\{\mathbf{S}, \tilde{\mathbf{S}}\}$ becomes rank deficient (i.e., drops rank). The other rows in $\bar{\mathbf{S}}$ remain linearly independent, since for distinct differential mode time delays, not more than one match is possible for a particular value of $\{\ell, \nu\}$. This implies that the rank of the $2M \times K$ matrix $\bar{\mathbf{S}}$ drops strictly by one, and is given by $\mathcal{R}\{\bar{\mathbf{S}}\} = 2M - 1$.

Since the number of modes $M \leq L$, and identifiability requires $K > 2L$, we have that $K > 2M$. This implies that the $2M \times K$ matrix $\bar{\mathbf{S}}$ has full rank $\mathcal{R}\{\bar{\mathbf{S}}\} = 2M$ under H_0, but has rank $2M - 1$ under H_1. These observations are summarized in Eqn. (13.72).

$$\begin{cases} H_0: \ \mathcal{R}\{\bar{\mathbf{S}}\} = 2M \\ H_1: \ \mathcal{R}\{\bar{\mathbf{S}}\} = 2M - 1 \end{cases} \tag{13.72}$$

As $\mathcal{R}\{\bar{\mathbf{S}}\} = 2M - 1$ under H_1, the Hermitian matrix $\bar{\mathbf{M}} = \bar{\mathbf{S}}\bar{\mathbf{S}}^\dagger$ is positive semi-definite, and a non-trivial solution for $\underline{\upsilon}$ exists to reduce $\|\bar{\mathbf{e}}\|^2 = \bar{\mathbf{e}}\bar{\mathbf{e}}^\dagger = \underline{\upsilon}^\dagger\bar{\mathbf{M}}\underline{\upsilon}$ to the minimum eigenvalue, which is equal to zero in this case. Hence, a non-trivial solution for $\underline{\upsilon}$ that yields $\bar{\mathbf{e}} = \mathbf{0}$ exists under H_1. Moreover, the non-trivial solution is given by a scaled version of the single eigenvector corresponding to the zero eigenvalue of $\bar{\mathbf{M}}$, which is unique to within a complex scale.

Specifically, the non-trivial solution for $\underline{\upsilon}$ in Eqn. (13.73) is given by a complex scale β, and the M-dimensional unit vectors $\{\mathbf{u}_i, \mathbf{u}_j\}$, which respectively select the two identical rows of $\bar{\mathbf{S}}$ but with the signs reversed.

$$\bar{\mathbf{e}} = \underline{\upsilon}^\dagger\bar{\mathbf{S}} = \mathbf{0} \iff \underline{\upsilon}^\dagger = [\mathbf{f}^\dagger, -\mathbf{g}^\dagger] = \beta[\mathbf{u}_i^\dagger, -\mathbf{u}_j^\dagger] \quad (\text{under } H_1) \tag{13.73}$$

The solution for $\underline{\upsilon}$ in Eqn. (13.73) implies that $\mathbf{f} = \mathbf{A}^\dagger\mathbf{w} = \beta\mathbf{u}_i$, and $\mathbf{g} = \mathbf{A}^\dagger\mathbf{r} = \beta\mathbf{u}_j$, where $\beta = e^{j\vartheta}/\|\mathbf{u}_i^\dagger\mathbf{S}\|$ satisfies the norm constraint in Eqn. (13.62) for an arbitrary rotation $e^{j\vartheta}$. As $\mathcal{R}\{\mathbf{A}\} = M < N$, this leads to infinitely many solutions for $\mathbf{w}$ and $\mathbf{r}$ in Eqn. (13.74). Here, $\mathbf{A}^+ = \mathbf{A}(\mathbf{A}^\dagger\mathbf{A})^{-1}$ is the Moore-Penrose pseudo-inverse of $\mathbf{A}$, $\mathbf{P}_{\mathbf{A}}^\perp = \mathbf{I} - \mathbf{A}^+\mathbf{A}^\dagger$ is the orthogonal projection matrix, and $\{\mathbf{q}_w, \mathbf{q}_r\}$ are any complex vectors in N-dimensional space. Importantly, the minimum norm solutions $\mathbf{w} = \beta\mathbf{A}^+\mathbf{u}_i$ and $\mathbf{r} = \beta\mathbf{A}^+\mathbf{u}_j$ in Eqn. (13.74) coincide with the clairvoyant signal-copy vectors that perfectly isolate modes $m = i$ and $m = j$, respectively, in Eqn. (13.31).

$$\mathbf{w} = \beta\mathbf{A}^+\mathbf{u}_i + \mathbf{P}_{\mathbf{A}}^\perp\mathbf{q}_w, \quad \mathbf{r} = \beta\mathbf{A}^+\mathbf{u}_j + \mathbf{P}_{\mathbf{A}}^\perp\mathbf{q}_r \tag{13.74}$$

For $i = M$, all solutions for $\mathbf{w}$ in Eqn. (13.74) exactly recover a complex scaled, time-delayed, and Doppler-shifted copy of the source waveform $\{s_k\}_{k=1}^K$ propagated by mode M at the processor output $\bar{z}_k$ in Eqn. (13.75). Similarly, all solutions for $\mathbf{r}$ in Eqn. (13.74) produce an output $\bar{y}_k = \mathbf{r}^\dagger\bar{\mathbf{u}}_k$ identical to $\bar{z}_k$, as expected for $\bar{e}_k = \bar{z}_k - \bar{y}_k = 0$. While $\mathbf{w}$ isolates mode i, $\mathbf{r}$ differs from $\mathbf{w}$, and isolates mode j. Application of $\mathbf{r}$ to the reference (instead of auxiliary) data for $j = 1$ yields $\bar{z}_k' = \mathbf{r}^\dagger\bar{\mathbf{x}}_k = \beta s_{k-\ell_1} e^{j2\pi\nu_1 k}$, such that the two weight vectors isolate the matched pair of modes.

$$\bar{z}_k = \mathbf{w}^\dagger\bar{\mathbf{x}}_k = \beta s_{k-\ell_M} e^{j2\pi\nu_M k} \tag{13.75}$$

Under the stated conditions, a number of observations may be made for the noiseless case:

- A non-trivial solution for $\underline{\upsilon}$ that yields a zero-error vector $\bar{\mathbf{e}}$ satisfying Eqn. (13.62) exists if and only if the delay-Doppler displacement $\{\ell, \nu\}$ identically matches the differential coordinates existing between a pair of modes (H_1). For mismatched $\{\ell, \nu\}$, such a solution does not exist (H_0).
- For M modes with distinct differential delay-Doppler coordinates, there are $\mathcal{L} = (M!/2)/(M-2)!$ different values of $\{\ell, \nu\}$ that can give rise to H_1, and the non-trivial solution for $\underline{\upsilon}$ satisfying Eqn. (13.62) for each particular value is unique up to a complex scale factor.
- For parameter values $\{\ell, \nu\}$ that match a particular pair of modes (i, j), the associated unique non-trivial solution for $\underline{\upsilon}$ satisfying $\bar{\mathbf{e}} = \mathbf{0}$ defines all possible solutions for the spatial filters $\{\mathbf{w}, \mathbf{r}\}$ that respectively isolate the matched pair of modes (i, j).

For the noiseless case, satisfying the criterion in Eqn. (13.62) therefore leads to spatial processing weight vector solutions that can exactly recover a complex scaled, time-delayed, and Doppler-shifted copy of the input sequence that is free of multipath. In the presence of noise, minimization of the error vector L_2-norm with respect to the spatial filters $\{\mathbf{w}, \mathbf{r}\}$, and delay-Doppler displacement $\{\ell, \nu\}$, provides a basis for GEMS to blindly estimate the source waveform.

13.3.1.2 Multiple Sources

The main assumptions specified in the single-source problem formulation are once again made for the multiple-source case, but modifications are required in three main areas. First, the number of sensors is assumed to be greater than the number of modes summed over the number of sources. As for the single-source case, the R mode wavefronts $\{\mathbf{a}_{mq}\}$ are assumed to be linearly independent, but otherwise arbitrary, such that $\mathbf{H}$ has full rank. These conditions are captured by Eqn. (13.76). The remaining two conditions, which relate to the source waveforms and distinct propagation mode delay-Doppler parameters, are described below. With these three generalizations, the previously described GEMS optimization criterion may be utilized for blind source and multipath separation.

$$\mathcal{R}\{\mathbf{H}\} = R < N \tag{13.76}$$

Identifiability requires the source waveforms to have linear complexity $P_q > 2L_q$, where $L_q = \max\{\ell_{mq}\}_{m=1}^{M_q}$ is the maximum channel impulse response duration for source q over all modes M_q. Recall that for a finite-length deterministic sequence $s_q(k)$, linear complexity is the smallest integer P_q for which there exist coefficients $\{\lambda_p\}_{p=1}^{P_q}$ that satisfy Eqn. (13.77). The sources are also assumed to emit different waveforms, where the term "different" implies that no two sources emit a time-delayed and Doppler-shifted version of a common signal. Besides meeting the linear complexity condition, and the sources emitting different waveforms, no other information is assumed about the signals $s_q(k)$. As the number of samples required to identify source q through an FIR channel of length

L_q must be greater than $K_q = 3L_q$ (Hua and Wax 1996), a total of $K > \max\{K_q\}_{q=1}^{Q}$ samples are assumed to be available such that all sources can be identified.

$$s_q(k) = -\sum_{p=1}^{P_q} \lambda_p \, s_q(k-p)\,, \quad k = 1, \ldots, K \tag{13.77}$$

Besides the relatively mild conditions assumed for the mode wavefronts $\mathbf{a}_{mq}$ and the source waveforms $s_q(k)$, both Q and M_q are assumed unknown along with the mode time-delays ℓ_{mq} and Doppler-shifts ν_{mq}. Similar to the single-source case, the differences between the mode parameter tuples $\underline{\rho}_{mq} = [\ell_{mq}, \nu_{mq}]$ are assumed to be distinct, as in Eqn. (13.78). This is reasonable for separated sources, since the differential time-delay *and* Doppler-shift between two propagation modes ($i \neq j$) of a particular source q will in general not be *identical* to that of another pair of modes ($i' \neq j'$) from the same or different source q' when attention is restricted to the relatively small number of dominant signal modes. Clearly, the condition $(i, j) \neq (i', j')$ is imposed in Eqn. (13.78) for $q = q'$.

$$\{\rho_{iq} - \rho_{jq}\}_{i,j=1}^{M_q} \neq \{\rho_{i'q'} - \rho_{j'q'}\}_{i',j'=1}^{M_{q'}} \tag{13.78}$$

The noiseless data $\mathbf{Hp}_k$ due to multiple sources has a similar mathematical form to the noiseless data $\mathbf{As}_k$ for a single-source. The mixing matrix $\mathbf{H}$ is a higher-dimension generalization of $\mathbf{A}$, while the stacked signal vector $\mathbf{p}_k$ behaves the same way as $\mathbf{s}_k$ in Section 13.3.1.1 under the stated assumptions. The assumption of different waveforms is required because sources that emit time-delayed and possibly Doppler-shifted copies of the same waveform, where the relative (inter-source) delay and Doppler shift values can reasonably be attributed to multipath, makes it difficult, if not impossible, to distinguish whether a waveform is a multipath component of a particular source, or is due to a different source.

Following a similar analysis to that in Section 13.3.1.1, it is readily shown that the spatial filter solutions $\{\mathbf{w}, \mathbf{r}\}$ that minimize the noiseless error vector to zero when $\{\ell, \nu\}$ matches the differential delay and Doppler between modes i and j of source q are given by Eqn. (13.79). Here $\mathbf{0}_q$ is a zero column vector of length M_q, and $\mathbf{u}_{mq}$ is a unit column vector of length M_q with unity in position $m \in [1, M_q]$. The terms $\mathbf{H}^+$ and $\mathbf{P}_{\mathbf{H}}^{\perp}$ denote the Moore-Penrose pseudo-inverse of $\mathbf{H}$ and the projection matrix orthogonal to the range space of $\mathbf{H}$, respectively. As before, $\mathbf{q}_w$ and $\mathbf{q}_r$ are any complex vectors in N-dimensional space.

$$\mathbf{w} = \beta\, \mathbf{H}^+ \begin{bmatrix} \mathbf{0}_1 \\ \vdots \\ \mathbf{u}_{iq} \\ \vdots \\ \mathbf{0}_Q \end{bmatrix} + \mathbf{P}_{\mathbf{H}}^{\perp}\mathbf{q}_w, \quad \mathbf{r} = \beta\, \mathbf{H}^+ \begin{bmatrix} \mathbf{0}_1 \\ \vdots \\ \mathbf{u}_{jq} \\ \vdots \\ \mathbf{0}_Q \end{bmatrix} + \mathbf{P}_{\mathbf{H}}^{\perp}\mathbf{q}_r \tag{13.79}$$

The key point is that spare degrees of freedom in the spatial filters are used to cancel unwanted (mismatched) modes from source q, as well as all other signal modes from the remaining $Q-1$ sources, such that the noiseless outputs $z_k = \mathbf{w}^{\dagger}\bar{\mathbf{x}}_k$ and $z'_k = \mathbf{r}^{\dagger}\bar{\mathbf{x}}_k$ contain only the waveforms carried by modes i and j of source q, respectively. This leads to the following conclusions for the multiple-source case.

- A non-trivial solution for $\underline{\upsilon}$ that yields a zero-error vector $\bar{\mathbf{e}}$ satisfying Eqn. (13.62) exists if and only if the delay-Doppler displacement $\{\ell, \nu\}$ identically matches the differential coordinates existing between a pair of modes from a source q. For mismatched $\{\ell, \nu\}$, such a solution does not exist.
- For the M_q modes of source q with distinct differential coordinates, there are $\mathcal{L}_q = (M_q!/2)/(M_q - 2)!$ different values of $\{\ell, \nu\}$ that can give rise to a zero-error vector $\bar{\mathbf{e}}$, and the non-trivial solution for $\underline{\upsilon}$ satisfying Eqn. (13.62) at each of these values is unique up to a complex scale factor.
- For a value of $\{\ell, \nu\}$ that matches a particular pair of modes (i, j) of source q, the associated unique non-trivial solution for $\underline{\upsilon}$ satisfying $\bar{\mathbf{e}} = \mathbf{0}$ defines all possible solutions for the spatial filters $\{\mathbf{w}, \mathbf{r}\}$ that respectively isolate the matched pair of modes (i, j) for source q.

In the noiseless case, the outputs of these spatial filters can exactly recover a complex-scaled, time-delayed, and possibly Doppler-shifted copy of the input sequence for source q free of contamination from multipath echoes and other source signals. In the presence of noise, minimization of the error vector L_2-norm $\|\bar{\mathbf{e}}\|^2$ with respect to non-trivial spatial filter solutions $\{\mathbf{w}, \mathbf{r}\}$, over a suitable domain of delay-Doppler displacements $\{\ell, \nu\}$, provides a basis for GEMS to blindly estimate all Q source waveforms.

13.3.1.3 ID Model

The focus has been on the coherently distributed (CD) signal model thus far, where the mode wavefronts are crinkled (i.e., non-planar) but are time-invariant over the processing interval. In practice, the shape of the mode wavefronts may change during the processing interval due to random variations in the diffuse scattering process. Considering a single source with $m = 1, \ldots, M$ modes, the ID multipath model represents time-varying mode wavefronts according to Eqn. (13.80), where $\mathbf{a}_m$ is a steady (non-fluctuating) component, and $\mathbf{Q}_m\underline{\varsigma}_m(k)$ is the changing (dynamic) component. The latter is assumed to be confined to a low-rank subspace $\mathbf{Q}_m \in \mathcal{C}^{N\times\mathcal{I}_m}$ of effective dimension $\mathcal{R}\{\mathbf{Q}_m\} = \mathcal{I}_m \ll N$.

$$\mathbf{a}_m(k) = \mathbf{a}_m + \mathbf{Q}_m\underline{\varsigma}_m(k) \tag{13.80}$$

By defining the $N \times \mathcal{I}$ matrix $\mathbf{B} = [\mathbf{Q}_1, \ldots, \mathbf{Q}_M]$, where $\mathcal{I} = \sum_{m=1}^{M} \mathcal{I}_m$, and the $\mathcal{I} \times M$ block diagonal matrix $\mathbf{D}_k = \text{diag}[\underline{\varsigma}_1(k), \ldots, \underline{\varsigma}_M(k)]$, it is possible to write the single ID source noiseless data snapshots $\bar{\mathbf{x}}_k$ in the form of Eqn. (13.81). Here, we have defined $\mathbf{C} = [\mathbf{A}, \mathbf{B}]$ as the augmented multipath mixing matrix, and $\mathbf{m}_k \in \mathcal{C}^{M+\mathcal{I}}$ as the stacked vector of $\{\mathbf{s}_k, \mathbf{i}_k\}$, where $\mathbf{i}_k = \mathbf{D}_k\mathbf{s}_k$ is a vector of $\mathcal{I}$ arbitrarily modulated versions of the source waveform that may be regarded as "interference" signals.

$$\bar{\mathbf{x}}_k = [\mathbf{A} + \mathbf{B}\mathbf{D}_k]\,\mathbf{s}_k = \mathbf{C}\mathbf{m}_k \tag{13.81}$$

For ID multipath signals, the time-varying mode wavefront components may be viewed as subspace interference. Providing $\mathbf{C}$ has full rank $M + \mathcal{I} < N$, and the $\mathcal{I}$ arbitrary modulation sequences are linearly independent for any displacement $\ell \in (0, 2L]$, the conclusions for the noiseless case also hold for the case of subspace interference due to an ID multipath signal. Under such conditions, it can be readily shown that the spatial filter solutions take the form of Eqn. (13.79), where the unit vector $\mathbf{u}_m$ has been defined previously, and $\mathbf{0}_\mathcal{I}$ is the zero vector of length $\mathcal{I}$. In this case, $\mathbf{w}$ and $\mathbf{r}$ preserve the undisturbed waveforms associated with the steady wavefronts $\mathbf{a}_i$ and $\mathbf{a}_j$, respectively, while spare DOFs are used to reject all other signal components, including those which carry the arbitrarily modulated versions of the source waveform. The robustness of GEMS to mode wavefront fluctuations over the processing interval will be illustrated using experimental data in Section 13.4.

$$\mathbf{w} = \beta\, \mathbf{C}^{+} \begin{bmatrix} \mathbf{u}_i \\ \mathbf{0}_{\mathcal{I}} \end{bmatrix} + \mathbf{P}_{\mathbf{C}}^{\perp}\mathbf{q}_w, \quad \mathbf{r} = \beta\, \mathbf{C}^{+} \begin{bmatrix} \mathbf{u}_j \\ \mathbf{0}_{\mathcal{I}} \end{bmatrix} + \mathbf{P}_{\mathbf{C}}^{\perp}\mathbf{q}_r \tag{13.82}$$

Mode wavefront fluctuations within the processing interval will expand the rank of the system by effectively increasing the number of unwanted signals. Under the stated assumptions, spare DOFs in the processor may be used to cancel these unwanted components, which can degrade the source waveform estimate.

Although GEMS is notionally based on the minimization of the error vector norm for the CD multipath model, robustness to the presence of time-varying wavefront distortions described by the ID multipath model can be achieved by increasing the number of sensors in the array. This mitigates the rank expansion (i.e., consumption of degrees of freedom) caused by random channel fluctuations over the processing interval. Similar arguments apply for the multiple-source case, where one or more signal modes may be described by the ID model.

13.3.2 Operational Procedure

This section describes an operational GEMS routine based on the optimization criterion described for the noiseless case. To simplify notation, the procedure is described for a single source using the CD model, although the same procedure applies for the multiple-source case and the ID model. In other words, the operational GEMS procedure may be used for both source and multipath separation, without modifying the technique for different models. The full receiver-space version of GEMS is presented here, but a rank-reduction transform (e.g., beam-space processing or truncated singular value decomposition) may be applied to the data as a preprocessing step to reduce dimensionality.

In the presence of full-rank noise, denoted by the $N \times K$ matrix $\mathbf{N} = [\mathbf{n}_1, \ldots, \mathbf{n}_K]$, we recall that the reference data matrix $\mathbf{X} \in \mathcal{C}^{N\times K}$ is given by Eqn. (13.83), where $\mathbf{A} = [\mathbf{a}_1, \ldots, \mathbf{a}_M]$ is the $N \times M$ mixing matrix, and $\mathbf{S} = [\mathbf{s}_1, \ldots, \mathbf{s}_K]$ is the $M \times K$ signal matrix.

$$\mathbf{X} = [\mathbf{x}_1, \ldots, \mathbf{x}_K] = \mathbf{AS} + \mathbf{N} \tag{13.83}$$

The auxiliary data vectors $\mathbf{u}_k$ are constructed as time-delayed and Doppler-shifted reference data vectors according to Eqn. (13.84). The delay and Doppler coordinates $\{\ell, \nu\}$ represent input parameters to the algorithm. The vector $\tilde{\mathbf{s}}_k = \mathbf{s}_{k-\ell}e^{j2\pi\nu k}$ is defined as before, while $\tilde{\mathbf{n}}_k = \mathbf{n}_{k-\ell}e^{j2\pi\nu k}$.

$$\mathbf{u}_k = \mathbf{x}_{k-\ell}e^{j2\pi\nu k} = \{\mathbf{A}\mathbf{s}_{k-\ell} + \mathbf{n}_{k-\ell}\}e^{j2\pi\nu k} = \mathbf{A}\tilde{\mathbf{s}}_k + \tilde{\mathbf{n}}_k \tag{13.84}$$

The auxiliary data matrix $\mathbf{U} \in \mathcal{C}^{N\times K}$ may be expressed in the form of Eqn. (13.85) by defining the $M \times K$ signal matrix $\tilde{\mathbf{S}} = [\tilde{\mathbf{s}}_1, \ldots, \tilde{\mathbf{s}}_K]$, and the $N \times K$ noise matrix $\tilde{\mathbf{N}} = [\tilde{\mathbf{n}}_1, \ldots, \tilde{\mathbf{n}}_K]$. Note that $\mathbf{U}$ is implicitly a function of $\{\ell, \nu\}$, but this dependence is momentarily dropped for notational convenience.

$$\mathbf{U} = [\mathbf{u}_1, \ldots, \mathbf{u}_K] = \mathbf{A}\tilde{\mathbf{S}} + \tilde{\mathbf{N}} \tag{13.85}$$

In the reference channel, the weight vector $\mathbf{w} \in \mathcal{C}^N$ processes the received data $\mathbf{X}$ to yield the output time-series $\mathbf{z} = [z_1, \ldots, z_K] = \mathbf{w}^\dagger \mathbf{X}$. Similarly, in the auxiliary channel, the weight vector $\mathbf{r} \in \mathcal{C}^N$ processes the data $\mathbf{U}$ to yield the output time-series $\mathbf{y} = [y_1, \ldots, y_K] = \mathbf{r}^\dagger \mathbf{U}$. The error vector $\mathbf{e} = [e_1, \ldots, e_K]$ is the difference between the reference and auxiliary outputs in Eqn. (13.86).

$$\mathbf{e} = \mathbf{z} - \mathbf{y} = \mathbf{w}^\dagger \mathbf{X} - \mathbf{r}^\dagger \mathbf{U} \tag{13.86}$$

Recall that the constraint ensuring a non-trivial solution in the noiseless case is $\|\mathbf{w}^\dagger \bar{\mathbf{X}}\|^2 = \|\mathbf{f}^\dagger \mathbf{S}\|^2 = 1$, where $\bar{\mathbf{X}}$ is the noiseless data matrix. In the presence of full-rank noise, $\bar{\mathbf{X}}$ is replaced by $\mathbf{X}$, such that the non-triviality constraint used for the operational procedure is given by Eqn. (13.87).

$$\|\mathbf{w}^\dagger \mathbf{X}\|^2 = \|\mathbf{f}^\dagger \mathbf{S} + \mathbf{w}^\dagger \mathbf{N}\|^2 = 1 \tag{13.87}$$

The GEMS algorithm is based on minimizing the L_2-norm of the error vector $\mathbf{e}$ in the quadratically constrained optimization problem of Eqn. (13.88). The dependence of the cost function $\epsilon(\ell, \nu)$ and auxiliary data matrix $\mathbf{U}(\ell, \nu)$ on the input delay and Doppler-shift settings $\{\ell, \nu\}$ is explicitly included in Eqn. (13.88).

$$\epsilon(\ell, \nu) = \min_{\mathbf{w},\mathbf{r}} \|\mathbf{w}^\dagger \mathbf{X} - \mathbf{r}^\dagger \mathbf{U}(\ell, \nu)\|^2 \quad \text{s.t.} \quad \|\mathbf{w}^\dagger \mathbf{X}\|^2 = 1 \tag{13.88}$$

For a particular (unnamed) delay and Doppler-shift setting, the objective function to be minimized $J(\mathbf{w}, \mathbf{r}) = \|\mathbf{e}\|^2$ may be expanded and expressed in the form of Eqn. (13.89), where the sample matrices are defined as $\mathbf{R} = \mathbf{X}\mathbf{X}^\dagger$, $\mathbf{F} = \mathbf{U}\mathbf{U}^\dagger$, and $\mathbf{G} = \mathbf{X}\mathbf{U}^\dagger$.

$$J(\mathbf{w}, \mathbf{r}) = (\mathbf{w}^\dagger \mathbf{X} - \mathbf{r}^\dagger \mathbf{U})(\mathbf{w}^\dagger \mathbf{X} - \mathbf{r}^\dagger \mathbf{U})^\dagger = \mathbf{w}^\dagger \mathbf{R}\mathbf{w} - \mathbf{r}^\dagger \mathbf{G}^\dagger \mathbf{w} - \mathbf{w}^\dagger \mathbf{G}\mathbf{r} + \mathbf{r}^\dagger \mathbf{F}\mathbf{r} \tag{13.89}$$

Similarly, the quadratic constraint $\mathcal{C}(\mathbf{w}) = \|\mathbf{w}^\dagger \mathbf{X}\|^2 = \mathbf{w}^\dagger \mathbf{X}\mathbf{X}^\dagger \mathbf{w}$ may be written in terms of the sample covariance matrix $\mathbf{R}$ in Eqn. (13.90).

$$\mathcal{C}(\mathbf{w}) = \mathbf{w}^\dagger \mathbf{R}\mathbf{w} \tag{13.90}$$

The optimization problem is to jointly find the minimizing weight vector arguments $\{\hat{\mathbf{w}}, \hat{\mathbf{r}}\}$ according to Eqn. (13.91), where the dependence of these weight vectors on the inputs $\{\ell, \nu\}$ is implicit.

$$\{\hat{\mathbf{w}}, \hat{\mathbf{r}}\} = \arg\min_{\mathbf{w},\mathbf{r}} J(\mathbf{w}, \mathbf{r}) \quad \text{s.t.} \quad \mathcal{C}(\mathbf{w}) = 1 \tag{13.91}$$

Differentiating $J(\mathbf{w}, \mathbf{r})$ with respect to the auxiliary weights $\mathbf{r}$, and setting the partial derivative to zero yields the minimizer $\hat{\mathbf{r}}$ in Eqn. (13.92). The matrix $\mathbf{F}$ is invertible in full-rank noise provided $K \geq N$.

$$\partial J(\mathbf{w}, \mathbf{r})/\partial \mathbf{r} = \mathbf{F}\mathbf{r} - \mathbf{G}^\dagger \mathbf{w} = 0 \;\Rightarrow\; \hat{\mathbf{r}} = \mathbf{F}^{-1}\mathbf{G}^\dagger \mathbf{w} \tag{13.92}$$

Substitution of the minimizer $\hat{\mathbf{r}} = \mathbf{F}^{-1}\mathbf{G}^\dagger \mathbf{w}$ for $\mathbf{r}$ in the objective function of Eqn. (13.89), and simplifying the terms, yields a cost function $\mathcal{J}(\mathbf{w}) = J(\mathbf{w}, \hat{\mathbf{r}})$ in terms of $\mathbf{w}$ only in Eqn. (13.93).

$$\mathcal{J}(\mathbf{w}) = \mathbf{w}^\dagger (\mathbf{R} - \mathbf{G}\mathbf{F}^{-1}\mathbf{G}^\dagger)\mathbf{w} \tag{13.93}$$

By defining $\mathbf{Q} = \mathbf{R} - \mathbf{G}\mathbf{F}^{-1}\mathbf{G}^\dagger$, the optimization problem in Eqn. (13.91) requires the minimization of a quadratic cost function in $\mathbf{w}$, subject to a quadratic equality constraint in $\mathbf{w}$, as in Eqn. (13.94).

$$\hat{\mathbf{w}} = \arg\min_{\mathbf{w}} \mathbf{w}^\dagger \mathbf{Q}\mathbf{w} \quad \text{s.t.} \quad \mathbf{w}^\dagger \mathbf{R}\mathbf{w} = 1 \tag{13.94}$$

The solution to this constrained optimization problem can be found by minimizing the function $\mathcal{H}(\mathbf{w}, \lambda)$ in Eqn. (13.95), where λ is a Lagrange multiplier.

$$\mathcal{H}(\mathbf{w}, \lambda) = \mathbf{w}^\dagger \mathbf{Q}\mathbf{w} + \lambda(1 - \mathbf{w}^\dagger \mathbf{R}\mathbf{w}) \tag{13.95}$$

Taking the derivative of Eqn. (13.95) with respect to $\mathbf{w}$ and equating to zero results in the generalized eigenvalue problem of Eqn. (13.96). The Lagrange multiplier λ is the generalized eigenvalue of the matrix pencil $\{\mathbf{Q}, \mathbf{R}\}$.

$$\partial\mathcal{H}(\mathbf{w}, \lambda)/\partial\mathbf{w} = \mathbf{Q}\mathbf{w} - \lambda\mathbf{R}\mathbf{w} = 0 \;\Rightarrow\; \mathbf{Q}\mathbf{w} = \lambda\mathbf{R}\mathbf{w} \tag{13.96}$$

As $\mathbf{R}$ and $\mathbf{Q}$ are positive-definite Hermitian matrices in full-rank noise (for $K \geq N$), it follows that for any generalized eigenvector $\mathbf{w}$, all generalized eigenvalues are real and positive from Eqn. (13.97).

$$\mathbf{w}^\dagger \mathbf{Q}\mathbf{w} = \lambda \mathbf{w}^\dagger \mathbf{R}\mathbf{w} \tag{13.97}$$

Application of the constraint $\mathbf{w}^\dagger\mathbf{R}\mathbf{w} = 1$ in Eqn. (13.97) leads to Eqn. (13.98). This shows that for any generalized eigenvector $\mathbf{w}$, the resulting cost $\mathcal{J}(\mathbf{w})$ is equal to the associated generalized eigenvalue λ.

$$\mathcal{J}(\mathbf{w}) = \mathbf{w}^\dagger \mathbf{Q}\mathbf{w} = \lambda \tag{13.98}$$

The solution $\hat{\mathbf{w}}$ is the generalized eigenvector corresponding to the smallest generalized eigenvalue $\lambda_{\min}$ of the matrix pencil $\{\mathbf{Q}, \mathbf{R}\}$. Multiplying Eqn. (13.96) by $\mathbf{Q}^{-1}$ we obtain Eqn. (13.99).

$$\mathbf{Q}^{-1}\mathbf{R}\mathbf{w} = \frac{1}{\lambda}\mathbf{w} \tag{13.99}$$

This is the characteristic equation for $\mathbf{Z} = \mathbf{Q}^{-1}\mathbf{R}$, where the minimum generalized eigenvalue $\lambda_{\min}$ corresponds to the *maximum* eigenvalue $(1/\lambda_{\min})$ of $\mathbf{Z}$. Hence $\hat{\mathbf{w}}$ takes the form of Eqn. (13.100), where the operator $\mathcal{P}\{\cdot\}$ returns the principal eigenvector of a matrix.

$$\mathbf{Z}\hat{\mathbf{w}} = \frac{1}{\lambda_{\min}}\hat{\mathbf{w}} \;\Rightarrow\; \hat{\mathbf{w}} \propto \mathcal{P}\{\mathbf{Z}\} = \mathbf{z}_1 \tag{13.100}$$

The scale of $\hat{\mathbf{w}}$ is determined by the constraint $\mathcal{C}(\hat{\mathbf{w}}) = \hat{\mathbf{w}}^\dagger\mathbf{R}\hat{\mathbf{w}} = 1$. Hence, the solution for a particular value of $\{\ell, \nu\}$ can be expressed in the closed-form expression of Eqn. (13.101). From Eqn. (13.92), the auxiliary weights minimizing the cost function are given by $\hat{\mathbf{r}} = \mathbf{F}^{-1}\mathbf{G}^\dagger\hat{\mathbf{w}}$.

$$\hat{\mathbf{w}} = \mathbf{z}_1(\mathbf{z}_1^\dagger \mathbf{R}\mathbf{z}_1)^{-1/2} \tag{13.101}$$

Scaling of the principal eigenvector by $(\mathbf{z}_1^\dagger\mathbf{R}\mathbf{z}_1)^{-1/2}$ to satisfy the quadratic constraint on the reference weights does not affect output SINR, but such normalization is important to meaningfully compare the cost $\epsilon(\ell, \nu) = \mathcal{J}(\hat{\mathbf{w}})$ at different values of $\{\ell, \nu\}$.

$$\epsilon(\ell, \nu) = \hat{\mathbf{w}}^\dagger \mathbf{Q}\hat{\mathbf{w}} = \lambda_{\min}(\mathbf{z}_1^\dagger \mathbf{R}\mathbf{z}_1)^{-1} \tag{13.102}$$

The coordinates of the deepest minimum of the cost function, denoted by $\{\hat{\ell}, \hat{\nu}\}$ in Eqn. (13.103), is expected at a delay and Doppler that matches the differential values between the two most dominant modes. As the SNR of these modes tends to infinity, the noiseless case is approached and $\epsilon(\ell, \nu) \to 0$ at the coordinates $\{\hat{\ell}, \hat{\nu}\}$ that satisfy this condition.

$$\{\hat{\ell}, \hat{\nu}\} = \arg\min_{\ell,\nu} \epsilon(\ell, \nu) \tag{13.103}$$

The GEMS spatial filter $\hat{\mathbf{w}}_{\mathcal{G}}$ is extracted as the weight vector solution at the coordinates of the global minimum of $\epsilon(\ell, \nu)$, i.e., $\hat{\mathbf{w}}_{\mathcal{G}} = \hat{\mathbf{w}}(\hat{\ell}, \hat{\nu})$. The source waveform estimate $\hat{s}_k$ is computed by processing the received data snapshots $\mathbf{x}_k$ with the GEMS spatial filter, as in Eqn. (13.104).

$$\hat{s}_k = \hat{\mathbf{w}}_{\mathcal{G}}^{\dagger}\mathbf{x}_k \tag{13.104}$$

For a single source with more than two dominant modes, a deep local minima is expected at delay-Doppler coordinates matching the differential values existing between each pair of modes. However, recovering $\hat{s}_k$ from the global minimum is expected to lead to the best waveform estimate, since this minimum is most likely to be formed on the two most dominant modes. This completely defines the operational GEMS procedure for the single-source case. Figure 13.8 illustrates the data-flow for the GEMS algorithm, while the following summarizes its underlying rationale.

- In the noiseless case, a non-trivial solution for $\underline{\upsilon}$ yielding $\bar{\mathbf{e}} = \underline{\upsilon}^{\dagger}\bar{\mathbf{S}} = \mathbf{0}$ in Eqn. (13.62) exists if and only if the delay-Doppler displacement $\{\ell, \nu\}$ matches the differential coordinates between a pair of modes (H_1). Under H_0, such a solution does not exist, and $\bar{\mathbf{e}}^{\dagger}\bar{\mathbf{e}} = \|\bar{\mathbf{e}}\|^2 > 0$ for all $\underline{\upsilon} \neq \mathbf{0}$.
- Under H_1, the non-trivial solution for $\underline{\upsilon}$ that satisfies $\bar{\mathbf{e}} = \underline{\upsilon}^{\dagger}\bar{\mathbf{S}} = \mathbf{0}$ in Eqn. (13.62) is unique up to a complex scale factor. This solution completely defines the set of all possible spatial filters $\{\mathbf{w}, \mathbf{r}\}$ that respectively isolate the matched pair of modes.
- In the absence of noise, the spatial filters $\{\mathbf{w}, \mathbf{r}\}$ spanned by this solution set will exactly recover a complex-scaled, time-delayed, and Doppler-shifted copy of the input sequence free of multipath which satisfies the processing objective.
- When noise is present, minimization of the error vector norm $\|\mathbf{e}\|^2$ in Eqn. (13.88) with respect to the spatial filters $\{\mathbf{w}, \mathbf{r}\}$, and delay-Doppler displacement $\{\ell, \nu\}$, provides a basis for GEMS to blindly estimate the source waveform.
- The GEMS spatial filter $\hat{\mathbf{w}}_{\mathcal{G}}$ optimizing this criterion function is used to derive the waveform estimate $\hat{s}_k$ in Eqn. (13.104). The auto-ambiguity function may be used to cue the GEMS procedure to the differential delay and Doppler between the two most dominant modes (as described shortly).
- While there are infinitely many spatial filters that can recover the source waveform in the *noiseless* case, the unique GEMS solution in the full-rank noise case attempts to reduce the noise contribution at the output as much as possible.

When multiple sources are present, it is necessary to interrogate more than one local minima in order to recover the different source waveforms. One approach is to interrogate all deep local minima of the GEMS cost function. Once the waveform estimates are

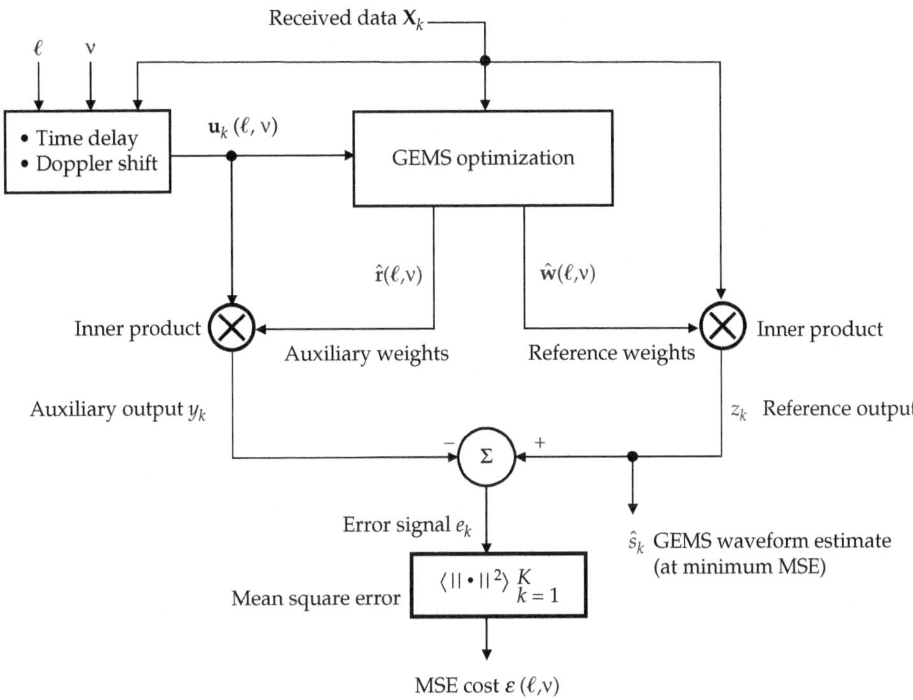

FIGURE 13.8 Flow-chart illustrating the GEMS algorithm. The procedure involves an optimization step that requires computing a closed-form solution for each input parameter setting over a bank of delays and Doppler shifts. Specifically, reference and auxiliary weight vectors are computed from a block of K input data vectors $\mathbf{x}_k$ for each input value of ℓ and ν as the solution of a generalized eigenvalue problem. Spatial processing is performed to generate the mean square error (MSE) cost function $\epsilon(\ell, \nu)$. In the single-source case, the GEMS waveform estimate may be extracted as the reference output at the delay-Doppler coordinates corresponding to the global minimum of the MSE cost function. In the multiple-source case, different waveform estimates may be extracted as reference outputs corresponding to the delay-Doppler coordinates of different deep local minima in the MSE cost function.

extracted from each deep local minima, measures based on cross correlation analysis may be adopted to distinguish between waveform estimates corresponding to different sources, or different modes of the same source. This method clearly assumes that the different sources do not emit time-delayed and Doppler-shifted versions of a common waveform, as stated previously. Practical application of GEMS to the single- and multiple-source scenarios will be illustrated using experimental data in Sections 13.4 and 13.5, respectively.

13.3.3 Computational Complexity

This section analyzes the computational complexity of the GEMS algorithm relative to that of the classic (DOA-based) signal-copy procedure using the MUSIC estimator, which serves as a benchmark for comparison. Numerical examples are provided for the single-source case, where only the global minimum of the cost function needs to be interrogated for waveform recovery. However, the computational complexity calculations may be scaled accordingly to reflect different scenarios, including the multiple-source case, where a number of deep local minima need to be interrogated. The analysis is presented in a manner that allows the computational complexity of GEMS to be readily calculated for other practical situations of interest not limited to the examples provided in this section.

As the coordinates of the global minimum $\{\hat{\ell}, \hat{\nu}\}$ of the cost function $\epsilon(\ell, \nu)$ are not known a priori, a number of options are available in an operational system:

- *Exhaustive Search*: Evaluate $\epsilon(\ell, \nu)$ over a bank of delay-Doppler bins restricted to plausible values of (13.69). The granularity of the bins may be set to the sampling period T_s in delay and the FFT bins spaced by T_o^{-1} in Doppler. In a time-invariant system, only a one-dimensional search in delay is required.
- *Localized Search*: For propagation channels with slowly-varying and smoothly evolving nominal mode parameter values (i.e., correlated changes in the nominal mode delay and Doppler coordinates), the search space may be narrowed for observation intervals subsequent to the first based on regions near the previously observed minima locations.
- *Point Search*: The auto-ambiguity function of the time-series output in a single receiver may be computed to estimate the locations of the mode differential delay-Doppler coordinates as the local peaks of this function, which may be used to cue the coordinates at which $\epsilon(\ell, \nu)$ is evaluated.

The first option can be computationally expensive, depending on the number of receivers N, the extent of delay-Doppler domain to be searched, and the grid resolution used. The faster (third) option is to compute the auto-ambiguity function of the reference receiver output x_k in Eqn. (13.105), where $K' = K - N_\ell$. It is well known that $|\chi(\ell, \nu)|$ exhibits local peaks at values of the differential delay-Doppler offset between pairs of modes (Zhang, Tao, and Ma 2004). The global maximum of the ambiguity function occurs at the origin, while the coordinates of the *next* highest peak is expected at the differential delay-Doppler offset between the two most dominant modes. The coordinates of this second highest local peak in $|\chi(\ell, \nu)|$ may be used as the input to GEMS (i.e., to cue the GEMS procedure), as will be shown in the next section.

$$\chi(\ell, \nu) = \sum_{k=1}^{K'} x_k x^*_{k+\ell} e^{-j2\pi\nu k/K'} \quad \begin{cases} \ell \in [0, N_\ell] \\ \nu = [-N_\nu/2, N_\nu/2] \end{cases} \tag{13.105}$$

The computational complexity analysis for the single-source case is based on the point-search option, where the ambiguity function surface is used to cue GEMS to the coordinates of the global minimum. This represents a computationally efficient implementation of GEMS, as only one point in the cost function needs to be evaluated and interrogated for waveform estimation. The computational complexity may be readily scaled to account for the number of points evaluated in the cost function, as well as the number of deep local minima interrogated for waveform estimation. In other words, the elements of the analysis for the point-search option may be used to deduce complexity for other implementations.

Processing Step	NCM	Study Case
Sample matrix $\mathbf{R}$ $[N \times N]$	$N^2 \times K$	3.2×10^7
Noise subspace $\mathbf{U}_n$ $[N \times N]$	$O(N^3)$	16,384
MUSIC spectrum $\mathbf{v}(\psi)^\dagger \mathbf{U}_n \mathbf{v}(\psi)$	$(N^2 + N) \times N_\theta \times N_\phi$	435,200
Mixing matrix $\mathbf{A}$ $[N \times M]$	–	–
Compute $\mathbf{B} = \mathbf{A}^\dagger \mathbf{A}$ $[M \times M]$	$M^2 \times N$	64
Invert $\mathbf{B}^{-1}$ $[M \times M]$	$O(M^3)$	32
Multiply $\mathbf{C} = \mathbf{B}^{-1}\mathbf{f}$ $[M \times 1]$	M^2	4
Weights $\mathbf{w} = \mathbf{AC}$ $[N \times 1]$	$N \times M$	32
Process $\hat{s}_k = \mathbf{w}^\dagger \mathbf{x}_k, \forall k \in [1, K]$	$N \times K$	2×10^6
Total Complexity	F_C	34,451,716

TABLE 13.1 Computational complexity of classic signal-copy procedure using MUSIC for DOA estimation. © Commonwealth of Australia 2011.

A standard figure of merit for computational complexity is the number of complex multiplications (NCM). The complexity of the classic signal-copy procedure using the MUSIC DOA estimator is shown in Table 13.1. The study case numbers reflect the practical system parameters of $N = 16$, $K = 125{,}000$, and $M = 2$, which are relevant to the experiment described in the next section. The MUSIC spectrum is calculated over N_θ azimuths and N_ϕ elevations with $N_\theta = N_\phi = 40$. An $N \times N$ matrix inversion or eigen-decomposition is assumed to have an NCM of $4N^3$ in all comparisons. Table 13.2 shows the complexity for GEMS. By eliminating unnecessary computations (Zhang et al. 2004),

Processing Step	NCM	Study Case
Ambiguity $\chi(\ell, \nu)$ $[N_\ell \times N_\nu]$	$f(K, N_\ell, N_\nu)$	10,254,258
Sample matrix $\mathbf{R}$ $[N \times N]$	$N^2 \times K$	3.2×10^7
Sample matrix $\mathbf{F}$ $[N \times N]$	$N^2 \times K$	3.2×10^7
Sample matrix $\mathbf{G}$ $[N \times N]$	$N^2 \times K$	3.2×10^7
Invert $\mathbf{F}^{-1}$ $[N \times N]$	$O(N^3)$	16,384
Compute $\mathbf{B} = \mathbf{F}^{-1}\mathbf{G}^\dagger$ $[N \times N]$	N^3	4,096
Multiply $\mathbf{C} = \mathbf{GB}$ $[N \times N]$	N^3	4,096
Form $\mathbf{Q} = \mathbf{R} - \mathbf{C}$ $[N \times N]$	–	–
Invert $\mathbf{Q}^{-1}$ $[N \times N]$	$O(N^3)$	16,384
Form $\mathbf{Z} = \mathbf{Q}^{-1}\mathbf{R}$ $[N \times N]$	N^3	4,096
Principal eigenvector $\mathbf{z}_1$ $[N \times 1]$	$O(N^3)$	16,384
Normalization $\beta = \sqrt{\mathbf{z}_1^\dagger \mathbf{R} \mathbf{z}_1}$	$(N^2 + N)$	272
GEMS weights $\mathbf{w} = \mathbf{z}_1/\beta$ $[N \times 1]$	N	16
Process $\hat{s}_k = \mathbf{w}^\dagger \mathbf{x}_k$ $\forall k \in [1, K]$	$N \times K$	2×10^6
Total Complexity	$F_{\mathcal{G}}$	108,315,986

TABLE 13.2 Computational complexity of the GEMS waveform estimation algorithm. © Commonwealth of Australia 2011.

and confining attention to the domain of uncertainty in $\{\ell, \nu\}$, the ambiguity function NCM is given by Eqn. (13.106), where an NCM of $(N/2)\log_2 N$ is assumed for an N-point FFT. The numbers $N_\ell = N_\nu = 40$ mirror the values used in the experimental results of the following section.

$$f(K, N_\ell, N_\nu) = N_\ell \times \{2K + (N_\nu/2)\log_2 N_\nu\} + 2K \tag{13.106}$$

The GAM signal-copy approach cited previously requires the solution of a generalized eigenvalue problem (GEP) of small dimension at *each* DOA grid point, so it is computationally more expensive than MUSIC. For comparisons, the MUSIC DOA-based signal-copy procedure of lower complexity will be used as a benchmark. The relative complexity $F_{\mathcal{R}} = F_{\mathcal{G}}/F_{\mathcal{C}}$ is defined as the NCM ratio of GEMS to that of the MUSIC DOA-based signal-copy procedure. As this figure varies with system parameters, $F_{\mathcal{R}}$ is plotted as a function of K using different values of N in Figure 13.9a, and as a function of N using different values of K in Figure 13.9b. The main observation is that the complexity of GEMS is higher, but of the same order of magnitude as the classic signal-copy method over a wide range of system parameters.

The performance of this cued GEMS procedure relative to the DOA and GAM signal-copy methods will be illustrated using experimental data with known truth of the source waveform in the following section. Such comparison enables the tradeoff between waveform estimation performance and computational complexity to be evaluated for these different techniques in a real-world system. The computational complexity of GEMS will clearly be higher for multiple sources. In this case, changes in the relative complexity $F_{\mathcal{R}}$ may be calculated by using the elements listed in Tables 13.1 and 13.2. An attractive feature of the GEMS algorithm is that the processing is highly parallelizable, such that the computational burden may be shared over multiple CPUs.

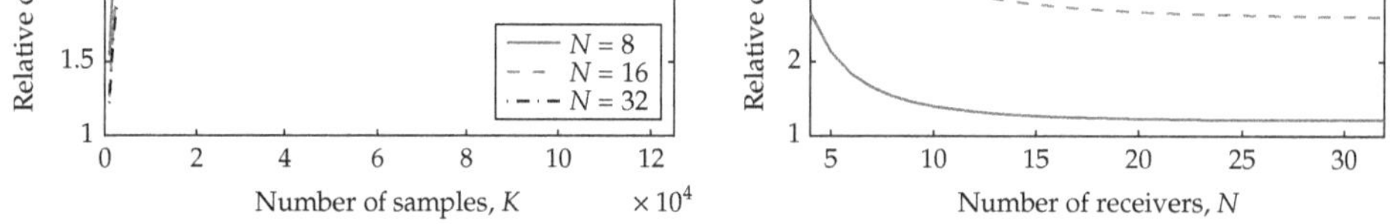

(a) Dependence on data length K.

(b) Dependence on array dimension N.

FIGURE 13.9 Relative complexity $F_{\mathcal{R}}$ as a function of data length K and number of receivers N. © Commonwealth of Australia 2011.

13.4 SIMO Experiment

The main reason signal-processing techniques such as GEMS are developed is that they may be gainfully employed in practice. Numerical results in a simulation environment can provide useful insights regarding performance when it comes to comparing different techniques under controlled conditions. However, the critical question often raised by signal-processing communities relates to the effectiveness and robustness of different techniques in real systems. Despite the value of simulation studies, what is of ultimate concern to a user is how well the newly proposed techniques perform in practice, and what additional capability they can provide in actual systems with respect to traditional methods.

Although many signal-processing techniques have been successfully developed and tested on the basis of computer simulations, the path chosen here is to assess and compare the performance of GEMS against two benchmark signal-copy methods using real data acquired by an experimental HF system. By evaluating performance on real data collected in the field, this contribution complements the many existing works on blind waveform estimation by redressing the lack of experimental results available on this topic in the open literature.

13.4.1 Data Collection

Experimental data were collected by a well-calibrated two-dimensional (L-shaped) HF antenna array composed of $N = 16$ vertical monopole elements, spaced 8 m apart on each arm, with a digital receiver per element. This system was previously described in Chapter 10. The signal of interest (SOI) was received via the ionosphere from a source located at a ground range of 1851 km. The source transmitted a repetitive linear frequency modulated continuous waveform with center frequency $f_c = 21.620$ MHz, bandwidth $B = 10$ kHz, and pulse repetition frequency $f_p = 62.5$ Hz. Knowledge of the transmitted signal provides ground-truth information for assessing the estimation performance of different techniques that do not assume any prior knowledge about the source waveform. The receiving system acquired in-phase and quadrature (I/Q) components of the down-converted and baseband-filtered signals at a sampling rate of $f_s = 62.5$ kHz. The data was collected continuously but processed as a sequence of coherent processing intervals (CPI) of $T_o = 2$-second duration.

By re-synthesizing the known reference signal, the channel-scattering function (CSF) resulting in the reference (first) receiver of the array is shown in Figure 13.10a as an intensity-modulated delay-Doppler display normalized by the maximum value. The received energy is dominated by $M = 2$ distinct propagation modes resolved at nominal time delays $\{\tau_1 = 6.54 \text{ ms}, \tau_2 = 6.88 \text{ ms}\}$. The Doppler spectra in Figure 13.10b are line plots taken from Figure 13.10a at time delays $\{\tau_1 = 6.54 \text{ ms}, \tau_2 = 6.88 \text{ ms}\}$ to show the relative strengths and Doppler shifts of the two modes. The mode SNRs are in excess of 50 dB, but the *SINR* of mode 1 is only 6 dB when mode 2 is considered as the interfering multipath signal to be removed for waveform estimation. The two propagation modes also have different Doppler shifts $\{f_1 = 0.3 \text{ Hz}, f_2 = 1.0 \text{ Hz}\}$ due to the different regular components of large-scale movement in the ionospheric layers responsible for signal reflection over the observation interval.

The normalized conventional spatial spectrum for the same CPI of data is shown in Figure 13.11a. The maximum value occurs at the azimuth and elevation $\{\theta = 134°$,

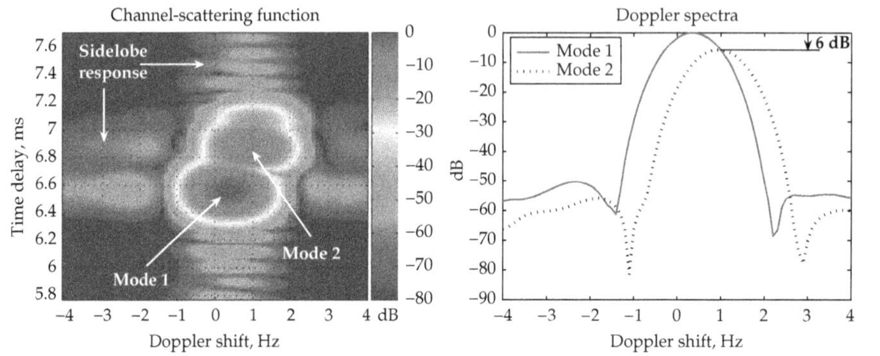

(a) Channel-scattering function (CSF) display. (b) Doppler line plots at the two mode time delays.

FIGURE 13.10 Location and relative strengths of the two dominant propagation modes resolved in the delay-Doppler plane. © Commonwealth of Australia 2011.

$\phi = 18^\circ\}$. The azimuth matches the known bearing of the source (134°), while the elevation (18°) is consistent with the angle expected based on geometrical considerations. However, the conventional spatial spectrum is not able to resolve the two modes in nominal DOA. Figure 13.11b shows the MUSIC spectrum for the same data assuming $M = 2$ signals. This spectrum is normalized by the maximum value, which also occurs at $\{\theta = 134^\circ, \phi = 18^\circ\}$.

The two dominant modes cannot be resolved by MUSIC, even when the number of assumed signals is increased to $M = 3$ and $M = 4$ (not shown here). The inability to resolve the two modes is not only due to their closely spaced nominal DOAs, but also the departures of the mode wavefronts from the plane-wave model. The performance of super-resolution DOA estimation techniques is very sensitive to model mismatch, and consequently, so are the waveform estimation (signal-copy) procedures based upon them.

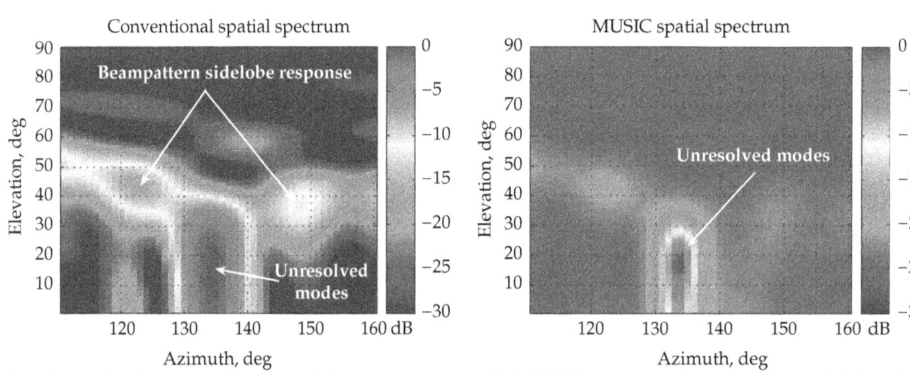

(a) Normalized conventional spatial spectrum. (b) MUSIC spectrum for same data with $M = 2$.

FIGURE 13.11 The two dominant modes cannot be resolved in nominal DOA by conventional beamforming and the MUSIC super-resolution technique. © Commonwealth of Australia 2011.

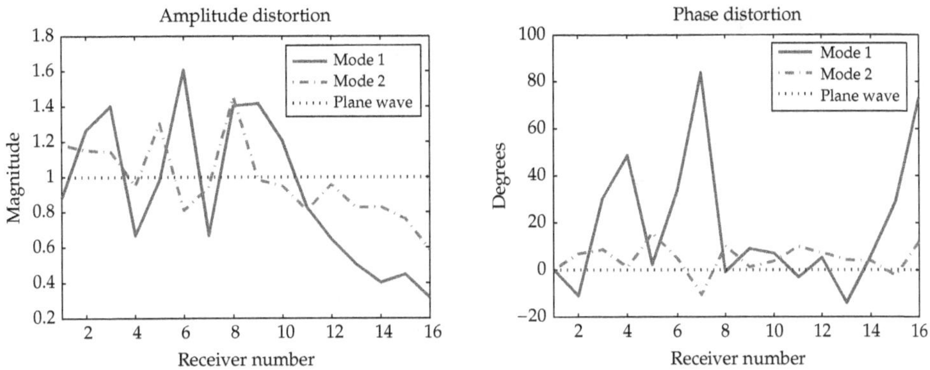

(a) Magnitude of wavefront distortions.

(b) Phase of wavefront distortions.

FIGURE 13.12 Complex-valued wavefront distortions received for modes 1 and 2 relative to the plane-wave model of best fit. © Commonwealth of Australia 2011.

The wavefront distortions may be observed by resolving the modes in the delay-Doppler plane across all $N = 16$ receivers using the known reference waveform, as for the CSF. The mode wavefronts are extracted as the data snapshots received in the delay-Doppler bins corresponding to the peak coordinates of each mode, given by $\{\tau_1 = 6.54$ ms, $f_1 = 0.3$ Hz$\}$ and $\{\tau_2 = 6.88$ ms, $f_2 = 1.0$ Hz$\}$. These snapshot vectors represent the mode wavefronts $\{\mathbf{a}_m\}_{m=1,2}$, which may be normalized to yield the mode spatial signatures $\{\mathbf{v}_m = \mathbf{v}(\psi_m) \odot \mathbf{d}_m\}_{m=1,2}$ defined previously. The plane waves of best fit in a least-squares sense were found to have nominal DOAs of $\psi_1 = [\theta_1 = 134°, \phi_1 = 17°]$ and $\psi_2 = [\theta_2 = 134°, \phi_2 = 18°]$. The multiplicative magnitude and phase distortions $\{\mathbf{d}_m\}_{m=1,2}$ imposed on each mode relative to these plane waves of best fit are plotted as a function of receiver number in Figures 13.12a and 13.12b, respectively.

Both the magnitude and phase deviate significantly with respect to the plane-wave model. In addition, the distortions are quite different for the two modes (i.e., propagation path dependent). The illustrated deviations from the presumed array steering vector manifold are the main reason why MUSIC fails to resolve the modes. Note that $\mathcal{F}\{\mathbf{v}_1, \mathbf{v}_2\} = 0.71$, while $\mathcal{F}\{\mathbf{v}(\psi_1), \mathbf{v}(\psi_2)\} = 0.98$, which implies that the mode spatial signatures are significantly further apart in N-dimensional space than the underlying plane waves of best fit. Hence, for techniques that do not rely on a manifold model parameterized by the signal DOA alone, the additional diversity in wavefront shape introduced by diffuse scattering provides greater opportunity to spatially resolve the modes. The nominal mode parameters listed in Table 13.3 were derived using the reference waveform, and represent ground-truth information for performance assessment in the analysis to follow.

Mode	Delay, ms	Doppler, Hz	Azimuth, deg.	Elevation, deg.
$m = 1$	6.54	0.3	134	17
$m = 2$	6.88	1.0	134	18

TABLE 13.3 Ground-truth information on nominal parameter values for propagation modes 1 and 2.

13.4.2 Signal-Copy Methods

In the case of spatially spread signals with non-planar wavefronts, a DOA estimation procedure can at best resolve and estimate the DOAs of the plane waves of best fit to the received spatial signatures. Although the nominal DOAs of the two dominant modes are not resolved by conventional beamforming or MUSIC, the plane waves of best fit extracted using the reference signal may be used to assess the performance of signal-copy procedures based on the plane-wave model. Reconstructing the mixing matrix using the plane waves known to be of best fit to the mode spatial signatures effectively provides an upper limit on the performance of any DOA-based signal-copy procedure for $M = 2$.

Setting $\hat{\mathbf{A}}_{doa} = [\mathbf{v}(\psi_1), \mathbf{v}(\psi_2)]$ and $\mathbf{u}_1 = [1, 0]^T$ in Eqn. (13.31) to preserve the first mode and eliminate the second mode yields the DOA-based signal-copy weight vector $\hat{\mathbf{w}}_{doa}$ according to Eqn. (13.31). Figure 13.13a shows that an SINR improvement of 8 dB can be achieved at the output of the spatial filter $\hat{\mathbf{w}}_{doa}$ with respect to that in the reference receiver. The choice to reject mode 2 in preference of mode 1 favors the DOA-based signal-copy procedure because this mode is more similar to a plane wave and is therefore canceled more effectively by the null at ψ_2.

A GAM technique was also implemented for performance comparisons. The GAM approach in Astely et al. (1998) models the wavefront of a distributed signal as $\mathbf{v}(\psi, \vartheta) = \mathbf{v}(\psi) + \vartheta\dot{\mathbf{v}}(\psi)$, where $\dot{\mathbf{v}}(\psi) = \partial\mathbf{v}(\psi)/\partial\psi$, and ϑ is a free complex scalar parameter. A MUSIC-like spectrum may also be computed using the GAM model, but this technique is also unable to resolve the two modes. The potential effectiveness of the GAM method may be similarly assessed on the mode wavefronts extracted using the reference waveform. The GAM vectors $\{\mathbf{v}(\psi_m, \vartheta_m)\}_{m=1,2}$ that provide the best least-squares fit to the extracted mode spatial signatures were computed, and the mixing matrix model $\hat{\mathbf{A}}_{gam} = [\mathbf{v}(\psi_1, \vartheta_1), \mathbf{v}(\psi_2, \vartheta_2)]$ was substituted for $\mathbf{A}$ in Eqn. (13.31) to derive the GAM-based signal-copy weight vector $\hat{\mathbf{w}}_{gam}$.

As shown in Figure 13.13b, the output of $\hat{\mathbf{w}}_{gam}$ provides a relative SINR improvement of 16 dB with respect to a single receiver. This is an additional 8 dB of SINR with respect to the traditional DOA-based signal-copy method. Such a substantial improvement testifies

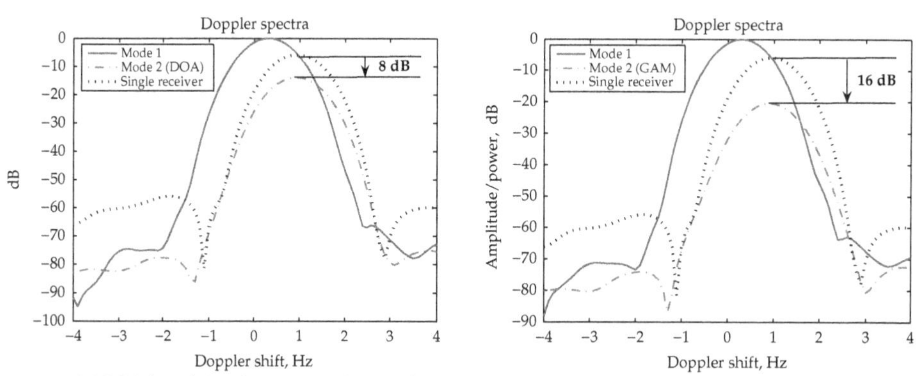

(a) DOA-based signal-copy procedure ≈8 dB.

(b) GAM-based signal-copy procedure ≈16 dB.

FIGURE 13.13 Doppler spectra showing improvement in output SINR resulting from the application of two benchmark signal-copy procedures relative to that in a single receiver.

to the potential value of the GAM approach in a practical application. However, the mode to be rejected is still more than 25 dB above the noise floor. The reason the GAM method cannot provide a higher level of rejection is that such a manifold model is valid only for small angular spreads, whereas the distortions encountered in this application are large, and hence not accurately represented by the GAM model.

13.4.3 Application of GEMS

The application of GEMS is described in two parts. The first illustrates the capability of GEMS to blindly recover an accurate estimate of the source waveform that is contaminated by multipath. The second illustrates the capability of GEMS to estimate the crinkled wavefronts of the incident signal modes. Unlike the DOA- and GAM-based signal-copy methods, GEMS is applied in a strictly blind manner, i.e., without use of the reference waveform to estimate the mode spatial signatures. This represents a truly operational implementation of the GEMS algorithm.

13.4.3.1 GEMS Waveform Estimation

Figure 13.14a shows the cost function $\epsilon(\ell, \nu)$ resulting for the GEMS algorithm, normalized by the maximum value over the search grid. The global minimum is located at {34 ms, 0.70 Hz}, which agrees well with the differential mode time delay $\tau_2 - \tau_1 = 34$ ms and Doppler shift $f_2 - f_1 = 0.7$ Hz using the values in Table 13.3. Figure 13.14b shows the normalized auto-ambiguity of the reference receiver output $|\chi(\ell, \nu)|/\chi(0, 0)$, where the delay-Doppler coordinates of the highest peak (excluding the maximum at the origin) match the location of the global minimum in Figure 13.14a to the grid resolution. The coordinates of this peak $\{\hat{\ell}, \hat{\nu}\}$ are used to compute the GEMS weight vector $\hat{\mathbf{w}}_{\mathcal{G}}$ for source waveform estimation.

Figure 13.15a shows that the SINR improvement at the output of $\hat{\mathbf{w}}_{\mathcal{G}}$ is over 40 dB with respect to a single receiver, and 25 dB better than the GAM signal-copy method. Importantly, this substantial improvement has been obtained *directly* from the data in a strictly blind manner by the GEMS algorithm, i.e., without resorting to the known

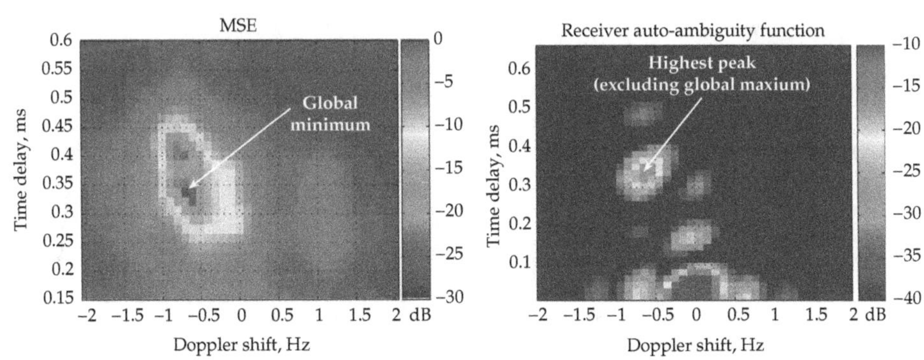

(a) GEMS cost function in delay and Doppler. (b) Auto-ambiguity function of single receiver.

FIGURE 13.14 GEMS cost function showing location of global minimum, and auto-ambiguity function of a single receiver showing location of the second highest peak. © Commonwealth of Australia 2011.

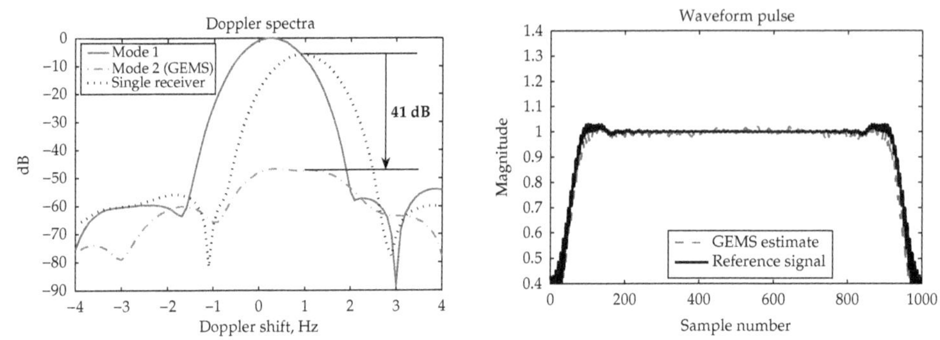

(a) SINR improvement for GEMS >40 dB.

(b) GEMS estimate of pulse shape.

FIGURE 13.15 SINR improvement for GEMS, and a comparison of the waveform pulse shape estimated by GEMS with the known reference over 1000 A/D samples (16 ms). © Commonwealth of Australia 2011.

reference signal for resolving the modes in delay and Doppler. Hence, unlike the DOA and GAM signal-copy procedures, the results achieved by GEMS are a real improvement attained in practice, as opposed to an upper limit on performance.

The impact on waveform recovery may be appreciated in Figures 13.15b and 13.16b, which compare the pulse shape of the known reference signal with those recovered by the three different estimation techniques. The GEMS estimate practically overlays the known reference in Figure 15.15b, while the benchmark signal-copy estimates exhibit signal envelope distortions due to residual multipath contamination in Figure 13.16b. Figure 13.16a shows the estimate resulting when the unit-norm non-triviality constraint $\|[\mathbf{w}^T, \mathbf{r}^T]\|^2 = 1$ is used instead of the GEMS constraint $\|\mathbf{w}^\dagger \mathbf{X}\|^2 = 1$ in Eqn. (13.88). The minimizing singular vector that arises for the former constraint attempts to cancel all signals and produces a very noisy estimate. This demonstrates the effectiveness of the GEMS constraint.

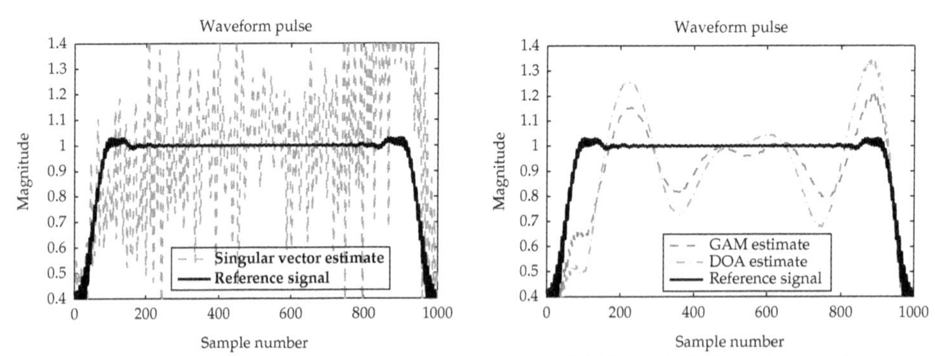

(a) Noisy singular vector estimate.

(b) DOA and GAM signal-copy estimates.

FIGURE 13.16 Comparison of the waveform pulse shape estimated using the singular vector method with those estimated using the DOA- and GAM-based signal-copy approaches. © Commonwealth of Australia 2011.

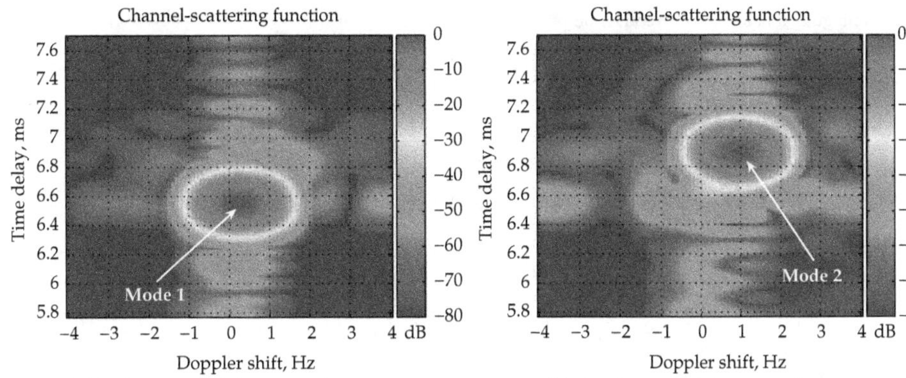

(a) Reference weight vector $\hat{\mathbf{w}}_G$ output.

(b) Auxiliary weight vector $\hat{\mathbf{r}}_G$ output.

FIGURE 13.17 CSF outputs showing that the GEMS weight vector $\hat{\mathbf{w}}_{\mathcal{G}}$ isolates mode 1 at the reference output, while $\hat{\mathbf{r}}_{\mathcal{G}} = \mathbf{F}^{-1}\mathbf{G}^{\dagger}\hat{\mathbf{w}}_{\mathcal{G}}$ isolates mode 2 at the auxiliary output. Compare these results to the CSF display for a single receiver in Figure 13.10, where both modes are present in the output. © Commonwealth of Australia 2011.

The reference $\hat{\mathbf{w}}_{\mathcal{G}}$ and auxiliary $\hat{\mathbf{r}}_{\mathcal{G}}$ GEMS weight vectors are expected to separate the two modes, such that the earlier and later arrivals are isolated in the reference and auxiliary outputs, respectively. The CSF computed for the reference and auxiliary GEMS outputs are shown in Figures 13.17a and 13.17b, respectively. A comparison of these two displays with the CSF resulting for a single receiver in Figure 13.10a shows that the two modes have been effectively isolated by the spatial filters $\hat{\mathbf{w}}_{\mathcal{G}}$ and $\hat{\mathbf{r}}_{\mathcal{G}}$. The earlier arrival appears at the output of $\hat{\mathbf{w}}_{\mathcal{G}}$, which is consistent with the expected performance of the technique. The removal of inter-mode contamination in each display of Figure 13.17 is quite remarkable, and confirms the practical effectiveness of GEMS for blind multipath separation.

An alternative visualization of the beneficial effect of multipath separation is presented in Figure 13.18, which shows the auto-ambiguity function (AF) of the GEMS waveform estimate compared to that of a single receiver output. The GEMS estimate in Figure 13.18b is consistent with the point-spread function of a repetitive linear FMCW waveform, while Figure 13.18a shows the deleterious effect of multipath in the single receiver output. Multipath causes significant "blurring" of the true point-spread function, and gives rise to a number of significant peaks that are well-separated from the main lobe. The poor quality of the resulting image in Figure 13.18a can significantly complicate the interpretation of the signal components present in the data based on the output of a cross-correlation receiver. With reference to these AF diagrams, the action of GEMS may be interpreted as the auto-focusing of a blurred image by virtue of multipath removal.

The single-source scenario is relevant to a number of practical applications, such as long-range HF communications and over-the-horizon passive radar (Fabrizio et al. 2009). In the latter case, an accurate estimate of the transmitted waveform from a source of opportunity is important for matched filtering (i.e., coherent range-Doppler processing). Such systems strive to minimize contamination due to multipath, such that matched filtering can be performed based on the source waveform point-spread function, as opposed to an inferior version that has been disturbed by multipath. The presence of multipath

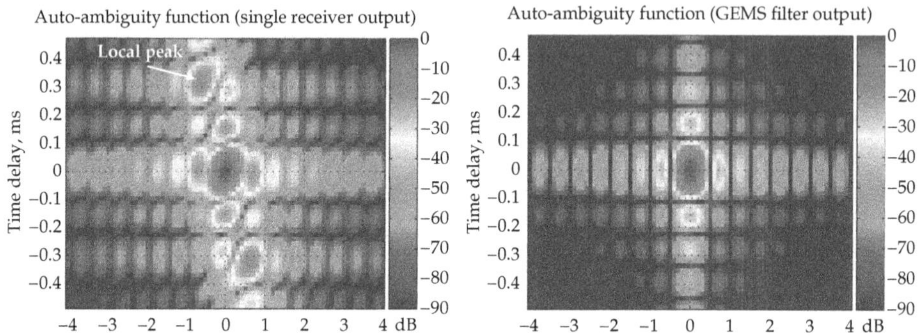

(a) Reference receiver output. (b) GEMS waveform estimate.

FIGURE 13.18 Auto-ambiguity function of a single receiver output and the GEMS waveform estimate. The former exhibits significant blurring due to multipath contamination, while the latter exhibits a well-focused point-spread function consistent with that of the transmitted source waveform. © Commonwealth of Australia 2011.

in the matching waveform may cause the detection of false targets in passive radar applications.

13.4.3.2 GEMS Wavefront Estimation

Once the mode waveforms have been separated using GEMS, it is possible to reconstruct an estimate of the signal matrix $\hat{\mathbf{S}}$ defined previously. In the case of $M = 2$ dominant modes, the rows of $\hat{\mathbf{S}} \in \mathcal{C}^{M\times K}$ are given by the outputs $z_k = \hat{\mathbf{w}}_G^\dagger \mathbf{x}_k$ and $z'_k = \hat{\mathbf{r}}_G^\dagger \mathbf{x}_k$ for $k = 1, \ldots, K$, where $\hat{\mathbf{w}}_G$ and $\hat{\mathbf{r}}_G$ are the weight vectors extracted from the global minimum of the GEMS cost function. Based on the data model $\mathbf{X} = \mathbf{AS} + \mathbf{N}$, and conditioned on the estimate $\hat{\mathbf{S}}$, the least-squares estimate of the mixing matrix containing the mode wavefronts is given by $\hat{\mathbf{A}}$ in Eqn. (13.107), where $\hat{\mathbf{S}}^+ = \hat{\mathbf{S}}^\dagger(\hat{\mathbf{S}}\hat{\mathbf{S}}^\dagger)^{-1}$.

$$\hat{\mathbf{A}} = [\hat{\mathbf{a}}_1, \ldots, \hat{\mathbf{a}}_M] = \mathbf{X}\hat{\mathbf{S}}^+ = \mathbf{A}\{\mathbf{S}\hat{\mathbf{S}}^\dagger(\hat{\mathbf{S}}\hat{\mathbf{S}}^\dagger)^{-1}\} + \mathbf{N}\hat{\mathbf{S}}^+ \tag{13.107}$$

Estimates of the mode spatial signatures, denoted by $\hat{\mathbf{v}}_m$, are obtained by normalizing the columns of $\hat{\mathbf{A}}$ as in Eqn. (13.108), such that $\hat{\mathbf{v}}_m^\dagger \hat{\mathbf{v}}_m = N$. For example, consider the second mode $m = 2$, which has a more planar wavefront. Figure 13.19 compares the magnitude of the spatial signature estimated by GEMS for this mode as a function of receiver number with the measurements based on the reference waveform. The latter may be interpreted as true measurements of the mode wavefront structure. The GEMS estimate matches these measurements quite well across all $N = 16$ receivers.

$$\hat{\mathbf{v}}_m = \frac{\hat{\mathbf{a}}_m\sqrt{N}}{\sqrt{\hat{\mathbf{a}}_m^\dagger \hat{\mathbf{a}}_m}} \tag{13.108}$$

Results for the unwrapped phase of the same mode are compared in Figure 13.20, where the first receiver is used as the phase reference. The phase progression is quasi-linear over

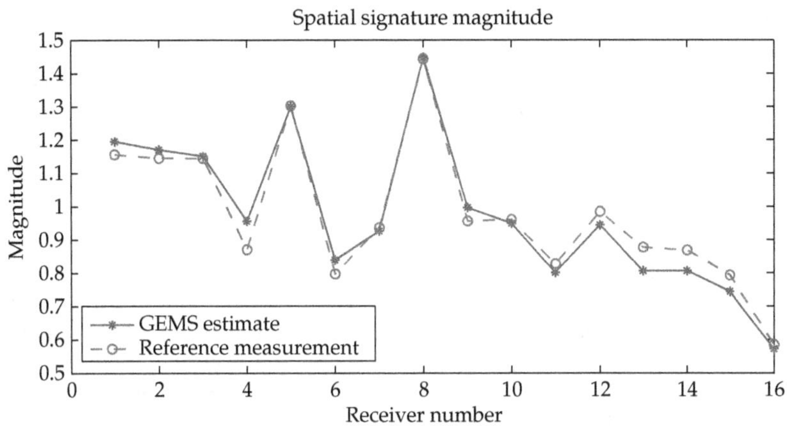

FIGURE 13.19 Receiver element comparison of the spatial signatures measured and estimated for mode 2 in magnitude. The reference measurements were extracted from the range-Doppler bin containing mode 2 after matched filtering is applied to each receiver using the known reference signal. The GEMS blind spatial signature estimate agrees well with the reference measurement. © Commonwealth of Australia 2011.

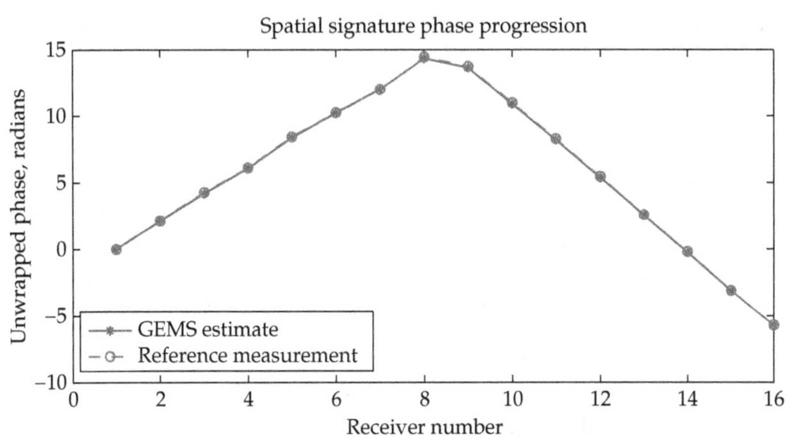

FIGURE 13.20 Receiver element comparison of the unwrapped phase across the two ULA arms of the L-shaped array for the spatial signatures estimated with GEMS and measured using the reference waveform. The phase estimates derived from GEMS agree well with reference measurements, and may be used to infer the nominal DOA of mode 2. © Commonwealth of Australia 2011.

each arm of the L-shaped array, where the receiver numbers $n = 1, \ldots, 8$ and $n = 9, \ldots, 16$ form ULAs at right angles to each other. The mean slope of each quasi-linear phase progression depends on the nominal DOA of mode 2 with respect to the boresight of each arm. The GEMS phase estimates also agree well with measurements using the reference waveform. These wavefront estimates are useful for inferring the nominal DOA of modes that cannot be resolved on the plane-wave manifold (such as in the case of MUSIC). As demonstrated in the final section of this chapter, nominal mode DOA information may be used in conjunction with an ionospheric model to geolocate an HF source using a single ULA.

13.4.3.3 CD and ID Wavefronts

Although GEMS is notionally based on the CD mode wavefront model, it was claimed in Section 13.3 that GEMS is robust to ID mode wavefronts, provided that sufficient spatial degrees of freedom are available. Earlier in this section, it was shown that GEMS blindly recovered accurate estimates of the transmitted source waveform and received mode wavefronts in a practical HF system. It is now of interest to investigate whether the two dominant modes present in the experimental data are better described by a CD or ID model. The wavefront planarity test proposed in Fabrizio et al. (2000a) may be applied to reveal the presence and characteristics of time-varying wavefront distortions over the processing interval for each mode.

Using the known reference waveform for range processing in each FMCW pulse, this test is based on the statistic defined by $\rho_m(t_p) \in [0, 1]$ in Eqn. (13.109), where $\mathbf{y}_m(t_p)$ is the array snapshot data vector extracted from range bin $g_m = c\tau_m/2$ containing mode m at pulse repetition interval $t_p = 1, \ldots, P$. In our case, we have that $P = 125$ for a 2-second CPI. Range processing allows the two dominant modes to be resolved into different group-range bins. This enables spatial signature variations to be analyzed over the different pulses in the CPI on a mode-separated basis.

$$\rho_m(t_p) = \max_{\psi} \left\{ \frac{|\mathbf{v}^\dagger(\psi)\mathbf{y}_m(t_p)|^2}{\mathbf{v}^\dagger(\psi)\mathbf{v}(\psi) \, . \, \mathbf{y}_m^\dagger(t_p)\mathbf{y}_m(t_p)} \right\} \tag{13.109}$$

The maximizing argument of Eqn. (13.109), denoted by $\psi_m(t_p)$, represents the DOA of the plane-wave model $\mathbf{v}(\psi)$ that best fits the received snapshot $\mathbf{y}_m(t_p)$ at pulse number t_p. The value of $\rho_m(t_p)$ is a measure of how well the wavefront that is received from mode m matches the best fitting plane-wave at pulse number t_p. A value near unity indicates an almost planar wavefront, while lower values indicate further deviation from the plane-wave manifold. Since the statistic $\rho_m(t_p)$ is invariant to the complex scale of $\mathbf{y}_m(t_p)$, it is agnostic to mode Doppler shifts. In other words, $\psi_m(t_p)$ and $\rho_m(t_p)$ are sensitive only to changes in the mode spatial signature. The CD model described previously is consistent with fixed values of $\psi_m(t_p)$ and $\rho_m(t_p)$ over the CPI, while variations with pulse number are expected for the ID model.

Figure 13.21 shows the profiles $\rho_m(t_p)$ resulting for the two modes $m = 1, 2$. The spatial signature of mode 1 is observed to deviate further from the plane-wave manifold than mode 2, which has a more planar spatial signature. The reasonably constant value of the profile for $m = 2$ indicates that this mode has a rather rigid wavefront over the CPI, i.e., more consistent with the CD signal model. On the other hand, the smooth variation of the profile observed for $m = 1$ indicates that the spatial signature of this mode is changing in a correlated manner over the CPI. This is more consistent with an (partially

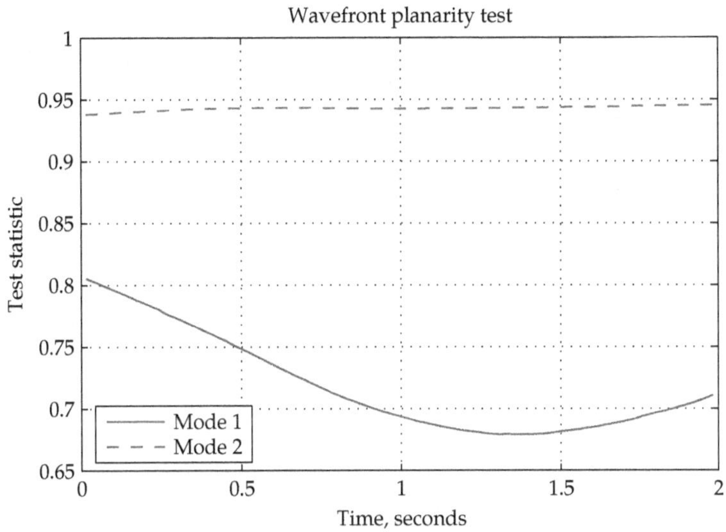

FIGURE 13.21 Example of wavefront planarity test statistic plotted for two simultaneously received modes over a 2-second CPI (125 PRI). The test indicates that mode 2 has a more planar and stable spatial signature over the CPI than mode 1. The spatial signature of the former is consistent with a CD model, while that of the latter exhibits greater temporal variation as is hence more aligned with the ID model.

correlated) ID signal model. In other words, the multipath environment is in this case best described by a mixture of CD and ID modes. Such results indicate that modes reflected from physically well-separated regions in the ionosphere can exhibit wavefront distortions that have quite different structures and dynamic characteristics. The previous results on waveform estimation using the same data demonstrate the robustness of GEMS to the presence of an ID mode.

13.5 MIMO Experiment

A situation that is often of practical interest to wireless communication and radar systems involves the presence of multiple sources on the same frequency channel (i.e., co-channel sources). The different sources are assumed to be spatially separated such that propagation to the system occurs via nonidentical multipath channels. In this case the sensor array receives a convolutive mixture of the transmitted source signals. This was previously called a finite impulse response multiple-input multiple-output (FIR-MIMO) system. In mobile communications, it is desirable to separate and recover copies of the signals transmitted from multiple sources, whereas in active radar, the presence of independent co-channel sources may represent interference to be removed by the system.

Figure 13.22 schematically illustrates an FIR-MIMO system for $Q = 2$ sources. In this example, one source corresponds to the radar signal, while the other represents interference. Incidental interference due to co-channel sources is a common problem in

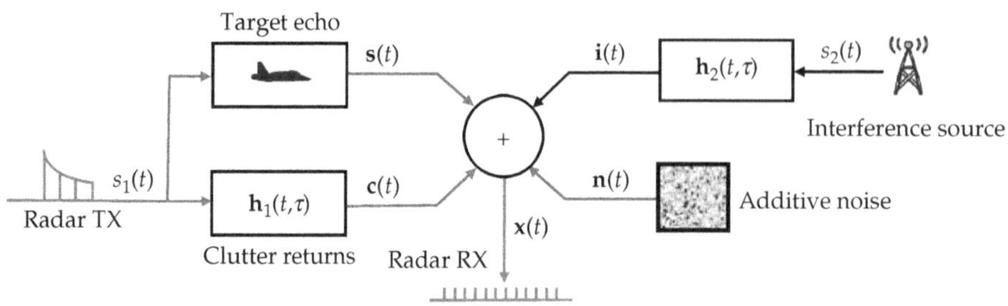

FIGURE 13.22 FIR-MIMO system illustrating the reception of two independent sources by an antenna array via different multipath channels. In this example, the first source $s_1(t)$ is a radar transmitter, which generates clutter and target echoes, while the second source $s_2(t)$ represents co-channel interference that is uncorrelated with the radar waveform. © Commonwealth of Australia 2011.

the HF environment. This is mainly due to the dense occupancy of the HF band, and the ability of the ionosphere to propagate HF signals over very long distances (particularly at night when there is no D-region absorption).

13.5.1 Data Collection

The L-shaped HF array described previously was used to simultaneously collect a number of different sources in the 62.5-kHz receiver bandwidth at a carrier frequency of 21.639 MHz. The spectrogram in Figure 13.23 shows a linear FMCW radar signal along side two AM radio signals acquired by a single receiver in the array. The different source transmissions occupy non-overlapping frequency channels to avoid mutual interference. For the purposes of this study, it is possible to superimpose two or more different signals onto a common center frequency. This creates a realistic co-channel source scenario for testing BSS techniques.

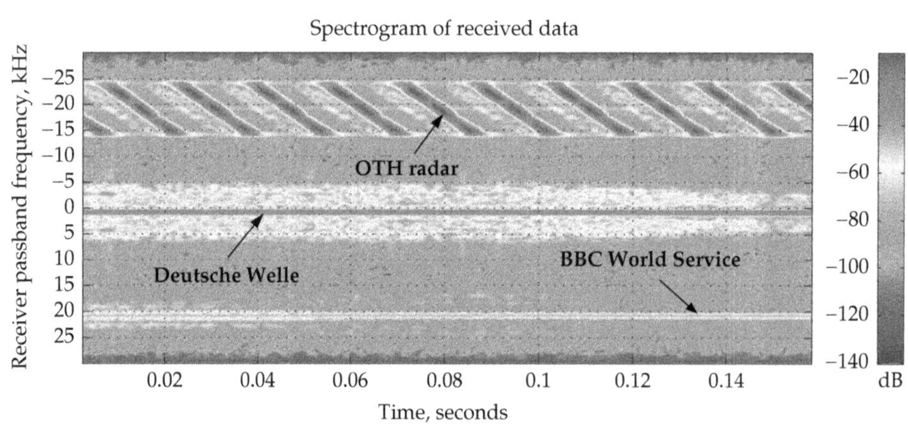

FIGURE 13.23 Sources in the receiver bandwidth include a linear FMCW radar signal, and two AM sources identified as BBC World Service and Deutsche Welle radio broadcasts. © Commonwealth of Australia 2011.

An FIR-MIMO system with $Q = 2$ sources was generated by digital down-conversion of the FMCW signal and BBC World Service broadcast to baseband and then superimposing the two received signals. This experimental technique avoids mutual interference, but effectively mimics the situation in which the two sources are present on the same frequency channel. Since the coherence bandwidth of the ionospheric channel is in the order of tens of kilohertz, this method faithfully replicates data that would arise for truly co-channel sources. GEMS and the classic signal-copy procedure are applied directly to the resulting AM-FM signal mixture.

For reference, the source emitting the linear FMCW was an HF radar, located at a ground range of 1851 km and bearing of 134° N with respect to the receiving system. The AM radio broadcast was from a BBC World Service station in Kranji, Singapore, located at a ground range of approximately 3400 km and bearing of 295° N from the receiver. As for the FIR-SIMO experiment, the data were collected continuously, but analyzed as a sequence of coherent processing intervals with 2-second duration.

13.5.2 Source and Multipath Separation

Figure 13.24 shows the MUSIC spatial spectrum for a particular CPI plotted as a function of azimuth θ and elevation ϕ assuming two signals. The DOA parameter vector is denoted by $\psi = [\theta, \phi]$. The two peaks occur at $\psi_1 = [134^\circ, 19^\circ]$ and $\psi_2 = [295^\circ, 16^\circ]$. The azimuth estimates match the known FM and AM source bearings. However, multiple propagation modes could not be resolved in elevation by MUSIC when more signals were assumed (not shown here). This is partly due to the close spacing of the modes in elevation, and the poor elevation angle resolution of a ground-deployed array at near grazing incidence.

Letting $\mathbf{w}_1^\dagger \mathbf{A} = \mathbf{u}_1$ where $\mathbf{A} = [\mathbf{v}(\psi_1), \mathbf{v}(\psi_2)]$ and $\mathbf{u}_1 = [1, 0]$ leads to $\mathbf{w}_1 = \mathbf{A}(\mathbf{A}^\dagger \mathbf{A})^{-1}\mathbf{u}_1^\dagger$ which yields the classic null-steering signal-copy weight vector for recovering the FM

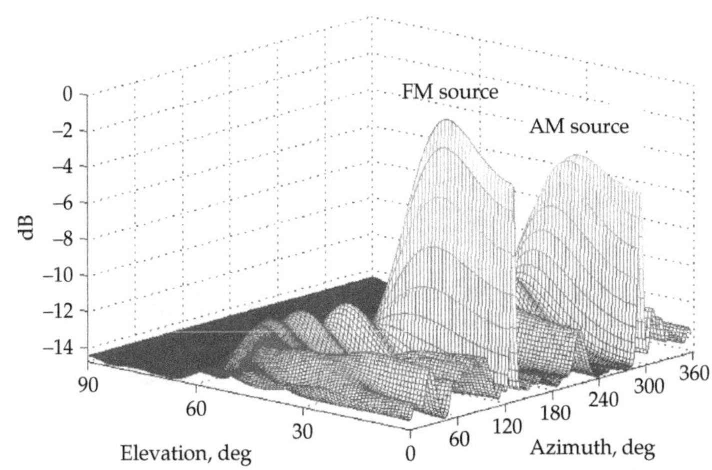

FIGURE 13.24 Two-dimensional MUSIC spectrum showing the azimuth and elevation of the FM and AM signals received by the L-shaped array. The different propagation modes of each HF source could not be resolved in elevation. © Commonwealth of Australia 2011.

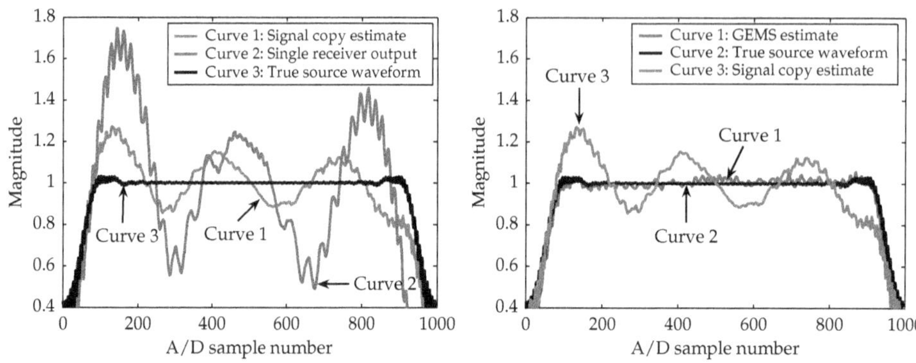

(a) Single receiver and signal-copy estimate. (b) GEMS estimate of waveform pulse shape.

FIGURE 13.25 Comparison of the true FM waveform pulse shape with the estimates derived from a single receiver output, the classic DOA-based signal-copy procedure, and GEMS. © Commonwealth of Australia 2011.

source waveform. The resulting estimate of this waveform $\tilde{s}_1(k) = \mathbf{w}_1^\dagger \mathbf{x}_k$ is compared with the true source waveform, and the output of a single receiver, in terms of the pulse shape in Figure 13.25a. Although the null has reduced the corrupting effect of the AM signal on the FM source waveform estimate, significant distortion remains due to multipath interference as well as residual AM signal energy. Figure 13.25b shall be referred to in a moment.

Figure 13.26 compares the GEMS cost function for the FIR-SIMO and FIR-MIMO data. Figure 13.26a is a replica of the cost function shown previously for the FM signal only in Figure 13.14a. Figure 13.26b, plotted using a different time-delay scale, shows the cost function when GEMS is applied to the AM-FM signal mixture. Two deep local minima

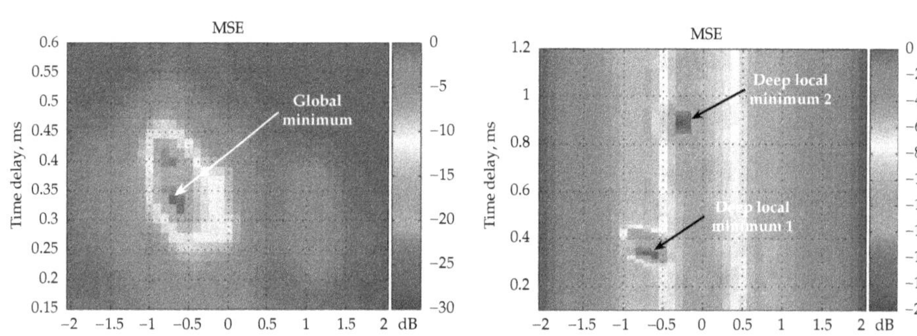

(a) FM signal only (FIR-SIMO system). (b) FM and AM signals (FIR-MIMO system).

FIGURE 13.26 Comparison of GEMS cost functions arising for the single-source (FM only) and multiple-source (FM and AM) cases. The right-hand figure is plotted with a different time-delay scale to capture the second minimum. © Commonwealth of Australia 2011.

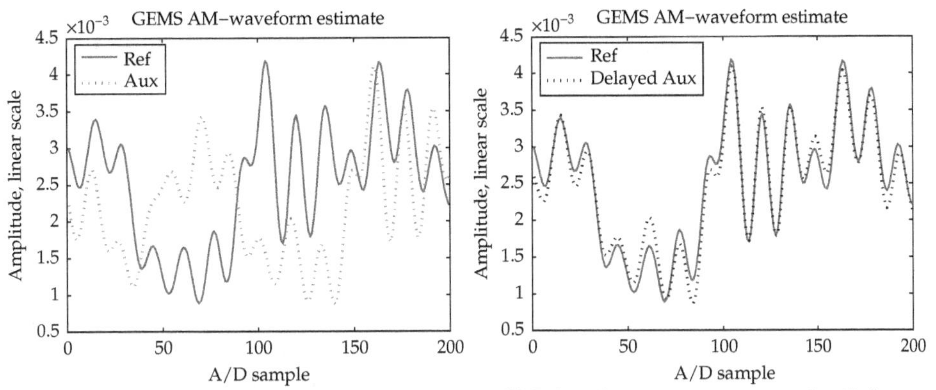

(a) Simultaneous Ref and Aux outputs. (b) Delayed Aux output compared to Ref output.

FIGURE 13.27 Magnitude of the reference and auxiliary GEMS waveform estimates for the AM source. In the figure on the right-hand side, the auxiliary output has been delayed by 0.88 ms (the location of the second minimum) to show the match between the two dominant propagation modes of the AM source.

$\{\hat{\ell}_q, \hat{\nu}_q\}_{q=1,2}$ are evident in the latter display. As expected, a minimum appears at the same coordinates as the global minimum in Figure 13.26a. This is labeled as "minimum 1" in Figure 13.26b, and is due to the FM source in the signal mixture. The application of GEMS to the FIR-MIMO system containing $Q = 2$ sources produces another deep local minimum, which is labeled as "minimum 2" in Figure 13.26b. This minimum is not related to the FM signal, but is formed due to the presence of the AM signal.

The GEMS weight vectors $\hat{\mathbf{w}}_{\mathcal{G}}(q)$ corresponding to the minima $\{\hat{\ell}_q, \hat{\nu}_q\}$ for $q = 1, 2$ were used to recover estimates of the associated source waveforms $\hat{s}_q(k) = \hat{\mathbf{w}}_{\mathcal{G}}(q)^{\dagger}\mathbf{x}_k$. Figure 13.25b compares the GEMS estimate $\hat{s}_1(k)$ and the DOA-based signal-copy estimate with the true source waveform. Clearly, the GEMS procedure yields a much more accurate waveform estimate than the classic signal-copy method. Indeed, the waveform recovered by GEMS exhibits minimal contamination and practically overlays the reference waveform.[3]

No ground-truth information is available to directly assess whether the GEMS waveform estimate $\hat{s}_2(k)$ has recovered a copy of the AM source signal. The magnitude envelope of the estimated AM signal is shown as the solid line in Figure 13.27a. The dashed line in Figure 13.27a shows the auxiliary output formed as $\bar{s}_2(k) = \hat{\mathbf{r}}_{\mathcal{G}}(2)^{\dagger}\mathbf{x}_k$, where $\hat{\mathbf{r}}_{\mathcal{G}}(2)$ is the auxiliary weight vector resulting at the minimum $\{\hat{\ell}_2, \hat{\nu}_2\}$. At first glance, the reference and auxiliary outputs appear to be unrelated in Figure 13.27a.

However, by delaying the auxiliary output a number of samples corresponding to the delay at which minimum 1 occurs (i.e., $\hat{\ell}_2$), it is possible to compare estimates of the AM source signal carried by different propagation modes. Figure 13.27b shows a remarkable agreement between the amplitude envelopes of $\hat{s}_2(k)$ and $\bar{s}_2(k - \hat{\ell}_2)$. It is evident that the two outputs are practically time-delayed replicas of one another, as expected for correct source and multipath separation. As these two outputs are derived from different spatial

[3]The transmitted FM signal is tapered in amplitude at the edges of the pulse to reduce out-of-band spectral emissions.

filters in the GEMS procedure, this result provides a high level of confidence that the AM source has also been estimated accurately in the presence of the FM source and multipath.

13.5.3 Radar Application

The question arises as to how the application of GEMS may provide a benefit in radar systems. To see this, consider a radar beam formed in the nominal direction of the AM interference source. A spectrogram of the resulting *beam* output is shown in Figure 13.28. The presence of the AM continuous wave (CW) signal and sideband components are clearly visible in the same frequency band as the FMCW radar signal. Two synthetic target echoes have also been injected, but are not distinguishable from the much more powerful AM and FMCW signals in Figure 13.28. These useful signals were injected at the receiver sample level (i.e., in the raw data prior to range processing and beamforming). The two synthetic targets have different ranges and Doppler shifts, but both are incident from the nominal direction of the AM "interference."

Figure 13.29a shows the range-Doppler map resulting from the conventional beam output of Figure 13.28. The two strong direct-wave clutter peaks indicated in this figure correspond to the two dominant modes of the FMCW radar signal, which are clearly visible near zero Doppler frequency in the range-Doppler display. The powerful CW component of the AM signal manifests itself as a vertical "stripe" that is well-localized in Doppler frequency, but spreads over all range bins after pulse compression. On the other hand, the energy contained in the AM signal sidebands (i.e., the modulation component) is distributed over the entire range-Doppler map. In this case, both the CW and sideband components of the AM source mask the injected targets.

The AM source waveform is estimated using GEMS, as described in the previous section. The only exception is that GEMS is applied to the data containing the two synthetic targets, in addition to the AM and FM source signals, as required in an operational radar

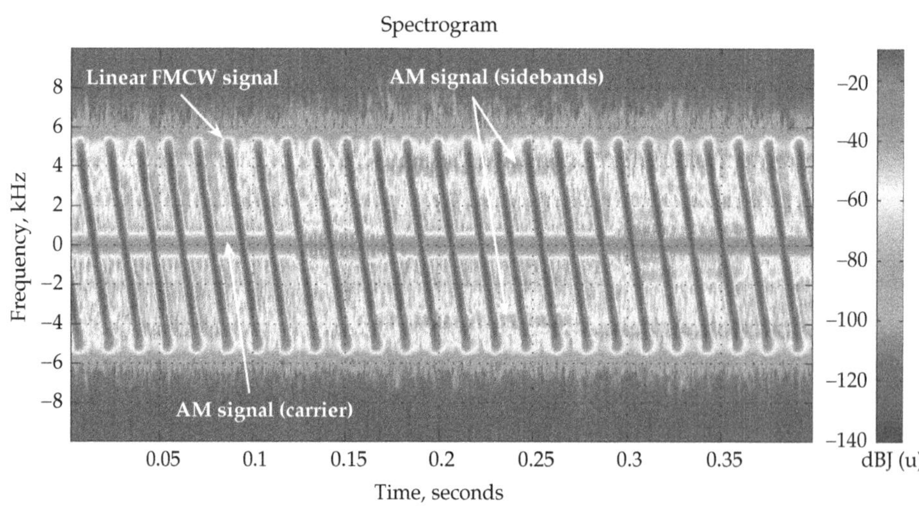

FIGURE 13.28 Spectrogram of the received AM source (BBC World Service) superimposed on the linear FMCW source from an OTH radar transmitter. © Commonwealth of Australia 2011.

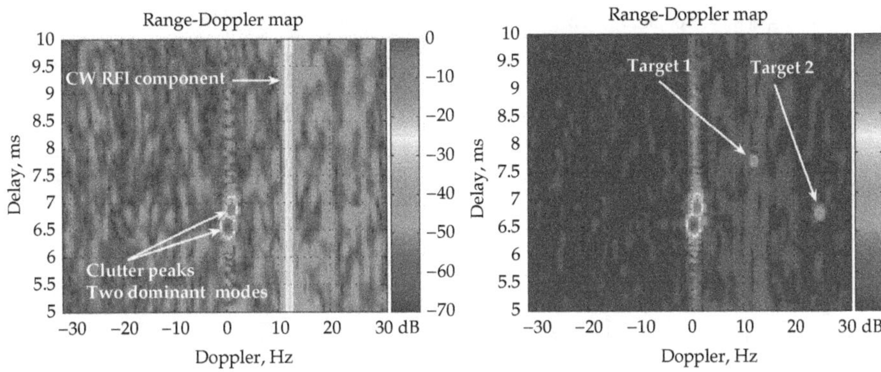

(a) Conventional range-Doppler map. (b) GEMS-filtered range-Doppler map.

FIGURE 13.29 Conventional and GEMS-filtered range-Doppler maps for a main-beam interference scenario, where two synthetic targets have been injected in the nominal direction of the AM source. © Commonwealth of Australia 2011.

context. Once the time-series of the interfering AM signal modes are blindly separated and estimated using GEMS, these signal estimates may be complex-weighted and linearly combined so as to provide the best least-squares fit to the conventional beam output time-series in the pre-range (decimated) A/D sample domain.

This weighted combination of the recovered AM signal modes represents an estimate of the multipath interference that is received in the main lobe of the conventional beam. This estimate of the received AM interference signal may then be subtracted from the conventional beam output in an operation that may be loosely referred to as "'waveform filtering." Once the interference estimate has been subtracted from the conventional beam output in the pre-range sample domain, the remaining signal may be range-Doppler processed in the standard manner. The result of such processing is shown in Figure 13.29b, where the two synthetic targets are clearly distinguishable.

Figure 13.30 shows the range-bin line spectra containing the two targets before and after the GEMS-based waveform filter is applied for main-beam cancelation (MBC). The CW component has been canceled by about 40 dB in Figure 13.30a, while the signal sideband energy has been reduced by about 20 dB in Figure 13.30b. This is more than sufficient to unmask the target echoes, which were not detected by conventional processing. Note that the range-Doppler map in Figure 13.29b is relatively free of undesirable artifacts or "side effects" of such processing. This practical example also highlights the immunity of the approach to target self-cancelation and target-copying effects, which were discussed in Chapters 10–12.

In summary, the main strength of the developed GEMS technique resides in the mild assumptions made with regard to the mode wavefronts and source waveforms, viewed collectively, relative to other spatial processing methods for blind signal separation. This feature of GEMS broadens its scope for practical applications, and also increases its robustness to instrumental and environmental uncertainties that are inevitably encountered in real-world systems. In this sense, the GEMS algorithm represents an advance on current blind spatial-processing techniques for the problem considered.

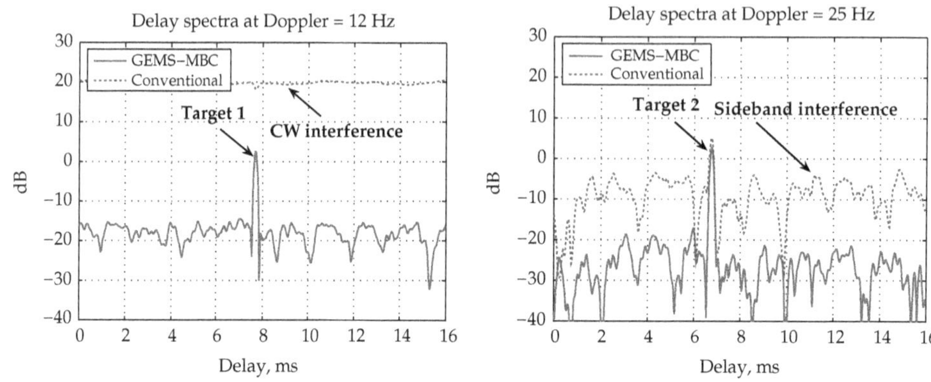

(a) Target 1 under AM signal carrier. (b) Target 2 under AM signal sidebands.

FIGURE 13.30 Range-bin spectra for the two synthetic targets before and after interference subtraction using the GEMS main-beam cancelation algorithm. © Commonwealth of Australia 2011.

13.6 Single-Site Geolocation

Thus far, relatively little attention has been paid to the use of channel information, which can also be blindly estimated along with the source waveform. The problem addressed in this section is that of HF single-site location (SSL), where it is required to estimate the geographic position of an uncooperative source located well beyond the line of sight from the signals received via the skywave propagation mode.

Much of the work in this area has focused around the use of two-dimensional apertures, which can unambiguously estimate both the azimuth and elevation angles of an incoming signal mode. A distinguishing aspect of the problem considered here is that geolocation of an unknown source is attempted using a receiving system based on a single ULA of antenna elements (with front-to-back directivity), where only the cone angle of an incoming signal mode can be measured.

A geolocation method that exploits multipath to resolve the "coning" ambiguity in a linear array is described and tested on real data in this section. Relative mode time-delay information extracted using the GEMS algorithm is also incorporated to make the approach robust in situations where the source is located near boresight. It is shown that by combining the cone-angle and time-delay information contained in all the multipath components opens up the possibility to perform meaningful HF-SSL using a linear array. The described method is also applicable to traditional HF-SSL systems based on two-dimensional apertures. However, the advantage in this case is reducing estimation errors, as opposed to resolving an ambiguity.

13.6.1 Background and Motivation

Figure 13.31 illustrates the classical HF-SSL concept, where a 2D aperture is used to measure the azimuth and elevation angles of the signal DOA. A real-time ionospheric model (RTIM) derived from a vertical incidence sounder (VIS) at the receiver site allows

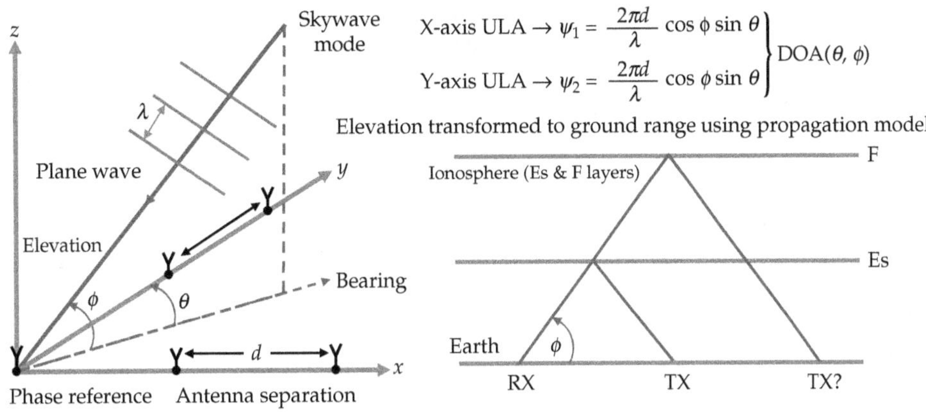

FIGURE 13.31 Classical HF-SSL concept illustrated for a 2D (L-shaped) array. Phase measurements made on both arms of the array yield two equations in two unknowns, such that the azimuth and elevation of the downcoming wave can be deduced. An ionospheric model is used to convert the elevation angle to a ground range based on the virtual height of reflection. © Commonwealth of Australia 2011.

the elevation angle to be converted to a ground range for source position estimation. Table 13.4 compares the main characteristics of various geolocation methods. Trade-offs between the number of sites required, system complexity at each site, and knowledge demanded of the ionosphere and signal is summarized. A detailed description of HF-DF systems is beyond the scope of this chapter, but may be found in the authoritative texts of McNamara (1991) and Gething (1991).

Geolocation Method	Number of Sites Needed	Type of System	Ionospheric Model	Source Waveform	Number of Modes
Multi-site TDOA	4 or more	Single-channel antenna element	Effectively independent	Knowledge required	1 or more at all sites
HF-DF network	2 or more	Multi-channel antenna array	Required	Arbitrary	1 or more at all sites
2D array SSL (e.g., SkyLOC)	1	Two-dimensional array	Required	Arbitrary	1 or more
1D array SSL (no system)	1	Linear array	Required	Arbitrary	2 or more

TABLE 13.4 Relative characteristics of different geolocation approaches. Reducing the number of sites tends to increase system complexity at each site, and increases reliance on knowledge of the ionosphere for propagation modeling. However, a redeeming feature of multi-channel array systems that use DOA as opposed to time-difference of arrival (TDOA) measurements for HF geolocation is that they are agnostic to the source waveform, which may be quite arbitrary. As far as HF-SSL systems are concerned, the last row signifies a reversal in the trend since system complexity is reduced from a 2D to 1D (linear) array. We shall see that it is possible to perform meaningful HF-SLL with a linear array provided that multipath is present, as indicated in the last column of the table. © Commonwealth of Australia 2011.

In particular, there currently appears to be no operational HF-SSL system based on a linear array. This is not surprising given that a linear array cannot estimate bearing and elevation independently due to the coning ambiguity. This section introduces and experimentally validates a geolocation method that exploits multipath to enable HF-SSL using a linear array. Such results have significant implications for implementing an HF-SSL capability on linear arrays existing in the field. Potential applications of this technology include search and rescue, enforcement of HF spectrum regulations, as well as in military situations McNamara (1991).

The key behind the approach is to simultaneously combine the cone angle and relative time-delay information contained in all of the signal modes received via skywave propagation from the source. This is in contrast to many current HF-SSL systems, which employ wavefront testing procedures (WFT) in order to produce valid estimates of the source position only at times of quasi uni-modal propagation (QUMP). This is tantamount to regarding multipath as a nuisance rather than an opportunity. Such systems not only miss the opportunity to use the additional information contained in the ensemble of paths, when considered jointly, but the severe pruning of data needed to satisfy QUMP conditions can also significantly limit the times at which measurements may be used.

For the purpose of this study, the HF-SSL system is assumed to be an antenna array connected to a multi-channel digital receiver system. Particular emphasis is on a ULA, but the described method is applicable to general array geometries with minor modification. In the ULA case, the antenna elements are assumed to have front-to-back directivity, such that the field of view is limited to $\pm 90^\circ$ from boresight. The source is assumed to be at relatively long range, such that propagation is via the skywave mode only. In other words, there is no reliance on the ground-wave mode, which is assumed to be highly attenuated. Ground waves can be useful for bearing estimation over short ranges, but its effective absence over long-range paths makes geolocation of an unknown emitter more challenging for a ULA.

The source of interest is assumed to emit an arbitrary waveform of finite bandwidth not restricted to a particular modulation format. The linear complexity of the waveform needs to exceed twice the maximum channel impulse response duration (as described previously for blind system identification in the context of identifiability). Importantly, propagation between the source and receiver is assumed to be via *more than one* ionospheric mode. Multipath propagation frequently occurs due to the E and F regions of the ionosphere, the presence of low- and high-angle rays, and magneto-ionic splitting that results in ordinary (o) and extraordinary (x) rays. Propagation via a single mode is the exception rather than the rule in this medium.

An RTIM is assumed to be available from a VIS at the receiver site. The quality of the ionospheric model may be improved by incorporating information from a network of spatially distributed ionosondes, or by using a backscatter sounder at the receiver site, for example. This is expected to be beneficial for long range paths, where the control point for the signal of interest is far from the receiver site. In this section, attention is restricted to path lengths less than 3000–4000 km. Range extents beyond this due to multi-hop ionospheric propagation will be considered in a future work.

13.6.2 Data Collection

Figure 13.32a shows the twin-dart antenna of the emitter to be geolocated in the experiment. This cooperative source was located near Broome, on the north-west coast of Australia. Figure 13.32b shows the receiving ULA of the east arm of the Jindalee

(a) Twin-dart antenna of the emitter near Broome.

(b) East arm of the JORN OTH radar receiving array.

FIGURE 13.32 Pictures of the twin-dart antenna used to emit the test signal from a site near Broome, Western Australia, and the east arm of the receiving uniform linear array of the JORN OTH radar at Laverton, Western Australia.

Operational Radar Network (JORN) OTH radar, located near Laverton in western Australia. This ULA has an aperture of 2970 m and consists of 480 twin-monopole (end-fire) elements with a digital receiver per element. The geometry of the experiment is illustrated in Figure 13.33a. We shall refer back to Figure 13.33b in due course.

The emitter is located at a ground distance of 1161 km from the receiver site with a great circle bearing of 1.1°N. The bearing of the source is −33.9° relative to the ULA boresight direction. Figure 13.34a shows an ionogram recorded by the VIS system at the receiver

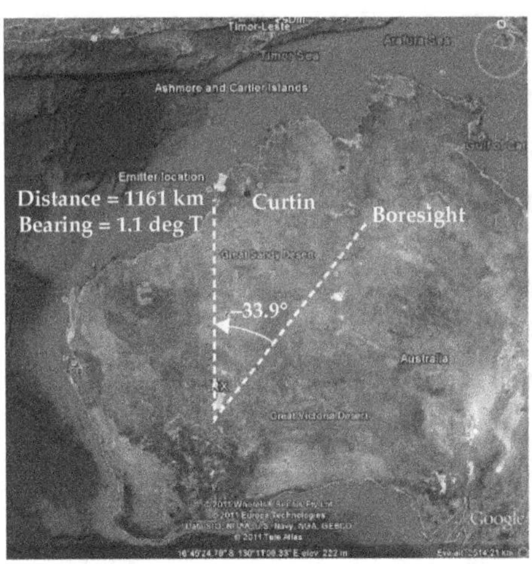

(a) Geographic view of experiment geometry.

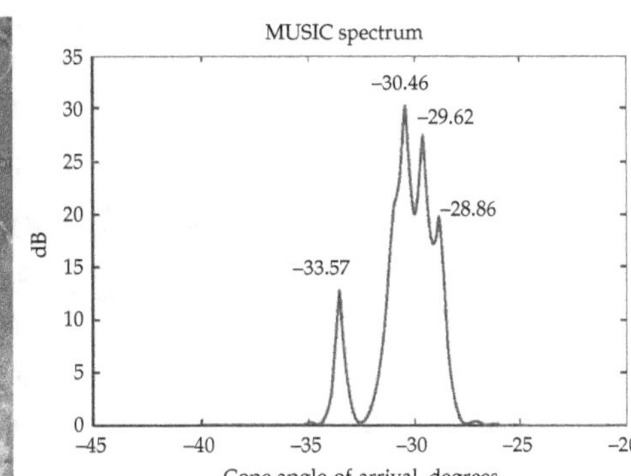

Mode	1Es	1F(I)	1Fo(h)	1Fx(h)
Mean	−33.55	−30.97	−29.79	−28.96
Std. Dev.	0.03	0.24	0.34	0.19

(b) Example MUSIC spectrum using 2.47 second of data.

FIGURE 13.33 Experiment geometry showing emitter ground range and bearing relative to the receiving ULA position and boresight direction. An example of MUSIC spectrum showing the estimated mode cone angles is also shown along with the mean and standard deviation of measurements averaged over a 1 minute interval of data.

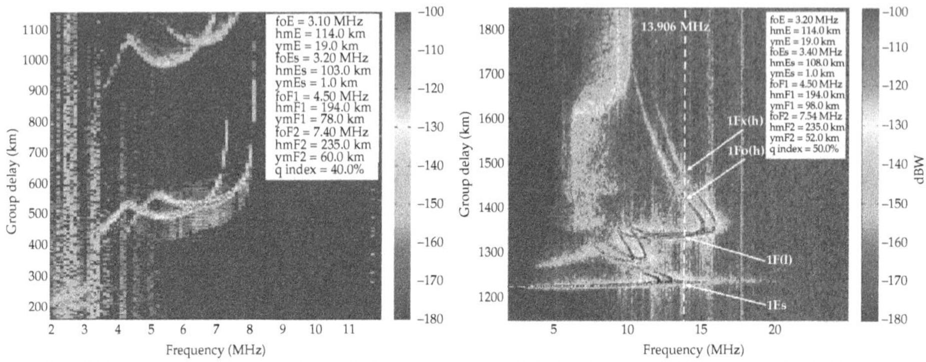

(a) VIS ionogram at Laverton (06:45:45 UT). (b) Curtin-Laverton OIS ionogram (06:37:53 UT).

FIGURE 13.34 VIS ionogram at Laverton (06:45:45 UT) and Curtin-Laverton OIS ionogram (06:37:53 UT). The QP ionospheric model parameters manually fitted to each ionogram are listed in the figures. © Commonwealth of Australia 2011.

site, while Figure 13.34b shows an oblique incidence sounder (OIS) ionogram recorded for the Curtin-Laverton path. The location of Curtin is indicated in Figure 13.33a. A multi-segment quasi-parabolic (QP) ionospheric profile model was manually fitted to the VIS and OIS ionograms. The fitted model parameters are indicated in the two figures showing the VIS and OIS ionograms, respectively.

The source emitted a narrowband waveform with flat spectral density. The carrier frequency was 13.906 MHz and the bandwidth was 8 kHz. A spectrogram of the waveform is shown in Figure 13.35. Although the source was cooperative, the signal waveform was assumed to be unknown. The received signal was digitally down-converted and decimated to a sampling frequency of 31.25 kHz. The array data were acquired in consecutive processing intervals of approximately 2.5-s duration. This experiment was performed on 18 August 2011 starting at 06:48 UT (14:48 local time).

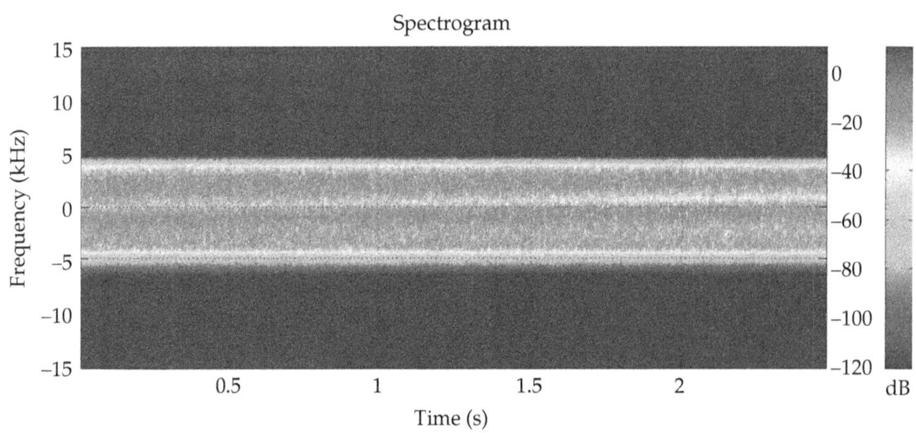

FIGURE 13.35 Spectrogram of the source signal with a bandlimited spectral density. No knowledge is assumed regarding the modulation format of the emitted waveform. © Commonwealth of Australia 2011.

Figure 13.34b shows that four ionospheric modes are resolved in the OIS trace for the Curtin-Laverton path at 13.906 MHz. These single-hop modes are sporadic-E "1Es," the F-layer low-angle ray "1F(l)," and the F-layer high-angle rays, namely, the ordinary wave "1Fo(h)" and extraordinary wave "1Fx(h)." The reader may refer to Chapter 2 for an explanation of low/high rays and o/x waves. The OIS path midpoint is about 86 km from the control point of the source-to-receiver path. This is considered to be a short distance on the spatial scale of ionospheric variability. However, the ionogram was recorded approximately 10 minutes prior to the source, which may be a significant interval on the temporal scale of ionospheric variability. On the other hand, the VIS was recorded less than 3 minutes earlier, but the control point for the source is about 586 km from the VIS.

13.6.3 Geolocation Method

The geolocation method is composed of three elements. The first makes use of phase measurements across the antenna elements in the array, from which the cone angles of the different signal modes can be estimated. A cost function that combines the phase measurements associated with the cone angles of all signal modes with predictions based on an ionospheric model may be used to obtain a position fix on the source. However, such a method is only applicable for sources with a bearing that is not near or at boresight. To address this issue, inter-mode TDOA information extracted using the GEMS technique is incorporated for range estimation, as the second component. The third component of the introduced geolocation method fuses the mode spatial phase and time-delay information to estimate the source position.

13.6.3.1 Phase Measurements

Define $\mathbf{x}_k(p) = [x_k(s), \ldots, x_k(s+N_s-1)]^T \in \mathcal{C}^{N_s}$ as the array snapshot vector recorded at time sample $k = 1, \ldots, K$ starting at receiver $s = 1, \ldots, N - N_s + 1$, where N_s and N are the number of elements in the subarray and array, respectively. To be clear, the subarray refers to a selected number $N_s < N$ of adjacent receivers in the full array. Compute $\mathbf{R}$ as the spatially smoothed sample spatial covariance matrix by forward-backward averaging in Eqn. (13.110), where $\mathbf{J}$ is the exchange matrix with ones on the anti-diagonal (Pillai 1989). The symbols T, $*$, and $\dagger$ denote the transpose, conjugate, and Hermitian (conjugate-transpose) operators, respectively.

$$\mathbf{R} = \frac{1}{K(N-N_s+1)} \sum_{s=1}^{N-N_s+1} \sum_{t=1}^{K} \mathbf{x}_k(s)\mathbf{x}_k(s)^\dagger + \mathbf{J}\mathbf{x}_k^*(s)\mathbf{x}_k^T(s)\mathbf{J} \tag{13.110}$$

The mode cone angles may be estimated using the MUSIC algorithm based on the eigen-decomposition of $\mathbf{R}$ in Eqn. (13.111), where $\mathbf{Q}_s$ and $\mathbf{Q}_n$ represent the signal and noise subspaces, respectively. Sub-aperture smoothing trades off robustness for correlated multipath arrivals against spatial resolution. Here, $N = 480$ and $N_s = 240$ was used as a compromise between the two competing objectives. The number of modes is determined as $M = 4$ in the following, so $\mathbf{Q}_n$ contains the $N_s - M$ eigenvectors with the smallest eigenvalues (Schmidt 1981).

$$\mathbf{R} = \mathbf{Q}_s\Lambda_s\mathbf{Q}_s^\dagger + \mathbf{Q}_n\Lambda_n\mathbf{Q}_n^\dagger \tag{13.111}$$

The MUSIC spectrum is computed as $p(\psi)$ in Eqn. (13.112), where $\mathbf{v}(\psi) = [1, e^{j\psi}, \ldots, e^{j(N-1)\psi}]^T$ is the ULA steering vector.[4] As illustrated in Figure 13.31, the phase $\psi = 2\pi d \sin\varphi/\lambda$, where $\sin\varphi = \cos\phi \sin\theta$ for a ULA on the x-axis, and φ is the cone angle. The MUSIC spectrum resulting for a 2.47-second interval of data is illustrated in Figure 13.33b. The estimated cone angles $\hat{\varphi}_m$ associated with the four ($m = 1, \ldots, M = 4$) dominant peaks are labeled in this figure.

$$p(\psi) = \frac{1}{\mathbf{v}(\psi)^\dagger \mathbf{Q}_n \mathbf{Q}_n^\dagger \mathbf{v}(\psi)} \tag{13.112}$$

This process may be repeated over successive processing intervals to obtain an average estimate of the mode cone angles. The table in Figure 13.33b shows the mean and standard deviation of the cone-angle estimates using one minute of data. Cone angles closer to boresight are associated with modes reflected from greater virtual heights. The estimated phases $\hat{\psi}_m = 2\pi d \sin\hat{\varphi}_m/\lambda$ can be arranged into the measurement vector $\underline{\hat{\psi}}$ in Eqn. (13.113), where $\hat{\psi}_1 < \hat{\psi}_2 < \ldots < \hat{\psi}_M$.

$$\underline{\hat{\psi}} = [\hat{\psi}_1, \ldots, \hat{\psi}_M]^T \tag{13.113}$$

For a hypothesized source ground range R, and great circle bearing θ, the ionospheric model can be used to predict a phase angle $\psi_m(R, \theta)$ for each propagation mode. These predictions were derived using analytical ray tracing and virtual ray path geometry principles assuming a spherically symmetric QP profile. The effect of ionospheric gradients or tilts on geolocation accuracy is less pronounced on long-range paths than it is on short-range ones. Neglecting tilts, the model predictions may be arranged into the vector $\underline{\psi}(R, \theta)$ in Eqn. (13.114), where $\psi_1(R, \theta) < \psi_2(R, \theta) < \ldots < \psi_M(R, \theta)$.

$$\underline{\psi}(R, \theta) = [\psi_1(R, \theta), \ldots, \psi_M(R, \theta)]^T \tag{13.114}$$

Even when ionospheric tilts can be ignored, and the cone angle furthest from the boresight direction (i.e., $\hat{\varphi}_1 = -33.55°$ in this case) is correctly associated with the sporadic-E propagation mode, it is still not possible to obtain a unique fix on the source position using a single mode from a linear array. This is because there is a continuous locus of source range-bearing pairs on the ground that give rise to the same cone angle. This ambiguity is illustrated for $\hat{\varphi}_1 = -33.55°$ in Figure 13.36a. The true source location is indicated by a black dot. Figures 13.36 and 13.37, in the same format, illustrate the ambiguities for the other estimated cone angles on a mode-by-mode basis.

The main point is that it is not possible to resolve these ambiguities when the phase measurements for each mode are considered separately and processed on an *independent* basis. These ambiguities can be resolved by *jointly* processing the phase measurements made on different modes, i.e., by exploiting multipath to estimate the source position. Specifically, the RMS phase error cost function in Eqn. (13.115) may be proposed,

[4]The angle ψ is defined here as the phase difference between adjacent antenna elements in the ULA for a plane wave incident from azimuth θ and elevation ϕ.

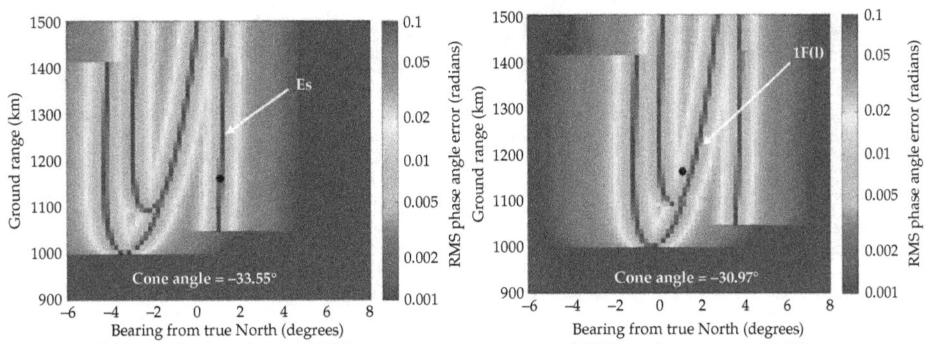

(a) Sporadic-E mode $\varphi = -33.55°$ (b) F-layer low ray $\varphi = -30.97°$

FIGURE 13.36 Loci of ambiguity in the range-bearing plane associated with the cone angles estimated for the sporadic-E and F-layer (low-ray) modes, denoted by the abbreviations 1Es and 1F(l), respectively. © Commonwealth of Australia 2011.

where $\varepsilon_\psi(R,\theta) = \underline{\psi}(R,\theta) - \hat{\underline{\psi}}$ is the error vector between the measurements and model.

$$c_\psi(R,\theta) = \sqrt{\varepsilon_\psi^T(R,\theta)\varepsilon_\psi(R,\theta)/M} \tag{13.115}$$

The source position may be estimated by evaluating $c_\psi(R,\theta)$ over a grid of candidate range-bearing tuples, and observing the coordinates of the minimum, i.e., where the phase measurements best fit the model predictions in a least-squares sense. The cost function evaluated using real data is shown in Figure 13.38a. The estimated and true source locations are also indicated in Figure 13.38a. In this example, the geolocation error is 25.8 km over a 1161 km ground distance path. Based on skywave propagation only, and the use of a single receive site, this is considered to be a new result for HF source geolocation with a ULA.

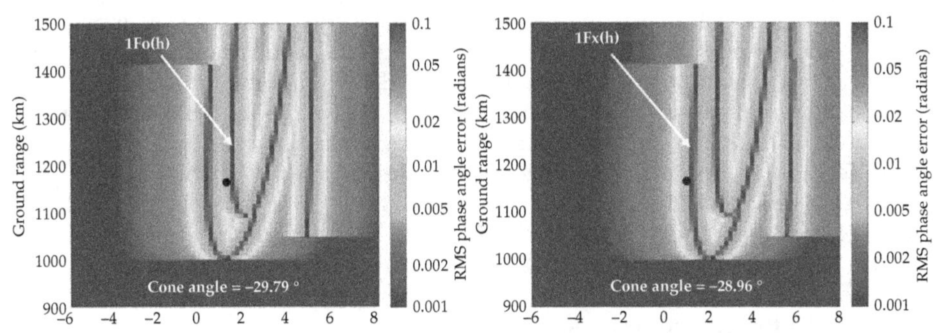

(a) Ordinary ray $\varphi = -29.79°$ (b) Extraordinary ray $\varphi = -28.96°$

FIGURE 13.37 Loci of ambiguity in the range-bearing plane corresponding to the cone angles estimated for the ordinary and extraordinary high angle rays in the F-region, denoted by the abbreviations 1Fo(h) and 1Fx(h), respectively. © Commonwealth of Australia 2011.

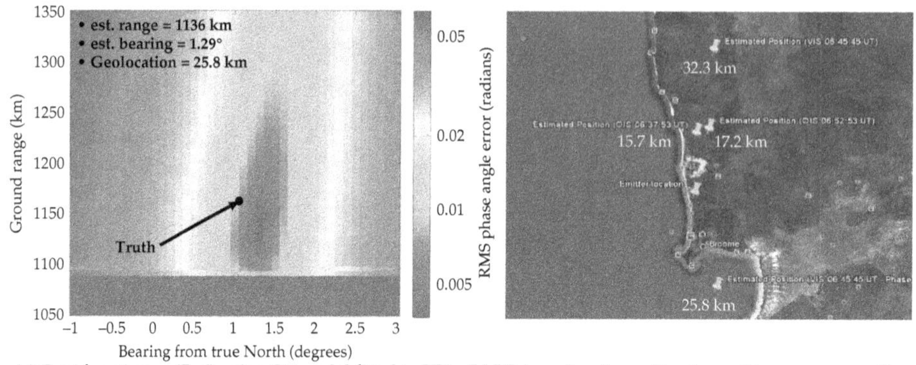

(a) Cost function $c_\psi(R, \theta)$ using QP model fitted to VIS. (b) ULA geolocation estimates and true source position.

FIGURE 13.38 Cost function $c_\psi(R, \theta)$ using the QP ionospheric model fitted to VIS ionogram, and geographic map showing the geolocation estimates relative to true source position. © Commonwealth of Australia 2011.

13.6.3.2 Delay Measurements

A potential problem with using $c_\psi(R, \theta)$ in isolation is that there is effectively no coning for a source near boresight. In this case, the cost function elongates in range such that only bearing estimation becomes possible. To overcome this limitation, it is proposed to incorporate an inter-mode TDOA method for range estimation. The relative mode time delays may be estimated by cross-correlation analysis when the source waveform has a known parametric form, e.g., linear FMCW. The problem is how to estimate the mode differential time delays for an unknown source that emits an arbitrary waveform. The previously introduced GEMS algorithm is used here to estimate the differential mode time delays.

Figure 13.39 shows the GEMS cost function computed using the same 2.47-second processing interval of data as in Figure 13.33. A total of $Q = \binom{M}{2}$ minima are expected,[5] excluding the trivial solution at the origin. The cost function exhibits $Q = 6$ minima, which corresponds to $M = 4$ modes, as predicted from the OIS ionogram in Figure 13.34b. The location of the minima are also listed in Figure 13.39. Importantly, GEMS exploits wavefront "crinkles" caused by diffuse scattering to resolve modes with similar or the same nominal DOAs. This enables GEMS to resolve sources near or at boresight.

The differential group-range estimates of the modes, denoted by $\hat{g}_q$, may be arranged into the measurement vector $\hat{\mathbf{g}} = [\hat{g}_1, \ldots, \hat{g}_Q]^T$, such that $\hat{g}_1 < \hat{g}_2 < \ldots < \hat{g}_Q$. Similarly, the ionospheric model may be used to predict the differential group ranges arising between different pairs of modes as a function of the hypothesized source ground range and bearing. The resulting model predictions, denoted by $g_q(R, \theta)$, are assembled in ascending order to form the vector $\mathbf{g}(R, \theta)$ in Eqn. (13.116).

$$\mathbf{g}(R, \theta) = [g_1(R, \theta), \ldots, g_Q(R, \theta)]^T \tag{13.116}$$

[5]The definition of Q stated in this section is not to be confused with its previous definition as the number of sources.

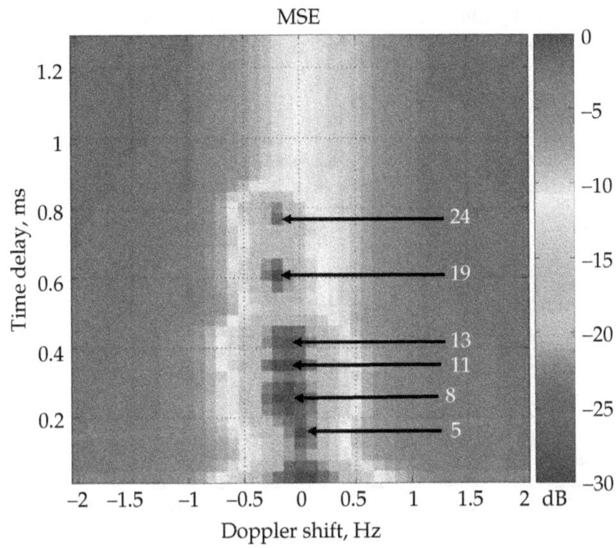

Sample number	5	8	11	13	19	24
Time delay, ms	0.160	0.256	0.352	0.416	0.608	0.768
Group range, km	48	77	106	125	182	230

FIGURE 13.39 GEMS cost function in the differential delay-Doppler plane. © Commonwealth of Australia 2011.

The RMS group-range cost function may then be computed as $c_g(R,\theta)$ in Eqn. (13.117), where the error vector is $\varepsilon_g(R,\theta) = \mathbf{g}(R,\theta) - \hat{\mathbf{g}}$. This function is very weakly dependent on bearing due to the impact of the Earth's magnetic field on signal propagation through the ionosphere. Figure 13.40 shows the representative range-only dependence of $c_g(R,\theta)$, which is more or less the same for all bearings.

$$c_g(R,\theta) = \sqrt{\varepsilon_g^T(R,\theta)\varepsilon_g(R,\theta)/Q} \tag{13.117}$$

The top panel in Figure 13.40 shows the differential group ranges measured using GEMS as horizontal dashed lines. The model predictions for each mode pair combination are plotted against source ground range as solid lines. The cost function $c_g(R,\theta)$ is shown as a function of source ground range in the bottom panel of Figure 13.40. The minimum occurs at a range of 1195 km, which corresponds to a range error of 34 km. This ground range estimate of the source location has been obtained using inter-mode TDOA information obtained at a single site for a far-field source that emits an unknown waveform. This is also regarded as a new result for HF-SSL.

13.6.3.3 Fused Measurements

The phase-only and group-range cost functions may be fused to take advantage of the relative benefits that each provides. A possible way to fuse the cost functions is to express the phase rms error $c_\psi(R,\theta)$ as a bearing rms error $c_\theta(R,\theta)$ using standard transformations,

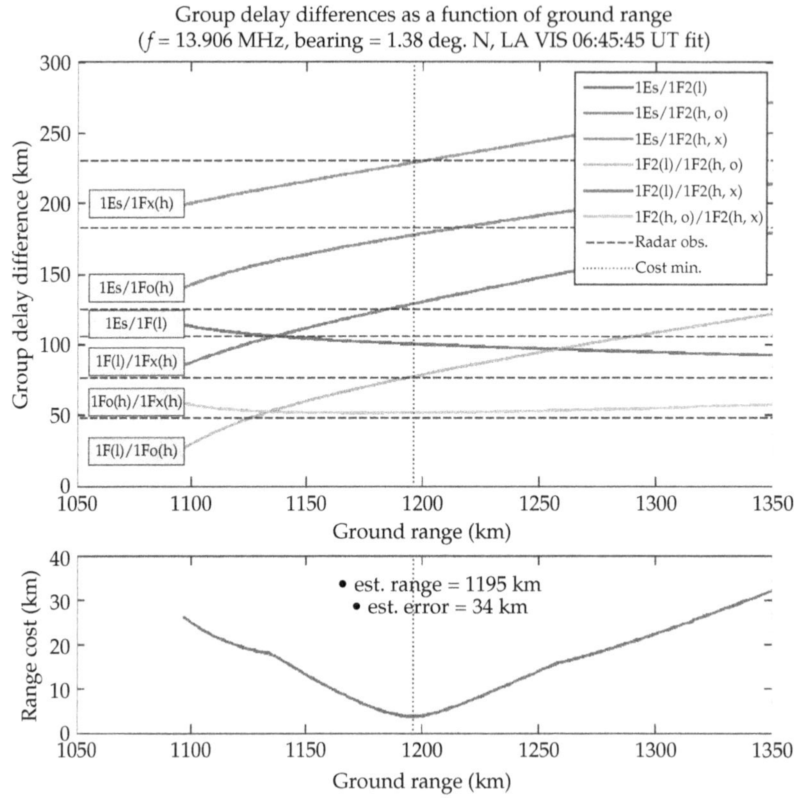

FIGURE 13.40 Cost function $c_g(R, \theta)$ used to estimate the source ground range.

and to compute the cross-range rms error as $Rc_\theta(R, \theta)$. The down-range rms error is well-approximated by $c_g(R, \theta)$. Hence, the rms Euclidean distance error may be approximated as $c(R, \theta)$ in Eqn. (13.118).

$$c(R, \theta) = \sqrt{R^2 c_\theta^2(R, \theta) + c_g^2(R, \theta)} \tag{13.118}$$

Figure 13.41a shows a plot of $c(R, \theta)$ with the estimated and true source positions indicated. When the QP model fitted to the VIS is used, the geolocation error is 32.3 km. Figure 13.41b shows that this error reduces to 15.7 km when the ionospheric model fitted to OIS record is used instead. Figure 13.38b shows the true and estimated source positions on a geographic map zoomed into the Broome area.

$$\{\hat{R}, \hat{\theta}\} = \arg \min_{R,\theta} c(R, \theta) \tag{13.119}$$

The estimate resulting for another OIS ionogram recorded later at 06:52 UT (with 17.2 km geolocation error) is also shown to indicate the robustness of the approach. It is emphasized that these accuracies should be interpreted as preliminary results only, rather

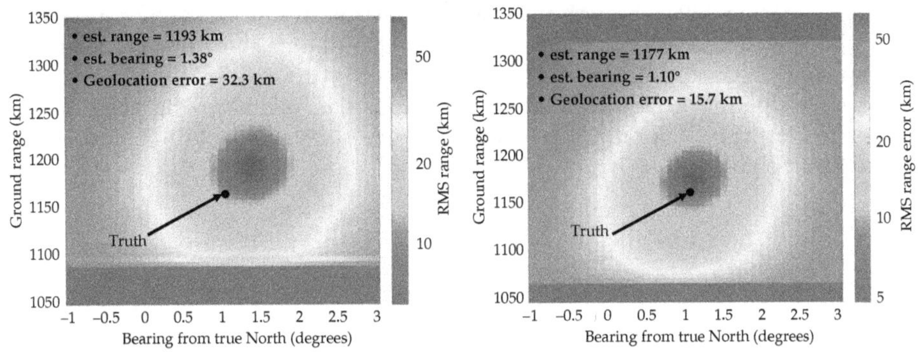

(a) Cost function for VIS ionospheric model.

(b) Cost function for OIS ionospheric model.

FIGURE 13.41 Fused cost function $c(R, \theta)$ based on QP ionospheric models fitted to the VIS and OIS ionograms. © Commonwealth of Australia 2011.

than upper bounds on achievable geolocation performance. It is envisaged that a number of refinements can be made to the geolocation technique to reduce source position estimation errors to around 10 km; see Fabrizio and Heitmann (2013). Some topics for future work are discussed below.

13.6.4 Summary and Future Work

An HF-SSL method for the geolocation of long range sources propagated exclusively via skywave paths has been proposed and experimentally validated on a ULA. The technique is essentially agnostic to the class of waveform emitted by the source and exploits the geometrical relationships satisfied by multipath propagation to resolve the inherent coning ambiguity of a ULA.

By incorporating all multipath measurements and the physical constraints associated with such propagation into the cost function, the introduced HF-SSL procedure allows the source position (and mode group ranges) to be estimated. This multipath-driven HF-SSL method represents a fundamentally new approach to HF geolocation.

Experimental trials confirm the validity of the approach, which combines both cone angle and TDOA information for all resolved propagation modes. The described approach is also applicable to two-dimensional receiver apertures with minor modification. In this case, the main advantage is expected to be estimation error reduction as opposed to resolving an ambiguity.

Besides the aforementioned practical applications of HF geolocation, the approach is also of interest for the inverse problem of estimating the signal virtual reflection heights when the position of an uncooperative source is known. The ability to estimate mode structure for a known path using emitters of opportunity that probe ionospheric control points in regions which cannot be monitored by dedicated sounders is desirable for coordinate registration in OTH radar systems. The estimation of ionospheric tilts when the source position is known is a subject of current investigation.

Future work involves performance analysis as a function of receive aperture length, mode signal-to-noise ratios, waveform bandwidth, and source distance. Extensions to the approach include the incorporation of known reference points (KRPs) when available, numerical ray tracing (NRT) through an RTIM derived from a network of ionosondes (for long ranges), ionospheric tilt correction (for short ranges), time averaging of estimates, and refinement of the cost function expression to account for ionospheric uncertainty, including the weighting of different mode contributions. It is anticipated that a multipath-driven estimation approach may also be useful in other systems, not limited to HF or electromagnetic signals.

PART IV

Appendixes and Bibliography

APPENDIX A

Sample ACS Distribution

Consider a time-continuous complex random process that is sampled every Δ seconds, and let $z(t)$ be a complex random variable that takes on the value of the process at different sampling instants, given by $t\Delta$ for $t = 0, 1, \ldots, P-1$.

$$z(t) = x(t) + jy(t) = m(t)e^{j\theta(t)} \tag{A.1}$$

In Eqn. (A.1), the scalars $x(t)$ and $y(t)$ are the real and imaginary parts of $z(t)$, respectively, while $m(t)$ and $\theta(t)$ are the associated magnitude and phase. It is assumed that $x(t)$ and $y(t)$ are *zero mean* Gaussian distributed random variables with the same variance σ^2. It is further assumed that $x(t)$ and $y(t)$ are independent stationary random processes with the same second order statistics, or auto-correlation function $r(\tau)$, given by Eqn. (A.2).

$$r_x(\tau) = E\{x(t)x(t+\tau)\}, \quad r_y(\tau) = E\{y(t)y(t+\tau)\}, \quad r_x(\tau) = r_y(\tau) = r(\tau) \tag{A.2}$$

For the complex random process $z(t)$, it can be shown that the magnitude $m(t)$ is Rayleigh distributed and the phase $\theta(t)$ is uniformly distributed over the interval $[-\pi, \pi)$, see Papoulis (1984). This general statistical model is suitable for the description of HF signals received at a single antenna sensor after reflection from the ionosphere, as experimentally verified by Watterson et al. (1970).

Under the above-mentioned assumptions, the maximum likelihood estimate of the auto-correlation function is the *unbiased* sample auto-correlation function. This estimate, denoted by $\hat{r}_z(\tau)$, is calculated as the average of a sum of lagged products, as in Eqn. (A.3), where P is the number of available samples of $z(t)$.

$$\hat{r}_z(\tau) = \frac{1}{P-\tau} \sum_{t=0}^{P-\tau-1} z(t)z^*(t+\tau) \tag{A.3}$$

The estimates $\hat{r}_z(\tau)$ for $\tau = 0, 1, \ldots, Q-1$, where $Q < P$, are collectively referred to as the sample auto-correlation sequence (ACS) Marple (1987). The estimator in Eqn. (A.3) is statistically consistent in the sense that the sample ACS $\hat{r}_z(\tau)$ tends to the statistically expected ACS $r_z(\tau) = E\{z(t)z^*(t+\tau)\}$ as the number of samples P tends to infinity.

By substituting Eqn. (A.1) into Eqn. (A.3), the sample ACS can be decomposed into real and imaginary parts, as in Eqn. (A.4).

$$\hat{r}_z(\tau) = [\hat{r}_x(\tau) + \hat{r}_y(\tau)] + j[\hat{r}_{yx}(\tau) - \hat{r}_{xy}(\tau)] \tag{A.4}$$

It is noted that $\hat{r}_x(\tau) = M\{x(t)x(t+\tau)\}$, $\hat{r}_y(\tau) = M\{y(t)y(t+\tau)\}$, $\hat{r}_{xy}(\tau) = M\{x(t)y(t+\tau)\}$, and $\hat{r}_{yx}(\tau) = M\{y(t)x(t+\tau)\}$, where $M\{\cdot\} = \frac{1}{P-\tau}\sum_{t=0}^{P-\tau-1}\{\cdot\}$. As the sample ACS estimates are unbiased, it is possible to define the following zero mean random variables.

$$\begin{aligned} e_x(\tau) &= \hat{r}_x(\tau) - r_x(\tau) \\ e_y(\tau) &= \hat{r}_y(\tau) - r_y(\tau) \\ e_{xy}(\tau) &= \hat{r}_{xy}(\tau) - r_{xy}(\tau) \\ e_{yx}(\tau) &= \hat{r}_{yx}(\tau) - r_{yx}(\tau) \end{aligned} \tag{A.5}$$

Note that $r_{xy}(\tau) = E\{x(t)y(t+\tau)\}$ and $r_{yx}(\tau) = E\{y(t)x(t+\tau)\}$ are both equal to zero as the real and imaginary parts of $z(t)$ are assumed to be independent. The asymptotic (large sample) variances of these random variables are given in Muirhead (1982). Specifically, they are given by Eqn. (A.6), where N is defined as the number of *independent* samples of $z(t)$ used to form the sample ACS according to Eqn. (A.3).

$$\begin{aligned} E\left\{e_x^2(\tau)\right\} &= E\left\{e_y^2(\tau)\right\} = [\sigma^4 + r^2(\tau)]/N \\ E\left\{e_{xy}^2(\tau)\right\} &= E\left\{e_{yx}^2(\tau)\right\} = \sigma^4/N \end{aligned} \tag{A.6}$$

In many practical situations, the samples recorded from a random process are correlated rather than independent. In this case, the substitution of P for N in Eqn. (A.6) will not yield the asymptotic (large sample) variance of the error terms in Eqn. (A.5). A notable exception for which this substitution is valid occurs when the random processes $x(t)$ and $y(t)$ are independent and identically distributed (IID) Gaussian *white noise* processes, wherein the consecutive samples are uncorrelated and hence statistically independent. The effect of correlation in the samples of a random process is to reduce the number of independent observations, which in turn increases the variance of the ACS estimates.

Fortunately, an *exact* result for the variances of the ACS estimation error terms in Eqn. (A.5) exists due to Bartlett (1946). As explained in Priestly (1981), the joint fourth-order cumulant in the expression derived by Bartlett (1946) vanishes for a Gaussian processes, so the exact sample ACS error variance is given by Eqn. (A.7).

$$\begin{aligned} E\{e_x(\tau)e_x(\tau+v)\} &= \frac{P}{(P-\tau)(P-\tau-v)} \sum_{m=-(P-\tau)+1}^{P-\tau-v-1} \\ &\left\{1 - \frac{\mu(m)+\tau+v}{P}\right\}[r(m)r(m+v) + r(m+\tau+v)r(m-\tau)] \end{aligned} \tag{A.7}$$

Note that $E\{e_y(\tau)e_y(\tau+v)\} = E\{e_x(\tau)e_x(\tau+v)\}$ as the two random processes $x(t)$ and $y(t)$ are assumed to be identically distributed. The term $\mu(m)$ is given by Eqn. (A.8).

$$\mu(m) = \begin{cases} m, & m > 0 \\ 0, & -v \le m \le 0 \\ -m-v, & -(P-\tau)+1 \le m \le -v \end{cases} \tag{A.8}$$

Observe that the cross terms $e_{xy}(\tau)$ and $e_{yx}(\tau)$ of the sample ACS in Eqn. (A.4) are derived as the average of a sum of lagged products between the samples of *two independent* processes, whereas the expression in Eqn. (A.7) has been derived for the average of a sum of lagged products between the samples of a *single* and potentially correlated random process. Consequently, the expression in Eqn. (A.7) cannot be used in the strictly presented form to obtain the variances of these cross terms.

If the processes $x(t)$ and $y(t)$ are of finite bandwidth, then $r(s) \rightarrow 0$ as $s \rightarrow \infty$. In other words, the samples of either random process are effectively uncorrelated, and hence statistically independent, when the separation between them is sufficiently large. Since $x(t)$ and $y(t)$ are IID Gaussian processes, it follows that $x(t+s)$ for $s \rightarrow \infty$ can be considered a valid realization of $y(t)$ for the purpose of evaluating the variances of the cross terms. Stated another way, the statistical properties of a sum of $P-\tau$ lagged products between the samples of $x(t)$ and $y(t+\tau)$ are equivalent to the statistical properties of the sum of $P-\tau$ lagged products between the samples of $x(t)$ and $x(t+s+\tau)$ for large s. Hence, substituting $P+s$ for P and $\tau+s$ for τ in Eqn. (A.7) yields the following expression for the variance of the cross terms as $s \rightarrow \infty$.

$$E\{e_{xy}(\tau)e_{xy}(\tau+v)\} = E\{e_{yx}(\tau)e_{yx}(\tau+v)\}$$

$$= \frac{P}{(P-\tau)(P-\tau-v)} \sum_{m=-(P-\tau)+1}^{P-\tau-v-1} \left\{1 - \frac{\mu(m)+\tau+v}{P}\right\} r(m)r(m+v) \tag{A.9}$$

Note that if the samples of $x(t)$ and $y(t)$ are white, such that $r(\tau) = \delta(\tau)\sigma^2$ and $N = P$ is appropriate, then as $P \rightarrow \infty$ the variances predicted by Eqn. (A.7) and Eqn. (A.9) for $v = 0$ coincide with the asymptotic variances reported by Muirhead (1982) in Eqn. (A.6). It may also be observed from Eqn. (A.7) and Eqn. (A.9) that for a colored Gaussian process the correlation between samples serves to increase the variance of the ACS estimates. This increase in variance is interpreted in terms of a decrease in the effective number of statistically independent samples used to estimate the ACS (Priestly 1981).

Before proceeding to determine the sample distributions for the magnitude and phase of the ACS estimation error, a few points are mentioned. A more compact expression for the variance of the *complex* ACS estimation error exists, as described in Thierren (1992), but this expression has not been used as nothing can be deduced about the variances of the real and imaginary parts of the ACS estimation error, which are required to evaluate the statistics of the sample ACS magnitude and phase. For Gaussian process that are stationary up to order four, the variances given by Eqn. (A.7) and Eqn. (A.9) are exact for any number of samples P, but the *distributions* of the error terms in Eqn. (A.5) *tend* to the normal density as the number of samples P tends to infinity. As typically more than $P = 10,000$ samples are used to estimate the ACS in the experimental analysis, a normal distribution for these random variables is assumed in the following. Although the theoretical expression for the variances will not be exact for non-Gaussian processes, they are known to provide a good approximation providing that $|\tau| \ll P$ (Thierren 1992).

The complex ACS estimation error $e_z(\tau)$ corresponding to Eqn. (A.4) can be written in the alternative form of Eqn. (A.10), where $a(\tau) = e_x(\tau) + e_y(\tau)$ and $b(\tau) = e_{yx}(\tau) - e_{xy}(\tau)$.

$$e_z(\tau) = \hat{r}_z(\tau) - r_z(\tau) = a(\tau) + jb(\tau) \tag{A.10}$$

The zero-mean random variables $e_x(\tau)$ and $e_y(\tau)$ have the same variance $\sigma_1^2(\tau)$ given by Eqn. (A.7) and are normally distributed in the asymptotic case. Similarly, the zero-mean normally distributed random variables $e_{xy}(\tau)$ and $e_{yx}(\tau)$ have a common variance $\sigma_2^2(\tau)$ given by Eqn. (A.9).

As shown in Muirhead (1982), the random variables $e_x(\tau)$ and $e_y(\tau)$ are uncorrelated such that $E\{e_x(\tau)e_y(\tau)\} = 0$. Hence, the random variable $a(\tau)$ is normally distributed with zero mean and a variance of $\sigma_a^2(\tau) = 2\sigma_1^2(\tau)$. The cross terms $e_{yx}(\tau)$ and $e_{xy}(\tau)$ are correlated, so the variance of the zero-mean normally distributed random variable $b(\tau)$ is equal to $\sigma_b(\tau) = 2[\sigma_2^2(\tau) - E\{e_{xy}(\tau)e_{yx}(\tau)\}]$. The random processes $x(t)$ and $y(t)$ are identically distributed, which implies that the random variables $e_{yx}(\tau)$ and $e_{xy}(-\tau)$ have the same *statistical* properties. The interchangeability of these two variables, as far as their statistical properties are concerned, is exploited to derive an expression for $E\{e_{xy}(\tau)e_{yx}(\tau)\}$ in Eqn. (A.11).

$$E\{e_{xy}(\tau)e_{yx}(\tau)\} = E\{e_{xy}(\tau)e_{xy}(-\tau)\} = E\{e_{xy}(\tau)e_{xy}(\tau+v)\}, \quad v = -2\tau \tag{A.11}$$

Using Eqn. (A.11), the quantity $E\{e_{xy}(\tau)e_{yx}(\tau)\}$ can be calculated by substituting $v = -2\tau$ in Eqn. (A.9). Note that for $\tau = 0$, the variance of $b(\tau)$ falls to zero, which implies that $e_z(0)$ does not have an imaginary component, or equivalently, that there is no phase error in this sample of the ACS. This is expected because both $\hat{r}_z(\tau)$ and $r_z(\tau)$ are real when $\tau = 0$.

It is shown in Papoulis (1984) that the joint probability density function $p(a, b)$ of the random variables a and b, with the dependence on τ being dropped for notational convenience, is given by Eqn. (A.12), where $\rho = E\{ab\}/(\sigma_a\sigma_b)$ is defined as the *correlation coefficient*.

$$p(a,b) = \frac{1}{2\pi\sigma_a\sigma_b\sqrt{1-\rho^2}} \exp\left\{\frac{-1}{2(1-\rho^2)}\left[\frac{a^2}{\sigma_a^2} - \frac{2\rho ab}{\sigma_a\sigma_b} + \frac{b^2}{\sigma_b^2}\right]\right\} \tag{A.12}$$

From Muirhead (1982), the normal random variables a and b are uncorrelated ($\rho = 0$) in this case because the error terms in the real part (e_x, e_y) are mutually uncorrelated with those in the imaginary part (e_{xy}, e_{yx}). The Cartesian random variables a and b may be related to the polar random variables (i.e., magnitude r and phase ϕ) by the transformations $a = r\cos\phi$ and $b = r\sin\phi$. The joint probability density function of the variables r and ϕ is given by the following expression,

$$f(r,\phi) = J(a,b)p(r\cos\phi, r\sin\phi) = rp(r\cos\phi, r\sin\phi) \tag{A.13}$$

where $J(a, b) = r$ is the Jacobian determinant of the transformation between Cartesian and polar coordinates Papoulis (1984). By setting $\rho = 0$ in Eqn. (A.12) and substituting the result into Eqn. (A.13), the polar coordinate joint density function $f(r, \phi)$ may be derived as,

$$f(r,\phi) = \frac{r}{2\pi\sigma_a\sigma_b} \exp\frac{-1}{2}\left[\frac{r^2\cos^2\phi}{\sigma_a^2} + \frac{r^2\sin^2\phi}{\sigma_b^2}\right] \tag{A.14}$$

where r and ϕ are the magnitude and phase of the complex ACS estimation error, respectively. The marginal densities $f_1(\phi)$ and $f_2(r)$ can then be derived from Eqn. (A.14). Specifically, the marginal density for the phase $f_1(\phi)$ is given by Eqn. (A.15).

$$f_1(\phi) = \int_{r=0}^{\infty} f(r,\phi)dr = \int_{r=0}^{\infty} Cr\exp\left[\frac{-Ar^2}{2}\right]dr \tag{A.15}$$

The constants C and A in Eqn. (A.15) are given by $C = (2\pi\sigma_a\sigma_b)^{-1}$ and $A = \cos\phi^2/\sigma_a^2 + \sin\phi^2/\sigma_b^2$. This integral is readily evaluated to derive the marginal density of the phase error in Eqn. (A.16).

$$f_1(\phi) = \left[\frac{-C}{A}\exp\left\{\frac{-Ar^2}{2}\right\}\right]_0^\infty = \frac{C}{A} = \frac{\sigma_a\sigma_b}{2\pi\left(\cos\phi^2\sigma_b^2 + \sin\phi^2\sigma_a^2\right)} \tag{A.16}$$

In the special case of $\sigma_a = \sigma_b$, the phase becomes uniformly distributed, $f_1(\phi) = 1/2\pi$, as expected. The marginal density for the magnitude $f_2(r)$ is given by,

$$f_2(r) = \int_{\phi=-\pi}^{\pi} f(r,\phi)\,d\phi = \int_{\phi=-\pi}^{\pi} F\exp\left(G\cos 2\phi\right)d\phi \tag{A.17}$$

where after some manipulation, the constants F and G can be written as:

$$F = \frac{r}{2\pi\sigma_a\sigma_b}\exp\left\{\frac{-r^2\left(\sigma_a^2+\sigma_b^2\right)}{4\sigma_a^2\sigma_b^2}\right\}, \quad G = \frac{-r^2\left(\sigma_b^2-\sigma_a^2\right)}{4\sigma_a^2\sigma_b^2} \tag{A.18}$$

The integral in Eqn. (A.17) can be evaluated by first forming a Taylor series expansion of the integrand,

$$f_2(r) = F\int_{\phi=-\pi}^{\pi}\sum_{n=0}^{\infty}\frac{G^n(\cos 2\phi)^n}{n!}d\phi \tag{A.19}$$

and then deriving an expression for the following generic term of this expansion.

$$\begin{aligned}\int_{\phi=-\pi}^{\pi}\cos^n 2\phi\,d\phi &= \int_{\phi=-\pi}^{\pi}\frac{1}{2^n}[e^{j2\phi}-e^{-j2\phi}]^n d\phi = \int_{\phi=-\pi}^{\pi}\frac{1}{2^n}\sum_{k=0}^{n}C_k^n e^{j2\phi k}e^{j2\phi(n-k)}d\phi \\ &= \frac{1}{2^n}C_{n/2}^n 2\pi = \frac{1}{2^n}\frac{n!}{(n/2)!(n/2)!}2\pi \quad (\text{for } n \text{ even})\end{aligned} \tag{A.20}$$

The odd terms in the Taylor series expansion do not contribute to the marginal density in Eqn. (A.19), which may be simplified to yield the expression in Eqn. (A.21)

$$f_2(r) = F\sum_{n=0}^{\infty}\frac{1}{2^n}C_{n/2}^n 2\pi\frac{G^n}{n!} \quad \text{for} \quad n = 0, 2, 4, \ldots, \tag{A.21}$$

If $\sigma_a = \sigma_b$, then we have that $F = \frac{r}{2\pi\sigma_a^2}\exp -r^2/2\sigma_a^2$ and $G = 0$. Substituting these values into Eqn. (A.21) yields the density,

$$f_2(r) = \frac{r}{\sigma_a^2}e^{-r^2/2\sigma_a^2} \tag{A.22}$$

which is, of course, the Rayleigh density. The marginal densities derived above allow us to construct statistical tests for validating or rejecting a hypothesised ACS model in terms of real and imaginary components, or magnitude and phase, from a sample ACS with a known level of confidence.

APPENDIX B

Space-Time Separability

For a general one-dimensional ARMA(p, q) model, the current output $\mathbf{c}_m^{[n]}(t)$ can be written as a linear combination of the current and past inputs $\underline{\epsilon}_m^{[n]}(t)$ that drive the process, as in Eqn. (B.1), where the complex scalars $\alpha_m(k)$ represent the linear combination coefficients. These coefficients are independent of the receiver number n in the described scalar-type generalized Watterson model (GWM), as the temporal second-order statistics that describe Doppler spread are assumed to be identical for all receivers in this model.

$$\mathbf{c}_m^{[n]}(t) = \sum_{k=0}^{\infty} \alpha_m(k)\underline{\epsilon}_m^{[n]}(t-k) \tag{B.1}$$

Similarly, the output $\mathbf{c}_m^{[n-j]}(t-i)$ of the process at a different receiver $n-j$ and time $t-i$ can be written as in Eqn. (B.2), with reference to the linear combination coefficients defined in Eqn. (B.1).

$$\mathbf{c}_m^{[n-j]}(t-i) = \sum_{k=0}^{\infty} \alpha_m(k+i)\underline{\epsilon}_m^{[n-j]}(t-i-k) \tag{B.2}$$

In accordance with the GWM, the innovative noise vector $\underline{\epsilon}_m(t)$ is temporally white, and the space-time correlation function of the inputs $\underline{\epsilon}_m^{[n]}(t)$ take the form of Eqn. (B.3), where $\rho_s(j)$ is the *spatial* ACS of the innovative noise vector, normalized such that $\rho(0) = 1$. In the GWM, this normalization follows from the definition of the scaling term ν_m, which is applied to the driving (scalar) white noise process $\gamma_{mn}(t)$ of unit variance to generate the realizations of $\underline{\epsilon}_m(t)$.

$$E\left\{\underline{\epsilon}_m^{[n]}(t)\underline{\epsilon}_m^{[n-j]*}(t-i)\right\} = \rho_s(j)\,\delta(i) \tag{B.3}$$

Using Eqn. (B.1) and Eqn. (B.3), the *temporal* ACS of the channel modulations is given by Eqn. (B.4). Note that the temporal ACS is independent of receiver number n. By definition, the term $E\{\underline{\epsilon}_m^{[n]}(t)\underline{\epsilon}_m^{[n]*}(t)\} = \rho_s(0) = 1$, so the temporal ACS may be expressed as a function of the linear combination coefficients. This function is denoted by the term $\rho_t(i) = \sum_{k=0}^{\infty} \alpha_m(k+i)\alpha_m^*(k)$ in Eqn. (B.4).

$$E\left\{\mathbf{c}_m^{[n]}(t)\mathbf{c}_m^{[n]*}(t-i)\right\} = E\left\{\underline{\epsilon}_m^{[n]}(t)\underline{\epsilon}_m^{[n]*}(t)\right\} \sum_{k=0}^{\infty} \alpha_m(k+i)\alpha_m^*(k) = \rho_t(i) \tag{B.4}$$

Due to the scaling term μ_m, which normalizes the power of the modulations in each receiver to unity, we also have that $\rho_t(0) = 1 = \sum_{k=0}^{\infty} |\alpha_m(k)|^2$. By inspection of the spatial ACS expression in Eqn. (B.5), this implies that the spatial ACS of the channel modulations is equal to the spatial ACS of the innovative noise vector.

$$E\left\{\mathbf{c}_m^{[n]}(t)\mathbf{c}_m^{[n-j]*}(t)\right\} = E\left\{\underline{\epsilon}_m^{[n]}(t)\underline{\epsilon}_m^{[n-j]*}(t)\right\}\sum_{k=0}^{\infty} |\alpha_m(k)|^2 = \rho_s(j) \tag{B.5}$$

Combining Eqn. (B.1) and Eqn. (B.2), the space-time correlation sequence $r(i, j)$ may be simplified to the form in Eqn. (B.6), since samples of the innovative noise vector at different times are uncorrelated.

$$r(i, j) = E\left\{\mathbf{c}_m^{[n]}(t)\mathbf{c}_m^{[n-j]*}(t-i)\right\} = E\left\{\underline{\epsilon}_m^{[n]}(t)\underline{\epsilon}_m^{[n-j]*}(t)\right\}\sum_{k=0}^{\infty} \alpha_m(k+i)\alpha_m^*(k) \tag{B.6}$$

From Eqn. (B.5), we have that $E\{\underline{\epsilon}_m^{[n]}(t)\underline{\epsilon}_m^{[n-j]*}(t)\} = E\{\mathbf{c}_m^{[n]}(t)\mathbf{c}_m^{[n-j]*}(t)\} = \rho_s(j)$, while from Eqn. (B.4), $\sum_{k=0}^{\infty} \alpha_m(k+i)\alpha_m^*(k) = E\{\mathbf{c}_m^{[n]}(t)\mathbf{c}_m^{[n]*}(t-i)\} = \rho_t(i)$. Hence, the space-time ACS $r(i, j)$ of the GWM is separable as the product of the temporal ACS $\rho_t(i)$ and the spatial ACS $\rho_s(j)$. The space-time ACS of the GWM channel modulations may therefore be represented in the form of Eqn. (B.7).

$$E\left\{\mathbf{c}_m^{[n]}(t)\mathbf{c}_m^{[n-j]*}(t-i)\right\} = E\left\{\mathbf{c}_m^{[n]}(t)\mathbf{c}_m^{[n-j]*}(t)\right\}E\left\{\mathbf{c}_m^{[n]}(t)\mathbf{c}_m^{[n]*}(t-i)\right\} = \rho_t(i)\rho_s(j) \tag{B.7}$$

For the special case of *first-order* AR temporal and spatial processes, the normalized ACS functions are given by $\rho_t(i) = \alpha^i$ and $\rho_s(j) = \beta^j$, where α and β are the correlation coefficients of the temporal and spatial AR(1) processes, respectively. As derived above, the space-time separability property of the GWM is not limited to this special case, and holds irrespective of the orders of the temporal and spatial AR processes.

APPENDIX C

Modal Decomposition

As described in Scharf (1991), the discrete-time impulse response of an ARMA $(M, M-1)$ process, denoted by $h(iT)$, can be expressed as the superposition of M modes, as in Eqn. (C.1), where T is the sampling interval (Nyquist rate of the ARMA process), and $h(iT) = 0$ for $i < 0$ (i.e., a causal system).

$$h(iT) = \sum_{m=1}^{M} c_m z_m^i \tag{C.1}$$

The modes z_m are constructed from the roots of the AR characteristic equation in Eqn. (C.2), where $a_0, a_1, \ldots, a_M$ are the AR process coefficients, and z_m are the poles of the ARMA transfer function, denoted by $H(z) = B(z)/A(z)$. The latter is given by the Z-transform of the impulse response $h(iT)$.

$$A(z) = a_0 z^M + a_1 z^{M-1} + \cdots + a_M = \prod_{m=1}^{M} (z - z_m) \tag{C.2}$$

The residues c_m in Eqn. (C.1) are constructed from the partial fraction expansion in Eqn. (C.3), evaluated at $z = z_m$, where the moving average characteristic polynomial is defined as $B(z) = \sum_{k=0}^{M-1} b_k z^{-k}$. The described modal decomposition of the impulse response is valid for the class of ARMA$(M, M-1)$ processes.

$$c_m = \left[\left(1 - z_m z^{-1}\right) \frac{B(z)}{A(z)} \right]_{z=z_m} \tag{C.3}$$

The power spectral density $S(z)$ of the ARMA$(M, M-1)$ process is given by the modulus squared of the transfer function $H(z)$ scaled by the sampling interval T, and is typically evaluated for frequencies in the interval $f \in [-1/2T, 1/2T)$, where $z = e^{j2\pi fT}$ is on the unit circle.

$$S(z) = T|H(z)|^2 = H(z)H^*(z) \tag{C.4}$$

The auto-correlation sequence (ACS) of the process $r(iT)$ is given by the inverse Fourier transform of the periodic power spectral density function $S(e^{j2\pi fT})$. Using Eqn. (C.4), and the convolution theorem, the ACS of the process can be expressed in terms of the modal impulse response function in Eqn. (C.5), where $h^*(-iT)$ is the inverse Fourier transform of $H^*(z)$ and the symbol $\oplus$ denotes convolution.

$$r(iT) = h(iT) \oplus h^*(-iT) = \sum_{j=0}^{\infty} h(j)h^*(j+i) \tag{C.5}$$

The sampling interval T has been omitted from the discrete convolution sum in Eqn. (C.5) for notational convenience. Substituting Eqn. (C.1) into Eqn. (C.5) and expanding the terms yields Eqn. (C.6). The geometric sum identity $\sum_{j=0}^{\infty}(z_{m'}z_m^*)^j = (1 - z_{m'}z_m^*)^{-1}$ for $\{m, m'\} \in [1, M]$ has been used to simplify the expansion. Note that the poles z_m are not outside the unit circle (i.e., $|z_m| \le 1$ for $m = 1, 2, \ldots, M$) in the assumed ACS model.

$$r(iT) = \sum_{m=1}^{M} g_m z_m^{*i}, \quad g_m = \sum_{m'=1}^{M} \frac{c_{m'}c_m^*}{1 - z_{m'}z_m^*} \tag{C.6}$$

From Eqn. (C.6), it is evident that the ACS of an ARMA(M, $M-1$) model can also be expressed as a modal decomposition, where the modes of $r(iT)$ are the complex conjugates of the modes of $h(iT)$. This result, which does not explicitly appear in Scharf (1991) or Marple (1987), can be exploited to *jointly* estimate the parameters g_m and z_m that provide a least squares fit to the sample ACS. Once the parameters g_m and z_m have been estimated from a finite length sample ACS, it is possible to extend the ACS model to an infinite number of lags to estimate the power spectral density using Eqn. (C.7).

$$S(z) = T \sum_{i=-\infty}^{\infty} r(iT)z^{-i} = \sum_{m=1}^{M} \frac{g_m}{|1 - z_m^* z^{-1}|^2} \tag{C.7}$$

Bibliography

Abed-Meraim, K., Qui, W., and Hua, Y.: 1997, Blind system indentification, *Proceedings of IEEE* **85**(8), 1310–1322.

Ablett, S. J. and Emery, D. J.: 2000, Waveform parameter advice: A clutter evaluation tool for high frequency surface wave radar, *Eighth International Conference on HF Radio Systems and Techniques* **474**, 213–217.

Aboutanios, E. and Mulgrew, B.: 2005, Assessment of the single data set detection algorithms under template mismatch, *Proceedings of the Fifth IEEE International Symposium on Signal Processing and Information Technology*, Athens, Greece, 269–274.

Aboutanios, E. and Mulgrew, B.: 2010, Hybrid detection approach for STAP in heterogeneous clutter, *IEEE Transactions on Aerospace and Electronic Systems* **46**(3), 1021–1033.

Abramovich, Y., Anderson, S., Lyudviga, Y., Spencer, N., Turcaj, P., and Hibble, B.: 2004, Space-time adaptive techniques for ionospheric clutter mitigation in HF surface wave radar systems, *IEEE International Conference on Radar Systems*, Toulouse, France.

Abramovich, Y. I.: 1981a, A controlled method for adaptive optimization of filters using the criterion of maximum SNR, *Radio Engineering and Electronic Physics* **26**(3), 87–95.

Abramovich, Y. I.: 1981b, A regularised method for adaptive optimisation of filters based on the maximum signal-to-interference ratio criterion, *Radioteknika i Electronika* **26**(3), 543. (In Russian.)

Abramovich, Y. I.: 1990, Analysis of a direct adaptive tuning method for interference compensation systems with auxiliary linear constraints, *Soviet Journal of Communications Technology and Electronics* **35**(1), 30–37. (Translated from *Radioteknika i Electronika*.)

Abramovich, Y. I.: 2000, Convergence analysis of linearly constrained SMI and LSMI adaptive algorithms, *Proceedings of ASSPCC*, Lake Louise, Canada, 255–259.

Abramovich, Y. I., Anderson, S. J., Frazer, G. J., and Solomon, I. S. D.: 1995, Measurement and interpretation of phase fluctuations in HF radar echoes propagated via sporadic E layers, *Proceedings of the IPS/DSTO Conference on Solar and Terrestrial Physics*, Adelaide, Australia, 152–155.

Abramovich, Y. I., Anderson, S. J., Gorokhov, A. Y., and Spencer, N. K.: 2004, Stochastically constrained spatial and spatio-temporal adaptive processing for nonstationary hot-clutter cancellation, In R.K. Klemm (ed.), *Applications of Space-Time Adaptive Processing*, London: Springer, 603–697.

Abramovich, Y. I., Anderson, S. J., and Solomon, I. S. D.: 1996, Adaptive ionospheric distortion correction techniques for HF skywave radar, *Proceedings of 1996 IEEE National Radar Conference*, Ann Arbor, Michigan, US, 267–272.

Abramovich, Y. I., Anderson, S. J., and Spencer, N.: 2000, Stochastic-constraints method in nonstationary hot-clutter cancellation, Part 2: Unsupervised training applications, *IEEE Transactions on Aerospace and Electronic Systems* **36**(1), 132–150.

Abramovich, Y. I., Mikhaylyukov, V. N., and Malyavin, I. P.: 1992, Stabilisation of autoregressive characteristics of spatial clutters while nonstationary spatial filtering, *Soviet Journal of Communications Technology and Electronics* **37**(2), 10–19. (Translated from *Radioteknika i Electronika*.)

Abramovich, Y. I., and Gorokhov, A. Y.: 1994, Adaptive OTHR signal extraction under nonstationary ionospheric propagation conditions, *IEEE International Radar Conference (RADAR 94)*, Paris, 420–425.

Abramovich, Y. I., Fabrizio, G. A., Anderson, S. J., Gray, D. A., and Turley, M. D.: 1997, Nonstationary HF interference rejection in adaptive arrays, *IEEE International Radar Conference (RADAR 97)*, Edinburgh, UK.

Abramovich, Y. I., Frazer, G. J., and Johnson, B. A.: 2011, Principles of mode-selective MIMO OTHR, *Accepted for Publication in IEEE Transactions Aerospace and Electronic Systems*.

Abramovich, Y. I., Gorokhov, A. Y., and Demeure, C.: 1996, Experimental verification of a generalized multivariate propagation model for ionospheric HF signals, *Proceedings of the European Signal Processing Conference (EUSIPCO-96)*, Trieste, Italy 3: 1853–1856.

Abramovich, Y. I., Gorokhov, A. Y., Mikhaylyukov, V. N., and Malyavin, I. P.: 1994, Exterior noise adaptive rejection for OTH radar implementations, *International Conference on Acoustics, Speech and Signal Processing (ICASSP-94)*, Adelaide, Australia, 105–107.

Abramovich, Y. I. and Kachur, V. G.: 1987, Effectiveness of adaptive tuning of correcting filters for cyclic convolution, *Soviet Journal of Communication Technology and Electronics* **32**(4), 180–186.

Abramovich, Y. I., Kachur, V. G., and Struchev, V. F.: 1984, Methods of digital channel correction in multichannel radar receivers, *Radio Engineering and Electronic Physics* **29**(9), 62–68.

Abramovich, Y. I., Mikhaylyukov, V. N., and Malyavin, I. P.: 1992a, Test of interference stationarity in adaptive filtering systems, *Soviet Journal of Communications Technology and Electronics* **37**(3), 1–10. (Translated from *Radioteknika i Electronika*.)

Abramovich, Y. I. and Nevrev, A. I.: 1981, An analysis of effectiveness of adaptive maximization of the signal-to-noise ratio which utilizes the inversion of the estimated correlation matrix, *Radio Engineering and Electronic Physics* **26**(12), 67–74.

Abramovich, Y. I., Spencer, N. K., Anderson, S. J., and Gorokhov, A. Y.: 1998, Stochastic-constraints method in nonstationary hot-clutter cancellation, Part 1: Fundamentals and supervised training applications, *IEEE Transactions on Aerospace and Electronic Systems* **34**(4), 1271–1292.

Abramovich, Y. I., Spencer, N. K., and Anderson, S. J.: 1999, Experimental trials on non-Gaussian environmental noise mitigation for surface-wave over-the-horizon radar by adaptive antenna array processing, *Proceedings of IEEE Signal Processing Workshop on Higher-Order Statistics*, Caesarea, Israel, 340–344.

Abramovich, Y. I., Spencer, N. K., and Anderson, S. J.: 2000, Experimental study of the spatial dynamics of environmental noise for a surface-wave OTHR application, *Eight International Conference on HF Radio Systems and Techniques*, Guildford, UK, 357–362.

Abramovich, Y. I., Spencer, N. K., Tarnavskii, S., and Anderson, S. J.: 2000, Experimental trials on environmental noise rejection by adaptive spatio-polarimetric processing for HF surface wave radar, *Proceedings of the International Radar Symposium (IRS-2000)*, Berlin, Germany.

Abramovich, Y. I., Turcaj, P., and Spencer, N. K.: 2008, Joint transient and CW external-noise mitigation in heavy clutter, *IEEE Radar Conference (RADARCON'08)*, Rome, Italy, 1–6.

Abramovich, Y. I., Yevstratov, F. F., and Mikhaylyukov, V. N.: 1993, Experimental investigation of efficiency of adaptive spatial unpremeditated noise compensation in HF radars for remote sea surface diagnostics, *Soviet Journal of Communications Technology and Electronics* **38**(10), 112–118.

Abramovich, Y., Mikhaylyukov, V., and Malyavin, I.: 1992b, Stabilisation of the autoregressive characteristics of spatial clutters in the case of nonstationary spatial filtering, *Soviet Journal of Communications Technology and Electronics* **37**, 10–19. (Translated from *Radioteknika i Electronika*.)

Abramowitz, M. and Stegan, I. A.: 1964, *Handbook of Mathematical Functions*, Dover Publications, Inc., Mineola, NY, US.

Adachi, F., Feeney, M., Williamson, A., and Parsons, J.: 1986, Cross-correlation between the envelopes of 900 MHz signals received at a mobile radio base station site, *IEE Proceedings Part F* **133**, 506–512.

Adve, R. S., Hale, T. B., and Wicks, M. C.: 1999, Transform domain localized processing using measured steering vectors and non-homogeneity detection, *Proceedings of the IEEE Radar Conference*, Waltham, Massachusetts, US, 285–290.

Adve, R. S., Hale, T. B., and Wicks, M. C.: 2000a, Practical joint domain localised adaptive processing in homogeneous and nonhomogeneous environments, parts 1 and 2, *IEE Proceedings—Radar, Sonar and Navigation* **147**(2), 57–74.

Adve, R. S., Hale, T. B., and Wicks, M. C.: 2000b, Practical joint domain localised adaptive processing in homogeneous and nonhomogeneous environments, part 1: Homogeneous environments, *IEE Proceedings—Radar, Sonar and Navigation* **147**(2), 57–65.

Adve, R. S., Hale, T. B., and Wicks, M. C.: 2000c, Practical joint domain localised adaptive processing in homogeneous and nonhomogeneous environments, part 2: Nonhomogeneous environments, *IEE Proceedings—Radar, Sonar and Navigation* **147**(2), 66–74.

Agee, B. G., Schell, S. V., and Gardner, W. A.: 1990, Spectral self-coherence restoral: A new approach to blind adaptive signal extraction using antenna arrays, *Proceedings of IEEE* **78**, 753–767.

Agy, V.: 1954, Geographical and temporal distribution of polar blackouts, *Journal of Geophysical Research* **59**, 499.

Ahearn, J. L., Curley, S. R., Headrick, J. M., and Trizna, D. B.: 1974, Tests of remote skywave measurement of oean surface conditions, *Proceedings of IEEE* **62**, 681–686.

Albert, A.: 1972, *Regression and the Moore-Penrose Pseudoinverse*, Academic Press, New York and London.

Alebastrov, V. A., Mal'tsev, A. T., Oros, V. M., Shlionskiy, A. G., and Yarko, O. I.: 1993, Some characteristics of echo signals, *Telecommunications and Radio Engineering* **48**, 92–95.

Ames, J.: 1964, Spatial properties of the amplitude fading of continuous HF radio waves, *Radio Science* **68D**(12), 1309–1318.

Anand, K., Mathew, G., and Reddy, V. U.: 1995, Blind separation of multiple co-channel BPSK signals arriving at an antenna array, *IEEE Signal Processing Letters* **2**, 176–178.

Anderson, D. N., Forbes, J. M., and Codrescu, M.: 1989, A fully analytical, low- and middle-latitude ionospheric model, *Journal of Geophysical Research* **94**, 1520–1524.

Anderson, R. H. and Krolik, J. L.: 2002, Track association for over-the-horizon radar with a statistical ionospheric model, *IEEE Transactions on Signal Processing* **50**, 2632–2643.

Anderson, S.: 2008a, Multiple scattering of HF skywave radar signals: Physics, interpretation and exploitation, *IEEE Radar Conference (RADARCON'08)*, Rome, Italy, 1–5.

Anderson, S.: 2008b, Prospects for tsunami detection and characterisation with HF skywave radar, *IEEE International Radar Conference (RADARCON'08)*, Rome, Italy, 639–645.

Anderson, S. J.: 1985, Simulation and modeling for the Jindalee over-the-horizon radar, *Mathematics and Computers in Simulation* **27**, 241–248.

Anderson, S. J.: 1986, Remote sensing with the Jindalee skywave radar, *IEEE Journal of Oceanic Engineering* **11**(2), 158–163.

Anderson, S. J.: 2004, The Doppler structure of diffusely-scattered skywave radar echoes, *IEEE International Radar Conference (RADAR-2004)*, Toulouse, France.

Anderson, S. J.: 2007, Optimising bistatic HF radar configurations for target and environmental signature discrimination, *Conference on Information, Decision, and Control*, Adelaide, Australia, 29–33.

Anderson, S. J. and Abramovich, Y. I.: 1998, A unified approach to detection, classification, and correction of ionospheric distortion in HF skywave radar systems, *Radio Science* **33**(4), 1055–1067.

Anderson, S. J., Abramovich, Y. I., and Boerner, W.-M.: 2002, Measuring polarization dynamics of the generalized HF skywave channel transfer function, *International Symposium on Antennas and Propagation (ISAP-2000)*, Japan, 1–4.

Anderson, S. J., Abramovich, Y. I., and Fabrizio, G. A.: 1997, Stochastic constraints in non-stationary hot clutter cancellation, *IEEE International Conference on Acoustics, Speech and Signal Processing (ICASSP-97)*, Munich, Germany, 3753–3756.

Anderson, S. J., Bates, B. D., and Tyler, M. A.: 1999, HF surface wave radar and its role in littoral warfare, *Journal of Battlefield Technology* **2**(3), 1–5.

Anderson, S. J., Edwards, P. J., Morrone, P., and Abramovich, Y. I.: 2003, Investigations with SECAR: A bistatic HF surface wave radar, *IEEE International Radar Conference (RADAR 2003)*, Adelaide, Australia, 717–722.

Anderson, S. J., Mei, F. J., and Peinan, J.: 2001, Enhanced OTHR ship detection via dual frequency operation, *Proceedings of China Institute of Electronics International Conference on Radar*, Beijing, 85–89.

Antonik, P., Schuman, H., Li, P., Melvin, W., and Wicks, M.: 1997, Knowledge-based space-time adaptive processing, *Proceedings of IEEE National Radar Conference*, Syracuse, NY, 372–377.

Applebaum, S. P.: 1976, Adaptive arrays, *IEEE Transactions on Antennas and Propagation* **24**(5), 585–598.

Appleton, E. V.: 1954, The anomalous equatorial belt in the F2-layer, *Journal of Atmospheric and Terrestrial Physics* **5**, 349.

Appleton, E. V. and Barnett, M. A. F.: 1925, Local reflection of wireless waves from the upper atmosphere, *Nature* **115**, 333–334.

Apaydin, G. and Sevgi, L.: 2010, Propagation modeling and path loss prediction tools for high frequency surface wave radars, *Turkish Journal of Electrical Engineering & Computer Sciences* **18**(3), 469–484.

Arato, M.: 1961, On the sufficient statistics for stationary Gaussian random processes, *Theory of Probability and its Applications* **6**, 199–201.

ARRL: 1991, *The ARRL Antenna Book*, The American Radio Relay League, Newington, CT.

Astely, D., Ottersten, B., and Swindelhurst, A.: 1998, A generalized array manifold model for wireless communication channels with local scattering, *IEE Proceedings—Radar, Sonar, and Navigation* **145**, 51–57.

Astely, D., Swindlehurst, A., and Ottersten, B.: 1999, Spatial signature estimation for uniform linear arrays with unknown receiver gains and phases, *IEEE Transactions on Signal Processing* **47**(8), 2128–2138.

Aurand, J. F.: 1987, An algorithm for computing the roots of a complex polynomial, *IEEE Transactions on Automatic Control* **32**, 164–166.

Bailey, D. K.: 1959, The effect of multipath distortion on the choice of operating frequencies for high-frequency communications circuits, *IRE Transactions on Antennas and Propagation* **7**, 398.

Baixiao, C., Duofang, C., Shouhong, Z., Hao, Z., and Maocang, L.: 2006, Experimental system and experimental results for coast-ship bi/multistatic ground-wave over-the-horizon radar, *International Conference on Radar (CIE'06)*, Shanghai, China, 1–5.

Baker, C. J., Griffiths, H. D., and Papoutsis, I.: 2005, Passive coherent location radar systems, part 2: Waveform properties, *IEE Proceedings—Radar, Sonar and Navigation* **152**(3), 160–168.

Balser, M. and Smith, W.: 1962, Some statistical properties of pulsed oblique HF ionospheric transmissions, *Journal of Research of the National Bureau of Standards, Section D* **66-D**, 721–730.

Bandiera, F., Besson, O., and Ricci, G.: 2008, An ABORT-like detector with improved mismatched signal rejection capabilities, *IEEE Transactions on Signal Processing* **56**(1), 14–25.

Bandiera, F., De Maio, A., De Nicola, S., Farina, A., Orlando, D., and Ricci, G.: 2010, Adaptive strategies for discrimination between mainlobe and sidelobe signals, *IEEE International Radar Conference (RADAR 2010)*, Washington D.C., US, 910–914.

Bandiera, F., De Maio, A., Greco, A., and Ricci, G.: 2007, Adaptive radar detection of distributed targets in homogeneous and partially homogeneous noise plus subspace interference, *IEEE Transactions on Signal Processing* **55**(4), 1223–1237.

Bandiera, F., Orlando, D., and Ricci, G.: 2006, CFAR detection of extended and multiple point-like targets without assignment of secondary data, *IEEE Transactions on Signal Processing* **13**(4), 240–243.

Baniak, I., Baker, G., Cunningham, A., and Martin, L.: 1999, Silent Sentry passive surveillance, *Aviation Week and Space Technology*.

Bar-Shalom, Y. and Fortmann, T. E.: 1988, *Tracking and Data Association*, Academic Press, New York.

Bar-Shalom, Y. and Tse, E.: 1975, Tracking in a cluttered environment with probabilistic data association, *Automatica* **11**(5), 451–460.

Barclay, L. E.: 2003, *Propagation of Radio Waves*, 2d. ed., The Institution of Electrical Engineers, London, UK.

Barghausen, A. L., Finney, J. W., Proctor, L. L., and Schulz, L. D.: 1969, Predicting long-term operational parameters of high-frequency sky-wave communications systems, *ESSA Technical Report ERL 110-ITS 78,* US Department of Commerce.

Barnes, R.: 1996, Automated propagation advice for OTHR ship detection, *IEEE Proceedings—Radar, Sonar, and Navigation* **143**, 53–63.

Barnes, R. I.: 1990, Modelling the horizontal structure of mid-latitude Es from its refraction effects on F-region echoes, *Journal of Atmospheric and Terrestrial Physics* **53**(1-2), 105–114.

Barnes, R. I.: 1992, Spread-Es structure producing apparent small scale structure in the F-region, *Journal of Atmospheric and Terrestrial Physics* **54**(3-4), 373–389.

Barnum, J., Maresca, J. J., and Serebreny, S.: 1977, High-resolution mapping of oceanic wind fields with skywave radar, *IEEE Transactions on Antennas and Propagation* **25**(1), 128–132.

Barnum, J. R.: 1973, Skywave polarization rotation in swept-frequency sea backscatter, *Radio Science* **8**(5),May, 411–423.

Barnum, J. R.: 1986, Ship detection with high-resolution HF skywave radar, *IEEE Journal of Oceanic Engineering* **11**(2), 196–209.

Barnum, J. R.: 1993, Long-range surveillance of private aircraft by OTH radar with CD applications, *Proceedings ONDCP/CTAC Tactical Technologies and Wide Area Surveillance Symposium*, Chicago, Illinois, 365–406.

Barnum, J. R. and Simpson, E.: 1997, Over-the-horizon radar sensitivity enhancement by impulsive noise excision, *IEEE National Radar Conference*, Syracuse, New York, US, 252–256.

Barnum, J. R. and Simpson, E. E.: 1998, Over-the-horizon radar target registration improvement by terrain feature localization, *Radio Science* **33**, 1067.

Barrick, D.: 2003, History, present status, and future directions of HF surface-wave radars in the US, *IEEE International Radar Conference*, Adelaide, Australia, 652–655.

Barrick, D. E.: 1971, Theory of HF and VHF propagation across the rough sea, parts 1 and 2, *Radio Science* **6**, 517–533.

Barrick, D. E.: 1972a, First order theory and analysis of MF/HF/VHF scatter from the sea, *IEEE Transactions on Antennas and Propagation* **20**, 2–10.

Barrick, D. E.: 1972b, Remote sensing of sea state by radar, Chapter 12 in *Remote Sensing of the Troposphere*, V. E. Derr (ed.), NOAA/Environmental Research Laboratories, Boulder, CO.

Barrick, D. E.: 1977, Extraction of wave parameters from measured HF radar sea-echo spectra, *Radio Science* **12**(3), 415.

Barrick, D. E., Headrick, J. M., Bogle, R. W., and Crombie, D. D.: 1974, Sea backscatter at HF: Interpretation and utilization of the echo, *Proceedings of IEEE* **62**, 673–680.

Barrick, D. and Snider, J.: 1977, The statistics of HF sea-echo Doppler spectra, *IEEE Journal of Oceanic Engineering* **2**(1), 19–28.

Barrick, E. and Evans, M. W.: 1976, Implementation of coastal current-mapping HF radar system, *Progress Report No.1 NOAA Technical Memorandum, 373-WPL-47*.

Barry, G. H.: 1971, A low-power vertical-incidence ionosonde, *IEEE Transactions* **GE–9**, 86.

Bartlett, M. S.: 1946, On the theoretical specification of the sampling properties of autocorrelated time series, *Journal of the Royal Statistical Society Supplement* **8**, 27–41.

Barton, D. K.: 1974, Low-angle radar tracking, *Proceedings of IEEE* **62**, 687–704.

Basler, R. P. and Scott, T. D.: 1973, Ionospheric structure from oblique-backscatter soundings, *Radio Science* **8**, 425.

Basler, R., Price, G., Tsunoda, R., and Wong, L.: 1988, Ionospheric distortion of HF signals, *Radio Science* **23**(4), 569–579.

Bates, H. F. and Hunsucker, R. D.: 1974, Quiet and disturbed electron-density profiles in the auroral zone ionosphere, *Radio Science* **9**, 455.

Bazin, V., Molinie, J. P., Munoz, J., Dorey, P., Saillant, S., Auffray, G., Rannou, V., and Lesturgie, M.: 2006, NOSTRADAMUS: An OTH radar, *IEEE Aerospace and Electronic Systems Magazine* **21**(10), 3–11.

Beex, A. A. and Scharf, L. L.: 1981, Covariance sequence approximation for parametric spectrum modelling, *IEEE Transactions on Acoustics, Speech and Signal Processing* **29**(5), 1042–1052.

Belouchrani, A., Abed-Meraim, K., Cardoso, J.-F., and Moulines, E.: 1997, A blind source separation technique using second-order statistics, *IEEE Transactions on Signal Processing* **45**, 434–444.

Bennett, J. A., Chen, J., and Dyson, P. L.: 1991, Analytic ray tracing for the study of HF magneto-ionic radio propagation in the ionosphere, *Applied Computational Electromagnetics Society Journal* **6**(1), 192–210.

Bennett, J. A. and Dyson, P. L.: 1986, The effect of small amplitude wave irregularities on radio wave observations of the ionosphere, *Radio Science* **21**, 375.

Bernhardt, P. A., Ganguli, G., Kelley, M. C., and Swartz, W. E.: 1995, Enhanced radar backscatter from space shuttle exhaust in the ionosphere, *Journal of Geophysical Research* **100**(23), 811–823,818.

Berry, L. A. and Chrisman, M. E.: 1966, A FORTRAN program for calculation of ground wave propagation over homogeneous spherical earth for dipole antennas, *National Bureau of Standards Report 9178*, Boulder, US.

Besson, O., Monakov, A., and Chalus, C.: 2004, Signal waveform estimation in the presence of uncertainties about the steering vector, *IEEE Transactions on Signal Processing* **52**(9), 2432–2440.

Besson, O. and Stoica, P.: 2000, Decoupled estimation of DOA and angular spread for a spatially distributed source, *IEEE Transactions on Signal Processing* **48**, 1872–1882.

Besson, O., Vincent, F., Stoica, P., and Gershman, A.: 2000, Approximate maximum likelihood estimators for array processing in multiplicative noise environments, *IEEE Transactions on Signal Processing* **48**, 2506–2518.

Bibl, K. and Reinisch, B. W.: 1978, The universal digital ionosonde, *Radio Science* **13**(13), 519–530.

Bilitza, D.: n.d., International reference ionosphere, *http://modelweb.gsfc.nasa.gov/ionos/iri.html*.

Black, Q. R., Wood, J. F., and Sherrill, W. M.: 1995, Mode angles of arrival in the 55 to 3500 km range, *Radio Science* **28**(1), 693–702.

Blackman, S. and Popoli, R.: 1999, *Design and Analysis of Modern Tracking Systems*, Artech House, Boston, Massachusetts.

Blake, T. M.: 2000, The detection and tracking of small fast boats using HF surface wave radar, *HF Radio Systems and Techniques, IEE Conference Publication* **474**, 219–224.

Blum, R. S. and McDonald, K. F.: 2000, Analysis of STAP algorithms for cases with mismatched steering and clutter statistics, *IEEE Transactions on Signal Processing* **48**(2), 301–310.

Bodonyi, J. and Pegram, T. W.: 1988, HF ground-wave radar: Theory and practice, *Fourth International Conference on HF Radio Systems and Techniques*, London, UK, 241–245.

Bogle, R. W., and Trizna, D. B.: 1976, Small boat HF radar cross sections, *Naval Research Laboratory Memo. Report*, NRL-MR-3322.

Bongioanni, C., Colone, F., and Lombardo, P.: 2008, Performance analysis of a multi-frequency FM based passive bistatic radar, *IEEE Radar Conference*, Rome, Italy.

Booker, H. G., Pasricka, P. K., and Powers, W. J.: 1986, Use of scintillation theory to explain frequency-spread on F-region ionograms, *Journal of Atmospheric and Terrestrial Physics* **48**, 327.

Booker, H., Ratcliffe, J., and Shinn, D.: 1950, Diffraction from an irregular screen with applications to ionospheric problems, *Philosophical Transactions of Royal Society* **A242**(856), 579–609.

Boroson, D. M.: 1980, Sample size considerations for adaptive arrays, *IEEE Transactions on Aerospace and Electronic Systems* **16**(4), 446–451.

Boswell, A.: 1995, Antennas for a 19 MHz groundwave radar, *Ninth International Conference on Antennas and Propagation (ICAP-95)* **1**, 463–464.

Boswell, A., Emery, D., and Bedford, M.: 2006, Performance of a tetrahedral antenna array in the HF band, *IET International Conference on Ionospheric Radio Systems and Techniques*, 49–53.

Bourdillon, A. and Delloue, J.: 1994, Phase correction of a HF multi-receiver antenna array using a radar transponder, *International Conference on Acoustics Speech and Signal Processing (ICASSP-94)*, Adelaide, Australia, 125–128.

Bourdillon, A., Delloue, J., and Parent, J.: 1989, Effects of geomagnetic pulsations on the Doppler shift of HF backscatter radar echoes, *Radio Science* **24**, 183–195.

Boutacoff, D. A.: 1985, Backscatter radar extends early warning times, *Defense Electronics* **17**, 71–83.

Bowhill, S.: 1961, Statistics of a radio wave diffracted by a random ionosphere, *Journal of Research of the National Bureau of Standards, Section D: Radio Propagation* **65D**(3), 275–292.

Bowman, G. G.: 1968, Movements of ionospheric irregularities and gravity waves, *Journal of Atmospheric and Terrestrial Physics* **30**, 721.

Boys, J.: 1968, Statistical variations in the apparent specular component of ionospherically reflected radio waves, *Radio Science* **3**(10), 984–990.

Bramley, E. N.: 1951, Diversity effects in spaced-aerial reception of ionospheric waves, *Proceedings of IEE, Part III* **98**, 19–25.

Bramley, E. N.: 1955, Some aspects of the rapid directional fluctuations of short radio waves reflected at the ionosphere, *Proceedings of the Institution of Electrical Engineers (London)* **122B**(4), 533–540.

Bremmer, H.: 1949, *Terrestrial Radio Waves: Theory of Propagation*, Elsevier Pub. Co., New York, US.

Bremmer, H.: 1958, Applications of operational calculus to ground wave propagation, particularly for long waves, *IRE Transactions on Antennas and Propagation* **6**(3), 267–272.

Brennan, L. E. and Reed, I. S.: 1973, Theory of adaptive radar, *IEEE Transactions on Aerospace and Electronic Systems* **9**(2), 237–252.

Bresler, Y. and Macovski, A.: 1986a, Exact maximum likelihood parameter estimation of superimposed exponential signals in noise, *IEEE Transactions on Acoustics, Speech and Signal Processing* **34**(5), 1081–1089.

Briggs, B. H. and Phillips, G. J.: 1950, A study of the horizontal irregularities of the ionosphere, *Proceedings of the Physical Society* **B63**, 907–923.

Brown, P. and Jones, J.: 1995, A determination of the strengths of the sporadic radio-meteor sources, *Earth, Moon and Planets* **68**, 223–245.

Bruzzone, S. and Kaveh, M.: 1984, Information tradeoffs in using the sample autocorrelation function in ARMA parameter estimation, *IEEE Transactions Acoustics Speech and Signal Processing* **32**(4), 701–715.

Budden, K. G.: 1985, *The Propagation of Radio Waves*, Cambridge University Press, UK.

Burke, G. J. and Poggio, A. J.: 1981, Numerical electromagnetic code (NEC)-method of moments, *NOSC Technical Document*, 116, California, US.

Burrus, C. S. and Parks, C. W.: 1985, *DFT/FFT and Convolution Algorithms*, John Wiley and Sons, New York.

Cadzow, J.: 1982, Spectral estimation: An overdetermined rational model equation approach, *Proceedings of IEEE* **70**, 907–938.

Cadzow, J. A., Baseghi, B., and Hsu, T.: 1983, Singular-value decomposition approach to time series modelling, *IEE Procedings Pt. F* **130**(3), 332–340.

Cai, L. and Wang, H.: 1990, On adaptive filtering with the CFAR feature and its performance sensitivity to non-Gaussian interference, *Proceedings of the 24th Annual Conference on Information Sciences and Systems*, Princeton, NJ, 558–563.

Cameron, A.: 1995, The Jindalee operational radar network: Its architecture and surveillance capability, *IEEE International Radar Conference (RADAR-95)*, Virginia, US, 692–697.

Capon, J.: 1969, High-resolution frequency-wavenumber spectrum analysis, *Proceedings of IEEE* **57**(8), 2408–2418.

Capraro, G. T., Farina, A., Griffiths, H., and Wicks, M.: 2006, Knowledge-based radar signal and data processing: A tutorial review, *IEEE Signal Processing Magazine*, **23**(1), 18–29.

Cardinali, R., Colone, F., Ferretti, C., and Lombardo, P.: 2007, Comparison of clutter and multipath cancellation techniques for passive radar, *IEEE International Radar Conference (RADAR 2007)*, Boston, MA, 469–474.

Cardoso, J.: 1998, Blind signal separation: Statistical principles, *Proceedings of IEEE* **86**(10), 2009–2025.

Cardoso, J.-F. and Souloumiac, A.: 1993, Blind beamforming for non-Gaussian signals, *Proceedings of the Institution of Electrical Engineers F* **140**(6), 362–370.

Carhoun, D. O., Kramer, J. D. R., and Rastogi, P.: 1995, Adaptive cancellation of atmospheric noise and ionospheric clutter for high frequency radar, *MITRE Report*, Bedford, Massachusetts.

Carlson, B. D.: 1988, Covariance matrix estimation errors and diagonal loading in adaptive arrays, *IEEE Transactions on Aerospace and Electronic Systems* **24**(4), 397–401.

Carrell, R. L.: 1961, The design of log-periodic dipole antennas, *IRE National Convention Record, Part 1*.

Carter, G. C.: 1971, *Coherence Estimation*, Naval Underwater Systems Centre, Newport Laboratory, Rhode Island.

CCIR: 1964, 1983, and 1988, World distribution and characteristics of atmospheric radio noise, *International Radio Consultative Committee (CCIR) Report 322, International Telecommunications Union*, editions 1964, 1983, and 1988.

CCIR: 1970, 12th Plenary Assembly, Volume II, part 2, New Delhi, India, 197–213.

Cervera, M.: 2010, Provision of pharlap raytracing toolbox, *Internal Communication, Defence Science and Technology Organisation (DSTO)*, Australia.

Cervera, M. A. and Elford, W. G.: 2004, The meteor response function: Theory and application to narrow beam MST radar, *Planet, Space, Science* **52**, 591–602.

Cervera, M. A., Holdsworth, D. A., Reid, I. M., and Tsutsumi, M.: 2004, Meteor radar response function: Application to the interpretation of meteor backscatter at medium frequency, *Journal of Geophysical Research, A11309* **109**, 591–602.

Checcacci, P. F. E.: 1975, Special Marconi issue, *Radio Science* **10**, 654.

Chen, J., Bennett, J. A., and Dyson, P. L.: 1992, Synthesis of oblique ionograms from vertical ionograms using quasi-parabolic segment models of the ionosphere, *Journal of Atmospheric and Terrestrial Physics* **54**(3-4), 323–331.

Chen, P., Melvin, W., and Wicks, M.: 1999, Screening among multivariate normal data, *Journal of Multivariate Analysis* **69**, 10–29.

Chen, W. S. and Reed, I. S.: 1991, A new CFAR detection test for radar, *Digital Signal Processing* **1**(4), 198–214.

Cheremisin, O. P.: 1982, Efficiency of adaptive algorithm with regularised sample covariance matrix. *Radio Engineering and Electronic Physics* **27**(10), 69–77.

Cherniakov, M., Nezlin, D., and Kubin, K.: 2002, Air target detection via bistatic radar based on LEOS communications system, *IEE Proceedings—Radar, Sonar and Navigation* **149**(1), 33–38.

Cherniakov, M., Zeng, T., and Plakidis, E.: 2003, Galileo signal-based bistatic system for avalanche prediction, *Proceedings of IGARSS-03*, Toulouse, France, 784–786.

Chunbo, L., Baixiao, C., Duofang, C., and Shouhong, Z.: 2006, Analysis of first-order sea clutter in a shipborne bistatic high frequency surface wave radar, *International Conference on Radar (CIE'06)*, Shanghai, China, 1–4.

Ciboci, J. W.: 1997, Over-the-horizon radar surveillance of airfields for counterdrug applications, *IEEE National Radar Conference*, Syracuse, New York, US, 178–181.

Ciboci, J. W.: 1998, Over-the-horizon radar surveillance of airfields for counterdrug applications, *IEEE Aerospace and Electronic Systems Magazine* **13**(1), 31–34.

Clancy, J. T., Bascom, H. F., and Hartnett, M. P.: 1999, Mitigation of range folded clutter by a nonrecurrent waveform, *IEEE Radar Conference*, Waltham, Massachusetts, US, 79–83.

Clark, M. and Scharf, L.: 1994, Two-dimensional modal analysis based on Maximum Likelihood, *IEEE Transactions on Signal Processing* **42**(6), 1443–1452.

Clark, R. H. and Tibble, D. V.: 1978, Measurement of the elevation angles of arrival of multi-component HF skywaves, *Proceedings of IEE* **125**, 17–24.

Clark, R. W.: 1971, *Sir Edward Appleton*, Pergamon Press, Oxford.

Cohen, L.: 1989, Time-frequency distributions—a review, *Proceedings of IEEE* **77**, 941–981.

Cohen, L.: 1995, *Time-Frequency Analysis*, Prentice-Hall Inc., New York, US.

Colegrove, S. B.: 2000, Project Jindalee: From bare bones to operational OTHR, *IEEE International Radar Conference*, Virginia, US, 825–830.

Colegrove, S. B., and Cheung, B.: 2002, A peak detector that picks more than peaks, *IEEE Radar Conference*, Long Beach, California, US, 167–171.

Colegrove, S. B. and Davey, S. J.: 2003, PDAF with multiple clutter regions and target models, *IEEE Transactions on Aerospace and Electronic Systems* **39**, 110–124.

Colegrove, S. B., Davey, S. J., and Cheung, B.: 2003, PDAF versus PMHT performance on OTH radar, *IEEE International Radar Conference*, Adelaide, Australia, 560–565.

Coleman, C. J.: 1993, A general purpose ionospheric ray-tracing procedure, *DSTO Technical Report SRL-0131-TR*.

Coleman, C. J.: 1997, The effect of an imperfectly conducting plane upon an incident electromagnetic field, *IEEE Transactions on Antennas and Propagation* **45**(10), 1445–1450.

Coleman, C. J.: 1998, A ray tracing formulation and its application to some problems in over-the-horizon radar, *Radio Science* **33**(4), 1187–1197.

Coleman, C. J.: 2000, The directionality of atmospheric noise and its impact upon an HF receiving system, *Proceedings of the 8th International Conference on HF Radio Systems and Techniques* **474**, 363–366.

Coleman, C. J.: 2002, A direction-sensitive model of atmospheric noise and its application to the analysis of HF receiving antennas, *Radio Science* **37**(3),110.

Coleman, C. J., Watson, R. A., and Yardley, H.: 2008, A practical bistatic passive radar system for use with DAB and DRM illuminators, *IEEE Radar Conference*, Rome, Italy, 1–6.

Colone, F., Cardinali, R., and Lombardo, P.: 2006, Cancellation of clutter and multipath in passive radar using a sequential approach, *IEEE National Radar Conference (RadarCon 2006)*, Verona, NY, 393–399.

Compton, R. T.: 1982, The effect of random steering vector errors in the Applebaum adaptive array, *IEEE Transactions on Aerospace and Electronic Systems* **18**(5), 392–400.

Compton, R. T.: 1988a, *Adaptive Antennas—Concepts and Performance*, Prentice-Hall, Englewood Cliffs, NJ.

Compton, R. T.: 1988b, The relationship between tapped delay-line and FFT processing in adaptive arrays, *IEEE Transactions on Aerospace and Electronic Systems* **36**(1), 15–26.

Conte, E., De Maio, A., Farina, A., and Foglia, G.: 2005, Design and analysis of a knowledge-based radar detector, *IEEE International Radar Conference (RADAR 2005)*, Arlington, Virginia, US, 387–392.

Conte, E., De Maio, A., and Ricci, G.: 2001, GLRT-based adaptive detection algorithms for range spread targets, *IEEE Transactions on Signal Processing* **49**(7), 1336–1348.

Conte, E., De Maio, A., and Ricci, G.: 2002, Recursive estimation of the covariance matrix of a compound-Gaussian process and its application to adaptive CFAR detection, *IEEE Transactions on Signal Processing* **50**(8), 1908–1915.

Conte, E., Lops, M., and Ricci, G.: 1995, Asymptotically optimum radar detection in compound-Gaussian clutter, *IEEE Transactions on Aerospace and Electronic Systems* **31**(2), 617–625.

Conte, E., Lops, M., and Ricci, G.: 1996, Adaptive matched filter detection in spherically invariant noise, *IEEE Signal Processing Letters* **3**(8), 248–250.

Cook, C. E. and Bonfeld, M.: 1967, *Radar Signals: An Introduction to Theory and Application*, Acedemic Press, New York, US.

Creekmore, J., Bronez, T., and Keizer, R.: 1993, Experimental results on high frequency direction finding with modern methods, *MILCOM*-93(1), 73–77.

Croft, T. A.: 1969, A review of oblique ray tracing and its application to the calculation of signal strength, In Jones, T.B. (ed.): *"Oblique Ionospheric Radiowave Propagation," AGARD Conference Proceeding No. 13. Advisory Group for Aerospace Research and Development*, North Atlantic Treaty Organisation, p. 137.

Croft, T. A.: 1972, Sky-wave backscatter: A means for observing our environment at great distances, *Reviews of Geophysics* **10**, 73.

Croft, T. A. and Hoogasian, H.: 1968, Exact ray calculations in a quasi-parabolic ionosphere with no magnetic field, *Radio Science* **3**, 69–74.

Crombie, D. D.: 1955, Doppler spectrum of the sea echo at 13.56 Mcs., *Nature* **175**, 681–682.

Daun, M. and Koch, W.: 2008, Multistatic target tracking for non-cooperative illumination by DAB/DVB-T, *IEEE Radar Conference (RadarCon 2008)*, Rome, Italy, 1–6.

Davey, S. J.: 2009, Track-before-detect for sensors with complex measurements, *12th International Conference on Information Fusion*, Seattle, WA, 618–625.

Davies, K.: 1990, *Ionospheric Radio*, Peter Peregrinus, London, UK.

Dawber, W. N., Pote, M. F., Turner, S. D., Graddon, J. M., Barker, D., Evans, G., and Wood, S. G.: 2006, Integrated antenna architecture for high frequency multifunction naval systems, *International Conference on Radar (CIE'06)*, Shanghai, China, 1–5.

De Maio, A., Fabrizio, G. A., Farina, A., Melvin, W. L., and Timmoneri, L.: 2007, Challenging issues in multichannel radar array processing, *IEEE International Radar Conference (RADAR 2007)*, Waltham, MA, 856–862.

de Prony, B. G. R.: 1795, Essai experimental et anlaytique: sur les lois de la dilatabilite de fluides elastiques et sur celles de la force expansive de la vapour de l'eau et de la vapeur de l'alkool, a differentes temperatures, *J. E. Polytech* **1**(2), 24–76.

Di Lallo, A., Farina, A., Fulcoli, R., Genovesi, P., Lalli, R., and Mancinelli, R.: 2008, Design, development and test on real data of an FM-based prototypical passive radar, *IEEE Radar Conference*, Rome, Italy, 1–6.

Dickel, G., Emery, D. J., and Money, D. G.: 2007, The architecture and operating characteristics of a multi-frequency HF surface wave radar, parts 1 and 2, *IEEE International Radar Conference*, Edinburgh, UK.

Dinger, R., Nelson, E., Anderson, S., Earl, F., and Tyler, M.: 1999, High-frequency radar cross section measurements of surrogate go-fast boats in Darwin, Australia, *SPAWAR System Center Technical Report* (1805).

Dogan, M. C. and Mendel, J. M.: 1994, Cumulant-based blind optimum beamforming, *IEEE Transactions on Aerospace and Electronic Systems* **30**, 722–741.

Ducharme, E. D., Petrie, L. E., and Eyfrig, R.: 1971, A method for predicting the F1 layer critical frequency, *Radio Science* **6**, 369–378.

DuHamel, R. H. and Isbell, D. E.: 1957, Broadband logarithmically periodic antenna structures, *IRE National Convention Record* **5**, 119–128.

Durbridge, L.: 2002, Spectral occupancy minimization for JFAS waveforms, *DSTO Quick Report (Internal Communication).*

Durbridge, L.: 2004, Emitters of opportunity radar, *Internal Communication, Defence Science and Technology Organization (DSTO)*, Adelaide, Australia.

Dyson, P. L. and Bennett, J. A.: 1988, A model of the vertical distribution of the electron concentration in the ionosphere and its application to oblique propagation studies, *Journal of Atmospheric and Terrestrial Physics* **50**(3), 251–262.

Dyson, P. L. and Bennett, J. A.: 1992, Exact ray path calculations using realistic ionospheres, *IEE Proceedings, Part H, Microwaves, Antennas and Propagation* **139**(5), 407–413.

Dzvonkovskaya, A., Gurgel, K.-W., Rohling, H., and Schlick, T.: 2008, Low power high frequency surface wave radar application for ship detection and tracking, *IEEE International Radar Conference*, Adelaide, Australia, 627–632.

Dzvonkovskaya, A. L.; Rohling, H.: 2006, CFAR target detection based on Gumbel distribution for HF radar, *International Radar Symposium (IRS-06)*, Krakow, Poland, 1–4.

Dzvonkovskaya, A. L., and Rohling, H.: 2007, HF radar ship detection and tracking using WERA system, *IET International Conference on Radar*, Edinburgh, UK, 1–5.

Earl, F.: 1987, The influence of receiver cross-modulation on attainable HF radar dynamic range, *IEEE Transactions on Instrumentation and Measurement* **36**, 776–782.

Earl, F.: 1991, An algorithm for the removal of radio frequency interference in ionospheric backscatter sounding, *Radio Science* **26**(3), 661–670.

Earl, G.: 1995a, HF radar receiving system image rejection requirements, *Sixth International Conference on Radio Receivers and Associated Systems*, University of Bath, UK, 128–132.

Earl, G. F.: 1991, Receiving system linearity requirements for HF radar, *IEEE Transactions in Instrumentation and Measurement* **40**(6), 1038–1041.

Earl, G. F.: 1995b, HF radar receiving system image rejection requirements, *Sixth International Conference HF Radio Systems and Techniques*, University of Bath, UK, 128–132.

Earl, G. F.: 1997, Consideration of reciprocal mixing in HF OTH radar design, *Seventh International Conference on HF Radio Systems and Techniques*, Nottingham, UK, 256–259.

Earl, G. F.: 1998, FMCW waveform generator requirements for ionospheric over-the-horizon radar, *Radio Science* **33**, 1069–1076.

Earl, G. F., Kerr, P. C., and Roberts, P. M.: 1991, OTH radar receiving system design using synoptic HF environmental database, *Fifth International Conference on HF Radio Systems and Techniques*, Edinburgh, UK, 48–53.

Earl, G. F. and Ward, B. D.: 1986, Frequency management support for remote sea-state sensing using the Jindalee sky-wave radar, *IEEE Journal of Oceanic Engineering* **11**, 164–172.

Earl, G. F. and Ward, B. D.: 1987, The frequency management system of the Jindalee over-the-horizon backscatter HF radar, *Radio Science* **22**, 275–291.

Earl, G. F. and Whitington, M. J.: 1999, HF radar ADC dynamic range requirements, *Third International Conference on Advanced A/D and D/A Conversion Techniques and Their Applications*, Glasgow, Scotland, 101–105.

Eather, R. H.: 1980, *Majestic Lights: The Aurora in Science, History, and the Arts*, American Geophysical Union, Washington, D.C.

Eckersley, P. P.: 1930, The calculation of the service area of braodcast stations, *Proceedings of IRE* **18**, 1160–1194.

Edjeou, A., Lemur, D., Bertel, L., and Cole, D.: 1994, First analysis of ionospheric tilts and their effects observed with a goniopolarimeter, *Sixth International Conference on HF Radio Systems and Techniques*, UK, 55–59.

Elfouhaily, T., Chapron, B., Katsaros, K., and Vandemark, D.: 1997, A unified directional spectrum for long and short wind-driven waves, *Journal of Geophysical Research* **102**, 15782–15796.

Elkins, T. J.: 1980, A model for high frequency radar auroral clutter, *RADC Report* **TR–80–122**.

Emery, D. J.: 2003, The choice of operating frequency in HF surface wave radar design, *Ninth International Conference on HF Radio Systems and Techniques* 493, 178–181.

Emery, D. J., Money, D. G., and Matthewson, P.: 2004, Target detection and tracking in a monostatic HF surface wave radar system, *IEEE International Radar Conference*, Toulouse, France.

Ertel, R., Cardieri, P., Sowerby, K., Rappaport, T., and Reed, J.: 1998, Overview of spatial channel models for antenna array communication systems, *IEEE Personal Communications Magazine* **5**, 10–22.

Essen, H.-H., Gurgel, K. W., and Schlick, T.: 2000, On the accuracy of current measurements by means of HF radar, *IEEE Journal of Oceanic Engineering* **25**(4), 472–480.

Evstratov, F. F., Kolossov, A. A., Kuzmin, A. A., Shustov, E. I., Yakunin, V. A., Abramovich, Y. I., and Alebastrov, V. A.: 1994, Over-the-horizon radiolocation in Russia and Ukraine (the history and achievements), *L'Onde Elecrique* **74**(3), 29–32.

Fabrizio, G. A.: 2000, Space-time characterisation and adaptive processing of ionospherically propagated HF signals, Ph.D. dissertation, Adelaide University, Australia.

Fabrizio, G. A.: 2008, High frequency over-the-horizon radar, *Tutorial presented at the 2008 IEEE Radar Conference*, Rome, Italy.

Fabrizio, G. A., Abramovich, Y. I., Gray, D. A., Anderson, S. J., and Turley, M. D.: 1998, Adaptive cancellation of nonstationary interference in HF antenna arrays, *IEE Proceedings—Radar, Sonar and Navigation* **145**(1), 19–24.

Fabrizio, G. A. and Crouch, C.: 2007, Eigenvalue shaping for robust adaptive array beamforming with finite sample support, *Proceedings of Tenth Australian Symposium on Antennas*, Sydney, Australia.

Fabrizio, G. A., Colone, F., Lombardo, P., and Farina, A.: 2008, Passive radar in the high frequency band, *IEEE Radar Conference*, Rome, 26–30.

Fabrizio, G. A., Colone, F., Lombardo, P., and Farina, A.: 2009, Adaptive beamforming for high-frequency over-the-horizon passive radar, *Proceedings of IET Radar, Sonar and Navigation* **3**(4), 384–405.

Fabrizio, G. A., De Maio, A., Farina, A., and Gini, F.: 2007, A generalized adaptive Doppler processing scheme for multi-channel radar systems, *International Radar Symposium (IRS 2007)*, Cologne, Germany.

Fabrizio, G. A. and Farina, A.: 2007a, GLRT-based adaptive Doppler processing for HF radar systems, *Proceedings of the International Conference on Acoustics, Speech, and Signal Processing, ICASSP-07*, Honolulu, Hawaii, 949–952.

Fabrizio, G. A. and Farina, A.: 2007b, A robust adaptive detection scheme for radar Doppler processing, *IEEE International Radar Conference (RADAR 2007)*, Edinburgh, UK.

Fabrizio, G. A. and Farina, A.: 2011a, An adaptive filtering algorithm for blind waveform estimation in diffuse multipath channels, *Proceedings of IET Radar, Sonar and Navigation* **5**(3), 322–330.

Fabrizio, G. A. and Farina, A.: 2011b, Exploiting multipath for blind source separation with sensor arrays, *Proceedings of IEEE International Conference on Acoustics, Speech, and Signal Processing*, Prague, Czech Republic, 2536–2539.

Fabrizio, G. A., Farina, A., and De Maio, A.: 2006, Knowledge-based adaptive processing for ship detection in OTH radar, *Proceedings of International Radar Symposium*, Krakow, Poland, 261–265.

Fabrizio, G. A., Farina, A., and Turley, M.: 2003a, Spatial adaptive subspace detection in OTH radar, *IEEE Transactions on Aerospace and Electronic Systems* **39**(4), 1407–1427.

Fabrizio, G. A., Farina, A., and Turley, M.: 2003b, Experimental performance of adaptive subspace detectors in over-the-horizon radar, *International Radar Symposium*, Dresden, Germany, 355–359.

Fabrizio, G. A., Frazer, G. J., and Turley, M. D.: 2006, STAP for clutter and interference cancellation in a HF radar system, *Proceedings of IEEE International Conference on Acoustics, Speech, and Signal Processing*, Toulouse, France, **4**, 1033–1036.

Fabrizio, G. A., Gershman, A. B., and Turley, M. D.: 2003, Non-stationary interference cancellation in HF surface wave radar, *IEEE International Radar Conference (RADAR 2003)*, Adelaide, Australia, 672–677.

Fabrizio, G. A., Gershman, A. B., and Turley, M. D.: 2004, Robust adaptive beamforming for HF surface wave over-the-horizon radar, *IEEE Transactions on Aerospace and Electronic Systems* **40**(2), 510–525.

Fabrizio, G. A., Gray, D. A., and Turley, M. D.: 1999, Using sources of opportunity to estimate digital compensation for receiver mismatch in HF arrays, *Fifth International Symposium on Signal Processing and Its Applications (ISSPA 99)*, Brisbane, Australia, **2**, 855–858.

Fabrizio, G. A., Gray, D., and Turley, M.: 2001, Using sources of opportunity to compensate for receiver mismatch in HF arrays, *IEEE Transactions on Aerospace and Electronic Systems* **37**(1), 310–316.

Fabrizio, G., Gray, D., and Turley, M.: 2003, Experimental evaluation of adaptive beamforming methods and interference models for high frequency over-the-horizon radar, *Multidimensional Systems and Signal Processing—Special Issue on Radar Signal Processing Techniques* **14**(1-3), 241–263.

Fabrizio, G. A., Gray, D. A., and Turley, M. D.: 2000, Parametric localisation of space-time distributed sources, *IEEE International Conference on Acoustics, Speech, and Signal Processing (ICASSP 2000)*, Istanbul, Turkey, 3097–3100.

Fabrizio, G. A., Gray, D. A., Turley, M. D., and Anderson, S. J.: 1998a, Adaptive correction of HF antenna arrays for nonstationary interference rejection, *Proceedings of the International Radar Symposium (IRS-98)*, Munich, Germany, 1165–1175.

Fabrizio, G. A., Gray, D. A., Turley, M. D., and Anderson, S. J.: 1998b, Modelling the spatial characteristics of ionospherically propagated HF signals, *Proceedings of the International Radar Symposium (IRS-98)*, Munich, Germany, 1187–1197.

Fabrizio, G. A. and Heitmann, A.: 2012, Single site geolocatiom method for a linear array, *IEEE Radar Conference*, Atlanta, 885–890.

Fabrizio, G. A. and Heitmann, A.: 2013, A multipath-driven approach to HF geolocation, *Signal Processing—Special Issue on Sensor Array Signal Processing*, http://dx.doi.org/10.1016/j.sigpro.2013.01.026i

Fabrizio, G. A. and Holdsworth, D. A.: 2008, Adaptive mitigation of spread clutter in high frequency surface-wave radar, *IEEE International Radar Conference (RADAR 2008)*, Adelaide, Australia, 235–240.

Fabrizio, G. A., Holdsworth, D. A., and Farina, A.: 2007, Experimental HF radar trial of real-time STAP, *Proceedings of IEEE Waveform Diversity and Design Conference*, Pisa, Italy, 316–320.

Fabrizio, G. A., Scharf, L. L., Farina, A., and Turley, M. D.: 2004a, Robust adaptive Doppler processing for HF surface-wave radar, *Defence Applications of Signal Processing (DASP) Workshop*, Salt Lake City, US.

Fabrizio, G. A., Scharf, L. L., Farina, A., and Turley, M. D.: 2004b, Ship detection with HF surface-wave radar using short integration times, *IEEE International Radar Conference (RADAR-2004)*, Toulouse, France.

Fabrizio, G. A. and Turley, M. D.: 2001, An advanced STAP implementation for surveillance radar systems, *Proceedings of the 11th IEEE Workshop on Statistical Signal Processing*, Singapore, 134–137.

Fabrizio, G. A. and Turley, M. D.: 2002, Adaptive spatial filtering with data-dependent training for improved radar signal detection and localisation in structured interference, *IEEE International Radar Conference (RADAR 2002)*, Edinburgh, UK, 497–501.

Fante, R. and Dhar, S.: 1990, A model for target detection with over-the-horizon radar, *IEEE Transactions on Aerospace and Electronic Systems* **26**, 68–83.

Fante, R. L. and Vacarro, J. J.: 1998, Cancellation of jammers and jammer multipath in a GPS receiver, *IEEE Aerospace and Electronic Systems Magazine* **13**, 25–28.

Fante, R. and Torres, J. A.: 1995, Cancellation of diffuse jammer multipath by an airborne adaptive radar, *IEEE Transactions on Aerospace and Electronic Systems* **31**(2), 805–820.

Farina, A.: 1992, *Antenna-Based Signal Processing Techniques for Radar Systems*, Artech House, Norwood, MA, US.

Farina, A., Fabrizio, G. A., Melvin, W. L., and Timmoneri, L.: 2004, Multichannel array processing in radar: State of the art, hot topics and way ahead (invited paper), *IEEE Sensor Array and Multichannel Signal Processing Workshop*, Sitges, Barcelona, Spain, 11–19.

Farina, A. and Russo, A.: 1986, Radar detection of correlated targets in clutter, *IEEE Transactions on Aerospace and Electronic Systems* **22**(5), 513–532.

Farley, D.: 1971, Radio wave scattering from the ionosphere, *Chapter 14, Methods of Experimental Physics*, Academic, New York, **9B**, 139–187.

Farley, D. T.: 1970, Incoherent scattering at radio frequencies, *Journal of Atmospheric and Terrestrial Physics* **32**, 693.

Featherstone, W., Strangeways, H. J., Darnell, M., and Mewes, H.: 1997, The application of super resolution direction finding to a variety of channel conditions in the HF band, *HF Radio Systems and Techniques*, Nottingham, UK, 306–310.

Fejer, B. G. and Kelley, M. C.: 1980, Ionospheric irregularities, *Reviews of Geophysics and Space Physics* **18**(2), 401–454.

Felgate, D. G. and Golley, M. G.: 1971, Ionospheric irregularities and movements observed with a large aerial array, *Journal of Atmospheric and Terrestrial Physics* **33**, 1353–1369.

Fenwick, R. B. and Villard, O. G. J.: 1963, A test of the importance of ionosphere-ionosphere reflections in long distance and around-the-world high-frequency propagation, *Journal of Geophysical Research* **68**, 5659.

Ferguson, B. G. and McNamara, L. F.: 1986, Calculation of HF absorption using the international reference ionosphere, *Journal of Atmospheric and Terrestrial Physics* **48**(1), 41–49.

Fernandez, D. M., Vesecky, J. F., and Teague, C.: 2006, Phase corrections of small-loop HF radar system receive arrays with ships of opportunity, *IEEE Journal of Oceanic Engineering* **31**(4), 919–921.

Fernandez, D. M., Vesecky, J., and Teague, C.: 2003, Calibration of HF radar systems with ships of opportunity, *IEEE International Geoscience and Remote Sensing Symposium*, New York, 4271–4273.

Ferraro, E. J. and Bucknam, J. N.: 1997, Improved over-the-horizon radar accuracy for the counter drug mission using coordinate registration enhancements, *IEEE National Radar Conference*, Syracuse, New York, US, 132–137.

Field, P. R. and Rishbeth, H.: 1997, The response of the ionospheric F2-layer to geomagnetic activity: An analysis of worldwide data, *Journal of Atmospheric and Solar-Terrestrial Physics* **59**, 163–180.

Flax, S. and O'Donnell, M.: 1988, Phase-aberration correction using signals from point reflectors and diffuse scatterers: Basic principles, *IEEE Transactions on Ultrasonics, Ferroelectrics, and Frequency Control* **35**(6), 758–767.

Fock, V. A.: 1945, Diffraction of radio waves around the earths surface, *Journal of Physics (Moscow)* **9**, 256–266.

Fowle, E. N., Key, E. L., Millar, R. I., and Sear, R. H.: 1979, The Enigma of the AN/FPS-95 OTH Radar (U), *MITRE Technical Report*, May 22, 1979 *(downgraded to unclassified March 23, 1993)*.

Fox, M. W. and McNamara, L. F.: 1988, Improved world-wide maps of monthly median foF2, *Journal of Atmospheric and Terrestrial Physics* **50**(12), 1077–1086.

Frazer, G. J.: 2001, High-resolution Doppler (Hoppler) processing for skywave radar, *Internal Communication*, Defence Science and Technology Organization (DSTO), Australia.

Frazer, G. J.: 2003, SkyLOS — OTH radar augmentation using a line-of-sight receiver system, *Discussion Paper DSTO-DP-0928, Defence Science and Technology Organization*, Australia.

Frazer, G. J.: 2007, Forward-based receiver augmentation for OTHR, *Proceedings of IEEE International Radar Conference 2007*, Waltham, MA, US.

Frazer, G. J., Abramovich, Y. I., and Johnson, B. A.: 2007, Spatially waveform diverse radar: Perspectives for high frequency OTHR, *IEEE Radar Conference*, Boston, Massachusetts, US, 385–390.

Frazer, G. J., Abramovich, Y. I., and Johnson, B. A.: 2009, Multiple-input multiple-output over-the-horizon radar: Experimental results, *Proceedings of IET Radar, Sonar, Navigation* **3**(4), 287–303.

Frazer, G. J. and Anderson, S. J.: 1999, Wigner-ville analysis of HF radar measurements of an accelerating target, *Fifth International Symposium on Signal Processing and its Applications (ISSPA-99)*, Brisbane, Australia, 317–320.

Frazer, G. J., Johnson, B. A., and Abramovich, Y. I.: 2007, Orthogonal waveform support in MIMO HF OTH radars, *Proceedings of IEEE Waveform Diversity and Design Conference*, Pisa, Italy, 423–427.

Frazer, G. J., Meehan, D., Abramovich, Y. I., and Johnson, B. A.: 2010, Mode-selective OTHR: A new cost-effective sensor for maritime domain awareness, *IEEE Radar Conference*, Washington, D.C., US, 935–940.

Friedlander, B. and Weiss, A.: 1994, Effects of model errors on waveform estimation using the MUSIC algorithm, *IEEE Transactions on Signal Processing* **42**(1), 147–155.

Frost, O. L.: 1972, An algorithm for linearly constrained adaptive array processing, *Proceedings of IEEE* **60**, 926–935.

Gabel, R. A., Kogan, S. M., and Rabideau, D. J.: 1999, Algorithms for mitigating terrain-scattered interference, *Electronics and Communication Engineering Journal* **11**(1), 49–56.

Gadwal, V. and Krolik, J.: 2003, A performance evaluation of autoregressive clutter mitigation methods for over-the-horizon radar, *Thirty-Seventh Asilomar Conference on Signals, Systems and Computers*, Pacific Grove, California, US, **1**, 937–941.

Games, R., Townes, S., and Williams, R.: 1991, Experimental results for adaptive sidelobe cancelellation techniques applied to a HF array, *Record of the 25th Asilomar Conference on Signals, Systems, Computers*, Pacific Grove, California, 153–159.

Gao, H. T., Zheng, X., and Li, J.: 2004, Adaptive anti-interference technique using subarrays in HF surface wave radar, *IEE Proceedings—Radar, Sonar and Navigation* **151**(2), 100–104.

Gebhard, L. A.: 1979, Evolution of naval radio-electronics and contributions of the Naval Research Laboratory, *Naval Research Laboratory Report*, Washington, D.C., New York, 8300.

Georges, T. M. and Harlan, J.: 1994, New horizons for over-the-horizon radar, *IEEE Antennas and Propagation Magazine* **36**, 14–24.

Georges, T. M., Harlan, J. A., Leben, R. R., and Lematta, R. A.: 1998, A test of ocean surface current mapping with over-the-horizon radar, *IEEE Transactions on Geoscience and Remote Sensing* **36**, 101–110.

Gerlach, K.: 1990, Implementation and convergence considerations of a linearly constrained adaptive array, *IEEE Transactions on Aerospace and Electronic Systems* **26**, 263–272.

Gersham, A., Mecklenbrauker, C., and Bohme, J.: 1997, Matrix fitting approach to direction-of-arrival estimation with imperfect spatial coherence of wavefronts, *IEEE Transactions on Signal Processing* **45**(7), 1894–1899.

Gershman, A. B., Nickel, U., and Bohme, J. F.: 1997, Adaptive beamforming algorithms with robustness against jammer motion, *IEEE Transactions on Signal Processing* **45**(7), 1878–1885.

Gershman, A., Turchin, V., and Zverev, V.: 1995, Experimental results of localization of moving underwater signal by adaptive beamforming, *IEEE Transactions on Signal Processing* **43**, 2249–2257.

Gething, P. J. D.: 1991, *Radio Direction Finding and Superresolution*, 2d ed., Peter Peregrinus Ltd, London, UK.

Gierull, C. H.: 1996, Performance analysis of fast projections of the Hung-Turner type for adaptive beamforming, *Signal Processing* (Eurosip special issue on subspace methods for detection and estimation), *Part 1* **50**, 17–28.

Gierull, C. H.: 1997, Statistical analysis of the eigenvector projection method for adaptive spatial filtering of interference, *IEE Proceedings Radar Sonar Navigation* **144**(2), 57–63.

Gillmor, C. S.: 1998, The history of the term "ionosphere," *Nature* **262**, 347.

Gini, F.: 1997, Sub-optimum coherent radar detection in a mixture of K-distributed and Gaussian clutter, *IEE Proceedings—Radar, Sonar and Navigation* **144**(1), 39–48.

Gini, F. and Farina, A.: 1999, Matched subspace CFAR detection of hovering helicopters, *IEEE Transactions on Aerospace and Electronic Systems* **35**(4), 1293–1304.

Gini, F. and Farina, A.: 2002a, Vector subspace detection in compound-Gaussian clutter, Part I: Survey and new results, *IEEE Transactions on Aerospace and Electronic Systems* **38**(4), 1295–1311.

Gini, F. and Farina, A.: 2002b, Vector subspace detection in compound-Gaussian clutter, Part II: Peformance analysis, *IEEE Transactions on Aerospace and Electronic Systems* **38**(4), 1312–1323.

Gini, F. and Greco, M.: 2002, Covariance matrix estimation for CFAR detection in heavy tailed clutter, *Signal Processing* **82**(10), 1495–1507.

Gini, F., Greco, M., and Farina, A.: 1999, Clairvoyant and adaptive signal detection in non-Gaussian clutter: A data-dependent threshold interpretation, *IEEE Transactions on Signal Processing* **47**(6), 1522–1531.

Gini, F. and Michels, J. H.: 1999, Performance analysis of two covariance matrix estimators in compound-Gaussian clutter, *IEEE Proceedings—Radar, Sonar, Navigation* **146**(3), 133–140.

Global Assimilative Ionospheric Model, JPL: n.d., *http://iono.jpl.nasa.gov/gaim/index.html*.

Goldberg, J. and Messer, H.: 1998, A polynomial rooting approach to the localisation of coherently scattered sources, *International Conference on Acoustics Speech and Signal Processing (ICASSP-98)*, Seattle, Washington, US, 2057–2060.

Goldstein, J. S. and Reed, I. S.: 1997, Reduced-rank adaptive filtering, *IEEE Transactions on Signal Processing* **45**(2), 492–496.

Gonen, E. and Mendel, J. M.: 1997, Applications of cumulants to array processing, Part III: Blind beamforming for coherent signals, *IEEE Transactions on Signal Processing* **45**(9), 2252–2264.

Goodman, N. R.: 1963, Statistical analysis based on a certain multivariate complex Gaussian distribution (an introduction), *Annals of Mathematical Statistics* **34**, 152–177.

Gooley, K. W.: 1995, Detecting and identifying spurious signals generated in radar receivers, *Sixth International Conference on Radio Receivers and Associated Systems*, 119–122.

Gorski, T., Fabrizio, G., Le Caillec, J., Kawalec, A., and Thomas, N.: 2008, Practical problems with covariance matrix estimation for adaptive MTI and space-time adaptive processing for target detection in HF surface wave radars, *International Radar Symposium (IRS 2008)*, Wroclaw, Poland, 1–4.

Goutelard, C.: 1990, The NOSTRADAMUS project: French OTH-B radar design studies, *Forty-Seventh AGARD Symposium on the Use or Reduction of Propagation and Noise Effects in Distributed Military Systems*, **CP-488**, Greece.

Goutelard, C.: 2000, STUDIO father of NOSTRADAMUS. Some considerations on the limits of detection possibilities of HF radars, *International Conference on HF Radio Systems and Techniques* (474), Guildford, UK, **474**, 199–205.

Goutelard, C., Caratori, J., and Barthes, L.: 1994, Studio—monostatic backscatter sounder father of Nostradamus French OTH radar project—new results, *International Conference on Acoustics Speech and Signal Processing (ICASSP-94)* **5**, 109–112.

Green, A. L.: 1946, Early history of the ionosphere, *A. W. A. Technical Review* **7**, 177.

Green, S. D., Kingsley, S. P., and Biddiscombe, J. A.: 1994, HF radar waveform design, *Sixth International Conference on HF Radio Systems and Techniques*, North Yorkshire, UK, 202–206.

Greenwald, R. A., Baker, K. B., Hutchins, R. A., and Hanuise, C.: 1985, An HF phased array radar for studying small-scale structure in the high latitude ionosphere, *Radio Science* **20**, 63–79.

Griffiths, H. D.: 2003, From a different perspective: Principles, practice and potential of bistatic radar, *IEEE International Radar Conference (RADAR 2003)*, Adelaide, Australia.

Griffiths, H. D.: 2013, The German WW2 HF radars ELEFANT and SEE-ELEFANT, *IEEE Aerospace and Electronic Systems Magazine* **28**(1), 4–12.

Griffiths, H. D. and Baker, C. J.: 2005, Measurement and analysis of ambiguity functions of passive radar transmissions, *IEEE International Radar Conference (RADAR 2005)*, Arlington, Virginia, US, 321–325.

Griffiths, H. D. and Long, N. R.: 1986, Television based bistatic radar, *IEE Proceedings Part F, Communications, Radar, and Signal Processing* **133**(7), 649–657.

Griffiths, H. D. and Willis, N. J.: 2010, Klein Heidelberg—the first modern bistatic radar system, *IEEE Transactions on Aerospace and Electronic Systems* **46**(4), 1571–1588.

Griffiths, J. D. and Baker, C. J.: 2005, Passive coherent location radar systems, Part 1: Performance prediction, *IEE Proceedings—Radar, Sonar and Navigation* **152**(3), 153–159.

Griffiths, L.: 1977, A comparison of quadrature and single-channel receiver processing in adaptive beamforming, *IEEE Transactions on Antennas and Propagation* **25**(2), 209–218.

Griffiths, L. J.: 1976, Time-domain adaptive beamforming of HF backscatter radar signals, *IEEE Transactions on Antennas and Propagation* **24**(5), 707–720.

Griffiths, L. J.: 1996, Linear constraints in hot clutter cancellation, *International Conference on Acoustics Speech and Signal Processing (ICASSP-96)*, Atlanta, **2**, 1181–1184.

Griffiths, L. J.: 1997, Linear constraints in pre-Doppler STAP processing, *International Conference on Acoustics Speech and Signal Processing (ICASSP-97)*, Munich, Germany, 5, 3481–3484.

Griffiths, L. J. and Buckley, K. M.: 1987, Quiescent pattern control in linearly constrained adaptive arrays, *IEEE Transactions on Acoustics, Speech and Signal Processing* **35**(7), 917–926.

Guerci, J. R.: 2003, *Space-Time Adaptive Processing for Airborne Radar*, Artech House, Norwood, MA.

Guerci, J. R., Goldstein, J. S., and Reed, I. S.: 2000, Optimal and adaptive reduced-rank STAP, *IEEE Transactions on Aerospace and Electronic Systems* **36**(2), 647–663.

Guest editorial reviewing recent progress in OTH radar technology, *Radio Science* **33**, July–August 1998.

Guinvarch, R., Lesturgie, M., Durand, R., and Cheraly, A.: 2006, On the use of HF surface wave radar in congested waters: Influence of masking effect on detection of small ships, *IEEE Journal of Oceanic Engineering* **31**(4), 894–903.

Gunner, A., Temple, M. A., and Claypoole, R. J.: 2003, Direct-path filtering of DAB waveform from PCL receiver target channel, *Electronics Letters* **39**(2), 1005–1007.

Guo, X., Ni, J.-L., and Liu, G.-S.: 2003, Ship detection with short coherent integration time in over-the-horizon radar, *IEEE International Radar Conference*, Adelaide, Australia, 667–671.

Guo, X., Sun, H., and Yeo, T. S.: 2005, Transient interference excision in over-the-horizon radar using adaptive time-frequency analysis, *IEEE Transactions on Geoscience and Remote Sensing* **43**(4), 722–735.

Guo, X., Sun, H., Yeo, T. S., and Lu, Y.: 2006, Lightning interference cancellation in high-frequency surface wave radar, *International Conference on Radar (CIE'06)*, Shanghai, China, 1–4.

Gurevich, A. V.: 1978, *Nonlinear Phenomena in the Ionosphere*. Springer-Verlag, New York.

Gurevich, A. V. and Tsedilina, E. E.: 1985, *Multiple Scattering of HF Radiowaves Propagating Across the Sea Surface*, Springer-Verlag, Berlin.

Gurgel, K. W., Antonischki, G., Essen, H.-H., and Schlick, T.: 1999, Wellen radar (WERA): A new ground-wave HF radar for ocean remote sensing, *Coastal Engineering Journal* **37**, 219–234.

Gurgel, K.-W., Barbin, Y., and Schlick, T.: 2007, Radio frequency interference suppression techniques in FMCW modulated HF radars, *OCEANS 2007 - Europe*, 1–4.

Gurgel, K. W. and Schlick, T.: 2005, HF radar wave measurements in the presence of ship echoes—problems and solutions, *Proceedings of OCEANS* **2**, 937–941.

Hall, H. M.: 1966, A new model of impulsive phenomena: Applications to atmospheric noise communications channels, *Technical Report 3412-8 and 7050-7, SU-SEL-66-052, Electronic Laboratory Stanford University*, Stanford, Calif.

Hall, H. M.: 1970, The generalized "T" model for impulsive noise, *Proceedings of International Symposium on Information Theory*, Noordwokjk, Netherlands.

Hall, M. P. M. and Barclay, L. W.: 1989, *Radiowave Propagation*, Peter Peregrinus, Ltd., London.

Hanson, W. B. and Moffett, R. J.: 1966, Ionization transport effects in the equatorial F region, *Journal of Geophysical Research* **71**, 5559.

Hanuise, C., Villain, J., Gresillon, D., Cabrit, B., Greenwald, R., and Baker, K.: 1993, Interpretation of HF radar ionospheric Doppler spectra by collective wave scattering theory, *Annales Geophysicae* **11**, 29–39.

Hargreaves, J. K.: 1979, *The Upper Atmosphere and Solar-Terrestrial Relations*, Van Nostrand Reinhold, New York.

Harris, F. J.: 1978, On the use of windows for harmonic analysis with the descrete Fourier transform, *Proceedings of IEEE* **66**(1), 51–83.

Hartnett, M. P., Clancy, J. T., and Denton Jr., R. J.: 1998, Utilization of a nonrecurrent waveform to mitigate range-folded spread Doppler clutter, *Radio Science* **33**(4), 1125.

Haselgrove, J.: 1963, The Hamilton ray path equations, *Journal of Atmospheric and Terrestrial Physics* **25**, 397–399.

Hashimoto, N. and Tokuda, M.: 1999, A Bayesian approach for estimation of directional wave spectra with HF radar, *Coastal Engineering Journal* **41**, 137–149.

Hasselmann, K.: 1961, On the non-linear energy transfer in a gravity-wave spectrum, Part 1: General theory, *Journal of Fluid Mechanics* **12**(4), 481–500.

Hasselmann, K., Ross, D. B., Muller, P., and Sell, W.: 1976, A parametric wave prediction model, *Journal of Physical Oceanography* **6**, 200–228.

Hatfield, V. E.: 1980, HF communications predictions 1978 (an economical up-to-date computer code, ambcom), *Solar Terrestrial Production Proceedings*, In F.R. Donnelley (ed.), *Prediction of Terrestrial Effects of Solar Activity*, National Oceanic and Atmospheric Administration, **4**, 1–15.

Hayden, E. C.: 1961, Instrumentation for propagation and direction finding measurements, *Journal of Research of the National Bureau of Standards* **65D**(3), 253.

Haykin, S.: 1994, *Blind Deconvolution*, Prentice-Hall, Englewood Cliffs, NJ.

Haykin, S.: 1996, *Adaptive Filter Theory*, 3d ed., Prentice-Hall, Upper Saddle River, NJ.

Hayward, S. D.: 1996, Adaptive beamforming for rapidly moving arrays, *Proceedings of the CIE International Conference on Radar* **8**(10), 480–483.

Hayward, S. D.: 1997, Effects of motion on adaptive arrays, *IEE Proceedings—Radar, Sonar and Navigation* **144**(1), 15–20.

Hazelgrove, J.: 1955, Ray theory and a new method for ray tracing, In *Physics of the Ionosphere, Physical Society*, London, UK, p. 355.

Headrick, J. M.: 1990a, HF over-the-horizon radar, Chapter 24 In *Radar Handbook*, M.I. Skolnik (ed.), 2d ed., McGraw-Hill, New York.

Headrick, J. M.: 1990b, Looking over the horizon (HF radar), *IEEE Spectrum* **27**(7), 36–39.

Headrick, J. M., Root, B. T., and Thomason, J. F.: 1995, RADARC model comparisons with Amchitka radar data, *Radio Science* **30**, 607–617.

Headrick, J. M. and Skolnik, M. I.: 1974, Over-the-horizon radar in the HF band, *Proceedings of IEEE* **62**, 664–673.

Headrick, J. M. and Thomason, J. F.: 1996, Naval applications of high frequency over-the-horizon radar, *Naval Engineers Journal* **108**(3), 353–362.

Headrick, J. M. and Thomason, J. F.: 1998, Applications of high-frequency radar, *Radio Science* **33**(4), 1045–1054.

Headrick, J. M., Thomason, J. F., Lucas, D. L., McCammon, S., Hanson, R., and Lloyd, J. L.: 1971, Virtual path tracing for HF radar including an ionospheric model, *Naval Research Laboratory Memo. Report*, **2226**.

Headrick, J. and Thomason, J.: 2008, The development of over-the-horizon radar at the Naval Research Laboratory, *IEEE Radar Conference (RADARCON'08)*, Rome, Italy, 1–5.

Herman, J. R.: 1966, Spread F and ionospheric F-region irregularities, *Reviews of Geophysics* **4**, 255.

Hickey, K., Khan, R., and Walsh, J.: 1995, Parametric estimation of ocean surface-currents with HF radar, *IEEE Journal of Ocean Engineeing* **20**(2), 139–144.

Hill, J. K.: 1979, Exact ray paths in a multisegment quasi-parabolic ionosphere, *Radio Science* **14**, 855–861.

Hisaki, Y.: 1996, Nonlinear inversion of the integral equation to estimate ocean wave spectra from HF radar, *Radio Science* **31**, 25–39.

Hocke, K. and Schlegel, K.: 1996, A review of atmospheric gravity waves and travelling ionospheric disturbances: 1982-1995, *Annales Geophysicae* **14**, 917–940.

Hoft, D. J. and Agi, F.: 1986, Solid state transmitters for modern radar applications, *Proceedings of the CIE International Radar Conference*, Beijing 775–781.

Holdsworth, D. A. and Fabrizio, G. A.: 2007, Investigation of multiple window spectrogram techniques for high frequency over the horizon radar, *Proceedings of Tenth Australian Symposium on Antennas*, Sydney, Australia.

Holdsworth, D. A. and Fabrizio, G. A.: 2008, HF interference mitigation using STAP with dynamic degrees of freedom allocation, *IEEE International Radar Conference (RADAR 2008)*, Adelaide, Australia, 317–322.

Horner, J., Kubik, K., Mojarrabi, B., Longstaff, I. D., Donskoi, E., and Cherniakov, M.: 2002, Passive bistatic radar sensing with LEOS based transmitters, *Proceedings of IGARSS-02*, Toronto, Canada 438–440.

Howland, P.: 1995, Passive tracking of airborne targets using only Doppler and DOA information, *Colloquium Digest/IEE, Institution of Electrical Engineers* **104**(7), 1–3.

Howland, P. E.: 1999, Target tracking using television-based bistatic radar, *IEE Proceedings – Radar, Sonar and Navigation* **146**(3), 166–174.

Howland, P. E. and Clutterbuck, C. F.: 1997, Estimation of target altitude in HF surface wave radar, *Seventh International Conference on HF Radio Systems and Techniques* **441**, 296–300.

Howland, P. E. and Cooper, D. C.: 1993, Use of the Wigner-Ville distribution to compensate for ionospheric layer movement in high-frequency skywave radar systems, *IEE Proceedings—Radar, Sonar, Navigation.* **140**(1), 29–36.

Howland, P. E., Maksimiuk, D., and Reitsma, G.: 2005, FM radio based bistatic radar, *IEE Proceedings — Radar, Sonar and Navigation* **152**(3), 107–115.

Hua, Y.: 1992, Estimating two-dimensional frequencies by matrix enhancement and matrix pencil, *IEEE Transactions on Signal Processing* **40**(9), 2267–2280.

Hua, Y.: 1996, Fast maximum likelihood for blind identification of multiple FIR channels, *IEEE Transactions on Signal Processing* **44**(3), 661–672.

Hua, Y., Sarkar, T. K., and Weiner, D. D.: 1991, An L-shaped array for estimating 2-D directions of wave arrival, *IEEE Transactions on Antennas and Propagation* **39**(2), 143–146.

Hua, Y. and Wax, M.: 1996, Strict identification of multiple FIR channels driven by an unknown arbitrary sequence, *IEEE Transactions on Signal Processing* **44**(3), 756–759.

Huang, C.-S., Kelley, M. C., and Hysell, D. L.: 1993, Nonlinear Rayleigh-Taylor instabilities, atmospheric gravity waves, and equatorial spread F, *Journal of Geophysical Research* **98**, 15,631–15,642.

Huba, J. D., Joyce, G., and Fedder, J. A.: 2000, Sami2 is another model of the ionosphere(SAMI2): A new low-latitude ionosphere model, *Journal of Geophysical Research* **105**, 23,035–23,053.

Hudnall, J. M. and Der, S. W.: 1993, HF-OTH radar performance results., *Naval Research Laboratory Technical Report. NRL/MR/5325-93-7326.*

Hudson, J. E.: 1981, *Adaptive Array Principles*, Peregrinus, Stevenage, UK.

Hughes, C. J. and Morris, D. W.: 1963, Phase characteristics of HF radio waves received after propagation by the ionosphere, *Proceedings of IEE* **110**, 1720–1734.

Hughes, D.: 1988, Tests verify OTH-B radar's ability to detect cruise missiles, *Aviation Week and Space Technology* **128**(12), 60–65.

Hughes, D. T. and McWhirter, J. G.: 1996, Sidelobe control in adaptive beamforming using a penalty function, *International Conference on Acoustics Speech and Signal Processing (ICASSP-96)*, pp. 200–203.

Hunsucker, R.: 1991, *Radio Techniques for Probing the Terrestrial Ionosphere*, Springer-Verlag, Berlin, Germany.

Hunsucker, R. D.: 1990, Atmospheric gravity waves and travelling ionospheric disburbances: Thirty years of research, *Ionospheric Effects Symposium (IES-90)*, Washington, DC.

International Telecommunications Union, R. I.-R.: 1994, The concept of transmission loss for radio links, *ITU-R Recommendation*, Geneva, 341–343.

International Telecommunications Union, R. I.-R.: 1999a, Radio noise, *ITU-R Recommendation*, Geneva, 372.

International Telecommunications Union, R. I.-R.: 1999b, World atlas of ground conductivities, *ITU-R Recommendation 832-2.*

International Telecommunications Union, R. I.-R.: 2003, Techniques for measurement of unwanted emissions of radar systems, *Recommendation ITU-R M.1177-3.*

International Telecommunications Union, R. I.-R.: 2006a, Out-of-band domain emission limits for primary radar systems, *Annex 8 of Recommendation ITU-R SM.1541-2.*

International Telecommunications Union, R. I.-R.: 2006b, Unwanted emissions in the out-of-band domain, *Recommendation ITU-R SM.1541-2.*

Issue, S.: 2005, Passive radar systems, *IEE Proceedings — Radar, Sonar and Navigation* **152**(3), 106–223.

Jangal, F., Saillant, S. and Helier, M.: 2007, Wavelets: A versatile tool for the high frequency surface wave radar, *IEEE Radar Conference*, Boston, Massachusetts, US, 497–502.

Japt.: 1968, Special issue on upper atmosphere winds, waves and ionospheric drifts, *Journal of Atmospheric and Terrestrial Physics* **30**, 657–1063.

Jarrott, R. K., Netherway, D. J., and Anderson, S. J.: 1989, Signal processing for ocean surveillance by HF skywave radar, *Australian Symposium on Signal Processing and Applications (ASSPA-89)*, Adelaide, Australia, 293–297.

Jeng, S., Lin, H., Xu, G., and Vogel, W.: 1995, Measurements of the spatial signature of an antenna array, *Proceeding of PIMRC*, Vol. 2, Toronto, Canada, 669–672.

Johnson, D. H. and Dudgeon, D. E.: 1993, *Array Signal Processing: Concepts and Techniques*, Prentice-Hall, Englewood Cliffs, NJ.

Johnson, J. R., Fenn, A. J., Aumann, H. M., and Willwerth, F. G.: 1991, An experimental adaptive nulling receiver utilising the sample matrix inversion algorithm with channel equlisation, *IEEE Transactions on Microwave Theory and Techniques* **39**(5), 798–808.

Johnson, R. C. and Jasik, H.: 1993, *Antenna Engineering Handbook, 3d ed.*, McGraw-Hill Book Company, New York.

Jones, J. and Brown, P.: 1993, Sporadic meteor radiant distributions: Orbital survey results, *Monthly Notices of the Royal Astronomical Society* **265**, 524–532.

Jones, R. M. and Stephenson, J. J.: 1975, A versatile three-dimensional ray tracing computer program for radio waves in the ionosphere, *Office of Telecomunications Reptort 75–76.*

Jouny, I. I. and Culpepper, E.: 1995, Modeling and mitigation of terrain scattered interference, *Proceedings of IEEE International Symposium on Antennas and Propagation* **1**, 455–457.

Juang, B., Perdue, R., and Thompson, D.: 1995, Deployable automatic speech recognition systems: Advances and challenges, *AT&T Technical Journal*, 45–55.

Kalson, S. Z.: 1992, An adaptive array detector with mismatched signal rejection, *IEEE Tranactions on Aerospace and Electronic Systems* **28**(1), 195–205.

Kalson, S. Z.: 1995, Adaptive array CFAR detection, *IEEE Transactions on Aerospace and Electronic Systems* **31**(2), 534–542.

Kay, S. M.: 1987, *Modern Spectral Estimation*, Prentice-Hall, Inc., Englewood Cliffs, NJ.

Kay, S. M.: 1993, *Fundamentals of Statistical Signal Processing: Estimation Theory*, Prentice-Hall, Inc., Englewood Cliffs, NJ.

Kay, S. M.: 1998, *Fundamentals of Statistical Signal Processing: Detection Theory*, Prentice-Hall, Inc., Englewood Cliffs, NJ.

Kazanci, O., Bilik, I., and Krolik, J.: 2007, Wavefront adaptive raymode processing for over-the-horizon HF radar clutter mitigation, *Conference Record of the Forty-First Asilomar Conference on Signals, Systems and Computers 2007 (ACSSC)*, 2191–2194.

Kelley, M. C.: 1989, *The Earth's Ionosphere*, Academic Press, Inc., San Diego, California.

Kelley, M. C.: 2009, *The Earth's Ionosphere: Plasma Physics and Electrodynamics*, Academic Press, 2d ed., Burlington, MA.

Kelley, M. C., Vlasov, M. N., Foster, J. C., and Coster, A. J.: 2004, A quantitative explanation for the phenomenon known as storm enhanced density, *Geophysical Research Letters* **31**, L19809.

Kelly, E. J.: 1986, An adaptive detection algorithm, *IEEE Transactions on Aerospace and Electronic Systems* **22**(1), 115–127.

Kelly, E. J.: 1989, Performance of an adaptive detection algorithm, *IEEE Transactions on Aerospace and Electronic Systems* **25**(2), 122–133.

Kelly, E. J. and Forsythe, K. M.: 1989, Adaptive detection and parameter estimation for multidimensional signal models, *Technical Report 848*, Massachusetts Institute of Technology Lincoln Laboratory.

Kelso, J. M.: 1964, *Radio Ray Propagation in the Ionosphere*, McGraw-Hill, New York.

Kent, G. S. and Wright, R. W. H.: 1968, Movements of ionospheric irregularities and ionospheric winds, *Journal of Atmospheric and Terrestrial Physics* **30**, 657.

Khan, R. H.: 1991, Ocean-clutter model for high-frequency radar, *IEEE Journal of Oceanic Engineering* **16**(2), 181.

Khan, R. H. and Mitchell, D. K.: 1991, Waveform analysis for high-frequency FMICW radar, *IEE Proceedings — Radar and Signal Processing* **138**(5), 411–419.

Khattatov, B., Murphy, M., Gnedin, M., Fuller-Rowell, T., and Yudin, V.: 2004, Advanced modeling of the ionosphere and upper atmosphere, *Environmental Research Technologies Report, A550924*.

Kim, A. G., Zumbrava, Z. F., Grozov, V. P., Kotovich, G. V., Mikhaylov, Y. S., and Oinats, A. V.: 2005, The correction technique for IRI model on the basis of oblique sounding data and simulation of ionospheric disturbance parameters, *Proceedings of the 28th URSI General Assembly*, New Delhi.

King, R. W. P.: 2003, Surface-wave radar and its application, *IEEE Transactions on Antennas and Propagation* **51**(10), 3000–3002.

Kinsman, B.: 1965, *Wind Waves*, Prentice-Hall, Inc., Englewood Cliffs, New Jersey.

Kirby, R. C.: 1982, Radio-wave propagation, *Section 18 in Electronics Engineer's Handbook,* D.G. Fink and D. Christiansen (eds.), **51**(5), 18-50–18-126.

Klemm, R.: 1998, *Space-Time Adaptive Processing — Principles and Applications*, IEE Publishers, London, UK.

Klemm, R.: 2002, *Principles of Space-Time Adaptive Processing, 2d ed.*, IEE, London, UK.

Klemm, R.: 2004, *Applications of Space-Time Adaptive Proceesing*, The Institution of Electrical Engineers, London, UK.

Knott, E., Shaeffer, J., and Tuley, M.: 1993, *Radar Cross Section,* 2d ed., Artech House, Boston, MA.

Kogon, S. M., Williams, D. B., and Holder, E. J.: 1996, Reduced-rank terrain scattered interference mitigation, *Proc. of IEEE international symposium on Phased array systems and technology, Boston, MA*, pp. 400–405.

Kogon, S. M., Williams, D. B., and Holder, E. J.: 1996, Beamspace techniques for hot clutter cancellation., *Proceedings of ICASSP-96,* Atlanta, **2**, 1177–1180.

Kogon, S. M., Williams, D. B., and Holder, E. J.: 1998, Exploiting coherent multipath for main beam jammer suppression, *IEE Proceedings—Radar, Sonar and Navigation* **145**(5), 303–308.

Kogon, S. M. and Zatman, M. A.: 2001, STAP adaptive weight training using phase and power selection criteria, *Record of thirty-fifth Asilomar conference on Signals, Systems and Computers*, Pacific Grove, CA, **1**, 98–102.

Kolosov, A. A. E.: 1984, *Fundamentals of Over-the-Horizon Radar,* (Translated from *Radio i Svyaz* by W. F. Barton, Artech House, Norwood, MA, 1987).

Kotaki, M. and Katoh, C.: 1984, The global distribution of thunderstorm activity observed by the ionosphere satellite (ISS-B), *Journal of Atmospheric and Terrestrial Physics* **45**, 833–850.

Kramer, J. D. R. and Williams, R. T.: 1994, High frequency atmospheric noise mitigation, *IEEE International Conference on Acoustics, Speech and Signal Processing (ICASSP-94)* **6**, 101–104.

Kraut, S. and Scharf, L. L.: 1999, The CFAR adaptive subspace detector is a scale-invariant GLRT, *IEEE Transactions on Signal Processing* **47**(8), 2538–2541.

Kraut, S. and Scharf, L. L.: 2001, Adaptive subspace detectors, *IEEE Transactions on Signal Processing* **49**(1), 1–16.

Kraut, S., Scharf, L. L., and Butler, R. W.: 2005, The adaptive coherence estimator: A uniformly most-powerful-invariant adaptive detection statistic, *IEEE Transactions on Signal Processing* **53**(2), 427–438.

Kreyszig, E.: 1988, *Advanced Engineering Mathematics,* 6th ed., John Wiley and Sons, New York, US.

Krim, H. and Viberg, M.: 1996, Two decades of array signal processing research: The parametric Approach, *IEEE Signal Processing Magazine* **13**(4), 67–94.

Krolik, J. I. and Anderson, R. H.: 1997, Maximum likelihood coordinate registration for over-the-horizon radar, *IEEE Transactions on Signal Processing* **45**, 945–959.

Kulpa, K. S. and Czekala, Z.: 2005, Masking effect and its removal in PCL radar, *IEE Proceedings — Radar, Sonar, and Navigation* **152**(3), 174–178.

Kumaresan, R., Scharf, L., and Shaw, A.: 1986, An algorithm for pole-zero modeling and spectral analysis, *IEEE Transactions on Acoustics Speech and Signal Processing* **34**(3), 637–640.

Kurashov, A. G. (Ed.): 1985, *Shortwave Antennas,* 2d ed. in Russian, *Radio i Svyaz.*

Kuschel, H., Heckenbach, J., Muller, S., and Appel, R.: 2008, On the potentials of passive, multistatic, low frequency radars to counter stealth and detect low flying targets, *IEEE Radar Conference,* Rome, Italy.

Lee, Y., Choi, J., Song, I., and Lee, S.: 1997, Distributed source modeling and direction-of-arrival estimation techniques, *IEEE Transactions on Signal Processing* **45**, 960–969.

Lees, M. L.: 1987, An overview of signal processing for over-the-horizon radar, *Proceedings of the International Symposium on Signal Processing and its Applications (ISSPA-87),* Adelaide, Australia.

Lemmon, J. J.: 2001, Wideband model of HF atmospheric radio noise, *Radio Science* **36**(6), 1385–1391.

Leong, H.: 1997, Adaptive nulling of skywave interference using horizontal dipole antennas in a coastal surveillance HF surface wave radar system, *IEEE International Radar Conference (RADAR'97)*, Edinburgh, UK, 26–30.

Leong, H.: 2000, A comparison of sidelobe cancellation techniques using auxiliary horizontal and vertical antennas in HF surface wave radar, *IEEE International Radar Conference*, Alexandria, Virginia, US, 672–677.

Leong, H.: 2002, Dependence of HF surface wave radar sea clutter on sea state, *IEEE Radar Conference 2002*, Long Beach, California, US, 56–60.

Leong, H.: 2006, The potential of bistatic HF surface wave radar system for the surveillance of water-entry area along coastline, *IEEE Radar Conference*, Verona, New York, US, p. 4.

Leong, H., Helleur, C., and Rey, M.: 2002, Ship detection and tracking using HF surface wave radar, *IEEE Radar Conference*, Long Beach, California, US, 61–65.

Leong, H. and Ponsford, A.: 2004, The advantage of dual-frequency operation in ship tracking by HF surface wave radar, *Proceedings of International Radar Conference*, Toulouse, France.

Leong, H. and Ponsford, A.: 2008, The effects of sea clutter on the performance of HF surface wave radar in ship detection, *IEEE Radar Conference (RADARCON'08)*, 1–6.

Leong, H. W.: 1999, Adaptive suppression of skywave interference in HF surface wave radar using auxiliary horizontal dipole antennas, *IEEE Pacific Rim Conference on Communications, Computers and Signal Processing*, British Columbia, Canada, 128–132.

Leong, H. W.: 2007, An estimation of radar cross sections of small vessels at HF, *IET International Conference on Radar Systems*, Edinburgh, UK, 1–4.

Leong, H. and Wilson, H.: 2006, An estimation and verification of vessel radar cross sections for high-frequency surface-wave radar, *IEEE Antennas and Propagation Magazine* **48**(2), 11–16.

Leontovich, M. and Fock, V. A.: 1946, Solution of the problem of electromagnetic waves along the Earth's surface by the method of parabolic equation, *Journal of Physics (USSR)* **10**, 13–24.

Levanon, N.: 1988, *Radar Principles*, John Wiley and Sons, New York.

Levanon, N. and Mozeson, E.: 2004, *Radar Signals*, John Wiley and Sons, Hoboken, NJ.

Lewis, B. L., Kretschmer, F. F., and Shelton, W. W.: 1986, *Aspects of Radar Signal Processing*, Artech House, Norwood, MA.

Li, B. and Yuan, Y.: 2006, A method for ship target extracting from broadened Bragg lines in bistatic shipborne SWR, *The 8th International Conference on Signal Processing* **4**, 16–20.

Li, J. and Stoica, P.: 1996, An adaptive filtering approach to spectral estimation and SAR filtering, *IEEE Transactions on Signal Processing* **44**(6), 1469–1483.

Li, J. and Stoica, P. E.: 2006, *Robust Adaptive Beamforming*, John Wiley and Sons, Hoboken, NJ.

Li, J., Stoica, P., and Zheng, D.: 1997, One-dimensional MODE algorithm for two dimensional frequency estimation, *Multidimensional Systems and Signal Processing* **8**, 449–468.

Li, L.-W.: 1998, High frequency over-the-horizon radar and ionospheric backscatter studies in China, *Radio Science* **33**(5), 1445–1448.

Li, S. T., Koyama, L. B., Schukantz, J. H., and Dinger, R. J.: 1995, EMC study of a shipboard HF surface wave radar, *IEEE International Symposium on Electromagnetic Compatibility*, Atlanta, Georgia, US, 406–410.

Liang, H. and Biyang, W.: 2006, Impulsive interference mitigation in high frequency radar, *First IEEE Conference on Industrial Electronics and Applications*, 2006, 1–4.

Lin, K. H., Yeh, K. C., Weng, A., and Yang, K. S.: 1995, Probing the ionosphere with an extremely narrow frequency separation, *Radio Science* **30**(3), 683–692.

Lipa, B.: 1977, Derivation of directional ocean-wave spectra by inversion of second order radar echoes, *Radio Science* **12**(4), 425–434.

Lipa, B. J. and Barrick, D. E.: 1983, Least-squares methods for the extraction of surface currents from CODAR crossed-loop data: Application at ARSLOE, *IEEE Journal of Oceanic Engineering* **8**(4), 226–253.

Lipa, B. J. and Barrick, D. E.: 1986, Extraction of sea-state from HF radar sea echo: Mathematical theory and modeling, *Radio Science* **21**(1), 81–100.

Liu, B., Chen, B., and Michels, J. H.: 2002, A GLRT for radar detection in the presence of compound-Gaussian clutter and additive white Gaussian noise, *IEEE Second Sensor Array and Multi-channel Signal Processing Workshop*, Washington, D.C.

Liu, D. and Waag, R.: 1995, A comparison of ultrasonic wavefront distortion and compensation in one-dimensional and two-dimensional apertures, *IEEE Transactions on Ultrasonics, Ferroelectrics, and Frequency Control* **42**(4), 726–733.

Liu, D. and Waag, R.: 1998, Estimation and correction of ultrasonic wavefront distortion using pulse-echo data received in a two-dimensional aperture, *IEEE Transactions on Ultrasonics, Ferroelectrics, and Frequency Control* **45**(2), 473–490.

Liu, H. and Xu, G.: 1996, A subspace method for signature waveform estimation in synchronous CDMA systems, *IEEE Transactions on Communications* **44**, 1346–1354.

Liu, H. and Zoltowski, M. D.: 1997, Blind equalization in antenna array CDMA systems, *IEEE Transactions on Signal Processing* **45**, 161–172.

Liu, Y.: 1996, Target detection and tracking with a high frequency ground wave over-the-horizon radar, *CIE International Radar Conference*, Beijing, China, 29–33.

Liu, Y., Xu, R., and Zhang, N.: 2003, Progress in HFSWR research at Harbin Institute of Technology, *IEEE International Radar Conference*, Adelaide, Australia, 522–528.

Lodge, O.: 1902, Mr. Marconi's results in day and night wireless telegraphy, *Nature* **66**, 222.

Lombardo, P., Colone, F., Bongioanni, C., Lauri, A., and Bucciarelli, T.: 2008, PBR activity at INFOCOM: Adaptive processing techniques and experimental results, *IEEE Radar Conference (RadarCon 2008)*, Rome, Italy.

Long, A. E. and Trizna, D. B.: 1973, Mapping of north Atlantic winds by HF radar sea backscatter interpretation, *IEEE Transactions on Antennas and Propagation* **21**, 680–685.

Lu, K., and Liu, X.: 2005, Enhanced visibility of maneuvering targets for high-frequency over-the-horizon radar, *IEEE Transactions on Antennas and Propagation* **53**(1), 404–411.

Lu, X., Kirlin, R. L., and Wang, J.: 2003, Temporal impulsive noise excision in the range-Doppler map of HF radar, *Proceedings of 2003 International Conference on Image Processing, ICIP*, **2**, 835–838.

Lu, X., Wang, J., Dizaji, R., Ding, Z., and Ponsford, A. M.: 2004, A switching constant false alarm rate technique for high frequency surface wave radar, *Canadian Conference on Electrical and Computer Engineering (CCECE)*, **4**, 2081–2084.

Lu, X., Wang, J., Ponsford, A. M., and Kirlin, R. L.: 2010, Impulsive noise excision and performance analysis, *IEEE Radar Conference*, Washington D.C., US, 1295–1300.

Lucas, D. L.: 1987, Ionospheric parameters used in predicting the performance of high frequency skywave circuits, *Interim Report on NRL Contract N00014-87-K-20009, Account 153-6943*, University of Colorado, Boulder.

Lucas, D. L. and Harper, J. D.: 1965, A numerical representation of CCIR report 322 high frequency (3-30 Mcs) atmospheric radio noise data, *National Bureau of Standards Note 318*.

Lucas, D. L. and Haydon, G. W.: 1966, Predicting statistical performance indexes for high frequency telecommunications systems, *ESSA Technical Report IER 1 ITSA 1*, US Department of Commerce.

Lucas, D. L., Lloyd, J. L., Headrick, J. M., and Thomason, J. F.: 1972, Computer techniques for planning and management of OTH radars, *Naval Research Laboratory Memo Report* **2500**.

Lucas, D., Pinson, G., and Pilon, R.: 1993, Some results of RADARC 2 equatorial spread Doppler clutter predicitions, *Proceedings of the Seventh International Ionospheric Effects Symposium*, Alexandria, Virginia, 2A5-1–2A5-8.

Lynch, J. T.: 1970, Aperture synthesis for HF radio signals propagated via the F-layer of the ionosphere, *Stanford Electronic Laboratory Technical Report 161 SU-SEL-70-066*, Stanford University.

Lyon, E. and French, M.: 1999, Detecting/tracking the A37-B aircraft with ROTHR, *Technical Memorandum to DoD Counterdrug Technology Development Program Office*.

Maclean, T. S. M. and Wu, Z.: 1993, *Radiowave Propagation over Ground*, Chapman and Hall, London.

Madden, J. M.: 1987, The adaptive suppression of interference in HF ground wave radar, *IEEE International Radar Conference*, London, UK, 98–102.

Maeda, R. and Inuki, H.: 1979, Radio disturbance warning issuance system, In R. F. Donnelly, (ed.): *Solar-Terrestrial Predictions Proceedings I*, US GPO, Washington, D.C. 20402 p. 223.

Malanowski, M. and Kulpa, K.: 2008, Digital beamforming for passive coherent location radar, *IEEE Radar Conference (RadarCon 2008)*, Rome, Italy.

Manolakis, D. G., Ingle, V. K., and Kogon, S. M.: 2000, *Statistical and Adaptive Signal Processing: Spectral Estimation, Signal Modeling, Adaptive Filtering and Array Processing*, McGraw-Hill, Boston, MA.

Maresca, J. W. and Barnum, J. R.: 1982, Theoretical limitation of the sea on the detection of low Doppler targets by over-the-horizon radar, *IEEE Transactions on Antennas and Propagation* **30**, 837–845.

Markowski, S., Hall, P., and Wylie, R.: 2010, *Defence Procurement and Industry Policy: A Small Country Perspective*, Routledge, Abingdon.

Marple, S. L.: 1987, *Digital Spectral Analysis with Applications*, Prentice-Hall—Signal Processing Series, Englewood Cliffs, New Jersey, US.

Marrone, P. and Edwards, P.: 2008, The case for bistatic HF surface wave radar, *IEEE International Radar Conference*, Adelaide, Australia, 633–638.

Maslin N.: 1987, *HF Communications—A Systems Approach*, Pitman Publishing, London.

Matsushita, S.: 1976, Geomagnetic disturbances and storms, In S. Matsushita, and W.H. Campbell, (eds.): *Physics of Geomagnetic Phenomena, II* **8**, 793–819.

Matsushita, S. and Campbell, W. H.: 1967, *Physics of Geomagnetic Phenomena I*, Academic Press, New York.

McKinley, D. W. R.: 1961, *Meteor Science and Engineering*, McGraw-Hill, New York.

McMillan, S. G.: 2011, OTH radar array calibration using LOS active returns, *M.S. Thesis, Department of Electrical and Electronic Engineering*, University of Adelaide, Australia.

McNamara, L. F.: 1973, Evening-type transequatorial propagation on Japan-Australia circuits, *Australian Journal of Physics* **26**, 521–543.

McNamara, L. F.: 1991, *The Ionosphere: Communications, Surveillance and Direction Finding*, Krieger Publishing Company, Malabar, Florida.

Melvin, W.: 2004, A STAP overview, *IEEE Aerospace and Electronic Sytems Magazine* **19**(1), 19–35.

Melvin, W. L.: 2000, Space-time adaptive radar performance in heterogeneous clutter, *IEEE Transactions on Aerospace and Electronic Systems* **36**(2), 621–633.

Melvin, W. L. and Scheer, J. A.: 2013, *Principles of Modern Radar: Advanced Techniques*, SciTech Publishing, Edison, NJ.

Melvin, W. L. and Wicks, M.: 1997, Improving practical space-time adaptive radar, *Proceedings of IEEE Radar Conference*, Syracuse, New York, US, 48–53.

Melvin, W., Wicks, M., Antonik, P., Salama, Y., Li, P., and Schuman, H.: 1998, Knowlege-based space-time adaptive processing for AEW radar, *IEEE Aerospace and Electronic Sytems Magazine* **13**(4), 37–42.

Mendillo, M.: 1973, A study of the relationship between geomagnetic storms and ionospheric disturbances at mid-latitudes, *Planetary and Space Science* **21**, 349.

Menelle, M., Auffray, G., and Jangal, F.: 2008, Full digital high frequency surface wave radar: French trials in the biscay bay, *IEEE International Radar Conference*, Adelaide, Australia, 224–229.

Meng, Y., Stoica, P., and Wong, K.: 1996, Estimation of the directions of arrival of spatially dispersed signals in array processing, *IEE Proceedings—Radar, Sonar and Navigation* **143**(1), 1–9.

Mewes, H.: 1997, Long term azimuth and elevation measurements in the HF band, *HF Radio Systems and Techniques*, Nottingham, UK, 344–348.

Michels, J.: 1997, Covariance matrix estimator performance in non-Gaussian clutter processes, *IEEE National Radar Conference*, Syracuse, NY.

Michels, J. H., Himed, B., and Rangaswamy, M.: 2000, Performance of STAP tests in Gaussian and compound-Gaussian clutter, *Digital Signal Processing* **10**(4), 309–324.

Michels, J. H., Himed, B., and Rangaswamy, M.: 2002, Performance of parametric and covariance based STAP tests in compound-Gaussian clutter, *Digital Signal Processing* **12**, 307–328.

Miles, J. W.: 1957, On the generation of surface waves by shear flows, *Journal of Fluid Mechanics* **3**(2), 185–204.

Miller, D. C. and Gibbs, J.: 1975, Ionospheric analysis and ionospheric modeling, *AFCRL Technical Report 75-549*.

Miller, T. W.: 1976, The transient response of adaptive arrays in TDMA systems, *Ohio State University Electroscience Laboratory Technical Report RADC-TR-76-390.*

Millington, G.: 1949, Ground-wave propagation over an inhomogeneous smooth earth, *Proceedings of IEE* **96**, 53–64.

Millington, G. and Isted, G. A.: 1950, Ground-wave propagation over an inhomogeneous smooth earth, Part 2: Experimental evidence and practical implications, *Proceedings of IEE* **97**, 209–222.

Millman, G. H. and Nelson, G.: 1980, Surface wave HF radar for over-the-horizon detection, *IEEE International Radar Conference (RADAR-80)*, 106–112.

Milsom, J. D.: 1997, HF groundwave radar equations, *Seventh International Conference on HF Radio Systems and Techniques*. **441**, 285–290.

Minkler, G. and Minkler, J.: 1990, *CFAR. The Principles of Automatic Radar Detection in Clutter*, Magellan Book Company, Baltimore, Maryland, US.

Mitra, A. P.: 1974, *Ionospheric Effects of Solar Flares*, Reidel Publishing Company, Dordrecht, Holland.

Money, D. G., Emery, D. J., Blake, T. M., Clutterbuck, C. F., and Ablett, S. J.: 2000, HF surface wave radar management techniques applied to surface craft detection, *IEEE International Radar Conference*, Virginia, US, 110–115.

Money, D. G., Emery, D. J., and Dickel, G.: 2007, Intelligent radar management techniques in high frequency surface wave radar, *3rd Institution of Engineering and Technology Seminar on Intelligent Sensor Management*, London, UK, 1–11.

Montbriand, L. E.: 1981, *The Dependence of HF Direction Finding Accuracy on Aperture Size*, Report 343, Communications Research Center, Ottawa, Canada.

Monzingo, R. and Miller, T.: 1980, *Introduction to Adaptive Arrays*, John Wiley and Sons, New York.

Moutray, R. E. and Ponsford, A. M.: 1995, IMS integrates surface wave radar with existing assets for continuous surveillance of the EEZ, *Conference Proceedings of IEEE Challenges of Our Changing Global Environment—OCEANS95* **2**, 696–702.

Moutray, R. E. and Ponsford, A. M.: 1997, Integrated maritime surveillance (IMS) for the Grand Banks, *IEEE Conference Proceedings of OCEANS'97* **2**, 981–986.

Moutray, R. E. and Ponsford, A. M.: 2003, Integrated maritime surveillance: Protecting national sovereignty, *IEEE International Radar Conference*, Adelaide, Australia, 385–388.

Moyle, D. E. and Warrrington, E. M.: 1997, Some superresolution DF measurements within the HF band, *Tenth International Conference on Antennas and Propagation*, Edinburgh, UK, 71–73.

Muggleton, L. M.: 1975, A method of predicting foE at any time and place, *Telecommunications Journal* **42**(7), 413–418.

Muirhead, R. J.: 1982, *Aspects of Multivariate Statistical Theory*, John Wiley and Sons, New York.

Nathanson, F. E.: 1969, *Radar Design Principles*, McGraw-Hill, New York, US.

Nathanson, F. E., Reilly, J. P., and Cohen, M. N.: 1999, *Radar Design Principles: Signal Processing and the Environment, 2d ed.*, SciTech Publishing, Mendham, NJ.

Neale, B. T.: 1985a, CH – the first operational radar, *The GEC Journal of Research* **3**(2), 73–83.

Netherway, D. J. and Ayliffe, J. K.: 1987, Impulsive noise suppression in HF radar, *IREE Digest of papers, 21st International Electronics Convention and Exhibition*, Sydney, Australia, 673–676.

Netherway, D. J. and Carson, C. T.: 1986, Impedance and scattering matrices of a wideband HF phased array, *Journal of Electronics Engineering*, Australia **6**, 29–39.

Netherway, D. J., Ewing, G. E., and Anderson, S. J.: 1989, Reduction of some environmental effects that degrade the performance of HF skywave radars, *Australian Symposium on Signal Processing and Applications (ASSPA-89)*, Adelaide, Australia, 288–292.

Newton, R. J., Dyson, P. L., and Bennett, J. A.: 1997, Analytic ray parameters for the quasi-cubic segment model of the ionosphere, *Radio Science* **32**, 567–578.

Ng, B. P., Er, M. H., and Kot, C.: 1994, Array gain/phase calibration techniques for adaptive beamforming and direction finding, *IEE Proceedings—Radar, Sonar and Navigation* **141**(1), 25–29.

Nickisch, L. J.: 1992, Non-uniform motion and extended media effects on the mutual coherence function: An analytic solution for spaced frequency, position, and time, *Radio Science* **27**(1).

Nickisch, L. J., St. John, G., Fridman, S. V., and Hausman, M.: 2011, HiCIRF: A high-fidelity HF channel simulation, *Proceedings of the 2011 Ionospheric Effects Symposium*.

Northey, B. J. and Whitham, P. S.: 2000, A comparison of DSTO and DERA HF background noise measuring systems with the international radio consultative committee (CCIR) model data, *DSTO Technical Report DSTO-TR-0855*.

Norton, K. A.: 1935, The propagation of radio waves over a plane earth, *Nature* **135**, 945–955.

Norton, K. A.: 1936, The propagation of radio waves over the surface of the earth and in the upper atmosphere, Part 1: Ground wave propagation from short antennas, *Proceedings of IRE* **24**, 1367–1387.

Norton, K. A.: 1937, The propagation of radio waves over the surface of the earth and in the upper atmosphere, Part 2: The propagation from vertical, horizontal, and loop antennas over a plane earth of finite conductivity, *Proceedings of IRE* **25**, 1203–1236.

Norton, K. A.: 1941, The calculation of ground wave field intensity over a finitely conducting spherical earth, *Proceedings of IRE* **29**, 623–639.

Norton, R. B.: 1969, The middle-latitude F-region during some severe ionospheric storms, *Proceedings of IEEE* **57**, 1147.

Nuttall, A.: 1976, Spectral analysis of a univariate process with bad data points, via maximum entropy and linear predictive techniques, *Naval Underwater Systems Center, Technical Document 5303, New London, CT* **1**.

Olkin, J. A., Nowlin, W. C., and Barnum, J. R.: 1997, Detection of ships using OTH radar with short integration times, *IEEE National Radar Conference*, Syracuse, New York, US, 1–6.

Ott, R. H.: 1992, Ring: An integral equation algorithm for HF/VHF propagation over irregular, inhomogenous terrain, *Radio Science* **27**(6), 867–882.

Ottersten, B., Stoica, P., and Roy, R.: 1998, Covariance matching estimation techniques for array signal processing applications, *Digital Signal Processing* **8**, 185–210.

Ottersten, B., Viberg, M., and Kailath, T.: 1992, Analysis of subspace fitting and ML techniques for parameter estimation from sensor array data, *IEEE Transactions on Signal Processing* **40**, 590–600.

Ottersten, B., Viberg, M., Stoica, P., and Nehorai, A.: 1993, Radar array processing, Analysis of Subspace Fitting and ML Techniques for Parameter Estimation from Sensor Array Data, in Radar Array Processing, Springer-Verlag, Berlin, 99–151.

Owsley, N.: 1985, Sonar array processing, In S. Haykin (ed.), *Array Signal Processing*, Prentice-Hall, Englewood Cliffs, NJ.

Papadias, C. and Paulraj, A.: 1997, A constant modulus algorithm for multiuser signal separation in presence of delay spread using antenna arrays, *IEEE Signal Processing Letters* **4**, 178–181.

Papazoglou, M. and Krolik, J. L.: 1999, Matched-field estimation of aircraft altitude from multiple over-the-horizon radar revisits, *IEEE Transactions on Signal Processing* **47**(4), 966–976.

Papoulis, A.: 1984, *Probability, Random Variables and Stochastic Processes*, McGraw-Hill, New York.

Parent, J. and Bourdillon, A.: 1987, A method to corrrect HF skywave backscattered signals for ionospheric frequency modulation, *IEEE Transactions on Antennas and Propagation* **36**, 127–135.

Parkinson, M. L.: 1997, Observations of the broadening and coherence of MF/lower HF surface-radar ocean echoes, *IEEE Journal of Ocean Engineering* **22**(2), 347–363.

Paul, A. K.: 1979, Radio pulse distortion in the ionosphere, *Journal of Geophysical Research* **46**, 15.

Paul, A. K. A.: 1977, Simplified inversion procedure for calculating electron density profiles from ionograms to use with minicomputers, *Radio Science* **12**, 119–122.

Paulraj, A. and Kailath, T.: 1988, Direction of arrival estimation by eigenstructure methods with imperfect spatial coherence of wavefronts, *Journal of the Acoustical Society of America* **83**, 1034–1040.

Paulraj, A. and Papadias, C.: 1997, Space-time processing for wireless communications, *IEEE Signal Processing Magazine* **14**, 49–83.

Pearce, T. H.: 1991, The application of digital receiver technology to HF radar, *Fifth International Conference on HF Radio Systems and Techniques*, Edinburgh, UK, 304–309.

Pearce, T. H.: 1997, Calibration of a large receiving array for HF radar, *International Conference HF Radio Systems and Techniques*, **3**, 260–264.

Pearce, T. H.: 1998, Receiving array design for over-the-horizon radar, *GEC Journal of Technology* **15**, 47–55.

Peckham, C. D., Haimovich, A. M., Ayoub, T. F., Goldstein, J. S., and Reed, I. S.: 2000, Reduced-rank STAP performance analysis, *IEEE Transactions on Aerospace and Electronic Systems* **36**(2), 664–676.

Pedersen, K., Mogensen, P., Fleury, B., Frederiksen, F., Olesen, K., and Larsen, S.: 1997, Analysis for time, azimuth and Doppler dispersion in outdoor radio channels, *Proceedings of ACTS Mobile Communications Summit*, Aalborg, Denmark.

Percival, D. J. and White, K. A. B.: 1997, Multipath track fusion for over-the-horizon radar, *Proceedings of SPIE 3163*, San Diego, California, US, 363–374.

Percival, D. J. and White, K. A. B.: 1998, Multihypothesis fusion of multipath over-the-horizon radar tracks, *Proceedings of SPIE 3373*, Orlando, Florida, US, 440–451.

Percival, D. J. and White, K. A. B.: 2001, Multipath coordinate registration and track fusion for over-the-horizon radar, In D.A. Cochran, B. Moran, and L. White (eds.), *Defence Applications of Signal Processing*, Elsevier, Amsterdam, 149–155.

Phillips, O. M.: 1966, *The Dynamics of the Upper Ocean*, The University Press, Cambridge, UK (2d and 3d eds., 1969 and 1977).

Pierson, W. J. and Moskowitz, L.: 1964, A proposed spectral form for fully developed wind seas based on the similarity theory of S.A. Kitaigordskii, *Journal of Geophysical Research* **69**(24), 5181–5190.

Piggot, W. R. and Rawer, K. E.: 1972, *World Data Center A for Solar-Terrestrial Physics, U.R.S.I Handbook of Ionogram Interpretation and Reduction*, 2nd Edition, Report UAG-23A.

Pillai, S. U.: 1989, *Array Signal Processing*, Springer-Verlag, New York.

Pillai, S. U., Kim, Y. L., and Guerci, J. R.: 2000, Generalized forward/backward subaperture smoothing techniques for sample starved STAP, *IEEE Transactions on Signal Processing* **48**(12), 3569–3574.

Pillai, U. and Kwon, B.: 1989, Forward/backward spatial smoothing techniques for coherent signal identification, *IEEE Transactions on Acoustics, Speech Signal Processing* **37**, 8–15.

Pilon, R. O. and Headrick, J. M.: 1986, Estimating the scattering coefficient of the ocean surface for high-frequency over-the-horizon radar, *Naval Research Laboratory Memo Report*, **5741**.

Podilchak, S. K., Leong, H., Solomon, R., and Antar, Y. M. M.: 2009, Radar cross-section modeling of marine vessels in practical oceanic environments for high-frequency surface-wave radar, *IEEE Radar Conference*, Pasadena, California, US, 1–6.

Ponsford, A. M.: 1993, A comparison between predicted and measured sea echo Doppler spectra for surface wave radar, *OCEANS Conference, Institute of Electrical and Electronics Engineers* **3**(1), 61–66.

Ponsford, A. M., Dizaji, R. M., and McKerracher, R.: 2003, HF surface wave radar operation in adverse conditions, *IEEE International Radar Conference*, Adelaide, Australia, 593–598.

Ponsford, A. M., Dizaji, R. M., and McKerracher, R.: 2005, Noise suppression system and method for phased-array based systems, *United States Patent 6867731*.

Ponsford, A. M., D'Souza, I. A., and Kirubarajan, T.: 2009, Surveillance of the 200 nautical mile EEZ using HFSWR in association with a spaced-based AIS interceptor, *IEEE Conference on Technologies for Homeland Security*, Waltham, Massachusetts, US, 87–92.

Ponsford, A., Sevgi, L., and Chan, H. C.: 2001, An integrated maritime surveillance system based on high-frequency surface-wave radars, Part 2: Operational status and system performance, *IEEE Antennas and Propagation Magazine* **43**(5), 52–63.

Ponsford, A. and Wang, J.: 2010, A review of high frequency surface wave radar for detection and tracking of ships,*Turk J. Elec. Eng. & Comp. Sci.* **18**(3), 409–428.

Poole, A. W. V. and Evans, G. P.: 1985, Advanced sounding: 2. First results from an advanced chirp ionosonde, *Radio Science* **20**, 1617–1623.

Porat, B. and Friedlander, B.: 1985, Aymptotic analysis of the bias of the modified Yule-Walker estimator, *IEEE Transactions on Automatic Control* **30**, 765–767.

Porat, B. and Friedlander, B.: 1991, Blind equalization of digital communication channels using higher order moments, *IEEE Transactions on Signal Processing* **39**(2), 522–526.

Poullin, D.: 2005, Passive detection using broadcasters (DAB, DVB) with CODFM modulation, *IEE Proceedings—Radar, Sonar and Navigation* **152**(3), 143–152.

Poullin, D.: 2008, Recent progress in passive coherent location (PCL) concepts and technique in France using DAB or FM broadcasters, *IEEE Radar Conference*, Rome, Italy, 1–5.

Praschifka, J., Durbridge, L. J., and Lane, J.: 2009, Investigation of target altitude estimation in skywave OTH radar using a high-resolution ionospheric sounder, *IEEE International Radar Conference*, Bordeaux, France, 1–6.

Priestly, M. B.: 1981, *Spectral Analysis and Time Series, Volume 1*, Academic Press, Harcourt Brace Publishers, San Diego, Calif.

Prolss, G. W.: 1995, Ionospheric F Region Storms, in *Handbook of Atmospheric Electrodynamics*, H. Volland (ed.), CRC Press, Boca Raton, FL.

Pulford, G. W.: 2004, OTHR multipath tracking with uncertain coordinate registration, *IEEE Transactions on Aerospace and Electronic Systems* **40**, 38–56.

Pulford, G. W. and Evans, R. J.: 1998, A multipath data association tracker for over-the-horizon radar, *IEEE Transactions on Aerospace and Electronic Systems* **34**(4), 1165–1183.

Pulsone, N. B. and Rader, M.: 2001, Adaptive beamformer orthogonal rejection test, *IEEE Transactions on Signal Processing* **49**(3), 521–529.

Pulsone, N. B. and Raghavan, R. S.: 1999, Analysis of an adaptive CFAR detector in non-Gaussian interference, *IEEE Transactions on Aerospace and Electronic Systems* **35**(3), 903–916.

Pulsone, N. B. and Zatman, M. A.: 2000, A computationally efficient two-step implementation of the GLRT, *IEEE Transactions on Signal Processing* **48**(3), 609–616.

Qian, S.: 2002, *Time-Frequency and Wavelet Transforms*, Prentice-Hall Inc., New York.

Qiang, Y., Jiao, L. C., and Bao, Z.: 2001, An approach to detecting the targets of aircraft and ship together by over-the-horizon radar, *CIE International Radar Conference*, Beijing, China, 95–99.

Qiao, X. L. and Jin, M.: 2001, Radio disturbance suppression for HF radar, *IEE Proceedings—Radar, Sonar and Navigation* **148**(2), 89–93.

Qiao, X. and Liu, Y.: 1990, Ship detection in heavy sea clutter echoes and man-made radio noise environment for an on-shore HF ground wave frequency agile radar, *IEEE International Radar Conference*, Arlington, Virginia, US, 34–37.

Raab, F. A., Asbeck, P., Cripps, S., Kenington, P. B., Popovic, Z. B., Pothecary, N., Sevic, J. F., and Sokal, N. O.: 2002, Power amplifiers and transmitters for RF and microwave, *IEEE Transactions on Microwave Theory and Techniques* **50**(3), 814–826.

Rabideau, D. J.: 1998, Multidimensional sidelobe target editing with applications of terrain scattered jamming cancellations, *Proceedings of Ninth IEEE Statistical Signal and Array Processing Workshop*, Portland, OR, 252–255.

Rabideau, D. J.: 2000, Clutter and jammer multipath cancellation in airborne adaptive radar, *IEEE Transactions on Aerospace and Electronic Systems* **36**(2), 565–583.

Rabideau, D. J. and Steinhardt, A. O.: 1999, Improved adaptive clutter cancellation through data-adaptive training, *IEEE Transactions on Aerospace and Electronic Systems* **35**(3), 879–891.

Radford, M. F.: 1983, High frequency antennas, In A.W. Rudge, K. Milne, A.D. Oliver, and P. Knight (eds.), *The Handbook of Antenna Design*, Peter Peregrinus Ltd, London, UK.

Raghavan, R. S., Pulsone, N. B., and McLaughlin, D. J.: 1999, Performance of the GLRT for adaptive vector subspace detection, *IEEE Transactions on Aerospace and Electronic Systems* **35**(3), 903–915.

Raich, R., Goldberg, J., and Messer, H.: 2000, Bearing estimation for a distributed source: Modeling, inherent accuracy limitations and algorithms, *IEEE Transactions on Signal Processing* **48**(2), 429–441.

Rao, N. N.: 1974, Inversion of sweep-frequency sky-wave backscatter leading edge for quasi-parabolic ionospheric layer parameters, *Radio Science* **9**, 845.

Ratcliffe, J. A.: 1959, *The Magneto-Ionic Theory and Its Applications to the Ionosphere: A Monograph*, Cambridge University Press, London, UK.

Ratcliffe, J. A.: 1967, The ionosphere and the engineer, *Proceedings of Institute of Electrical Engineers*, London **114**(1).

Ratcliffe, J. A.: 1972, *An Introduction to the Ionosphere and Magnetosphere*, Cambridge University Press, London, UK.

Rawer, K. and Bilitza, D.: 1989, Electron density profile description in the international reference ionosphere, *Journal of Atmospheric and Terrestrial Physics* **51**(9-10), 781–790.

Reed, I. S., Mallet, J. D., and Brennan, L. E.: 1974, Rapid convergence rate in adaptive arrays, *IEEE Transactions on Aerospace and Electronic Systems* **10**(6), 853–863.

Regimbal, R.: 1965, A transmitter-spectrum synthesis technique for EMI prediction, *IEEE Transactions on Electromagnetic Compatibility* **7**, 125–141.

Reid, G. C.: 1967, Ionospheric disturbances, In S. Matsushita, and W.H. Campbell, (eds.): *Physics of Geomagnetic Phenomena II*, Academic Press, New York, p. 627.

Reinisch, B. W., Galkin, I. A., Khmyrov, G. M., et al.: 2009, New digisonde for research and monitoring applications, *Radio Science* **44**, 1–15.

Reinisch, B. W. and Huang, X.: 1983, Automatic scaling of electron density profiles from digital ionograms, Part 3: Processing of bottomside ionograms, *Radio Science* **18**(3), 477–492.

Rice, D. W.: 1971, The CRC high frequency direction finding research facility, *Conference Digest, 1971 International Electrical and Electronics Conference, Toronto*, IEEE, New York.

Rice, D. W.: 1973, Phase characteristics of ionospherically-propagated radio waves, *Nature Physical Science* **244**, 86–88.

Rice, D. W.: 1976, High resolution measurement of time delay and angle of arrival over a 911 km HF path, Blackband, W.T. (ed.), *Radio Systems and the Ionosphere, Advisory Group for Aerospace Research and Revelopment (AGARD) Conference Proceedings No. 173*, p. 33.

Rice, D. W.: 1982, HF direction finding by wave front testing in a fading signal environment, *Radio Science* **17**(4), 827–836.

Rice, S. O.: 1951, Reflection of electromagnetic waves from slightly rough surfaces, In M. Kline (ed.), *Theory of Electromagnetic Waves*, Interscience Publishers, New York, 351–378.

Richards, M. A.: 2005, *Fundamentals of Radar Signal Processing*, McGraw-Hill Professional, US.

Richards, W. L. and Scheer, J. A.: 2010, *Principles of Modern Radar: Basic Principles*, SciTech Publishing, Edison, NJ.

Richmond, C. D.: 1996, A note on non-Gaussian adaptive array detection and signal parameter estimation, *IEEE Signal Processing Letters* **3**, 251–252.

Riddolls, R. J.: 2006, Implementation of method for operating multiple high frequency surface wave radars on a common carrier frequency, *IEEE Conference on Radar*, Verona, New York, US, p. 4.

Rifkin, R.: 1994, Analysis of CFAR performace in Weibull clutter, *IEEE Transactions on Aerospace and Electronic Systems* **30**(2), 315–329.

Rihaczek, A. W.: 1971, Radar waveform selection—a simplified approach, *IEEE Transactions on Aerospace and Electronic Systems* **7**(3), 1078–1086.

Rihaczek, A. W.: 1985, *Principles of High Resolution Radar*, Peninsula Publishing, Los Altos, California, US.

Ringelstein, J., Gershman, A., and Bohme, J.: 1999, Sensor array processing for random inhomogeneous media, *Advanced Signal Processing Algorithms, Architectures and Implementations (SPIE-99)* **IX-3807**, 267–276.

Ringer, M. A. and Frazer, G. J.: 1999, Waveform analysis of transmissions of opportunity for passive radar, *International Symposium on Signal Processing and its Applications (ISSPA-99)* **2**, 511–514.

Rishbeth, H.: 1991, F-region storms and thermospheric dynamics, *Journal of Geomagnetism and Geoelectricity* **43**, 513–524.

Rishbeth, H. and Field, P. R.: 1997, Latitudinal and solar-cycle patterns in the response of the ionosphere F2-layer to geomagnetic activity, *Advances in Space Research* **20**(9), 1689–1692.

Ristic, B., Arulampalam, S., and Gordon, N.: 2004, *Beyond the Kalman Filter: Particle Filters for Tracking Applications*, Artech House, Boston, MA.

Ritcey, J. A.: 1986, Performance analysis of the censored mean-level detector, *IEEE Transactions on Aerospace and Electronic Systems* **22**(4), 443–454.

Ritcey, J. A.: 1989, Performance of max-mean-level detector with and without censoring, *IEEE Transactions on Aerospace and Electronic Systems* **25**(2), 213–223.

Robey, F. C., Fuhrmann, D. R., Kelly, E. J., and Nitzberg, R.: 1992, A CFAR adaptive matched filter detector, *IEEE Transactions on Aerospace and Electronic Systems* **28**(1), 208–216.

Robinson, L. A.: 1989, Summary of RCS measurements at SRI International—1982-1988, *Final Report, SRI Project 3786 for Lincoln Laboratory.*

Rohling, H.: 1983, Radar CFAR thresholding in clutter and multiple target situations, *IEEE Transactions on Aerospace and Electronic Systems* **19**(4), 608–621.

Roman, J. R., Rangaswamy, M., Davis, D. W., Zhang, Q., Himed, B., and Michels, J.: 2000, Parametric adaptive matched filter for airborne radar applications, *IEEE Transactions on Aerospace and Electronic Systems* **36**(2), 667–692.

Root, B.: 1998, HF radar ship detection through clutter cancellation, *IEEE National Radar Conference*, Dallas, US, 281–286.

Root, B. T. and Headrick, J. M.: 1993, Comparison of RADARC high-frequency radar performance prediction model and ROTHR Amchitka data, *Naval Research Laboratory Memo Report NRL/MR/5320-93-7181*.

Rotheram, S.: 1981a, Ground wave propagation, part 1: Theory for short distances, *Proceedings of the IEE, Part F: Communications, Radar and Signal Processing* **128**(5), 275–284.

Rotheram, S.: 1981b, Ground wave propagation, parts 1 and 2, *IEEE Proceedings Part F* **128**, 275–295.

Roy, R. and Kailath, T.: 1989, ESPRIT—estimation of signal parameters via rotational invariance techniques, *IEEE Tranactions on Acoustics, Speech, Signal Processing* **37**, 984–995.

Ruck, G., Barrick, D., Stuart, W., and Krichbaum, D.: 1970, *Radar Cross Section Handbook*, Volumes I and II, Plenum Press, New York.

Rumi, G. C.: 1975, Around-the-world propagation, *Radio Science* **10**, 711.

Rush, C. M., Pokempner, M., Anderson, D. N., Perry, J., Stewart, F. G., and Reasoner, R.: 1984, Maps of foF2 derived from observations and theoretical data, *Radio Science* **19**, 1083.

Rutten, M. G., Gordon, N. J., and Percival, D. J.: 2003, Track fusion in over-the-horizon radar networks, *Proceedings of Sixth International Conference on Information Fusion* **1**, 334–341.

Rutten, M. G. and Percival, D. J.: 2001, Joint ionospheric and track target state estimation for multipath OTHR track fusion, *SPIE Conference on Signal and Data Processing of Small Targets*, San Diego, US, 118–129.

Rutten, M. G. and Percival, D. J.: 2002, Fusion of multipath tracks for a network of over-the-horizon radars, *Final Program and Abstracts, Information, Decision and Control*, Adelaide, Australia, 193–198.

Sacchi, M., Ulrych, T., and Walker, C.: 1998, Interpolation and extrapolation using a high-resolution discrete fourier transform, *IEEE Transactions on Signal Processing* **46**(1), 31–38.

Sacchini, J., Steedly, W., and Moses, R.: 1993, Two-dimensional prony modelling and parameter estimation, *IEEE Transactions on Signal Processing* **41**(11), 3127–3137.

Sahr, J. D. and Meyer, M.: 2004, Opportunities for passive VHF radar studies of plasma irregularities in the equatorial E and F regions, *Journal of Atmospheric and Solar Terrestrial Physics* **66**, 1675–1681.

Saillant, S., Auffray, G., and Dorey, P.: 2003, Exploitation of elevation angle control for a 2-D HF skywave radar, *IEEE International Radar Conference*, Adelaide, Australia, 662–666.

Sailors, D. B.: 1995, Discrepancy in the international radio consultative committee report 322-3 radio noise model: The probable cause, *Radio Science* **30**, 713–728.

Saini, R. and Cherniakov, M.: 2005, DTV signal ambiguity function analysis for radar application, *IEE Proceedings—Radar, Sonar and Navigation* **152**(3), 133–142.

Saini, R., Cherniakov, M., and Lenive, V.: 2003, Direct path interference suppression in bistatic system: DTV based radar, *IEEE International Radar Conference (RADAR 2003)*, Adelaide, Australia, 309–314.

Sangston, K., Gini, F., Greco, M., and Farina, A.: 1999, Structures for radar detection in compound-Gaussian clutter, *IEEE Transactions on Aerospace and Electronic Systems* **35**(2), 445–458.

Sangston, K. I. and Gerlach, K.: 1994, Coherent detection of radar targets in a non-Gaussian background, *IEEE Transactions on Aerospace and Electronic Systems* **30**(2), 330–340.

Sarkar, T. K., Wang, H., and Park, S.: 2001, A deterministic least-squares approach to space-time adaptive processing (STAP), *IEEE Transactions on Antennas and Propagation* **49**(1), 91–103.

Sarma, A. and Tufts, D. W.: 2001, Robust adaptive threshold for control of false alarms, *IEEE Signal Processing Letters* **8**(9), 261–263.

Sarunic, P. and Rutten, M. G.: 2001, Over-the-horizon radar multipath track fusion incorporating track history, *Proceedings of ISIF*, Paris, France, 13–19.

Sato, Y.: 1975, A method of self-recovering equalization for multilevel amplitude-modulation systems, *IEEE Transactions on Communications* **23**(6), 679–682.

Scharf, L.: 1991, *Statistical Signal Processing, Detection, Estimation and Time Series Analysis*, Addison-Wesley Publishing Company, US.

Scharf, L. L. and Friedlander, B.: 1994, Matched subspace detectors, *IEEE Transactions on Signal Processing* **42**(8), 2146–2157.

Scharf, L. L. and Mc Whorter, L. T.: 1996, Adaptive matched subspace detectors and adaptive coherence, *Proceedings of 30th Asilomar Conference on Signals, Systems, and Computers*, Pacific Grove, Calif., US.

Schmidt, R. O.: 1979, Multiple emitter location and signal parameter estimation, *Proceedings of the RADC Spectral Estimation Workshop*, Rome, NY, 243–258.

Schmidt, R. O.: 1981, A signal subspace approach to multiple emitter location and spectral estimation, Ph.D. Dissertation, Stanford University, Stanford, Calif.

Schmidt, R. O.: 1986, Multiple emitter location and signal parameter estimation, *IEEE Transactions on Antennas and Propagation* **34**, 276–280.

Schunk, R. W., and Nagy, A. F.: 2009, *Ionospheres: Physics, Plasma Physics and Chemistry*, 2d ed., Cambridge University Press, Cambridge, UK.

Secan, J. A.: 2004, WBMOD ionospheric scintillation model, an abbreviated user's guide, *NWRA-CR-94 R172/Rev 7*, NorthWest Research Associates, Inc., Bellevue, WA.

Secan, J. A., Bussey, R. M., Fremouw, E. J., and Basu, S.: 1995, An improved model of equatorial scintillation, *Radio Science* **30**, 607–617.

Seliktar, Y., Williams, D. B., and Holder, E. J.: 2000, Beam-augmented STAP for joint clutter and jammer multipath mitigation, *IEE Proceedings—Radar, Sonar and Navigation* **147**(5), 225–232.

Sennitt, A. G. and Kuperus, B.: 1997, *World Radio TV Handbook*, Glen Heffernan Billboard Books, Amsterdam, Netherlands.

Sevgi, L.: 2001, Target reflectivity and RCS interactions in integrated maritime surveillance systems based on surface-wave high-frequency radars, *IEEE Antennas and Propagation Magazine* **43**(1), 36–51.

Sevgi, L.: 2003, *Complex Electromagnetic Problems and Numerical Simulation Approaches*, IEEE Press, Hoboken, NJ.

Sevgi, L., Ponsford, A., and Chan, H. C.: 2001, An integrated maritime surveillance system based on high-frequency surface-wave radars, Part 1: Theoretical background and numerical simulations, *IEEE Antennas and Propagation Magazine* **43**(5), 28–43.

Shan, T. J., Wax, M., and Kailath, T.: 1985, On spatial smoothing for direction of arrival estimation of coherent signals, *IEEE Transactions on Acoustics Speech and Signal Processing* **33**(4), 806–811.

Shearman, E. D. R.: 1983, Propagation and scattering in MF/HF ground-wave radar, *IEE Proceedings F* **130**(7), 579–590.

Shepherd, A. and Lomax, J.: 1967, Frequency spread in ionospheric radio propagation, *IEEE Transactions on Communication Technology* **15**(2), 268–275.

Sherill, W. and Smith, G.: 1977, Direcional dispersion of sporadic-E modes between 9 and 14 MHz, *Radio Science* **12**(5), 773–778.

Shimazaki, T.: 1962, A statistical study of occurrence probability of spread F at high latitudes, *Journal of Geophysical Research* **67**, 4617.

Shivaprasad, A. P.: 1971, Some pulse characteristics of atmospheric radio noise bursts at 3 MHz, *Journal of Atmospheric and Terrestrial Physics* **33**(10), 1607–1608.

Shnidman, D. A.: 2003, Expanded swerling target models, *IEEE Transactions on Aerospace and Electronic Systems* **39**(3), 1059–1069.

Shunjun, W. and Yingjun, L.: 1995, Adaptive channel equalisation for space-time adaptive processing, *IEEE International Radar Conference (RADAR-95)*, Alexandria, Virginia, US, 624–628.

Silberstein, R.: 1959, The origin of the current nomenclature of the ionospheric layers, *Journal of Atmospheric and Terrestrial Physics* **13**, 382.

Sinnott, D.: 1986, The Jindalee over-the-horizon radar system, *Conference on Air Power in the Defence of Australia*, Australian National University, Research School of Pacific Studies, Strategic and Defence Studies Centre, Canberra, Australia.

Sinnott, D. H.: 1987, Jindalee - DSTO's over-the-horizon radar project, *IREE, Digest of papers, 21st International Electronics Convention and Exhibition*, Sydney, Australia, 661–664.

Sinnott, D. H.: 1988, The development of over-the-horizon radar in Australia, *DSTO Bicentennial History Series—Australian Government Publishing Service.*

Sinnott, D. H. and Haack, G. R.: 1983, The use of overlapped subarray techniques in simultaneous receive beam arrays, *Proceedings of the Antenna Applications Symposium, University of Illinois*, Moticello, Illinois, US, 21–37.

Six, M., Parent, J., Bourdillon, A., and Delloue, J.: 1996, A new multibeam receiving equipment for the Valensole skywave HF radar: Description and applications, *IEEE Transactions on Geoscience and Remote Sensing* **34**(3), 708–719.

Skolnik, M.: 2008a, *Radar Handbook,* 3d ed., McGraw-Hill, New York, US.

Skolnik, M.: 2008b, HF over-the-horizon radar, J. M. Headrick and S. J. Anderson (Chapter 20), *Radar Handbook,* 3d ed., McGraw-Hill, New York, US.

Skolnik, M.: 2008c, Electronic counter-countermeasures, A. Farina (Chapter 22), *Radar Handbook,* 3d ed., McGraw-Hill, New York, US.

Slock, D. T. and Padias, C. B.: 1995, Further results on blind identification and equalisation of multiple FIR channels, *International Conference on Acoustics Speech and Signal Processing (ICASSP-95)*, Detroit, Michigan, US.

Smith, E. K.: 1957, Worldwide occurrence of sporadic E, *Circular 582, National Bureau of Standards.*

Smith, E. K. and Matsushita, S. E.: 1962, *Ionospheric Sporadic E*, Macmillan, New York.

Sojka, J. J.: 1989, Global scale, physical models of the F region ionosphere, *Reviews of Geophysics* **27**(3), 371–403.

Solomon, I. S. D.: 1998, Over-the-horizon radar array calibration, Ph.D. Dissertation, The University of Adelaide, Adelaide, Australia.

Solomon, I. S. D., Gray, D. A., Abramovich, Y. I., and Anderson, S. J.: 1998, Over-the-horizon radar array calibration using echoes from ionised meteor trails, *IEE Proceedings—Radar, Sonar and Navigation* **145**(3), 173–180.

Solomon, I. S. D., Gray, D. A., Abramovich, Y. I., and Anderson, S. J.: 1999, Receiver array calibration using disparate sources, *IEEE Transactions on Antennas Propagation* **47**, 496–505.

Solomon, R. C., Leong, H., and Antar, Y. M. M.: 2008, Forward scattering effects in RCS of complex targets in the 3–20 MHz high frequency range, *International Symposium of the IEEE Antennas and Propagation Society (AP-S)*, San Diego, California, US, 1–4.

Sommerfeld, A.: 1909, The propagation of waves in wireless telegraphy, *Annals of Physics* **28**, 665–736.

Somov, V. G., Leusenko, V. A., Tyapkin, V. N., and Shaidurov, G. Y.: 2003, Effect of nonlinear and focusing ionospheric properties on qualitative characteristics of radar in the decametric wave band, *Journal of Communications Technology and Electronics* **48**, 850–858.

Spaulding, A. D. and Washburn, J. S.: 1985, Atmospheric radio noise: Worldwide levels and other characteristics, *NTIA Report 85-173, National Telecommunications and Information Administration.*

Steel Yard OTH Radar: 2008, *http://www.globalsecurity.org/wmd/world/russia/steelyard.htm.*

Steinhardt, A. O. and Van Veen, B. D.: 1989, Adaptive beamforming, *International Journal of Adaptive Control and Signal Processing* **3**, 253–281.

Stimson, G. W.: 1998, *Introduction to Airborne Radar*, 2d ed., SciTech Publishing, Mendham, New Jersey, US.

Stoica, P., Besson, O., and Gershman, A.: 2001, Direction-of-arrival estimation of an amplitude-distorted wavefront, *IEEE Transactions on Signal Processing* **49**, 269–276.

Stoica, P. and Nehorai, A.: 1989, MUSIC, Maximum Likelihood and Cramer-Rao lower bound: Further results and comparisons, *IEEE Transactions on Acoustics Speech and Signal Processing* **37**, 720–741.

Stoica, P. and Sharman, K.: 1990a, Maximum likelihood methods for direction-of-arrival estimation, *IEEE Transactions on Acoustics, Speech, Signal Processing* **38**, 1132–1143.

Stoica, P. and Sharman, K. C.: 1990b, Novel eigen analysis method for direction estimation, *IEE Proceedings, Part F* **137**(1), 19–26.

Stone, L. D., Barlow, C. A., and Corwin, T. L.: 1999, *Bayesian Multiple Target Tracking*, Artech House, Boston, MA.

Stove, A.: 1992, Linear FMCW radar techniques, *IEE Proceedings F* **139**(5), 343–350.

Strausberger, D. J., Garber, F. D., Chamberlain, N. F., and Walton, E. K.: 1992, Modeling and performance of HF/OTH radar target classification systems, *IEEE Transactions on Aerospace and Electronic Systems* **28**(2), 396–403.

Summers, B.: 1995, Radar range processing, *Patent G01S1/34, International publication number WO 95/32437.*

Sweeney, L. E.: 1970, Spatial properties of ionospheric radio propagation as determined with half degree azimuthal resolution, Ph.D. Dissertation, Stanford University, Calif.

Swerling, P.: 1960, Probability of detection for fluctuating targets, *IEEE Transactions on Information Theory* **6**, 269–308.

Swerling, P.: 1997, Probability of detection for some additional fluctuating target cases, *Reprinted in IEEE Transactions on Aerospace and Electronic Systems* **33**(2), 698–709.

Swindlehurst, A.: 1998, Time delay and spatial signature estimation using known asynchronous signals, *IEEE Transactions on Signal Processing* **46**(2), 449–462.

Swindlehurst, A. and Kailath, T.: 1992, A performance analysis of subspace-based methods in the presence of model errors, Part I: The MUSIC algorithm, *IEEE Transactions on Signal Processing* **40**, 1758–1773.

Swingler, D. and Walker, R.: 1988, Linear-predictive extrapolation for narrowband spectral estimation, *Proceedings of IEEE* **76**(9), 1249–1251.

Talwar, S., Viberg, M., and Paulraj, A.: 1994, Blind estimation of multiple co-channel digital signals using an antenna array, *IEEE Signal Processing Letters* **1**, 29–31.

Tan, D., Sun, H., Lui, Y., Lesturgie, M., and Chan, H.: 2005, Passive radar using global system for mobile communication signal: Theory, implementation and measurements, *IEE Proceedings—Radar, Sonar and Navigation* **152**(3), 116–123.

Tarran, C. J.: 1997, Operational HF DF systems employing real-time super resolution processing, *HF Radio Systems and Techniques*, Nottingham, UK, 311–319.

Taylor, A. H. and Hulbert, E. O.: 1926, The propagation of radio waves over the earth, *Physical Review* **27**(2), 189–215.

Taylor, J. E.: 1902, Characteristics of electric earth-current disturbances, and their origin, *Proceedings of the Royal Society of London*, **71** (467-476), 225–227.

Teague, C. C., Tyler, G. L., and Stewart, R. H.: 1975, The radar cross-section of the sea at 1.95 MHz: Comparison of in-situ and radar determinations, *Radio Science* **10**, 847–852.

Teitelbaum, R.: 1991, A flexible processor for a digital adaptive array radar, *IEEE National Radar Conference (RadarCon-91)*, Los Angeles, US, 103–107.

Teters, L. R., Lloyd, J. L., Haydon, G. W., and Lucas, D. L.: 1983, Estimating the performance of telecommunication systems using the ionospheric transmission channel—ionospheric communications analysis and prediction program users manual, *National Telecommunications and Information Administration Report 83–127*.

Thayaparan, T. and Kennedy, S.: 2004, Detection of a maneuvering air target in sea-clutter using joint time-frequency analysis techniques, *IEE Proceedings—Radar, Sonar and Navigation* **151**, 19–30.

Thayaparan, T. and MacDougall, J.: 2005, Evaluation of ionospheric sporadic-E clutter in an Arctic environment for the assessment of high frequency surface wave radar surveillance, *IEEE Transactions on Geoscience and Remote Sensing* **43**(5), 1180–1188.

Thierren, C. W.: 1992, *Discrete Random Signals and Statistical Signal Processing*, Prentice-Hall, Inc., Englewood Cliffs, New Jersey.

Thomas, J. M., Baker, C. J., and Griffiths, H. D.: 2007, DRM signals for HF passive bistatic radar, *IEEE International Radar Conference (RADAR 2007)*, Edinburgh, UK.

Thomas, J. M., Griffiths, H. D., and Baker, C. J.: 2006, Ambiguity function analysis of Digital Radio Mondiale signals for HF passive bistatic radar, *Electronics Letters* **42**(25).

Thomas, L.: 1970, F2-region disturbances associated with major magnetic storms, *Journal of Atmospheric and Terrestrial Physics* **18**, 917.

Thomas, L. and Venables, F. H.: 1966, The onset of the F-region disturbance at middle latitudes during magnetc storms, *Journal of Atmospheric and Terrestrial Physics* **28**, 599.

Thomas, R. M., Whitham, P. S., and Elford, W. G.: 1988, Response of high frequency radar to meteor backscatter, *Journal of Atmospheric and Terrestrial Physics* **50**, 703–724.

Thomason, J. F.: 2003, Development of over-the-horizon radar in the United States, *IEEE International Radar Conference (RADAR 2003)*, Adelaide, Australia, 599–601.

Thomason, J., Skaggs, G., and Lloyd, J.: 1979, A global ionospheric model, *Naval Research Laboratory Report*(8321).

Titheridge, J. E.: 1988, The real height analysis of ionograms: A generalized formulation, *Radio Science* **23**(5), 831–839.

Tong, L. and Perreau, S.: 1998, Multichannel blind identification: From subspace to maximum likelihood methods, *Proceedings of IEEE* **86**(10), 1951–1968.

Tong, L., Xu, G., and Kailath, T.: 1994, Blind identification and equalization based on second-order statistics: A time domain approach, *IEEE Transactions on Information Theory* **40**, 340–349.

Treharne, R. F.: 1967, Vertical triangulation using skywaves, *Proceedings of the Institution of Radio and Electronics Engineers*, Australia, **28**, 419–423.

Treichler, J. and Agee, B.: 1983, A new approach to multipath correction of constant modulus signals, *IEEE Transactions on Acoustics, Speech, Signal Processing* **31**(2), 459–472.

Trenkle, F.: 1979, *Die Deutschen Funkmeverfahren bis 1945* (in German), Motorbuch-Verlag, Stuttgart.

Trizna, D. B.: 1982, Estimation of the sea surface radar cross section at HF from second-order Doppler spectrum characteristics, *Naval Research Laboratory Memo Report* (8579), 916.

Trizna, D. B. and Headrick, J. M.: 1961, Ionospheric effects on HF over-the-horizon radar, In J.M. Goodman, (ed.), *Proceedings Effect of the Ionosphere on Radiowave Systems, ONR/AFGL-sponsored*, 262–272.

Trizna, D., Moore, J., Headrick, J., and Bogle, R.: 1977, Directional sea spectrum determination using HF Doppler radar techniques, *IEEE Transactions on Antennas and Propagation* **25**(1), 4–11.

Trost, T.: 1979, Electron concentrations in the E and upper D region at Arecibo, *Journal of Geophysical Research* **84**(A6), 2736–2742.

Trump, T. and Ottersten, B.: 1996, Estimation of nominal direction of arrival and angular spread using an array of sensors, *Signal Processing* **50**(9), 57–69.

Tufts, D. and Melissinos, C.: 1986, Simple, effective computation of the principle eigenvectors and their eigenvalues and application to high-resolution of frequencies, *IEEE Transactions on Acoustics Speech and Signal Processing* **34**, 1046–1053.

Turley, M. D.: 1997, Hybrid CFAR techniques for HF radar, *Proceedings of International Radar Conference*, 36–40.

Turley, M. D.: 1999, Skywave radar spatial adaptive processing with quiescent pattern control, *Fifth International Symposium on Signal Processing and Its Applications (ISSPA-99)*, Brisbane, Australia, 337–340.

Turley, M. D.: 2003, Impulsive noise rejection in HF radar using a linear prediction technique, *Proceedings of the International Radar Conference*, Adelaide, Australia, 358–362.

Turley, M. D.: 2006, FMCW radar waveforms in the HF band, *Technical note presented at ITU-R JRG 1A-1C-8B meeting*.

Turley, M. D.: 2008, Signal processing techniques for maritime surveillance with skywave radar, *IEEE International Radar Conference (RADAR 2008)*, Adelaide, Australia.

Turley, M. D.: 2009, Bandwidth formula for linear FMCW radar waveforms, *IEEE International Radar Conference (RADAR 2009)*.

Turley, M. D. and Voight, S.: 1992, The use of a hybrid AR/classical spectral analysis technique with application to HF radar, *International Symposium on Signal Processing and its Applications (ISSPA-99)*, Gold Coast, Australia.

Turley, M. and Lees, M. L.: 1987, An adaptive impulsive noise suppressor for FMCW radar, *Proceedings of IREECON'87, 21st International Electronic Convention and Exhibition*, Sydney, Australia, 665–668.

Tyler, G. L., Teague, C. C., Stewart, R. H., Peterson, A. M., Munk, W. H., and Joy, J. W.: 1974, Wave directional spectra from synthetic aperture observations of radio scatter, *Deep-Sea Research* **21**, 989–1016.

Uman, M. A.: 1987, *The Lightning Discharge*, Academic, San Diego, Calif.

Valaee, S., Champagne, B., and Kabal, P.: 1995, Parametric localisation of distributed sources, *IEEE Transactions on Signal Processing* **43**(9), 2144–2153.

van der Pol, B. and Bremmer, H.: 1937a, The diffraction of electromagnetic waves from an electrical point source round a finitely conducting sphere – Part I, *Philosophical Magazine Series 7* **24**, 141–176.

van der Pol, B. and Bremmer, H.: 1937b, The diffraction of electromagnetic waves from an electrical point source round a finitely conducting sphere – Part II, *Philosophical Magazine Series 7* **24**, 825–864.

van der Pol, B. and Bremmer, H.: 1938, The propagation of radio waves over a finitely conducting spherical earth, *Philosophical Magazine Series 7* **25**, 817–834.

van der Pol, B. and Bremmer, H.: 1939, Further note on the propagation of radio waves over a finitely conducting spherical earth, *Philosophical Magazine Series 7* **26**, 261–265.

van der Veen, A.-J.: 1998, Algebraic methods for deterministic blind beamforming, *Proceedings of IEEE* **86**(10), 1987–2008.

van der Veen, A. and Paulraj, A.: 1996, An analytical constant modulus algorithm, *IEEE Transactions on Signal Processing* **44**, 1136–1155.

Van Trees, H. L.: 2002, *Detection, Estimation, and Modulation Theory*, John Wiley and Sons, New York, US.

Van Veen, B. D.: 1991, Minimum variance beamforming with soft response constraints, *IEEE Transacations on Signal Processing* **39**(9), 1964–1972.

Van Veen, B. D. and Buckley, K. M.: 1988, Beamforming: A versatile approach to spatial filtering, *IEEE Acoustics Speech and Signal Processing Magazine* **5**(2), 4–24.

Viberg, M. and Ottersten, B.: 1991, Sensor array processing based on subspace fitting, *IEEE Transactions on Signal Processing* **39**(5), 1110–1121.

Viberg, M., Ottersten, B., and Kailath, T.: 1991, Detection and estimation in sensor arrays using weighted subspace fitting, *IEEE Transactions on Signal Processing* **39**, 2436–2449.

Villain, J., Andre, R., Hanuise, C., and Gresillon, D.: 1996, Observation of high latitude ionosphere by HF radars: Interpretation in terms of collective wave scattering and characterization of turbulence, *Journal of Atmospheric and Terrestrial Physics* **58**(8), 943–958.

Villars, F. and Weisskopf, V.: 1955, On the scattering of radio waves by turbulent fluctuations in the ionosphere, *Proceedings of the IRE* **43**, 1232–1239.

Vizinho, A. and Wyatt, L. R.: 2001, Evaluation of the use of the modified-covariance method in HF radar ocean measurement, *IEEE Journal of Oceanic Engineering* **26**(4) 832–840.

Vorobyov, S., Gershman, A., and Luo, Z.: 2003, Robust adaptive beamforming using worst-case performance optimization: A solution to the signal mismatch problem, *IEEE Transactions on Signal Processing* **51**, 313–324.

Vozzo, A.: 1995, Performance assessment of receivers used in HF radar, *Sixth International Conference on Radio Receivers and Associated Systems*, Bath, UK, 133–138.

Wait, J. R.: 1956, Radiation from a vertical antenna over a curved stratified ground, *Journal of Research of the National Bureau of Standards* **56**(4), 237–244.

Wait, J. R.: 1964, *Electromagnetic Surface Waves, Advances in Radio Research*, Academic Press, New York.

Walton, E. K. and Young, J. D.: 1984, The Ohio State University compact radar cross section measurement range, *IEEE Transactions on Antennas and Propagation* **32**, 1218–1223.

Wan, X., Xiong, X., Cheng, F., and Ke, H.: 2007, Experimental investigation of directional characteristics for ionospheric clutter in HF surface wave radar, *IET Proceedings—Radar, Sonar and Navigation* **1**(2), 124–130.

Wang, E., Wang, J., and Ponsford, A. M.: 2011, An adaptive hierarchal CFAR for optimal target detection in mixed clutter environments, *IEEE Radar Conference*, Kansas, Missouri, US, 543–547.

Wang, G., Xia, X.-G., Root, B. T., Chen, V. C., Zhang, Y., and Amin, M.: 2003, Manoeuvring target detection in over-the-horizon radar using adaptive clutter rejection and adaptive chirplet transform, *IEE Proceedings—Radar, Sonar, Navigation* **150**(4), 292–298.

Wang, J., Dizaji, R., and Ponsford, A. M.: 2004, Analysis of clutter distribution in bistatic high frequency surface wave radar, *Canadian Conference on Electrical and Computer Engineering (CCECE)*, Niagara Falls, 1301–1304.

Ward, J.: 1994, Space-time adaptive processing for airborne radar, *MIT Lincoln Laboratory TR 1015, ESC-TR-94-109.*

Warrington, E. M.: 1995, Measurements of the direction of arrival of HF skywave signals by means of a wide aperture array and super resolution direction finding algorithms, *IEE Proceedings on Microwaves, Antennas and Propagation* **142**(2), 136–144.

Warrington, E. M., Nasyrov, I., Stocker, A. J., and Jacobsen, B.: 2003, Measurements of the delay, Doppler and directional characteristics of obliquely propagating HF signals over several northerly paths and a comparison with vertical ionosonde and HF radar observations, *Ninth International Conference on HF Radio Systems and Techniques,* **493**, 159–164.

Warrington, E. M., Thomas, E. C., and Jones, T. B.: 1990, Measurements on the wavefronts of ionospherically-propagated HF radio waves made with a large aperture antenna array, *Proceedings of IEE*, **137H**(1), 25–30.

Washburn, T. W. and Sweeney, L. E.: 1976, An on-line adaptive beamforming capability for HF backscatter radar, *IEEE Transactions on Antennas and Propagation* **24**(5), 721–732.

Watson, G. N.: 1919, The transmission of electric waves round the earth, *Proceedings of the Royal Society of London, Series A* **95**, 546–563.

Watson-Watt, R. A.: 1929, Weather and wireless, *Quarterly Journal of the Royal Meteorological Society* **55**(273).

Watterson, C. C., Juroshek, J. R., and Bensema, W. D.: 1970, Experimental confirmation of an HF channel model, *IEEE Transactions on Communications* **18**(6), 792–803.

Watts, J. M. and Davies, K.: 1960, Rapid frequency analysis of fading radio signals, *Journal of Geophysical Research* **65**, 2295.

Wax, M. and Kailath, T.: 1985, Detection of signals by information theoretic criteria, *IEEE Transactions on Acoustics Speech and Signal Processing* **33**, 387–392.

Wei, W., Yingning, P., Taifan, Q., and Yongtan, L.: 1999, HF OTHR target detection and estimation subsystem, *IEEE Aerospace and Electronic Systems Magazine* **14**(4), 39–45.

Weiner, D. D., Capraro, G. T., and Wicks, M. C.: 1998, An approach for utilizing known terrain and land feature data in estimation of the clutter covariance matrix, *Proceedings of the IEEE Radar Conference (RADARCON)*, Dallas, TX, 381–386.

Weiss, A. and Friedlander, B.: 1996, Almost blind steering vector estimation using second-order moments, *IEEE Transactions on Signal Processing* **44**(4), 1024–1027.

Weiss, M.: 1982, Analysis of some modified cell-averaging CFAR processors in multipl-target situations, *IEEE Transactions on Aerospace and Electronic Systems* **18**(1), 102–114.

Wenyu, Z. and Peinan, J.: 1994, Signal processing of skywave OTH-B radar, *IEEE International Conference on Acoustics, Speech and Signal Processing (ICASSP'94)* **6**, 117–120.

Wenyu, Z. and Xu, M.: 1991, Bistatic FMCW OTH-B experimental radar, *International Conference on Radar ICR-91*, China Institute of Electronics, 138–141.

Whale, H. A.: 1969, *Effects of Ionospheric Scattering on Very-Long-Distance Radio Communication*, Plenum Press, New York.

Whale, H. and Gardiner, C.: 1966, The effect of a specular component on the correlation between the signals received on spaced antennas, *Radio Science* **1**(5), 557–570.

Whitehead, J. D.: 1962, Distribution of echo amplitudes from an undulating surface, *Journal of Atmospheric and Terrestrial Physics* **24**, 715.

Whitehead, J. D.: 1989, Recent work on mid-latitude and equatorial sporadic E, *Journal of Atmospheric and Terrestrial Physics* **51**, 401.

Wicks, M. C., Melvin, W. L., and Chen, P.: 1997, An efficient architecture for nonhomogeneity detection in space-time adaptive processing airborne early warning radar, *Proceedings of Radar 97*, Edinburgh, UK, 295–299.

Wicks, M. C., Rangaswamy, M., Adve, R., and Hale, T. B.: 2006, Space-time adaptive processing: A knowledge-based perspective for airborne radar, *IEEE Signal Processing Magazine* **23**(1), 51–65.

Widrow, B., Mantey, P., Griffiths, L., and Goode, B.: 1967, Adaptive antenna systems, *Proceedings IEEE* **55**(12), 2143–2159.

Willis, N. J.: 1991, *Bistatic Radar*, Artech House, Norwood, MA.

Willis, N. J. and Griffiths, H. D.: 2007, *Advances in Bistatic Radar*, SciTech, North Carolina, US.

Wilson, H. and Leong, H.: 2003, An estimation and verification of vessel radar-cross-sections for HF surface wave radar, *IEEE International Radar Conference*, Adelaide, Australia, 711–716.

Wood, L. and Treitel, S.: 1975, Seismic signal processing, *Proceedings of the IEEE* **63**, 649–661.

Wu, G.-X. and Deng, W.-B.: 2006, HF radar target identification based on optimized multi-frequency features, *International Conference on Radar (CIE'06)*, Shanghai, China, 1–4.

Wu, Q. and Wong, K. M.: 1996, Blind adaptive beamforming for cyclostationary signals, *IEEE Transactions on Signal Processessing* **44**(11), 2757–2767.

Wyatt, L. R.: 1990, A relaxation method for integral inversion applied to HF radar measurement of the ocean wave directional spectrum, *International Journal of Remote Sensing* **11**, 1481–1494.

Wyatt, L. R., Green, J. J., Middleditch, A., Moorhead, M. D., Howarth, J., Holt, M., and Keogh, S.: 2006, Operational wave, current, and wind measurements with the Pisces HF radar, *IEEE Journal of Oceanic Engineering* **31**(4), 819–834.

Wylder, J.: 1987, The frontier for sensor technology, *Signal* **41**, 73–76.

Xianrong, W., Feng, C., and Hengyu, K.: 2006a, Experimental trials on ionospheric clutter suppression for high-frequency surface wave radar, *IEE Proceedings—Radar, Sonar and Navigation* **153**(1), 23–29.

Xianrong, W., Feng, C., and Hengyu, K.: 2006b, Main beam cochannel interference suppression by range adaptive processing for HFSWR, *IEEE Signal Processing Letters* **13**(1), 29–32.

Xianrong, W., Xinlong, X., and Hengyu, K.: 2006, Ionospheric clutter suppression in HF surface wave radar OSMAR, *7th International Symposium on Antennas, Propagation and EM Theory*, Guilin, China, 1–6.

Xiaodong, T., Yunjie, H., and Wenyu, Z.: 2001, Skywave over-the-horizon backscatter radar, *CIE International Radar Conference*, Beijing, China, 90–94.

Xie, J., Yuan, Y., and Liu, Y.: 1998, Super-resolution processing for HF surface wave radar based on pre-whitened MUSIC, *IEEE Journal of Oceanic Engineering* **23**(4), 313–321.

Xie, J., Yuan, Y., and Liu, Y.: 2001, Experimental analysis of sea clutter in shipborne HFSWR, *IEE Proceedings—Radar, Sonar and Navigation* **148**(2), 67–71.

Xingpeng, M., Yongtan, L., and Weibo, D.: 2004, Radio disturbance of high frequency surface wave radar, *Electronics Letters* **40**(3), 202–203.

Xingpeng, M., Yongtan, L., Weibo, D., Changjun, Y., and Zilong, M.: 2004, Sky wave interference of high-frequency surface wave radar, *Electronics Letters* **40**(15), 968–969.

Xiongbin, W., Feng, C., Zijie, Y., and Hengyu, K.: 2006, Broad beam HFSWR array calibration using sea echoes, *International Conference on Radar, (CIE'06)*, Shanghai, China, 1–3.

Xu, G. and Kailath, T.: 1992, Direction-of-arrival estimation via exploitation of cyclostationarity: A combination of temporal and spatial processing, *IEEE Transactions on Signal Processing* **40**(7), 1775–1786.

Yakunin, V. A., Evstratov, F. F., Shustov, E. L., Alebastrov, V. A., and Abramovich, Y. I.: 1994, Thirty years of eastern OTH radars: History, achievements and forecast, *L'Onde Electrique* **74**(3), 29–32.

Yao, K.: 1973, A representation theorem and its applications to spherically invariant random processes, *IEEE Transactions on Information Theory* **19**, 600–608.

Yasotharan, A. and Thayaparan, T.: 2006, Time-frequency method for detecting an accelerating target in sea clutter, *IEEE Transactions on Aerospace and Electronic Systems* **42**(4), 1289–1310.

Yeh, K. C. and Liu, C. H.: 1972, *Theory of Ionospheric Waves*, Academic Press, New York.

Yellin, D. and Porat, B.: 1993, Blind identification of FIR systems excited by discrete-alphabet inputs, *IEEE Transactions on Signal Processing* **41**, 1331–1339.

Yen, J.: 1985, Image reconstruction in synthesis radio telescope arrays, S. Haykin (ed.), *Array Signal Processing*, Prentice-Hall, Englewood Cliffs, NJ.

Yuen, N. and Friedlander, B.: 1996, Performance analysis of blind signal copy using fourth order cumulants, *Journal of Adaptive Control and Signal Processing* **10**(2), 239–266.

Zatman, M.: 1996, Forwards-backwards averaging for adaptive beamforming and STAP, *Proceedings of IEEE International Conference on Acoustics, Speech and Signal Processing*, Atlanta, GA, **5**, 2630–2633.

Zatman, M. A. and Strangeways, H. J.: 1994, Bearing measurements on multi-moded signals using single snapshot data. *IEEE Antennas and Propagation Society International Symposium Digest* **3**(4), Seattle, Washington, 1934–1937.

Zenneck, J.: 1907, Uberdie fortpflanzungebenerelectromagnetische wellenlangseinerebenen leiterflcheund ihre beziehungzurdrachtlosen telegraphie, *Annalender Physik* **23**, 846–866.

Zetterberg, P. and Ottersten, B.: 1995, The spectrum efficiency of a base station antenna array system for spatially selective transmission, *IEEE Transactions on Vehicular Technology* **44**(3), 651–660.

Zhang, D. and Liu, X.: 2004, Range sidelobes suppression for wideband randomly discontinuous spectra OTH-HF radar signal, *IEEE Radar Conference*, Pennsylvania, US, 577–581.

Zhang, W.-Q., Tao, R., and Ma, Y.-F.: 2004, Fast computation of the ambiguity function, *Proc. International Conference on Signal Processing* **3**, 2124–2127.

Zhang, Y., Amin, M. G., and Frazer, G. J.: 2003, High-resolution time-frequency distributions for maneuvering target detection in over-the-horizon radars, *IEE Proceedings—Radar, Sonar and Navigation* **150**, 299–304.

Zhang, Z., Yuan, Y., and Meng, X.: 2001, HF shipborne over-the-horizon surface wave radar background clutter statistics, *CIE International Radar Conference*, 100–104.

Zhou, H., Wen, B., and Wu, S.: 2005, Dense radio frequency interference suppression in HF radars, *IEEE Signal Processing Letters* **12**(5), 361–364.

Zhou, W. and Jiao, P.: 1994, Signal processing of skywave OTH-B radar, *Proceedings of ICASSP-94* **VI**, Adelaide, Australia, 117–120.

Zywicki, D. J., Melvin, W. L., Showman, G. A., and Guerci, J. R.: 2003, STAP performance in site-specific clutter environments, *Proceedings of 2003 IEEE Aerospace Conference*, Big Sky, Montana.

Index

D

E

J

S

T

U

www.ingramcontent.com/pod-product-compliance
Lightning Source LLC
LaVergne TN
LVHW080102011026
849160LV00013B/1015